EXCITONS

MODERN PROBLEMS IN CONDENSED MATTER SCIENCES

Volume 2

Series editors

V.M. AGRANOVICH

Moscow, USSR

A.A. MARADUDIN

Irvine, California, USA

NORTH-HOLLAND PUBLISHING COMPANY
AMSTERDAM · NEW YORK · OXFORD

EXCITONS

Volume editors

E.I. RASHBA
Moscow, USSR

M.D. STURGE
Murray Hill, NJ, USA

1982

NORTH-HOLLAND PUBLISHING COMPANY
AMSTERDAM · NEW YORK · OXFORD

ISBN 0444 86202 1

PUBLISHERS:
NORTH-HOLLAND PUBLISHING COMPANY
AMSTERDAM · NEW YORK · OXFORD

SOLE DISTRIBUTORS FOR THE USA AND CANADA:
ELSEVIER SCIENCE PUBLISHING COMPANY, INC.
52 VANDERBILT AVENUE
NEW YORK, N.Y. 10017

Library of Congress Cataloging in Publication Data
Main entry under title:

Excitons.

(Modern problems in condensed matter sciences)
Complements the vol. of the same title published in 1979.
Bibliography: p.
Includes indexes.
1. Exciton theory. I. Rashba, E. I. (Emmanuel I.), 1927- . II. Sturge, M. D. (Michael Dudley), 1931- . III. Series.
QC176.8.E9E952 530.4'1 82-3464
ISBN 0-444-86202-1 AACR2

PRINTED IN THE NETHERLANDS

MODERN PROBLEMS IN CONDENSED MATTER SCIENCES

Vol. 1. SURFACE POLARITONS
V. M. Agranovich and D. L. Mills, *editors*

Vol. 2. EXCITONS
E. I. Rashba and M. D. Sturge, *editors*

Vol. 3. ELECTRONIC EXCITATION ENERGY TRANSFER IN CONDENSED MATTER
V. M. Agranovich and M. D. Galanin

Vol. 4. SPECTROSCOPY AND EXCITATION DYNAMICS OF CONDENSED MOLECULAR SYSTEMS
V. M. Agranovich and R. M. Hochstrasser, *editors*

Oh, how many of them there
are in the fields!
But each flowers in its
own way—
In this is the highest achievement
of a flower!

Matsuo Bashó
1644–1694

PREFACE TO THE SERIES

"Modern Problems in Condensed Matter Sciences" is a series of contributed volumes and monographs on condensed matter science that is published by North-Holland Publishing Company. This vast area of physics is developing rapidly at the present time, and the numerous fundamental results in it define to a significant degree the face of contemporary science. This being so, it is clear that the most important results and directions for future developments can only be covered by an international group of authors working in cooperation.

Both Soviet and Western scholars are taking part in the series, and each contributed volume has, correspondingly, two editors. Furthermore, it is intended that the volumes in the series will be published subsequently in Russian by the publishing house "Nauka".

The idea for the series and for its present structure was born during discussions that took place in the USSR and the USA between the former President of North-Holland Publishing Company, Drs. W.H. Wimmers, and the General Editors.

The establishment of this series of books, which should become a distinguished encyclopedia of condensed matter science, is not the only important outcome of these discussions. A significant development is also the emergence of a rather interesting and fruitful form of collaboration among scholars from different countries. We are deeply convinced that such international collaboration in the spheres of science and art, as well as other socially useful spheres of human activity, will assist in the establishment of a climate of confidence and peace.

The General Editors of the Series,

V.M. Agranovich A.A. Maradudin

Ya. I. Frenkel

PREFACE

50 Years of exciton spectroscopy

The outstanding achievement of the new quantum mechanical theory of solids in the late 1920's was the band theory of crystals, i.e., the Bloch scheme describing the independent motion of charge-carrying electrons and holes. In 1931 Frenkel went beyond the scope of this scheme in advancing the concept of excitons: electron excitation waves which, do not carry electric current and involve correlated motion of electrons and holes. While some absorption bands in crystal spectra, which turned out subsequently to be exciton bands, were experimentally observed in the 1920's, the firm identification of exciton spectra required the joint efforts of experimentalists and theorists over many years. The first successes were achieved for alkali halides in the 1930's, for molecular crystals in the 1940's and for semiconductors in the 1950's. Excitons have now been found in all the basic types of non-metallic crystals as well as in certain rare earth metals and intermetallics; their manifestation in numerous physical phenomena is established, and the methods of the investigation of excitons have been diversified and refined.

As it reaches its 50th anniversary, the spectroscopy of excitons has become a highly diversified field of science. It is three-faced like Hecate – there being three basic types of excitons (Frenkel, charge transfer and Wannier excitons). The interactions of excitons with each other and with other quasi-particles drastically affect all aspects of the phenomena. Furthermore, the subject is intimately connected with many other fields of physics. Therefore, an attempt to expound the physics of excitons from first principles, treating all its branches in proportion, would lead to a book which would at most merely repeat material already elucidated in other review papers and textbooks, and in which current investigations could only be treated to a very limited extent. In short, the well-known aphorism of Koz'ma Prutkov "Nobody can embrace the unembraceable" is entirely applicable to excitons.

For this reason this contributed volume does not try to be either Lehrbuch or Handbuch, textbook or encyclopedia. Our plan has been to have separate chapters covering modern achievements in the most active branches of exciton physics, where the most fundamental experimental and theoretical advances are being made. We hope that with this structure, the book will be of use to experts wishing to familiarize themselves with progress in adjacent fields. To facilitate the use of this book by newcomers to the field, the "Introduction" (chapter 1) introduces the basic ideas of exciton physics and sets forth the range of problems dealt with in the subsequent chapters. The Introduction also contains some brief historical remarks; these do not pretend to be complete or authoritative, but aim merely to outline the development of the field, incidentally drawing the reader's attention to points sometimes overlooked.

We did our best to minimize the overlap of this book with the topical reviews available. General surveys of problems already well represented on contemporary literature (e.g., molecular excitons and bound excitons in semiconductors) have not been included (see the Bibliography). We have also excluded surface excitons, magneto-spectroscopy of excitons and exciton condensation into electron–hole droplets, since surveys of these subjects will appear in other volumes of this series.

The authors of review papers are usually inclined to emphasize those problems with which they are personally involved and which, therefore, especially inspire them. This is inevitable, and has the desirable result that different papers on similar problems acquire an individual colouring. However, an international series such as this provides a particularly suitable vehicle for the elucidation of those areas where workers in various laboratories and in different countries have taken different points of view. We asked our contributors to make this their aim as far as possible. The reader will judge how far we have succeeded in this.

The Bibliography at the end of the book gives a selection of recent reviews and books in the field of Excitons. These have been chosen with the aim of directing the reader to authoritative treatments of those parts of the field we were unable to include in our book, and of drawing his attention to points of view which differ from those of our contributors.

The original concept of the exciton was developed by the great Soviet physicist Yakov I. Frenkel (1894–1952) during his stay in the USA. It seems appropriate that this joint publication should go to press in the 50th anniversary year of Frenkel's pioneer paper.

We are grateful to Miss Michelle Torchia for her help with the index.

E.I. Rashba
Moscow, USSR

M.D. Sturge
Murray Hill, NJ, USA

CONTENTS

CHAPTER 1

Introduction

M.D. STURGE

Bell Laboratories
Murray Hill, New Jersey 07974
U.S.A.

Excitons
Edited by
E.I. Rashba and M.D. Sturge

Contents

1. Frenkel and charge-transfer excitons

The concept* of the exciton as a quantum of excitation propagating in an insulating crystal originated with Frenkel (1931) and Peierls (1932). Frenkel considered N identical atoms (or molecules) in a crystal, in which one atom is raised to an excited state of energy E_1. In the absence of interatomic coupling, the wave function of the crystal is $\Phi_j = A\phi_1\phi_2 \cdots \psi_j \cdots \phi_N$, where ϕ_i represents the ground state of atom i, ψ_j its excited state, and A is the antisymmetrization operator. Clearly Φ_j is N-fold degenerate since the excitation can be on any atom. Introduction of inter-atomic coupling lifts this degeneracy. The problem is analogous to that of N coupled oscillators and has a well-known solution: a band of N normal modes with Bloch-type wave functions,

$$\Phi_{\boldsymbol{K}}(\boldsymbol{R}) = U_{\boldsymbol{K}}(\boldsymbol{R})\exp(\mathrm{i}\boldsymbol{K}\cdot\boldsymbol{R}), \tag{1}$$

where $U_{\boldsymbol{K}}(\boldsymbol{R})$ has the periodicity of the lattice. In the tight-binding approximation appropriate here, eq. (1) can be written:

$$\Phi_{\boldsymbol{K}}(\boldsymbol{R}) = N\,\Sigma_j \exp(\mathrm{i}\boldsymbol{K}\cdot\boldsymbol{R}_j)\Phi_j(R - R_j), \tag{2}$$

where $\boldsymbol{R}_j$ is the position vector of atom j. Equation (1) represents the motion of a particle, which we call the exciton, with quasi-momentum $\hbar K$. The corresponding energy levels form a band, centered (to a first approximation) on E_1. For nearest-neighbor interactions, the bandwidth is proportional to the strength of the inter-atomic transfer matrix element, whose sign determines if the point $K = 0$ is at the top or bottom of the band. More complicated interactions lead to more complicated band structure, and the band mimimum can be anywhere in the zone.

The Frenkel exciton differs from an electron in that it is neutral and carries no current: it does however, carry energy. It also differs in that K must lie within the first Brillouin zone: the extended zone scheme has no meaning. In this respect it is analogous to a phonon. As in the phonon case, if there are n translationally inequivalent atoms or molecules per unit cell, there is a Φ_j for each, and consequently n branches to the exciton spectrum for each atomic excited state. The separation between the different $K = 0$ excitons arising from a single parent state is called the

* The name "exciton" came later (Frenkel 1936).

Davydov splitting and its observation (Prikhot'ko 1944, 1949, Broude et al. 1951, Davydov 1948, 1951) was the first evidence for the existence of Frenkel excitons in molecular crystals (although previous observations of sharp absorption lines by Kronenberger and Pringsheim (1926) and Obreimov and de Haas (1928) were subsequently interpreted as being due to excitons). That this "exciton" is in fact a mobile entity capable of moving from point to point in the crystal was demonstrated rather directly by Simpson (1956).

Frenkel's model is appropriate for molecular excitons (see chs. 15 and 16) and for the d-shell excitations of transition-metal insulators (see ch. 14). Organic molecules have excited states of two types: (spin) singlet and triplets. Transitions from the ground state to singlets are generally allowed; they dominate the absorption spectrum, and give short-lived emission (fluorescence). A triplet state is usually the lowest excited state of the molecule; transitions to it are forbidden, and it shows up in long-lived phosphorescence. A molecule in an allowed excited state E_1 (i.e. one to which electric dipole transitions from the ground state are allowed) has an effective dipole moment $\boldsymbol{P}_1 \exp(\mathrm{i}Et/\hbar)$. Its classical dipole interaction potential with another dipole $\boldsymbol{P}_2$ of the same frequency at a distance $\boldsymbol{R}$ is

$$V_{\mathrm{dd}} = R^{-3}\boldsymbol{P}_1 \cdot \boldsymbol{P}_2 - 3R^{-5}(\boldsymbol{P}_1 \cdot \boldsymbol{R})(\boldsymbol{P}_2 \cdot \boldsymbol{R}).$$

This interaction can induce an upward transition in the second molecule, de-exciting the first; thus the exciton can move and an exciton band is formed. The bandwidth is $\sim P^2\langle R^{-3}\rangle$; for $P \sim 1$ Debye, $\langle R\rangle \sim 4$ Å, this is $\sim 10^{-2}$ eV, which is large enough to see but much less than the usual separation between molecular excited states. Note that this mechanism does not depend on overlap of wave functions and has a long range. Higher order multipole interactions can give bandwidths of a similar order of magnitude.

Since the triplet transition is spin forbidden, the dipolar and higher multipole interactions are very weak; but, because the excited state wave functions of neighboring molecules overlap slightly, there is a small exchange matrix element. However, a very small perturbation is enough to suppress it.

The d-shell excitations of transition metal ions and the f-shell excitations of rare-earth ions are electric-dipole forbidden, and, while they can be made allowed in a solid by non-centrosymmetric crystal fields and phonon interactions, the dipolar interaction is rarely important. However exchange (or, more usually, superexchange, due to overlap of the wave functions of two metal ions with those of an intervening non-magnetic ion) is very important. The off-diagonal matrix elements which lead to transfer of excitation are (for spin-allowed transitions) of the same order as the diagonal ones which lead to magnetic order; thus for a transition-metal oxide crystal ordering at (say) 100 K, we would expect exciton bandwidths ~0.01 eV. This is rarely observed; probably because of self-trapping, which is a consequence of the exciton–phonon interaction and is discussed in ch. 13. Rare-earth ions have smaller exchange but also less phonon coupling. In rare-earth metals and intermetallic compounds, $E(K)$ relations for f-shell Frenkel excitons have been established by neutron scattering (see bibliography).

In his original paper on "Excitation Waves" Frenkel (1931) drew heavily on the analogy between electronic and vibrational excitation. This was natural at the time, since the Born–von Karman theory of lattice vibrations was well-established and familiar to Frenkel's readers, while the exact formulation of the band theory of solids was in its infancy. In ch. 18 of this book, Belousov shows that in molecular crystals this analogy is more than a historical curiosity. The internal vibrations of neighboring molecules couple weakly and their excitation propagates just as a Frenkel exciton does.

Frenkel's model is inadequate for most insulators and semiconductors, because it ignores the fact that the exciton has an internal structure of its own. The excited atom can be regarded as having an electron in a normally empty state, and a hole in a normally filled one. Viewed in this way, the concept can be generalized to the case where the electron lies on a *different* atom to the hole (Slater and Shockley 1936, Overhauser 1956). In molecular spectroscopy, such a configuration is called a "charge-transfer" state. Excitons in alkali halides and other strongly ionic insulators involve such transfer to nearest neighbors. Such excitons are sometimes loosely referred to as Frenkel excitons, but it seems best to retain the original definition with its limitation to excitation on a single atom or molecule, and to call these "charge-transfer" excitons. They are the main subject of ch. 12.

It has been suggested (Little 1964, Ginzberg 1964, 1968, Abrikosov 1978) that certain one-dimensional (1-D) or inhomogeneous 2-D systems might be superconducting at relatively high temperatures, exciton exchange taking the place of phonon exchange in the BCS theory. Current theoretical opinion (Bardeen 1978) seems to be that this is unlikely to occur in a homogeneous system. In an inhomogeneous system the situation is confused both theoretically (Allender et al. 1973 a, b, Inkson and Anderson 1973) and experimentally (Brandt et al. 1978, Chu et al. 1978, Wilson 1978).

2. Wannier excitons

2.1. Elementary effective mass model

Because the exciton is a two-particle system, the general problem of calculating its energy levels and wave functions is, in principle, much more complicated than the corresponding electron problem. However, in the

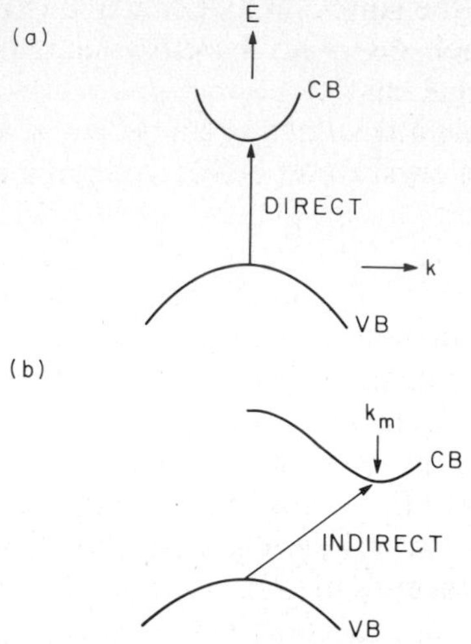

Fig. 1. Direct (a) and indirect (b) transitions between valence band (VB) and conduction band (CB) in insulators. The VB maximum does not have to be at the zone center, though in semiconductors it usually is.

limit (opposite to the Frenkel case) where the electron and hole are separated by many inter-atomic spacings, we can use the effective mass approximation (EMA)* and the problem simplifies enormously, as was first shown by Wannier (1937).

The essential point is that, in ordinary single particle band theory, the lowest excited state of an insulator consists of an electron at the absolute minimum of the (normally empty) conduction band and a hole at the absolute maximum of the (normally filled) valence band. (The generalization to relative minima is obvious). This state is at an energy E_g, the bandgap, above the ground state. Near these extrema, the kinetic energies of the electron and the hole are positive quadratic functions of the change in momentum $\hbar(\boldsymbol{k} - \boldsymbol{k}_m)$, where $\hbar \boldsymbol{k}_m$ is the momentum at the extremum. Figure 1a illustrates a simple isotropic non-degenerate band extremum at the zone center (i.e. $k_m = 0$) for which the kinetic energy is

$$E_k = \hbar^2 k^2 / 2m^*, \tag{3}$$

where m^* is the effective mass (m_e^* for the electron m_h^* for the hole). There is no known case in nature to which this simple theory strictly applies. The generalization to anisotropic, degenerate and near degenerate bands, and to the "indirect" case illustrated in fig. 1b, where $\boldsymbol{k}_m$ is not the same for electron and hole, is obvious in principle though complicated in practice

* For the theoretical justification of the EMA see Kohn (1957), Sham and Rice (1966).

(Cho 1979). However, the simple model described here exhibits most of the essential features which distinguish the exciton, and is an excellent approximation in some real cases.

The electron and hole attract each other. If they are so far apart that the atomic structure of the crystal can be ignored, their mutual potential can be written*:

$$V(\boldsymbol{\rho}) = -e^2/\kappa\rho, \tag{4}$$

where $\boldsymbol{\rho}$ is the electron–hole separation and κ is the macroscopic (long wavelength) dielectric constant. For weak exciton–phonon coupling, κ is (approximately) the dielectric constant for the frequency corresponding to the exciton binding energy E_x. This includes the lattice contribution, if, as in most semiconductors, E_x is small relative to the optical phonon energy. In the opposite limit the high frequency dielectric constant, κ_∞, is appropriate. For a more accurate treatment, see Pollman and Büttner (1975, 1977).

A Coulomb potential like eq. 4 gives rise to an infinite series of discrete hydrogenic bound states, with energies given by

$$E_n(\boldsymbol{k}) = -\frac{e^4\mu}{2\hbar^2\kappa^2 n^2} + \frac{\hbar^2 K^2}{2M}, \tag{5}$$

where $n = 1, 2, \ldots, \infty$, $\mu^{-1} = (m_e^*)^{-1} + (m_h^*)^{-1}$, $M = m_e^* + m_h^*$, $\hbar\boldsymbol{K}$ is the momentum of the center of mass, and E_n is measured from the band edge E_g. In the presence of exciton–phonon coupling, the relevant masses are (approximately) the "bare" electron and hole effective masses, *not* the polaron masses (Kane 1978).

The continuum (unbound) states of the electron and hole are also profoundly modified by the Coulomb interaction, which thus has a marked effect on the band-to-band optical absorption spectrum (see the next section).

Equation 5 and its generalizations include the N internal states of the exciton, (N = number of unit cells in the crystal) which the Frenkel formulation could not. Furthermore, all the parameters in eq. 5 can in principle be determined by experiments which do not involve the exciton, such as cyclotron resonance.

Pursuing the analogy with the H atom, we see that the wave function of the $n = 1$ exciton is (disregarding the normalization factor)

$$\psi(\boldsymbol{R}, \boldsymbol{\rho}) = \exp(i\boldsymbol{k}\cdot\boldsymbol{R} - \rho/a_x) \tag{6}$$

where $\boldsymbol{R}$ is the center of mass coordinate, and the most probable value of ρ

* Wannier (1937) omitted the factor κ^{-1}, which does not appear in a first-principles single-particle derivation of the EMA. It was added heuristically by Mott (1938) (hence the common appellation "Wannier–Mott" exciton), and was rigorously justified by Sham and Rice (1966).

is a_x, the "exciton Bohr radius", given by

$$a_x = \hbar^2 \kappa / \mu e^2. \tag{7}$$

For excited states $\exp(-\rho/a_x)$ in eq. 6 is replaced by the appropriate hydrogenic wave function $F_{nlm}(\rho/a_x)$.

The binding energy of the $n = 1$ exciton, relative to the free electron–hole pair at the band edge, often called the "exciton Rydberg", is

$$E_x = E(0) = \frac{\mu e^4}{2\hbar^2\kappa^2} = \frac{e^2}{2\kappa a_x} = \frac{\hbar^2}{2\mu a_x^2}. \tag{8}$$

Equation 8 provides a convenient way of estimating a_x from E_x if the effective value of κ or of μ is unknown.

E_x can be very small; for example in InSb, for which $\mu = 0.015\ m_0$ (m_0 is the free-electron mass) and $\kappa = 18$, $E_x = 0.6$ meV, $a_x = 600$ Å. Such an exciton is clearly only going to be stable at very low temperatures. At the other extreme, solid Kr has $\kappa_\infty = 1.8$, $\mu \sim 0.5$, giving $E_x \sim 2$ eV, $a_x \sim 2$ Å. Since the inter-atomic spacing is 3.9 Å, the Frenkel limit is reached in this case and the lowest excitons are close to the corresponding atomic states (Baldini 1962); nevertheless higher exciton states are seen in the spectrum and the Wannier model still gives a remarkably good account of the entire exciton spectrum (Fowler 1963, see also ch. 12). For most semiconductors, noble metal halides and chalcogenides, and for at least the excited exciton states of alkali halides, the model works quite well; while serious deviations occur for charge transfer excitons they are usually ascribed to the effects of the exciton–phonon interaction rather than to failure of the EMA.

Equation 6 is useful, but on an atomic scale it is misleading, since it treats the electron and hole as if they were truly free particles whose wave functions are plane waves. In fact, of course, the wave functions are Bloch functions, and

$$\psi(\boldsymbol{K}, \boldsymbol{R}, \rho) = \sum_{k_e k_n} G_{nlm}(k_e, k_h)\psi_{k_e}(r_e)\psi_{k_h}(r_h), \tag{9}$$

where $\boldsymbol{K} = \boldsymbol{k}_e + \boldsymbol{k}_n$ and $G_{nlm}(k_e k_h)$ is the Fourier transform of $\exp(i\boldsymbol{K}\cdot\boldsymbol{R})F_{nlm}(r_e, r_h)$.

Equation 9 takes into account the modulation of the wave function near the atom cores, allows for the case of the "indirect" exciton where the electron and hole extrema are at different k, and can readily be generalized to the case of degenerate bands (Dresselhaus 1956) and to include the effect of electron–hole exchange (Cho 1976). For a review see Cho (1979). These complications introduce extra splittings and deviations from the simple Rydberg formula (5).

The results of the detailed calculations, in which all the material parameters are known from independent measurements (such as cyclotron resonance) have been compared with experimental exciton energy levels in several semiconductors. Excellent agreement is obtained in most cases (Baldereschi and Lipari 1971, Lipari and Altarelli 1977, Altarelli and Lipari 1977, Timusk et al. 1978). Disagreement can usually be attributed to experimental difficulties rather than to failure of the model. Accurate band parameters are only available for a few semiconductors. In an indirect semiconductor the experimental exciton binding energy E_x is often in error due to uncertainties in the precise band gap. For example, the optical properties of GaP are the most thoroughly studied of all the indirect gap semiconductors, yet for 11 years the accepted value of E_x (Dean and Thomas 1966) was a factor of two low (Carter et al. 1977, Sturge et al. 1978). Recently it has been shown that when the "camel's back" structure of the conduction band minimum in GaP is taken into account, good agreement can be obtained with the complex exciton absorption data, including the effect of applied stress (Capizzi et al. 1977, Humphreys et al. 1978, Glinskii et al. 1979).

Excitons can be associated with any critical point in the band structure, the criterion for their existence being that $\nabla_k E_e = \nabla_k E_h$; i.e. that the electron and hole have equal group velocity. Saddle-point excitons are discussed by Phillips (1966).

2.2. Exciton complexes

The analogy between the Wannier exciton and the hydrogen atom was carried a step further by Lampert (1958) and Moskalenko (1958). They pointed out the possibility of exciton complexes analogous to the hydrogen or positronium molecules (the "biexciton") and to the H_2^+ ion ("trions", of either charge).

These complexes should always exist at low temperatures and high exciton density, but it has proved difficult to establish their existence because of the previously unexpected appearance, in most semiconductors, of the electron–hole liquid (Keldysh 1968, Asnin et al. 1968) which is the subject of a separate volume in this series. Definitive evidence for free biexcitons has only recently been obtained (see chs. 9 and 11); the existence of trions is still problematical.

When bound to an impurity, exciton complexes are more stable and, because of their sharp line spectra, much easier to identify. Not only biexcitons but multi-exciton complexes in indirect semiconductors have a well-established spectroscopy, which is described in ch. 10.

Exciton complexes (biexcitons in particular) also exist in the Frenkel case: see chs. 14 and 18.

3. *Optical effects of excitons in pure crystals*

The exciton is a short lived excited state of the crystal, typically a few eV above the ground state, so that virtually the only way of studying it is by optical spectroscopy.

In this section we consider, again on a very elementary level, how excitons show up in optical spectra.

3.1. Allowed transitions

The exciton energy level scheme is an infinity of overlapping continua; nevertheless, its optical transitions create excitons of well-defined energy and can therefore be sharp. This is a consequence of momentum conservation. Since the ground state of the crystal has $K = 0$, a photon of energy $\hbar\omega$ can only create an exciton with $K = n_r\omega/c$ where n_r is the refractive index. For $\hbar\omega \sim \mathrm{eV}$, $\omega/c \sim 10^5\,\mathrm{cm}^{-1}$, so that for $m \sim m_0$, $\hbar^2K^2/2M \sim 10^{-5}\,\mathrm{eV}$. Thus the only excitons which can be directly created by a photon have a well-defined energy, and their kinetic energy is negligible. Transitions involving the creation of free electron–hole pairs, on the other hand, give continuous spectra for $E > E_g$, since momentum conservation only puts one condition on $\boldsymbol{k}_e$ and $\boldsymbol{k}_h$, leaving the relative motion of the electron and hole free to be specified by energy conservation. For the exciton, the requirement that the electron and hole stay together eliminates this degree of freedom. Thus, from the Wannier point of view, the existence of sharp optical transitions in a pure crystal is evidence for the existence of excitons.

There are, of course, many selection rules governing the optical transitions other than $K = n_r\omega/c$, which is peculiar to excitons. An allowed transition, excitonic or not, must be direct, i.e. vertical on the $E(\boldsymbol{k})$ diagram. Since excitons are associated with extrema, these extrema must have the same $\boldsymbol{k}_m$ (see fig. 1a). For an allowed transition, the interband transition matrix element $p_{cv} = \langle u_0^c(\boldsymbol{r}_e)|e\boldsymbol{p}|u_0^v(\boldsymbol{r}_h)\rangle \neq 0$, where $e\boldsymbol{p}$ is the electric dipole operator, and the u's are the periodic parts of the conduction and valence band Bloch functions at $k = k_m$; and $F_{nlm}(0) \neq 0$, since the electron and hole coincide at the moment of their creation. Thus only transitions to hydrogenic S states are allowed (Elliot 1957) as in the analogous case of positronium (Sakharov 1948).

While the most dramatic effect of the electron–hole interaction is the creation of sharp exciton states, it also has a profound effect on the intensity of band-to-band transitions (Elliott 1957). By encouraging the electron and hole to stay together, it greatly enhances the transition probability in the vicinity of the band edge, converting the $(E - E_g)^{1/2}$ dependence, expected for non-interacting particles, into a step function, which merges into the absorption due to discrete levels.

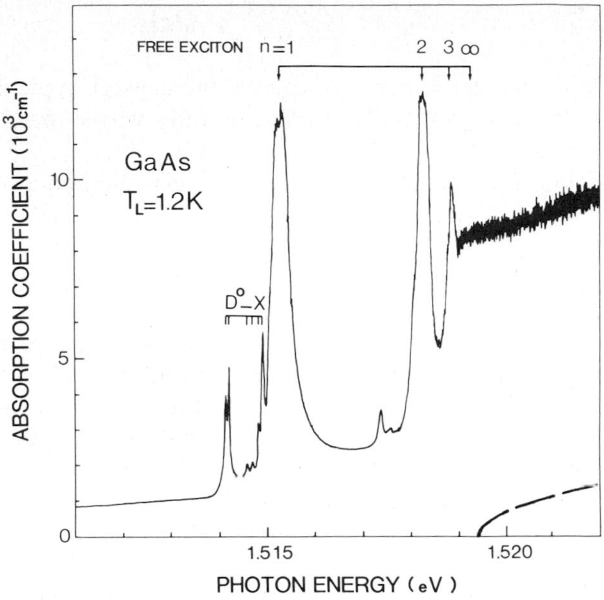

Fig. 2. Absorption spectrum at 1.2 K of ultra-pure GaAs near the band edge (Weisbuch 1977). The $n = 1, 2, 3$ free exciton peaks are shown; also the bandgap E_g, determined by extrapolation to $n = \infty$, and impurity lines (D_0X) from the excitons bound to $\sim 10^{15}\ \mathrm{cm}^{-3}$ donors. (The rise at high energy is due to substrate absorption.) The dashed line shows the $(E - E_g)^{1/2}$ behavior expected in the absence of electron–hole interaction (the absolute magnitude is chosen to fit the absorption far from the band edge).

Figure 2 shows the absorption spectrum of a very thin, very pure epitaxial crystal of GaAs. Allowed transitions creating $n = 1$, 2 and 3 excitons are clearly seen, followed by the envelope of $n > 3$ excited states which leads smoothly into the continuum. The difference between the data and the dashed line, calculated neglecting the electron–hole interaction, illustrates the importance of taking excitonic effects into account.

3.2. Forbidden transitions

Transitions forbidden by the above selection rules become weakly allowed under various perturbations. Such transitions have played an important historical role, since they are often weak enough to be resolved in samples of reasonable thickness. The $\boldsymbol{k}_m = 0$ selection rule, forbidding indirect transitions, is relaxed by the electron–phonon coupling, since a phonon can take up the unwanted momentum. A detailed account of the weak coupling case, as it applies to indirect semiconductors like Ge and Si, is given by McLean (1960). The absorption spectrum is a continuum (since $\boldsymbol{K}$ can now take on any value) but there are discontinuities in the absorption derivative which allow determination of the exciton and phonon energies.

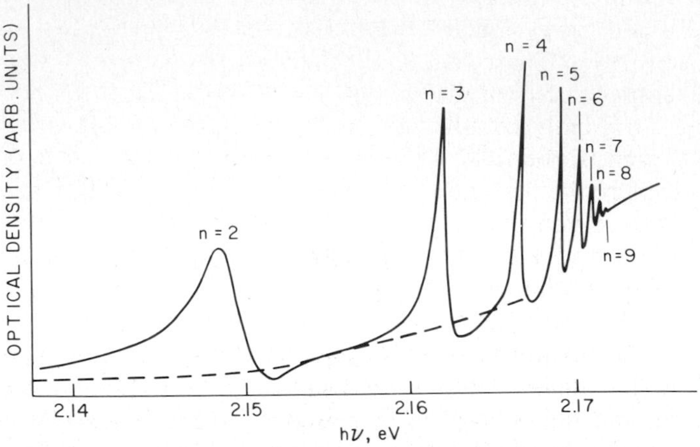

Fig. 3. Absorption spectrum of the "yellow" exciton in Cu_2O at 1.8 K. (Shindo et al. 1974). Since transitions are allowed only to p states of this exciton, the $n = 1$ exciton is forbidden. The asymmetry of the lines is due to interference with the background continuum associated with a lower lying band edge (Phillips 1966).

The $\boldsymbol{K}$ and $\boldsymbol{k}_m$ selection rules are consequences of translational symmetry. The hope of seeing distinguishable excitonic effects in amorphous materials might therefore be thought rather dim*. Most crystals have some degree of rotational as well as translational symmetry; for example, semiconductors are usually cubic, belonging to the O_h or T_d point group, or hexagonal. The zone center has the full point group symmetry, so that $k = 0$ Bloch functions transform as, and are usually labelled by, a representation Γ_i of that group**. Other points in the zone usually have lower symmetry.

According to the symmetry of the u_k, an electric dipole transition may or may not be allowed. If it is forbidden, transitions to exciton states with $l \neq 0$ may still be weakly allowed. For example, if the conduction and valence band edges of a centro-symmetric crystal have the same parity, the transition to S states is electric-dipole forbidden, but to hydrogenic p states it will be weakly allowed, since the overall final state has p-like symmetry. The classic case of the "yellow" exciton in Cu_2O, shown in fig. 3, is an example of this. Note the weak absorption compared to GaAs and the absence of a transition to the 1s hydrogenic state.

Wannier excitons were first identified in the optical absorption spectra of

* See, however, Mott (1977), Bonch–Bruevich and Iskra (1975) and sect. 4.3.

** See the Appendix A to Lax (1974) for the various labelling schemes common in the literature.

Cu_2O (Hayashi 1952, Hayashi and Katsuki 1952*, Gross et al. 1952, 1953) and of Cu and heavy metal halides (Nikitine et al. 1954, 1956, Gross and Kaplyansky 1956). There had in fact been many previous observations of excitons in the spectra of alkali halides, (Hilsch and Pohl 1928, 1929, 1930) but these are greatly complicated by the electron–phonon interaction and have only come to be fully understood recently (see ch. 12). Cu_2O and the Cu halides were the first inorganics to show "no-phonon" exciton lines: i.e. transitions in which excitons are created without the participation of a phonon. For a review of this early work, see Gross (1962), Nikitine (1969).

Unfortunately, the crystals available at that time were of doubtful quality, and it was open to sceptics to attribute the observed spectra to impurities and stoichiometric defects (Kokhahenko 1952, 1954, Dresselhaus 1956). The first unassailable evidence for Wannier excitons was obtained by Roberts and his colleagues (Macfarlane et al. 1958, McLean 1960); a sharp absorption line was observed just below the direct band gap in very thin samples of Ge known to be of extremely high purity. Subsequent work on the absorption and reflectance of Ge, Si, and other semiconductors, with and without a magnetic field, has demonstrated that excitonic effects always dominate the low temperature spectra of pure semiconducting crystals near band edges, direct or indirect.

3.3. Excitonic polaritons

The fact that an allowed exciton must be p-like, i.e. an orbital triplet transforming like a polar vector, has another very important consequence, as was first pointed out by Pekar (1957).

The exciton has a polarization $\boldsymbol{P} \exp(\mathrm{i}\boldsymbol{K} \cdot \boldsymbol{R})$ associated with it, and, like an optic phonon, can be longitudinal ($\boldsymbol{P} \| \boldsymbol{K}$) or transverse ($\boldsymbol{P} \perp \boldsymbol{K}$). Only the transverse exciton can interact with an electromagnetic field. This (dipolar) interaction splits the exciton into longitudinal and transverse branches and produces non-analytic behavior as $K \to 0$ (since at $K = 0$ the distinction between longitude and transverse ceases to exist). The effect of this interaction is illustrated in fig. 4, which is a schematic $E(K)$ diagram for the longitudinal (L) and transverse (T) excitons, and for the photon (the dashed "light line"). As in the analogous case of the optical phonon (Tolpygo 1950, Huang 1951, Henry and Hopfield 1965) the photon and the transverse exciton mix in the crossover region, losing their identity in a combined particle called the "polariton". Figure 4 differs from the corresponding phonon diagram in one vitally important way: because of the kinetic energy term $\hbar^2K^2/2M$, the exciton branches curve upwards, so that the well-known

* Hayashi and Katsuki (1950) had previously observed sharp lines in Cu_2O which were subsequently attributed to excitons.

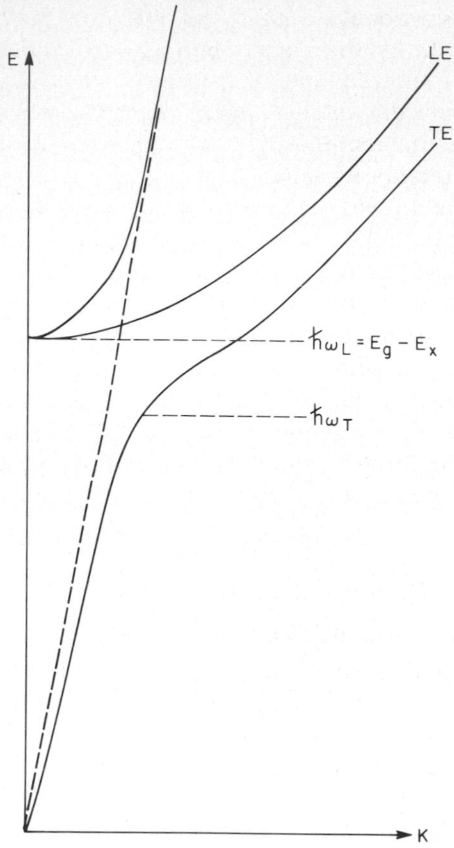

Fig. 4. $E(\boldsymbol{K})$ diagram for an allowed exciton polariton. ---$E = \hbar Kc/n_r$, where the contribution to n_r from the exciton itself is excluded. The size of this contribution and of the corresponding longitudinal–transverse splitting, $\hbar(\omega_L - \omega_T)$, is greatly exaggerated for clarity.

"forbidden band" between the LO and TO phonons does not exist in the exciton case. The curvature leads to a k-dependent non-local dielectric function, a phenomenon known as "spatial dispersion"* (see ch. 2).

The advent of the tunable laser as a spectroscopic tool has made possible the direct exploration of this $E(K)$ diagram by k-vector spectroscopy. This is two-photon spectroscopy (non-linear absorption or Raman or Brillouin scattering) in which ΔE and $\Delta \boldsymbol{K}$ can be varied independently; this work is described in ch. 3. Other observable polariton effects are discussed in chs. 2 and 4.

While k-vector spectroscopy has established that an $E(\boldsymbol{K})$ diagram

* Strictly this should be called "wave-vector dispersion" to parallel "frequency dispersion".

exists, it says nothing about the phase coherence of the excitons apart from the lower limit given by the linewidths. Hopfield and Thomas (1960) and Gross et al. (1961) showed that in the non-centrosymmetric crystal CdS, the absorption spectrum in a magnetic field $\boldsymbol{H}$ depends on whether $\boldsymbol{K}$ (determined by the incident light direction) is parallel or anti-parallel to $\boldsymbol{H}$. Thus the exciton retains its initial $\boldsymbol{K}$ for $\gtrsim 10^{-11}$ s (the inverse linewidth). Thomas and Hopfield (1961) measured the exciton velocity, using crossed electric and magnetic fields. Recently Chase et al. (1979) established very elegantly the existence of $\boldsymbol{K}$ for biexcitons in CuCl. These were created by two-photon absorption in two different ways: with plane-polarized light in which two photons of opposite circular polarization and equal $\boldsymbol{K}$ combine to produce a biexciton with wave vector $\boldsymbol{K}_2 = 2\boldsymbol{K}$; and with two counter-propagating circularly polarized beams, in which the two photons have opposite $\boldsymbol{K}$ to give $\boldsymbol{K}_2 = 0$. A luminescent transition forbidden at $\boldsymbol{K}_2 = 0$ is observed only in the former configuration, demonstrating that the biexciton indeed has a well-defined $\boldsymbol{K}_2$ for the order of its lifetime ($\sim 10^{-9}$ s).

3.4. *Effect of external fields: electric, magnetic and strain*

Because of its large size the Wannier exciton is extremely polarizable. Consequently an applied electric or magnetic field has a pronounced effect on its energy and wave functions, as does crystal strain. From general symmetry considerations it is possible to make useful predictions of the splitting of an exciton associated with a particular band extremum. Magneto-optics, piezo-reflectance and electro-reflectance have become powerful tools for the experimental elucidation of band structure and excitonic effects in semiconductors (Cardona 1969, Aspnes 1980). The splitting of optical transitions by applied stress and by magnetic fields has proved to be particularly useful in the identification of bound excitons (see next section and Dean and Herbert 1979). The quantitative theory of electric field effects turns out to be extremely difficult, and is reviewed in ch. 7 (for reviews of magnetic and stress effects, see the bibliography).

4. *Impurity effects: bound excitons*

So far we have considered only perfect crystals. Real crystals contain impurities and defects, and even in a semiconductor such as GaAs, which, because of its technical importance is available with an impurity content of less than 10^{15} cm^{-3}, these play a crucial role. In particular, as we shall see, excitons bound to impurities tend to dominate the optical emission spectra at low temperatures.

4.1. Bound Frenkel and charge transfer excitons

We consider an impurity or defect at site j which has an excited state at $E_1' = E_1 + \Delta$. If $|\Delta|$ is large enough (for the precise meaning of "large enough" see sect. 2.2 of ch. 18, also Sommer and Jortner 1969) a localized exciton state is produced which is centered on the impurity site and whose energy is outside the exciton band. If $\Delta < 0$ this state is below the band and a finite activation energy is required for the exciton to escape: hence the name "bound" exciton*. A "free" exciton, created by absorption into the exciton band of the host crystal, moves from site to site in the host until it either decays radiatively or is trapped at an impurity or defect site, where (at low temperatures) it will remain until it decays, emitting a photon characteristic of the bound, not of the free exciton.

The exciton motion may be diffusive (total loss of phase memory after each intersite transfer) or ballistic (perfect phase coherence: no scattering) or, more usually, something in between. In either case, typical intersite jump times for an allowed molecular exciton are $\lesssim 10^{-13}$ s if the localizing effects of electron–phonon coupling and disorder can be neglected. Since radiative lifetimes are $\gtrsim 10^{-9}$ s, an impurity present at a concentration of only 10^{-4} has a very good chance of capturing the exciton and thereby appearing in the emission spectrum, while only making a negligible contribution to the absorption (Broude et al. 1959).

In certain cases the impurity center may be a "reaction center", i.e., one whose excitation initiates a chemical reaction. The host then acts as an "antenna", collecting light energy and tunneling it to the reaction center. This effect is of great importance in biology; it is the initial step in photosynthesis and is the subject of ch. 17.

The existence of this energy transfer to luminescent or reaction centers is not, in itself, evidence for the existence of mobile excitons. Direct transfer of excitation, from a "donor"** which has absorbed a photon to an emitting "acceptor" also occurs, and careful study of the kinetics is necessary to distinguish these very different processes. Since the time scale is usually in the picosecond range, such study has only recently become technically feasible and is still subject to serious pitfalls.

Even if the energy difference $|\Delta|$ is not large enough to produce a localized state, an impurity can still have a marked effect on the optical spectrum of the exciton by breaking translational symmetry and hence the $\boldsymbol{K} = 0$ selection rule. Because of this, in the vicinity of the impurity an

* Knox (1963) remarks that in the Frenkel case "bound exciton" is simply another name for an excited state of the trapping center; however, this state is not necessarily localized on the center itself, but spreads into the surrounding crystal, even in the Frenkel case.

**This usage of "donor" and "acceptor" is not to be confused with the semiconductor usage (see note next page).

exciton of any $\boldsymbol{K}$ can be created by a photon; the use of this fact to determine the exciton density of states is described in ch. 18.

Probably the first identification of a bound exciton was by Seitz (1954) who assigned the α and β absorption bands of irradiated alkali halides to excitons bound to defects. Excitons bound to impurities in silver halides were reported by Tsukashi and Kanzaki (1971).

4.2. Bound Wannier excitons

Turning to the Wannier exciton, we ask if corresponding impurity bound states can exist for it. Lampert (1958) pointed out the analogy between the hydrogen molecule and an exciton interacting with a shallow neutral donor or acceptor*, which can, like the exciton, be described by the hydrogenic model. Just as the hydrogen atoms bind to each other in the H_2 molecule, so should the exciton be bound to the neutral impurity. Calculation of the binding energy has proved difficult, since the Born–Oppenheimer approximation, which separates electronic from nuclear motion in the H_2 case, cannot be used here. However, it has been established (Hopfield 1964, Herbert 1977) that the exciton can be bound by a neutral donor or acceptor for all possible m_e/m_h. Since the recognition of bound excitons in the emission spectrum of Si doped with As by Haynes (1960)**, hundreds of cases have been identified.

The bound Wannier exciton dominates the emission even more than in the Frenkel case; especially in indirect gap semiconductors, even though they are usually much purer than molecular crystals. The large exciton radius a_x leads to a correspondingly large cross section σ ($\sim a_x^2 \sim 10^{-12}\,\mathrm{cm}^2$) for capture by an impurity. A free exciton at 2 K has a velocity $v \sim 10^6\,\mathrm{cm/s}$, so for an impurity concentration $N \sim 10^{15}\,\mathrm{cm}^{-3}$ its lifetime before capture is expected to be $(N\sigma v)^{-1} \sim 10^{-9}\,\mathrm{s}$. This is comparable to the radiative lifetime in a direct gap semiconductor and very much shorter than that in an indirect one.

The large size of the bound Wannier exciton has another important consequence: a "giant" oscillator strength, since all the free exciton dipole strength within a volume $\sim a_x^3$ is concentrated in one line (Rashba and Gurgeneshvili 1962). The short lifetime thus predicted (~ 0.5 ns in CdS) was observed by Henry and Nassau (1970). The strong bound exciton line in fig. 2, which arises from only $\sim 10^{15}\,\mathrm{cm}^{-3}$ donors, is also illustrative of this effect. A qualitatively similar result is obtained in the Frenkel case if the

* In semiconductor language, a "donor" is a neutral impurity atom, such as P in Si, which can easily give up an electron to the conduction band; an "acceptor" (e.g. B in Si) can accept one from the valence band, creating a hole.

** Bound exciton emission from semiconductors had been seen earlier, but not identified as such. See, for example, Grillot et al. (1956), Arkhangelskaya and Feofilov (1956).

bound exciton is close to the bottom of the free exciton band (Rashba 1957) since, in such a case, the exciton is not strictly localized on the impurity but spreads over many of the neighboring atoms.

In indirect semiconductors, the radiative lifetime is long, since the transition to the ground state is forbidden, nevertheless short lifetimes ($\sim 10^{-8}$ s) and low luminescent efficiencies are found for donor and acceptor bound excitons. This is the result of non-radiative Auger decay; since there are three particles present, two can readily recombine and give up the band-gap energy to the kinetic energy of the third.

Lampert also predicted that excitons would bind to ionized impurities. Such bound excitons have been reported in II–VI compounds (see Reynolds and Collins 1981) and in SiC*. The theoretical conditions for binding are discussed by Dean and Herbert (1979).

Oddly enough, the bound Wannier exciton which is most closely analogous to the bound Frenkel exciton discussed above was not predicted theoretically, and its experimental discovery by Aten and Haanstra (1964) and Thomas et al. (1965) came as a surprise. This is the exciton bound to an "isoelectronic trap" (a more precise name would be "isovalent center"), for example, N substituting for P in GaP or O for Te in ZnTe. Since the valency of the impurity atom is the same as that of the host atom it replaces, there is no long range Coulomb field and hydrogenic bound states do not exist. The binding arises from the short range potential; in the case cited the impurity is strongly electronegative, i.e. attractive to electrons, relative to the host. If this potential is strong enough a bound state is formed; even if it is not, its short range makes it very effective in breaking the $\boldsymbol{K}$ and $\boldsymbol{k}_{\mathrm{m}}$ selection rules, making possible the optical creation and decay of indirect excitons without the participation of phonons. Furthermore, there is no third particle, so that Auger recombination is not possible; consequently the luminescent efficiency is usually high.

The observable behavior of bound excitons in semiconductors, particularly indirect ones, is multifarious, and the literature correspondingly, multitudinous. Fortunately, the subject has recently been covered in a masterly review by Dean and Herbert (1979), and for this reason we omit it from this volume.

The generalization that impurity effects dominate the low temperature emission spectra except in material of extreme purity does not always apply. "Self-trapping" (see ch. 13) by preventing diffusion of the exciton, permits it to decay in its own way. Under intense excitation, the impurity centers in a semiconductor can be saturated, and intrinsic effects take over. Of course all sorts of high level effects will also appear: biexcitons,

* Choyke et al. (1974). The α-band of irradiated alkali halides is thought to arise from an exciton bound to an ionized vacancy (Seitz 1954, Schulman and Compton 1962).

electron–hole droplets, stimulated emission, etc., and it is difficult to study the isolated free exciton this way.

In systems of low dimensionality (1-D or 2-D) intrinsic exciton luminescence is easier to obtain; for example, in 2-D naphthalene (Shpak and Sheka 1960). This may be a consequence of localization of the intrinsic exciton by disorder (see sect. 4.3). If this is the correct explanation, the recent observation of intrinsic exciton luminescence 2 K in a two-dimensional *semiconducting* system (Weisbuch et al. 1981) is of great interest. This system is a thin (~200 Å) layer of GaAs sandwiched between two layers of the wider band gap $Ga_{1-x}Al_xAs$. Motion normal to the layer is quantized, so that in the lowest state motion is two-dimensional. Although no purer than bulk GaAs, which shows strong bound excitons and no intrinsic luminescence at 2 K, such a layer emits at exactly the lowest free exciton energy. Furthermore the spin-polarization (see ch. 6) is just that predicted for a free, but not for a bound, exciton.

4.3. Disorder

If the impurity content is sufficiently large, transfer of excitation directly between bound states becomes possible without activation into the free exciton band. Since impurities are randomly distributed the problem is one of diffusion in a random lattice, treated first classically by Broadbent and Hammersley (1957) and quantum mechanically by Anderson (1958), and subsequently the subject of a vast literature (see, for example, Kopelman 1976, Mott and Davis 1979). A similar problem arises if the host itself is disordered. The results depend critically on dimensionality. Many molecular crystals are essentially one-dimensional, the molecules forming chains between which the interaction is too weak to matter. In a one-dimensional system any deviation from perfect order, however weak, is sufficient to produce localization at 0 K (Berezinski 1973); some real systems showing such behavior are described in ch. 15. In three dimensions localized and delocalized states are theoretically possible. The criteria which determine which will occur in any particular case are still a subject of debate, and are beyond the scope of this book. Theoretically, if dipolar (long range) interactions are the principal cause of intermolecular transfer, as is usually the case for allowed excitons, localization cannot occur at 0 K (Anderson 1958); however, this has yet to be established experimentally.

Another important theoretical prediction, still under debate, is the existence of a "mobility edge" in systems with short range interactions for which Anderson localization is theoretically possible (Mott 1966, Cohen et al. 1969, Economou and Cohen 1972). Disorder spreads a sharp state out into a continuum, and gives a sharp band edge a tail. Mott argues that states sufficiently far down in the tail are localized, while those in the band are not; that a localized state cannot be degenerate with a delocalized one; hence, that there must be a well-defined energy, the "mobility edge", separating the two. There is no generally accepted experimental evidence for its existence in any disordered system; certainly not one involving excitons, although this is a very active area of research. Most effort has been directed towards elucidating energy transfer processes in disordered molecular systems. The luminescence of dilute probe molecules (acceptors) is only strongly excited when energy transfer is possible amongst the randomly distributed donor molecules. The most convincing results have been obtained in the 2-D system naphthalene, containing isotopic impurities where the excitons are strongly localized and their motion can be accounted for in terms of classical percolation theory (Kopelman 1976, 1980, Argyrakis and Kopelman 1978). These will be discussed in another volume in this series (Kopelman, unpublished). (Another view is put forward in chapter 15 of the present book.) The new technique of transient grating spectroscopy (Eichler 1977, 1978, Dlott et al. 1978) also shows promise of demonstrating exciton motion and localization directly.

Semiconducting alloys, in which the lattice sites are occupied randomly by different (though similar) constituents, would appear to be promising subjects for the study of disorder effects. However, no uncontrovertible results have been reported so far, although the possible observation of an exciton localized by alloy fluctuations in $GaAs_{1-x}P_x$ has been reported by Lai and Klein (1980). Most work on alloy semiconductors was concentrated on the fact that their bandgaps can be varied continuously by changing the composition. This fact, of obvious technical importance, is very useful in research, particularly on impurity states, since it provides an extra parameter to vary. Recent work in this area is reviewed in ch. 8.

5. *Interaction of excitons with phonons and other crystal excitations*

5.1. The deformation potential and the Fröhlich interactions

The title of Frenkel's original (1931) paper is "On the transformation of light into heat by solids" and the bulk of it is devoted to what we would now call the exciton–phonon interaction. His main point was that the excited states of isolated atoms cannot dispose of their energy non-radiatively, while it was obvious from experiment that those of solids could. Excitons can be strongly coupled to phonons, since their energy is a function of the inter-atomic interaction, which is very sensitive to lattice distortion. This "deformation potential" coupling is very large for Wannier excitons with radii a_x of a few inter-atomic spacings (e.g. those in rare gases and alkali halides). It becomes progressively weaker as a_x increases, since coupling is only possible to phonons with $k < \sim a_x^{-1}$, and the number of these varies as a_x^3*.

Another type of exciton–phonon coupling, very important in alkali halides and dominant for Wannier excitons in polar semiconductors, comes from the Fröhlich interaction. A long wavelength LO phonon in a polar crystal produces an electric field proportional to the effective ionic charge. As remarked in sect. 2, the large size of a Wannier exciton makes it highly polarizable, so the LO phonons couple strongly to it. The optical consequences of this coupling are discussed in ch. 5.

A free or large radius bound exciton can only couple to phonons of well-defined $\boldsymbol{k}(\boldsymbol{k} = 0$ in the case of a direct exciton, and the momentum-conserving phonon with $\boldsymbol{k} = \boldsymbol{k}_m$ in the indirect case). Consequently the "phonon sidebands" (transitions in which a phonon is created or destroyed with the exciton) tend to be sharp, and there is a well-defined "no-phonon" line, such as those shown in fig. 2. A small exciton localized in a region of the order of lattice spacing by disorder, impurities or self-trapping (see below) can couple strongly to all phonons and consequently has broad bands. Non-radiative transitions, in which all the excitation energy is dissipated as phonons before the exciton has time to radiate, are then possible and often dominant, as Frenkel realized (whereas for large excitons, capture by impurities and subsequent Auger decay is the dominant loss mechanism). It is probable, though not yet established, that in the absence of self-trapping almost all non-radiative exciton decay is extrinsic.

5.2. *Dressed excitons: self trapping*

When the exciton–phonon coupling is strong there is another effect, theoretically more dramatic, though experimentally difficult to detect. This is the "dressing" of the exciton by phonons, leading in the strong coupling limit to "self-trapping". This is easiest to understand in the Frenkel model. If the local state Φ_j from which the exciton is made up (eq. (2)) is strongly coupled to phonons, its energy will be substantially reduced by a local lattice distortion. Thus the lowest energy excitation is "dressed" by this distortion which must be dragged along with it if it is to move through the crystal. (This is analogous to the "small polaron", which is a dressed electron.) The transfer matrix element is multiplied by vibrational overlap factor $\langle \chi_j \mid \chi_{j+1} \rangle$, where χ_j is the vibrational wave function corresponding to the lowest distorted state on site j. This factor, and hence the exciton bandwidth, is roughly proportional (at 0 K) to $\exp(-E_D/\hbar\bar{\omega})$, where $\bar{\omega}$ is the

* Self-trapping (see sect. 5.3 and ch. 13) or binding to a short range potential (as in an isoelectronic trap) can introduce large $\boldsymbol{k}$ components into the exciton wave function, and thereby permit coupling to large $\boldsymbol{k}$ phonons.

mean phonon frequency and E_D the energy gained by local distortion (Holstein 1959). Thus, if the distortion is large, the bandwidth can become so small that the exciton cannot move at all, it is "self-trapped". This phenomenon is of great importance in the spectra of alkali halides and rare gases, and is treated fully in chs. 12 and 13. It is also present in the magnetic crystals discussed in ch. 14. It is primarily responsible for the success in these crystals of ordinary crystal field theory, which neglects excitonic effects and treats atomic excited states, even as regards their phonon interactions, as localized impurity-like states.

Similar results, analogous to the "large polaron", are obtained for the Wannier exciton. The Fröhlich interaction is usually dominant in this case.

In semiconductors the exciton–phonon interaction is usually not strong enough to produce self-trapping; nevertheless it can still produce "bound" but mobile exciton–phonon complexes (Liang and Yoffe 1968, Toyozawa and Hermanson 1968, Levinson and Rashba 1973) which may themselves bind to impurities (Manchon and Dean 1970). Such states, bound and virtual, are of great importance in the temperature-dependent lineshape ("Urbach–Martienssen tail") of exciton transitions (Sumi and Toyozawa 1971).

5.3. Other exciton interactions

In a magnetic crystal below its Curie temperature another quantum of excitation exists, namely the magnon or spin wave. (This itself can be regarded as a special sort of exciton; see ch. 14). An atomic excited state, particularly if it has a different spin from the ground state, represents a very substantial perturbation from the magnon's point of view. Hence exciton–magnon coupling can be expected to be very strong.

At high exciton densities exciton–exciton scattering becomes very important (Adamowski and Bednarek 1979). This can give rise to processes in which one exciton is raised to a higher excited state, or ionized, while the other decays, producing a photon of correspondingly lower energy (Büttner 1973, Bille 1973). It is probably an important process in the thermalization of excitons but this has been little studied. It limits the lifetime of polaritons and hence the effectiveness of non-linear absorption (Fröhlich et al. 1970) as a tool in k-vector spectroscopy (see ch. 3).

Even less studied is the interaction of excitons with electrons and holes (Kravchenko et al. 1979, Weisbuch 1978, Elkomoss et al. 1979). This is probably important in the early stages of formation of electron–hole drops, but is difficult to study in isolation, since at temperatures low enough for the exciton to exist, electrons and holes are likely to be bound to impurities. However, in such semiconductors as GaAs and InSb, whose electrons have very low mass, even a small number of donors ($\sim 10^{14}$ in InSb) overlap sufficiently to produce an impurity band, in which the electron remains (more or less) free at 0 K. These could perhaps form model systems for the experimental study of the exciton–electron interaction.

6. Conclusion

Work on Wannier excitons has in the past tended to proceed almost in ignorance of that on Frenkel excitons, and vice versa. This is in part an unfortunate consequence of the artificial educational and semantic barriers which often separate chemists from physicists. However, there have recently been encouraging signs of a reapproachement; the field of excitons, vast as it is, is coming to be perceived once again as a unity. For example, the difficult theoretical and experimental problems associated with exciton localization and transport engage the interests of both solid state physicists and organic chemists, and each is discovering that he has much to learn from the other. Self-trapping and the formation of exciton–phonon bound states are now recognized phenomena in all branches of exciton physics. Several new experimental techniques, stemming principally from recent advances in laser technology, are proving valuable in the study of all types of excitons. Selectively excited luminescence, resonant light scattering, optical orientation, picosecond spectroscopy and transient gratings come to mind. These experiments have in turn stimulated new theoretical work.

One of the aims of this book is to emphasize this underlying unity of the subject. The editors hope that it will encourage specialists in particular areas of exciton physics to learn from, and contribute to, other areas where the questions at issue may be similar, though sometimes couched in unfamiliar terms.

Acknowledgement

I am most grateful to Dr. E.I. Rashba and Dr. E.O. Kane for helpful comments.

References

Abrikosov, A.A., 1978, J. Less Common Metals, **62**, 451.
Adamowski, J. and S. Bednarek, 1979, Solid State Electron. **22**, 33.
Allender, D., J. Bray and J. Bardeen, 1973a, Phys. Rev. **B7**, 1973; 1973b, ibid **B8**, 4433.
Altarelli, M. and N.O. Lipari, 1977, Phys. Rev. **B15**, 4898.
Anderson, P.W., 1958, Phys. Rev. **109**, 1492.
Argyrakis, P. and R. Kopelman, 1978, J. Theor. Biol. **73**, 205.

Arkhangelskeia, V.A. and P.P. Feofilov, 1956, J. Phys. Rad. **17**, 824.
Asnin, V.M. and A.A. Rogachev, 1968, Zh. Eksp. Teor. Fiz. Pis'ma Red. **7**, 464; JETP Lett. **7**, 360.
Aspnes, D.E., 1980, in: Optical Properties of Solids, ed. M. Balkanski (North-Holland, Amsterdam) p. 109.
Aten, A.C. and J.H. Haanstra, 1964, Phys. Lett. **11**, 97.
Baldereschi, A. and N.O. Lipari, 1971, Phys. Rev. **B3**, 439.
Baldini, G., 1962, Phys. Rev. **128**, 1562.
Bardeen, J., 1978, J. Less Common Metals, **62**, 447.
Berezinski, V., 1973, Zh. Eksp. Teor. Fiz. **65**, 1251; Sov. Phys. JETP **38**, 620.
Bille, J., 1973, Festkörperprobleme, **13**, 111.
Bonch-Bruevich, V.L. and V.D. Iskra, 1975, Phys. Status Solidi, **B68**, 369.
Brandt, N.B., S.V. Kurshinikov, A.P. Rusakov and V.M. Semenov, 1978, Zh. Eksp. Teor. Fiz. Pis'ma Red. **27**, 37; JETP Lett. **27**, 33.
Broadbent, S.R. and J.M. Hammersley, 1957. Proc. Cambridge Phil. Soc. **53**, 629.
Broude, V.L., 1961, Usp. Fiz. Nauk. **74**, 577; Sov. Phys. Usp. **4**, 514.
Broude, V.L., V.S. Medvedev and A.F. Prikhot'ko, 1951, Zh. Eksp. Teor. Fiz. **21**, 665.
Broude, V.L., A.F. Prikhot'ko and E.I. Rashba, 1959, Usp. Fiz. Nauk. **67**, 99; Sov. Phys. Usp. **2**, 38.
Büttner, 1973, Festkörperprobleme **13**, 145.
Cardona, M., 1969, Optical Modulation Spectroscopy of Solids (Academic, New York).
Carter, A.C., P.J. Dean, M.S. Skolnick and R.A. Stradling, 1977, J. Phys. C. **10**, 5111.
Chase, L.L., H.N. Peyghambrian, G. Grynberg and A. Mysyrowicz, 1979, Phys. Rev. Lett. **42**, 1231.
Cho, K., 1976, Phys. Rev. **B14**, 4463.
Cho, K., 1979, Top. Curr. Phys. **14**, (Excitons) ed. K. Cho (Springer, Berlin) p. 15.
Choyke, W.J., L. Patrick and P.J. Dean, 1974, Phys. Rev. **B10**, 2554.
Chu, C.W., A.P. Rusakov, S. Huang, S. Early, T.H. Geballe and C.Y. Huang, 1978, Phys. Rev. **B18**, 2116.
Cohen, M.H., H. Fritsche and S.R. Ovshinsky, 1969, Phys. Rev. Lett. **22**, 1065.
Davydov, A.S., 1948, Zh. Eksp. Teor. Fiz. **18**, 210.
Davydov, A.S., 1951. Tr. Inst. Fiz. Akad. Nauk. Ukr. S.S.R. **3**, 36.
Dean, P.J. and D.G. Thomas, 1966, Phys. Rev. **150**, 691.
Dean, P.J. and D.C. Herbert, 1979, Top. Curr. Phys. **14** (Excitons) ed. K. Cho (Springer, Berlin) p. 55.
Dlott, D.D., M.D. Fayer, J. Salcedo and A.E. Siegman, 1978, Springer Ser. Chem. Phys. **4** (Picosecond Phenomena) eds. C.V. Shank et al. (Springer, Berlin) p. 240.
Dresselhaus, G., 1956, J. Phys. Chem. Solids, **1**, 15.
Economou, E.N. and M.H. Cohen, 1972, Phys. Rev. **B5** 2931.
Eichler, H.J., 1977, Optica Acta, **24**, 621.
Eichler, H.J., 1978, Festkörperprobleme, **18**, 241.
Elkomoss, S.G. and G. Munschy, 1979, J. Phys. Chem. Solids, **40**, 471.
Elliott, R.J., 1957, Phys. Rev. **108**, 1384.
Fowler, W. Beall, 1963, Phys. Rev. **132**, 1591.
Frenkel, J.I., 1931, Phys. Rev. **37**, 17; ibid, 1276.
Frenkel, J.I., 1936, Phys. Z. Sowjet Union, **9**, 158.
Fröhlich, D., E. Mohler and P. Wiesner, 1971, Phys. Rev. Lett. **26**, 554.
Ginzburg, V.L., 1964, Zh. Eksp. Teor. Fiz. **47**, 2316; Sov. Phys. JETP **20**, 1549.
Ginzburg, V.L., 1968, Contemp. Phys. **9**, 355.
Grillot, E., M. Grillot, P. Pesteil and A. Zmerli, 1956, Comptes Rend. Acad. Sci. Paris, **242**, 1794.
Gross, E.F., 1962, Usp. Fiz. Nauk **76**, 433; Sov. Phys. Usp. **5**, 195.

Gross, E.F. and A.A. Kaplyansky, 1956, J. Techn. Phys. USSR, **25**, 2061.
Gross, E.F. and N.A. Karryeff, 1952, Dokl. Akad. Nauk, USSR, **84**, 261, 471.
Gross, E.F. and B.P. Zakharchenya, 1953, ibid **90**, 745.
Gross, E.F., B.P. Zakharchenya and O.V. Konstantinov, 1961, Fiz. Tver. Tela **3**, 305; Sov. Phys. Solid State, **3**, 221.
Hayashi, M., 1952, J. Fac. Sci. Hokkaido Univ. (II) **4**, 107.
Hayashi, M. and K. Katsuki, 1950, J. Phys. Soc. Japan **5**, 381.
Hayashi, M. and K. Katsuki, 1952, J. Phys. Soc. Japan, **7**, 599.
Haynes, J.R., 1960, Phys. Rev. Lett. **4**, 361.
Henry, C.H. and J.J. Hopfield, 1965, Phys. Rev. Lett. **15**, 964.
Henry, C.H. and K. Nassau, 1970, Phys. Rev. **B1**, 1828.
Herbert, D.C., 1977, J. Phys. C **10** L, 131.
Hilsch, R. and R.W. Pohl, 1928, Z. Phys. **48**, 384.
Hilsch, R. and R.W. Pohl, 1929, ibid, **57**, 145.
Hilsch, R. and R.W. Pohl, 1930, ibid, **59**, 145.
Holstein, T., 1959, Ann. Phys. **8**, 343.
Hopfield, J.J., 1964, Proc. 7th Int. Conf. Phys. Semicond., Paris, ed. Dunod, p. 725.
Hopfield, J.J. and D.G. Thomas, 1960, Phys. Rev. Lett. **4**, 357.
Huang, Kun, 1951, Proc. Roy. Soc. (London) **206**, 352.
Inkson, J.C. and P.W. Anderson, 1973, Phys. Rev. **B8**, 4429.
Kane, E.O., 1978, Phys. Rev. **B18**, 6849.
Keldysh, L.V., 1968, Proc. 9th Int. Conf. Phys. Semicond., Moscow, p. 1303.
Knox, R., 1963, Theory of Excitons (Academic, New York).
Kohn, W., 1957, Adv. Solid State Phys. **5**, 257, eds. F. Seitz and D. Turnbull (Academic, New York).
Kokhahenko, P.N., 1952, Comptes Rend. Acad. Sci. USSR **85**, 543.
Kokhahenko, P.N., 1954, Zh. Eksp. Teor. Fiz. **26**, 121.
Kopelman, R., 1976, in: Top. Appl. Phys. **15** (Radiationless Processes) ed. F.K. Fong (Springer, Berlin) p. 298.
Kopelman, R., 1980, Organic Coatings and Plastics Chem. **42**, 371.
Kopelman, R., unpublished. To appear in "Molecular Spectroscopy" eds., V.M. Agranovich and R.M. Hochstrasser (this series).
Kravchenko, A.F., V.V. Nazintzev, A.P. Savchenko and A.S. Terekhov, 1979, Fiz. Tver. Tela. **21** 1551; Sov. Phys. Solid State, **21**, 894.
Kronenberger, A. and P. Pringsheim, 1926, Z. Phys. **40**, 75.
Lai, S. and M.V. Klein, 1980, Phys. Rev. Lett. **44**, 1087.
Lampert, M.A., 1958, Phys. Rev. Lett. **1**, 450.
Lax, M., 1974, Symmetry Principles in Solid State and Molecular Physics (Wiley, New York).
Levinson, I.B. and E.I. Rashba, 1973, Reps. Prog. Phys. **36**, 1499.
Liang, W.Y. and A.D. Yoffe, 1968, Phys. Rev. Lett. **20**, 59.
Lipari, N.O. and M. Altarelli, 1977, Phys. Rev. **B15**, 4883.
Little, W.A., 1964, Phys. Rev. **134A**, 1416.
Macfarlane, G.G., T.P. McLean, J. Quarrington and V. Roberts, 1958, Proc. Phys. Soc. (London) **71**, 863.
Manchon, D.D. and P.J. Dean, 1970, Proc. 12th Int. Conf. Physics Semicond., eds. S.P. Keller et al., (U.S.A.E.C., Oak Ridge), p. 720.
McLean, T.P., 1960, Progress in Semiconductors, eds. A.F. Gibson et al. (Hayward, London) **5**, 55.
Moskalenko, S.A., 1958, Opt. Spekt. **5**, 147.
Mott, N.F., 1938, Trans. Faraday Soc. **34**, 500.
Mott, N.F., 1966, Phil. Mag. **13**, 899.
Mott, N.F., 1977, Adv. Phys. **26**, 363.

Mott, N.F. and E.A. Davis, 1979, Electronic Properties of Non-crystalline materials, 2nd. ed, (Univ. Press, Oxford).
Nikitine, S., 1969, Optical Properties of Solids, eds. S. Nudelman and S.S. Mitra (Plenum, New York) p. 197.
Nikitine, S. and G. Perny, 1956, J. Phys. Rad. **17**, 1017.
Nikitine, S., G. Perny and M. Sieskind, 1954, J. Phys. Rad. **15**, 518.
Nikitine, S., M. Sieskind, L. Couture and G. Perny, 1954, Comptes Rend. Acad. Sci. Paris **238**, 1786.
Novikov, B.V., 1975, in: Excitons at high density, eds. H. Haken and S. Nikitine (Springer, Berlin).
Obreimov, I.V. and W.J. de Haas, 1928, Proc. Acad. Sci. Amsterdam **31**, 353.
Obreimov, I.V. and W.J. de Haas, 1929, ibid. **32**, 1324.
Overhauser, A.W., 1956, Phys. Rev. **101**, 1702.
Peierls, R., 1932, Ann. Phys. (Leipzig) **13**, 905.
Pekar, S., 1957, Zh. Eksp. Teor. Fiz. **33**, 1022; Sov. Phys. JETP **6**, 785.
Phillips, J.C., 1966, Adv. Solid State Phys. **18**, 56, eds. F. Seitz and D. Turnbull (Academic, New York).
Pollman, J. and H. Buttner, 1975, Solid State Commun, **17**, 1171.
Pollman, J. and H. Buttner, 1977, Phys. Rev. **B16**, 4480.
Prikhot'ko, A.F., 1944, J. Phys. USSR **8**, 157.
Prikhot'ko, A.F., 1949, Zh. Eksp. Teor. Fiz. **19**, 383.
Rashba, E.I., 1957, Opt. Spekt. **2**, 568.
Rashba, E.I., 1973, Reps. Prog. Phys. **36**, 1499.
Rashba, E.I. and G.E. Gurgeneshvili, 1962, Fiz. Tver. Tela. **4**, 1029; Sov. Phys. Solid State **4**, 759.
Reynolds, D.C. and T.C. Collins, 1981, Excitons, their properties and uses (Academic, New York).
Sakharov, A.D., 1948, Zh. Eksp. Teor. Fiz. **18**, 631.
Schulman, J.H. and W.J. Compton, 1962, Color centers in Solids (Pergamon, Oxford).
Seitz, F., 1954, Revs. Mod. Phys. **26**, 7.
Sham, L.J. and T.M. Rice, 1966, Phys. Rev. **144**, 708.
Shindo, K., T. Goto and T. Anzai, 1974, J. Phys. Soc. Japan, **36**, 753.
Shpak, M.T. and E.F. Sheka, 1960, Opt. i Spekt **8**, 66; Opt. Spectry **8**, 31.
Simpson, O.J., 1956, Proc. Roy. Soc. (London) **A238**, 402.
Slater, J.C. and W. Shockley, 1436, Phys. Rev. Rev. **50**, 705.
Sommer, B.J. and J. Jortner, 1969, J. Chem. Phys. **50**, 822.
Sturge, M.D., A.T. Vink and F.P.J. Kuijpers, 1978, Appl. Phys. Lett. **34**, 1.
Sumi, H. and Y. Toyozawa, 1971, J. Phys. Soc. Japan, **31**, 342.
Thomas, D.G. and J.J. Hopfield, 1961, Phys. Rev. **124**, 657.
Thomas, D.G. and J.J. Hopfield, 1962, Phys. Rev. **128**, 2135.
Thomas, D.G., J.J. Hopfield and C.J. Frosch, 1965, Phys. Rev. Lett. **5**, 857.
Timusk, T., H. Navarro, N.O. Lipari and M. Altarelli, 1978, Solid State Comm. **25**, 217.
Tolpygo, K.B., 1950, Zh. Eksp. Teor. Fiz. **20**, 497.
Toyozawa, Y. and J. Hermanson, 1968, Phys. Rev. Lett. **21**, 1637.
Tsukasi, M. and H. Kanzaki, 1971, J. Phys. Soc. Japan, **30**, 1243.
Wannier, G.H., 1937, Phys. Rev. **52**, 191.
Weisbuch, C., 1977, Contribution to the study of optical pumping in III–V semiconductors, Thèse de doctorat d'Etat, Univ. Paris 7.
Weisbuch, C., 1978, Solid State Electron. **21**, 179.
Weisbuch, C., R.C. Miller, R. Dingle, A.C. Gossard and W. Wiegmann, 1981, Solid State Comm. **37**, 219.
Wilson, J.A., 1978, Phil. Mag. **B38**, 427.

CHAPTER 2

Electrodynamic and Non-Local Optical Effects Mediated by Exciton Polaritons

JOSEPH L. BIRMAN*

Department of Physics
City College of the City University of New York
New York, N.Y. 10031
U.S.A.

* Fellow of the John Simon Guggenheim Memorial Foundation 1980–81.

Excitons
Edited by
E.I. Rashba and M.D. Sturge

Contents

1. Introduction

1.1 Background

In 1957 Pekar proposed that finite wavevector effects in the dispersion equation for the total energy of an exciton could play a decisive role in the optics of crystalline condensed matter systems. As far as the writer is aware this was the first, albeit controversial, proposal in our modern epoch to analyze the optical effects due to wavevector dependence of the dielectric response function. Such finite wavevector effects may have been known in other branches of contemporary physics e.g. plasma physics; physics of the ionsphere, but in crystalline solid-state matter physics it seems to have been Pekar who first convincingly called the attention of solid-state scientists to these spatial-dispersion effects in the optical region of excitonic resonances. There is an implication in Pekar's early papers that the ever-fertile scientific mind of L.D. Landau suggested this study. Very shortly thereafter, Ginzburg (1958), Agranovich and Rukhadze (1958) and other Soviet scientists made important and deep contributions to this subject. The strong stream of Soviet scientific contributions – theoretical and experimental – continues to the present. An excellent and comprehensive monograph by Agranovich and Ginzburg (1979, 1981) recently appeared, containing a full bibliography, and bringing up to date the earlier review of these authors (Agranovich and Ginzburg 1964, 1966).

The attention of the American scientific community was first drawn to coupled-wave or polariton effects on optical absorption, due to mixing of a photon and a stationary or infinite mass exciton, by Hopfield (1958). Earlier treatments of mixed waves had been given, *inter alia* by Huang and Rhys (1951) and Born and Huang (1952) but these papers were generally interpreted as mainly relating to phonon–photon mixing*. In 1963, Hopfield and Thomas forcefully called attention to the possible existence of spatial dispersion effects in the optics of crystals. The latter paper conveyed, in a transparently clear way, much of the relevant physics of spatially dispersive media. It introduced the resonance oscillator model (now known at the "dielectric approximation" or the cut-off translationally invariant

* The 1951 reference was only recently made available to the writer by Prof. Huang – it does treat the exciton–photon mixing and anticipates some later work. Tolpygo (1950) discussed phonon polaritons (this reference was recently made available by Prof. E.I. Rashba).

model), suggested an exciton-free "dead layer" due to surface repulsion of excitons, gave illustrative quantitative applications, and explained many optical effects observed by high precision spectroscopy in CdS. At the 1966 Kyoto Semiconductor Conference, Hopfield (1966) reviewed the status of work on spatial dispersion.

It has been pointed out by Pekar (1962) and Agranovich and Ginzburg (1962) that an earlier recognition of the importance of wavevector dependent phenomena in optics goes back at least to Lorentz (1879) who, more than 100 years ago, developed a molecular optics theory of optical rotatary power – gyrotropy – by taking account of wavelength dependent terms in the molecular polarizability. It would be no surprise if precursor theories existed even to that of Lorentz since optical rotation was known well before Lorentz's time.

The lengthy interval of 6 years separating the beginnings of investigations of spatial dispersion effects by Soviet and American scientists is only one unfortunate reflection of the difficulty of scientific and personal communication characteristic of the epoch in the 1950's when Soviet scientific journals and monographs were available with difficulty in America (and the West) and conversely. In addition there was practically no face to face meeting of Soviet and American scientists. Let us work toward the positive goal of free flow of scientists and scientific information, and the negative goal of non-recurrence of that period, when scientific progress occurred despite such impediments. The present article is a contribution to the ongoing work and the ongoing dialogue between scientists on an important subject in modern optics.

1.2. Content of this article

The scientific content of this article is strictly limited to a review of recent work on the structure of the elementary excitation known as the exciton polariton in a bounded crystal, and to novel optical consequences of that structure. The article is highly selective and does not claim to be comprehensive. A selection of citations to literature has been given – an extensive citation is in the monograph by Agranovich and Ginzburg (1979).

In the most immediate and accessible form, these exciton-polariton effects manifest themselves in an altered phenomenology: the constitutive relation between induced polarization $\boldsymbol{P}(\boldsymbol{r}, \omega)$ at field point $\boldsymbol{r}$ frequency component ω, and inducing electric field $\boldsymbol{E}(\boldsymbol{r}', \omega)$ is no longer spatially local. A brief discussion is given in sect. 2 of the dielectric response function in bounded medium. Three model non-local dielectric susceptibilities are presented. These are extremely useful models and can represent major physical effects due to non-locality and both for Frenkel and for Wannier excitons. Although the general linear response theory by which to compute the susceptibility is well known (Pines 1964) we do not possess reliable, general theoretical algorithms by which to compute accurate analytical dielectric susceptibilities for a given finite crystal with its own peculiarities: chemical composition, boundary surfaces, imperfections, etc.

In sect. 3 we discuss the structure of the "physical" polariton mode as this exists in a model semi-infinite bounded crystal. This physical mode is a "correct" linear combination of exciton–polaritons. Each exciton polariton is itself a mixture of exciton and photon. Determination of the correct linear combination of these polaritons can be reduced to a boundary value problem so we are immediately led to the problem of boundary conditions in finite systems, and perforce to the problem of additional boundary conditions or "abc". Of the many interesting and controversial problems on spatial dispersion stimulated by Pekar's (1957) article, that of the abc has been one of the longest lived and most active, with discussion continuing through the time of writing. Our emphasis will be: on the logical necessity for boundary conditions; on the clearest possible statement of the problem as perceived by the workers in the field; on mathematically consistent solutions of the problem. In particular we will show how, given a model susceptibility function for the system, a consistent set of abc can be obtained without further assumptions.

In the presence of divergent theoretical views such as on the abc problem it could be expected that some decisive experiment would be performed. The experimental situation is reviewed in accompanying articles (chs. 3 and 4 of this volume) by E. Koteles and E.L. Ivchenko.

In the writer's opinion the most likely definitive work will require a combination of elastic (reflectivity, transmittant) and inelastic (resonant Brillouin and Raman) scattering experiments on the *same* crystal, together with an improved theory including delicate interference effects between different inelastic scattering channels.

In sect. 4 we present a theorist's view of novel optical consequences of the existence of spatial dispersion. These exciton–polariton spectroscopic effects include brief theoretical background for some of the results on reflectivity and Brillouin scattering. Also included are proposals for new lines of experimental investigations into transient optical spectra of spatially dispersive media. A predicted excitonic precursor and other associated fast phenomena (time scale about 1 ps = 10^{-12} s) should stimulate research for these transient elastic effects. They can give a complementary view of the physical polariton, by probing a larger region in frequency space of the physical polariton dispersion than heretofore accessible. In this fashion direct experimental study of group velocity, energy velocity and signal velocities of excitonic polaritons may now be possible.

2. Response functions in spatially dispersive media

In this section we give a brief discussion of response functions in spatially dispersive media. Our objective is to obtain realistic and analytic model expressions for the non-local dielectric susceptibility to be used in the investigation of the electrodynamics and optics of such media. The subject here is excitonic polaritons so all the models will refer to phenomena dominated by excitonic resonances. Within that subject a distinction can still be made as to whether the excitons are tightly bound (Frenkel type) or delocalized (Wannier type). Different susceptibilities result depending on which type of exciton is our major concern.

The simplest "universal" model susceptibility which is now known as the "resonance oscillator" or "dielectric approximation" has been an important testing ground for theory. Probably most of the qualitative physics of spatially dispersive media can be understood on the basis of this model. This form applies for an infinite homogeneous media where translational invariance forces response functions to behave as functions only of separation $(\boldsymbol{r} - \boldsymbol{r}')$. For finite media the boundary breaks the symmetry and forces other types of terms to be present. We shall present three model susceptibilities which can be, and have been, quantitatively used in theories, and which contain major physics, namely: the dielectric approximation and two others.

2.1. The signature of spatial dispersion

A concise description of the origin of excitonic spatial dispersion effects in insulators is that they are due to the "motion" of electronic elementary excitations in the system. This motion is characterized by transport of kinetic energy: $\hbar^2 k^2/2M$ where $\boldsymbol{p} = \hbar \boldsymbol{k}$ is the exciton quasi-momentum and M the total exciton mass (the sum of electron and hole effective band masses: $M = m_e^* + m_h^*$; anisotropic effects need to be included where appropriate). The motion of an excitation of finite mass is the peculiarity taken into account in spatial dispersion. When $M \to \infty$ the excitation is localized and non-spatially dispersive. Classical optical phenomena arise if only the frequency dispersion of the response function is considered.

The "signature" of spatial dispersion is the linear, non-local relationship between the response and the applied field. In our case this manifests itself in the integral constitutive relationship for the polarization:

$$\boldsymbol{P}(\boldsymbol{r}, \omega) = \hat{\chi}\boldsymbol{E} \equiv \int \chi(\boldsymbol{r}, \boldsymbol{r}', \omega)\boldsymbol{E}(\boldsymbol{r}', \omega)\, \mathrm{d}\boldsymbol{r}'. \tag{1}$$

The integral operator is sometimes abbreviated $\hat{\chi}$ as shown. Physically the essential point is that the polarization $\boldsymbol{P}(\boldsymbol{r}, \omega)$ at a point $\boldsymbol{r}$ in the medium is determined by the exciting field $\boldsymbol{E}(\boldsymbol{r}, \omega)$ in a neighborhood whose dimension is the range of $\chi(\boldsymbol{r}, \boldsymbol{r}', \omega)$. In eq. (1) and the following we neglect anisotropy, which must be included as required in making comparisons with experiment.

A useful mechanical model to picture what is occurring was given by Hopfield and Thomas (1963) in terms of coupled localized simple harmonic oscillations. The localized oscillators represent excitons (localized excitations) or in another context an assembly of excited two-level atoms (Ginzburg 1960, 1964, Silin and Rukhadze 1961) whose coupling provides the delocalization or motion by means of which the kinetic energy and mass of the excitation enter the theory. This pictorialization of the distant coupling which produces non-locality can be quite useful.

For our purposes it is still more apt to refer to the results of linear response theory (see below) and interpret the source of non-locality in excitonic insulators differently. The non-local part of the susceptibility is seen to be determined by matrix elements of a current operator $\hat{j}(\boldsymbol{r})$, using a basis of complete many-body eigenstates of the system denoted $|q\rangle$. The relevant matrix elements are of the form

$$\langle 0|\boldsymbol{j}(\boldsymbol{r})|q\rangle\langle q|\boldsymbol{j}(\boldsymbol{r}')|0\rangle \tag{2}$$

and denote the electronic excitation of the system at $\boldsymbol{r}'$ from its ground state $|0\rangle$, to the state $|q\rangle$, followed by deexcitation at $\boldsymbol{r}$ from $|q\rangle$ to $|0\rangle$. If the eigenstate $|q\rangle$ of the entire system is extended or delocalized then the deexcitation of the system $|q\rangle \to |0\rangle$ can occur at point $\boldsymbol{r}$ spatially separated from $\boldsymbol{r}'$. The propagation of excitation in the excited state $|q\rangle$ is the ultimate source of non-locality. Another manifestation of the propagation of excitation or delocalization of excitation is that the eigenenergy of the system, in some state $|q\rangle$ will include a part ω_q^0 which can be identified as "internal" energy of the excitation and a part corresponding to kinetic energy of motion of the excitation. The energy can be represented taking $\hbar = 1$, as

$$\omega_q = \omega_q^0 + q^2/2M, \tag{3}$$

accounting for translational motion. Both the matrix element and the denominator are required in a theory of non-local susceptibility. As shown below various simplifications may be made and an approximate representation of the susceptibility obtained.

2.2. The susceptibility

According to linear response theory (Pines 1964) the general expression for the non-local susceptibility of a medium due to electronic excitations is

$$\chi_{ij}(\boldsymbol{r}, \boldsymbol{r}', \omega) = (\chi_0 - \omega_p^2/4\pi\omega^2)\delta(\boldsymbol{r}-\boldsymbol{r}')\theta(z, z')\delta_{ij} + (c^2/\omega^2)\sum_{\nu}\{\langle 0|j_i(\boldsymbol{r})|\nu\rangle\langle|j_j(\boldsymbol{r}')|0\rangle/(\omega_{\nu 0}+\omega+i\eta) + \langle 0|j_i(\boldsymbol{r}')|\nu\rangle\langle\nu|j_j(\boldsymbol{r})|0\rangle/(\omega_{\nu 0}-\omega-i\eta)\} \quad (4)$$

where $\langle 0|$, is the ground state eigenfunction of the many-body system; $\langle\nu|$ is a complete set of many-body eigenstates characterized by quantum number ν, $\omega_{\nu 0}$ is the excitation energy $0\to\nu$; χ_0 is some background (possibly frequency dependent) dielectric coefficient and ω_p is the electronic plasma frequency. We reserve the symbol $\theta(z, z')$ to account for possible boundary or spatially inhomogeneous effects. Cartesian indices are denoted (ij), but will be ignored in the simple geometry used in this paper; an isotropic model will be used.

2.2.1. *Local optics*

For an infinite homogeneous medium take $\theta(z, z') = 1$. The simplest ansatz for the susceptibility will be to treat only the isotropic medium and to consider the contribution to $\chi(\boldsymbol{r}, \boldsymbol{r}', \omega)$ from a single excited eigenstate $|\nu\rangle = |q\rangle$. Further, taking state $|q\rangle$ to have constant energy corresponding to the discrete eigenenergy of a "1s" exciton $\omega_\nu = \omega_q^0 = \omega_{1s}$ and taking the ground state energy as reference (zero) we have $\omega_{\nu 0} = \omega_0$ ($\hbar = 1$). For the matrix elements of the current operator it is natural to assume in this model extreme locality so that,

$$\langle 0|j(\boldsymbol{r})|q\rangle\langle q|j(\boldsymbol{r}')|0\rangle = a^2\omega_0^2\delta(\boldsymbol{r}-\boldsymbol{r}'). \quad (5)$$

The susceptibility takes the form familiar in local optics, with only frequency dispersion.

$$\chi(\omega) = \chi_b + a^2\omega_0^2/(\omega_0^2 - \omega^2 - i\omega\Gamma), \quad (6)$$

here Γ is the effective damping coefficient. This is the "one resonance" oscillator model. This model might be taken to represent a localized exciton, i.e., the case of *no* overlap or motion of exciton from one site to another. In effect the exciton mass is infinite: $M\to\infty$. This one resonance model is the conventional model for a frequency dispersive-dielectric medium (Jackson 1975, Landau and Lifshitz 1960); in the usual theory the origin of the resonance oscillator is not specified and might be an electronic, or phonon, or other excitation.

2.2.2. *Spatial dispersion: infinite homogeneous medium*

Continuing with the infinite homogeneous medium $\theta(z, z') = 1$. Again we restrict attention to the contribution from a single excitation $|\nu\rangle$ to the sum. Now let this be an exciton with finite mass: $M \neq \infty$. In this case we take

($\hbar = 1$),

$$\omega_{\nu 0} = \omega_q = \omega_0 + bq^2, \tag{7}$$

with $b = (1/2M)$. The simplest approximation is to replace the current matrix elements by some constant matrix element (in effect as will be seen below this uses a $|\boldsymbol{q}| \to 0$ approximation for the matrix elements) to obtain the susceptibility

$$\chi(\boldsymbol{q}, \omega) = [\chi_b + a^2\omega_0^2/(\omega_0^2 + bq^2 - \omega^2 - \mathrm{i}\omega\Gamma)]. \tag{8}$$

In this expression $\boldsymbol{q}$ is to be taken as the wavevector of the excitation. This model dielectric function was introduced by Hopfield and Thomas (1963). It has been widely used and is often denoted as the "dielectric approximation". Note that this expression is assumed to apply to the unbounded medium as we can easily see by taking the Fourier transform. Writing (Birman and Sein (1972)):

$$\chi(\boldsymbol{r}, \omega) = (2\pi)^{-3/2} \int \chi(\boldsymbol{q}, \omega) \exp(\mathrm{i}\boldsymbol{q} \cdot \boldsymbol{r})\, \mathrm{d}\boldsymbol{q}, \tag{9}$$

we find,

$$\chi(\boldsymbol{r}) = \chi_0\delta(\boldsymbol{r}) + \chi_1 G_+(\boldsymbol{r}, \omega), \tag{10}$$

$$G_+(\boldsymbol{r}, \omega) \equiv \exp(\mathrm{i}k_+R)/R, \quad R \equiv |\boldsymbol{r}|. \tag{11}$$

The auxiliary complex quantity k_+ is defined by reexpressing $\chi(\boldsymbol{q}, \omega)$ as

$$\chi(\boldsymbol{q}, \omega) = (2\pi)^{-3/2}[\chi_0 + 4\pi\chi_1/(q^2 - k_+^2)]. \tag{12}$$

Hence k_+ is a complex wavenumber, giving the location in the complex plane, of the simple poles of $\chi(\boldsymbol{q}, \omega)$ at $q = \pm k_+$. We can write k_+ explicitly as

$$k_+ \equiv (|C(\omega)|/b)^{1/2} \exp[\mathrm{i}(\theta + \pi)], \tag{13}$$

where

$$|C(\omega)| = [(\omega_0^2 - \omega^2)^2 + \omega^2\Gamma^2]^{1/2},$$
$$\tan\theta = -\omega\Gamma/(\omega_0^2 - \omega^2); \quad -\pi \leqslant \theta < 0, \tag{14}$$

and we note that $G_+(|\boldsymbol{r}|, \omega)$ is a Green function, since

$$(\nabla^2 + k_+^2) G_+(|\boldsymbol{r}|, \omega) = 2\pi\delta(|\boldsymbol{r}|). \tag{15}$$

This "dielectric approximation" to the susceptibility has the merit of being a simple, analytical expression which surely contains some of the physics of spatially dispersive media when the exciting laser frequency is close to an exciton resonance (as, for example, one of the discrete members of a Balmer series for a particular exciton). It presumably applies best well inside the crystal away from the surface where evanescent waves have decayed and surface-related phenomena are unimportant.

In order to emphasize that point we can write the constitutive equation as

$$\boldsymbol{P}(\boldsymbol{q}, \omega) = \chi(\boldsymbol{q}, \omega)\boldsymbol{E}(\boldsymbol{q}, \omega) \tag{16}$$

for the polarization at wavevector $\boldsymbol{q}$, frequency ω, induced by electric field $\boldsymbol{E}(\boldsymbol{q}, \omega)$. Taking the Fourier transform of this expression over the unbounded (infinite) crystal we have

$$\boldsymbol{P}(\boldsymbol{r}, \omega) = (2\pi)^{-3/2} \int \chi(\boldsymbol{r} - \boldsymbol{r}', \omega)\boldsymbol{E}(\boldsymbol{r}', \omega)\, \mathrm{d}\boldsymbol{r}'. \tag{17}$$

Two points are apparent here. First of all the spatial non-locality of the response. Secondly the translational invariance of the susceptibility. An infinite, unbounded, system is homogeneous and its response function can be expected to be translationally invariant. By contrast, for a system with a broken spatial symmetry, caused for example by a boundary, translational invariance would not hold. The terminology "non-local" which was first used by Hopfield (1966) in the context of exciton-polariton optics has much merit if one observes that near resonance $\omega \to \omega_0$ the range of interaction can be very large (even infinite if damping is neglected). Consequently the departure of $\chi(\boldsymbol{r}, \boldsymbol{r}', \omega)$ from proportionality to $\delta(\boldsymbol{r} - \boldsymbol{r}')$ signals the non-locality. Direct examination of the integral constitutive equation (17) justifies the terminology "non-local", as distinguished from the usual classical "local" optics, and we shall adhere to this usage*.

2.2.3. Spatially dispersive bounded media

Several susceptibilities have been proposed to model the optical response of bounded spatially dispersive media. In the present article, a selection will be made for illustrative purposes. Each of the specific models has merits and shortcomings.

2.2.3.1 Frenkel (tightly bound) excitons. A simple way to treat exciton motion (non-localization) in the presence of a boundary is to treat the exciton center of mass motion (for a specified internal state) as comprising a plane wave traveling toward the boundary plus a reflected wave (Hopfield and Thomas 1963, Zeyher et al. 1972, Zeyher and Birman 1974). The wavevector is taken as $\boldsymbol{q} = (q_x, q_y, q_z)$.

As before we take the simplest geometry with the crystal boundary being the plane $z = 0$. Let the reflection coefficient for the excitation (exciton) at the boundary be $R(q_z)$; it depends only on the normal component of

* The recent comment by van Kronendonk and Sipe (1977) questioning this usage seems inappropriate. Their comment is based on the "local" equation of motion which $\boldsymbol{P}(\boldsymbol{r}, t)$ satisfies. (*Vide infra* eq. (53)). However that equation involves spatial derivatives of $\boldsymbol{P}(\boldsymbol{r}, t)$ as well as the usual local oscillator terms $\ddot{P}(\boldsymbol{r}, t) - \omega_0^2 P(\boldsymbol{r}, t)$, and the spatial derivatives imply a "non-point field" interaction.

wavevector, q_z. Let us assume the matrix element in the susceptibility to be given by:

$$(\omega_q/8\pi e^2)^{-1/2}\langle 0|j(\boldsymbol{r})|q\rangle = \omega_q\phi_q(\boldsymbol{r}), \tag{18}$$

with,

$$\phi_q(\boldsymbol{r}) = [\exp(\mathrm{i}q_z z) + R(q_z)\exp(-\mathrm{i}q_z z)]\exp[\mathrm{i}(q_x x + q_y y)]\theta(z). \tag{19}$$

The susceptibility arising from this assumption can be shown to be of the form (Zeyher et al. 1972):

$$\chi(\boldsymbol{r}, \boldsymbol{r}', \omega) = [\chi_0\delta(\boldsymbol{r}-\boldsymbol{r}') + \chi^+(\boldsymbol{r}-\boldsymbol{r}') + \chi^-(\boldsymbol{\eta}-\boldsymbol{\eta}', z+z')], \tag{20}$$

where

$$\chi^+ = [\omega_p^2/(2\pi)^3]\int \mathrm{d}^3q \frac{\exp \mathrm{i}\boldsymbol{q}\cdot(\boldsymbol{r}-\boldsymbol{r}')}{[\omega_q^2-(\omega+\mathrm{i}\eta)^2]}, \tag{21}$$

$$\chi^- = [\omega_p^2/(2\pi)^3]\int \mathrm{d}^3q \frac{R(q_z)\exp \mathrm{i}[\boldsymbol{q}_\parallel\cdot(\boldsymbol{\eta}-\boldsymbol{\eta}') + q_z(z+z')]}{[\omega_q^2-(\omega+\mathrm{i}\eta)^2]}, \tag{22}$$

with $\eta \equiv x\hat{i} + y\hat{j}$, and $\boldsymbol{q}_\parallel$ the component of $\boldsymbol{q}$ parallel to the surface. $R(q_z)$ is an effective reflection coefficient for the exciton. The term χ^+ has exactly the form of the non-local susceptibility in the infinite system while the term χ^-, accounting for the boundary, takes exciton reflection into account. To make the analogy still closer we can use the expression given above in eq. (7) for ω_q so that

$$\chi^+ = \chi_1 G_+(\boldsymbol{r}-\boldsymbol{r}'; \omega). \tag{23}$$

If we make an added assumption of total reflection of the exciton on the boundary, taking $R(q_z) \sim R(0) = -1$,

$$\chi^- = -\chi_1 G_+(\boldsymbol{\eta}-\boldsymbol{\eta}', z+z'; \omega). \tag{24}$$

Note that in both cases the *integrand* in χ^+ and χ^- has a simple pole at $q = \pm k_+$. We can call $\chi^+(\boldsymbol{r}-\boldsymbol{r}'; \omega)$ the propagating part of the polarization – the justification for this will follow below.

The assumption which has been made in this model calculation is that the exciton reflects coherently at the boundary. That is, electron and hole retain their relative-motion eigenstates during the process of interaction with the boundary. In effect this assumes that electron and hole act "rigidly" – inducing an *image* exciton across the boundary, and then interacting with it to produce the characteristic image term which depends on $(z+z')$ in the susceptibility term χ^-. This ansatz is most applicable to the tight-binding or Frenkel excitons where electron and hole are in close proximity.

To go beyond this straightforward physical ansatz of coherent interaction requires a much more elaborate theory. Such a theory was given by Hizhnyakov et al. (1975) based on a detailed microscopic model and they obtained a susceptibility whose form is in principle more complicated than eq. (20). However, it does appear that most of the physics of the tight-binding exciton is given in eq. (20). It does not seem that any detailed applications of this more elaborate model to optical properties of media have been published.

2.2.3.2. Wannier (delocalized) exciton. A model susceptibility which takes into account the effects of a Wannier exciton moving in a bounded medium requires a completely self-consistent solution of the Schrödinger equation for the exciton in presence of the boundary. Such a problem has, it seems, not yet been solved. As a physical approximation we may consider the solution of the effective mass approximation (EMA) Hamiltonian for the exciton in the presence of an infinite potential barrier for electron and hole at $z = 0$. It was shown by Ting et al. (1975) that the resultant exciton eigenfunction has the form

$$f(z) = \sum_{q_z} [\phi_{q_z}(0) \cos q_z z - \phi_{q_z}(2z) \cos(\alpha q_z z) + g_{q_z}(z)], \tag{25}$$

where $\phi_q(r) \sim \exp(-r/a_0)$ is the "1s" bound state exciton eigenfunction and $\alpha \equiv (m_e^* - m_h^*)/(m_e^* + m_h^*)$. The first term results from bulk contribution, the second and third terms are surface-related evanescent terms. The important result is that while the bulk exciton center-of-mass eigenfunction does *not* have a node at the surface $z = 0$, the total eigenfunction $f(z) \to 0$ as $z \to 0$ owing to cancellation of the first terms (bulk and surface) in the square bracket (the correction term $g(z) \to 0$).

For the calculation of the bulk susceptibility we take the current matrix element (18) in the Wannier model:

$$\omega_p \phi_q(r) = \omega_p \cos q_z z \exp[i(q_x x + q_y y)]\theta(z), \tag{26}$$

where

$$\theta(z) = 1, \quad z > 0, \quad \theta(z) = 0, \quad z \leq 0.$$

This should be contrasted with the previous ansatz (24) with $R(q_z) \sim -1$, in which the center-of-mass eigenfunction of the bulk exciton has a node at $z \to 0$ ($\sin q_z z \to 0$). Substituting this matrix element into the general expression for the susceptibility we can evaluate the bulk non-local susceptibility in the Wannier case. *Formally* this simply results in taking $R \sim +1$ in the previous expression, and we then have:

$$\chi(\boldsymbol{r}, \boldsymbol{r}'; \omega) = [\chi_0 \delta(\boldsymbol{r} - \boldsymbol{r}') + \chi_1 G_+(\boldsymbol{r} - \boldsymbol{r}'; \omega) + \chi_1 G_+(\boldsymbol{\eta} - \boldsymbol{\eta}', z + z'; \omega)]. \tag{27}$$

Expressed in Fourier transform, the two last terms have simple poles at $q_z = \pm k_+$ as before.

Comparing eqs (20) and (27), we note that formally these two different susceptibilities, originating in quite different physical assumptions (Frenkel vs. Wannier exciton), have remarkably similar structure, differing only in the sign of the last term (± 1 respectively). These two expressions (20) and (27) and the "dielectric approximation" can be combined into the general form (74) below, which can go over to each case depending on the value of one parameter $R(0)$, whose microscopic significance is of a "reflection coefficient".

2.2.3.3. Recent work. Some recent discussions of the motion of Wannier exciton in bounded media have appeared, and we cite two. Mattis and Beni (1978) have analyzed a soluble one-dimensional model taking electron and hole masses equal, replacing the Coulomb attraction term in the effective mass Hamiltonian by an attractive delta function and retaining the "hard wall" (infinitely repulsive barrier). They obtain eigenstates and eigenenergies. It would be interesting to have the form of susceptibility corresponding to their eigensolutions but this has not been reported yet. Zeyher (1980) has examined a microscopic model Hamiltonian, equivalent to the effective mass approximation in lowest approximation, for electron and hole motion in bounded medium. He discusses the solution in terms of the next correction to the effective mass approximation via the two-body Bethe–Salpeter equation. No closed form for the non-local susceptibility was given.

2.2.4. Some generalities

In view of the evident lack of a single unique all-embracing form for the spatially dispersive dielectric function, and the introduction of many specific physical and mathematical models the question naturally arises as to whether there are any general properties of $\chi(\boldsymbol{r}, \boldsymbol{r}'; \omega)$ or its Fourier transform $\chi(\boldsymbol{q}, \boldsymbol{q}'; \omega)$ which could serve for the analysis of electrodynamic and optical properties. This question cannot yet be answered in sufficient generality.

As a response function, the susceptibility $\chi(\boldsymbol{r}, \boldsymbol{r}'; \omega)$ must obey all the analyticity properties imposed by causality and, when appropriate, relativistic causality. Consider for example the spatially homogeneous response function $\chi(\boldsymbol{q}, \omega)$. For fixed $\boldsymbol{q}$, this function can be treated as a function of complex variable ω and $\chi(\boldsymbol{q}, \omega)$ must be analytic in the upper half (complex) ω plane. It must obey the Kramers–Kronig relations (at fixed $\boldsymbol{k}$) in terms of which Im $\chi(\boldsymbol{q}, \omega)$ and Re $\chi(\boldsymbol{q}, \omega)$ are Hilbert transforms of each other. These relations follow from the analyticity of $\chi(\boldsymbol{q}, \omega)$ which is given by Cauchy's theorem as

$$[\chi(\boldsymbol{q}, \omega) - 1] = \frac{1}{\pi \mathrm{i}} \fint_{-\infty}^{\infty} \frac{[\chi(\boldsymbol{q}, \omega') - 1]}{\omega' - \omega} \mathrm{d}\omega', \quad (\boldsymbol{q} \text{ fixed}) \tag{28}$$

where the barred integral refers to principal value and $\chi(\omega')$ (suppressing $\boldsymbol{q}$) is taken on the real axis (Landau and Lifschitz 1960).

An important paper by Leontovich (1961) treats the general dispersion relations in spatially dispersive media. If we restrict our attention, as before, to the translationally invariant case, Leontovich's criterion for causality of $\chi(\boldsymbol{q}, \omega)$ is

$$\chi(\boldsymbol{q}, \omega) = (1/\mathrm{i}\pi) \fint_{\infty}^{+\infty} \chi(\boldsymbol{q} + (v/c)(\omega' - \omega), \omega) \, \mathrm{d}\omega'/(\omega' - \omega). \tag{29}$$

This important condition was obtained by treating $(\boldsymbol{q}, \omega)$ as a relativistic 4-vector in the material medium, and using relativistic covariance. It is evidently of general importance that $\chi(\boldsymbol{q}, \omega)$ should be causal, in particular when considering transient optical effects (*vide infra*) such as signal propagation; thus it is essential not to introduce spurious model effects by using causality-violating susceptibilities. Frankel and Birman (1977) verified that the model non-local susceptibilities which we discuss in detail in this paper do satisfy the Leontovich condition for propagating solutions.

A general discussion of dispersion relations and causality in the context of non-local response functions was given by Martin (1967, 1968) and more recently by Kirzhnits (1976); but these seem not to have been fully utilized in analysis of electrodynamics or optics of spatially dispersive media. One noteworthy paper concerning application of dispersion relations to evaluation of the complex refractive indices of a crystal taking spatial dispersion into account was given by Ginzburg and Meiman (1964).

3. Structure of the "physical" polariton mode

In this section we discuss the structure of the physical polariton which is the correct linear combination of elementary exciton polaritons. Each exciton polariton is a coupled exciton and photon excitation, the coupling depending on frequency. The correct linear combination arises from satisfying mathematically correct boundary conditions for "launching" the physical polariton into the medium under the aegis of an external photon coupled in. We give the solution of this problem – the abc problem – by several mathematically correct methods.

We consider only the "perfect crystal" case, where scattering can only occur at the surface. The effect of impurity scattering has been considered by Hopfield (1968, 1969) and by Sugakov (1968).

3.1. Necessity for boundary conditions

We need to first set out the problem clearly in words. In spatially dispersive media the dielectric susceptibility (or linear response function) is non-local, as described in the introduction. Consequently the relation between total polarization $\boldsymbol{P}(\boldsymbol{r})$ induced at field point $\boldsymbol{r}$, and the applied macroscopic field $\boldsymbol{E}(\boldsymbol{r})$ is a linear integral relation as in eq. (1). Fourier transforming this equation gives in general

$$\boldsymbol{P}(\boldsymbol{q}, \omega) = \int \chi(\boldsymbol{q}, \boldsymbol{q}', \omega)\boldsymbol{E}(\boldsymbol{q}'\omega)\, \mathrm{d}\boldsymbol{q}'. \tag{30}$$

In the case that one is only concerned with long wavelength phenomena and can neglect "umklapp" effects, the non-local susceptibility $\chi(\boldsymbol{q}, \boldsymbol{q}', \omega)$ will only depend on the wavevector $\boldsymbol{q}$ and the constitutive relation becomes:

$$\boldsymbol{P}(\boldsymbol{q}, \omega) = \chi(\boldsymbol{q}, \omega)\boldsymbol{E}(\boldsymbol{q}, \omega). \tag{31}$$

The corresponding dielectric coefficient is

$$\epsilon(\boldsymbol{q}, \omega) = 1 + 4\pi\chi(\boldsymbol{q}, \omega). \tag{32}$$

The essential point here is the wavevector dependence in $\epsilon(\boldsymbol{q}, \omega)$ in addition to the frequency dependence or the usual frequency dispersion of local optics. In a physical system (semiconductor crystal, plasma, etc.) whose constitutive relation involves an essential $\boldsymbol{q}$ dependence, new phenomena occur, by contrast to the situation where ϵ is merely frequency dependent. One new effect first predicted by Pekar (1957) and Ginzburg (1958) is the existence of new or "additional" propagating waves. These propagating waves are "new" or "additional" in the sense that the counterpart material system described by the dielectric coefficient $\epsilon(\boldsymbol{q} = 0, \omega)$ has fewer propagating "normal" waves at the same frequency.

In any geometry, presence of spatial dispersion permits added waves at frequency ω. Each such wave is a frequency dependent mixture of exciton and photon portions. To anticipate the later discussion we remark that each wave satisfies Maxwell equations with harmonic time dependence:

$$\boldsymbol{\nabla} \times \boldsymbol{\nabla} \times \boldsymbol{E} = k_0^2\boldsymbol{D} = k_0^2\epsilon(\boldsymbol{q}, \omega)\boldsymbol{E}, \tag{33}$$

or if we assume plane-wave solutions with wave vector $\boldsymbol{q}$:

$$-\boldsymbol{q} \times \boldsymbol{q} \times \boldsymbol{E} = k_0^2\boldsymbol{D} = k_0^2\epsilon(\boldsymbol{q}, \omega)\boldsymbol{E}. \tag{34}$$

This expression gives rise to transverse and longitudinal waves. The transverse wave propagation vector $\boldsymbol{q}$ satisfies

$$q_j^2 = k_0^2\epsilon(\boldsymbol{q}_j, \omega), \quad (k_0 = \omega/c) \quad j = 1, \ldots n \tag{35}$$

where n is the number of solutions.

The functional relation $q_j(\omega)$ is the dispersion relation of the coupled wave. This coupled wave (part photon, part exciton) is the "exciton polariton". Recall that each propagating solution to eq. (6) gives a "branch" $q_j(\omega)$ labelled by fixed j. For each such $q_j(\omega)$ and $\boldsymbol{E}$, a $\boldsymbol{D}$ (or $\boldsymbol{P}$) field exists. Examination of the field (say $\boldsymbol{P}_j$ corresponding to wave vector q_j, frequency ω) shows that it is composed of a mixture of photon plus exciton part (Hopfield, 1958) or in the analogous phonon problem, photon plus phonon (Huang and Rhys 1951, Born and Huang 1952). At fixed ω several degenerate exciton polariton excitations may exist – each specified by fixed j, and representing a distinct entity at different k.

If the system is infinite (no boundary) then *any* linear combination of these distinct exciton polaritons can be taken as a propagating excitation in the crystal.

When the system is bounded, the situation is completely changed. The vacuum or external region is the source of incident radiation, and also the region in which scattered radiation exists. In a typical problem radiation enters through the boundary, producing the excitation launched into the crystal and scattered (reflected) radiation propagating back in the vacuum. Hence the boundary provides a region on which exterior and interior fields must be properly matched.

The proper matching of exterior and interior fields produces a "physical" excitation throughout all of space. In the *exterior* region this is the externally controllable incident field, joined at the boundary to the reflected exterior field. In the *interior* (medium) region this physical excitation, propagating into the medium is a coupled polariton.

This coupled propagating mode, consisting of the correct linear combination of the polariton modes which are individually solutions of Maxwell's equation, is what we call the "physical polariton" mode in the medium.

Central to any quantitative investigation of the kinematic or dynamic processes of scattering of physical polaritons is quantitative information about their structure. The structure of the physical polaritons is strictly determined by the solution of both the individual dispersion equations for each separate polariton mode (branch) giving $q_j(\omega)$, and by the solution of the boundary condition problem which provides the correct coupling of interior and exterior solutions to produce the physical polaritons. A *sine qua non* is the determination of all the boundary conditions required in order to join exterior and interior solutions.

It is because of the importance of obtaining knowledge of the structure of the physical polariton that considerable effort has been devoted to the determination of the correct set of boundary conditions. We shall therefore attempt to give a selective review and perspective on the problem of boundary conditions in what follows in this section.

3.2. *The problem of boundary conditions*

For the present we follow simple traditional usage in electrodynamics and assume that the inhomogeneous system consists of a semi-infinite medium in the region $z > 0$ terminated by a plane surface $z = 0$ with vacuum located in the region $z < 0$. The medium is homogeneous right up to the boundary. In this model system the dielectric coefficient for $z < 0$ is $\epsilon = 1$ (vacuum) while for $z > 0$, the dielectric coefficient is $\epsilon(\boldsymbol{r}, \boldsymbol{r}', \omega)$. Physically we can assume that the dielectric coefficient changes rapidly but smoothly from the interior value $\epsilon(\boldsymbol{r}, \boldsymbol{r}', \omega)$ to $\epsilon = 1$. On the wavelength scale involved here it may be reasonable at times to neglect the smooth variation and take

$$\epsilon_{\text{sys}}(\boldsymbol{r}, \boldsymbol{r}'\omega) = \epsilon(\boldsymbol{r}, \boldsymbol{r}'\omega)\theta(z) + \theta(-z),$$

which produces a surface discontinuity in ϵ at the boundary plane $z = 0$.

The usual Maxwell constructions (Stratton 1941, Born and Wolf 1958) by which boundary conditions on the field are obtained, begins with Maxwell equations:

$$(1)\quad \nabla \times \boldsymbol{E} = -\frac{1}{c}\frac{\partial \boldsymbol{B}}{\partial t}; \qquad (2)\quad \nabla \times \boldsymbol{H} = \left(\frac{1}{c}\frac{\partial D}{\partial t} + \frac{4\pi}{c}\boldsymbol{J}\right). \tag{36}$$

Using Stokes theorem for a rectangular path C which cuts the surface of discontinuity S we have

$$\oint \boldsymbol{E} \cdot \mathrm{d}l = -1/c \int_S (\partial \boldsymbol{B}/\partial t) \cdot \hat{n}\, \mathrm{d}a, \tag{37}$$

from which by passing to the limit we obtain:

$$\hat{n} \times (\boldsymbol{E}_2 - \boldsymbol{E}_1) = 0 \qquad \text{or} \qquad \boldsymbol{E}_{t_1} = \boldsymbol{E}_{t_2}, \tag{38}$$

since $\partial \boldsymbol{B}/\partial t$ is finite on S. Likewise with $\partial \boldsymbol{D}/\partial t$ and $\boldsymbol{J}$ finite on S,

$$\hat{n} \times (\boldsymbol{H}_2 - \boldsymbol{H}_1) = 0 \qquad \text{or} \qquad \boldsymbol{H}_{t_1} = \boldsymbol{H}_{t_2}. \tag{39}$$

The other Maxwell boundary conditions are derived from

$$(3)\quad \nabla \cdot \boldsymbol{B} = 0; \qquad (4)\quad \nabla \cdot \boldsymbol{D} = 4\pi\rho, \tag{40}$$

and application of the divergence theorem using a "pillbox" (small cylindrical volume) whose area is A, e.g.:

$$\oint \boldsymbol{B} \cdot \hat{n} \, da = 0. \tag{41}$$

Letting the volume shrink to zero we find

$$\hat{n} \cdot (\boldsymbol{B}_2 - \boldsymbol{B}_1) = 0, \qquad \text{or} \qquad \boldsymbol{B}_{n_1} = \boldsymbol{B}_{n_2}, \tag{42}$$

and from (4)

$$\hat{n} \cdot (\boldsymbol{D}_2 - \boldsymbol{D}_1) = 0, \qquad \text{or} \qquad \boldsymbol{D}_{n1} = \boldsymbol{D}_{n2}, \tag{43}$$

in the absence of free surface charge density on S.

The four conditions (38) (39) (42) (43) are the *Maxwell boundary conditions.* It is worth emphasizing that the constitutive relations are not used in deriving these Maxwell boundary conditions but only the field equations.

For local electrodynamics or optics there is no problem of boundary conditions. Recall the "model problem". Take the crystal to have dielectric coefficient $\epsilon(\omega)$, and $\mu = 1$. The crystal surface is the plane $z = 0$. Let a plane-polarized electromagnetic wave travelling in vacuum in the $+z$ direction be incident normally on this surface and take the electric field as $E_0 \exp(ik_0 z)$. The reflected wave is $E_0' \exp(-ik_0 z)$. Here $k_0 = (\omega/c)$. Call the transmitted wave $E_1 \exp(ik_1 z)$. Correspondingly the magnetic fields are $k_0 E_0 \exp(ik_0 z)$, $-k_0 E_0' \exp(-ik_0 z)$ and $k_1 E_1 \exp(ik_1 z)$. In order to completely determine all fields we need to determine the *two* ratios E_0'/E_0 and E_1/E_0, it being assumed that E_0 is independently controllable. The continuity of $\boldsymbol{E}_t$ and $\boldsymbol{H}_t$ across the boundary $z = 0$ give *two* boundary conditions

$$E_0 + E_0' = E_1, \tag{44}$$

$$k_0(E_0 - E_0') = k_1 E_1, \tag{45}$$

from which we obtain the Fresnel equations for normal incidence:

$$(E_1/E_0) = \frac{1}{2}\left(\frac{1}{1 + (k_1/k_0)}\right); \qquad (E_0'/E_0) = \frac{1}{2}\left(\frac{1 - k_1/k_0}{1 + k_1/k_0}\right). \tag{46}$$

Thus the conventional Maxwell boundary conditions exactly suffice to determine the two unknown ratios of field amplitudes. Within the dielectric medium we obtain the propagating transverse electric displacement field $\boldsymbol{D}(\boldsymbol{r}, t)$ or $\boldsymbol{D}(\boldsymbol{k}, \omega)$ in the conventional way from Maxwell's equations. The solution of the equation (36) above, making the plane wave ansatz, and using the constitutive equation $\boldsymbol{D} = \epsilon(\omega)\boldsymbol{E}$, requires:

$$(q/k_0)^2 = \epsilon(\omega), \quad \text{or} \quad q(\omega) = \pm k_0 \epsilon(\omega)^{1/2}, \tag{47}$$

which produces a single propagating wave in the medium at each frequency:

$$D_1(q, \omega) = \epsilon(\omega) E_1(q, \omega), \tag{48}$$

or

$$4\pi P_1(q, \omega) = [\epsilon(\omega) - 1] E_1(q, \omega), \tag{49}$$

where $q(\omega)$ is given by eq. (47). With *one* propagating wave in the medium, plus incident and reflected waves exterior to it, two Maxwell boundary conditions exactly suffice to determine the two needed amplitude ratios.

For non-local media an entirely new situation obtains. The dispersion equation for transverse wave propagation is now (eq. (35)), i.e.,

$$q(\omega)^2 = k_0^2 \epsilon(q, \omega), \tag{50}$$

where $\epsilon(q, \omega)$ is a non-trivial function of q. Consequently the algebraic equation for solutions q_i will be higher than second degree, with solutions $q_1(\omega), q_2(\omega), \ldots, q_j(\omega) \ldots$. In the single oscillator model with an exciton of finite mass, there will be two roots (q_1, q_2) of the dispersion equation

$$(q/k_0)^2 = \epsilon_0 + 4\pi F/[C(\omega) + bq^2], \tag{51}$$

(each with $\pm$ sign). Consequently *two* propagating fields exist in the medium at the frequency ω. For example

$$E_1 = \exp(iq_1 z - i\omega t); \quad E_2 = \exp(iq_2 z - i\omega t), \tag{52}$$

and also *two* propagating fields with $-q_1$ and $-q_2$ (propagating in the negative z direction). For the model problem of a normally incident plane wave from vacuum, we consider only propagation in $+z$ direction in the medium as in eq. (52) plus incident and reflected waves. Hence there are the following amplitudes in question: E_0, E_0', E_1, E_2, i.e., four amplitudes (one of which is externally controllable) or, *three* ratios. But the Maxwell boundary conditions give only *two* ratios. The problem of determining the additional unknown ratio can be solved if one is presented with one more boundary condition. This need for supplementary boundary conditions in order to fully determine the structure of the wave field *inside* the medium has given rise to the topic of "additional boundary conditions" or the abc problem. We emphasize several points.

The dispersion relations for waves in the medium are independent of boundary conditions. The equation for transverse waves $q^2 = k_0^2 \epsilon(q, \omega)$ holds as a strict consequence of Maxwell's equations and the constitutive relations in the medium. Insofar as we are only concerned with these dispersion surfaces $q_j(\omega)$ for waves propagating in an infinite spatially dispersive medium, the boundary is irrelevant. But, in a bounded medium the totality of boundary conditions are essential in order to construct the "physical polariton". In quantum mechanical calculations the physical polariton wave field is quantized and these quantum states are the initial, intermediate, and final eigenstates of a scattering problem (Brenig et al. 1972).

3.3. *A perspective on the history of the abc problem and its solution*

Simultaneously with the prediction of the existence of additional waves in spatially dispersive media the need for more than the usual Maxwell boundary conditions was realized (Pekar 1957, Ginzburg 1958), and strenuous theoretical efforts were made to solve this problem.

In his first paper Pekar (1957) discussed a quantum mechanical approach to obtaining additional boundary conditions. This utilized ultimately an equation of motion for the exciton polarization $\boldsymbol{P}_e(\boldsymbol{r}, t)$. Schematically (avoiding tensorial indices) it was proposed that in the medium $\boldsymbol{P}_e(\boldsymbol{r}, t)$ satisfies the equation of motion

$$(-\omega_0^2 + \partial^2/\partial t^2 + b\nabla^2 + \gamma\partial/\partial t)\boldsymbol{P}_e(\boldsymbol{r}, t) = \alpha\boldsymbol{E}(\boldsymbol{r}, t), \tag{53}$$

with $-\omega_0^2$, $b = 1/M$, γ, α, being coefficients and the macroscopic field $\boldsymbol{E}(\boldsymbol{r}, t)$ acting as a driving force on the polarization. For harmonic plane wave propagation in the medium, this equation immediately leads to the macroscopic constitutive relation for the exciton polarization ($-\mathrm{i}\nabla \to q$):

$$\boldsymbol{P}_e(\boldsymbol{q}, \omega) = \chi_e(\boldsymbol{q}, \omega)\boldsymbol{E}(\boldsymbol{q}, \omega), \tag{54}$$

with

$$\chi_e(\boldsymbol{q}, \omega) \equiv \alpha/(\omega_0^2 + bq^2 - \omega^2 - \mathrm{i}\gamma\omega), \tag{55}$$

or if we add a background term *not* connected with this oscillator the total susceptibility is

$$\chi(\boldsymbol{q}, \omega) = \chi_0 + \chi_e(\boldsymbol{q}, \omega). \tag{56}$$

It is helpful to emphasize that eq. (53) is an equation of motion which is "semi-macroscopic". The equation is obtained as a long wavelength approximation to an exciton Schrödinger equation with dipole moment (or polarization) defined via the exciton eigenfunction in a bounded medium. The *total* macroscopic polarization $\boldsymbol{P}$ is identified as this excitonic polarization subject to eq. (53) plus a background χ_0. The equation (53) was solved subject to the boundary condition

$$\boldsymbol{P}_e(\boldsymbol{r}, t)\,|_{\Sigma} = 0 \tag{57}$$

since the polarization vanishes outside the medium. Pekar then asserted that this boundary condition should apply to the exciton polarization and be taken as the abc needed to complete the solution of Maxwell's equations in the presence of spatial dispersion.

We now sharpen this point. In this approach it is taken that: the macroscopic susceptibility in the bounded medium is $\chi(\boldsymbol{q}, \omega)$ as in eq. (56), and the supplementary boundary condition appropriate is $P_e|_\Sigma = 0$. Although this is a natural and appealing procedure, this boundary condition is *not* a mathematical consequence of the assumed susceptibility.

In a paper published soon after, Ginzburg (1958) proposed a phenomenological approach, in which a spatially dispersive susceptibility was postulated for the bounded medium by writing $\epsilon(\boldsymbol{q}, \omega)$ in a power series up to bilinear terms in $\boldsymbol{q}$. Two other supplementary boundary conditions were proposed, one of these being

$$D'|_\Sigma \equiv (\boldsymbol{D} - \epsilon \boldsymbol{E})|_\Sigma = 0, \tag{58}$$

where ϵ is the nonspatially-dispersive (q independent) part of $\epsilon(\boldsymbol{q}, \omega)$. In this paper Ginzburg (1958) stated "within the framework of the phenomenological approach employed here additional boundary conditions can be obtained only by introducing certain assumptions".

In a comprehensive review article Agranovich and Ginzburg (1962) further analyzed boundary conditions starting from essentially the same constitutive relation as in the Ginzburg (1958) paper for the exciton polarization,

$$-\omega^2 \boldsymbol{P}_e + \beta \boldsymbol{P}_e + \gamma \nabla \boldsymbol{P}_e + \alpha \nabla^2 \boldsymbol{P}_e = \lambda \boldsymbol{E}, \tag{59}$$

(we ignore tensorial indices) and they observe that the general form of boundary condition at Σ should be

$$(-\omega^2 + \beta)\boldsymbol{P}_e + \gamma(\nabla \boldsymbol{P}_e) = \lambda \boldsymbol{E} \qquad \text{on } \Sigma, \tag{60}$$

with all fields evaluated on Σ. A somewhat more general phenomenological boundary condition was given in the book by Agranovich and Ginzburg (1966) in which they wrote (here put schematically):

$$(-\omega^2 + \beta)\boldsymbol{P}_e + \gamma \nabla \boldsymbol{P}_e = \Lambda \boldsymbol{E} + \Lambda^{(1)} \nabla \boldsymbol{E} \qquad \text{on } \Sigma, \tag{61}$$

all fields at Σ.

In both papers they remark on the frequency-independence of the coefficients β, γ, Λ, $\Lambda^{(1)}$ and propose for their susceptibility a limiting (long wavelength) abc:

$$\boldsymbol{P}_e + \Gamma \boldsymbol{E} = 0 \qquad \text{at } \Sigma, \tag{62}$$

with Γ frequency independent.

At about this time Hopfield and Thomas (1963) took up the abc problem. Following somewhat along the lines of Pekar's (1957) paper they introduced the single oscillator resonance type dielectric function as in eq. (53) *and* analyzed a quantum-mechanical model which would give a Schrödinger eigenfunction for the exciton. From this eigenfunction they deduced the needed abc, using a classical correspondence. They recovered the Pekar abc for the resonance oscillator model. Thus, these authors, like Pekar, took the single resonance oscillator model for the non-local susceptibility of the bounded medium along with the Pekar abc. To allow for the fact that a Wannier exciton has a finite radius, and hence must be affected by the presence of the surface even when its center of mass is some distance away, a "dead layer" or exciton-free local region was postulated, and this three layer geometry was invoked in their analysis of data.

From these published papers the impression was conveyed in the literature that there was an "abc" problem to be solved by following either of two lines. The first is to set up a quantum mechanical exciton model, solve for the Schrödinger eigenfunction then use this eigenfunction to deduce the boundary condition on the macroscopic polarization to be used in conjunction with a non-local dielectric function. The second is to postulate a non-local dielectric function and "introduce certain assumptions". While the first procedure is of course in principle correct – one should solve the coupled Schrödinger–Maxwell problem self-consistently – in practice no such self-consistent solution was given for the bounded non-local medium. The actual abc obtained by workers following this line were gotten by injecting some sort of plausibility argument into the analysis at a penultimate point and not by a mathematically clear procedure directly from the assumed macroscopic susceptibility without added assumptions or microscopic models. The belief, current at that time, that extra assumptions or models *were* needed is clearly seen in the excellent review given by Hopfield (1966) p. 82–84 and discussion following that article.

A major part of the residual controversy and uncertainty about the abc problem evidently relates to which of these approaches is used by a particular author, and different or even contradictory assumptions made by different workers. Evidently the solution of the combined Schrödinger–Maxwell equations is the *desideratum*, and will continue to give rise to investigations in the future.

The second phenomenological approach however is capable of being "well posed" *without* added assumptions and we turn to it. We postulate that the macroscopic Maxwell equations hold *and* that an *a priori* prescribed macroscopic constitutive relation holds between $\boldsymbol{P}(\boldsymbol{r}, \omega)$ and $\boldsymbol{E}(\boldsymbol{r}, \omega)$ in the *bounded* medium. With these hypotheses only, *no* additional assumptions are needed to solve the electrodynamics. Up to 1969–70 this seems not to have been explicitly realized in the literature.

To the writer's knowledge, the first correct solution of this "well posed" problem of obtaining abc was given by his (then) doctoral student John J. Sein (1969, 1970). Sein's thesis topic had been an investigation of the

Ewald–Oseen extinction theorem of classical (local) optics (Born 1933, Rosenfeld 1951, Born and Wolf 1958).

The present writer suggested to Sein that on physical grounds an extinction theorem must apply also to non-local media. By carefully working through an integral formulation of optics which he developed (which is not quite equivalent to the usual formulation owing to different assumptions on lack of singularities of current sources) and by correctly interpreting the result, Sein was led both to the correct extinction theorem for non-local optics *and* to the correct solution of the abc problem. It should be remarked that Sein's abc include the explicit solutions for both transverse and longitudinal polariton modes for the single oscillator model (Sein 1969,1970). Most of the work at this time concerned the simple model geometry (normal incidence-transverse waves) which, as Hopfield and Thomas (1963) pointed out, contains all the essentials of the problem, but Sein's work dealt with the vectorial electrodynamics of bounded non-local media, *in extenso.* As will be discussed later, Sein also solved the spatially dispersive multioscillator model by his integral equation method (Sein 1969, 1970).

In addition to giving the correct solution of the abc problem for his model problem invoking the resonance oscillator (dielectric approximation) susceptibility, Sein also opened a new view of the extinction theorem, as a "cooperative" (SALTUS) effect due to all the dipoles in the crystal, and not only those at the surface. This interpretation marked a new departure and has led to considerable deepening and further elaboration of the understanding of the extinction theorem by Sein (1970, 1975), de Goede and Mazur (1972) and Wolf (1973).

The integral equation method was reexpressed in the framework of a polarization approach and the extinction theorem and abc obtained again by Birman and Sein (1972). In this approach the role of the local field becomes apparent (it is somewhat obscure in Sein's original formulation), and can be separately studied: an initial report was given on the effect of modifying the "local field factor" by Kraft and Birman (1974). Because of the clarity and power of the integral equation approach we shall use it in this article – this method deserves wider use in other problems as was earlier emphasized by Born and Wolf (1958).

Agarwal et al. (1971, 1973) reformulated the problem of the electrodynamics of non-local media in a differential equation approach extended to higher order owing to the non-local (integral) constitutive relation. Their analysis recovered Sein's results but proved (rather than assumed) the expansion of the fields in plane waves, in the model geometry. They also exhibited and elaborated upon the full vectorial nature of all fields and analyzed various geometries and polarizations. Maradudin and Mills (1973) developed a differential equation formulation which is in essence equivalent to that of Agarwal et al. (1973). They also recovered Sein's abc results for the susceptibility (8) and (10). That paper contains an important extension of the theory to surface polaritons and will be discussed in another volume of this series.

Another successful approach to the abc problem was meanwhile developed by Zeyher et al. (1972) in the course of their investigation of the non-translationally invariant susceptibility.

They examined the condition for self-consistency of propagation of polarization waves in non-local medium, and by taking advantage of the structure of $\chi(\boldsymbol{q}, \omega)$ in the complex $\boldsymbol{q}$ plane they were able to obtain the abc essentially from an extinction theorem in a differential approach, rather than the integral equation method which Sein used.

Skettrup (1973) has published an interesting variational approach to the abc problem based on extremizing the Lagrangian of a matter system in non-local medium allowing for "corner" (Weierstrass–Erdmann) conditions of discontinuity. This work recovered Pekar's abc for the susceptibility (8). Since Pekar's abc are not equivalent to the correct Sein abc for the prescribed susceptibility (8), some further investigation of this variational method is indicated.

In the following paragraphs some salient aspects of each method will be presented. It is worth again, emphasizing that the abc problem is in reality the problem of the structure of the physical polariton in a realistic, bounded medium with non-local susceptibility. As such it is as crucial a problem for the physics of these excitations as that of determining the dispersion $q_j(\omega)$ for each branch of propagating polariton mode.

3.4. The integral formulation and solution of the abc problem

3.4.1. Integral formulation

The integral formulation of electrodynamics is a powerful method of treating electromagnetic phenomena in material media (Born 1933, Rosenfeld 1955, Born and Wolf 1958). We assume the material system is comprised of polarizable entities each with dipole moment $\boldsymbol{p}_0$, or dipolar "pseudo-oscillators", distributed on the crystal lattice. These entities may be taken as a model for the exciton.

Let the jth dipole be located at site $\boldsymbol{R}_j$, let $R_{jl} \equiv |\boldsymbol{R}_j - \boldsymbol{R}_l|$. The total electric field at site $\boldsymbol{R}_j$ when an incident electromagnetic field $\boldsymbol{E}^{(i)}(\boldsymbol{r}, t)$ is incident upon the medium is

$$\boldsymbol{E}'_j = \boldsymbol{E}^i(\boldsymbol{r}_j, t) + \sum_l{}' \boldsymbol{E}_{jl}, \tag{63}$$

where $\boldsymbol{E}_{jl}$ is the field at site j due to the excited dipole at site l; the primed sum excludes the self-term $l = j$. The field $\boldsymbol{E}_{jl}$ is

$$\boldsymbol{E}_{il} = \underset{l}{\nabla} \times \underset{l}{\nabla} \times \frac{\boldsymbol{p}_0(t - R_{jl}/c)}{R_{ji}}. \tag{64}$$

Assume it is physically and mathematically possible to "smooth" the dipole distribution so that a dipolar density $\boldsymbol{P}(\boldsymbol{r}, t)$ can be defined at every space-time point $(\boldsymbol{r}, t)$. Passing to the limit we have for the field at the site $(\boldsymbol{r}, t)$:

$$\boldsymbol{E}'(\boldsymbol{r}, t) = \boldsymbol{E}^{(i)}(\boldsymbol{r}, t) + \int_{\sigma(\boldsymbol{r})}^{\Sigma} \nabla \times \nabla \times \left[\frac{\boldsymbol{P}(\boldsymbol{r}', t - R/c)}{R}\right] \mathrm{d}\boldsymbol{r}', \tag{65}$$

where $R \equiv |\boldsymbol{r}' - \boldsymbol{r}|$, and the total polarization is $\boldsymbol{P}(\boldsymbol{r}, t)$ (dipolar density per unit volume). The field $\boldsymbol{E}'(\boldsymbol{r}, t)$ which is the field at $(\boldsymbol{r}, t)$ due to all dipolar oscillators except that at $(\boldsymbol{r}, t)$ is to be identified as the Lorentz local field $\boldsymbol{E}_L(\boldsymbol{r}, t)$. Note that integration is over that volume of the crystal occupied by the matter and bounded by Σ, omitting a small region $\sigma(\boldsymbol{r})$ about the point $\boldsymbol{r}$. Assume the field is monochromatic and time harmonic:

$$\boldsymbol{P}(\boldsymbol{r}, t) = \boldsymbol{P}(\boldsymbol{r}, \omega) \exp(-\mathrm{i}\omega t), \tag{66}$$

then

$$\frac{\boldsymbol{P}(\boldsymbol{r}', t - R/c)}{R} = \boldsymbol{P}(\boldsymbol{r}', \omega) G_0(R), \tag{67}$$

where

$$G_0(R) \equiv [\exp(\mathrm{i}k_0 R)]/R, \quad k_0 = \omega/c, \tag{68}$$

and

$$\nabla^2 G_0(R) + k_0^2 G_0(R) = 4\pi\delta(R). \tag{69}$$

A separate assumption now invokes the Lorentz–Lorenz relation

$$\boldsymbol{E}_L(\boldsymbol{r}, t) = \boldsymbol{E}(\boldsymbol{r}, t) + \tfrac{4}{3}\pi \boldsymbol{P}(\boldsymbol{r}, t) \tag{70}$$

to eliminate the local ("microscopic") field $\boldsymbol{E}_L(\boldsymbol{r}, t)$ in favor of the Maxwell (macroscopic) field $\boldsymbol{E}(\boldsymbol{r}, t)$. Inserting this expression, and carrying out the interchange of $\nabla \times \nabla \times$ and integral in the usual manner (Rosenfeld 1955) gives

$$\boldsymbol{E}(\boldsymbol{r}, \omega) + 4\pi \boldsymbol{P}(\boldsymbol{r}, \omega) = \boldsymbol{E}^{i}(\boldsymbol{r}, \omega) + \nabla \times \nabla \times \int_{\sigma(\boldsymbol{r})}^{\Sigma} \boldsymbol{P}(\boldsymbol{r}', \omega) G_0(R)\, \mathrm{d}\boldsymbol{r}'. \tag{71}$$

At this point we convert as many terms as possible into $\boldsymbol{P}(\boldsymbol{r}, \omega)$ by applying the integral operator $\hat{\chi}$ (see eq. (1)) and integrating appropriately over the medium. Omitting arguments we have

$$\boldsymbol{P} + 4\pi\hat{\chi}\boldsymbol{P} = \hat{\chi}\boldsymbol{E}^{(i)} + \hat{\chi}\nabla\times\nabla\times\int_{\sigma(r)}^{\Sigma}\boldsymbol{P}(\boldsymbol{r}'\omega)G_0(R)\,\mathrm{d}\boldsymbol{r}'. \tag{72}$$

The electrodynamics of bounded spatially dispersive material media is contained in this equation, including a complete set of boundary conditions. Both longitudinal *and* transverse polarized solutions can arise from the solution of this equation.

We shall restrict attention to the transverse polarized case in the interest of simplicity. All our results have also been obtained for longitudinal polarization (Sein 1969, 1970, Birman and Sein 1972, Birman and Zeyher 1974); thus for general wave vector.

We now assume an incident monochromatic plane wave, denoted $\boldsymbol{E}(\boldsymbol{r}, \omega)$, impinging on the medium which is bounded by plane $z = 0$, and satisfying the vacuum wave equation, gives rise to several transverse propagating (possibly damped) plane polarization waves denoted $\boldsymbol{P}_j(\boldsymbol{r}, \omega)$. These waves satisfy the Helmholtz equations respectively:

$$\nabla^2\begin{pmatrix}\boldsymbol{E}^{(i)}(\boldsymbol{r}, \omega)\\ \boldsymbol{P}_j(\boldsymbol{r}, \omega)\end{pmatrix} = -\begin{pmatrix}k_0^2\\ k_j^2\end{pmatrix}\begin{pmatrix}\boldsymbol{E}^{(i)}(\boldsymbol{r}, \omega)\\ \boldsymbol{P}_j(\boldsymbol{r}, \omega)\end{pmatrix}, \tag{73}$$

where $k_0 \equiv \omega/c$ is the vacuum wavenumber, k_j the unknown wavenumber of the propagating polarization. Hereafter we use $k_j(\omega)$ for wavenumber.

Substituting this ansatz into the general equation, and carrying out straightforward mathematical manipulations we shall find the conditions under which solutions of the proposed type can exist. We must now specify the susceptibility explicitly.

3.4.2. Three model susceptibilities

We shall now introduce a general analytical form for the three non-local susceptibilities given previously as eqs. (10), (20) and (27). Introducing a parameter $R(0)$, these can be written in compact form as:

$$\chi(\boldsymbol{r}, \boldsymbol{r}, \omega) = [\chi_0\delta(\boldsymbol{r}-\boldsymbol{r}') + \chi_1 G_+(R) + R(0)\chi_1 G_+(\xi)]\theta(z)\theta(z'), \tag{74}$$

where, as before

$$R \equiv |\boldsymbol{r}-\boldsymbol{r}'|,$$

$$\xi \equiv |(\boldsymbol{\eta}-\boldsymbol{\eta}')^2 + (z+z')^2|. \tag{75}$$

The parameter $R(0)$ takes three values in these analytical models. Recall that $R(0)$ is, formally, the effective reflection coefficient of the exciton at the surface Σ.

The three cases are:

$$R(0) = -1 \qquad \text{(Frenkel)} \tag{76}$$

this is the Frenkel tight binding case (Zeyher et al. 1972);

$$R(0) = +1 \qquad \text{(Wannier)} \tag{77}$$

is the Wannier-like case (Ting et al. 1975); finally, taking

$$R(0) = 0 \qquad \text{(infinite crystal)} \tag{78}$$

gives the single oscillator (translationally invariant) or dielectric approximation model (Hopfield and Thomas 1963). The third case evidently does *not* conserve probability since $|R(0)| \neq 1$ (see Bishop and Maradudin, 1976).

3.4.3. *The solution*

The integral equation method was applied (Birman and Zeyher 1974) to the models with $R(0) = -1$ and $R(0) = 0$, but it is obviously trivial to formally permit $R(0) = +1$ in the identical framework. Proceeding as in that reference we insert eqs. (73) and (74) into eq. (72) and solve to find:

(*A*) *The extinction theorem.* The incident wave is "extinguished" in the medium and the propagating "refracted" waves travel at phase velocity c/n_j where $n_j \equiv k_j/k_0$. We can write the extinction theorem in terms of the amplitudes of the (Fourier component) of propagating polarizations. In all cases using susceptibilities (74), this theorem is

$$\boldsymbol{E}^{(i)}(\boldsymbol{k}, \omega) = 2\pi[k_0 + \sum_{j=1}^{2} k_j \boldsymbol{P}_j(\boldsymbol{k}_j, \omega)/(k_j^2 - k_0^2)]. \tag{79}$$

If we now use the relationship between $\boldsymbol{P}_j(\boldsymbol{k}_j, \omega)$ and $\boldsymbol{E}_j(\boldsymbol{k}_j, \omega)$, namely:

$$\boldsymbol{P}_j(\boldsymbol{k}_j, \omega) = \boldsymbol{E}_j(\boldsymbol{k}_j, \omega)\chi(\boldsymbol{k}_j, \omega), \tag{80}$$

this can be transformed to:

$$\boldsymbol{E}^i(\boldsymbol{k}, \omega) - \tfrac{1}{2}\sum_j (n_j + 1)\boldsymbol{E}_j(\boldsymbol{k}_j, \omega) = 0. \tag{81}$$

(*B*) *The dispersion equation.* In *all* cases the wavenumber k_j of the propagating modes satisfies the dispersion equation

$$(k_j^2/k_0^2) = 1 + [(4\pi)/(2\pi)^{3/2}](\chi_0 + 4\pi\chi_1)/(k_j^2 - k_+^2), \tag{82}$$

where k_+ is given in eqs. (13) and (14). This equation can be recognized as $(k_j^2/k_0^2) = \epsilon_p(k_j, \omega)$ where $\epsilon_p(k_j, \omega)$ is the Fourier transform of the propagating, or translationally invariant, part of the dielectric function, evaluated at $q = k_j$. In "$\boldsymbol{r}$" space this is the portion depending on $|\boldsymbol{r} - \boldsymbol{r}'|$ only.

We pause to note that neither the extinction theorem nor the dispersion equation depends on the reflection coefficient parameter $R(0)$: both depend only on the "propagating part" of the susceptibility.

(*C*) *The additional boundary conditions.* Turning to the abc, the formalism of the integral equation method produces, as a necessary condition for a solution to be possible, certain specific constraints (Birman and Zeyher 1974) which we now write out in detail for an *arbitrary* bounding surface Σ. The two coupled constraints or abc are:

$$\chi_1 \sum_j (1 + R(0))\boldsymbol{S}_+(\boldsymbol{P}_j)[(k_j^2 - k_0^2)(k_j^2 - k_+^2)]^{-1} = 0, \tag{83}$$

and

$$-(4\pi)^2\chi_1 R(0) \sum_j \boldsymbol{P}_j(\boldsymbol{r})[(k_j^2 - k_0^2)(k_j^2 - k_+^2)]^{-1} = 0. \tag{84}$$

Here

$$\boldsymbol{P}_j(\boldsymbol{r}) = \boldsymbol{P}_j \exp(\mathrm{i}\boldsymbol{k}_j \cdot \boldsymbol{r} - \mathrm{i}\omega t) \tag{85}$$

is the propagating polarization at wavenumber $\hat{k}_j$, and $\boldsymbol{S}_+(\boldsymbol{P}_j)$ is the surface integral

$$\boldsymbol{S}_+(\boldsymbol{P}_j) \equiv \int_\Sigma (\boldsymbol{P}_j G_+(R)_n - G_+(k)\boldsymbol{P}_{jn})\,\mathrm{d}S, \tag{86}$$

with $G_+(R)_n$ and $\boldsymbol{P}_{jn}$ being the *normal* derivative of $G_+(R)$ (defined in eq. (11)) and $\boldsymbol{P}_j(\boldsymbol{r})$ respectively, over the surface Σ. Equations (83) and (84) are the abc.

In the special case of a plane surface (xy-plane, $z = 0$), which is the model geometry used here we can reduce these integrals as discussed in Birman and Sein (1972) to obtain a simpler expression for the abc. We have

$$(1 + R(0)) \sum_{j=1}^{2} [(k_j^2 - k_0^2)(k_j - k_+)]^{-1}\boldsymbol{P}_j(\boldsymbol{k}_j, \omega) = 0, \tag{87}$$

and

$$R(0) \sum_{j=1}^{2} [(k_j^2 - k_0^2)(k_j^2 - k_+^2)]^{-1}\boldsymbol{P}_j(\boldsymbol{k}_j, \omega) = 0. \tag{88}$$

Equations (87) and (88) can be combined to give

$$\sum_j \left(\frac{1}{(k_j - k_+)} + \frac{R(0)}{(k_j + k_+)}\right) \boldsymbol{E}_j(\boldsymbol{k}_j, \omega) = 0. \tag{89}$$

This general expression for the abc was first given by Zeyher et al. (1972), eq. (3.10) (see eq. (117) below). We discuss here the model cases $R(0) = \pm 1, 0$; Halevi and Fuchs (1978) applied this expression to other cases.

For comparison with Ivchenko (ch. 4) note that for plane wave propagation a factor in eq. (87) can be rewritten using the substitution

$$(1 + k_j/k_+) \rightarrow \left(1 - \frac{\mathrm{i}}{k_+}\frac{\partial}{\partial z}\right)\Bigg|_\Sigma \tag{90}$$

where the r.h.s. of eq. (90) implies that the derivative is to be evaluated over the boundary surface Σ. In the present paper we consider plane boundary Σ but the formalism applies with evident modification to arbitrary surface Σ.

For $R(0)=0$, which is the translationally invariant susceptibility or dielectric approximation, we obtain the rigorous abc (using the above)

$$\sum_j (k_j + k_+)[(k_j^2 - k_0^2)]^{-1}\boldsymbol{P}_j(\boldsymbol{k}, \omega) = 0, \tag{91}$$

which was first obtained by Sein (1969, 1970) and Birman and Sein (1972). Using the convention of eq. (90) this can be rewritten as:

$$\left(1 - \frac{\mathrm{i}}{k_+}\frac{\partial}{\partial z}\right)\sum_{j=1} [(k_j^2 - k_0^2)(k_j^2 - k_+^2)]^{-1}\boldsymbol{P}_j(\boldsymbol{k}_j, \omega)\Big|_{\Sigma} = 0, \tag{92}$$

with the derivative operating on $\boldsymbol{P}_j(\boldsymbol{r})$, defined in eq. (84), and the result evaluated on the surface Σ.

For $R(0)=-1$, we have the Frenkel or tight binding exciton case for which the abc is

$$\sum_j [(k_j^2 - k_0^2)(k_j^2 - k_+)]^{-1}\boldsymbol{P}_j(\boldsymbol{k}_j, \omega) = 0 \qquad \text{(Frenkel)}. \tag{93}$$

This can be rewritten:

$$(\boldsymbol{P}_1(\boldsymbol{k}_1, \omega) - \chi_0\boldsymbol{E}_1(\boldsymbol{k}_1, \omega)) + (\boldsymbol{P}_2(\boldsymbol{k}_2, \omega) - \chi_0\boldsymbol{E}_2(\boldsymbol{k}_2, \omega)) = 0. \tag{94}$$

This abc was first obtained by Zeyher et al. (1972) from the analysis of the *macroscopic* complex susceptibility (this will be discussed in sect. 3.5 below). The abc (94) is identical to that proposed by Pekar, eq. (59), but there is a crucial difference. The abc (94) applies to the susceptibility (74) with $R(0)=-1$, which is *not* Pekar's susceptibility (56). In fact, eq. (56) corresponds to the dielectric approximation, $R(0)=0$, for which the correct abc is eq. (91) or eq. (92). The reader should not be confused by this situation. The literature from 1958–1970 is replete with instances of a susceptibility being postulated (or derived) for the bounded crystal, and mathematically inconsistent set of abc then being inserted, instead of a consistent macroscopic electrodynamics being developed.

For $R(0)=+1$ we have the Wannier exciton case, and the abc

$$\frac{\partial}{\partial z}\sum_j [(k_j^2 - k_0^2)(k_j^2 - k_+^2)]\boldsymbol{P}_j(\boldsymbol{r})\,|_{\Sigma} = 0, \qquad \text{(Wannier)} \tag{95}$$

where again the derivatives are evaluated on Σ. This abc was first obtained by Ting *et al.* (1975).

In conclusion: rigorous macroscopic electrodynamics in the integral equation framework has been shown to give a complete and unambiguous solution of the abc problem. To each model susceptibility there corresponds, without further assumption or microscopic theory, one unique abc. Johnson and Rimbey (1976) have shown that in some cases, the symmetry of $\chi(\boldsymbol{r}, \boldsymbol{r}', \omega)$ under reflection in a plane boundary can give abc; a similar observation was made earlier in an unrelated problem of reflectivity off a plane surface in metal optics by Kliewer and Fuchs (1968). These observations are restricted to plane surfaces.

The general result from the integral equation formalism (83) and (84) does not depend on the surface being plane since the surface integral is carried out over the physical surface Σ which may have no relationship whatsoever to crystal symmetry or to possible mirror plane. Similar general results are obtained in the other formalisms discussed below in sects. 3.5 and 3.6.

The integral equation approach was applied to bounded optically active – gyrotropic–nonlocal media by Puri and Birman (1981).

3.4.4. *Multi-oscillator model and abc*

Sein (1969, 1970) also studied a "multi-oscillator" model consisting of N discrete excitonic levels each one with a possibly distinct kinetic energy or mass term. The form of susceptibility appropriate for the multioscillator model is

$$\chi(\boldsymbol{k}, \omega) = \left[\chi_b + \sum_{s=1}^{N} \frac{\chi_{1s}}{(k^2 - k_{+s}^2)}\right], \tag{96}$$

where s labels the oscillator, $\chi_{1s} \equiv (\pi a^2 \omega_{0s}^2)/b_s(2\pi)^{1/2}$ with $b_s \equiv (1/m_s)$ the inverse mass of this oscillator, ω_{0s} the resonance frequency of the oscillator, and k_{+s} is the identical expression as eqs. (13) and (14) with b_s replacing b, m_s replacing m, ω_{0s} replacing ω_0. This will be recognized as a representation of $\chi(\boldsymbol{k}, \omega)$ as a meromorphic function and as such is of greater generality than merely applicable to excitons. The constitutive relationship expressed in configuration space is

$$\boldsymbol{P}(\boldsymbol{r}, \omega) = \chi_0 \boldsymbol{E}(\boldsymbol{r}, \omega) + \sum_{s=1}^{N} \chi_{1s} \int G_{+s}(\boldsymbol{r} - \boldsymbol{r}')\boldsymbol{E}(\boldsymbol{r}', \omega)\, \mathrm{d}\boldsymbol{r}'. \tag{97}$$

The entire program carries through just as before and generalizes previous results for $\boldsymbol{R}(0) = 0$. We find:

(*A*) *The extinction theorem.* In this case the incident wave is extinguished by the superposition of contributions from $(N + 1)$ waves:

$$\boldsymbol{E}^{(i)}(\boldsymbol{k}, \omega) - \tfrac{1}{2} \sum_{j=1}^{N+1} (n_j + 1)\boldsymbol{E}_j(\boldsymbol{k}_j, \omega) = 0. \tag{98}$$

(*B*) *The dispersion equation.* The $(N + 1)$ propagating modes have wavenumbers determined as solutions of the dispersion equation:

$$(k_j^2/k_0^2) = 1 + [(4\pi)/(2\pi)^{3/2}]\left[\chi_0 + \sum_{s=1}^{N} \frac{4\pi\chi_{1s}}{(k_j^2 - k_{+s}^2)}\right]. \tag{99}$$

(*C*) *The additional boundary conditions.* Each oscillator – each value of s – gives rise to one additional boundary condition in the case of normal incidence, transverse propagation: or a total of N abc. Each of them is of form:

$$\left(1 - \frac{i}{k_{+s}} \frac{\partial}{\partial z}\right) \sum_{j=1}^{N+1} [(k_j^2 - k_0^2)(k_j^2 - k_{+s}^2)]^{-1} \boldsymbol{P}_j(k_j, \omega)\Big|_{\Sigma} = 0 \quad s = 1, \ldots N. \tag{100}$$

Worth noting is that in this case the "model experiment" (normal incidence reflectivity) comprises: *one* incident field amplitude $E_0(\boldsymbol{k}, \omega)$, *one* reflected field amplitude $\boldsymbol{E}_0'(\boldsymbol{k}, \omega)$, and $(N + 1)$ propagating amplitudes. Consequently we have to determine $(N + 2)$ ratios: E_0'/E_0, E_j/E_0. The totality of boundary conditions available comprises 2 Maxwell b.c. plus N abc. All boundary conditions needed are available rigorously from the theory!

Recently, Cho (1978) reported some investigations of a multi-oscillator model. His proposed abc were the extended Pekar conditions $\boldsymbol{P}_{\text{TOT}}(\Sigma) = 0$, where $\boldsymbol{P}_{\text{TOT}}(\Sigma)$ here refers to the total excitonic polarization evaluated on Σ. This abc is not a correct consequence of the multi-oscillator susceptibility.

Two remarks now seem appropriate. Evidently the more general model susceptibility (74) can be generalized to include N discrete levels.

$$\chi(\boldsymbol{r}, \boldsymbol{r}', \omega) = [(\chi_0\delta(\boldsymbol{r} - \boldsymbol{r}') + \sum_s^N \chi_{1s}G_{+s}(\boldsymbol{R}) + \sum_s^N R_s(0)\chi_{1s}G_{+s}(\xi)]\theta(z)\theta(z'). \tag{101}$$

Although we do not write down the explicit abc for all cases, they are easy to do and further investigations of this type could prove fruitful.

No systematic comparison of a multilevel oscillator model with experiment seems to have been reported.

3.5. *Susceptibility in the complex k plane and solution of the abc problem*

Using a method based upon the structure of the Fourier transform of $\chi(\boldsymbol{r}, \boldsymbol{r}', \omega)$ Zeyher et al. (1972) first showed how to obtain all needed abc by a direct procedure which will now be briefly recapitulated. The non-local constitutive relation, for normal incidence, can be written in scalar form (suppressing cartesian indices) since all fields can be taken colinear, as

$$P(\boldsymbol{r}) = \int \chi(\boldsymbol{r}, \boldsymbol{r}')E(\boldsymbol{r}')\, d\boldsymbol{r}', \tag{102}$$

with the susceptibility tensor given in eq. (74). We shall seek propagating solutions of this equation in the medium in the case of the simple model geometry: normal incidence plane polarized fields.

With incident field ($z<0$) travelling as $E_0\exp(ik_0z)$, reflected field ($z<0$) as $E_0'\exp(-ik_0z)$, we shall assume inside the medium

$$E(\boldsymbol{r})=\sum_j E_j\exp(ik_jz),$$

$$P(\boldsymbol{r})=\sum_j P_j\exp(ik_jz),\qquad z>0, \tag{103}$$

the right-running solutions are expressed as Fourier series with as yet undetermined propagation numbers k_j, and number of waves $i=1\ldots N$.

From Maxwell's equation for transverse wave propagation,

$$-\nabla^2E(\boldsymbol{r})=k_0^2(E(\boldsymbol{r})+4\pi\hat{\chi}E(\boldsymbol{r})), \tag{104}$$

or writing out the kernel

$$-(\nabla^2+k_0^2)E(\boldsymbol{r})=4\pi k_0^2\int\chi(\boldsymbol{r},\boldsymbol{r}')E(\boldsymbol{r}')\,\mathrm{d}\boldsymbol{r}'$$

$$=4\pi k_0^2P(\boldsymbol{r}). \tag{105}$$

Left and right hand sides of this equation must be consistent, of course.

Let us now examine more closely the polarization $P(\boldsymbol{r})$ which can arise from an assumed propagating field, by considering:

$$\int\chi(\boldsymbol{r},\boldsymbol{r}')E(\boldsymbol{r}')\,\mathrm{d}\boldsymbol{r}'. \tag{106}$$

Using the general expression for $\chi(\boldsymbol{r},\boldsymbol{r}')$ we consider the region of a single excitation. Near this frequency we write (changing a sum to an integral).

$$\chi(\boldsymbol{r},\boldsymbol{r}')=\chi_0\delta(\boldsymbol{r}-\boldsymbol{r}')+\frac{\omega_{\mathrm{p}}}{2\omega^2}\int\Bigg[\frac{\omega_q\phi_q(\boldsymbol{r})\phi_q^*(\boldsymbol{r}')}{\omega_0+bq^2+\omega+\mathrm{i}\eta}$$

$$+\frac{\omega_q\phi_q^*(\boldsymbol{r})\phi_q(\boldsymbol{r}')}{\omega_0+bq^2-(\omega+\mathrm{i}\eta)}\Bigg]\frac{\mathrm{d}^3q}{(2\pi)^3}, \tag{107}$$

where $\phi_q(\boldsymbol{r})$ is defined in eq. (19). Consider one term in the $\boldsymbol{q}$ integral and write

$$b(q^2+\omega/b+\omega_0/b+\mathrm{i}\eta)=b(q^2+\lambda^2),$$

$$\lambda^2\equiv(\omega/b+\omega_0/b+\mathrm{i}\eta). \tag{108}$$

The first term then takes the form:

$$\frac{\omega_q\phi_q(\boldsymbol{r})\phi_q^*(\boldsymbol{r}')}{2\lambda b}\left(\frac{1}{q-\lambda}-\frac{1}{q+\lambda}\right), \tag{109}$$

with poles at $q = \pm\lambda$. Recall that

$$q = (q_x^2 + q_y^2 + q_z^2)^{1/2}. \tag{110}$$

Now we can substitute for the current matrix element the expressions (18) and (19), and for the $\boldsymbol{E}$ field eq. (100). We obtain a contribution to $\boldsymbol{P}(\boldsymbol{r})$ which will have originated in the local part of the susceptibility

$$\sum_j (4\pi\chi_0 - \omega_p^2/\omega^2) E_j \exp(\mathrm{i}k_j z), \tag{111}$$

plus a contribution of the form

$$\sum_i \left(\frac{\omega_p^2}{2\omega^2}\right) E_i \int \frac{\mathrm{d}^3 q}{(2\pi)^3} \left(\frac{1}{2\lambda b}\right) \left[\frac{1}{(q-\lambda)} - \frac{1}{(q+\lambda_+)}\right] \phi_q(\boldsymbol{r}) \times \left[\int \mathrm{d}^3 r' \phi_q^*(\boldsymbol{r}') \exp(\mathrm{i}k_j z') + \text{c.c.}\right]. \tag{112}$$

Doing the integral over $\boldsymbol{r}'$ in the second (square) bracket we obtain

$$[\ ] = \left(\frac{\mathrm{i}}{k_j - q_z + \mathrm{i}\eta} + \frac{\mathrm{i}R^*(q_z)}{k_j + q_z + \mathrm{i}\eta}\right) \delta(q_x)\delta(q_y). \tag{113}$$

It now remains to perform the $\boldsymbol{q}$ integration which gives directly three contributions to $P(\boldsymbol{r})$. Thus $P(\boldsymbol{r})$ has the form:

$$P(\boldsymbol{r}) = \sum_j{}' P_j \exp(\mathrm{i}k_j z) + P_+ \exp(\mathrm{i}k_+ z). \tag{114}$$

Now note that the k_j are as yet not determined. The P_j are coefficients given as

$$P_j = \chi_P(k_j, \omega) E_j, \tag{115}$$

where, as before χ_p is the propagating susceptibility.

Now return to the Maxwell equation (105). Substituting the expansion for $E(\boldsymbol{r})$ on the left, the expression for P on the right, we obtain, as in the integral equation method several groups of terms, each characterized by a propagation wavenumber (k_j, or k_+) or plane-wave factor $\exp(\mathrm{i}k_j z)$ or $\exp(\mathrm{i}k_+ z)$ respectively. When independent terms are separately equated we find: (a) Dispersion equation for k_j:

$$k_j^2/k_0^2 = 1 + 4\pi\chi_p(k_j, \omega), \tag{116}$$

where χ_p is the "propagating" part of the susceptibility as in eq. (82). We also find (b): the abc. From the terms at propagation number k_+ the abc:

$$P_+ \equiv -\sum_j r_+ \left(\frac{1}{k_j - k_+} + \frac{R(k_+)}{k_j + k_+}\right) = 0, \tag{117}$$

where $R(k_+)$ is the value of the "reflection coefficient" $R(q_z)$ at the wavenumber k_+, and r_+ the residue (a non-zero constant) which factors out of the equation, to give the abc.

If as before we choose $R(k_+) \sim R(0) = -1, 0, +1$, respectively, we immediately recover all three cases of abc (92), (94) and (95) as obtained in the integral equation formulation. Note that the present method admits more general reflection coefficients $R(q_z)$ with exciton-wavenumber dependent phase shifts. By previous definition of $R(q_z)$ we should have $|R(q_z)| = 1$ but if exciton attenuation at the surface occurs then $|R(q_z)| < 1$. Recently Halevy and Fuchs (1979) have discussed treating $R(0)$ as a phenomenological parameter and have extended (74) to the anisotropic case.

3.6. Differential form of Maxwell equations and abc

Two treatments of the electrodynamics of spatially dispersive media which utilize a differential form of Maxwell equations were given at about the same time, by Agarwal, Pattanayak, and Wolf, (1971, 1973) and by Maradudin and Mills (1973). Both these theories used the one resonance oscillator ("dielectric approximation") model for the susceptibility of the bounded medium.

We shall briefly outline these methods. Maradudin and Mills begin with the Maxwell equation:

$$\nabla \times \nabla \times \boldsymbol{E}(\boldsymbol{r}, \omega) = k_0^2[\boldsymbol{E}(\boldsymbol{r}, \omega) + 4\pi \boldsymbol{P}(\boldsymbol{r}, \omega)]. \tag{118}$$

As previously let us consider transverse wave propagation so that $\nabla \times \nabla \times = \nabla(\nabla \cdot) - \nabla^2 = -\nabla^2$ and

$$-(\nabla^2 + k_0^2)\boldsymbol{E}(\boldsymbol{r}, \omega) = 4\pi k_0^2 \boldsymbol{P}(\boldsymbol{r}, \omega). \tag{119}$$

Now

$$\boldsymbol{P}(\boldsymbol{r}, \omega) = \hat{\chi}\boldsymbol{E}(\boldsymbol{r}, \omega), \tag{120}$$

and writing this out using the explicit form for the susceptibility corresponding to the models (74) with $R(0) = 0$, we have

$$\boldsymbol{P}(\boldsymbol{r}, \omega) = \int \mathrm{d}\boldsymbol{r}'[\chi_0 \delta(\boldsymbol{r} - \boldsymbol{r}') + \chi_1 G_+(\boldsymbol{r} - \boldsymbol{r}', \omega)]\boldsymbol{E}(\boldsymbol{r}'). \tag{121}$$

As was noted in sect. 1 (see eq. (15)), the object G_+ is the Green function of Helmholtz operator:

$$(\nabla^2 + k_+^2)G_+(\boldsymbol{r} - \boldsymbol{r}') = -4\pi\delta(\boldsymbol{r} - \boldsymbol{r}'). \tag{122}$$

Comparing these equations and the Maxwell equation which we now rewrite as

$$-(\nabla^2 + k_0^2 + 4\pi k_0^2 \chi_0)\boldsymbol{E}(\boldsymbol{r}, \omega) = k_0^2 \chi_1 \int [G_+(\boldsymbol{r} - \boldsymbol{r}', \omega)]\boldsymbol{E}(\boldsymbol{r}')\, \mathrm{d}\boldsymbol{r}', \tag{123}$$

it is clear that the integral on the right-hand side can be eliminated if we apply the Helmholtz operator $(\nabla_r^2 + k_+^2)$ where the differentiation is with respect to the free coordinate $\boldsymbol{r}$. Thus

$$-(\nabla_r^2 + k_+^2)(\nabla^2 + k_0^2 + 4\pi k_0^2\chi_0)\boldsymbol{E}(\boldsymbol{r}, \omega) = 4\pi k_0^2\chi_1\boldsymbol{E}(\boldsymbol{r}, \omega). \tag{124}$$

Clearly the second order integro-differential equation has been converted to a fourth order partial differential equation. If we assume there are plane-wave solutions with wavevector $\boldsymbol{k}$, then for the electric and magnetic fields $\nabla^2 \to -k^2$ and the waves obey the dispersion equation:

$$(k^2 - k_+^2)(-k^2 + k_0^2(1 + 4\pi\chi_0)) = 4\pi k_0^2\chi_1, \tag{125}$$

or

$$-k^2 + k_0^2\epsilon_0 = \frac{4\pi k_0^2\chi_1}{(k^2 - k_+^2)}, \tag{126}$$

which is

$$k^2 = k_0^2\epsilon(\boldsymbol{k}, \omega), \tag{127}$$

where, as previously,

$$\epsilon(\boldsymbol{k}, \omega) = \epsilon_0 + \frac{4\pi\chi_1}{(k^2 - k_+^2)}. \tag{128}$$

This is the dispersion equation for volume propagating modes. Just as previously, it only involves the propagating part of the susceptibility.

To find the abc we now construct the propagating solutions for the electric field. These are

$$\boldsymbol{E}(\boldsymbol{r}) = E_1 \exp(ik_1 z) + E_2 \exp(ik_2 z), \tag{129}$$

where k_1 and k_2 are the wavenumbers gotten as the solutions of the dispersion equation. Now follow a procedure like that previously given in eqs. (106)–(117). If this electric field is to be a propagating solution it must satisfy the integro-differential equation (123). Substituting into this equation and integrating we find that left and right side of these equations will be consistent if an extra term propagating with wavenumber k_+ vanishes. This gives the abc of Sein (1969, 1970) as before

$$E_1/(k_1 - k_+) + E_2/(k_2 - k_+) = 0. \tag{130}$$

This differential equation approach has the merit of giving directly the required dispersion and abc. It can also be extended to deal with surface related problems. Since the incident electric field does not appear in any part of the theory, the extinction theorem does not occur in the form (81).

The treatment of Agarwal et al. (1971, 1973) analyzes the general electromagnetic mode structure in bounded spatially dispersive media. The particular geometries treated in this paper include the half space and the finite slab. The paper contains *inter-alia* a rigorous proof that in the half-space plane waves (homogeneous and inhomogeneous) are eigen-modes of the Maxwell equations and a rather complete discussion of longitudinal and transverse modes. The proof given here provides an alternate rigorous justification of the plane-wave ansatz made by Sein (1969), Birman and Sein (1972), Zeyher et al. (1972), and Maradudin and Mills (1973). The latter papers supported their assumption by constructing such solutions. Agarwal et al. (1971) used as an analytical tool, the powerful angular spectrum, or Weyl, representation of the fields in a slab, or a half space. The method has been used in studies of gyrotropic media – Pattanayak et al. (1980), Goos–Hänschen shift (Birman et al. 1982) and other problems. We shall, in the interest of completeness, give a sketch of the principal mathematical statement which is the Weyl representation of the Green Function $G_+(|\boldsymbol{r}-\boldsymbol{r}'|, \omega)$. Take

$$G_+(|\boldsymbol{r}-\boldsymbol{r}'|) = \frac{\mathrm{i}}{(2\pi)} \int\int_{-\infty}^{\infty} \exp[\mathrm{i}(u(x-x') + v(y-y') + \omega_+|z-z'|)]\, \mathrm{d}u\, \mathrm{d}v, \tag{131}$$

with

$$\omega_+ \equiv (k_+^2 - u^2 - v^2)^{1/2}, \tag{132}$$

where k_+^2 is defined as in eqs. (13) and (14), and in the square root take $\operatorname{Re}\omega_+ > 0$, $\operatorname{Im}\omega_+ > 0$. The electric field in a slab $0 < z < d$ is expressed as a two dimensional Fourier integral with respect to (x, y),

$$\boldsymbol{E}(\boldsymbol{r}, \omega) = \int\int_{-\infty}^{\infty} \mathbf{E}(u, v, z, \omega) \exp[\mathrm{i}(ux + vy)]\, \mathrm{d}u\, \mathrm{d}v. \tag{133}$$

Using Maxwell's equations, an equation for $\boldsymbol{E}(\boldsymbol{r}, \omega)$ can be derived which has both longitudinal and transverse solutions. Restricting attention to the transverse case ($\boldsymbol{\nabla}\cdot\boldsymbol{E} = 0$) gives

$$(u^2 + v^2 - k_0^2\epsilon_0)E - \frac{\partial^2 \boldsymbol{E}}{\partial z^2} = \mathrm{i}\frac{\chi_1 k_0^2}{2} \int_0^d \frac{\exp(\mathrm{i}\omega_+|z-z'|)}{\omega_+} \boldsymbol{E}(u, v, z', \omega)\, \mathrm{d}z'. \tag{134}$$

Since the object

$$g_+(|z-z'|) \equiv \frac{\exp(\mathrm{i}\omega_+|z-z'|)}{\omega_+}, \tag{135}$$

is the Green Function of a "one-dimensional Helmholtz operator",

$$(\partial^2/\partial z^2 + \omega_+^2) g_+(|z-z'|) = 2\pi\mathrm{i}\delta(z-z'), \tag{136}$$

the integro-differential equation can be reduced to a higher order differential equation as previously. The electric field then obeys the fourth order system:

$$\partial^4 \mathbf{E}/\partial z^4 + (k_0^2\epsilon_0 - u^2 - v^2 + \omega^2)\partial^2 \mathbf{E}/\partial z^2 + [\omega_+^2(k_0^2\epsilon_0 - u^2 - v^2) - \chi_1 k_0^2]\mathbf{E} = 0. \tag{137}$$

Analysis of the solutions of this equation, subject to the requirement that the solution of the differential equation must solve the initial integro-differential equation, gives the dispersion relation and the abc of the dielectric approximation as before.

As Agarwal et al. (1973) emphasized, the abc are "mode coupling parameters", giving the correct linear combination of the polariton waves which satisfy Maxwell equations, i.e. the "physical" polaritons.

3.7. *Penultimate comment on abc's*

Why penultimate? Clearly the approach of finding a fully self-consistent microscopic theory of the solution of coupled Schrödinger and Maxwell equations in a bounded medium should be pursued. In this review we have given a deliberately selective and partial account, emphasizing the macroscopic aspects.

Many authors, Soviet and American, have followed Pekar's scenario of setting up a microscopic model equation of motion for the elementary polarization oscillators (like eq. (53)) and solving it subject to certain boundary conditions on the oscillator. This yields the susceptibility. The susceptibility and the same boundary conditions are injected into Maxwell macroscopic equations. The procedure is unacceptable at the macroscopic level since, as we emphasized, each susceptibility $\chi(\boldsymbol{r}, \boldsymbol{r}', \omega)$ comes with all information to produce the needed boundary conditions over an arbitrary surface Σ.

The acceptable, or well posed problem, as we discussed it here is: to solve Maxwell's macroscopic equations in a given medium subject to a prescribed non-local constitutive relation, in the medium bounded by surface Σ. The methods for definitive solution of this problem have been presented here, permitting attention now to be preferentially given to the microscopic self-consistent problem.

4. *Exciton-polariton spectroscopy*

As set out in the introduction, one of the main purposes of this article is to give an introduction to electrodynamic and optical consequences of the finite mass exciton effects leading to spatial dispersion. In this section we will briefly touch on several types of optical consequences of non-locality which occur when the exciting laser frequency ω_L is in the region of the exciton resonance: reflectivity (elastic scattering); resonant Brillouin (inelastic) scattering; transient and precursor (optical) effects. Other novel optical effects are being investigated by the present writer and collaborators (see Birman and Pattanayak 1979, Puri and Birman 1981, Pattanayak et al. 1980 and others). Since the experimental status of the first two of these topics will be thoroughly reviewed in the accompanying articles by Ivenchko and Koteles, respectively, we shall only make some brief remarks on them here.

4.1. *Reflectivity anomalies*

For a local medium, characterized by frequency dependent dielectric coefficient $\epsilon(\omega)$, the wavenumber of transverse waves satisfies the equation $k = (\omega/c)\epsilon(\omega)^{1/2}$. In such a local medium, in the resonance region, the single oscillator model (6) or (8) with $b = 0$ gives the traditional description of optical effects. Consider the case of no damping $\Gamma = 0$. Then in the frequency region $\omega_0 < \omega < \omega_\ell$, where the longitudinal frequency is defined by $\epsilon(\omega_\ell) = 0$, there is a "stop band", where k is pure imaginary and homogeneous plane waves do not propagate in the medium. Small damping, or $\Gamma \sim 0$ produces complex k, with a small real part of k, but does not change this conclusion significantly in $\omega_0 < \omega < \omega_\ell$, which is a region of near-total reflectivity. Since each ω corresponds to a single k, in the frequency regions where $\mathrm{Re}\, k \gg \mathrm{Im}\, k$ propagation or transmission in the crystal will occur, and the reflectivity will diminish.

In the presence of spatial dispersion, the transverse-wave dispersion equation $k^2 = (\omega/c)^2\epsilon(k, \omega)$ admits additional propagating solutions and the stop band is eliminated. A decrease in magnitude of reflectivity is to be expected compared to the local case, and is observed experimentally in the region $\omega_0 < \omega < \omega_\ell$. Just at frequency $\omega_L = \omega_\ell$ two propagating modes exist. The reflectivity as function of frequency $\rho(\omega)$ is expected to show a dip or some further structure at that frequency.

As fully discussed in Ivchenko's accompanying article three major features occur experimentally (on CdS, one of the more popular materials for investigation): larger damping constants Γ seem needed to account for reflectivity than are measured *on the same sample* in transmission experiments; a "spike" appears near ω_ℓ; the spike near ω_ℓ depends markedly upon polarization of the radiation with respect to the crystal axis. A fourth feature is the generally lower magnitude of the reflectivities than expected from the "classical" theory with the same parameters (except $b = 0$).

Some comments may be in order here. The accepted explanation of the spike requires introduction of an exciton-free "dead layer" of some thickness l, and physically this results in multiple reflection - interference producing the spike. Thus the spike at ω_ℓ seems not to be a simple consequence of the opening of a new channel (2 polaritons) at ω_ℓ in spatially dispersive media, but results from fairly complicated cancellations and reinforcements of different contributions of polaritons in the medium and in the "dead layer" (Ivchenko, chapter 4 of this volume). There is a certain irony in this since the "spike" was originally taken as giving strong experimental support to existence of spatial dispersion effects (Hopfield and Thomas 1963, Hopfield 1966). In fact, by suitably parameterizing the theory using several intrinsic parameters: oscillator strength, α, exciton mass M, damping coefficient Γ, background dielectric constant ϵ_0, resonance frequency ω_0 and dead-layer width λ, one can fit the reflectivity (in some cases) neglecting spatial dispersion altogether (Sein 1969). In recent work, introduction of an exciton-free dead layer with prescribed thickness l is treated as a useful model to account for rough surface and similar effects which markedly decrease exciton lifetime near the surface (Ivchenko 1980). However, with so many parameters at one's disposal, an "outsider" may legitimately question how compelling the experimental evidence for spatial dispersion is, based on reflectivity.

In those cases where identical, independently measured, parameters are used on the same samples for reflectivity, transmittivity and other properties one can feel some confidence that ability to "fit the reflectivity data" gives support to the existence of spatial dispersion effects. Certainly, in the earlier literature (1957–1977) fitting reflectivity data was "faute de mieux" taken as adequate support for spatial dispersion effects, but at present much attention has shifted to the inelastic resonant Brillouin scattering studies which can directly reveal many kinematic and dynamic features of the exciton polariton.

4.2. Resonance Brillouin scattering spectroscopy

The most important experimental information needed for confirmation of excitonic-polariton optical effects is the dispersion of the polaritons or the determination of $\omega(k)$ for each branch of propagating excitation *and* the composition of the physical polariton modes (as determined by the abc). To this end a theory of resonance Brillouin scattering was developed by Brenig et al. (1972), and an experimental search for these effects was proposed. This method has become a very powerful branch of the spectroscopy of solids and is being used to determine not only the basic exciton and photon parameters, mass M, damping constant Γ for exciton decay, resonance energy $\hbar\omega_0$ for the exciton, as well as the oscillator strength for exciton–photon coupling, but also the exciton–phonon coupling interaction which produces the scattering interaction. All these questions are discussed in greater detail by Koteles in chapter 3 of this volume, by Yu (1979), and by Ulbrich and Weisbuch (1978). We now give a brief review of the existing theory.

In order to probe the exciton polariton dispersion by an inelastic scattering process, consider that initial and final polariton states have energy $\omega(\boldsymbol{k})$, $\omega'(\boldsymbol{k}')$ and quasi-momentum $\boldsymbol{k}$, $\boldsymbol{k}'$ respectively. For example, state $\omega(\boldsymbol{k})$ may be created by the incident laser photon ω_L. Further, let $\Omega(\boldsymbol{q})$ be the dispersion relation of an acoustic phonon: $\Omega(\boldsymbol{q}) = C_s\boldsymbol{q}$ where q is the quasi-momentum, C_s the sound velocity. Then energy and quasi-momentum will be conserved if in the transition from initial to final states (assumed to be a Stokes or phonon-production process),

$$\omega(\boldsymbol{k}) - \omega'(\boldsymbol{k}') = \Omega(\boldsymbol{q}) = C_s\boldsymbol{q}, \tag{138}$$

and

$$\boldsymbol{k} - \boldsymbol{k}' = \boldsymbol{q}. \tag{139}$$

Figures illustrating these transitions are given in Koteles' article (ch. 3, fig. 3). It is very useful to think of the linear dispersion (138) of the acoustic phonon as being a "ruler" which originates at the initial polariton state, and terminates at the final polariton state to which a transition can occur. For $\omega < \omega_\ell$ only one (essentially undamped) photonic polariton propagates.

For experimental reasons the scattering experiment in the resonance regime is performed in a backscattering geometry so that $\boldsymbol{k}$ is for example in $+z$, $\boldsymbol{k}'$ in $-z$ directions. For visualization one reflects the multibranch excitation-dispersion curve about the $\boldsymbol{k} = 0$ axis. Now in the region $\omega \ll \omega_0$ the curve $\omega(\boldsymbol{k}) \sim (c/n)|\boldsymbol{k}|$ where $n = \epsilon_0^{1/2}$ is a "background refractive index" which in the non-resonant frequency regime has relatively small frequency dispersion. In this regime, the photon-like excitation has essentially a vertical slope (the polariton travels at essentially infinite velocity compared to the phonon) and the "ruler" joins points which lie on parallel, essentially vertical lines. Thus $(\omega - \omega') = \Omega$ will be essentially independent of the frequency of the exciting laser. This non-resonant Brillouin scattering is well known (Cummins and Schoen 1972).

As ω_L increases to ω_0 and then to ω_ℓ, qualitative differences occur owing to the (non-linear) dispersive properties of the polariton dispersion $\omega(\boldsymbol{k})$. At the 'end of the outer branch curve the tuning condition has significantly changed: the momentum transfer $\boldsymbol{q}$ increases ($|\boldsymbol{q}| \sim 2k$) and since the phonon dispersion is strictly linear the energy shift $\Omega(\boldsymbol{q})$ increases. This is the first qualitative effect which can be appreciated just from inspection of the forward and reflected dispersion curves: strong dispersion of the frequency shift as ω_L approaches ω_0, which has been observed even prior to the full observation of all resonance Brillouin scattering effects (Bruce and Cummins 1977).

The major class of phenomena which are of interest concern the full determination of mode dispersion and structure by, for example, Stokes scattering experiments for $\omega_L > \omega_\ell$ when the full two branch structure (coupled waves and abc) needs to be invoked. A quantum-mechanical and a classical theory have been developed which turn out, with minor differences, to be the same. In the presence of the acoustic phonon the (non-local) susceptibility will be modulated by the phonon of wavevector $\boldsymbol{q}$, as

$$\chi(\boldsymbol{r}, \boldsymbol{r}') \to \chi(\boldsymbol{r}, \boldsymbol{r}') + \delta\chi(\boldsymbol{r}, \boldsymbol{r}'). \tag{140}$$

The modulated susceptibility term $\delta\chi \sim \exp[\mathrm{i}(\boldsymbol{q} \cdot \boldsymbol{r} + \Omega t)]$ and thus the source term in Maxwell's equations for the scattered wave at frequency $\omega' \equiv Ck_0'$ is $4\pi k_0^2(\delta\chi)\boldsymbol{E}(\boldsymbol{r}, \omega)$ where $\boldsymbol{E}(\boldsymbol{r}, \omega)$ is the incident field at frequency ω. Then the equation for the scattered field $\boldsymbol{E}'$ is:

$$-(\nabla^2 + k_0'^2(1 + 4\pi\chi_{k_0'}))E' = 4\pi k_0'^2(\delta\chi E_{k_0})_{k_0'}. \tag{141}$$

Using a deformation potential approximation we write

$$\delta\hat{\chi} = \frac{\partial\hat{\chi}}{\partial n}\delta n = -\hat{\chi}C\hat{\chi}\delta n \tag{142}$$

where C is a deformation potential constant, independent of frequency, δn is the density fluctuation due to the phonon, $\hat{\chi}$ is the non-local susceptibility operator (1) with $\chi_0 = 0$. Formally solving eq. (141) gives for thc scattered field

$$E' = -\frac{4\pi k_0'^2}{[\nabla^2 + k_0'^2(1 + 4\pi\chi_{k_0'})]}\chi_{k_0'}C(\delta n P_{k_0})_{k_0'}. \tag{143}$$

Using a complete set of eigenfunctions $E_{k_0}^{\mathrm{INC}}(\boldsymbol{r})$ of the denominator, with incoming boundary conditions, we can write the scattered field as

$$E'_{k_0} = [(k_0'^2)/R]\exp(\mathrm{i}k_0'R)T_{k_0'k_0}, \tag{144}$$

showing the usual spherical wave scattering amplitude $\exp(\mathrm{i}k_0'R)/R$ at distance R, times a t-matrix element. The incident and scattered polariton fields can now be written using $P_{k_0}^{\mathrm{INC}} = \chi E_{k_0}^{\mathrm{INC}}$,

$$P_{k_0} = P_1\exp(\mathrm{i}k_1 z) + P_2\exp(\mathrm{i}k_2 z),$$

$$P_{k_0}^{\mathrm{INC}} = P_1'\exp(-\mathrm{i}k_1' z) + P_2'\exp(-\mathrm{i}k_2' z), \tag{145}$$

where k_i and k_i' solve the dispersion equation. The t-matrix elements become:

$$T_{k_0'k_0} = CF\sum_{ij} P_i P_j' \int_{(\infty)} \mathrm{d}t \int_0^\infty \mathrm{d}z[\exp \mathrm{i}((k_i + k_j')z - \mathrm{i}(\omega - \omega')t)], \tag{146}$$

where F is the cross-sectional area.

The differential scattering cross section is then:

$$d^2\sigma/d\Omega\, d\omega' = \frac{d^2 I_s/d\Omega\, d\omega'}{I_I} = \frac{c}{8\pi} k_0'^4 \langle |T_{k_0 k_0'}|^2 \rangle \tag{147}$$

where a thermal average $\langle\ \rangle$ must be taken, and I_I is the incident energy flux $I_I = (c/8\pi)|E_{IN}|^2$. Carrying out the integral transforms and the thermal average gives the final expression for Stokes plus anti-Stokes parts of the scattering. The Stokes part is:

$$d^2\sigma/d\Omega\, d\omega' = \frac{k_0'^4}{4\rho C_s^2} c^2 \sum_{ij=1}^{2} |P_i|^2 |P_j|^2 (F/\pi C_s) \times \left[\frac{(1+n(\Omega))\hbar}{[k_i + k_j' - \Omega/C_s]^2 + (\kappa_i + \kappa'_j)^2}\right], \tag{148}$$

where Re $k_i = k_i$, Im $k_i = \kappa_i$, ρ is the density, C_s the sound velocity, $n(\Omega)$ is the Bose–Einstein distribution at Ω, and $\Omega = (\omega - \omega')$.

In writing eq. (148), interference terms in the amplitudes have been neglected. Interference terms would be of the type: $(\pm)\ P_1 P_2^* P_1' P_2'^* \times \exp[i(k_1 - k_2)(k_1' - k_2')]$ and if $\mathrm{Re}(k_1 - k_2) \gg \mathrm{Im}(k_1 - k_2)$, and likewise for scattered states, these terms may be neglected. However, very close to resonance this condition needs to be carefully studied to determine if it is quantitatively applicable. At the time of writing there has not been any experimental study of these interference effects, which in principle may make significant contributions to the lineshape and intensity near ω_0 or ω_ℓ. In our discussion we neglect these terms. Thus the expression (148) only includes superposition of intensities of separate channel scatterings. In the classical limit $[1 + n(\Omega)]h|\Omega| \to kT$ and the scattered light gives a Lorentzian lineshape with maximum at $\Omega = c|k_i + k_j'|$ and linewidth $\Delta = 2(\kappa_i + \kappa'_j/k_i + k_j')$. At the time of writing this article, there seems to have been no reports of lineshape measurements. Such measurements could provide an important test of the available theory. Tilley (1980) has recently presented a theory of resonant Brillouin scattering of exciton polaritons which agrees in most respects with that of Brenig et al. (1972); his theory also gives explicit instances for which non-Lorentzian lineshapes can occur, and it would be important to determine whether this can be observed. One caution needs mention, however: Tilley (1980, eq. (5.7) and following paragraph) does not take proper account of the abc. The susceptibility employed in that paper (eqs. (4.6) and (4.7)) is the "dielectric approximation" for which the Sein abc eq. (92) is correct, rather than the ansatz used by Tilley. Hence it is preferable to view Tilley's results as indicating types of possible behavior rather than quantitative predictions for particular media. Using eq. (148) we can remark that the total scattering cross section is here decomposed into a sum of partial cross sections from each possible channel. In the approximation of neglecting interference the factors $|P_i|^2$, $|P_j'|^2$ are probabilities of occupancy of state i, j, with polarization P_i, P'_j. The frequency dependence of P_1, P_2 (incident waves) depends strongly on the mode structure through the abc, since different abc produce different frequency variation of the relative compositions of the two waves in the physical mode.

Yu (1979a, 1979b) has obtained a simplified form of eq. (148) by integrating the cross section over ω', which gives the following expression when F, the cross sectional area, is taken equal to one:

$$\frac{(1+n(\Omega))}{8\pi^2 h^2}(\omega_s/c)[|M_{if}|^2|P_i|^2|P_f|^2/v_i v_s(\kappa_i+\kappa_s)], \tag{149}$$

where v_i and v_s are energy velocities of the polaritons i and f in the resonance regime. He has shown that for exciton damping $\Gamma < \Gamma_C$(critical) the different abc produce qualitatively different frequency-dependent efficiencies. Thus the experimental determination of abc may be facilitated.

Also it should be noted that in the discussion given here no account was taken of non-deformation (Fröhlich) exciton–phonon coupling; owing to the k dependence in the matrix element this produces pronounced enhancement of large momentum transfers and is important for piezoelectric phonons. Details are discussed in the reviews of Koteles (this volume), Yu (1979), and Weisbuch and Ulbrich (1978).

A more serious matter is that to the writer's knowledge all the analyses of experiments so far carried out, use a "factorized" form of the scattering cross section to interpret the results:

$$\partial^2\sigma/\partial\Omega\partial\omega' = (\partial^2\sigma/\partial\Omega\partial\omega')_\infty T(\omega)T(\omega')\frac{F}{\kappa(\omega)+\kappa(\omega')}, \tag{150}$$

where $(\ )_\infty$ means the scattering cross section calculated for interbranch scattering in infinite volume, $T(\omega)$ and $T(\omega')$ are transmission factors for incident/scattered polaritons and $F/[\kappa(\omega)+\kappa(\omega')]$ is an effective scattering volume. Zeyher et al. (1974) have shown that if spatial dispersion is present this factorization is *not* mathematically justifiable. In fact to obtain eq. (150) from the more general expressions seems to require neglect of spatial dispersion effects entirely. However since the form (150) is most convenient to use in practice, it would evidently be most desirable to establish under what conditions (perhaps "weak spatial dispersion") an exact scattering cross section expression (as in Zeyher et al. 1974, eq. A.10) goes over into eq. (150).

A first measurement of linewidth as a function of laser frequency was reported by Wicksted et al. (1981); this is not yet sufficiently precise to enable comparison with theory to be made.

4.3. Transient exciton-polariton spectroscopy

The spectroscopic experiments discussed in sects. 4.1 and 4.2 may be considered as steady state optical spectroscopy. Transient optical experiments are another tool by which the optical consequences of spatial dispersion can be investigated. We briefly discuss four types of transient optical experiments which have been proposed and analyzed theoretically. One purpose of this part of the article is to stimulate additional experimental and theoretical work on these lines. One major new effect arising in spatially dispersive media is the occurrence of a new exciton precursor in addition to the classical Sommerfeld and Brillouin precursors. Another important effect due to spatial dispersion is the absence of a stop band: this produces a major change in the signal velocity of a propagating pulse. The transient transmittivity of a thin slab will show an anomalous enhancement for some characteristic frequency. A fourth new effect, which should be observable in transient reflectivity, offers the possibility of separating local and non-local components and locating ω_0 and ω_ℓ in a transient measurement.

Recent discussions of transient effects in local optics were given by Skrotskaya et al. (1969), Gitterman and Gitterman (1976) and Vainshtein (1976).

4.3.1. Excitonic, Sommerfeld and Brillouin precursors

The first extension of the theory of transient optical response – the local optics precursor theory of Sommerfeld (1914) and Brillouin (1914) (1960) to spatially dispersive non-local media was given by Birman and Frankel (1975) (1977). This work investigated the time development of the response of the non-local medium to the imposition of a truncated monochromatic laser (frequency ω_L) applied as the incident wave to a plane surface. It follows from the physical interpretation of the extinction theorem that when a wave (with sharp front in time and space) enters the medium, the leading edge of the electromagnetic wave propagates with the vacuum velocity c since the elementary dipoles of the medium have had no opportunity to respond and build up the refractive index (propagation number k). Some time behind the leading edge of the wave there will be a kind of "ringing" effect as the elementary dipole oscillators begin their forced motion under the influence of the incident field. This will manifest itself for some time period as a series of oscillatory disturbances in the amplitude detected by an observer at (z, t). After the transient interval, the signal arrives consisting of the predominant response $f(z, t) \sim \sin(\omega_L t - \phi)$ where ϕ is some phase lag. The transient or precursor regime is of interest here. The time regime of interest after application of the laser is approximately $z/c < t < z/v_s(\omega_L)$ where $v_s(\omega_L)$ is the "signal velocity" at the laser frequency ω_L, which is a measure of the speed of arrival of the main amplitude at ω_L.

In the model, semi-infinite, geometry we consider that at $t = 0$ a monochromatic laser of frequency ω_L, unit amplitude, is applied at normal incidence at the front surface $z = 0$, or $f(0, t) = (\sin \omega_L t)\theta(t)$. Inside the medium at point z, time t we determine the amplitude $f(z, t)$ of the disturbance measured by an observer. Generalizing the integral representation technique of Sommerfeld (1914) to the case of spatial dispersion where two transverse waves propagate, the desired amplitude is written

$$f(z, t) = \frac{\text{Re}}{2\pi} \oint \sum_{j=1}^{2} \frac{A_j(\omega) \exp[ik_j(\omega)z - i\omega t]}{(\omega - \omega_L)} \, d\omega, \tag{151}$$

Here the propagation wavenumbers satisfy the usual dispersion equation $k_j^2 = k_0^2(1 + 4\pi\chi_p(k_j, \omega))$ where χ_p is the (Fourier transform) of the propagating part of the susceptibility and $k_0 = \omega/c$. The coefficients $A_j(\omega)$ are determined from the relevant abc and are the mode coupling coefficients. The integral is to be performed on a closed contour in the complex ω plane containing the real axis $(-\infty, \infty)$ on one side. When $t < 0$, $z = 0$, closing the contour in the upper half plane gives zero; when $t > 0$ closing in the lower half plane gives at $z = 0$ two contributions from the residues,

$$f(0^+, t) = [A_1(\omega_L) + A_2(\omega_L)] \sin \omega_L t, \tag{152}$$

which is a field existing just inside the boundary after the pulse has penetrated.

The amplitude of the field at $t > 0$, arbitrary z, requires careful evaluation of the integral. It will be recalled that $k_j(\omega)$ is the solution of a fourth degree algebraic equation so that $k_j(\omega)$ as a function of complex ω has 2 branch cuts in the lower half complex ω plane, and a branch cut in the upper half plane. The singularity in the upper half plane is disconcerting but careful investigation shows that it makes no contribution (for $t < 0$) to the integral owing to a cancellation of the two terms in the sum (151). This is as it should be since the propagating susceptibility χ_p does in fact satisfy Leontovich's generalized analyticity condition (Frankel and Birman 1977), so the theory must be causal and no net contribution from singularities in the upper half plane can occur.

In order to evaluate the integral (151) in detail it is necessary to specify the abc, since the initial condition is the values of A_1 and A_2 which are abc – and frequency – dependent. By contrast the $k_j(\omega)$ only depends on the propagating part of the susceptibility. The analysis can be carried out using standard asymptotic methods and we cite the results:

At (z, t) for $t < z/c$ no disturbance is received:

$$f(z, t) = 0, \quad t < z/c. \tag{153}$$

Then the disturbance can be written to asymptotic accuracy

$$f(z, t) = \sum_{\alpha} f_\alpha(z, t)\theta(t - t_\alpha); \quad \frac{z}{c} < t < \frac{z}{v_s(\omega_L)}, \tag{154}$$

where the $f(z, t)$ is the sum of three partially overlapping contributions $f_\alpha(z, t)$ each starting after some delay time t_α. The asymptotic evaluation has assumed that each of the coefficients $A_j(\omega)$ is slowly varying around the saddle point and the standard method of steepest descents was used. The delay times t_α are given as:

$$t_{SP} = \sqrt{\epsilon_0}\, z/c,$$

$$t_{BP} = \sqrt{\epsilon_0}(z/c)(1 + a^2/2\epsilon_0\omega_0^2),$$

$$t_{EP} = z/\sqrt{b}, \tag{155}$$

where a and b are defined in eq. (8).

The three contributions labelled by α are:

The (new) exciton precursor (EP);
The Sommerfeld precursor (SP);
The Brillouin precursor (BP).

The quantities $f_\alpha(z, t)$ can be expressed in terms of sums of Bessel plus Airy functions and can be approximately given as

$$f_\alpha(z, t) = B_\alpha(\omega_\alpha, z) \cdot (t - t_\alpha)^{1/4} \exp[-\Gamma_\alpha(t - t_\alpha)] \cos(\omega_\alpha(t - t_\alpha) + \pi/4)$$

$$\alpha = \text{SP, EP}, \tag{156}$$

and

$$f_\alpha(z, t) = B_\alpha(\omega_\alpha, z)(t - t_\alpha)^{-1/4} \cos[\omega_\alpha(t - t_\alpha)],$$

$$\alpha = \text{BP}. \tag{157}$$

The ω_α are (time varying) saddle point frequencies, which arise as follows: ω_{EP} and ω_{SP} arise from saddle points on the two high frequency propagating branches for $\omega > \omega_\ell$, while ω_{BP} arise from the saddle point on the low frequency branch $\omega < \omega_0$ of the dispersion curves. Coefficients $B_\alpha(\omega_\alpha z)$ and Γ_α are amplitude and damping constants.

The observer at (z, t) will receive three partially overlapping precursor wave packets. Each such packet represents a time varying disturbance containing a packet of frequencies. Roughly speaking the packet in the new exciton precursor contains high frequencies sliding down the *outer* polariton branch for $\omega > \omega_\ell$, and centered near ω_ℓ^+, the Sommerfeld precursor contains high frequencies sliding down the *inner* polariton branch for $\omega > \omega_\ell$ centered near ω_ℓ^+. The Brillouin precursor packet originates on the low frequency ($\omega < \omega_0$) branch with main frequency content just below ω_L.

Frankel and Birman (1977) gave an illustration of the time scale, amplitude, duration and content of each of the packets which is reproduced in table 1. Parameters used were typical of CdS which may not be an optimum material for this experiment but delay times t_α, amplitudes and other quantities may be obtained for other materials from the expressions given in that reference by appropriately substituting parameter values.

Table 1

Illustrative values giving certain characteristics of the precursor packets. Major frequencies refer to a region where the relative amplitude changes by a factor $\leqslant 10$. The parameters chosen are appropriate to CdS: $\epsilon_0 = 8$, $\hbar\omega_0 = 2.5528$ eV, $M^* = 0.9M_e$. z is taken as 0.1 cm and Γ is taken equal to 0. The amplitude is taken relative to incoming signal amplitude, normalized to 1. $\omega_\ell = \omega_0(1 + 4\pi\alpha/\epsilon_0)^{1/2}$, ω_1 is taken here to be 2×10^{15} Hz. (Frankel and Birman 1977)

Precursor	Delay time t_α (10^{-12} s)	Major precursor frequencies	Relative amplitude	Duration (s)
SP	9.428	$\omega_\ell < \omega < 1.23\omega_\ell$	2.0×10^{-3}	7.35×10^{-15}
BP	9.535	$0.8\omega_1 < \omega < \omega_1$	1.4×10^{-5}	4×10^{-8}
EP	1.42×10^3	$\omega_\ell < \omega < 1.23\omega_\ell$	2.3×10^{-4}	7×10^{-10}

4.3.2. *Signal, energy, and group velocities*

The group velocity is the most familiar and most used object in discussion of wave propagation due to a packet. In a dispersive medium $v_g(\omega) = (dk(\omega)/d\omega)^{-1} = |\nabla_k\omega(k)|$. However in a region near the resonance, group velocity may not be a useful construct (Brillouin 1960). It will be recalled that, in the local optics case, in the region $\omega_0 < \omega < \omega_\ell$, k can be complex so that the derivative $(d\omega/dk)$ loses any immediate significance. Often a quantity $(d\omega/dk^R)$ is taken as the group velocity, where k^R is the real part of k but this is of dubious significance, especially since this quantity can become larger in magnitude than the vacuum velocity c.

The energy velocity v_E is defined as $|v_E| = |\bar{S}/\bar{u}|$ where $\bar{S}$ is the time averaged Poynting vector $\boldsymbol{S} = (c/8\pi)(\boldsymbol{E}\times\boldsymbol{H}^*)$ and u is the energy density $u = 1/8\pi(E^2 + H^2)$. The energy velocity has been discussed by Loudon (1970) who shows that in a single oscillator local model it can be written

$$(c/v_E) = n + c\kappa^I/(\Gamma/2), \tag{158}$$

where $k = (\omega/c)\epsilon^{1/2}$; Γ is the damping constant and $\sqrt{\epsilon} = n + ic\kappa^I/\omega$.

The signal velocity v_s is defined in a more indirect way and represents the first arrival of the laser signal at frequency ω_L, after the precursor scenario of the previous part has ended. Recalling the integral expression for $f(z, t)$ given in eq. (151) we note that at the pole $(\omega = \omega_L)$, $f(z, t)$ will oscillate as $\sin(\omega_L t - \phi)$, with some amplitude $|f(z, t)_{\text{pole}}|$. On the other hand the integral can be evaluated asymptotically by saddle point method, to get $f(z, t)_{\text{sp}}$. The location of the saddlepoint depends implicitly on time, and at that time when

$$|f(z, t)_{\text{pole}}| = |f(z, t)_{\text{sp}}|, \tag{159}$$

the signal is defined to have arrived (Baerwald 1930, Brillouin 1960). An analysis of the signal velocity in spatially dispersive media was given in Frankel and Birman (1975, 1977)*. In the case of large z it can be shown that

$$(v_s/c)^{-1} \sim c\kappa^{I}/(\Gamma/2), \tag{160}$$

so that $v_s \approx V_E$ if $\mathrm{Re}\,\epsilon^{1/2} \ll c\kappa^{I}/\Gamma$. In general there seems no simple relation between v_s and V_E. A plot of $(v_s/c)^{-1}$ as function of ω in region $\omega_0 < \omega < \omega_\ell$ shows, for CdS parameters, nearly a factor of 5 increase in signal velocity at $\omega \sim \omega_L$ in the presence of spatial dispersion.

Recently Ulbrich and Fehrenbach (1979) and Masumoto et al. (1979) measured the velocity of propagation of an optical pulse whose carrier frequency is close to resonance, in spatially dispersive GaAs, and CuCl respectively. Both groups compared their measurements with "group velocities" $v_g(\omega)$. As mentioned earlier the group velocity was computed from dispersion curves $\omega(k_j)$ for the exciton polaritons, in the sense of identifying the derivative $(\mathrm{d}\omega/\mathrm{d}k^R)$ as $v_g(\omega)$. Both groups report very good agreement between their measurements and the group velocity $v_g(\omega)_j \equiv (\mathrm{d}\omega/\mathrm{d}k_j^R)$ where $\omega(k_j^R)$ is the dispersion for the most "photon-like" branch of polariton. Careful examination of the measurements shows some interesting and (at the time of writing) puzzling features. For GaAs, for $\omega_L < \omega_0$, measured velocities are somewhat smaller than $v_g(\omega)_j$; for $\omega_0 < \omega_L < \omega_\ell$ (in the "pseudo-gap" or resonance regime) measured velocities are larger (by as much as a factor ten) than computed; for $\omega_\ell < \omega_L$, measured velocities jump from "outer" to "inner" branch of the polariton dispersion and agree well with computed $v_g(\omega)_j$. Puri and Birman (1981) computed group, energy, and signal velocities for exciton polaritons in GaAs and compared the dispersion (frequency dependence) of each of these near resonance. They modified the expressions of Bishop and Maradudin (1976), and dropped certain cross terms in energy velocity, in making this comparison. The situation for $\omega_0 < \omega_L$ remains puzzling. In CuCl experimental measurement and theoretical calculation of $v_g(\omega)_j$ appear to be in excellent agreement (allowing for the cross-over at $\omega = \omega_L$) in the entire frequency range explored**.

In the accompanying article by Ivenchko (ch. 4) a careful discussion is given of the measurement of exciton-polariton dispersion curves (and thus the group velocity $v_g(\omega)$) by interference experiments in thin plates and wedges. Agreement with calculated dispersion curves is excellent.

It is not surprising that static measurements such as reflectivity and

* Figure 2 of Frankel and Birman (1977) has $b \neq 0$ and $b = 0$ curves mislabelled: $b = 0$ is dashed curve, $b \neq 0$ is solid curve with $b = 1/M$.

** It has been shown by Garrett and McCumber (1970) that in a lossy local medium the peak of a Gaussian pulse always propagates with the velocity v_g, whatever the system parameters. Work is in progress (Puri and Birman 1982) investigating the theory of Gaussian pulse propagation in spatially dispersive media.

interference give agreement between theory and experiment; but further investigation of the range of validity of various "velocities" for coupled-wave spatial dispersion optics seems called for.

4.3.3. *Transient transmittivity*

In an interesting extension of these proposals Johnson (1978) analyzed the transient transmittivity of a finite slab of a spatially dispersive medium for laser frequency ω_L close to a frequency ω^* where the polariton group velocity v_g has a minimum $(dv_g/d\omega)_{\omega^*} = 0$. The merit of Johnson's suggestion is that it attempts to overcome the problem of small amplitudes in the predicted precursors (EP, SP, BP). Speaking very roughly the polariton waves "bunch up" at ω^* rather than passing quickly through the medium as in the precursor experiment. By adjusting the parameters of the system, e.g., of thickness of slab, and considering a time scale· ~2–3 ps for a CdS-like medium, the transmitted intensity follows an expression like

$$I(t) \sim e^{-\beta a} a^{-2/3} |Ai(\alpha t - T_0)|^2, \qquad (161)$$

where Ai is the Airy function and α, β are coefficients, a is the slab thickness, $T_0 = a/v_g(\omega^*)$. Careful study of this expression shows that the prediction is that the transmitted intensity will oscillate in time due to "beating" of the two branches of polaritons against one another giving an oscillatory transmittivity before the entire field decays exponentially.

Time scale and intensities for this experiment seem, for CdS, to be more promising than for the precursor experiments, but we emphasize the importance of attempting these experiments as well on other materials exhibiting spatial dispersion due to exciton polaritons, such as GaAs, CuCl, or others.

4.3.4. *Transient reflectivity*

Another type of experiment which can reveal novel transient effects due to non-locality has been analyzed by (Pattanayak et al. 1981). This case considers the transient reflectivity from a half space non-local medium with a normally incident pulse, in the usual model geometry. The merit of this type of experiment is that no attenuation occurs as in the case of the electromagnetic fields passing through the medium since the reflected wave is in vacuum.

Although the most interesting case is for an incident pulse of duration T (see below, eq. (169)) carrier frequency ω_L, we first analyze the reflectivity from a plane surface with incident "chopped" laser applied to the boundary plane. For convenience in the calculation in this case put the boundary at plane $z = L$, medium at $z > L$, vacuum at $z < L$. At $t = 0^+$, $z = L$, the laser is applied. The incident laser amplitude can be represented as $f(0, t) = (\sin \omega_L t)\theta(t)$ where $\theta(t)$ is the Heaviside function.

For the reflected wave we again will use an integral representation and write the amplitude of the electric field at $z = 0$, arbitrary time as

$$f_R(0, t) = \frac{1}{2\pi} \oint_{-\infty}^{\infty} \frac{R(\omega)}{\omega - \omega_L} \exp\left[-i\omega\left(t - \frac{2L}{c}\right)\right] d\omega, \tag{162}$$

where $R(\omega)$ is the frequency-dependent reflectivity of the non-local medium at frequency ω, and the contour is to be closed as before. A small change is that we place the boundary plane at $z = L$ for convenience in the calculation, and this brings a phase factor $\exp(2i\omega L/c)$. The transient reflectivity scenario begins after $t = 2L/c$ before which no amplitude is recieved.

Careful analysis of the integral (162) in the complex plane requires finding singularities of the function in the integrand $R(\omega)$ in addition to the simple pole at $\omega = \omega_L$. First remark that the local case $b = 0$ was carefully analyzed by Elert (1930), who discussed the entire scenario starting with precursor reflectivity immediately after $t = 2L/c$ and continuing through the regime at which the signal was established. Elert plotted the time-dependent scenario (Elert 1930, fig. 10) which reveals the existence of a first Sommerfeld precursor, a second Brillouin precursor and finally the signal arrival. Amplitudes of these precursors were estimated by Elert at about 10^{-2} of the main signal using parameters then available for media.

For the incident "chopped laser" which we discuss first, the time period of interest is the interval *between* the first arrival of a measurable amplitude (due to the Sommerfeld precursor effect in reflectivity) and the establishment of the steady-state reflectivity at laser frequency ω_L. The transient amplitude rises smoothly from zero at $t = 2L/c$, oscillates with an envelope which decays in time and joins smoothly (after the signal arrival time) to be steady-state reflectivity. For the case of a square pulse (see eq. (169) below) our particular interest will be the transient amplitude between $t = (T + 2L/c)$, corresponding to the trailing edge of the pulse, and the time $t > (T + 2L/c + t_c)$ when the transient has disappeared and the reflected amplitude has fallen to zero. The reflected amplitude in this time interval first oscillates and then decays to zero showing a cross-over behavior described below.

In the analysis of Pattanayak et al. (1981) for non-local media the model susceptibility (74) and the relevant abc (94) were used. The reflectivity $R(\omega)$ can be expressed in conventional form as:

$$R(\omega) = (1 - \bar{n}(\omega)/(1 + \bar{n}(\omega)), \tag{163}$$

where $\bar{n} \equiv (n_1 n_2 + \epsilon_0)/(n_1 + n_2)$ and $n_j \equiv (k_j/k_0)$ are the effective refractive indices of the two waves, ϵ_0 the background dielectric coefficient as before.

The structure of singularities in $R(\omega)$ shows 6 branch point singularities (two in the upper half plane, which produce no causality, owing to cancellation just as in the case studied by Frankel and Birman (1977)).

Apart from the steady state contribution to the reflectivity at laser frequency ω_L originating in the pole contribution to the integral, there is transient reflectivity due to contributions from branch-line singularities in the integrand. The transient reflectivity has been evaluated asymptotically; in addition a second approximation was made for the purpose of gaining additional physical insight into the results. A dimensionless parameter $\delta \equiv (h\omega_0/Mc^2)^{1/2} \sim 10^{-3}$ is defined and the integrals representing transient reflectivity were expanded in powers of δ. Keeping terms linear in δ, it turns out that a clear separation is possible in terms of whether or not the integral vanishes when $\delta \to 0$ (infinite mass) where this theory should revert to Elert's case of transient reflectivity in local optics ($M \to \infty$, $\delta \to 0$). Terms which are non-vanishing represent a "local" contribution, while those terms vanishing are the non-local contribution. Worth noting is that for $\delta \neq 0$ both types of terms depend on δ, as could be expected from the mixed nature of the physical polariton.

Hence the total reflected field can be written:

$$f_R(0, t) = E^s(\tau) + E_L^T(\tau, \delta) + E_{NL}^T(\tau, \delta), \tag{164}$$

where τ is the retarded time $\tau \equiv (t - 2L/c)$;

$$E^S(\tau) = |R(\omega_L)| \sin(\omega_L\tau - \phi(\omega_L)), \tag{165}$$

which is the steady state signal reflectivity at frequency ω_L with $R(\omega_L)$ the modulus of the ordinary reflection coefficient at ω_L, $\phi(\omega_L)$ is the plane; E_L^T is the local part of the transient response with

$$\lim_{\delta\to 0} E_L^T(\tau, \delta) \neq 0, \tag{166}$$

and E_{NL}^T is the non-local part of the transient response

$$\lim_{\delta\to 0} E_{NL}^T(\tau, \delta) = 0. \tag{167}$$

For the model system, certain analytical results of interest can be obtained. For $\omega_L \approx \omega_0$ ("on-resonance" case) we define a characteristic time $t_c \equiv (\beta\delta\omega_0)^{-1}$ with $\beta \equiv (4\pi\alpha_0)^{1/2}$. Then the non-local contribution to the transient reflectivity is approximately

$$E_{NL}^T(\tau, \delta) \sim [\exp(-\tau/\tau_0)]/\tau \quad \text{with} \quad \tau_0 \equiv 2/\Gamma. \tag{168}$$

For CdS, $t_c \sim 1$ ps. It turns out that for the non-local reflectivity the form of decay changes from exponential to τ^{-1} around $\tau = t_c$. Numerical analysis of the integrals confirms this and suggests similar cross-over behavior both on resonance and off resonance. The intensity of the non-local contribution to transient reflectivity is approximately 10^{-2} that of the signal and 1% discrimination seems possible for modern detection systems. Since the transient as defined in this work contains many frequencies while the signal is monochromatic some optical "filter" which would remove the reflected wave at ω_L may also enable discrimination of transient and signal.

The "local" part of the transient reflectivity nonetheless depends on exciton mass parameter δ, and away from exact resonance $\omega_L = \omega_0$ the magnitude of local part is about 10^{-2} of the non-local transient reflectivity (i.e. about 10^{-4} of the signal).

Some very interesting frequency-dependent structure appears in the local part of the reflectivity at fixed time $\tau < \tau_c$. As the laser frequency sweeps through (but close to) the resonance region $\omega_L \sim \omega_0$, the local part shows resonant enhancement both at ω_0 (the bare transverse frequency) and at ω_ℓ.

All the cited theoretical results have been obtained assuming that the exciting field (applied at $z = 0^+$) is the chopped laser $f(0^+, t) = (\sin \omega_L t)\theta(t)$. A transient reflectivity experiment which seems more easily implementable will occur for an incident pulse of duration T, carrier frequency (laser) ω_L:

$$f0, t) = (\sin \omega_L t)\theta(t)[1 - \theta(t - T)]. \tag{169}$$

The analysis was carried out using an integral representation which modified eq. (162). The physical entity to be studied is the time decay of reflectivity, after the incident or exciting pulse is terminated, i.e. $t > (T + 2L/c)$, measured at $z = 0$. The envelope of the amplitude of this transient decays to zero, showing cross-over from exponential to decay after about 1 ps for CdS. Other spectral and temporal features of the transients are identical to what was described for the transient build-up of reflectivity: especially at resonance when $\omega_L = \omega_0$, the transient is enhanced and (according to the theory) shows clearly the cross-over in decay (Pattanayak et al. 1981, Agrawal et al. 1982).

4.3.5. *Summary*

It is to be hoped that the measurements of Ulbrich and Fehrenbach (1979) and Masumoto et al. (1979, 1980) on velocity of propagation of optical pulses in non-local media, near resonance will be followed by experimental investigations on other members of the "family" of transient optical effects in non-local media.

It will be recalled that the initial theoretical work on these effects for classical, local media, is nearly 70 years old going back to the beautiful papers of Sommerfeld (1914) and Brillouin (1914), on the precursors, or forerunners; Baerwald (1930) on signals; and Elert (1930) on reflectivity. The first observation of the structure of electromagnetic precursors was by

Palocz and Pleshko (1969) and this was not on optical systems but in the microwave region.

The challenge of experimental discovery of new aspects of transient optical response, including linear and non-linear, local and non-local, remains an important frontier for modern optical investigations.

Acknowledgements

All my coworkers on problems of theory of spatial dispersion and non-local optics have contributed to my understanding of this fascinating phenomenon. In the writing of this article I have particularly benefited from recent discussions and joint collaborative work with Mr. Ashok Puri, Dr. D.N. Pattanayak, and Dr. E. Koteles, some of which is incorporated into this article. They may not share all my opinions but they have been of considerable help in bringing this review to completion. I acknowledge with thanks the courtesy of Drs. Koteles, Ivchenko and Permagorov in making preprints of their papers in this volume available to me. I also express my sincere thanks to Dr. M. D. Sturge and to Prof. E. I. Rashba for their scientific and editorial help in getting this manuscript ready for publication.

With Prof. H.Z. Cummins and Prof. R.R. Alfano, I have discussed many problems of bringing theoretical results into confrontation with real world of experiments and vice versa.

My thanks to Mrs. E. De Crescenzo for her expert and tireless work in reducing my script to a legible manuscript.

This paper was supported by research grants from National Science Foundation DMR78–12399, Army Research Office, DAAG29-79-G-0400 and PSC-BHE grant # RF-13404. I also thank the Institute des Hautes Etudes Sciéntifique, Bures sur Yvette, France; the Lady Davis Fellowship Trust and Department of Physics Technion, Haifa, Israel; and the Science Research Council and Department of Theoretical Physics, University of Oxford, U.K., for support and hospitality when this article was completed.

References

Agarwal, G.S., D.N. Pattanayak and E. Wolf, 1971, Phys. Rev. Lett. **27**, 1022.

Agarwal, G.S., D.N. Pattanayak and E. Wolf, 1973, Phys. Rev. **B8**, 4768.

Agranovich, V.M. and V.L. Ginzburg, 1962, UFN **76**, 643; **77**, 663; (1962) Sov. Phys. Uspekhi **5**, 323, 675.

Agranovich, V.M. and V.L. Ginzburg, 1964, 1966, Spatial Dispersion in Crystal Optics and the Theory of Excitons (Wiley, New York).

Agranovich, V.M. and V.L. Ginzburg, 1979, Spatial Dispersion in Crystal Optics and the Theory of Excitons, 2nd ed. (Nauka, Moscow); Engl. transl. (Springer, Berlin) (1981).

Agranovich, V.M. and A.A. Rukhadze, 1958, JETP **35**, 982; (1959) Soviet Phys. JETP **7**, 685.

Agrawal, G., J.L. Birman, D.N. Pattanayak, A. Puri, 1982, Phys. Rev. B (to appear).

Baerwald, H.G., 1930, Ann. der Phys. [5]**6**, 295.
Birman, J.L. and M. Frankel, 1975, Optics Commun. **13**, 303.
Birman, J.L. and D.N. Pattanayak, 1979, in: Light Scattering in Solids, eds. J.L. Birman, H.Z. Cummins, K.K. Rebane (Plenum Press, New York).
Birman, J.L., A. Puri and D.N. Pattanayak, 1982 (unpublished).
Birman, J.L. and J.J. Sein, 1972, Phys. Rev. **B6**, 2482.
Birman, J.L. and R. Zeyher, 1972, Phys. Rev. **B6**, 4617.
Birman, J.L. and R. Zeyher, 1974, in: Polaritons, eds. E. Burstein and F. De Martini (Pergamon Press, London, New York).
Bishop, M. and A.A. Maradudin, 1976, Phys. Rev. **B14**, 3384.
Born, M., 1932, Optik (Springer, Berlin) reprinted 1972.
Born, M., and K. Huang, 1952, Dynamical Theory of Crystal Lattices (University Press, Oxford).
Born, M. and E. Wolf, 1958, Principles of Optics (Pergamon, Oxford).
Brillouin, L., 1914, Ann. Phys. [4], **44**, 203; (1960) Wave Propagation and Group Velocity (Academic Press).
Bruce, R.H. and H.Z. Cummins, 1977, Phys. Rev. **B16**, 4462.
Cho, K., 1978, Solid State Commun. **27**, 305.
Cummins, H.Z. and P.E. Schoen, 1972, in: Laser Handbook Vol. 2, eds. F.T. Arrechi and E.O. Schultz (North-Holland, Amsterdam).
De Goede, J. and P. Mazur, 1972, Phys. Rev. **B6**, 2482.
Elert, D., 1930, Ann. Phys. [5] **7**, 65.
Frankel, M. and J.L. Birman, 1977, Phys. Rev. **A15**, 2000.
Garrett, C.G.B. and D.E. McCumber, 1970, Phys. Rev. **A1**, 305.
Ginzburg, V.L., 1958, JETP **34**, 1593; (1958) Sov. Phys. JETP **7**, 813.
Ginzburg, V.L., 1960, 1964, Propagation of Electro-Magnetic Waves in a Plasma (Pergamon, London).
Ginzburg, V.L. and N.N. Meiman, 1964, JETP **46**, 243; Sov. Phys. JETP **19**, 169.
Gitterman, E. and M. Gitterman, 1976, Phys. Rev. **A13**, 763.
Halevy, P. and K. Fuchs, 1979, Inst. Phys. Conf. Series, **43**, 863: Proc. 14th Int'l. Conf. Phys. Semiconductors (Edinburgh 1978).
Hizhnyakov, V.V., A.A. Maradudin and D.L. Mills, 1975, Phys. Rev. **B11**, 3149.
Hopfield, J.J., 1958, Phys. Rev. **112**, 1555.
Hopfield, J.J., 1966, J. Phys. Soc. Japan, **21** (Supplement) 77.
Hopfield, J.J., 1969, Phys. Rev. **182**, 945.
Hopfield, J.J. and D.G. Thomas, 1963, Phys. Rev. **132**, 563.
Huang, K. and A. Rhys, 1951, Chinese J. Phys. **8**, 208 (in English).
Ivchenko, I.L. 1981, Chapter 4 of this volume.
Jackson, J.S., 1975, Classical Electrodynamics (Wiley, New York).
Johnson, D.L. and P.R. Rimberg, 1976, Phys. Rev. **B14**, 2398.
Johnson, D.L., 1978, Phys. Rev. Lett. **41**, 417.
Kirzhnits, D.A., 1976, UFN **119**, 357; Sov. Phys. Usp. **19**, 530.
Kraft, D. and J.L. Birman, 1974, in: Polaritons, eds. E. Burstein and F. de Martino (Pergamon, London/New York).
Koteles, E., 1981, Chapter 3 of this volume.
Landau, L.D. and E.M. Lifschitz, 1960, Electrodynamics of Continuous Media (Pergamon Press).
Leontovich, M.A., 1960, JETP **40**, 907; (1961) Sov. Phys. JETP **13**, 634.
Lorentz, H.A., 1879, Verh. d.k. Akad. Wet. Amsterdam **18**.
Loudon, R., 1970, *J. Phys.* **A3**, 233.

Maradudin, A.A. and D.L. Mills, 1973, Phys. Rev. **B7**, 2787.

Martin, P.C., 1967, Phys. Rev. **161**, 143; 1968, in: Many Body Physics (Gordon and Breach, New York) ch. 2.

Masumoto, Y., Y. Unuma, Y. Tanaka and S. Shionoya, 1979, J. Phys. Soc. Japan **47**, 1844; (1980) Proc. of the 15th International Conference on Semiconductors, Kyoto, Japan (to be published).

Mattis, D. and G. Beni, 1978, Phys. Rev. **B18**, 3816.

Pattanayak, D.N., G. Agrawal and J.L. Birman, 1981, Phys. Rev. Lett. **46**, 174.

Pekar, S.I., 1957, JETP **33**, 1022; Sov. Phys. JETP **6**, 785.

Pekar, S.I., 1962, FTT **4**, 1301; Sov. Phys. Solid State **4**, 953.

Permogorov, S.A., 1982, this volume, ch. 5.

Pines, D., 1964, Elementary Excitations in Solids (Benjamin, New York).

Pleshko, P. and I. Palocz, 1969, Phys. Rev. Lett. **22**, 1201.

Puri, A. and J.L. Birman, 1981, Phys. Rev. Lett. **47**, 173.

Puri, A. and J.L. Birman, 1981a, Opt. Commun. **37**, 81.

Puri, A. and J.L. Birman, 1982 (to be published).

Rosenfeld, L., 1951, Theory of Electrons (North-Holland, Amsterdam).

Sein, J.J., 1969, Ph.D. Dissertation, New York University, Available from University Microfilm, Ann Arbor, MI.

Sein, J.J., 1970, Phys. Lett. **32A**, 141.

Sein, J.J., 1970, Optics Commun. **2**, 170.

Sein, J.J., 1975, Optics Commun. **14**, 157.

Silin, V.P. and A.A. Rukhadze, 1961, Electrodynamic Properties of Plasma and Plasma-Like media (Gasatomizdat).

Skettrup, T., 1973, Phys. Status Solidi (b) **60**, 695.

Skrotskaya, E.G., A.N. Makhlin, V.A. Kashen and G.V. Skrotskii, 1969, JETP **56**, 220; Sov. Phys. JETP **29**, 123.

Sommerfeld, A., 1914, Ann. Phys. **44**, 177.

Sugakov, V.I., 1968, Opt. Spektrosk. **24**, 477; Opt. Spectrosc. (USSR) **24**, 253.

Stratton, J.A., 1941, Electromagnetic Theory (McGraw Hill, New York).

Thomas, D.G. and J.J. Hopfield, 1968, Phys. Rev. **175**, 1020.

Tilley, D.R., 1980, J. Phys. C. Solid State Physics **13**, 781.

Ting, C.S, M. Frankel and J.L. Birman, 1975, Solid State Commun. **17**, 1285.

Tolpygo, K.B. 1950, Zh. E.T.F. **20**, 497 (in Russian).

Ulbrich, R.G. and G.W. Fehrenbach, 1979, Phys. Rev. Lett. **43**, 963.

Ulbrich, R.G. and C. Weisbuch, 1978, in: Festkörperprobleme XVIII (Vieweg, Braunschweig) 217.

Vainshtein, L.A., 1976, UFN, **118**, 339; Sov. Phys. Uspekhi **19**, 189.

Van Kronendonk, J. and J.E. Sipe, 1977, in: Progress in Optics XV, ed. E. Wolf (North-Holland, Amsterdam).

Wicksted, J., M. Matsushida and H.Z. Cummins, 1981, Solid State Commun. **38**, 777.

Wolf, E., 1973, in: Coherence and Quantum Optics, eds. L. Mandel and E. Wolf (Plenum, New York).

Yu, P., 1979, in: Excitons, ed. K. Cho (Springer, Berlin) Chap. 6.

Yu, P., 1979b, Solid State Commun. **32**, 29.

Zeyher, R., 1980, Proc. European Physical Society Conference, Amsterdam (to be published).

Zeyher, R. and J.L. Birman, 1974, in: Polaritons, eds. E. Burstein and F. de Martini (Pergamon, London, New York).

Zeyher, R., J.L. Birman and W. Brenig, 1972, Phys. Rev. **B6**, 4613.

Zeyher, R., C.S. Ting and J.L. Birman, 1974, Phys. Rev. **B10**, 1725.

CHAPTER 3

Investigation of Exiton-Polariton Dispersion using Laser Techniques

EMIL S. KOTELES

Advance Technology Laboratory
GTE Laboratories, Inc.
Waltham, MA 02254
U.S.A.

Excitons
Edited by
E.I. Rashba and M.D. Sturge

Contents

1. Introduction

From the time that the concept of exciton–photon coupling was first formulated over two decades ago by Pekar (1958) and Hopfield (1958), there have been numerous attempts to experimentally verify the most interesting prediction of the model: that for certain frequencies, two or more similarly-polarized modes can propagate simultaneously in a solid. This effect arises from the wavevector dependence of the exciton energy ("spatial dispersion"). Since a detailed theoretical description of exciton–photon coupling leading to spatial dispersion effects is provided in other chapters of this book (in particular, see E.L. Ivchenko and J.L. Birman) the theoretical discussion here will be limited to a short review of exciton-polariton dispersion in general but with some elaboration of systems which have been experimentally investigated.

Inherent in the concept of additional propagating modes is the problem of the additional boundary condition (ABC). It is fair to say that, to date, the experimental evidence is insufficient to confidently choose the correct ABC from among the many candidates. Although there was hope that some of the experiments described below in sect. 3 could resolve the question, so far conclusive results are lacking. In this chapter discussion of this problem will be omitted due to shortage of space. We will concentrate instead on the search for experimental proof of the existence of two or more simultaneously propagating modes and will describe experiments which attempt to accurately measure the parameters of exciton-polariton dispersion curves. Only bulk free exciton-polaritons will be discussed. Limitations of space prevent this review from being exhaustive. We hope to point out some of the exciting work which has been performed in this rapidly developing field and to indicate some interesting problems which remain to be solved.

The early theoretical work stimulated a great deal of experimental activity. At first the exciton resonance was probed almost entirely with standard optical techniques. The broadband reflection, transmission and luminescence spectra were studied in great detail in order to accumulate evidence of the polariton nature of the exciton. Detailed discussion of much of this work is available in ch. 4 by E.L. Ivchenko and will not be repeated here. In general, these experiments attempted to seek out subtle differences between measured spectra and those calculated using coupled exciton–photon models. Although abundant circumstantial evidence of spatial dispersion was gathered, the sensitivity of these techniques to sample quality effects precluded precise determination of parameter values. For example, reflectivity experiments depend critically on surface quality and the past history of the sample. The overall shape of the exciton-polariton reflection spectra is a sensitive function of the presence of the so-called exciton-free layer (e.g., see Evangelisti et al. 1974). The width of this layer can be altered by a number of techniques, including temperature cycling and annealing. Since reflectivity spectra are sensitive not only to exciton-polariton parameters but also to ABCs, meaningful reproducible results can only be obtained from experiments performed on well-characterized samples.

Transmission experiments require extremely thin crystals (e.g., ~0.1 to 1 μm for CdS) since the exciton absorption is so strong*. These are difficult (if not impossible) to handle without adding an unknown amount of strain. This would adversely affect the sample quality and probably modify the exciton-polariton structure in some manner difficult to predict. In addition, it is not clear that optical effects observed in such thin samples are necessarily characteristic of the bulk.

Broadband exciton-polariton luminescence spectra are also sensitive to sample quality, especially the presence of defects, impurities, etc. In addition, their interpretation is subject to the uncertainties attendant with the ABC problem.

To reiterate, classical broadband optical experiments performed on dipole-active excitons yielded abundant evidence of the polariton nature of the system. However, they have not proven precise or clear-cut enough to provide accurate exciton-polariton parameter values or to answer the ABC riddle.

On the other hand, the existence of two or more simultaneously propagating modes has been demonstrated in a more direct and straightforward manner with innovative techniques made possible by the development of tunable lasers. The measurements probe the bulk, not the surface, of a crystal, and the positions, not the intensities, of the experimental features (e.g., scattering peaks) are the relevant factors in the study of the kinematics of exciton polaritons. Thus restrictions on sample and surface quality can be relaxed to a large extent. These experiments will be discussed in sect. 3. In the next section, a short review will be presented of the different exciton-polariton systems which have been studied to date using these new techniques.

* Exciton absorption is a function of crystal quality which controls polariton decay. Only in extremely good crystals is transmission possible in thicker samples.

2. Exciton-polariton structure

To a first approximation, a free exciton can be thought of as an electron–hole pair in an insulator or semiconductor coupled by the Coulomb interaction. It possesses a finite effective mass (equal to the sum of the electron and hole masses) and is free to travel throughout the crystal. Its properties, in particular symmetry and mass, are related to the characteristics of the conduction and valence bands associated with the constituent particles.

Exciton polaritons are composite quasi-particles formed by the coupling of light and dipole-active excitons. In fig. 1a, the dashed lines represent the

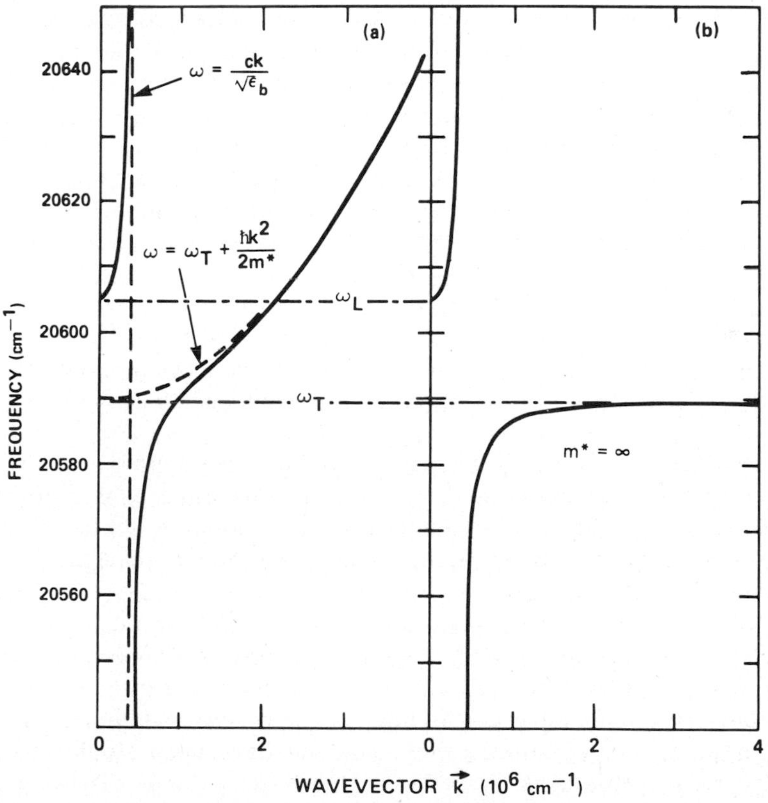

Fig. 1 (a) Schematic diagram of photon and exciton frequency–wavevector dispersion curves: in the absence of coupling (dashed lines) and in the exciton-polariton picture (solid lines). Parameter values applicable to the A exciton in CdS (table 2) are used. (b) Exciton-polariton dispersion curve in the absence of spatial dispersion.

energy vs. wavevector dispersion curves of an uncoupled photon (light-line), $\omega = ck/\sqrt{\epsilon_b}$, and a single exciton, which is a parabolic band in the simplest case ($\omega_e(k) = \omega_T + \hbar k^2/2m^*$). ω is the photon frequency, k is the wavevector, ϵ_b is the background dielectric constant which contains contributions from all interactions except the exciton in question, ω_T is the exciton frequency at $k = 0$, m^* is the exciton effective mass, $\hbar$ is Planck's constant/2π and c is the vacuum velocity of light. If the exciton is dipole active and couples with the photon, the dispersion curve of the resultant exciton polariton is given by the following implicit equation for the frequency dependence of the dielectric function:

$$\frac{c^2k^2}{\omega^2} = \epsilon_b + \sum \frac{4\pi\beta(k)}{1 - \omega^2/\omega_e^2(k)}, \tag{1}$$

$4\pi\beta(k)$ is the k-dependent oscillator strength. Exciton-polariton damping is neglected in this discussion since it generally has a negligible effect on the shape of the exciton-polariton dispersion curve. For a single parabolic exciton band, the dispersion curve of the exciton polariton is given by the solid line in fig. 1a. Note that above a certain energy, ω_L, two similarly-polarized modes can propagate simultaneously. From eq. (1), when $k = 0$, $4\pi\beta(0)$ is simply related to the zone center longitudinal–transverse exciton splitting; $4\pi\beta(0) \simeq 2\epsilon_b\omega_{LT}/\omega_T$ where $\omega_{LT} = \omega_L - \omega_T$. Far below resonance, ω_T, the exciton polariton is fundamentally photon-like. As the frequency increases, the polariton takes on more exciton character until at and above ω_T it is primarily exciton-like. Above ω_L a new inner (or upper) polariton branch arises which quickly takes on predominantly photon character. These modes exist simultaneously because the effective exciton mass, m^*, is finite, and so the exciton dispersion is wavevector dependent. If the mass were infinite (i.e., in the absence of spatial dispersion effects) only one mode would exist at each polariton frequency and a gap would open between ω_L and ω_T in which propagating modes could no longer be supported (see fig. 1b). Experimental tests of spatial dispersion effects in exciton-polariton dispersion curves require, as a minimum, evidence of the curvature of the lower polariton branch (or a modification of the curvature of the upper branch). Existence of an upper branch is insufficient proof.

The most studied system, the A exciton-polariton in CdS (composed of a conduction-band electron and a hole from the top valence band) possesses this simple structure, and experimental data from many other excitons have been interpreted in the light of this model (e.g., GaAs, CdSe, HgI_2). However, many excitons possess more complex structures. Examples of complex exciton polaritons are described briefly below.

2.1. *Degenerate valence band effects*

In zincblende materials which possess cubic symmetry, the two uppermost valence bands are degenerate at the Γ point ($k = 0$). When holes from these valence bands are coupled with conduction band electrons, "heavy" and "light" excitons are formed, each of which is able to couple with light (Kane 1975, Fishman 1978). The result is a three branch exciton-polariton dispersion curve (fig. 2). Thus, in certain frequency regions, one external photon can generate three similarly polarized, simultaneously propagating exciton-polariton modes in the crystal. Experimental confirmation of this model has recently been achieved in ZnSe by Sermage and Fishman (1979, 1981) and will be discussed later.

2.2. *k-Linear term effects*

Exciton complexity may also result from non-quadratic terms in the exciton energy. In CdS, a wurtzite structure crystal, terms linear in wave-vector in the conduction and second valence bands modify the energy of the B exciton so that it is non-parabolic in k when $\mathbf{k} \perp \mathbf{c}$ (Mahan and Hopfield 1964).

$$\hbar\omega_{\mathrm{e}}^{\pm}(k) = \hbar\omega_{\mathrm{T}}^{\mathrm{B}} + \frac{\hbar^2 k^2}{2m_{\perp}} + \frac{\Delta_{\mathrm{i}}}{2} \pm \left[\left(\frac{\Delta_{\mathrm{i}}}{2}\right)^2 + (\phi k)^2\right]^{1/2}, \tag{2}$$

ϕ is the coefficient of the k-linear term in the exciton energy, $m_{\perp} = m_{\mathrm{e}\perp} + m_{\mathrm{h}\perp}$ is the transverse effective exciton mass for $\mathbf{k} \perp \mathbf{c}$ and $\hbar\omega_{\mathrm{T}}^{\mathrm{B}}$ is the

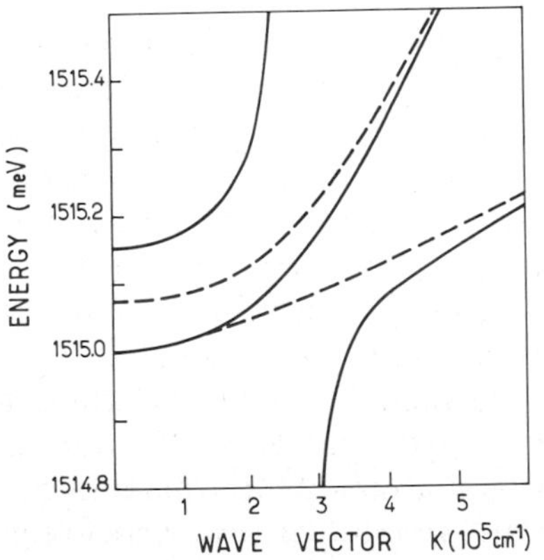

Fig. 2. Calculated dispersion curves of excitons (dashed lines) and exciton polaritons (solid lines) in GaAs. (From Fishman, 1978.)

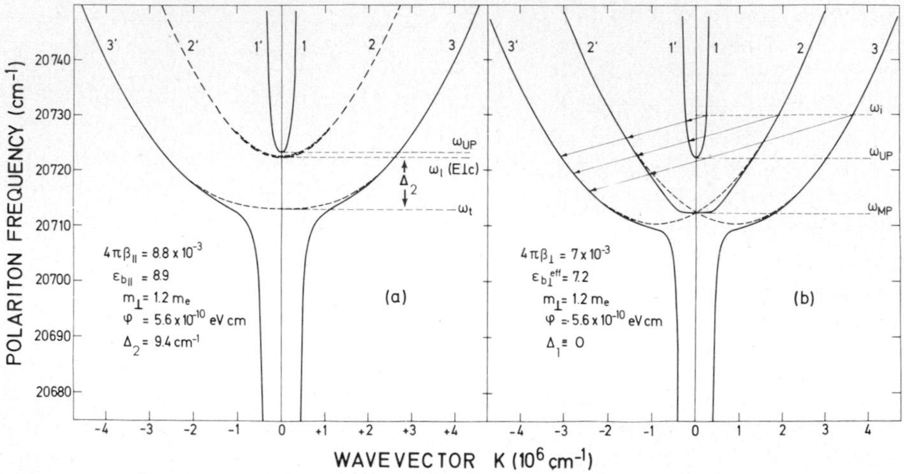

Fig. 3. Calculated dispersion curves of excitons (dashed lines) and exciton polaritons (solid lines) in the vicinity of the B exciton in CdS; $\boldsymbol{k} \perp \boldsymbol{c}$. (a) The exciting light is polarized parallel to the c-axis; $\boldsymbol{E} \| \boldsymbol{c}$. (b) $\boldsymbol{E} \perp \boldsymbol{c}$. Arrows depict possible single LA phonon scattering transitions. (From Koteles and Winterling, 1980.)

transverse exciton energy at $k = 0$. The k-linear term, ϕk, produces a mixing among the ground states of the B exciton, Γ_1, Γ_2, Γ_{5T}, Γ_{5L}. Only states $\Gamma_1(\boldsymbol{E} \| \boldsymbol{c})$ and $\Gamma_{5T}(\boldsymbol{E} \perp \boldsymbol{c})$ are dipole active ($\boldsymbol{c}$ is the optic axis). $\Delta_{i=1,2}$ is the energy separation of the coupled states at $k \to 0$. $\Delta_1 \simeq 0(\Gamma_2$ and $\Gamma_{5T})$ and $\Delta_2 > 0(\Gamma_1$ and $\Gamma_{5L})$. The strongest mixing of oscillator strengths between coupled states occurs if they are degenerate at $k = 0$. Thus for $\boldsymbol{E} \perp \boldsymbol{c}$, both hybrid states will have equal and k-independent oscillator strengths and each will couple to the photon. In this case, since $\Delta_1 = 0$, eq. (2) reduces to

$$\hbar\omega_e^{\pm}(k) = \hbar\omega_T^B + \frac{\hbar^2 k^2}{2m_\perp} \pm \phi k. \tag{3}$$

These exciton dispersion curves are represented by the dashed lines in fig. 3(b). Coupling between these dipole-active excitons and photons is accomplished mathematically by substituting eq. (3) into eq. (1). The exciton-polariton dispersion curves are shown as solid lines in fig. 3(b). On the other hand, for $\boldsymbol{E} \| \boldsymbol{c}$, $\Delta_2 > 0$ and there is only weak mixing between the coupled states. One hybrid state is dominated by Γ_1 and so retains most of the oscillator strength and is the only state which couples strongly with light. Figure 3(a) illustrates the exciton (dashed lines) and exciton-polariton (solid lines) dispersion curves in this case. Thus, experimentally, it is possible to change from a situation in which the exciton polariton has only two active branches to one in which three branches are present simply by rotating a polarizer. Note that in both cases the values of the parameters of the exciton dispersion are identical, with the exception of Δ_1. Recently experimental results from resonant Brillouin scattering (Koteles and Winterling 1980) have verified this model.

In wurtzite crystals an additional complication in the interpretation of data arises due to the relatively large anisotropy inherent in the crystal structure. This will be discussed in greater detail later.

3. Experiments

3.1. Introduction

During the last few years, rapid progress has been realized in studies of the kinematics of exciton polaritons. Provided that a few basic experimental requirements can be fulfilled, it is now possible to routinely, and accurately, determine the dispersion curves of simple, or complex, exciton polaritons. The experimental criteria are adequate sample quality (i.e., relatively pure single crystals) and the availability of suitable optical probes. High quality crystals of some materials, in particular, CdS, have been extensively studied in the past using conventional techniques. But the explosive increase in our knowledge of exciton-polariton properties is the result of the advent of new optical sources coupled with innovative experimental techniques.

The instrument which is most responsible for this situation is the tunable dye laser. It is now possible to probe a crystal excitation with an intense source which is sharply defined in energy and/or time and which can be precisely tuned into and through coincidence with an exciton resonance. By restricting the exciting beam to a narrow frequency interval, it is possible to populate a well-defined portion of the exciton-polariton dispersion curve and to directly observe its decay via acoustic and/or optical phonon scattering (resonant light scattering). When the sum of two incident photon energies is tuned to approximate equality with the exciton energy, polaritons in the upper branch can be probed through non-linear effects. By varying the angle between two intersecting beams, the wavevector of the resultant polariton is tunable. Thus its dispersion curve, near $k \to 0$, can be measured. If the exciting photons are in the vicinity of one-half of the biexciton frequency, molecular excitons may be directly created if the incident beam is sufficiently intense. By observing the decay products of these biexcitons as functions of incident frequency and angles of incidence and observation, it is possible to derive the exciton-polariton dispersion curve using iterative calculations. Finally, the group velocity of exciton polaritons may be measured directly by observing the transit time of picosecond pulses of light through thin crystals. In the region of the "bottleneck," exciton-polariton velocities vary by several orders of magnitude within a narrow frequency interval. These experiments provide dramatic proof that in the vicinity of a dipole-active exciton, the normal modes of energy propagation are no longer photons but polaritons, coupled exciton–photon quasi-particles.

3.2. k-Vector spectroscopy

In non-linear materials, when two beams from intense monochromatic sources overlap, a strong polarization, quadratic in the electric field amplitude, is induced in the solid. This non-linear effect generates significant radiation at the sum frequency ($\omega_s = \omega_1 + \omega_2$). The wavevector of this light is defined by the k vectors of the incident beams, $\boldsymbol{k}_1$, $\boldsymbol{k}_2$, and θ, the angle between them (i.e., the vector sum). Optimum transfer of intensity from incident to sum frequencies takes place when phase matching is achieved. This occurs when the sum energy lies on the dispersion curve relating to the propagation of energy inside the solid (i.e., energy and wavevector are conserved). As the incident frequency is scanned, the phase matched condition is detectable as either a peak in the magnitude of the sum signal (Haueisen and Mahr 1971) or as a maximum in the absorption of the incident light (Fröhlich et al. 1971). In either case, by varying the angle between the incoming beams (and thus changing the resultant wave-vector, $\boldsymbol{k}_s$) it is possible to study the dispersion of the sum frequency over the range $0 \leq k_s \leq |k_1| + |k_2|$. When the sum frequency is in the vicinity of an exciton resonance, only the upper polariton and longitudinal exciton branches may be probed. The lower polariton branch has wavevectors consistently larger than $|k_1| + |k_2|$ as a result of exciton-polariton dispersion. Thus the lower branch is inaccessible since the conditions for phasematching (i.e., $\omega_s = \omega_1 + \omega_2$ and $\boldsymbol{k}_s = \boldsymbol{k}_1 + \boldsymbol{k}_2$, where ω_s, $\boldsymbol{k}_s$ is a state on an exciton-polariton branch) can never be achieved. Presently k-vector spectroscopy is the most accurate method for measuring the dispersion of the upper polariton. It can also be used to determine the "dispersion" of the longitudinal exciton near the zone center. However, in this k-vector range, the longitudinal exciton dispersion is generally featureless; i.e., its energy is independent of wavevector, within experimental error.

CuCl was the first material whose exciton-polariton dispersion curve was probed using k-vector spectroscopy (Fröhlich et al. 1971). The absorption spectrum of the crystal was measured as a function of a sum excitation energy produced by two laser beams of energy $\hbar\omega_1$ and $\hbar\omega_2$. Two relatively sharp absorption bands were observed (fig. 4). These defined the energies of the exciton-polariton modes. As the angle between the two incident beams was varied from 180° to 60°, one band remained stationary while the

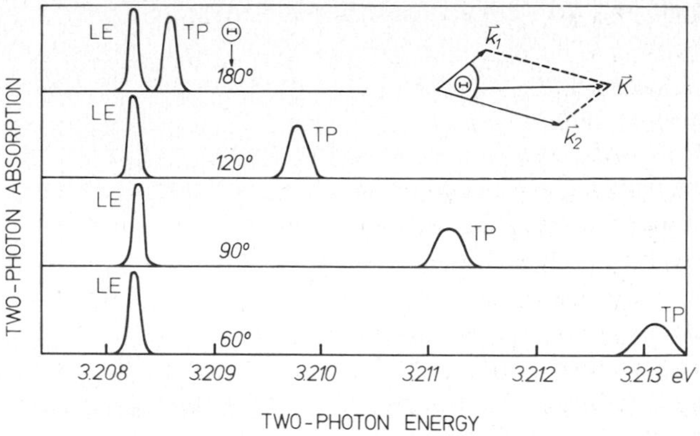

Fig. 4. Two-photon absorption peaks in CuCl as a function of the angle, θ, between the incident beams. LE: longitudinal exciton; TP: transverse exciton polariton (upper branch). Inset: experimental geometry. (From Fröhlich et al. 1971.)

position of the other increased in energy. The resultant wavevector was calculated using the known wavevectors of the incident beams and the angle between them. The dispersion curves of the resultant exciton polaritons are plotted in fig. 5; the curve whose energy increases with $\boldsymbol{k}$ is attributed to the upper branch of the transverse polariton (TP) while the other, which is stationary in energy, is due to the longitudinal exciton (LE).

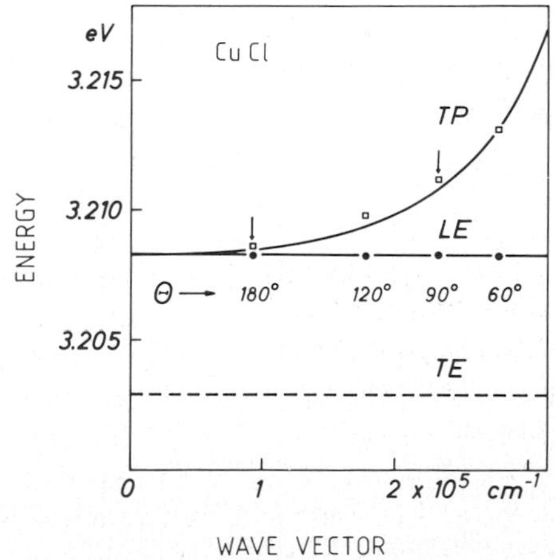

Fig. 5. Measured (circles and squares) and calculated (lines) dispersion curves of the upper exciton-polariton branch (TP) and the transverse (TE) and longitudinal exciton (LE) in CuCl. (From Fröhlich et al. 1971.)

The flatness of the LE curve was thought to arise from a rather large exciton mass in CuCl ($m^* \sim 13\, m_e$). Large exciton masses increase the difficulty of distinguishing effects of spatial dispersion on the shape of the upper polariton branch.

An identical dispersion curve for the upper polariton in CuCl was obtained by measuring second harmonic generation in the region of the first exciton absorption band (Haueisen and Mahr 1971). In this case the two incident beams were identical and the generation of second harmonic light ($\omega_s = 2\omega_1$) was used as a probe to indicate phase matching. As these authors point out, second harmonic generation (SHG) and two-photon absorption are complementary processes. For, once the incident photons are inside the crystal, they "combine" to form the sum exciton-polariton state. If this polariton reaches the surface of the crystal and escapes, it is detected as second harmonic radiation. Since these polaritons couple strongly to phonons, the experimental geometry is designed so that SHG polaritons are generated close to crystal surfaces to increase the chances of their being detected. On the other hand, if the polariton is scattered before reaching the surface, the depletion of the incident beams is observed as two-photon absorption.

The effect of spatial dispersion ($m^* \neq \infty$) on the upper branch of the

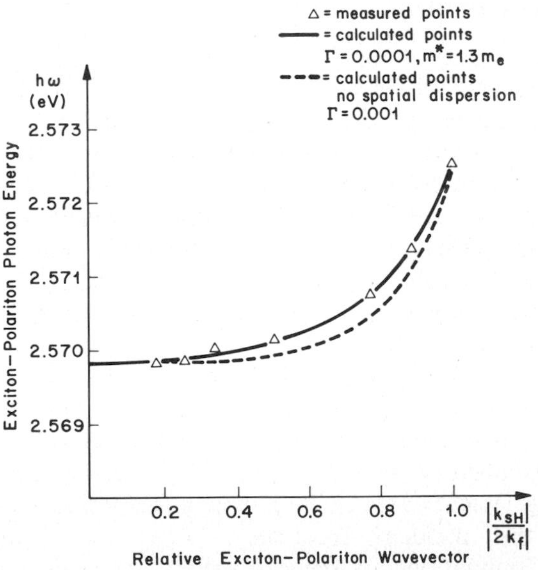

Fig. 6. Measured (triangles) and calculated (lines) dispersion curves of the upper branch of the B exciton-polariton in CdS for $\boldsymbol{E} \| \boldsymbol{c}$. Solid line, $m^* = 1.3\, m_e$; dashed line, $m^* = \infty$. (From Jackel and Mahr, 1978.)

polariton has recently been measured in the B exciton in CdS using k-vector spectroscopy (Jackel and Mahr 1978). The best fit with the experimental points was obtained when spatial dispersion, utilizing a simple two-branch model with $m^* = 1.3\, m_e$, was included in the theory (fig. 6). However, it is known that an energy term linear in k must be included in the calculation of the dispersion curve of the B exciton in CdS (sect. 2.2). For $\boldsymbol{E} \| c$ this additional term increases the magnitude of the fitted effective mass above that present in the normal two-branch model. Although these measurements clearly show the effect of a finite mass on the dispersion of the upper polariton, the accuracy of this technique appears inadequate to detect the difference between the simple two-branch model for the B exciton and the more sophisticated theory incorporating k-linear terms. The consequences of this more complicated model can be observed directly in resonant Brillouin scattering experiments (sect. 3.5.2.2).

The use of k-vector spectroscopy to determine the dispersion curves of upper branch polaritons has been limited, so far, to CuCl and the B exciton in CdS. Other experiments however (Levine et al. 1975), indicate that phase-matched two-photon absorption also occurs in the vicinity of the A excitons in CdS and ZnO. The technique should be applicable to these and many other materials.

3.3. Hyper-Raman scattering

If a high enough density of exciton polaritons is generated in a crystal, coupling between pairs of these quasi-particlcs is likely. If the resulting biexciton (or molecular exciton) is created in a virtual state, it quickly relaxes. The decay products are simply polaritons on one or more branches of the normal exciton-polariton dispersion curve. By investigating the behavior of these resultant states, it is possible to derive their dispersion curves in a semi-empirical manner. This technique is dealt with at greater length in ch. 11 in this volume (J.B. Grun, B. Hönerlage and R. Levy), so this discussion will only review the method and concentrate on some results relevant to this chapter.

Several years ago sharp emission peaks were observed when the frequency of a high intensity laser beam was tuned into the vicinity of one-half the biexciton frequency of CuCl (Nagasawa et al. 1976). This phenomenon was called two-photon resonance Raman scattering, or hyper-Raman scattering (HRS) since these peaks scaled (albeit in a non-linear manner) with the incident frequency. Theoretical papers quickly established the foundations of the phenomenon (Inoue and Hanamura 1976, Henneberger and Voigt 1976, Bechstedt and Henneberger 1977, Henneberger et al. 1977). The earliest work was primarily concerned with identifying the mechanism, and it soon became evident that a polariton model was necessary in order to explain the details of the frequency behavior of the decay peaks (Henneberger and Voigt 1976, Itoh et al. 1977, Hönerlage et al. 1977, Vu Duy Phach et al. 1978). The usual exciton-polariton model employing parameters found in the literature was used to identify and predict the behavior of these anomalous peaks. However, the inverse procedure was soon developed; i.e., the use of the measured behavior of these peaks to derive the values of the parameters of the dispersion curves of the exciton polariton (Hönerlage et al. 1978).

The experiment consists of irradiating a single crystal at low temperature with a high intensity dye laser beam. When the incident frequency, ω_i, is tuned to the vicinity of one-half of the biexciton frequency (which is less than the exciton resonance due to the binding energy of the biexciton) biexcitons with an energy $2\omega_i$ and wavevector $2k_i$ are resonantly excited by the simultaneous absorption of two photons. When the incident frequency is slightly detuned from this resonance energy, only virtual biexcitons are generated which quickly decay into two exciton polaritons. Energy and momentum conservation laws govern these decay processes and, in conjunction with the exciton-polariton dispersion curve, uniquely define the dependence of the emission spectra on the angles of excitation and observation. Extracting the dispersion curve from the experimental data necessitates self-consistent computation.

The biexciton can decay into several states. In the case of a crystal which possesses a simple two-branch exciton-polariton dispersion curve, such as CuCl, the pair decay products may be (1) two polaritons on the lower branch, (2) one polariton on the lower and one on the upper branch, or (3) one polariton on the lower and one on the longitudinal branch (Hönerlage et al. 1978). Generally one of the polaritons is photon-like and one exciton-like. The latter interact so strongly with phonons and impurities that they have a small probability of reaching the surface of the crystal, transforming into an external photon, and being observed. However, the photon-like polariton is readily observed. Invoking energy conservation, the frequency of its partner polariton can be deduced. By varying the incident frequency and the angles of the exciting light and of the direction of observation, it is possible to "tune" the pair of exciton polaritons through various regions of the dispersion curve. Although the energies of the decay products are directly measurable, their wavevectors are not. Thus extracting the exciton-polariton dispersion curve from this data requires an involved iterative calculation: first an exciton-polariton dispersion model with appropriate values for its parameters must be assumed; then the energies of decay exciton-polaritons must be calculated as a function of incident frequency, incident angle, and observation angle, employing energy and momentum conservation laws. If these calculated values do not satisfactorily agree with the experimental data, the polariton

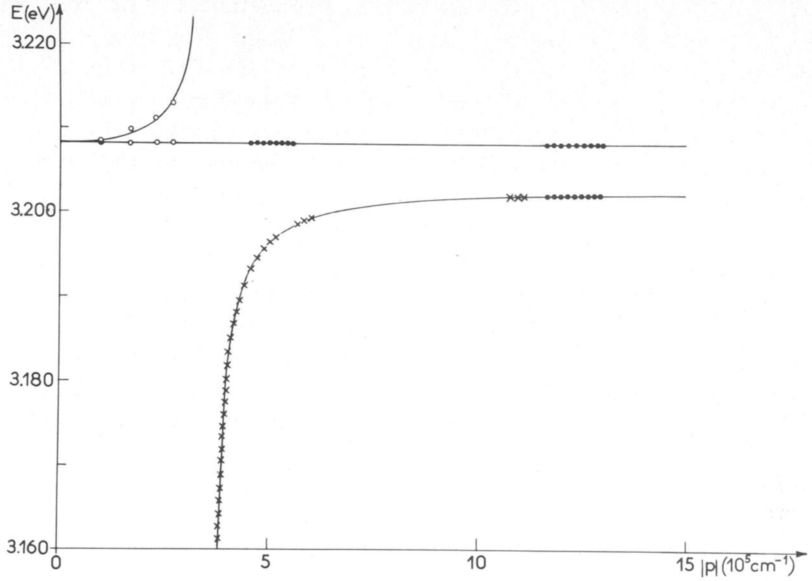

Fig. 7. Exciton-polariton dispersion curves in CuCl derived from hyper-Raman scattering experiments. Solid line, calculated; crosses, direct observation; solid circles, indirect results (using energy conservation); open circles, from Fröhlich et al. (1971). (From Hönerlage et al. 1978.)

energies must be recalculated, utilizing a slightly different set of parameters. This process is repeated until agreement is reached. Fortunately, one of the parameters, the energy of the longitudinal exciton, may be obtained in a more direct manner. If the exciton mass is sufficiently large, then, near $\boldsymbol{k} \rightarrow 0$, the energy of this exciton is essentially constant. A plot of "partner" polariton emission peak energy versus biexciton energy ($2\omega_i$) yields a straight line with unity slope and intercept equal to the longitudinal exciton energy.

Using hyper-Raman scattering, Hönerlage et al. (1978) have measured the longitudinal and lower branches of the exciton-polariton dispersion curve of CuCl (fig. 7). They obtained the same value for the energy of the longitudinal exciton as Fröhlich et al. (1971) but measured a rather smaller exciton mass ($m^* \simeq (2.5 \pm 0.3)\, m_e$ compared with $m^* \simeq 13\, m_e$). Even with such a small mass, effects of spatial dispersion on the curvature of the lower branch were not observed due to the small range of the wavevector ($<1.3 \times 10^6\ \mathrm{cm}^{-1}$). An increase in $\boldsymbol{k}$ of a factor of at least two or three is necessary for the upward curvature of the lower branch to be discernable.

This has recently been achieved by employing two different tunable laser frequencies to excite the sample (Mita et al. 1980). One laser frequency, ω_1,

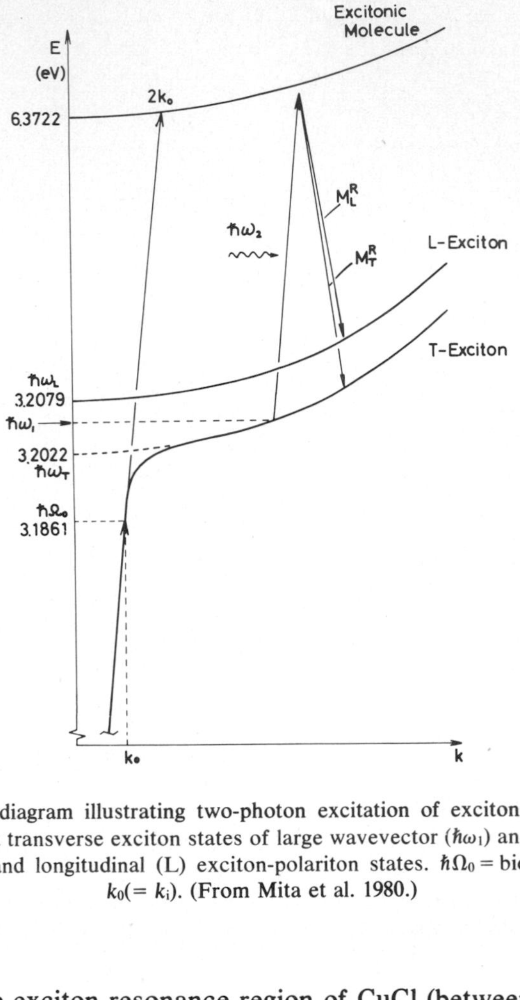

Fig. 8. Schematic diagram illustrating two-photon excitation of excitonic molecules (biexcitons) beginning at transverse exciton states of large wavevector ($\hbar\omega_1$) and subsequent decay to transverse (T) and longitudinal (L) exciton-polariton states. $\hbar\Omega_0$ = biexciton energy/2 at $k_0(= k_i)$. (From Mita et al. 1980.)

was fixed in the exciton resonance region of CuCl (between ω_T and ω_L) in order to create exciton-polaritons possessing large wave-vectors. The second laser, ω_2, was tuned so that the sum, $\omega_1 + \omega_2$, was approximately equal to the biexciton frequency. Under these conditions two hyper-Raman lines, related to biexciton decay to large wavevector states on the longitudinal and transverse exciton-polariton branches, were observed (fig. 8). The standard technique described previously generates biexcitons with wavevectors $k \simeq 2k_i$ and probes final states with wavevectors between k_i and $3k_i$ depending on the scattering angle. In this case, biexcitons with $\sim 4k_i < k < 15k_i$ are formed and the spatial dispersion of the exciton polariton has been investigated as far as $k \simeq 6.2 \times 10^6\,\mathrm{cm}^{-1}$, i.e., to about 10% of the Brillouin zone boundary value (fig. 9). Surprisingly, the measured

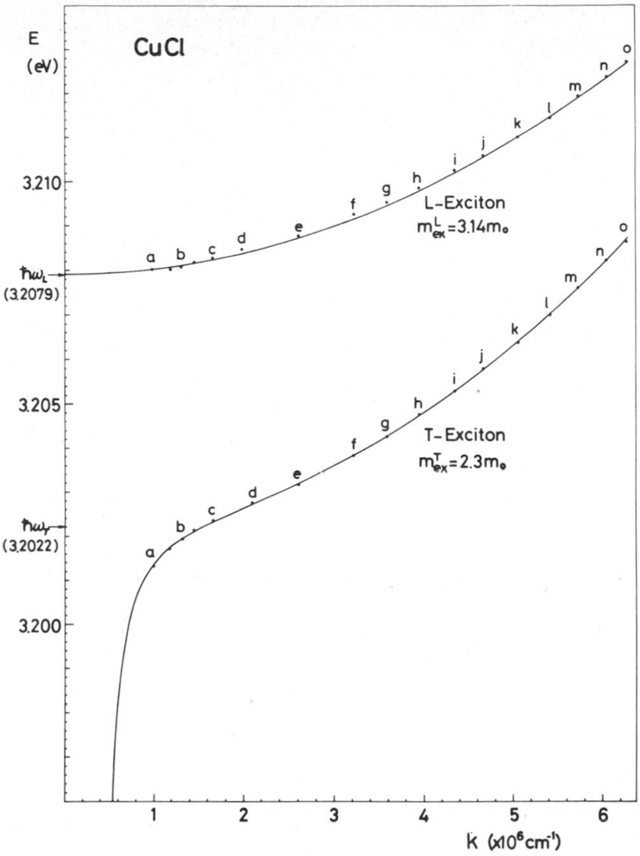

Fig. 9. Calculated (solid lines) and measured (labeled points) exciton-polariton dispersion curves for CuCl illustrating wide range of dispersion curve probed, using processes illustrated in fig. 8. (From Mita et al. 1980.)

effective longitudinal exciton mass ($(3.1 \pm 0.1)\,m_e$) was found to be significantly larger than the effective transverse exciton mass ($(2.3 \pm 0.1)\,m_e$).

The usual experimental geometry limits the technique to investigations of the lower and longitudinal polariton branches. However, if biexcitons whose wavevector is less than $2\boldsymbol{k}_i$ are generated, then decay to the upper polariton branch is possible. In order to reduce $\boldsymbol{k}$ below $2\boldsymbol{k}_i$, it is necessary to introduce an angle between the two interacting exciton polaritons in a manner analogous to the technique of k-vector spectroscopy (sect. 3.2). Suzuki et al. (1978) accomplished this by splitting their dye-laser beam and impinging their thin CuCl sample from front and back. By varying the angle

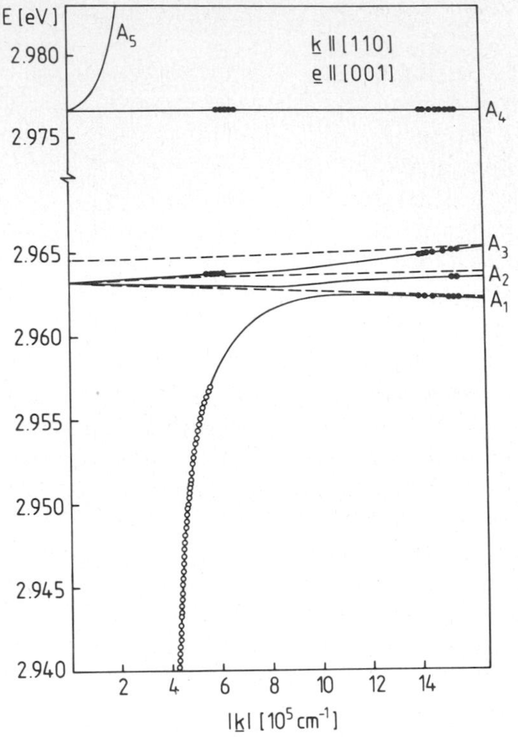

Fig. 10. Calculated exciton (dashed lines) and exciton-polariton (solid lines) dispersion curves in CuBr. Open and closed circles denote semi-empirical points. (From Bivas et al. 1979.)

between the two beams, they were able to scan $\boldsymbol{k}$ from $0.67\boldsymbol{k}_i$ to $0.39\boldsymbol{k}_i$. They observed an extra emission peak which they attributed to upper branch polaritons populated by the decay of biexcitons. They fitted their experimental points with a theory developed in the absence of spatial dispersion (i.e., $m^* = \infty$) (Itoh and Suzuki 1978).

Hyper-Raman scattering has recently been employed to investigate the exciton-polariton dispersion curves of CuBr (Bivas et al. 1979). In contrast to CuCl, which has a simple exciton structure, degeneracy of the valence bands in CuBr leads to multi-component exciton-polaritons. This is more typical of zincblende semiconductors. In this case, emission peaks attributed to exciton-polariton states on four of five possible branches were observed. Using the self-consistent calculation procedure described previously, the parameters of the exciton-polariton dispersion curve were derived. Figure 10 illustrates the fit between the calculated dispersion curves and the semi-experimental points. More details are available in ch. 11.

It is clear that hyper-Raman scattering is a useful tool in the study of the dispersion curves of exciton polaritons. In contrast with resonant Brillouin scattering (RBS – see sect. 3.5.2), dipole-inactive branches (such as the longitudinal exciton) may be probed. However, the large incident intensities required to generate virtual biexcitons may, simultaneously, distort the normal exciton-polariton dispersion curve being investigated. It has recently been shown that even intermediate intensities can significantly renormalize the exciton-polariton dispersion curve in CuCl in the vicinity of the two-photon resonance (Itoh et al. 1978, May et al. 1979). This effect is expected to be stronger in CdS (Hönerlage and Rössler 1980). In addition, the iterative process required to derive the parameters of the exciton-polariton dispersion requires careful study. It is important to show that the derived set of parameters is not only reasonable but also unique. A recent investigation of CdS using hyper-Raman scattering demonstrates the importance of these questions (Schrey et al. 1979).

Table 1 lists the parameters of the A exciton-polariton in CdS as derived from hyper-Raman scattering (HRS), from resonant Brillouin scattering (RBS) and using other experimental techniques. It is evident that discrepancies exist between the values derived using hyper-Raman scattering and those obtained from other techniques (the apparent agreement in the exciton masses is illusory; the values used in the HRS studies were taken from the literature). The major problem lies in the background dielectric constant, ϵ_b, as is clearly evident in fig. 11. The only agreement between the dispersion curve calculated using the parameter values derived

Table 1
Experimental parameters of the A exciton-polariton in CdS

Technique	Resonant Brillouin scattering[a]	Hyper Raman scattering[b]	Other determinations[d]
ϵ_b	9.3	5.4 ± 0.2	8.0–8.5[e]
m^*	$0.89\ m_e$	$0.9\ m_e$[c]	$0.94\ m_e$
ω_T	$20589.5\ \text{cm}^{-1}$	$20585.6 \pm 0.8\ \text{cm}^{-1}$	$20589.5\ \text{cm}^{-1}$
ω_{LT}	$15.4\ \text{cm}^{-1}$	$21.0\ \text{cm}^{-1}$	$15.0\ \text{cm}^{-1}$

[a] Winterling and Koteles (1977b).
[b] Schrey et al. (1979).
[c] Hopfield and Thomas (1961).
[d] Voigt et al. (1976).
[e] Bruce and Cummins (1977).

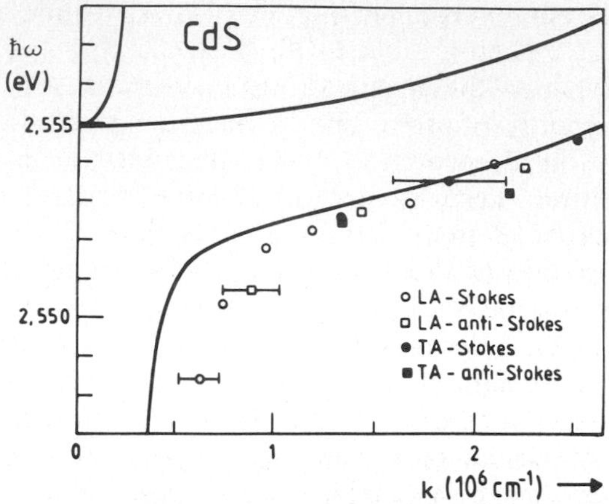

Fig. 11. Measured (points from RBS – Winterling and Koteles, 1977b) and calculated (solid lines – using parameters derived from HRS) exciton-polariton dispersion curves for the A exciton in CdS. (From Schrey et al. 1979.)

from HRS and the experimental points measured by RBS is in the region where the polariton is exciton-like. This is not surprising, as the important parameter here is the exciton mass which, as mentioned previously, is fortuitously identical for RBS (measured) and HRS (taken from the literature). In the region of maximum disagreement below the exciton resonance, ω_T, the dominant parameter is the background dielectric constant, ϵ_b. This can be measured directly using Brillouin scattering. For incident frequencies, $\omega_i \ll \omega_T$, the Brillouin shift, depends directly on ϵ_b and v, the acoustic phonon (sound) velocity (see eq. (8) below). v has been measured to four significant figures using ultrasonic techniques (Gerlach 1967). The values of ϵ_b obtained from Brillouin scattering results ($\epsilon_b \simeq 8.5$ measured $40 \rightarrow 100\ \mathrm{cm}^{-1}$ below ω_T (Bruce and Cummins 1977) and ($\epsilon_b \simeq 9.3$ measured $20 \rightarrow 30\ \mathrm{cm}^{-1}$ below ω_T (Winterling and Koteles 1977b)) are significantly larger than the value derived indirectly using HRS ($\epsilon_b = 5.4$ measured $80 \rightarrow 240\ \mathrm{cm}^{-1}$ below ω_T). The precise cause of this anomalously low value of ϵ_b is as yet unknown but could be related to renormalization effects resulting from the intense laser beams necessary for this technique.

3.4. Time-resolved spectroscopy

Exciton polaritons are the normal modes of energy propagation in crystals with dipole-active excitons. The velocity of their propagation, v_g, may be calculated directly from the slope of their dispersion curve. For a simple harmonic oscillator model of the exciton polariton, and in the absence of damping, there is one energy near the exciton resonance at which v_g is a minimum. Even if damping is included, generally v_g will decrease several orders of magnitude as the polariton energy is increased from below ω_T, The velocity of energy propagation can be experimentally determined by means of time resolved spectroscopy. This provides a sensitive test of the overall shape of the exciton-polariton dispersion curve and is another dramatic direct proof of the existence of these mixed propagating modes.

The first observation of the strong variation of the group velocity of an exciton polariton in the vicinity of an exciton resonance was in CuCl (Segawa et al. 1978). These workers measured the transit time of a 25 picoseconds (ps) wide light pulse through a 1.7 mm thick sample. They also determined the delay time of two satellites of the laser peak attributed to decay products of virtual biexcitons (see sect. 3.3). In interpreting their data they assumed that these satellites were formed in a narrow layer near the surface of the crystal and that their passage through the bulk of the sample was not significantly affected by inelastic collison processes. The measured differences between the transit times of the fastest and slowest pulses (700 ps) was attributed to the differences in frequency of the

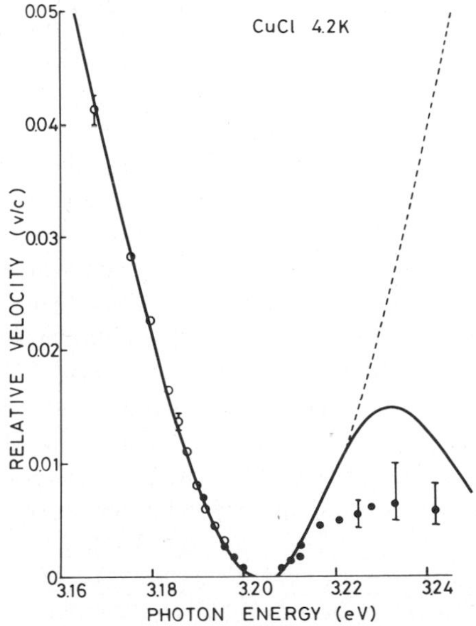

Fig. 12. Relative exciton-polariton group velocities in CuCl as a function of incident photon energy. $\hbar\omega_L \simeq 3.209$ eV. Experimental points are indicated by open and closed circles. Calculated curves: dashed lines, one-oscillator model; solid lines, two-oscillator model. (From Segawa et al. 1979.)

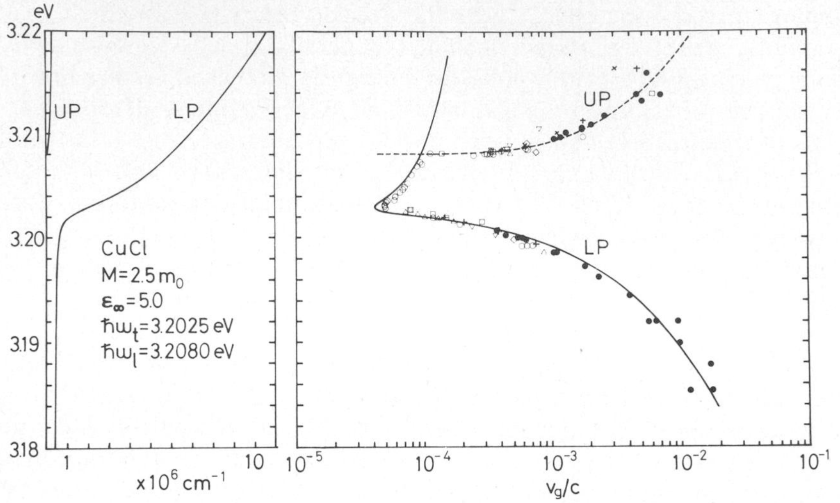

Fig. 13. The dispersion curve of CuCl (left) and the group velocity of exciton polaritons as a function of incident frequency (right). Different symbols denote different samples. The solid and dashed lines are calculated group velocities for the lower (LP) and upper (UP) exciton-polariton branches respectively. (From Masumoto et al. 1979.)

respective exciton polaritons. Absolute velocities were not determined but they obtained a good fit between measured relative delay times and times calculated using a one-oscillator exciton-polariton model neglecting spatial dispersion. Only the lower branch polariton was probed. In a later paper (Segawa et al. 1979) more precise results of the relative group velocities of both the lower and upper branch polaritons were obtained. Over the range investigated, the decrease in v_g for the lower branch was measured to be about a factor of 60 as the incident laser frequency approached the transverse Z_3 exciton in CuCl (fig. 12). However, exactly at resonance, the crystal was opaque. The calculated velocities were in good agreement with the experimental values for the lower branch polariton. However, the model yielded velocities for the upper polariton which were much larger than those measured. The discrepancy was reduced somewhat by including effects of the higher energy Z_{12} exciton.

Recently, another group, using very thin samples, was able to track the group velocity through the resonance regime of CuCl (Masumoto et al. 1979). They measured a minimum in group velocity at the inflection point of the lower branch of the exciton polariton and, at a somewhat higher energy, the onset of energy propagation on the upper polariton branch (fig. 13). Over a narrow frequency interval near ω_L two peaks were observed (fig. 14). These correspond to exciton polaritons simultaneously propagat-

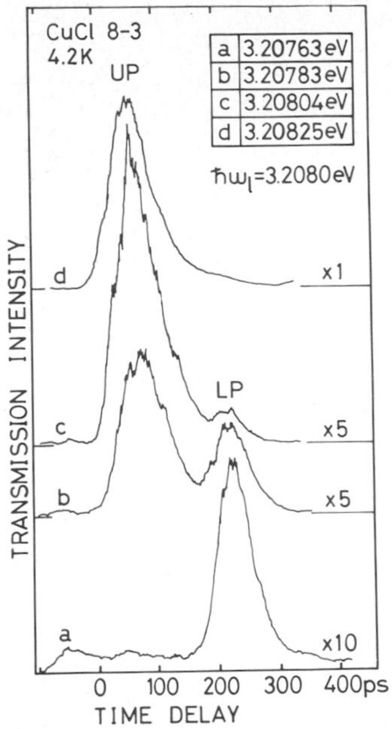

Fig. 14. Transient response of light pulses (FWHM = 0.45 meV) when the incident energy $\sim\hbar\omega_L$. UP and LP label peaks due to propagation on the upper and lower exciton-polariton branches respectively. (From Masumoto et al. 1979.)

ing on the upper and lower branches and provide dramatic proof of the existence of two branches. By precisely measuring the ratio of the intensities of these two modes as a function of incident frequency, it should be possible to derive information about ABCs. These workers extended the utility of this experiment by attempting to determine exciton-polariton parameter values from their results. Assuming a two band model ($Z_{1,2}$ and Z_3 excitons–fairly widely separated– 536 cm^{-1}), best agreement was reached when $m^* = 2.1\, m_e$. This value is somewhat lower than that derived from hyper-Raman experiments (see sect. 3.3).

Time-resolved spectroscopy has also been applied to GaAs, a more difficult case than CuCl since the resonance region is extremely narrow (Ulbrich and Fehrenbach 1979). The accuracy of the measurement was increased by splitting a single dye laser beam into two and using one to probe the sample and the other as a variable delay reference (fig. 15a). The simultaneous arrival of the two beams at the detector was determined by recording the second harmonic signal generated in a $LiNO_3$ crystal by the overlapping beams. The maximum delay time was about 30 ps while the laser width was 12 ps (fig. 15b). Good agreement was achieved between the

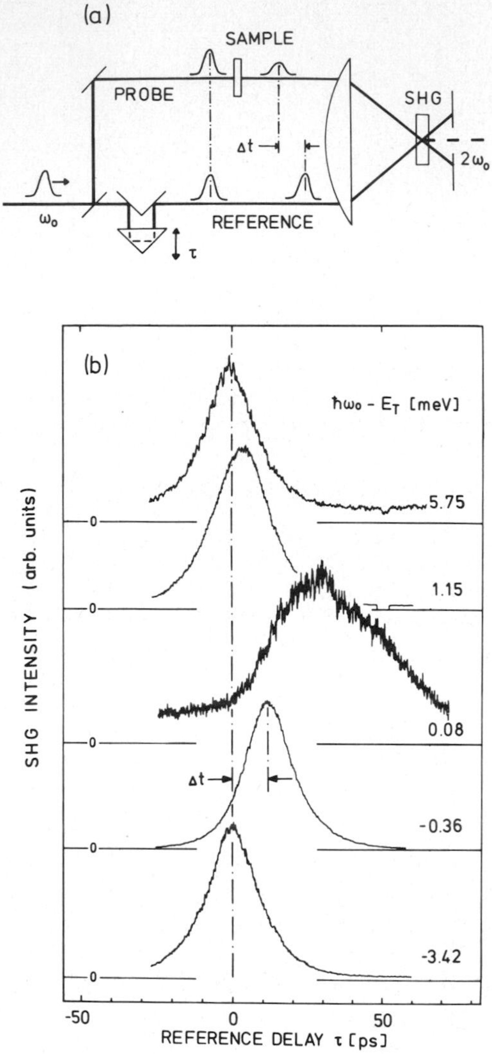

Fig. 15. (a) Experimental arrangement for the measurement of the propagation delay of picosecond light pulses transmitted through a thin sample. (b) The cross-correlation (SHG) signal for five different incident frequencies, ω_0, in the vicinity of the transverse exciton, E_T, of GaAs at 1.3 K. (From Ulbrich and Fehrenbach, 1979).

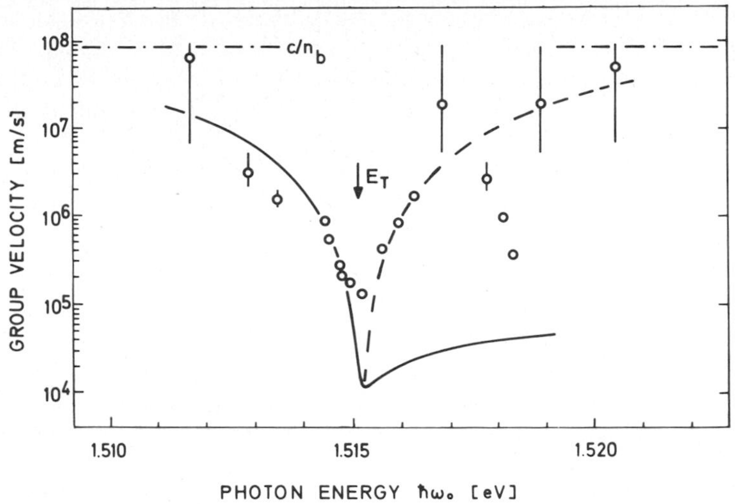

Fig. 16. Measured (open circles) and calculated (solid line, lower branch; dashed line, upper branch) exciton-polariton group velocities in GaAs. c/n_b denotes the phase velocity due to the background dielectric constant. The anomaly at 1.518 eV is due to the first excited state of the exciton. (From Ulbrich and Fehrenbach, 1979.)

experimental points and theoretical curves calculated using a one-oscillator exciton-polariton model (fig. 16). The agreement illustrates (1) that the light exciton branch required by a rigorous theory is not strongly populated (in agreement with RBS results – see sect. 3.5.2.2) and (2) that above $E_T + \hbar\omega_T$, energy propagation is primarily by upper branch polaritons in agreement with the CuCl results above. Near E_T, the magnitude of the minimum of v_g was determined to be about 2 orders of magnitude larger than predicted. The cause of the discrepancy is not clear but may be due to frequency dependent damping effects distorting the experimental spectra. It must be remembered that with such short pulses (12 ps) the uncertainty in energy due to the Heisenberg uncertainty principle (0.4 cm^{-1}) is significant compared with the width of the resonance ($\omega_{LT} = 0.6\,cm^{-1}$).

3.5. Resonant light scattering

3.5.1. Introduction

Since the introduction of tunable lasers, there has been a veritable explosion of investigations of resonant light scattering in solids. Until very recently, these experiments consisted of measuring the intensity of photons scattered inelastically by phonons or other excitations in a solid as a function of incident frequency. Resonant enhancement of the cross sections was studied as the laser frequency was tuned to the vicinity of some electronic level present in the crystal. Resonant Raman (optical phonon) and Brillouin (acoustic phonon) scattering offered a convenient, precise means to investigate energy levels in a solid.

However, in this section, the interest is not in the magnitudes of the *intensities* of scattered light but rather in the magnitudes of the Brillouin (and sometimes Raman) *shifts*. In standard light scattering experiments, only scattering intensities are studied as the incident frequency is scanned. The frequency shifts remain constant (Raman) or vary linearly with incident frequency (Brillouin). When the laser is tuned to an exciton-polariton resonance, however, the magnitudes of both the scattered intensity and the Brillouin shifts vary dramatically. The intensity and the width of a Brillouin peak depends on a large number of complicated factors; the nature and strength of the mechanism coupling exciton polaritons and acoustic phonons, the density of exciton-polariton states, additional boundary conditions, scattering of exciton polaritons by impurities, etc. It is not straightforward to predict the frequency dependence of each one of these factors, let alone their product, especially near an exciton resonance. On the other hand, the variation of Brillouin shifts with incident frequency depends only on the kinematics of the exciton polariton, e.g., its dispersion curve which, to a good approximation, is independent of damping and the other intensity-modifying factors mentioned above. At present resonant Brillouin scattering is the most direct and precise method available for determining the values of the parameters of the exciton-polariton dispersion curve. The emphasis in this section will be on the kinematics of exciton polaritons as determined by resonant light scattering. Occasional forays, however, will be made into exciton-phonon coupling theory in order to try to understand seemingly anomalous data.

3.5.2. Resonant Brillouin scattering

3.5.2.1. Introduction. In 1972, Brenig, Zeyher and Birman suggested that a convenient method of experimentally investigating additional boundary conditions (ABC) would be to measure the intensities of Brillouin scattering peaks as a function of ω_i in the vicinity of an exciton resonance. The intensities of these peaks near resonance depend not only on population factors, polariton–phonon coupling mechanisms, etc., but also, and quite sensitively, on the relevant ABC at the crystal–vacuum boundary. In addition, they pointed out that, as a consequence of the strongly dispersing shape of the exciton polariton at resonance, the magnitude of the Brillouin *shifts* would undergo a strong, distinctive dispersion. Further, due to the possibility of populating two simultaneously propagating modes when

$\omega_i \geqslant \omega_L$, the Stokes–anti-Stokes doublet usually seen in Brillouin scattering would be supplemented by six additional peaks.

Inherently, resonant Brillouin scattering (RBS) must be studied close to an intense dipole-active exciton which renders the crystal effectively opaque (the absorption constant is typically $\sim 10^5\ \text{cm}^{-1}$ at resonance). Thus the backscattering geometry is mandatory. Even so, it was not readily apparent that RBS experiments could succeed. Early attempts (Pine 1972) indicated that as the band gap energy of CdS was thermally tuned to a fixed laser frequency, the cross section for acoustic phonon backscattering increased proportionally to the square root of the absorption coefficient. This sub-linear dependence implied that the resonant enhancement of the cross section would be overcome by the increase in resonant absorption as ω_T was tuned into resonance with ω_i. This is indeed the case if the tuning is accomplished thermally since the phonon-assisted absorption increases more rapidly with temperature than the Brillouin cross section (Bruce and Cummins 1977). However, if the sample is kept at a low temperature, and the incident laser frequency is brought into resonance through the use of a suitable tunable dye laser, Brillouin peaks can be readily observed above the background in pure samples.

The kinematics of the Brillouin scattering process are based on the conservation of energy (frequency) and momentum (wavevector),

$$\omega_s = \omega_i \pm \omega_q, \tag{4}$$

$$\boldsymbol{k}_s = \boldsymbol{k}_i \pm \boldsymbol{q}, \tag{5}$$

between the incident (i) and scattered (s) photons and an acoustic phonon (q). The minus and plus signs indicate Stokes (phonon emission) and anti-Stokes (phonon absorption) processes respectively. For the polariton, $k_{i,s} = n_{i,s}\omega_{i,s}/c$ where $n_{i,s}$ is the refractive index for the respective waves and c is the vacuum velocity of light. For the acoustic phonon, $q = \omega_q/v$, where v is the sound velocity near the Brillouin zone center. Substituting these expressions into the momentum conservation law and assuming the backscattering geometry, the exact expression for the Brillouin shift ($\Delta\omega^{\pm}$) is

$$\omega_q = \Delta\omega^{\pm} = \pm \frac{v\omega_i}{c}(n_s + n_i)\left(1 \mp \frac{n_s v}{c}\right)^{-1}. \tag{6}$$

Since $v \ll c$, this reduces to

$$\Delta\omega^{\pm} = \pm \frac{v\omega_i}{c}(n_s + n_i). \tag{7}$$

If ω_i is in the transparent region (i.e., $\omega_i \ll \omega_T$) then $n_s \simeq n_i = \sqrt{\epsilon_b}$ and

$$\Delta\omega^{\pm} = \pm \frac{2v\omega_i\sqrt{\epsilon_b}}{c}, \tag{8}$$

where ϵ_b is the background dielectric constant. For example, referring to fig. 17a, consider an exciton polariton created at point A on the dispersion curve, and travelling in the forward direction which scatters, via a single LA phonon into state B^+ (anti-Stokes) or B^- (Stokes) travelling in the backward direction (i.e., 180° from exciton-polariton A). The total frequency shift ($\Delta\omega^+ = B^+ - A$, $\Delta\omega^- = B^- - A$) is given by eq. (8). In this region, the magnitudes of the Stokes and anti-Stokes Brillouin shifts are approximately equal and both vary linearly with ω_i (see lower portions of fig. 17(b). However, as the incident frequency is increased to the neighborhood of the exciton resonance, this simple relationship for the Brillouin shifts is no longer valid. It must be replaced by a more rigorous expression containing the non-linear behavior of the dispersion curve of the exciton polariton. In practice it is difficult to obtain an explicit relationship equivalent to eq. (8). It is relatively easy, however, to program the problem on a computer.

For example, consider Stokes scattering involving a one oscillator model for the exciton polariton. The problem consists of finding the points of intersection of the exciton-polariton dispersion curve and the acoustic phonon dispersion curve taking, as an initial condition, that one of the points (the coordinates of the incident exciton polariton, ω_i) is common to both curves. Solving these two equations simultaneously reduces the problem to a fourth order polynomial. When the initial frequency is less than ω_L, two of the solutions of this polynomial are imaginary and only one real $\Delta\omega^-$ is found. When $\omega_i > \omega_L$, all four solutions are real and there are four real $\Delta\omega^-$'s corresponding to two intrabranch ($1 \to 1'$, $2 \to 2'$) and two interbranch ($1 \to 2'$, $2 \to 1'$) transitions (fig. 17(a)).

In interpreting *Brillouin shift* dispersion curves (fig. 17(b)), the acoustic phonon sound velocity near the zone center is assumed independent of wavevector and is taken as known. For many crystals, it has been determined to four significant figures in ultrasonic experiments (e.g., Gerlach 1967). Inelastic neutron scattering experiments indicate that v is constant out to one-quarter or more of the Brillouin zone for many materials. For the one-oscillator exciton-polariton model, the relevant parameters are ϵ_b, the background dielectric constant, ω_T, the transverse exciton energy, m^*, the exciton effective mass and $4\pi\beta$, the exciton-polariton oscillator strength. Although all these parameters contribute to the overall shape of the $\Delta\omega^\pm$ dispersion curve, each may be determined almost uniquely, i.e., without reference to the others. The value of ω_T is determined by observing at which frequency $\Delta\omega^\pm$ begins its rapid increase (fig. 17(b)). The "elbow" of the ($2 \to 2'$) curve is relatively insensitive to ϵ_b, m^* and $4\pi\beta$. The exact shape of the ($2 \to 2'$) curve for $\omega_i > \omega_T$ depends primarily on m^*, and it is quite independent of the values of $4\pi\beta$, ω_T, or ϵ_b. For $\omega_i < \omega_T$, $\Delta\omega^\pm$ is directly related to ϵ_b through eq. (8). Finally, ω_L, the onset frequency of

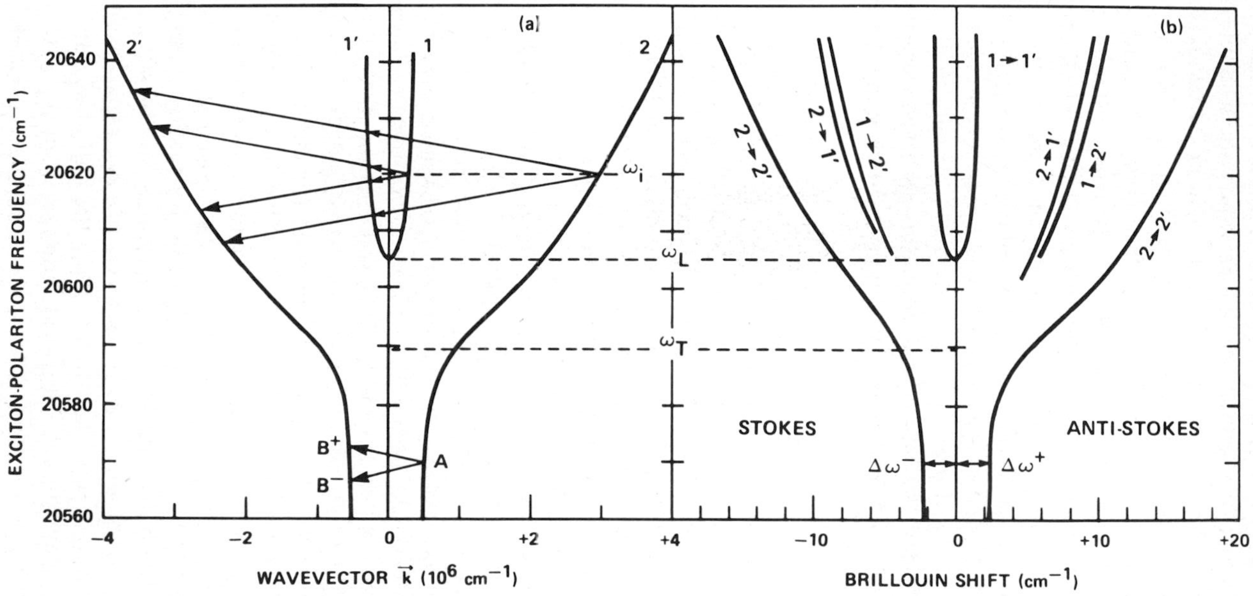

Fig. 17. (a) Exciton-polariton dispersion in the vicinity of the A exciton in CdS. The arrows illustrate the dynamics of Stokes (downward) and anti-Stokes (upward) Brillouin backscattering for incident frequencies A and ω_i. $\boldsymbol{k}$ and $-\boldsymbol{k}$ denote directions in reciprocal space 180° apart. For $\boldsymbol{k} \perp \boldsymbol{c}$ in CdS, the dispersion curves are isotropic. (b) Calculated Brillouin shift magnitudes for LA phonon scattering as a function of exciting frequency based on the exciton-polariton dispersion curve plotted in part (a). The notation $(n \rightarrow m')$ labels scattering from an initial state on branch n to a final state on branch m'. In this and all further figures, branches are labeled in increasing sequential order beginning at the innermost branch unless otherwise noted.

upper polariton branch, can be determined by observing at which frequency additional interbranch transitions (e.g., $1 \rightarrow 2'$) are first evident. The difference, $\omega_L - \omega_T$, is directly related to $4\pi\beta$, the oscillator strength. Thus accurate values of each of these parameters can be determined in a straightforward, unambiguous manner.

An additional advantage of RBS is that only low excitation intensities are required. There is little danger of altering the optical properties of the crystal in some unforeseen manner with an intense laser beam. However, only dipole-active exciton-polariton branches may be populated. Thus transitions involving states with weak or non-existent oscillator strengths (such as the longitudinal exciton in the example we have been following) are not observed. Also, with RBS (and especially with two-phonon Raman scattering) it is possible to probe the exciton-polariton dispersion curve at wavevectors which are a significant fraction of Brillouin zone boundary values.

Resonant Brillouin scattering experiments are performed at low temperatures to sharpen the exciton resonance and to reduce phonon related broadband background emission. To observe intense Brillouin peaks over a wide range, competing relaxation processes must be reduced to a minimum. Even small concentrations of impurities can produce strong impurity emission peaks which can overlap exciton resonances. Generally, the best available, pure, single crystals are investigated. A rule-of-thumb is to search out samples in which the free-exciton luminescence at low temperatures is strong and well resolved from the usually intense bound exciton peaks. RBS can sometimes be observed in poorer quality samples near the exciton resonance, but usually only weakly.

Brillouin scattering experiments in the transparent region of crystals ($\omega_i \ll \omega_T$) result in weak peaks very close to the exciting frequency. The largest Brillouin shifts are normally $1 \rightarrow 2\,\mathrm{cm}^{-1}$ (~ 0.1 meV). Their observation necessitates special optical systems, such as Fabry–Perot interferometers, requiring relatively sophisticated measurement techniques and frequency-stable sources. In contrast, RBS peaks are resonantly enhanced so that their intensities can be as large as 10^{-2} of the elastically scattered light (Ulbrich and Weisbuch 1977) and their frequency shifts may be an order of magnitude or more larger than that measured off-resonance (Winterling and Koteles 1982). Thus they may be readily detected with standard spectroscopic techniques such as those employed in Raman scattering. In fact, all the work reported so far has been carried out using double grating spectrometers. The lower resolution of these instruments permits the use of less highly stabilized sources. However, in order to secure more precise information on peak widths (related to polariton life times), the exact magnitude of small Brillouin shifts (e.g., $1 \rightarrow 1'$) etc., it will be necessary to study RBS with interferometric techniques. A monochromator or some other pre-filter will be needed to eliminate adjacent background emission.

Coupling between acoustic phonons and exciton polaritons is usually accomplished via the deformation potential interaction. In simple terms, lattice vibrations modulate the index of refraction of the solid which, in turn, modulates the propagating excitation. The selection rules for this coupling are well known and can be readily calculated for each crystal structure (Cummings and Schoen 1972). In certain materials, Fröhlich coupling may also be very important, especially near resonance. (For a review of Fröhlich interaction induced resonant Raman scattering by longitudinal optical phonons, see Martin and Falicov (1975).) This coupling can also act between the longitudinal electric fields of LA and piezoelectrically active TA phonons and the electric fields of exciton polaritons. It is q dependent and thus especially strong near resonance where the wavevectors of scattered phonons can become quite large. This interaction is sometimes called "forbidden" since it can produce a large effect in some geometries in which the normal, q independent, deformation potential interaction is not allowed. As the main emphasis of this article is on kinematics, the details of these coupling mechanisms will not be explored. However, it is important to be aware of them as they govern which particular scattering processes are dominant in practice, as will be seen later.

The first results of resonant Brillouin scattering experiments employing tunable dye lasers were reported in 1977 on zincblende GaAs (Ulbrich and Weisbuch 1977) and wurtzite CdS (Winterling and Koteles 1977a), the classic material for exciton-polariton studies. In the following discussion RBS experiments will be grouped, somewhat arbitrarily, according to the crystal structure of the material studied. Table 2 summarizes the values of the parameters of exciton-polariton dispersion curves which have been derived, so far, through the use of resonant Brillouin scattering.

3.5.2.2. Cubic zincblende structure. While investigating the resonant fluorescence of exciton-polaritons traveling in the [100] direction in GaAs, a III–V compound semiconductor which crystallizes in the zincblende structure (T_d), Ulbrich and Weisbuch noticed strong narrow peaks in the emission spectra close to and approximately symmetrical about the incident laser frequency, ω_i. When ω_i was tuned into the vicinity of the GaAs exciton resonance, these peaks increased in intensity and were joined by other narrow peaks. They soon had identified this structure with the resonant Brillouin peaks predicted earlier by Brenig et al. in 1972 (Ulbrich and Weisbuch 1977). Below the exciton resonance, the Brillouin shift of the symmetric doublet was observed to approach an asymptotic value as ω_i was decreased and, taking ϵ_b from the literature, they calculated an

Table 2
Exciton-polariton parameter values derived from resonant Brillouin scattering

	Direction of $\boldsymbol{k}$	ϵ_b	ω_T (cm^{-1})	m^* (m_e)	ω_{LT} (cm^{-1})	Additional Information
Zincblende						
GaAs[1]	[100]	12.6	12219	0.7	0.6	m^* for heavy excitons
CdTe[1]	[110]	10.4	12868	2.4	3.2	
ZnSe[2]	[100]	8.7	22602	$m^*_H = 1.11\, m_e$ $m^*_L = 0.38\, m_e$	11.7 ± 0.4	Δ(exchange energy) $= (-0.8 \pm 0.8)\ cm^{-1}$ m^*_H = heavy exciton mass m^*_L = light exciton mass
	[110]	8.7	22604	$m^*_H = 1.95\, m_e$ $m^*_L = 0.37\, m_e$	11.7 ± 0.4	
Wurtzite						
CdS						
A exciton	$\perp \boldsymbol{c}$[3]	9.3	20589.5	0.89	15.4	
	$\parallel \boldsymbol{c}$[4]	9.3	20585	2.85	15.4	$m^* = 2.7\, m_e$[5]
B exciton	$\perp \boldsymbol{c}$, $\boldsymbol{E} \perp \boldsymbol{c}$[6]	7.2	20712.3	1.2	10.1	ϕ(k-linear term coefficient) $= 5.6 \times 10^{-10}$ eVcm
	$\perp \boldsymbol{c}$, $\boldsymbol{E} \parallel \boldsymbol{c}$[6]	8.9	20713.0	1.2	10.2	$\phi = 5.6 \times 10^{-10}$ eVcm
	$\parallel \boldsymbol{c}$, $\boldsymbol{E} \perp \boldsymbol{c}$[4]	7.3	20713.5	0.74	9.0	
CdSe[7] (A exciton)	$\perp \boldsymbol{c}$	8.4	14713 ± 1	0.40 ± 0.05	4 ± 1	
	$\parallel \boldsymbol{c}$			1.3		
HgI_2[8]	[001]	6.8*	18837	1.2 ± 0.1	43*	*taken from literature
	[201]	6.8*	18837	0.75	43*	
	[100]	6.8*	18837	0.68 ± 0.1	43*	

[1] Ulbrich and Weisbuch (1978).
[2] Sermage and Fishman (1981).
[3] Winterling and Koteles (1977b).
[4] Winterling and Koteles (1982).
[5] Yu and Evangelisti (1978).
[6] Koteles and Winterling (1980).
[7] Hermann and Yu (1980).
[8] Goto and Nishina (1979).

acoustic sound velocity consistent with that of the longitudinal acoustic phonon. They concluded that exciton polaritons in GaAs couple predominantly to LA phonons. This is certainly true for polaritons travelling in the [100] direction in zincblende structure crystals but not true in general, as will be seen later.

The dispersion of their experimentally measured Brillouin shifts (ΔE) as

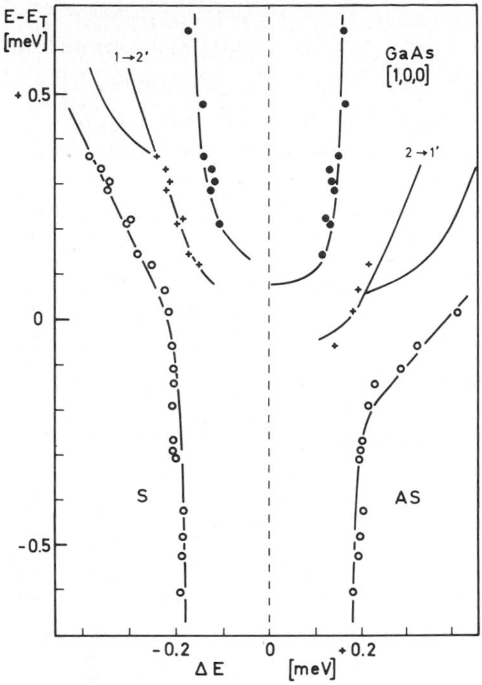

Fig. 18. Measured (points) and calculated (solid lines) Brillouin shifts, ΔE, as a function of relative incident light energy ($E - E_T$) for $k \parallel [1\,0\,0]$ in GaAs. (From Ulbrich and Weisbuch 1978.)

a function of the difference between the incident frequency and the exciton resonance ($E - E_T$) is illustrated in fig. 18. The asymmetry of the Stokes and anti-Stokes Brillouin shifts is clearly evident and the magnitudes of the shifts increase by a factor of two in the vicinity of the resonance. The solid lines were calculated assuming single LA scattering and a two-oscillator exciton model (see sect. 2.1). The best fit was achieved when $m^* = 0.7\,m_e$ (table 2). This value of the mass is close to that expected for the heavy exciton. Transitions involving the light exciton mass, which gives rise to a third exciton-polariton branch situated between the upper and lower branches, were not observed.

RBS experiments were also performed on GaAs in the [100] and [111] directions. The Brillouin shift dispersion curves could be fitted with the same exciton-polariton parameters as for [100] except that the masses were slightly larger (Ulbrich and Weisbuch 1978). In addition, in the [110] direction, Brillouin peaks attributable to TA phonon scattering were observed although "forbidden" by deformation potential selection rules (see sect. 3.5.2.2).

Another cubic zincblende material which has been investigated with RBS is the II–VI compound semiconductor CdTe (Ulbrich and Weisbuch 1978). At resonance many peaks were observed since the experiment was performed in the [110] direction and both TA and LA scattering occurred. Good agreement with the measured dispersion curves was reached by assuming a single harmonic oscillator model. The derived parameter values are given in table 2. It is not yet clear why scattering involving the third exciton-polariton branch has not been observed in GaAs or CdTe.

However, recently the three branch model for the exciton polariton in cubic zincblende has been experimentally confirmed by RBS in ZnSe (Sermage and Fishman 1979). The experiment was performed near the [100] direction in the vicinity of the $n = 1$ exciton. As ω_i was scanned through the resonance, a number of narrow, intense Brillouin peaks were observed. In the [100] direction only LA phonon scattering is expected. These new peaks were attributed to *intrabranch* Brillouin scattering involving three different polariton branches. Only a few *interbranch* scattering peaks were seen and over only a short frequency interval. This result was somewhat surprising. Six interbranch Stokes–Brillouin shifts are predicted for each type of acoustic phonon scattering when three exciton-polariton modes can be excited (see, for example, fig. 3). More recently, this interpretation has been significantly modified by new results (Sermage and Fishman 1981). Evidence acquired by additional RBS experiments performed with $\boldsymbol{k} \parallel [110]$ as well as $\boldsymbol{k} \parallel [100]$ indicate that some peaks previously considered as due to Brillouin scattering are probably non-wavevector conserving scattering due to impurities. Now, with $\boldsymbol{k} \parallel [100]$, as many as seven LA Stokes–Brillouin peaks were observed for a single incident frequency. The dispersion of these Brillouin peaks as a function of incident energy is presented in fig. 19. The presence of a weak TA peak (with an intensity two orders of magnitude smaller than LA scattering) is believed to indicate that the exciton-polariton propagation direction is not precisely parallel to [100]. The parameter values used to calculate the theoretical curves in fig. 19 are listed in table 2, along with values derived from measurements taken with $\boldsymbol{k} \parallel [110]$. The agreement is quite good.

Before proceeding, it is worthwhile to point out that all of the RBS data on cubic zincblende crystals can be interpreted by invoking scattering by *single* acoustic phonons.

3.5.2.3. Wurtzite structure. The first RBS experiments on a wurtzite structure material, CdS (C_{6v}) were performed shortly before the initial results on GaAs were reported (Bruce and Cummins 1977, Winterling and Koteles 1977a). CdS is a II–VI compound semiconductor whose optical properties have been investigated in hundreds of papers over the last three decades. It

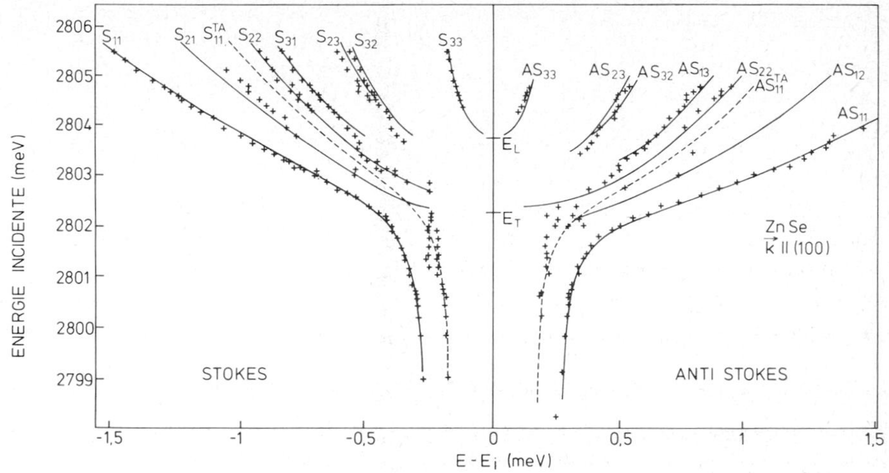

Fig. 19. Experimental (crosses) and calculated (solid and dashed lines) Brillouin shifts ($E - E_i$) as a function of incident energy in the vicinity of the $n = 1$ exciton in ZnSe. The Brillouin shift dispersion curves are labeled S_{if} (Stokes) or AS_{if} (anti-Stokes). S_{if} denotes scattering between the initial (i) and final (f) polariton branch and i, f = 1, 2, 3 (counting from the outer branch). (From Sermage and Fishman, 1981.)

was chosen for the initial experiments since it is a well characterized material, is available in high-quality single crystal platelet form, and is the crystal used as the example in the pioneering theoretical paper postulating RBS (Brenig et al. 1972). The theoretical calculations indicated that the Brillouin peaks at resonance would be broad and thus difficult to distinguish above the strong free-exciton luminescence background. While a 5-pass Fabry–Perot interferometer thought necessary for the experiment was in the process of construction, a preliminary emission spectrum was taken with a double-grating spectrometer in order to establish the background spectra. Surprisingly, the first scans revealed strong, sharp, resonant Brillouin peaks when ω_i was in the vicinity of the A exciton and work commenced immediately using the grating instrument.

Most of the samples investigated in the work discussed below were thin (~30 μm) single crystal platelets (typically 2×3 mm) grown by vapor phase epitaxy. Under these growth conditions, the c-axis usually lies in the plane of the crystal. Experiments were performed on as-grown surfaces with $\boldsymbol{k} \perp c$ unless otherwise indicated.

The luminescence spectra of a number of good-looking (i.e., thin, smooth-surfaced platelets) samples irradiated by band-gap laser light were measured. The crystals were glued to a copper holder with a single drop of diluted GE varnish to minimize strain and placed in an optical flow-through

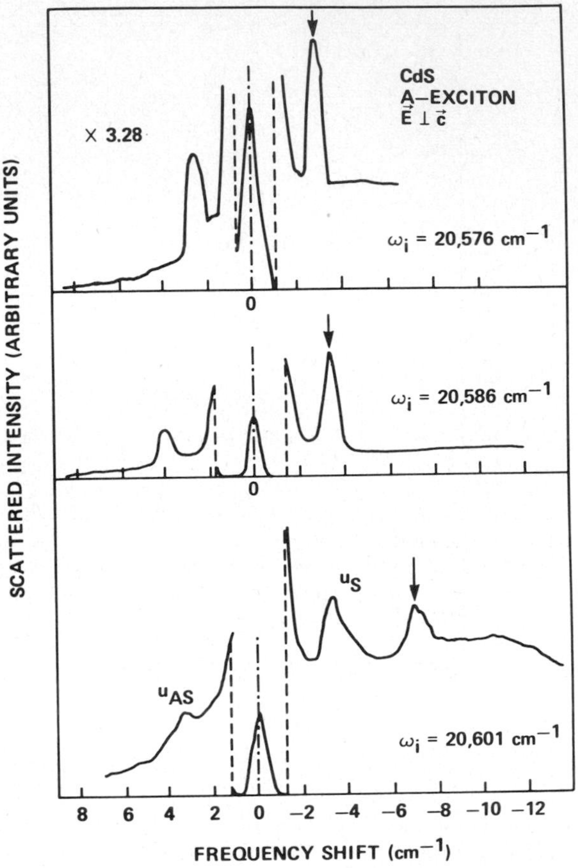

Fig. 20. Experimental Brillouin scattering spectra excited by three incident frequencies, ω_i, in the vicinity of the A exciton-polariton in CdS. The central portion of the spectra, delimited by the dashed vertical lines, contains the Rayleigh elastically scattered peak and is experimentally attenuated by a factor of 3×10^3. (From Winterling and Koteles, 1977a.)

helium dewar. Those platelets possessing strong free-exciton luminescence, well resolved from bound exciton lines, were selected for RBS experiments.

Typical Brillouin spectra, for ω_i in the vicinity of the A exciton frequency in CdS, $\omega_T = 20589.5\ \text{cm}^{-1}$, are presented in fig. 20. Brillouin peaks due to LA(2→2′) Stokes transitions are labeled with an arrow. The magnitude of the Brillouin shift increases from $2.2\ \text{cm}^{-1}$ below resonance to about $7\ \text{cm}^{-1}$ above resonance in this figure. The dispersion of the measured LA Brillouin shifts as a function of ω_i is shown in fig. 21. Again,

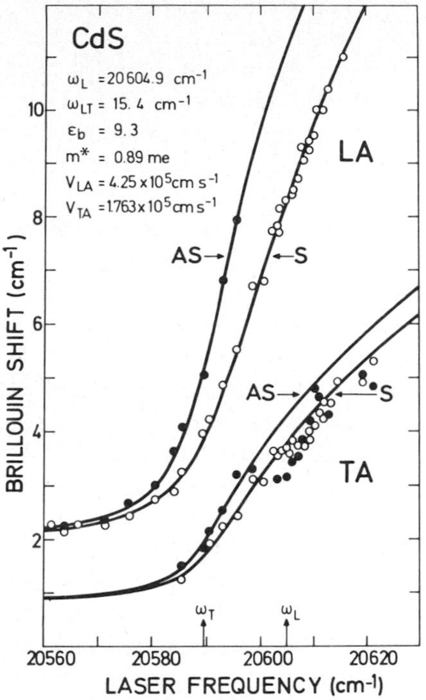

Fig. 21. Experimental Stokes (open circles) and anti-Stokes (closed circles) Brillouin shifts as a function of the incident laser frequency in the vicinity of the A exciton-polariton in CdS. The solid curves were calculated assuming a single harmonic oscillator model for the exciton with the parameter values given in the figure. (From Winterling and Koteles, 1977b.)

the rapid increase in the shift when $\omega_i \simeq \omega_T$ and the asymmetry between the Stokes (S) and anti-Stokes (AS) shifts are clearly evident. The solid curves were calculated employing a single harmonic oscillator model for the A exciton and adjusting the parameters as discussed previously. The figure illustrates excellent agreement between experiment and theory.

In fig. 20, when $\omega_i = 20601\ \mathrm{cm}^{-1}$, extra peaks (labeled U_S and U_{AS}) were observed. These could not be due to interbranch scattering as ω_i was still below the frequency at which the upper polariton branch begins. The dispersion of the Brillouin shifts of these "unknown" peaks is given in the lower part of fig. 21. The solid curves were calculated using the exciton-polariton parameters derived by fitting the LA Brillouin shift dispersion and replacing the LA phonon velocity with the TA phonon velocity. The excellent agreement attained indicates that these "unknown" peaks result from Brillouin scattering involving TA phonons in spite of the fact that such interactions are "forbidden" by deformation potential coupling selection rules. Physically, sheer strains associated with transverse acoustic phonons modulate only the component of the dipole moment which is parallel to the wavevector, $\boldsymbol{q}$, of the phonon. Thus, radiation in the backward direction, $(-\boldsymbol{q})$, is impossible. This argument applies to both deformation-potential coupling and to coupling through the first-order electrooptic effect. However, there is another interaction which can couple phonons and photons, the Fröhlich interaction. In Raman scattering, this coupling proceeds via the macroscopic longitudinal electric field of the LO phonon and can be present even in geometries in which deformation-potential coupling is not possible. Thus, it is commonly termed "forbidden" coupling. It occurs only for finite wavevector and its strength, for small $\boldsymbol{q}$, varies as q^2. The importance of this interaction for Brillouin scattering will become clear when an additional fact is added. The slow transverse acoustic phonon in CdS possesses a *longitudinal* electric field as a consequence of the action of the piezoelectric effect. Thus, in analogy with Fröhlich LO scattering, "forbidden" coupling between TA phonons and the A exciton-polariton in CdS is possible. This interaction is resonantly enhanced when $\omega_i \simeq \omega_T$, both due to terms in the denominator of the coupling matrix element which tend to zero at resonance and to the increase in q of the phonon due to the dispersion of the exciton polariton near ω_T (Winterling et al. 1977). This coupling mechanism is operative not only in CdS but in any piezoelectrically-active crystal. For example, some TA phonons traveling in the [110] direction in GaAs are piezoelectrically active and thus are observable in RBS due to the action of Fröhlich coupling (Ulbrich and Weisbuch 1978). On the other hand, TA phonons in the [100] direction in GaAs are not piezoelectrically active and thus TA Brillouin scattering is not observed (Ulbrich and Weisbuch 1977). This coupling mechanism is also responsible for the presence of two-phonon peaks observed in the resonant Brillouin and Raman scattering spectra of CdS. The narrowness and strength of these two-phonon scattering peaks is a consequence of the anisotropy of the wurtzite structure. Thus, it is a good idea to briefly discuss the effects of crystal anisotropy of this coupling mechanism on some of the other properties of the solid. The phonon velocity, effective mass, and the square of the longitudinal electric field of piezoelectrically-active acoustic phonons in CdS are plotted as functions of the angle, ϕ, between $\boldsymbol{c}$ and the propagation direction in fig. 22. Although phonons are purely transverse or longitudinal only along certain crystallographic directions (i.e., $\boldsymbol{q} \parallel \boldsymbol{c}$ or $\boldsymbol{q} \perp \boldsymbol{c}$), the curves are labeled "TA" and "LA" for simplicity. For $\theta = 90°$ (i.e., $\boldsymbol{q} \perp \boldsymbol{c}$) there is a maximum in the magnitude of the piezoelectrically induced longitudinal electric field associated with the TA phonon. Thus Fröhlich coupling is possible and, in fact, produces "forbidden" TA back scattering as discussed above. But, there is another, stronger maximum at 32° and maxima also for LA phonons at 0°

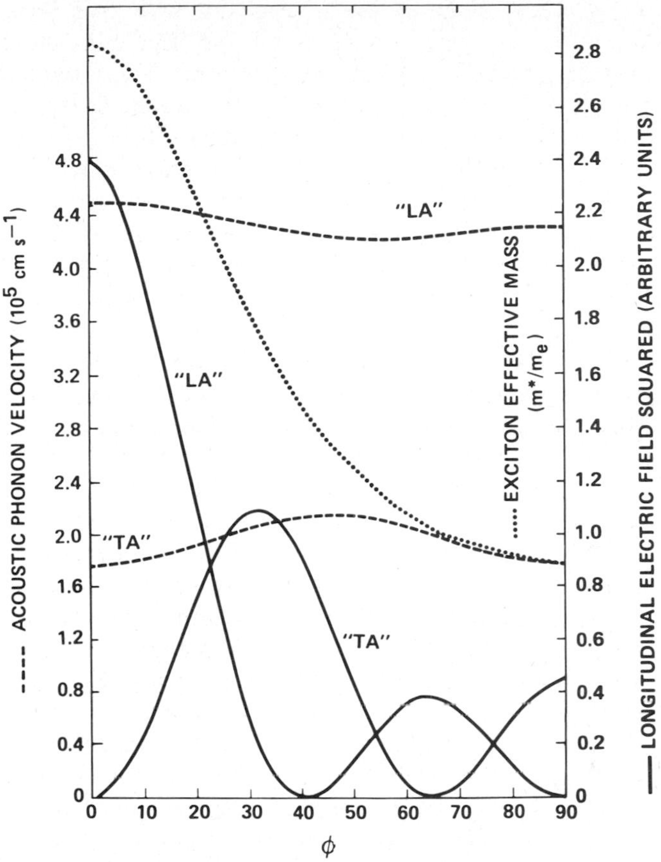

Fig. 22. Anisotropy of several properties of CdS. ϕ is the angle between the optic axis, $\boldsymbol{c}$, and $\boldsymbol{k}$. Solid curves, E_L^2 (longitudinal electric field associated with piezoelectrically-active acoustic phonons)^2; dashed curves, v (acoustic phonon velocities); dotted curve, $m^* = (\sin^2 \phi/m_\perp^* + \cos^2 \phi/m_\parallel^*)^{-1}$ (exciton effective mass). "TA" and "LA" denote the quasi-transverse and quasi-longitudinal acoustic phonon modes.

and 64°. Thus Fröhlich coupling is possible for LA phonons traveling parallel to $\boldsymbol{c}$ or at $\theta = 64°$ and for TA phonons at $\theta = 32°$ and 90°. The magnitude of the Brillouin shift which will result will depend on the dispersion of the acoustic phonon (i.e., its velocity) and on the dispersion of the exciton polariton. For a given frequency greater than ω_T, the wavevector of the lower branch polariton is primarily determined by the exciton effective mass, which, as fig. 22 shows, is also highly anisotropic. Thus, if RBS back scattering experiences were performed as a function of angle, ϕ, the exact Brillouin shift dispersion curves measured would

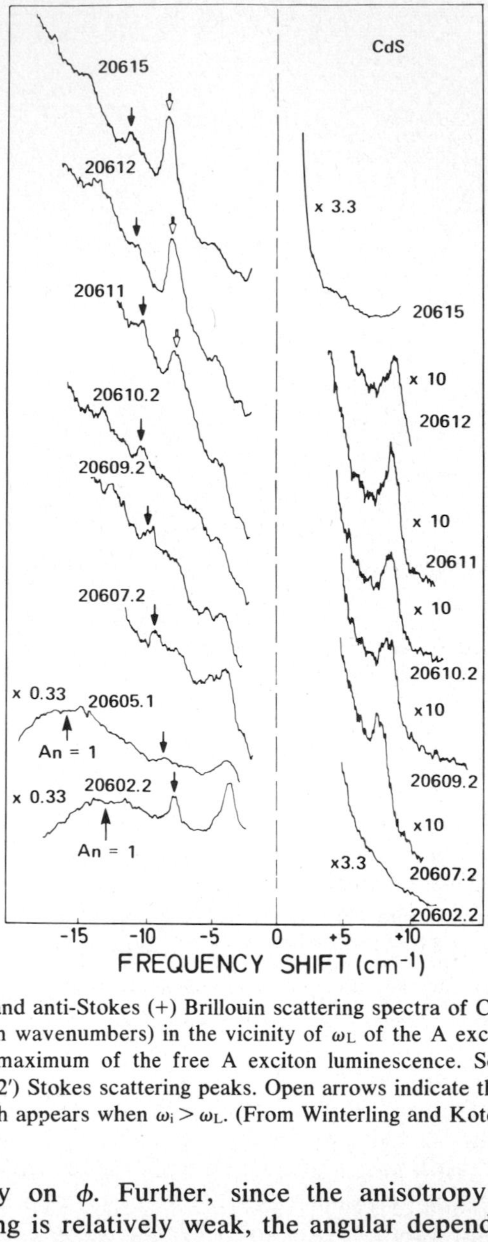

Fig. 23. Stokes (−) and anti-Stokes (+) Brillouin scattering spectra of CdS at several incident frequencies (given in wavenumbers) in the vicinity of ω_L of the A exciton-polariton in CdS. $A_{n=1}$ identifies the maximum of the free A exciton luminescence. Solid arrows mark the position of LA (2→2′) Stokes scattering peaks. Open arrows indicate the two-phonon Stokes peak which appears when $\omega_i > \omega_L$. (From Winterling and Koteles, 1977b.)

depend critically on ϕ. Further, since the anisotropy of deformation-potential coupling is relatively weak, the angular dependence of the scattering intensities of TA and LA phonons is strongly correlated with the Fröhlich interaction anisotropy. Thus, there will be enhanced Brillouin scattering in directions in which the longitudinal electric fields of the acoustic phonons have maximum values. This effect is so strong in the

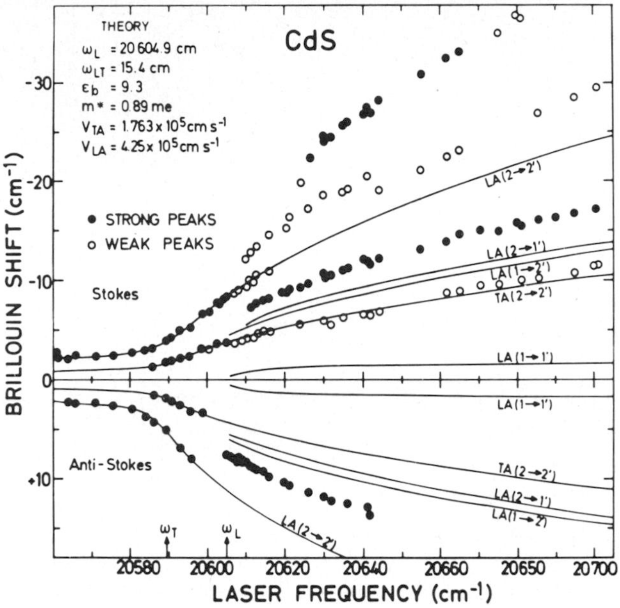

Fig. 24. Open and closed circles denote experimental Stokes (−) and anti-Stokes (+) Brillouin shifts as a function of incident frequency in the vicinity of the A exciton-polariton in CdS. The solid curves were calculated assuming a single oscillator exciton-polariton model, single-phonon scattering and the parameter values listed in the figure. (From Winterling and Koteles, 1977.)

region of the A exciton in CdS for $\omega_i > \omega_L$ that it produces dominating features in the scattering spectra even when $\boldsymbol{k} \perp \boldsymbol{c}$.

Figure 23 illustrates that, as ω_i was tuned through ω_L, LA intrabranch scattering $(2 \to 2')$ (labeled with solid downward pointing arrows in the figure) weakened and was quickly replaced by strong, narrow peaks (open arrows). The Brillouin shifts of some of these new peaks are much larger than the maximum possible single phonon scattering peak (LA$(2 \to 2')$) as is evident in fig. 24. They have been identified with two-phonon scattering processes (Winterling and Koteles 1977b, Yu and Evangelisti 1978). It is interesting that for $\omega_i > \omega_L$ (fig. 23) a second order process completely dominates a first order process. The Fröhlich interaction dominates "allowed" scattering since: (1) the phonons participating in strong two-phonon scattering travel in directions inside the crystal in which their longitudinal electric fields and thus their scattering cross sections are at maxima and (2) their wavevectors are large.

Figure 25(a) is a schematic diagram illustrating a typical two-phonon Brillouin scattering process. When $\omega_i > \omega_L$, the upper polariton branch (1)

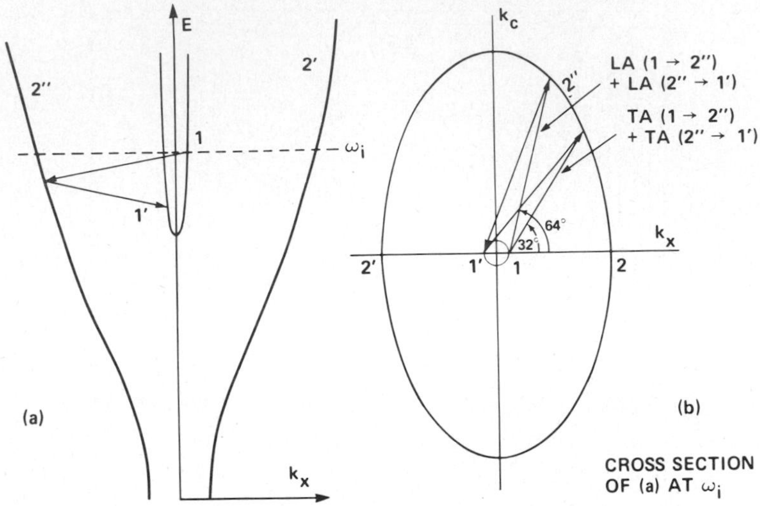

Fig. 25. (a) A exciton-polariton dispersion curves in CdS. Arrows illustrate a two-phonon Brillouin scattering process ($1 \rightarrow 2'' \rightarrow 1'$). (b) Cross section of exciton-polariton dispersion curve at frequency ω_i. Note that while unprimed and singly-primed states are always $\perp \boldsymbol{c}$ and 180° from each other, doubly-primed intermediate states are in k-space directions in which Fröhlich coupling is enhanced.

is more heavily populated than the lower branch (2). Thus only transitions ($1 \rightarrow 1'$) and ($1 \rightarrow 2'$) are predicted to be intense. The latter transition quickly weakens as ω_i is increased further since the lower branch in this region is exciton-like and polaritons on this branch have a small probability of reaching the crystal surface and escaping. On the other hand, branch 1 polaritons are photon-like and have a good chance of escaping and being observed. Consider, however, a scattering process which connects branch 1 with branch 2″ employing a phonon which travels in a direction in the crystal in which Fröhlich coupling is strong (e.g. $\phi = 32°$ for TA phonons). Many polaritons would accumulate on branch 2″, but they would be difficult to observe outside the crystal since branch 2″ is exciton-like. However, the reverse process ($2'' \rightarrow 1'$) is equally strong and would leave the polariton in a photon-like final state for which the crystal is essentially transparent. Thus, two acoustic phonons can connect the same initial (1) and final (1′) states as in the case of strong single phonon scattering ($1 \rightarrow 1'$). Figure 25(b) is a cross section of the anisotropic exciton-polariton dispersion curve of the A exciton in CdS, illustrating two of the probable two-phonon Brillouin scattering processes. The exact shape of the resultant two-phonon scattering peaks is an involved, but straightforward, calculation involving the anisotropic parameters given in fig. 22. Yu and Evangelisti (1978) have

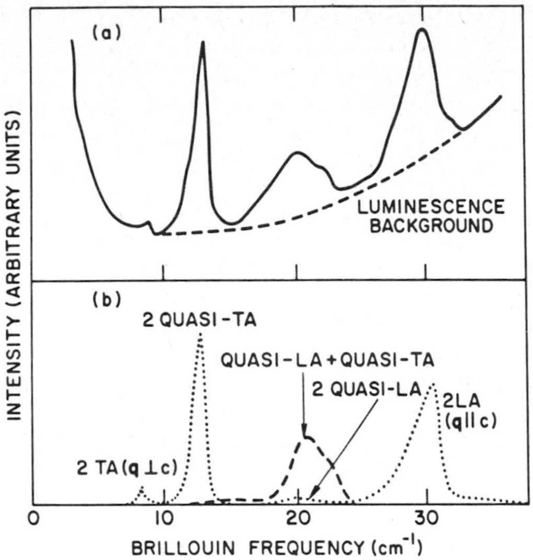

Fig. 26. Experimental (a) and calculated (b) two-phonon Brillouin spectra of CdS at an incident frequency of 20655 cm^{-1}. (From Yu and Evangelisti, 1978.)

found good agreement between calculated and experimental spectra (fig. 26). In fact, using the exciton mass parallel to c as an adjustable parameter, they were able to derive a value for it $(2.7 \pm 0.2)\, m_e$, which is in good agreement with that measured directly with single phonon RBS $(2.85 \pm 0.1)\, m_e$ (Winterling and Koteles 1982). It is worth noting in passing that the excellent agreement between experimental and calculated Brillouin shift dispersion curves for two-phonon scattering when $\omega_i \gg \omega_L$ implies that the basic assumptions of the exciton model are valid. Even for $k \sim \frac{1}{5}$ Brillouin zone boundary value (e.g., at $\omega_i \simeq 20680\ cm^{-1}$) the A exciton in CdS can be represented by a simple harmonic oscillator assuming that the dispersions of acoustic phonons remain linear that far into the Brillouin zone. Two-phonon resonant scattering is the most sensitive technique presently available for accurately probing the exciton-polariton dispersion curve out to large wavevector values. Two-phonon Raman scattering, in which scattering by an acoustic phonon precedes LO phonon scattering, proceeds via a similar two-step process. This has been observed previously in Cu_2O (Yu and Shen 1974) and other materials and has also been seen recently near the A exciton in CdS. In the latter case, however, features attributable to the polariton nature of the exciton were discerned for the first time (sect. 3.5.3).

It is obvious that Fröhlich coupling is of fundamental importance in interpreting Brillouin and Raman spectra when the incident laser frequency

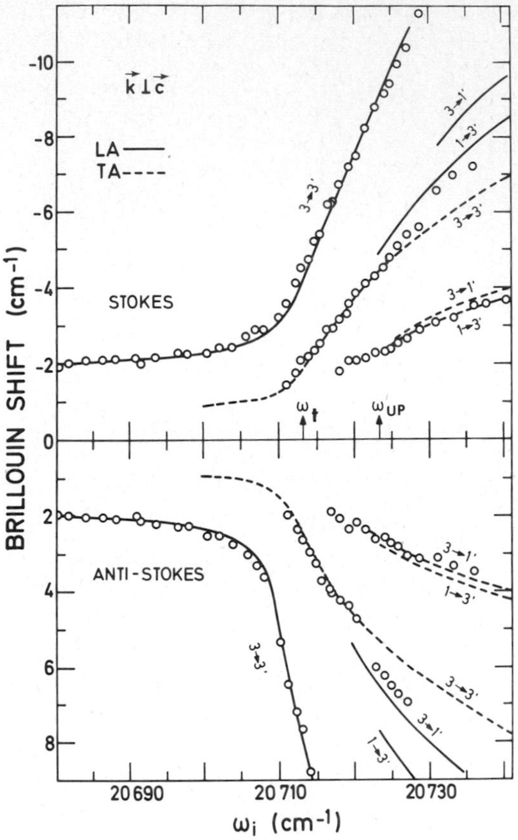

Fig. 27. Experimental (circles) and calculated (solid and dashed lines) Brillouin shifts as a function of incident frequency, ω_i, in the vicinity of the B exciton-polariton in CdS. $\boldsymbol{E} \parallel \boldsymbol{c}$ for both incident and scattered light and $\boldsymbol{k} \perp \boldsymbol{c}$. (From Koteles and Winterling, 1980.)

is tuned to the vicinity of the A exciton in CdS. It is curious therefore that this mechanism appears to play a minor role in resonant Brillouin scattering in the vicinity of the B exciton. All of the experimental data can be interpreted in terms of single-phonon scattering, even though "forbidden" TA scattering is observed (Koteles and Winterling 1979a, 1980).

The geometry for RBS experiments performed in the vicinity of the B exciton in CdS was arranged so that the incident and back scattered polariton always traveled perpendicular to $\boldsymbol{c}$. In one set of experiments the incident (E_i) and scattered (E_s) light were polarized parallel to the c-axis ($\boldsymbol{E}_i \parallel \boldsymbol{E}_s \parallel \boldsymbol{c}$) and, in another set, $\boldsymbol{E}_i \parallel \boldsymbol{E}_s \perp \boldsymbol{c}$. The measured dispersion of the Brillouin shifts for the two cases are given in figs. 27 and 28 respectively.

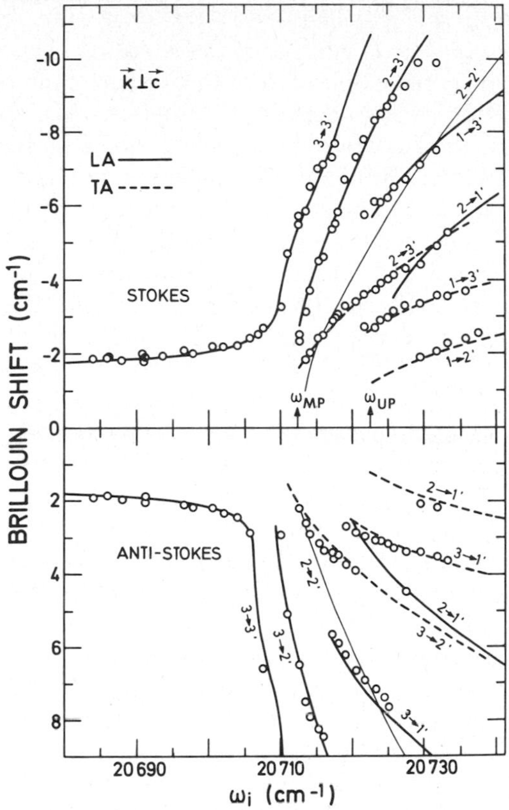

Fig. 28. Experimental (circles) and calculated (solid and dashed lines) Brillouin shifts as a function of incident frequency, ω_i, in the vicinity of the B exciton-polariton in CdS. $\boldsymbol{E} \perp \boldsymbol{c}$ for both incident and scattered light and $\boldsymbol{k} \perp \boldsymbol{c}$. (From Koteles and Winterling, 1980.)

Even in the absence of a model, it is obvious by glancing at these figures that there are fundamental differences between these two sets of results. For $\boldsymbol{E} \parallel \boldsymbol{c}$, only a few dispersing Brillouin peaks were observed, indicative of a relatively simple exciton-polariton dispersion curve. For $\boldsymbol{E} \perp \boldsymbol{c}$, however, the number of peaks increased and even peaks which were present in the other polarization (i.e., $3 \rightarrow 3'$ for $\omega_i < \omega_T$) exhibited significantly different behavior when ω_i was in the region of $\omega_T = \omega_{MP}$.

The exciton-polariton model invoked to interpret these results is due to Mahan and Hopfield (1964) and has been discussed in detail earlier (see sect. 2.2). Basically, the polarization dependence of the exciton-polariton dispersion curves results from the fact that an exciton energy term linear in wavevector couples (1) states degenerate at $k \rightarrow 0$ when $\boldsymbol{E} \perp \boldsymbol{c}$, producing strong mixing and (2) widely-separated states when $\boldsymbol{E}\|\boldsymbol{c}$, leading only to a weak perturbation of the exciton-dispersion curve. Thus, for $\boldsymbol{E}\|\boldsymbol{c}$, the exciton-polariton dispersion curve has the appearance of, and can be replaced by, a simple two-branch model (Koteles and Winterling 1979a). For $\boldsymbol{E}\perp\boldsymbol{c}$, however, the complete three-branch model is necessary in order to fit the experimental Brillouin shift dispersion. The values of the parameters employed to obtain the good agreement shown in fig. 26 are given in table 2. The derived value of the coefficient of the k-linear term, ϕ, is close to that obtained by fitting reflection experiments (Mahan and Hopfield 1964) and is related to the coefficients of the k-linear terms of the Γ_7 conduction and second valence bands (C^e and C^h respectively) by:

$$\phi = (C^h m_{h\perp} - C^e m_{e\perp})/m_\perp, \qquad (9)$$

where $\perp$ signifies effective masses for $\boldsymbol{k}\perp\boldsymbol{c}$. C^e has recently been measured by spin-flip Raman scattering ($C^e = 1.6\times10^{-10}$ eVcm (Romestain et al. 1977). Using eq. (9) and the known values of the effective masses, $C^h = (6.7\pm0.7)\times10^{-10}$ eVcm. If C^h and C^e arise from spin–orbit coupling (Hopfield 1961) then $|C^h/C^e|$ should be $\sim10^2$, not 4 as measured. This large discrepancy suggests that the physical origin of the observed k-linear terms requires further study.

Using the exciton parameters ($m_\perp$, ϕ and $\omega_T^B = \omega_{MP}$) derived by fitting the Brillouin shift dispersion curves for $\boldsymbol{E}\perp\boldsymbol{c}$, it is possible to fit the results for $\boldsymbol{E}\|\boldsymbol{c}$ by slightly modifying the "polariton" parameters, i.e., $\epsilon_{b\|}$ and $4\pi\beta_\|$ (fig. 27). As expected, $\epsilon_{b\|} = 8.9 \simeq \epsilon_b$ in the vicinity of the A exciton. $\epsilon_{b\perp}^{eff}$, however, is significantly smaller than ϵ_b since the A exciton is active in this polarization and thus can modify the background dielectric constant near the B exciton.

The complexity of the Brillouin scattering spectra for $\boldsymbol{E}\perp\boldsymbol{c}$ is a consequence of the three active exciton-polariton branches. When $\omega_i > \omega_{UP}(=\omega_L)$, there are nine Stokes–Brillouin transitions possible for each incident frequency compared with four in the two-branch model (see fig. 3). Many of these are observed (fig. 28), but others are not, for varying reasons. Some transitions are overlapped, within experimental uncertainty, by other, stronger, transitions (e.g., Stokes $3\rightarrow2'$ shifts have the same magnitude as $2\rightarrow3'$); some shifts are too small to be resolvable with the double grating spectrometer (i.e., $1\rightarrow1'$); by the time some transitions become possible, the density-of-states of the initial or final state is so small that the Brillouin peak is very weak (e.g., Stokes $3\rightarrow1'$). But, notwithstanding the above arguments, both Stokes and anti-Stokes LA($2\rightarrow2'$) should be strong, at least near ω_{MP}. In fact, neither is observed. This is a consequence of the nature of the matrix elements coupling the initial and final polariton states (Winterling and Koteles 1980).

To date, the only other wurzite structure crystal which has been investigated using RBS is CdSe (Hermann and Yu 1978, 1980). The quality of their samples made it difficult to obtain clear-cut Brillouin shift dispersion curves over as wide a range as is possible in CdS. For $\omega_i > \omega_L$ in undoped samples and, in addition, for all ω_i in doped samples, non-wavevector conserving scattering processes involving defects were found to dominate the normal wavevector conserving Brillouin processes. Nevertheless, they were able to derive values for the parameters of the A exciton in CdSe (see table 2). At resonance, in their pure samples, they observed both "allowed" LA and "forbidden" TA back scattering as in the case of CdS. Unlike CdS, they did not observe any two-phonon scattering processes.

3.5.2.4. Other crystal structures. The A exciton region of the layer semiconductor red HgI_2 has recently been investigated using RBS (Goto and Nishina 1979). The crystal structure of this compound (D_{4h}) is highly

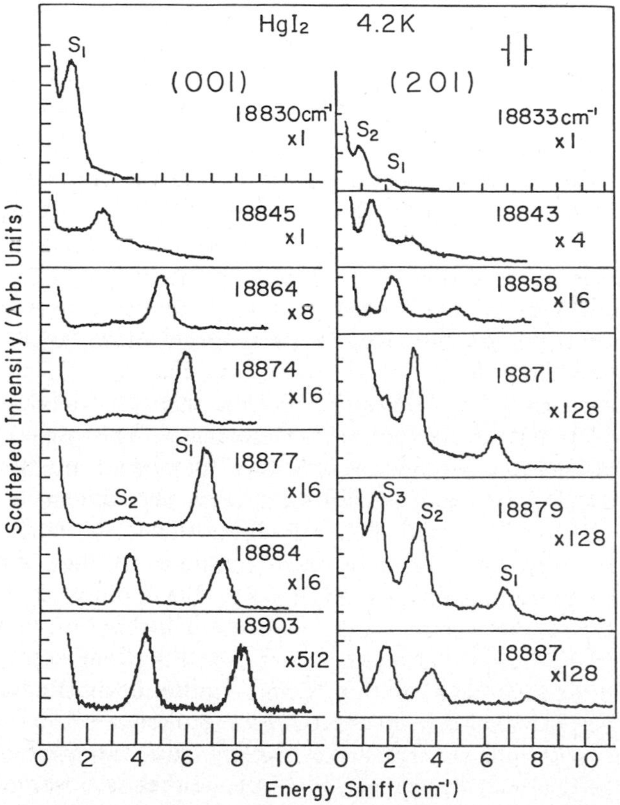

Fig. 29. Experimental Stokes–Brillouin spectra in the vicinity of the A exciton-polariton in red HgI_2 for several different frequencies (given in wavenumbers) and for two incident directions; $\boldsymbol{k}\parallel[001]$ and $\boldsymbol{k}\parallel[201]$. (From Goto and Nishina, 1979.)

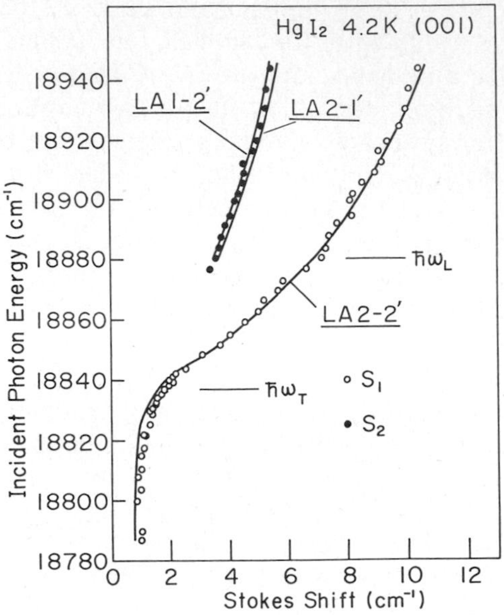

Fig. 30. Experimental (circles) and calculated (solid lines) Brillouin–Stokes shifts for $\boldsymbol{k} \parallel [001]$ in the vicinity of the A exciton-polariton in red HgI_2 as a function of incident photon energy. (From Goto and Nishina, 1979.)

anisotropic. The magnitude of the effective exciton mass was measured in two directions in the crystal in order to determine the mass anisotropy. The angular dependence of the exciton mass has a functional relationship similar to that of wurzite crystals (see fig. 22). The A exciton is formed by the combination of a Γ_6^+ conduction electron and a Γ_7^- valence hole and is active only for light polarized perpendicular to the uniaxial direction. Measurements were performed on as-grown surfaces in the [001] and [201] directions. Typical Brillouin scattering spectra are shown in fig. 29. For [001], only LA scattering was observed whereas for [201], both LA and TA scattering are permitted. Plots of the dispersion of the Brillouin shifts versus ω_i are presented in figs. 30 and 31. The data were fitted with a single-oscillator exciton model and, even though some of the parameters were taken from the literature (e.g., $\omega_L = 18{,}880\ \text{cm}^{-1}$ and $\epsilon_b = 6.8$) there is substantial agreement between theory and experiment. The derived values of the effective exciton masses are $m_\perp = (0.68 \pm 0.1)m_e$ and $m_\parallel = (1.2 \pm 0.1)\ m_e$. Although this is a layer compound, its effective exciton mass anisotropy ($m_\parallel/m_\perp = 1.76$) is less than that of the A excitons in CdS ($m_\parallel/m_\perp = 3.2$) or CdSe (3.25). It should be noted that all of the observed Brillouin peaks were interpreted by assuming only single-phonon scattering.

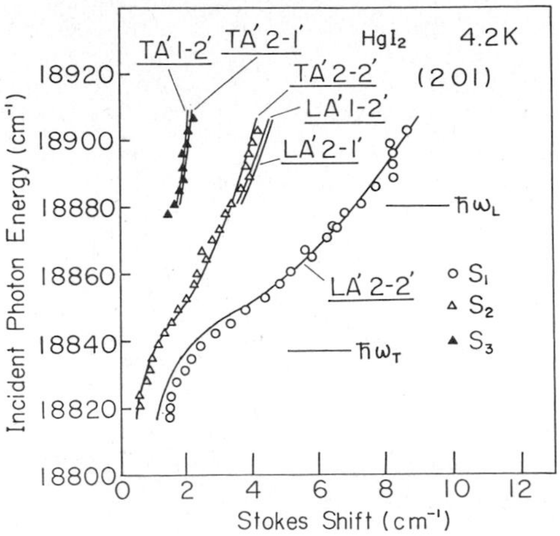

Fig. 31. Experimental (circles and triangles) and calculated (solid lines) Brillouin–Stokes shifts for $\boldsymbol{k} \parallel [201]$ in the vicinity of the A exciton-polariton in red HgI_2 as a function of incident photon energy. (From Goto and Nishina, 1979.)

3.5.3. Resonant Raman scattering

It was shown previously that if strong Fröhlich coupling between phonons and exciton polaritons is present, narrow two-phonon Brillouin peaks are possible when $\omega_i > \omega_L$. Such scattering was observed in the vicinity of the A exciton in CdS. When the "forbidden" Raman spectrum was investigated in the same region, analogous dispersing Raman peaks were found (Koteles and Winterling 1978, 1979b). The Raman shifts of these lines were observed to increase ("Stokes") or decrease ("anti-Stokes") as ω_i was scanned between the A and the B exciton (see fig. 32). These peaks were attributed to a two-step cascade process in which exciton-polariton scattering by an acoustic phonon was immediately followed by LO phonon scattering. This is illustrated in fig. 33.

Such a mechanism was invoked earlier (Permogorov and Travnikov 1976) to account for a broadening of the LO Raman peak in the vicinity of the exciton resonance. Unfortunately experimental resolution did not permit a detailed investigation of the LO sidebands. As in the case of two-phonon Brillouin scattering, the intermediate state (branch 2″) is in some direction in k space in which Fröhlich coupling is enhanced. This experiment is of some interest as several features of the results require the dual propagating modes (spatial dispersion) of the A exciton-polariton model. These are:

1) Dispersing two-phonon modes are first observed when $\omega_i \simeq \omega_{UP} (= \omega_L)$, not when $\omega_i \simeq \omega_T$ as would be the case if the exciton model

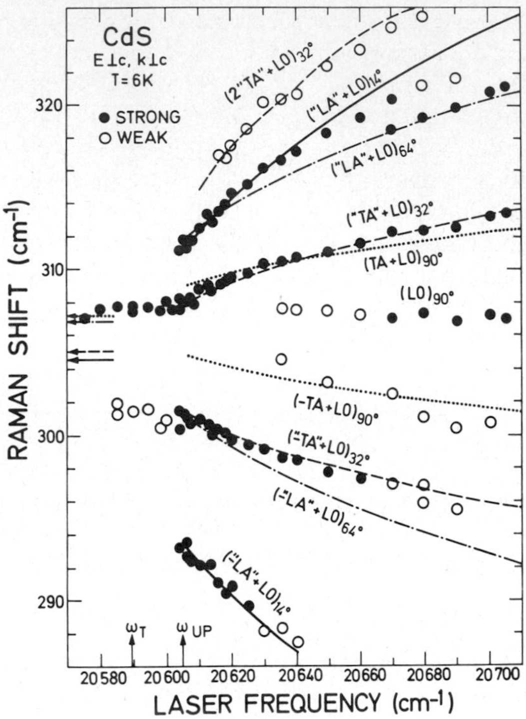

Fig. 32. Experimental (circles) and calculated (lines) two-phonon Raman shifts as a function of incident frequency in the vicinity of the A exciton-polariton in CdS. The numbers beside the brackets refer to the angles, ϕ, between the phonon wavevector and the c-axis. The arrows at the left border indicate LO frequencies for $q \to 0$ and $\phi = 14°, 32°, 64°$, and $90°$, in sequence of increasing frequency. (From Koteles and Winterling, 1979b.)

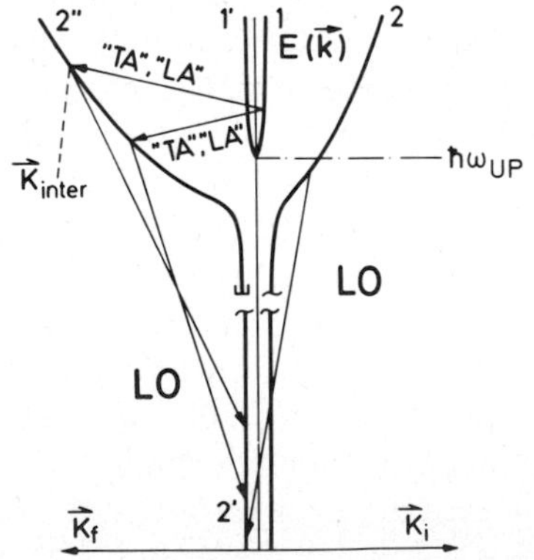

Fig. 33. Schematic exciton-polariton dispersion curves in the vicinity of the A exciton-polariton in CdS. Possible one- (on right) and two- (on left) phonon Raman scattering processes are illustrated by arrows. $\boldsymbol{k}_{\mathrm{i}}$, $\boldsymbol{k}_{\mathrm{inter}}$ and $\boldsymbol{k}_{\mathrm{f}}$ denote the wavevectors of incident, intermediate and final state exciton polaritons respectively. (From Koteles and Winterling, 1979b.)

were applicable (e.g., such a model was used to interpret two-phonon Raman scattering involving dipole-forbidden exciton states in Cu_2O (Yu and Shen 1974)). As with two-phonon Brillouin scattering, two-phonon Raman scattering is intense when both the initial and final states are photon-like. In the vicinity of the exciton resonance, it is the upper polariton which is primarily photon-like and, once this branch can be populated ($\omega_i > \omega_{UP}$), two-phonon processes become dominant.

2) Immediately above ω_{UP}, single LO scattering loses intensity and cannot be distinguished above the background. The two-phonon "side-bands" dominate the spectrum (fig. 34). The sudden, dramatic weakening of

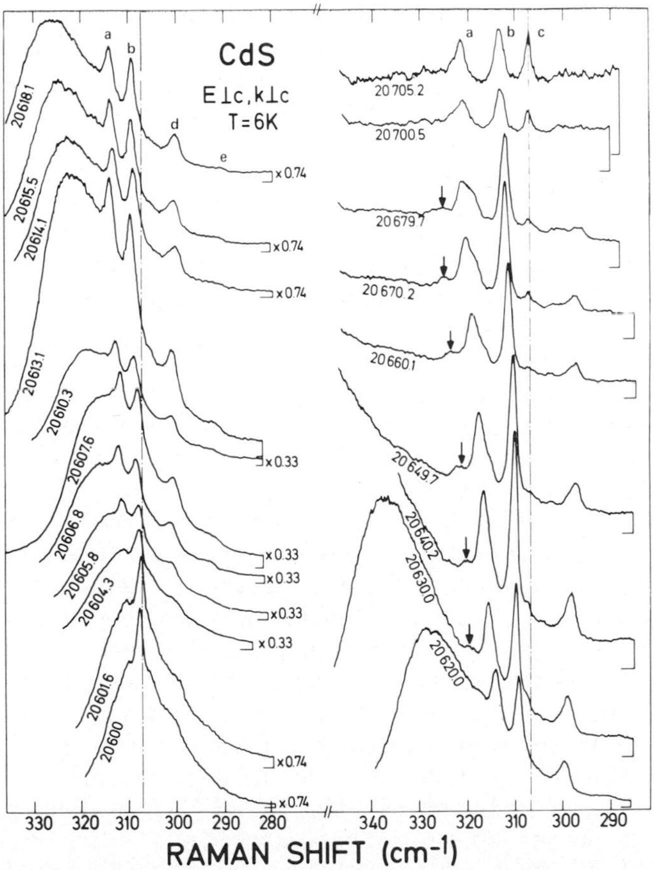

Fig. 34. Secondary emission spectra of CdS excited by several different incident frequencies (given in wavenumbers). Referring to fig. 32, the labeled peaks are: (a) ("LA" + LO)$_{14^\circ}$ or ("LA" + LO)$_{64^\circ}$; (b) ("TA" + LO)$_{32^\circ}$; (c) (LO)$_{90^\circ}$; (d) (−"TA" + LO)$_{32^\circ}$ and (e) (−"LA" + LO)$_{14^\circ}$. The dash-dotted line marks the location of the LO peak when $\omega_i \ll \omega_T$. The arrows identify a peak due to a three phonon process. (From Koteles and Winterling, 1979b.)

this scattering peak is inexplicable in an exciton model. It is consistent with the two-branch exciton-polariton model. At ω_{UP}, a new LO scattering process ($1 \rightarrow 2'$) becomes possible. However, the wavevector of the LO phonon scattering via this channel is five times smaller than the q of the previous channel ($2 \rightarrow 2'$). Thus, since branch 1 is preferentially populated when $\omega_i > \omega_{UP}$, and since the magnitude of the Fröhlich coupling matrix varies as q^2, the intensity of LO scattering should decrease by a factor of 25 as ω_i is tuned through ω_{UP}.

The excellent overall agreement in fig. 32 between the calculated curves and the experimental data points supports the assumptions of the model; a parabolic exciton band, a linear dispersion for acoustic phonons and q-independent LO frequencies. Again, as with two-phonon Brillouin scattering, the exciton-polariton dispersion curve is probed far out into the Brillouin zone (as far as 20% of the distance to the zone edge).

The results of Raman-scattering experiments in the vicinity of the B exciton in CdS are consistent with and just as puzzling as the results of two-phonon Brillouin scattering. While the two-phonon Raman peaks associated with the A exciton could be followed past the B exciton, no two-phonon Raman peaks involving the B exciton were ever observed.

In CdS, when $\omega_i > \omega_L$ for the A exciton, the first step of the two-phonon Raman scattering process is scattering by an acoustical phonon. LO scattering followed by acoustic phonon scattering is not likely since the second step is far removed from resonance and the phonon wavevectors are small. However, if ω_i were increased to $\omega_L + \omega_{LO}$, the second scattering step would occur near resonance. Then two-phonon Raman scattering in which an LO phonon is scattered from branch 1 to 2″ and is followed by acoustic phonon scattering from 2″ to 1′ should be possible. This has recently been observed in ZnTe (Oka and Cardona 1979). As shown in fig. 35, two-phonon Raman peaks are first observed when $\hbar\omega_i \simeq E_L$ (i.e., $\hbar\omega_L$) as with CdS. They are attributed to acoustic phonon scattering in reciprocal space directions in which Fröhlich coupling is strong followed by LO phonon scattering (e.g., TA + LO). When ω_i is increased to $\omega_T + \omega_{LO}$, new dispersing peaks appear and the original peaks disappear. The new features are due to LO phonon scattering followed by acoustic phonon scattering (e.g., LO + TA). Unlike CdS, at no time are only two-phonon peaks present in the scattering spectrum of ZnTe. The single LO scattering peak always dominates. This reflects the fact that the piezoelectric coupling constant in ZnTe is about eight times smaller than that of CdS. In addition, since the LO intensity doesn't weaken dramatically as ω_i is scanned through ω_L, single LO scattering is not entirely due to Fröhlich scattering. Other, non-wavevector concerning scattering processes may also be important (e.g., impurity scattering).

Even earlier, the frequency dependence of the intensity of single LO

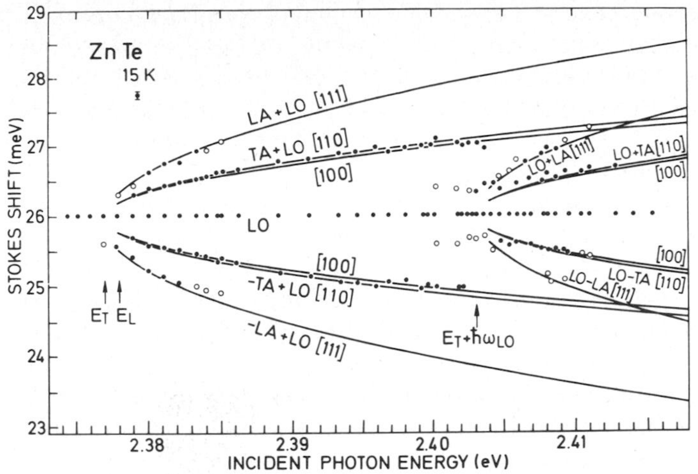

Fig. 35. Experimental (circles) and calculated (solid lines) two-phonon Raman shifts as a function of incident photon energy in the vicinity of the exciton polariton in ZnTe. The square brackets give the direction of propagation of the acoustic phonons. (From Oka and Cardona, 1979.)

phonon scattering in the vicinity of the A exciton in CdS was interpreted in terms of an exciton-polariton model (Gross et al. 1971, Permogorov and Travnikov 1971). The cross section for "forbidden" scattering varies as the square of the phonon wavevector which, in turn, increases rapidly in the vicinity of an exciton resonance due to the exciton-polariton dispersion. The net result is an enhancement in 1LO scattering intensity. By measuring this increase as a function of exciting frequency, it is possible, in theory, to determine the parameters of the exciton-polariton dispersion. In practice, the accuracy of this technique depends on narrow, intense exciting sources and extremely pure samples so that non-wavevector conserving scattering processes can be safely neglected. The discrepancies between parameter values derived from RBS and this technique illustrate the problems inherent in experiments which are based on intensity measurements.

4. *Future work*

It is clear from this brief review that tunable laser sources have added new scope to investigations of exciton-polariton dispersion. In the short time since the techniques described above have come into use, less than a dozen, mainly well-characterized, exciton-polariton systems have been investigated. As new dyes or techniques (such as the use of frequency-doubling crystals, F-center lasers, injection lasers) extend the frequency range of tunable lasers, more exotic materials will undoubtedly be explored. For example, it has been suggested (Cardona 1963) that some II–VI compound semiconductors may possess large negative hole masses. If this is the case, the resulting negative exciton masses will produce a new class of exciton-polariton dispersion curves. The curvature in the region of the exciton resonance will be downward, not upward, and a gap will open in which propagating modes cannot be supported (Hopfield 1966). It will be an extremely interesting system to investigate. Other potential exciton-polariton systems worthy of study are those in semiconductor superlattices, organic semiconductors, and many of the materials discussed in this book.

Laser techniques also provide new, promising approaches to the long-standing ABC problem. In particular, RBS and time-dependent spectroscopy have special attractions. Information concerning ABCs is present in the intensities of the measured peaks, not in their position. Thus ABC information can be readily separated from data about the values of the parameters of the exciton-polariton dispersion. Since this separation is not possible in broadband spectroscopic techniques, data analysis is more complicated and uncertain. Unfortunately, the frequency dependence of the intensities of the Brillouin peaks in RBS is more involved than originally anticipated. For example, Fröhlich coupling between acoustic phonons and excitons, the exciton-free layer at the crystal surface and exciton-impurity scattering were omitted in the original theory (Brenig et al. 1972). Recent experimental and theoretical work demonstrate the significance of these factors (Yu 1979, Yu and Evangelistic 1979) although the question of the relative importance of, for instance, impurity versus acoustic phonon scattering at resonance, has yet to be resolved (Ulbrich and Weisbuch 1978). Similar problems are present in the interpretation of the intensities measured in time-dependent spectroscopy experiments. The relative populations of exciton polaritons traveling simultaneously in the upper and lower branches depend, not only on the ABC at the surfaces, but also on the history of the interactions of the pulses as they travel through the crystal. In particular, exciton polaritons on the upper branch, being photon-like, will interact less strongly with acoustic phonons than exciton polaritons in the exciton-like lower branch, thus modifying the intensity ratio of these pulses (sect. 3.4). These problems can be resolved through further experimental and theoretical work and should lead to definitive statements about ABCs.

Additional questions which require answers are: how important to the Brillouin intensity are the natures of the branches connected by acoustic phonon scattering? why is the role of piezoelectric Fröhlich coupling dominant to the extent of producing two-phonon Brillouin scattering only near the A exciton in CdS? can the subtle influence of dampening be detected in the kinematics of exciton-polariton dispersion? and others.

Finally these laser techniques provide a convenient and precise method of monitoring changes in exciton-polariton structure produced by external parameters such as magnetic fields, electric fields, strain, etc. These possibilities have yet to be exploited.

Acknowledgments

It is a pleasure to acknowledge that the work on CdS described in this article was the fruit of an enjoyable collaboration with Dr. Gerhard Winterling at the Max-Planck Institute in Stuttgart, West Germany. I would also like to thank Professors Joseph Birman and Herman Cummins of the City College of New York for stimulating discussions.

Notes added in proof:

Since the completion of this chapter, there have been several developments in this rapidly growing field which require brief discussion in order to bring this work up to date.

Results from hyper-Raman scattering in ZnSe have recently been used to challenge the three-branch exciton-polariton model employed to interpret the resonant Brillouin scattering results given in sect. 3.5.2.2 (Nozue et al. 1981). The technique has also been utilized to study exciton-polariton dispersion curves of CdS and ZnO in strong magnetic fields (Kurtze et al. 1980). Finally, quantitative evidence of large renormalization effects on the exciton-polariton dispersion curve of CdS when a strong exciting laser beam is tuned to half the biexciton energy has been presented (Kempf et al. 1981). This is in agreement with the proposal put forth in sect. 3.3 in order to resolve the discrepancy between hyper-Raman and resonant Brillouin scattering results.

The technique of time-resolved spectroscopy (sect. 3.4) has recently been extended to investigate exciton-polariton dispersion in CdSe (Itoh et al. 1981) and CdS (Segawa et al. 1980). The derived values of the parameters are consistent with those obtained from resonant Brillouin scattering.

There have been several important developments in the utilization of resonant Brillouin scattering to investigate exciton polaritons in semiconductors. The first investigation of CuBr using RBS has yielded results inconsistent with those derived from hyper-Raman scattering (Vu et al. 1981). The authors believe that the ability of RBS to probe as much as an order of magnitude deeper into the Brillouin zone than HRS gives RBS an advantage in investigating fine details of the dispersion relations of exciton polaritons. On the other hand, recent theoretical considerations suggest that misinterpretation of the many observed Brillouin peaks may be the cause of the discrepancy (Cho and Yamane 1981). The postulated selection rule, which in certain situations would forbid backscattering between exciton-polariton states with almost pure spins if spin-flip is involved, could also explain the absence of $2 \rightarrow 2'$ scattering in the B exciton in CdS (sect. 3.5.2.3). Two groups have recently coupled Fabry–Perot interferometers with grating spectrometers in order to improve the resolution of RBS spectra. As a consequence, LA ($1 \rightarrow 1'$) scattering has been observed for the first time in CdS (Wicksted et al. 1981). The half-width of the TA ($2 \rightarrow 2'$) Brillouin peak was observed to increase rapidly and monotonically in the vicinity of the resonance to a value of $0.12\ cm^{-1}$ before the peak disappeared. It did not peak between ω_T and ω_L as expected. This behavior was confirmed by another group (Flynn and Geschwinn 1981) who reported much smaller maximum linewidths ($1.3 \times 10^{-2}\ cm^{-1}$ for LA and $8.3 \times 10^{-3}\ cm^{-1}$ for TA scattering). Finally, the first example of utilizing a magnetic field in conjunction with RBS has recently been reported (Broser et al. 1980). The magnetic field mixes allowed and forbidden states changing the two branch A exciton-polariton in CdS into a three branch system at high fields.

References

Akopyan, I., B. Novikov, S. Permogorov, and V. Travnikov, 1975, Phys. Status Solidi (b) **70**, 353.

Bechstedt, F. and F. Henneberger, 1977, Phys. Status Solidi (b) **81**, 211.

Bivas, A., Vu Duy Phach, B. Hönerlage, U. Rössler and J.B. Grun, 1979, Phys. Rev. **B20**, 3442.

Brenig, W., R. Zeyher and J.L. Birman, 1972, Phys. Rev. **B6**, 4617.

Broser, I., M. Rosenzweig, R. Broser, E. Beckmann and E. Birkicht, 1980, J. Phys. Soc. Japan **49** Suppl. A., 401.

Bruce, R.H. and H.Z. Cummins, 1977, Phys. Rev. **B16**, 4462.

Cardona, M., 1963, J. Phys. Chem. Solids, **24**, 1543.

Cho, K. and M. Yamane, 1981, Solid State Commun. **40**, 121.

Cummins, H.Z. and P.E. Schoen, 1972, Laser Handbook, Vol. 2, eds. Arecchi and Schultz, Dubois (North-Holland, Amsterdam) 1029.

Evangelisti, F., A. Frova and F. Patella, 1974, Phys. Rev. **B10**, 4253.

Fishman, G., 1978, Solid State Commun. **27**, 1097.

Flynn, E.J. and S. Geschwind, 1981, Bull. Amer. Phys. Soc. **26**, 488.

Fröhlich, D., E. Mohler and P. Wiesner, 1971, Phys. Rev. Lett. **26**, 554.

Gerlich, D., 1967, J. Phys. Chem. Solids, **28**, 2575.

Goto, T. and Y. Nishina, 1979, Solid State Commun. **31**, 751.

Gross, E.F., S.A. Permogorov, V.V. Travnikov and A.V. Selkin, 1971, Sov. Phys. Solid State, **13**, 578.

Haueisen, D.C. and H. Mahr, 1971, Phys. Lett. **36A**, 433.

Henneberger, F. and J. Voigt, 1976, Phys. Status Solidi (b) **76**, 313.

Henneberger, F., K. Henneberger and J. Voigt, 1977, Phys. Status Solidi (b) **83**, 439.

Hermann, C. and P.Y. Yu, 1978, Solid State Commun. **28**, 313.

Hermann, C. and P.Y. Yu, 1980, Phys. Rev. **B21**, 3675.

Hönerlage, B. and U. Rössler, 1980, Private communication.

Hönerlage, B., Vu Duy Phach, A. Bivas and E. Ostertag, 1977, Phys. Status Solidi (b) **83**, K101.
Hönerlage, B., A. Bivas, and Vu Duy Phach, 1978, Phys. Rev. Lett. **41**, 49.
Hopfield, J.J., 1958, Phys. Rev. **112**, 1555.
Hopfield, J.J., 1961, J. Appl. Phys. Suppl. **32**, 2277.
Hopfield, J.J., 1966, J. Phys. Soc. Japan, **21**, Suppl., 77.
Hopfield, J.J. and D.G. Thomas, 1961, Phys. Rev. **122**, 35.
Inoue, M. and E. Hanamura, 1976, J. Phys. Soc. Japan **41**, 1273.
Itoh, T. and T. Suzuki, 1978, J. Phys. Soc. Japan **45**, 1939.
Itoh, T., T. Suzuki and M. Ueta, 1977, J. Phys. Soc. Japan, **42**, 1069.
Itoh, T., T. Suzuki and M. Ueta, 1978, J. Phys. Soc. Japan, **44**, 345.
Itoh, T., P. Lavallard, S. Reydellet and C. Benoit à la Guillaume, 1981, Solid State Commun. **37**, 925.
Jackel, J. and H. Mahr, 1978, Phys. Rev. **B17**, 3387.
Kane, E.O., 1975, Phys. Rev. **B11**, 3850.
Kempf, K., G. Schmieder, G. Kurtze and C. Klingshirn, 1981, Phys. Status Solidi (b) **107**, 297.
Koteles, E.S. and G. Winterling, 1978, in: Physics of Semiconductors 1978, ed. B.L.H. Wilson (Institute of Physics, London) 481.
Koteles, E.S. and G. Winterling, 1979a, J. Lumin. **18/19**, 267.
Koteles, E.S. and G. Winterling, 1979b, Phys. Rev. **B20**, 628.
Koteles, E.S. and G. Winterling, 1980, Phys. Rev. Lett. **44**, 948.
Kurtze, G., W. Maier, K. Kempf, G. Schmieder, H. Schrey, C. Klingshirn, B. Hönerlage and U. Rössler, 1980, J. Phys. Soc. Japan **49**, Suppl. A., 559.
Levine, B.F., R.C. Miller and W.A. Nordland, 1975, Phys. Rev. **B12**, 4512.
Mahan, G.D. and J.J. Hopfield, 1964, Phys. Rev. **135**, A428.
Martin, R.M. and L.M. Falicov, 1975, in: Light Scattering in Solids, ed. M. Cardona (Springer, Berlin) 79.
Masumoto, Y., Y. Unuma, Y. Tanaka and S. Shionoya, 1979, J. Phys. Soc. Japan, **47**, 1844.
May, V., K. Henneberger, R. Enderlein and F. Henneberger, 1979, J. Lumin. **18/19**, 563.
Mita, T., K. Sotome and M. Ueta, 1980, J. Phys. Soc. Japan, **48**, 496, Solid State Commun. **33**, 1135.
Nagasawa, N., T. Mita and M. Ueta, 1976, J. Phys. Soc. Japan, **41**, 929.
Nozue, Y., M. Itoh and K. Cho, 1981, J. Phys. Soc. Japan **50**, 889.
Oka, Y. and M. Cardona, 1979, Solid State Commun. **30**, 447.
Pekar, S.I., 1958, Sov. Phys. JETP **6**, 785.
Permogorov, S.A. and V.V. Travnikov, 1971, Sov. Phys. Solid State **13**, 586.
Permogorov, S. and V. Travnikov, 1976, Phys. Status Solidi (b) **78**, 389.
Pine, A.S., 1972, Phys. Rev. **B5**, 3003.
Romestain, R., S. Geschwind and G. Devlin, 1977, Phys. Rev. Lett. **39**, 1583.
Sakuma, F., H. Fukutani and G. Kuwabara, 1978, J. Phys. Soc. Japan, **45**, 1349.
Schrey, H., V.G. Lyssenko, C. Klingshirn and B. Hönerlage, 1979, Phys. Rev. **B20**, 5267.
Segawa, Y., Y. Aoyagi, K. Azuma and S. Namba, 1978, Solid State Commun. **28**, 853.
Segawa, Y., Y. Aoyagi and S. Namba, 1979, Solid State Commun. **32**, 229.
Segawa, Y., Y. Aoyagi, T. Baba and S. Namba, 1980, J. Phys. Soc. Japan **49**, Suppl. A, 389.
Sermage, B. and G. Fishman, 1979, Phys. Rev. Lett. **43**, 1043.
Sermage, B. and G. Fishman, 1981, Phys. Rev. **B23**, 5107.
Suzuki, T., T. Itoh and M. Ueta, 1978, P. Phys. Soc. Japan, **44**, 347.
Ulbrich, R.G. and G.W. Fehrenbach, 1979, Phys. Rev. Lett. **43**, 963.
Ulbrich, R.G. and C. Weisbuch, 1977, Phys. Rev. Lett. **38**, 865.
Ulbrich, R.G. and C. Weisbuch, 1978, Festkörperprobleme XVIII – Advances in Solid State Physics, ed. J. Treusch (Vieweg, Braunschweig) 217.

Voigt, J., M. Senoner and I. Rückmann, 1976, Phys. Status Solidi (b) **75**, 213.
Vu, Duy Phach, A. Bivas, B. Hönerlage and J.B. Grün, 1978, Phys. Status Solidi (b) **86**, 159.
Vu, Duy-Phach, Y. Oka and M. Cardona, 1981, Phys. Rev. **B24**, 765.
Wicksted, J., M. Matsushita and H.Z. Cummins, 1981, Solid State Commun. **38**, 777.
Winterling, G. and E.S. Koteles, 1977a, Solid State Commun. **23**, 95.
Winterling, G. and E.S. Koteles, 1977b, Lattice Dynamics, Paris 1977, ed. M. Balkanski (Flammarion Sciences, Paris) 170.
Winterling, G. and E.S. Koteles, 1980, unpublished results.
Winterling, G. and E.S. Koteles, 1982, to be published.
Winterling, G., E.S. Koteles and M. Cardona, 1977, Phys. Rev. Lett. **39**, 1286.
Yu, P.Y., 1979, Solid State Commun. **32**, 29.
Yu, P.Y. and F. Evangelisti, 1978, Solid State Commun. **27**, 87.
Yu, P.Y. and F. Evangelisti, 1979, Phys. Rev. Lett. **42**, 1642.
Yu, P.Y. and Y.R. Shen, 1974, Phys. Rev. Lett. **32**, 939.

CHAPTER 4

Spatial Dispersion Effects in the Exciton Resonance Region

E.L. IVCHENKO

A.F. Ioffe Physico-Technical Institute
Leningrad
USSR

Excitons
Edited by
E.I. Rashba and M.D. Sturge

Contents

1. Introduction

Additional propagating light waves in solids near exciton absorption peaks were first predicted by Pekar in 1957. For many years the spatial dispersion effects and optical phenomena (reflectivity, absorption, light scattering, photoluminescence) in the vicinity of excitonic resonances have been attracting attention. At present the existence of additional waves due to the finite exciton effective mass has been established in various independent experiments. Nevertheless, the interest in optical studies of spatial dispersion effects is still considerable.

In the present paper we discuss effects of the finite exciton mass on light reflectance and transmission in semiconducting crystals. In semiconductors the effects have been experimentally investigated rather well. A particularly detailed comparison of experimental and theoretical results has been carried out for II–VI compounds where spatial dispersion is pronounced. Consequently, it is not surprising that the majority of spatial dispersion effects are illustrated here using the results obtained on CdS and CdSe crystals.

2. Polariton dispersion relation

In the crystal optics spatial dispersion refers to the dependence of the dielectric tensor $\boldsymbol{\epsilon}(\omega, \boldsymbol{k})$ on the wave vector $\boldsymbol{k}$. Well away from resonance absorption lines, the tensor $\boldsymbol{\epsilon}(\omega, \boldsymbol{k})$ can be expanded in powers of $\boldsymbol{k}$ and the spatial dispersion effects are determined by the small parameter $ka \sim a/\lambda$, where a is the interatomic distance and λ is the light wavelength inside the crystal (Agranovich and Ginzburg 1966). Hence, under nonresonant conditions it is in practice sufficient to retain only the first few terms in the dielectric tensor expansion. On the other hand, in the exciton resonance region the components $\epsilon_{ij}(\omega, \boldsymbol{k})$ are in general complicated functions of ω and $\boldsymbol{k}$ and their resonant behaviour prevents the dielectric tensor from being usefully expanded in powers of $\boldsymbol{k}$.

For an isolated excitonic resonance in a cubic crystal, the displacement field $\boldsymbol{D}$ and the electric field $\boldsymbol{E}$ are connected by the relation

$$\boldsymbol{D} = \epsilon_0 \boldsymbol{E} + 4\pi \boldsymbol{P}, \qquad (1)$$

where $\boldsymbol{P}$ is the contribution of the exciton under consideration to the dielectric polarization, ϵ_0 is the background dielectric constant, i.e., the contribution to the dielectric constant from other excitons and other crystal excitations. The frequency dependence of ϵ_0 may usually be ignored as long as we are interested only in frequencies very close to the isolated exciton frequency. For the simple model of a single electric-dipole-active exciton resonance with the isotropic effective mass, the differential equation of motion of $\boldsymbol{P}$ is (Hopfield and Thomas 1963)

$$\left(\frac{\partial^2}{\partial t^2} + \omega_0^2 - \frac{\hbar\omega_0}{M}\nabla^2 + \Gamma\frac{\partial}{\partial t}\right)\boldsymbol{P}(\boldsymbol{r}, t) = \omega_0^2\alpha_0\boldsymbol{E}(\boldsymbol{r}, t), \tag{2}$$

where ω_0 is the resonance frequency, M is the exciton effective mass, Γ is the exciton damping, $\alpha_0 = 2d^2/\hbar\omega_0$, d is the dipole matrix element for optical excitation of the exciton. In this case the dielectric tensor is

$$\epsilon_{ij}(\omega, \boldsymbol{k}) = \epsilon(\omega, k)\delta_{ij},$$

where

$$\epsilon(\omega, k) = \epsilon_0 + \frac{4\pi\alpha_0\omega_0^2}{\omega_0^2 - \omega^2 + (\hbar k^2/M)\,\omega_0 - \mathrm{i}\omega\Gamma}. \tag{3}$$

Throughout this chapter, we consider the resonance frequency region, so that

$$|\omega - \omega_0| \ll \omega_0. \tag{4}$$

Then eq. (3) may be reduced to the form

$$\epsilon(\omega, k) = \epsilon_0 + \frac{2\pi\alpha_0\omega_0}{\omega_0(k) - \omega - \mathrm{i}\Gamma/2}, \tag{5}$$

where $\hbar\omega_0(k) = \hbar\omega_0 + \hbar^2k^2/2M$ is the exciton band energy.

The excitonic contribution to the polarization $\boldsymbol{P}(\boldsymbol{r}, t)$ can be represented as

$$P_\alpha(\boldsymbol{r}, t) = 2d \;\mathrm{Re}\; C_\alpha(\boldsymbol{r}, t), \tag{6}$$

where $C_\alpha(\boldsymbol{r}, t)$ determines the wavefunction of the exciton

$$\Psi^{(\mathrm{ex})}(\boldsymbol{r}, t) = \sum_\alpha C_\alpha(\boldsymbol{r}, t)\varphi_\alpha. \tag{7}$$

Here $\boldsymbol{r}$ is the center of mass motion of a Mott–Wannier exciton, $\varphi_\alpha(\alpha = x, y, z)$ are the exciton wavefunctions at $\boldsymbol{k} = 0$, transforming like x, y, z (the representation Γ_5 of the point group $\mathrm{T_d}$ or the representation Γ_4^- of the group $\mathrm{O_h}$). Equations (3) and (5) are valid in the frequency region $\hbar|\omega - \omega_0| \ll \Delta\mathscr{E}$, $\Delta\mathscr{E}$ being the energy difference between the given and any other excitonic levels.

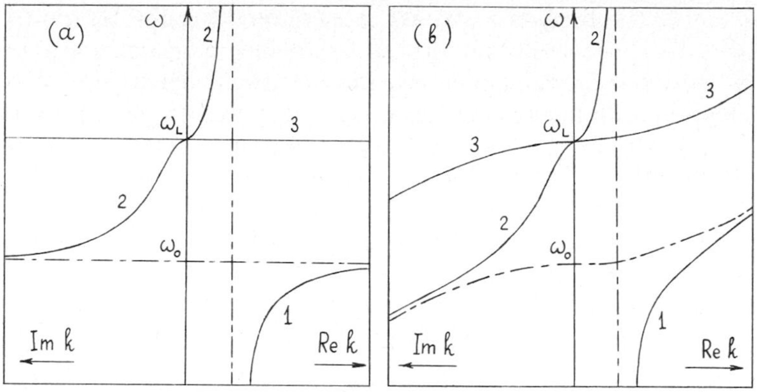

Fig. 1. The frequency–wavevector dispersion relation for the normal modes in the absence (a) and in the presence (b) of spatial dispersion. No damping is included. The normal mode wavevectors are in either case either purely real or purely imaginary and are plotted to the right or left accordingly. The dashed-and-dotted lines show the dispersion relation for bare photons and excitons ($\omega_{LT} = 0$): (1) and (2)-transverse modes, (3)-longitudinal mode.

Equation (2) plus the Maxwell equations constitute the complete set of equations one needs to find normal modes of the crystal in the resonance frequency region. For the isotropic case, there exist two types of modes: transverse (with $\boldsymbol{P}, \boldsymbol{E} \perp \boldsymbol{k}$) and longitudinal (with $\boldsymbol{P}, \boldsymbol{E} \parallel \boldsymbol{k}$). The modes satisfy the dispersion equations

$$\begin{aligned} \epsilon(\omega, k) &= (ck/\omega)^2 \quad \text{(transverse mode)}, \\ \epsilon(\omega, k) &= 0 \quad \text{(longitudinal mode)}. \end{aligned} \tag{8}$$

Figure 1 shows the dispersion curves of the normal modes in the vicinity of the exciton resonance. The exciton–photon interaction renormalizes the energy spectra of photons and mechanical excitons leading to coupled exciton–photon modes – exciton polaritons (Born and Huang 1954, Tolpygo 1950). In fig. 1 the branches 1 and 2 are doubly degenerate and correspond to transverse exciton polaritons, the branch 3 is nondegenerate and corresponds to longitudinal exciton states. Figure 1(a) presents the polariton dispersion relation in the absence of spatial dispersion, i.e., for the case where the dielectric constant depends on the frequency, but not on the wavevector. The curves calculated in the presence of spatial dispersion (M is finite) are plotted in fig. 1(b). In a spatially dispersive medium additional electromagnetic waves can propagate: according to fig. 1(b) for a given ω and a given propagation direction there exist five waves. Neglecting the exciton damping, the wavevectors of the waves with frequencies above ω_L

are real (ω_L is the frequency of the longitudinal exciton at $k = 0$). For frequencies $\omega < \omega_L$ the wavevectors corresponding to branch 1 are real and the wavevectors of the remaining three waves are imaginary.

The longitudinal–transverse splitting $\omega_{LT} = (\omega_L - \omega_0)$ is given by

$$\omega_{LT} = \frac{2\pi\alpha_0\omega_0}{\epsilon_0} = \frac{4\pi d^2}{\hbar\epsilon_0}. \tag{9}$$

At large values of k, so that $k \gg \sqrt{\epsilon_0}k_0$ ($k_0 = \omega/c$), the energy difference between the longitudinal exciton and transverse exciton polariton 1 equals ω_{LT}. At small values of k, so that $|k| \ll \sqrt{\epsilon_0}k_0$, the dispersion relation for exciton polaritons 2 can be written as follows:

$$\omega_2(k) \cong \omega_L + \hbar k^2/2\mu, \tag{10}$$

where

$$\frac{1}{\mu} = \frac{1}{M} + \frac{1}{\mu_0}, \qquad \mu_0 = \frac{\hbar\epsilon_0}{2\omega_{LT}}\left(\frac{\omega_0}{c}\right)^2. \tag{11}$$

Using eqs. (5) and (8), one can find the refractive indices $n_r = ck_r/\omega$ ($r = 1, 2, 3$) corresponding to exciton polaritons 1 and 2 and longitudinal excitons ($r = 3$)(Pekar 1957)

$$n_{1,2}^2 = \frac{1}{2}\epsilon_0\left\{1 + \frac{M}{\mu_0}\frac{\omega - \omega_0}{\omega_{LT}} \pm \left[\left(1 - \frac{M}{\mu_0}\frac{\omega - \omega_0}{\omega_{LT}}\right)^2 + 4\frac{M}{\mu_0}\right]^{1/2}\right\}, \tag{12}$$

$$n_3^2 = \frac{2M(\omega - \omega_L)}{\hbar}\left(\frac{c}{\omega_0}\right)^2.$$

To take account of the exciton damping one has to perform in eq. (12) the transformation $\omega \to \omega + \mathrm{i}\Gamma/2$. It should be noted that the refractive indices n_1, n_2, n_3 are connected by the relation (Permogorov et al. 1972):

$$n_3^2 = n_1^2 n_2^2/\epsilon_0. \tag{13}$$

3. *Additional boundary conditions and energy transfer by exciton polaritons*

As mentioned above, if spatial dispersion is present the incident light wave with frequency close to ω_0 will excite more than two waves in the crystal when it strikes the crystal surface. Therefore in order to calculate optical spectra in this case, unlike in classical optics, additional boundary conditions (ABC) are required beyond the usual Maxwell boundary conditions,

$$\boldsymbol{E}_{tg}(-0) = \boldsymbol{E}_{tg}(+0), \qquad \boldsymbol{H}_{tg}(-0) = \boldsymbol{H}_{tg}(+0), \tag{14}$$

where $\boldsymbol{E}_{tg}(\pm 0)$, $\boldsymbol{H}_{tg}(\pm 0)$ are the tangential components of the electric and magnetic field vectors, $\boldsymbol{E}$ and $\boldsymbol{H}$, on the right- and left-hand side of the plane $z = 0$. For the case of an electric-dipole-active exciton in a cubic crystal three additional scalar conditions (or one vector condition) should be provided.

The simplest form of ABC is

$$\boldsymbol{P}|_{z=0} = 0, \tag{15}$$

where the z-axis is directed along the normal to the crystal surface. The ABC in the form (15) have been proposed in the first works on the theory of additional light waves near exciton absorption lines (Pekar 1957, 1958). They correspond to a node of the wavefunction of the mechanical exciton on the crystal boundary: $C_\alpha(\boldsymbol{r}, t)|_{z=0} = 0$. Taking into account that the amplitudes of the polarization $\boldsymbol{P}_\mathrm{r}$ and the electric field $\boldsymbol{E}_\mathrm{r}$ (r = 1, 2, 3) satisfy the condition

$$\boldsymbol{P} = \frac{n^2 - \epsilon_0}{4\pi}\boldsymbol{E} \quad \text{(transverse waves, r = 1, 2)}, \tag{16}$$

$$\boldsymbol{P} = -\frac{\epsilon_0}{4\pi}\boldsymbol{E} \quad \text{(longitudinal waves, r = 3)}.$$

Equation (15) can be reduced to the form with the electric field amplitudes being present only,

$$(n_1^2 - \epsilon_0)\boldsymbol{E}_1 + (n_2^2 - \epsilon_0)\boldsymbol{E}_2 - \epsilon_0\boldsymbol{E}_3 = 0. \tag{17}$$

The form of ABC depends essentially on the behaviour of the potential $U(z)$ for the exciton near the crystal boundary. At present a theory taking into account the detailed mechanism by which Mott–Wannier excitons interact with the surface has not been completely developed. In order to simplify the problem Hopfield and Thomas (1963) introduced an exciton-free surface layer (the so-called "dead-layer" model). In this model the potential $U(z)$ is replaced by an infinite potential barrier a finite distance l inside the crystal, i.e.,

$$U(z) = \begin{cases} 0 & (l < z < +\infty), \\ +\infty & (0 < z < l). \end{cases} \tag{18}$$

The spatial region $0 < z < l$ is characterized by a frequency-independent dielectric constant which is assumed to be about the same as ϵ_0. The usual Maxwell boundary conditions are applied at the boundary "vacuum–deadlayer", and beyond the Maxwell boundary conditions the Pekar ABC (15) are applied at $z = l$. Physically, the "dead-layer" thickness should be of the order of the exciton Bohr radius a_B. It should be noted that attempts have been made to reveal the origin of the "dead-layer" with the model of an image force (Skaistys and Sugakov 1974, Sakoda 1976).

At present the ABC

$$\left(P+\frac{c}{\omega}\gamma\frac{\partial P}{\partial z}\right)_{z=0}=0 \tag{19}$$

are frequently used, where the dimensionless parameter γ is generally supposed to depend on the light frequency and the angle of incidence, φ. Zeyher et al. (1972) showed the intimate connection between ABC for exciton polaritons and boundary conditions in the solution of the Schrödinger equation for a mechanical exciton. For a bounded crystal well away from the boundary, $z \gg a_B$, the wavefunction of the mechanical exciton is a linear combination of the two plane waves

$$c_k(\boldsymbol{r},\omega) \sim \theta(z)[\exp(\mathrm{i}k_z z) + r_e \exp(-\mathrm{i}k_z z)]\exp(\mathrm{i}\boldsymbol{k}_\perp \cdot \boldsymbol{r}_\perp), \tag{20}$$

where r_e is the exciton reflection coefficient, $\theta(z)$ is the usual step function. The parameters γ and r_e are interrelated (Zeyher et al. 1972, Halevi and Fuchs 1978)

$$\gamma = -\mathrm{i}\frac{k_0}{k_{ez}}\frac{1+r_e^*}{1-r_e^*}, \tag{21}$$

where $k_0 = \omega/c$ is the vacuum wavevector of the wave, $k_{ez} = (k_e^2 - k_\perp^2)^{1/2}$, $k_e^2 = 2M(\omega-\omega_0)/\hbar$, $k_\perp^2 = k_0^2 \sin^2\varphi$. If loss processes are neglected, one must have $|r_e| = 1$ and r_e may be conveniently written in the form $r_e = -\exp(\mathrm{i}2\delta)$, where the phase δ is a real function of frequency. The relationship between γ and δ is

$$\gamma = (\omega/2ck_{ez})\operatorname{tg}\delta. \tag{22}$$

Attempts to analyze the problem for the case in which both the conduction electron and the valence hole experience an infinite potential barrier at $z=0$, so that the condition $\Psi^{(ex)}(\boldsymbol{r}_1, \boldsymbol{r}_2) = 0$ for $z_1 = 0$ or $z_2 = 0$ has to be satisfied, have been undertaken by Ting et al. (1975) and Konstantinov et al. (1975).

For the isotropic case, the vector functions $\boldsymbol{D}(\boldsymbol{r},\omega)$ and $\boldsymbol{E}(\boldsymbol{r},\omega)$ are connected by the integral relation with a scalar kernel $\epsilon(\boldsymbol{r},\boldsymbol{r}',\omega)$. In an unbounded medium $\epsilon(\boldsymbol{r},\boldsymbol{r}',\omega)$ depends on $\boldsymbol{r}$ and $\boldsymbol{r}'$ through their difference. In a bounded medium, for $\boldsymbol{r}$ and $\boldsymbol{r}'$ near the boundary, the function $\epsilon(\boldsymbol{r},\boldsymbol{r}',\omega)$ depends on spatial coordinates $\boldsymbol{r}$ and $\boldsymbol{r}'$ separately. For example, if r_e is real, then the relationship between $\boldsymbol{D}$ and $\boldsymbol{E}$ becomes (Zeyher et al. 1972, Halevi and Fuchs 1978):

$$\boldsymbol{D}(\boldsymbol{r},\omega) = \int_{z'>0} \mathrm{d}\boldsymbol{r}'\epsilon(z-z',\omega)\boldsymbol{E}(\boldsymbol{r}',\omega) + r_e \int_{z'>0} \mathrm{d}\boldsymbol{r}'\epsilon(z+z',\omega)\boldsymbol{E}(\boldsymbol{r}',\omega). \tag{23}$$

Here $\epsilon(z \pm z',\omega)$ is equal to $\epsilon(\boldsymbol{r}_1,\boldsymbol{r}_2,\omega)$ for the unbounded crystal at $\boldsymbol{r}_1$ and $\boldsymbol{r}_2$ which satisfy the condition $z_1 - z_2 = z \pm z'$, $\boldsymbol{r}_{1\perp} - \boldsymbol{r}_{2\perp} = \boldsymbol{r}_\perp - \boldsymbol{r}'_\perp$.

Various authors have discussed the assumption that for a semi-infinite crystal $\epsilon(\boldsymbol{r}, \boldsymbol{r}', \omega)$ depends on $(\boldsymbol{r} - \boldsymbol{r}')$ as well as in the case of an infinitely extended medium (the so-called dielectric model; Birman and Sein 1972, Maradudin and Mills 1973, Agarwal et al. 1974). This assumption is equivalent to the condition $r_e = 0$ in eq. (21) or eq. (23) and leads to the ABC which reduce at normal incidence to

$$\left[\boldsymbol{P} - (\mathrm{i}/k_e)\frac{\partial \boldsymbol{P}}{\partial z}\right]_{z=0} = 0. \tag{24}$$

It is clear that, if $|r_e| < 1$, loss occurs on the boundary. Note that in this case, according to eq. (21), the imaginary part of γ is non-zero, $\mathrm{Im}\,\gamma < 0$.

Let us now analyze the energy transfer through a crystal boundary (Agranovich and Ginzburg 1966, Permogorov and Selkin 1973, Bishop and Maradudin 1976, Selkin 1977). The energy flux density $\boldsymbol{S}$ of a radiation field in the vicinity of an isolated isotropic exciton is given by

$$\boldsymbol{S} = \frac{c}{4\pi}(\boldsymbol{E} \times \boldsymbol{H}) - \frac{2\pi\hbar}{\omega_{\mathrm{LT}}\epsilon_0 M}\sum_\alpha \dot{P}_\alpha \frac{\partial P_\alpha}{\partial \boldsymbol{r}}. \tag{25}$$

One has to keep in mind, however, that there is an arbitrariness in the definition of $\boldsymbol{S}$ and we can add to the right-hand side of eq. (25) the time derivative or curl of any vector function. In other words, the only uniquely defined quantity is S_z averaged over the time interval $T = 2\pi/\omega$ and over a plane perpendicular to the z-axis.

According to eq. (14) the first term in the right-hand side of eq. (25) is continuous at the boundary. In the case of ABC given in the general form (19) the second term is continuous only for real γ. Hence, the energy flux is always continuous at $z = 0$ for the Pekar ABC. In the case of $\mathrm{Im}\,\gamma < 0$ energy is dissipated on the boundary, whereas in the case $\mathrm{Im}\,\gamma > 0$ energy is generated at $z = 0$.

Above we considered only homogeneous ABC. Agranovich and Ginzburg (1966) proposed the more general case of non-homogeneous ABC which connect at $z = 0$ the polarization and its first spatial derivative with the electric field. Akhmediev and Yatsishen (1978) showed that non-homogeneity of the boundary conditions for polarization together with the requirement of the continuous energy flux result in non-homogeneity of the Maxwell boundary conditions. For the case of normal incidence they proposed a set of boundary conditions in the following form

$$\left(\boldsymbol{P} + \frac{c}{\omega}\gamma\frac{\partial \boldsymbol{P}}{\partial z}\right)_{z=0} = \omega_0^2\alpha_0\gamma\eta\boldsymbol{E}(+0),$$
$$\boldsymbol{E}(+0) = \boldsymbol{E}(-0), \qquad \boldsymbol{H}(+0) - \boldsymbol{H}(-0) = 4\pi\mathrm{i}\eta\beta\boldsymbol{P}(+0), \tag{26}$$

where $\beta = \hbar\omega_0 k_0^2/M$, constants γ and η are real. If $\eta = 0$, eqs. (26) reduce to the set of homogeneous boundary conditions given by eqs. (14) and (19).

Near the crystal boundary, z being of the order of a_B, the exciton wavefunction differs from eq. (20). So far we supposed $ka_B \ll 1$, in which case this difference may be neglected and optical spectra can be calculated using the proper choice of ABC. If however $ka_B \gtrsim 1$, the approximation of ABC is not applicable, one is required to resort to a microscopic theory and the calculation of optical spectra becomes annoyingly difficult. Kiselev (1979) computed the reflectivity in the exciton resonance region for various surface potentials $U(z)$ using the adiabatic approximation. In this case the real potential is replaced by a function of a staircase form. At the crystal boundary the Pekar ABC are applied, at the points of discontinuity of the approximated potential the polarization and its first spatial derivative are assumed to be continuous. The method has been used in order to calculate reflection spectra of semiconductors with Schottky barrier on the boundary (Kiselev 1979).

4. *Reflection and transmission spectra of bulk crystals*

Reflection and transmission spectra in the vicinity of an excitonic resonance show a set of anomalies and provide considerable information about excitonic states in solids. Spatial dispersion produces striking effects on these spectra. As stated above, due to the non-infinite exciton mass one of the modes in the interval between ω_L and ω_0 is a propagating mode characterized in the absence of dissipation ($\Gamma \to 0$) by a real value of the wavevector (branch 1 in fig. 1(b)). Consequently, the incident wave will excite this mode and the reflectivity in the interval $\omega_0 < \omega < \omega_L$ will differ from 100% even for vanishing value of the exciton damping. This is one of the principal differences between the structure in the reflectivity due to excitons and stopbands in the reflectivity spectra due to phonons. In the present section we discuss peculiarities of optical spectra of bulk crystals caused by spatial dispersion, i.e. by the finite effective mass.

4.1. Normal incidence

We first consider radiation normally incident on the material from the vacuum. In this case the transverse waves are excited only (branches 1 and 2 in fig. 1(b)). For the plane-polarized radiation the Maxwell boundary conditions and the Pekar ABC become

$$E_0 + E_R = E_1 + E_2, \qquad E_0 - E_R = n_1 E_1 + n_2 E_2,$$
$$(n_1^2 - \epsilon_0) E_1 + (n_2^2 - \epsilon_0) E_2 = 0. \tag{27}$$

Here E_0 and E_R are the electric field amplitudes of the incident and reflected waves respectively, E_1 and E_2 are the amplitudes of the two transmitted waves 1 and 2. The set of equations (27) can be readily solved and the reflection coefficient is given by

$$R = |r|^2, \qquad r \equiv \frac{E_R}{E_0} = \frac{1-\bar{n}}{1+\bar{n}}, \qquad \bar{n} = \frac{n_1 n_2 + \epsilon_0}{n_1 + n_2}. \tag{28}$$

This result has been obtained by Pekar in 1957. Far enough away from the resonance frequency ω_0, where $|n_2| \gg \epsilon_0$, n_1 (long-wavelength limit, $\omega < \omega_0$) or $|n_1| \gg \epsilon_0$, n_2 (short-wavelength limit, $\omega > \omega_0$), eq. (28) reduces to the well-known expression $R = |(1-n)/(1+n)|^2$, where n is the refractive index corresponding to the photon-like parts of polariton dispersion curves (branch 1 for $\omega < \omega_0$ and branch 2 for $\omega > \omega_L$).

Shortly after the theoretical prediction of additional electromagnetic waves in solids, several attempts to verify the theory experimentally were made (Brodin and Pekar 1960, 1960a, Gorban' and Timofeev 1961, Brodin and Strashnikova 1962). The first analysis of the experimental reflectivity measurements using a spatial dispersion approach was carried out by Hopfield and Thomas (1963). Figure 2 shows the experimentally observed reflection curves in the range of the exciton resonances A(n = 1), B(n = 1) and A(n = 2) in CdS crystals. The spectra were measured for different but

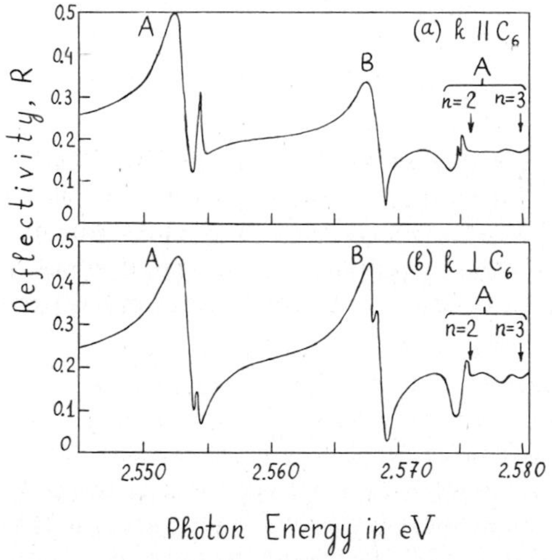

Fig. 2. The measured normal-incidence reflection spectra for excitons A and B in a CdS crystal at 1.6 K. (a) refers to $\boldsymbol{k} \parallel \boldsymbol{c}$, (b) refers to $\boldsymbol{k} \perp \boldsymbol{c}$. The position of the $n = 2$ and $n = 3$ states of A exciton has been marked. The large peaks labelled A and B are the $n = 1$ states of excitons A and B. Notice the differences for these two classically equivalent geometries (both have $\boldsymbol{E} \perp \boldsymbol{c}$). (After Hopfield and Thomas 1963.)

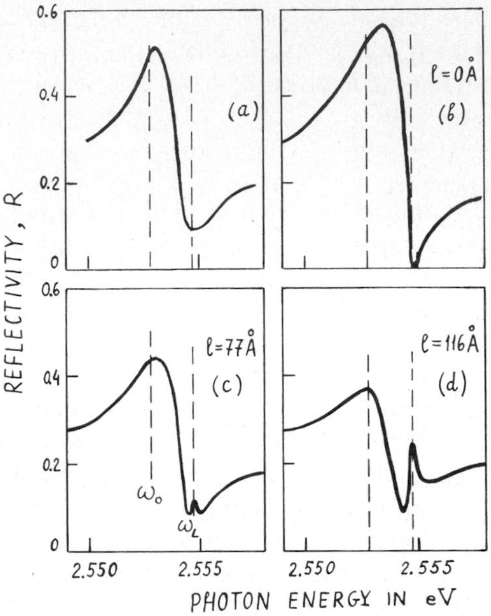

Fig. 3. The calculated normal-incidence reflection curves. (a) The reflectivity of a classical oscillator ($\hbar\Gamma = 10^{-3}$ eV, $\epsilon_0 = 8.1$, $4\pi\alpha_0 = 0.0094$). (b) The calculated curve taking spatial dispersion into account. The parameters are $\hbar\omega_0 = 2.5528$ eV, $M = 0.9\, m_0$, m_0 being the free electron mass, $4\pi\alpha_0 = 0.0125$, $\epsilon_0 = 8$, $\Gamma = 0$, $l = 0$. (c) Same as (b) but with a "dead-layer" of thickness $l = 77$ Å. (d) Same as (b) but for $l = 116$ Å. The longitudinal and transverse exciton energies at $k = 0$ are indicated by the vertical dashed lines.

classically equivalent geometries: (a) $\boldsymbol{E} \perp \boldsymbol{c}$, $\boldsymbol{k} \parallel \boldsymbol{c}$ and (b) $\boldsymbol{E} \perp \boldsymbol{c}$, $\boldsymbol{k} \perp \boldsymbol{c}$, where $\boldsymbol{c}$ is the unit vector along the hexagonal axis (c-axis). In spite of the anisotropy of the material, the theory developed for an isotropic exciton can be applied for the two geometries. One has to keep in mind, however, that the exciton mass in eqs. (3) and (5) is $M_\parallel$ for $\boldsymbol{k} \parallel \boldsymbol{c}$ and $M_\perp$ for $\boldsymbol{k} \perp \boldsymbol{c}$, where $M_\parallel$ and $M_\perp$ are the longitudinal and transverse masses of the exciton. In this section we confine ourselves to the discussion of optical spectra of the A($n = 1$) exciton. The reflectivity in the spectral region of the B($n = 1$) exciton resonance will be considered in sect. 6.

The structure in the reflectivity spectra shown in fig. 2 differs markedly from that expected from the model of a classical oscillator which considers only the frequency dependence of the dielectric constant in the exciton region. The calculated reflectivity of a classical oscillator is shown in fig. 3(a). Three quandaries present themselves. First, the damping which quantitatively fits the measured reflectivity is $\hbar\Gamma = 1$ meV, whereas it follows from transmission experiments that $\hbar\Gamma < 0.1$ meV. Second, there occurs near the reflectivity minimum a subsidiary sharp peak ("spike"). Third, the two reflectivity curves in fig. 2 should be the same in the classical theory, whereas the extra peak at the longitudinal frequency of the A($n = 1$) exciton in fig. 2(b) markedly differs from that in fig. 2(a). Figure 3(b) shows the calculated reflectivity taking spatial dispersion into account, using the Pekar ABC and neglecting the exciton damping. The parameters used are given in the caption to the figure. Figures 3(c) and 3(d) show the effect of the "dead-layer" on the calculated reflectivity: (c) refers to $l = 77$ Å and (d) refers to $l = 116$ Å.

Taking into account the exciton-free layer of a thickness l, the effective refractive index $\bar{n}$ in eq. (28) is given by

$$\bar{n} = n\,\frac{(\bar{n}_0 + n)\exp(-2\mathrm{i}k_0 nl) - n + \bar{n}_0}{(\bar{n}_0 + n)\exp(-2\mathrm{i}k_0 nl) + n - \bar{n}_0}, \tag{29}$$

where n is the refractive index of the "dead-layer", $\bar{n}_0 \equiv \bar{n}(l = 0)$. The important result is that the presence of the "dead-layer" can explain the anomalous structure of the reflection curves at the longitudinal frequency ω_{L}. Physically, the reason for the appearance of the "spike" is the interference of the wave reflected from the crystal surface and from the interface between the "dead-layer" and the volume. In order to demonstrate this result mathematically, let us consider the frequency dependence of R in the vicinity of ω_{L}. According to eqs. (10) and (12) in the frequency region $|\omega - \omega_{\mathrm{L}}| \ll \omega_{\mathrm{LT}}$ for $\Gamma \to 0$ the refractive indices n_1 and n_2 can be written approximately as follows

$$n_1 = [\epsilon_0(1 + M/\mu_0)]^{1/2}, \qquad n_2 = A(\omega - \omega_{\mathrm{L}})^{1/2}, \tag{30}$$

where $A = (2\mu/\hbar k_0^2)^{1/2}$. Using the fact that n_2 is real for $\omega > \omega_{\mathrm{L}}$ and imaginary for $\omega < \omega_{\mathrm{L}}$, one can find the reflectivity R in the following approximate form

$$R = \frac{[1 - a - A(\omega - \omega_{\mathrm{L}})^{1/2}]^2 + b^2}{[1 + a + A(\omega - \omega_{\mathrm{L}})^{1/2}]^2 + b^2} \quad (\omega > \omega_{\mathrm{L}}),$$

$$R = \frac{(1 - a)^2 + [b + A(\omega_{\mathrm{L}} - \omega)^{1/2}]^2}{(1 + a)^2 + [b + A(\omega_{\mathrm{L}} - \omega)^{1/2}]^2} \quad (\omega < \omega_{\mathrm{L}}), \tag{31}$$

where $a = \epsilon_0/n_1$ (in CdS $a < 1$), $b = -n \operatorname{tg}\Phi$, $\Phi = k_0 nl$. Equation (31) is valid if $|\omega - \omega_{\mathrm{L}}| \ll \omega_{\mathrm{LT}}$, $n_1^2 \gg 1$ and $|\operatorname{tg}\Phi| \leq 1$. Then for $\operatorname{tg}\Phi > 0$, when $b < 0$, R in the vicinity of ω_{L} is indeed smaller than R at $\omega = \omega_{\mathrm{L}}$. In the absence of a surface barrier ($l = 0$), according to eq. (31) the reflection coefficient exhibits only one minimum ($R_{\min} = 0$) at the frequency $\omega_{\min} = \omega_{\mathrm{L}} + [(1 - a)/A]^2$ which satisfies the condition

$$n_2^2(\omega_{\min}) = (1 - a)^2. \tag{32}$$

Observe that for CdS $(\omega_{\min} - \omega_{\mathrm{L}}) \ll \omega_{\mathrm{LT}}$ (fig. 3(b)).

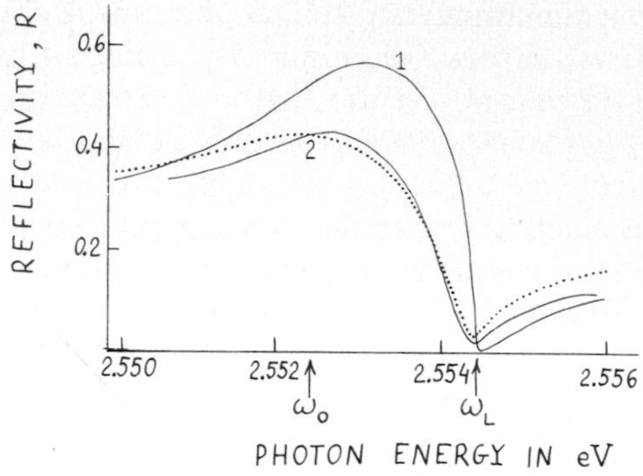

Fig. 4. The reflection spectra of CdS at normal incidence. The dotted curve is the experiment in the geometry $\boldsymbol{k} \perp \boldsymbol{c}$, $\boldsymbol{E} \perp \boldsymbol{c}$ at $T = 1.8\,\mathrm{K}$; the curve 1 is the theory using the Pekar ABC ($\hbar\omega_0 = 2.5524\,\mathrm{eV}$, $\hbar\omega_{\mathrm{LT}} = 2\,\mathrm{meV}$, $\epsilon_{0\perp} = 8.3$, $M = 0.9\,m_0$) (Permogorov et al. 1972). The curve 2 was calculated for the ABC (19) with $\gamma = 0.02 + \mathrm{i}0.03$ (Konstantinov and Saifullaev 1978).

The difference between fig. 2(a) and fig. 2(b) can be understood as well in terms of the effect of spatial dispersion and an exciton-free layer assuming the layer thickness l to depend upon the exciton mass.

The reflection spectra have been also calculated for the ABC (19) (Akhmediev and Yatsishen 1976, Konstantinov and Saifullaev 1978). The best fit with experiment for the frequency-independent parameter γ takes place at $\gamma = 0.02 + \mathrm{i}0.03$ (fig. 4). As stated above, for complex values of γ the energy flux is discontinuous. Moreover, for $\mathrm{Im}\,\gamma > 0$ energy is generated on the boundary. In order to satisfy the energy conservation law, Akhmediev and Yatsishen (1978) explored the non-homogeneous boundary conditions (26). In this case the best fitting parameters are $\gamma = 0.055$ and $\xi = 100$, where $\xi = 4\pi\omega_0^2\alpha_0\eta$ and η is defined in eq. (26).

The exciton reflectance line shape strongly depends on the surface conditions of a crystal. Reflectivity spectra depend markedly on extrinsic effects, such as surface impurities, surface fields, additional light, electron and ion bombardment, heating, etc. (see, for example, Permogorov et al. 1972, Davydova et al. 1973, 1974, Evangelisti et al. 1972, 1974, 1974a, Patella et al. 1976, Benemanskaya et al. 1975, 1977, Kiselev 1979, 1979a). For example, a subsidiary peak near ω_{L} was absent in the CdS crystals investigated by Permogorov et al. (1972), but the "spike" appeared if the samples were illuminated at low temperatures with radiation of frequencies above the fundamental absorption edge. When the samples were heated to room temperature and cooled again to liquid-helium temperature, the "spike" disappeared. Many changes in optical spectra can be understood in the "dead-layer" model taking into account the variation of l in the samples undergoing some surface treatment.

The exciton damping strongly influences the reflectivity in the vicinity of ω_L. Increase of Γ washes out the "spike". At large Γ the "spike" disappears. For very large values of Γ, so that

$$\Gamma > 4\omega_{LT}(\mu_0/M)^{1/2},$$

the light propagating in the crystal can in practice be described by one normal mode with a complex refractive index which is close to that calculated in the absence of spatial dispersion (Davydov and Myasnikov 1974).

4.2. *Oblique incidence*

If at oblique incidence the electric field vector is perpendicular to the plane of incidence (s polarization) then, as at normal incidence, only transverse waves are excited inside the crystal. For oblique incidence of p-polarized light both transverse and longitudinal waves are excited (fig. 5).

It is convenient to write the reflection coefficient as above in the form

$$R = \left|\frac{1-\bar{n}}{1+\bar{n}}\right|^2.$$

For ABC in the general form (19) one can obtain in the case of p-polarized radiation (Halevi and Fuchs 1978):

$$\bar{n}_p = \cos\varphi \,\frac{(\epsilon_1/n_{1z})(2,3)+(\epsilon_2/n_{2z})(3,1)}{(1,2)+(2,3)+(3,1)}, \tag{33}$$

where

$$\epsilon_r = \epsilon(\omega, k_x, k_{rz}), \quad (l, m) \equiv a_l b_m - a_m b_l,$$
$$a_r = (1 + i\gamma n_{rz})(n_r^2 - \epsilon_0), \quad b_r = \alpha_r a_r, \quad n_{rz} = (n_r^2 - \sin^2\varphi)^{1/2},$$
$$\alpha_r = -n_x/n_{rz}\ (r = 1, 2), \quad \alpha_3 = n_{3z}/n_x, \quad n_x = \sin\varphi.$$

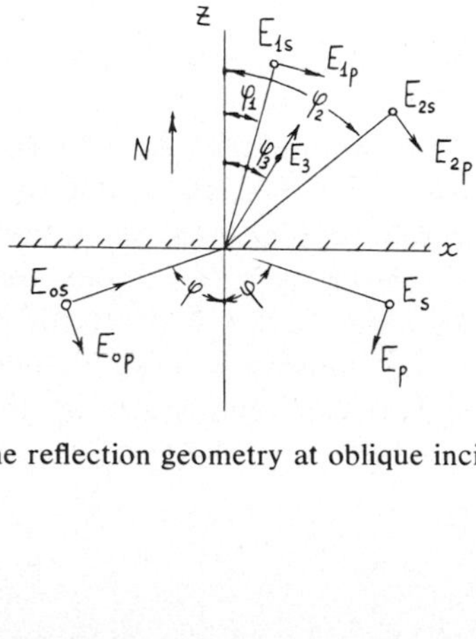

Fig. 5. The reflection geometry at oblique incidence.

The analogous formula in the case of s-polarized radiation reads

$$\bar{n}_s = \frac{a_1 n_{2z} - a_2 n_{1z}}{\cos\varphi(a_1 - a_2)}. \tag{34}$$

For the Pekar ABC eqs. (33) and (34) reduce to the corresponding formulae derived by Pekar (1958) and Permogorov et al. (1972, 1973). In particular, at $\gamma = 0$ eq. (34) reduces to

$$\bar{n}_s = \frac{\epsilon_0 - \sin^2\varphi + n_{1z} n_{2z}}{\cos\varphi(n_{1z} + n_{2z})}. \tag{35}$$

If $\cos\varphi > a = \epsilon_0/n_1(\omega_L)$, then the minimum value of the reflectivity R in the s component is zero and the minimum occurs approximately at the point ω_{min} which is determined by the condition (Permogorov et al. 1973)

$$n_2^2(\omega_{min}) = \sin^2\varphi + (a - \cos\varphi)^2. \tag{36}$$

At normal incidence eq. (36) reduces to eq. (32). In the absence of spatial dispersion, when $a = 0$, eq. (36) transforms to $n_2^2(\omega_{min}) = 1$ and does not depend on φ. If $\cos\varphi \leqslant a$ (in the case of CdS this occurs for $\varphi > 60$–$65°$) the minimum value of R_{ss} is no longer zero and ω_{min} is determined by the condition

$$n_2^2(\omega_{min}) = \sin^2\varphi. \tag{37}$$

Thus spatial dispersion produces observable effects on the angular dependence of reflection spectra.

Reflectivity experiments at non-normal incidence near the A($n = 1$) exciton resonance of CdS have been performed in the configurations CYP ($\boldsymbol{c} \parallel y$, p component), CXS ($\boldsymbol{c} \parallel x$, s component) and CXP ($\boldsymbol{c} \parallel x$, p component) (Permogorov et al. 1972, 1973, Stössel and Wagner 1978, Broser et al. 1978). In the cases CXS and CXP the excitons are excited with effective masses

$$M_r = (M_\perp^{-1} \cos^2\varphi_r + M_\parallel^{-1} \sin^2\varphi_r)^{-1} \quad (r = 1, 2, 3).$$

However, the anisotropy of the exciton effective mass is unimportant in the analysis of the reflection spectra. This can be understood taking into account that in fact the refractive index n_2 does not depend on the effective masses if $M_{\perp,\parallel} \gg \mu_0$. On the other hand, in the frequency region where the refractive index n_1 strongly depends on M, $|n_1| \gg 1$ and the exciton polariton of the branch 1 therefore propagates almost perpendicular to the c-axis ($\varphi_1 \simeq 0$). Hence, the frequency ω_{min} defined by eq. (36) depends via the parameter a only upon the transverse effective mass $M_\perp$.

Figure 6 shows the reflectivity spectra obtained in the geometry CXS for

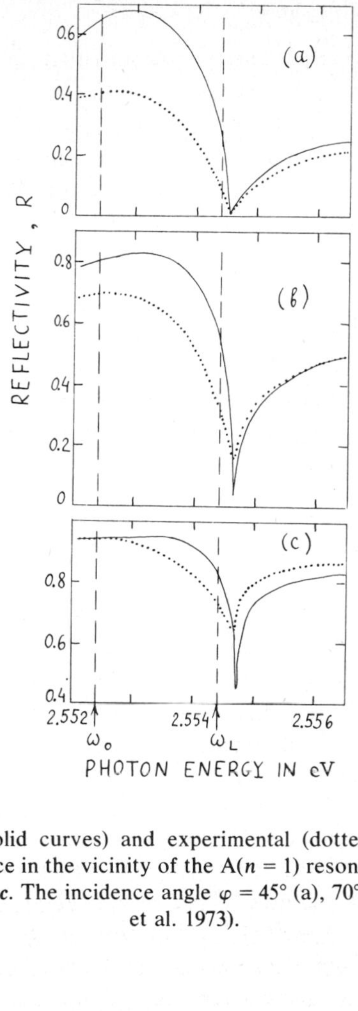

Fig. 6. The theoretical (solid curves) and experimental (dotted curves) reflection spectra obtained at oblique incidence in the vicinity of the A($n = 1$) resonance in CdS ($T = 1.8$ K). The geometry used is $\boldsymbol{E} \perp \boldsymbol{c}$, $\boldsymbol{k} \perp \boldsymbol{c}$. The incidence angle $\varphi = 45°$ (a), 70° (b) and 85° (c) (Permogorov et al. 1973).

three angles of incidence. In agreement with the theoretical predictions ω_{min} depends on φ and for $\cos \varphi < a$ $R_{ss,min} = 0$, while for $\cos \varphi > a$ the value of $R_{ss,min}$ is non-zero and increases with increasing φ.

Konstantinov and Saifullaev (1976) calculated the reflectivity and transmissivity of a slender optical wedge. The angle of the wedge being small enough, the transmitted radiation is split into two beams due to the difference between the refractive indices n_1 and n_2. The calculations show that an increase of γ in eq. (19) from 0 to 1 leads to a decrease of three to four orders of magnitude in the intensity of the beam corresponding to the refraction of the exciton polariton 1.

4.3. Phase change on reflection

The phase change $\Delta\Phi$ on reflection is equal to the phase of the amplitude reflection coefficient,

$$r = E_R/E_0 = |r| \exp(i\Delta\Phi). \tag{38}$$

Taking into account an exciton-free layer the coefficient r at normal incidence is given by eq. (37) and $\Delta\Phi$ satisfies the condition (Komarov et al. 1978):

$$\text{tg}\,\Delta\Phi = \frac{\rho_{23}(1 - r_{12}^2)\sin(\Phi_{23} + 2\Phi)}{r_{12}(1 + \rho_{23}^2) + \rho_{23}(1 + r_{12}^2)\cos(\Phi_{23} + 2\Phi)}. \tag{39}$$

Here $r_{12} = (1 - n)/(1 + n)$ is the amplitude reflection coefficient for the boundary "vacuum - deadlayer", $\Phi = k_0 nl$, $\rho_{23}\exp(i\Phi_{23})$ is the amplitude reflection coefficient for the boundary "deadlayer - spatially dispersive medium":

$$\rho_{23} = \frac{(n - n')^2 + (n'')^2}{(n + n')^2 + (n'')^2}, \qquad \text{tg}\,\Phi_{23} = \frac{2nn''}{(n')^2 + (n'')^2 - n^2}, \tag{40}$$

n' and n'' being the real and imaginary parts of $n(l = 0) = (n_1 n_2 + \epsilon_0)/(n_1 + n_2)$.

The spatial dispersion effects on the phase of the reflected light wave have been investigated in CdS (Solovjev and Babinskii 1976, Komarov et al. 1978) and β-AgJ (Mashlyatina et al. 1979). The experiments in CdS were performed at normal incidence in the configuration $\boldsymbol{k} \| z$, $\boldsymbol{c} \| y$, the angle between the electric field vector and the c-axis was equal to 45°. The phase difference $(\Phi_\perp - \Phi_\|)$ between the components $E_\perp$ and $E_\|$, transverse and longitudinal with respect to the c-axis, was measured. The optical dipole transitions to the A(n = 1) exciton states are forbidden for $\boldsymbol{E} \| \boldsymbol{c}$. Hence, the phase change $\Delta\Phi_\|$ of the $E_\|$ component on reflection equals π and does not depend on frequency. Since $\Phi_\perp - \Phi_\| = \Delta\Phi_\perp - \Delta\Phi_\|$, the measured phase difference differs from $\Delta\Phi_\perp$ only by $(-\pi)$. The frequency dependence of $\Delta\Phi_\perp$ is markedly affected by spatial dispersion. The measured spectra of the phase change $\Delta\Phi_\perp$ and the reflectivity $R_\perp$ for the $E_\perp$ component obtained at $T = 4.2$ and 77 K are shown in fig. 7. In the non-resonant region, where exciton–photon coupling and spatial dispersion effects are negligible, $\Delta\Phi_\perp$ is close to π in agreement with the prediction of the classical optics. The calculated spectra of $\Delta\Phi_\perp$ and $R_\perp$, neglecting spatial dispersion, qualitatively disagree with the experiment performed at 4.2 K and quantitatively differs from the experimental data obtained at 77 K. In contrast with this, the calculated spectra, taking into account spatial dispersion and the effect of a "dead-layer", satisfactorily reproduce the experimental curves. According to fig. 7 some discrepancies remain only for the spectrum of $R_\perp$ at $T = 4.2$ K.

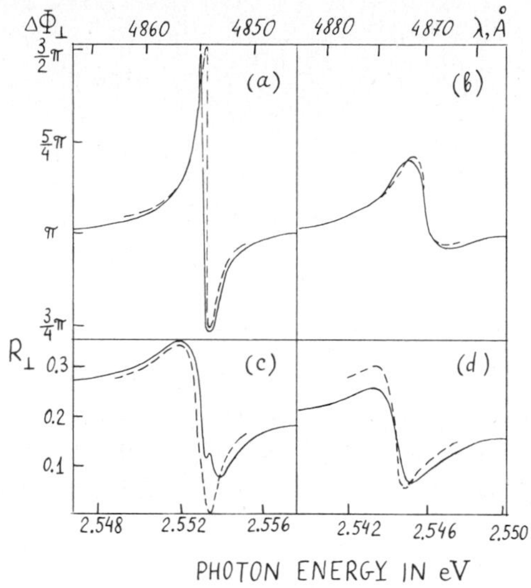

Fig. 7. The normal-incidence spectra of phase change $\Delta\Phi_\perp$ (a), (b) and reflection coefficient $R_\perp$ (c), (d) in CdS near the resonance A(n = 1). The spectra were measured at T = 4.2 K (a), (c) and 77 K (b), (d). Solid curves are experimental, dashed curves are calculated taking into account spatial dispersion for $\Gamma(77°) = 8.6\ \mathrm{cm}^{-1}$, $l(77°) = 76.5$ Å, $\Gamma(4.2°) = 2.74\ \mathrm{cm}^{-1}$, $l(4.2°) =$ 83.5 Å. Other parameters coincide with those used by Pekar and Strashnikova (1975). (After Komarov et al. 1978.)

5. Interference spectra of thin crystals

In the previous section it was shown that the structure of optical spectra near excitonic resonances can be naturally understood using a spatial dispersion approach. However, the existence of additional light waves was only demonstrated indirectly in the above-mentioned experiments. The direct verification of the theory of additional waves has been achieved by reflectivity and transmissivity measurements in thin CdS and CdSe crystals (Kiselev et al. 1973, 1974, 1975) and by observation of resonant Brillouin scattering of exciton polaritons in GaAs (Ulbrich and Weisbuch 1977) and CdS (Winterling and Koteles 1977).

Figure 8(b) shows the normal-incidence reflection spectrum of a thin CdSe crystal ($\boldsymbol{k} \parallel \boldsymbol{c}$, $\boldsymbol{E} \perp \boldsymbol{c}$). On the low-frequency side of ω_L in the frequency interval of about 15 cm^{-1} a very distinct set of lines can be seen. This set is due to the Fabry–Perot interference of exciton polaritons 1 (branch 1 in fig. 8(a)). On the high-frequency side of ω_L two series of interference lines occur. The first series corresponds to the Fabry–Perot interference of exciton polaritons 2, the lines of this series are widely spaced. The second

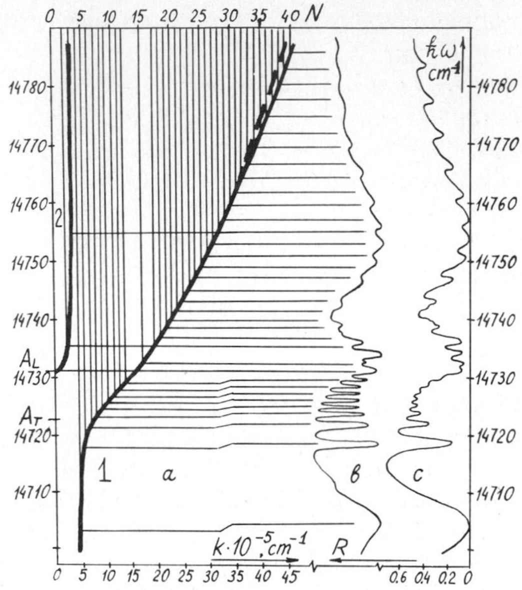

Fig. 8. The reflection spectra of a thin CdSe crystal. (a) Experimental curves for dispersion of exciton polaritons A(n = 1) – solid lines 1, 2; theoretical dispersion – dashed line. (b) Microphotometer trace of the reflectivity at T = 4.2 K, $\mathbf{E} \perp \mathbf{c}$, d = 0.356 μm. (c) Computed spectrum. (Kiselev et al. 1974.)

series, superimposed on the first one, consists of much more closely-spaced peaks and is due to the mutual interference of exciton polaritons 1 and 2. The dispersion branches 1 and 2 and the theoretical curve (c) in fig. 8 are calculated taking the parameters

$$\hbar\omega_0 = 14\,723\ \text{cm}^{-1}, \quad \hbar\omega_{\text{LT}} = 7.5\ \text{cm}^{-1}, \quad M_{\|} = 0.58\ m_0, \quad \epsilon_{0\perp} = 10.8,$$
$$\hbar\Gamma = 1\ \text{cm}^{-1}.$$

The best fitting thickness was found to be $d = 0.356\ \mu$m. In further works by Kiselev et al. (1975a, 1977) more detailed comparison of the experimental data with the theory was carried out and the values of $M_{\|}$ and $\epsilon_{0\perp}$ were corrected: $M_{\|} = 0.415\ m_0$, $\epsilon_{0\perp} = 8.45$. In addition, the order of interference is indicated in fig. 8. The frequency corresponding to N was computed by virtue of the Fabry–Perot condition

$$2n'_r d = \lambda_0 N, \tag{41}$$

where n'_r is the real part of the refractive index n_r (r = 1, 2), $\lambda_0 = 2\pi c/\omega$ is the light wavelength in vacuum. The shift of singularities of the spectrum in fig. 8 with respect to the frequencies defined by eq. (41) is caused by the influence of the normal mode 2 (for $\omega < \omega_L$ it is spatially damped) and by the effect of exciton damping.

An interesting feature of the spectrum above ω_L is that the structure has an approximately doubled period in comparison with the set of frequencies defined by eq. (41) for $r = 1$. This is because the light wave can pass once (forward, say) as a polariton 1 and once (say, back after reflection from the back surface of the crystal) as a polariton 2.

With the crystal thickness d increasing the oscillations become more frequent and their amplitude decreases. The condition for the superimposed oscillations to be registered is as follows

$$d \lesssim L_1, \tag{42}$$

where $L_1 = (n_1'' k_0)^{-1}$ is the mean free path of an exciton polariton 1. It should be noted that optical spectra of thin crystals have been observed by other experimental groups (e.g. Brodin and Strashnikova 1962, Reynolds et al. 1968, Voigt and Mauersberger 1973). However, due to Kiselev, Razbirin and Uraltsev these studies have become thorough and systematic.

The interference structure of optical spectra is very sensitive to the form of ABC (Kiselev et al. 1975a, Bishop 1976). For example, with the increasing of the parameter γ in the ABC (19) the amplitude of the additional oscillations for $\omega > \omega_L$ rapidly decreases. The reason for this is quite obvious since different ABC result in different ratios E_2/E_1, where E_1 and E_2 are the amplitudes of the electric field corresponding to polariton modes 1 and 2. According to eq. (19) at normal incidence for a semi-infinite crystal we obtain

$$\frac{E_1}{E_2} = -\frac{1 + i\gamma n_2}{1 + i\gamma n_1}\frac{n_2^2 - \epsilon_0}{n_1^2 - \epsilon_0}. \tag{43}$$

The modulus of the first multiplier in eq. (43) varies from 1 to $|n_2/n_1| \ll 1$ with increasing γ from 0 to $+\infty$. Hence, for $|\gamma| \gg 1$ the role of wave 1 is diminished and the fringes due to the mutual interference of waves 1 and 2 disappear.

Figure 9 shows the reflection spectra of three CdSe crystals differing slightly in thickness ($d \simeq 0.8\ \mu$m). One can see that, beyond frequency intervals with pronounced additional structure, there are intervals (in the vicinity of $\lambda = 6772$ Å and in the region $\lambda = 6758$–6755 Å) where such structure is absent. This can be explained if one assumes that γ in the ABC (19) depends on frequency. The curves (d) and (e) in fig. 9 are computed respectively for the Pekar ABC and taking into account the frequency dependence of γ. In the case (e) the approximation was used (Kiselev et al. 1977a),

$$\delta(\omega)/k_0 = [2.67(\omega - \omega_0) - 53.9]\ \text{Å},$$

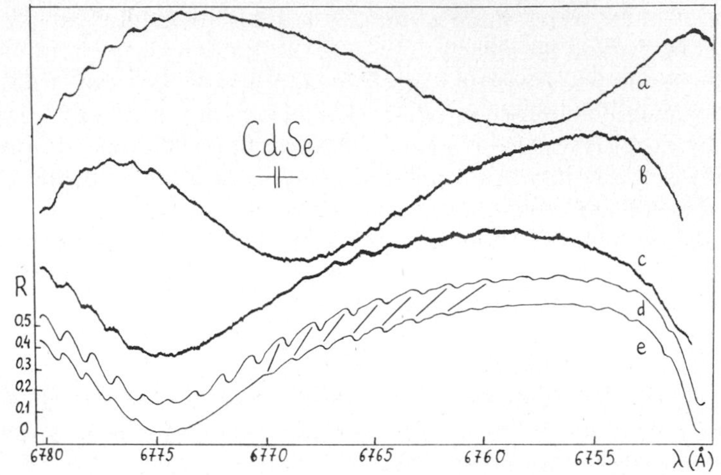

Fig. 9. The reflectivity spectra of thin CdS crystals in a region between the $n = 1$ and $n = 2$ states of the A exciton. (a), (b), (c) – experiment at $T = 1.6$ K, $\mathbf{E} \perp \mathbf{c}$; (d) – computed spectrum for $\delta = 0$; (e) – computed spectrum taking into account the frequency dependence of δ (Kiselev et al. 1977a).

where the phase δ of the reflected exciton wave is related with γ by eq. (21) and the frequency is in cm^{-1}. If $\delta = \pm\pi N$, then $\gamma = 0$ and the fine structure is seen quite well. If $\delta = \pm\pi(2N + 1)/2$, so that $\gamma = \pm\infty$, then the structure disappears. Thus the interference spectra of thin crystals are very informative and permit one to make a choice among various ABC.

Finally, we note that the theoretical description of the optical oscillations in thin crystals can be also formulated in terms of the size quantization of a mechanical exciton (Kiselev et al. 1977). This is of course equivalent to the wave interference method discussed above.

6. *Fine structure of excitonic levels and optical spectra*

So far we have concentrated efforts on effects which can be described within the simple model of an isolated isotropic electric-dipole-active exciton. We now turn to more complicated cases of exciton band structure and discuss the effect of the band structure peculiarities upon the reflectivity and transmissivity.

6.1. The general scheme

We begin with the general scheme for calculation of optical spectra taking into account the fine structure of excitonic levels (Ivchenko et al. 1978,

Ivchenko and Selkin 1979, Cho 1978). Let a group of excitonic levels be isolated from other excitonic states. The crystal symmetry is assumed to be arbitrary. By analogy with eq. (7), we express the exciton wavefunction as an expansion in the excitonic eigenfunctions φ_ν at $k = 0$, where $\nu = 1, \ldots, s$ is the index of the exciton states under consideration. Including the interaction of the macroscopic electric field $\boldsymbol{E}$ with the crystal and applying the Fourier transformation to the time-dependent Schrödinger equation we obtain the following set of equations for the Fourier components $C_\nu(\boldsymbol{k}, \omega)$

$$\sum_{\nu'} [H_{\nu\nu'}(\boldsymbol{k}) - \hbar\omega\delta_{\nu\nu'}]C_{\nu'}(\boldsymbol{k}, \omega) = \boldsymbol{d}_\nu(\boldsymbol{k}) \cdot \boldsymbol{E}(\boldsymbol{k}, \omega). \tag{44}$$

Here $\boldsymbol{E}(\boldsymbol{k}, \omega)$ is the electric field amplitude, $\boldsymbol{d}_\nu(\boldsymbol{k}) = \langle \nu | \hat{\boldsymbol{d}}(\boldsymbol{k}) | 0 \rangle$ is the dipole matrix element for optical transition to the excitonic state ν and, for simplicity, the exciton damping is neglected. The effective Hamiltonian $H_{\nu\nu'}(\boldsymbol{k})$ of a mechanical exciton contains the kinetic energy of the exciton in the effective-mass approximation, including terms linear in wavevector (Rashba and Sheka 1959, Casella 1960, Hopfield and Thomas 1961, Sheka 1965, Pikus and Bir 1973), the short-range exchange interaction and the terms due to external perturbations, e.g. stress, electric or magnetic field (Bir and Pikus 1974, Cho 1978, Röseler and Henneberger 1979, and references therein). Since the right-hand side of eq. (44) includes both the transverse and longitudinal components of the macroscopic electric field, one should not include in $H_{\nu\nu'}(\boldsymbol{k})$ the long-range exchange (or annihilation) interaction (Agranovich and Ginzburg 1966). The relationship between the Fourier components of the excitonic polarization $\boldsymbol{P}$ and the coefficients C_ν is given by

$$\boldsymbol{P}(\boldsymbol{k}, \omega) = \sum_{\nu} [\boldsymbol{d}^*_\nu(\boldsymbol{k})C_\nu(\boldsymbol{k}, \omega) + \boldsymbol{d}_\nu(-\boldsymbol{k})C^*_\nu(-\boldsymbol{k}, -\omega)]. \tag{45}$$

The displacement vector $\boldsymbol{D}$ is defined by eq. (1) in which the scalar ϵ_0 should in general be replaced by the background dielectric tensor $\boldsymbol{\epsilon}_0$.

For the simple model of an isotropic exciton considered above we have

$$H_{\alpha\alpha'}(\boldsymbol{k}) = \left(\hbar\omega_0 + \frac{\hbar^2 k^2}{2M}\right)\delta_{\alpha\alpha'}, \quad \boldsymbol{d}_\alpha(\boldsymbol{k}) \cdot \boldsymbol{E}(\boldsymbol{k}, \omega) = \mathrm{d}E_\alpha(\boldsymbol{k}, \omega). \tag{46}$$

One can easily show that in this case eq. (44) reduces to eq. (2).

For the case of s isolated excitonic states, s additional boundary conditions are needed for a complete description of the electrodynamics. The straightforward generalization of the Pekar ABC (15) is that the wavefunction of a mechanical exciton vanishes at the boundary surface

$$C(\boldsymbol{r}, \omega)|_{z=0} = 0 \quad (\nu = 1, \ldots, s). \tag{47}$$

To take into account the surface potential, a "dead-layer" of the effective thickness l can be included with the ABC (47) applied at $z = l$.

In order to illustrate the general approach we next consider the effect of linear in $\boldsymbol{k}$ terms and the influence of an external magnetic field on optical properties of crystals with a wurtzite structure, such as CdS and CdSe (C_{6v} point symmetry). Due to the exchange interaction the ground state of A-series exciton ($\Gamma_7 \times \Gamma_9$) in these crystals is split into two terms Γ_5 and Γ_6, while the ground state B($n = 1$) or C($n = 1$) (exciton $\Gamma_7 \times \Gamma_7$) is split into three terms Γ_5, Γ_1 and Γ_2. As a set of basic states, we choose the states Γ_{5x}, Γ_{5y}, Γ_{61} and Γ_{62} (A-exciton), transforming like (x, y), $(x^2 - y^2, -2xy)$, and the states Γ_{5x}, Γ_{5y}, Γ_1 and Γ_2 (B- or C-exciton), transforming like (x, y), z and J_z, where z is along the principal axis and J_z is the z component of any axial vector.

6.2. *Effect of linear in k terms*

Using general symmetry considerations, one can obtain the contribution to $H_{\nu\nu'}(\boldsymbol{k})$ linear in $\boldsymbol{k}$ for the A and B exciton in the above-mentioned bases (Hopfield and Thomas 1961, Pikus and Bir 1973)

$$H_k = \begin{bmatrix} 0 & D \\ D^+ & 0 \end{bmatrix}, \qquad D^{(A)} = \mathrm{i}\beta_A \begin{bmatrix} k_x & -k_y \\ -k_y & -k_x \end{bmatrix},$$

$$D^{(B)} = \mathrm{i}\begin{bmatrix} \beta_1 k_x, & -\beta_2 k_y \\ \beta_1 k_y, & \beta_2 k_x \end{bmatrix}. \tag{48}$$

The wurtzite valence band structure near $k = 0$ may be approximated by introducing a (111) strain into a zincblende valence band (the so-called quasicubic model). It then follows that the constant β_A is quite small and, in fact, terms linear in $\boldsymbol{k}$ are unimportant for A-series exciton. So we shall analyze the effect of the linear terms for B-series exciton. Note that in the quasicubic model $\beta_1 \simeq -\beta_2$. Applying to eq. (44) the inverse Fourier transformation $C_\nu(\boldsymbol{k}, \omega) \to C_\nu(\boldsymbol{r}, \omega)$ and defining for convenience the quantity

$$Q(\boldsymbol{r}, \omega) = d_\perp[C_{\Gamma_2}(\boldsymbol{r}, \omega) + C^*_{\Gamma_2}(\boldsymbol{r}, -\omega)] \tag{49}$$

for the B($n = 1$) exciton we can reduce eq. (44) to the following form (Ivchenko et al. 1978, Ivchenko and Selkin 1978)

$$[\mathscr{E}_{\Gamma_5}(-\mathrm{i}\nabla) - \hbar\omega]\boldsymbol{P}_\perp + \beta_1 \frac{d_\perp}{d_\parallel} \nabla_\perp P_z + \beta_2(\boldsymbol{c} \times \nabla Q) = d_\perp^2 \boldsymbol{E}_\perp, \tag{50}$$

$$[\mathscr{E}_{\Gamma_1}(-\mathrm{i}\nabla) - \hbar\omega]P_z - \beta_1 \frac{d_\parallel}{d_\perp} \operatorname{div} \boldsymbol{P}_\perp = d_\parallel^2 E_z, \tag{51}$$

$$[\mathscr{E}_{\Gamma_2}(-\mathrm{i}\nabla) - \hbar\omega]Q - \beta_2 \operatorname{rot}_z \boldsymbol{P}_\perp = 0. \tag{52}$$

Here

$$\mathscr{E}_\nu(\boldsymbol{k}) = \mathscr{E}_\nu(0) + \frac{\hbar^2 k_\perp^2}{2M_\perp} + \frac{\hbar^2 k_z^2}{2M_\parallel},$$

$M_\perp$ and $M_\parallel$ being the transverse and longitudinal effective masses of the exciton, $\boldsymbol{c}$ is a unit vector directed along the c-axis, $d_\perp$ and $d_\parallel$ are the dipole matrix elements corresponding to the states Γ_5 and Γ_1, respectively. For simplicity, we ignore the wavevector dependence of $\boldsymbol{d}_\nu$, neglecting therefore the quadrupole corrections, i.e., linear terms in the expansion of $\boldsymbol{d}_\nu(\boldsymbol{k})$ with respect to $\boldsymbol{k}$. In this case the relationship between the coefficients C_ν and the excitonic contribution to the dielectric polarization is quite simple

$$P_{x,y}(\boldsymbol{r}, \omega) = d_\perp[C_{\Gamma_{5x,y}}(\boldsymbol{r}, \omega) + C^*_{\Gamma_{5x,y}}(\boldsymbol{r}, -\omega)],$$

$$P_z(\boldsymbol{r}, \omega) = d_\parallel[C_{\Gamma_1}(\boldsymbol{r}, \omega) + C^*_{\Gamma_1}(\boldsymbol{r}, -\omega)], \tag{53}$$

which is equivalent to eq. (7) when $d_\perp = d_\parallel$. The energy flux density can be written in the case under consideration as follows

$$\boldsymbol{S}(\boldsymbol{r}, t) = \frac{c}{4\pi}(\boldsymbol{E} \times \boldsymbol{H}) - \frac{\beta_1}{d_\perp d_\parallel}\dot{P}_z\boldsymbol{P}_\perp + \frac{\beta_2}{d_\perp^2}(\boldsymbol{P}_\perp \times \boldsymbol{c})Q$$

$$- \sum_{i=x,y}\frac{\hbar^2}{d_\perp^2}\dot{P}_i\hat{\mu}\nabla P_i - \frac{\hbar^2}{d_\parallel^2}\dot{P}_z\hat{\mu}\nabla P_z - \frac{\hbar^2}{d_\perp^2}\dot{Q}\hat{\mu}Q, \tag{54}$$

where $(\hat{\mu}\nabla) = (1/2M_{ii})\nabla_i$, $M_{zz} = M_\parallel$, $M_{xx} = M_{yy} = M_\perp$. One can easily verify that for the boundary conditions (14) and (47) the normal component of the energy flux density is continuous at the boundary. If we set $d_\perp = d_\parallel = d$, $M_\parallel = M_\perp = M$, $\beta_1 = \beta_2 = 0$ and omit the last term in eq. (54), then eq. (54) will reduce to eq. (25).

We restrict the consideration here to the case $\boldsymbol{k} \perp \boldsymbol{c}$ which corresponds to the geometries used experimentally. Let us suppose that the c-axis lies in the boundary plane and choose coordinates x, y, z in such a way that $z \parallel \boldsymbol{c}$ and y is parallel to the normal to the boundary plane. Then the ABC (47) are applied at $y = 0$ or $y = l$.

Of course the ABC in the form (47) are not the only ones which satisfy the energy conservation law at the boundary. Another set of ABC with this property is

$$Q|_{y=0} = 0,$$

$$\left\{\boldsymbol{p}_\perp + u_\perp\left[\frac{\hbar^2}{2M_\perp d_\perp^2}(\boldsymbol{N}\cdot\nabla)\boldsymbol{p}_\perp - p\frac{\beta_1}{d_\perp d_\parallel}\boldsymbol{N}p_z\right]\right\}_{y=0} = 0,$$

$$\left\{p_z + u_\parallel\left[\frac{\hbar^2}{2M_\perp d_\parallel^2}(\boldsymbol{N}\cdot\nabla)p_z + (1-p)\frac{\beta_1}{d_\perp d_\parallel}(\boldsymbol{N}\cdot\boldsymbol{p})\right]\right\}_{y=0} = 0, \tag{55}$$

where $u_\perp$, $u_\parallel$ and p are real coefficients. If the quadrupole corrections are added to the matrix elements $\boldsymbol{d}_\nu$, then it is necessary to include in eq. (55) the terms linear in the electric field $\boldsymbol{E}$.

Mahan and Hopfield (1964) were the first to study optical effects of exciton energy terms linear in wavevector. They examined the influence of

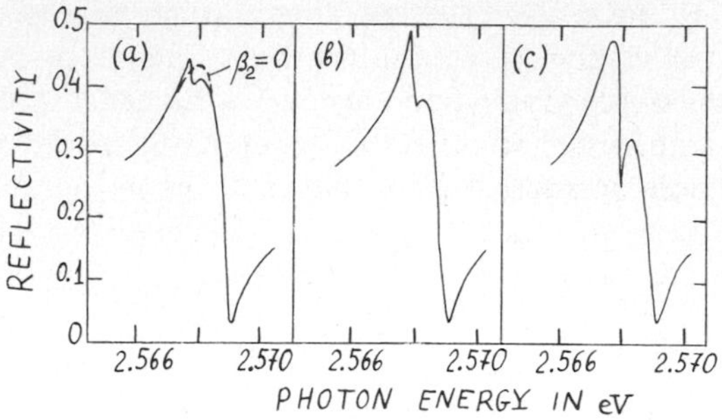

Fig. 10. The theoretical effect of increasing splitting linear in k on the reflection spectrum at normal incidence ($\boldsymbol{k} \perp \boldsymbol{c}$, $\boldsymbol{E} \perp \boldsymbol{c}$). (a) $\beta_2 = 0$ (dashed line) and $\beta_2 = 0.3 \times 10^{-9}$ eV cm (solid line); (b) $\beta_2 = 0.5 \times 10^{-9}$ eV cm; (c) $\beta_2 = 0.8 \times 10^{-9}$ eV cm. The middle value appears to give the most reasonable fit to the experimental data (Mahan and Hopfield 1964).

such terms in the attempt to understand the origin of some anomalous structure of the reflection curves measured in the vicinity of B excitons in CdS at normal incidence (see fig. 2). According to fig. 2(b) the geometry $\boldsymbol{k} \perp \boldsymbol{c}$, $\boldsymbol{E} \perp \boldsymbol{c}$ shows the extra shoulder near the transverse frequency ω_0 (the main reflectivity peak). This anomaly is not caused by an exciton-free layer, otherwise the subsidiary peak should fall at the frequency ω_L, i.e. near the reflectivity minimum. Of particular importance is the fact that in the geometries $\boldsymbol{k} \parallel \boldsymbol{c}$, $\boldsymbol{E} \perp \boldsymbol{c}$ (fig. 2(a)) and $\boldsymbol{k} \perp \boldsymbol{c}$, $\boldsymbol{E} \parallel \boldsymbol{c}$ the structure is not observed. The spectra shown in fig. 10 were calculated for various values of β_2. Other parameters used are $\hbar\omega_0 = 2.5679$ eV, $M_\perp = 1.3\, m_0$, $\alpha_0 = \omega_{LT}^{\perp}\epsilon_{0\perp}/2\pi\omega_0 = 0.65 \times 10^{-3}$, $l = 70$ Å, $\hbar\Gamma = 7.5 \times 10^{-5}$ eV. The value of Γ was chosen large enough to wash out the "spike" at ω_L. The splitting of the B-exciton states due to the short-range exchange interaction was neglected. The important result is that the presence of energy terms linear in $\boldsymbol{k}$ can explain the anomalous features of the reflectivity. The value of $\beta_2 = 0.5 \times 10^{-9}$ eV cm gives the best fit to the experimental data.

6.3. Optical activity of wurtzite crystals

Crystals with a wurtzite structure (C_{6v} symmetry) are optically active. By definition, the symmetry of optically active crystals allows linear terms in the expansion of the dielectric tensor in powers of $\boldsymbol{k}$ (Agranovich and Ginzburg 1966):

$$\epsilon_{\lambda\mu}(\omega, \boldsymbol{k}) = \epsilon_{0,\lambda\mu}(\omega) + i\gamma_{\lambda\mu\nu}(\omega)k_\nu + \cdots . \tag{56}$$

Optical activity is usually associated with the rotation of the polarization plane of linearly polarized light wave propagating through the optically active medium. However, in crystals of symmetry groups C_{3v}, C_{4v} and C_{6v} the terms linear in $\boldsymbol{k}$ give rise to no optical rotation for any propagation direction in spite of the fact that they are allowed in the expansion (56) (Fedorov 1959, Fedorov et al. 1962, Bokut' and Serdyukov 1971). In crystals of the three symmetries the tensor $\epsilon(\omega, \boldsymbol{k})$ to the first order in $\boldsymbol{k}$ is given by

$$\|\epsilon_{\lambda\mu}(\omega, \boldsymbol{k})\| = \begin{bmatrix} \epsilon_{0\perp} & 0 & \mathrm{i}\bar{\gamma}k_x \\ 0 & \epsilon_{0\perp} & \mathrm{i}\bar{\gamma}k_y \\ -\mathrm{i}\bar{\gamma}k_x & -\mathrm{i}\bar{\gamma}k_y & \epsilon_{0\|} \end{bmatrix}. \tag{57}$$

In such crystals, as in non-active uniaxial crystals, the ordinary (transverse) and extraordinary (mixed) modes can propagate. For the extraordinary waves the electric field vector lies in a plane containing both the c-axis and the wavevector $\boldsymbol{k}$. Unlike non-active crystals, in an optically active wurtzite crystal the extraordinary mode has an elliptic polarization, the plane of the polarization ellipse coinciding with the $(\boldsymbol{c}, \boldsymbol{k})$ plane (fig. 11). According to eq. (57) for the wave propagating perpendicular to the c-axis $(\boldsymbol{k} \perp \boldsymbol{c})$ the ratio

$$E_\ell/E_\mathrm{t} = -\mathrm{i}\bar{\gamma}k/\epsilon_{0\perp}.$$

Such ellipticity, directly connected with optical activity, can be observed in reflectivity experiments at oblique incidence. Let the c-axis be perpendicular to the plane of incidence (fig. 11); then at non-normal incidence of the s-polarized light, beyond the s component, the orthogonal (p-polarized) component will be present in the reflected wave. This is because in an optically active wurtzite crystal the longitudinal component $\boldsymbol{E}_\ell$ of the

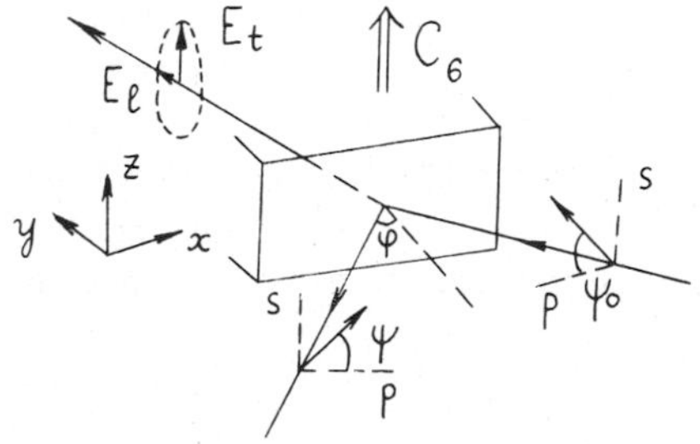

Fig. 11. Geometry of the reflectance spectrum of CdS in the spectral region of the B(n = 1) exciton resonance.

mixed mode has a non-zero projection on the reflecting face. At normal incidence ($\varphi = 0$) this projection is zero and the reflectance in the orthogonal configuration vanishes. In analogous fashion, the p-polarized incident light can induce the s component on reflection.

Far away from resonance absorption lines the intensity of the orthogonal component is proportional to $(a/\lambda)^2 \ll 1$. Hence in nonresonant conditions its value is negligibly small. However, in the resonant polariton region the effects due to optical activity must markedly increase (as well as other spatial dispersion effects).

The special case of optical activity discussed above has been first observed for the B exciton in CdS crystals in the geometry of fig. 11 (Ivchenko et al. 1978, 1978a). Theoretically the problem of optical activity in this case reduces to the calculation of the oblique-incidence reflectivity taking into account linear in $\boldsymbol{k}$ terms in the effective exciton Hamiltonian. Ivchenko and Selkin (1979) have analyzed the effect of the constants β_1 and β_2 on optical activity. Far below the resonance, the contribution from the B(n = 1)-exciton states to $\bar{\gamma}$ is given approximately by

$$\bar{\gamma}(\omega) = -\beta_1 \frac{(\epsilon_{0\perp}\epsilon_{0\|}\omega_{\mathrm{LT}}^{\perp}\omega_{\mathrm{LT}}^{\|})^{1/2}}{\hbar(\omega-\omega_0)^2}. \tag{58}$$

Here $\hbar\omega_{\mathrm{LT}}^{\perp} = 4\pi d_{\perp}^2/\epsilon_{0\perp}$ and $\hbar\omega_{\mathrm{LT}}^{\|} = 4\pi d_{\|}^2/\epsilon_{0\|}$ are the longitudinal–transverse splittings for the states Γ_5 and Γ_1, respectively, ω_0 is the resonance frequency at $k = 0$ neglecting the exchange interaction. According to eq. (58) optical activity is associated with the constant β_1. The latter result is general and holds in resonant conditions when the expansion (57) is not valid. According to eqs. (50)–(52) the terms corresponding to β_1 mix the transverse exciton polariton Γ_1 with the longitudinal exciton Γ_5. Only this mixing leads to the elliptical polarization of the mixed modes and, consequently, to the reflected light signal in the orthogonal configurations s → p or p → s.

Figure 12 shows the reflection spectra R_{pp}, R_{ss}, R_{sp} and R_{ps}, in the configurations p → p, s → s, p → s and s → p, respectively, measured in the vicinity of the B(n = 1) resonance in CdS for three angles of incidence at $T = 2$ K. For small incidence angles the intensity of the orthogonal component is small and comparable with the background, the latter probably being due to the contribution from diffuse reflectance. The reflectivities R_{ps} and R_{sp} exhibit a maximum approximately at $\varphi = 45°$. From the time inversion symmetry it follows that

$$R(\boldsymbol{k}, \boldsymbol{e}|\tilde{\boldsymbol{k}}, \tilde{\boldsymbol{e}}) = R(-\tilde{\boldsymbol{k}}, \tilde{\boldsymbol{e}}^*|-\boldsymbol{k}, \boldsymbol{e}^*), \tag{59}$$

where $\boldsymbol{k}$ and $\boldsymbol{e}$ ($\tilde{\boldsymbol{k}}$ and $\tilde{\boldsymbol{e}}$) are the wavevector and polarization vector of the incident (reflected) wave. According to eq. (59) in the geometry of fig. 11 the reflection coefficients R_{ps} and R_{sp} must coincide. This is in fact the case

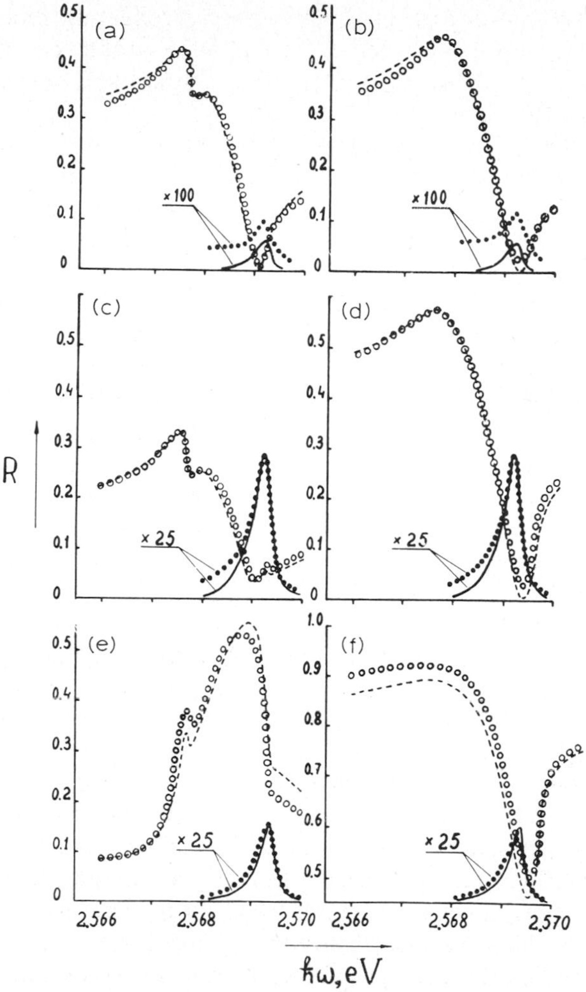

Fig. 12. Experimental and theoretical reflection spectra of R_{pp}, R_{sp} (a), (c), (e) and R_{ss}, R_{ps} (b), (d), (f) in the vicinity of the B(n = 1) resonance at the incidence angles 8° (a), (b); 45° (c), (d); 81.5° (e), (f). (○) – experimental spectra of R_{pp} and R_{ss}, (●) – experimental spectra of R_{sp} and R_{ps}. Dashed lines are theoretical spectra of R_{pp} and R_{ss}, solid lines are theoretical spectra of R_{sp} and R_{ps} (Ivchenko and Selkin 1979).

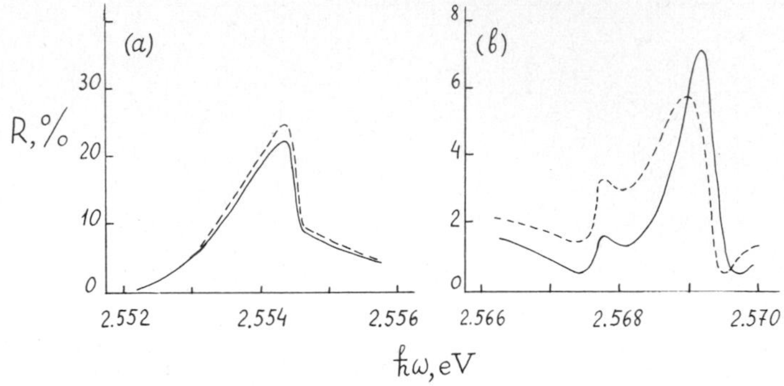

Fig. 13. The reflection spectra of CdS ($T = 2$ K) near the resonances A($n = 1$) (a) and B($n = 1$) (b) in the geometry (45°, 45°, 45°) (solid curves) and (135°, 45°, 135°) (dashed lines). (After Ivchenko et al. 1978b.)

experimentally. The dashed curves in fig. 12 are calculated for $|\beta_1|\omega_0/c = 0.075$ meV and $|\beta_2|\omega_0/c = 0.06$ meV.

The coefficients R_{sp} and R_{ps} do not depend on the sign of β_1. The sign of β_1 affects the difference between the coefficients $R(\psi_0, \varphi, \psi)$ and $R(\psi_0 + 90°, \varphi, \psi + 90°)$, if $(\psi_0 + \psi) \neq \pm l\pi$ $(l = 0, 1, \ldots)$. Here φ is the angle of incidence, the angles ψ_0 and ψ are indicated in fig. 11. It can be shown that the difference is an odd function of β_1. Figure 13 shows the reflection spectra $R(45°, 45°, 45°)$ and $R(135°, 45°, 135°)$ of a CdS crystal. In the absence of optical activity the two spectra would coincide. However, one can see that near the B($n = 1$) resonance the spectra markedly differ from each other. This is an additional manifestation of optical activity in CdS.

Reflection spectra of β-AgJ crystals (C_{6v} symmetry) in the resonant frequency region have been measured by Mashlyatina et al. (1978). The calculation, performed in the framework of the theory taking into account linear in $\boldsymbol{k}$ terms, is in agreement with the experimental data.

The effect of terms linear in wavevector on optical spectra of D_3 symmetry crystals near phonon and exciton resonances have been theoretically investigated by Bishop and Maradudin (1977). The calculation was performed for the ABC (24). It was shown that the reflection coefficients of D_3 symmetry crystals for the left- and right-handed circular polarization differ from each other.

6.4. *Magnetic field effect*

For the A($n = 1$) exciton the linear Zeeman interaction in the terms of the basis Γ_{5x}, Γ_{5y}, Γ_{61} and Γ_{62} is given by (Venghaus et al. 1977):

$$H_H^{(A)} = \frac{\hbar}{2}\begin{bmatrix} 0 & i\Omega_\parallel & i\begin{pmatrix}\Omega_y & \Omega_x \\ \Omega_x & -\Omega_y\end{pmatrix} \\ -i\Omega_\parallel & 0 & \\ -i\begin{pmatrix}\Omega_y & \Omega_x \\ \Omega_x & -\Omega_y\end{pmatrix} & & \begin{matrix} 0 & i\Omega'_\parallel \\ -i\Omega'_\parallel & 0 \end{matrix} \end{bmatrix}. \tag{60}$$

Here

$$\hbar\Omega_\parallel = (g_{e\parallel} - g_{h\parallel})\mu_0 H_z, \quad \hbar\Omega'_\parallel = (g_{e\parallel} + g_{h\parallel})\mu_0 H_z, \quad \hbar\boldsymbol{\Omega}_\perp = g_{e\perp}\mu_0 \boldsymbol{H}_\perp,$$

$\boldsymbol{H}$ is the magnetic field vector, μ_0 is the Bohr magneton, $g_{e\parallel}$ and $g_{e\perp}$ are the g factors for electrons and $g_{h\parallel}$ is the longitudinal component of the hole g factor ($g_{h\perp} = 0$). Figure 14 shows schematically the dispersion relation for exciton polaritons in a magnetic field $\boldsymbol{H} \parallel \boldsymbol{c}$ for $\boldsymbol{k} \perp \boldsymbol{c}$ in the case $\omega_{LT} > \Omega_\parallel$. In this case the magnetic field mixes transverse and longitudinal waves. At $k \simeq 0$, the exciton polaritons 2 and 3 are circularly polarized in such a way that the electric field vector lies in a plane containing the vectors $\boldsymbol{H}$ (or $\boldsymbol{c}$) and $\boldsymbol{k}$. Now at normal incidence all three modes 1, 2 and 3 are excited. In the absence of spatial dispersion ($M_\perp \to \infty$) and for zero damping the reflection coefficient in the two regions

$$\omega'_0 + \tfrac{1}{2}(\omega_{LT} \pm \sqrt{\omega_{LT}^2 + \Omega_\parallel^2}) < \omega < \omega'_L \pm \tfrac{1}{2}\Omega_\parallel$$

would be equal to 100%.

Figure 15 shows the experimental reflection spectra in the geometry $\boldsymbol{H} \parallel \boldsymbol{c}$, $\boldsymbol{k} \perp \boldsymbol{c}$, $\boldsymbol{E} \perp \boldsymbol{c}$. With increasing magnetic field the whole structure shifts to higher frequencies and the additional reflectance minimum appears. The

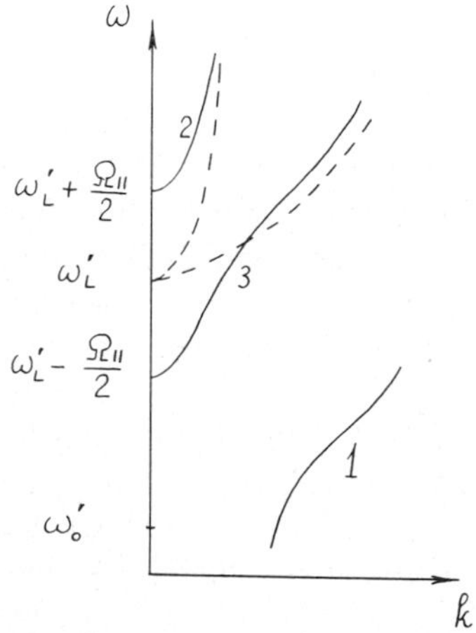

Fig. 14. The polariton dispersion curves in a magnetic field $\boldsymbol{H} \parallel \boldsymbol{c}$ for $\boldsymbol{k} \perp \boldsymbol{c}$. The dashed lines show the dispersion relation for the branches 2 and 3 neglecting the Zeeman splitting of excitonic states but taking into account the diamagnetic shift of the excitonic level ($\omega'_0 \neq \omega_0$ if $H \neq 0$).

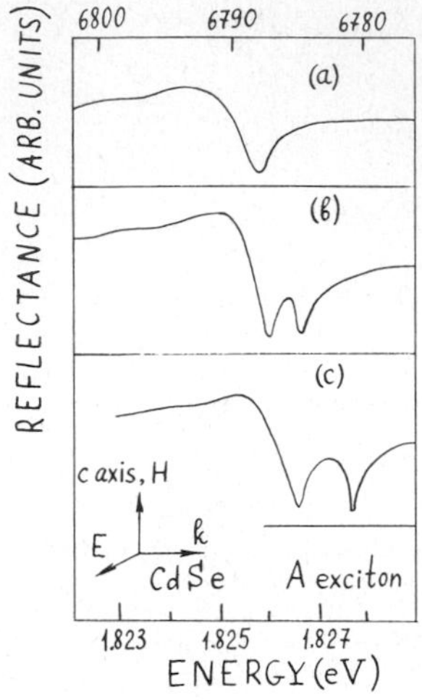

Fig. 15. Magnetoreflectance of CdSe for $\boldsymbol{H} \parallel \boldsymbol{c}$, $\boldsymbol{E} \perp \boldsymbol{c}$ and $\boldsymbol{k} \perp \boldsymbol{c}$ at 4 K. (a) $H = 0$; (b) $H = 6$ T; (c) $H = 10$ T (Venghaus et al. 1977).

transmission spectra of thin CdSe crystals in the same geometry $\boldsymbol{H} \parallel \boldsymbol{c}$, $\boldsymbol{k} \perp \boldsymbol{c}$, $\boldsymbol{E} \perp \boldsymbol{c}$ have been measured by Kochereshko et al. (1978). In the presence of a magnetic field two minima are observed in the spectra. They correspond to the two extrema $(\omega_L' \pm \frac{1}{2}\Omega_\parallel)$ in fig. 14.

7. The long-wavelength limit

As mentioned above, at frequencies far away from resonance the dielectric tensor can be expanded in powers of $\boldsymbol{k}$. In this case the phenomenological approach taking into account spatial dispersion corrections can be used for the calculation of optical spectra (Bokut' and Serdyukov 1971, Agranovich and Yudson 1972, 1973). The important point is that in order to satisfy the energy conservation law at the boundary it is necessary to use non-homogeneous Maxwell boundary conditions,

$$\boldsymbol{E}_{\mathrm{tg}}(-0) = \boldsymbol{E}_{\mathrm{tg}}(+0) + \tilde{\boldsymbol{E}}, \quad \boldsymbol{H}_{\mathrm{tg}}(-0) = \boldsymbol{H}_{\mathrm{tg}}(+0) + \tilde{\boldsymbol{H}}. \tag{61}$$

Here the vectors $\tilde{\boldsymbol{E}}$ and $\tilde{\boldsymbol{H}}$ are linear combinations of the spatial derivatives of the fields $\boldsymbol{E}$ and $\boldsymbol{H}$.

The physical meaning of additional terms in eq. (61) can be understood if we analyze the long-wavelength limit, where the expansion (56) is valid but, on the other hand, the inequality $(\omega_0 - \omega) \ll \omega_0$ still holds and eq. (44) is applicable.

For convenience, we first consider the model of an isolated isotropic exciton for the normal incidence. According to eq. (3), far below the resonance frequency ω_0 the dielectric constant is given by

$$\epsilon(\omega, k) = \epsilon(\omega) + Dk^2,$$

$$\epsilon(\omega) = \epsilon_0\left(1 + \frac{\omega_{LT}}{\omega_0 - \omega}\right), \qquad D = -\frac{\omega_{LT}\epsilon_0}{(\omega_0 - \omega)^2}\frac{\hbar}{2M}. \tag{62}$$

At normal incidence two modes 1 and 2 are excited in the crystal. One of the modes (branch 1 in fig. 1) is a propagating mode, slightly damped for non-zero exciton damping. The other mode is spatially damped, the wave-vector of this mode being purely imaginary. Using a set of ABC, we can express the amplitude of the wave 2, E_2, in terms of that of the wave 1, E_1, and substitute this into the right-hand sides of the Maxwell boundary conditions (27). Then these boundary conditions reduce to eq. (61) where the additional terms $\tilde{\boldsymbol{E}}$ and $\tilde{\boldsymbol{H}}$ arise from the terms E_2 and n_2E_2 in eq. (27). For the Pekar ABC in the case under consideration $\tilde{\boldsymbol{E}}$ and $\tilde{\boldsymbol{H}}$ become

$$\tilde{\boldsymbol{E}} = [D/\epsilon(\omega)]\nabla^2\boldsymbol{E}, \qquad \tilde{\boldsymbol{H}} = |n_2|[Dk_0/\epsilon(\omega)](\boldsymbol{N} \times \text{rot}\, \boldsymbol{H}), \tag{63}$$

where $\boldsymbol{E}$ and $\boldsymbol{H}$ are the electric and magnetic field vectors of the propagating wave 1. Thus, the additional terms in eq. (61) arise after eliminating the amplitudes of strongly damped additional light waves from the Maxwell boundary conditions.

For C_{6v} symmetry crystals in the geometry of fig. 11 the components of $\tilde{\boldsymbol{E}}$ and $\tilde{\boldsymbol{H}}$ can be written to first order in k in the following form

$$\tilde{E}_z = 0, \qquad \tilde{E}_x = q\frac{\bar{\gamma}}{\epsilon_{0\perp}}\frac{\partial E_z}{\partial x},$$

$$\tilde{H}_z = 0, \qquad \tilde{H}_x = (1-q)\frac{\bar{\gamma}}{\epsilon_{0\perp}}\frac{\partial H_z}{\partial x}, \tag{64}$$

where q is a real coefficient. Its value depends on the form of ABC. For the ABC $\boldsymbol{P}(0) = 0$, $Q(0) = 0$, it equals $\frac{1}{2}$; if $P_z(0) = 0$, $\partial \boldsymbol{P}_\perp(0)/\partial y = 0$, i.e. $u_\parallel = 0$ and $u_\perp \to \infty$ in eqs. (55), then $q = 1$.

The amplitude reflection coefficients in the orthogonal configurations $s \to p$ or $p \to s$ are given by

$$r_{ps} = r_{sp} =$$

$$(1-q)\frac{i\bar{\gamma}k_0 \sin 2\varphi}{((\epsilon_{0\parallel} - \sin^2\varphi)^{1/2} + \cos\varphi)((\epsilon_{0\perp} - \sin^2\varphi)^{1/2} + \epsilon_{0\perp}\cos\varphi)}. \tag{65}$$

According to eq. (65), r_{ps} and r_{sp} are zero at normal incidence and the orthogonal reflectance exhibits a maximum at $\varphi \simeq 45°$.

Acknowledgement

The author wishes to thank V.A. Kiselev, G.E. Pikus and A.V. Selkin for their careful reading of the manuscript and for helpful discussions.

References

Agarwal, G.S., D.N. Pattanayak and E. Wolf, 1974, Phys. Rev. **B10**, 1447.

Agranovich, V.M. and V.L. Ginzburg, 1966, Spatial Dispersion in Crystal Optics and the Theory of Excitons (Wiley, New York).

Agranovich, V.M. and V.I. Yudson, 1972, Optics Commun. **5**, 422.

Agranovich, V.M. and V.I. Yudson, 1973, Opt. Commun. **9**, 58.

Akhmediev, N.N. and V.V. Yatsishen, 1976, Fiz. tverd. Tela **18**, 1679 (Sov. Phys. Solid State, **18**, 975).

Akhmediev, N.N. and V.V. Yatsishen, 1978, Solid State Commun. **27**, 357.

Benemanskaya, G.V., B.V. Novikov and A.E. Cherednichenko, 1975, Zh. eksper. teor. Fiz., Pisma, **21**, 650 (JETP Lett. 21, 307 (1975)).

Benemanskaya, G.V., B.V. Novikov and A.E. Cherednichenko, 1977, Fiz. tverd. Tela **19**, 1389 (Sov. Phys.-Solid State **19**, 806 (1977)).

Bir, G.L. and G.E. Pikus, 1974, Symmetry and Strain-Induced Effects in Semiconductors (Keter Publishing House, Jerusalem) Sc. 40.

Birman, J.L. and J.J. Sein, 1972, Phys. Rev. **B6**, 2482.

Bishop, M.F., 1976, Solid State Commun. **20**, 779.

Bishop, M.F. and A.A. Maradudin, 1976, Phys. Rev. **B14**, 3384.

Bishop, M.F. and A.A. Maradudin, 1977, Solid State Commun. **23**, 507.

Bokut', B.V. and A.N. Serdyukov, 1971, Zh. eksper. teor. Fiz. **61**, 1808 (Sov. Phys.-JETP **34**, 962 (1972)).

Born, M. and Huang K., 1954, Dynamical Theory of Crystal Lattices (Oxford University Press, London).

Brodin, M.S. and S.I. Pekar, 1960, Zh. eksper. teor. Fiz. **38**, 74 (Sov. Phys.-JETP **11**, 55 (1960)).

Brodin, M.S. and S.I. Pekar, 1960a, Zh. eksper. teor. Fiz. **38**, 1910 (Sov. Phys.-JETP **11**, 1373 (1960)).

Brodin, M.S. and M.I. Strashnikova, 1962, Fiz. tverd. Tela **4**, 2454 (Sov. Phys.-Solid State **4**, 1798 (1963)).

Broser, I., M. Rozenzweig, R. Broser, M. Richard and E. Birkicht, 1978, Phys. Stat. Sol. (b) **90**, 77.

Casella, R.C., 1960, Phys. Rev. Lett. **5**, 371.

Cho, K., 1978, Solid State Commun. **27**, 305.

Davydov, A.S. and E.N. Myasnikov, 1974, Phys. Stat. Sol. (b) **63**, 325.

Davydova, N.A. and E.N. Myasnikov and M.I. Strashnikova, 1973, Fiz. tverd. Tela **15**, 3332 (Sov. Phys.-Solid State **15**, 2217 (1974)).

Davydova, N.A., E.N. Myasnikov and M.I. Strashnikova, 1974, Fiz. tverd. Tela **16**, 1173 (Sov. Phys.-Solid State **16**, 752 (1974)).

Evangelisti, F., A. Frova and J.U. Fischbuch, 1972, Phys. Rev. Lett. **29**, 1001.

Evangelisti, F., A. Frova and J.U. Fischbuch, 1974, Phys. Rev. **B9**, 1516.

Evangelisti, F., A. Frova and F. Patella, 1974, Phys. Rev. **B10**, 4253.

Fedorov, F.I., 1959, Optika i Spectroskopiya **6**, 377 (Opt. Spectrosc. **6**, 237).

Fedorov, F.I., B.V. Bokut' and A.F. Konstantinova, 1962, Kristallografiya **7**, 910 (Sov. Phys. Cryst. **7**, 748).
Gorban', I.S. and V.B. Timofeev, 1961, Doklady Akad. Nauk SSSR **141**, 791 (Sov. Phys. Doklady **6**, 871).
Halevi, P. and R. Fuchs, 1978, Proc. XIV Int. Conf. Phys. Semicond. (Edinburgh, 1978) p. 863.
Hopfield, J.J. and D.G. Thomas, 1961, Phys. Rev. **122**, 35.
Hopfield, J.J. and D.G. Thomas, 1963, Phys. Rev. **132**, 563.
Ivchenko, E.L., S.A. Permogorov and A.V. Selkin, 1978, Solid State Commun. **28**, 345.
Ivchenko, E.L., S.A. Permogorov and A.V. Selkin, 1978a, Zh. eksper. teor. Fiz., Pisma **27**, 27 (JETP Lett. **27**, 24 (1978)).
Ivchenko, E.L., S.A. Permogorov and A.V. Selkin, 1978b, Zh. eksper. teor. Fiz., Pisma **28**, 649 (JETP Lett. **28**, 599 (1978)).
Ivchenko, E.L. and A.V. Selkin, 1979, Zh. eksper. teor. Fiz. **76**, 1837 (Sov. Phys. JETP **49**).
Kiselev, V.A., 1978, Fiz. tverd, Tela **20**, 2173 (Sov. Phys. Solid State **20**, 1255).
Kiselev, V.A., 1979, Zh. eksper. teor. Fiz., Pisma **29**, 369 (JETP Lett. 29, 332 (1979)).
Kiselev, V.A., 1979a, Fiz. tverd. Tela **21**, 1069 (Sov. Phys. Solid State, **21**, 621).
Kiselev, V.A., B.S. Razbirin and I.N. Uraltsev, 1973, Zh. eksper. teor. Fiz., Pisma **18**, 504 (JETP Lett. **18**, 296 (1973)).
Kiselev, V.A., B.S. Razbirin and I.N. Uraltsev, 1974, Proc. XII Int. Conf. Phys. Semicond. (Stuttgart, 1974) p. 996.
Kiselev, V.A., B.S. Razbirin, I.N. Uraltsev and V.P. Kochereshko, 1975, Fiz. tverd. Tela **17**, 640 (Sov. Phys.-Solid State **17**, 418 (1975)).
Kiselev, V.A., B.S. Razbirin and I.N. Uraltsev, 1975a, Phys. Stat. Sol. (b) **72**, 161.
Kiselev, V.A., I.V. Makarenko, B.S. Razbirin and I.N. Uraltsev, 1977, Fiz. tverd. Tela **19**, 1348 (Sov. Phys.-Solid State 19, 1374 (1977)).
Kiselev, V.A., I.V. Makarenko, B.S. Razbirin and I.N. Uraltsev, 1977a, Zh. eksper. teor. Fiz., Pisma **26**, 352 (JETP Lett. **26**, 234 (1977)).
Kochershko, V.P., B.S. Razbirin and I.N. Uraltsev, 1978, Zh. eksper. teor. Fiz., Pisma **27**, 285 (JETP Lett. **27**, 266 (1978)).
Komarov, A.V., S.M. Ryabchenko and M.I. Strashnikova, 1978, Zh. eksper. teor. Fiz. **74**, 251 (Sov. Phys.-JETP **47**, 128 (1978)).
Konstantinov, O.V. and Sh. R. Saifullaev, 1976, Fiz. tverd. Tela **18**, 3433 (Sov. Phys. Solid State, **18**, 1998).
Konstantinov, O.V. and Sh. R. Saifullaev, 1978, Fiz. tverd. Tela **20**, 1745 (Sov. Phys. Solid State, **20**, 1010).
Konstantinov, O.V., M.M. Panakhov and Sh. R. Saifullaev, 1975, Fiz. tverd. Tela **17**, 3551 (Sov. Phys.-Solid State **17**, 2315 (1975)).
Mahan, G.D. and J.J. Hopfield, 1964, Phys. Rev. **A135**, 428.
Maradudin, A.A. and D.L. Mills, 1973, Phys. Rev. **B7**, 2787.
Mashlyatina, T.M., D.S. Nedzvetskii and A.V. Selkin, 1978, Zh. eksper. teor. Fiz., Pisma **27**, 573 (JETP Lett. **27**, 539 (1978)).
Mashlyatina, T.M., D.S. Nedzvetskii and L.E. Solov'ev, 1979, Fiz. tverd. Tela **21**, 2040 (Sov. Phys. Solid State, **21**, 1169).
Patella, F., F. Evangelisti and M. Capizzi, 1976, Solid State Commun. **20**, 23.
Pekar, S.I., 1957, Zh. eksper. teor. Fiz. **33**, 1022 (Sov. Phys.-JETP **6**, 785 (1958)).
Pekar, S.I., 1958, Zh. eksper. teor. Fiz. **34**, 1176 (Sov. Phys.-JETP **7**, 813 (1958)).
Pekar, S.I. and M.I. Strashnikova, 1975, Zh. eksper. teor. Fiz. **68**, 2047 (Sov. Phys.-JETP **41**, 1024 (1975)).
Permogorov, S.A. and A.V. Selkin, 1973, Fiz. tverd. Tela **15**, 3025 (Sov. Phys.-Solid State **15**, 2015 (1974)).
Permogorov, S.A., V.V. Travnikov and A.V. Selkin, 1972, Fiz. tverd. Tela **14**, 3642 (Sov. Phys.-Solid State **14**, 3051 (1973)).

Permogorov, S.A., A.V. Selkin and V.V. Travnikov, 1973, Fiz. tverd. Tela **15**, 1822 (Sov. Phys.-Solid State **15**, 1215 (1973)).
Pikus, G.E. and G.L. Bir, 1973, Fiz. i tekhn. poluprovodnikov **7**, 119 (Sov. Phys. Semicond. **7**, 81).
Rashba, E.I., 1961, Proc. Int. Conf. Semicond. Phys. (Prague, 1960) p. 45.
Rashba, E.I. and V.I. Sheka, 1959, Fiz. tverd. Tela, collection No. 2, 162.
Reynolds, D.S., R.N. Euwema and T.C. Collins, 1968, Proc. IX Int. Conf. Phys. Semicond. (Moscow, 1968) vol. 1, p. 210.
Röseler, J. and K. Henneberger, 1979, Phys. Stat. Sol. (b) **93**, 213.
Sakoda, S., 1976, J. Phys. Soc. Japan **40**, 152.
Selkin, A., 1977, Phys. Stat. Sol. (b) **83**, 47.
Sheka, V.I., 1965, Fiz. tverd. Tela **7**, 1783 (Sov. Phys.-Solid State **7**, 1437 (1965)).
Skaistys, E. and V.I. Sugakov, 1974, Lit. fiz. sbornic **14**, 297.
Solov'ev, L.E. and A.V. Babinskii, 1976, Zh. eksper. teor. Fiz., Pisma **23**, 291 (JETP Lett. **23**, 263 (1976)).
Stössel, W. and H.J. Wagner, 1978, Phys. Stat. Sol. (b) **89**, 403.
Ting, C.S., M.J. Frankel and J.L. Birman, 1975, Solid State Commun. **17**, 1285.
Tolpygo, K.B., 1950, Zh. eksper. teor. Fiz. **20**, 497.
Ulbrich, R. and C. Weisbuch, 1977, Phys. Rev. Lett. **38**. 865.
Venghaus, H., S. Suga and K. Cho, 1977, Phys. Rev. **B16**, 4419.
Voigt, J., and G. Mauerberger, 1973, Phys. Stat. Sol. (b) **60**, 679.
Winterling, G., and E. Koteles, 1977, Solid State Commun. **23**, 95.
Zeyher, R., J.L. Birman and W. Brenig, 1972, Phys. Rev. **B6**, 4613.

CHAPTER 5

Optical Emission due to Exciton Scattering by LO Phonons in Semiconductors

S. PERMOGOROV

A.F. Ioffe Physical-Technical Institute
Leningrad, 194021
USSR

Excitons
Edited by
E.I. Rashba and M.D. Sturge

Contents

1. *Introduction*

In this paper we shall discuss those secondary emission processes in semiconductors which are essentially based on exciton scattering by longitudinal optical (LO) phonons. Free excitons serve as the intermediate states (real or virtual) for the most of the optical processes in the spectral range of the fundamental absorption edge. Energy relaxation of these intermediate states takes place through phonon scattering. In polar semiconductors the excitons are most strongly scattered by LO phonons.

Among the emission processes due to exciton–phonon interaction we shall first discuss the phonon-assisted luminescence of free excitons. An important property of the corresponding emission lines is that their spectral shape reflects the kinetic energy distribution of free excitons. Thus the information on the thermalization of exciton system and its temperature can be obtained.

Another optical phenomenon, also involving the exciton intermediate states, is preresonant Raman light scattering. Since in Raman experiments various excitation and scattering geometries can be used, detailed information on the exciton–phonon interaction can be obtained from Raman spectra.

Finally, we shall discuss the LO relaxation of crystal electronic excitations which are created by light absorption in the fundamental region. This relaxation process leads to the appearance of multiphonon LO Raman scattering in the fundamental absorption. On the other hand, the same process of hot exciton relaxation, as can be concluded from the oscillatory structure in the luminescence and photoconductivity excitation spectra, forms a very important mechanism for the transformation of the optical energy absorbed by the crystal.

2. *Polar interaction of excitons with LO phonons in semiconductors*

A characteristic feature of the optical processes in the spectral region of the exciton resonance in polar semiconductors is the predominance of interaction with LO phonons as compared with phonons of other types. It can be concluded that the interaction of the principal intermediate states for these processes (i.e. free excitons) with phonons is mainly due to the phonon's macroscopic electric field and can be described as Fröhlich polar intraband scattering (Fröhlich 1954). For other mechanisms of exciton–phonon interaction (e.g. for deformation potential scattering), both LO and TO phonons can be observed in the optical spectra with comparable strength. In what follows we shall restrict our analysis of the experimental data to the Fröhlich model of polar exciton–LO phonon interaction. It will be shown that this model well accounts for the main features of the observed phenomena.

Even in the strongly ionic semiconductors, such as ZnO and CdS, Fröhlich exciton–phonon polar interaction can be regarded as "weak" since the corresponding coupling constant is usually smaller than unity. However, optical processes including up to ten LO phonons can be observed in the exciton region. The unusually high strength of multiphonon processes in this case is connected with the existence of multiple resonances for the intermediate scattering states (Martin and Falicov 1975). Such resonances in the scattering cross section can appear when the Nth order multiphonon process can be described as a successive repetition of several one-phonon interactions with the inclusion of $(N-1)$ additional intermediate states. An exact resonance for all these intermediate states can be achieved in principle only for Raman scattering in the energy region of fundamental absorption. However, multiple resonances are important in preresonant scattering also. When the exciting frequency approaches the exciton resonance, the intensity of multiphonon processes increases the faster, the greater the number of preresonant intermediate states, i.e. the higher the scattering order. As a result, multiphonon lines of high order can be observed in the resonant Raman scattering with unexpectedly high intensity. The model of successive multiphonon interaction has been confirmed experimentally by Oka and Kushida (1972) in the study of the frequency dependence of preresonant Raman 2LO scattering in CdS and ZnO. They have found that the resonant enhancement of the scattering cross section in this case is a stronger function of frequency than that expected for the exciton–phonon interaction quadratic in phonon amplitudes.

In order to understand the main peculiarities of the emission processes caused by LO scattering of excitons it is necessary to consider the explicit dependence of the exciton–phonon scattering matrix element on the scattering wavevector $\boldsymbol{q}$ (Bulyanitsa 1970). If an exciton with wavevector $\boldsymbol{K}_1$ and wavefunction ψ_{λ_1} is scattered from the quantum state λ_1 into the quantum state λ_2 with wavevector $\boldsymbol{K}_2$ and wavefunction ψ_{λ_2} by the LO phonon with wavevector $\boldsymbol{q} = \boldsymbol{K}_1 - \boldsymbol{K}_2$, the matrix element $H_{\mathrm{EL}}^{\lambda_1\lambda_2}$ for such scattering is proportional to

$$H_{EL}^{\lambda_1\lambda_2} \propto \frac{g\Omega_{LO}}{a_0 q}\langle\psi_{\lambda_2}, (\exp(i\mathbf{q}_c\mathbf{r}) - \exp(-i\mathbf{q}_v\mathbf{r}))\psi_{\lambda_1}\rangle, \tag{1}$$

where g is the Fröhlich exciton–phonon coupling constant given by

$$g^2 = \frac{1}{\hbar\Omega_{LO}}\frac{e^2}{2\epsilon_0 u}\left(\frac{\epsilon_0}{\epsilon_\infty} - 1\right) = \left(\frac{R}{\hbar\Omega_{LO}}\right)^{1/2}\left(\frac{\Omega_{LO}^2}{\Omega_{TO}^2} - 1\right). \tag{2}$$

Here $\hbar\Omega_{LO(TO)}$ is the energy of LO(TO) phonon; R is the exciton binding energy, $R = e^2/2\epsilon_0 a_0$; e is the free electron charge; ϵ_0 and ϵ_∞ are the static and high frequency dielectric constants; a_0 is the exciton Bohr radius;

$$u = \left(\frac{\hbar}{2\mu\Omega_{LO}}\right)^{1/2}; \qquad \mathbf{q}_{c,v} = \frac{\mu}{m_{c,v}}\mathbf{q}, \tag{3}$$

where $m_{c,v}$ are the carrier effective masses for the bands from which the exciton is formed, and $\mu = m_c m_v/(m_c + m_v)$ is the exciton reduced mass.

The behaviour of the matrix element (1) depends critically on the symmetry of the exciton states λ_1 and λ_2. For states of the same parity (e.g. scattering within the same band 1S–1S or scattering 1S–2S, 3S etc.) the exciton–phonon scattering has forbidden character, i.e. $H_{EL}^{\lambda_1\lambda_2} \to 0$ as $q \to 0$. As it was shown by Anselm and Firsov (1956) the matrix element for the 1S–1S scattering is given by

$$H_{EL}^{1S1S} \propto \frac{g\Omega_{LO}}{a_0 q}\left\{\frac{1}{[1 + (a_0 q_c/2)^2]^2} - \frac{1}{[1 + (a_0 q_v/2)^2]^2}\right\}. \tag{4}$$

For small $q (a_0 q \ll 1)$ the matrix element H_{EL}^{1S1S} can be approximated as

$$H_{EL}^{1S1S} \propto \left(\frac{m_v - m_c}{m_v + m_c}\right) a_0 q. \tag{5}$$

It reaches a maximum at $a_0 q \sim 1$ and rapidly decreases at larger q. All the matrix elements for scattering between states of the same parity nS–nS, nP–nP, etc. have a similar dependence on q. However, the maximum value of the matrix element in these cases is reached at the reciprocal radii of the corresponding exciton states a_n,

$$q \sim 1/na_n; \qquad a_n = na_0. \tag{6}$$

In the case of equal carrier masses ($m_v = m_c$) one-phonon Fröhlich exciton scattering is totally forbidden.

For scattering between the exciton states of different parity (e.g. 1S–2P) the matrix element $H_{EL}^{\lambda_1\lambda_2}$ has a nonzero value at $q = 0$, and the scattering can be regarded as allowed. However, in this case $H_{EL}^{\lambda_1\lambda_2}$ also decreases rapidly at $q > 1/a_0$. The dependence of $H_{EL}^{\lambda_1\lambda_2}$ on q is shown in fig. 1 for several possible cases.

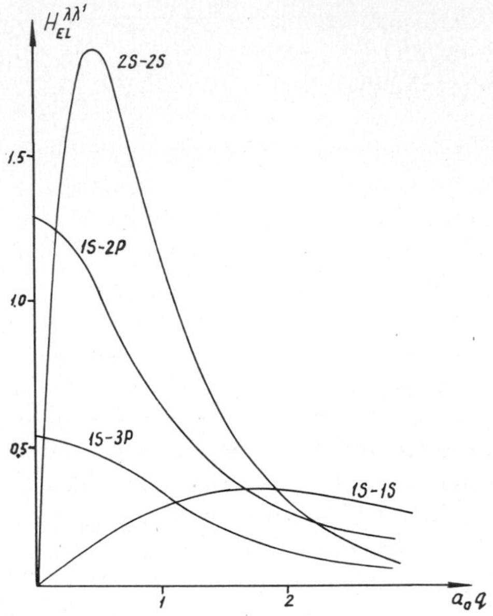

Fig. 1. The dependence of Fröhlich exciton–phonon scattering matrix elements $H_{EL}^{\lambda\lambda'}$ on the reduced phonon wavevector $a_0 q$ (a_0 is the exciton Bohr radius) as calculated in accordance with Bulyanitsa (1970) for the scattering between the different exciton states λ and λ'. The relation of the effective masses $m_v/m_c = 6$ corresponds to the case of CdS crystals.

3. *Multiphonon luminescence of free excitons*

Under the continuous optical excitation a stationary population of the exciton states can be created in the crystal. Due to the free motion in the crystal and the interaction with the crystal lattice the gas of free excitons is spread over some region of kinetic energies. However, usually only the lowest $n = 1S$ exciton band is populated at low temperatures.

The luminescent emission of free excitons can take place either in resonance with the exciton absorption line (a so called zero-phonon luminescence) or can be shifted in energy due to the simultaneous creation of phonons (phonon-assisted luminescence). In the resonant exciton emission only excitons with small wavevectors of the order of the photon wavevector $K \approx 0$ can participate. In semiconductors with direct allowed interband transitions the resonant exciton luminescence can be properly described only when strong exciton–photon coupling is taken into account (Toyozawa 1959, Agranovich 1960, Bonnot and Benoit a la Guillaume 1970, Weiher and Tait 1972). Due to this coupling mixed exciton–photon states (polaritons) are formed in the region of exciton resonance (Hopfield 1958).

The polariton emission spectrum is strongly influenced by spatial dispersion effects (Permogorov and Selkin 1973), by the energy distribution function in the region of the polariton anticrossing (Sumi 1977) and also by the spatial distribution of polaritons inside the crystal (Weisbuch and Ulbrich 1979). When the polariton concept is used, the resonant exciton luminescence cannot be considered, strictly speaking, as a zero-phonon process. First, the energy distribution in the region of polariton bottleneck is determined mainly by radiation transfer and by interaction with acoustic phonons (Wiesner and Heim 1975). Furthermore, an essential part of the exciton resonant emission is the result of inter-branch polariton scattering by acoustic phonons. The complex problem of resonant polariton emission is far beyond the scope of the present paper and will not be discussed further. A part of it, namely the polariton scattering by acoustic phonons, is reviewed by E. Koteles in this book (ch. 3).

Phonon-assisted luminescence of free excitons in polar semiconductors is mainly due to the creation of LO phonons. The elementary luminescence process for the exciton state with initial energy E_i and wavevector K_i obeys the following conservation laws:

$$E_i - \hbar\omega_s = \sum_n^N \hbar\Omega_n, \tag{7}$$

$$\mathbf{K}_i - \mathbf{K}_s = \sum_n^N q_n, \tag{8}$$

where $\hbar\omega_s$ and $K_s \approx 0$ are the energy and the wavevector of the emitted photon; $\hbar\Omega_n$ and q_n are energies and wavevectors of LO phonons, and N is the number of created LO phonons, which will be called the order of the emission process.

It follows from the wavevector conservation law (8) that excitons with any value of K_i can participate in phonon-assisted luminescence. Since the energy of LO phonons has a weak dependence on wavevector q, the spectrum of emitted photons $\hbar\omega_s$ is simply related to the initial distribution of exciton energies E_i.

Figure 2 shows the spectrum of free exciton luminescence in CdS crystals at $T = 30$ K due to creation of 1, 2, 3 and 4 LO phonons. The shape of the phonon-assisted lines in this spectrum corresponds with good accuracy to the equilibrium Maxwell distribution of the exciton kinetic energies

$$F(\epsilon) \propto \epsilon^{1/2} \exp(-\epsilon/kT), \tag{9}$$

where $\epsilon = E_i - E_0$ is the kinetic energy of the exciton in a band with a bottom energy E_0. The first factor in (9) is the density of states in the parabolic exciton band neglecting some minor corrections caused by

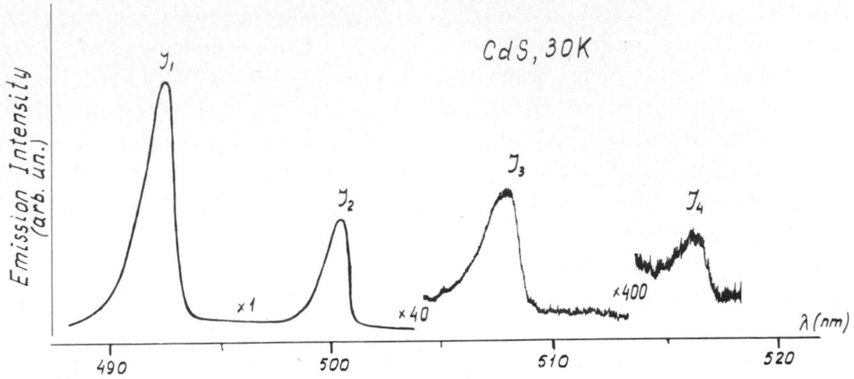

Fig. 2. The spectrum of LO phonon assisted exciton luminescence in CdS crystals at $T = 30$ K. The lines J_1, J_2, J_3 and J_4 correspond to emission processes with the simultaneous creation of 1, 2, 3 and 4LO phonons. Note the changes in gain (Klochikhin et al. 1976a).

polariton dispersion. It has been shown (Tait et al. 1967) that these corrections are essential only at very low temperatures, when $kT \approx \Delta E_{LT}$ (ΔE_{LT} is the energy of exciton longitudinal–transverse splitting, typically 1–2 meV). The second factor in (9) reflects the exciton statistics, which can be taken to be Boltzmann at low exciton densities. The temperature T in the exponent of (9) corresponds to the crystal lattice temperature, demonstrating the existence of thermal equilibrium in the exciton system. Equilibrium distribution of exciton kinetic energies with the temperature of crystal lattice has been also observed in many other semiconducting materials. A review of early studies of phonon-assisted exciton luminescence in semiconductors was given by Gross et al. (1971).

In the case of thermal equilibrium in the exciton system, the spectral shape of phonon-assisted lines can be described by the general formula

$$J_N(\epsilon) = \epsilon^{1/2} \exp(-\epsilon/kT) W_N(\epsilon), \tag{10}$$

where $W_N(\epsilon)$ stands for the probability of Nth order phonon-assisted annihilation for the exciton with the kinetic energy ϵ.

The dependence of W_N on ϵ can be found by fitting the experimental spectra with eq. (10). However, a more suitable method is the study of the temperature dependence of the linewidths and the temperature dependence of the relative integrated intensities for the lines of different order. If W_N is a simple power function of the energy $W_N \propto \epsilon^L$, the halfwidth of eq. (10) depends linearly on temperature with a slope unambiguously related to the exponent L. The shift Δ of the line maximum from its low-energy threshold ($\epsilon = 0$) also increases linearly with temperature:

$$\Delta = (L + \tfrac{1}{2})kT. \tag{11}$$

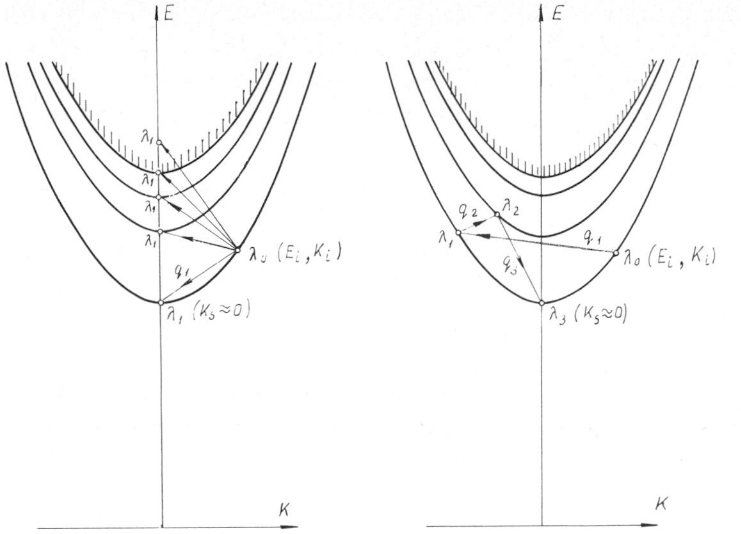

Fig. 3. A schematic representation of exciton scattering between intermediate states which leads to the appearance of the phonon-assisted exciton luminescence in the framework of perturbation theory. The point λ_0 represents the initial state of a thermalized exciton (E_i, K_i). For the case of one-phonon luminescence (left part) the different possible intermediate states λ_1 ($K_s \approx 0$) in the discrete and continuum exciton spectrum are shown. The right part shows the three-phonon scattering through the intermediate states λ_1, λ_2 and λ_3. For the calculation of the luminescence probability (13) summation over the quantum numbers of intermediate states λ_n and integration over the phonon wavevectors q_n is necessary.

If $W_{N_1} \propto \epsilon^{L_1}$ and $W_{N_2} \propto \epsilon^{L_2}$, the line integrated intensities I_N are related by

$$S = I_{N_1}/I_{N_2} \propto T^{(L_1-L_2)}. \quad (12)$$

These simple relations also permit us to find $W_N(\epsilon)$ in a more general form.

A theoretical expression for $W_N(\epsilon)$ can be obtained from the microscopic treatment of the exciton emission due to the scattering by LO phonons. In fact this process is inverse to the phonon-assisted direct exciton absorption which was theoretically studied by Thomas et al. (1960) and Segall (1966, 1967). In accordance with perturbation theory one should consider the scattering of the initial exciton state $\lambda_0 = 1S$ (E_i, K_i) through the intermediate states of discrete and continuum exciton spectrum $\lambda_1, \lambda_2, \ldots, \lambda_N$ with the emission of N LO phonons. Radiative annihilation of the exciton takes place from the last intermediate state λ_N with appropriate wavevector $K_{\lambda_N} = K_s \approx 0$. A schematic representation of the phonon-assisted exciton emission is given in fig. 3.

A general expression for the scattering cross section can be written as

$$W_N = \int d^3q_1\, d^3q_2 \ldots d^3q_{N-1} |A_N|^2, \tag{13}$$

$$A_N = \sum_{\lambda_1,\ldots,\lambda_N} \frac{H_{ER}^{\lambda_N} H_{EL}^{\lambda_N\lambda_{N-1}} \cdots H_{EL}^{\lambda_1\lambda_0}}{(\hbar\omega_s - E_{\lambda_N})(E_i - (N-1)\hbar\Omega - E_{\lambda_{N-1}}) \ldots (E_i - \hbar\Omega - E_{\lambda_1})}, \tag{14}$$

where $H_{EL}^{\lambda\lambda'}$ are the Fröhlich scattering matrix elements between the exciton states $\lambda(E_\lambda)$ and $\lambda'(E_{\lambda'})$ discussed in the previous section; $H_{ER}^{\lambda_N}$ is the matrix element for the radiative transition from the last intermediate state $\lambda_N(E_{\lambda_N}, K_s \approx 0)$ to the crystal ground state. Summation in eq. (14) is over all states of discrete exciton bands $\lambda_n(K_n)$ and over the continuum exciton spectrum. For the calculation of the scattering cross section (13) the possible permutations of the phonon wavevectors q_n in the scattering amplitude (14) should be taken into account. As a result, the cross section (13) consists of $N!$ topologically nonequivalent terms. Integration in (13) is over the $(N-1)$ independent phonon wavevectors q_n. The expression for the scattering cross section (13), which corresponds to the emission probability $W_N(\epsilon)$, can be used for the analysis of the phonon-assisted line shapes $J_N(\epsilon)$.

The experimental studies of exciton luminescence in CdS (Gross et al. 1971), ZnO (Weiher and Tait 1968) and CdSe (Abramov et al. 1970) have shown that the spectral shape of 1 LO luminescence line can, with sufficient accuracy, be approximated by:

$$J_1(\epsilon) \propto \epsilon^{3/2} \exp(-\epsilon/kT), \tag{15}$$

which shows that $W_1(\epsilon) \propto \epsilon$. This result implies that the main intermediate state for the process of 1 LO luminescence is $\lambda_1 = 1S$ $(K \approx 0)$.

In the case of 1 LO scattering the probability (13) can be rewritten as

$$W_1 \propto \left| \sum_{\lambda_1} \frac{\psi_{\lambda_1}(0) H_{EL}}{(E_i - \hbar\Omega - E_{\lambda_1})} \right|^2, \tag{16}$$

where $|\psi_{\lambda_1}(0)|^2$ is the oscillator strength for the optical transition from the state λ_1 in dipole approximation. For allowed transitions $|\psi_{\lambda_1}(0)|^2$ is maximal for 1S state. The energy denominator in (16) at low temperatures has also the minimal value for $\lambda_1 = 1S$. The square of the scattering matrix element $|H_{EL}^{1S1S}|^2$ is proportional to q^2 for small q. Taking into account wavevector conservation (8) and the parabolic dispersion of the exciton band one can obtain

$$|H_{EL}^{1S1S}|^2 \propto q_i^2 \approx K_i^2 \propto \epsilon. \tag{17}$$

Since for the small exciton kinetic energies ($\epsilon \ll \hbar\Omega$) the energy denominator in (16) has a weak dependence on the energy, the total energy dependence of W_1 for $1S-1S$ scattering can be approximated by $W_1(\epsilon) \propto \epsilon$, which is in accordance with the experiment. However, at high temperatures or for small $\hbar\Omega$ the energy denominator in (16) can markedly change the energy dependence of W_1.

While scattering through 1S state dominates the 1LO process, exact calculations of the exciton–phonon absorption (Segall 1966) have shown that the inclusion of other intermediate states improves the quantitative agreement with the experiment.

The experimental study of the multiphonon ($N > 1$) free exciton luminescence lines in CdS and CdSe (Klochikhin et al. 1976a) has shown that the spectral shape of these lines in general form can be described as:

$$J_N \propto B_N \epsilon^{1/2} \exp(-\epsilon/kT)[1 + \gamma_N \epsilon + \cdots]. \qquad (18)$$

Here the first term in square brackets represents the part of emission probability which is independent of the exciton kinetic energy, whereas the second term gives the dependence of the emission probability on the exciton kinetic energy ϵ, or the square of exciton wavevector K_i^2. Within the experimental precision no terms with higher powers in ϵ have been detected.

For the two-phonon exciton annihilation it has been found that W_2 is practically independent of the exciton kinetic energy (Gross et al. 1971), which means $\gamma_2 = 0$ in (18). This independence, as was shown by Klochikhin et al. (1976a), is caused by the inclusion of sequences of intermediate states with alternating parity (e.g. 1S – 2P – 1S) in the 2LO scattering process. The large contribution of such sequences to the scattering probability is a consequence of the high oscillator strength for the last intermediate state (1S) and the allowed character of 1S–2P and 2P–1S Fröhlich scattering. In a rigorous calculation of emission probability other nS–nP scattering processes should be also taken into account.

Since the matrix elements for allowed scattering do not depend strongly on the scattering wavevector in the region of small q, the probability of the two-phonon annihilation can be considered to be independent of wavevector and energy ($W_2(\epsilon) = \text{const}$). The energy denominators in (14) for the 2 LO process have an even weaker dependence on K_i than in the first order process.

The fact that the two-phonon annihilation probability is independent of the exciton kinetic energy is of great experimental importance. First of all, the integrated intensity of the corresponding emission line is proportional to the exciton concentration in the crystal and can be used for its measurement. The integrated intensity of the 2LO line (I_2) is also very convenient for the normalization of intensities of other emission lines, and we shall take $B_2 = 1$. Secondly, the shape of 2LO line directly reproduces the energy distribution of excitons, which is very important for the study of the nonequilibrium distributions.

Table 1
Relative integrated intensities of phonon-assisted luminescence lines (I_N) in the low temperature exciton luminescence spectra of CdS, CdSe, ZnO, LiD and LiH crystals. All the intensities are normalized to the second order intensity I_2. Two last columns list the shape parameters γ_3 and γ_4 of eq. (18) for the $N = 3$ and $N = 4$ lines, respectively. In accordance with eq. (19) the γ_N values are measured in reciprocal Kelvins (1/K).

	I_1	I_2	I_3	I_4	γ_3	γ_4
CdS [a]	0.7	1	0.015	0.0017	0.034	<0.005
CdSe [a]	1.2	1	0.016	0.0017	0.041	<0.005
ZnO [b]	0.1	1	0.06	0.015	0.023	0.014
LiD [c]	0.68	1	0.015	0.011	—	—
LiH [d]	0.41	1	0.0065	0.0033	—	—

[a] Klochikhin et al. (1976a) $T = 4.2$ K.
[b] Original data, $T = 4.2$ K.
[c] O'Connel-Bronin and Plekhanov (1979) $T = 78$ K.
[d] Plekhanov and O'Connel-Bronin (1978) $T = 78$ K.

Table 1 presents the relative intensities of the phonon-assisted exciton luminescence lines in the low temperature spectra of CdS, CdSe, ZnO and LiH crystals. Although lithium hydride is a wide gap insulator and cannot be considered as a semiconductor, the strength of the exciton–phonon coupling in this material is surprisingly close to that in II–VI compounds. As a result, the multiphonon exciton luminescence spectra of LiH exhibit a close similarity to those of II–VI semiconductors. Two last columns of table 1 shows the available data on γ_N parameters. If we assume that $B_2 = 1$ and $\gamma_2 = 0$ in eq. (18) for the 2LO line, the relative temperature dependence of the integrated intensities I_N for $N > 2$ will be given by,

$$S_N = I_N/I_2 = B_N(1 + \tfrac{3}{2}\gamma_N T), \qquad (19)$$

which makes it possible to measure the γ_N values. By the definition of eq. (19) the γ_N values are measured in reciprocal Kelvin degrees $1/K$.

As has been mentioned earlier, the 1LO exciton luminescence has a strongly forbidden character with emission probability proportional to the exciton kinetic energy. Now we shall briefly discuss how the dependence on the exciton kinetic energy enters the emission probability for the multiphonon processes.

If the scattering amplitude (14) contains only the matrix elements for allowed Fröhlich scattering, its contribution to the emission probability will have negligible dependence on the wavevector and hence on the energy of the initial state. Such situation apparently occurs for the 2LO emission process. Due to inclusion of the allowed scattering, the intensity of the second order line I_2 is over a wide temperature range, of the same order as that of the first order line I_1, and even exceeds it at low temperatures.

The inclusion of one or more forbidden scattering matrix elements into the amplitude (14) will give it a strong wavevector dependence. However, in the emission probability this dependence will be smoothed to a great extent after the integration over the phonon wavevectors q_n in eq. (13). Nevertheless, since the conservation law for the q_n (eq. (8)) holds, the resulting probability will preserve the dependence on $\boldsymbol{K}_i - \boldsymbol{K}_s \approx \boldsymbol{K}_i$. This relatively weak dependence can be clearly detected in the spectral shape of the multiphonon lines at elevated temperatures.

It can be seen from table 1, that for all the crystals studied γ_3 is higher than γ_4. From simple considerations it can be seen that for even-order scattering processes, sequences of intermediate states with alternating parity are possible including the last intermediate state $\lambda_N = 1S(K \approx 0)$ which has high oscillator strength. Such sequences cannot be formed in odd-order processes. As a result, the relative contribution of the wavevector dependent forbidden scattering is greater in the latter case, which increases the corresponding γ_N values.

The intensities of the multiphonon luminescence lines strongly decrease with increasing order. The dependence of the relative intensities on order is caused mostly by the increasing denominators in eq. (14) for the multiphonon processes, and also by the increasing power of the coupling constant g^{2N}. The decrease is partly compensated by the increase of the number of permutation terms in the scattering cross section (13). However, no simple dependence of the intensity distribution on the order N can be established, unlike the case of the phonon repetitions in the vibronic spectra of localized centers.

The general distribution of the relative intensities in the phonon-assisted exciton luminescence spectra of table 1 is similar for the different compounds. Nevertheless, characteristic differences in the intensity distribution can be noted. This shows that the multiphonon exciton luminescence spectra can be used, in principle, for the determination of the crystal parameters relevant to the Fröhlich scattering. However, it should be mentioned that the theoretical evaluation of $W_N(\epsilon)$ for the multiphonon processes is very difficult since multiple summation over the intermediate states λ_n and multiple integration over the phonon wavevectors q_n are involved.

4. Resonant Raman scattering below the exciton resonance

A characteristic feature of the resonant Raman scattering is a sharp selective enhancement of certain phonon lines in comparison with other phonons due to inclusion of new scattering mechanisms. In polar crystals

Table 2
Relative integrated intensities of preresonant Raman scattering lines (I_N) in CdS and LiD crystals. All the intensities are normalized on the second order intensity I_2. Δ is the shift of the exciting energy $\hbar\omega_i$ below the exciton resonance E_0.

	I_1	I_2	I_3	I_4
CdS [a]				
$\Delta = 12$ meV	0.064	1	0.013	0.0016
LiD [b]				
$\Delta = 40$ meV	0.04	1	0.015	0.011

[a] Klochikhin et al. (1976a).
[b] O'Connel-Bronin and Plekhanov (1979).

with well developed exciton structure in the fundamental absorption edge this selective enhancement is usually observed for LO scattering. The steep increase of LO scattering intensity near the exciton resonance has been observed in CdS and ZnO (Oka and Kushida 1972, Martin and Damen 1971, Callendar et al. 1973), GaSe (Reydelet et al. 1976), HgI_2 (Goto and Nishina 1978), etc. It can be concluded from these experimental results that the main intermediate states for the resonant Raman scattering are the free excitons which are scattered by LO phonons through the Fröhlich interaction.

If the energy shift of the exciting laser line from the exciton resonance is of order of, or less than the exciton binding energy R, the multiphonon LO scattering processes can be observed with relatively high intensity. For example, in the case of CdS and LiD crystals under the proper resonant excitation the Raman scattering processes with the creation of up to 4LO phonons were detected. As it can be seen from table 2 the distribution of the multiphonon Raman intensities is similar to that in the multiphonon exciton luminescence spectrum. From a general point of view this fact illustrates that exciton luminescence is the limiting case of resonant Raman scattering. In the microscopic approach this fact shows that the same scattering mechanisms and intermediate states are important in both phenomena.

The energy and wavevector conservation laws for the Raman scattering have the same formulation, as in the case of phonon-assisted exciton luminescence (7) and (8). However, the initial state for the Raman scattering is the photon with the wavevector $K_i \approx 0$ and the energy $\hbar\omega_i$, smaller than the energy of the exciton resonance E_0. Due to this fact, in order to calculate the Raman scattering amplitude within the framework of the perturbation theory one should consider an additional virtual transition into the intermediate exciton state λ_0, $K_i \approx 0$. The amplitude for the Nth order Raman scattering in this case can be written as:

$$A_N = \sum_{\lambda_0, \lambda_1, \ldots, \lambda_N} \frac{H_{ER}^{\lambda_N} H_{EL}^{\lambda_N \lambda_{N-1}} \ldots H_{EL}^{\lambda_1 \lambda_0}}{(\hbar\omega_s - E_{\lambda_N})(\hbar\omega_i - (N-1)\hbar\Omega - E_{\lambda_{N-1}}) \ldots (\hbar\omega_i - \hbar\Omega - E_{\lambda_1})} \times \frac{H_{ER}^{\lambda_0}}{(\hbar\omega_i - E_{\lambda_0})}. \tag{20}$$

This amplitude, as compared to the exciton luminescence amplitude (14) includes the additional summation over the intermediate states λ_0, $K_i \approx 0$, the additional matrix element $H_{ER}^{\lambda_0}$ for the optical transition into the λ_0 state and the additional energy denominator $(\hbar\omega_i - E_{\lambda_0})$, which accounts for the energy deficit for such transition.

The highest intensity line in the resonant Raman scattering spectrum is the second order line 2LO. As in the case of phonon-assisted luminescence, the high intensity of this line comes from intermediate states of alternating parity, e.g. 1S–2P–1S. The small relative intensity of the 1LO line is due to the forbidden character of the first order scattering and the small value of the wavevector $\boldsymbol{q}_1 = \boldsymbol{K}_i - \boldsymbol{K}_s$ transferred to the phonon. In the case of Raman scattering q_1 is of the order of the photon wavevector and is much smaller than the thermal wavevector K_i involved in the first order luminescence.

The dependence of the 1LO Raman cross section on the magnitude of q has been directly tested by comparing the 1LO intensities in forward and backward scattering geometries. In CdS at $T = 2$ K under the 4880 Å excitation the corresponding change of q_1 is from $2 \times 10^4\,\mathrm{cm}^{-1}$ to $8 \times 10^5\,\mathrm{cm}^{-1}$, respectively. It has been found experimentally (Klochikhin et al. 1976b) that the 1LO scattering intensity in the forward direction σ_f is considerably less than the backward scattering intensity σ_b (fig. 4). For forbidden Fröhlich scattering it is expected (Martin and Damen 1971) that the value σ_f/σ_b after the reflection corrections should be equal to $(q_f/q_b)^2$. However, the measured value far exceeds the expected one, indicating that 1LO Raman scattering cross section has a contribution independent of the wavevector q_1. Moreover, it has proved to be dependent on the excitation frequency, sharply decreasing as $\hbar\omega_i$ approaches the exciton absorption line $n = 1$S (fig. 5).

The observed behaviour of the 1LO scattering cross section in the region of small wavevectors is a result of "size" resonance of the phonon wavelength $\tilde{\lambda} = 2\pi/q_1$ with the radii of exciton intermediate states a_n. The matrix element of the Fröhlich forbidden exciton–phonon scattering $H_{EL}^{\lambda_n \lambda_n}$ has a maximum value at $q_1 \sim 1/n^2 a_0$. Since the phonon wavevector q_1 in 1LO scattering is fixed by the momentum conservation law (8) and is small, the main contribution to the scattering cross section will come from the

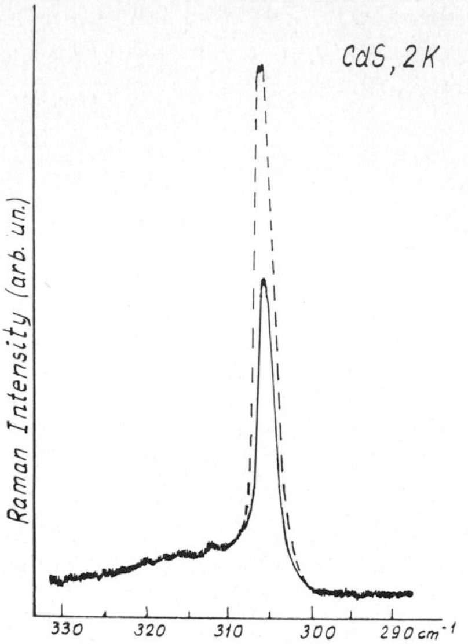

Fig. 4. A comparison of the 1 LO Raman scattering spectra in CdS crystals taken at $T = 2$ K under 4880 Å excitation in the forward X(YY)X (solid line) and backward X(YY)$\bar{X}$ (dashed line) directions (Klochikhin et al. 1976b).

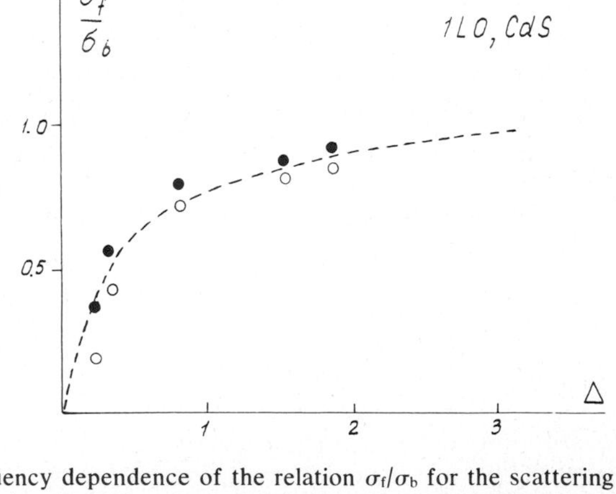

Fig. 5. The frequency dependence of the relation σ_f/σ_b for the scattering intensities of 1 LO Raman cross section in CdS crystals in the forward (σ_f) and backward (σ_b) directions in X(YY)X geometry. Δ is the shift of the excitation energy $\hbar\omega_i$ from the $n = 1S$ exciton resonance E_0 in units of the LO phonon energy $\hbar\Omega$, $\Delta = (E_0 - \hbar\omega_i)/\hbar\Omega$. Circles are the experimental data corrected (open) and uncorrected (filled) for reflection inside the crystal. The dashed line is calculated taking into account the energy dependence of the relative contribution of intermediate states with different quantum numbers (Klochikhin et al. 1976b).

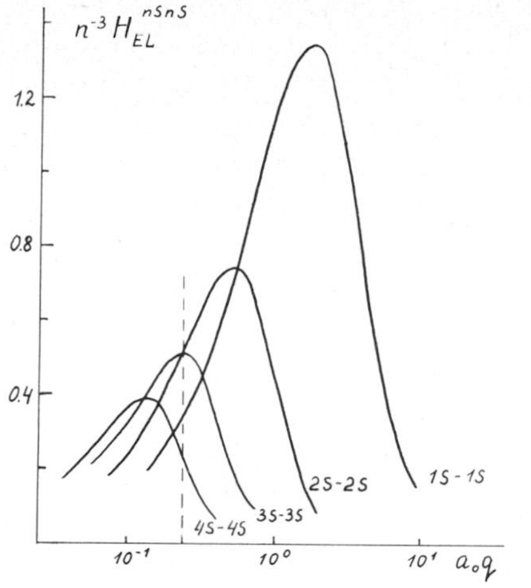

Fig. 6. The dependence of the diagonal exciton–phonon scattering matrix elements H_{EL}^{nSnS} (scattering inside the same exciton band) on the reduced phonon wavevector a_0q (semilogarithmic plot). The additional factor n^{-3} accounts for the decrease of the oscillator strength with increasing quantum number n. The vertical dashed line marks the wavevector q_l for 4880 Å back scattering in CdS at 2 K. It can be seen that the main contribution to the back-scattering cross section in this case comes from 2S–2S and 3S–3S intermediate states.

intermediate states with large quantum numbers n_0, which are selected by the value of q_l. It has been shown (Klochikhin et al. 1976b) that the decrease of the oscillator strength in the exciton–photon matrix element $H_{ER}^{\lambda_n} \propto n^{-3}$ is to a great extent compensated by the increase of the maximum value of $H_{EL}^{\lambda_n\lambda_n}$ (fig. 6). For example, in the case of 4880 Å Raman scattering in CdS at 2 K, $n_0 = 2$–3 for back scattering and $n_0 = 10$–15 for forward scattering.

As a result, the dependence of the 1 LO cross section on q_l in preresonant Raman scattering is not given by the q dependence of any particular matrix element $H_{EL}^{\lambda\lambda'}(q)$ but is to a great extent determined by the redistribution of intermediate states, provided that the exciting frequency is not too close in resonance with one of the exciton levels ($\hbar\omega_i - E_0 \gtrsim R$). The above considerations account for the relatively weak q dependence of 1 LO Raman cross section observed experimentally. The exact wavevector and frequency dependence of the 1 LO preresonance Raman cross section has been recently calculated also by Lottici and Razzetti (1978).

When the exciting frequency is approaching the $n = 1S$ exciton resonance, the contribution of this intermediate state to the scattering amplitude increases sharply due to the decrease of the last energy denominator in eq. (20). Since for the 1S–1S scattering the q_l value is well below the region of H_{EL}^{1S1S} maximum, the scattering cross section acquires a stronger dependence on q_1. The study of 1 LO scattering inside the $n = 1S$ polariton resonance in CdS crystals (Permogorov and Travnikov 1971) has shown that in this case the scattering cross section can be taken to be strictly proportional to q^2. The scattering wavevector was in this case monitored by the shift of the excitation frequency $\hbar\omega_i$ in the region of strong polariton dispersion $K_i(\hbar\omega_i)$.

The strong decrease of the Fröhlich exciton–phonon interaction in the region $qa_0 \gg 1$, limits the phonon wavevector in multiphonon processes. In both forbidden and allowed scattering processes, only the phonons close to the center of the Brillouin zone take part. As a result, the multiphonon Raman lines in resonant scattering are relatively narrow. In the case of anisotropic (uniaxial) crystals, as was shown by Klein and Colwell (1971), the main contribution to the multiphonon linewidths comes from the dependence of the LO energy at $K = 0$ on the direction of phonon propagation (directional dispersion).

5. Multiphonon Raman scattering in the region of fundamental absorption

The secondary emission spectra of semiconducting crystals under monochromatic excitation in the region of fundamental absorption often exhibit a series of equidistant lines shifted to the lower energies from the exciting line by the energies of 2, 3, 4 etc. LO phonons (Bendow et al. 1970). The line at position 1 LO in perfect crystals is much weaker than the 2 LO line. Other lines have comparable relative intensities, and show pronounced enhancement in the vicinity of the exciton resonance (fig. 7).

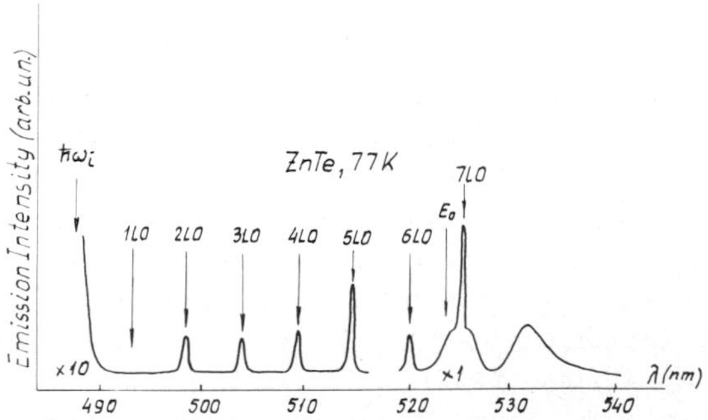

Fig. 7. Multiphonon resonance Raman scattering in the region of fundamental absorption of ZnTe crystals at 77 K. Excitation with Ar^+ laser, $\lambda = 4880$ Å. An arrow E_0 shows the position of the $n = 1S$ exciton resonance. The gain in the left part of the spectrum is ten times higher than in the right one. The first-order line 1 LO is too weak to be seen in the spectrum.

The width of the multiphonon lines does not increase with increasing order, and usually does not exceed several cm^{-1}, showing that only long-wave LO phonons take part in this emission process. On the basis of its spectral properties (positions and widths) these lines can apparently be assigned to multiphonon Raman scattering. The preferential interaction with LO phonons, the restriction of phonon wavevectors to small values, and the resonance with exciton transitions indicate that this emission process is also dominated by Fröhlich scattering through exciton intermediate states.

The amplitude of the Nth order Raman scattering in this case is given by the same general expression (20) which has been used for the preresonant scattering. However, the energy denominators of (20), in distinction to the preresonant scattering or phonon-assisted exciton luminescence, do not increase with the increasing scattering order. On the contrary, most of these denominators can be small simultaneously, which corresponds to "multiple resonance" for the intermediate states (Martin and Falicov 1975). As a result, the multiphonon scattering lines of different orders can have comparable intensities.

In the region of fundamental absorption the band states of both the discrete and continuum exciton spectra should be considered as the scattering intermediate states. However, as a first approximation models taking into account only one type of intermediate states have been used for the cross-section calculations and the qualitative interpretation of the results.

The successive LO scattering through the intermediate states of discrete exciton bands has been considered by Gross et al. (1973), Bir et al. (1976) and Klochikhin et al. (1976a). In this model an exact resonance for all the intermediate states, except for the first and the last, occurs which corresponds to the scattering through the real intermediate states. The intensity of the Nth order line for such scattering is given by

$$I_N \propto A_1(\hbar\omega_i)\frac{A_2(\hbar\omega_s)}{\Gamma(\hbar\omega_s+\hbar\Omega)}\prod_{n=1}^{N-2}\frac{\Gamma_0(\hbar\omega_i-n\hbar\Omega)}{\Gamma(\hbar\omega_i-n\hbar\Omega)}, \tag{21}$$

where $A_1(\hbar\omega_i)$ is the probability of the indirect exciton absorption at energy $\hbar\omega_i$; $A_2(\hbar\omega_s)$ is the probability of 1 LO radiative annihilation for the exciton with the energy $(\hbar\omega_s+\hbar\Omega)$; $\Gamma(E)$ is the reciprocal total lifetime for the exciton with the energy E and $\Gamma_0(E)$ is the probability of exciton LO scattering inside the band with the lowering of energy by $\hbar\Omega$. The ratio Γ_0/Γ represents the "quantum yield" of the exciton relaxation. In accordance with estimates of Aristova et al. (1978) it decreases only slowly with increasing exciton energy.

The qualitative interpretation of eq. (21) is that in the case of real intermediate states multiphonon Raman scattering is equivalent to hot exciton luminescence (Klein 1973) and can be treated as absorption followed by emission. However, the careful examination of the scattering amplitude (20) by Bir et al. (1976) and Ivchenko et al. (1978) has shown that for the 2LO scattering there exists an additional interference contribution to the scattering cross section which doubles the intensity of the 2LO line and cannot be understood within the hot luminescence model. Although this interference contribution vanishes for $N \geq 3$, it can be concluded that the general approach of Raman scattering to the LO-assisted secondary emission in the region of fundamental absorption is more complete than the hot luminescence model, even when the scattering takes place through real intermediate states.

The multiphonon scattering model, taking into account only the intermediate states of the continuum exciton spectrum (or free electron–hole pairs), has been used by Martin (1974), Zeyher (1975) and Abdumalikov and Klochikhin (1977). Due to the two-particle nature of the intermediate state spectrum, the scattering process in this case cannot be described as light absorption followed by emission since strong interference of different contributions to the scattering cross section takes place. As a result, even for the most simple 2LO process the scattering cross section through the pair states has been evaluated only numerically (Abdumalikov and Klochikhin 1977).

It should be noted that 1LO scattering is forbidden for both models of intermediate states since it involves the scattering of an exciton or an electron–hole pair by a small wavevector $\boldsymbol{q}_1 = \boldsymbol{K}_i - \boldsymbol{K}_s$.

It follows from the experimental results that the real situation rather corresponds to the coexistence of both discrete and continuum exciton intermediate states with the relative contribution being dependent on the energy region. The frequency dependence of the multiphonon LO scattering intensity in CdS (Permogorov and Travnikov 1976) and GaSe (Reydellet et al. 1976, Camassel et al. 1976) shows that near the fundamental absorption edge the discrete $n = 1S$ exciton intermediate states are the most important. However, at higher energies the contribution of the continuum spectrum predominates. Figure 8 compares the experimental frequency dependence of 2LO cross section in ZnTe crystals with the theoretical predictions of the two models. The insert of the figure shows that the contribution of the discrete exciton states decreases with energy much faster than that of the electron–hole pair states. This situation is possible when the scattering starts through the pair states and then switches to the discrete exciton states. The multiphonon emission processes incorporating at last stages the discrete excitons have been observed for 4LO scattering in GaSe (Camassel et al. 1976) and 9LO scattering in CdS (Leite et al. 1969).

Multiphonon Raman scattering in the region of the fundamental absorptin has its counterpart in the oscillatory structure of the luminescence or photoconductivity excitation spectra. Such structure in both these

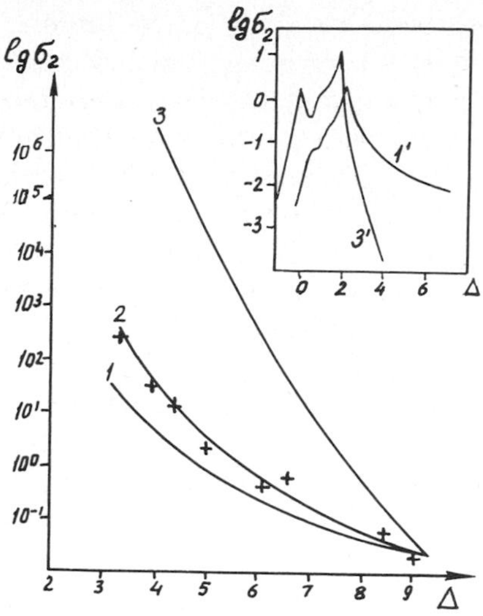

Fig. 8. The frequency dependence of the 2 LO Raman scattering cross section σ_2 in the region of the fundamental absorption of ZnTe crystals (semilogarithmic plot). Δ is the shift of the excitation energy $\hbar\omega_i$ from the lowest exciton resonance $n = 1S(E_0)$ in the units of LO phonon energy $\hbar\Omega$, $\Delta = (\hbar\omega_i - E_0)/\hbar\Omega$. Crosses are the experimental points. It can be seen that the slope of the experimental curve (2) is close to that for the theoretical curve (1), calculated for the model of electron–hole pair intermediate states. Curve (3) is calculated with the help of eq. (21) within a model which takes into account only $n = 1S$ discrete exciton intermediate states, assuming $\Gamma_0/\Gamma = 1$. The insert shows the 2 LO Raman cross section for electron–hole pair (1′) and $n = 1S$ discrete exciton (3′) intermediate states calculated in the same units for the parameters of ZnTe crystals (Klochikhin et al. 1978a).

phenomena results from the same process of LO relaxation of optically excited electronic intermediate states. The oscillatory structure with period $\hbar\Omega_{LO}$ have been observed in the excitation spectra of most of the polar semiconductors (for a review see, e.g., Permogorov 1975).

The maxima in the excitation spectra correspond to the creation of a nonequilibrium population in some chosen points of the exciton bands through the fast relaxation processes with the emission of several LO phonons. Since additional relaxation via the acoustic phonons can be involved in the generation of luminescence or photoconductivity, the maxima in the excitation spectra are usually broader than the multiphonon Raman scattering lines (Nakamura and Weisbuch 1978).

The appearance of oscillating structure in the excitation spectra is connected with the incomplete establishment of thermal equilibrium in the crystal excited state and is favoured by the short lifetime of the intrinsic excitations of the crystal. However, the same relaxation processes are operative in samples with arbitrary exciton lifetime. The only difference between the multiphonon Raman scattering and the fast relaxation processes in luminescence or photoconductivity is that in the former process the final state of LO scattering is a photon with $K_s \approx 0$, whereas in the latter cases the final states may be excitons with $K_s \neq 0$. However the same intermediate states and scattering mechanisms are involved in all three processes.

The multiphonon Raman scattering lines in the region of fundamental absorption represent the most nonequilibrium component of the secondary emission spectrum. The simultaneous population of real states in the exciton bands due to fast LO relaxation leads to the appearance first of nonthermalized and then of thermalized luminescence. The temporal evolution of the exciton secondary emission spectrum with the increase of the free exciton lifetime has been studied experimentally in CdS crystals (Permogorov and Travnikov 1979).

It should be mentioned that under linearly polarized optical excitation the LO-assisted secondary emission is strongly polarized, which is the result of the optical alignment of the scattering intermediate states (see ch. 6 by E. Pikus and E. Ivchenko in this book). Optical alignment has been observed in both excitation (Bonnot et al. 1974) and multiphonon Raman (Permogorov et al. 1975) spectra. This indicates that the spin correlation of electron and hole in the intermediate states is retained throughout the relaxation process. On this basis, the intermediate states of LO scattering even at high energies should be considered as continuum exciton states rather than as free electron–hole pairs. When lowering the energy through the LO emission these continuum states can be transformed into the discrete exciton states.

The oscillatory structure in many cases constitutes the substantial part of the integrated area of the excitation spectrum. It can be concluded that the LO relaxation of hot excitons through the continuum and discrete states is an important mechanism for the transformation of the optical energy absorbed by the crystal in the region of fundamental absorption. Study of the oscillatory structure in excitation spectra shows (Permogorov 1975) that this mechanism is efficient in the spectral region up to 10 LO above the fundamental absorption edge.

6. *Exciton scattering in crystals with defects*

A characteristic feature of the Fröhlich exciton scattering by LO phonons is the forbidden character of the first order processes. As a result the intensity of 1 LO line in the Raman spectra of perfect crystals is much lower than that of 2 LO line, both in the fundamental absorption and the transparency region. However, in crystals with defects ("defect crystals") the relative intensity of 1 LO scattering increases drastically. Such an increase has been observed in the fundamental region of CdSe samples with the mechanical defects introduced by surface polishing (Gross et al. 1973). In CdS samples doped with concentrations of the order of $10^{17}\,cm^{-3}$ Ni the enhancement of 1 LO relative intensity in the preresonant scattering is more than 30 times (Permogorov and Reznitsky 1976). In what follows we shall briefly discuss some mechanisms by which defects can increase the 1 LO scattering intensity.

(1) If the absorption spectrum of the crystal contains the strong impurity absorption bands, resonance Raman scattering through localized impurity intermediate states can be observed. The intensity of 1 LO scattering through the localized states will depend on the electron–phonon coupling of the impurity center and can exceed the intensity of forbidden intrinsic 1 LO scattering. In semiconductor crystals resonance can take place with the impurity bound exciton transitions, for which "giant" oscillator strengths are predicted (Rashba 1974). Resonant enhancement of the forbidden 1 LO scattering cross section in the region of bound exciton lines has been observed by Damen and Shah (1971) in CdS crystals under tunable laser excitation. No enhancement of the allowed 2LO cross section was found in these experiments. A detailed discussion of the resonant bound exciton LO scattering was recently given by Klochikhin et al. (1978b).

(2) The breakdown of the crystal translational symmetry by the defects relaxes the wavevector conservation law (8) in the Raman scattering. As a result, the magnitude and direction of the phonon wavevector in the first order process are no longer strictly related to the wavevectors of the exciting and emitted light. Figure 9 shows 1 LO scattering spectra of CdS samples doped with different concentrations of Ni impurity in the $Z(YY)\bar{Z}$ geometry, which corresponds to the excitation of A_1(LO) phonons with wavevectors along the optical axis c ($\boldsymbol{q} \parallel c$). As can be seen, with increasing impurity concentration the wavevector forbidden E_1(LO) $\boldsymbol{q} \perp c$ scattering gradually appears in the spectrum. The general increase of 1 LO integrated intensity with the impurity concentration is partly accounted for by the inclusion in the first order scattering of phonons with $q \sim 1/a_0$ for which the scattering probability is a maximum.

(3) The intensity of 1 LO preresonant scattering in defect samples is even more strongly affected by the inclusion of the allowed Fröhlich exciton–phonon interaction. The breakdown of the wavevector conservation by the defect can be regarded as an additional elastic scattering, with the inclusion

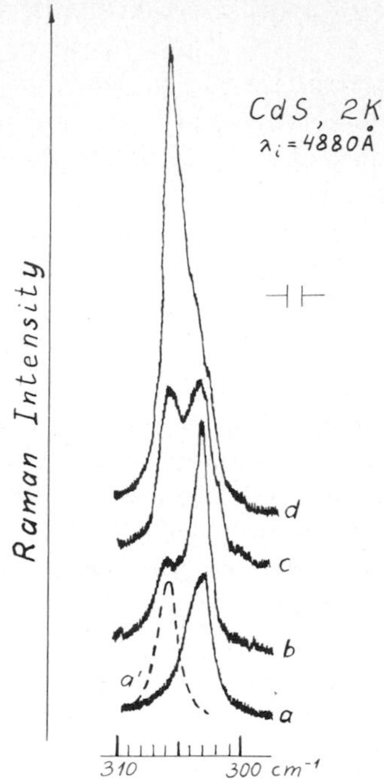

Fig. 9. Resonant Raman 1 LO scattering spectrum in CdS samples doped with Ni impurity with different concentrations. The impurity concentration increases from (a) to (d). Scattering geometry $Z(YY)\bar{Z}$ corresponds to the excitation of $A_1(LO)$ phonons in the perfect crystals. With increasing impurity concentration, the $E_1(LO)$ component appears in the spectrum. The dashed line is the $X(YY)\bar{X}$ $E_1(LO)$ spectrum in the perfect crystal (Permogorov and Reznitsky 1976).

of an extra intermediate state. As to the number of the intermediate states, one-phonon scattering in the defect crystal is equivalent to the two-phonon scattering in a perfect crystal (Klochikhin and Plyukhin 1975). Since the 2P exciton band can be incorporated as an additional intermediate state, the allowed Fröhlich scattering through the states of the alternating parity enters 1 LO cross section.

(4) When the exciting light energy $\hbar\omega_i$ is above the threshold exciton energy E_0 elastic defect scattering induces the transitions to real exciton intermediate states with $K \neq 0$. These intermediate states give the main contribution to the defect 1LO scattering cross section in the region of the fundamental absorption (Gogolin and Rashba 1976). The increase of the scattering order and excitation of the real intermediate states were taken into account by Klochikhin et al. (1976c) when calculating the frequency

dependence of 1LO Raman cross section in disordered (Zn, Cd) Te solid solutions.

It should be mentioned that all the discussed mechanisms of defect scattering strongly increase the intensity of the forbidden 1LO scattering. At the same time the allowed 2LO scattering is not influenced so strongly by the defects.

Crystal defects should also have a strong influence on the phonon-assisted first order exciton luminescence. However, the study of the exciton luminescence in defect crystals is usually hindered by the quenching of exciton emission by defects. Nevertheless, in the case of ZnO samples with internal strains a quasi-equilibrium exciton kinetic energy distribution has been observed in luminescence (Verbin et al. 1977). It has been found that in ZnO containing defects, the relative intensity of 1LO exciton luminescence line I_1 is approximately 20 times higher than that in the perfect samples at $T = 2$ K. The analysis of the 1LO phonon-assisted spectral shape shows that the probability of this emission process does not depend on the exciton kinetic energy in defect samples. Such anomalous behaviour of the 1 LO exciton luminescence can be easily understood as a result of defect-induced allowed Fröhlich scattering through the states with alternating parity.

7. *Conclusion*

The experimental material presented in this paper shows that scattering of excitons by LO phonons constitutes an essential part of many emission processes in the region of the fundamental edge. An important point is that strong interaction with LO phonons is a characteristic of intrinsic electronic excitations in polar semiconductors.

A detailed study of multiphonon exciton luminescence and Raman scattering shows that the main features of these spectra are well described by a rather general model which takes into account Fröhlich scattering through hydrogen-like exciton intermediate states. This simple model is valid for a wide class of ionic crystals. On the other hand, the quantitative characteristics of multiphonon spectra depend on the crystal parameters. Further elaboration of the theory may lead to a method of determining the crystal parameters relevant to the exciton–phonon interaction.

Multiphonon LO Raman scattering in the region of fundamental absorption and oscillatory structure of luminescence and photoconductivity excitation spectra reveals the existence of energy relaxation processes in which the photo-created electron–hole pair loses its energy as a whole, without separation into independent carriers. This relaxation mechanism, which can be regarded as hot-exciton relaxation, is operative over a wide energy region, of order 10 LO energies above the fundamental edge. This region contains a substantial part of the integrated intensity of the photoconductivity and luminescence excitation spectra. So it can be concluded that the hot-exciton relaxation with emission of LO phonons plays an important role in transformation of crystal energy after optical excitation.

Acknowledgements

The author is very grateful to Dr. A. Klochikhin and Dr. A. Reznitsky for many helpful discussions on the nature of the exciton Fröhlich scattering. Dr. A. Reznitsky has also performed most of the numerical calculations used in this paper.

References

Abdumalikov, A.A. and A.A. Klochikhin, 1977, Phys. Stat. Sol. (b)**80**, 43.
Abramov, V.A., S.A. Permogorov, B.S. Razbirin and A.I. Ekimov, 1970, Phys. Status Solidi **42**, 627.
Agranovich, V.M., 1960, Uspekhi Fiz. Nauk. **71**, 141.
Anselm, A.I. and Yu. A. Firsov, 1956, J. Eksper. Teor. Fiz. **30**, 719 (transl: Sov. Phys. Uspekhi **3**, 564).
Aristova, A.K., C. Trallero Giner, I.G. Lang and S.T. Pavlov, 1978, Phys. Status Solidi (b)**85**, 351.
Bendow, B., J.L. Birman, A.K. Ganguly, T.C. Damen, R.C.C. Leite and J.F. Scott 1970, Opt. Commun. **1**, 267.
Bir, G.L., E.L. Ivchenko and G.E. Pikus, 1976, Izv. Akad. Nauk USSR, ser. fiz. **40**, 1866 (Bull. Acad. Sci. USSR, Phys. Rev. **40**, No. 9, 81).
Bonnot, A. and C. Benoit a la Guillaume, 1970, Phys. Rev. Lett. **24**, 1235.
Bonnot, A., R. Planel and C. Benoit a la Guillaume, 1974, Phys. Rev. **B9**, 690.
Bulyanitsa, D.S., 1970, Fiz. Tekhn. Polupr. **4**, 1273 (transl: Sov. Phys. Semicond. **4**, 1081).
Callender, R.H., S.S. Sussmann, M. Seders and R.K. Chang, 1973, Phys. Rev. **B7**, 3788.
Camassel, J., T.C. Chiang, Y.R. Shen, J.P. Voitchovsky and N.M. Amer, 1976, Solid State Commun. **19**, 483.
Damen, T.C. and J. Shah, 1971, Phys. Rev. Lett. **27**, 1506.
Fröhlich, H., 1954, Adv. Phys. **3**, 325.
Gogolin, A.A. and E.I. Rashba, 1976, Solid State Commun. **19**, 1177.
Goto, T. and Y. Nishina, 1978, Phys. Rev. **B17**, 4565.
Gross, E., S. Permogorov and B. Razbirin, 1966, J. Phys. Chem. Sol. **27**, 1647.
Gross, E.F., S.A. Permogorov and B.S. Razbirin, 1971, Uspekhi Fiz. Nauk **103**, 431 (transl: Sov. Phys. Uspekhi, **14**, 104).
Gross, E., S. Permogorov, Ya. Morozenko and B. Kharlamov, 1973, Phys. Status Solidi (b)**59**, 551.
Hopfield, J.J., 1958, **112**, 1955.
Ivchenko, E.L., I.G. Lang and S.T. Pavlov, 1978, Phys. Status Solidi (b)**85**, 81.
Klein, M.V., 1973, Phys. Rev. **B8**, 919.
Klein, M.V. and P.J. Colwell, 1971, Proc. 2nd Int. Conf. Light Scattering in Solids, ed. M. Balkanski (Flammarion Sciences, Paris) p. 65.

Klochikhin, A.A. and A.G. Plyukhin, 1975, Pis'ma JETF **21**, 267 (transl: Sov. Phys. JETP Lett. **21**, 122).
Klochikhin, A.A., S.A. Permogorov and A.N. Reznitsky, 1976a, J. Eksper. Teor. Fiz. **71**, 2230 (transl: Sov. Phys. JETP **44**, 1176).
Klochikhin, A.A., S.A. Permogorov and A.N. Reznitsky, 1976b, Fiz. Tverd. Tela **18**, 2239 (transl: Sov. Phys. Solid State, **18**, 1304).
Klochikhin, A.A., A.G. Plukhin, L.G. Suslina and E.B. Shadrin, 1976c, Fiz. Tverd. Tela **18**, 1909 (transl: Sov. Phys. Solid State, **18**, 1112).
Klochikhin, A.A., Ya.V. Morozenko and S.A. Permogorov, 1978a, Fiz. Tverd. Tela **20**, 3557 (transl: Sov. Phys. Solid State, **20**, 2057).
Klochikhin, A.A., A.G. Plyukhin, L.G. Suslina and D.L. Fedorov, 1978b, Fiz. Tekhn. Polupr. **12**, 2365 (transl: Sov. Phys. Semicond. **12**, 1406).
Leite, R.C.C., J.F. Scott and T.C. Damen, 1969, Phys. Rev. Lett. **22**, 780.
Lottici, P.P. and C. Razzetti, 1978, Solid State Commun. **25**, 427.
Martin, R.M., 1974, Phys. Rev. **B10**, 2620.
Martin, R.M. and T.C. Damen, 1971, Phys. Rev. Lett. **26**, 86.
Martin, R.M. and L.M. Falicov, 1975, Resonant Raman Scattering, in: Light Scattering in Solids, ed. M. Cardona (Springer Verlag, Berlin, Heidelberg, New York), ch. 3.
Nakamura, A. and C. Weisbuch, 1978, Solid State Electron. **21**, 1331.
O'Connel-Bronin, A.A. and V.G. Plekhanov, 1979, Phys. Status Solidi (b)**95**, 75.
Oka, Y. and T. Kushida, 1972, J. Phys. Soc. Jap. **33**, 372.
Permogorov, S.A., 1975, Phys. Status Solidi (b)**68**, 9.
Permogorov, S. and A. Reznitsky, 1976, Solid State Commun. **18**, 781.
Permogorov, S.A. and A.V. Selkin, 1973, Fiz. Tverd. Tela **15**, 3025 (transl: Sov. Phys. Solid State **15**, 2015).
Permogorov, S.A. and V.V. Travnikov, 1971, Fiz. Tverd. Tela **13**, 709 (transl: Sov. Phys. Solid State **13**, 586).
Permogorov, S. and V. Travnikov, 1976, Phys. Status Solidi (b)**78**, 389.
Permogorov, S. and V. Travnikov, 1979, Solid State Commun. **29**, 615.
Permogorov, S.A., Ya.V. Morozenko and B.A. Kazennov, 1975, Fiz. Tverd. Tela **17**, 2970 (transl: Sov. Phys. Solid State **17**, 1974).
Plekhanov, V.G. and A.A. O'Connel-Bronin, 1978, Fiz. Tverd. Tela **20**, 2078 (transl: Sov. Phys. Solid State **20**, 1200).
Reydellet, J., M. Balkanski and J.M. Besson, 1976, J. de Physique, **37**, L219.
Rashba, E.I., 1974, Fiz. Tekhn. Polupr. **8**, 1241 (transl: Sov. Phys. Semicond. **8**, 807).
Segall, B., 1966, Phys. Rev. **150**, 734.
Segall, B., 1967, Phys. Rev. **163**, 769.
Sumi, H., 1977, J. Phys. Soc. Jap. **41**, 526.
Tait, W.C., D.A. Campbell, J.R. Packard and R.L. Weiher, 1967, in: Proc. Int. Conf. II–VI Semicond. Compounds, ed. D.G. Thomas (Benjamin, New York) p. 370.
Thomas, D.G., J.J. Hopfield and M. Power, 1960, Phys. Rev. **119**, 570.
Toyozawa, Y., 1959, Progr. Theor. Phys. Suppl. **12**, 111.
Verbin, S.Yu., S.A. Permogorov, A.N. Reznitzky and A.N. Starukhin, 1977, Fiz. Tverd. Tela **19**, 3468 (transl: Sov. Phys. Solid State **19** 2028).
Weiher, R.L. and W.C. Tait, 1968, Phys. Rev. **166**, 791.
Weiher, R.L. and W.C. Tait, 1972, Phys. Rev. **B5**, 623.
Weisbuch, C. and R.G. Ulbrich, 1979, J. Luminescence **18/19**, 27.
Wiesner, P. and U. Heim, 1975, Phys. Rev. **B11**, 3071.
Zeyher, R., 1975, Solid State Commun. **16**, 49.

CHAPTER 6

Optical Orientation and Polarized Luminescence of Excitons in Semiconductors

G.E. PIKUS and E.L. IVCHENKO

A.F. Ioffe Physico-technical Institute
USSR Academy of Sciences, Leningrad
USSR

Excitons
Edited by
E.I. Rashba and M.D. Sturge

Contents

1. *Introduction*

Optical orientation methods were first extensively used to study atomic states in gases and localized states in solids. In recent years, the methods have been applied to interband transitions in semiconductors. These methods are based on the fact that, under optical pumping in gases and solids by circularly or linearly polarized light, states with a preferential orientation of spins (angular momenta) or with a definite direction of oscillating electric-dipole moment are excited. The former phenomenon is known as optical spin orientation, while the latter one is frequently called optical alignment.

Optical orientation manifests itself in polarization of the recombination light. Hence, a convenient way to study optical orientation is to measure the polarization of the photoluminescence. Measurements of the change of polarization due to an external magnetic field (Hanle effect) or due to a rf field in a magnetic field (optical detection of electron spin resonance) considerably extend the capabilities of optical orientation methods. Moreover, the hyperfine coupling between spin polarized excitations and the nuclei of the lattice leads to dynamical nuclear orientation which can be observed by means of nuclear magnetic resonance.

In the first experiments on optical orientation in semiconductors, spin orientation of free carriers was obtained (Lampel 1968, Parsons 1969, Ekimov and Safarov 1970, Zakharchenya et al. 1971). Optical orientation of excitons was first achieved in CdSe by Gross et al. (1971). Then the effect was observed for excitons in various compounds: InP (Weisbuch and Lampel 1972, Weisbuch et al. 1974, Weisbuch and Hermann 1975), CdS (Bonnot et al. 1972), GaSe (Veshchunov et al. 1972, Razbirin et al. 1975, Minami et al. 1976), GaAs (Weisbuch et al. 1974, Fishman et al. 1974, Weisbuch and Fishman 1976), CdTe (Fishman et al. 1974) and ZnTe (Oka and Kushida 1977, 1978, Permogorov and Morozenko 1979). Optical alignment effects predicted for excitons in semiconductors by Bir and Pikus (1972a, 1972b) has been measured in CdS (Bonnot et al. 1974, Permogorov et al. 1975), CdSe (Permogorov et al. 1975, Nawrocki et al. 1976) and GaSe (Razbirin et al. 1975, 1976, Gamarts et al. 1977).

The optical orientation technique is a powerful tool for studying the fine structure of excitonic levels and for measurement of exciton lifetimes and spin relaxation times. Optical orientation of excitons bound to impurities helps one to determine the nature and charge of the impurities. If an exciton is trapped on a neutral donor only the hole can retain its orientation in the complex. If the valence band is nondegenerate, similar situation takes place for exciton bound to a neutral acceptor: in this case only the electron can be oriented.

It follows then that excitons bound to neutral impurities cannot be aligned, i.e., the radiation due to excitons trapped on neutral impurities will be unpolarized under linearly polarized excitation (Bonnot et al. 1974). Optical alignment can be achieved for free excitons or excitons bound to ionized impurities or isoelectronic centers. Optical pumping experiments permit one to clarify formation processes of excitonic complexes, namely, to trace whether it occurs by trapping of a free exciton or by successive electron and hole trapping (see e.g. Weisbuch and Lampel 1972). Optical orientation methods give information concerning the lifetime of intermediate states. For example, the optical detection of electron spin resonance performed on the line due to bound exciton recombination in pure GaAs and InP showed that in this case the time taken to form an excitonic complex is longer than the complex lifetime (Weisbuch et al. 1974). Experimental data on optical orientation of free excitons will be discussed in detail in sects. 4 and 5.

The condition for the optical orientation effect to be readily observed is that the spin relaxation time, τ_s, is not too small relative to the lifetime, τ_0, of a photocreated excitation. If $\tau_s \ll \tau_0$ then the excitation almost completely loses the preferential orientation during the lifetime. In this case a more convenient method is to observe the thermal orientation of spins in a magnetic field (D'yakonov and Perel' 1972, Dzhioev et al. 1973). For $\tau_s \ll \tau_0$, in the steady-state regime the populations of the Zeeman levels are close to their equilibrium values and the polarization of the photoluminescence is determined by the way the photoexcited levels split in the magnetic field. In direct-gap semiconductors the condition $\tau_s \gg \tau_0$ is usually satisfied and optical orientation methods are efficient in this case. On the other hand, for semiconductors with an indirect forbidden gap, so that the lower conduction band and upper valence band extrema lie in different points of Brillouin zone, the lifetime, τ_0, as a rule, is considerably longer than τ_s. Hence, in this case the thermal orientation of spins in a magnetic field is used. The thermal orientation of indirect excitons in Ge and Si has been observed by Asnin et al. (1976, 1979) and Altukhov et al. (1978).

In the present review we discuss the main results of theoretical and experimental studies on optical orientation of excitons in direct-gap III–V, II–VI and III–VI semiconductors and on thermal orientation of excitons in Ge and Si in a magnetic field. In addition, we give the necessary information about the properties of excitonic states in these crystals, since the specific features of the effects of interest are ultimately determined by the fine structure of excitonic levels and selection rules.

2. Structure of excitonic levels and selection rules for direct excitons

If allowance is made for free carrier spin, then in the effective mass theory the ground state of an exciton is degenerate, even in the case of simple bands. The ground-state degeneracy is equal to the product of the conduction- and valence-band degeneracies at $k = 0$. The wavefunctions of the ground state transform according to the representation $D_{\text{exc}} = D_c \times D_v$, where D_c and D_v are the representations according to which the wavefunctions of an electron and a hole at the extremum transform. In general, the representation D_{exc} is reducible and may be decomposed into irreducible representations. The electron–hole exchange interaction partially removes the degeneracy and splits the ground state of an exciton into corresponding irreducible representations.

In this section we briefly recall the structure of excitonic levels and selection rules for direct excitons in III–V, II–V and III–VI compounds.

2.1. III–V and II–VI cubic crystals (T_d symmetry)

The band structure near $k = 0$ for a cubic direct-gap semiconductor is shown schematically in fig. 1. Without taking account of the electron spin, the conduction band Γ_6 is nondegenerate. The valence bands Γ_8 and Γ_7 issue from the threefold-degenerate band Γ_{15} as a result of spin–orbit splitting. The ground state of the direct exciton $\Gamma_6 \times \Gamma_8$ is eightfold degenerate. The eight eigenfunctions are characterized by spin indices $\pm\frac{1}{2}$ of the electron and $\pm\frac{3}{2}$, $\pm\frac{1}{2}$ of the hole. The exchange interaction splits the state $\Gamma_6 \times \Gamma_8$ into three terms

$$\Gamma_6 \times \Gamma_8 = \Gamma_{12} + \Gamma_{15} + \Gamma_{25}. \tag{1}$$

The Hamiltonian describing the splitting of the excitonic level is determined by two constants

$$H_{\text{ex}} = \tfrac{1}{4}\Delta(\boldsymbol{J} \cdot \boldsymbol{\sigma}) + \Delta_1(J_x^3\sigma_x + J_y^3\sigma_y + J_z^3\sigma_z), \tag{2}$$

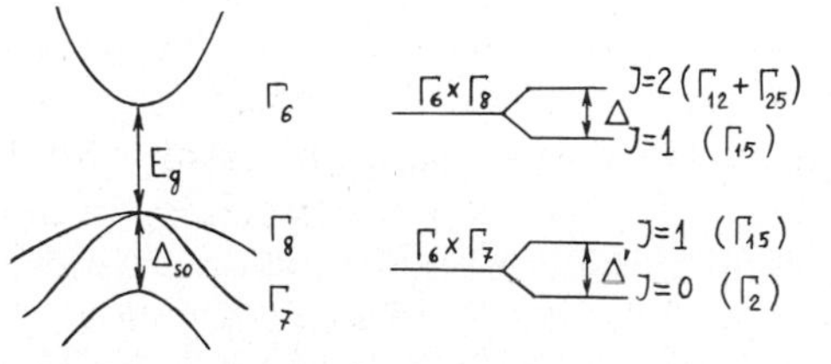

Fig. 1. The band structure of a cubic direct-gap semiconductor and the exchange splitting for $\Gamma_6 \times \Gamma_8$ and $\Gamma_6 \times \Gamma_7$ excitons.

where $\sigma_\alpha (\alpha = x, y, z)$ are the Pauli matrices in the basis of the electron wavefunctions $|\pm\frac{1}{2}\rangle$, J_α are the matrices of the hole angular momentum operator in the basis of the hole functions $|\pm\frac{3}{2}\rangle$, $|\pm\frac{1}{2}\rangle$, x, y, z are the principal crystal axes. The triplet level Γ_{15} corresponds to the total angular momentum $J = 1$, the component of angular momentum along a polar axis being $m = 1, 0, -1$. The constant Δ_1 is determined by the contribution to the Γ_8 states from other bands Γ_{12}, Γ_{25} and Γ_{15}, due to spin–orbit mixing. In the spherical approximation $\Delta_1 = 0$ and the exchange interaction does not remove the additional degeneracy of the terms Γ_{12} and Γ_{25}. These terms correspond to $J = 2$ with $m = \pm 2, \pm 1, 0$ and are shifted by Δ with respect to the term Γ_{15} (fig. 1). The selection rules for the electric-dipole-active Γ_{15} exciton are analogous to those for optical transitions $1s \to 2p$ in a hydrogen atom. The states with $J = 2$ are electric-dipole-inactive.

The ground state of the $\Gamma_6 \times \Gamma_7$ exciton corresponding to the split-off valence band Γ_7 is fourfold degenerate. The exchange interaction splits it into the electric-dipole-active level Γ_{15} with $J = 1$ $(m = \pm 1, 0)$ and the inactive singlet level Γ_2. The exchange splitting Δ' as Δ is proportional to the modulus squared, $|\Psi(\boldsymbol{r})|^2_{r\to 0}$, of the exciton envelope wavefunction at $r = 0$, where $\boldsymbol{r} = \boldsymbol{r}_e - \boldsymbol{r}_h$, $\boldsymbol{r}_e$ and $\boldsymbol{r}_h$ being the positions of the electron and hole, respectively. If the electron effective mass in the conduction band is much smaller than the hole effective mass, then the values of $\Psi(0)$ for $\Gamma_6 \times \Gamma_7$ and $\Gamma_6 \times \Gamma_8$ excitons coincide and $\Delta' = \frac{3}{4}\Delta$. The spin wavefunctions corresponding to the states Γ_{15} are as follows

$$\Phi_{1,1} = \varphi_{1/2}\psi_{1/2}, \qquad \Phi_{-1,-1} = \varphi_{-1/2}\psi_{-1/2}, \qquad (3)$$
$$\Phi_{1,0} = (1/\sqrt{2})(\varphi_{1/2}\psi_{-1/2} + \varphi_{-1/2}\psi_{1/2}),$$

while the spin wavefunction of the state Γ_2 is

$$\Phi_{0,0} = (1/\sqrt{2})(\varphi_{1/2}\psi_{-1/2} - \varphi_{-1/2}\psi_{1/2}). \qquad (4)$$

Here φ_s and ψ_s are electron and hole Bloch functions. In this basis nonzero matrix elements of the current density operator satisfy the relations

$$\langle 1, -1|j_-|0\rangle = -\langle 1, 1|j_+|0\rangle = \langle 1, 0|j_z|0\rangle = j_0, \qquad (5)$$

where

$$j_\pm = (1/\sqrt{2})(j_x \pm i j_y).$$

2.2. *II–VI wurtzite-type crystals (C_{6v} symmetry)*

In II–VI crystals with a wurtzite structure the conduction band is simple, whereas the valence band consists of three close-lying bands Γ_9, Γ_7 and Γ_7 (fig. 2). For the so-called quasicubic model the wurtzite valence band structure near $k = 0$ can be approximated by introducing a (111) strain into

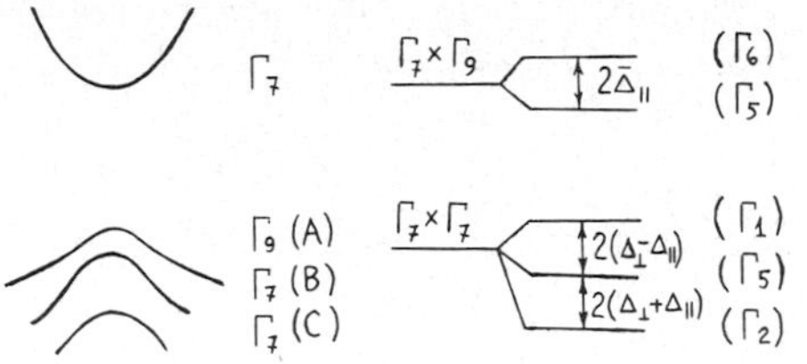

Fig. 2. The band structure of wurtzite-type crystals and the exchange splitting of the ground state of $\Gamma_7\times\Gamma_9$ (A) and $\Gamma_7\times\Gamma_7$ (B) and (C) excitons.

the valence band of a cubic crystal. In this approximation the three bands Γ_9, Γ_7 and Γ_7 arise from the band Γ_{15} due to spin–orbit and crystal splitting. Consequently, three series of excitons are observed in wurtzite-type crystals: $A(\Gamma_7\times\Gamma_9)$, $B(\Gamma_7\times\Gamma_7)$ and $C(\Gamma_7\times\Gamma_7)$. The ground state $n = 1$ of each of these excitons is fourfold degenerate. The exchange interaction splits the $A(n = 1)$ level into two terms, Γ_5 and Γ_6, and the $B(n = 1)$ or $C(n = 1)$ level into three terms Γ_1, Γ_2 and Γ_5. If the electron effective mass is much smaller than the hole effective masses, then all the exchange splitting constants can be expressed in terms of one independent constant (Bir and Pikus 1974). The states Γ_1 and Γ_2 correspond to the component of angular momentum $m_z = 0$, the term Γ_5 corresponds to $m_z = \pm1$ and for the term Γ_6, $m_z = \pm2$, where z is along the hexagonal axis (or the c-axis). Optical transitions are allowed into the state $\Gamma_1(\langle\Gamma_1|j_z|0\rangle \neq 0)$ and into the states Γ_5 with nonzero matrix elements

$$\langle\Gamma_5, -1|j_-|0\rangle = -\langle\Gamma_5, 1|j_+|0\rangle. \tag{6}$$

2.3. *III–VI crystals*

We shall discuss in detail optical orientation of excitons in GaSe. It is known that gallium selenide is crystallized in three modifications: $\beta(D_{6h}$ symmetry), $\gamma(C_{3v})$ and $\epsilon(D_{3h})$. All the modifications have similar band structures. We shall therefore consider only the ϵ modification. The lower conduction band and the upper valence band at the point Γ are simple and correspond to the representations $\Gamma_4(A_2^-)$ and $\Gamma_1(A_1^+)$. Taking spin into account, the representation Γ_4 transforms into Γ_8 and Γ_1 transforms into Γ_7. It should be noted that spin–orbit interaction noticeably mixes wavefunctions of the band Γ_1 with wavefunctions of the nearest valence band Γ_5 (fig. 3).

The ground state of the exciton is split into three terms: $\Gamma_7\times\Gamma_8 = \Gamma_3+\Gamma_4+\Gamma_6$. The exchange interaction is described by the Hamiltonian

$$H_{ex} = -\tfrac{1}{2}(\tfrac{1}{2}\Delta_1\boldsymbol{\sigma}_e\cdot\boldsymbol{\sigma}_h + \Delta\sigma_{ez}\sigma_{hz}), \tag{7}$$

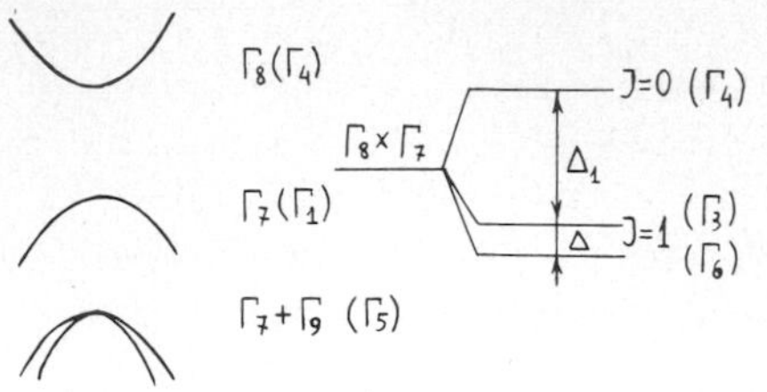

Fig. 3. Energy bands near $\boldsymbol{k} = 0$ for GaSe and the exchange splitting of the $\Gamma_8 \times \Gamma_7$ exciton.

where $\sigma_{e\alpha}$ and $\sigma_{h\alpha}$ ($\alpha = x, y, z$) are the Pauli matrices for electrons and holes. In the absence of spin–orbit mixing of the bands Γ_1 and Γ_5, $\Delta = 0$ and the terms Γ_3 and Γ_6 have the same energy. They correspond to the total spin $S = 1$ and the component $m_z = \pm 1, 0$ of the total spin angular momentum $\boldsymbol{S} = \frac{1}{2}(\boldsymbol{\sigma}_e + \boldsymbol{\sigma}_h)$ along the c-axis. The upper singlet level Γ_4 corresponds to a total spin $S = 0$ and is shifted by Δ_1 from the terms Γ_3 and Γ_6. According to Mooser and Schlüter (1973) the splitting Δ_1 is about 2 meV.

Neglecting the inter-valence-band mixing, direct optical transitions are allowed only to the state Γ_4 for the polarization $\boldsymbol{E} \parallel z$. When the spin–orbit mixing is taken into account the optical transitions are allowed to the state Γ_4, for which only the matrix element $\langle \Gamma_4 | j_z | 0 \rangle$ differs from zero, and to the states Γ_6, for which the nonzero matrix elements are

$$\langle \Gamma_6, -1 | j_- | 0 \rangle = \langle \Gamma_6, 1 | j_+ | 0 \rangle. \tag{8}$$

For excitons bound to charged centers or isoelectronic traps, both the exchange interaction and selection rules are determined by the same formulae as for free excitons. It should be noted, however, that in the case of bound excitons the contribution to the exchange splitting constants comes not only from the short-range exchange interaction, but also from the long-range dipolar interaction (Pikus and Bir 1972). For free excitons the latter interaction results in the longitudinal–transverse splitting of excitonic states (see sect. 5).

3. *Optical orientation of bound excitons*

The effect of optical orientation has been observed both on free excitons and on excitons bound to impurities. In the lowest energy level of an exciton bound to a neutral donor, the two like electrons of the complex have antiparallel spins. If the valence band is orbitally nondegenerate, then similarly the ground state of an exciton bound to a neutral acceptor is characterized by antiparallel spins of the two like holes. Consequently, in the former case only the hole can be oriented,* while in the latter case, when a free exciton is trapped on a neutral acceptor, only the electron spin orientation survives (Gross et al. 1971, Bonnot et al. 1972). For both cases the effect of a transverse magnetic field is similar to the Hanle effect for free carriers. We therefore consider here the more characteristic case of excitons bound to charged centers or to isoelectronic traps.

3.1. *The phenomenological theory*

In the present section we use the density matrix formalism to describe optical orientation of excitons. This phenomenological method enables one to treat in a compact manner the various problems concerning bound excitons. For free excitons the method is applicable when the density matrix can be written as a product of a matrix which depends only on spin indices, and the distribution function, which depends only on the wave vector $\boldsymbol{k}$. This is the case if both the lifetime, τ_0, and the spin relaxation time, τ_s, of the exciton considerably exceed the momentum scattering time, τ_p. If this condition is not satisfied, then it is necessary to resort to a microscopic theory, which will be considered in the following sections.

Let us now define the exciton density matrix. If the exciton state may be described by the wavefunction $\Phi = \Sigma_m C_m \Phi_m$, where Φ_m is a set of basic functions, then the density matrix is defined as follows

$$\rho_{mm'} = C_m C^*_{m'}. \tag{9}$$

In a mixed state eq. (9) should be averaged over the statistical ensemble. Physically, the diagonal component ρ_{mm} of the density matrix determines the probability of finding the exciton in the basic state m, while the off-diagonal component $\rho_{mm'}(m \neq m')$ characterizes the degree of coherence between the basic states m and m'. In a pure state described by eq. (9) the diagonal and off-diagonal components are interrelated, so that $|\rho_{mm'}|^2 = \rho_{mm}\rho_{m'm'}$. For mixed states the inequality $|\rho_{mm'}|^2 < \rho_{mm}\rho_{m'm'}$ holds.

The polarization tensor of the exciton luminescence is connected with $\rho_{mm'}$ as follows

$$d_{\alpha\beta} \propto \sum_{mm'} j^{\alpha *}_m j^{\beta}_{m'} \rho_{mm'}, \tag{10}$$

where $j^{\alpha}_m = \langle m|j_\alpha|0\rangle$ is the matrix element of the current density operator, which corresponds to optical excitation of the state m.

* Fishman et al. (1974a) showed that, in an exciton created through binding of a polarized electron and unpolarized hole, the strong electron–hole exchange interaction leads to spin orientation of the hole. If the exciton is trapped on a neutral donor and the hole retains its polarization, the ($D^\circ X$) line will be polarized. In this case, for $\Gamma_6 \times \Gamma_8$ excitons in a cubic crystal, the signs of circular polarization of the ($D^\circ X$) and free exciton lines will be opposite.

The density matrix satisfies the equation

$$\frac{\partial \rho}{\partial t} = \left(\frac{\partial \rho}{\partial t}\right)_{\text{rec}} + \left(\frac{\partial \rho}{\partial t}\right)_{\text{sr}} - \frac{\mathrm{i}}{\hbar}[H_{\text{ex}}\rho] - \frac{\mathrm{i}}{\hbar}[H'\rho] + G. \tag{11}$$

Hereafter we impose the normalization condition of ρ to the total number of excitons, $N = \mathrm{Sp}\,\rho$. In a steady-state regime, $\partial\rho/\partial t = 0$. The first two terms on the right-hand side of eq. (11) describe the variation of the density matrix due to decay (recombination) and spin relaxation processes, respectively, the third term takes account of the exchange interaction H_{ex}; the forth term determines the effect of external fields (stress, magnetic field etc.). The matrix G gives the generation rate of the density-matrix components. The form of G essentially depends on a method of exciton generation. We consider here two types of exciton formation. The first one is the direct creation of excitons in the resonant excitation conditions when the incident photon energy is close to the exciton resonance energy. The second type is the formation of excitons from the binding of photo-created electrons and holes.

In the case of resonant excitation of excitons the generation matrix G is given by

$$G_{mm'} \propto \sum_{\alpha\beta} j_m^{\alpha} j_{m'}^{\beta *} d_{\alpha\beta}^0, \tag{12}$$

where $d_{\alpha\beta}^0 = e_\alpha e_\beta^*$ is the polarization tensor of the incident light, $\boldsymbol{e}$ being the polarization unit vector.

If excitons are created through binding of free electrons and holes, the matrix G is determined by the spin orientation of free carriers. If the probability of binding of an electron and a hole does not depend on their spin indices, G can be written as a direct product

$$G = G_{\text{e}} \times G_{\text{h}}, \tag{13}$$

where the matrices G_{e} and G_{h} describe the spin orientation of the electron and the hole immediately after their binding into an exciton.

In order to make transparent the main features of optical orientation of excitons, in the present section, as well as in the following one, we first consider the simplest case of the $\Gamma_6 \times \Gamma_7$ exciton in cubic crystal and then discuss briefly the effects arising in cases of more complicated structures of excitonic levels.

It follows from eqs. (5) and (10) that, for a given density matrix of the $\Gamma_6 \times \Gamma_7$ exciton, the degree of circular polarization, P_{circ}, of the exciton luminescence can be written in the form

$$P_{\text{circ}} = \frac{\rho_{11} - \rho_{-1,-1}}{\rho_{11} + \rho_{-1,-1}}. \tag{14}$$

Here the angular momentum component m is quantized in the propagation direction of the light. As is well known, the linear polarization of light is uniquely defined, either by the degrees of linear polarization, $P'_{\rm lin}$ and $P''_{\rm lin}$, referred respectively to the rectangular axes x, y and to the axes x', y', such that x' makes the angle 45° with x, or by the total degree of linear polarization, $P_{\rm lin}$, and the angle, θ, between the x-axis and the plane of polarization. The relationship between $P'_{\rm lin}$, $P''_{\rm lin}$ and $P_{\rm lin}$, θ is as follows

$$P_{\rm lin} = (P'^2_{\rm lin} + P''^2_{\rm lin})^{1/2}, \qquad \theta = \tfrac{1}{2}\,\mathrm{arctg}\,(P''_{\rm lin}/P'_{\rm lin}). \tag{15}$$

If the phases of exciton states $|\pm 1\rangle$ are chosen in such a way that the selection rules (5) are valid for the axes x, y, then $P'_{\rm lin}$ and $P''_{\rm lin}$ are given by

$$P'_{\rm lin} = -\frac{2\,\mathrm{Re}\,\rho_{1,-1}}{\rho_{11} + \rho_{-1,-1}}, \quad P''_{\rm lin} = \frac{2\,\mathrm{Im}\,\rho_{1,-1}}{\rho_{11} + \rho_{-1,-1}}. \tag{16}$$

Equations (14)–(16) provide a complete set of parameters characterizing the polarization of light propagating in a given direction. The total intensity, I, of the light emitted in the same direction is proportional to the sum $(\rho_{11} + \rho_{-1,-1})$. It should be mentioned that eqs. (14) and (16) are valid as well for the case of the $\Gamma_6 \times \Gamma_8$ exciton (in the corresponding basis).

According to eqs. (5) and (12) under resonant absorption of circularly polarized light σ_+, so that $e_- = (e_x - \mathrm{i}e_y)/\sqrt{2} = 1$ and $e_+ = (e_x + \mathrm{i}e_y)/\sqrt{2} = 0$, only the exciton state $m = 1$ is excited and the only nonzero component of G is G_{11}. Thus in this case the exciton angular momenta are oriented. On the other hand, the plane-polarized light excites the exciton state with a well-defined direction of the oscillating electric-dipole moment $\boldsymbol{d} \parallel \boldsymbol{e}$. This state can be described by the wavefunction

$$\begin{aligned}\Phi &= \frac{1}{\sqrt{2}}(\Phi_{1,1}\mathrm{e}^{\mathrm{i}\theta} - \Phi_{-1,-1}\,\mathrm{e}^{-\mathrm{i}\theta}) \\ &= \frac{1}{\sqrt{2}}(\varphi_{1/2}\psi_{1/2}\,\mathrm{e}^{\mathrm{i}\theta} - \varphi_{-1/2}\psi_{-1/2}\,\mathrm{e}^{-\mathrm{i}\theta}),\end{aligned} \tag{17}$$

where θ is the angle between the x-axis and the polarization plane of the incident light. Hence, under absorption of linearly polarized light the non-zero components of G are

$$G_{11} = G_{-1,-1} = \tfrac{1}{2}G_0, \qquad G_{1,-1} = G^*_{-1,1} = -\tfrac{1}{2}\,\mathrm{e}^{2\mathrm{i}\theta}G_0. \tag{18}$$

It is interesting to note that in the state defined by eq. (17) or eq. (18) no orientation can be assigned to the electron or the hole alone. In fact, for this state the expectation values of electronic and hole spins, defined as $s_{\rm e,h} = \mathrm{Sp}(\boldsymbol{\sigma}_{\rm e,h}\rho)/2\,\mathrm{Sp}\,\rho$, are zero. Thus, excitons can be not only optically oriented, like free carriers, upon excitation by circularly polarized light, but can also become aligned when excited by linearly polarized light, in the similar fashion as is observed for atoms in gases* (Lombard 1967, 1969, Nedelec et al. 1967). The term "optical orientation of excitons" is used here to include both orientation of exciton angular momenta and alignment of exciton oscillating dipole moments under optical excitation.

When the excitons are formed by the binding of free electrons and holes, the matrix G for the $\Gamma_6 \times \Gamma_7$ exciton can be written as

$$G = \tfrac{1}{4}(1 + 2\boldsymbol{s}_e^0 \cdot \boldsymbol{\sigma}_e)(1 + 2\boldsymbol{s}_h^0 \boldsymbol{\sigma}_h), \tag{19}$$

where $\boldsymbol{s}_e^0$ and $\boldsymbol{s}_h^0$ are the expectation values of the electron and hole spin at the moment of capture. Since optical orientation of free carriers occurs only for circular polarization of the incident light, in this case for linearly polarized excitation no alignment takes place and the exciton luminescence is unpolarized.

Now we turn to the terms $(\partial\rho/\partial t)_{rec}$ and $(\partial\rho/\partial t)_{sr}$ in eq. (11). The term $(\partial\rho/\partial t)_{rec}$, which describes the radiative and nonradiative exciton recombination, is diagonal in the basis of eigenstates of the exchange interaction Hamiltonian

$$\left(\frac{\partial \rho_{mm'}}{\partial t}\right)_{rec} = -\frac{\rho_{mm'}}{\tau_{mm'}}, \qquad \frac{1}{\tau_{mm'}} = \frac{1}{2}\left(\frac{1}{\tau_m} + \frac{1}{\tau_{m'}}\right), \tag{20}$$

where τ_m is the lifetime of the state m. For the states belonging to the same irreducible representation the lifetimes are identical. It is apparent that in the case of the $\Gamma_6 \times \Gamma_7$ or $\Gamma_6 \times \Gamma_8$ exciton in a cubic crystal the lifetime τ_0 of the electric-dipole-active level Γ_{15} is shorter than those of the inactive states Γ_2, Γ_{12}, or Γ_{25}.

The spin-relaxation term $(\partial\rho/\partial t)_{sr}$ can be written in general as

$$\left(\frac{\partial \rho_{mm'}}{\partial t}\right)_{sr} = \sum_{nn'} W_{mm'nn'}\rho_{nn'}. \tag{21}$$

By using symmetry considerations one can find a set of linearly independent components of W and express all the others in terms of this set. To determine such a set it is convenient to form from the components $\rho_{mm'}$ linear combinations $\rho_i^{(\nu,l)}$ which transform according to the irreducible representation D_ν of the crystal point group ($i = 1, 2, \ldots, n_\nu$, where n_ν is the dimensionality of the representation). The index l specifies linearly independent sets belonging to the same representation. It is clear that only

* It should be mentioned that the linear polarization of luminescence due to free carriers can be also observed. However, in this case the linear polarization is connected not with the coherence between the electron and hole spins discussed above, but with the anisotropic momentum distribution of the photo-created electrons in a semiconductor with a degenerate valence band (Zemskii et al. 1976, Dymnikov et al. 1976) or with the anisotropic population of valleys in a many-valley semiconductor (Areshev et al. 1976).

components $\rho_i^{(\nu,l)}$ corresponding to the same representation D_ν and the same index i can be interrelated. Hence, in the new basis eqs. (21) becomes

$$\left(\frac{\partial \rho_i^{(\nu,l)}}{\partial t}\right)_{\rm sr} = \sum_{l'} W_{ll'}^{(\nu)} \rho_i^{(\nu,l')}. \tag{22}$$

The operator (21) conserves the total number of excitons, i.e. the component $\rho^{(1,1)} = \mathrm{Sp}\,\rho$ that transforms according to the identity representation. This implies that in eq. (22) $W_{1l'}^{(1)} \equiv 0$. It should be noted that eq. (22) is valid provided the influence of external fields upon spin relaxation processes is negligible.

We shall illustrate the procedure by considering the spin relaxation term in a wurtzite-type crystal for large exchange splitting (Bir and Pikus 1973). In this case the electric-dipole active level Γ_5 can be treated apart. The density matrix components $\rho_{mm'}(m,\ m' = \pm 1)$ transform according to the representation $\Gamma_5 \times \Gamma_5 = \Gamma_1 + \Gamma_2 + \Gamma_6$. Combinations transforming according to the representations Γ_1, Γ_2 and Γ_6 are as follows

$$\rho^{(1)} = \rho_{11} + \rho_{-1,-1} \quad (\Gamma_1), \qquad \rho^{(2)} = \rho_{11} - \rho_{-1,-1} \quad (\Gamma_2),$$

$$\rho_1^{(6)} = \rho_{1,-1}, \qquad \rho_2^{(6)} = \rho_{-1,1} \quad (\Gamma_6). \tag{23}$$

Consequently, eq. (22) reduces to

$$\left(\frac{\partial \rho^{(2)}}{\partial t}\right)_{\rm sr} = -\frac{\rho^{(2)}}{\tau_{s1}}, \qquad \left(\frac{\partial \rho_i^{(6)}}{\partial t}\right)_{\rm sr} = -\frac{\rho_i^{(6)}}{\tau_{s2}} \quad (i = 1, 2). \tag{24}$$

It follows then from eqs. (10)–(12), (24) that under resonant excitation in the steady-state regime

$$P_{\rm circ} = P_{\rm circ}^0 \frac{T_1}{\tau_0}, \qquad P_{\rm lin} = P_{\rm lin}^0 \frac{T_2}{\tau_0}, \tag{25}$$

where τ_0 is the lifetime, $T_{1,2}^{-1} = \tau_0^{-1} + \tau_{s1,2}^{-1}$, $P_{\rm circ}^0$ and $P_{\rm lin}^0$ are the degrees of circular and linear polarization of the incident radiation. Thus, $P_{\rm circ}$ depends on the spin relaxation time τ_{s1}, while $P_{\rm lin}$ is governed by τ_{s2}.

Now we discuss a specific mechanism of exciton spin relaxation in cubic crystals with a degenerate valence band (Pikus and Bir 1974). In such crystals the hole spin relaxation time τ_s is very short. If the two-phonon processes of hole spin disorientation prevail and the exchange splitting $\Delta \ll kT$, the leading term in eq. (21) is determined only by one independent parameter

$$\left(\frac{\partial \rho_{mm',nn'}}{\partial t}\right)_{\rm sr} = -\frac{1}{\tau_s}\left(\rho_{mm',nn'} - \frac{1}{4}\delta_{mm'} \sum_{m''} \rho_{m''m'',nn'}\right), \tag{26}$$

m, m' being the spin indices of the hole and n, n' being those of the electron. The operator (26) conserves the total number of excitons, $N =$

Sp ρ, and the average electron spin, $s_e = \mathrm{Sp}(\boldsymbol{\sigma}_e\rho)/2\,\mathrm{Sp}\,\rho$. If the dominant mechanism of the electron spin relaxation is the electron–hole exchange interaction, the electron spin relaxation time which determines s_e is given by

$$\frac{1}{\tau_{es}} = \frac{1}{T}\frac{5(\Delta T/\hbar)^2}{8+3(\Delta T/\hbar)^2}, \tag{27}$$

where $T^{-1} = \tau_0^{-1} + \tau_s^{-1}$, the lifetime τ_0 is supposed to be identical for the states Γ_{15}, Γ_{12} and Γ_{25}. If the lifetime for the level Γ_{15}, τ_1, differs from that of the levels Γ_{12} and Γ_{25}, τ_2, but $\tau_s \ll \tau_{1,2}$, eq. (27) is also valid, except that τ_0 should be replaced by the average exciton lifetime defined as

$$\frac{1}{\bar{\tau}} = \frac{1}{8}\left(\frac{3}{\tau_1} + \frac{5}{\tau_2}\right). \tag{28}$$

In the case when the excitons are formed by the binding of free carriers, so that only the electrons are oriented, one obtains

$$P_{circ} = -\,s_e^0\frac{1+T/\tau_{es}}{1+\bar{\tau}/\tau_{es}}. \tag{29}$$

If the photon energy lies in the interval between E_g and $E_g + \Delta_{so}$ and the depolarization of photoelectrons does not take place before exciton formation, the value of s_e^0 is close to $\frac{1}{4}$.

For the $\Gamma_6 \times \Gamma_7$ exciton, which involves an electron from the conduction band Γ_6 and a hole from the split-off valence band Γ_7, the hole spin relaxation time, τ_{hs}, is in general comparable with the electron spin relaxation time, τ_{es}. If, however, for any reason the electron spin disorientation is related to the electron–hole exchange interaction only, the leading term in $(\partial\rho/\partial t)_{sr}$ for the $\Gamma_6 \times \Gamma_7$ exciton is given by eq. (26), where the factor $\frac{1}{4}$ is replaced by $\frac{1}{2}$. Then, τ_{es} is found to be as follows

$$\frac{1}{\tau_{es}} = \frac{1}{2T}\frac{(\Delta' T/\hbar)^2}{1+(\Delta' T/\hbar)^2}. \tag{30}$$

In resonant excitation conditions for this case we obtain

$$P_{circ} = \frac{P_{circ}^0}{1+\tau_0/\tau_{es}}, \qquad P_{lin} = \frac{P_{lin}^0}{1+\tau_0/2\tau_s}. \tag{31}$$

From eq. (31) one can see that in the circular configuration optical orientation is governed by τ_{es}, whereas in the linear configuration it depends on τ_s.

3.2. Magnetic field effects for bound excitons

3.2.1. $\Gamma_6 \times \Gamma_7$ and $\Gamma_6 \times \Gamma_8$ excitons in a cubic crystal

The effect of a magnetic field on optical orientation is intimately connected with the Zeeman splitting of exciton levels. We first consider the variation

Table 1
Variation of polarization, P, and intensity, I, of the exciton radiation with magnetic field for $\Gamma_6 \times \Gamma_7$ exciton at $\Delta' = 0$.

	$\boldsymbol{H} \parallel z$		$\boldsymbol{H} \parallel x$		
	lin, $\boldsymbol{e} \parallel x$	circ	lin, $\boldsymbol{e} \parallel \boldsymbol{H}$	lin, $\boldsymbol{e} \perp \boldsymbol{H}$	lin, $\boldsymbol{e} \parallel x'$ $(x, \hat{}\, x') = 45°$
P_{circ}	0	$\frac{L_e + L_h}{2L(H_\perp)}$	0	0	0
P'_{lin}	L_+	$\frac{L_- - L_+}{4L(H_\perp)}$	1	−1	$\frac{L_- - L_+}{4L(H_\perp)}$
P''_{lin}	$\Omega^+ \tau_0 L_+$	0	0	0	$\frac{L_e + L_h}{2L(H_\perp)}$
$I(H)/I_0$	1	$L(H_\perp)$	$\frac{1}{2}(1 + L_-)$	$\frac{1}{2}(1 + L_+)$	$L(H_\perp)$

Here

$I_0 = I(0)$, $\hbar\Omega_{e,h} = g_{e,h}\mu_0 H$, $\Omega^{\pm} = \Omega_e \pm \Omega_h$,
$L_{e,h} = 1/[1 + (\Omega_{e,h}\tau_0)^2]$, $L_{\pm} = 1/[1 + (\Omega^{\pm}\tau_0)^2]$, $L(H_\perp) = \frac{1}{4}(2 + L_+ + L_-)$.

of luminescence polarization with increasing magnetic field for the $\Gamma_6 \times \Gamma_7$ exciton in a cubic crystal.

The linear Zeeman interaction is described by the Hamiltonian

$$H_H = \tfrac{1}{2}\mu_0(g_e \boldsymbol{\sigma}_e \cdot \boldsymbol{H} + g_h \boldsymbol{\sigma}_h \cdot \boldsymbol{H}), \tag{32}$$

where μ_0 is the Bohr magneton, g_e and g_h are the effective g factors of an electron and a hole. For simplicity, we assume that the lifetimes of the levels Γ_2 and Γ_{15} are the same and neglect the exciton spin relaxation. Of course, spin relaxation processes can be easily taken into account by making use of the general expression (22). To determine the magnetic-field dependence of the polarization ratio, P, as well as that of the total intensity in a given direction, I, it is necessary to solve eq. (11) for the density matrix substituting the Hamiltonian (32) into eq. (11) and using the expressions (5), (12) or (13) for G. Once ρ is determined, we can readily find P by using eqs. (14) and (16).

Table 1 gives the dependences of P_{circ}, P'_{lin}, P''_{lin} and I on the value and direction of the magnetic field under resonant excitation at small exchange splitting $\Delta' < \hbar/\tau_0$. The z-axis is in the propagation direction of the incident light which coincides with the direction of observation of exciton luminescence.

In order to obtain the expressions for the four quantities, P_{circ}, P'_{lin}, P''_{lin} and $I(H)/I_0$, at large exchange splitting,

$$\Delta' \gg \hbar/\tau_0, \quad \Delta' \gg \hbar\Omega_{e,h}, \tag{33}$$

it suffices to perform the transformations

$$\tfrac{1}{2}(L_e + L_h) \to L_+, \qquad L_- \to 1, \qquad L(H_\perp) \to \tfrac{1}{4}(3 + L_+). \tag{34}$$

One can see from table 1 and eq. (15), that, under linearly polarized excitation in a longitudinal magnetic field $\boldsymbol{H} \| z$, the plane of polarization of the exciton radiation is rotated around the direction of magnetic field through the angle $\theta = \frac{1}{2}\,\mathrm{arctg}\,\Omega_\| \tau_0$ ($\Omega_\| = \Omega^+$), while the degree of linear polarization, $P_{\mathrm{lin}} = (P'^2_{\mathrm{lin}} + P''^2_{\mathrm{lin}})^{1/2}$, decreases by the factor of $[1 + (\Omega_\| \tau_0)^2]^{1/2}$. If, for example, the initial light is polarized along the x-axis ($\boldsymbol{e} \| x$), then we have

$$P'_{\mathrm{lin}}(H_\|) = \frac{1}{1 + (\Omega_\| \tau_0)^2}, \qquad P''_{\mathrm{lin}}(H_\|) = \frac{\Omega_\| \tau_0}{1 + (\Omega_\| \tau_0)^2}. \tag{35}$$

It should be noted that these equations are valid for any value of Δ'. The longitudinal magnetic field does not affect the circular polarization in the considered above approximations, i.e. so long as $(\partial\rho/\partial t)_{\mathrm{sr}} = 0$.

In the transverse magnetic field $\boldsymbol{H} \| x$ for $\boldsymbol{e} \| x$ or $\boldsymbol{e} \| y$ excitation the linear polarization is the same and the intensity falls by a factor of two. At large exchange splitting the intensity decreases only for $\boldsymbol{e} \perp \boldsymbol{H}$. In the configuration $\boldsymbol{H} \| x$, $\boldsymbol{e} \| x'$, the angle between x and x' being 45°, $I(H)$ and $P_{\mathrm{lin}}(H)$ decrease and the plane of polarization is rotated. Under circularly polarized excitation a transverse field leads to the decrease of P_{circ} and gives rise to a linear polarization of the radiation. For large Δ' and $\boldsymbol{H} \| x$,

$$P_{\mathrm{circ}}(H_\perp) = \frac{1}{1 + \frac{3}{4}(\Omega_\perp \tau_0)^2}, \qquad P'_{\mathrm{lin}}(\mathrm{H}_\perp) = \frac{(\Omega_\perp \tau_0)^2}{4 + 3(\Omega_\perp \tau_0)^2}. \tag{36}$$

where $\Omega_\perp = \Omega^+$. In the strong-field limit, $\Omega_\perp \tau_0 \gg 1$, P'_{lin} approaches $\frac{1}{3}$. The same functions describe the variation of $P''_{\mathrm{lin}}(H_\perp)$ and $P'_{\mathrm{lin}}(H_\perp)$, respectively, in the linear configuration $\boldsymbol{e} \| x'$, $\boldsymbol{H} \| x$. According to table 1 the transverse magnetic field leads to a linear polarization of the radiation even under excitation by unpolarized light. The dependence $P'_{\mathrm{lin}}(H_\perp)$ is in this case the same as under circularly polarized excitation.

Physically, the formulae above can be derived by using the following intuitive approach. For large exchange splitting, the $\Gamma_6 \times \Gamma_7$ exciton may be regarded as a classical oscillating dipole. Optical pumping in the configuration $\boldsymbol{e} \| x$ excites the electric-dipole moment $\boldsymbol{d}_0$ of the oscillator oriented along the x-axis. In a longitudinal magnetic field this dipole will rotate around the z-axis with angular frequency $\Omega_\|/2$. Hence, for the exciton created at $t = 0$ the components d_x, d_y, $d_{x'}$ and $d_{y'}$ at time $t > 0$ are

$$d_x(t) = d_0 \cos(\tfrac{1}{2}\Omega_\| t), \qquad d_y(t) = d_0 \sin(\tfrac{1}{2}\Omega_\| t),$$

$$d_{x'}(t) = d_0 \cos(\tfrac{1}{2}\Omega_\| t - \tfrac{1}{4}\pi), \qquad d_{y'}(t) = d_0 \sin(\tfrac{1}{2}\Omega_\| t - \tfrac{1}{4}\pi).$$

The degrees of linear polarization, $P'_{\rm lin}$ and $P''_{\rm lin}$, can be written in the form

$$P'_{\rm lin} = \frac{\overline{d_x^2} - \overline{d_y^2}}{\overline{d_x^2} + \overline{d_y^2}}, \qquad P''_{\rm lin} = \frac{\overline{d_{x'}^2} - \overline{d_{y'}^2}}{\overline{d_{x'}^2} + \overline{d_{y'}^2}}, \tag{37}$$

where

$$\overline{d_\alpha^2} = \frac{\int_0^\infty \exp(-t/\tau_0) d_\alpha^2(t)\,{\rm d}t}{\int_0^\infty \exp(-t/\tau_0)\,{\rm d}t}. \tag{38}$$

Substitution of $(d_x^2 - d_y^2) = d_0^2 \cos \Omega_\| t$, $(d_{x'}^2 - d_{y'}^2) = d_0^2 \sin \Omega_\| t$ and $(d_x^2 + d_y^2) = (d_{x'}^2 + d_{y'}^2) = d_0^2$ into eq. (37) and integration over t gives the same expressions for $P'_{\rm lin}$ and $P''_{\rm lin}$ as in eq. (35). Similarly, one can derive eq. (36) taking into account the fact that in the transverse magnetic field the dipole moment will rotate around the field vector with the angular frequency $\frac{1}{2}\Omega_\perp$, while the angular momentum excited by circular polarized light will precess around the field vector with the frequency $\Omega_\perp$.

In analogous fashion, one can obtain the functions given in table 1, if use is made of the fact that for $\Delta' = 0$ the electron and hole spins precess independently with the angular frequencies $\Omega_{\rm e}$ and $\Omega_{\rm h}$.

For the $\Gamma_6 \times \Gamma_8$ exciton in a cubic crystal, besides the appearance of a linear polarization in a transverse magnetic field, under unpolarized excitation a circular polarization of the exciton luminescence may be induced by a longitudinal magnetic field. The latter effect, called recombination orientation of excitons, arises in the case when the lifetimes of the electric-dipole-active states Γ_{15} and inactive states Γ_{12} and Γ_{25} are different (Pikus and Bir 1974). Optical orientation of $\Gamma_6 \times \Gamma_8$ excitons is rather sensitive to strain (Pikus and Bir 1974a). For small exchange splitting Δ, a small uniaxial strain results in an increase and anisotropy of the electron spin relaxation time. With increasing strain, when the strain-induced splitting of the excitonic states $\Delta_{\rm str} \sim \hbar/\tau_{\rm s}$, the hole spin relaxation time, $\tau_{\rm s}$, rapidly increases. This leads at $\Delta\tau_{\rm s} \ll \hbar$ to the decrease in $\tau_{\rm es}$. Hence, the electron spin relaxation time, as a function of stress, exhibits a maximum. For large exchange splitting so that $\Delta\tau_{\rm s} > \hbar$, even a small transverse strain can almost completely suppress the polarization. Internal random strains (and electric fields) can therefore prevent the observation of optical orientation of $\Gamma_6 \times \Gamma_8$ excitons with a large exchange splitting. The effect of internal strains may be diminished by applying an external longitudinal strain or a strong longitudinal magnetic field.

3.2.2. $\Gamma_7 \times \Gamma_9$ *and* $\Gamma_7 \times \Gamma_7$ *excitons in a wurtzite-type crystal*

The effect of a magnetic field on optical orientation of excitons in wurtzite-type crystals has been theoretically studied by Bir and Pikus (1972a, 1972b, 1973). Throughout the present paper, we assume that for uniaxial crystals, both the incident and radiated light propagate along the c-axis of a crystal.

Then for a wurtzite crystal the variation of polarization in a longitudinal magnetic field is described by eq. (35), just as for the $\Gamma_6 \times \Gamma_7$ exciton in a cubic crystal. However, the behaviour of polarization under the influence of a transverse magnetic field is quite different. For the $A(\Gamma_7 \times \Gamma_9)$ exciton, as well as for a hole in the A valence band (Γ_9), the Zeeman effect in the transverse magnetic field is entirely absent. In this case, therefore, the transverse magnetic field gives rise to no depolarizing effect, provided that $g_\perp^e \mu_0 H_\perp \ll |E_{\Gamma_5} - E_{\Gamma_6}| = 2\bar{\Delta}_\parallel$, i.e. unless the field mixes strongly the terms Γ_5 and Γ_6. For the level Γ_5 of the $\Gamma_7 \times \Gamma_7$ exciton (B or C), the Zeeman splitting, $\Delta E_\perp$, is quadratic in $H_\perp$. It follows that under resonant excitation of the $\Gamma_7 \times \Gamma_7$ exciton by circularly polarized light, the transverse magnetic field depolarizes the circular component and induces a linear component of the exciton radiation (Bir and Pikus 1973), namely,

$$P_{\text{circ}}(H_\perp) = \frac{P^0_{\text{circ}}}{1+K^2}, \quad P'_{\text{lin}} = 0, \quad P''_{\text{lin}} = \frac{P^0_{\text{circ}} K}{1+K^2} \left(\frac{T_2}{T_1}\right)^{1/2}. \tag{39}$$

Here $K = (T_1 T_2)^{-1/2} \Delta E_\perp / \hbar$, T_1 and T_2 are the total lifetimes of orientation and alignment introduced in the previous section (see eq. (25)), the x-axis is directed along the field vector $\boldsymbol{H}_\perp$. It should be mentioned that the depolarization curve $P_{\text{circ}}(H_\perp)$ does not follow a classical Lorentzian since the splitting $\Delta E_\perp$ is proportional to $H_\perp^2$. The angle between the vector $\boldsymbol{H}_\perp$ and the direction of the field-induced linear polarization equals 45°. Conversely, upon resonant excitation by linearly polarized light in the configuration $\boldsymbol{H}_\perp \| x$, $\boldsymbol{e} \| x'$, the transverse magnetic field induces a circular component of the radiation.

3.2.3. $\Gamma_8 \times \Gamma_7$ *excitons in GaSe*

Similar partial transformation of circular polarization into linear polarization and vice versa can take place in III–VI compounds, e.g. GaSe. For GaSe, the splitting of the $\Gamma_8 \times \Gamma_7$ exciton level in the magnetic field $\boldsymbol{H}$ is described by the Hamiltonian

$$H_H = \tfrac{1}{2}\mu_0[(g_\parallel^e \sigma_z^e + g_\parallel^h \sigma_z^h) H_z + (g_\perp^e \boldsymbol{\sigma}_\perp^e + g_\perp^h \boldsymbol{\sigma}_\perp^h) \cdot \boldsymbol{H}_\perp]. \tag{40}$$

Table 2 gives the variation of polarization and intensity of the $\Gamma_8 \times \Gamma_7$ exciton radiation with a magnetic field for $\Delta_1 \gg \hbar/\tau_0$ (recall that this is the case in GaSe). One can see that for the $\Gamma_8 \times \Gamma_7$ exciton in GaSe, like the $\Gamma_6 \times \Gamma_7$ exciton in a cubic crystal, in a transverse magnetic field $\boldsymbol{H} \| x$ a linear polarization P'_{lin} is nonzero under excitation by circularly polarized, linearly polarized in the configuration $\boldsymbol{e} \| x'$ and unpolarized light. For these three cases the dependence $P'_{\text{lin}}(H_\perp)$ is the same. With increasing magnetic field strength P'_{lin} varies from zero to $(-\tfrac{1}{3})$. According to table 2 the transformation of the circular component into the linear component, P''_{lin}, occurs only for $\Delta \neq 0$. If $\Delta\tau_0 \gg \hbar$, then the value of P''_{lin} goes through a

Table 2
Variation of polarization, P, and intensity, I, of the exciton radiation with magnetic field for $\Gamma_8 \times \Gamma_7$ exciton in GaSe ($\Delta_1 \gg \hbar/\tau_0, \Delta$).

	$\boldsymbol{H} \parallel z$		$\boldsymbol{H} \parallel x$		
	lin, $\boldsymbol{e} \parallel x$	circ	lin, $\boldsymbol{e} \parallel H$	lin, $\boldsymbol{e} \perp H$	lin, $\boldsymbol{e} \parallel x'$ $(x, \hat{\ } x') = 45°$
P_{circ}	0	$u_1(H_\perp)$	0	0	$u_2(H_\perp)$
P'_{lin}	$L_\parallel$	$-\dfrac{1-L_\perp}{1+L_\perp}$	1	-1	$-\dfrac{1-L_\perp}{1+L_\perp}$
P''_{lin}	$\Omega_\parallel \tau_0 L_\parallel$	$-u_2(H_\perp)$	0	0	$u_1(H_\perp)$
$I(H)/I_0$	1	$\frac{1}{2}(1+L_\perp)$	$L_\perp$	1	$\frac{1}{2}(1+L_\perp)$

Here

$$I_0 = I(0), \quad \hbar\Omega_\parallel = (g_\parallel^e + g_\parallel^h)\mu_0 H_\parallel, \quad \hbar\Omega_\perp = (g_\perp^e + g_\perp^h)\mu_0 H_\perp,$$

$$L_\parallel = \frac{1}{1+(\Omega_\parallel\tau_0)^2}, \quad L_\perp = 1 - \frac{1}{2}\frac{(\Omega_\perp\tau_0)^2}{1+(\Omega_\perp^2+\bar{\Delta}^2)\tau_0^2}, \quad \bar{\Delta} = \Delta/\hbar,$$

$$u_1(H_\perp) = \frac{2[1+\tau_0^2(\Omega_\perp^2/4+\bar{\Delta}^2)]}{(1+L_\perp)\{[1+(\Omega_\perp\tau_0/2)^2]^2+(\bar{\Delta}\tau_0)^2\}},$$

$$u_2(H_\perp) = \frac{2\bar{\Delta}\tau_0(\Omega_\perp\tau_0/2)^2}{(1+L_\perp)\{[1+(\Omega_\perp\tau_0/2)^2]^2+(\bar{\Delta}\tau_0)^2\}}.$$

maximum at $\Omega_\perp = 2(\Delta/\hbar\tau_0)^{1/2}$ and the maximum value is about 0.5. If $\Delta\tau_0 \ll \hbar$, P''_{lin} has a maximum value of $\Delta\tau_0/4\hbar$ at $\Omega_\perp \approx 2/\tau_0$.

For $\Delta = 0$, the functions presented in table 2 can be obtained by looking upon the exciton as an oscillating dipole that rotates in a magnetic field. However, due to the particular symmetries of the conduction and valence bands of GaSe, in a transverse magnetic field $\boldsymbol{H}_\perp$ the effective dipole moment rotates not around the vector $\boldsymbol{H}_\perp$, as in the case of a classical dipole, but around the direction perpendicular both to $\boldsymbol{H}_\perp$ and to the c-axis. Moreover, the longitudinal component d_z which appears during the precession does not, in fact, lead to dielectric polarization of the medium and does not excite radiation. Bearing this in mind, we are now in a position to interpret the results presented in table 2.

The linearly polarized light with $\boldsymbol{e} \parallel y$ excites the dipole moment $\boldsymbol{d}_0 \parallel y$ which does not precess in a transverse magnetic field $\boldsymbol{H}_\perp \parallel x$. It follows therefore that in GaSe in the configuration $\boldsymbol{e} \parallel y$, $\boldsymbol{H} \parallel x$ neither the intensity nor the polarization of the exciton radiation changes in a magnetic field. In the configuration $\boldsymbol{e} \parallel x$, $\boldsymbol{H} \parallel x$ we have $d_x(t) = d_0 \cos(\frac{1}{2}\Omega_\perp t)$ and $d_y = 0$. Hence, in this case the polarization of the emitted light is also unaffected, but the intensity decreases with increasing $H_\perp$,

$$I_x(H_\perp) \propto \overline{d_x^2} = \frac{d_0^2}{\tau_0}\int_0^\infty \exp(-t/\tau_0)\cos^2\frac{\Omega_\perp t}{2}\,\mathrm{d}t$$

$$= \frac{1}{2}d_0^2\left[1 + \frac{1}{1+(\Omega_\perp\tau_0)^2}\right]. \tag{41}$$

In a strong field I_x falls by the factor of two. Hence, under excitation by unpolarized light the total intensity decreases in the transverse magnetic field as

$$\frac{I(H_\perp)}{I(0)} = \frac{1}{4}\left[3 + \frac{1}{1+(\Omega_\perp\tau_0)^2}\right]. \tag{42}$$

As $\overline{d_x^2}$ decreases and $\overline{d_y^2}$ is constant the radiation acquires the linear polarization in the coordinate system attached to the vector $\boldsymbol{H}_\perp$,

$$P'_{\text{lin}} = \frac{\overline{d_x^2} - \overline{d_y^2}}{\overline{d_x^2} + \overline{d_y^2}} = -\frac{(\Omega_\perp\tau_0)^2}{4+3(\Omega_\perp\tau_0)^2}. \tag{43}$$

It is clear that eqs. (42) and (43) are also valid for excitation by circularly polarized light or light that is linearly polarized at 45° to the direction of $\boldsymbol{H}_\perp$.

Because of the $H_\perp$ dependence of the intensity, the degrees of polarization, P_{circ} and P''_{lin}, as functions of H, differ from classical Lorentzians and are given by

$$\frac{P_{\text{circ}}(H_\perp)}{P^0_{\text{circ}}} = \frac{P''_{\text{lin}}(H_\perp)}{P''^0_{\text{lin}}} = \frac{I_0}{I(H_\perp)}\frac{1}{1+(\Omega_\perp\tau_0)^2}. \tag{44}$$

In the above we have assumed that the exciton lifetimes of the electric-dipole-active state Γ_6, τ_1, and of the inactive Γ_3 state, τ_2, are equal. In this case the linear polarization, P'_{lin}, does not appear in the transverse field if the exciton is formed from an independently excited free electron and hole. For different times $\tau_1 \neq \tau_2$, however, a transverse magnetic field induces linear polarization even when excitons are created from nonoriented electrons and holes. In this case:

$$P'_{\text{lin}} = \frac{[1-(\tau_1/\tau_2)^2]\Omega_\perp^2}{8(\Delta/\hbar)^2(\tau_1/\tau_2) + \Omega_\perp^2[3+4(\tau_1/\tau_2)+(\tau_1/\tau_2)^2]}. \tag{45}$$

If $\hbar\Omega_\perp \gg \Delta(8\tau_1/3\tau_2)^{1/2}$ and the radiative channel of the exciton recombination prevails over nonradiative one, so that $\tau_1 \ll \tau_2$, then $P'_{\text{lin}} = \frac{1}{3}$.

4. *Optical orientation of free excitons under resonant excitation*

4.1. Secondary emission in terms of light scattering

As mentioned above, the phenomenological theory is applicable for free excitons only if the exciton lifetime is much longer than the exciton momentum relaxation time. Otherwise it is necessary to apply a microscopic theory that treats the secondary radiation as the result of light scattering processes with excitons acting as intermediate states, and that takes into account multiple scattering of the excitons by impurities and phonons. Before we proceed to discuss optical orientation of free excitons, it is instructive to demonstrate the equivalence of the general microscopic approach and the phenomenological theory for the simple case of the $\Gamma_6 \times \Gamma_7$ exciton in a cubic crystal bound to a charged center.

By employing the well-known formula of quantum mechanics, the intensity, $I(\omega)$, of the bound exciton radiation under resonant monochromatic excitation can be written in the form

$$I(\omega) \propto \delta(\omega - \omega_0) \frac{1}{2\pi} \sum_{\alpha} \left| \sum_m \frac{j_m^{\alpha *}(j_m \cdot e)}{\omega_0 - \omega_m + i\Gamma} \right|^2 I^0(\omega_0). \tag{46}$$

Here I^0, e and ω_0 are respectively the intensity, polarization vector and frequency of the incident light, m is the index of an excitonic state, $\hbar\omega_m$ is the exciton energy in the state m, $\Gamma = 1/(2\tau_0)$ is the exciton damping. Equation (46) is similar to the result obtained in the theory of resonant fluorescence of atoms. By analogy with eq. (46), the polarization tensor is given by

$$d_{\alpha\beta}(\omega) \propto \delta(\omega - \omega_0) \frac{1}{2\pi} \left(\sum_m \frac{|j_m^{\alpha}|^2}{\omega_0 - \omega_m + i\Gamma} \right) \left(\sum_{m'} \frac{|j_{m'}^{\beta}|^2}{\omega_0 - \omega_{m'} + i\Gamma} \right)^* d_{\alpha\beta}^0(\omega_0). \tag{47}$$

Here the basis vectors, e_α, of light polarization are conveniently chosen in such a way that the transition to each excitonic state is allowed for one polarization component only, i.e. $j_m^{\alpha} j_m^{\beta *} = \delta_{\alpha\beta} |j_m^{\alpha}|^2$. The form of the basis depends on the direction of the magnetic field. For a longitudinal magnetic field, these are the unit vectors of σ_+ and σ_- circular polarization, $e_\pm$, while in the case of a transverse field $H \| x$ the basic vectors are e_x and e_y.

According to eq. (46) the frequency of the scattered light coincides with the excitation frequency whatever the value of Γ. This result can be understood from energy conservation: since the initial and final states of the crystal are the same, the energies of the outcoming and incoming photons must be identical.

Let us show that eq. (47) reduces to eq. (35) in the limiting case, when the linewidth of the incident light $\Delta\omega_0 \gg \Gamma$ and the scattered radiation is detected in a frequency range $\Delta\omega \gg \Gamma$. In order to calculate the polarization tensor, $d_{\alpha\beta}$, which is actually measured in these conditions, we have to integrate the expression (47) for $d_{\alpha\beta}(\omega)$ over ω and ω_0. For $\Delta\omega$, $\Delta\omega_0 \gg \Gamma$ the limits of integration can be extended to $\pm\infty$. On performing the ω and ω_0 integrations, we find

$$d_{\alpha\beta} \propto d^0_{\alpha\beta}\tau_0 \sum_{mm'} \frac{|j^\alpha_m|^2 |j^\beta_{m'}|^2}{1 + \mathrm{i}\omega_{mm'}\tau_0}, \tag{48}$$

where $\omega_{mm'} = \omega_m - \omega_{m'}$. Then, by making use of the selection rules (5), for optical excitation of the $\Gamma_6 \times \Gamma_7$ exciton by circularly or linearly polarized light, we obtain the expressions for P_{circ}, P'_{lin} and P''_{lin} which coincide with eqs. (35) and (36).

4.2. Allowance for multiple scattering of free excitons

According to table 2 in a longitudinal magnetic field the degree of linear polarization, P'_{lin}, is a monotonic function of $H_\parallel$. However, in GaSe crystals the linear polarization reverses sign with increasing magnetic field strength (see fig. 4). A similar change in sign is observed for optical alignment of atoms in gases under cascade excitation, in which an atom is first excited into the upper level 1 and then goes to the lower excited state 2. If at $t = 0$ the excited atoms are in the state 1, the number of atoms in the state 2, as well as the intensity of the radiation due to transitions from the state 2, varies with time as

$$f(t) = \frac{\tau_2}{\tau_2 - \tau_1}[\exp(-t/\tau_2) - \exp(-t/\tau_1)], \tag{49}$$

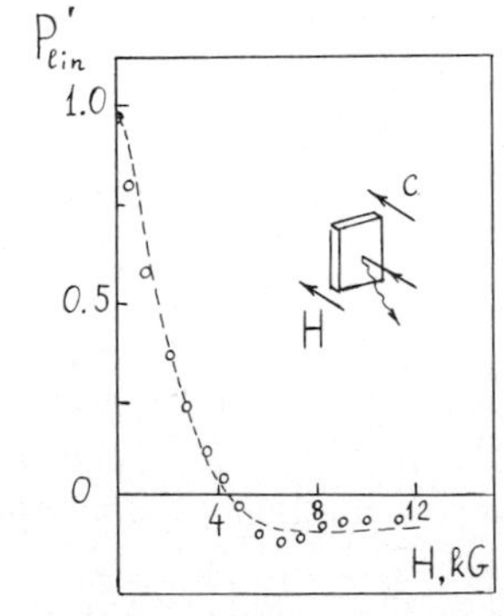

Fig. 4. The degree of linear polarization, P'_{lin}, as a function of the longitudinal magnetic field for GaSe. The dashed curve is calculated from eq. (52). (After Gamarts et al. 1977.)

where τ_1 and τ_2 are the lifetimes of the levels 1 and 2, respectively. In this case the expressions (37) and (38) for $P'_{\rm lin}$ and $P''_{\rm lin}$ are also valid if in the integrand of eq. (38) the exponential $\exp(-t/\tau_0)$ is replaced by the function (49). Then, after integration over t we obtain for $P'_{\rm lin}$

$$P'_{\rm lin}(H_\parallel) = \frac{1 - \Omega_\parallel^2 \tau_1 \tau_2}{(1 + \Omega_\parallel^2 \tau_1^2)(1 + \Omega_\parallel^2 \tau_2^2)}. \tag{50}$$

This function changes sign at $\Omega_\parallel = (\tau_1 \tau_2)^{-1/2}$.

Under resonant excitation of the $n = 1$ excitonic level the principal quantum number of the photo-created excitons, n, does not change. Hence, in order to understand the anomalous behaviour of the Hanle effect in GaSe, it is necessary to modify the cascade picture discussed above. Razbirin et al. (1976) associated the inversion of polarization with the fact that under resonant excitation conditions excitons are created in a state with the wavevector $\boldsymbol{k} = \boldsymbol{q}_0$, $\boldsymbol{q}_0$ being the wavevector of the initial light, while the radiation (observed in the back-scattering geometry) is due to the emission of the excitons with the wavevector $\boldsymbol{k}' \approx -\boldsymbol{q}_0$ which differs from $\boldsymbol{k}$. Therefore, to radiate light in the direction of observation the photo-created exciton has to be scattered, at least once, in the crystal. By analogy with the cascade excitation of atoms, the excitonic state $\boldsymbol{k} = \boldsymbol{q}_0$ is, in fact, the first step and the states with $\boldsymbol{k}' \approx -\boldsymbol{q}_0$ form the final step of the effective cascade. Assume that the excitons take part only in elastic collisions and that the scattering cross section do not depend on the scattering angle. Then it is readily shown that the probability to find an exciton in the state $\boldsymbol{k} \neq \boldsymbol{q}_0$ is proportional to the function (49) in which τ_2 and τ_1 should be replaced by τ_0 and $\tau = \tau_0 \tau_{\rm p}/(\tau_0 + \tau_{\rm p})$ respectively, where τ_0 is the exciton lifetime and $\tau_{\rm p}$ is the exciton momentum scattering time. For τ_0 of the same order as $\tau_{\rm p}$, the observed anomalies can be explained.

The initial idea of relating the inversion of polarization in a magnetic field to the relaxation of exciton momentum has proved to be relevant. However, it follows from a rigorous microscopic approach that the simple formula which was deduced by substituting the function (49) into eq. (38) instead of the exponential is incorrect. The microscopic theory of resonant light scattering developed by Ivchenko et al. (1977) takes into account elastic scattering of excitons which appear as intermediate states of the material system in the light scattering process. It is important to emphasize that multiple scattering of excitons should be included. In other words, it is necessary to sum the contributions from N-fold scattering over all $N (N = 1, 2, \ldots)$. In practice, only $N \lesssim \tau_0/\tau_{\rm p}$ are important, the ratio $\tau_0/\tau_{\rm p}$ being the mean number of collisions during the exciton lifetime. For elastic scattering of excitons by impurities, the frequencies of the scattered and initial light coincide as in the previous subsection. Moreover, in this case the contribution to the polarization tensor, $d_{\alpha\beta}(\omega)$, from the N-fold scattering

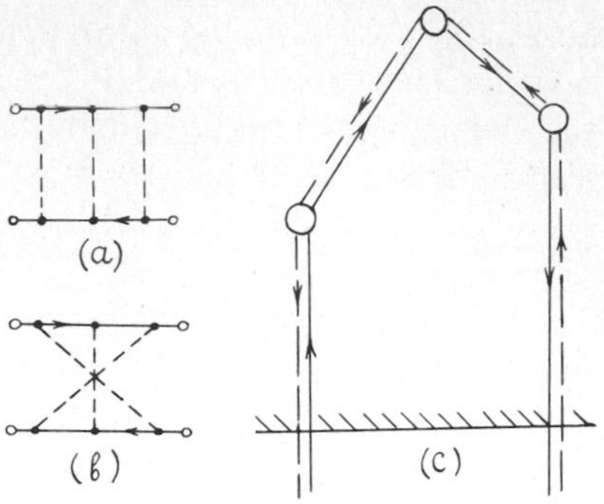

Fig. 5. The normal (a) and anomalous (b) diagrams contributing to the threefold scattering. (c) The interference of two backscattering processes.

can be written in the form (47) provided the sum

$$\sum_m |j_m^\alpha|^2/(\omega_0 - \omega_m + \mathrm{i}\Gamma)$$

is replaced by

$$\sum_m \frac{j_m^\alpha d_{-q_0,k_N} d_{k_N,k_{N-1}} \cdots d_{k_2,k_1} d_{k_1,q_0} j_m^\alpha}{\Omega_{-q_0m}\Omega_{k_Nm}\Omega_{k_{N-1}m} \cdots \Omega_{k_2m}\Omega_{k_1m}\Omega_{q_0m}}, \tag{51}$$

where $d_{k',k}$ is the matrix element of exciton scattering from the state $\boldsymbol{k}$ into the state $\boldsymbol{k}'$,

$$\Omega_{km} = \omega_0 - \omega_{km} + \mathrm{i}\Gamma, \qquad 2\Gamma = \tau^{-1} = \tau_0^{-1} + \tau_\mathrm{p}^{-1}, \qquad \omega_{km} = \omega_m + \hbar k^2/2M_\mathrm{exc},$$

$\hbar\omega_m$ is the energy of the exciton in the state m at $k = 0$, M_exc is the exciton effective mass. Figure 5(a) shows graphically the process described by the expression (51). In addition to this set of diagrams, for $N \geqslant 2$ diagrams of fig. 5(b) should be taken into account. In the back-scattering configuration, the terms corresponding to the diagrams depicted in fig. 5(a) and (b), are identical. As a result, the intensity of the back-scattered radiation due to twofold, threefold, . . . , scattering is doubled. The additional contribution is of general character and is due to the fact that, on backward scattering, the phase shift of the wave scattered by the impurity centers $1, 2, \ldots, N$ coincides exactly with that of the wave scattered in reverse order, i.e., scattered in succession by the impurity centers N, $N-1, \ldots, 1$ (see fig. 5(c)). Thus, the anomalous diagram of fig. 5(b) is a result of the interferences of the two rays. On scattering in a direction making the angle $\theta > \lambda/l_\mathrm{D}$ with the vector $(-\boldsymbol{q}_0)$, the phases of the two rays are different and the extra contribution is negligible. Here λ is the wavelength of the light inside the crystal and l_D is the diffusion length of the exciton.

Since the polarization of the contribution due to the N-fold scattering is a function of N and the intensity of the backward scattering is doubled beginning with $N=2$, the interference contribution influences on the polarization of the exciton radiation.

4.3. *Faraday rotation, birefringence and dichroism in a magnetic field*

The variation of the polarization in a magnetic field can arise not only from the variation of the exciton density matrix. The measured degree of polarization can also vary because the light polarization changes as the light propagates in the crystal before the exciton creation and after the exciton annihilation. Consider, for example, linearly polarized excitation at the resonance frequency in a longitudinal magnetic field. In this case, the polarization rotation angles due to the Faraday effect and due to the rotation of the exciton dipole moment are determined by the products $\Omega_{\|}\tau$ and $\Omega_{\|}\tau_0$, respectively, and have opposite signs. Thus, for comparable values of τ_0 and τ_p it is necessary to take account of the variation of light polarization along the direction of the light propagation in the crystal.

In the experiment by Razbirin et al. (1976) under excitation by a broad line $\Delta\omega \gg 2\Gamma = \tau^{-1}$ the exciton radiation linewidth $\delta\omega \gg 2\Gamma$. This can be understood in terms of inhomogeneous broadening which arises from the spatial fluctuations of the exciton eigenfrequency, ω_m, induced, for example, by random strains. If the deviation of ω_m depends only on the transverse coordinates, x and y, the polarization of the radiation is insensitive to the inhomogeneous broadening. On the contrary, if ω_m varies markedly over the light absorption length, the polarization rotation angle due to the Faraday effect is determined by the ratio $\Omega_{\|}/\delta\omega$ instead of the product $\Omega_{\|}\tau$. Since $\delta\omega \gg \tau^{-1}$, the change of polarization caused by the light propagation in the crystal can be neglected in the latter case.

4.4. *Hanle effect in Faraday and Voigt geometries*

It follows from the microscopic theory which takes into account both the interference effect and the Faraday rotation, that in a longitudinal magnetic field P'_{lin} and P''_{lin} are given by

$$P'_{\text{lin}}(H_{\|}) = \Phi(\Omega_{\|}, \tau_0, \tau)/(1 - \tfrac{1}{2}\tau/\tau_0), \tag{52}$$

where

$$\Phi(\Omega, \tau_0, \tau) = \frac{1}{1+\Omega^2\tau^2}\left(\frac{1-\Omega^2\tau_0\tau}{1+\Omega^2\tau_0^2} - \frac{\tau}{2\tau_0}\frac{1-\Omega^2\tau^2}{1+\Omega^2\tau^2}\right),$$

and

$$P''_{\text{lin}}(H_{\|}) = \frac{1}{(1-\frac{1}{2}\tau/\tau_0)(1+\Omega_{\|}^2\tau^2)}\left[\frac{\Omega_{\|}(\tau_0+\tau)}{1+\Omega_{\|}^2\tau_0^2} - \frac{\Omega_{\|}\tau^2}{\tau_0(1+\Omega_{\|}^2\tau^2)}\right]. \tag{53}$$

The second term in the right-hand part of eq. (52) or (53) comes from the interference in the back-scattering configuration, as well as the factor $1/(1-\frac{1}{2}\tau/\tau_0)$. If this term is omitted and $(1-\frac{1}{2}\tau/\tau_0)$ is replaced by unity, then eq. (52) reduces to eq. (50) obtained for the cascade model. However, this correspondence is accidental. Indeed, in the case of the inhomogeneous broadening, when the Faraday effect is negligible, the function $P'_{\text{lin}}(H_{\|})$ differs from eq. (52) and does not reduce to eq. (50) if the interference effect is ignored. In the limiting case $\tau \ll \tau_0$, the results of the microscopic theory reduce, as expected, to the results of the phenomenological theory presented in table 2. Figure 4 shows the depolarization curve $P'_{\text{lin}}(H_{\|})$ measured in GaSe in a longitudinal magnetic field. The dashed line is calculated by making use of eq. (52).

In sect. 4 we discussed the partial transformation of circular polarization into linear polarization and vice versa in a transverse magnetic field. However, experiment shows that for free excitons in GaSe this polarization transformation is absent within experimental uncertainty. According to table 2 this means that the value of $\Delta\tau/\hbar$ is very small. The absence of the effect can be explained by assuming that in GaSe $\Delta \leqslant 0.01$ meV in agreement with Anno and Nishina (1979). Thus, hereafter we set $\Delta = 0$, neglecting the splitting between the states Γ_3 and Γ_6 of the free exciton in GaSe. As stated above, in this case the results of the phenomenological theory can be obtained in terms of an effective oscillating dipole. The microscopic theory leads to more complicated functions than those of eqs. (41)–(44). The important result is the qualitative difference between the behaviour of the intensity $I_x(H_\perp)$ for two limiting cases, namely: in the presence (case 1) and in the absence (case 2) of birefringence and dichroism due to the field. In the case 1 for $\boldsymbol{H} \| x$ the polarization tensor components are given by

$$d_{xx} = d_{xx}^0\left\{\left(1-\frac{\tau}{2\tau_0}\right)\left[1-\frac{1}{2(1+\Omega_\perp^2\tau^2)^{1/2}}\right]\right.$$
$$\left. + [1-\tfrac{1}{2}(1+\Omega_\perp^2\tau^2)^{1/2}]\,\Phi(\Omega_\perp, \tau_0, \tau)\right\}, \tag{54}$$
$$d_{yy} = d_{yy}^0(1-\tfrac{1}{2}\tau/\tau_0), \qquad d_{xy} = d_{yx}^* = d_{xy}^0\Phi(\tfrac{1}{2}\Omega_\perp, \tau_0, \tau),$$

where $\Phi(\Omega, \tau_0, \tau)$ is given by eq. (52). The theoretical curve 1 in fig. 6 gives the ratio $I_x(H_\perp)/I_x(0)$ for this case. The curve 2 corresponds to case 2, i.e., when the inhomogeneous broadening over the light absorption length is important. With increasing magnetic field curve 1 exhibits minimum and

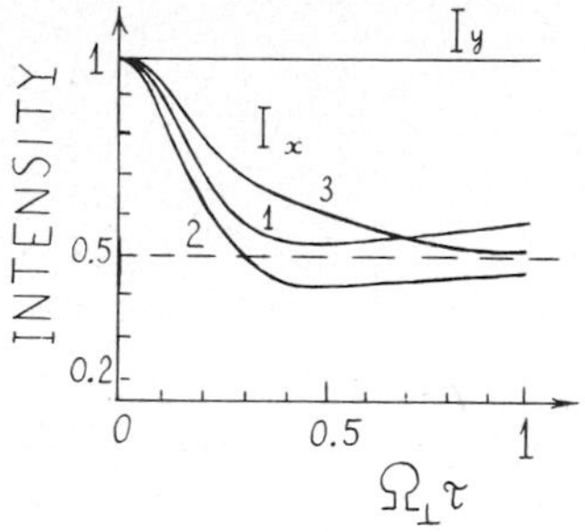

Fig. 6. The intensities I_x and I_y of exciton radiation under unpolarized resonant excitation in GaSe in a transverse magnetic field. The curves are calculated taking into account magnetic field dependence of light absorption coefficient (1), for large inhomogeneous broadening (2) and using the phenomenological theory (3).

then tends to unity, whereas curve 2 approaches $\frac{1}{2}$. From comparison with experiment (see fig. 7) it follows that the experimental data are in agreement with curve 1 and disagree both with curve 2 and with curve 3, obtained within the framework of the phenomenological theory.

Although the functions $P_{\text{circ}}(H_\perp)$ and $P''_{\text{lin}}(H_\perp)$ given by the microscopic theory differ markedly from those derived from the phenomenological theory, they remain identical to each other. This theoretical prediction is in agreement with the experimental data shown in figs. 8 and 9. The dashed lines are calculated curves using eqs. (52)–(54) with $\tau_0 = 1.6 \times 10^{-11}$ s and $\tau_p = 0.5 \times 10^{-11}$ s.

Thus, measurements of the degree of polarization as the function of the magnetic field in the resonant excitation conditions yield separately τ_0 and τ_p and, in addition, provide information concerning the character of inhomogeneous broadening.

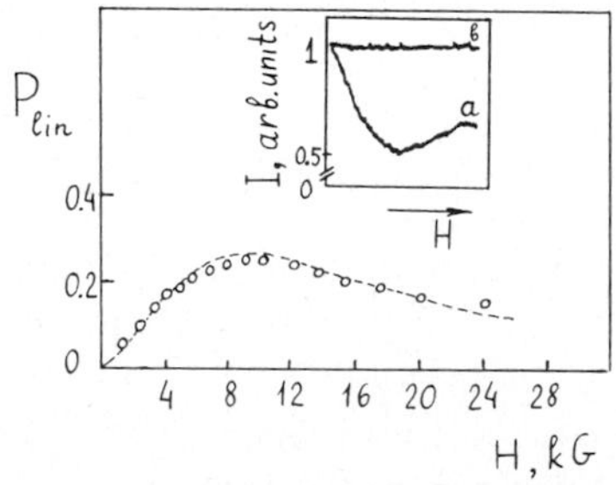

Fig. 7. The degree of linear polarization as a function of the transverse magnetic field under unpolarized resonant excitation. Points represent experimental data for GaSe. The dashed line is calculated by using eq. (54). Insert shows the variation of the intensities I_y(a) and I_x(b) (Gamarts et al. 1977).

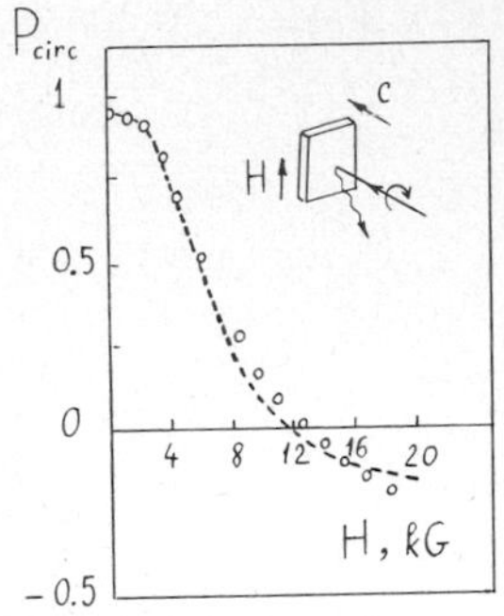

Fig. 8. Variation of the circular polarization ratio with transverse magnetic field for GaSe. The dashed curve is calculated by making use of eq. (54) (Gamarts et al. 1977).

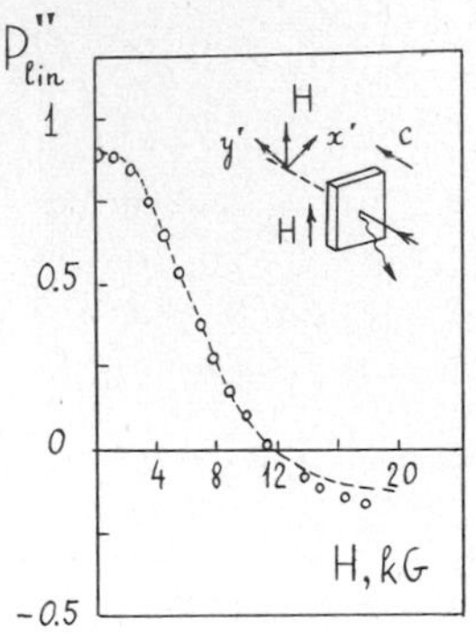

Fig. 9. The degree of linear polarization, $P^{\parallel}_{\rm lin}$, as a function of the transverse magnetic field under linearly polarized excitation of a GaSe crystal. The angle between the direction of $\boldsymbol{H}_{\perp}$ and the polarization plane of the incident light is equal to 45°. (After Gamarts et al. 1977.)

4.5. Validity of the microscopic theory

Let us now consider the criteria for the validity of the microscopic theory discussed above. First of all, they include the conditions under which the contributions from exciton scattering processes of any order N can be summed up. We assume the exciton wavevector to be a well-defined quantity. This implies that the exciton momentum uncertainty $\hbar l^{-1}$, where l is the mean free path of the exciton, is small in comparison with the excited momentum, i.e. in comparison with $\hbar q_0$. Since l is the product of the exciton velocity, $v_{q_0} = \hbar q_0/M_{\rm exc}$, and the total momentum lifetime, τ, this condition can be reduced to

$$\omega_{q_0}\tau \gg 1, \qquad (55)$$

where $\hbar\omega_{q_0} = \hbar^2 q_0^2/2M_{\rm exc}$ is the exciton kinetic energy.

We suppose also that the exciton–photon interaction is sufficiently weak that polariton effects are negligible. This approximation is valid if $l \ll l_{\rm rad}$, where $l_{\rm rad} = \Gamma/q_0\omega_{\rm LT}$ is the mean free path of the exciton with respect to radiative recombination or, equivalently, the light absorption length at the resonance frequency, $\hbar\omega_{\rm LT}$ is the longitudinal–transverse splitting. The condition $l \ll l_{\rm rad}$ can be rewritten in the form

$$\omega_{\rm LT}\tau \ll (\omega_{q_0}\tau)^{-1}. \qquad (56)$$

We neglect the exciton diffusion. This is permissible if $l_{\rm rad}$ exceeds the diffusion length of the exciton, $l_D = v_{q_0}(\tau\tau_0)^{1/2}$, or equivalently, if

$$\omega_{LT}\tau \ll (\omega_{q_0}\sqrt{\tau\tau_0})^{-1}. \tag{57}$$

In addition, above we have ignored the reemission processes, i.e. the emission of the excitons excited by the scattered light. This is valid if the radiative lifetime of the exciton, $\tau_{rad} = v_{q_0}^{-1} l_{rad}$ is much larger than the nonradiative lifetime τ_0, i.e.

$$\omega_{LT}\tau \ll (\omega_{q_0}\tau_0)^{-1}. \tag{58}$$

It is clear that among the conditions (56)–(58) the third one is the most stringent. Indeed, the inequalities (56) and (57) are certainly satisfied if the condition (58) is valid.

Moreover, for simplicity in the present section we neglected the spin relaxation of excitons and considered the exciton scattering by impurities in the Born approximation.

Estimates show that for GaSe all the above-mentioned conditions are satisfied quite well except the criterion (55) for the validity of the kinetic equation. According to Gamarts et al. (1977) the value of $\omega_{q_0}\tau$ is of the order of unity.

4.6. Reabsorption and reemission of excitonic radiation

In this subsection we shall analyze the effect of reemission on optical orientation of free excitons. Except for the inequality (58), all the other conditions of subsect. 4.5 are assumed to be valid. In this case it is necessary to take into account the multiple reabsorption and reemission of the excitonic radiation. Under resonant excitation conditions the absorption of an incoming photon, the multiple scattering of the excited exciton and, finally, the emission of an outgoing photon can be regarded as a single process of light scattering. The mean number of scattering processes in which the radiation takes part before it reaches the crystal boundary depends on the quantum efficiency of the single scattering process, $\tilde{\omega}_0 = \tau_0/(\tau_0 + \tau_{rad})$, which is usually called the albedo.

For the case of the $\Gamma_6 \times \Gamma_7$ exciton in a cubic crystal, the matrix that connects the Stokes parameters of the radiation scattered in a volume element of the crystal and the Stokes parameters of the incident radiation is a multiple of the analogous matrix for Rayleigh scattering. Thus, the problem is equivalent to problems of radiation transport in scattering mediums (see e.g. Chandrasekhar 1953, Lenable 1970). It is important to remark, that, in contrast to the problems of astrophysics and geophysics, it is necessary in the solid state physics to make allowance for the reflection of light from the crystal boundary. In fact the radiation can leave the crystal only if it is incident on the boundary within a small solid angle

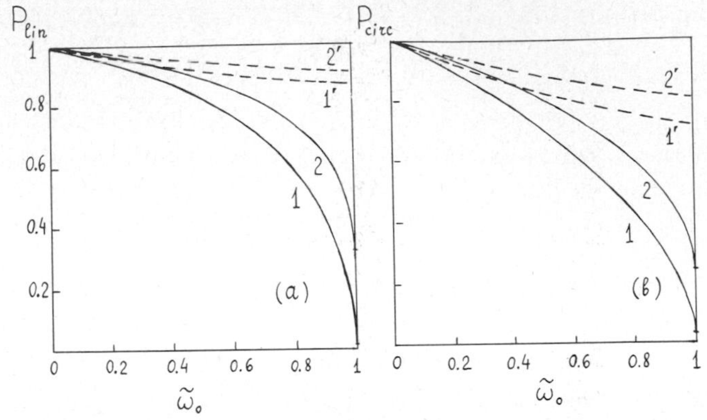

Fig. 10. The albedo dependence of the degree of linear (a) and circular (b) polarization taking into account light reabsorption and reemission processes. The curves 1 and 2 are calculated, respectively, taking account of and neglecting the light reflection from the boundary. The similar curves 1′, 2′ are calculated considering only one- and two-fold light scattering (Ivchenko and Yuldashev 1979).

$\Delta\Omega = 4\pi \sin^2(\bar{\theta}/2)$ about the normal to the boundary plane, where $\bar{\theta} = \arcsin \epsilon^{-1/2}$ is the angle of the total reflection and ϵ is the dielectric constant (in semiconductors $\epsilon \gg 1$). If θ exceeds $\bar{\theta}$, all the light is reflected back into the crystal.

For $\tau_{rad} \lesssim \tau_0$, the allowance for reflection leads to dramatic changes in the polarization of the secondary radiation. Figure 10 shows the calculated $\tilde{\omega}_0$ dependences of the degree of linear polarization under linearly polarized excitation, P_{lin}, and the degree of circular polarization under circularly polarized excitation, P_{circ}. For $\tilde{\omega}_0 \ll 1$, so that $\tau_0 \ll \tau_{rad}$, exciton creation by the scattered light can be neglected and the radiation is completely polarized in the absence of a magnetic field and of spin relaxation processes. Each light scattering process results in the partial depolarization of the light. Consequently, for $\tilde{\omega}_0 \neq 0$ the values of P_{lin} and P_{circ} are smaller than 100%. For a conservative medium ($\tilde{\omega}_0 = 1$) neglecting the reflection on the boundary, P_{lin} and P_{circ} are equal to 32% and 24%, respectively, in agreement with Chandrasekhar (1950). It follows from fig. 10 that, for a conservative medium with $\epsilon = 10$, P_{lin} and P_{circ} are smaller than 4% if allowance for reflection is made. This occurs for two reasons. First, the polarization of the radiation changes, in general, upon reflection. Second, the multiple light reflection from the boundary increases the mean number of light scattering processes.

Since under resonant excitation in GaSe at low temperatures the polarization of exciton radiation is close to 100%, we conclude that the criterion (58), as well as the condition (57) or (56), is satisfied in this case.

5. Optical orientation of hot excitons

5.1. Multiphonon secondary emission

It has been found for a number of semiconductors that under excitation with light of frequency ω_0 in the region of the fundamental absorption edge the secondary radiation contains a series of narrow lines at frequencies $(\omega_0 - N\omega_{LO})$, where ω_{LO} is the longitudinal optical-phonon frequency and $N = 2,3,\ldots$. The polarization properties of these lines have been experimentally investigated on CdS, CdSe, ZnTe and ZnSe for linearly and circularly polarized exciting light (Bonnot et al. 1974, Permogorov et al. 1975, Minami et al. 1976, Permogorov and Morozenko 1979). It has been established that the polarization of the observed overtones decreases as N increases. In this section we shall discuss some possible mechanisms of the overtone depolarization.

In terms of a consistent quantum-mechanical description, the multiphonon lines of the secondary radiation should be viewed as a result of resonant light scattering with excitons and electron–hole pairs acting as intermediate states and interacting with LO phonons (see, e.g., Bir et al. 1976, Klochikhin et al. 1978, Ivchenko et al. 1978 and references therein, and Permogorov, ch. 5 of this volume). If exciton–phonon coupling constants are small then the contribution of free electron–hole pairs to the multiphonon spectrum is of higher order in these constants (Zeyher 1975, Bir et al. 1976) and the main contribution comes from excitons, i.e., the bound states of an electron and a hole. In this case the secondary radiation can be described as hot luminescence of nonthermalized excitons arising from the following cascade process: the formation of a hot exciton by indirect excitation involving an LO phonon, the emission of $(N-2)$ LO phonons by the exciton and the indirect radiative annihilation of the exciton accompanied by emission of an additional LO phonon. This picture holds for any value of N except $N = 2$. For the $(\omega_0 - 2\omega_{LO})$ overtone there exists an additional interference contribution (Bir et al. 1976, Ivchenko et al. 1977a, 1978a). In the back-scattering geometry this contribution doubles the radiation intensity without affecting the polarization. In contrast to the elastic multiple scattering considered in subsect. 4.2, here the interference contribution for $N > 2$ is small.

If exciton–phonon coupling is not weak, important intermediate states are free electron–hole pairs modified by the Coulomb interaction (Klochikhin et al. 1978).

5.2. Heavy–light exciton splitting

For hot excitons there exist specific spin-relaxation mechanisms due to the interaction between the angular momentum and the wavevector $\boldsymbol{k}$ of the exciton. As an example, let us analyze optical orientation of hot excitons in a cubic crystal with the degenerate valence band Γ_8. The $\Gamma_6 \times \Gamma_8$ exciton dispersion relation consists of two branches. They are called the branches of heavy and light excitons, because, for hot excitons of sufficiently high kinetic energy exceeding the exciton binding energy, a heavy exciton is formed from an electron and a heavy hole and a light exciton is a bound state of an electron and a light hole. The oscillating electric-dipole-moment of a heavy exciton lies in the plane perpendicular to the vector $\boldsymbol{k}$, while that of a light exciton is predominantly directed along $\boldsymbol{k}$. Therefore the change in direction of the exciton wavevector in the process of phonon scattering leads to decrease in the exciton orientation or alignment. The polarization can be also partially lost during the indirect creation or annihilation of the exciton.

The N dependences of the polarization of LO-phonon overtones, $P_{\text{lin}}^{(N)}$ and $P_{\text{circ}}^{(N)}$, have been calculated by Bir et al. (1976) and Ivchenko et al. (1978). The assumptions used are as follows. The heavy–light exciton splitting of hot excitons is assumed to be much larger than the uncertainty in the exciton energy, $\hbar/\tau$, where τ is the exciton lifetime at a corresponding stage of the cascade. In this case the off-diagonal components of the exciton density matrix taken between the light- and heavy-exciton states vanish within the exciton lifetime at any stage and can be neglected. This results in decrease of the correlation between the angular momenta of the electron and the hole. The other assumptions are

$$m_e \ll m_L,\ m_H; \qquad (ka_B)^2 \gg [m_e^{-1}(m_L^{-1} - m_H^{-1})^{-1}], \tag{59}$$

where m_e is the electron effective mass, m_L and m_H are the effective masses of the light and heavy holes, respectively, a_B is the Bohr radius of the exciton that corresponds here to the mass m_e.

Neglecting the momentum dependence of the electron–phonon interaction as well as that of the LO-phonon frequency, the polarizations of the overtones $[\omega_0 - (N+1)\omega_{LO}]$ are given by

$$P_{\text{lin}}^{(N+1)} = P_{\text{lin}}^{0} \frac{3 + 13 \times 3^N}{1 + 11 \times 3^N + 4 \times 5^N}, \qquad P_{\text{circ}}^{(N+1)} = P_{\text{circ}}^{0} \frac{10 + 5 \times 3^N + 5^N}{1 + 11 \times 3^N + 4 \times 5^N}, \tag{60}$$

where P_{lin}^{0} and P_{circ}^{0} correspond to the polarization of the exciting light. The degree of linear polarization falls to zero as the number of emitted LO phonons increases. At the same time, the degree of circular polarization tends to 25% for large values of N. The latter result can be understood by taking into account the fact that the spin-relaxation mechanism under consideration leads to disorientation of the hole angular momentum, while the mean electron spin is conserved for this mechanism.

Allowance for the electron–hole exchange interaction results in a decrease of the degree of the electron orientation. The exchange interaction splits the states of a heavy hot exciton into two terms with the angular momentum components $m_z = \pm 1$ and $m_z = \pm 2$, where the z-axis is along the vector $\boldsymbol{k}$. The states of a light hot exciton are split into three terms. If the exchange splittings exceed the value of $\hbar/\tau$, then the corresponding off-diagonal components of the exciton density matrix vanish. In this case the rate of depolarization essentially depends on the balance between the light- and heavy-exciton contributions to the secondary radiation. The degree of circular polarization of the radiation due to heavy excitons can be written in the form

$$P_{\rm circ}^{(N+1)} = P_{\rm circ}^{0} \frac{5[3^N + 3(7/3)^{N-1}]}{1 + 11 \times 3^N + 4 \times 5^N}. \tag{61}$$

One can readily verify that $P_{\rm circ}^{(N)}$ vanishes for large N. As mentioned above, formulae (60) and (61) are obtained neglecting the momentum dependence of the electron–phonon interaction. In the case where small-angle scattering of excitons plays a leading role, for $N \geq 3$ the depolarization is diminished. The polarization of the $(\omega_0 - 2\omega_{\rm LO})$ line does not change if the allowance is made for the momentum dependence of the electron–phonon interaction.

The results obtained for cubic crystals are applicable for wurtzite-type crystals if the kinetic energies of the hot excitons exceed the crystal splitting of the valence bands at $\boldsymbol{k} = 0$. The values of $P_{\rm circ}^{(N)}$ and $P_{\rm lin}^{(N)}$ calculated from eqs. (60) and (61) are shown in fig. 11. The crosses represent the results of polarization measurements for an undoped sample of CdS. In the case of CdS some criteria for validity of the theory are not satisfied. Nevertheless, good agreement is obtained between the experimental and theoretical values.

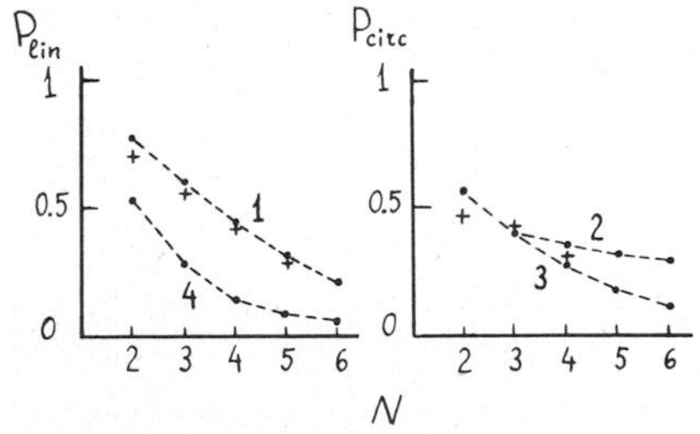

Fig. 11. The dependence of the polarization degree of LO phonon replica upon the number of emitted phonons. The points 1, 2 are calculated neglecting the exchange and longitudinal–transverse splitting (eq. (60)); the points 3 are calculated for large exchange splitting; the points 4 correspond to large longitudinal–transverse splitting (Ivchenko et al. 1978). Crosses are experimental data obtained by Permogorov et al. (1975) on CdS under excitation with $\lambda = 4579$ Å at 77 K.

5.3. Longitudinal–transverse splitting

Bonnot et al. (1974) proposed an exciton spin-relaxation mechanism connected with the longitudinal–transverse splitting of the exciton states. The mechanism can be efficient even in semiconductors with simple bands. For the $\Gamma_6 \times \Gamma_7$ exciton in a cubic crystal the splitting is described by the Hamiltonian

$$H_{\mathrm{LT}}\left(\frac{\boldsymbol{k}}{k}\right) = -\hbar\omega_{\mathrm{LT}}\left[\left(\boldsymbol{J}\cdot\frac{\boldsymbol{k}}{k}\right)^2 - 1\right], \tag{62}$$

where J_α is the matrix of the operator of the angular momentum component ($\alpha = x, y, z$) in the basis $J = 1, m_z = \pm 1{,}0$ of the representation Γ_{15}.

Due to the longitudinal–transverse (or annihilation) interaction (62) between the electron and hole in an exciton, the exciton dispersion curve is split into two branches, or, to be more precise, the longitudinal branch is shifted by $\hbar\omega_{\mathrm{LT}}$ with respect to the branch of transverse excitons. Note that the interaction (62) does not alter the dipole inactive state Γ_2 of the $\Gamma_6 \times \Gamma_7$ exciton. For a longitudinal exciton the electric-dipole moment $\boldsymbol{d}$ is directed along $\boldsymbol{k}$. For the doubly degenerate states of the transverse exciton, the vector $\boldsymbol{d}$ is perpendicular to $\boldsymbol{k}$. For linearly polarized excitation we shall analyze the behaviour of the vector $\boldsymbol{d}$, while for circular polarization it is more convenient to discuss the behaviour of the exciton angular momentum.

We first consider the case of large longitudinal–transverse splitting, so that $\omega_{\mathrm{LT}} \gg \tau^{-1} = \tau_0^{-1} + \tau_{\mathrm{p}}^{-1}$, and neglect any other spin relaxation processes. For $\omega_{\mathrm{LT}}\tau \gg 1$, the coherence between the states of longitudinal (L) and transverse (T) excitons vanishes and exciton scattering into the states L and T can be considered separately. It is clear that, for the exciton scattering from the state with the wavevector $\boldsymbol{k}_0$ and the dipole moment $\boldsymbol{d}$ into the state (L, $\boldsymbol{k}$), only the component of $\boldsymbol{d}$ along the direction of $\boldsymbol{k}$ is conserved. For the transition to the state (T, $\boldsymbol{k}$), only the component of $\boldsymbol{d}$ perpendicular to $\boldsymbol{k}$ is conserved. In addition, the former transition leads to complete disorientation of the exciton angular momentum, while for the latter one the projection of the angular momentum on $\boldsymbol{k}$ is conserved. It follows from this that the polarization of hot exciton radiation will decrease with increasing number of involved LO phonons. If the cross section for exciton scattering by LO phonons is independent of angle, then the results are as follows:

$$P_{\mathrm{lin}}^{(N+1)} = P_{\mathrm{lin}}^{0}\frac{3^{N+1}}{3^N + 2\times 5^N}, \qquad P_{\mathrm{circ}}^{(N+1)} = P_{\mathrm{circ}}^{0}\frac{5^N}{3^{N-1}(3^N + 2\times 5^N)}. \tag{63}$$

One can observe that $P_{\mathrm{circ}}^{(N+1)}$ falls more quickly than $P_{\mathrm{lin}}^{(N)}$.

Scattering time of excitons by LO phonons is very short. But when the kinetic energy, ϵ_{kin}, of the hot excitons becomes smaller than $\hbar\omega_{LO}$, the role of LO phonons in establishing the steady-state population of excitonic states is diminished. Hence, in the energy region $\epsilon_{kin} < \hbar\omega_{LO}$ the energy distribution of hot excitons is governed by exciton nonradiative decay (trapping on impurities) and quasi-elastic scattering (by acoustic phonons). For large ω_{LT} the orientation and alignment of the hot excitons with $\epsilon_{kin} < \hbar\omega_{LO}$ will decrease rapidly with the exciton energy decreasing. The energy distribution function of these excitons can be found from measurements of secondary radiation spectra arising due to LO and 2LO phonon-assisted radiative annihilation of the excitons (Gross et al. 1966, Permogorov and Travnikov 1979, Permogorov, ch. 5 of this volume). It is evident that, if ω_{LT} is large, the long-wavelength tail of the spectra will be completely depolarized.

If $\omega_{LT} \ll \tau^{-1}$ and $\tau_p \ll \tau_0$, then optical orientation of hot excitons can be described by the average density matrix, $\bar{\rho}$, that does not depend on $\boldsymbol{k}$.

In this case we can treat the depolarizing effect of the longitudinal–transverse splitting by analogy with the treatment of the free-electron spin relaxation due to k^3 splitting of spin states in the conduction band for III–V semiconductors (D'yakonov and Perel' 1971, 1971a). The equation for $\bar{\rho}$ can be written in the form

$$\frac{\bar{\rho}}{\tau_0} + \frac{\mathrm{i}}{\hbar}[H'\bar{\rho}] + \frac{\tau_p'}{\hbar^2}\overline{[H_{LT}[H_{LT}\bar{\rho}]]} = G. \tag{64}$$

Here, by definition,

$$\bar{F} = \int_{4\pi} F\,\frac{d\Omega}{4\pi}, \qquad \frac{1}{\tau_p'} = \frac{3}{4}\int_0^{\pi} W(\theta)\sin^3\theta\, d\theta,$$

where $W(\theta)/2\pi$ is the probability for exciton scattering by impurities per unit time per unit solid angle in the direction θ. If $W(\theta) = W = \text{const}$, then $\tau_p' = W^{-1}$ and coincides with the momentum relaxation time τ_p. The last term in the left-hand part of eq. (64) describes the spin relaxation connected with the interaction (62). For simplicity, we neglect in eq. (64) the contribution from the scattering by acoustic phonons.

As in sect. 2, it is convenient to introduce the components $\bar{\rho}_i^{(\nu)}$ which transform according to irreducible representations. Then the spin relaxation term of eq. (64) can be reduced to

$$-\left(\frac{\partial \bar{\rho}_i^{(\nu)}}{\partial t}\right)_{sr} = \frac{\bar{\rho}_i^{(\nu)}}{\tau_{s\nu}}. \tag{65}$$

In the spherical approximation the components of $\bar{\rho}$ transform according to the representations $2D_0^+$, $3D_1^+$ and D_2^+ of the orthogonal group. The components $\mathrm{Sp}(\bar{\rho}J_\alpha)$ form a basis for the representation D_1^+; the components

$\mathrm{Sp}(\bar{\rho}R_j)$, R_j being $(J_x^2 - J_y^2)$, $(J_y^2 - J_z^2)$, $(J_\alpha J_\beta + J_\beta J_\alpha)$ $(\alpha \neq \beta)$, transform according to D_2^+. One can show that for the two sets of components the spin relaxation times are given by

$$1/\tau_{s1} = \tfrac{2}{3}\omega_{LT}^2\tau_p', \qquad 1/\tau_{s2} = \tfrac{2}{5}\omega_{LT}^2\tau_p', \tag{66}$$

respectively. It is interesting to note that as the collision rate increases the spin relaxation is diminished (so-called "motional narrowing"). For the off-diagonal components of $\bar{\rho}$ taken between the states Γ_2 and Γ_{15}, the spin relaxation time is equal to τ_{s1}. For the polarization of the hot exciton secondary radiation we obtain

$$P_{\mathrm{lin}} = P_{\mathrm{lin}}^0 \frac{1}{1 + \frac{2}{3}\tau_0/\tau_{s2}}, \qquad P_{\mathrm{circ}} = P_{\mathrm{circ}}^0 \frac{1 + \tau_0/\tau_{s2}}{(1 + \tau_0/\tau_{s1})(1 + \frac{2}{3}\tau_0/\tau_{s2})}. \tag{67}$$

5.4. *Polariton effects*

If one of the frequencies $\omega_N = \omega_0 - N\omega_{LO}$ $(N = 0, 1 \ldots)$ lies in the exciton resonance region, it is necessary to take account of exciton–photon interaction. If the conditions (56) and (57) of the weak exciton–photon coupling are satisfied, the theory discussed in sect. 4 can be readily extended to describe the intensity and polarization of the line ω_N. In this subsection we consider the opposite limit of the strong exciton–photon coupling, $\omega_{LT} \gg \tau^{-1}, (\omega_{q_0}\tau^2)^{-1}$, so that the exciton–photon interaction leads to the formation of coupled modes, or exciton polaritons (Pekar 1957, Hopfield 1958, Agranovich 1959; see also chs. 2–4 of this volume). For the Γ_{15} exciton in a cubic crystal, the exciton–polariton dispersion relation is shown in fig. 12. The doubly degenerate branches 1 and 2 correspond to transverse waves, while nondegenerate branch 3 is the dispersion relation of longitudinal waves. In the frequency region $(\omega - \omega_T) \gg \omega_{LT}$, the branches 1 and 3 are

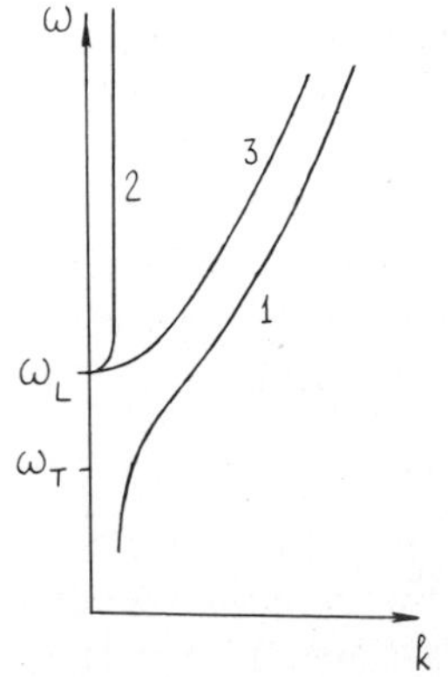

Fig. 12. The exciton-polariton dispersion relation for the Γ_{15} exciton in a cubic crystal.

shifted with respect to each other by the value of the longitudinal–transverse splitting, ω_{LT}.

For the Γ_5 exciton states in wurtzite-type crystals with the wave vector $\boldsymbol{k} \parallel \boldsymbol{c}$ the exciton polariton spectrum consists of two branches of transverse waves which are similar to the branches 1 and 2 in fig. 12. Longitudinal waves are absent in this case, because, in contrast to the threefold degenerate term Γ_{15}, the term Γ_5 is doubly degenerate. If the angle between the direction of $\boldsymbol{k}$ and $\boldsymbol{c}$ differs from 0 or π, then the degeneracy of branches 1 and 2 is removed and each of them splits into two branches of transverse and mixed modes.

In crystals with the wurtzite structure polariton effects have been found to be pronounced. For example, in CdSe crystals $\hbar\omega_{LT}$ equals $7.5\ \mathrm{cm}^{-1}$ (or 0.92 meV) and considerably exceeds the exciton damping $\hbar\Gamma$ for hot excitons with $\epsilon_{kin} < \hbar\omega_{LO}$ (Kiselev et al. 1974).

It follows then that within the exciton–polariton "bottleneck" the spin relaxation time is of the order of the momentum scattering time. Hence, in optical orientation experiments the polarization of exciton-polariton radiation will be, basically, connected with those polaritons which are created with $\boldsymbol{k} \parallel \boldsymbol{c}$ and reach the surface, before being scattered by acoustic phonons or impurities.

The effect of magnetic field on optical orientation of exciton polaritons has been observed by Nawrocki et al. (1976). The experiment was performed on CdSe crystals excited $2\hbar\omega_{LO}$ above the "bottleneck" region. In this case the secondary radiation spectrum in the excitonic region can be divided into two parts. The first part is almost independent of the exciting energy. The second one is a sharp peak at $(\omega_0 - 2\omega_{LO})$ superimposed over this "permanent" spectrum. The peak arises due to nonequilibrium polaritons created after a rapid LO-phonon cascade, i.e. by successive emission of two LO phonons, as described in subsect. 5.1.

Under linearly polarized excitation, the "permanent" spectrum was found to be unpolarized, whereas the superimposed peak was strongly polarized, the degree of polarization lying between 40 and 80%, depending on the exciting energy, $\hbar\omega_0$. The polaritons are created with rather well-defined energy, $\hbar(\omega_0 - 2\omega_{LO})$, but the initial direction of their wavevector is arbitrary. The peak radiation corresponds to the polaritons which propagate in a small solid angle about the normal to the boundary plane. The mean lifetime of such polaritons is given by (Planel et al. 1977)

$$T^{-1} = v_g/d + \tau_p^{-1}. \tag{68}$$

Here $d = \alpha^{-1}$ is the depth of absorption of incoming photons (α is the light absorption coefficient), v_g is the polariton group velocity which depends on the energy, $\hbar\omega$, and the polariton branch number, τ_p is the momentum scattering time. As strong polarization is observed, we may conclude that

τ_p is due not to impurities, but to acoustic phonon scattering. In fact, scattering by impurities, if important, would lead to the depolarization of the peak radiation.

It follows from eq. (35) that in the case under consideration the depolarizing effect of a longitudinal magnetic field is determined by the product $\Omega_{\parallel}^{pl}T(\omega)$, where $\hbar\Omega_{\parallel}^{pl}$ is the Zeeman splitting between the states $m_z = \pm 1$. According to eqs. (15) and (35) the product $\Omega_{\parallel}^{pl}T$ can be expressed in terms of the polarization rotation angle, θ, or the degree of linear polarization, P_{lin}, as follows:

$$\Omega_{\parallel}^{pl}T = \mathrm{tg}\, 2\theta,$$

$$\Omega_{\parallel}^{pl}T = [(P_{lin}(0)/P_{lin}(H_{\parallel}))^2 - 1]^{1/2}. \tag{69}$$

The effect of a longitudinal magnetic field may be considered in terms of the Faraday rotation of the polariton polarization plane as the polariton propagates from the point of its creation to the crystal boundary. The contribution to the radiation intensity from polaritons which are excited at the distance z from the crystal surface is proportional to $\exp[-(\alpha + \alpha_p)z]$, $\alpha_p = (v_g\tau_p)^{-1}$ being the polariton absorption coefficient. The average distance that polariton travels before reaching the crystal surface is $\bar{z} = v_g T$. The rotation angle for a polariton created in the space point z is given by $\theta(z) = (q_+ - q_-)z/2$, where $q_\pm$ are the wavevectors of the right- and left-handed circular polarized polaritons. On performing the z integration of the polarization tensor, we obtain

$$P'_{lin}(H_{\parallel}) = P'_{lin}(0)/[1 + (q_+ - q_-)^2(\alpha + \alpha_p)^{-2}],$$
$$P''_{lin}(H_{\parallel}) = P'_{lin}(0)(q_+ - q_-)(\alpha + \alpha_p)^{-1}/[1 + (q_+ - q_-)^2(\alpha + \alpha_p)^{-2}]. \tag{70}$$

Because of the z dependence of θ, the magnetic field leads to the partial depolarization of the radiation. Since $\Omega_{\parallel}^{pl} = (q_+ - q_-)v_g$, we have $(q_+ - q_-)/(\alpha + \alpha_p) = \Omega_{\parallel}^{pl}T$ and eq. (70) reduces to eq. (35).

Figure 13 shows the value of $\Omega_{\parallel}^{pl}T$ obtained from the measured values of θ and $P_{lin}(0)/P_{lin}(H_{\parallel})$ by making use of eq. (69). The theoretical predictions are well verified for a field smaller than 50 kG. In the "bottleneck" region $\hbar\Omega_{\parallel}^{pl}$ is close to the Zeeman splitting of exciton states, $\hbar\Omega_{\parallel}$. For the exciton g-value of 0.69, the lifetime T was found to be 6×10^{-12} s. The sign of rotation was in agreement with the sign of the exciton g-factor. The discrepancy between experimental results and theory for high fields can be explained taking into account that, for $H > 50$ kG, the Zeeman splitting of the term Γ_5 is comparable with the exchange splitting, 0.12 meV, between the Γ_5 and Γ_6 terms of the A exciton in CdSe.

While deriving eqs. (68)–(70), we assumed that the dominant contribution to polariton scattering comes from acoustic phonons and that each scattering leads to the polariton transfer out from the narrow energy interval of

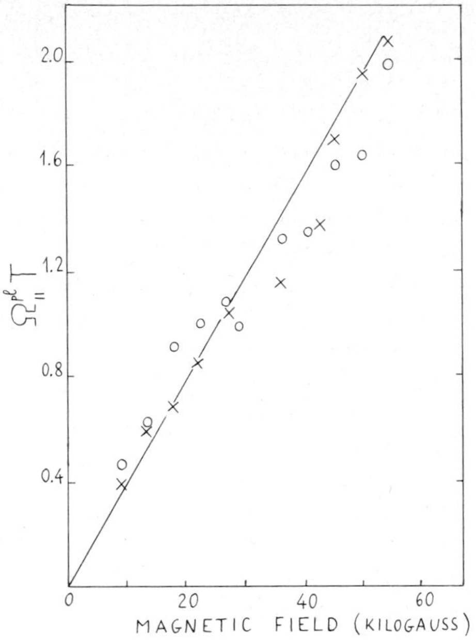

Fig. 13. The quantity $\Omega_{\parallel}^{pl}T$ obtained from measurements of the rotation angle (crosses) and the rate of linear polarization (circles). The solid line is calculated for $g = 0.69$ and $T = 6 \times 10^{-12}$ s (Nawrocki et al. 1976).

the two-phonon peak. In the general case it is necessary, by analogy with subsect. 4.6, to develop a theory of polariton transport that takes into account interbranch scattering and reflection from the boundary. The general transport theory is required as well for the description of polariton radiation when excitons are formed by the binding of free carriers.

In conclusion, it should be noted that the polariton effects have been found to be important also in optical orientation experiments on zinc-blende-type crystals, like GaAs (Weisbuch and Fishman 1976).

6. *Polarized luminescence of excitons in Ge and Si in a magnetic field*

As noted in the Introduction, for indirect excitons formed by an electron and a hole which belong to extrema lying at different points of the Brillouin zone, the exciton lifetimes are usually very long, since their recombination is accompanied by transfer of a large amount of momentum to a phonon or impurity center. For this reason, so far, optical orientation of excitons or free carriers has not been observed in indirect-gap semiconductors by measuring the polarization of the recombination radiation. Observation of the optical orientation of electrons by optical pumping in Si was performed only through induced polarization of the nuclei, using a rather sensitive method of nuclear magnetic resonance. This was the method employed by Lampel in 1968 for the first observation of optical orientation of free carriers in semiconductors. On the other hand, the smallness of the spin relaxation time τ_s, in comparison to the indirect-exciton lifetime, τ_0, permits one to observe easily the thermal spin orientation in magnetic field. For $\tau_s \ll \tau_0$, the magnetically split spin states of photocarriers are thermally populated and, hence, from measurements of the polarization of the photoluminescence one can determine the exciton level structure and the level splitting in a magnetic field. Long lifetimes and high degree of degeneracy of electron states in many-valley semiconductors favour forming bound multiexciton complexes (BMEC). Thus, investigating the polarization of the BMEC radiation one can clarify the BMEC energy structure.

Because of lower symmetry of indirect excitons and of higher degeneracy, the exciton level structure in Ge and Si is more complicated than in direct-gap crystals. In addition, the selection rules depend on the type of a phonon or impurity involved in an indirect process. Moreover, to obtain sufficiently high degrees of the light polarization, it is usually necessary to apply far stronger magnetic fields than in the case of Hanle effect studies. In such fields the diamagnetic effects may be important. The exchange splitting in Ge and Si is at most 10^{-4} eV and does not directly manifest itself in experiment, although for Si the exchange interaction is an efficient mechanism of the electron spin relaxation in an exciton. We therefore ignore the exchange splitting in Ge and Si.

In this section, we briefly discuss the level structure for free and bound excitons in Ge and Si and for BMEC in Si, describe the splitting of the levels in a magnetic field and consider the properties of the thermal orientation of excitons in Ge and Si.

6.1. *Energy structure of excitonic states and selection rules for indirect excitons in Ge and Si*

6.1.1. *Wavefunctions and energy levels*

In Ge and Si indirect excitons are formed by holes in the valence band $\Gamma_8^+(\Gamma'_{25})$ and electrons in the conduction band $L_6(L_1)$ in Ge or $\Delta_6(\Delta_1)$ in Si (in the brackets the representations without taking account of the electron spin are indicated). The wavefunction of the exciton ground state, $\Psi_{l,s}(\boldsymbol{r}_e, \boldsymbol{r}_h)$ ($l = \pm\frac{3}{2}, \pm\frac{1}{2}$ and $s = \pm\frac{1}{2}$), can be represented as a product of a function F_l and the Bloch function at the bottom of the conduction band, φ_s^{e}. The function F_l is a superposition of the products of the exciton envelope functions, $f^l_m(\boldsymbol{r})$ ($\boldsymbol{r} = \boldsymbol{r}_{\mathrm{e}} - \boldsymbol{r}_{\mathrm{h}}$), and the hole Bloch functions, ψ^{h}_m, so that

$$\Psi_{l,s}(\boldsymbol{r}_{\mathrm{e}}, \boldsymbol{r}_{\mathrm{h}}) = F_l\varphi_s^{\mathrm{e}} = \sum_m f^l_m(\boldsymbol{r})\psi^{\mathrm{h}}_m\varphi^{\mathrm{e}}_s. \tag{71}$$

The function φ_s^{e} corresponds to the definite component $s_z = s$ ($s = \pm\frac{1}{2}$) of the electron spin along the principal axis of the corresponding extremum, and the function F_l corresponds to the component $M_z = l$ of the hole total angular momentum. For the ground state, $l = \pm\frac{3}{2}, \pm\frac{1}{2}$. The "slow" functions $f^l_m(\boldsymbol{r})$ are superpositions of functions transforming according to the irreducible representations which occur in the direct product $\Gamma_8^+ \times \Gamma_8^+ = \Gamma_1 + \Gamma_2 + \Gamma_{12} + 2\Gamma'_{15} + 2\Gamma'_{25}$. The component transforming according to the identity representation, Γ_1, contributes to the functions $f^l_m(\boldsymbol{r})$ with $l = m$ only. Consequently, at $r = 0$ only the functions f^l_l are nonzero and the value of $f^l_l(0)$ is independent of l. This fact is quite significant. Since the probability of the free exciton excitation or recombination is determined by the values of $f^l_m(\boldsymbol{r})$ at $r = 0$, the selection rules corresponding to the exciton state (l, s) remain the same as for the electron–hole pair with electron spin s and hole angular momentum $m = l$. Since the symmetry of the point L or Δ is lower than that of the point Γ, the ground state of the exciton is split in accordance with the splitting of the state Γ_8^+ into the terms $L_4 + L_5$, L_6 for Ge and Δ_6, Δ_7 for Si. Hereinafter we shall refer to this splitting as the crystal splitting. The crystal splitting is described by the Hamiltonian

$$H_{\mathrm{cr}} = -\tfrac{1}{2}\Delta_{\mathrm{cr}}(J_z^2 - \tfrac{5}{4}). \tag{72}$$

Here J_α are the angular-momentum matrices in the basis of the "hole" functions $F_l(l = \pm\frac{3}{2}, \pm\frac{1}{2})$, the z-axis is directed along the principal axis of an extremum L or Δ. In Ge the value of the crystal splitting is about -1 meV, in Si $\Delta_{\mathrm{cr}} \approx -0.3$ meV. For a negative value of Δ_{cr} the lower states belong to $l = \pm\frac{3}{2}$ and the higher states correspond to $l = \pm\frac{1}{2}$. The fourfold degeneracy of each excitonic branch is not removed for the nonzero wavevector of the exciton, $\boldsymbol{k}$. However, the splitting between the two branches depends on the value and direction of $\boldsymbol{k}$. The Hamiltonian $H(\boldsymbol{k})$ which determines together with H_{cr} the exciton spectrum has the form

$$H(\boldsymbol{k}) = \frac{\hbar^2 k_z^2}{2M_\parallel} + \frac{\hbar^2 k_\perp^2}{2M_\perp} + \frac{\bar{\gamma}}{m_0}[(\boldsymbol{J}\cdot\boldsymbol{k})^2 - \tfrac{5}{4}k^2]. \tag{73}$$

The last term on the right-hand side of eq. (73) describes the quadratic (in k) splitting of the exciton branches and is written in the spherical approximation. In general it is determined by six linearly-independent constants for Ge and four constants for Si (Frova et al. 1975, Bir and Pikus 1976, Lipari and Altarelli 1976).

6.1.2. Selection rules for Ge

As stated above, the selection rules for the free exciton excitation coincide with the selection rules for the excitation of the electron–hole pair with the same spin indices. For indirect transitions some momentum-conserving process is necessary during the exciton recombination. The momentum-conserving process can be either the absorption and emission of a phonon or interaction with an impurity center. In Ge and the indirect transition is accompanied by the virtual scattering of a Γ'_{25} hole into one of L bands or of a L_1 electron into Γ bands. In Si the virtual scattering from the Γ'_{25} band to Δ bands or from the Δ_1 band to Γ bands takes place. If several channels markedly contribute to the matrix element of the second-order process, it is necessary to take into account the interference of the channels. However, there is no need to consider separately the contributions from various channels, because by using the method of invariants one can derive a general expression for the matrix element from the symmetry considerations (Asnin et al. 1976, Pikus 1977). The selection rules for the recombination of an electron from the band L_6 and a hole from the band Γ_8^+ obtained by the method of invariants are summarized in table 3(a). For Ge at $k = 0$ the allowed indirect transitions are those due to the longitudinal acoustic (LA) phonon, u_{LA}, corresponding to the representation L'_2, and to the transverse optical (TO) phonons, u_x and u_y, corresponding to the representation L'_3. For the no-phonon (NP) line related to the scattering by a substitutional impurity atom, the selection rules are determined by the symmetry of the scattering potential, $V(\boldsymbol{r})$, which transforms according to the representation Γ_1 of the group T_d. The potential V can be written as the sum of two potentials, V^+ and V^-, accordingly even and odd with respect to the operation of spatial inversion which is contained in the group $L = C_{3v} \times i$ but does not belong to T_d. The functions V^+ and V^- transform according to the representations L_1 and $L'_2 L'_1$, respectively. At $k = 0$ the transitions are allowed only for the scattering by the potential V^-.

The selection rules of table 3(a) are obtained neglecting the admixture to the states L_6 and Γ_8^+ of states from other bands due to the spin–orbit interaction. In this approximation radiative recombination of the $(\frac{3}{2}, \frac{1}{2})$ and $(-\frac{3}{2}, -\frac{1}{2})$ excitons is forbidden. The possible admixture of exciton excited states to the terms $l = \pm\frac{3}{2}$ and $l = \pm\frac{1}{2}$ due to the crystal splitting is ignored in table 3(a) as well. The values of l and s indicated in the table correspond to the angular-momentum components along the principal axis of a given ellipsoid. Note that in table 3(a) all the constants are real.

In Ge the energy difference between the extrema L_1 and Γ'_2 is much smaller than the difference between the extremum L_1 or Γ'_{25} and any other extremum. Consequently, for the LA and NP lines the main contribution comes from the transitions via the band Γ'_2. In this case $\eta = \lambda$, $\bar{\eta} = \bar{\lambda}$ and

Table 3
Selection rules for indirect excitons in Ge and Si

(a) Germanium ($\Gamma_8^+ \to L_6$).

l, s phonon	$3/2, -1/2$	$-3/2, 1/2$	$1/2, 1/2$	$-1/2, -1/2$	$+1/2, -1/2$	$-1/2, -1/2$
LA	$-\eta e_- u_{LA}$	$\eta e_+ u_{LA}$	$\sqrt{1/3}\eta e_- u_{LA}$	$-\sqrt{1/3}\eta e_+ u_{LA}$	$\sqrt{2/3}\lambda e_z u_{LA}$	$-\sqrt{2/3}\lambda e_z u_{LA}$
TO	$-(\alpha e_z u_- + \gamma e_+ u_+)$	$\alpha e_z u_+ - \gamma e_- u_-$	$\sqrt{1/3}(\alpha e_z u_- + \gamma e_+ u_+)$	$-\sqrt{1/3}(\alpha e_z u_+ + \gamma e_- u_-)$	$-\sqrt{2/3}\beta(e_+ u_- + e_- u_+)$	$\sqrt{2/3}\beta(e_+ u_- + e_- u_+)$
NP	$-\bar{\eta} e_-$	$\bar{\eta} e_+$	$\eta e_- /\sqrt{3}$	$-\bar{\eta} e_+ /\sqrt{3}$	$\sqrt{2/3}\bar{\lambda} e_z$	$-\sqrt{2/3}\bar{\lambda} e_z$

(b) Silicon ($\Gamma_8^+ \to \Delta_6$).

l, s phonon	$3/2, -1/2$	$-3/2, 1/2$	$1/2, 1/2$	$-1/2, -1/2$	$1/2, -1/2$	$-1/2, 1/2$
LA	$-\gamma e_+ u_{LA}$	$-\gamma e_- u_{LA}$	$\gamma e_+ u_{LA}/\sqrt{3}$	$\gamma e_- u_{LA}/\sqrt{3}$	0	0
LO	$-\eta e_- u_{LO}$	$\eta e_+ u_{LO}$	$\eta e_- u_{LO}/\sqrt{3}$	$-\eta e_+ u_{LO}/\sqrt{3}$	$\sqrt{2/3}\lambda e_z u_{LO}$	$-\sqrt{2/3}\lambda e_z u_{LO}$
TO TA	$-\alpha e_z u_-$	$\alpha e_z u_+$	$\alpha e_z u_- /\sqrt{3}$	$-\alpha e_z u_+ /\sqrt{3}$	$-\sqrt{2/3}\beta(e_+ u_- + e_- u_+)$	$\sqrt{2/3}\beta(e_+ u_- + e_- u_+)$
NP	$-(\bar{\gamma} e_+ + \bar{\eta} e_-)$	$-(\bar{\gamma} e_- - \bar{\eta} e_+)$	$(\bar{\gamma} e_+ + \bar{\eta} e_-)/\sqrt{3}$	$(\bar{\gamma} e_- - \bar{\eta} e_+)/\sqrt{3}$	$\sqrt{2/3}\bar{\lambda} e_z$	$-\sqrt{2/3}\bar{\lambda} e_z$

where

$$a_\pm = \mp(a_x \pm i a_y)/\sqrt{2}.$$

the selection rules are the same as for the $\Gamma_8^+ \times \Gamma_2'$ direct exciton. The transitions via the band L_3' contribute only to the constants η and $\bar{\eta}$; the transitions via the band L_2' contribute only to λ and $\bar{\lambda}$; for transitions via Γ_{15} only, one obtains $\lambda = -2\eta$. For a TO-phonon-assisted process, the transitions via the band Γ_2' are forbidden, the transitions via the band L_3' contribute only to β and γ, the transitions via the band L_2' contribute only to α, and in the case of transitions via the band Γ_{15} one can show that $\alpha = \beta = \frac{1}{2}\gamma$.

6.1.3. *Selection rules for Si*

Table 3(b) shows the selection rules for the recombination of a Δ_6 electron with a Γ_8^+ hole in Si. The indirect transitions are allowed for all phonons, namely, the LA phonon, u_{LA}, corresponding to the representation Δ_1, the LO phonon, u_{LO} (the representation Δ_2'), the TO, TA phonons, u_x, u_y (Δ_5). In table 3(b) the functions u_x and u_y are assumed to transform like yz and zx. In order to determine the selection rules for the NP line, it is convenient to write the impurity potential, $V(\boldsymbol{r})$, as the sum of two potentials, V^+ and V^-, which are even and odd respective to the operation C_{4z} that is present in the group $\Delta(C_{4v})$ and absent in T_d. The functions V^+ and V^- transform according to the representations Δ_1 and Δ_2', respectively. Therefore, the selection rules for the NP line are governed by three constants. In table 3(b) all the constants are real.

In Si the transitions via different bands, give comparable contributions. For the transitions via the band Γ_{15}', non-zero constants are γ and $\alpha = \beta$; the transitions via Δ_5 contribute to γ, η and β; for transitions via Γ_2', $\lambda = \eta$ and the remaining three constants vanish.

6.2. *Polarized luminescence of free excitons in Ge and Si in a magnetic field*

6.2.1. *Splitting of excitonic levels in a magnetic field*

As mentioned above, in considering the thermal orientation it is necessary, beyond the paramagnetic splitting, to take account of the diamagnetic splitting of excitonic states. If the electron–hole exchange interaction is unimportant the paramagnetic splitting is determined separately by the values of the electron and hole angular momentum, s and l. In Ge, because of the large anisotropy of the electron g factor, the splitting of the electron states depends on the angle between the magnetic field vector and the principal axis of a given extremum, z, and is described by the Hamiltonian

$$H_e^p(\boldsymbol{H}) = \mu_0[g_\parallel s_z H_\parallel + g_\perp(\boldsymbol{s}_\perp \cdot \boldsymbol{H}_\perp)]. \tag{74}$$

Here s_z, $\boldsymbol{s}_\perp$ and $H_\parallel$, $\boldsymbol{H}_\perp$ are the components of $\boldsymbol{s}$ and $\boldsymbol{H}$ which are parallel or perpendicular with respect to the z-axis. For Si, $g_{\parallel} \approx g_{\perp}$. The paramagnetic splitting of hole states is also anisotropic and is determined by the Hamiltonian

$$H_{\rm h}^{\rm p}(\boldsymbol{H}) = \mu_0[g_1(\boldsymbol{J} \cdot \boldsymbol{H}) + g_2 \sum_{\alpha} J_{\alpha}^3 H_{\alpha}], \tag{75}$$

where the index α labels the crystal principal axes.

The diamagnetic effects, in addition to the uniform shift of excitonic levels, lead to relative shifts of excitonic levels corresponding to different valleys. These shifts also depend on the angle between $\boldsymbol{H}$ and the z-axis, θ, and are described by the Hamiltonian

$$H_{\rm e}^{\rm d}(\boldsymbol{H}) = \lambda_2(3H_{\parallel}^2 - H^2) = \lambda_2 H^2(3\cos^2\theta - 1). \tag{76}$$

For Ge and Si, where the transverse electron mass, $m_{\perp}$, is smaller than the longitudinal electron mass, $m_{\parallel}$, the constant λ_2 is positive. Moreover, the diamagnetic effects give rise to the splitting of hole states which is determined in the spherical approximation by the Hamiltonian

$$H_{\rm h}^{\rm d}(\boldsymbol{H}) = \bar{\lambda}_3[(\boldsymbol{J} \cdot \boldsymbol{H})^2 - \tfrac{5}{4}H^2]. \tag{77}$$

Thus, the diamagnetic effects lead to the splitting between the states with $l = \pm\frac{3}{2}$ and $l = \pm\frac{1}{2}$. For Ge, $\bar{\lambda}_3 > 0$, i.e., the diamagnetic splitting shifts the states with $l = \pm\frac{3}{2}$ to higher energies and thus reduces the effect of the crystal splitting.

Figure 14 shows the splitting of excitonic levels in a magnetic field for excitons corresponding to one valley at $\boldsymbol{H} \parallel z$. For other directions of $\boldsymbol{H}$ the picture is more complicated. Taking into account the anisotropy of the band Γ_8^+, the diamagnetic splitting (77), as well as the k^2 splitting (73), is governed by six constants for Ge and by four constants for Si.

Equations (75)–(77) are applicable as long as the Landau level spacing, $\hbar\omega_H$, for free electrons or holes is smaller than the exciton binding energy, $E_{\rm B}$. In the opposite limiting case, $\hbar\omega_H \gg E_{\rm B}$, excitons are formed by electrons and holes belonging to separate Landau subbands.

6.2.2. *Polarization of exciton radiation in Ge*

The polarization of the LA line due to free excitons in Ge has been investigated by Asnin et al. (1976). Figure 15 shows the dependence $P_{\rm circ}(H)$ for three geometries: $\boldsymbol{H} \parallel (100)$, $\boldsymbol{H} \parallel (110)$ and $\boldsymbol{H} \parallel (111)$. The direction of the light propagation coincides with the direction of $\boldsymbol{H}$. For $\boldsymbol{H} \parallel (100)$ all the extrema are equivalent and equally populated. The direction of the preferential orientation of electron spins differs from the magnetic field direction because of the difference between the longitudinal and transverse g factors for electrons, $g_{\parallel}$ and $g_{\perp}$ (for Ge, $g_{\parallel} = 0.90$ and $g_{\perp} = 1.92$). Hence, the saturation value of the polarization rate is less than 100%.

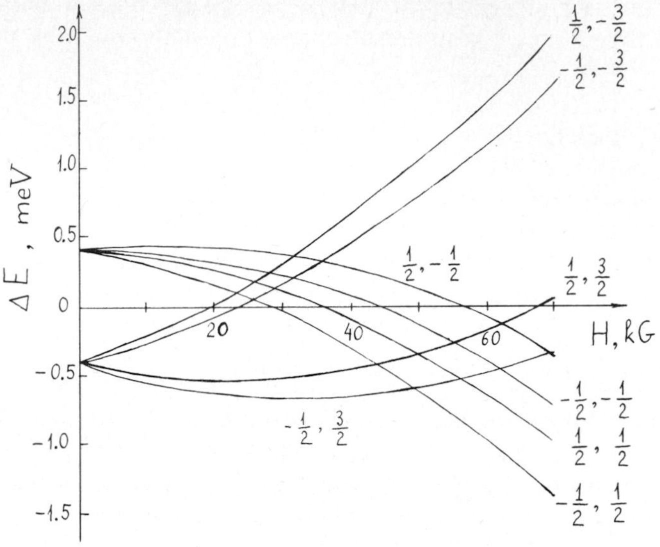

Fig. 14. The level splitting in magnetic field for an exciton, corresponding to one valley ⟨111⟩ in Ge. $\boldsymbol{H}\|(111)$. (for $\Delta_{cr} = -0{,}8$ meV, $\bar{\lambda}_3 = 2{,}54 \times 10^{-4}$ meV/(kG)^2, $g_1 = -1{,}6$; $g_2 = 0$, $g_\| = 0.90$).

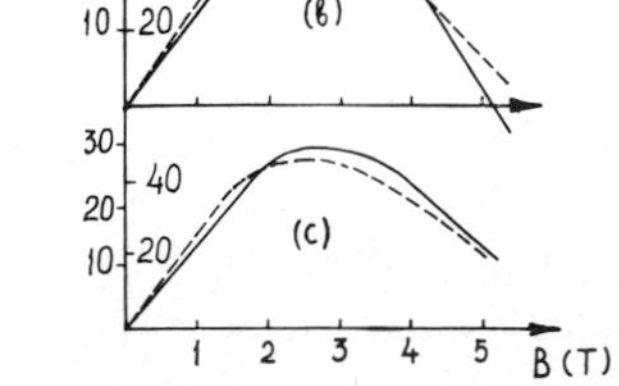

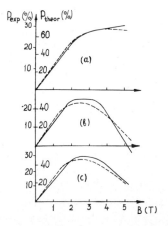

Fig. 15. The magnetic field dependence of the degree of circular polarization for the LA line due to the free exciton emission in Ge for (a) $\boldsymbol{H}\|(100)$, (b) $\boldsymbol{H}\|(110)$ and (c) $\boldsymbol{H}\|(111)$ at $T = 4.2$ K. The dashed lines are calculated for the following parameters: $g_\| = 0.90$, $g_\perp = 1.92$, $g_1 = -1.6$, $g_2 = 0$; $\Delta_{cr} = -0.8$ meV, $\lambda_2 = 2 \times 10^{-4}$ meV/kG^2, $\bar{\lambda}_3 = 1{,}5\ 10^{-4}$ meV/kG^2 (a) and $\bar{\lambda}_3 = 2.54 \times 10^{-4}$ meV/kG^2(b) and (c). (After Asnin et al. 1976.)

For $\boldsymbol{H} \parallel (110)$, the increase in magnetic field strength leads to predominant occupation of the extrema $\langle\bar{1}11\rangle$ and $\langle 1\bar{1}1\rangle$ which lie by the value of $2\lambda_2 H^2$ below the extrema $\langle 11\bar{1}\rangle$ and $\langle 111\rangle$. One can see from table 3 that, for the extrema $\langle\bar{1}11\rangle$ and $\langle 1\bar{1}1\rangle$, the radiation due to the lower levels $l = \pm\frac{3}{2}$, and emitted in the direction (110) perpendicular to the z-axis of the extrema is linearly polarized. Thus, with increasing population of these levels the degree of circular polarization decreases.

For $\boldsymbol{H} \parallel (111)$, at strong magnetic fields the extremum $\langle 111\rangle$ shifts by $(\frac{8}{3})\lambda_2 H^2$ upwards relative to the remaining three extrema. As the excitons transfer to other extrema from this one with increasing magnetic field, the degree of circular polarization decreases, as in the previous case.

The difference in the behaviour of $P_{\text{circ}}(H)$ for different $\boldsymbol{H}$ directions can thus be qualitatively understood in terms of the intervalley diamagnetic splitting. However, in order to obtain quantitative agreement with experiment, it is necessary in addition to take into account the splitting (77) of hole states which compensates the crystal splitting and leads to an increase in population of the terms $l = \pm\frac{1}{2}$. The predominant occupation of these states in a strong magnetic field leads, in agreement with the experimental data, to the noticeable decrease in P_{circ} for $\boldsymbol{H} \parallel (111)$ and to the sign reversal of P_{circ} for $\boldsymbol{H} \parallel (110)$. In fig. 15 the dashed lines are theoretical curves calculated in accordance with table 3 for $\eta = \lambda$. In this case P_{circ} is determined only by the mean values of electron and hole angular-momentum components along the direction of the light propagation, z, as follows

$$P_{\text{circ}} = \frac{2\langle J_z\rangle + 3\langle s_z\rangle + 4(\langle \boldsymbol{s}\rangle \cdot \langle \boldsymbol{J}J_z\rangle)}{\frac{3}{4} + \langle J_z^2\rangle + (\langle \boldsymbol{s}\rangle \cdot \langle \boldsymbol{J}\rangle) + 4\langle s_z\rangle\langle J_z\rangle - 4(\langle \boldsymbol{s}\rangle\langle \boldsymbol{J}J_z^2\rangle)}. \tag{78}$$

The calculation was performed neglecting the k^2 splitting. In order to take into account the anisotropy of the diamagnetic splitting, the value of the constant $\bar{\lambda}_3$ was adjusted to agree with experiment for each direction of $\boldsymbol{H}$. The values of $\bar{\lambda}_3$ have been found to be different for different $\boldsymbol{H}$ directions. One can see from fig. 15 that the theory explains the behaviour of $P_{\text{circ}}(H)$ observed experimentally.

For $g_2 = 0$ in a weak magnetic field P_{circ} is independent of the magnetic field direction and, for $\Delta_{\text{cr}} \gg kT$, is determined by

$$P_{\text{circ}}(H) = \tfrac{1}{4}(g_\parallel - 3g_1)\mu_0 H/kT. \tag{79}$$

From the comparison of the theoretical and experimental curves in this region one can find the depolarizing factor due to multiple reflections from sample boundaries. Because of the depolarizing factor, the P_{circ} scales for the theoretical and experimental curves in fig. 15 are different.

For $H \gtrsim 50$ kG, the Landau level spacing for electrons in Ge is comparable to the exciton binding energy. This should lead to the increase in the relative shifts of the valleys. Moreover, the deformation of electron wavefunctions in the magnetic field should change the direction of the spin quantization axis and the value of the crystal splitting (72).

6.2.3. *Polarization of exciton radiation in Si*

Due to the relatively small value of the exciton radius, the diamagnetic shift and diamagnetic splitting in silicon can be neglected in magnetic fields up to 60–70 kG. For Si, the electron g factor is isotropic, $g_\parallel = g_\perp = g = 2$. Consequently, if the crystal and k^2 splittings are negligible and g_2 in eq. (75) is assumed to vanish, then both the electron and hole angular momenta are oriented along the magnetic field vector and the energy of the (l, s) state is

$$E_{l,s} = \mu_0 H (g_1 l + g s), \tag{80}$$

where l and s are the angular-momentum components along $\boldsymbol{H}$ for a hole ($l = \pm\frac{3}{2}, \pm\frac{1}{2}$) and an electron ($s = \pm\frac{1}{2}$), respectively.

The degree of circular polarization of the luminescent light propagating in the $\boldsymbol{H}$ direction is given by

$$P_{\text{circ}}(H) = \sum_{l,s} d_{l,s}^{-} \exp\left(-\frac{E_{l,s}}{kT}\right) \Big/ \sum_{l,s} d_{l,s}^{+} \exp\left(-\frac{E_{l,s}}{kT}\right). \tag{81}$$

For the LA and TO, TA replicas the values of $d_{l,s}^{\pm}$ averaged over all six extrema Δ depend on the direction of the light propagation in the crystal. Table 4 gives the values of $d_{l,s}^{\pm}$ for $z \parallel (111)$. In weak magnetic fields the

Table 4
Relative intensities of exciton emission lines in Si for $\boldsymbol{H} \parallel (111)$,
$d^- = I_+ - I_-$, $d^+ = I_+ + I_-$.

<table>
<tr><th rowspan="2">l,s</th><th colspan="2">LA line</th><th colspan="2">LO line</th><th colspan="2">TO, TA lines</th></tr>
<tr><th>d^-</th><th>d^+</th><th>d^-</th><th>d^+</th><th>d^-</th><th>d^+</th></tr>
<tr><td>3/2, −1/2</td><td>−3</td><td rowspan="2">3</td><td>$-9\Phi_{LO}$</td><td rowspan="2">$3(2-\Phi_{LO})$</td><td>$-9\Phi_T$</td><td rowspan="2">$3(2+\Phi_T)$</td></tr>
<tr><td>−3/2, 1/2</td><td>3</td><td>$9\Phi_{LO}$</td><td>$9\Phi_T$</td></tr>
<tr><td>1/2, 1/2</td><td>−1</td><td rowspan="2">1</td><td>$-3\Phi_{LO}$</td><td rowspan="2">$2-\Phi_{LO}$</td><td>$-3\Phi_T$</td><td rowspan="2">$2+\Phi_T$</td></tr>
<tr><td>−1/2, 1/2</td><td>1</td><td>$3\Phi_{LO}$</td><td>$3\Phi_T$</td></tr>
<tr><td>1/2, −1/2
−1/2, 1/2</td><td>0</td><td>4</td><td>0</td><td>$4(1+\Phi_{LO})$</td><td>0</td><td>$4(1-\Phi_T)$</td></tr>
</table>

$$\Phi_{LO} = -\frac{\eta(\eta + 2\lambda)}{2\eta^2 + \lambda^2}, \qquad \Phi_T = \frac{\alpha\beta}{\alpha^2 + \beta^2}.$$

polarization of the radiation is independent of the orientation of $\boldsymbol{H}$ and is determined by

$$P^N_{\rm circ}(H) = \Phi_N(\langle j_z\rangle - \langle s_z\rangle) = -\frac{1}{4}\Phi_N \frac{\mu_0 H}{kT}(5g_1 - g). \tag{82}$$

Here

$$\Phi_{\rm LA} = \tfrac{1}{2}, \qquad \Phi_{\rm LO} = -\eta(\eta + 2\lambda)/(2\eta^2 + \lambda^2), \qquad \Phi_{\rm T} = \alpha\beta/(\alpha^2 + \beta^2).$$

The polarization of the TO line in the magnetic field in Si has been measured by Altukhov et al. (1978). The measured dependence $P_{\rm circ}(H)$ shown in fig. 15 agrees with the curve calculated using eqs. (80) and (81) for $g = 1.2$, $g = 2$ and $\Phi_{\rm TO} = 0.4$. The value of $\Phi_{\rm TO}$ corresponds to the ratio $\alpha/\beta = 0.5$, which is in agreement with the value of α/β obtained by Alkeev et al. (1976) from measurements of the relative intensities and polarization of the TO line in a strained Si crystal.

Equation (81) is derived assuming that the equilibrium populations are achieved in the Zeeman sublevels of an exciton. This is true if the spin relaxation times of electrons, $\tau_{\rm es}$, and holes, $\tau_{\rm hs}$, are much shorter than the exciton lifetime, τ_0. This condition is well satisfied for holes in Ge and Si and for electrons in Ge. However for Si, due to its weak spin–orbit coupling, the usual spin-relaxation mechanisms for electrons are unimportant and for the electron in an exciton the dominant mechanism of spin relaxation is the exchange interaction with the hole. According to eq. (27) in this case, $\tau_{\rm es}^{-1} = \frac{5}{8}(\Delta/\hbar)^2\tau_{\rm hs}$. If $\tau_{\rm es}$ is comparable with or longer than τ_0, then the exponent $\exp(-g_e\mu_0H/kT)$ in eq. (81) should be replaced by the factor

$$[\exp(-g_e\mu_0H/kT) + \tau_{\rm es}/\tau_0]/(1 + \tau_{\rm es}/\tau_0),$$

which for $(\tau_{\rm es}/\tau_0) \gg \exp(-g_e\mu_0H/kT)$ reduces to $(1 + \tau_0/\tau_{\rm es})^{-1}$.

In Si the polarization of the luminescence in a magnetic field is determined predominantly by the hole spin orientation and, in fact, the nonequilibrium population of the electron spin sublevels has no influence on the polarization of the exciton radiation. However, the degree of the electron spin orientation can be determined from measurements of the polarization of the luminescence due to electron–hole drops (EHD) resulting from the condensation of free excitons at high excitation levels (Altukhov et al. 1978). The polarization of the EHD luminescence was found to exceed considerably the equilibrium value and to have the sign that corresponds to electron spin orientation and is inverse as compared to the sign of $P_{\rm circ}$ for the exciton line. The nonequilibrium orientation of electron spins in EHD can be explained in the following way. In EHD, due to the decrease of the electron–hole exchange interaction in comparison with that in a free exciton state, the spin relaxation time for electrons increases considerably. Consequently, the electrons captured by EHD conserve the spin orientation acquired when they were bound in excitons. On the contrary, the holes rapidly lose their spin orientation during the lifetime in EHD. This leads to the sign reversal of the light polarization. The existence of nonequilibrium electron-spin orientation shows that the evaporation of free excitons from the surface of a drop is unimportant. Altukhov et al. (1978) have observed that at high excitations, the degree of the electron spin orientation in EHD decreases with further increase of the excitation intensity, i.e., with the decrease of the exciton lifetime which is limited in this case by the rate of free exciton condensation into drops. This means that under these conditions the equilibrium value of the electron spin orientation in an exciton is not achieved during the exciton lifetime. At low excitation levels the exciton lifetime is longer than the spin relaxation time due to the electron–hole exchange interaction, τ_{es}, and the populations of the electron spin sublevels in a free exciton are close to their equilibrium values.

6.3. Polarized luminescence of bound excitons and multi-exciton complexes in Ge and Si

In Ge and Si, in addition to free exciton lines the luminescence spectra reveal lines arising from the radiative recombination of an exciton bound to a neutral donor (ND) or to a neutral acceptor (NA). At higher excitation levels, a number of new lines appear in Si below the bound exciton lines. The new lines are due to the radiative decay of electron–hole pairs in bound multiexciton complexes (BMEC) made up of a number of excitons bound to one impurity center (Kaminskii et al. 1970; see also Thewalt, ch. 10 of this volume). The observation of similar recombination lines in Ge has been reported by Martin (1974).

The wavefunctions and energy structure of bound excitons have some specific features which distinguish them from free excitons. These features result in a substantial change of the polarization of the bound exciton radiation and in the modification of the dependence of the polarization degree upon the number of excitons bound in a BMEC. This can be understood taking into account, firstly, that a complex of an exciton bound to a neutral impurity center or a BMEC contains two or more like particles, electrons and/or holes. Due to the correlation between the spins of the like particles, the splitting of energy levels in a magnetic field for bound excitons differs from that for free excitons. Secondly, the short-range contribution to the attractive impurity potential gives rise to a valley–orbit splitting of the electronic states in excitons bound to a ND, which leads to the mixing of the states from different valleys. Consequently, not only occupations of the corresponding states change, but the selection rules for NP processes become different for different levels. This influences both the value and sign of the light polarization. Finally, because of the difference between the wavefunctions, the selection rules for phonon-assisted transitions of free and bound excitons are also somewhat different.

6.3.1. *Structure of energy levels of bound excitons and multi-exciton complexes*

Many features of the recombination luminescence attributed to bound excitons and BMEC can be understood in terms of a shell model (Kirzcenov 1977, 1977a). In this model the wavefunction of a collection of electrons and holes bound to an impurity center is represented by a antisymmetrized product of single-particle wavefunctions. In other words, the electrons and holes in bound excitons and BMEC are supposed to fill, in accordance with the Pauli principle, the lowest levels which correspond to the lowest donor levels for electrons and acceptor levels for holes. It should be noted that, of course, the energy of a given electron or hole level depends on the number of particles bound in the complex and differs from the corresponding energy in a ND or NA.

The valley–orbit interaction splits the ground state of a donor into two terms, $\Gamma_1(A_1)$ and $\Gamma_5(F_2)$, for Ge and into three terms, $\Gamma_1(A)$, $\Gamma_3(E)$ and $\Gamma_5(F_2)$, for Si. Experimental results show that the valley symmetric state Γ_1 is lowest in energy and the Γ_3 and Γ_5 states have in Si practically the same energy. For an exciton bound to a ND, it is energetically favourable for both electrons to occupy the singlet state Γ_1, so that their total spin $S = 0$. As successive excitons are added to the complex, the additional electrons occupy the states which are next lowest in energy after the Γ_1 state (see fig. 16).

For excitons bound to NA, the valley–orbit splitting is small and lies within the experimental uncertainty. The ground state of a hole at an acceptor center belongs to the spin $\frac{3}{2}$ quartet $\Gamma_8(G')$. Hence, the ground state for two holes in an exciton bound to a NA corresponds to a two-hole antisymmetric spin function with the total angular momentum $J = 0\,(\Gamma_1)$ or $J = 2$. The crystal field interaction splits the latter state into two terms, $\Gamma_3(E)$ and $\Gamma_5(F_2)$. For three holes the ground state belongs to the $J = \frac{3}{2}$ quartet Γ_8. For four holes the Γ_8 shell is completely filled and the ground state belongs to the singlet Γ_1. For complexes which contain more than four holes, the additional holes occupy excited acceptor states (fig. 16). The crystal field interaction can remove the degeneracy of states with the hole angular momentum $J > 0$. When this splitting is small in comparison with the valley–orbit splitting, as for many-exciton complexes bound to ND, it can manifest itself only if electrons occupy degenerate states Γ_3 or Γ_5. In the shell model it is assumed that the energy of a BMEC is insensitive whether two electrons belong to one and only valley or to different valleys.

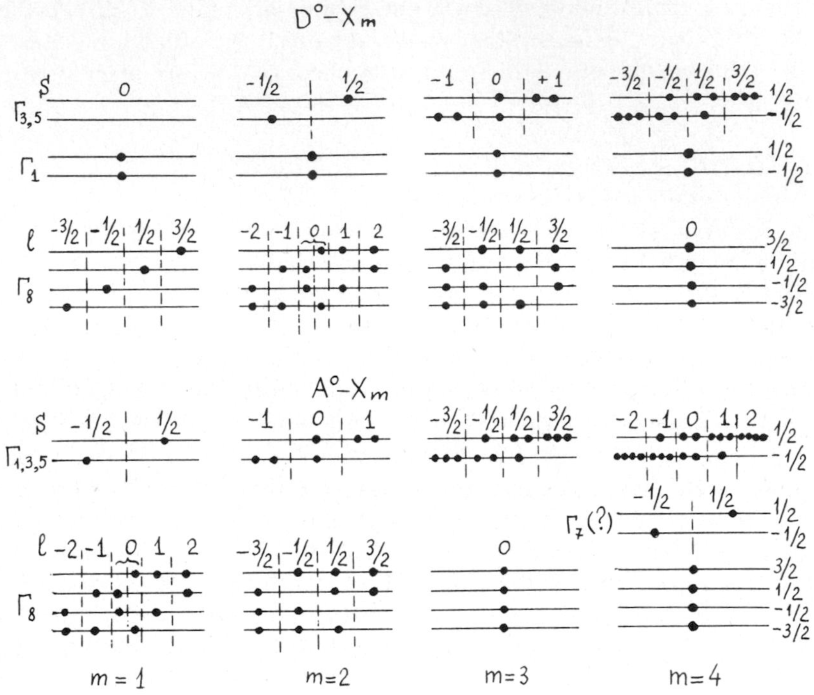

Fig. 16. Scheme showing the filling of donor and acceptor BMEC levels.

The 'exchange-valley' splitting, i.e. the energy difference between these two states should cause the additional splitting of the terms Γ_3 and Γ_5 if they are occupied by two or more electrons. Recently Mayer and Lightowlers (1979, 1979a, 1979b) have observed the fine structure of the excited state of excitons bound to ND (As, P) in Ge which can be explained in terms of the above-mentioned crystal and exchange-valley splittings.

6.3.2. *Selection rules for bound excitons in Si*

As discussed in subsect. 6.1, the matrix element for an indirect transition of a free exciton is proportional to the value of F_l at $\boldsymbol{r} = \boldsymbol{r}_e - \boldsymbol{r}_h = 0$ and is therefore determined by the functions f_m^l in eq. (71) with $l = m$ which contain the component transforming according to the representation Γ_1. For phonon-assisted transitions of bound excitons, the matrix element is proportional to the integral $\int \mathrm{d}\boldsymbol{r} f_e(\boldsymbol{r}) f_h(\boldsymbol{r})$, where f_e and f_h are the envelope functions for electrons and holes. Thus the contribution to the matrix element comes not only from $f_l^l(0) = f_0(0)$, but also from any other functions f^l_m which contain components transforming according to the identity representation L_1 or Δ_1 of the group of the wavevector at point L for Ge or Δ for Si, respectively. These are the functions corresponding to the representation Γ_1 of the group Γ and contributing to f^l_m with $m = l$ and one of the functions corresponding to the representation Γ'_{25} for Ge or Γ_{12} for Si. Now the recombination rates for transitions involving a hole with $l = \pm\frac{3}{2}$ and $l = \pm\frac{1}{2}$ do not coincide and the selection rules for the LO, LA and TO, TA lines differ from those summarized in table 3 by the factors $(1+\delta)$ for $l = \pm\frac{3}{2}$ and $(1-\delta)$ for $l = \pm\frac{1}{2}$ where the constant $\delta \ll 1$. Note that a large crystal splitting can lead to the same change in the selection rules for free excitons. Neglecting these corrections, in Si the values of $d^{\pm}_{\bar{l},s}$ for the lines above (averaged over all three states of the term Γ_5 or two states of the term Γ_3) are the same for all terms and coincide with those of table 4.

Since a NP process is accompanied by the transfer of a large value of momentum to an impurity center, the matrix element for such a process is determined by the short-range part of the impurity potential stretching over the interatomic distance. Hence, the transition matrix element is proportional to the product $f_e(0)f_h(0)$ and the result is that the selection rules for the NP line due to bound excitons are determined by table 3 as well. However, in contrast to phonon-assisted lines, in order to calculate the total probability of the transition in this case, one has to sum up not the probabilities, but matrix elements corresponding to various valleys. Consequently, the values of $d^{\pm}$ are different for different states. Moreover, for the term Γ_3 in Si the values of $d^{\pm}$ depend on the light propagation direction, z. Table 5 gives $d^{\pm}_{\bar{l},s}$ for Si with $\boldsymbol{H} \parallel (111)$. As in table 4, the values of $d^{\pm}_{\bar{l},s}$ are averaged over two Γ_3 states or three Γ_5 states. One can observe that the selection rules for the Γ_1 state coincide with those for the $\Gamma_8^+ \times \Gamma'_2$ direct exciton. In addition table 5 presents the values of $d^{\pm}$ for free excitons which are the averages of $d^{\pm}$ over all three states.

6.3.3. Polarization of radiation due to bound excitons and multi-exciton complexes in Si

The behaviour of BMEC in Si in a magnetic field is determined by the values of the total angular-momentum component of electrons, S, and holes, L, as well as by the electron distribution over the Γ_1 and $\Gamma_{3,5}$ shells. If both the g factor for holes, g_1, and g factor for electrons are identical for all states and independent of the number of excitons bound in a complex and if in addition $g_2 = 0$, then the variation of the BMEC energy in a magnetic field is determined only by the values S and L, so that

$$\Delta E_{S,L} = (gS + g_1 L)\mu_0 H. \tag{83}$$

Here the axis of spin quantization is parallel to $\boldsymbol{H}$. In the single particle shell model the splitting of the Zeeman components of luminescence lines

Table 5
Relative line intensities for NP transitions in Si (summed up over all the states Γ_3 or Γ_5, in the case of Γ_3 states for $\boldsymbol{H}\parallel(111)$).

	$\Gamma_1(A_1)$		$\Gamma_3(E)$		$\Gamma_5(F_2)$		free exciton	
$l,\ s$	d_-	d_+	d_-	d_+	d_-	d_+	d_-	d_+
$\frac{3/2,\ -1/2}{-3/2,\ 1/2}$	$\frac{3}{-3}$	3	$\frac{-3\theta_3}{3\theta_3}$	$3\theta_3$	$\frac{-3\theta_5}{3\theta_5}$	$3\theta_5$	$\frac{-9\Phi_{NP}}{9\Phi_{NP}}$	$3(2-\Phi_{NP})$
$\frac{1/2,\ 1/2}{-1/2,\ -1/2}$	$\frac{1}{-1}$	1	$\frac{-\theta_3}{\theta_3}$	θ_3	$\frac{-\theta_5}{\theta_5}$	θ_5	$\frac{-3\Phi_{NP}}{3\Phi_{NP}}$	$2-\Phi_{NP}$
$\frac{1/2,\ -1/2}{-1/2,\ 1/2}$	0	0	0	$4\theta_3$	0	$4\theta_5$	0	$4(1+\Phi_{NP})$

$$\theta_3 = \frac{(\bar{\eta}-\bar{\lambda})^2}{(2\bar{\eta}+\bar{\lambda})^2}, \quad \theta_5 = \frac{3\bar{\gamma}^2}{(2\bar{\eta}+\bar{\lambda})^2}, \quad \Phi_{NP} = \frac{1-Z}{1+Z}, \quad Z = \theta_3 + \theta_5.$$

and their polarization are related only to the values of angular-momentum components, s and l, of the electron and hole involved in the indirect transition. The shift of the corresponding component in a magnetic field is

$$\Delta\omega_{s,l} = (gs + g_1 l)\mu_0 H/\hbar. \tag{84}$$

It follows then that each line of the BMEC luminescence in Si splits into six components in a magnetic field. The component spacings are determined by eq. (84) and the polarizations are given by tables 4 and 5. The number of excitons in a BMEC, and the populations of the Γ_1 and $\Gamma_{3,5}$ shells in the initial and final states, influence only the relative intensity of each component, $I_{s,l}$. Then the degree of polarization of the line on the whole can be written as

$$P_{\text{circ}} = \sum_{s,l} d^-_{l,s} I_{l,s} \Big/ \sum_{s,l} d^+_{l,s} I_{l,s}. \tag{85}$$

In order to compute the intensity $I_{l,s}$, it is convenient to define two quantities, g^{Sm}_{sr} and g^{Lm}_{l}. By definition, g^{Sm}_{sr} is the number of electrons with the spin s in the r shell (Γ_1 or $\Gamma_{3,5}$) for the BMEC formed by m excitons with the total electron spin S. g^{Lm}_{l} is the number of holes with the angular momentum l for the BMEC with the total hole angular momentum L. In accordance with the scheme of the filling of the BMEC levels shown in fig. 16, g_l^L can be either 0 or 1, while $g_{s,r}^S$ can vary from 0 to g_r, where g_r is the degeneracy of the r shell (but g_r is at most the number of electrons in a BMEC). For the Γ_1, Γ_3 and Γ_5 shells, $g_r = 1, 2, 3$, respectively. The only exception is the complex which contains two holes with $L = 0$. For the terms $J = 2$, $J_z = L = 0$ and $J = L = 0$, $g_l^0 = \frac{1}{2}$ for all values of $l = \pm\frac{3}{2}, \pm\frac{1}{2}$. If these terms have the same energy, then instead of them one can consider two other states shown in fig. 16.

According to eq. (83) the relative intensity of the (l, s) line arising from the recombination of an electron in the r shell of the m complex is given by

$$I_{sl}^{rm} = \sum_S G_{Sm} g_{sr}^{Sm} \exp(-g\mu_0 HS/kT) \sum_L g_l^{Lm} \exp(-g_1\mu_0 HL/kT), \tag{86}$$

where G_{Sm} is the statistical factor of the state S, m. Kirczenow (1977, 1977a) put forward the suggestion that the dominant lines of the donor BMEC luminescence are due to processes in which a Γ_1 electron recombines with a Γ_8 hole, so that the complex is left in an excited final state. These lines are known as the α_m lines, where m labels the number of excitons bound to a ND.

In the TO-phonon region another series of sharp lines is observed. This series is attributed to the transitions which go from ground state to ground state and thus involve the recombination of a $\Gamma_{3,5}$ electron. The lines of this series are called the β_{m-1} lines (Kirczenow 1977, 1977a). No corresponding structure was seen for the NP processes which means that the constants θ_3 and θ_5 in table 5 should be very small. The explanation of the α and β line series has been justified both by measurements of relative line intensities as functions of the excitation power (Thewalt 1977, 1977a) and by studies of the line behaviour in a transverse magnetic field and under a uniaxial strain (Kulakovskii et al. 1979, Kaminskii and Pokrovskii 1979, Thewalt 1978). For the α lines associated with the recombination of an electron from the filled Γ_1 shell in a BMEC, $g_{s\Gamma_1}^{Sm} = 1$ for all S and s and electron spins cannot be oriented in this case. Hence, the polarization of the α lines is determined by the spin orientation of holes only and I_{sl}^{rm} depends only on the index l.

The polarization of the TO-phonon and NP replicas in a longitudinal magnetic field has been studied for both donor and acceptor BMEC (Altukhov et al. 1977, 1978, 1980, 1980a, 1980b). Figure 17 shows the magnetic-field dependence of the degree of polarization for the neutral donor BMEC lines in P-doped Si. One can observe that the polarization degree for the α_1, α_2, α_3 lines decreases with increasing the number of bound excitons, m. The TO phonon and NP lines have opposite signs of the polarization. This is connected with different signs of d^- corresponding to the term Γ_1 for the TO-phonon and NP replicas (see tables 4, 5). The α_4

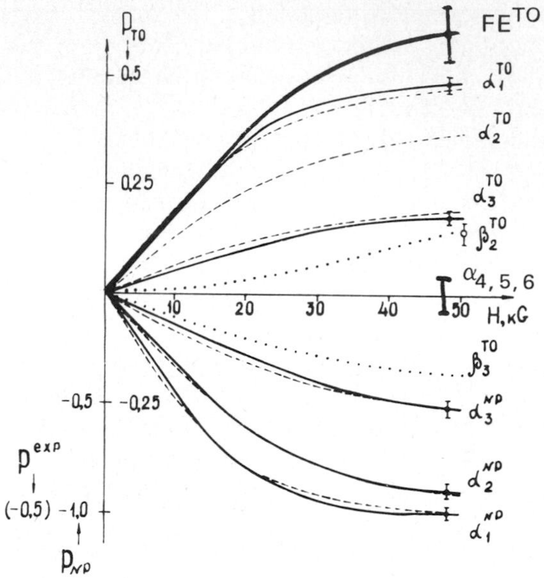

Fig. 17. The magnetic field dependence of the degree of polarization for the TO-phonon and NP lines due to multi-exciton complexes bound to ND (P) and for the free exciton lines. (After Altukhov et al. 1980, 1980b.)

lines of these replicas are unpolarized, because they correspond to the BMEC state with the full Γ_8 hole shell. The α_5 and α_6 lines are also unpolarized. This implies that the α_5, α_6 lines, as well as the α_4 line, are due to the recombination of a hole from the full Γ_8 shell with an electron from the full Γ_1 shell. The fifth and sixth holes of the $m = 5$ and $m = 6$ complexes occupy a higher shell. For Si, the next nearest in energy state is the doubly degenerate level Γ_7. The luminescence lines associated with the recombination of a hole from the Γ_7 shell were not seen in experiments by Altukhov et al. (1980, 1980a, 1980b).

For the β_1 and β_2 lines of the TO replica associated with the recombination of a $\Gamma_{3,5}$ electron the sign of the circular polarization is mainly determined by the spin orientation of holes and coincides with that for the α_2 and α_3 lines. For the line corresponding to $m = 4$, the sign of the polarization is opposite, since the holes of the full Γ_8 shell are unoriented and the thermal orientation of electrons in a magnetic field leads to the opposite sign of the polarization.

Figure 18 shows the magnetic field dependence of the polarization degree for the neutral acceptor BMEC luminescence in B-doped Si. For the acceptor BMEC, the valley–orbit splitting is small and, as a rule, the α and β lines are not resolved. Nevertheless in this case for $m = 1$ the luminescence spectra reveal some fine structure which is related to the splitting

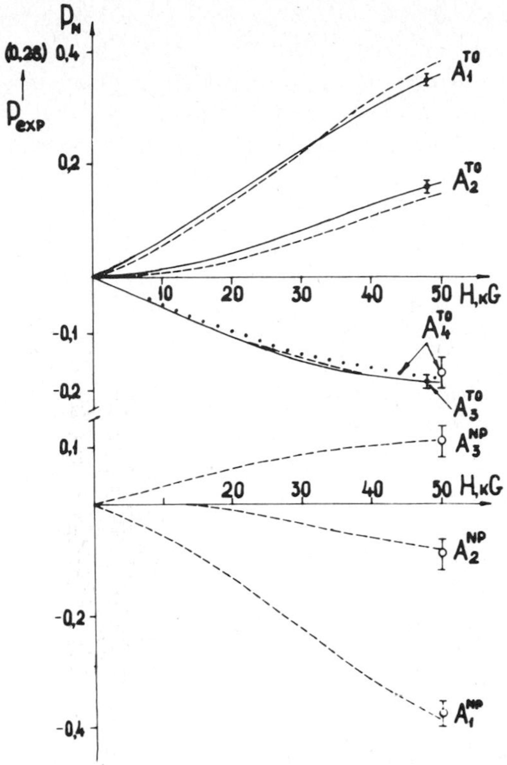

Fig. 18. The magnetic field dependence of the degree of polarization for the TO-phonon and NP lines due to multi-exciton complexes bound to NA (B). (After Altukhov et al. 1978, 1980a, 1980b.)

between the $J = 0$ and $J = 2$ states. In fig. 18 lines labelled A_m correspond to the recombination of a $\Gamma_{1,3,5}$ electron with a Γ_8 hole. The A_1 line is due to transitions from both the $J = 0$ and $J = 2$ states. The sign of the polarization of the A_1 and A_2 lines is determined by the hole spin orientation. For the line A_3, corresponding to the recombination of a hole from the full Γ_8 shell, the sign of the polarization is opposite, as for the β_3 line in fig. 17, and is determined by the electron spin orientation. The coincidence of the circular polarization directions for the A_3 and A_4 lines indicates that the A_4 line, as well as the $\alpha_{5,6}$ lines in fig. 17, is due to the recombination of a hole from the "inner" Γ_8 shell. In fig. 18, as in fig. 17, the NP and TO-phonon lines differ in the sign of polarization. One can see from tables 4 and 5 that the polarizations of the NP lines due to the recombination of an electron from the Γ_1 and $\Gamma_{3,5}$ shells are opposite. Therefore, the polarization of the A_m^{NP} lines is determined by the populations of these states and by the value of the constant Z in table 5. In the absence of the valley–orbit splitting, when electrons fill the $\Gamma_{1,3,5}$ states in accordance with the statistical factor of each state, the values of $d^{\pm}_{l,s}$ in eq. (85) for NP lines are determined, as for free excitons, by the last column in table 5. The sign of the polarization for the NP lines corresponding to fig. 18 shows that, in spite of the larger statistical factor for the $\Gamma_{3,5}$ states, the main contribution to the emission comes from the transitions involving a Γ_1 electron and, hence, $Z \ll 1$. In figs. 17 and 18 the dashed lines are theoretical curves calculated using eqs. (85) and (86) and tables 4 and 5. The values of $g = 2$ and $g_1 = 1.2$ were used. For neutral donor BMEC with $m = 2$, the splitting between the $J = 0$ and $J = 2$ states was neglected. For neutral acceptor BMEC, the similar splitting was assumed to be equal to -0.1 meV (the $J = 0$ level is low). In addition, for the latter case the valley–orbit splitting was ignored and the value of 0.2 was taken for the constant Z. One can see that the theoretical results obtained for the above-mentioned assumptions are in good agreement with the experimental data. For convenience, fig. 17 shows also the magnetic field dependence of P_{circ} for free excitons. Observe that saturation values of P_{circ} for free and bound excitons differ somewhat. Probably, this occurs because of the above mentioned difference between the selection rules for free and bound excitons. In order to explain the observed discrepancy it is necessary to assume that for bound excitons the constant δ introduced in subsect. 6.3.2 is about 0.2.

References

Agranovich, V.M., 1959, Zh. eksper. teor. Fiz. **37**, 430 (Sov. Phys.-JETP 10, 307 (1960)).

Alkeev, N.V., A.S. Kaminskii and Ya. E. Pokrovskii, 1976, Fiz. tverd. Tela **18**, 713 (Sov. Phys. Solid State, **18**, 410).

Altukhov, P.D., K.N.El'tsov, G.E. Pikus and A.A. Rogachev, 1977, Zh. eksper. teor. Fiz., Pisma **26**, 468 (JETP Lett. 26, 337 (1977)).

Altukhov, P.D., G.E. Pikus and A.A. Rogachev, 1978. Fiz. tverd. Tela, **20**, 489 (Sov. Phys. Solid State, **20**, 283).

Altukhov, P.D., K.N. El'tsov, G.E. Pikus and A.A. Rogachev, 1980, Fiz. tverd. Tela **22**, 239 (Sov. Phys. Solid State, **22**, 140).

Altukhov, P.D., K.N. El'tsov, G.E. Pikus and A.A. Rogachev, 1980a, Fiz. tverd. Tela **22**, 599 (Sov. Phys. Solid State, **22**, 350).

Altukhov, P.D., N.S. Averkiev, K.N. El'tsov, C.E. Pikus, A.A. Rogachev and E.M. Ushakova, 1980b, Fiz. tverd. Tela **22**, 1980 (Sov. Phys. Solid State, **22**, 1154).

Anno, H. and Y. Nishina, 1979, Solid State Commun. **29**, 439.

Areshev, I.P., A.M. Danishevskii, C.F. Kochegarov and V.K. Subashiev, 1976, Zh. eksper. teor. Fiz., Pisma 24, 594 (JETP Lett. **24**, 552 (1976)).

Asnin, V.M., Yu. N. Lomasov and A.A. Rogachev, 1975, Zh. techn. Fiz., Pisma, **1**, 596 (Sov. Tech. Phys. Lett. **1**, 268).

Asnin, V.M., G.L. Bir, Yu. N. Lomasov, G.E. Pikus and A.A. Rogachev, 1976, Zh. eksper. teor. Fiz. **71**, 1600 (Sov. Phys.-JETP **44**, 838 (1976)).

Bir, G.L. and G.E. Pikus, 1972, Simmetriya i deformatsionnye effekty v poluprovodnikach (Nauka, Moscow); 1974, Symmetry and Strain-Induced Effects in Semiconductors (A Halsted Press Book, Jerusalem, London).

Bir, G.L. and G.E. Pikus, 1972a, Zh. eksper, teor. Fiz. Pisma, **15**, 730 (JETP Lett. **15**, 516 (1972)).

Bir, G.L. and G.E. Pikus, 1972b, Proc. XI Int. Conf. Phys. Semicond. (Warsaw, 1972) p. 1341.
Bir, G.L. and G.E. Pikus, 1973, Zh. eksper. teor. Fiz. **64**, 2210 (Sov. Phys.-JETP **37**, 1116 (1973)).
Bir, G.L. and G.E. Pikus, 1976, Fiz. tverd. Tela, **18**, 220 (Sov. Phys. Solid State, **18**, 127).
Bir, G.L., E.L. Ivchenko and G.E. Pikus, 1976, Izv. Akad. Nauk SSSR (ser. fiz.) **40**, 1866 (Bull. Acad. Sci. USSR, Phys. ser. **40**, No. 9, 81).
Bonnot, A., R. Planel, C. Benoît á la Guillaume and G. Lampel, 1972, Proc. XI Int. Conf. Phys. Semicond. (Warsaw, 1972) p. 1334.
Bonnot, A., R. Planel and C. Benoît á la Guillaume, 1974, Phys. Rev. **B9**, 690.
Chandrasekhar, S., 1950, Radiative Transfer (Oxford University Press, London).
D'yakonov, M.I. and V.I. Perel', 1971, Zh. eksper. teor. Fiz. **60**, 1954 (Sov. Phys.-JETP **33**, 1053 (1971)).
D'yakonov, M.I. and V.I. Perel', 1971a, Fiz. tverd. Tela, **13**, 3681 (Sov. Phys.-Solid State 13, 3023 (1972)).
D'yakonov, M.I and V.I. Perel', 1972, Fiz. tverd. Tela, **14**, 1452 (Sov. Phys.-Solid State **14**, 1245 (1972)).
Dymnikov, V.D., M.I. D'yakonov and V.I. Perel', 1976, Zh. eksper. teor. Fiz. **71**, 2373 (Sov. Phys.-JETP **44**, 1252 (1976)).
Dzhioev, R.I., B.P. Zakharchenya and V.G. Fleisher, 1973, Zh. eksper. teor. Fiz., Pisma, **17**, 244 (JETP Lett. **17**, 174 (1973)).
Ekimov, A.I. and V.I. Safarov, 1970, Zh. eksper. teor. Fiz., Pisma, **12**, 293 (JETP Lett. **12**, 198 (1970)).
Fishman, G., C. Hermann and G. Lampel, 1974a, J. Phys. (Paris) **35**, C3–13.
Fishman, G., C. Hermann, G. Lampel and C. Weisbuch, 1974b, J. Phys. (Paris) **35**, C3–7.
Frova, A., G.A. Thomas, R.E. Miller and E.O. Kane, 1975, Phys. Rev. Lett. **34**, 1572.
Gamarts, E.M., E.L. Ivchenko, M.I. Karaman, V.P. Mushinskii, G.E. Pikus, B.S. Razbirin and A.N. Starukhin, 1977, Zh. eksper. teor. Fiz. **73**, 1113 (Sov. Phys.-JETP **46**, 590 (1977)).
Gross, E.F., A.I. Ekimov, B.S. Razbirin and V.I. Safarov, 1971, Zh. eksper. teor. Fiz., Pisma, **14**, 108 (JETP Lett. **14**, 108 (1971)).
Gross, E., S. Permogorov and B. Razbirin, 1966, J. Phys. Chem. Solids **27**, 1647.
Hopfield, J.J., 1958, Phys. Rev. **112**, 1555.
Ivchenko, E.L., G.E. Pikus, B.S. Razbirin and A.I. Starukhin, 1977, Zh. eksper. teor. Fiz. **72**, 2230 (Sov. Phys.-JETP **45**, 1172 (1977)).
Ivchenko, E.L., G.E. Pikus and L.V. Takunov, 1978, Fiz. tverd. Tela **20**, 2598 (Sov. Phys. Solid State, **20**, 1502).
Ivchenko, E.L., I.G. Lang and S.T. Pavlov, 1977a, Fiz. tverd. Tela **19**, 1751 (Sov. Phys.-Solid State **19**, 1610 (1977)).
Ivchenko, E.L., I.G. Lang and S.T. Pavlov, 1978a, Phys. Status Solidi (b) **85**, 81.
Ivchenko, E.L. and N.Kh. Yuldashev, 1979, Fiz. tverd. Tela, **21**, 2182 (Sov. Phys.-Solid State, **21**, 1255).
Kaminskii, A.S., Ya. E. Pokrovskii and N.V. Alkeev, 1970, Zh. eksper. teor. Fiz. **69**, 1937 (Sov. Phys.-JETP **32**, 1048 (1971)).
Kaminskii, A.S. and Ya. E. Pokrovskii, 1979, Zh. eksper. teor. Fiz. **76**, 1727 (Sov. Phys. JETP **49**).
Kirczenow, G., 1977, Solid State Commun. **21**, 713.
Kirczenow, 1977a, Canad. J. Phys. **55**, 1787.
Kiselev, V.A., B.S. Razbirin and I.N. Uraltsev, 1974, Proc. XII Int. Conf. Phys. Semicond. (Stuttgart, 1974) p. 996.
Klochikhin, A.A., Ya. V. Morozenko and S.A. Permogorov, 1978, Fiz. tverd. Tela **20**, 3557 (Sov. Phys. Solid State, **20**, 2057).
Kulakovskii, V.D., A.V. Malevnik and V.B. Timofeev, 1979, Zh. eksper. teor. Fiz. **76**, 272 (Sov. Phys.-JETP **49**, 139 (1979)).
Lampel, G., 1968, Phys. Rev. Lett. **20**, 491.

Lenable, J., 1970, J. Quant. Spectrosc. Radiat. Transfer **10**, 533.
Lipari, N.O. and M. Altarelli, 1976, Solid State Commun. **18**, 951.
Lipari, N.O. and A. Baldareschi, 1978, Solid State Commun. **25**, 665.
Lombard, M., 1967, Compt. Rend. **265**, B 191.
Lombard, M., 1969, J. de Phys. **30**, 631.
Martin, R.W., 1974, Solid State Commun. **14**, 369.
Mayer, A.E. and E.C. Lightowlers, 1979, J. Phys. C: Solid State Phys. **12**, L507.
Mayer, A.E. and E.C. Lightowlers, 1979a, J. Phys. C: Solid State Phys. **12**, L539.
Mayer, A.E. and E.C. Lightowlers, 1979b, J. Phys. C: Solid State Phys. **12**, L945.
Minami, F., Y. Oka and T. Kushida, 1976, J. Phys. Soc. Japan, **41**, 100.
Mooser, E. and M. Schlüter, 1973, Nuovo Cimento **18B**, 164.
Nawrocki, M., R. Planel and C. Benoìt á la Guillaume, 1976, Phys. Rev. Lett. **36**, 1343.
Nedelec, O., P. Baltayen and A. Orizet, 1967, Compt. Rend. **265**, B542.
Oka, Y. and T. Kushida, 1977, Nuovo Cimento, **39B**, 483.
Oka, Y. and T. Kushida, 1978, Proc. XIV Int. Conf. Phys. Semicond. (Edinburgh, 1978) p. 863.
Parsons, R.R., 1969, Phys. Rev. Lett. **23**, 1152.
Pekar, S.I., 1957, Zh. eksper. teor. Fiz. **33**, 1022 (Sov. Phys.-JETP **6**, 785 (1958)).
Permogorov, S.A. and Ya. V. Morozenko, 1979, Fiz. tverd. Tela, **21**, 784 (Sov. Phys. Solid State, **21**, 458).
Permogorov, S. and V. Travnikov, 1979, Solid State Commun. **29**, 615.
Permogorov, S.A., Ya. V. Morozenko and B.A. Kazennov, 1975, Fiz. tverd. Tela, **17**, 2970 (Sov. Phys.-Solid State **17**, 1974 (1975)).
Pikus, G.E., 1977, Fiz. tverd. Tela, **19**, 1653 (Sov. Phys.-Solid State, **19**, 965 (1977)).
Pikus, G.E. and G.L. Bir, 1972, Zh. eksper. teor. Fiz. **62**, 324 (Sov. Phys.-JETP **35**, 174 (1972)).
Pikus, G.E. and G.L. Bir, 1974, Zh. eksper. teor. Fiz. **67**, 788 (Sov. Phys.-JETP **40**, 390 (1975)).
Pikus, G.E. and G.L. Bir, 1974a, Fiz. tverd. Tela, **16**, 2701 (Sov. Phys.-Solid State **16**, 1746 (1975)).
Planel, R., M. Nawrocki and C. Benoît á la Guillaume, 1977, II Nuovo Cimento, **39B**, 519.
Razbirin, B.S., V.P. Mushinskii, M.I. Karaman, A.N. Starukhin and E.M. Gamarts, 1975, Zh. eksper. teor. Fiz., Pisma **22**, 203 (JETP Lett. **22**, 94 (1975)).
Razbirin, B.S., V.P. Mushinskii, M.I. Karaman, A.N. Starukhin and E.M. Gamarts, 1976, Izv. Akad. Nauk SSSR (ser. fiz.) **40**, 1872 (Bull. Acad. Sci. USSR, Phys. Sov. **40**, No. 9, 88).
Thewalt, M.L.W., 1977, Solid State Commun. **21**, 937.
Thewalt, M.L.W., 1977a, Canad. J. Phys. **55**, 1463.
Thewalt, M.L.W., 1978, Proc. XIV Int. Conf. Phys. Semicond. (Edinburgh, 1978) p. 605.
Varshalovich, D.A., A.I. Moskalev and V.K. Khersonskii, 1975, Kvantovaya teoriya uglovogo momenta (Nauka, Leningrad).
Veshchunov, Yu.P., B.P. Zakharchenya and E.M. Leonov, 1972, Fiz. tverd. Tela **14**, 2678 (Sov. Phys.-Solid State, **14**, 2312 (1973)).
Weisbuch, C. and G. Fishman, 1976, J. Lumines. **12/13**, 219.
Weisbuch, C. and C. Hermann, 1975, Solid State Commun. **16**, 659.
Weisbuch, C. and G. Lampel, 1972, Proc. XI Int. Conf. Phys. Semicond. (Warsaw, 1972) p. 1327.
Weisbuch, C., C. Hermann and G. Fishman, 1974, Proc. XII Int. Conf. Phys. Semicond. (Stuttgart, 1974) p. 761.
Zakharchenya, B.P., V.G. Fleisher, R.I. Dzhioev and Yu.P. Veshchunov, 1971, Zh. eksper. teor. Fiz., Pisma **13**, 195 (JETP Lett. **13**, 137 (1971)).
Zemskii, V.I., B.P. Zakharchenya and D.N. Mirlin, 1976, Zh. eksper. teor. Fiz., Pisma, **24**, 96 (JETP Lett. **24**, 82 (1976)).
Zeyher, R., 1975, Solid State Commun. **16**, 49.

CHAPTER 7

Exciton Electrooptics

A.G. ARONOV and A.S. IOSELEVICH

Leningrad Nuclear Physics Institute
Gatchina, Leningrad, 188350
USSR

Excitons
Edited by
E.I. Rashba and M.D. Sturge

Contents

Introduction

Experimental research on exciton electrooptics has its source in the pioneering work of Gross et al. (1954, 1957), in which the Stark effect and ionization of the exciton in Cu_2O were investigated. These experiments copied the well-known Stark-effect experiments in gases and demonstrated the similarity of the exciton to the hydrogen atom. Franz (1958) and Keldysh (1958) have independently predicted the effect of an electric field on the absorption edge in semiconductors – the effect, known now as the Franz–Keldysh effect. During the past twenty years many papers have appeared in which this effect was investigated theoretically and experimentally. Now it is widely used as a method of band-structure analysis in semiconductors, and a new field has arisen in semiconductor physics – electrooptics.

In the early work the Coulomb interaction between an electron and a hole was not taken into account, but very soon it became clear that these phenomena (exciton electroabsorption and the Franz–Keldysh effect) are closely related and, are in fact different aspects of the same effect. The computer calculations performed by Ralph (1968), Dow and Redfield (1970), Blossey (1970, 1971), Dow et al. (1971), Weinstein et al. (1971) and Lao et al. (1971) made it possible to evaluate the exciton effects, and confirmed, the important role of the exciton in the optical spectrum in an electric field. Merkulov and Perel' (1973, 1974a) built up the theory of exciton electroabsorption in the gap and in the vicinity of the exciton ground state, Aronov and Ioselevich (1978a) developed the theory of exciton electroabsorption above the absorption edge and near the excited states.

One of the most significant experimental applications of electrooptics is the determination of critical points of the crystal spectrum (van Hove singularities) by means of the electroreflectance method. This method turned to be much more sensitive and more explicit than the reflectance method in the absence of an external field, because the electric field affects the reflectance most drastically just near the van Hove singularities. The electroreflectance spectra of a wide group of materials (such as Ge, Si, GaAs, CdS, etc.) were studied. Since there exist excellent reviews by Seraphin (1970), by Cardona (1969) and by Aspnes and Bottka (1970), treating this class of phenomena, we do not consider the electroreflectance technique in detail.

A group of phenomena related to electroreflectance, are the effects of nonhomogeneous surface fields on the motion of the exciton centre of mass. Experimental and theoretical consideration of this problem (Permogorov et al. 1972, Evangelisti et al. 1972, Kiselev 1978, 1979) have proved the sensitivity of reflectance and electroreflectance to the details of surface barrier.

Another field of investigation is the study of the polarization dependence of the electroabsorption in cubic crystals (a theory of this effect was developed by Keldysh et al. (1969)), and of anisotropic electrooptical effects (Pockels effect and Kerr effect). A theory of anisotropic effects was developed by two groups of authors (Bagaev et al. 1966, Aronov and Pikus 1968) disregarding exciton effects, and was generalized by Aronov and Ioselevich (1978b) who took exciton effects into account. Experimental measurements of the spectral dependence of these effects confirmed the theoretical predictions.

In this review paper we aim to discuss the physical and theoretical foundations of the Franz–Keldysh effect, considering the exciton effects in semiconductors with simple bands and with degenerate ones.

In sect. 1 we consider the physics of electroabsorption without exciton effects; in sect. 2 exciton electroabsorption in the case of simple bands. In sects. 3 and 4 we discuss electroabsorption in semiconductors with degenerate valence bands without, and with exciton effects. In sects. 5 and 6 forbidden and indirect transitions in an electric field are discussed, and in sect. 7 we consider dispersion effects in an electric field. In the last section (sect. 8) we discuss the effects of nonhomogeneous surface fields on the exciton electroreflectance spectra.

For lack of space we cannot give a detailed analysis of the experimental work and experimental results are used only as illustrative material.

1. *Franz–Keldysh effect*

The Franz–Keldysh effect is the effect of an electric field on the absorption coefficient near the absorption edge. In fig. 1 the energy bands and wave functions in an electric field are shown. The electron potential energy is $-Fz$ ($\boldsymbol{F} \parallel z$). If the photon energy $\hbar\omega$ is less than the energy gap E_g the electron can transit from the valence band to the conduction band only by tunneling through the triangular barrier. The height of the barrier is $E_g - \hbar\omega = -\Delta$ and its width is $-\Delta/F$. Therefore, the absorption coefficient α has an exponential form for $\Delta < 0$ because of the exponential character of the tunneling transition probability (Keldysh 1958, Landau and Lifschitz 1963):

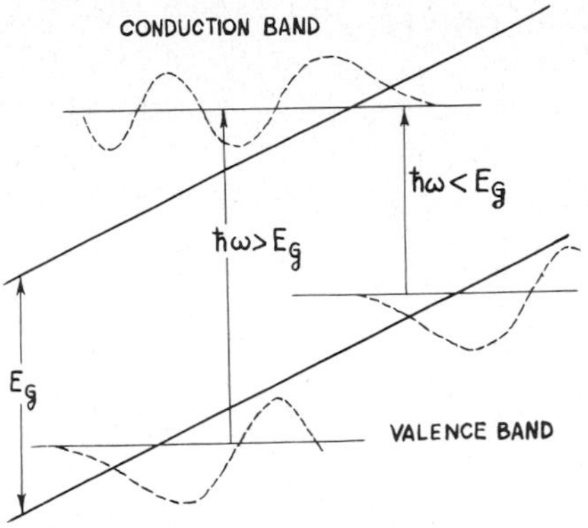

Fig. 1. Energy bands and wave functions in an electric field.

$$\alpha \propto \exp\left[-\frac{4\sqrt{2m^*}}{3\hbar F}|\Delta|^{3/2}\right].$$

Here m^* is the reduced effective mass of an electron and a hole, $1/m^* = 1/m_c + 1/m_h$, m_c being the electron effective mass, m_h being the hole effective mass.

The physical nature of the absorption at $\hbar\omega > E_g$ is connected with another phenomenon, and we shall discuss it in the end of this section.

The general expression for the dielectric permeability without respect to the spatial dispersion has the form (see, for example, Konstantinov and Perel' 1959, Aronov and Pikus 1968):

$$\epsilon_{\alpha\beta}(\omega) = \frac{4\pi}{\hbar\omega^2 V}\sum_{m\neq n} F_m^{(0)}\left\{\frac{J^\alpha_{mn}(0)J^\beta_{nm}(0)}{\omega-\omega_{mn}+i\delta} - \frac{J^\alpha_{nm}(0)J^\beta_{mn}(0)}{\omega+\omega_{mn}+i\delta}\right.$$
$$\left. + \frac{J^\alpha_{mn}(0)J^\beta_{nm}(0)+J^\alpha_{nm}(0)J^\beta_{mn}(0)}{\omega_{mn}}\right\}, \qquad (1.1)$$

where V is a crystal volume, $\hbar\omega_{mn} = \epsilon_m - \epsilon_n$, ϵ_m is the electron energy of the mth state, $F_m^{(0)} = 1$ for filled states and $F_m^{(0)} = 0$ for the empty states,

$$J^\alpha_{mn}(0) = \int \mathrm{d}^3\boldsymbol{x} J^\alpha_{mn}(\boldsymbol{x}). \qquad (1.2)$$

$J^\alpha(\boldsymbol{x})$ is a current operator. The interband current operator for the non-degenerate bands has a simple form (Bir and Pikus, 1972):

$$J^\alpha = e\rho_2 s\sigma_\alpha, \qquad (1.3)$$

where σ_α are the Pauli matrices,

$$\rho_2 = \begin{pmatrix} 0 & -\mathrm{i}I \\ \mathrm{i}I & 0 \end{pmatrix},$$

I is a unit two-by-two matrix, $s = |P_{cv}|/m$, P_{cv} is the matrix element of $\hat{\boldsymbol{p}}$ between the Bloch functions of the conduction and valence bands of crystal momentum $\boldsymbol{k} = 0$.

Substitute eq. (1.3) into eq. (1.1) and calculate a trace over the spin indices. Then the expression $\epsilon_{\alpha\beta}(\omega)$ can be rewritten with the aid of the Green function for electron–hole relative motion

$$\epsilon_{\alpha\alpha}(\omega) = \frac{8\pi e^2 s^2}{\hbar\omega^2}\{G^{(-)}_{-\hbar\omega-E_g}(0,0) + G^{(+)}_{\Delta}(0,0) - 2G^{(+)}_{-E_g}(0,0)\}, \tag{1.4}$$

where,

$$G^{(\pm)}_E(\boldsymbol{r}_1, \boldsymbol{r}_2) = -\hbar \sum_n \frac{\psi_n(\boldsymbol{r}_1)\psi^*_n(\boldsymbol{r}_2)}{E - \epsilon_n \pm \mathrm{i}\delta}, \tag{1.5}$$

$\psi_n(\boldsymbol{r})$ is the wave function for relative motion of an exciton, ϵ_n is an energy eigenvalue. From eq. (1.4) we can get the Elliot formula for the absorption coefficient (for example, see Bir and Pikus (1972)):

$$\mathrm{Im}\,\epsilon_{\alpha\alpha} = \frac{8\pi^2 e^2 s^2}{\hbar\omega^2}\sum_n |\psi_n(0)|^2 \delta(\hbar\omega - E_g - \epsilon_n). \tag{1.6}$$

The wave function for relative motion without the Coulomb interaction is determined by the Schrödinger equation:

$$\left(-\frac{\hbar^2}{2m^*}\nabla^2 + Fz\right)\psi(\boldsymbol{r}) = \epsilon\psi(\boldsymbol{r}). \tag{1.7}$$

The solution of this equation is well known (Landau and Lifschitz, 1963),

$$\psi(\boldsymbol{r}) = \frac{1}{\sqrt{\pi}}\exp(\mathrm{i}\boldsymbol{k}_\perp \boldsymbol{r}_\perp)\left[\frac{(2m^*)^2}{\hbar^4 F}\right]^{1/6} Ai\left[\left(\frac{2m^*F}{\hbar^2}\right)^{1/3}\left(z + \frac{\epsilon - \hbar^2 \boldsymbol{k}_\perp^2/2m^*}{F}\right)\right]. \tag{1.8}$$

Here $Ai(z) = (1/\sqrt{\pi})\int_0^\infty du\,\cos(\frac{1}{3}u^3 + uz)$ is the Airy function. The wave function is normalized to a δ function of energy ϵ; $\boldsymbol{k}_\perp, \boldsymbol{r}_\perp$ are the wave vector and space coordinate perpendicular to the electric field. Substituting eq. (1.8) into eq. (1.5) and integrating over $\boldsymbol{k}_\perp$ and ϵ at $\boldsymbol{r}_1 = \boldsymbol{r}_2 = 0$ we get

$$G^{(\pm)}_E(0,0) = -\frac{(2m^*)^{3/2}}{4\pi^2\hbar^2}\theta^{1/2}\int_{-\infty}^{+\infty}\frac{d\epsilon}{E - \epsilon \pm \mathrm{i}\delta}\int_{-\epsilon/\theta}^{\infty} dz Ai^2(z), \tag{1.9}$$

where $\theta^3 = (\hbar F)^2/2m^*$. If $\omega > 0$, the calculation of the absorption only requires $G^{(+)}(0,0)$,

$$\text{Im}\,\epsilon_{\alpha\alpha} = \frac{2e^2s^2}{\hbar^3\omega^2}\theta^{1/2}(2m^*)^{3/2}\int_{-\Delta/\theta}^{\infty} dz Ai^2(z).$$

Carrying out integration by parts and using the equality $zAi(z) = Ai''(z)$ we obtain (Tharmalingham 1963, Callaway 1963, 1964):

$$\text{Im}\,\epsilon_{\alpha\alpha} = \frac{2e^2s^2}{\hbar^3\omega^2}\theta^{1/2}(2m^*)^{3/2}\left\{Ai'^2\left(-\frac{\Delta}{\theta}\right) + \frac{\Delta}{\theta}Ai^2\left(-\frac{\Delta}{\theta}\right)\right\}. \tag{1.10}$$

Equation (1.10) describes the Franz–Keldysh effect at $\Delta > 0$ as well as at $\Delta < 0$ (fig. 2). If $\Delta < 0$ and $|\Delta| \gg \theta$ the asymptotic limit of the Airy function is

$$Ai(z) \approx \frac{1}{2z^{1/4}}\exp(-\tfrac{2}{3}z^{3/2})\left(1 - \frac{5}{48z^{3/2}}\right), \quad (z \gg 1)$$

and therefore,

$$\text{Im}\,\epsilon_{\alpha\alpha} = \frac{e^2s^2m^*}{2\hbar^2\omega^2}\frac{F}{|\Delta|}\exp\left[-\frac{4\sqrt{2m^*}}{3}\frac{|\Delta|^{3/2}}{\hbar F}\right]. \tag{1.11}$$

According to eq. (1.11) $\text{Im}\,\epsilon_{\alpha\alpha}$ is defined by the quasi-classical exponential tunneling factor. If $\Delta > 0$ and $\Delta \gg \theta$ the Airy function has the asymptotic form

$$Ai(z) \approx \frac{1}{|z|^{1/4}}\left\{\cos(\tfrac{2}{3}|z|^{3/2} - \tfrac{1}{4}\pi) + \frac{5}{48|z|^{3/2}}\sin(\tfrac{2}{3}|z|^{3/2} - \tfrac{1}{4}\pi)\right\}, \quad (z < 0) \tag{1.12}$$

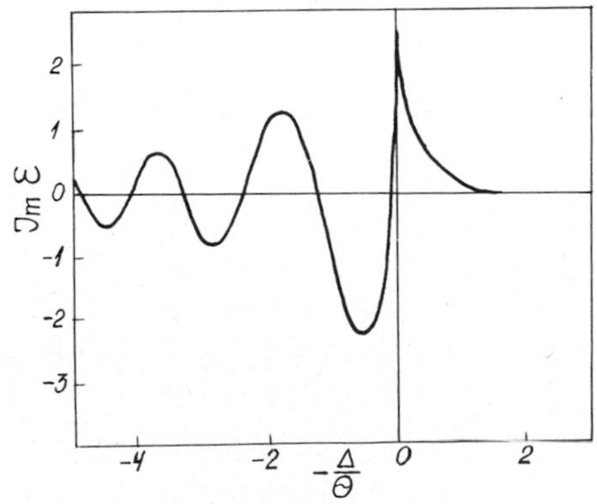

Fig. 2. The change of light absorption in an electric field without exciton effects, according to Aspnes (1966).

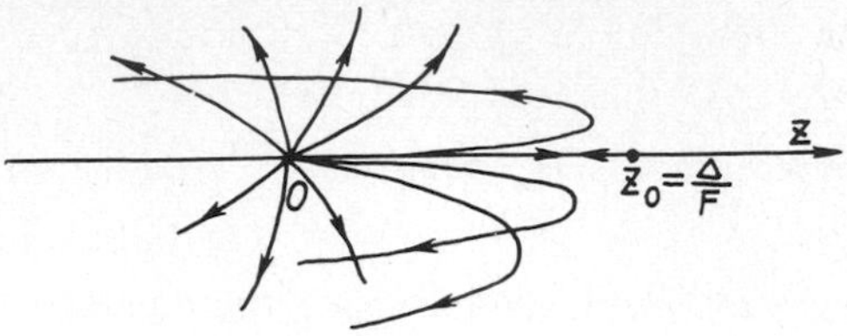

Fig. 3a. The particle trajectory in an electric field at different values of $k_\perp$, $k^4/\tilde{F} \gg 1$ (Coulomb distortion of the trajectory is neglected).

and

$$\operatorname{Im} \epsilon_{\alpha\alpha} = \frac{2e^2s^2}{\hbar^3\omega^2}(2m^*)^{3/2}\left[\Delta^{1/2} - \frac{\theta^{3/2}}{4\Delta}\cos\frac{4}{3}\left(\frac{\Delta}{\theta}\right)^{3/2}\right]. \tag{1.13}$$

We see from eq. (1.13) that the absorption oscillates as a function of the light frequency for $\Delta > 0$. Let us discuss the physical nature of the oscillations. All absorption is determined by the value $|\psi(0)|^2$, and the exciton (or the free electron–hole pair) wave function is determined by eq. (1.7) for the particle, which has mass m^* and energy ϵ. This particle is created at the point $r = 0$, from which point the wave propagates. For $\Delta \gg \theta$ the motion is quasi-classical. The particle trajectories are shown in the fig. 3a for different transverse momenta $k_\perp$. The particle moving in the direction opposite to the field direction travels a distance $z_0 = \Delta/F$ and is then reflected. At $r = 0$ the reflected wave interferes with the incident wave. The quasi-classical action is extreme for this trajectory. The actions for a neighbouring trajectory is

$$S = \sqrt{2m^*}\int_0^{z_0} dz\sqrt{\Delta - Fz - \frac{\hbar k_\perp^2}{2m^*}} = \tfrac{4}{3}\hbar\left(\frac{\Delta}{\theta}\right)^{3/2}(1-\varphi^2)^{3/2},$$

where $\varphi^2 = \hbar^2k_\perp^2/2m^*\Delta$. At small $k_\perp$, φ is the angle of the departure from the

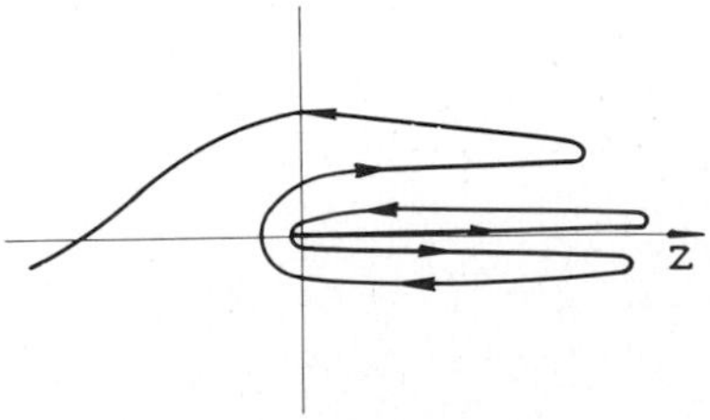

Fig. 3b. The particle trajectory in both electric and Coulomb fields, $k \ll 1$, $k^4/\tilde{F} \ll 1$ (the trajectory is distorted drastically by the Coulomb potential).

origin. Let us consider the solid angle Ω, formed by the trajectories, for which the action differs from the extreme action by an amount of the order of $\hbar$:

$$\Omega \sim \varphi^2_{\max} \sim (\theta/\Delta)^{3/2}.$$

All waves which depart in this solid angle interfere at the point $r = 0$. If ψ_0 is the incident wave amplitude, then

$$|\psi(0)|^2 \simeq |\psi_0 + \Omega\psi_0 \exp(2iS/\hbar)|^2 \propto 1 + C\left(\frac{\theta}{\Delta}\right)^{3/2} \cos\frac{4}{3}\left(\frac{\Delta}{\theta}\right)^{3/2},$$

(C being of the order of unity), and the absorption coefficient oscillates as a function of the light frequency.

2. *Exciton electroabsorption (simple energy bands)*

The main qualitative phenomena are described by the simple theory (sect. 1), but as noted above, it is necessary to take excitons into account for the semi-quantitative and even qualitative comparison of theory and experiment. Numerical calculations have shown that exciton effects lead to strong enhancement of the effect, to a change of the oscillation period, and to other phenomena (fig. 4).

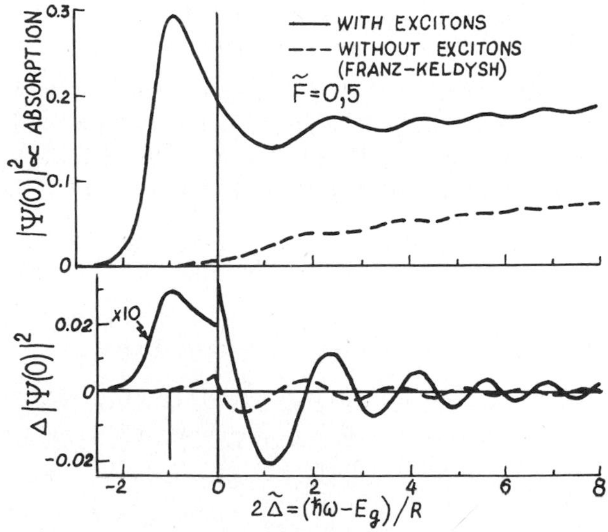

Fig. 4. Absorption coefficient and its changing in an electric field with and without respect to an exciton effect, according to Dow et al. (1971).

The principal results of the theory developed by Merkulov and Perel' (1973, 1974) and by Aronov and Ioselevich (1978a) can be summarized briefly as follows: at large energy deficits $|\Delta| \gg \theta$ the Coulomb interaction alters the pre-exponential factor in the asymptotic expression for the Franz–Keldysh effect (see eq. (1.11)) both as a result of the change of the wave function at short distances and as a result of the lowering of the potential barrier. At photon energies corresponding to the production of the exciton in a bound state, the absorption has a Lorentzian form. The width of the peak is determined by the probability of ionization by the electric field.

In the continuous spectrum the absorption is changed due to the appearance of the Sommerfeld factor, but the relative strength of the oscillating part is not changed by the Coulomb interaction. Moreover, the oscillation period is changed due to the change of the phase of the wave function both at short and large distances. The first contribution does not depend on the electric field. Near the absorption threshold the oscillations are transformed into a sawtooth curve.

In this section we shall follow the method used in the author's paper (Aronov and Ioselevich 1978a) because this method is traditional and allows us to obtain results throughout the spectrum, excluding a narrow region near the threshold.

In the effective mass approximation the wave function of an electron–hole pair in an electric field and in the Coulomb potential $-e^2/\epsilon_0 r$ is described by the Schrödinger equation,

$$\left(-\frac{\hbar^2}{2m^*}\nabla^2 - \frac{e^2}{\epsilon_0 r} + Fz\right)\psi(\boldsymbol{r}) = \Delta\psi(\boldsymbol{r}), \tag{2.1}$$

where ϵ_0 is the static dielectric permeability. Let us introduce Coulomb units: $\tilde{\Delta} = \Delta/2R$, $\tilde{F} = Fa/2R$, $R = m^*e^4/2\epsilon_0^2\hbar^2$ is the exciton binding energy and $a = \hbar^2\epsilon_0/m^*e^2$ is the exciton radius. It is well known that the variables in eq. (2.1) separate in parabolic coordinates ξ, η, φ, connected with Cartesian coordinates by the relations

$$\begin{gathered} x = \sqrt{\xi\eta}\cos\varphi, \quad y = \sqrt{\xi\eta}\sin\varphi, \quad z = \tfrac{1}{2}(\xi - \eta), \\ r = (x^2 + y^2 + z^2)^{1/2} = \tfrac{1}{2}(\xi + \eta), \quad \mathrm{d}^3r = \tfrac{1}{4}(\xi + \eta)\,\mathrm{d}\xi\,\mathrm{d}\eta\,\mathrm{d}\varphi. \end{gathered} \tag{2.2}$$

The optical absorption coefficient receives contributions only from states with azimuthal quantum number $m = 0$. Equation (2.1) reduces to two equations (Landau and Lifschitz 1963):

$$\chi_1'' + \left(\frac{\tilde{\Delta}}{2} + \frac{\beta_1}{\xi} + \frac{1}{4\xi^2} - \frac{\tilde{F}}{4}\xi\right)\chi_1 = 0, \quad \chi_2'' + \left(\frac{\tilde{\Delta}}{2} + \frac{\beta_2}{\eta} + \frac{1}{4\eta^2} + \frac{\tilde{F}}{4}\eta\right)\chi_2 = 0, \tag{2.3}$$

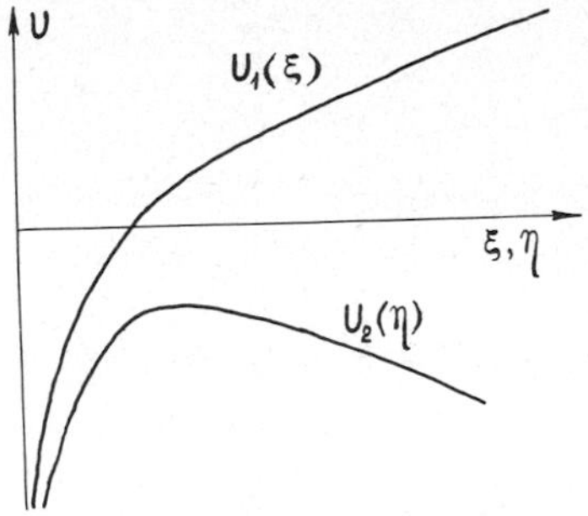

Fig. 5. The effective potentials $U_1(\xi)$, $U_2(\eta)$ for finite and infinite motions respectively.

where

$$\psi(\boldsymbol{r}) = f_1(\xi)f_2(\eta) = \frac{\chi_1(\xi)}{\sqrt{\xi}}\frac{\chi_2(\eta)}{\sqrt{\eta}}, \tag{2.4}$$

and $\beta_1 + \beta_2 = 1$.

The effective potentials $U_1(\xi)$ and $U_2(\eta)$ are shown in fig. 5. For $\tilde{\Delta} > 0$ the motion along η is infinite, and along ξ it is finite; for $\tilde{\Delta} < 0$ the motion along both ξ and η is finite if the energy $|\tilde{\Delta}|$ is large enough.

2.1. Optical absorption by an exciton in the continuous spectrum

To calculate the optical absorption far from the threshold let us investigate the quasi-classical solutions of eqs. (2.3) and (2.4). To solve these equations we use the fact that at short distances we can neglect the electric field. If the influence of the electric field becomes substantial only in the region where the motion is quasi-classical, then it is possible to match the quasi-classical Coulomb function with the quasi-classical function in which the electric field is taken into account. The condition for the applicability of the quasi-classical approximation is

$$\frac{U_2'(\eta)}{k^3(\eta)} = \frac{\beta_2/2\eta^2 + 1/2\eta^3 - \frac{1}{4}\tilde{F}}{(\frac{1}{4}k^2 + \beta^2/\eta + 1/4\eta^2 + \frac{1}{4}\tilde{F}\eta)^{3/2}} \ll 1, \tag{2.5}$$

where $\tilde{\Delta} = k^2/2$.

It is seen from eq. (2.3) that the influence of the field can be neglected if

$$\frac{\tilde{F}\eta}{4} \ll \max\left\{\frac{1}{\eta^2}, \frac{\beta_2}{\eta}, \frac{k^2}{4}\right\}. \tag{2.6}$$

It will be shown below that the significant values are $\beta_2 \simeq k$. Then eqs. (2.5) and (2.6) lead to the following condition for the applicability of the quasi-classical method,

$$k^3/\tilde{F} \gg 1, \tag{2.7}$$

or in dimensional variables we have,

$$(\Delta/\theta)^{3/2} \gg 1.$$

This inequality is the only restriction of this theory. It requires that the electric field be quasi-classical and imposes no limitations on the Coulomb energy.

The normalization conditions for the functions of the infinite and finite motion $\chi_2(\eta, \beta_2)$ and $\chi_1(\xi, \beta_1)$ are:

$$\int_0^\infty \mathrm{d}\eta \chi_{2,\tilde{\Delta}'}(\eta, \beta_2')\chi^*_{2,\tilde{\Delta}}(\eta, \beta_2) = (2/\pi)\delta(\tilde{\Delta} - \tilde{\Delta}')\delta_{\beta_2\beta_2'}, \tag{2.8}$$

$$\int_0^\infty \frac{\mathrm{d}\xi}{\xi}\chi_1^2(\xi, \beta_1) = 1. \tag{2.9}$$

Under the condition of eq. (2.6) we have (Landau and Lifschitz 1963):

$$f_2(\eta, \beta_2) = f_2(0, \beta_2)\exp(-\mathrm{i}k\eta/2)F(\tfrac{1}{2} + \mathrm{i}\beta_2/k, 1, \mathrm{i}k\eta). \tag{2.10}$$

If, furthermore, $k\eta \gg 1$ then,

$$f_2(\eta, \beta_2) = f_2(0, \beta_2)\left\{\frac{2}{\pi k\eta}\left[1 + \exp\left(-\frac{2\pi\beta_2}{k}\right)\right]\right\}^{1/2} \times \cos\left[\frac{k\eta}{2} + \frac{\beta_2}{k}\ln k\eta - \frac{\pi}{4} - \delta\left(\frac{\beta_2}{k}\right)\right], \tag{2.11}$$

where $\delta(z) = \arg\Gamma(1/2 + \mathrm{i}z)$.

The function, eq. (2.11), has quasi-classical form. Hence in the region $\eta > (k/\tilde{F})^{1/2}$, where it is necessary to take the first term into account it continuously transforms into the quasi-classical function in an electric field. The normalization condition, eq. (2.8), yields

$$|f_2(0, \beta_2)|^2 = \frac{1}{\pi[1 + \exp(-2\pi\beta_2/k)]}. \tag{2.12}$$

It is seen from eq. (2.12) that $|f_2(0, \beta_2)|^2$ does not depend at all on the electric field; for, owing to the smallness of the above – barrier reflection, the function for infinite motion can be normalized by calculating the probability flux at any point, where the quasi-classical condition, eq. (2.5), is satisfied, including a point where the electric field is no longer of significance.

For the finite motion the conditions (2.5) and (2.6) also lead to the inequality (2.7). Therefore, by analogy with the preceding case, we can represent $f_1(\xi, \beta_1)$ in the region $1 \ll k\xi \ll (k^3/\tilde{F})^{1/2}$ in the form:

$$f_1(\xi, \beta_1) = f_1(0, \beta_1)\left[1 + \exp\left(-\frac{2\pi\beta_1}{k}\right)\right]^{1/2} \frac{1}{\sqrt{\pi\xi k(\xi)}} \cos S(\xi), \tag{2.13}$$

$$S(\xi) = \frac{k\bar{\xi}}{2} + \frac{\beta_1}{k} \ln k\bar{\xi} - \frac{\pi}{4} + \int_{\bar{\xi}}^{\xi} k(\xi')\, d\xi' - \delta\left(\frac{\beta_1}{k}\right), \tag{2.14}$$

where $k(\xi) = (k^2/4 + \beta_1/\xi - \tilde{F}\xi/4)^{1/2}$ with $\bar{\xi}$ an arbitrary value of ξ from the interval $1/k \ll \xi \ll (k/\tilde{F})^{1/2}$. The condition for quantization in the potential $U_1(\xi)$ is

$$S(\xi_2) = \pi(l + \tfrac{1}{4}), \tag{2.15}$$

where $\xi_2 = k^2/\tilde{F}$ is the right-hand classical turning point.

Far from the edge, as we shall show, an important contribution to the action $S(\xi_2)$ is made by values $\beta_1 \sim k$. Then eqs. (2.14) and (2.15) yield

$$\pi l = \frac{k^3}{3\tilde{F}} + \frac{\beta_1}{k} \ln \frac{4k^3}{\tilde{F}} - \delta\left(\frac{\beta_1}{k}\right) - \frac{\pi}{2}. \tag{2.16}$$

We shall need the quasi-classical state density,

$$\nu(\beta_1) = \frac{\partial l}{\partial \beta_1} = \frac{1}{\pi k}\left\{\ln \frac{4k^3}{\tilde{F}} - \frac{1}{2}\left[\psi\left(\frac{1}{2} + i\frac{\beta_1}{k}\right) + \psi\left(\frac{1}{2} - i\frac{\beta_1}{k}\right)\right]\right\}, \tag{2.17}$$

where

$$\psi(z) = \frac{d}{dz} \ln \Gamma(z).$$

To calculate $f_1^2(0, \beta_1)$ we use the normalization condition (2.9), and break up entire region $0 \leqslant \xi \leqslant \xi_2$ of integration with respect to ξ into two regions: (1) $0 \leqslant \xi \leqslant \bar{\xi}$ and (2) $\bar{\xi} \leqslant \xi \leqslant \xi_2$. In the integration over the first region it is necessary, by virtue of the condition (2.6), to use the exact Coulomb function, while in the second region it is necessary to use the quasi-classical function in the electric and Coulomb fields. As a result of simple calculations using eq. (2.17) for $\nu(\beta_1)$ we obtain,

$$f_1^2(0, \beta_1) = \frac{1}{\nu(\beta_1)} \frac{1}{1 + \exp(-2\pi\beta_1/k)}. \tag{2.18}$$

We note that eq. (2.18), or more accurately its connection with the state density, would be trivial if only the quasi-classical regions were to contribute to the normalization integral and to the action. In our case, however, the nonquasi-classical region also makes a substantial contribution (the second term in eq. (2.17)).

To find Im $\epsilon_{\alpha\alpha}$ we must know $|\psi(0)|^2$,

$$|\psi(0)|^2 = \sum_{\beta_1+\beta_2=1} f_1^2(0, \beta_1) f_2^2(0, \beta_2). \tag{2.19}$$

To calculate the sum we use the Poisson formula,

$$|\psi(0)|^2 = \int_{-\infty}^{+\infty} d\beta_1 \nu(\beta_1) f_1^2(0, \beta_1) f_2^2(0, 1-\beta_1)$$

$$+2\,\mathrm{Re} \sum_{m=1}^{\infty} \int_{-\infty}^{+\infty} d\beta_1 \nu(\beta_1) \exp[2\pi\, iml(\beta_1)] f_1^2(0, \beta_1) f_2^2(0, 1-\beta_1). \tag{2.20}$$

The integration in eq. (2.20) is carried out not from the end point of the spectrum $\beta_1^{(0)} = -k^4/16\tilde{F}$, but from $-\infty$, since allowance for the fact that $\beta_1^{(0)}$ is finite leads to a correction of the order of $\exp(-k^3/\tilde{F})$, which is smaller than the error in the quasiclassical approach.

Substituting in eq. (2.20) the explicit eqs. (2.12), (2.18) and (2.16) we obtain,

$$|\psi(0)|^2 = \frac{1}{\pi[1-\exp(-2\pi/k)]} - \frac{k}{\pi^2} \mathrm{Re} \exp\left(i\frac{2k^3}{3\tilde{F}}\right) \int_{-\infty}^{+\infty} dt\, \Gamma^2(\tfrac{1}{2} - it)$$

$$\times \exp\left[\left(\pi + 2i \ln \frac{2k^3}{\tilde{F}}\right) t\right] \left\{1 + \exp\left[2\pi\left(t - \frac{i}{k}\right)\right]\right\}^{-1}. \tag{2.21}$$

Going around the point 0 in the upper half of the complex t-plane along a semicircle of radius $1 \ll R \ll k^3/\tilde{F}$ and neglecting, relative to the parameter $(k^3/\tilde{F})^{-1}$, both the integral along the semicircle and the residues of all the poles of the integrand that lie inside the contour, except the one closest to the real axis, we obtain,

$$\mathrm{Im}\, \epsilon_{\alpha\alpha} = \frac{\alpha_0}{\pi[1-\exp(-2\pi/k)]}$$

$$\times \left\{1 - \frac{\tilde{F}}{2k^3} \cos\left[\frac{2}{3}\frac{k^3}{\tilde{F}} + \frac{2}{k} \ln \frac{4k^3}{\tilde{F}} + 2\arg\Gamma\left(1 - \frac{i}{k}\right)\right]\right\}, \tag{2.22}$$

or in dimensional variables we have,

$$\mathrm{Im}\, \epsilon_{\alpha\alpha} = \frac{2e^2 s^2 (2m^*)^{3/2}}{\hbar^3 \omega^2} \frac{2\pi R^{1/2}}{1-\exp(-2\pi\sqrt{R/\Delta})} \left\{1 - \frac{1}{4}\left(\frac{\theta}{\Delta}\right)^{3/2} \cos\left(\frac{4}{3}\left(\frac{\Delta}{\theta}\right)^{3/2} + \right.\right.$$

$$\left.\left. + 2\left(\frac{R}{\Delta}\right)^{1/2} \ln 8\left(\frac{\Delta}{\theta}\right)^{3/2} + 2\arg\Gamma\left[1 - i\left(\frac{R}{\Delta}\right)^{1/2}\right]\right)\right\}. \tag{2.23}$$

2.2. Optical absorption near the threshold

Let us consider the absorption near the threshold when $\Delta \ll R$, assuming that $\Delta \gg \theta$. We shall show that in this case the principal parameter on the theory is $\Delta^2/\theta^{3/2}R^{1/2}$ (or in dimensionless variables, $k^4/\tilde{F}$). For $k^4/\tilde{F} \gg 1$ the results obtained for $k \ll 1$ are matched with the results obtained above for $k \gtrsim 1$.

First, let us note that eq. (2.12) for $|f_2(\eta)|^2$ was obtained without any restriction on the value of k. For the finite motion the form of expression for $|f_1(0)|^2$ is likewise preserved. Only the expression for the density of state changes. In fact the quantization condition, eq. (2.16), was derived under the condition $\beta_1 \simeq k$, which was verified in the course of the integration in eq. (2.21). In the case $k \ll 1$ the dominant values in the integration are $\beta_1 \simeq 1$ so that the quantization condition must be derived anew. For $\xi \ll \beta_1/k^2$, $(\beta_1/\tilde{F})^{1/2}$ the wave function in the Coulomb field is (Landau and Lifschitz 1963):

$$f_1(\xi) = f_1(0)J_0(2\sqrt{\beta_1\xi}), \tag{2.24}$$

where $J_0(x)$ is the Bessel function. If, furthermore, $\xi \gg 1/\beta_1$, then we can write the asymptotic form of eq. (2.24):

$$f_1(\xi) = \frac{f_1(0)}{\sqrt{\pi}}(\beta_1\xi)^{-1/4}\cos\left(2\sqrt{\beta_1\xi} - \frac{\pi}{4}\right). \tag{2.24a}$$

It is seen that eq. (2.24a) is the quasi-classical wave function with

$$k(\xi) = \left(\frac{\beta_1}{\xi}\right)^{1/2}, \qquad S(\xi) = \int_0^\xi k(\xi')\,\mathrm{d}\xi'.$$

Proceeding further to values of ξ where the electric field comes into play, and matching eq. (2.24a) with quasi-classical function in this region, we find that

$$S(\xi) = \int_0^\xi (\tfrac{1}{4}k^2 + \beta_1/\xi - \tfrac{1}{4}\tilde{F}\xi)^{1/2}\,\mathrm{d}\xi. \tag{2.25}$$

The quantization condition takes the form

$$S(\xi_2) = \pi(l + \tfrac{1}{2}),$$

where

$$\xi_2 = \frac{k^2}{2\tilde{F}}\left(1 + \sqrt{1 + \frac{16\beta_1\tilde{F}}{k^4}}\right)$$

is the right-hand turning point. After simple calculations we get for $S(\xi_2)$:

$$S(\xi_2) = \frac{k^3}{3\tilde{F}}\left(\frac{1+\sqrt{1+x}}{2}\right)^{3/2}(1+a)^{1/2}\left[aK\left(\frac{1}{\sqrt{1+a}}\right)+(1-a)E\left(\frac{1}{\sqrt{1+a}}\right)\right] \tag{2.26}$$

here

$$a = \frac{\sqrt{1+x}-1}{\sqrt{1+x}+1}, \quad x = \frac{16\tilde{F}\beta_1}{k^4} = \frac{\beta_1}{\gamma k} = x_0\beta_1,$$

$K(x)$, $E(x)$ are complete elliptic integrals of the first and second kind, respectively. The density of state has the form:

$$\nu(\beta_1) = \frac{\partial l}{\partial \beta_1} = \frac{1}{\pi}\frac{\partial S(\xi_2)}{\partial \beta_1} = \frac{2}{\pi k}\frac{1}{(1+x)^{1/4}}K\left(\frac{1}{\sqrt{1+a}}\right). \tag{2.27}$$

The condition $\beta_1 \gg k$ is equivalent to the condition $x\gamma \gg 1$. By direct calculation we can verify that the normalization integral, as before, is proportional to $\nu(\beta_1)$. Therefore, in accordance with eq. (2.20),

$$|\psi(0)|^2 = \frac{1}{\pi[1-\exp(-2\pi/k)]} + \sum_{m=1}^{\infty}(-1)^m \frac{2}{\pi x_0}\operatorname{Re}\int_{-\infty}^{+\infty} dx \exp\left[i\frac{32}{3}\gamma m\varphi(x)\right]$$
$$\times [1+\exp(-2\pi\gamma x)]^{-1}\{1+\exp[-2\pi\gamma(x_0-x)]\}^{-1} \tag{2.28}$$

where

$$\varphi(x) = (3\tilde{F}/k^3)S(x), \tag{2.29}$$

$$\varphi(x) \simeq 1 + 3/16x(\ln 64/x + 1), \quad \text{at } x \ll 1 \tag{2.30}$$

$$\varphi(x) \simeq \frac{1}{2}K\left(\frac{1}{\sqrt{2}}\right)x^{3/4} + \frac{3}{2}\left[E\left(\frac{1}{\sqrt{2}}\right) - \frac{1}{\sqrt{2}}K\left(\frac{1}{\sqrt{2}}\right)\right]x^{1/4}, \quad \text{at } x \gg 1. \tag{2.30a}$$

Thus in the upper half-plane the integrand in eq. (2.28) attenuates exponentially, and in the calculation of the integral we can close the contour in the upper half-plane. It turns out that there are two sets of poles:

$$x_l = (i/\gamma)(l+\tfrac{1}{2}) \quad \text{and} \quad x_n = x_0 + (i/\gamma)(n+\tfrac{1}{2}), \quad l, n = 0, 1, 2 \ldots .$$

The residues of the poles of the first set are small in the parameter $k \ll 1$ relative to the corresponding residues of the second set, and can be neglected, so that the expression under the summation sign with respect to m takes the form

$$I_m = (-1)^m \frac{2k}{\pi}\sum_{n=0}^{\infty}\exp\left[i\frac{32}{3}\gamma m\varphi\left(x_0 + \frac{i}{\gamma}\left(n+\frac{1}{2}\right)\right)\right]. \tag{2.31}$$

At small k the expression in the exponential can be expanded in the form:

$$\varphi\left[x_0+\frac{i}{\gamma}\left(n+\frac{1}{2}\right)\right]=\varphi(x_0)+\frac{3i}{8\gamma}\frac{K((1+a)^{-1/2})(n+\frac{1}{2})}{(1+x)^{1/4}}. \tag{2.32}$$

Substituting eqs. (2.32) and (2.31) into eq. (2.28) and changing the order of the summation we obtain,

$$|\psi(0)|^2=\frac{1}{\pi}\left\{1-2k\sum_{n=0}^{\infty}\operatorname{Im}\frac{1}{1+\exp[-i\Theta+(2n+1)q]}\right\}, \tag{2.33}$$

where

$$\Theta=\tfrac{32}{3}\gamma\varphi(x_0),\quad q=2(1+x_0)^{-1/4}K((1+a)^{-1/2}).$$

Let us consider first the case $x_0=16\tilde{F}/k^4 \ll 1$. Using the asymptotic expressions for the complete elliptic integrals, we obtain:

$$\Theta=\tfrac{2}{3}k^3/\tilde{F}+2/k\ln 4k^3/\tilde{F}+2/k(\ln k+1);\quad q=\ln 4k^3/\tilde{F}. \tag{2.34}$$

Since $e^q \gg 1$ it is sufficient to retain in the sum only the term with $n=0$, and the remaining terms will be small in terms of the parameter $\tilde{F}/4k^4=x_0/64 \ll 1$. Moreover, using the asymptotic expression for $\Gamma(1-i/k)$ in eq. (2.22) we verify that the asymptotic forms of eq. (2.33) at $x_0 \ll 1$ and eq. (2.22) at $k \ll 1$ coincide. Thus, eq. (2.22) describes the light absorption at arbitrary k under the condition $k^4/\tilde{F} \gg 1$.

If $x_0 \gg 1$, then $q \ll 1$, so that the sum over n can be replaced by an integral. As a result we have

$$|\psi(0)|^2=\frac{1}{\pi}\left\{1+\frac{\tilde{F}^{1/4}}{K(1/\sqrt{2})}\operatorname{arc\,tg}\left[\frac{(1-q)\sin\Theta}{1+(1-q)\cos\Theta}\right]\right\}. \tag{2.35}$$

If $\Theta-\pi(2j+1) \gg q$ $(j=0,1,2\ldots)$ then

$$\operatorname{arc\,tg}\left[\frac{(1-q)\sin\Theta}{1+(1-q)\cos\Theta}\right]\simeq[(\Theta/2\pi+\tfrac{1}{2})-\tfrac{1}{2}]\pi,$$

but if $\Theta-\pi(2j+1) \lesssim q \ll 1$, then

$$\operatorname{arc\,tg}\left\{\frac{(1-q)\sin\Theta}{1+(1-q)\cos\theta}\right\}=\frac{\pi(2j-1)-\Theta}{q}.$$

Here $[z]$ is the non-integral part of z. Thus, at small k the oscillations in the Franz–Keldysh effect, with allowance for the Coulomb interaction, are transformed into a sawtooth curve.

At large $k \gtrsim 1$ the particle on being reflected by the electric field barrier turns back to the origin, passes it by, and, being accelerated by field, goes away from the origin (fig. 3a), contributing to the interference only once. For $k \ll 1$, a particle, passing the origin, is strongly affected by the Coulomb field and passes the origin along a highly eccentric hyperbolic orbit, being turned in a direction, almost opposite to the field direction (fig. 3b), so that it repeats its motion in almost the same manner many times. Such a repeated motion enhances the higher harmonics.

A similar phenomenon is evidently responsible for the appearance of the "electric-field-induced resonances above the ionization threshold", observed by Freeman et al. (1978) in an atomic beam of Rb, which behaves as a one-electron atom. Freeman et al. (1978) calculated the action numerically, with results similar to eq. (2.25), and obtained resonances which agree with experimental results.

2.3. *Optical absorption in the gap*

The change of the absorption in an electric field, including the Coulomb interaction of the electron and hole was investigated by Merkulov and Perel' (1973) at large photon-energy deficits and by Merkulov (1974) near the resonance with the ground level of the exciton.

When finite motion is considered at sufficiently large energy deficits ($\kappa^3/\tilde{F} \gg 1$) and far from resonance, we can neglect the electric field completely and use the wave function of the electron in an unperturbed Coulomb potential (Landau and Lifschitz 1963),

$$f_1(\xi, n_1) = f_1(0)\exp\left(-\frac{\kappa\xi}{2}\right)F(-n_1, 1, \kappa\xi), \quad n_1 = 0, 1, 2 \dots. \tag{2.36}$$

The quantum number is $\beta_1 = \kappa(n_1 + \frac{1}{2})$ and $\frac{1}{2}\kappa^2 = |\Delta|$. The normalization condition yields,

$$f_1^2(0) = \kappa. \tag{2.37}$$

In the region $\eta \ll (\beta_2/\tilde{F})^{1/2}$ we can neglect the electric field also for infinite motion, and the wave function then takes the form:

$$f_2(\eta) = f_2(0)\exp(-\tfrac{1}{2}\kappa\eta)F(-\tfrac{1}{2} + \beta_2/\kappa, 1, \kappa\eta), \tag{2.38}$$

for $\kappa\eta \gg 1$ we have

$$f_2(\eta) = \frac{f_2(0)}{(\kappa\eta)^{1/2}}\frac{\exp(\frac{1}{2}\kappa\eta - (\beta_2/\kappa)\ln\kappa\eta)}{\Gamma(\frac{1}{2} - \beta_2/\kappa)}, \tag{2.39}$$

and the motion in this region is quasi-classical. Just as in the preceding subsection, the wave function (2.39) can be matched with the quasi-classical function in the region where the electric field comes into play. This function is, as usual, of the form

$$f_2(\eta) = f_2(0)[2\eta|k(\eta)|]^{-1/2}\exp(S(\eta)).$$

The complete "below the barrier" action up to the right hand turning point is equal to

$$S(\eta_2) = \frac{\kappa\eta}{2} - \frac{\beta_2}{\kappa}\ln\kappa\bar{\eta} + \int_{\bar{\eta}}^{\eta_2} d\eta\sqrt{\frac{\kappa^2}{4} - \frac{\tilde{F}\eta}{4} - \frac{\beta_2}{\eta}} \simeq \frac{\kappa^3}{3\tilde{F}} - \frac{\beta_2}{\kappa}\ln\frac{4\kappa^3}{\tilde{F}}, \tag{2.40}$$

where $\eta_2 = \kappa^2/\tilde{F}$ is the right-hand turning point and $\bar{\eta}$ is chosen so that $\kappa\bar{\eta} \gg 1$ but $\tilde{F}\bar{\eta}/4 \ll \max\{\beta_2/\bar{\eta}, \kappa^2/4\}$. To the right of the turning point we obtain the following quasi-classical function:

$$f_2(\eta) = \frac{f_2(0)}{(2\eta)^{1/2}} \frac{\exp S(\eta_2)}{\Gamma(\frac{1}{2} - \beta_2/\kappa)} \frac{2\cos S(\eta)}{|k(\eta)|^{1/2}},$$

but according to the flux normalization condition the coefficient of $|k(\eta)|^{-1/2}\cos S(\eta)$ should be equal to $1/\pi$. Therefore

$$f_2^2(0) = \frac{1}{2\pi^2}\exp[-2S(\eta_2)]\Gamma^2\left(\frac{1}{2} - \frac{\beta_2}{\kappa}\right),$$
$$\beta_2/\kappa = 1/\kappa - \beta_1/\kappa = 1/\kappa - (n_1 + \tfrac{1}{2}). \tag{2.41}$$

Thus, using eqs. (2.37), (2.40) and (2.41), we obtain

$$|\psi(0)|^2 = \sum_{n_1=0}^{\infty} f_1^2(0) f_2^2(0) = \frac{\kappa}{2\pi^2}\sum_{n_1=0}^{\infty}\Gamma^2\left(n_1 + 1 - \frac{1}{\kappa}\right)$$
$$\times \exp\left\{-\frac{2}{3}\frac{\kappa^3}{\tilde{F}} + \frac{2}{\kappa}\ln\frac{4\kappa^3}{\tilde{F}} - (2n_1 + 1)\ln\frac{4\kappa^3}{\tilde{F}}\right\}. \tag{2.42}$$

In the lowest approximation $\tilde{F}/4\kappa^3 \ll 1$, we rctain in the sum only one term with $n_1 = 0$. The result is

$$|\psi(0)|^2 = \frac{\tilde{F}}{8\pi^2\kappa^2}\Gamma^2\left(1 - \frac{1}{\kappa}\right)\exp\left(-\frac{2}{3}\frac{\kappa^3}{\tilde{F}} + \frac{2}{\kappa}\ln\frac{4\kappa^3}{\tilde{F}}\right). \tag{2.43}$$

This expression is the result of Merkulov and Perel' (1973), which they obtained by another method, assuming $\kappa^2/\tilde{F} \ll 1$. In dimensional variables we have

$$\operatorname{Im}\epsilon_{\alpha\alpha} = \frac{e^2 s^2 m^*}{2\hbar^3\omega^2}\frac{F}{|\Delta|}\Gamma^2\left(1 - \sqrt{\frac{R}{|\Delta|}}\right)\exp\left\{-\frac{4}{3}\left(\frac{|\Delta|}{\theta}\right)^{3/2} + 2\sqrt{\frac{R}{|\Delta|}}\ln 8\left(\frac{|\Delta|}{\theta}\right)^{3/2}\right\}. \tag{2.43a}$$

In fig. 6 the comparison of the absorption spectrum defined by the eq. (2.43a) and the numerical calculations by Dow and Redfield (1970) is shown. We can see that the agreement is good.

Equation (2.43) is incorrect when the photon frequency is very close to the energy of a transition to any bound state of the exciton. Near resonance it is necessary, first, to take into account the Stark effect, and, second, one cannot neglect the damped exponential in the quasi-classical

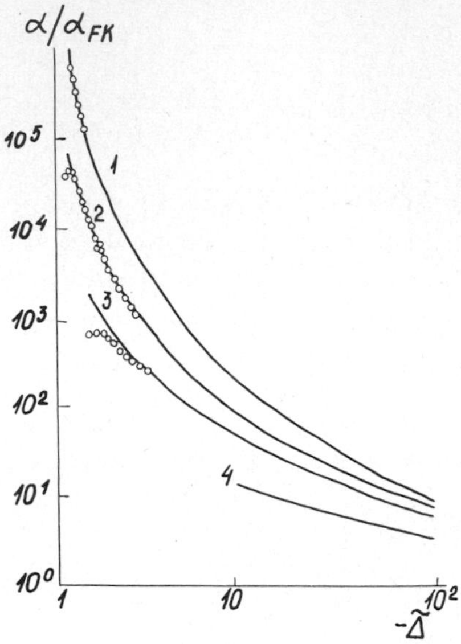

Fig. 6. Comparison of the numerical results of Dow and Redfield (1970) with eq. (2.43a) according to Merkulov (1974a). α_{FK} is the absorption coefficient without exciton effects: (1) – $\tilde{F} = 0.05$; (2) – $\tilde{F} = 0.2$; (3) – $\tilde{F} = 0.5$; (4) – $\tilde{F} = 5$.

Coulomb wave function, since the coefficient of the growing exponential vanishes at resonance. Indeed, although the wave function of the finite motion remains the same as before and is described by eqs. (2.36) and (2.37), while the wave function in the Coulomb region $f_2(\eta)$ is described as before by eq. (2.38), the form of $f_2(\eta)$ changes near resonance at $\eta\kappa \gg 1$,

$$f_2(\eta) = f_2(0)\frac{(-1)^{n_2}}{\sqrt{\kappa\eta}}\left\{\frac{1}{n_2!}\,e^{-S(\eta)} - n_2!\,\frac{y-\Delta\beta_{n_2}}{\kappa}\,e^{S(\eta)}\right\}, \qquad (2.44)$$

where $\beta_1 = \kappa(n_1+\frac{1}{2}) + \Delta\beta_{n_1}$, $\beta_2 = \kappa(n_2+\frac{1}{2}) + y$, $\Delta\beta_{n_1}$ is the perturbation-theory correction due to the Stark effect (Landau and Lifshitz 1963).

The quasi-classical asymptotic form of the Coulomb function (2.44) can be matched with the quasi-classical function in an electric field first below the barrier, which in turn is matched with the function to the right of the right-hand turning point. Once matched, the wave function in this region takes the form,

$$f_2(\eta) = \frac{2f_2(0)(-1)^{n_2}}{[2\eta k(\eta)]^{1/2}}\left\{\frac{1}{2n_2!}\,e^{-S(\eta_2)}\cos\left[S(\eta)-\frac{\pi}{4}\right] - \frac{y-\Delta\beta_{n_2}}{\kappa}\,n_2!\cos\left[S(\eta)+\frac{\pi}{4}\right]e^{S(\eta_2)}\right\}. \qquad (2.45)$$

where $S(\eta_2)$ is determined by eq. (2.40). Using the normalization condition, we obtain:

$$|\psi(0)|^2 = \frac{2\kappa}{\pi^2}\frac{\Gamma}{\Gamma^2 + [2(y - \Delta\beta_2)/\kappa]^2},$$
$$\Gamma = \frac{1}{(n_2!)^2}\exp[-2S(\eta_2)]. \tag{2.46}$$

Thus, the absorption coefficient has a Lorentzian shape with a width equal to the ionization probability of the given level. Near the resonances, eq. (2.46) must also be transformed by substituting the explicit expression for $\Delta\beta_2$ and by using the connection between the parabolic quantum numbers and the radial quantum number N: $N = n_1 + n_2 + 1$. After simple transformations we obtain

$$\mathrm{Im}\,\epsilon_{\alpha\alpha} \propto |\psi(0)|^2 = \frac{\kappa}{\pi^2 N^3}\frac{\Gamma_{n_1n_2}}{\Gamma^2_{n_1n_2} + (\Delta - E_{n_1n_2})^2}, \tag{2.47}$$

$$\Gamma_{n_1n_2} = \frac{1}{2N^3}\frac{1}{(n_2!)^2}\left(\frac{4}{\tilde{F}N^3}\right)^{2n_2+1}\exp\left\{-\frac{2}{3\tilde{F}N^3} + 3(N-1) - 6n_2\right\},$$

$$E_{n_1n_2} = -\frac{1}{2N^2} + \frac{3}{2}\tilde{F}N(n_1 - n_2) - \frac{\tilde{F}^2}{16}N^2[17N^2 - 3(n_1 - n_2)^2 + 19]. \tag{2.48}$$

Figure 7 shows the schematic form of the spectrum. It is seen that the linewidth increases with increasing number of the level in the multiplet. The latter is natural, in as much as for large values of the parabolic quantum number n_2 the wave functions are shifted relative to the field, and their tunneling probability is therefore larger.

Equation (2.48) coincides with the ionization probability of the hydrogen atom in an electric field (Lanczos 1930, Damburg and Kolosov 1977).

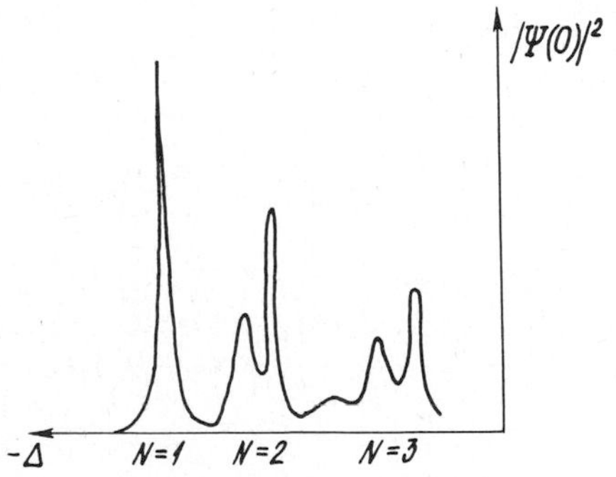

Fig. 7. The absorption spectrum in the gap in the vicinity of exciton absorption lines.

2.4. Green function approach to electrooptics

To conclude this section we derive some of our results again, using another method, which operates with the relative-motion Green function directly. We shall apply this method in the next section, so it is useful to demonstrate its advantages in the case of a simple zone structure, before using it in more complicated circumstances.

This method is based on the well-known identity concerning Green functions in the r–t representation (Feynman and Hibbs 1965)

$$G(\boldsymbol{r}_2 t_2, \boldsymbol{r}_1 t_1) = \int \mathrm{d}^3\boldsymbol{\rho}\, G(\boldsymbol{r}_2 t_2, \boldsymbol{\rho}\tau) G(\boldsymbol{\rho}\tau, \boldsymbol{r}_1 t_1), \quad t_1 \leqslant \tau \leqslant t_2, \tag{2.49}$$

which expresses the fact that the amplitude of transition from point $\boldsymbol{r}_1 t_1$ to point $\boldsymbol{r}_2 t_2 (\boldsymbol{r}_1 t_1 \to \boldsymbol{r}_2 t_2)$ equals to the sum over all the intermediate points $\boldsymbol{\rho}$ of products of the transition amplitude $(\boldsymbol{r}_1, t_1 \to \boldsymbol{\rho}, \tau)$ and that $(\boldsymbol{\rho}, \tau \to \boldsymbol{r}_2, t_2)$.

Using the identify (2.49) twice we can write:

$$\begin{aligned} G_\Delta(0,0) &= \int \mathrm{d}t G(0,0,t)\, \mathrm{e}^{\mathrm{i}\Delta t} \\ &= \int \mathrm{d}t G(0, \boldsymbol{r}_1, \tau_1) G(\boldsymbol{r}_1, \boldsymbol{r}_2, t-\tau_1-\tau_2) G(\boldsymbol{r}_2 0, \tau_2)\, \mathrm{e}^{\mathrm{i}\Delta t}\, \mathrm{d}^3\boldsymbol{r}_1\, \mathrm{d}^3\boldsymbol{r}_2 \\ &= \int G_{\epsilon_1}(0, \boldsymbol{r}_1) G_\Delta(\boldsymbol{r}_1, \boldsymbol{r}_2) G_{\epsilon_2}(0, \boldsymbol{r}_2) \exp[\mathrm{i}(\Delta-\epsilon_1)\tau_1 \\ &\quad + \mathrm{i}(\Delta-\epsilon_2)\tau_2] \frac{\mathrm{d}\epsilon_1\, \mathrm{d}\epsilon_2}{(2\pi)^2}\, \mathrm{d}^3\boldsymbol{r}_1\, \mathrm{d}^3\boldsymbol{r}_2. \end{aligned} \tag{2.50}$$

In eq. (2.50) we have introduced G_ϵ by means of the inverse Fourier transformation. At short distances the electric field can be neglected, so (if $k^4/\tilde{F} \gg 1$) we can choose such moments τ_1, τ_2, that Green functions $G(0, \boldsymbol{r}_1, \tau_1)$ and $G(0, \boldsymbol{r}_2, \tau_2)$ correspond to the motion in Coulomb potential alone. Furthermore at the final points $\boldsymbol{r}_1$ and $\boldsymbol{r}_2$ the motion is already quasi-classical. Then $G_\Delta(\boldsymbol{r}_2, \boldsymbol{r}_1)$ describes quasi-classical motion in an electric field. Coulomb interaction is rather weak in the $\boldsymbol{r}_1$–$\boldsymbol{r}_2$ region, so its contribution is expressed by a relatively small correction to the classical action. Thus, we obtain

$$\begin{aligned} G_\Delta^{\mathrm{c},F}(0,0) = \int & G_{\epsilon_1}^{\mathrm{c}}(0, \boldsymbol{r}_1) G_\Delta^{\mathrm{c},F}(\boldsymbol{r}_1, \boldsymbol{r}_2) G_{\epsilon_2}^{\mathrm{c}}(\boldsymbol{r}_2, 0) \\ & \times \exp[\mathrm{i}(\Delta-\epsilon_1)\tau_1 + \mathrm{i}(\Delta-\epsilon_2)\tau_2] \frac{\mathrm{d}\epsilon_1\, \mathrm{d}\epsilon_2}{(2\pi)^2}\, \mathrm{d}^3\boldsymbol{r}_1\, \mathrm{d}^3\boldsymbol{r}_2, \end{aligned} \tag{2.51}$$

G_ϵ^{c} being the Coulomb Green function, $G_\Delta^{\mathrm{c},F}$ being the quasi-classical Green function in an electric field and relatively weak Coulomb field. The exact exciton Green function in the absence of external field has the form (Baz' et al. 1971):

$$G_\Delta(\boldsymbol{r}_1, \boldsymbol{r}_2) = \frac{1}{2\pi\kappa} \frac{\Gamma(1-1/\kappa)}{|\boldsymbol{r}_1 - \boldsymbol{r}_2|} \left(\frac{\partial}{\partial y} - \frac{\partial}{\partial x}\right) W_{1/\kappa,1/2}(\kappa x) M_{1/\kappa,1/2}(\kappa y). \tag{2.52}$$

Here $x = |r_1| + |r_2| + |r_2 - r_1|$, $y = |r_1| + |r_2| - |r_1 - r_2|$; $W_{\alpha\beta}(x)$ and $M_{\alpha\beta}(x)$ are Whittaker functions (Gradstein and Ryzhik 1962). If $r_2 \to 0$

$$G_\Delta(r, 0) = G_\Delta(0, r) = \frac{1}{2\pi r}\Gamma\left(1 - \frac{1}{\kappa}\right)W_{1/\kappa, 1/2}(2\kappa r). \tag{2.53}$$

Note that eq. (2.53) enables one to obtain a closed expression for an interband complex dielectric permeability. Substituting eq. (2.53) into eq. (1.4) we write (in dimensional variables) (Aronov and Ioselevich 1978):

$$\epsilon_{\alpha\alpha}(\omega) = \frac{4e^2s^2m^*}{\hbar^2\omega^2}[2\kappa_g f_0(\kappa_g a) - \kappa_+ f_0(\kappa_+ a) - \kappa_- f_0(\kappa_- a)]. \tag{2.54}$$

Here

$$\hbar\kappa_g = \sqrt{2m^*E_g}, \qquad \hbar\kappa_\pm = \sqrt{2m^*(E_g \pm \hbar\omega)},$$

and

$$f_0(\kappa a) = 1 + \frac{2}{\kappa a}\left[\ln \kappa a + \psi\left(1 - \frac{1}{\kappa a}\right)\right],$$

$\psi(z)$ is a logarithmic derivative of the Γ function. In eq. (2.54) the logarithmic term arises from the Coulomb effects in the continuous spectrum, and the last term with $\psi(z)$ describes the dispersion near absorption lines. If $\kappa_- a \gg 1$ then $f_0 \approx 1$ and we have the usual expression for the dielectric permeability in the free-electron approximation (see, e.g. Bir and Pikus 1972). Above the absorption edge (for $\Delta > 0$):

$$\epsilon_{\alpha\alpha}(\omega) = \frac{4e^2s^2m^*}{\hbar^2\omega^2}\left\{2\kappa_g - \kappa_+ + \frac{2}{a}\left[\ln\frac{\kappa_g^2}{\kappa_+ k} + 2\psi\left(1 - \frac{1}{\kappa_g a}\right) - \psi\left(1 - \frac{1}{\kappa_+ a}\right) - \right.\right.$$
$$\left.\left. - \operatorname{Re}\psi\left(1 - \frac{1}{ka}\right)\right] + \frac{2\pi i}{a}\left[1 - \exp\left(-\frac{2\pi}{ka}\right)\right]^{-1}\right\}. \tag{2.55}$$

Thus, the imaginary part of $\epsilon_{\alpha\alpha}(\omega)$ is described by the well-known Sakharov–Elliot (Sakharov 1948, Elliot 1957) formula for the exciton absorption ($\hbar k = (2m^*\Delta)^{1/2}$).

Return now to the calculation of the absorption in an electric field. If $k_{\epsilon_1} r_1 \gg 1$ the following asymptotic limit of a Coulomb Green function (2.53) can be used:

$$G^c_{\epsilon_1}(0, r_1) = C(\epsilon_1)\exp[i\delta S_c(0, r_1; \epsilon_1)]G^{(0)}_{\epsilon_1}(0, r_1). \tag{2.56}$$

Here $G^{(0)}_{\epsilon_1} = (1/2\pi r_1)\exp(ik_{\epsilon_1} r_1)$ is the free-motion Green function, and $\delta S_c(0, r_1; \epsilon_1) = (1/k_{\epsilon_1})\ln 2k_{\epsilon_1} r_1$ is the Coulomb correction to the action, $C(\epsilon) = \Gamma(1 - i/k_\epsilon)\exp(\pi/2k_\epsilon)$. If $\epsilon > 0$ then

$$C(\epsilon) = \frac{\psi^c(0)}{\psi^{(0)}(0)}\exp(i\delta_c), \tag{2.57}$$

$\psi^{c}(0)$ and $\psi^{(0)}(0)$ being the magnitudes of the S-like wave function at the origin for a particle in the Coulomb potential and for a free particle respectively, δ_c being a Coulomb phase. We have already noted that the motion is quasi–classical everywhere in the trajectory interval r_1–r_2 so we can use the general quasi-classical expression for $G(r_2t_2; r_1t_1)$ (Feynman and Hibbs 1965, Baz' et al. 1971):

$$G(r_2t_2; r_1t_1) = \frac{\theta(t_2-t_1)}{[2\pi i(t_2-t_1)]^{3/2}} \exp[iS(r_2t_2, r_1t_1)], \tag{2.58}$$

$$\theta(x)=1 \quad (x>0), \qquad \theta(x)=0 \quad (x<0),$$

S being the classical action. It consists of two parts: the first one is the action in an electric field, and the second one is the Coulomb correction: $S = S_F + \delta S_c$. It is known from classical mechanics that the correction to the action due to a perturbation potential $V(\boldsymbol{r})$ has a form $\delta S = -\int V(\boldsymbol{r}(t))\,dt$, $\boldsymbol{r}(t)$ being the unperturbed trajectory. Actually, the action is the integral of Lagrangian along the trajectory, the correction to a potential being taken with opposite sign. In our case:

$$\delta S_c = \int_{t_1}^{t_2} dt/r(t).$$

The classical trajectory is defined by following equation: $z(t) = kt - \tilde{F}t^2/2$ and boundary conditions $z(t_{1,2}) = r_{1,2}$ or $t_1 = r_1/k$, $t_2 = 2k/\tilde{F} - r_2/k$. The integration leads to

$$\delta S_c = 2/k \ln(2k^3/\tilde{F}) - 1/k \ln(4k^2 r_1 r_2). \tag{2.59}$$

Substituting these results into eq. (2.51) we get:

$$\begin{aligned}\delta G_\Delta^{c,F}(0,0) = \int C(\epsilon_1)C(\epsilon_2) \exp\{i[\delta S_c(0,\boldsymbol{r}_1,\epsilon_1) + \delta S_c^F(\boldsymbol{r}_1,\boldsymbol{r}_2,\Delta) + \delta S_c(\boldsymbol{r}_2,0,\epsilon_2)]\}\\ \times G_{\epsilon_1}^{(0)}(0,\boldsymbol{r}_1) G_\Delta^F(\boldsymbol{r}_1,\boldsymbol{r}_2) G_{\epsilon_2}^{(0)}(\boldsymbol{r}_2,0) \exp[i(\Delta-\epsilon_1)\tau_1\\ + i(\Delta-\epsilon_2)\tau_2] \frac{d\epsilon_1\, d\epsilon_2}{(2\pi)^2} d^3\boldsymbol{r}_1\, d^3\boldsymbol{r}_2.\end{aligned} \tag{2.60}$$

All the factors in the first line of eq. (2.60) are smooth functions of coordinates and energies, compared with those in the last part. In carrying out the integration we may therefore treat the factors in the first line as constants. Note that the integral of the factors in the last part looks like the Green function of a particle in an electric field alone (without Coulomb interaction). After the integration over ϵ_1, ϵ_2 a factor $\delta(\epsilon_1-\Delta)\delta(\epsilon_2-\Delta)$ appears, thus, in eqs. (2.59) and (2.60) we have to substitute $\epsilon_1=\epsilon_2=\Delta$. Thus, the dependence on $\boldsymbol{r}_1, \boldsymbol{r}_2$ of the "smooth" factors disappears and $G_\Delta^{c,F}$

takes the following form:

$$\delta G_{\Delta}^{c,F}(0,0) = C^2(\Delta)\delta G_{\Delta}^{F}(0,0)\exp\left[i\frac{2}{k}\ln\frac{4k^3}{\tilde{F}}\right]. \tag{2.61}$$

The Green function in an electric field is (Baz' et al. 1971):

$$G_{\Delta}^{F}(0,0) = \int_0^{\infty} dt \frac{1}{(2\pi i t)^{3/2}}\exp\left[-i\frac{\tilde{F}^2 t^3}{24} + i\frac{k^2 t}{2}\right], \tag{2.62}$$

δG_{Δ}^{F} is a quasi-classical part of G_{Δ}^{F}, so it corresponds to a contribution to eq. (2.62) which may be obtained with the aid of the method of steepest descent. The steepest descent point is $t_0 = 2k/\tilde{F}$. Expanding the exponential to the second order in $(t - t_0)$ in the vicinity of t_0 we get the following expression for the field-dependent correction to the Green function:

$$\delta G_{\Delta}^{c,F} = i\frac{\tilde{F}}{k^3}\frac{1}{1-\exp(-2\pi/k)}\exp\left\{i\left[\frac{2k^3}{3\tilde{F}} + \frac{2}{k}\ln\frac{4k^3}{\tilde{F}} + 2\arg\Gamma\left(1-\frac{i}{k}\right)\right]\right\}. \tag{2.63}$$

Note that eq. (2.63) describes the real part of the Green function as well as the imaginary one and thus, according to eq. (1.4), both real and imaginary parts of the dielectric permeability may be written without a Kramers–Krönig analysis. The formula obtained is valid for $\Delta > 0$, but the result for $\Delta < 0$ can be obtained analogously.

3. *Franz–Keldysh effect (degenerate energy bands)**

The simple model of nondegenerate bands used above is inapplicable to such important semiconductors as Ge, Si, GaAs and many others, which have a degenerate valence band. The presence of degeneracy leads to some specific effects, first studied by Keldysh et al. (1969).

The electron states in the valence band are formed by atomic p-like states, and are three-fold degenerate (six-fold degenerate with electron spin). Spin–orbit interaction splits this multiplet into a quartet and a doublet, the latter being of lower energy. Hence the valence band maximum is four-fold degenerate and is described by a total angular momentum $J = 3/2$. Except in Si, the spin–orbit splitting is much greater than an exciton Rydberg, so that in considering fundamental absorption near the threshold one may neglect the split-off band corresponding to $J = 1/2$.

* *Note added in proof.*
The electrooptical effects in semiconductors with degenerate energy bands were examined narrowly in a recent paper by Aronov and Ioselevich (1981).

In the effective-mass approximation the hole Hamiltonian is:

$$\hat{H}_v = (\hbar^2/2m)[(\gamma_1 + \tfrac{5}{2}\gamma_2)k^2\hat{I} - 2\gamma_2(k_x^2\hat{J}_x^2 + k_y^2\hat{J}_y^2 + k_z^2\hat{J}_z^2) - 2\gamma_3(k_xk_y\{\hat{J}_x, \hat{J}_y\} + k_yk_z\{\hat{J}_y, \hat{J}_z\} + k_zk_x\{\hat{J}_z, \hat{J}_x\})], \quad (3.1)$$

(see, e.g. Bir and Pikus, 1972) γ_i being Luttinger constants, k being the hole wavevector, m the free electron mass, $\hat{J}_i$ the components of an angular momentum 3/2 operator, $\{\hat{J}_i, \hat{J}_j\} = \hat{J}_i\hat{J}_j + \hat{J}_j\hat{J}_i$, and $\hat{I}$ a unit four-by-four matrix.

The $\hat{H}$ operator has two eigenvalues, each of them is two-fold degenerate;

$$\epsilon_{h,\ell}(\boldsymbol{k}) = \frac{\hbar^2k^2}{2m}\left[\gamma_1 \mp 2g\left(\frac{\boldsymbol{k}}{|\boldsymbol{k}|}\right)|\gamma_2|\right]. \quad (3.2)$$

The upper sign corresponds to the heavy hole, the lower to the light one, and

$$g(\boldsymbol{n}) = \left[1 + 3\frac{\gamma_3^2 - \gamma_2^2}{\gamma_2^2}(n_x^2n_y^2 + n_y^2n_z^2 + n_z^2n_x^2)\right]^{1/2}. \quad (3.3)$$

Because the conduction band is simple, the electron–hole pair Hamiltonian $\hat{H}_0$ coincides with $\hat{H}_v$ with γ_1 changed to $\gamma_1' = \gamma_1 + m/m_c$.

In the gauge $\varphi = 0$, $\mathbf{A}(t) = -c/e\boldsymbol{F}(t)$ for an electric field, and the pair Hamiltonian including Coulomb interaction has the form:

$$\hat{H} = \hat{H}_0(\hbar\boldsymbol{k} - \boldsymbol{F}t) - (e^2/\epsilon_0 r)\hat{I}. \quad (3.4)$$

In order to obtain a formula for dielectric permeability, we derive a Green function of the Schrödinger equation with the Hamiltonian given by eq. (3.4) corresponding to energy Δ. The complex dielectric permeability is related to the Green function by a following relation:

$$\epsilon_{\alpha\beta}(\omega) = (4\pi e^2/\hbar\omega^2)v_{\sigma j}^{*\alpha}G_{\Delta}^{jj'}(0, 0)v_{\sigma j'}^{\beta} \quad (3.5)$$

$v_{\sigma j}^{\alpha}$ being the matrix element of the component of velocity between the Bloch function ($k = 0$) of an electron state of spin σ in the conduction band and that of a hole state of spin j in the valence band. If the spin-quantization axis is z, the $v_{\sigma j}^{\alpha}$ matrices are:

$$v_x = \frac{s}{2}\begin{Vmatrix} 0 & -1 & 0 & -\sqrt{3} \\ i\sqrt{3} & 0 & i & 0 \end{Vmatrix}, \quad v_y = \frac{s}{2}\begin{Vmatrix} 0 & -i & 0 & i\sqrt{3} \\ -\sqrt{3} & 0 & 1 & 0 \end{Vmatrix},$$

$$v_z = \frac{s}{2}\begin{Vmatrix} 0 & 0 & 2i & 0 \\ 0 & 2 & 0 & 0 \end{Vmatrix},$$

(Bir and Pikus 1972). $s = (2/3)^{1/2}P_{cv}/m$, P_{cv} being the interband matrix element of $\hat{P}$.

Let us first discuss the pair-motion of a positive energy ($\Delta > 0$); later we shall consider tunneling motion ($\Delta < 0$). Just as in the case of simple bands, the Green function $G_\Delta(0, 0)$ consists of two parts: the first one G^0, corresponding to motions at short distances of the order of a wavelength, where the electric field may be neglected; the second one G^F corresponding to classical motion along a trajectory returning to the origin. G^F consists of two contributions from the two different sorts of holes.

The returning hole trajectories of both sorts are straight lines, but in general they are not parallel to the field direction (as the electron trajectories are). The reason is that the Newton equation $\hbar\dot{\boldsymbol{k}} = \boldsymbol{F}$ imposes the condition $\boldsymbol{k} \parallel \boldsymbol{F}$ for the trajectory to return to the origin, but $v_i = m_{ij}^{-1}(\boldsymbol{k})\hbar k_j$. Since m_{ij}^{-1} is a tensor the velocity direction does not coincide with the field direction, unless either spherical symmetry holds ($\gamma_2 = \gamma_3$), or the field is directed along one of the symmetry axes [001], [011], [111].

Thus, there are two contributions to G^F, differing in pre-exponential factors as well as in classical actions. Consequently, there are two types of Franz–Keldysh oscillations which interfere. Their different period is due to the different effective masses of light and heavy holes.

The treatment of tunneling motion is analogous. There are two exponentially small contributions, the first one related to a light hole tunneling, the second one to heavy hole tunneling. For large $|\Delta|$ only the light hole contributes to an absorption because of its smaller tunnel action. Tunneling motion was firstly treated by Keldysh et al. (1969), the theory was later improved by Merkulov (1974b) who included the Coulomb potential.

In this section we neglect Coulomb interaction. It will be considered later. To calculate G^0, we turn the electric field off. Then momentum is conserved and the Green function in k–E representation has the standard form:

$$\hat{G}(\boldsymbol{k}) = -\hbar[\Delta - \hat{H}_0(\boldsymbol{k}) + \mathrm{i}\delta]^{-1}. \tag{3.6}$$

To remove the divergence of $G_\Delta(0,0)$ which, as is clear from eq. (3.4), vanishes in the final result, we subtract $G_{\Delta=0}(0,0)$. By inverse Fourier transformation of eq. (3.6) we obtain:

$$\hat{G}_\Delta(0,0) - \hat{G}_0(0,0) = -\hbar \int \frac{\mathrm{d}^3\boldsymbol{k}}{(2\pi)^3}[(\Delta - \hat{H}(\boldsymbol{k}) + \mathrm{i}\delta)^{-1} + \hat{H}^{-1}(\boldsymbol{k})].$$

The operator structure of $\hat{H}(\boldsymbol{k})$ depends on the direction of $\boldsymbol{k}$, but not on its modulus. Hence, in integrating over $|\boldsymbol{k}|$ we may treat $H(\boldsymbol{k})$ as a numerical function. Actually, for any particular direction of $\boldsymbol{k}$ we can diagonalize the integrand and then integrate over $|\boldsymbol{k}|$ for each eigenvalue separately. After that, we return to the initial representation. Furthermore, because of the homogeneity of the Hamiltonian, we may write:

$$\hat{H}(\boldsymbol{k}) = \hbar^2k^2\hat{H}(\boldsymbol{n}), \qquad \boldsymbol{n} = \boldsymbol{k}/|\boldsymbol{k}|,$$

$$\hat{G}_\Delta(0,0) - \hat{G}_0(0,0) = -\frac{\Delta}{2\pi^2\hbar}\int \frac{\mathrm{d}\Omega_n}{4\pi}\hat{H}^{-2}(\boldsymbol{n})\int_0^\infty \mathrm{d}k[(\Delta + \mathrm{i}\delta)\hat{H}^{-1}(\boldsymbol{n}) - \hbar^2k^2]^{-1}$$

$$= \mathrm{i}\frac{\Delta^{1/2}}{4\pi\hbar^2}\int \frac{\mathrm{d}\Omega_n}{4\pi}H^{-3/2}(\boldsymbol{n}). \tag{3.7}$$

For the last integration, note that the result must be a cubic invariant. However, we have no free vector from which to construct a cubic invariant (which might be similar to the Hamiltonian). Hence, the only possible form of G^0 is a unit matrix multiplied by a scalar function which we obtain by means of calculating the trace of the integral,

$$\hat{G}_\Delta(0,0) - \hat{G}_0(0,0) = \mathrm{i}\frac{\Delta^{1/2}}{2\pi\hbar^2}\hat{I}\frac{1}{8}\int\frac{\mathrm{d}\Omega_n}{4\pi}\,\mathrm{Tr}\,\hat{H}^{-3/2}(\boldsymbol{n}) = \mathrm{i}\frac{(2\Delta)^{1/2}M^{3/2}}{2\pi\hbar^2}\hat{I}, \tag{3.8}$$

where

$$M^{3/2} = \frac{1}{2}\int\frac{\mathrm{d}\Omega_n}{4\pi}[m_{\parallel\ell}^{3/2}(\boldsymbol{n}) + m_{\parallel\mathrm{h}}^{3/2}(\boldsymbol{n})],$$

$$m_{\parallel\mathrm{h},\ell} = m[\gamma_1' \mp 2|\gamma_2|g(\boldsymbol{n})]^{-1}. \tag{3.9}$$

We have used eq. (3.2) for the eigenvalues. Substituting eq. (3.8) into eq. (3.4), we get:

$$\epsilon_{\alpha\beta} = \delta_{\alpha\beta}\mathrm{i}\frac{2e^2s^2}{\hbar^3\omega^2}M^{3/2}(2\Delta)^{1/2}. \tag{3.10}$$

This is the well known expression for the interband contribution to the dielectric permeability (see, e.g. Bir and Pikus 1972).

To compute G^{F} we have to solve the equation for the Green function in an electric field:

$$\hbar\frac{\partial\hat{G}}{\partial t} + \mathrm{i}\hat{H}_0(\hbar\boldsymbol{k} - \boldsymbol{F}t)\hat{G} = \hbar\hat{I}\delta(t). \tag{3.11}$$

Consider an unitary transformation $\hat{S}(t)$ which diagonalizes the Hamiltonian $\hat{H}_0(\hbar\boldsymbol{k} - \boldsymbol{F}t)$,

$$\hat{S}\hat{H}_0\hat{S}^{-1} = \hat{H}_0', \qquad \hat{S}\hat{G}\hat{S}^{-1} = \hat{G}'.$$

$\hat{H}_0'$ is a diagonal operator, but $\hat{G}'$ in general, does not have to be diagonal. Let us rewrite eq. (3.11) in this time-dependent representation:

$$-\hat{S}\dot{\hat{S}}^{-1}\hat{G}' - \hat{G}'\hat{S}\dot{\hat{S}}^{-1} + \dot{\hat{G}}' + 1/\hbar H_0'G' = \hat{I}\delta(t). \tag{3.12}$$

In the quasi-classical approximation we may neglect the first two terms. Then a solution of eq. (3.12) may be easily obtained because H_0' is diagonal:

$$G'(t) = \exp\left[-\frac{\mathrm{i}}{\hbar}\int_0^t H_0'(\tau)\,\mathrm{d}\tau\right]. \tag{3.13}$$

Hence, the approximation made in eq. (3.12) makes it possible to replace the operator H_0 in eq. (3.14) by each of its eigenvalues and thus obtain the corresponding eigenvalues of G. In other words, we neglect the interband transitions between the light-hole and the heavy-hole bands, and treat each one independently. Such an approximation is quite natural in the quasi-classical limit; the only troubling fact is the presence of turning points on a "returning" classical trajectory. In the vicinity of such a point the motion cannot be treated as quasi-classical. However, we shall see that the region where eq. (3.13) is not valid is quite small and does not contribute to G.

Fourier transforming eq. (3.13) we get the Green function in the $\boldsymbol{r}$–E representation:

$$G_{\mathrm{F}}^{nn}(\boldsymbol{r}=0,\Delta)=\int\frac{\mathrm{d}^3\boldsymbol{k}}{(2\pi)^3}\int \mathrm{d}t\,\exp\left\{\frac{\mathrm{i}\Delta t}{\hbar}-\frac{\mathrm{i}}{\hbar}\int_0^t \hat{H}'_{nn}(\tau)\,\mathrm{d}\tau\right\}. \tag{3.14}$$

For the integration it is convenient to use a variable $q_i = k_i - F_i t/2\hbar$ and expand the exponential in powers of q_i:

$$\int_0^t H'_{nn}(\tau)\,\mathrm{d}\tau=\frac{F^2t^3}{24m_{\|}^{n}(\boldsymbol{e})}+\frac{t}{2}\frac{\partial^2 H'_{nn}}{\partial k_i\partial k_j}\bigg|_{q=0} q_iq_j, \tag{3.15}$$

$\boldsymbol{e}$ being a unit vector, parallel to the electric field. The integration over q_i becomes quite simple if the quadratic form

$$\frac{\partial^2 H_{nn}}{\partial k_i\partial k_j}\bigg|_{q=0} q_iq_j$$

is diagonalized in advance. Then, after integration over $\boldsymbol{q}$, a factor

$$\left[\det\frac{\partial^2 H'_{nn}}{\hbar^2\partial k_i\partial k_j}\bigg|_{q=0}\right]^{-1/2}$$

appears in the pre-exponential instead of $m^{*3/2}$ which appears in the case of simple bands.

The integration over t is carried out quite similarly to that in eq. (2.62). For the field-dependent correction to the Green function we find

$$\delta G_{\mathrm{F}}^{nn}(0,0,\Delta)=\mathrm{i}\frac{F}{4\pi\Delta\hbar}m_{\perp n}(\boldsymbol{e})\exp\left[\mathrm{i}\frac{4\sqrt{2}}{3}\frac{\Delta^{3/2}m_{\|n}^{1/2}(\boldsymbol{e})}{\hbar F}\right]. \tag{3.16}$$

Here

$$\begin{aligned} m_{\perp}^{\ell,\mathrm{h}} &= \left[m_{\|}^{\ell,\mathrm{h}}(\boldsymbol{e})\det\frac{\partial^2 H^{\ell,\mathrm{h}}}{\hbar^2\partial k_i\partial k_j}\right]^{-1/2}\\ &=\frac{m}{|\gamma_2|}\left[\frac{a\pm 2g(\boldsymbol{e})}{a\pm(2+c^2)g^{-1}(\boldsymbol{e})}\right]^{1/2}\\ &\quad\times[a^2+c^2+8\pm(c^2+4)g^{-3}(\boldsymbol{e})(a\pm 3g(\boldsymbol{e}))]^{-1/2},\\ a&=\gamma_1'/|\gamma_2|,\qquad c^2=3(\gamma_3^2-\gamma_2^2)/\gamma_2^2. \end{aligned} \tag{3.16a}$$

The calculation of the determinant is described in the Appendix.

Let us now evaluate the contribution of nonquasi-classical terms in eq. (3.12). The operator structure of $\hat{S}$ as well as that of $\hat{H}_0$ is determined by the direction $\boldsymbol{n}$, $S = S(\boldsymbol{n})$, $\boldsymbol{n} = \hbar\boldsymbol{k} - \boldsymbol{F}t/|\hbar\boldsymbol{k} - \boldsymbol{F}t|$,

$$\dot{\hat{S}} = \dot{\boldsymbol{n}}\frac{\partial \hat{S}(\boldsymbol{n})}{\partial \boldsymbol{n}} \propto |\hbar\boldsymbol{k} - \boldsymbol{F}t|^{-3}((\hbar\boldsymbol{k} - \boldsymbol{F}t)\cdot[\boldsymbol{F}\times\boldsymbol{k}]).$$

Thus, $\dot{\hat{S}}$ is small everywhere except for the narrow region where $|\hbar\boldsymbol{k} - \boldsymbol{F}t| \lesssim |\hbar\boldsymbol{k}\times\boldsymbol{e}|$. We shall now evaluate the contribution of this region. The condition $|\hbar\boldsymbol{k} - \boldsymbol{F}t| \sim \hbar k_\perp$ is valid during a short time interval $\tau \sim \hbar k_\perp/F$. On the other hand, in the same time interval $\hat{H}(\tau) \sim \hbar^2 k_\perp^2/m$, hence $\delta G/G \sim \tau H/\hbar \sim \hbar^2 k_\perp^3/mF$. But how large are the $k_\perp$ involved? To answer this question we appeal to eq. (3.15). We see then that when one integrates the exponent over $\boldsymbol{q}$, such $\boldsymbol{q}$ (and $k_\perp$) are important for which the second term in eq. (3.15) is of the order of $\hbar$, hence $k_\perp \sim (m/t\hbar)^{1/2} \sim (mF/k\hbar^2)^{1/2}$ and $\delta G/G \sim (mF/\hbar^2 k^3)^{1/2} \ll 1$. Thus, the time interval τ appears to be too small to contribute to the variation of the Green function. In other words, the region, where the perturbation represented by two first terms in eq. (3.12) is not small is much less than a wavelength. So the perturbation is ineffective.

By means of the $\hat{S}$-operator and the eigenvalues of $\hat{G}$ we can rewrite the Green function in the initial representation:

$$\delta G_\Delta^{\mathrm{F}}(0,0) = \mathrm{i}\frac{F}{8\pi\Delta\hbar}\hat{S}^{-1}(\boldsymbol{e})\left\| m_{\perp n}(\boldsymbol{e})\exp\left[\mathrm{i}\frac{4\sqrt{2}}{3}\frac{\Delta^{3/2}m_{\|n}^{1/2}(\boldsymbol{e})}{\hbar F}\right]\right\|\hat{S}(\boldsymbol{e}). \tag{3.17}$$

Under the spherical approximation the masses do not depend on $\boldsymbol{e}$, $m_{\perp n} = m_{\|n}$, $M^{3/2} = \frac{1}{2}(m_\ell^{3/2} + m_\mathrm{h}^{3/2})$. If $\Delta < 0$:

$$\delta G_\Delta^{\mathrm{F}}(0,0) = \mathrm{i}\frac{F}{8\pi\hbar|\Delta|}S^{-1}(\boldsymbol{e})\left\| m_{\perp n}(\boldsymbol{e})\exp\left[-\frac{4\sqrt{2}}{3}\frac{|\Delta|^{3/2}m_{\|n}^{1/2}(\boldsymbol{e})}{\hbar F}\right]\right\|\hat{S}(\boldsymbol{e}). \tag{3.18}$$

The correction is exponentially small, as is expected for the Franz–Keldysh effect. For sufficiently large negative Δ only the light hole contributes to the result.

Equation (3.17) leads to a following correction to the dielectric permeability:

$$\begin{aligned}\delta\epsilon_{\alpha\beta}^{\mathrm{F}}(\omega) &= -\frac{e^2F}{\hbar^2\omega^2\Delta}\sum_{\sigma,j,j',n} v_{\sigma j}^{*\alpha}S_{jn}^{-1}m_{\perp n}(\boldsymbol{e})\exp\left[\mathrm{i}\frac{4\sqrt{2}}{3}\frac{\Delta^{3/2}m_{\|n}^{1/2}(\boldsymbol{e})}{\hbar F}\right]S_{nj'}v_{j'\sigma}^{\beta}\\ &= -\frac{e^2Fs^2}{\hbar^2\omega^2\Delta}\sum_\nu \Theta_{\alpha\beta}^\nu(\boldsymbol{e})m_{\perp\nu}(\boldsymbol{e})\exp\left\{\mathrm{i}\frac{4\sqrt{2}}{3}\frac{\Delta^{3/2}m_{\|\nu}^{1/2}(\boldsymbol{e})}{\hbar F}\right\},\end{aligned} \tag{3.19}$$

where

$$\Theta_{\alpha\beta}^\nu(\boldsymbol{e}) = \frac{1}{s^2}\sum_{\sigma jj'n} v_{\sigma j}^{*\alpha}S_{jn\nu}^{-1}(\boldsymbol{e})S_{n\nu j'}(\boldsymbol{e})v_{j'\sigma}^{\beta}.$$

The summation runs over two states n_ν, corresponding to the same ν (ν

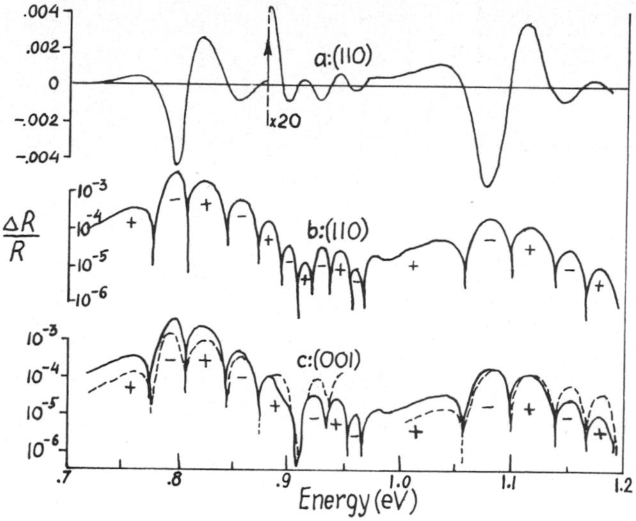

Fig. 8. The oscillations of electroreflectance in Ge according to Handler et al. (1969) at different light polarizations: (a) and (b) – $\boldsymbol{n} \parallel [110]$, (c) – $\boldsymbol{n} \parallel [001]$. Curve (a) is plotted linearly; curves (b) and (c) are plotted semi-logarithmically. Solid curves are experimental; the dashed line in (c) is a least squares computer fit using one-electron theory. Electric field along [110] axis, $E_g = 0.799$ eV, $E_g + \Delta_{SO} = 1.1$ eV. (+) or (−) signs are related to the signs of ΔR.

denotes the type of hole). Rather tedious computations lead to the following result (Bir and Pikus 1972):

$$\Theta^{\ell,h}_{\alpha,\beta} = \frac{1}{2|\gamma_2|g(\boldsymbol{e})}[(2|\gamma_2|g(\boldsymbol{e}) \mp \gamma_2)\delta_{\alpha\beta} \pm 3\gamma_3 e_\alpha e_\beta \pm 3(\gamma_2 - \gamma_3) \sum_{\lambda,\delta} \rho_{\alpha\beta\lambda\delta} e_\lambda e_\delta], \tag{3.20}$$

$\rho_{\alpha\beta\lambda\delta}$ being a fourth order cubic symmetry tensor. If x, y, z are the principal axes, $\rho_{xxxx} = \rho_{yyyy} = \rho_{zzzz} = 1$, all other components being zero. Equations (3.19), (3.20) and (3.10) determine the dielectric permeability completely.

Handler et al. (1969) observed about 11 oscillations in the electroreflectance spectrum which represents the sum of light-hole and heavy-hole contributions (fig. 8). In fig. 8 we see a clear polarization dependence of electroreflectance in the region of fundamental absorption edge. For photon energies in the vicinity of $E_g + \Delta_{so}$ (Δ_{so} being a spin–orbit splitting) polarization dependence was not observed. This fact is in agreement with theory because the corresponding valence band is nondegenerate. In this region ($\hbar\omega \approx E_g + \Delta_{so}$) the theory of sect. 1 should hold.

Recently Ovsyuk and Sinyukov (1978) observed an even richer spectrum

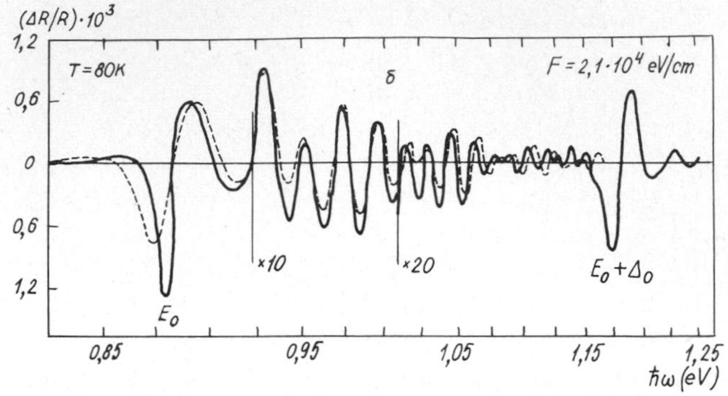

Fig. 9. The oscillations of electroreflectance in Ge according to Ovsyuk and Sinyukov (1978) $\boldsymbol{e} \parallel [111]$, $\boldsymbol{n} \parallel [100]$. The solid curve is experimental; the dashed line is a theoretical one-electron spectrum with reduced masses: $m_h = 0.196$ m, $m_\ell = 0.0354$ m, $E_g = 0.872$ eV. $\Gamma = 7$ meV is the broadening parameter, $T = 80$ K.

of oscillations (fig. 9). They describe the entire spectrum by a sum of two types of oscillations including the Coulomb interaction, damping and nonparabolicity of the light hole and electron bands.

Let us discuss the polarization dependence of absorption in detail now. For simplicity we make the spherical approximation ($\gamma_2 = \gamma_3$). If $\gamma_2 > 0$ then

$$\epsilon_{\alpha\beta}(\omega) = \frac{2ie^2s^2}{\hbar^3\omega^2}(2\Delta)^{1/2}M^{3/2}\left\{\delta_{\alpha\beta} - \frac{\hbar F}{4(2M\Delta)^{3/2}}\left[m_\ell(\delta_{\alpha\beta} + 3e_\alpha e_\beta)\right.\right.$$
$$\left.\left.\times \exp\left(i\frac{4\sqrt{2}}{3}\frac{\Delta^{3/2}m_\ell^{1/2}}{\hbar F}\right) + 3m_h(\delta_{\alpha\beta} - e_\alpha e_\beta)\exp\left(i\frac{4\sqrt{2}}{3}\frac{\Delta^{3/2}m_h^{1/2}}{\hbar F}\right)\right]\right\}. \tag{3.21}$$

Thus if the polarization vector is parallel to the electric field then only light holes contribute to oscillations. If they are oriented perpendicular to each other the ratio of contributions of light and heavy holes is $m_\ell/3m_h$. If $\gamma_2 < 0$ then light and heavy holes change their places: for parallel orientation only heavy holes contribute to oscillations, and for perpendicular orientation the ratio of light holes contribution to that of heavy holes is $3m_\ell/m_h$.

The interesting effects in the absorption in the gap were investigated by Keldysh et al. (1969). As said above only the light hole contributes to the absorption in the gap:

$$\operatorname{Im} \epsilon_{\alpha\beta} = \frac{e^2s^2}{\hbar^2\omega^2}\frac{F}{8|\Delta|}\frac{m_{\perp\ell}(\boldsymbol{e})}{g(\boldsymbol{e})}\exp\left[-\frac{4\sqrt{2}}{3}\frac{|\Delta|^{3/2}m_{\parallel\ell}^{1/2}(\boldsymbol{e})}{\hbar F}\right]$$
$$\times\left\{(2g(\boldsymbol{e}) - \operatorname{sign}\gamma_2)\delta_{\alpha\beta} + 3\frac{\gamma_3}{|\gamma_2|}e_\alpha e_\beta\right.$$
$$\left. + 3\left[\operatorname{sign}\gamma_2 - \frac{\gamma_3}{|\gamma_2|}\right]\sum_{\lambda\delta}\rho_{\alpha\beta\lambda\delta}e_\lambda e_\delta\right\}. \tag{3.22}$$

In the spherical approximation the ratio $\operatorname{Im} \epsilon_{\parallel}/\operatorname{Im} \epsilon_{\perp} = 4$ or $\gamma_2 > 0$ and $\operatorname{Im} \epsilon_{\parallel} = 0$ for $\gamma_2 < 0$. The degree of polarization is defined as $P = W_{\parallel} - W_{\perp}/W_{\parallel} + W_{\perp}$, $W_{\parallel,\perp}$ being the absorption per unit volume for the parallel and perpendicular polarizations respectively. Then for $\gamma_2 > 0$, $P = \frac{3}{2}$ and for $\gamma_2 < 0$, $P = -1$.

Now we shall discuss physical sources of the polarization dependence of light absorption in an electric field, following Keldysh et al. (1969).

Consider the polarization of an elementary absorption or recombination act. If $\gamma_2 > 0$ (as it is e.g. in GaAs) and the momentum is directed along the principal axis, the heavy hole wave function transforms as $x + iy$ under symmetry operations, and the light hole wave function transforms as a linear combination of $x + iy$ and z, such that the squared modulus of z-like amplitude is four times larger than that of x- or y-like amplitude.

When an electron of $k_z = 0$ and a heavy hole recombine, the radiation pattern is that of a dipole rotating in the xy plane. The ratio of polarizations is:

$$|P_z^{\mathrm{h}}|^2 : |P_x^{\mathrm{h}}|^2 : |P_y^{\mathrm{h}}|^2 = 0:1:1.$$

Recombination of an electron and a light hole gives dipole radiation polarized chiefly in the z direction:

$$|P_z^{\ell}|^2 : |P_x^{\ell}|^2 : |P_y^{\ell}|^2 = 4:1:1.$$

The total radiation arising from both transitions is not polarized. The polarization is a consequence of an anisotropic carrier distribution. Light holes tunnel more efficiently than heavy ones, and the smaller the component of momentum perpendicular to the field the easier the tunneling. Since only transitions between states of momenta parallel to the field direction are involved in electroabsorption and diagonal tunneling, the light holes dominate. The degree of polarization (for a perpendicular orientation) is:

$$P = \frac{|P_z^{\ell}|^2 - |P_x^{\ell}|^2}{|P_z^{\ell}|^2 + |P_x^{\ell}|^2} = \frac{4-1}{4+1} = 60\%.$$

Alferov et al. (1969) observed polarization effects in diagonal tunneling in p–n junctions made of GaAs and $\mathrm{pAl}_x\mathrm{Ga}_{1-x}\mathrm{As}$–nGaAs heterojunctions. They found that the radiation caused by diagonal tunneling is polarized strongly (fig. 10). The polarization degree was about 40–50%.

Equations (3.19) and (3.21) show that polarization experiments enable one to determine not only γ_2 and γ_3 moduli but their signs too. Furthermore, the angular dependence of electroabsorption will be strong if the nonsphericity of the valence band is included because of the exponential dependence on the hole mass.

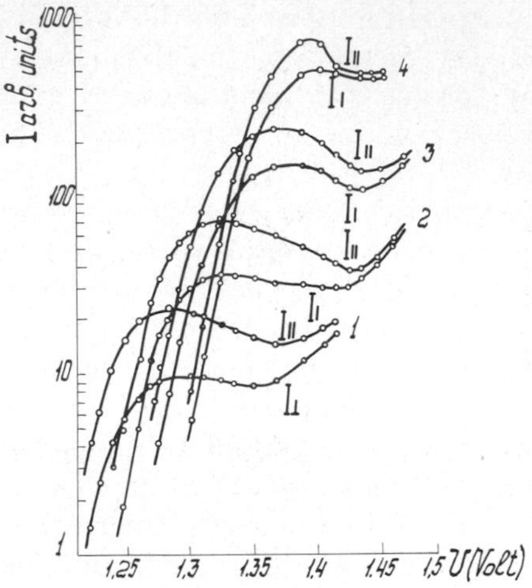

Fig. 10. Dependence of emission intensity versus the bias voltage for both light polarizations ($\boldsymbol{n} \parallel \boldsymbol{e}$ and $\boldsymbol{n} \perp \boldsymbol{e}$) for different frequencies of light, according to Alferov et al. (1969), $\boldsymbol{e} \parallel [111]$. $T = 77$ K, $\hbar\omega$: (1) – 1.232 eV, (2) – 1.263 eV, (3) – 1.296 eV, (4) – 1.334 eV.

4. *Exciton effects in degenerate bands*

The inclusion of electron–hole Coulomb interaction in semiconductors with a degenerate valence band appears to be a rather complicated problem. In fact, analytical expressions for the Green functions or wave functions have not been found. This fact means that one cannot write down a pre-exponential related to a wave function distortion at short distances where quantum effects are important. (In the simple band this is the Sommerfeld factor.) However, in the spherical approximation the Green function in nonzero electric field can be expressed in terms of that in zero electric field, which, while it can not be derived analytically, can be determined from the experimental absorption data in zero field or from numerical calculations. Thus the field dependence of the dielectric permeability is obtainable. A particularly simple result can be found for the modification of Franz–Keldysh oscillations ($\Delta > 0$).

This inclusion of anisotropy together with the Coulomb interaction does not lead, as a rule, to any new effects beyond those described in sect. 3, where exciton effects were neglected. An exception is the case of very small positive energies: $0 < \Delta \ll R$, where the Coulomb interaction enhances the anistropy of Franz–Keldysh effect. The experimental observation of this enhancement is evidently difficult, for it arises in a narrow region of frequencies and for very small electric fields, leading to very small amplitudes of oscillation.

Now we have to derive $\hat{G}^{c}_{\epsilon}(0, \boldsymbol{r})$ which (in the spherical approximation) we can express in terms of the total angular momentum eigenfunctions,

$$\hat{G}^{(c)nn'}_{\epsilon}(0, \boldsymbol{r}) = -\int d\epsilon' \sum_{\mathcal{F},\mathcal{F}_z} \frac{\psi^{n}_{\mathcal{F},\mathcal{F}_z,\epsilon'}(\boldsymbol{r})\psi^{*n'}_{\mathcal{F},\mathcal{F}_z,\epsilon'}(0)}{\epsilon - \epsilon' + i\delta}, \tag{4.1}$$

$\mathcal{F}$ being the total angular momentum, and $\mathcal{F}_z$ its z projection. It is convenient to choose the z-axis along the $\boldsymbol{r}$ vector. Then $\mathcal{F}_z = J_z$ since the projection of the orbital angular momentum onto the propagation direction is zero. In fact all the spherical harmonics with $m \neq 0$ are zero at $\theta = 0$. Thus:

$$\psi^{n}_{\mathcal{F},\mathcal{F}_z,\epsilon}(\boldsymbol{r}) = \delta_{n\mathcal{F}_z}\psi^{n}_{\mathcal{F},n,\epsilon}(\boldsymbol{r}). \tag{4.2}$$

If $\boldsymbol{r} = 0$, the projections of $\mathcal{F}$ and $\boldsymbol{J}$ onto the arbitrary axis are equal so $\mathcal{F} = \boldsymbol{J}$ and,

$$\psi^{n}_{\mathcal{F},\mathcal{F}_z,\epsilon}(0) = \delta_{\mathcal{F},3/2}\delta_{\mathcal{F}_z,n}\psi_{3/2,\epsilon}(0). \tag{4.3}$$

Using eqs. (4.2) and (4.3) we obtain (in the representation quantized along $\boldsymbol{r}$):

$$G^{(c)nn}_{\epsilon}(0, \boldsymbol{r}) = -\int d\epsilon' \frac{\psi^{n}_{3/2,n,\epsilon'}(\boldsymbol{r})\psi^{*n}_{3/2,n,\epsilon'}(0)}{\epsilon - \epsilon' + i\delta}. \tag{4.4}$$

For positive energy $\epsilon > 0$ and $r \to \infty$:

$$\psi^{(c)n}_{3/2,n,\epsilon}(\boldsymbol{r}) = \psi^{(0)n}_{3/2,n,\epsilon}(\boldsymbol{r}) \exp[i\varphi_n(r, \epsilon)], \tag{4.5}$$

$\psi^{(0)}$ being the wave function in zero Coulomb field, and $\varphi_n(r, \epsilon)$ being the extra Coulomb phase consisting, as in the nondegenerate band, of two parts:

$$\varphi_n(r, \epsilon) = \delta_n(\epsilon) + \frac{e^2}{\hbar\epsilon_0}\left(\frac{2m_n}{\epsilon}\right)^{1/2} \ln\left(\frac{2r\sqrt{2m_n\epsilon}}{\hbar}\right).$$

The first part comes from the quantum region and cannot be computed exactly in the presence of degeneracy. The second one is quasi-classical. $\psi^{(0)}(\boldsymbol{r})$ is a rapidly varying function of energy compared with all other factors under the integration in eq. (4.4). Hence we can take all factors except $\psi^{(0)}(\boldsymbol{r})$ and $(\epsilon - \epsilon' + i\delta)^{-1}$ outside the integral and take $\epsilon' = \epsilon$. Thus:

$$G^{(c)nn}_{\epsilon}(0, \boldsymbol{r}) = \frac{\psi^{*}_{3/2,\epsilon}(0)}{\psi^{*(0)}_{3/2,\epsilon}(0)} \exp[i\varphi_n(r, \epsilon)]G^{(0)nn}_{\epsilon}(0, \boldsymbol{r}). \tag{4.6}$$

The identity analogous to eq. (2.50) is valid:

$$G^{Fnn}_{\Delta}(0, 0) = \int G^{nn}_{\epsilon}(0, \boldsymbol{r})G^{Fnn}_{\Delta}(\boldsymbol{r}, \boldsymbol{r}')G^{nn}_{\epsilon'}(\boldsymbol{r}', 0) \times \exp\left[\frac{i(\Delta - \epsilon)\tau + i(\Delta - \epsilon')\tau'}{\hbar}\right] d^3r\, d^3r' \frac{d\epsilon\, d\epsilon'}{(2\pi\hbar)^2}. \tag{4.7}$$

All the Green functions in eq. (4.7) are diagonal. The remaining calculations are similar to the derivation of eq. (2.63). The result is:

$$\delta\hat{G}_{\Delta}^{(c,F)nn} = \frac{\operatorname{Im} G_{\Delta}^{(c)}(0,0)}{\operatorname{Im} G_{\Delta}^{(0)}(0,0)}\delta G_{\Delta}^{(F)nn}(0,0)\exp[i\varphi_{\Delta}^{n}(0,0)], \tag{4.8}$$

or

$$\hat{G}_{\Delta}^{(c,F)}(0,0) = \hat{G}_{\Delta}^{(c)}(0,0) - \frac{iF\hbar}{2(2M\Delta)^{3/2}}\operatorname{Im} G_{\Delta}^{(c)}(0,0)\hat{S}^{-1}(\boldsymbol{e})\|m_n\,e^{i\Phi_n}\|\hat{S}(\boldsymbol{e}), \quad (4.9)$$

$$\Phi_n = \frac{4\sqrt{2}}{3}\frac{\Delta^{3/2}m_n^{1/2}}{\hbar F} + \frac{e^2}{\hbar\epsilon_0}\left(\frac{2m_n}{\Delta}\right)^{1/2}\ln\frac{8\sqrt{2}m_n^{1/2}\Delta^{3/2}}{\hbar F} + 2\delta_n(\Delta). \quad (4.9)$$

The expression (3.21) for dielectric permeability also changes: a factor $\operatorname{Im} G_{\Delta}(0,0)$ and an extra phase appear:

$$\epsilon_{\alpha\beta}(\omega) = \epsilon_0(\omega)\delta_{\alpha\beta} - \frac{iF\hbar}{4(2M\Delta)^{3/2}}\operatorname{Im}\epsilon_0(\omega)\{[(2-\operatorname{sign}\gamma_2)\delta_{\alpha\beta} + 3\operatorname{sign}\gamma_2 e_\alpha e_\beta]$$

$$\times m_\ell\,e^{i\Phi_\ell} + [(2+\operatorname{sign}\gamma_2)\delta_{\alpha\beta} - 3e_\alpha e_\beta\operatorname{sign}\gamma_2]m_h\,e^{i\Phi_h}\}, \tag{4.10}$$

$\epsilon_0(\omega)$ being the dielectric permeability in zero electric field. Neither the polarization dependence nor the relative magnitude of oscillations change.

The reason that this result is relatively simple is as follows: Franz–Keldysh oscillations are due to those electron–hole pairs which separate and then return to each other when reflected by the electric field. In this process the distance between them is so great (much greater than a wavelength) that if there were no electric field it would be impossible for them to meet once more, although it will be possible, due to quantum effects, if the distance is of order or less than a wavelength. Thus, the probability of such a separation equals the probability of absorption in the absence of the field. If, moreover, the motion of the particles near the origin is isotropic, the probability of separation in the same angle interval varies in proportion to the total probability. These facts explain the form of the pre-exponent in eq. (4.10). The phase modification was already discussed above.

For negative energies the situation is a little more complicated. Equation (4.5), connecting Coulomb wave function with the free one, is valid in this case, too, but now $\delta(\Delta)$ has an imaginary part and thus affects not only the phase but also the pre-exponent. However the field dependence remains clear,

$$\hat{G}_{\Delta}(0,0) = G_{|\Delta|}^{F=0}(0,0) + \frac{iF}{4\pi|\Delta|\hbar}\hat{S}^{-1}(\boldsymbol{e})\|m_n C_n(\Delta)\,e^{-\Phi_n}\|\hat{S}(\boldsymbol{e}),$$

$$\Phi_n = \frac{4\sqrt{2}}{3}\frac{|\Delta|^{3/2}m_n^{1/2}}{\hbar F} - \frac{e^2}{\hbar\epsilon_o}\left(\frac{2m_n}{|\Delta|}\right)^{1/2}\ln\frac{8\sqrt{2}|\Delta|^{3/2}m_n^{1/2}}{\hbar F}, \tag{4.11}$$

the $C_n(\Delta)$ being some unknown functions of energy. If the exponential is large enough the heavy hole contribution may be neglected:

$$\operatorname{Im} \epsilon_{\alpha\beta} = \frac{e^2 s^2}{\hbar^2 \omega^2} \frac{F m_\ell}{8|\Delta|} C_\ell(\Delta)[(2 - \operatorname{sign} \gamma_2)\delta_{\alpha\beta} + 3 \operatorname{sign} \gamma_2 e_\alpha e_\beta] \, e^{-\Phi_\ell}. \tag{4.12}$$

In the case of a simple band $C(\Delta) = \Gamma^2(1 - 1/\kappa)$. In any case $C(\Delta)$ must have singularities at the bound state energies. For all energies $C(\Delta) \geqslant 1$ and if $|\Delta| \gg R$ then $C(\Delta) \approx 1$. We see that although the Coulomb interaction enhances the absorption, the polarization dependence does not change. (This insensitivity is due to the spherical approximation.)

Let us discuss now the effects of nonsphericity, which lead to curvature of radial trajectories in the Coulomb field. The Coulomb force is always directed along the line connecting the particle and a Coulomb centre, but for the tensorial effective mass the acceleration is not directed along the force. So radial motion is impossible (except along symmetry axes). On the other hand, it is clear that in the region where Coulomb potential is smaller than the kinetic energy this curvature can be neglected. So the question is: how much can the angular distribution of separating particles be distorted while they are in the region where the Coulomb potential is important? In order to make clear the physical nature of effects we consider a small anisotropy ($\gamma_2 \approx \gamma_3$). The wave function at the origin (at the distances of the order of the exciton Bohr radius) is then approximately isotropic. Leaving this region, the particle begins to propagate quasi-classically and its trajectory is curved. It ceases to curve when the Coulomb potential becomes smaller than the kinetic energy. The relevant radius is $r_0 = e^2/\epsilon_0\Delta$, so a curvature is noticeable if $a \ll r_0$ or $\Delta \ll R$. The angular distribution appears to be distorted in such a way that most trajectories are directed along particular axes ([001] for light holes and [111] for heavy ones). Thus, the Franz–Keldysh oscillations are enhanced when the electric field is directed along one of these axes and decrease for other directions of field.

5. *Forbidden exciton electroabsorption*

As was noted above, the first experiments in which the exciton electroabsorption was observed were made by Gross et al. (1954, 1957) in Cu_2O. In Cu_2O the transitions between the bands at $k = 0$ are forbidden. Later on Brams and Cardona (1968) obtained the spectrum of the direct forbidden exciton and the spectrum of the phonon-assisted forbidden exciton at different light polarizations. Hence, it is interesting to discuss forbidden exciton electroabsorption. The detailed theory of forbidden exciton electroabsorption is not published but such theory may be easily built up similar to that in sect. 2. Moreover, Areshev (1979), using the method developed by Aronov and Ioselevich (1978a) built up the theory of two-photon electroabsorption, which is closely connected with the theory of the forbidden transitions. In this review paper we shall confine ourselves to the qualitative description of this phenomenon.

For forbidden optical transitions at $\tilde{F} = 0$ an exciton may be excited only into a p-state, so the transitions into the $N = 1$ state are forbidden. The electric field mixes s and p-states and the transitions into the exciton state $N = 1$ become allowed. The forbidden lines appear for $\boldsymbol{n} \parallel \boldsymbol{e}$.

In an electric field the level $N = 2$ is split into one two-fold degenerate level $m = \pm 1$ $(0 \leq |m| \leq N - 1)$ and two nondegenerate levels $m = 0$, $n_1 = 0, 1$. If the light polarization vector $\boldsymbol{n}$ is parallel to $\boldsymbol{e}$, the states with $m = 0$ are excited and two lines are observed (fig. 11). The absorption line has a Lorentzian form in an electric field, similar to a direct allowed transition. The width of the lines is equal to the ionization probability of the corresponding states. (If we take into account other broadening mechanisms, the total width determines the line shape.) The width of level $m = 0$ $\Gamma_{N,n_1}^{m=0}$ is described by the eq. (2.48). The width of the exciton line $m = \pm 1$ was obtained by Areshev (1979) and, naturally, it agrees with the ionization probability of the corresponding state of the hydrogen atom (Damburg and Kolosov 1977):

$$\Gamma_{N,n_2}^{m=\pm 1} = \frac{4}{N^3 \tilde{F}\, \mathrm{e}^3 (1 + n_2)} \Gamma_{N,n_2}^{m=0}. \tag{5.1}$$

(Here e is the base of logarithms, not electron charge.)

Merkulov and Perel' (1973, 1974) have shown that the optical absorption in the gap $(\Delta < 0)$ can be interpreted as the tail of the exciton ground state resonance. Similarly, the absorption for forbidden transitions in the gap is the tail of exciton absorption line $N = 2$ and has an exponential form, the pre-exponential factor being $\Gamma^2(2 - 1/\kappa)$. Moreover, the absorption for $\boldsymbol{e} \perp \boldsymbol{n}$ is much less than that for $\boldsymbol{e} \parallel \boldsymbol{n}$,

$$\frac{\alpha_\perp}{\alpha_\parallel} \sim \frac{\tilde{F}}{\kappa^3} \ll 1, \tag{5.2}$$

i.e., strong polarization dependence of the light absorption should be

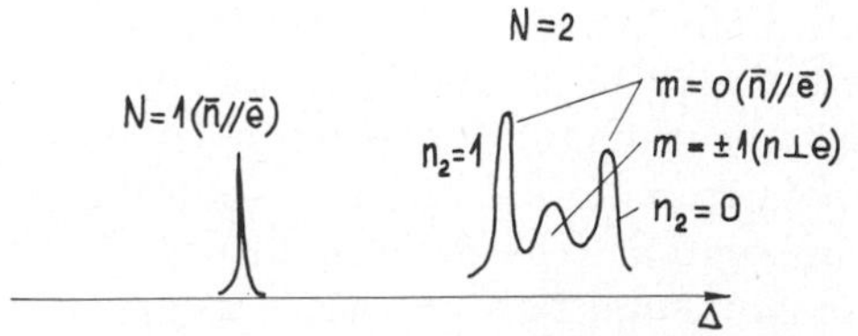

Fig. 11. Schematic electroabsorption spectrum for direct forbidden transitions.

observed. In the continuous spectrum the Franz–Keldysh oscillations are changed by the Coulomb interaction: at short distances an additional term $2\arg\Gamma(2-\mathrm{i}/k)$ and at large distances $(2/k)\ln 4k^3/\tilde{F}$, the quasi-classical phase, appear in the argument of the cosine. The amplitude ratio of the oscillating terms for $\boldsymbol{n}\parallel\boldsymbol{e}$ and $\boldsymbol{n}\perp\boldsymbol{e}$ is $k^3/\tilde{F}\gg 1$, and the oscillations at $\boldsymbol{n}\perp\boldsymbol{e}$ are strongly damped. We note that Dinges and Frohlich (1976) observed the two-photon electroabsorption spectrum of an exciton in β-AgI, consisting of the two lines corresponding to $N=2$.

6. Indirect exciton electroabsorption

In semiconductors such as Ge and Si the fundamental absorption edge is connected with indirect phonon-assisted transitions. In Si, where the extrema of the conduction band are at Δ points of the Brillouin zone, these transitions are allowed for all types of phonons. In germanium, in which the extrema of the conduction band lie on the boundary of the Brillouin zone, LA- and TO-phonon assisted transitions are allowed and TA-phonon assisted transitions are forbidden. Chester and Wendland (1965a) and Frova et al. (1965, 1966) investigated the differential absorption spectra in Ge and Si of allowed and forbidden transitions. Aronov et al. (1971) investigated the line shape of the indirect transitions in Ge in an electric field in detail. The spectra of the allowed indirect transitions in Ge and Si are shown in figs. 12a and 12b and the results for the forbidden transition (TA-phonon) in Ge are shown in fig. 12c.

Let us note that the form of the differential absorption spectrum (i.e. the variation of the absorption spectrum due to an electric field) are significantly different for allowed and forbidden transitions. For allowed transitions it is a curve with a high maximum and a deep minimum, followed by small oscillations. For forbidden transitions it has just one large maximum with small oscillations. This fact does not depend on the type of the assisting phonon and, therefore may be used for identification of the type of transiton. We shall demonstrate below that an absorption for allowed transitions is a derivative of that for forbidden transitions with respect to Δ. The one-electron theory of the indirect electroabsorption was developed by other authors (Penchina 1965, Chester and Fritsche 1965b). The numerical calculations taking exciton effects into account were performed by Lao et al. (1971). These calculations have shown that inclusion of exciton effects makes it possible to explain the form of the observed spectra even if the band structure anisotropy is not taken into account.

The absorption coefficient for indirect allowed transitions α_{ind} is,

$$\alpha_{\text{ind}} = C\begin{Bmatrix} n(\omega_{\text{ph}}) \\ 1+n(\omega_{\text{ph}}) \end{Bmatrix} \int_{-\infty}^{\Delta_\pm} \mathrm{d}\epsilon(\Delta_\pm-\epsilon)^{1/2}\,\mathrm{Im}\,G_\epsilon^+(0,0), \tag{6.1}$$

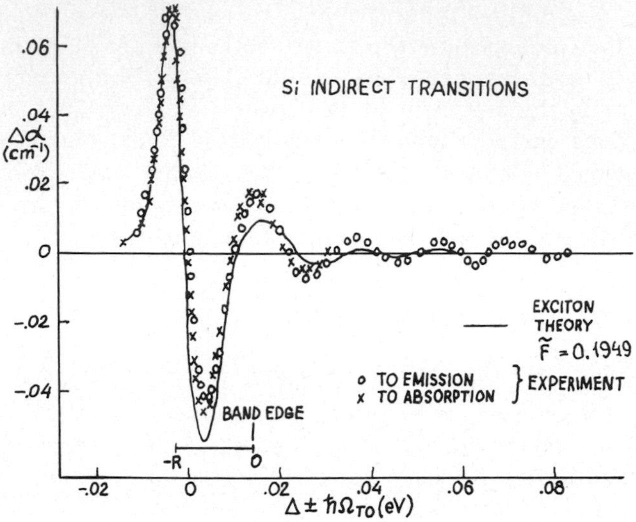

Fig. 12a. Electroabsorption spectrum for indirect allowed transitions with the emission and the absorption of the transverse optical phonon (TO) according to Frova et al. (1966). The TO absorption curve has been multiplied by 7.5 for coincidence at the main positive peak. $T = 23$ K, $F = 1.4 \times 10^4$ eV/cm. The solid curve is the result of the numerical calculations (Lao et al. 1971).

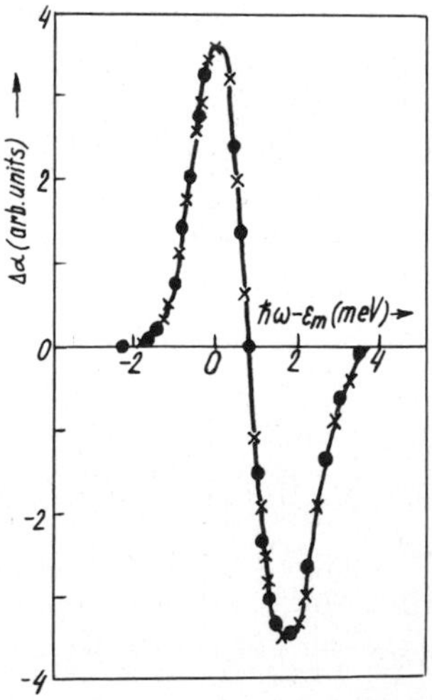

Fig. 12b. Electroabsorption spectrum for indirect allowed transitions with emission of TO (×) and LA (·) phonons according to Aronov et al. (1971) $F = 500$ eV cm^{-1}.

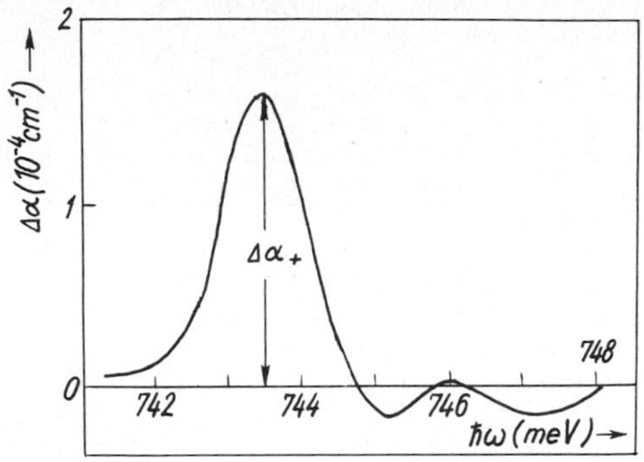

Fig. 12c. Electroabsorption spectrum for indirect forbidden transitions with emission of a TA phonon according to Aronov et al. (1971) $F = 300\ \mathrm{eV\ cm^{-1}}$.

where $\Delta_\pm = \hbar\omega \pm \hbar\omega_{ph} - E_g$, ω_{ph} being the phonon frequency, $n(\omega_{ph})$ being the Planck function, C being a constant which contains the interband matrix element of momentum and the electron–phonon interaction constant. In fact eq. (6.1) was used previously by Lao et al. (1971).

As was noted above, the isotropic approximation fits well with experimental data. Using eq. (6.1) we can get the closed expression for the indirect absorption coefficient. Substituting eq. (2.54) in eq. (6.1) and integrating over ϵ we obtain

$$\alpha_{ind} = C\left\{\begin{matrix} n(\omega_{ph}) \\ 1+n(\omega_{ph})\end{matrix}\right\}\frac{(2m^*)^{3/2}}{32}\left[\Delta_\pm + \frac{16}{3}\Delta_\pm^{3/2}R^{1/2} + \frac{4\pi^2}{3}\Delta_\pm R\right.$$

$$\left. + \frac{8\pi^2}{45}R^2 + 16R^2\sum_{n=1}^{\infty}\frac{1}{n^4}\left(1+\frac{\Delta_\pm}{R}n^2\right)^{1/2}\right], \quad \text{at } \Delta_\pm > 0,$$

$$\alpha_{ind} = C\left\{\begin{matrix} n(\omega_{ph}) \\ 1+n(\omega_{ph})\end{matrix}\right\}(2m^*)^{3/2}R^2\sum_{n=1}^{\infty}\frac{1}{n^4}\left(1+\frac{\Delta_\pm}{R}n^2\right)^{1/2}\theta\left(1+\frac{\Delta_\pm}{R}n^2\right),$$

at $\Delta_\pm < 0$.

To calculate the light absorption in an electric field we use the expression for Green function $\mathrm{Im}\, G_E(0,0) \propto \alpha$, which was found in sect. 2.

In the vicinity of the threshold the absorption coefficient is given by the following expression

$$\alpha_{ind} \propto \int_{-\infty}^{\Delta_\pm} d\epsilon (\Delta_\pm - \epsilon)^{1/2} \frac{\Gamma_{n_1 n_2}}{\Gamma_{n_1 n_2}^2 + (\epsilon - E_{n_1 n_2})^2}. \tag{6.2}$$

Integrating over ϵ and taking into account the fact that the important values of ϵ is $\Delta_{\pm} \simeq E_{n_1 n_2}$ we get

$$\alpha_{\text{ind}} \propto \{[(\Delta_{\pm} - E_{n_1 n_2})^2 + \Gamma^2_{n_1 n_2}]^{1/2} + \Delta_{\pm} - E_{n_1 n_2}\}^{1/2}. \tag{6.2a}$$

All the changes in the absorption are connected with Stark splitting and ionization broadening. Note that $\Gamma_{n_1 n_2}$ is the total width including the exciton–phonon broadening.

In the continuous spectrum change of optical absorption due to an electric field is ($k^3_{\pm}/\tilde{F} \gg 1$):

$$\delta\alpha_{\text{ind}} \propto \Delta^{3/2}_{\pm}\left[1 - \exp\left(-\frac{2\pi}{k_{\pm}}\right)\right]^{-1}\left(\frac{\tilde{F}}{k^3_{\pm}}\right)^{5/2} \cos\left\{\frac{2}{3}\frac{k^3_{\pm}}{\tilde{F}} + \frac{2}{k_{\pm}}\ln\frac{4k^3_{\pm}}{\tilde{F}} + 2\arg\Gamma\left(1 - \frac{\mathrm{i}}{k_{\pm}}\right) + \frac{\pi}{4}\right\}, \tag{6.3}$$

where $k_{\pm}$ in dimensional variables has the form,

$$k_{\pm} = (\Delta_{\pm}/R)^{1/2}.$$

Similarly to eq. (6.1) we can get the expression of $\alpha^{\text{f}}_{\text{ind}}$ for the forbidden indirect transitions

$$\alpha^{\text{f}}_{\text{ind}} = C'\left\{\begin{matrix} n(\omega_{\text{ph}}) \\ 1 + n(\omega_{\text{ph}}) \end{matrix}\right\} \int_{-\infty}^{\Delta_{\pm}} \mathrm{d}\epsilon (\Delta_{\pm} - \epsilon)^{3/2} \operatorname{Im} G^{+}_{\epsilon}(0, 0). \tag{6.4}$$

We shall not write out the full expression for the absorption coefficient in an electric field in this case, since it can easily be obtained in a way similar to the allowed indirect transitions. However, we note that $\alpha_{\text{ind}} \propto (\partial/\partial\Delta_{\pm})\alpha^{\text{f}}_{\text{ind}}$, so that the spectrum of the forbidden indirect transitions can be obtained by the integration of the spectrum of the allowed indirect transitions. Figures 12b and 12c show that the experimental data agree with this relation.

7. *Anisotropic electrooptical effects*

Let us discuss two anisotropic electrooptical effects.

(1) The linear electrooptical effect in the crystals without inversion symmetry (Pockels effect):

$$\delta\epsilon_{\alpha\beta} = a_{\alpha\beta\gamma}F_{\gamma}. \tag{7.1}$$

In a cubic crystal (the crystal classes T_d and T) the tensor $\alpha_{\alpha\beta\gamma}$ has only one linearly independent component a_{xyz}. The Pockels effect has been observed in some semiconductors, e.g. in GaAs (Bagaev et al. 1966, Walters 1966), in GaP (Nelson and Reinhart 1964), in ZnS (Sham 1936, Poulett 1955, Namba 1959, 1961, Mayers and Powell 1966), in ZnTe (Seiker and Jost 1966).

(2) The even electrooptical effect (Kerr effect):

$$\delta\epsilon_{\alpha\beta} = b_{\alpha\beta\gamma\lambda}F_\gamma F_\lambda. \tag{7.2}$$

The Kerr effect in the crystal with inversion symmetry can arise from the following:

(i) the anisotropy of the effective masses,
(ii) the nonsphericity of the valence bands,
(iii) the nonparabolicity of the energy bands.

The theory of the interband Kerr effect was built up by Aronov and Pikus (1968). In this paper we shall not discuss the Kerr effect but the interband Pockels effect only. The theory of this effect was developed by Keldysh (Bagaev et al. 1966) and Aronov and Pikus (1968) without exciton effects. Later on Aronov and Ioselevich (1978) developed the theory including exciton effects.

If the energy bands are simple and the extrema are at the centre of the Brillouin zone, the linear electrooptical effect is due to the terms in the interband current operator which are linear in $\boldsymbol{k}$:

$$J_\alpha = es\rho_2\sigma_\alpha + e\lambda\rho_1(\sigma_{\alpha+1}k_{\alpha+2} + \sigma_{\alpha+2}k_{\alpha+1}), \tag{7.3}$$

where,

$$\rho_1 = \begin{pmatrix} 0 & I \\ I & 0 \end{pmatrix}, \qquad \rho_2 = \begin{pmatrix} 0 & -\mathrm{i}I \\ \mathrm{i}I & 0 \end{pmatrix}$$

are 4×4 matrices.

To calculate $\epsilon_{xy}(\omega)$ it is necessary to substitute the second term in eq. (7.3) instead of one of the current operators in eq. (1.1) for the dielectric permeability. Calculating the trace over spin variables and transforming eq. (7.3) to the $\boldsymbol{r}$ representation we find that the resonance term has the form,

$$\epsilon_{xy}(\omega) = \frac{16\pi e^2 s\lambda}{\omega^2}\frac{\partial}{\partial z}G_\Delta^+(\boldsymbol{r}, 0)\,|_{r\to0}. \tag{7.4}$$

In zero order ($F = 0$) $\epsilon_{xy} = 0$ and it is necessary to calculate $(\partial/\partial z)G_\Delta^+(\boldsymbol{r}, 0)\,|_{r\to0}$ to first order in $\boldsymbol{F}$. We have

$$\frac{\partial}{\partial z}G_\Delta(\boldsymbol{r}, 0)|_{r\to0} = -F\int \mathrm{d}^3\boldsymbol{r}'\frac{\partial G_\Delta^0(\boldsymbol{r}, \boldsymbol{r}')}{\partial z}z'G_\Delta^0(\boldsymbol{r}', 0)\,|_{r\to0}. \tag{7.5}$$

Using the expressions for the exciton Green function (2.53) and (2.54) and making direct but cumbersome transformations, we get:

$$\frac{\partial}{\partial z}G_\Delta(\boldsymbol{r}, 0)\,|_{r\to0} = -\frac{m^{*2}F}{12\pi\kappa}\Gamma^2\left(1-\frac{1}{\kappa a}\right)\left[3J_0\left(\frac{1}{\kappa a}\right) - \frac{1}{\kappa a}J_1\left(\frac{1}{\kappa a}\right)\right], \tag{7.6}$$

where

$$J_n(\mu) = \int_0^\infty dzz^n W^2_{\mu,1/2}(z), \tag{7.7}$$

and $W_{\alpha,\beta}(z)$ is the Whittaker function (Ryzhik and Gradstein 1963). The calculation of the integrals $J_n(\mu)$ is given in the original paper (Aronov and Ioselevich 1978). Finally, we can write the dielectric permeability in the form:

$$\epsilon_{xy} = -\frac{4e^2m^{*2}s\lambda}{\omega^2}F_z\frac{f(\kappa_- a)}{\kappa_-}. \tag{7.8}$$

where

$$f(\kappa a) = 1 + \frac{5}{3}\frac{1}{\kappa a} - \frac{1}{(\kappa a)^2} - \frac{2}{(\kappa a)^3} + \frac{2}{(\kappa a)^2}\left[1 - \frac{1}{(\kappa a)^2}\right]\psi^1\left(1 - \frac{1}{\kappa a}\right), \tag{7.9}$$

and

$$\psi'\left(1 - \frac{1}{\kappa a}\right) = \sum_{N=1}^{\infty}\left(N - \frac{1}{\kappa a}\right)^{-2}. \tag{7.9a}$$

From eqs. (7.8) and (7.9) we see that in the vicinity of the exciton level $N = 1$, $\epsilon_{xy}(\omega)$ has the form of a standard anomalous dispersion but near the lines $N > 1$, $\epsilon_{xy}(\omega)$ has poles of second order. This is connected with the fact that while the exciton level $N = 1$ is not split in an electric field, and the levels $N > 1$ have a linear Stark effect. At $\kappa_- a \gg 1$, $f \to 1$ and $\epsilon_{xy}(\omega)$ is transformed into the expression for $\epsilon_{xy}(\omega)$ without an exciton effect (Bagaev et al. 1966, Aronov and Pikus 1968). The spectral form of the Pockels effect near the absorption line in the Stark multiplet is

$$\frac{\epsilon_{xy}(\omega)}{\epsilon_{xx}(\omega)} = -\frac{2}{a}\frac{\lambda}{s}\frac{n_1 - n_2}{N}. \tag{7.10}$$

We see that $\epsilon_{xy}(\omega)$ does not depend on an electric field in this case, and the spectral dependence has a form analogous to $\epsilon_{xx}(\omega)$. The width of the region where eq. (7.10) holds increases with electric field.

8. *Influence of surface field on electroreflectance*

As noted in the introduction the electroreflectance has wide applications as a method of analysis of electron spectra in solids (Seraphin 1964). This method is based on the fact that an electric field has the strongest influence on the optical properties of solids near absorption thresholds. So if an alternating electric field is applied to a sample the reflectance coefficient is modulated and the electroreflectance spectrum shows a characteristic structure near such thresholds.

The electric field may be applied along the surface (transverse electroreflectance TER) or perpendicular to it by means of the field effect (surface bearrier electroreflectance SBER). The disadvantage of the first method is that its application is limited to insulators and heavily compensated semiconductors because of significant heating of the samples. The second method has no such limitation. In principle in the TER method a homogeneous electric field can be oriented in an arbitrary direction in the surface) so that the polarization dependence can be studied. These advantages make TER method preferable for the verification of subtle details of the theoretical predictions. However, it must be noted that the presence of a surface field (as is common in semiconductors) which depends on temperature and on the intensity of the light, leads to a difference between the total field direction and that of the external field near the surface, so an additional analysis of surface effects is needed. On the other hand, the SEBR method provides experimentalists with larger (though non-homogeneous) fields, increasing the sensitivity. Hence, SBER is more convenient if one is only interested in the positions of van Hove singularities. The experimental technique and the results obtained from electroreflectance spectra are presented in the reviews by Seraphin (1964) and Cardona (1969).

In the interpretation of experiments it must be taken into account that the reflectance and electroreflectance spectra near the exciton band (where the most drastic effects arise) strongly depend on the surface conditions (Evangelisti et al. 1972, Permogorov et al. 1972). The nature of this fact may be described by the "dead layer" model (Hopfield and Thomas 1963). There are two reasons at least for the existence of a "dead" layer:

(1) The exciton is not able to come closer to the surface than to its radius, which is finite. This is equivalent to a repulsive potential preventing the exciton from coming too close to the surface. (2) The strong electric field of the surface barrier dissociates the excitons, diminishing the exciton absorption. The real part of the dielectric permeability changes, becoming independent of frequency in the exciton region. The depth of the "dead" layer is determined by the condition $\tilde{F} \sim 1$ and may be of the order of several microns. So two boundaries appear: an external sample boundary and an internal boundary, outside which excitons can not exist. The dielectric permeability changes at each boundary so that light is reflected by them and interference occurs. By varying the surface potential by means of the field effect one can vary the "dead" layer depth. This leads to oscillations of reflectance as a function of bias voltage (at a fixed wavelength of light). A detailed discussion of the influence of the surface barrier and the additional boundary conditions on the electroreflectance

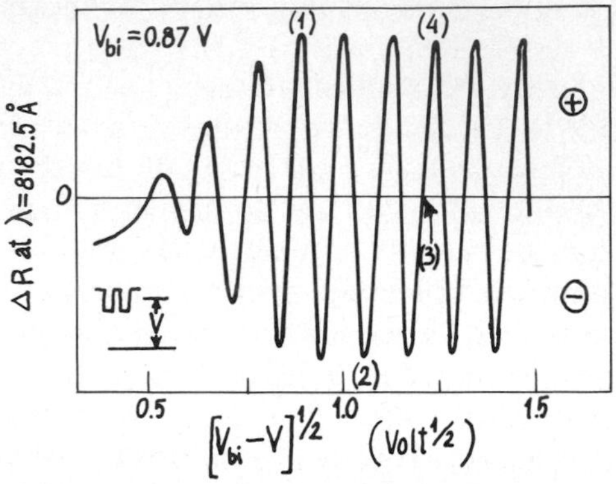

Fig. 13a. Electroreflectance interference effects in the GaAs structure (Evangelisti et al. 1972). The value of the built-in barrier $V_{bi} = 0.87$ V. The data correspond to the wavelength $\lambda = 8182.5$ Å, related to the peak in the plot of fig. 13b.

spectra may be found in the article by Ivchenko (ch. 4 of the present volume).

In the plot of fig. 13a the variation of reflectance R with the bias voltage, applied to Au–GaAs structure is shown (Evangelisti et al. 1972). The horizontal axis gives $(V_{bi} - V)^{1/2}$, proportional to the "dead" layer depth. In

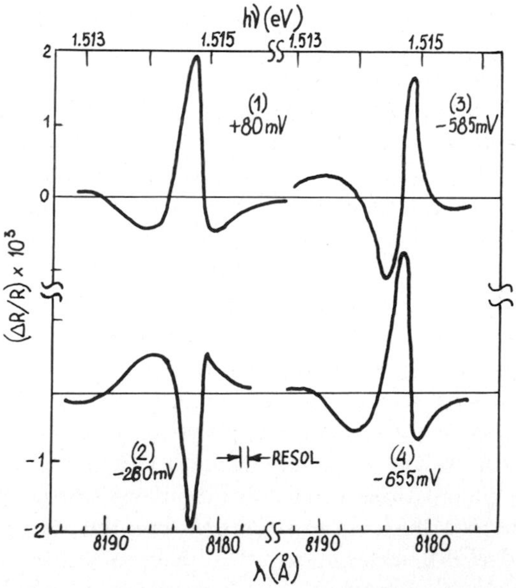

Fig. 13b. Electroreflectance in the region of $n = 1$ exciton level for different values of the voltage across the Schottky barrier. Curves (1) and (4) correspond to maxima in the plot of fig. 13a, curve (2) to a minimum, and curve (3) to zero.

fig. 13b the spectral dependence of reflectance at different bias voltages is shown. Curve 1 corresponds to a maximum in the plot of fig. 13a, curve (2) – to a minimum, curve (3) – to a null, curve (4) – to a maximum again. We see that despite the significant difference in the applied fields the curves (1) and (4) are almost the same, curve (2) is out of phase and curve (3) has zeros where the others have extrema and vice versa.

Kiselev (1978) has computed a reflectance for a variety of barrier forms numerically. These computations show a strong dependence of the spectra on the details of potential (such as the tails, the presence of a potential well, etc.). It was shown also that the inversion of the reflectance spectrum can be explained as being due to the repulsive potential, imitating the "dead" layer effect, and to the spatial dispersion of the dielectric permeability. A "dead" layer leads to the appearance of an additional sharp peak at the longitudinal exciton frequency the so called "spike" (see ch. 4 by Ivchenko in the present volume). The nonhomogeneous field is concentrated near the surface in a layer which is much thicker than an exciton radius (as in the experiments of Evangelisti et al. (1972)) and causes the following variation of the exciton potential energy:

$$W = -\tfrac{1}{2}\alpha e^2 F^2(x).$$

α being the polarizability of an exciton in given state; the x-axis is directed perpendicular to surface. The presence of such a potential well enhances the exciton reflection band in spite of the greater exciton damping in an electric field.

Thus, the electroreflectance spectra turn out to have a strong dependence on the state of the surface. It is clear now that these effects must be allowed for in the analysis of the spectra, not only near the fundamental absorption edge but near the other thresholds as well.

Conclusion

We will summarize the principal results obtained so far, and some important problems still awaiting the solution.

(1) The present-day theory describes the following exciton effects in electrooptics: the modification of electroabsorption and electroreflectance in the gap, in the exciton region and in the continuous spectrum for different types of transitions (direct and indirect, allowed and forbidden).

(2) The theory describes the exciton effects in electrooptics of semiconductors with degenerate valence band and the polarization effects in cubic crystals. Note that the magnitude of the exciton electroabsorption above the threshold is related simply to the exciton absorption in the zero electric field.

(3) The theory explains the mechanism responsible for the interband anisotropic electrooptical effects in cubic crystals (Pockels effect, Kerr effect) and the frequency dispersion of these effects near the absorption edge, taking exciton effects into account.

(4) The theory qualitatively describes the principal features of the influence of the sample surface on the electroreflectance spectra in the exciton region.

The results of experiments on a wide variety of semiconductors agree with the theoretical predictions and, we have tried to illustrate this agreement throughout the article. The principal experimental facts are named below.

(1) Enhancement of the Franz–Keldysh effect, the increase in the oscillations period, and polarization effects in both electroabsorption and diagonal tunneling have been observed.

(2) The spectral dependence of the linear electrooptical effect was studied.

(3) The influence of inhomogeneous surface fields on the electroreflectance was studied for the different materials. Experiments have shown the important role of the form of the surface barrier.

There are, however, a certain number of unsolved problems in electrooptics, so far, both theoretical and experimental.

(1) The influence of electron and hole damping on the electroabsorption and electroreflectance spectra is an interesting though rather complicated problem. Damping was considered phenomenologically by Aspnes (1974). If the decay of the exciton bound state is exponential, then the total width of the line is the sum of the broadening due to the ionization in an electric field and the broadening due to all other mechanisms of damping. The total width has to be inserted instead of Γ in eq. (2.47) of sect. 2. Inclusion of damping in the continuous spectrum leads to an exponential decrease of oscillation amplitude with the energy Δ, which is expressed in the factor $\exp[-\theta^{-3/2}\int_0^{\Delta}\Gamma(\epsilon)\epsilon^{-1/2}\,d\epsilon]$, where $\Gamma(\epsilon)=\Gamma_e(\epsilon m^*/m_e)+\Gamma_h(\epsilon m^*/m_h)$ and $\Gamma_{eh}(E)$ the electron and hole damping, respectively, at energy E. For $\Gamma=\text{const}$, this expression agrees with that derived by Aspnes (1974). Hence, the study of the light frequency dependence of the oscillation amplitude can provide information on the damping of electrons and holes and on its energy dependence. However such an information may only be obtainable if the electric field is sufficiently homogeneous and the number of observed oscillations is large.

(2) A theory of electrooptical effects at saddle-type van Hove singularities (M_2), including Coulomb interactions, does not exist at present and the experimental results can not be used for the evaluation of exciton effects.

(3) Although a complete one-to-one reconstruction of the surface properties by means of electroreflectance is evidently not possible, some

conclusions concerning the exciton motion and damping near the surface can be made. Further investigations in this field are, in our opinion, of great importance.

Acknowledgement

The authors are grateful to Prof. G.E. Pikus and Dr. E.L. Ivchenko for helpful discussions.

Appendix

We compute here $m_\perp(\boldsymbol{e})$ in the pre-exponent. We shall carry out all the calculations for a light hole (plus sign in eq. (3.2) of sect. 3). The result for heavy holes differs only in sign of g function. Differentiating eq. (3.2) with respect to $\boldsymbol{k}$ we find:

$$m_{ij}^{-1} = \left.\frac{\partial^2 \epsilon_e(\boldsymbol{k})}{\partial k_i \partial k_j}\right|_{\boldsymbol{k}=\boldsymbol{e}} = \frac{|\gamma_2|}{m}[a\delta_{ij} - g^{-3}(\boldsymbol{e})e_i e_j(2+c^2(1-e_i^2))(2+c^2(1-e_j^2)) + g^{-1}(\boldsymbol{e})(\delta_{ij}(c^2(1-3e_i^2)+2)+2e_ie_j(2+c^2))]. \tag{A1}$$

The resulting expression for $D = \det \|m_{ij}^{-1}(\boldsymbol{e})\|$ must possess cubic symmetry and can depend on $\boldsymbol{e}$ only through a cubical invariant $\Lambda = e_x^2e_y^2 + e_y^2e_z^2 + e_z^2e_x^2$. For this reason we are able to simplify our task. Let us assume $e_z = 0$, and compute D which depends on $\Lambda = e_x^2e_y^2$ in this case. In order to write a general expression valid for arbitrary $e_z \neq 0$ we have to substitute $e_x^2e_y^2 + e_y^2e_z^2 + e_z^2e_x^2$ for $e_x^2e_y^2$.

So, let $e_z = 0$

$$m_{zz}^{-1} = \frac{|\gamma_2|}{m}[a + g^{-1}(2+c^2)], \quad g = (1+c^2\Lambda)^{1/2},$$

$$m_{zx}^{-1} = m_{zy}^{-1} = 0, \qquad m_{xy}^{-1} = \frac{|\gamma_2|}{m}[(4+c^2)\Lambda^{1/2}(g^{-1}-g^{-3})],$$

$$m_{xx(yy)}^{-1} = \frac{|\gamma_2|}{m}\left[a + \frac{1}{2}g^{-3}(c^2+4) \pm \frac{e_x^2 - e_y^2}{2}(4g^{-1} + (c^2+4)g^{-3})\right],$$

$$D = m_{zz}^{-1}(m_{xx}^{-1}m_{yy}^{-1} - (m_{xy}^{-1})^2)$$
$$= \left(\frac{|\gamma_2|}{m}\right)^3 [a + g^{-1}(2+c^2)][a^2 + 8 + c^2 + (c^2+4)g^{-3}(a+3g)].$$

We have used the fact that $[(e_x^2 - e_y^2)/2]^2 = \frac{1}{4} - \Lambda$ and expressed Λ in terms of g. Now we can write a final result:

$$m_{\perp}^{\ell,h}(e) = (m_{\parallel}(e)D)^{-1/2}$$
$$= \frac{m}{|\gamma_2|}\left[\frac{a \pm 2g(e)}{a \pm (2+c^2)g^{-1}(e)}\right]^{1/2} [a^2 + c^2 + 8 \pm (c^2+4)g^{-3}(e)$$
$$\times (a \pm 3g(e))]^{-1/2}. \qquad \text{(A2)}$$

References

Alferov, J.I., D.Z. Garbuzov, E.P. Morozov and E.L. Portnoy, 1969, Fiz. Tekh. Polupr. **3**, 1054 (Sov. Phys.-Semiconductors **3**, 878).

Areshev, I.P., 1979, Fiz. Tverd. Tela **21**, 765 (Sov. Phys.-Solid State **21**, 447).

Aronov, A.G. and G.E. Pikus, 1968, Fiz. Tverd. Tela **10**, 825 (Sov. Phys.-Solid State **10**, 648).

Aronov, A.G. and A.S. Ioselevich, 1978a, Zh. Eksp. Teor. Fiz. **74**, 1043 (Sov. Phys. JETP **47**, 548).

Aronov, A.G. and A.S. Ioselevich, 1978b, Fiz. Tverd. Tela **20**, 2615 (Sov. Phys.-Solid State, **20**, 1511).

Aronov, A.G. and A.S. Ioselevich, 1981, Zh. ETF **81**, 336.

Aronov, A.G., I.V. Mochan and V.I. Zemskiy, 1971, Phys. Status Solidi (b)**45**, 395.

Aspnes, D.E., 1966, Phys. Rev. **147**, 554.

Aspnes, D.E., 1974, Phys. Rev. **B10**, 4228.

Aspnes, D.E. and N. Bottka, 1970, in: Semiconductors and Semimetals, eds., R.K. Willardson and A.C. Beer (Academic, New York) vol. VI.

Bagaev, V.S., Y.N. Berozashvili and L.V. Keldysh, 1966, ZhETF Pis'ma Red. **4**, 364; (JETP Lett. **4**, 246).

Baz', A.I., Ya.B. Zeldovich and A.M. Perelomov, 1971. Rasseyanie reaktsii i raspadi v nerelyativistskoi kvantovoy mekhanike (Scattering Reactions and Decays in Nonrelativistic Quantum Mechanics) (Nauka, Glavnaya redakzia fiz.-mat. literaturi, Moscow).

Bir, G.L. and G.E. Pikus. 1972, Symmetriya i deformationnie effecti v poluprovodnikah (Nauka, Glavnaja redakzia fiz.-mat. literaturi, Moscow). Bir G.L., G.E. Pikus, Symmetry and strain induced effects in semiconductors (Keter Publishing House Ltd., Jerusalem, 1974, Halsted Press).

Blossey, D.E., 1970, Phys. Rev. **B2**, 8976.

Blossey, D.F., 1971, Phys. Rev. **B3**, 1382.

Brahms, S. and M. Cardona, 1968, Solid State Commun. **6**, 733.

Callaway, J., 1963, Phys. Rev. **130**, 549.

Callaway, J., 1964, Phys. Rev. **134**, 898.

Cardona, M., 1969, Modulation Spectroscopy (Academic, New York).

Chester, M. and L. Fritsche, 1965, Phys. Rev. 139A, 518.

Chester, M. and P. M. Wendland, 1965, Phys. Rev. **140A**, 1384.

Damburg, R.J. and V.V. Kolosov, 1977, Phys. Lett. **61A**, 233.

Dinges, R. and D. Fröhlich, 1976, Solid State Commun. **19**, 61.

Dow, J.D. and D. Redfield, 1970, Phys. Rev. **B1**, 3351.

Dow, J.D., B.Y. Lao and S.A. Newman, 1971, Phys. Rev. **B3**, 2571.

Elliot, R.J., 1957, Phys. Rev. **108**, 1384.

Evangelisti, F., A. Frova and J.U. Fischbach, 1972, Phys. Rev. Lett. **26**, 1001.

Feynman, R.P. and A.R. Hibbs, 1965, Quantum Mechanics and Path Integrals (McGraw-Hill, New York).

Franz, W., 1958, Z. Naturforsch. **13a**, 487.
Freeman, R.R., N.P. Economou, L.C. Bjorklund and K.T. Lu, 1978, Phys. Rev. Lett. **41**, 1463.
Frova, A. and P. Handler, 1965, Phys. Rev. **137**, 1857.
Frova, A., P. Handler, F.A. Germano and D.E. Aspnes, 1966, Phys. Rev. **145**, 575.
Gradshtein, I.S. and I.M. Ryzhik, 1963, Tablitzi integralov summ, ryadov i proizvedenii (Fizmatgiz, Moscow) (Tables of integrals, series and products, Academic Press, New York 1965).
Gross, E.F., 1957, Usp. Fiz. Nauk. **63**, 575.
Gross, E.F., B.P. Zakharchenya and N.M. Reinov, 1954, Dokl. Akad. Nauk. **97**, 57.
Handler, P., S. Jasperson and S. Koeppen, 1969, Phys. Rev. Lett. **23**, 1387.
Hopfield, J.J., D.G. Thomas, 1963, Phys. Rev. **132**, 563.
Keldysh, L.V., 1958, Zh. Eksp. Teor. Fiz. **34**, 1138 (Sov. Phys. JETP **7**, 788).
Keldysh, L.V., O.V. Konstantinov and V.I. Perel', 1969, Fiz. Tekh. Polupr. **3**, 1042 (Sov. Phys.-Semiconductors **3**, 876).
Kiselev, V.A., 1978, Fiz. Tverd. Tela **20**, 2173 (Sov. Phys. Solid State **20**, 1255).
Kiselev, V.A., 1979, ZhETF Pis. Red. **29**, 369 (Sov. Phys. JETP Lett. **29**, 332).
Konstantinov, O.V. and V.I. Perel', 1959, Zh. Eksp. Teor. Fiz. **37**, 786 (Sov. Phys. JETP **37**, 560).
Lanczos, C., 1930, Zeit. für Phys. **62**, 518.
Lanczos, C., 1931, Zeit. für Phys. **68**, 204.
Landau, L.D. and E.M. Lifschitz, 1963, Kvantovaya Mekhanika (Fizmatgiz, Moscow) (L.D. Landau and E.M. Lifschitz, 1959, Quantum Mechanics, (Pergamon, New York)).
Lao, B.Y., J.D. Dow and F.C. Weinstein, 1971, Phys. Rev. **B4**, 4424.
Merkulov, I.A., 1974a, Zh. Eksp. Teor. Fiz. **66**, 2314 (Sov. Phys. JETP **39**, 1140).
Merkulov, I.A., 1974b, Fiz. Tekh. Polupr. **8**, 2094 (Sov. Phys.-Semiconductors **8**, 1374).
Merkulov, I.A. and V.I. Perel', 1973, Phys. Lett. **45A**, 83.
Myers, R.A. and C.G. Powell, 1966, Appl. Phys. Lett. **9**, 326.
Namba, S., 1959, J. Appl. Phys. Japan **28**, 432.
Namba, S., 1961, J. Opt. Soc. Amer. **51**, 76.
Nelson, D.F. and F.K. Reinhart, 1964, Appl. Phys. Lett. **5**, 148.
Ovsyuk, N.N. and M.P. Sinyukov, 1978, Zh. Eksp. Teor. Fiz. **75**, 1075 (Sov. Phys. JETP **48**, 542).
Penchina, C.M., 1965, Phys. Rev. **138**, A924.
Permogorov, S.A., V.V. Travnikov and A.V. Sel'kin, 1972, Fiz. Tverd. Tela **14**, 3642 (Sov. Phys.-Solid State **14**, 3051).
Poulett, H., 1955, J. Phys. Rad. **16**, 257.
Ralph, H.I., 1968, J. Phys. **C1**, 378.
Sakharov, A.D., 1948, Zh. Eksp. Teor. Fiz. **18**, 631.
Seiker, T.R. and J.M. Jost, 1966, J. Opt. Soc. Amer. **56**, 132.
Seraphin, B.O., 1964, in: Physics of Semiconductors, Proceedings of the Seventh International Conf., ed. M. Hulin (Dunod, Paris) p. 165.
Seraphin, B.O., 1970, in: Semiconductors and Semimetals, eds. R.K. Willardson and A.C. Beer (Academic, New York).
Sham, C., 1936, Ann. Phys. **25**, 309.
Tharmalingam, K., 1963, Phys. Rev. **130**, 2204.
Walters, W.L., 1966, J. Appl. Phys. **37**, 916.
Weinstein, F.C., J.D. Dow and B.Y. Lao, 1971, Phys. Rev. **B4**, 3502.

CHAPTER 8

Excitons in Semiconductor Alloys

R.J. NELSON

Bell Laboratories
Murray Hill, New Jersey 07974
U.S.A.

Excitons
Edited by
E.I. Rashba and M.D. Sturge

Contents

1. Introduction

Possibly the most important feature of semiconductor alloys is that the band structure may be varied over a wide range by altering the crystal composition. Numerous technical applications have been pursued including optoelectronic devices such as visible light emitting diodes and semiconductor lasers (Casey 1978). In addition to the technological importance of semiconductor alloys there has also been considerable interest in the study of alloys for scientific reasons. An ideal semiconductor alloy would be a perfect crystal with the constituent atoms randomly distributed on the proper lattice sites. Even though an alloy semiconductor might have a perfect lattice geometry, as revealed by X-ray diffraction, the translational symmetry of the potential energy is violated due to the random distribution. The Bloch theorem and the suitability of the wave vector ($\boldsymbol{k}$) as a quantum number are no longer strictly valid. It is common, however, to treat the non-periodic part of the potential as a small perturbation since the mean free paths of electrons and holes are considerably larger than atomic radii. As a result, the major concepts of crystal band theory are often applied to semiconductor alloys. Thus, the selection rules which apply to transitions between energy states in perfectly periodic crystals also apply in this treatment to semiconductor alloys. Manifestations of the random atomic placement or of short-range correlations are observed in some cases, however, as described below.

Transitions involving both free and bound excitons are observed in semiconductor alloys in spite of the random contribution to the potential energy. Narrow free-exciton peaks have been observed for example in the absorption spectra of direct bandgap $GaAs_{1-x}P_x$ (Nelson 1976a). Because of the large Bohr radius of excitons in direct bandgap III–V alloys, potential fluctuations are apparently averaged out and do not produce significant broadening for the direct exciton. Bound as well as indirect excitons, on the other hand, may have a much smaller radius and stronger broadening may be expected. This is, in fact observed for the exciton bound to the nitrogen isoelectronic trap in alloys such as $In_{1-x}Ga_xP:N$ (Nelson 1976b), although the alloy broadening is not as large as was believed from earlier work.

In this review we concentrate on recent experimental observations of exciton transitions in III–V semiconductor alloys because they have been investigated most completely. An excellent review of earlier work has been presented by Pikhtin (1977). The first section is devoted to free-exciton transitions in both direct- and indirect-bandgap alloys. The second section deals with bound excitons. Emphasis is placed on excitons bound to isoelectronic traps such as nitrogen in $In_{1-x}Ga_xP$. The early literature on nitrogen-doped alloys has been reviewed by Craford and Holonyak (1976). We therefore concentrate on the recent developments beginning with the new interpretation of Nelson and Holonyak (1976b) for the origin of the major luminescence and absorption band in these materials. This work led to a re-examination of both experimental and theoretical aspects of nitrogen-doped alloys as reviewed here.

2. Free excitons in semiconductor alloys

The fundamental absorption edge is associated either with direct vertical transitions for alloys whose absolute valence band maximum and conduction band minimum occur at the same point in $\boldsymbol{k}$ space (usually at the center of the Brillouin zone, see fig. 1a) or with transitions in indirect bandgap material (see fig. 1b) involving both a photon and phonon. The fundamental absorption edge shifts along with the bandgap variation as the crystal composition is varied in semiconductor alloys. It is well known that in heavily doped semiconductors the density of states near the band edges is perturbed by the fluctuating impurity potential producing band tailing (Bonch-Bruevich 1966). It might also be expected that random disorder or short-range correlations and associated potential fluctuations should distort the absorption edge and broaden the exciton transitions in alloys.

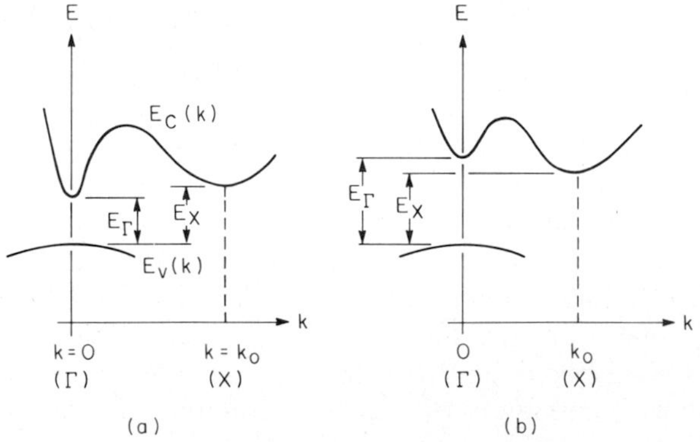

Fig. 1. Representative semiconductor band structure showing the upper valence band (E_v) and lowest conduction band (E_c). Part (a) shows a direct bandgap semiconductor. Part (b) shows an indirect bandgap semiconductor.

However, the potential fluctuations caused by differences of the pseudopotentials of the respective atoms making up the alloy are usually sufficiently small and localized to produce minor broadening effects in comparison with the Coulomb potential associated with ionized impurities.

2.1. Direct bandgap material

Fluctuations in crystal composition, internal stresses or high impurity levels often prevent the observation of free exciton transitions in imperfect alloy crystals. The first observations of direct-exciton transitions in the absorption spectra of III–V alloys were for high quality $In_{1-x}Ga_xP$ (see fig. 2) grown by liquid phase epitaxy (Nelson 1976b) and for $Al_xGa_{1-x}As$ (Dingle 1977). The observation of the exciton in $In_{1-x}Ga_xP$ led to a precise determination of the position of the direct bandgap energy as a function of x for this alloy, resolving a long controversy over the crystal composition at which the bandgap changed from direct to indirect (Bachrach 1971, Onton 1971a, Stringfellow 1972, Alibert 1972, Macksey 1973, Merle 1974). A similar situation was resolved by Dingle's work on

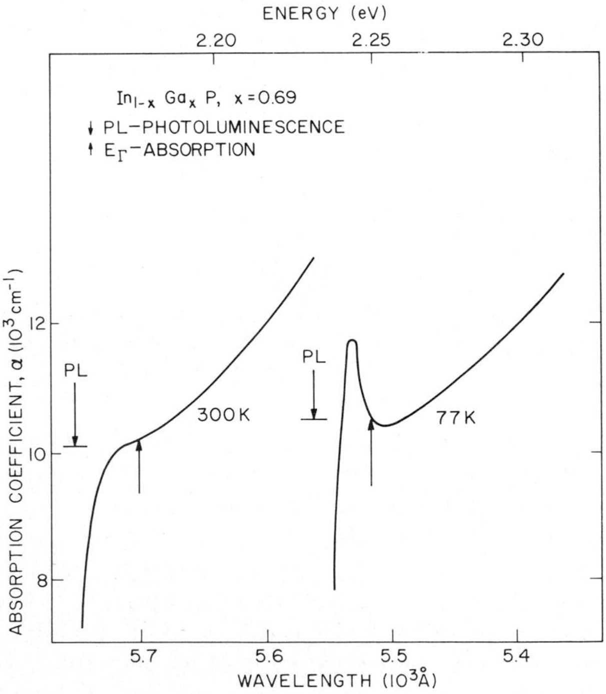

Fig. 2. Absorption spectra of a direct bandgap sample of $In_{1-x}Ga_xP$, $x = 0.69$ at 77 K and 300 K. The direct exciton peak is clearly observed in the 77 K data. (After Nelson 1976b.)

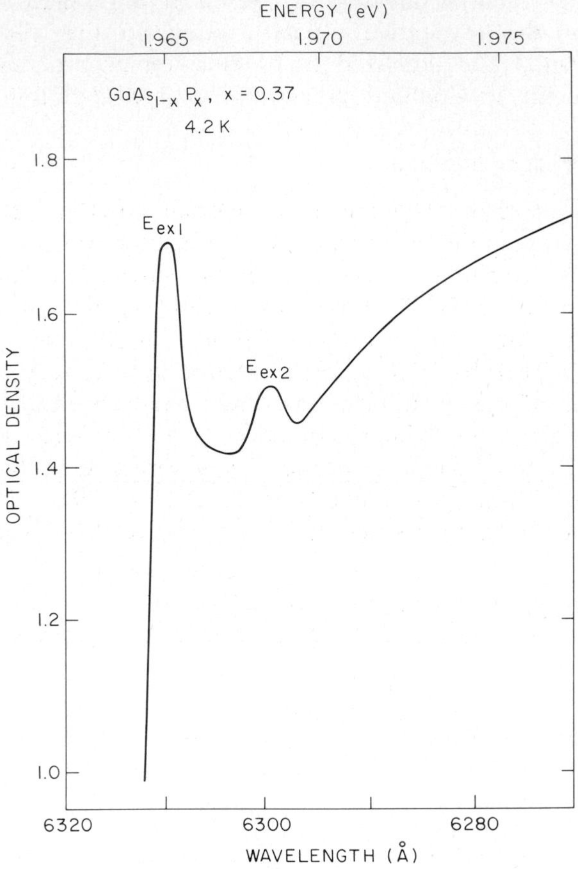

Fig. 3. Absorption spectra of a direct bandgap $GaAs_{1-x}P_x$ ($x = 0.37$) sample at 4.2 K. The absorption peaks associated with the $n = 1$ and $n = 2$ states of the exciton are observed. (After Nelson 1976a.)

$Al_{1-x}Ga_xAs$. As described in the next section, the precise determination of the direct bandgap also led to a re-evaluation of the interpretation of the near bandgap luminescence processes in nitrogen-doped alloys. Subsequently both the ground state ($n = 1$) and the first excited state ($n = 2$) of the free exciton were observed with peak widths less than 1 meV (see fig. 3) for high purity $GaAs_{0.63}P_{0.37}$ at 4.2 K (Nelson 1976a). The narrow width of the exciton transitions in fig. 3 is a result of the large Bohr radius of the direct-exciton in III–V materials. Typically the Bohr radius is approximately ten times the dimension of a unit cell. Potential fluctuations are sufficiently averaged to prevent significant broadening. Both the energy gap and the exciton effective mass can be obtained from a fit to the observed absorption spectra using Elliott's expression (Elliott 1957) for the absorption coefficient. The values of the exciton effective mass thus obtained for direct-gap alloys are in good agreement with the values calculated from the hydrogenic model (Nelson 1976a, b).

2.2. Indirect bandgap material

The band structure of indirect-gap semiconductors is such that the absolute conduction band minima and the absolute valence band maximum do not occur at the same point in the Brillouin zone. Transitions between the bands are allowed only in second order and occur via intermediate virtual states. In a perfect crystal without impurities, these transitions require the participation of a phonon, either in emission or absorption. The absence of periodicity in semiconductor alloys, however, allows indirect optical transitions (including excitons) without phonon involvement (often called quasidirect transitions). An interesting contrast between direct and indirect free excitons is the apparent effect of alloy broadening on the transitions involving these states. As mentioned previously, direct excitons have been observed with narrow line widths in alloys. Fluctuation broadening on the other hand is believed to be responsible for the experimentally observed half-width (~20 meV) of the free exciton luminescence band in indirect gap $GaAs_{0.12}P_{0.88}$ (Craford 1972). The radius of indirect excitons is considerably smaller than for direct excitons because of the difference in direct and indirect conduction band effective masses in III–V materials. The indirect exciton is more localized and may therefore be more sensitive to alloy fluctuations.

A theoretical treatment of band edge smearing caused by composition fluctuations has been given (Baranovskii 1978). According to this calculation the exciton broadening is given by

$$E_0 = \frac{1}{178}\frac{\alpha^4 M^3 x^2 (1-x)^2}{\hbar^6 N^2},$$

where $\alpha \equiv dE/dx$, the rate of change of bandgap with composition, M is the exciton mass, x is the composition and N is the number of atom sites per unit volume. For typical III–V alloys, α is larger for the direct bandgap region than for the indirect gap region. However, the indirect exciton mass is larger than that of the direct exciton. As a result the predicted alloy broadening is typically somewhat smaller for indirect excitons than for direct excitons. The predicted broadening for the direct exciton in $GaAs_{1-x}P_x$ is 0.01 meV which is less than the observed width shown in fig. 3 (strain or impurity effects may be responsible for the observed width). For indirect gap $GaAs_{1-x}P_x$ the calculated broadening is $\sim 3 \times 10^{-3}$ meV which is considerably less than the reported 20 meV line width of the indirect exciton in this material (Craford 1972). It is often found, however, that the low temperature luminescence width is greater than the width observed in the exciton absorption spectrum even for no-phonon transitions. This is probably because of thermalization into tail states. Broadening of exciton states in $Zn_xCd_{1-x}Te$, $Zn_xCd_{1-x}S$, and $CdS_{1-x}Se_x$ has also been examined (Suslina 1978, Goede 1979) with good agreement found with the calculated width given above.

Absorption due to quasidirect transitions was first observed in $GaAs_{1-x}P_x$ (Dean 1969). Further work (Onton 1971b, Pikhtin 1972, 1973) considered the effect of band structure changes and alloy scattering on the strength of the zero-phonon transitions. Calculations suggest that the absorption coefficient α should show a square root dependence on energy and should be directly proportional to the concentration of scattering sites. An inverse dependence on the square of the difference (Δ) between the direct–indirect conduction bandgaps is also expected. Thus, $\alpha(\hbar) = A_0x(1-x)(\hbar - E_{gx})^{1/2}/\Delta^2$, where A_0 is nearly a constant, x is the crystal composition and E_{gx} is the exciton band gap. Good agreement with experiment is found for $GaAs_{1-x}P_x$ (Pikhtin 1973) although for large values of x it is found that $\alpha \propto (\hbar\omega - E_{gx})^{1/4}$.

At high excitation levels an electron–hole plasma has been observed in indirect $Ga_{1-x}Al_xAs$ at low temperatures (Cohen 1980). The observed luminescence is associated with no-phonon transitions as a result of scattering by potential fluctuations as well as weaker phonon assisted transitions. In contrast, electron–hole plasmas observed in other (binary or elemental) indirect semiconductors show only the phonon-assisted transitions. A phase separation is observed between the plasma and the free exciton in $Ga_{1-x}Al_xAs$.

3. Bound excitons

Excitons bound to impurities are represented by localized states formed from the allowed states in a perfect crystal. Hence, the changes in the band structure of semiconductor alloys result in changes in the characteristics of the impurity states as described by the virtual crystal approximation. In addition, the random distribution of the component atoms should have an effect on these states. The degree of localization of the exciton determines in large part the extent of this alloy disorder effect. Shallow impurity states are characterized by large Bohr radii and are adequately described in the effective mass approximation. Deep centers with non-Coulomb potentials produce a strong localization of the exciton wave function in the central cell. Alloy disorder may result in significant broadening of transitions associated with these states. In addition, random potential fluctuations may bind excitons at low temperatures.

3.1. Excitons bound to Coulomb impurities and to composition fluctuations

A study of the near bandedge luminescence and the symmetry of the phonon coupling in indirect $Ga_{1-x}Al_xAs$ has suggested that the highest energy transition shown in fig. 4 is associated with the decay of shallow bound excitons associated with group V site donors (Dingle 1979). In this study it was pointed out that alloy disorder scattering could possibly lead to phonon coupling similar to that observed. Recent studies of $GaAs_{1-x}P_x$ have indicated that a high energy transition observed only at low excitation levels could be associated with excitons bound to fluctuations in the alloy potential (Lai 1980). The photoluminescence peak (M_0^x) reported by Lai and Klein is observed approximately 20 meV above the exciton bound to a neutral donor and approximately 10 meV below the free exciton peak. No known extrinsic bound exciton is believed to have the characteristics of this level. It is believed that the M_0^x peak is associated with an effective mass level owing to the strength of the LA^x phonon sideband which is not observed for a deep center like the N-isoelectronic trap. The observed luminescence decay time for M_0^x is 10^2–10^3 times longer than that typically observed for excitons bound to neutral donor or acceptors, which, being three particle systems, decay rapidly by Auger recombination. Excitons bound to charged donors or acceptors are not expected to occur in III–V compounds on theoretical grounds. Hence, the suggestion has been made that this transition is due to excitons bound to fluctuations in the potential caused by local fluctuations in the composition. Further work is required to completely rule out impurity or strain effects.

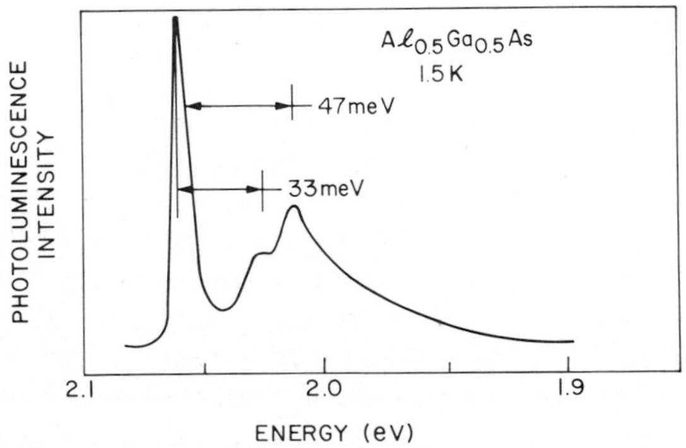

Fig. 4. Photoluminescence spectra of an indirect bandgap sample of $Al_{0.5}Ga_{0.5}As$ at 1.5 K (After Dingle 1979).

3.2. Excitons bound to isoelectronic impurities

The discovery in 1964 that substitutional isoelectronic impurities could produce bound exciton states in semiconductors (Aten 1964, Thomas 1965) has generated a great interest in the properties of these isoelectronic traps for both scientific and technological reasons. The best known example of this trap is nitrogen in GaP and the alloys $In_{1-x}Ga_xP$ and $GaAs_{1-x}P_x$. Now more than 15 years after its discovery in GaP:N, experimental evidence of the nature of the N trap is still being accumulated for both GaP and III–V alloys. Properties of nitrogen-doped alloys were first reported in 1969 (Dean 1969). For several years it was believed that the behavior of the N trap in alloys was essentially similar to that of GaP:N. Luminescence transitions attributed to both isolated N atoms (A line, see fig. 5) as well as ith nearest neighbor pairs of N atoms (called NN_i pairs, see fig. 5) were thought to occur in both the binary (GaP) and the alloys. The binding energy of the exciton bound to the single N impurity was believed to be only slightly larger in the alloys than in the binary. Alloy disorder was believed to cause significant broadening, (>50 meV) of the NN pair states. A review of the early literature on N-doped alloys was given by Craford and Holonyak (Craford 1976). In 1975, however, it was found that the N trap in $In_{1-x}Ga_xP$ does not exhibit the NN pair luminescence transitions (Nelson 1976b). The broad luminescence band previously attributed to alloy broadened NN pairs was shown to be associated with a phonon side band of a single excitonic energy level bound to a single N atom. Both the binding energy and the phonon coupling were found to increase sharply in the alloy. This work led to a renewed interest in both the experimental and theoretical aspects of nitrogen-doped alloys. Subsequent work showed that a similar interpretation was appropriate also for $GaAs_{1-x}P_x$:N (Nelson 1976c, Wolford 1976a). It was also shown that an additional exciton level associated with the direct conduction band edge is bound to the N trap over a small composition range near the direct–indirect transition (Nelson

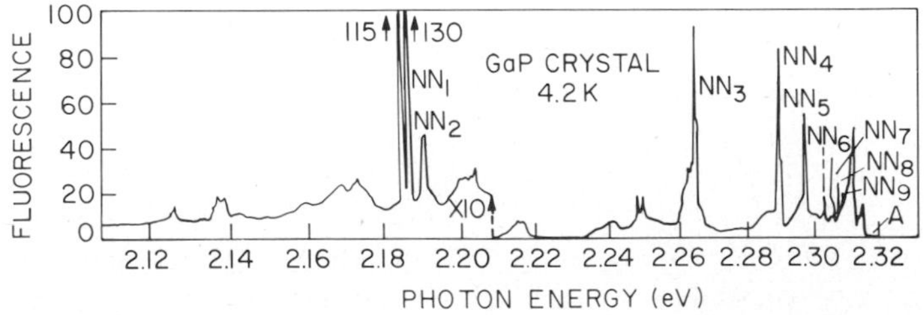

Fig. 5. Photoluminescence spectra of heavily doped ($N \sim 2 \times 10^{18}\,cm^{-3}$) GaP:N showing the transitions associated with single N atoms (A line) and pairs of N atoms (NN_i). (After Thomas 1967.)

1976c, 1976d, Wolford 1976b). While our understanding of the N trap continues to evolve, the technological importance of $GaAs_{1-x}P_x$:N and GaP:N for efficient light emitting diodes is well known (Craford 1976).

Theoretical models of the nitrogen isoelectronic trap differ in their interpretation of the importance of the difference in the covalent radii and the electronegativities of the impurity atom (N) and the host atom (P in GaP) it replaces. In one approach (Faulkner 1968) the fact that N is electronegative relative to P is believed to provide an attractive short-range potential for electrons. An electron can be bound if this potential is sufficiently strong. The Coulomb potential of a bound electron can then bind a hole. An electropositive impurity (e.g. Bi) would bind a hole first according to this model. Allen's theory (Allen 1968, 1971) relies instead on the strain fields which arise from the size difference between N and P to explain the binding of an exciton in the absence of bound single-particle states. It is difficult to distinguish between these models owing to the fact that in neither case do the calculated binding energies show good agreement with experiment. In addition, the electron bound state for isolated N atoms predicted by the Faulkner model has not been experimentally observed in GaP:N although it has been observed for NN_i pairs (Cohen, 1976).

The importance of changes in band structure in indirect-gap semiconductor alloys is illustrated by the "band structure enhancement" of the radiative recombination probability (without phonon participation) of an exciton bound to the nitrogen isoelectronic trap in $GaAs_xP_{1-x}$. The intensity of this transition is proportional to the square of the amplitude of the electron wave function $\psi(\boldsymbol{k})$ near $\boldsymbol{k} = 0$ (see fig. 6). Since a reduction in the separation between the impurity level and the direct minimum produces an enhanced contribution to the $\boldsymbol{k} = 0$ component of the impurity wave function, an increase in the radiative recombination probability should be observed for such a reduction. Calculations (see fig. 7) and observations of this effect have been reported by Craford (1972) and other workers (Campbell 1974, Chevallier 1976, Nelson 1976e).

Figure 8 shows the energies of observed nitrogen related transitions on a bandgap vs. composition diagram of $In_{1-x}Ga_xP$. The transitions along the line labeled N_X were originally believed to be associated with NN pairs broadened as a result of local alloy fluctuations. The transitions labeled N_Γ were believed to be associated with single N atoms (i.e., the A line as in GaP:N). The first indication that the luminescence transitions in N-doped alloys could not be simply interpreted in terms of the A line and alloy broadened NN pairs came when Garbuzov and coworkers (Garbuzov 1975) examined the luminescence spectra of $In_{1-x}Ga_xP$:N for large x. They discovered a spectral peak labeled A_0 in fig. 9 which they attributed to transitions at single N impurity sites which have a different local potential

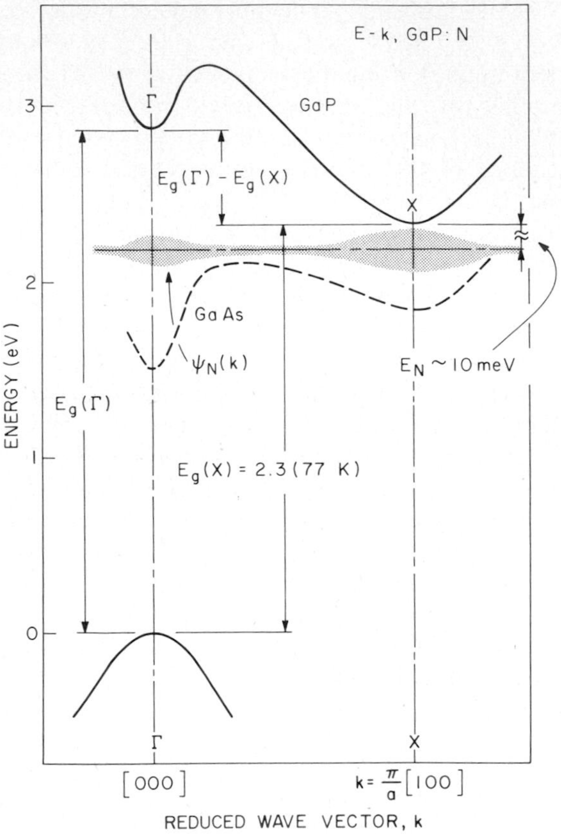

Fig. 6. *E–k* diagram of GaP:N with the conduction band of GaAs shown dotted for reference. The bound state of the nitrogen isoelectronic trap lies ~10 meV below the conduction band minima at X. As the mole fraction of As is increased in semiconductor alloys such as $GaAs_{1-x}P_x$ and the Γ minimum approaches the energy of the N bound state in $GaAs_{1-x}P_x$:N, the modulus of the bound state wave function at $\boldsymbol{k} = 0(|\psi_N(\boldsymbol{k} = 0)|^2)$ increases. (After Holonyak 1973.)

owing to the presence of In atoms. This peak is observed in addition to the A line transition associated with recombination at unperturbed N sites. Figure 9 also shows the development of the broad N luminescence band for increasing In content in the alloy.

A study of $In_{1-x}Ga_xP$:N over a wide composition range led to the first suggestion that the broad luminescence band previously attributed to NN pairs in III–V alloy semiconductors is actually associated with a single N impurity level (Nelson 1976b). These N-trap transitions were shown to be strongly coupled to the lattice with associated phonon sidebands and a

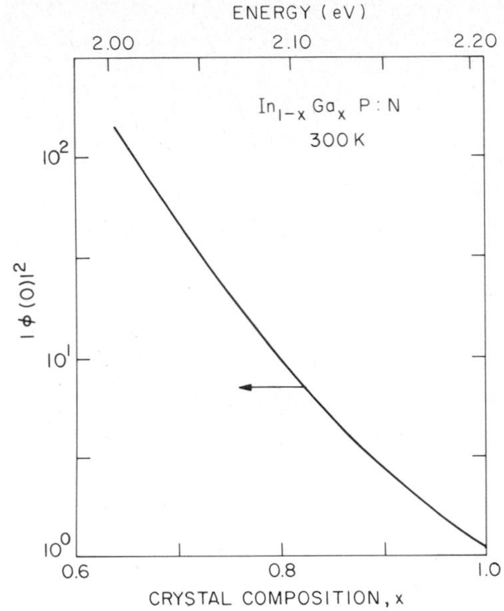

Fig. 7. Modulus of the wave function $|\phi(\mathbf{k}=0)|^2$ of the trapped electron of $In_{1-x}Ga_xP$:N as a function of crystal composition. The increase in $|\phi(\mathbf{k}=0)|^2$ for decreasing x results in an increased radiative recombination rate. (After Nelson 1976e.)

Stokes shift between absorption and emission. One major difficulty with earlier work on nitrogen-doped alloys was that samples of sufficient purity could not be grown for complete identification of optical transitions in undoped and intentionally doped samples. The purity of the samples grown by Nelson and Holonyak was such that, in the direct bandgap region, the exciton peak could be clearly seen in the absorption spectra (see fig. 3). This allowed the clear identification of the N-related luminescence transitions.

Figure 10 shows the luminescence spectra of four $In_{0.17}Ga_{0.83}P$ samples including one undoped and three intentionally doped crystals. In all cases, the excitation level is greater than that necessary to saturate the intensity of the observed transitions. Transitions associated with the Te donor and the N trap are identified. The observation of only one broad transition (labeled N_X) in the N-doped sample conflicts with the early identification (Macksey 1973) of the broad band with NN pairs and a higher transition (A line) with single N atoms. As shown in fig. 9, the higher energy transitions are associated with donor impurities. In the earlier work, however, crystals could not be grown without N doping for comparison owing to the N incorporation from the silica ampoules used in the early growth procedure.

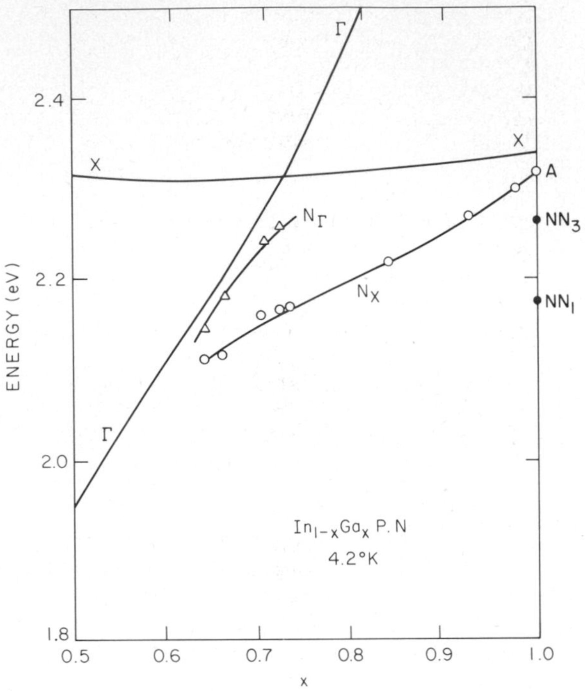

Fig. 8. Energies (4.2 K) of the N_X and N_Γ bound-state luminescence transitions in $In_{1-x}Ga_xP$:N. Energies of the Γ and X bandgaps as well as the A, NN_3 and NN_1 levels in GaP are also shown. (After Nelson 1976g.)

Evidence that the broad N_X band is associated with a single transition rather than NN pair states of various energies is shown in fig. 11 (Nelson 1976a). Spectra (a)–(c) show the luminescence spectra of GaP:N and $x = 0.98$ and 0.94 $In_{1-x}Ga_xP$:N samples respectively. NN pair transitions are clearly seen in the GaP:N sample. The development of the broad N_X band in the alloy is shown in (b) and (c). The luminescence decay time measurements of fig. 11(a) for the GaP:N sample clearly shows three distinct decay times for each of the major transitions (A, NN_3 and NN_1) and their phonon replicas. In contrast, there is only a slight change in decay time for the transitions observed in the $x = 0.98$ and $x = 0.94$ samples. In the GaP:N case the different decay times are indicative of different centers with decay times longer than the minority-carrier lifetime. For the alloy samples, the single decay time suggests that either the minority-carrier lifetime is longer than that of the N-trap levels or that there is only one N-impurity level. The decrease in the observed decay time near the band edge where free carrier emission is dominant at 77 K indicates that the

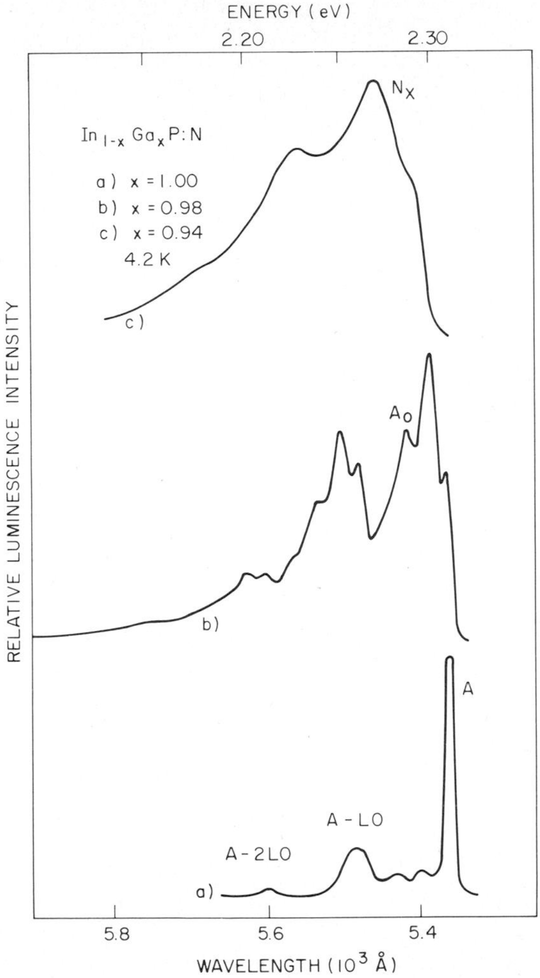

Fig. 9. Photoluminescence spectra of GaP:N, $In_{0.02}Ga_{0.98}P$:N and $In_{0.06}Ga_{0.94}P$:N at 4.2 K for samples grown by liquid phase epitaxy with estimated nitrogen concentrations of approximately 5×10^{16} cm^{-3} (R.J. Nelson, unpublished).

minority-carrier lifetime is shorter than the N-trap decay times. One must therefore conclude that the observed N-trap luminescence in the alloy is associated with a single energy level and its phonon sideband. Furthermore, this broad N-trap luminescence is observed in samples with nitrogen concentrations $<8 \times 10^{16}$ cm^{-3} where the pair concentration would be too low to be observed, confirming that NN pairs are not involved.

If the broad luminescence N_X band is associated with no-phonon tran-

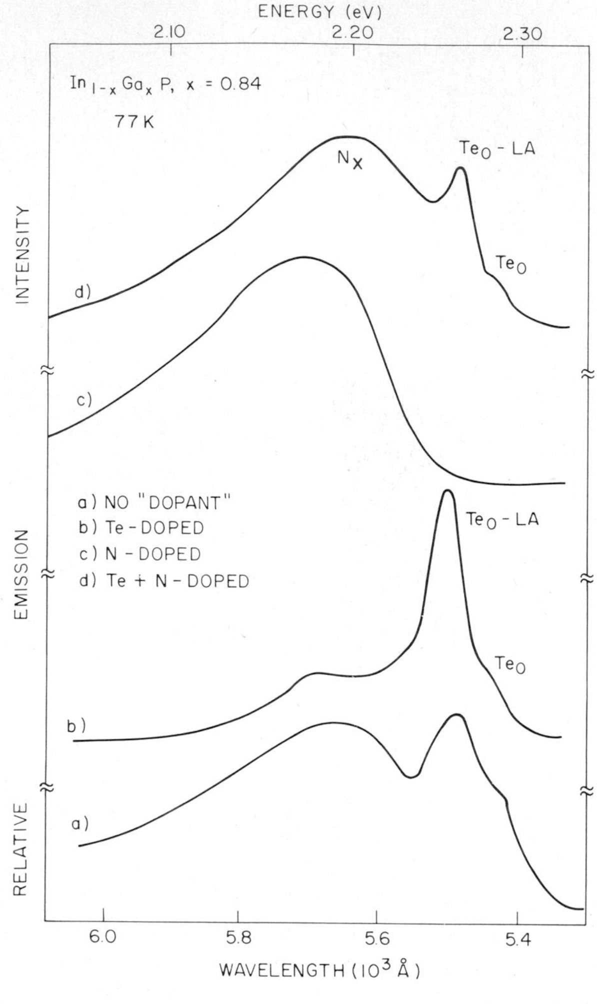

Fig. 10. Photoluminescence spectra of $In_{0.16}Ga_{0.84}P$ grown by LPE with (a) no-intentional dopant, (b) Te doping, (c) nitrogen doping and (d) Te and N doping. (After Nelson 1976b.)

sitions at single N atoms with a phonon sideband then a Stokes shift should be observed. This is indeed the case as shown in fig. 12. The temperature dependence of the N-band width is well described by a calculation (Nelson 1976a) using the standard configuration-coordinate model (Keil 1965, Dishman 1971) for impurities strongly coupled to the lattice. In addition to this Stokes shift, an additional alloying induced shift between no-phonon emission and absorption due to thermalization has recently been reported by Mariette (1979).

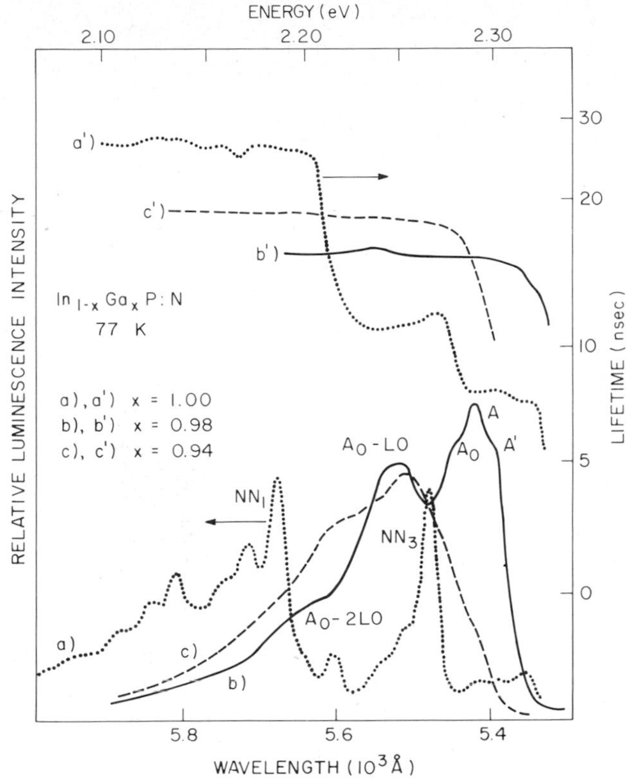

Fig. 11. Photoluminescence spectra and luminescence time decay spectra of $In_{1-x}Ga_xP$:N for $x = 1.0$, 0.98 and 0.94. (After Nelson 1976b.)

The general behavior of the N trap in alloys was qualitatively explained (Nelson 1976b) in the following way. As the In concentration increases, the N impurity sites are perturbed by the local strain fields in the alloy. For $x < 0.95$ most of the N atoms would have at least one In atom on a nearest or next nearest column III lattice site. To estimate the binding energy E' of an exciton at this perturbed N site, the strain interaction model previously used to account for the NN-pair binding energies in GaP:N (Allen 1968) was used. It is assumed that E' is not sensitive to impurity separation, $\boldsymbol{r}$, (for small $\boldsymbol{r}$) but only to the difference in the impurity host radii, δR, since in GaP, $E_{NN_1} \approx E_{NN_2}$. For an In atom on the nearest or next-nearest column III site, the exciton binding was estimated to be $E' \approx (\delta R'/\delta R)E_{NN_1}$ where $\delta R'$ refers to the difference in the In–Ga radii and δR to the N–P difference. The energy thus obtained is 62 meV compared with an observed binding energy of 40 meV for the N exciton at $x = 0.94$. For higher values

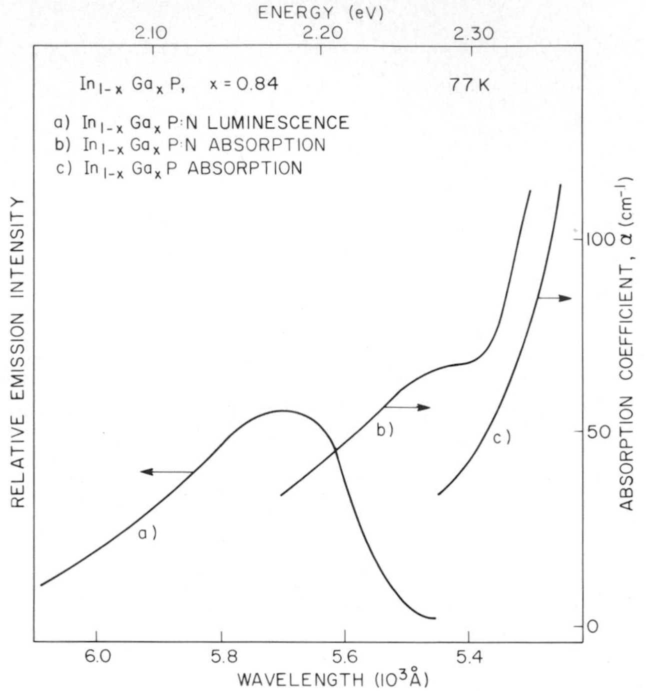

Fig. 12. Luminescence and absorption spectra of $In_{0.16}Ga_{0.84}P$:N demonstrating the Stokes shift between absorption and luminescence. (After Nelson 1976b.)

of x the average In–N separation is larger, and the binding energy is expected to be smaller as observed. For smaller values of x, the binding energy continues to increase as a result of strain, as well as through the influence of other conduction band minima and changes in the host crystal composition. The simple mismatch model described here would not be expected to apply rigorously in this region of lower x values.

As the binding energy of the N trap increases the N-luminescence band is observed to broaden owing to increased phonon coupling. In the direct-bandgap region of x, however, the decreasing energy difference between the conduction band minimum Γ and the N state enhances the no-phonon emission ($\bar{k} = 0$) on the high energy side of the N band producing the form observed in fig. 13 (Nelson 1976b, d).

The discovery that single N impurities not NN pairs are responsible for the broad luminescence band in $In_{1-x}Ga_xP$:N prompted the suggestion (Nelson 1976b) that other III–V alloys including $GaAs_{1-x}P_x$:N should have the same type of luminescence band. This was indeed shown to be the case in later reports of studies on $GaAs_{1-x}P_x$:N (Nelson 1976c, d, Wolford

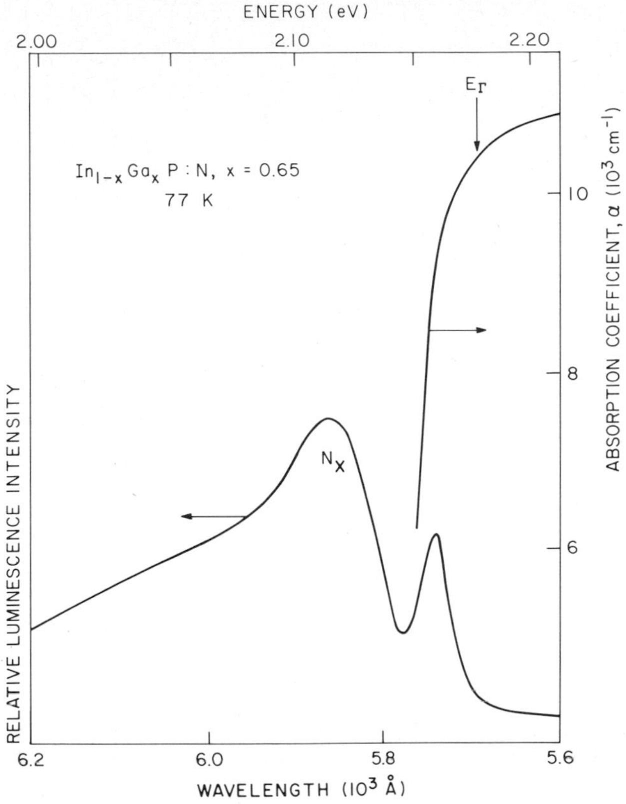

Fig. 13. Luminescence and absorption spectra of $In_{1-x}Ga_xP$:N showing the enhanced no-phonon component of the N_X band in the direct composition region. (After Nelson 1976b.)

1976a). An energy vs. composition diagram is shown in fig. 14 for $GaAs_{1-x}P_x$:N. A Stokes shift between the luminescence and absorption of the N band (labeled N_X in fig. 15) was demonstrated for $GaAs_{1-x}P_x$:N (Nelson 1976c). The luminescence decay time was also found to be constant for the N_X band in $GaAs_{1-x}P_x$:N as shown in fig. 16 (Nelson 1976f). As in $In_{1-x}Ga_xP$:N described earlier, this is evidence that a single transition causes the broad luminescence band, not a continuum of NN pair states. Further substantiation for this view came in ion-implantation work where the dosage of implanted N atoms could be varied over a wide range (Wolford 1976a). Problems associated with implantation damage and subsequent annealing appear to have been minimized in this work. It is difficult however to determine the extent of the influence of non-radiative centers on the results of such a study. Figure 17 shows the photolumines-

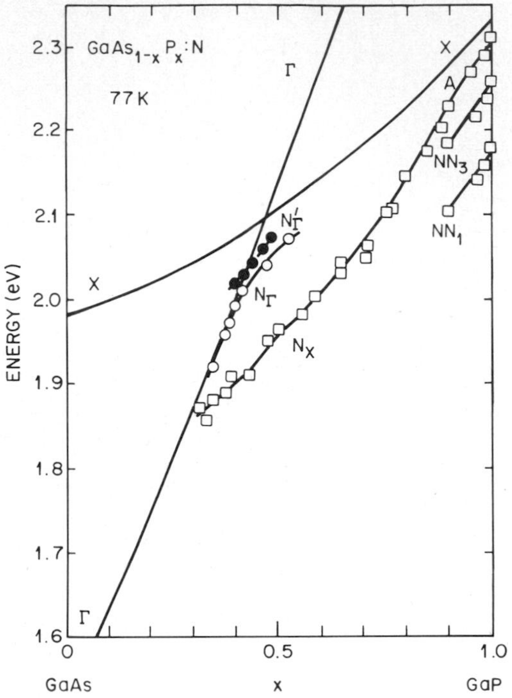

Fig. 14. Energies of the N_X, N_Γ and N_Γ transitions in $GaAs_{1-x}P_x$:N plotted as a function of composition x. Also shown are the Γ and X bandgaps and the positions of the A, NN_3 and NN_1 transitions in GaP.

cence spectra of $GaAs_{0.13}P_{0.87}$ N implanted to the indicated peak concentration and annealed for 30 min. at 950°C. For comparison the spectra of N-free material is also shown. The luminescence spectra observed in the implanted samples for $N > 5 \times 10^{17}\,cm^{-3}$ is observed to be independent of the N concentration supporting the association of the N_X band luminescence with isolated N atoms not pairs. The concentration of NN_1 pairs in the sample with $N = 5 \times 10^{17}\,cm^{-3}$ is approximately $10^{14}\,cm^{-3}$ which is probably too small to be observed in luminescence. The role of implantation damage in these results has not been fully explored.

While the shape of the N-luminescence band observed in $GaAs_xP_{1-x}$:N can be fitted theoretically using strong phonon coupling at single nitrogen sites (Wolford 1979a, b) a similar fit assuming alloy broadened NN pairs has also been demonstrated (Gail 1977). It appears that the luminescence time decay data described above is the most persuasive evidence ruling out alloy broadened NN pair states.

The interpretation of the origin of the broad N_X band in terms of an

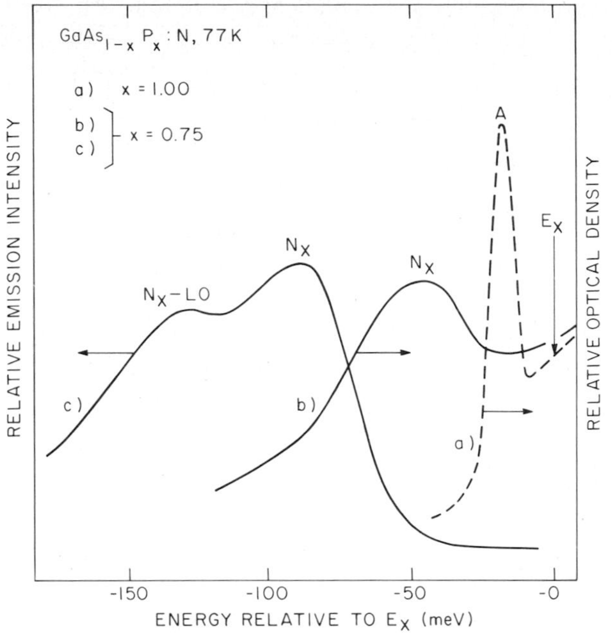

Fig. 15. Luminescence and absorption spectra of $GaAs_{0.25}P_{0.75}$:N demonstrating the Stokes shift between the luminescence and absorption peaks. The narrower absorption peak (A) observed in GaP:N is also shown. (After Nelson 1976c.)

excitonic bound state at a single N atom with strong phonon coupling led to a re-examination of the N related luminescence observed near the band edge in the direct bandgap region. These transitions were originally thought to be due to recombination at single N atoms (A line), NN_3 pairs or above gap resonant states. It is difficult to determine the origin and nature of the observed transitions because of uncertainties in crystal composition, radiative efficiencies, doping levels and nitrogen concentrations. A study of ion-implanted $GaAs_{1-x}P_x$:N in the direct bandgap region indicated that NN_3 pairs were not involved and suggested that the shallow N state was associated with the Γ conduction band minimum (Wolford 1976b). A typical luminescence spectrum in this composition region is shown in fig. 18 which reproduces data from early work (Dupuis 1973) with the transitions labeled as in later work (Nelson 1976c).

High-pressure studies of $GaAs_{1-x}P_x$:N allowed the examination of the effect of continuous changes in the band structure without the uncertainties associated with comparison of crystals of different compositions. Figures 19 and 20 show the intensity decrease of N_Γ and the energy shifts of band edge, N_Γ and N_X emission as a $GaAs_{0.58}P_{0.42}$:N sample is subjected to

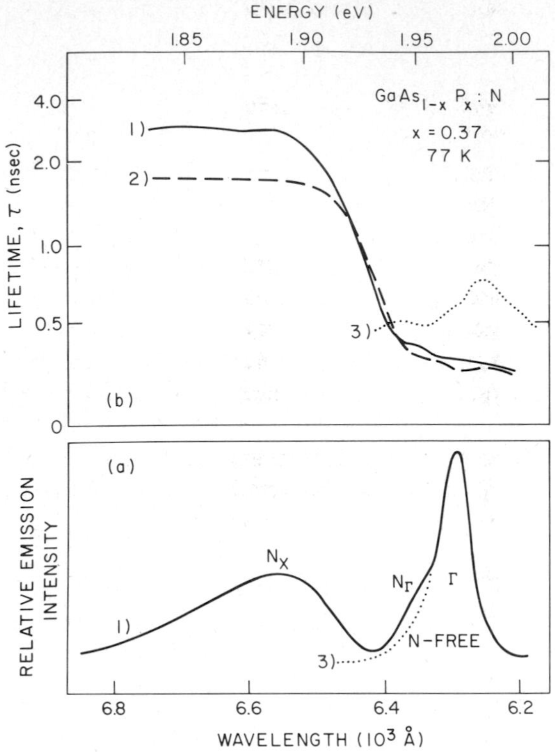

Fig. 16. Luminescence and luminescence time decay spectra of N-doped and N-free $GaAs_{0.43}P_{0.57}$ demonstrating the constant decay time for the broad N_X band. (After Nelson 1976e.)

pressures up to 6.5 kbar. Since the effective composition change with pressure is $dx/dp \sim 1.1\%$/kbar (Nelson 1976c), a composition range of $\sim 7\%$ can be effectively scanned with one sample. The band edge emission at low pressure shows the characteristic increase in energy observed for the Γ conduction band minimum (see fig. 20). At high pressure, the band edge luminescence shows the behavior characteristic of the X minima. The lowest energy N_X transition is observed to be nearly independent of pressure. The higher energy N_Γ state is observed to follow the Γ minimum for $p < 4$ kbar then bends over and follows X for $p \geq 5$ kbar. The inset of fig. 19 shows the pressure variation of the relative intensity of the N_Γ emission. The observed decrease suggests a similar decrease in the radiative transition probability for this state. This appears to explain the fact that the N_Γ transition has not been observed in $GaAs_{1-x}P_x$:N crystals with compositions $x > 0.53$. The pressure data clearly shows that the N_Γ state is

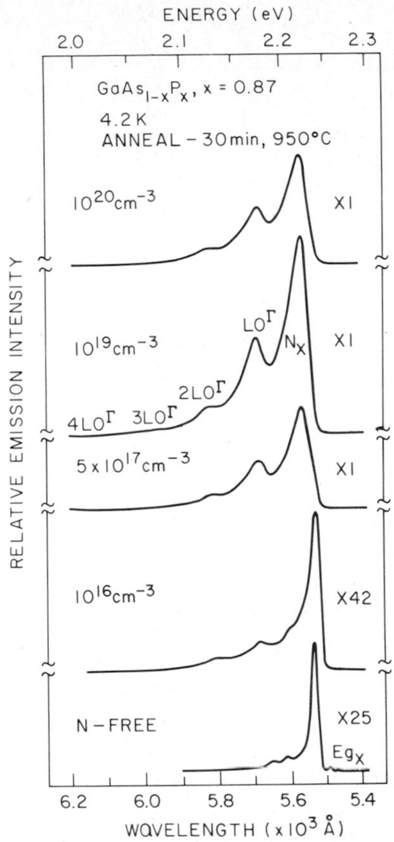

Fig. 17. Photoluminescence spectra of $GaAs_{1-x}P_x$ ($x = 0.87$) N implanted to the peak concentrations indicated and annealed for 30 min at 950°C. (After Wolford 1976a.)

associated with the lowest conduction band minimum, whereas earlier data on samples of different compositions (Dupuis 1973) were subject to conflicting interpretations.

The N_Γ transition has also been observed in $In_{1-x}Ga_xP$:N (Nelson 1976g) as shown in fig. 21. Comparison of the luminescence and absorption spectra clearly shows that the luminescence transition labeled Γ can be attributed to intrinsic recombination and not to an above gap resonant line associated with N (Macksey 1973). The binding energy of N_Γ is nearly the same as in $GaAs_{1-x}P_x$:N. The N_Γ state has not been observed in $Al_{1-x}Ga_xAs$:N (Makita 1977, Wolford 1979a).

Various attempts have been made to theoretically model the N trap in alloy semiconductors. Most of these treatments are only semi-

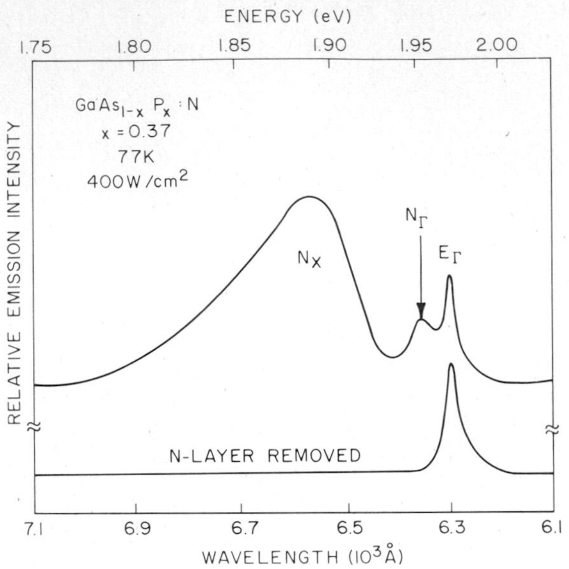

Fig. 18. Photoluminescence spectra of N-doped and N-free $GaAs_{0.37}P_{0.63}$. (After Nelson 1976c.)

phenomenological at best. A severely restricted Koster–Slater calculation has been employed by Kleiman et al. (Kleiman 1976–1978) and Hsu et al. (1977). Kleiman includes a short-range isoelectronic potential, V_s, which induces a state (n, see fig. 22a) delocalized in momentum space, in addition to a long-range disorder and strain-induced potential, V_e, which by itself introduces one state (n_Γ) associated with Γ and one (n_X) associated with X. The eigenstates corresponding to $V_e + V_s$ (denoted by N'_Γ, N_Γ and N_X in order of decreasing energy as shown in fig. 22b) contain admixtures of the n, n_Γ and n_X states. There is strong splitting of the energies in the regions near $X \sim X_{N\Gamma}$, and $X \sim X_{\Gamma X}$ as shown in fig. 22b. A one band-one site Koster–Slater approximation is used for V_s. The long-range potential is described by a spherical square well. Reasonable agreement is found with the experimental variations of the N-trap binding energies.

According to Kleiman's treatment three nitrogen-related bound states are predicted to exist in the region near the direct–indirect crossover. Data has been presented which shows evidence of the possible existence of this state (labeled N'_Γ) $GaAs_{1-x}P_x$:N as shown in fig. 23 (Holonyak 1977). There is some controversy over the existence of this state. Wolford (Wolford 1977) has reported photoluminescence data on an ion-implanted sample of $GaAs_{1-x}P_x$:N at nearly the same composition as that of fig. 23. Only one clearly defined peak was reported by these workers (see fig. 17 in Wolford

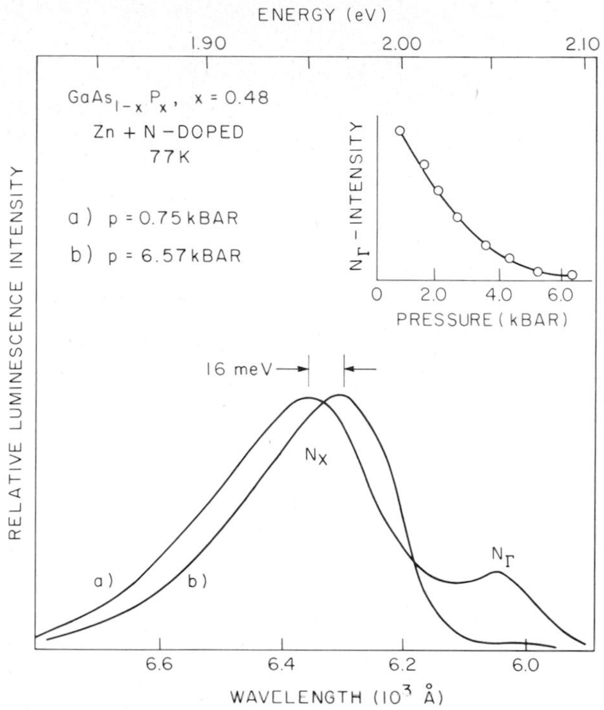

Fig. 19. Variation of the luminescence spectrum of $GaAs_{0.48}P_{0.52}$:N with pressure. The decrease of the intensity of the N_Γ transition with pressure is shown in the inset. (After Nelson 1976d.)

1977) in the vicinity of the N'_Γ and Γ transitions shown in fig. 23. Further studies may be required to resolve this issue, although it is tempting at this point to have a higher level of confidence in the data of Holonyak (fig. 23) which is taken on high-quality vapor-phase-epitaxy $GaAs_{1-x}P_x$:N rather than on $GaAs_{1-x}P_x$ samples which have been ion bombarded with N and subsequently annealed. A considerable amount of lattice damage and a high concentration of non-radiative centers may interfere with the unambiguous interpretation of data on these implanted samples*.

* Thermal line broadening is probably not responsible for the different spectra obtained by these two groups as measurements were made at 4.2 K in each case. It should be noted, however, that the line widths of the Γ and N_Γ transitions reported by Holonyak are smaller than that observed in the implanted samples of Wolford. The samples used by Holonyak are sufficiently thin that cavity modes could not cause spectral variations at the wavelength separation of the Γ and N'_Γ peaks of fig. 23. In this regard it may be useful to point out that the data of Wolford are taken on "bulk samples" consisting of a GaP or GaAs substrate, a graded composition (and lattice constant) region and finally a $GaAs_{1-x}P_x$ layer. This type of sample configuration is likely to be highly strained. The samples used by Holonyak et al. are typically ~1 μm thick with constant composition.

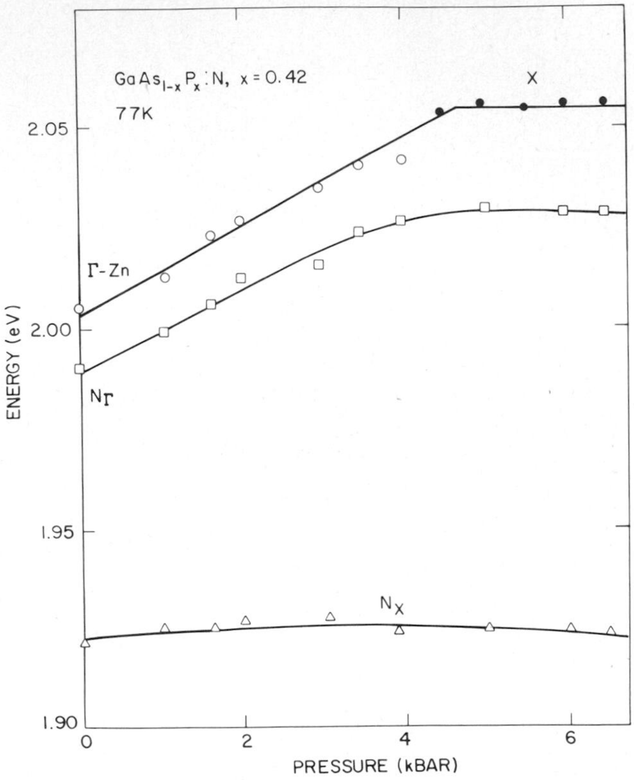

Fig. 20. Variation of the energy of the N_X, N_Γ and Γ–Zn transitions in $GaAs_{0.42}P_{0.48}$:N with pressure showing that the N_X level is independent of the Γ–N_X separation. The N_Γ level is observed to follow the Γ minima. (After Nelson 1976d.)

Another theoretical view point has been presented by Hsu (1977). This work is also a semi-phenomenological model which treats the nitrogen trap using a Koster–Slater calculation which includes an extended potential that affects only those electrons contacting the nitrogen impurity cell and its 12 nearest-neighbor primitive cells. This potential tends to overestimate the short-range parts of the deformation potential. Good agreement has been found with experiment for the N_X and N_Γ states using four adjustable parameters. This theory predicts only two bound nitrogen states and therefore does not treat the observed evolution of N_Γ into $N_{\Gamma-X}$ near $x = 0.5$.

Jaros and Ross (Jaros 1979) have treated the nitrogen impurity using local pseudo-potentials in a multiband calculation employing the virtual crystal approximation. The virtual crystal model neglects the role of alloy

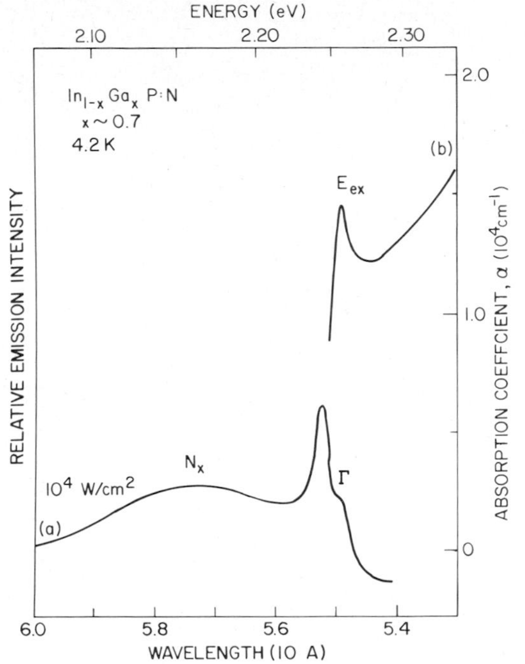

Fig. 21. Photoluminescence and absorption spectra (4.2 K) of N-doped $In_{1-x}Ga_xP$, $x \approx 0.70$. Curve (a) shows the two nitrogen bound state transitions N_X and N_Γ along with the intrinsic recombination Γ. The free-exciton absorption peak, typical of direct-gap semiconductors of low impurity concentration, is shown in curve (b) and confirms the Γ peak assignment of curve (a). (After Nelson 1976g.)

disorder. Local lattice distortion is included although the magnitude of this distortion is unknown. Only one bound state is predicted using this model. This calculation could be altered by a strong change in the local symmetry possibly as a result of lattice relaxation. It is clear that proper theoretical treatment of the nitrogen problem in III–V alloys is a formidable task with much work to be done.

4. *Conclusion*

Exciton transitions in semiconductor alloys have been the subject of considerable study, particularly in III–V materials. Direct free-exciton transitions, which have been observed in high-purity crystals, have not been found to be significantly perturbed by alloy disorder. Bound excitons, particularly those associated with the nitrogen isoelectronic trap have been

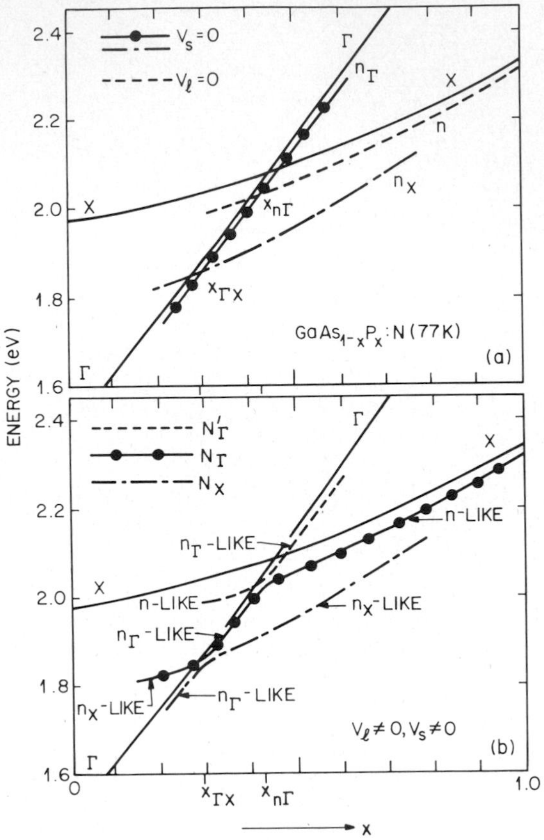

Fig. 22. (a) Schematic illustration of the composition dependence of bound-state energies produced by the long-range (V_ℓ) and short-range (V_s) nitrogen potentials before they are combined. (b) Corresponding diagram for bound states produced by combining V_ℓ and V_s. (After Kleiman 1976.)

found to be strongly affected by disorder and band structure change. New theories of this interesting impurity state are likely to appear in the future which should add to our understanding of isoelectronic traps and semiconductor alloys.

Acknowledgement

The author would like to thank Prof. N. Holonyak, Jr., for his important contributions and discussions relating to nitrogen-doped semiconductor alloys.

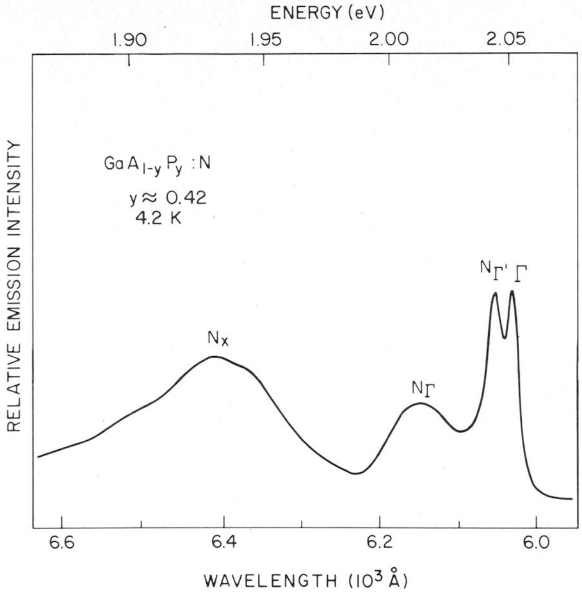

Fig. 23. Photoluminescence spectra obtained at 4.2 K on a $GaAs_{0.42}P_{0.58}$:N sample. At low excitation the N_X, N_Γ, N_Γ' and Γ emission peaks are evident. (After Holonyak 1977.)

References

Alibert, C., G. Bordure, A. Laugier and J. Chevallier, 1972, Phys. Rev. B6, 1301.

Allen, J.W., 1968, J. Phys. C. **1**, 1136.

Allen, J.W., 1971, J. Phys. C: Solid State Phys., **4**, 1936.

Aten, A.C., J.H. Haanstra and H. de Vries, 1965, Phillips Res. Rept. **20**, 395.

Bachrach, R.Z. and B.W. Hakki, 1971, J. Appl. Phys. **42**, 5102.

Baranovskii, S.D. and A.L. Efros, 1978. Sov. Phys.-Semicond. **12**, 1328.

Bonch-Bruevich, V.L., 1966, Semiconductors and Semimetals I, eds., R.K. Willardson and A.C. Beer (Academic, New York) p. 101.

Campbell, J.C., N. Holonyak Jr., M.G. Craford and D.L. Keune, 1974, J. Appl. Phys. **45**, 4543.

Casey, H.C., Jr and M.B. Panish, 1978, Heterostructure Lasers (Academic Press, New York).

Chevallier, J., H. Mariette, D. Diguet and G. Poiblaud, 1976, Appl. Phys. Lett. **28**, 1976.

Cohen, E. and M.D. Sturge, 1977, Phys. Rev. **B15**, 1039.

Cohen, E., M.D. Sturge, M.L. Olmstead and R.A. Logan, 1980, Phys. Rev. **B22**, 771.

Craford, M.G. and N. Holonyak, Jr., 1976, The Optical Properties of the Nitrogen Isoelectronic Trap in $GaAs_{1-x}P_x$, in: Optical Properties of Solids: New Developments, ed., B.O. Seraphin (North-Holland, Amsterdam) 187.

Craford, M.G., R.W. Shaw, A.H. Herzog and W.O. Groves, 1972, J. Appl. Phys. **43**, 4075.

Dean, P.J., G. Kaminsky and R.B. Zetterstrom, 1969, Phys. Rev. **181**, 1149.

Dingle, R., R.A. Logan and J.R. Arthur, Jr., 1977, Gallium Arsenide and Related Compounds (Institute of Physics Conf. Series **33a**, London) 210.

Dingle, R., R.A. Logan and R.J. Nelson, 1979, Solid State Commun. **29**, 171.

Dishman, J.M. and M. DiDomenico, Jr., 1971, Phys. Rev. **B4**, 2621.

Dupuis, R.D., N. Holonyak Jr., M.G. Craford, D. Finn and W.O. Groves, 1973, Solid State Commun. **12**, 489.
Elliott, R.J., 1957, Phys. Rev. **108**, 1384.
Faulkner, R.A., 1968, Phys. Rev. **175**, 991.
Gail, M., T. Görög and A. Keresztury, 1977, Solid State Commun. **21**, 491.
Garbuzov, D.Z., S.G. Konnikov, P.S. Kopier and V.A. Mishurnyi, 1975, Sov. Phys.-Semicond. **8**, 998.
Goede, O., D. Henning and L. John, 1979, Phys. Status Solidi **B96**, 671.
Holonyak, N. Jr., J.C. Campbell, M.H. Lee, J.T. Verdeyen, W.L. Johnson, M.G. Craford and D. Finn, 1973, J. Appl. Phys. **44**, 5517.
Holonyak, N. Jr., R.J. Nelson, J.J. Coleman, P.D. Wright, D. Finn, W.O. Groves and D.L. Keune, 1977, J. Appl. Phys. **48**, 1963.
Hsu, W.Y., J.D. Dow, D.J. Wolford and B.G. Streetman, 1977, Phys. Rev. **B16**, 1597.
Jaros, M. and S. Brand, 1979, J. Phys. C: Solid State Phys. **12**, 525.
Keil, T.H., 1965, Phys. Rev. **140**, A601.
Kleiman, G.G., 1977, Phys. Rev. **B15**, 802.
Kleiman, G.G. and M.F. Decker, 1978, Phys. Rev. **B17**, 924.
Kleiman, G.G., R.J. Nelson, N. Holonyak Jr. and J.J. Coleman, 1976, Phys. Rev. Lett. **37**, 375.
Lai, S. and M.V. Klein, 1980, Phys. Rev. Lett. **44**, 1087.
Macksey, H., N. Holonyak Jr., R.D. Dupris, J.C. Campbell and G.W. Zack, 1973, J. Appl. Phys. **44**, 1333.
Makita, Y. and S. Gonda, 1977, J. Appl. Phys. **48**, 1628.
Mariette, H. and J. Chevallier, 1979, Solid State Commun. **29**, 263.
Merle, P., D. Auvergne, H. Mathieu and J. Chevallier, 1977, Phys. Rev. **B15**, 2032.
Nelson, R.J. and N. Holonyak, Jr., 1976b, J. Phys. Chem. Solids **37**, 629.
Nelson, R.J. and N. Holonyak Jr., 1976e, J. Appl. Phys. **47**, 1704.
Nelson, R.J. and N. Holonyak, Jr., 1976g, Solid State Commun. **20**, 549.
Nelson, R.J., N. Holonyak Jr. and W.O. Groves, 1976a, Phys. Rev. **B13**, 5415.
Nelson, R.J., N. Holonyak Jr., J.J. Coleman, D. Lazarus, W.O. Groves, D.L. Keune, M.G. Craford, D.J. Wolford and B.G. Streetman, 1976c, Phys. Rev. **14**, 685.
Nelson, R.J., N. Holonyak Jr., J.J. Coleman, D. Lazarus, D.L. Keune, W.O. Groves and M. G. Craford, 1976d, Phys. Rev. **14**, 3511.
Nelson, R.J., N. Holonyak Jr., W.O. Groves and D.C. Keune, 1976f, J. Appl. Phys. **47**, 3625.
Onton, A., 1971b, Phys. Rev. **B4**, 4449.
Onton, A., M.R. Lorenz and W. Reuter, 1971a, J. Appl. Phys. **42**, 3420.
Pikhtin, A.N., 1977, Sov. Phys.-Semicond. **11**, 245.
Pikhtin, A.N., V.N. Razbegaer and D.A. Yaskov, 1972, Phys. Stat. Sol. (b) **50**, 717.
Pikhtin, A.N., V.N. Razbegaer and D.A. Yaskov, 1973, Sov. Phys. Semicond. **7**, 337.
Stringfellow, G.B., P.F. Lindquist and R.A. Burmeister, 1972, J. Electron. Mater. **1**, 437.
Suslina, L.G., A.G. Plyukhin, D.C. Fedoror and A.G. Areshkin, 1978, Sov. Phys. Semicond. **12**, 1331.
Thomas, D.G., J.J. Hopfield and C.J. Frosch, 1965, Phys. Rev. Lett. **15**, 857.
Wolford, D.J., B.G. Streetman, R.J. Nelson and N. Holonyak, Jr., 1976a, Solid State Commun. **19**, 741.
Wolford, D.J., B.G. Streetman, W.Y. Hsu, J.D. Dow, R.J. Nelson and N. Holonyak, Jr., 1976b, Phys. Rev. Lett. **36**, 1400.
Wolford, D.J., R.E. Anderson and B.G. Streetman, 1977, J. Appl. Phys. **48**, 2442.
Wolford, D.G., W.Y. Hsu, J.D. Dow and B.G. Streetman, 1979a, J. Luminescence **18/19**, 863.
Wolford, D.G., B.G. Streetman, S. Lai and M.V. Klein, 1979b, Solid State Commun. **32**, 51.

CHAPTER 9

Free Many Particle Electron–Hole Complexes in an Indirect Gap Semiconductor

V.B. TIMOFEEV

Institute of Solid State Physics
Academy of Science of the USSR
142432 Chernogolovka
USSR

Dedicated to the Memory of
Professor Vladimir L. Broude

Excitons
Edited by
E.I. Rashba and M.D. Sturge

Contents

1. Introduction

At low temperatures free electrons and holes, regardless how they are created in a semiconductor, are bound into excitons by Coulomb forces. In the limit of low nonequilibrium density n, when the dimensionless parameter r_s satisfies the inequality $r_s \equiv (3/4\pi n)^{1/3}/a_{ex} \gg 1$ (where a_{ex} is the Bohr radius of the exciton orbit) the exciton gas may be considered as a weakly interacting Boltzmann gas. By analogy with "polyelectronic" complexes first considered by Wheeler (1946), Lampert (1958) suggested that in a nonequilibrium electron–hole gas in a semiconductor free multiparticle complexes more complicated than the exciton can exist, namely, neutral excitonic molecules (bound states of two excitons, often called biexcitons) and excitonic ions (bound states of an exciton with an electron or a hole). The possibility of the existence of biexcitons in semiconductors was independently pointed out by Moskalenko (1958). The stability of such free complexes was later confirmed by reliable variational calculations.

Since the first work by Lampert many attempts have been made to discover these free multiparticle complexes, in particular the biexcitons, by spectroscopic means. The progress in experimental investigation of this problem is most evident during the last few years both in direct gap semiconductors (see, for example, the review article by Hanamura and Haug (1977) and by Grun in this volume) and in indirect gap semiconductors.

In contrast to ordinary gases the electron–hole (e–h) gas in a semiconductor is in principle a nonequilibrium system due to recombination processes. For example, in direct gap semiconductors the electron–hole recombination times are extremely short (usually of the order of nanoseconds). As a result the formation and decay of complexes takes place under highly nonequilibrium conditions which means the absence of equilibrium both with respect to the lattice temperature and between the components of the e–h system itself. A number of experimental methods have been proposed to study the energy spectrum of biexcitons which have proved to be effective in such nonequilibrium conditions. From practical point of view the most attractive of these methods are the resonant two-photon excitation of biexcitons (Hanamura 1973) and the stimulated one-photon transformation of an exciton into a biexciton (Gogolin and Rashba 1973). The effectiveness of these processes in direct gap semiconductors is due to extremely large oscillator strengths of corresponding optical transitions. To the same class of phenomena belongs the two-photon resonant (with respect to the biexciton state) Raman scattering (Ueta and Nagasawa 1975). The application of these methods to indirect gap semiconductors is less effective because of very small probabilities of optical transitions. On the other hand, indirect gap semiconductors have much longer recombination times (e.g. in Ge and Si these times are of the order of microseconds). Therefore one can expect that the chemical reactions of binding into complexes and the decay of the latter will occur under quasi-equilibrium conditions at low temperatures even at relatively small e–h gas densities. Under such conditions important information about the properties of free complexes can be obtained immediately from their emission spectra.

Because the strongest pair correlations of internal motion in the complex are between the electron and hole, the most probable radiative decay consists of the creation of a photon and a corresponding "recoil" particle. In indirect gap semiconductors such processes are accompanied by the emission of a phonon to conserve momentum. Experimental efforts were first concentrated on the search for new lines in the emission spectra corresponding to the radiative decay of excitonic molecules. The experimental activity in this direction was greatly stimulated by the well-known paper by Haynes (1966), who discovered a new radiative channel in an exciton gas of sufficiently high density in Si and attributed it to excitonic molecules. However, later it was established that this phenomenon was connected with the condensation of electrons and holes Pokrovski'i et al. 1969, 1972). For direct gap semiconductors the most convincing evidence for the biexciton radiative decay was found in CuCl and CuBr crystals, where according to variational calculations, the excitonic molecules have large binding energies. It should, however, be born in mind that compound semiconductors in contrast to atomic semiconductors such as Ge and Si contain, as a rule, relatively high concentrations of residual shallow electrically active impurity centers. At low temperatures excitons are easily bound forming immobile bound exciton complexes (BE) (Lampert 1958, Haynes 1960). Such immobile complexes in direct gap semiconductors typically have extremely large oscillator strengths for optical transitions (Rashba et al. 1962, Rashba 1975) and at low temperatures they provide the most effective radiative channel.

Since the spectral positions of the emission lines of bound and free complexes emission lines are very close it is difficult to identify free complexes in the spectra of radiative recombination in direct gap semiconductors.

This difficulty caused by the dominance of BE radiation in the emission spectra is absent in the indirect gap elemental semiconductors Ge and Si when of high purity. Modern technologies make it possible to grow Ge and Si crystals with extremely small concentrations of electrically active impurities, of the order of 10^{10} cm^{-3} (Ge) and 10^{12} cm^{-3} (Si) or even less. However, in Ge and Si another problem arises, namely, the condensation into an electron–hole liquid (EHL) (Keldysh, 1968).

In the region of parameters where such a phase transition exists the gaseous phase density is restricted by the condensation threshold. In Ge and Si and, in other indirect gap semiconductors free multiparticle complexes are less stable than the EHL. Consequently, at low temperatures the partial fraction of excitonic molecules and excitonic ions in the nonequilibrium e–h gas is small even near the condensation threshold. In the high-temperature region of the gas–liquid coexistence curve a new complication arises caused by the ionization of weakly bound states in the gaseous phase as a result of thermal processes and screening of the Coulomb interaction.

The conditions for the experimental detection of free multiparticle complexes in indirect gap semiconductors become more favourable when the degeneracy of the conduction and valence bands is lifted. The stability of complexes is insensitive to such degeneracy whereas the EHL binding energy decrease substantially when the degeneracy is lowered. Under such conditions it becomes possible to increase considerably the partial fraction of biexcitons in the saturated e–h gas. In our opinion up to now the free complexes have been most comprehensively studied in Si crystals, and in the present review we shall mainly use experimental results concerning this indirect gap material.

2. *The binding energy of excitonic molecules*

Despite the direct analogy between the Wannier–Mott exciton and the hydrogen atom one should bear in mind that in all known semiconductors the electron-to-hole ratio of effective masses is much greater than the corresponding ratio for free electron to proton, which is about 1/1840. In this connection first of all the question arose whether the excitonic molecule (EM) is stable with respect to the decay into two excitons for any values of the mass ratio parameter $\sigma = m_e/m_h$.

When the electron and hole effective masses do not differ considerably the four body problem of EM stability lacks a small parameter for a series expansion. Consequently in this case the widely used (particularly in molecular spectroscopy) adiabatic approximation is inapplicable. In other words at $m_e \approx m_h$ the zero oscillation amplitude is of the order of the EM dimension itself and consequently the concept of "nuclear equilibrium position" becomes meaningless.

In the limit $\sigma = 1$ the problem of EM stability is identical to the problem for the positronium molecule which was first solved by Hylleraas and Ore (1947). From variational calculations they found that the positronium molecule binding energy is only about 0,017 of the positronium atom of Rydberg. Such a small binding energy as compared with the H_2 molecule results from the large contribution of the kinetic energy of intramolecular motion due to the small difference between the masses m_e and m_h.

The EM binding energy Δ_{EM} as a function of the parameter σ was first calculated by Sharma (1968a, b) who used the variational method of James and Coolidge (1933) with a 5 parameter trial wave function. According to his calculations the excitonic molecule is stable everywhere except the interval $0.2 \leqslant \sigma \leqslant 0.4$. This result, it turned out later, was wrong and was corrected by Wehner (1969) and Adamowski et al. (1971). It was shown in these papers that if the EM is stable in the limit $\sigma = 1$ then it should be stable within the whole region $0 \leqslant \sigma \leqslant 1$ ($\partial^2\Delta_{EM}/\partial\sigma^2 > 0$ for the whole range $\sigma \in [0, 1]$).

The EM binding energy for semiconductors with various band structures was calculated by Akimoto and Hanamura (1972a, b) and Brinkman et al. (1973).

In variational calculations the wave function is usually chosen in the form of a product of the Hylleraas–Ore (H–O) function and the envelope function for the holes:

$$\psi = [\psi_{H\text{-}O}(r)/S(R)]F(R), \tag{1}$$

where $S^2(R) \equiv \int \psi^2_{H\text{-}O}(r)\, d\tau_r$, and R represents the inter-hole distance. In the limit $\sigma \to 0$, substituting $F(R)$ in eq. (1) for the δ function one obtains the trial function which was previously used by Inui (1938, 1941) for variational calculations of the H_2 molecule. In this limit, variational calculations yield $\Delta_{EM}/Ry^{ex} \approx 0.3$ (Ry^{ex} is the exciton Rydberg) whereas the exact result is 0.35. In the other limiting case when $\sigma = 1$ one has $F(R) = 1$ and the wave function (1) coincides with the (H–O) function for the positronium molecule. The dependence of the EM ground state energy on the effective mass ratio, $E(\sigma)$, is then found using standard variational procedure by minimizing the Hamiltonian expectation values (1) with respect to the parameters in the trial function (1). The EM binding energy relative to the dissociation into two free excitons, $\Delta_{EM} = E - 2Ry^{ex}$ is determined. Figure 1 shows the dependence of the EM binding energy (in exciton Rydberg units) on the parameter σ in the interval $0 \leqslant \sigma \leqslant 1$ calculated by Akimoto and Hanamura (1972). A similar result was also obtained by Brinkman et al. (1973). For $\sigma > 1$ the abscissa is to be read as σ^{-1} by symmetry. It is seen from fig. 1 that the EM should be stable at any mass ratio and the ratio Δ_{EM}/Ry^{ex} is a monotone, rapidly decreasing function of σ in the range $0 \leqslant \sigma \leqslant 1$.

It is interesting to know how the size of the EM depends on the

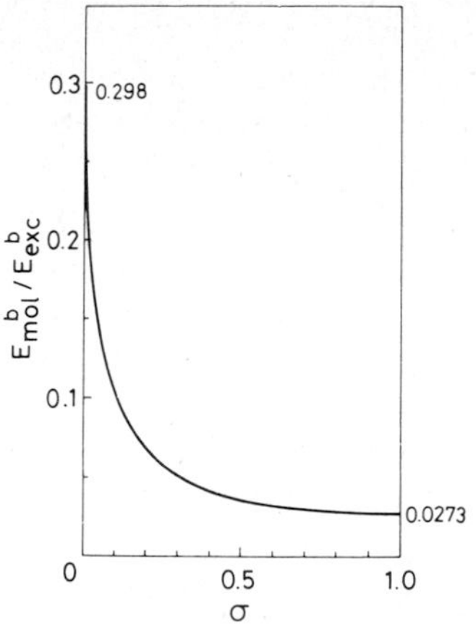

Fig. 1. The binding energy of the excitonic molecules in units of exciton Rydberg for different electron to hole mass ratios. (From Akimoto and Hanamura 1972.)

parameter σ. For $\sigma = 0$ the inter-hole distance is fixed. With increasing σ the kinetic energy of the holes increases leading to their delocalization and to the growth of the mean inter-hole distance $\langle R\rangle$, which is calculated from the formula

$$\langle R\rangle = \int \psi^2 R \, d\tau_r \Big/ \int \psi^2 \, d\tau_r. \tag{2}$$

The evaluation of $\langle R\rangle$ as a function of the parameter σ was performed by Akimoto and Hanamura (1973) and the result is shown in fig. 2 from which it is seen that the ratio $\langle R\rangle/a_{ex}$ increases monotonically from 1.44 to 3.47 in the interval $0 \leq \sigma \leq 1$. Thus the EM is a more loose formation than the hydrogen molecule. It is also interesting to note that $\langle R\rangle$ changes much more slowly than Δ_{EM}.

Less information is available about the effects of anisotropy of the effective mass spectrum on the magnitude of the EM binding energy. Brinkman et al. (1973) and Adamowski et al. (1978) have found that with increasing anisotropy parameter $\gamma \equiv m^e_\parallel/m^e_\perp$ (or $m^h_\parallel/m^h_\perp$) the binding energy of the excitonic molecule decreases.

3. *The gas of excitons and excitonic molecules and the electron–hole liquid in uniaxially strained silicon crystals*

In indirect gap semiconductors the main obstacle hampering the experimental detection of excitonic molecules from emission spectra is the smallness of the EM binding energy which, as a rule, is less than the EHL

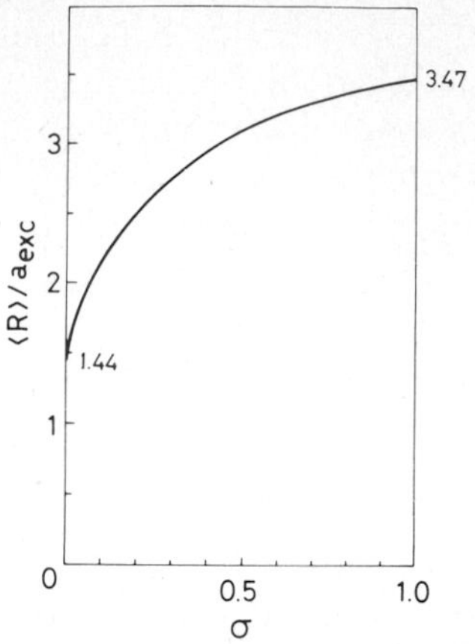

Fig. 2. The ratio of the mean inter-hole distance of the excitonic molecules to the Bohr radius of the exciton for different electron to hole mass ratios. (From Akimoto and Hanamura 1972.)

binding energy. In such semiconductors the e–h gas density in the region of gas–EHL coexistence is always restricted by the condensation threshold (Hensel et al., 1977). In unstrained Ge and Si at low temperatures the partial fraction of EM in the e–h gas along the gas–liquid phase boundary is almost two orders of magnitude less than that of excitons. The density of e–h gas saturated vapours can be substantially increased by increasing the electronic temperature, however this entails other complications because of thermodissociation processes. The conditions for the experimental observation of biexcitons become more favourable when the indirect gap semiconductor is uniaxially strained so that the EHL binding energy considerably decreases. It is well known that uniaxial elastic deformation lifts the degeneracy of the bands in Ge and Si. As a result of this the mean kinetic energy per pair in the EHL increases whereas the binding energy and the equilibrium pair density in the condensed phase decrease (Bagaev et al. 1969, Benoit a la Guillaume et al. 1972). We note that under the same conditions as was shown by Bir and Pikus (1974) the binding energy of the exciton (and, apparently, of the EM) are not altered significantly. The change of the EHL binding energy is accompanied by considerable changes in the partial composition of saturated vapour along the gas–liquid phase boundary. We first consider the change of the EHL binding energy when the energy spectrum is altered by the lifting of the band degeneracy by directional deformation in Si crystals.

Kulakovskii et al. (1978a, b) have investigated the recombination–radiation spectra of Si deformed along the axes ⟨100⟩, ⟨110⟩ and ⟨111⟩ (i.e. Si (2–1), Si(4–1) and Si(6–1)) where the first figure denotes the number of the lowest split-off electron valleys, and the second is the multiplicity of the valence band degeneracy at the extremum (with spin degeneracy neglected). At pressures $p < p_{cr}$, a continuous restructuring of the EHL recombination spectrum takes place as the deformation is increased. Above the pressure p_{cr}, when the splitting of the bands $\Delta_{c,v}$ exceeds the corresponding values of the Fermi energies of the electron and holes in the EHL, the recombination spectrum of the EHL acquires a canonical form for each given compression direction and is independent of the applied pressure. Figure 3 shows the recombination spectra of Si(6–1), Si(4–1) and Si(2–1) measured at $T = 1.8$ K and under pulsed laser excitation. For comparison the figure also shows the spectrum of undeformed Si(6–2). Each spectrum contains a wide band L corresponding to the EHL emission and emission lines FE connected with the gas of free excitons. Figure 3 demonstrates qualitatively that as the "multivalley character" is decreased the EHL band becomes narrower and its maximum and "violet" edge lies closer to the exciton line. In the strained crystals the condensation threshold increases noticeably.

The equilibrium concentration of e–h pairs in the EHL in indirect gap semiconductors is determined by analyzing the shape of the recombination spectrum, using the following expression,

$$I(E) \propto \int^{E} D_e(E')D_h(E'-E)f_e(E')f_h(E'-E)\,dE', \qquad (3)$$

where $D(E')$, $D(E'-E)$ are the densities of states in the electron and hole bands, f_e and f_h are the distribution functions of electrons and holes. The energy of the recombination photon is

$$\hbar\omega = E + E^{i}_{gap} - E_{xc} - \hbar\Omega, \qquad (4)$$

where E^{i}_{gap} is the width of the indirect gap, E_{xc} is the sum of the exchange and correlation energies, and $\hbar\Omega$ is the energy of the emitted phonon. In this expression no account is taken of the dependence of the transition matrix element on energy, nor of the corrections to the density of states due to interactions between the particles. Hence at a known electron temperature the density of e–h pairs in the condensed phase, n_0, is the single fit parameter in expression (3).

The approximation to the EHL spectrum shape given by expression (3) is shown in fig. 3. Within the framework of this description the sum of the Fermi energies of electron and holes, $E^{F}_{e}(n_0)$ and $E^{F}_{h}(n_0)$ is the width of the band L, and the EHL binding energy φ is determined as the distance between the position of chemical potential $\tilde{\mu}$ and the boundary of the exciton–phonon spectrum.

Kulakovskii et al. (1978a) have found that at maximum lifting of degeneracy in the bands, i.e. in Si(2–1), the EHL binding energy undergoes

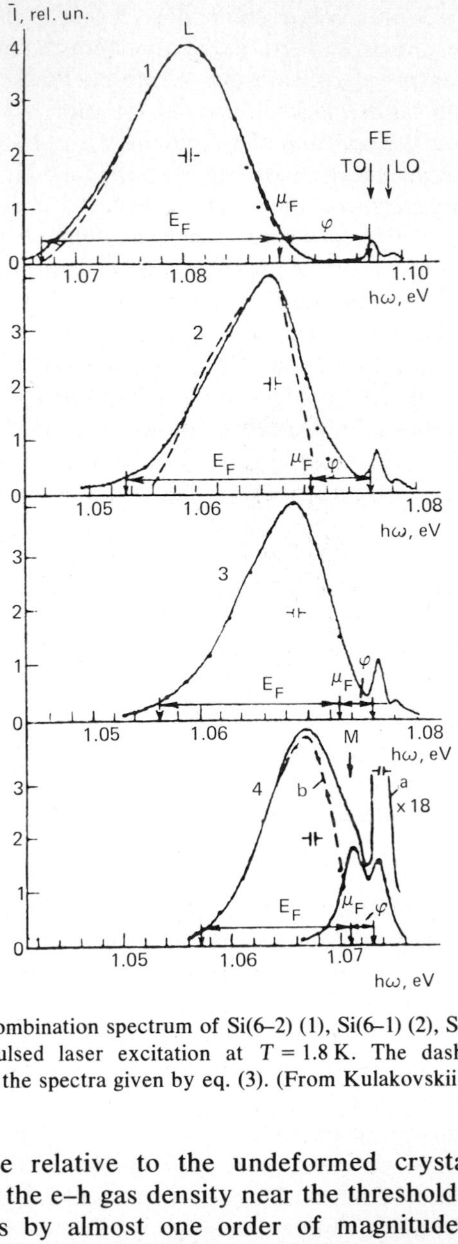

Fig. 3. The radiative recombination spectrum of Si(6–2) (1), Si(6–1) (2), Si(4–1) (3) and Si(2–1) (4) under 30 kW/cm^2 pulsed laser excitation at $T = 1.8$ K. The dashed lines show the approximation of the spectra given by eq. (3). (From Kulakovskii et al. 1978a.)

a fourfold decrease relative to the undeformed crystal Si(6–2). Consequently in Si(2–1) the e–h gas density near the threshold of condensation into EHL increases by almost one order of magnitude compared with Si(6–2). Qualitatively this is seen from the comparison of the Si(6–2) and Si(2–1) spectra shown in fig. 3, which were measured at equal generation levels of nonequilibrium carriers. With decreasing EHL binding energy φ the exciton line intensity increases. At the same time an increase of the background emission between the EHL and exciton bands is observed. However only in the spectra of Si(2–1) does this background exhibit a clearly pronounced spectral feature: a separate new line M due to the emission of excitonic molecules.

The appearance of the M line is clear from the time evolution of the emission spectra of Si(2–1) under pulsed excitation illustrated by fig. 4. At the instant of pulsed excitation of e–h pairs with a concentration of about $n_{e,h} \approx 10^{17}\,cm^{-3}$ the EHL emission (L band) predominates. The free exciton emission line FE is also clearly seen in the spectra. Besides the L and FE lines a new line M appears in the spectra when the applied uniaxial stress p is greater than 20 kg/mm^2. The spacing between the maxima of the lines M and FE does not change with increasing pressure (up to the destruction of the crystal at $p \approx 90$–100 kg/mm^2) and equals 2 meV. After the decay of the EHL emission band the shape of the M line remains unchanged and its decay time is $\tau_M \approx 0{,}5\tau_{FE}$.

The line M was first discovered in the emission spectra of Si by

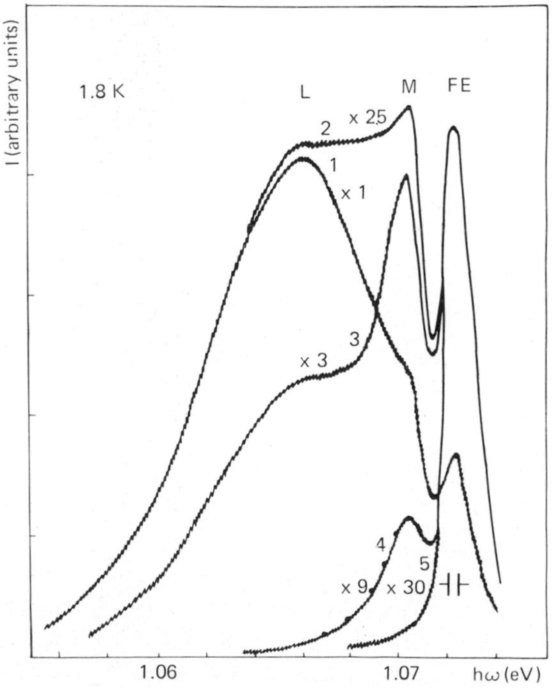

Fig. 4. Kinetics of the emission spectrum of the Si(2–1) under pulsed excitation (30 kW/cm^2) and $T = 1.8$ K. Spectra 1–5 correspond respectively to delays of 0.05, 0.25, 0.35, 0.7, and 1.1 μs relative to the exciting pulse. (From Kulakovskii et al. 1978a.)

Kulakovskii and Timofeev (1977) and was attributed to the indirect radiative decay of excitonic molecules in conformity with the conservation law,

$$E_p^{\mathrm{EM}} = E_q^{\mathrm{ex}} + \hbar\Omega_{p-q} + \hbar\omega, \tag{5}$$

where E_p^{EM} and E_q^{ex} are the energies of the biexciton and exciton, respectively, and $\hbar\omega$ is the energy of the emitted photon. The molecular origin of the M line was confirmed by experiments performed under stationary volume excitation. Figure 5 shows the evolution of the emission spectrum of Si(2–1) at $p = 50\,\mathrm{kg/mm^2}$ in the region of the exciton structure under

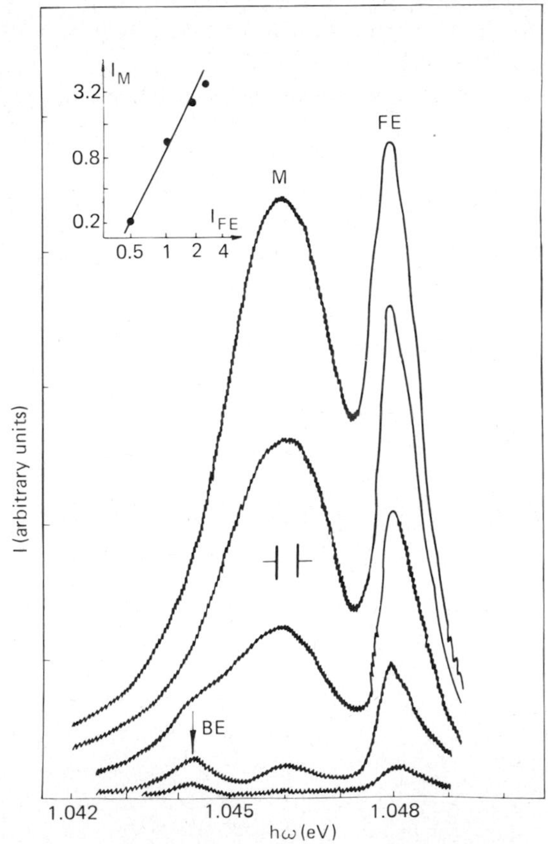

Fig. 5. Variation of the emission spectra of Si(2–1) ($p \approx 45\,\mathrm{kg/mm^2}$, $T = 1.8\,\mathrm{K}$) with changing excitation density as n is increased from 10^{14} to $10^{15}\,\mathrm{cm^{-3}}$. The insert shows the $J_{\mathrm{M}}(J_{\mathrm{FE}})$ dependence in a logarithmic scale, the straight line corresponds to $J_{\mathrm{M}} - J_{\mathrm{FE}}^2$. (From Kulakovskii et al. 1978a.)

stationary volume laser excitation ($\lambda = 1.064$ m) and at $T = 1.8$ K. By the use of such a source it is possible to obtain volume excitation in Si(2–1) crystals starting with a deformation $p \gtrsim 20$ kg/mm^2. At minimum excitation densities the spectrum contains only the free exciton emission line and the line BE of the exciton bound to a neutral acceptor (boron atoms). With increasing pumping the intensity of the BE line is saturated. The M line then appears between the BE and FE emission bands and with increasing exciton concentration its intensity increases quadratically with exciton line intensity (inset in fig. 5). This quadratic dependence is one of the main confirmations of the molecular origin of the M band. The intensity ratio of the lines FE and M ceases to depend on excitation density at $T = 1.8$ K and at concentrations $n_{e,h} \approx 10^{15}$ cm^{-2}, when condensation of the exciton and biexciton gas into EHL sets in. It follows from these experiments performed under the conditions of uniaxial deformation, that only in Si(2–1) are the partial pressures of excitons and biexcitons of the same order of magnitude at densities $n_{e,h}$ corresponding to the onset of condensation into EHL.

If one judges from the known variational estimates of the EM binding energy and from the experimental values of the EHL binding energy, a similar situation should exist in Ge crystals. In unstressed Ge near the threshold of condensation into EHL the density of excitons must exceed that of biexcitons by more than two orders of magnitude. A situation favourable for the detection of biexcitons from emission spectra should arise in Ge crystals under the conditions of uniaxial stress along the direction ⟨III⟩ (Ge(1–1), i.e. at maximum lifting of the degeneracy in the electron and hole bands).

4. *The rate of indirect annihilation and the shape of excitonic molecule emission lines*

In the case considered of indirect electronic transitions, the radiative annihilation of an excitonic molecule results into creation of a photon, phonon and exciton. Therefore the corresponding line in the luminescence must have a finite width even if the EM is initially at rest. This is due to the fact that a part of the EM energy is given up to the created "recoil" exciton to increase its kinetic energy. It is these "recoil" processes that determine the essential differences of the shape of the EM emission spectrum in the case of indirect annihilation as compared to their emission spectrum in direct gap semiconductors.

In order to estimate the width of the EM emission line at $T \to 0$ for the case of indirect transitions we use the fact that the characteristic dimension of the EM wave function is of the order of the exciton radius α_{ex} (fig. 2).

Having this in mind it is quite natural to assume that the matrix element $M(\boldsymbol{P} \mid \boldsymbol{q}, \boldsymbol{Q})$ of the corresponding radiative transition will differ from zero at $|\boldsymbol{q}| \lesssim a_{ex}^{-1}$. Therefore the typical "recoil" energy of the exciton, which determines the linewidth, is $\hbar^2/2m_{ex}a_{ex}^2$. Hence the width Γ of the biexciton emission line is of the order of

$$\Gamma \sim m^*/m_{ex}\mathrm{Ry}^{ex}, \tag{6}$$

where m^* and m_{ex} are the reduced and translational masses of the exciton. According to eq. (6) the EM emission linewidth in Si(2–1) should be $\Gamma \approx 3$ meV, and in Ge(1–1) about 1.5 meV.

For the calculation of the biexciton-emission lineshape it is necessary to know the dipole matrix element of the corresponding transition. The indirect EM annihilation amplitude was calculated by Kulakovskii et al. (1978a). The matrix element of the corresponding transition was found in standard fashion by second-order perturbation theory (interaction with phonons and electromagnetic fields) and after simple transformations it was reduced to the form,

$$M(\boldsymbol{P} \mid \boldsymbol{Q}, \boldsymbol{q}) \propto \int \mathrm{d}\boldsymbol{r}\, \mathrm{d}\boldsymbol{r}_e\, \mathrm{d}\boldsymbol{r}_h\, \psi^{\boldsymbol{P}}(\boldsymbol{r}, \boldsymbol{r}_e \mid \boldsymbol{r}, \boldsymbol{r}_h)\, \mathrm{e}^{-\mathrm{i}\boldsymbol{Q}\boldsymbol{r}}\, \psi^{*\boldsymbol{q}}(\boldsymbol{r}_e \mid \boldsymbol{r}_h), \tag{7}$$

where $\psi^{\boldsymbol{P}}$, $\mathrm{e}^{-\mathrm{i}\boldsymbol{Q}\boldsymbol{r}}$ and $\psi^{\boldsymbol{q}}$ are the biexciton, phonon, and exciton wave functions. In the derivation it was assumed that the spin function of the electrons and holes in the biexciton is antisymmetric under the permutation of both the electrons and the holes while the coordinate function is symmetric.

Expression (7) is a generalization of the well-known formula of Elliott (1957) and Sakharov (1948) for the annihilation amplitude which is proportional to $\psi_{ex}(0)$ of the exciton (or the positronium atom) and also has a simple physical interpretation. In fact, expression (7) is evidently the probability amplitude of finding one of the electrons and one of the holes at the same point in the biexciton while the other electron and hole form an exciton with a wave function $\psi^{\boldsymbol{q}}$.

It follows from the translational invariance of the exciton and biexciton in the effective-mass method, that $M(\boldsymbol{P} \mid \boldsymbol{Q}, \boldsymbol{q})$ contains $\delta(\boldsymbol{P}-\boldsymbol{Q}-\boldsymbol{q})$ in conformity with the quasimomentum-conservation law. In addition, since the internal motion in the exciton and the EM is independent of the center of gravity, the function $M(\boldsymbol{P} \mid \boldsymbol{Q}, \boldsymbol{q})$ after separating out the δ function, will depend only on one linear combination of the remaining two independent vectors,

$$M(\boldsymbol{P} \mid \boldsymbol{Q}, \boldsymbol{q}) = \delta(\boldsymbol{P}-\boldsymbol{Q}-\boldsymbol{q})N(2\boldsymbol{q}-\boldsymbol{P}). \tag{8}$$

To determine M and N explicitly it is necessary to know the wave function of the ground state of the EM. There is no exact solution of this problem, therefore Kulakovskii et al. (1978a) have evaluated M using the variational approximation of Brinkman et al. (1973), which is the most suitable for the biexciton in Si(2–1). Substitution of the expression for the variational wave function of the biexciton and also of the expressions for the exciton and phonon wave functions into (7) and numerical integration yields the function N. The values of $N(q)$ thus calculated in the region $|\boldsymbol{q}|a_{ex} \lesssim 2{,}5$ are shown in fig. 6. As expected, the function N is concentrated in the region $|\boldsymbol{q}|a_{ex} < 2$.

If phonon dispersion is neglected, then the probability for the emission of a photon of energy $\hbar\omega$ is

$$W(\hbar\omega) \propto \int \mathrm{d}\boldsymbol{P}\, \mathrm{d}\boldsymbol{q}\, \exp(-P^2/8m_{ex}kT) N^2(|2\boldsymbol{q} - \boldsymbol{P}|) \times \delta(E_{\boldsymbol{P}}^{EM} - E_{\boldsymbol{q}}^{ex} - \hbar\Omega - \hbar\omega). \tag{9}$$

At sufficiently low temperatures, such that the mean thermal momentum $\langle P \rangle$ of the EM satisfies the inequality $\langle P \rangle a_{ex} \ll 1$, one can see from (9) that the line shape is given by the expression

$$W(\hbar\omega) \propto |q| N^2(2q), \tag{10}$$

with $q^2/4m_{ex} = E_{\boldsymbol{q}}^{ex} - \Delta_{EM} - \hbar\Omega - \hbar\omega$. Figure 7 shows the fit of the experimental shape of the EM emission line in Si(2–1) at $T = 1.8$ K with the aid

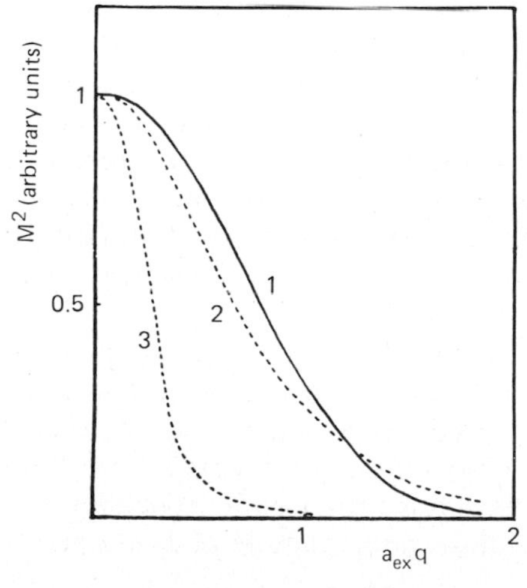

Fig. 6. The dependence of the matrix element of the excitonic molecule radiative decay transition on the parameter $a_{ex}q$. Curve (1) is calculated by Kulakovskii et al. (1978a) with the use of the variational wave function of Brinkmann et al. (1973), curves (2) and (3) are calculated with the use of Cho's approximation (1973) and correspond to $a_{EM} = a_{EX}$ and $a_{EM} = 3a_{EX}$, respectively.

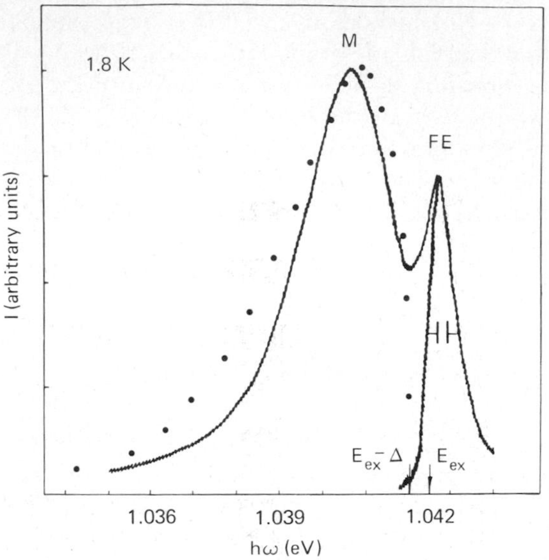

Fig. 7. Comparison of the calculated luminescence spectrum of EM (points) with the experimental one. The figure shows also the emission spectrum of excitons (FE) obtained at low excitation density. $T = 1.8$ K. (From Kulakovskii et al. 1978a.)

of the theoretical relation (10). The numerical values of the function N correspond to fig. 6. To obtain the result of fig. 7 the following was assumed: an exciton binding energy of 12 meV, $a_{ex} = 52.5$ Å, and a EM binding energy for Si(2–1), corresponding to the trial function of Brinkman et al. (1973), $\Delta_{EM} = 0.4$ meV. The agreement between theory and experiment seems quite satisfactory, especially if one remembers that no free parameters were used in the approximation.

It is important to emphasize that both the line shape and the binding energy are uniquely determined by the EM wave function and hence are not independent quantities. Therefore it is impossible to determine the EM binding energy accurately on the basis of the analysis of the line shape. One could try to determine Δ_{EM} from the position of the high-frequency boundary of the emission spectrum, however the experimental determination of this boundary in the case of biexcitons is not sufficiently exact.

Previously Cho (1973) has proposed a model-derived expression for the indirect biexciton-annihilation amplitude of the form

$$M(\boldsymbol{P} \mid \boldsymbol{q}) \propto [(\tfrac{1}{2}\boldsymbol{P} - \boldsymbol{q})^2 + a_{EM}^{-2}]^{-2}. \qquad (11)$$

It turned out, however, that the model expression (11) describes satisfactorily only the long-wave wing of the experimental profile of the EM line, which corresponds, as is seen from (11), to a large momentum transfer. The same conclusion can be drawn from a comparison of expression (11) with the numerical calculation of the function $N(q)$. Thus the estimates of the EM binding energy obtained by Pelant et al. (1976) and by Gourley and Wolfe (1978) by fitting the experimental profile with a model expression for the matrix element $M(\mathbf{P} \mid \mathbf{Q}, \mathbf{q})$ seem not to be exact. For example, in AgBr crystals Pelant et al. (1976) used expression (11) to fit the indirect EM emission profile and obtained an anomalously large value of the binding energy, $\Delta_{EM} \approx \frac{1}{2}\mathrm{Ry}^{ex}$, which has no simple explanation within the framework of existing theoretical ideas. Apparently, this anomaly is caused by the procedure used for fitting the experimental profile.

The EM binding energy in an indirect gap semiconductor can be estimated experimentally in a thermodynamic way if quasi-equilibrium between the components of the gas phase, namely the FE/EM equilibrium, holds. Under conditions of thermodynamic equilibrium the concentrations of excitons and biexcitons, n_{ex} and n_{EM}, are related in the following way

$$n_{EM} = n_{ex}^2(4\pi\hbar^2/kTm_{ex})^{3/2}g_{EM}/g_{ex}^2)\exp(\Delta_{EM}/kT), \tag{12}$$

where g_{EM}, g_{ex} are the statistical weights. Then the EM binding energy can be determined from the temperature dependence of the ratio of FE line to M line intensities, using relation (12). Such measurements were made by Kulakovskii et al. (1978b) for Si(2–1) in a temperature range 2 to 4.2 K and at concentrations $n_{e,h} \approx 10^{15}\,\mathrm{cm}^{-3}$ and the results are shown in fig. 8. These authors used volume excitation with the aid of a Nd:YAG laser and thus avoided the difficulties arising from temperature changes of the spatial distribution of the FE/EM gas and usually encountered when surface excitation is used. They found for the EM binding energy in Si(2–1) a value of $\Delta_{EM} = 0.55 \pm 0.15$ meV, or $\Delta_{EM} \approx 0.04\,\mathrm{Ry}^{ex}$. If in this experiment the quasi-equilibrium in the gas phase was incomplete, then the true value of Δ_{EM} might be somewhat greater.

One can judge the state of FE/EM equilibrium from studies of the kinetics of the emission spectra under the conditions of pulsed excitation. If radiation is emitted mainly through the channel of excitons and EM (EHL emission being absent) then at quasi-equilibrium the nonstationary density of EM will decay according to the law $n_{EM} \propto n_{ex}^2 \propto \exp(-2t/\tau_{exc})$, i.e. twice as fast as the exciton density. Studying the decay kinetics of excitons and biexcitons in Si(2–1) with 50 ns time resolution Kulakovskii et al. (1978a) found $\tau_{EM} = 0.2\,\mu$s and $\tau_{FE} = 0.38\,\mu$s at 2 K. The kinetics of the exciton and EM emission lines under pulsed excitation of $n_{e,h} \approx 10^{17}\,\mathrm{cm}^{-3}$, as well as the decay of the EHL line, are shown in fig. 9. It is seen that 0.5 μs after the exciting pulse, when the drops of EHL have evaporated the decay of the FE and EM line intensities is described by an exponential law. With increasing temperature the respective times also increase at $T = 10$ K, they are equal to $\tau_{EM} \approx 1\,\mu$s and $\tau_{FE} = 1.9\,\mu$s. Thus for Si(2–1) in the investigated

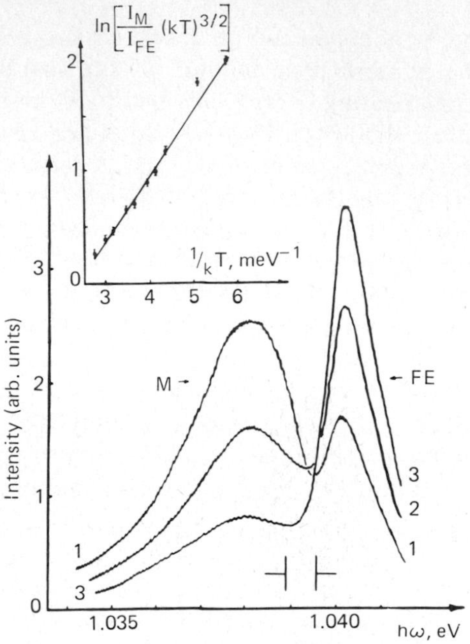

Fig. 8. Spectra 1, 2, 3 of Si(2–1) under stationary excitation and $T = 2$, 3.1, and 4.2 K respectively. (From Kulakovskii et al. 1978b.)

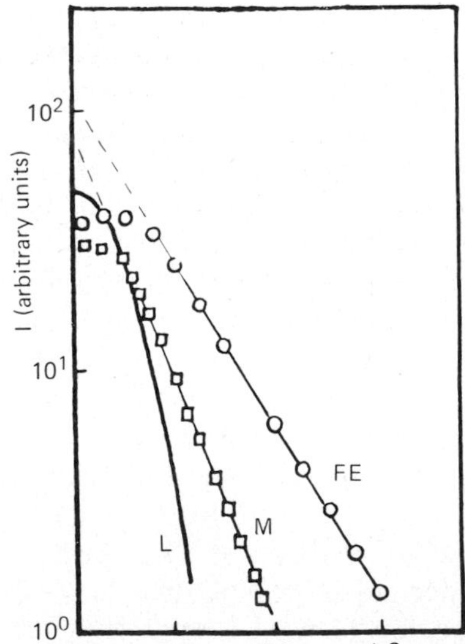

Fig. 9. Dependence of the emission line intensities of excitons (FE), excitonic molecules (M), and EHL(L) on the delay relative to the excitation pulse in Si(2–1) at $T = 2$ K. (From Kulakovskii et al. 1978a.)

temperature interval $\tau_{EM} \approx \frac{1}{2}\tau_{ex}$, indicative of the vicinity to FE/EM quasi-equilibrium, when the mean e–h concentration, $n_{e,h}$, is about 10^{15} cm^{-3}.

It is interesting to compare the experimental value of Δ_{EM} for Si(2–1) with the values resulting from variational calculations. Brinkman et al. (1973), Akimoto and Hanamura (1972a, b) and Forney and Baldereschi (1978) have found that $\Delta_{EM} \approx 0.03$ Ry^{ex}. Huang (1973) obtained a considerably larger value $\Delta_{EM} \approx 0.11$ Ry^{ex}. Note, that in the above variational calculations of the binding energy no account was taken of anisotropy effects, which are possibly a significant case of silicon.

Recently radiative decay of the excitonic molecule has been found in the emission spectra of uniaxially compressed Ge by Kukushkin et al. (1980). It is demonstrated in this paper that the electron–hole liquid (EHL) stability in Ge remains large (~0.28 Ry, Ry is excitonic Rydberg) with respect to the excitonic molecule binding energy ($\Delta_M \approx 0.03$ Ry) even with a high stress along ⟨111⟩ axis, which lifts orbital degeneracy of the bands completely. Because of this stability, the partial fraction of the molecules, relative to the free exciton (FE), in the gas phase in Ge ⟨111⟩ at low temperature is too small for the EM emission to be detectable in the spectra (see fig. 10, upper part). The EHL binding energy should decrease to ~0.18 Ry in Ge compressed along a nonsymmetrical direction close (5°) to the ⟨001⟩ axis (e.g. ⟨1, 1, 16⟩), due to the decrease in the hole effective mass density of states. Kukushkin et al. have found that in this case the partial fractions of excitons and molecules in the gas phase at the coexistence boundary with EHL at $T = 2$ K are of the same order of magnitude. Under such conditions the EM radiative decay line is visible in the emission spectra (M line) (see fig. 10, lower part). The shape of this line has been analyzed and the EM binding energy has been estimated equal to $\Delta_M \approx (0.15 \pm 0.1)$ meV by the thermodynamic method. It is interesting to mention that for the same experimental values of kT/Ry and I_M/I_{FE}, and with normalized scales (Ry(Si/Ry(Ge) ≈ 4.8), the shapes of the EM emission lines for Si(2–1) and Ge⟨1, 1, 16⟩ completely coincide (see curve 9 of fig. 10).

5. *Gas of excitons and excitonic molecules in a parabolic potential well*

The gas of excitons and EM exhibits interesting features when the nonequilibrium e–h system is confined within a three-dimensional spherically symmetric potential well described by a potential $V(r) = \alpha r^2$, where α is a force constant. Gourley and Wolfe (1978) have shown that it is possible to realize such conditions by means of inhomogeneous stress applied along one of the main crystallographic directions. The center of the parabolic potential well is then localized inside the sample where the stress is

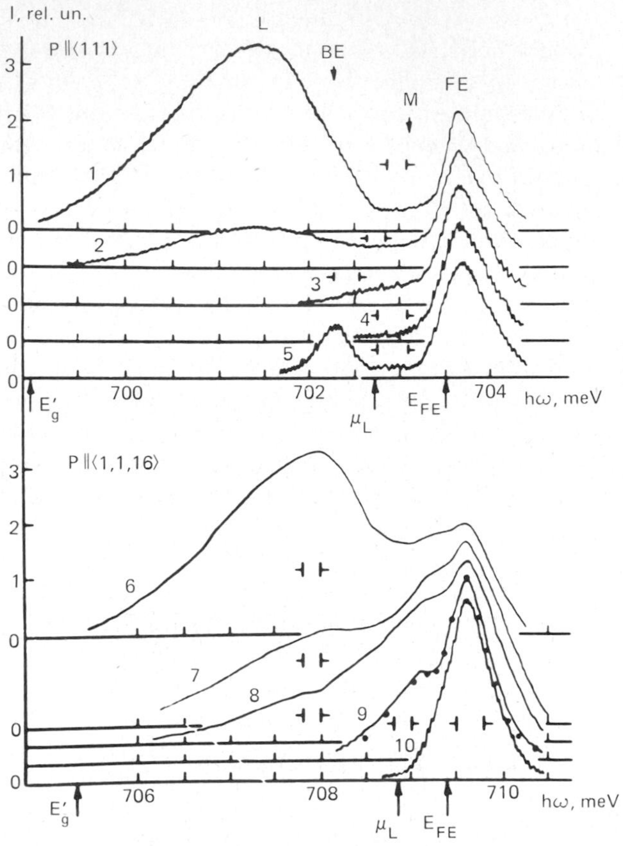

Fig. 10. LA-phonon assisted spectra of Ge stressed along ⟨111⟩ (≈130 MPa) and ⟨1, 1, 16⟩ (≈200 MPa) at $T = 1.8$ K for different pumping powers P (W/cm^2). Spectra (1)–(10) correspond to P = 5, 1, 0.3, 0.03, 0.03, 50, 20, 8, 2, 0.03. Spectra (1)–(4), (6)–(10) correspond to the pure samples with $N_i = N_D + N_A \approx 5 \times 10^{11}$ cm^{-3}, spectrum 5 to a sample with $N_i = 5 \times 10^{12}$ cm^{-3}. E'_g, μ_L and E_{FE} shown by arrows indicate the boundary of the band, EHL chemical potential and free exciton level. The dots at curve (9) correspond to the shape of EM and FE lines for Si(2–1) taken at $T = 9$ K. (From Kukushkin et al. 1980.)

maximum. Gourley and Wolfe accomplished this by stressing Si monocrystals, cut in the form of rectangular parallelepipeds, along the axis ⟨100⟩ with a rounded steel plunger. In this case the stress maximum was about 50 kg/mm^2 at 0.3 mm inside the surface to which the external stress was applied. Earlier this strain-confinement technique was used by Wolfe et al. (1975a, b) to detect giant electron–hole drops (so-called γ-EHD) in Ge crystals.

The nonequilibrium e–h pairs created by laser light near the surface or in the volume of the sample are driven by the stress to the center of the potential well, where the stress is maximum. Figure 11 is a photograph showing the migration of nonequilibrium carriers in Si crystals to the center of the potential well where the excitons and EM are condensed into EHL at $T < T_c$. In nonuniformly stressed Si crystals, due to the long exciton lifetimes the nonequilibrium e–h system, comes to a quasi-equilibrium state in the vicinity of the potential well center. In this case the spatial distribution of the e–h gas component (excitons, biexcitons, free electrons and holes) is determined by the shape of the potential well. It follows from the paper by Gourley and Wolfe (1978), that the local density of free particles confined to three-dimensional parabolic potential well is given by the expression:

$$n_i(r) = n_i(0)\exp(-\alpha_i r^2/kT), \qquad (13)$$

where the subscript i refers to the type of particle (excitons, EM etc.). Here $n_i(0) = (\alpha_i/\pi kT)^{3/2} N_i$, where N_i is the total number of particles in the

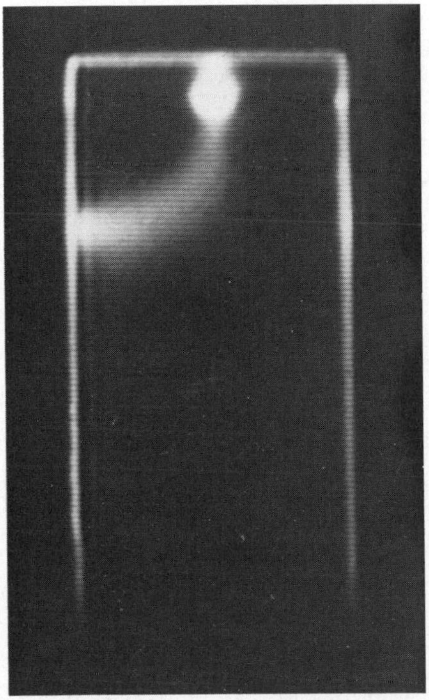

Fig. 11. Side view of the exciton luminescence from a 4 mm × 4 mm × 1.8 mm Si crystal at T = 15 K, showing explicitly the migration of excitons from the excited surface to the potential well. (From Gourley and Wolfe 1978.)

gas which occupies the effective volume $V_i = (\pi kT/\alpha_i)^{3/2}$. According to eq. (13) the spatial profile of the recombination–radiation spot of the ith component of the gas confined to the potential well has a Gaussian form $I(r_x) = I_0 \exp(-\alpha_i r_x^2/kT)$. The full width at half-maximum (FWHM), ΔH, of this distribution is

$$\Delta H = (2.77kT/\alpha_i)^{1/2}. \tag{14}$$

Thus, the width of the spatial distribution of gas particles of the ith kind is independent of the gas density and consequently of the excitation level, and is a function only of temperature. In fig. 12 FWHM of the spatial distribution of the exciton gas emission in nonuniformly stressed Si(2–1)

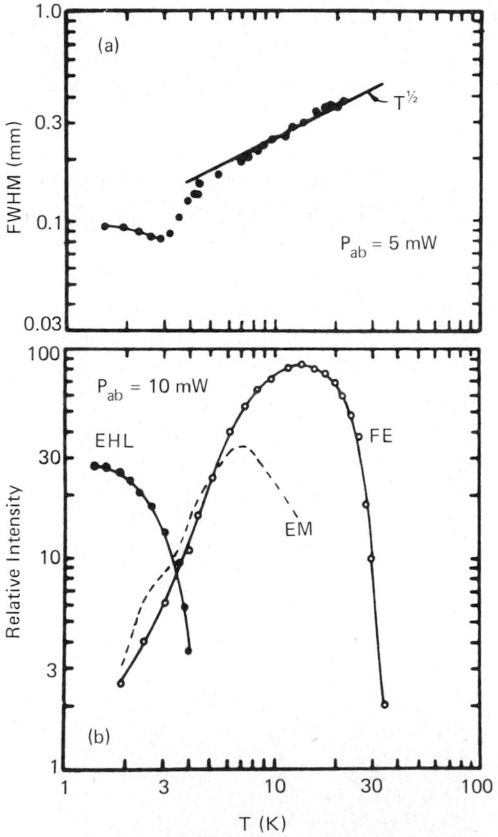

Fig. 12. (a) Spatial extent of the luminescence in Si from the potential well vs T. Above $T = 5$ K, ideal gas behaviour is observed. Below $T = 4$ K, the FE gas has condensed into an EHL. (b) Relative intensities of the three phases obtained with I (meV) spectral resolution. (From Gourley and Wolfe 1978.)

crystals is plotted versus temperature. Within the temperature range 22–5 K the width is proportional to $T^{1/2}$, in agreement with eq. (14). It is also seen from fig. 11 that this temperature dependence ceases to hold at $T \leqslant 4$ K, when under the excitation level used in the experiment condensation of excitons into drops of EHL occurs. As the EM, unlike the exciton, consist of two e–h pairs, it is natural to expect the force constants to be related by $\alpha_{EM} = 2\alpha_{FE}$. Therefore at a fixed temperature the FWHM of the spatial distribution of the EM gas is narrower than that of the exciton gas, namely, $\Delta H_{FE} = \sqrt{2}\Delta H_{EM}$. The measured spatial profiles of the radiation from the gas of excitons and excitonic molecules confined to a potential well confirm this conclusion (fig. 13).

Furthermore, the density of states $D(E)$ of the free ideal gas confined to a three-dimensional parabolic potential well suffers substantial modification and is given by $D(E) \propto E^2$. As a result of this the shape of the exciton–phonon luminescence spectrum is well described by the expression

$$I(E) \propto E^2 \exp(-E/kT). \tag{15}$$

Figure 14 shows the result of fitting the spectrum by eq. (15). For comparison the function $E^{1/2} \exp(-E/kT)$ which usually describes the shape of the exciton–phonon luminescence in the case of indirect allowed transitions in unstressed crystals, is plotted.

The EM binding energy in nonuniformly stressed Si(2–1) crystals, when the nonequilibrium e–h gas is spatially confined by a parabolic potential well, was extracted from the temperature dependence of the FE/EM emission–intensity ratio. Under these conditions Gourley and Wolfe (1979) have obtained for the intensity ratio $R \equiv I^2_{FE}/I_{EM}$ in the case of ther-

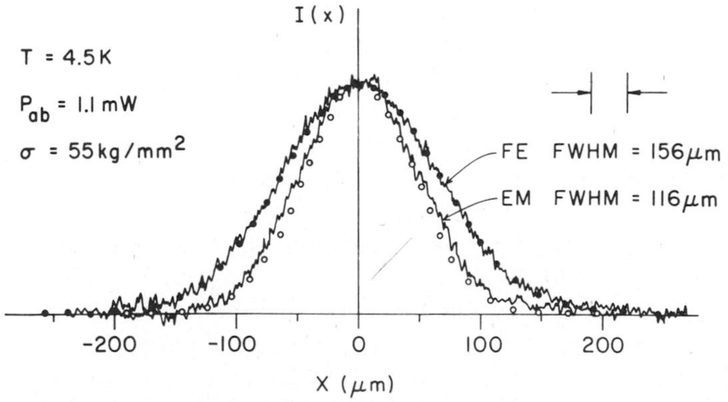

Fig. 13. Spatial profiles of the EM and FE distributions obtained by scanning the image of the Si crystal across the entrance slit of a monochromator set at the respective wavelength. (From Gourley and Wolfe 1979.)

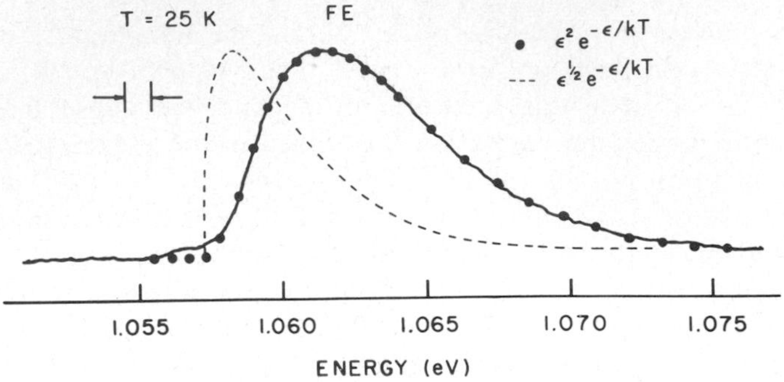

Fig. 14. High temperature FE line shape showing the excellent fit of the simple fitting function with $D(\epsilon) \propto \epsilon^2$ (From Gourley and Wolfe 1979.)

modynamic equilibrium the following expression:

$$R(T) = \beta \frac{g_{FE}^2}{g_{EM}} \left[\frac{1}{2\hbar^2} \frac{m_{ex}^2 \alpha_{EM}}{m_{EM} \alpha_{FF}^2}\right]^{3/2} (kT)^3 \exp\left(-\frac{\Delta_{EM}}{kT}\right), \tag{16}$$

where β is proportional to the FE to EM ratio of radiative probabilities, which is assumed to be independent of temperature. The ratio $R(T)$ does not depend on the total number of e–h pairs in the potential well, nor therefore on the excitation level. The temperature dependence of $R(T)$ given by eq. (16) differs from that in the case of free particles by a factor $T^{3/2}$. In nonuniformly stressed Si(2–1) crystals the EM binding energy was estimated by Gourley and Wolfe (1979) from the FE to M intensity ratio measured in the temperature range 3.5–10 K. The EM binding energy found is $\Delta_{EM} \approx 1.5$ meV, which is more than twice the value found in the same way for uniformly stressed Si(2–1) crystals. The cause of such discrepancy remains unclear.

In conclusion to this section it should be emphasized that under quasi-equilibrium in the potential well the distribution of the EM gas and the spatial extent of this distribution are insensitive to the excitation level and to the related density gradient. The latter is unavoidable under surface excitation of unstressed or uniformly stressed crystals, and the distribution itself depends on the excitation level in this case, which is particularly critical when the EM binding energy is estimated in a thermodynamic way. This seems to provide one of the main advantages of the strain-confinement method in studying the radiative annihilation of the EM gas in comparison with the case of unstressed or uniformly stressed crystals where there is no potential well. On the other hand, it should be born in mind that the resulting gas distribution will depend on the shape of the potential well. In particular, eq. (16) used to estimate the EM binding energy is valid only for the case of a parabolic well. To our opinion, it remains an open question whether the shape of the potential well can be exactly determined from emission spectra, if the well is produced by nonuniform stress, as well as to what extent such a potential well may be regarded as uniform. The investigation of this problem is of undoubted interest.

6. *Excitonic molecules in silicon crystals with different degrees of degeneracies of the electron valleys; "hot" excitonic molecules*

It is interesting to find out how the EM stability will be affected if the degeneracy of electronic valleys and their relative orientation in k space are changed. Such changes can be accomplished with the aid of uniform stress along various crystallographic axes. Remember that according to Bir and Pikus (1974) this does not lead to significant changes in the EM binding energy. In the case of a "cold" EM in Si(2–1) two electrons, in conformity with the Pauli principle, may either be in one valley X(100) (a state of the type XX), or in two lowest different valleys but lying on the same stress axis (a state of the type $X\bar{X}$). With the stress applied in other directions, e.g. along ⟨110⟩ or ⟨111⟩, as well as in unstressed Si crystals, the two electrons of the EM may belong to valleys with mutually perpendicular principal axes (states of the type XY). By analogy with the so-called D^- centers in Ge and Si considered by Gershenson et al. (1971) and Narita et al. (1976, 1977) one could assume that the binding energy for an excitonic molecule with XY electrons will be greater than for one with XX or $X\bar{X}$ electrons. This assumption is based on the expected reduction of the Coulomb repulsion of XY electrons due to the anisotropy of their effective masses.

With this purpose Kulakovskii and Timofeev (1980a) have investigated the emission spectra of "cold" EM in Si(2–1) and Si(4–1), as well as of "hot" molecules in Si(2–1). The "hot" biexciton discovered by Kaminskii and Pokrovskii (1979) implies a situation with one of the EM electrons in the upper split-off valley.

Figure 15 displays the emission spectrum of Si stressed along ⟨100⟩ at $T = 2$ K and excitation density $n_{e,h} \approx 4 \times 10^{15}$ cm^{-3}, in which the lines M of "cold" and M_n of "hot" biexcitons are distinctly seen. First of all, making use of the impact ionization method, it was examined whether these lines correspond to weakly bound states, because other radiative processes, e.g. exciton–exciton collisions and electron–hole plasma radiation, may contribute to the spectral region of interest. Figure 15 shows how the spectra are affected by the application of a constant electric field to heat the

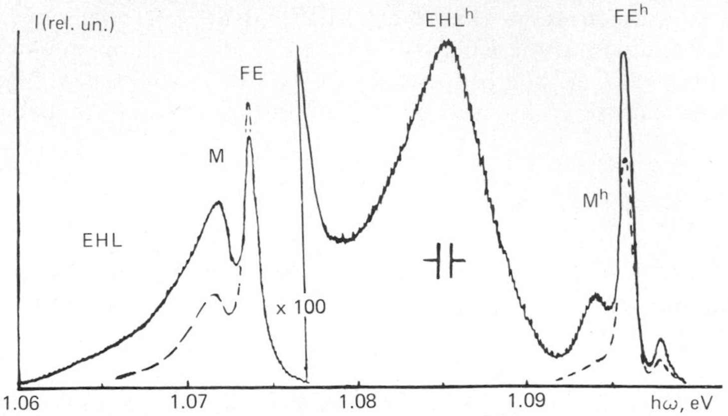

Fig. 15. The emission spectra of "cold" and "hot" excitons (FE), excitonic molecules (M), and EHL of Si stressed along the ⟨100⟩ axis at excitation level $n \approx 4 \times 10^{15}\ \mathrm{cm}^{-3}$ and $T = 2$ K. The dotted curve is measured with electric field $E \approx 10$ V/cm applied. (From Kulakovskii and Timofeev 1980a.)

carriers. It is seen that the intensity of the biexciton line, M, decreases when a field $E \approx 10$ V/cm is applied, whereas the intensity of the exciton emission line, FE, increases (due to impact ionization of EM). The effect of the electric field is more pronounced in the case of "hot" excitons and "hot" EM, as the emission line of the latter disappears from the spectrum at smaller fields. This is due to the decrease of the inter-valley relaxation time as a result of carrier heating by the electric field. Thus these impact ionization experiments prove explicitly that the lines M and M_h correspond to weakly bound states.

A similar biexciton line M is found in the spectra of Si(4–1), too (fig. 16). However its intensity, under some other conditions, is approximately an order of magnitude less than in Si(2–1). This is due to the larger EHL binding energy in Si(4–1), as compared to Si(2–1). Consequently the density of the saturated e–h gas in these crystals is several times less than in Si(2–1). The intensity of this line increases quadratically with the FE intensity. The shape of this line coincides with the contour of the emission line of "cold" EM in Si(2–1) (shown by a dashed line in fig. 16).

A more detailed investigation has been carried out of the "hot" EM line M_h in Si stressed along ⟨100⟩. It appeared that the shape of the line M_h also coincides with the shape of the "cold" biexciton line M and its intensity is $I_n \propto I_{FE} \cdot I^h_{FE}$ (i.e. is proportional to the product of "cold" and "hot" FE intensities). It follows uniquely from this result that this line is due to recombination of the "hot" exciton in an EM consisting of one "cold" and one "hot" exciton.

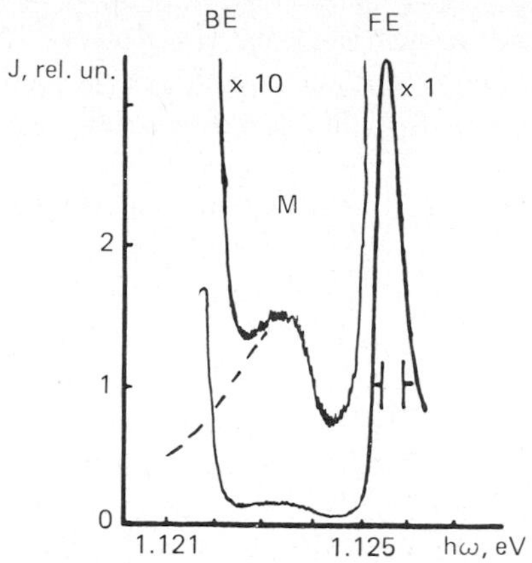

Fig. 16. The TA-phonon assisted component of the emission spectrum of Si stressed along the ⟨110⟩ axis at 2 K. The "cold" EM emission line of Si(2–1) is shown by dotted line for comparison. (From Kulakovskii and Timofeev 1980a.)

Consider now the binding energies of biexcitons in three different cases: when the electronic states in the molecule are of the type XX or $X\bar{X}$ ("cold" EM in Si(2–1)), of the type XX, $X\bar{X}$ or X ("cold" EM in Si(4–1)) and, only of the type XY ("hot" EM in Si(2–1)). To determine the binding energy Kulakovskii and Timofeev (1980a) used eq. (12) relating the EM and FE densities, which is valid under the conditions of thermodynamic equilibrium in the gas phase. To compare the EM binding energies in Si(2–1) and Si(4–1) the intensity ratio I_{EM}/I_{FE} was compared in these crystals at equal FE emission-line intensities, measured in the same experimental geometry. In addition, equal diffusion depths of excitons in Si(2–1) and Si(4–1) were assumed. From such a comparison it was found that the EM binding energy in Si(4–1) is approximately 0.1 meV less than in Si(2–1).

It is easier to estimate the ratio of "hot" to "cold" EM binding energies in Si stressed along ⟨100⟩ because "hot" and "cold" excitons and biexcitons are excited within the same volume, then the diffusion length does not come into question in this case. For the ratio $R = I^{h}_{EM} I_{FE} / I^{h}_{FE} I_{EM}$ we have:

$$R = \frac{1}{2}\left(\frac{m^{h}_{EM}}{m_{EM}}\right)^{3/2} \frac{g^{h}_{EM}}{g_{EM}} \left(\frac{g_{FE}}{g^{h}_{FE}}\right)^{2} \exp\frac{\Delta^{h}_{EM} - \Delta_{EM}}{kT}. \tag{17}$$

In Si(2–1), where there are two lower and four split-off electron ellipsoids, the statistical weights are $g_{FE} = 8$, $g^{h}_{FE} = 16$, $g_{EM} = 16$ and $g^{h}_{EM} = 32$. When the experimental values for the FE, FE^{h}, EM and EM^{h} line intensities were substituted into eq. (17), it was established that the binding energy of the EM with XY electronic states is 0.5 meV less than of those with XX or $X\bar{X}$ states. Thus it turned out unexpectedly that the EM binding energy in Si is insensitive to the degeneracy of the electron valleys, and, contrary to the existing belief, even shows a tendency to decrease in the case of mutually perpendicular orientation of the electron ellipsoids or when the number of the latter is increased.

It is extremely important to know the stability of the biexciton in unstressed Si(6–2). This is so, particularly in connection with the hope of discovering more complicated, free multiparticle complexes containing more than two e–h pairs. Wang and Kittel (1972) considered a multiparticle complex in a model semiconductor with large degeneracy of the electron and hole bands and with an infinite hole mass ($\sigma \to 0$). They found that in the case of eight-fold degeneracy eight-exciton complexes with a binding energy about $6\,\mathrm{Ry}^{\mathrm{ex}}$ should be highly stable. The model considered is remote from a real semiconducting structure. However, if we use the analogy with the multiparticle bound-exciton complex, the structure of which is well described within the framework of the shell model (Kirczenow 1977, Thewalt 1977, see also Ch. 10 of this volume) we may expect that in Ge and Si, where the valence band is four-fold degenerate, four-exciton complexes with electrons and holes in the singlet state will be stable. Note, that a three-exciton complex in these crystals will be unstable because 3 electrons (3 holes) cannot form a spin-singlet state.

Up to now, only Thewalt and Rostworowski (1978) have reported the observation of the biexciton emission line in unstressed Si(6–2) crystals. The EM binding energy was estimated in this paper at $\Delta_{\mathrm{EM}} \approx 1.3$ meV. As this estimate was based on the analysis of the spectrum shape and was not obtained by the thermodynamic method, it is rather difficult so far to say anything about its exactness. In addition, it was found in this paper that in Si(6–2) the EM to FE ratio of line intensities has a maximum value of $I_{\mathrm{EM}}/I_{\mathrm{FE}} \approx 1/20$, whereas from the data of Kulakovskii and Timofeev (1980a) it follows that at the same temperature ($T = 2$ K) even close to the EHL condensation threshold the background radiation in the region of the M line does not exceed 1/50 of I_{FE}. Thus the problem of biexciton stability in Si(6–2) as well as the possibility, in principle, of the existence of more complicated free multiparticle complexes in these crystals needs further investigation.

7. *Excitonic molecules in a magnetic field*

By virtue of the small binding energy the excitonic molecules in semiconductors appear to provide a convenient model for experimental studies of molecular properties in magnetic field. For example, in the case of EM in Si(2–1) which is close to the model of positronium molecules, fields of the order of 80 kOe are sufficient to make the diamagnetic-shift energy $\delta E_{\mathrm{dia}}^{\mathrm{ex}}$ of the free exciton and the paramagnetic-splitting energy $g_{e,h} \cdot \mu_B \cdot H$ of the electron and hole spin states in the exciton comparable to the EM binding energy. The same situation for the hydrogen molecule could be expected at astronomical magnetic fields of the order of 10^5 kOe and 10^7 kOe, respectively.

In this connection the problem arises of biexciton stability under the effect of such magnetic fields. It seemed, a priori, that the biexciton with its approximately equal effective masses m_e and m_h, and its loose formation compared with the hydrogen molecule, should have a magnetic susceptibility considerably greater than twice the exciton susceptibility. In fact, the mean interparticle distances $\langle r_{ee}\rangle$, $\langle r_{eh}\rangle$, $\langle r_{hh}\rangle$ in the EM calculated with the aid of the most convenient variational wave function for $m_e \approx m_h$ are of the order of $3a_{ex}$ (fig. 2). Therefore, by analogy with atoms, where according to Langevin the diamagnetic susceptibility is proportional to the square of the distance between the electron and the nucleus, it could be expected that $\chi_{EM} \sim 10\chi_{ex}$ where χ_{ex} is the diamagnetic susceptibility of the exciton. In this case it could appear that the stability of the EM decreases in magnetic field. On the other hand, the smallness of the EM binding energy as compared to the exciton Rydberg is an indication to the fact that the electron–hole correlations in the biexciton are of almost the same order as in the exciton and, consequently, χ_{EM} will not differ significantly from $2\chi_{ex}$.

The experimental observations of Kulakovskii et al. (1978b, 1979b) have shown that in Si(2–1) the EM emission line is present in the spectra up to magnetic fields $H \approx 100$ kOe, with the FE to EM intensity ratio almost unchanged. We first consider the diamagnetic properties of EM.

The Hamiltonian of the EM in magnetic field is

$$\hat{H} = \frac{1}{2m}\left[\left(\boldsymbol{p}_1 + \frac{e}{c}\boldsymbol{A}_1\right)^2 + \left(\boldsymbol{P}_a - \frac{e}{c}\boldsymbol{A}_a\right)^2 + \left(\boldsymbol{P}_2 + \frac{e}{c}\boldsymbol{A}_2\right)^2 + \left(\boldsymbol{P}_b - \frac{e}{c}\boldsymbol{A}_b\right)^2\right] + V_{\text{Coul}}, \tag{18}$$

where $\boldsymbol{r}_1$, $\boldsymbol{r}_2$ and $\boldsymbol{r}_a$, $\boldsymbol{r}_b$ are the coordinates of electrons and holes, respectively, and $\boldsymbol{A}_r = \frac{1}{2}\boldsymbol{Hr}$. Here no account is taken of the interaction of spins with magnetic field, for it is assumed that this does not affect the orbital motion. Remember that in semiconductors with nondegenerate bands two electrons and two holes form spin singlets, in conformity with the Pauli principle. Further it will be always assumed that the Zeeman energies are much less than the Coulomb interaction energy.

The motion of the centre of gravity of the molecule can be separated from the internal motion by the method of Lamb (1952), namely, the operator of the EM total momentum (which is a motion integral) is written in the form:

$$\hat{\boldsymbol{K}} = \boldsymbol{P}_1 + \boldsymbol{P}_2 + \boldsymbol{P}_a + \boldsymbol{P}_b + \frac{e}{c}[\boldsymbol{A}_a + \boldsymbol{A}_b - \boldsymbol{A}_1 - \boldsymbol{A}_2]. \tag{19}$$

If reduced coordinates,

$$z = \boldsymbol{r}_1 - \boldsymbol{r}_a, \qquad \zeta = \boldsymbol{r}_2 - \boldsymbol{r}_b, \qquad \boldsymbol{R} = \tfrac{1}{2}(\boldsymbol{r}_1 + \boldsymbol{r}_2 - \boldsymbol{r}_a - \boldsymbol{r}_b)$$
$$\boldsymbol{q} = \tfrac{1}{4}(\boldsymbol{r}_1 + \boldsymbol{r}_2 + \boldsymbol{r}_a + \boldsymbol{r}_b), \tag{20}$$

are introduced, then the eigenfunction of the K operator corresponding to the eigenvalue K is given by

$$\phi_{\boldsymbol{K}} = \exp\left\{\frac{i}{\hbar}\left(\boldsymbol{K} + \frac{e}{c}\,\boldsymbol{A}_{z+\zeta}\right)q\right\}\psi_{\boldsymbol{K}}(\boldsymbol{R}, z, \zeta). \tag{21}$$

Here $\psi_{\boldsymbol{K}}$, as was shown by Edelstein (1979), is the eigenfunction of the Hamiltonian $H_{\boldsymbol{K}}$:

$$\hat{H}_{\boldsymbol{K}} = \frac{\boldsymbol{K}^2}{8m} + \frac{e}{mc}\,\boldsymbol{K}\boldsymbol{A}_{z+\zeta} + \hat{H}_0 + \hat{H}_1 + \hat{H}_2, \tag{22}$$

where

$$\hat{H}_0 = \frac{1}{2}\,\boldsymbol{P}_{\boldsymbol{R}}^2 + \frac{1}{m}(\boldsymbol{P}_z^2 + \boldsymbol{P}_\zeta^2) + V_{\text{Coul}}$$

is the Hamiltonian of the molecule in the centre-of-gravity system with no magnetic field applied. Edelstein (1979) has found that an energy-correction term proportional to the square of the magnetic field comes from $\hat{H}_1$ in the second order perturbation theory (van Vleck orbital paramagnetism) and from $\hat{H}_2$ in the first order perturbation theory (Langevin diamagnetism).

Consider first the contribution of $\hat{H}_2$ to the magnetic susceptibility χ_{EM}. It follows from the symmetry of the ψ function of the ground state $\hat{H}_0$ with respect to permutation of two electrons that the Langevin correction to the energy is given by,

$$\delta E_{\text{EM}}^{\text{L}} = \frac{e^2}{mc^2}\left[\tfrac{3}{4}(\langle A_z^2\rangle + \langle A_\zeta^2\rangle) + \tfrac{1}{2}\langle \boldsymbol{A}_z\boldsymbol{A}_\zeta\rangle\right]. \tag{23}$$

It is interesting to compare $\delta E_{\text{EM}}^{\text{L}}$ with the Langevin formula for the diamagnetic shift of the atom $\delta E_{\text{at}}^{\text{L}}$. We have

$$\delta E_{\text{at}}^{\text{L}} = \frac{e^2}{8c^2}\sum_i \frac{\langle 0|[\boldsymbol{H}\boldsymbol{r}_i]|0\rangle}{m_i}, \tag{24}$$

where $|0\rangle$ is the ground state vector of the atom. It is seen that $\delta E_{\text{EM}}^{\text{L}}$ does not reduce to the mean square of some distance, as happens in the case of atoms. The calculation of integrals in eq. (23) using the variational function suggested by Brinkman et al. (1973) yields

$$\delta E_{\text{EM}}^{\text{L}} = 11.4\,\frac{e^2}{6mc^2}\,H^2 a_{\text{ex}}^2. \tag{25}$$

As is known the diamagnetic energy-shift for one exciton is

$$\delta E_{ex}^{dia} = 3\frac{e^2}{6mc^2}H^2a_{ex}^2. \tag{26}$$

The van Vleck paramagnetic shift is $\delta E^{VV} < 0$. Hence for the EM to FE ratio of magnetic susceptibilities we obtain the inequality

$$\chi_{EM}/2\chi_{ex} < 1.9. \tag{27}$$

Thus it appears that the EM diamagnetic susceptibility calculated by Edelstein (1979) is not very far from twice the FE susceptibility, although the used variational function did not contain explicitly the wave functions of the two excitons. It should be born in mind that the accuracy of eq. (27) corresponds to that of the trial wave function.

The diamagnetic properties of the EM were investigated experimentally by Kulakovskii et al. (1979b). Since the excitonic state is the final state of EM radiative decay the properties of excitons in magnetic field were simultaneously studied. From the Zeeman splitting the g factors of the electron and hole were determined. In the case of holes this factor, as expected, was found to be anisotropic. In addition, the susceptibility of the exciton was estimated from the magnitude of the diamagnetic spectral shift. It appeared that the diamagnetic shift of the exciton, δE_{ex}^{dia}, is anisotropic too. This is mainly due to the strong anisotropy of the electron effective mass (in Si(2–1) $m_{h\|} \approx 0.8m_{h\perp}$, whereas $m_{e\|} \approx 5m_{e\perp}$). It was possible to follow the anisotropy of the diamagnetic shift for the geometry $H\|p\|$ $\langle 100\rangle$ in the case of "cold" and "hot" excitons which have electron ellipsoids with the long axis parallel or perpendicular to the magnetic field respectively. If we use the relation $\chi = H^{-1}\,\partial E/\partial H$ for the diamagnetic susceptibility, then from the experimentally observed diamagnetic shifts we obtain for the "cold" exciton $\chi_c \approx 2.8\times 10^{-25}$ erg/Oe^2 and for the "hot" exciton $\chi_h \approx 1.3\times 10^{-25}$ erg/Oe^2. These values may be compared to the diamagnetic susceptibility of the hydrogen atom $\chi_H = 1.3\times 10^{-30}$ erg/Oe^2.

Since Si(2–1) has two equivalent valleys in the conduction band the formation is possible of biexcitons with the total spin 0 and 1. Therefore the lowest EM state is split by the magnetic field into a triplet,

$$E_{S_z,0}^{EM}(H) = E^{EM}(0) + \tfrac{1}{2}g_eS_z\mu_BH + \delta E_{EM}^{dia}, \tag{28}$$

where $S_z = S_{z_1} + S_{z_2} = 0, \pm 1$. In a magnetic field the energy of an exciton with electron and hole spins S_z and j_z, respectively, is

$$E_{S_z,j_z}^{ex}(H) = E^{ex}(0) + (g_eS_z + g_hj_z)\mu_BH + \delta E_{ex}^{dia}. \tag{29}$$

Hence for the EM binding energy in a magnetic field H we obtain

$$\Delta_{EM}(H) = 2E_{-1/2,-1/2}^{ex} - E_{-1,0}^{EM} = \Delta_{EM}(0) - g_h\mu_BH + 2\delta E_{ex}^{dia} - \delta E_{EM}^{dia}. \tag{30}$$

Besides, for uniaxially stressed Si(2–1) crystals it should be taken into account that the spin-relaxation times are comparable with the exciton lifetime (Kulakovskii et al. 1979a). Therefore, even when the paramagnetic splitting $g\mu_B H \approx 3kT$ the population of different spin states is of the same order. Because of this, in Si(2–1) with magnetic field applied they are lower in energy.

Those biexciton states in which the holes form a spin singlet are lowered in energy. Then the EM binding energy is given by the expression

$$\Delta_{EM}(H) = -(E^{EM}_{S_{z_1}+S_{z_2},0} - E^{ex}_{S_{z_1},1/2} - E^{ex}_{S_{z_1},-1/2}) = \Delta_{EM}(0) + 2\delta E^{dia}_{ex} - \delta E^{dia}_{EM}. \quad (31)$$

The FE and EM emission spectrum of Si(2–1) in a field of about $H \approx 80$ kOe is shown in fig. 17 together with the scheme of expected transitions. In a magnetic field the emission lines are strongly broadened, so that it is not possible to resolve the structure of separate Zeeman components. Therefore the change of the EM binding energy in the magnetic field can be estimated only from the change of the I_{EM} to I_{ex} ratio of integrated intensities under the assumption of quasi-equilibrium between excitons and biexcitons, $I_{EM}/I_{ex} \propto \exp(\Delta_{EM}/kT)$. Examining this ratio over the whole range $H \leqslant 80$ kOe both for $\bar{H}\|\bar{p}\|\langle 100\rangle$ and for $\bar{H}\perp\bar{p}\|\langle 100\rangle$ it was found that the EM diamagnetic shift does not exceed 3 excitonic shifts, in agreement with the above calculation. This experimental upper limit of the EM to FE ratio of susceptibilities is a direct indication of very strong electron–hole correlations in the biexciton, analogous to the positronium molecule.

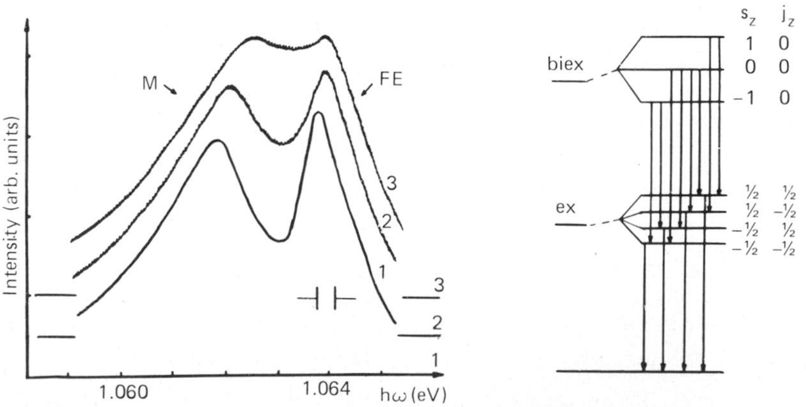

Fig. 17. Emission spectra of excitons and excitonic molecules in Si(2–1) at $T = 2$ K without (1) and with (2), (3) magnetic field $H = 8$ T (Voigt configuration). (2) $\boldsymbol{H} \perp \boldsymbol{E}$, (3) $\boldsymbol{H} \parallel \boldsymbol{E}$. The right side figure shows the optical transitions between split excitonic molecule levels and exciton levels when magnetic field is applied. (From Kulakovskii et al. 1979b.)

It is interesting to mention the case of a biexciton in extremely strong magnetic field when the contribution to the ground state energy caused by the magnetic field is much greater than the energy of Coulomb interaction. The stability of the EM in such magnetic fields, according to the theoretical analysis of Chui (1974) should increase as $\ln^2 H$. In this limit of magnetic field the electron and hole spins are aligned parallel to the field and both equal to 1. EM with such a spin orientation are completely unstable at $H = 0$. We note, in this connection, that the experimental data of Kulakovskii et al. (1979b) do not exclude a monotonic increase of the EM binding energy in Si(2–1) with increasing magnetic field.

8. *Excitonic trions*

By analogy with the localized states D^- and A^+ in semiconductors one can suppose that at low temperatures stable, three-particle complexes or trions may be present in the nonequilibrium e–h gas, namely, FE ions $X^-(e_2h)$ and EM ions $X_2^+(eh_2)$. These complexes can be regarded as bound states of an exciton with an electron or a hole. The idea of such complexes was proposed by Lampert (1958) who predicted their stability from the analogy with the ions of hydrogen atom and molecule, H^- and H_2^+.

Munschy and Stebe (1974, 1975) and Insepov and Norman (1975) have carried out a variational calculation of the trion binding energy. An isotropic model was used for the calculation with no account for spin states and band degeneracy. The stability of trions was investigated as a function of the mass–ratio parameter $\sigma = m_e/m_h$ $(0 \leqslant \sigma \leqslant 1)$. It follows from variational calculations, the result of which is presented in fig. 18 that the trions (e_2h) and (eh_2) should be stable at any σ. The calculations did not invoke band degeneracy, anisotropy of the effective mass spectrum and interaction of spin states. If these effects were taken into account the dependence $\Delta_T(\sigma)$ could be modified, however this would not hardly change the main result already obtained, namely, that $\Delta_T > 0$, $\partial^2\Delta_T/\partial\sigma^2 > 0$ over the entire range $\sigma \in [0, \infty]$.

Up to now two reports are known where the model of excitonic trions has been used to interpret some of the observed peculiarities connected with the high-density e–h gas, in the emission spectra of Ge (Thomas and Rice 1977) and in the cyclotron-resonance spectra of uniaxially stressed Si (Kawabata et al. 1977).

First we consider the features to be expected in the emission spectra of trions. It can be affirmed a priori that the emission spectrum of trions in indirect gap semiconductors, as in the biexciton case, must have a finite width even when the complex is at rest, i.e. as $T \to 0$. This is due to the fact that in an indirect optical transition a part of the trion energy will be

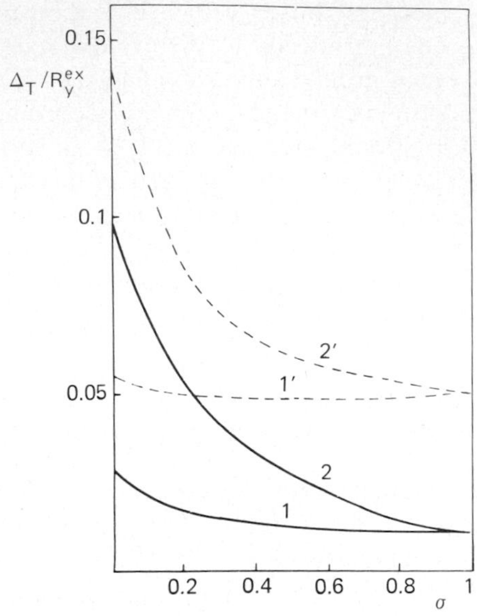

Fig. 18. The binding energy of the trions (e_2h) (1) and (eh_2) (2) in units of the exciton Rydberg for different electron-to-hole mass ratios. The solid lines are calculated by Insepov and Norman (1975), the dashed lines by Stebe and Munschy (1975).

consumed in imparting kinetic energy to the "recoil" particle, namely, to the electron in the case of a (e_2h) complex, and to the hole in the case of (eh_2). Compared to the EM emission line, the emission band of (e_2h) or (eh_2) trions should be broader by a factor $(1+m_e/m_h)$ or $(1+m_h/m_e)$, respectively (see eq. (6)). Therefore the full width of the trion emission spectrum is much greater than its binding energy (in Ge and Si more than by an order of magnitude). As to the line shape of the trion spectrum, this, apparently, should not differ considerably from that of the biexciton line, especially in its low-energy part which corresponds to large transfer momenta. The main obstacle for the experimental detection of trions from emission spectra consists in the fact that the spectral positions of the trion and EM emissions are very close. Furthermore, the partial fraction of trions in the nonequilibrium e–h gas is less than that of biexcitons because of the smaller binding energy. This additionally hampers the extraction of the trion contribution from the resulting spectrum.

Thomas and Rice (1977) and Thomas et al. (1978) have studied the emission spectra of indirect excitons in Ge at temperatures above the critical temperature of condensation into EHL ($T > T_c$). With increasing density of the e–h gas they observed a broadening of the exciton–phonon emission spectrum (the TA component) towards lower energies. The authors attribute this low-energy tail to trions with a small fraction of biexcitons; the partial fraction of these increases with increasing density. However, the EM and trion contributions to the resulting spectrum could not be separated; moreover, it was not possible to isolate the trion emission from the exciton band, in contrast to the situation in uniaxially stressed silicon. It should be born in mind that at densities exceeding the density of the exciton-plasma transition the emission from the electron–hole plasma sets in just in the spectral region where EM and trion emission is expected. Zhydkov and Pokrovskii (1979) have used the method of impact ionization of weakly bound states (of the biexciton type) and find that the observed broadening (or the low-energy tail) of the spectrum in uniaxially stressed Ge could be caused by EHP emission rather than by EM or trions. Thus, there have yet to be reliable observations of trion emission by means of optical spectroscopy.

Now consider attempts to detect trions in the cyclotron-resonance (CR) absorption spectra. Since the trion is a free charged complex the application of an external magnetic field leads to its cyclotron motion. The complex will move as a whole around the magnetic field direction if the Zeeman energy of the complex is much less than its binding energy. It is reasonable in this connection to use lower cyclotron frequencies without violating, however, the general condition $\omega_c\tau \gg 1$ necessary for CR observations.

Kawabata et al. (1977) studied CR in the high-density nonequilibrium e–h gas in uniaxially stressed Si crystals. The uniaxial stress was applied along the direction of the external magnetic field in order to simplify the expected picture of the CR spectrum for excitonic ions. In addition, under the conditions of uniaxial stress it was possible to avoid the complications arising from the condensation into EHL. The measurements were performed in the Voigt configuration. Figure 19(a), (b), (c) displays the CR spectra measured in Si(2–1), Si(4–1) and Si(6–1) at $T = 1,7$ K high excitation levels. In all stress configurations two new resonance absorption peaks are observed (denoted by the indices X_1 and X_2) which are found in the region of stronger magnetic fields as compared to the canonical CR spectrum of free electrons and holes (the lines e and h in the figure). These new peaks disappear from the spectra at $T = 4.2$ K and their intensity grows more rapidly with increasing excitation level then the intensity of the lines e and h. If the peaks X_1 and X_2 are associated with the cyclotron resonance absorption of the e_2h and eh_2 trions, then it is possible from their positions in the spectrum to determine the cyclotron masses of the trions and to compare them with the expected values, $(2m_e + m_h)$ and $(m_e + 2m_h)$ respectively. It appears that in all stress configurations the cyclotron masses of the supposed complexes e_2h and eh_2 found from CR spectra are approximately by 15–20% larger than expected. One of the possible origins

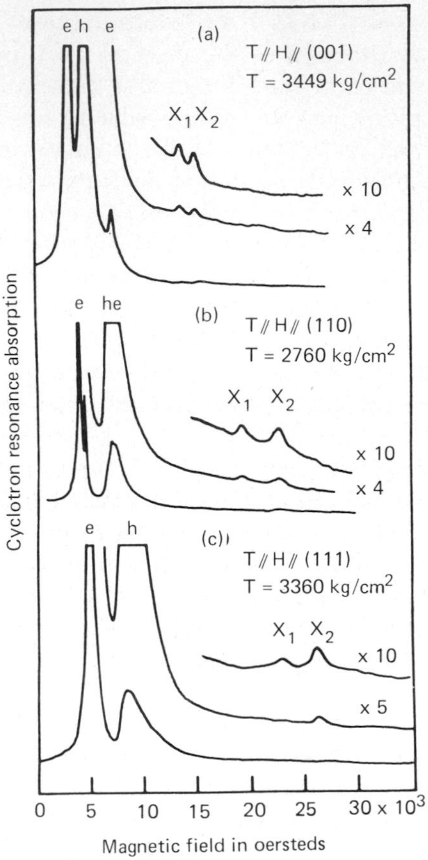

Fig. 19. The cyclotron resonance absorption spectra of stressed Si at 0.7 K under light excitation in three different configurations of stress (a), (b), (c). In this spectra, e and h denote the cyclotron resonance of free electron and the split band hole. (From Kawabata et al., 1977.)

of this discrepancy (Kawabata et al. 1977) is the fact that the excitonic states $M_J = \pm 1/2$ and $M_J = \pm 3/2$ remain mixed under relatively small uniaxial stresses used.

Here attention should be called to two circumstances which so far preclude the possibility of unique interpretation of the new peaks detected in the CR spectra of uniaxially stressed Si within the framework of the excitonic-ion model. The first circumstance is the spin resonance of free excitons which could be expected in the same spectral region. Secondly, for not very large uniaxial stresses and under intense optical pumping the cyclotron resonance of "hot" holes in both split-off band may be observed (Kulakovskii et al. 1980b). Due to the interaction between the split-off bands the dispersion law $E(k)$ for the holes strongly deviates from quadratic in an energy range of the order of the stress splitting. Therefore the CR spectrum of "hot" holes in the split-off bands differs from that of holes in the points of zero slope. It should be stressed, however, that despite the above apprehensions the investigation of CR spectra seems to be one of the most promising methods for the search and study of free, charged multiparticle complexes in semiconductors.

9. Gas of excitons and excitonic molecules along the equilibrium boundary with electron–hole liquid

In the indirect gap semiconductors Ge and Si, the exciton gas at low temperatures and sufficiently high densities may condense into EHL (Keldysh 1968). This condensation phenomenon is a phase transition of the first kind. The region of coexistence of the gas and liquid phases in the nonequilibrium e–h system is determined by the phase diagram and is usually described in coordinates n, T. From the experimental point of view the gas–EHL phase diagrams have been most thoroughly investigated in germanium (Thomas et al. 1974a, b, 1979) and in silicon (Shah et al. 1977 and Dite et al. 1977). By now the properties of the EHL itself are well known and it has been established that up to the critical region it can be described as a two-component Fermi liquid (Rice 1977, Hensel et al. 1977).

The interest of the phenomenon of condensation of a dielectric exciton gas into a metallic EHL becomes still greater in connection with the problem of the metal–insulator transition. It should be recalled that, taking mercury as an example, Landau and Zeldovitch (1943) have discussed the possibility, in principle, of the separate occurrence of two phase transitions, namely, gas–liquid and metal–insulator in a single system, and of two critical regions corresponding to these transitions. However hitherto these two transitions have not been observed separately in the same system.

The nonequilibrium e–h system in a semiconductor at low temperatures may be considered as a convenient model for studying this problem. It was expected, on the basis of critical-density estimates, that in such a system the two-phase transitions might occur separately (Rice 1974). The possibility of two critical points existing on the gas–EHL phase diagram was discussed theoretically in the papers by Insepov et al. (1972, 1976), Rice (1974), Sander (1976), Ebeling et al. (1976) and Thomas (1977). Here two criteria can be pointed out, which are used to estimate the critical density $n_c(r_s^c)$ corresponding to the ionization collapse of excitons, when the dielectric gas of excitons transforms into a completely ionized e–h plasma (exciton–plasma (E–P) transition). One of the criteria (known as Mott criterion) is based on the numerical solution of a two-particle Schrödinger equation with a statically screened Coulomb potential $V(r) = (e^2/\kappa r)\exp(-r/r_0)$ and establishes that the FE binding energy vanishes at

$$r_0(\bar{n}) \approx 0.84 a_{ex}, \tag{32}$$

where $r_0(\bar{n})$ is the static screening radius.

Approximately the same result was obtained recently by Zimmermann et al. (1978) who took into account the effects of dynamic screening of Coulomb interaction. In a model semiconductor with simple bands and isotropic and equal effective masses $m_e = m_h$, according to such a criterion of plasma screening, the E–P transition at $T = 0$ K occurs at critical densities corresponding to $r_s^{cp} = 6.8$ (Sander and Fairobent, 1976). This density is more than two orders of magnitude less than the critical-density value for the gas–EHL phase transition ($r_s^0 \sim 1$). Note that in the case of mercury and other liquid metals the critical-density estimates for the transitions metal–insulator and gas–liquid are of the same order of magnitude (Mott 1974). Within the framework of the other approach the critical density of the E–P transition is estimated from the "collapse", due to dielectric screening of Coulomb interaction of the energy gap in the spectrum of one-particle collective excitations in the high-density exciton system (Bisti and Silin, 1978). In this approximation the excitons remain stable at larger densities, up to $r_s^{cd} \approx 1.8$, i.e. until the wave functions overlap. It remained an open question which of these criteria is appropriate to the experiment in the case of an e–h system in a semiconductor and whether the E–P transition will have the character of a phase transition and show peculiarities along the gas–liquid phase boundary.

The most convenient object for the experimental investigation of the problems appeared to be Si crystals uniaxially stressed along the axis ⟨100⟩, Si(2–1). As has already been pointed out in sect. 3, the gas–EHL phase boundary is shifted towards higher densities by more than an order of magnitude as compared to unstressed Si crystals. As a result of this at low temperatures the FE and EM concentrations in such a dense e–h gas are of the same order of magnitude. This enabled Kulakovskii et al. (1980c), by measuring the radiative recombination spectra and photoconductivity, to follow experimentally the partial composition of the gas phase (excitons, biexcitons, free electrons and holes) in Si(2–1) up to densities corresponding to $r_s \approx 2$ to 2.5. The composition of the gas phase was analyzed along the equilibrium boundary. The density n_g of saturated vapours of the gas phase, at which condensation into EHL occurred, was determined from threshold measurements of the FE and EHL line intensities. The absolute values of n_g were estimated with the aid of special calibrational measurements of $\bar{n}_{e,h}$ with accuracy to a factor of 2. In determining the FE and EM concentrations from the intensity ratio of their lines it was taken into account that the EM to FE ratio of radiative probabilities calculated with the use of the variational function of Brinkman et al. (1973) is equal to 2.3. Also, the free carrier concentration was estimated independently by means of photoconductivity measurements.

Kulakovskii et al. (1980c) have established that with increasing density up to $r_s \approx (2.7 \mp 0.3)$ and $T = 12.5$ K the gas phase consists mainly of FE and EM, the fraction of free carriers (and, apparently, of trions) being more than an order of magnitude smaller. The partial composition of the gas phase in Si(2–1) is seen from fig. 20, which shows the gas–EHL phase diagram.

The E–P transition in Si(2–1) was studied by investigating the smoothing of the FE and EM discrete spectra into the structureless spectrum of the e–h plasma at higher excitation levels and at temperatures $T > T_c$. No critical density n_c of the E–P transition was seen, but a gradual change over the range $T = 13$ to 21 K with densities $n_c = (1.5–3) \times 10^{17}\ \text{cm}^{-3}$ ($r_s = 2–2.5$).

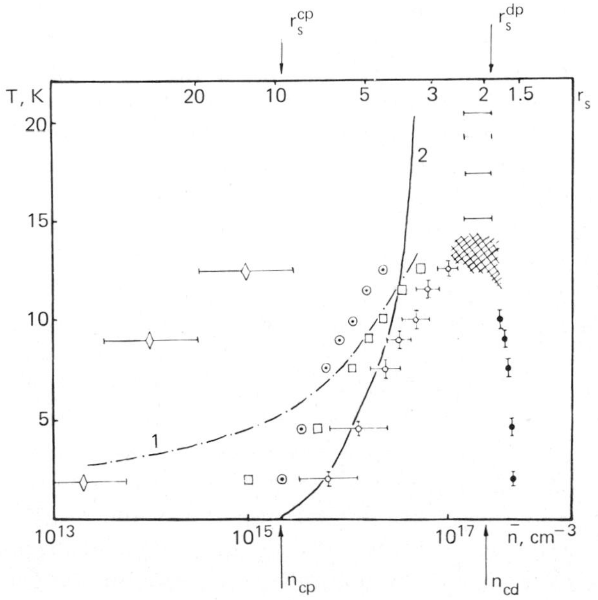

Fig. 20. Gas–liquid phase diagram for Si(2–1). (●) density of EHL, (○) onset density of the gas phase (□) excitons, (⊙) excitonic molecule, (◇) free carriers). The critical region is dashed. Curve (1) corresponds to the equilibrium density of the exciton gas at the phase boundary and (2) is the line calculated for the Mott transition. The region of exciton-plasma transformation is shown by the short horizontal lines at $T \geqslant T_c$. Arrows indicate the magnitude of the critical density of the exciton-plasma transition calculated in terms of plasma – (r_s^{cp}), and dielectric – (r_s^{cd}) screening of Coulomb interaction. (From Kulakovskii et al. 1980c.)

Thus, in Si(2–1) the E–P change-over occurs close to the critical region of the gas–EHL transition (in fig. 20 the critical region is shaded). It was possible to observe the E–P transition in Si(2–1) at lower temperatures (approximately half) than in Si(6–2) (Dite et al. 1977) because of considerably higher gas-phase densities which can be realized in these crystals. We recall that on the basis of the plasma-screening model the ionizational breakup of excitons was expected to occur abruptly and to be accompanied by a splitting of the excited volume into regions occupied by weakly and highly ionized exciton gas (Insepov et al. 1972, 1976, Ebeling et al. 1976). However in Si(2–1), as well as in Si(6–2) (Shah et al. 1977, Dite et al. 1977), and in Ge(4–2) (Thomas 1978) the observed transformation of the discrete excitonic spectrum into a plasma one is smooth (diffuse) and does not exhibit features characteristic of a first order phase transition.

Thus, the above analysis of the partial composition of the e–h gas along the phase boundary with the EHL shows that the excitonic (and molecular) states in Si(2–1) retain their individuality up to densities corresponding to $r_s \approx 2.5$. It is interesting to compare these experimental observations with the existing estimates for the E–P transition. According to the Mott criterion (eq. 32) based on the plasma-screening approach the critical density n_{cp} of the E–P transition in Si(2–1) at $T = 0$ K corresponds to $r_s^{cp} \approx 10$ (indicated by an arrow in fig. 20). It is seen that at $T \to 0$ K this value of the critical density is just within the two-phase region, i.e., the E–P transition is preceded by the condensation of the exciton gas into EHL. With increasing temperature the effectiveness of the screening decreases, and the critical values n_{cp} are shifted to higher densities. It is easy to calculate the line of the Mott transition for Si(2–1) in the temperature range of interest (2–20 K) using formula (32) and the following expression for the static screening radius,

$$r_{scr}^{-2} = \sum_{e,h} \frac{e^2 (m_{e,h})^{3/2} (2kT)^{1/2}}{\kappa \pi \hbar^3} F_{-1/2}(E_{e,h}^F / kT), \tag{33}$$

which is valid for arbitrary degeneracy of the electron and hole gas. In eq. (33) $F_{-1/2}(E_{e,h}^F/kT)$ is the Fermi integral. The calculated Mott density is shown in fig. 20 (curve 2). It is seen that at $T \approx 8$ K the threshold density of the e–h gas consisting mainly of FE and EM considerably exceeds the critical density of the E–P transition calculated from the Mott criterion. It means that the plasma-screening approximation is inadequate for the state in this region. This is not surprising since the critical-density values obtained in this approximation fall into the region where the e–h gas cannot be regarded as weakly interacting. It is quite natural to expect that at such densities and temperatures its state should be considerably affected by electron–hole correlations which are not taken into account within the plasma-screening approach. Of great interest, in this connection, are the estimates obtained by Bisti and Silin (1978) in the approximation of dielectric screening of Coulomb interaction. It was found that the dielectric gap in the spectrum of one-particle excitations of the exciton system becomes vanishingly small at $r_s^{cd} \approx 1.8$ which is close to the experimental data (shown by an arrow in fig. 20). In connection it should be noted that in direct gap semiconductors the discrete excitonic structure is smoothed out and practically disappears at densities corresponding to $r_s \approx 2$ (Frova et al. 1977, Lyssenko and Revenko 1978).

The gas–EHL phase diagram in Ge was studied recently by Thomas et al. (1978) with the purpose of considering the possible existence of two critical regions, namely, gas–liquid and metal-insulator. The authors have carried out spectroscopic measurements on the gas side of the phase boundary but did not find evidence for a metal-insulator phase transition separate from the gas–liquid transition. However, they did find that the shape of the phase boundary near the critical region deviates from that observed in the case of simple fluids. Thomas et al. (1979) refer to these deviations as a "Mott distortion" of the phase diagram believing that the gradual ionization of the dielectric FE and EM gas affects the shape of the phase diagram. As to the partial composition of the gas phase, this remains an open question in the case of Ge since for the time being there are no reliable observations of excitonic molecules and trions along the equilibrium boundary with the liquid in these crystals.

10. Conclusion

Thus, we have seen that in the class of indirect gap semiconductors the subject of most comprehensive studies has been up to now the four-particle complex, or the excitonic molecule in uniaxially stressed silicon crystals. Since the average masses of electrons and holes in Si(2–1) are close to each other the discovery of excitonic molecules in these crystals may be regarded as experimental evidence for the stability of a free complex analogous in structure to the positronium molecule. By virtue of its long electron–hole recombination times Si(2–1) appears, at present, to be most convenient for the search for Bose–Einstein condensation of the EM gas, since the main question here is whether it is possible, in principle, to cool the saturated EM gas at sufficiently high density down to the critical temperature corresponding to such a phase transition.

There are a number of other problems related to the subject of the present review which still wait for their solution. Among these we may mention the problem of the binding energy of four-exciton complexes in indirect gap semiconductors like Si and Ge. So far no reliable variational calculations of the binding energy of such complicated electron–hole structures are known, they are only expected by analogy with bound multi-exciton complexes. Nor is the effect of effective-mass anisotropy and complicated band structure on the stability of free complexes known. For example, it may be expected that, compared to the isotropic model, the stability of the complexes will be greater in semiconducting structures with strong and approximately equal anisotropy of electron and hole effective masses when the directions of the principal axes of the electron and hole ellipsoids coincide. Finally, of a certain interest is the problem of stability of multiparticle complexes in polar indirect gap semiconductors because their binding energy may be considerably modified by the strong interaction with optical phonons.

Acknowledgements

The author wishes to express his sincere gratitude to his colleagues Dr. A.F. Dite, Dr. V.M. Edelstein and Dr. V.D. Kulakovskii for their discussions. Thanks are also due to Prof. E.I. Rashba for a careful reading of the manuscript and some useful comments.

References

Adamowski, J., S. Bednarek and M. Suffczynski, 1971, Solid State Commun. **9**, 2037.
Adamowski, J., Bednarek and M. Suffczynski, 1978, J. Phys. **C11**, 4515.
Akimoto, A. and E. Hanamura, 1972a, Solid State Commun. **10**, 253.
Akimoto, A. and E. Hanamura, 1972b, J. Phys. Soc. Japan, **33**, 1357.
Bagaev, V.S., T.I. Galkina, O.V. Gogolin and L.V. Keldish, 1969, Pis'ma Red. Zh. Eksp. Teor. Fiz. **10**, 309 (1969, JETP Lett. **10**, 195).
Benoit à la Guillaume, C., M. Voos and F. Salvon, 1972, Phys. Rev. **B5**, 3079.
Bir, G.L. and G.E. Pikus, 1974, Symmetry and Strain-Induced Effects in Semiconductors (Keter Publish House Jerusalem) ch. 7.
Bisti, V.E. and A.P. Silin, 1978, Fiz. Tverd. Tela (Sov. Phys. Solid State) **20**, 1850.
Brinkman, W.F., T.M. Rice and J.B. Bell, 1973, Phys. Rev. **B8**, 1570.
Cho, K., 1973, Opt. Commun. **8**, 412.
Chui, S.T., 1974, Phys. Rev. **B9**, 3438.
Dite, A.F., V.D. Kulakovskii and V.B. Timofeev, 1977, Zh. Eksp. Teor. Fiz. **72**, 1156 (1977, Sov. Phys. JETP **45**, 604).
Ebeling, W., W. Kraeft and D. Kremp, 1976, Ergebnisse der Plasmaphysik und der Gaselektronik, ed., R. Rompe and M. Steenback (Academie-Verlag, Berlin) ch. 5.
Edelstein, V.M., 1979, Zh. Eksp. Teor. Fiz. **77**, 760.
Elliott, R.J., 1957, Phys. Rev. **108**, 1384.
Forney, J.J. and A. Baldereschi, 1978. Proc. 14th Intern. Conf. Physics Semic. Edinburgh, ed., B.L.H. Wilson (The Institute of Physics Bristol and London) p. 635.
Frova, A., P. Schmid, A. Grisel and F. Levy, 1977, Sol. State Commun. **23**, 45.

Gershenson, E.M., G.M. Gol'tsman and A.P. Mel'nikov, 1971, Pis'ma Red. Zh. Eksp. Teor. Fiz. **14**, 281 (1971, JETP Lett. **14**, 185).
Gourley, P.L., J.P. Wolfe, 1978. Phys. Rev. Lett. **40**, 526.
Gourley, P.L. and J.P. Wolfe, 1979, Phys. Rev., **B20**, 3319.
Gogolin, A.A. and E.I. Rashba, 1973, Pis'ma Red. Zh. Eksp. Teor. Fiz. **17**, 690 (1973, JETP Lett. **17**, 478).
Hanamura, E., 1973, Solid State Commun. **12**, 951.
Hanamura, E. and H. Haug, 1977, Phys. Rep. **33C**, 210.
Haynes, J.R., 1960, Phys. Rev. Lett. **4**, 450.
Haynes, J.R., 1966, Phys. Rev. Lett. **17**, 860.
Hensel, J.C., T.G. Phillips and G.A. Thomas, 1977, Solid State Phys. **32**, (Academic Press, New York, San Francisco, London) p. 88.
Huang, W.T., 1973, Phys. Status Solidi **60(b)**, 309.
Hylleraas, E.A. and A. Ore, 1947, Phys. Rev. **71**, 493.
Insepov, S.A. and G.E. Norman, 1972, Zh. Eksp. Teor. Fiz. **62**, 2290 (1972, Sov. Phys. JETP **35**, 1198).
Insepov, Z.A. and G.E. Norman, 1975, Zh. Eksp. Teor. Fiz. **69**, 1321 (1975, Sov. Phys. JETP **42**, 474).
Insepov, Z.A., G.E. Norman and L.Yu. Shurova 1976, Zh. Eksp. Teor. Fiz. **71**, 1960 (1976, Sov. Phys. JETP **44**, 1028).
Inui, T., 1938, Proc. Phys.—Math. Soc. Japan, **20**, 770;
Inui, T., 1941, Proc. Phys.—Math. Soc. Japan, **23**, 992.
James, H.M. and Coolidge, 1933, J. Chem. Phys. **1**, 335.
Kaminskii, A.S. and Ya.E. Pokrovskii, 1979, Zh. Eksp. Teor. Fiz. **76**, 1727 (Sov. Phys. JETP **49**).
Kawabata, T., K. Muro and S. Narita, 1977, Solid State Commun. **23**, 267.
Keldysh, L.V., 1968, Proc. Int. Conf. Phys. Semicond. 9th, Moscow, Nauka, Leningrad, p. 1303.
Keldysh, L.V., 1971, Excitons in Semicond. (Nauka Moscow) ch. 2.
Kirczenow, G., 1977, Sol. State Commun. **21**, 713.
Kukushkin I.V., Kulakovsky V.D. and Timofeev V.B. 1980, Pis'ma Zh. Eksp. Teor. Fiz. **32**, 304 (1980, JETP Lett. **32**, 280).
Kulakovskii, V.D. and V.B. Timofeev, 1977, Pis'ma Red. Zh. Eksp. Teor. Fiz. **25**, 487 (1977, JETP Lett. **25**, 458).
Kulakovskii, V.D. and V.B. Timofeev, 1980a, Solid State Commun. **33**, 1187.
Kulakovskii, V.D., V.B. Timofeev and V.M. Edelstein, 1978a, Zh. Eksp. Teor. Fiz. **74**, 372. (1978, Sov. Phys. JETP **47**, 193).
Kulakovskii, V.D., V.B. Timofeev and V.M. Edelstein, 1978b, Proc. 14th Intern. Conf. Phys. Semicond, Edinburgh, ed., B.L.H. Wilson (The Institute of Physics Bristol and London) p. 381.
Kulakovskii, V.D., A.V. Malyavkin and V.B. Timofeev, 1979a, Zh. Eksp. Teor. Fiz. **76**, 272 (Sov. Phys. JETP **49**, 139).
Kulakovskii, V.D., A.V. Malyavkin and V.B. Timofeev, 1979b, Zh. Eksp. Teor. Fiz. **77**, 752.
Kulakovskii, V.D., V.A. Tulin and V.B. Timofeev, 1980b, Pis'ma Red. Zh. Eksp. Teor. Fiz. **31**, 22 (1980, JETP Lett. **31**, 20).
Kulakovskii,V.D., I.M. Kukushkin and V.B. Timofeev, 1980c. Zh. Eksp. Teor. Fiz. **78**, 392.
Lamb, W.E. Jr., 1952, Phys. Rev. **85**, 259.
Lampert, M.A., 1958, Phys. Rev. Lett. **1**, 450.
Landau, L.D. and Ja.B. Zeldovich, 1943, Acta Phys. Chem. USSR **18**, 194.
Lyssenko, V.G. and V.I. Revenko, 1978, Fiz. Tverd. Tela **20**, 2144. (Sov. Phys.-Solid State **20**, 1238).

Moskalenko, S.A., 1958, Opt. Spectrosc. USSR. **5**, 147.
Mott, N.F., 1967, Advanc. Phys. **16**, 49.
Mott, N.F., 1974, Metal–Insulator Transitions (Barnes a. Noble, New York).
Munschy, G. and B. Stebe, 1974, Phys. Status Solidi (b), **64**, 213.
Munschy, G. and B. Stebe, 1975, Solid State Commun. **17**, 1051.
Narita, S. and M. Taniguchi, 1976, Phys. Rev. Lett. **36**, 913.
Narita, S. and M. Taniguchi, 1977, J. Phys. Soc. Japan, **43**, 1262.
Pelant, I., A. Mysyrowicz and C. Benoit a'la Guillaume, 1976, Phys. Rev. Lett. **37**, 1708.
Pokrovskii, Ya.E. and K.I. Svistunova, 1969, Pis'ma Red. Zh. Eksp. Teor. Fiz. **9**, 435 (1969, JETP Lett. **9**, 261).
Pokrovskii, Ya.E., 1972, Phys. Status Solidi **A11**, 385.
Rashba, E.I., 1975, Springer Tracts in Modern Phys. **73**, 150.
Rashba, E.I. and G.E. Gurgenishvili, 1962, Fiz. Tverd. Tela **4**, 1029 (1962, Sov. Phys. -Solid State **4**, 759).
Rice, T.M., 1974, Proc. XII Int. Conf. Phys. Semic., Stuttgart, ed., M.H. Pilkuhn (Teubner, Stuttgart, 1974) p. 23.
Rice, T.M., 1977, Solid State Phys. (Academic Press, New York, San Francisco, London) **32**, 1.
Saharov, A.D., 1948, Zh. Eksp. Teor. Fiz. **18**, 631.
Sander, L.M. and D.K. Fairobent, 1976, Solid State Commun. **20**, 631.
Shah, J., M. Combescot and A.H. Dayem, 1977, Phys. Rev. Lett. **38**, 1497.
Sharma, R.R., 1968a, Phys. Rev. **170**, 770.
Sharma, R.R., 1968b, Phys. Rev. **171**, 36.
Thewalt, M.L.W., 1977, Canad. J. Phys. **55**, 1463.
Thewalt, M.L.W. and R. Rostworovskii, 1978, Solid State Commun. **25**, 991.
Thomas, G.A., 1977, 11 Nuovo Cimento **39**, 561.
Thomas, G.A. and T.M. Rice, 1977, Solid State Commun. **23**, 359.
Thomas, G.A., T.M. Rice and J.C. Hensel, 1974a, Proc. Int. Conf. Phys. Semic., XII (Teubner, Stuttgart) p. 105.
Thomas, G.A., T.M. Rice and J.C. Hensel, 1974b, Phys. Rev. Lett. **33**, 219.
Thomas, G.A., J.B. Mock and M. Capizzi, 1978, Phys. Rev. **B18**, 4250.
Ueta, M. and N. Nagasawa, 1975, Proc. Oji Semin. Tomakomai, Japan, 1976, eds., M. Ueta and Y. Nishina (Springer-Verlag) p. 1.
Wang, J.S.Y. and C. Kittel, 1972, Phys. Lett. **A42**, 189.
Wehner, R.K., 1969, Solid State Commun. **7**, 457.
Wheeler, J.A., 1946, Ann. N.Y. Acad. Sci. **48**, 219.
Wolfe, J.P., R.S. Markiewicz, C. Kittel and C.D. Jeffries, 1975a, Phys. Rev. Lett. **34**, 275.
Wolfe, J.P., W.L. Hansen, E.E. Haller, R.S. Markiewicz, C. Kittel and C.D. Jeffries, 1975b, Phys. Rev. Lett. **34**, 1292.
Wolfe, J.P., R.S. Markiewicz, S.M. Kelso, J.E. Furneaux and C.D. Jeffries, 1978, Phys. Rev. **B18**, 1479.
Zimmermann, R., K. Kilimann, W.D. Kraeft, D. Kremp, G. Röpke, 1978, Phys. Status Solidi (b) **90**, 175.
Zhydkov A.E. and Ya.E. Pokrovskii, 1979, Pis'ma Red. Zh. Eksp. Teor. Fiz. **30**, 499.

CHAPTER 10

Bound Multiexciton Complexes

M.L.W. THEWALT*

IBM Thomas J. Watson Research Center
Yorktown Heights, New York 10598
U.S.A.

*Present address: Physics Department, Simon Fraser University, Burnaby, British Columbia, Canada

Excitons
Edited by
E.I. Rashba and M.D. Sturge

Contents

1. Introduction

Between 1931 and 1958 several authors predicted the existence of free excitons (FE's), biexcitons, bound excitons (BE's), negative donor ions and positive acceptor ions as relatively stable excited states of semiconductors (Frenkel 1931, Wannier 1937, Mott 1938, Lampert 1958). These predictions resulted from the simple analogies which could be drawn between states composed of effective-mass-like particles in semiconductors and atomic or molecular hydrogenic and positronium systems whose properties were already well known. In all the above cases the predictions significantly preceded any experimental evidence of the existence of the postulated states. This was not the case for bound multiexciton complexes (BMEC's, sometimes also referred to as multiple bound excitons, MBE's), which are localized complexes containing more than one electron–hole pair bound to an impurity site in the semiconductor, and in the foregoing analogy can be thought of as exotic "atoms" having as a nucleus a single proton (or anti-proton) which is surrounded by neutralizing clouds of both electrons and positrons. In this chapter we will give a brief review of the history of the BMEC field and then describe the current understanding of their behavior in more detail.

Luminescence resulting from the recombination of an electron with a hole in BMEC's typically results in a series of sharp lines lying at energies below that of the associated BE luminescence line. Such spectra were first observed in the mid-1960's by Haynes in his studies of the photoluminescence of lightly doped Si, but were apparently not reported for lack of a suitable explanation (Dean and Herbert 1979). Kaminskii, Pokrovskii and co-workers (1970, 1971, 1972) were the first to publish spectra showing the new line series and to propose the basic BMEC explanation. One of their early spectra showing FE, BE and BMEC luminescence in B-doped Si is given in fig. 1. They reasoned that since the new lines were impurity specific and did not show the kinetic energy broadening typical of the FE, the luminescent centers must be localized on impurity sites. Also, since the new lines showed a stronger dependence on excitation density than did the BE or the FE, they proposed that the centers must be composed of more than one electron–hole pair. These authors were also the first to suggest that BMEC's might provide the nucleation centers for electron–hole droplets, a subject which continues to be of considerable interest.

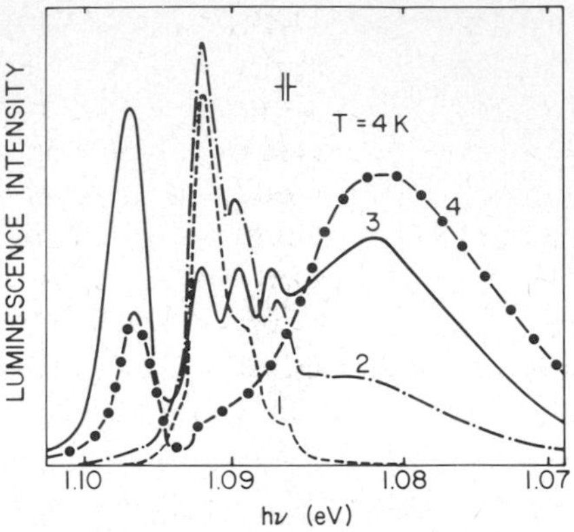

Fig. 1. One of the earliest Si photoluminescence spectra showing TO phonon assisted BMEC recombination. From left to right the luminescence lines are the FE, the B-associated BE and $m = 2$ and 3 BMEC's and the (much broader) electron-hole droplet. The spectra are labelled 1 through 4 in the order of increasing excitation density vs. dopant concentration, as shown below. From Kaminski and Pokrovskii (1970): (1) – $N_B = 5 \times 10^{14}\,\mathrm{cm}^{-3}$, $P = 40$ mW, (2) – $N_B = 5 \times 10^{14}\,\mathrm{cm}^{-3}$, $P = 160$ mW, (3) – $N_B = 5 \times 10^{12}\,\mathrm{cm}^{-3}$, $P = 50$ mW, (4) – $N_B = 3 \times 10^{12}\,\mathrm{cm}^{-3}$, $P = 160$ mW.

The detection of BMEC recombination luminescence remains to this day the only way in which these complexes can be studied, since the extremely small oscillator strengths of the BMEC transitions have unfortunately precluded either absorption or excitation spectroscopy. Later studies of BMEC luminescence benefited from the availability of excitation sources capable of achieving high energy densities, such as well collimated laser or electron beams, and also from the replacement of photoconductive detectors with the much more sensitive photo-multiplier tubes having S1, or more recently, quaternary InGaAsP photocathodes. Si remains by far the most heavily studied host even though the spectral range of its luminescence is less convenient from a detection standpoint than those of the wider band-gap semiconductors such as SiC or GaP which also show BMEC luminescence. This preference results not from any fundamental differences in the materials but rather from the highly advanced state of Si technology, which makes readily available samples lightly doped ($\leq 1 \times 10^{15}\,\mathrm{cm}^{-3}$) with a wide variety of donor and acceptor impurities, and showing a very high degree of doping specificity. Since the BE and BMEC spectra associated with a *single* impurity species can be quite complex, it is this doping specificity which gives Si its greatest advantage.

Later studies of the new line series by Sauer (1973) and Kosai and Gershenzon (1974) further supported the BMEC model, and in a 1974 review Sauer concluded that of all the suggested models only that based on BMEC was not ruled out by the then-available data. There was however a problem with the apparently rather large binding energies of some of the complexes (we will define the binding energy to be the energy required to remove a zero kinetic energy FE from a complex in its ground state, leaving behind a ground state complex smaller by one electron–hole pair, where the "ground state" of a complex is taken to mean the lowest total energy configuration of the electrons and holes). This problem resulted from a rather natural misinterpretation of the spectra in all the early papers caused by the very regular spacings of the luminescence lines, which for increasing m (m is an index used to give the number of electron–hole pairs bound in the complex which is the initial state of a given transition – for the BE, $m = 1$) converged smoothly to an energy well below the electron-hole-droplet binding energy, as is shown in fig. 2. This regularity of the line spacings for BMEC's associated with different impurities led to the general assumption that all the transitions were of the same type, namely ground

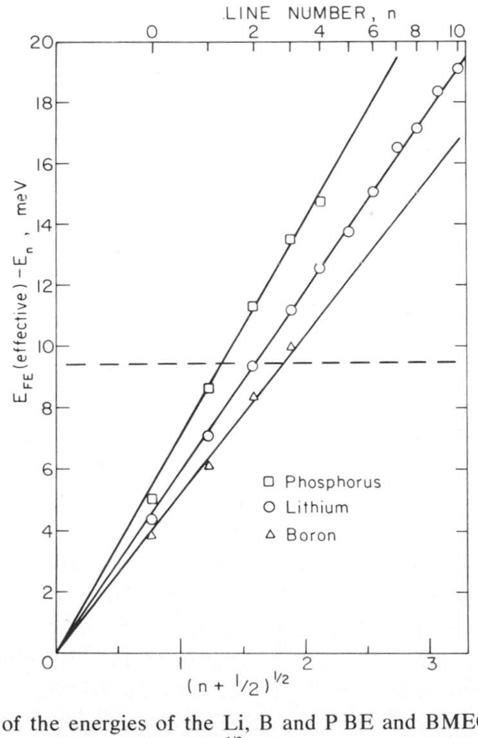

Fig. 2. An early plot of the energies of the Li, B and P BE and BMEC lines relative to the effective FE energy as a function of $(n + 1/2)^{1/2}$, where $n = m - 1$. At that time (1974) only the main Li^m line series was known for Si:Li, and only the α^m series for Si:P. The dotted line marks the electron–hole droplet binding energy. The apparent regularity of the line spacings led to the then prevalent erroneous assumption that all of the transitions were of the same basic type, namely ground state to ground state. From Kosai and Gershenzon (1974).

state to ground state. Given this assumption the binding energy of the mth complex equals the energy difference between the mth line and the low energy threshold of the FE, which for all impurities considerably exceeded the electron–hole droplet binding energy in the limit of large m. This was of course unexpected, since in the large m limit there should be no difference between BMEC and electron–hole droplet binding energies.

Sauer and Weber (1976) then discovered that the Zeeman spectra of the BMEC lines in Si:P showed exactly the same splittings as did the BE line, the only difference being in the thermalization of the components. They found similarly uniformity in the Zeeman spectra of BE and BMEC in Si:B as well as in a stress-split spectrum of Si:P. On the basis of the relative simplicity of these results they categorically ruled out any model based on BMEC as an explanation of the new line series, although no alternate model was apparent.

It was soon realized that this blanket rejection was premature. The first specifically structured BMEC model which sought to explain the new results was the BE-polyexciton complex model proposed by Morgan (1976). In this model the electron–hole pair which recombines is in all cases in a BE state, the energy of which could be perturbed by polyexciton complexes of different size loosely bound to the BE by van der Waals interactions. This model had to be eliminated after several studies of Si doped with Al or Ga (which are more fully described in sect. 2.4) convincingly demonstrated that the *final* states of the $m = 2$ transitions were BE states (Thewalt 1977a, 1977b, Lightowlers and Henry 1977, Lyon et al. 1978a).

In a study of BMEC luminescence in β-SiC Dean et al. (1976a, 1976b) argued that the stress and Zeeman splittings could readily be explained by a model in which the electrons and holes filled degenerate single particle levels in shells, provided that two reasonable assumptions were made. These were that the applied perturbation overwhelmed any residual electron–electron, electron–hole or hole–hole couplings (the well-known Paschen–Back limit), and that all electrons (holes) in both the initial and final states had the same g values. The Zeeman and stress behavior could then be determined from a consideration of the possible values of J for the initial and final states, along with the usual electric-dipole selection rules. The lowest electron shell was taken to have the full twelve-fold degeneracy of the conduction band, and the electrons filled it by pairing with opposing spins. The first hole shell was taken to have the four-fold degeneracy of the $j = 3/2$ valence band edge. This scheme for explaining the Zeeman splittings of the donor BE and $m = 2$ donor BMEC is outlined in sections (b) and (c) of fig. 3. The assumption of ground state to ground state transitions for all the lines, and the resulting large binding energies, were retained.

Although Dean et al. (1976a) were the first to introduce the concept of

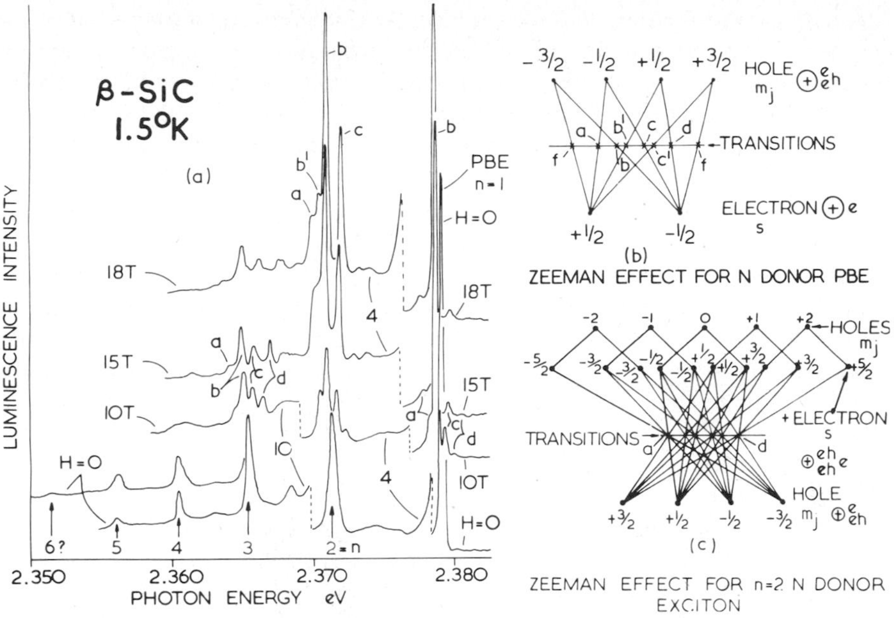

Fig. 3. (a) Zeeman spectra of the SiC:N luminescence lines. (b) and (c) show how the BE and $m = 2$ BMEC Zeeman components observed for SiC:N (and also for Si:P (Saver and Weber 1976)) could be explained by the model introduced by Dean et al. (1976b).

the electrons and holes in BMEC filling single particle levels in shells, its full implications were realized by Kirczenow 1977a, 1977b in his shell model of the BE and BMEC structure. In the shell model the wave function of the complex is represented by the properly antisymmetrized product of single particle wave functions classified according to their transformation properties under the tetrahedral point group T_d (all the impurities in Si reviewed here have point group symmetry T_d). As in the model of Dean et al. (1977a, 1977b), the lowest hole shell is taken to have the four-fold degeneracy of the valence band edge, and in the shell model is labelled Γ_8. In the shell model the twelve-fold degeneracy of the conduction bands can be lifted by the valley-orbit interaction, and the 1S-like lowest electron shell can thus be split into the Γ_1, Γ_3 and Γ_5 valley-orbit states already well known for the neutral donors (Kohn and Luttinger 1955), having degeneracies of 2, 4 and 6, respectively. It was assumed that the ordering of the valley-orbit electron states in donor complexes would be the same as the ordering of the states in the corresponding isolated donors, and that because of the electrostatic repulsion between the electrons and acceptor ions, these valley-orbit splittings would be much less important for acceptor complexes. The shell model also removed the problem of the large binding energies since it predicted that because of the exclusion principle complexes having m larger than some specific value (which can be readily deduced for the various impurities) would have electrons and/or holes in more than one shell. Such complexes can then decay either into ground state or excited state configurations, and the low energy luminescence lines are taken to result from decays into excited final states.

As will be described in more detail in the following sections, the shell model had great success in explaining the then existing data and, after a period of considerable controversy over both experimental results and interpretations (Thewalt 1979), also accounted for almost all of the more recent spectroscopic data. One of the remaining problems is the discrepancy between the spectroscopically determined binding energies and those obtained from measurements (Lyon et al. 1978b, 1978c) of the thermodynamic properties of these systems. This will be outlined in sect. 6. There has recently been some interest in calculating the ground state and excited state energies of BMEC's – a most challenging task from a theoretical standpoint – which is summarized in sect. 7.

Since the preponderance of BMEC data has been obtained for various dopants in Si, we limit ourselves strictly to the Si case when discussing the properties of BMEC in sects. 2 through 6. Results for the other semiconductors are given in sect. 8, while the rather different but conceptually related case of the biexciton bound to the isoelectronic N_P center in GaP is treated in sect. 9.

2. *Unperturbed spectra*

2.1. *Substitutional donors*

The most heavily studied substitutional donor in Si is P; a typical Si:P spectrum is shown in fig. 4. Prior to the introduction of the shell model only the main α^m line series had been observed, and only one BE level was known. The α line series can be observed both in the no-phonon (NP) region as well as in the replicas corresponding to the emission of TO and TA momentum-conserving phonons. Kirczenow's (1977a, 1977b) shell model level diagram for the substitutional donor BE and BMEC's is shown in fig. 5. The lowest electron shell is the two-fold degenerate Γ_1, while Γ_3 and Γ_5 (which are only slightly split in isolated donors) are assumed to be degenerate in the complexes and to form the next electron shell, $\Gamma_{3,5}$. The well known α transitions were all taken to result from the recombination of a Γ_1 electron with a Γ_8 hole, and thus of all the α transitions only the BE line, α^1, left its final state in a ground state configuration. The strong NP transitions of all the α lines were also explained, since Γ_1 electrons have a large amplitude in the central cell region, and in donor complexes it is the

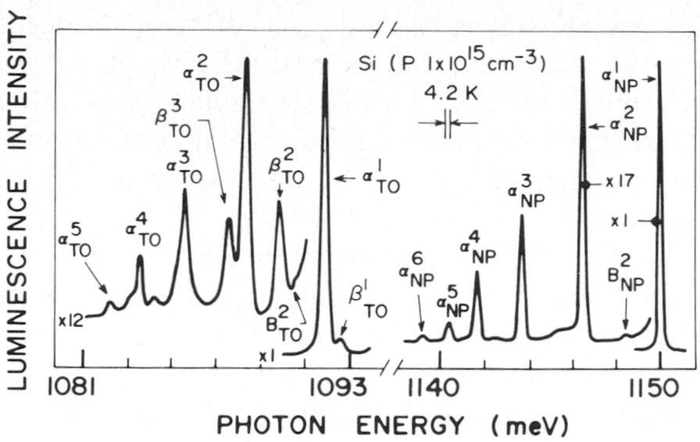

Fig. 4. The Si:P photoluminescence spectrum at 4.2 K as it is currently understood. The α lines are seen both in the NP and the phonon-assisted regions, while the β lines appear only in the phonon replicas. Some B contamination resulting in a B^2 line is evident. From Thewalt (1977b).

scattering of an electron by the short-range impurity potential which leads to NP luminescence.

The level diagram shown in fig. 5 also predicted a number of new lines. Each BMEC can also decay by the recombination of a $\Gamma_{3,5}$ electron, giving rise to a β line series which is of the ground state to ground state type. Since $\Gamma_{3,5}$ electrons have zero amplitude in the central cell, the β lines were expected to occur only in the phonon-assisted replicas. Note that in the

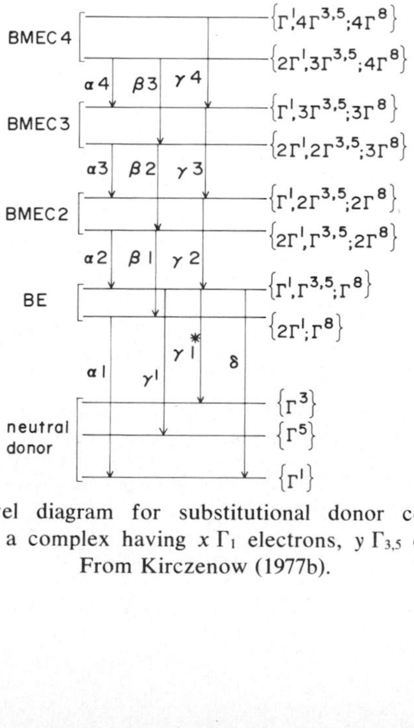

Fig. 5. Shell model level diagram for substitutional donor complexes up to $m = 4$. $\{x\,\Gamma_1, y\,\Gamma_{3,5}; z\,\Gamma_8\}$ denotes a complex having $x\,\Gamma_1$ electrons, $y\,\Gamma_{3,5}$ electrons and $z\,\Gamma_8$ holes. From Kirczenow (1977b).

shell model β^{m-1} and α^m have the same initial state, the mth BMEC, and that it is the energy difference between the FE edge and the β^{m-1} line which gives its binding energy. The β line series was observed experimentally soon after being predicted in the shell model (Thewalt 1977c).

Other transitions, originating from excited initial states, were also predicted on the basis of fig. 5. Of these, the BE δ, γ^1 and γ^{1*} lines have been observed in elevated-temperature photoluminescence spectra, as shown in fig. 6. The δ line, which involves the recombination of a $\Gamma_{3,5}$ electron, could only be observed in phonon-assisted replicas, as expected (Thewalt 1977c). The δ transition had actually been observed before the advent of the shell model, in the absorption spectra of Nishino et al. (1977), but these authors interpreted the line as a FE transition. This error was later rectified in the higher resolution absorption spectra of Henry and Lightowlers (1977). The two γ^1 lines originate from the same excited BE state as does δ, but result from recombination of the Γ_1 electron. The γ^1 transitions thus leave the donor in 1S valley-orbit excited states, and have strong NP transitions. No transitions originating from the excited states of BMEC's have yet been observed, probably as a result of the drastic decrease in BMEC luminescence at the elevated temperatures necessary for these states to be significantly populated. The energies of all the observed transitions in Si:P

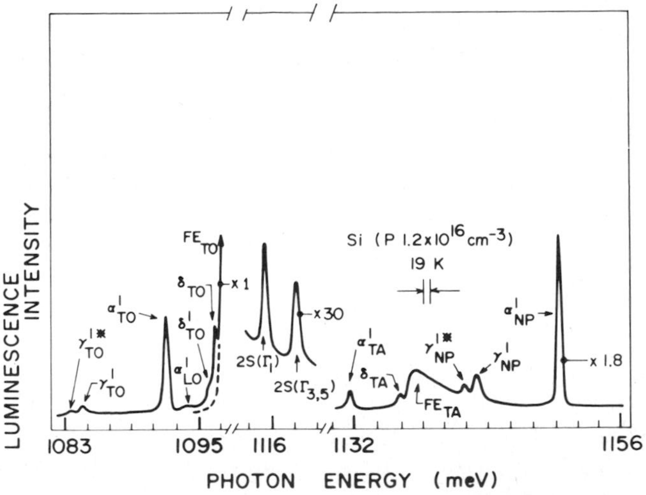

Fig. 6. High temperature (19 K) photoluminescence spectrum of Si:P. The δ lines is seen only in the phonon replicas while the γ^1 lines also appear in the NP region. 2S Γ_1 is a two-electron transition of a ground state BE resulting in a donor in the 2S Γ_1 excited state, while 2S $\Gamma_{3,5}$ originates from the BE excited state and leaves the donor in the 2S $\Gamma_{3,5}$ state. From Thewalt (1977b).

are in good agreement with fig. 5 and the known 1S valley-orbit splittings of P (Thewalt 1977b).

Using the shell model transition scheme and the photon energies of the transitions shown in fig. 4 (which are tabulated by Thewalt (1977b)) one finds that for Si:P the $m = 2$ complex is less tightly bound than the BE, while the $m = 3$ and 4 complexes are progressively more tightly bound. This can be qualitatively explained by analogy to the shell models of atomic or nuclear physics. One expects a very stable complex, and therefore a local maximum in the binding energy, for values of m which give closed outer electron or hole shells. Since the Γ_1 electron shell is closed at $m = 1$, the relative stability of the BE compared to the $m = 2$ complex can be understood. Similarly, the $m = 4$ complex has a closed Γ_8 hole shell and would be expected to be very stable. The $m = 5$ and 6 complexes should have smaller binding energies than the $m = 4$, but no data is available since the $m = 5$ and 6 ground state to ground state transitions have not been observed. These lines are presumably obscured by the more intense $m < 5$ luminescence.

Another result of the stability of complexes having closed outer electron or hole shells might be a reduced cross section for FE capture, which would cause a pattern in the relative intensities of the luminescence lines. Such patterns are indeed clearly visible in the BE and BMEC spectra associated with different impurities. In the Si:P spectrum shown in fig. 4 the drops in relative intensity between $m = 1$ and 2 and between $m = 4$ and 5 are very evident. Although the relative intensities of the luminescence lines originating from different complexes will of course change with changing excitation density, this overall pattern remains and is in fact typical of the spectra of all the substitutional donors. The excitation density dependence of the relative intensities of the lines can be used to test the shell model prediction that β^{m-1} and α^m have the same initial state. Henry and Lightowlers (1978) verified that the β^2/α^3 and β^3/α^4 intensity ratios did not change with the excitation level, supporting the shell model.

Fine structure has been observed in some of the Si:P NP α transitions by Parsons (1977) who used much higher resolving power than that of the other BMEC photoluminescence studies. The α^1, α^4, α^5 and α^6 lines were found to be without structure, while α^2 consisted of three strong components and one weak one, and α^3 had a doublet structure. Parsons explained these results by augmenting the original shell model scheme with small splittings resulting from interparticle interactions, which were described in a j–j coupling model.

These luminescence studies have also been extended to substitutional donors other than P. Lightowlers et al. (1977) have observed BE transitions similar to those just described for Si:P in Si doped with Sb, As and Bi, while Elliott and McGill (1978) reported on both BE and BMEC lumines-

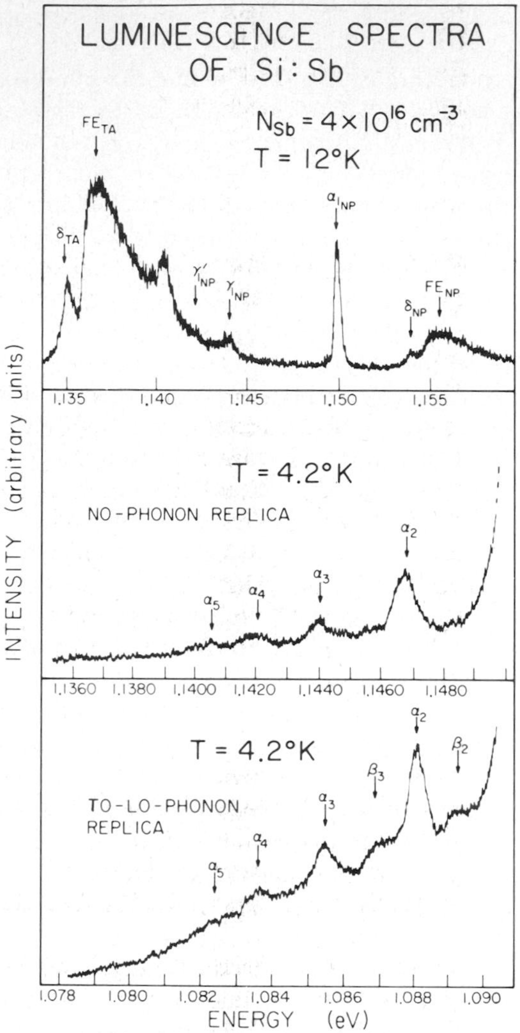

Fig. 7. Photoluminescence spectra of Si:Sb at 12 K and 4.2 K showing the α, β, δ and γ lines. From Elliott and McGill (1978).

cence in Si:Sb and Si:As, as shown in figs. 7 and 8. Both groups noted that, unlike the Si:P case, for these other substitutional donors the δ to γ^1 splittings did not exactly match the known donor 1S valley-orbit splittings, but suggested that the problem might lie in problems with the identification of the δ transition(s). Even the Si:P spectrum reveals an unidentified BE excited state, labelled δ' in fig. 6. This line was also observed in absorption spectra (Henry and Lightowlers 1977).

Elliott and McGill (1978) noted that there was no strong correlation

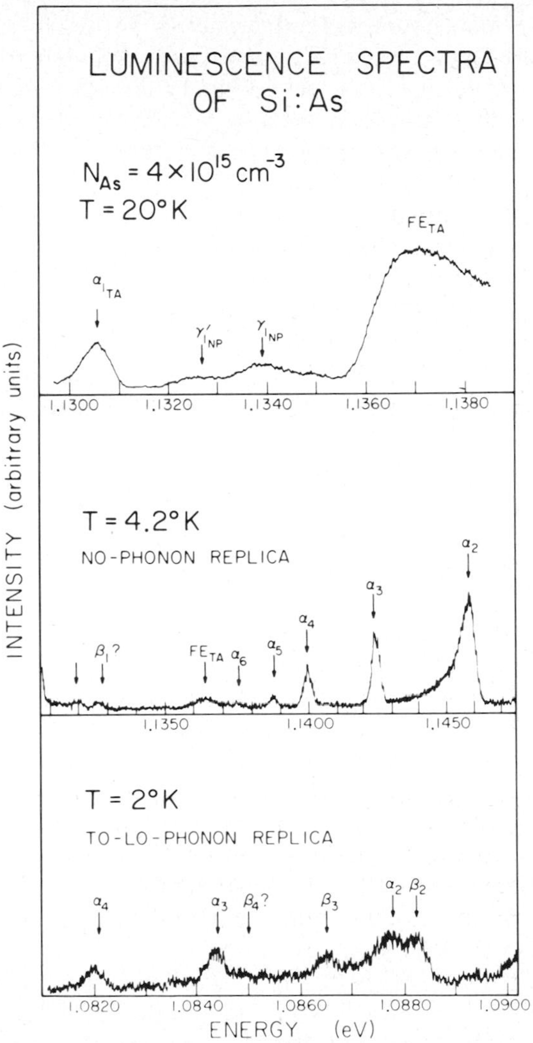

Fig. 8. Photoluminescence spectra of Si:As at 20, 4.2 and 2 K showing the α, β and γ lines. From Elliott and McGill (1978).

between the size of the valley-orbit splitting of a complex and the size of the splitting of the substitutional donor to which that complex was bound. The explanation of this effect must await realistic energy level calculations. More easily understood is the observation (Elliot and McGill 1978) that the size of the valley-orbit splitting decreases greatly in going from the donor to the BE, and is almost constant for $m \geq 1$. This follows directly from the shell model, since the magnitude of the interaction of a Γ_1 electron with the central cell potential should depend primarily on the number of Γ_1 electrons, and not on the $\Gamma_{3,5}$ electrons which are excluded from that region. The valley-orbit splitting of the BE is much less than that of the donor because the second Γ_1 electron must reduce the probability of either Γ_1 electron being in the central cell region, but for $m > 1$ the number of Γ_1 electrons remains equal to two.

The previously described shell-closing effects can also be observed in the Si:Sb and Si:As spectra. The $m = 2$ binding energies, as interpreted in the shell model are in both cases less than that of the BE (Elliott and McGill 1978). The large decreases in luminescence intensity in going from $m = 1$ to $m = 2$ and $m = 4$ to $m = 5$ are also present. Very similar BMEC spectra for Si:Sb have recently been reported by Kaminskii et al. (1978). There have been no reports of BMEC luminescence from Si doped with the deepest substitutional donor, Bi. This could result either from the expected short lifetime of the Bi BE due to strong Auger recombination, or from a very small, $m = 2$, binding energy. For the other substitutional donors the $m = 2$ binding energy is found to decrease as the donor binding energy (and therefore the BE binding energy) increases.

2.2. Interstitial Li donor

The Si:Li photoluminescence spectrum was first studied by Kosai and Gershenzon (1974). After a considerable hiatus, Lyon et al. (1978b) and Thewalt (1978) published higher resolution spectra which revealed many new lines. Although the data in these two studies were very similar, there were considerable differences in interpretation which resulted from the use of a shell model scheme by Thewalt, and a thermodynamically determined $m = 2$ binding energy (discussed in more detail in sect. 6) by Lyon et al. Here we will follow the notation and shell model interpretation introduced by Thewalt (1978).

The Si:Li spectrum provided an important test of the usefulness of the shell model, since interstitial Li is the only donor in Si having an inverted ordering of the 1S valley-orbit states, with the $\Gamma_{3,5}$ level lying 1.83 ± 0.07 meV below Γ_1. One would therefore expect a BMEC spectrum very different from that of the substitutional donors since for Li complexes the first electron shell only becomes full at $m = 9$. This is borne out by the experimental results, some of which are shown in fig. 9. The large intensity drop between $m = 1$ and $m = 2$ which is typical of the substitutional donor spectra is absent for Si:Li, while the drop between $m = 4$ and $m = 5$ resulting from the closure of the Γ_8 hole shell is still clearly seen. Both of these effects are also evident in the spectra of Lyon et al. (1978b) as well as in the earlier (Kosai and Gershenzon 1974) work. The main Li^1 through Li^4 lines are thus interpreted as being ground state to ground state transitions,

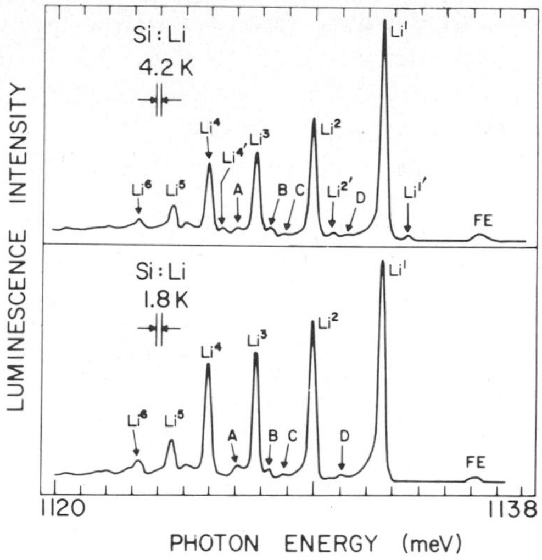

Fig. 9. Si : Li photoluminescence spectra taken at 4.2 K and 1.8 K showing the disappearance of the $Li^{m'}$ excited state lines at the lower temperature. From Thewalt (1978).

while the Li^5 and Li^6 lines (and also the much weaker lines at even lower energies) have excited final states, since these transitions involve the recombination of inner-shell holes.

Comparing the 4.2 K and 1.9 K spectra in fig. 9 we see a thermally activated line series, $Li^{m'}$, with the $Li^{3'}$ line almost obscured by a line labelled B which does not vanish at low temperatures. The $Li^{m'}$ series is interpreted as being due to transitions from complexes in valley-orbit excited states, in which a $\Gamma_{3,5}$ electron is promoted to the Γ_1 shell. This is outlined in the shell model level diagram given in fig. 10. An excited state to excited state transition, labelled Li^{1*} in figs. 10 and 11, has been observed for the BE, and the $Li^{1'}$ to L^{1*} splitting was found to agree with the known Li 1S valley-orbit splitting. Similar transitions would also be expected for the BMEC's, but these lie too near the main Li^m lines to be resolved. We should note that all of the Li lines have vanishingly small NP intensities, and the Si : Li spectrum must therefore be recorded in a phonon-assisted replica, with the concomitant phonon broadening (Thewalt et al. 1976).

Two additional BE excited states are revealed in fig. 11. Thewalt (1978) proposed that $Li^{1''}$ must be due to a BE having an excited hole state, since no replica of this line could be found which left the Li donor in any excited state. As before, the nature of the second hole shell is not known and it is simply labelled Γ_x. This interpretation was strengthened by the discovery of the $Li^{1'''}$ transition by Henry and Lightowlers (1979). Since the $Li^{1'''}$ to $Li^{1''}$ spacing was close to that of $Li^{1'}$ to Li^1 they interpreted the $Li^{1'''}$ line as being

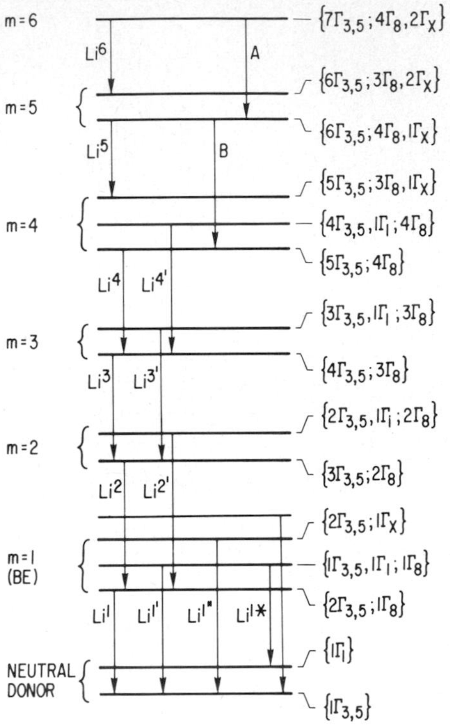

Fig. 10. A shell model level diagram for complexes associated with the interstital donor Li. The labelling of the complexes is similar to the scheme used in fig. 4. Only excited states which are needed to explain the observed transitions are shown. The identifications of the A and B lines are tentative. The unlabelled highest energy BE excited state is $\{1\Gamma_{3,5}, 1\,\Gamma_1; 1\,\Gamma_x\}$ and according to Henry and Lightowlers (1979) the transition from it to the Li ground state is $Li^{1'''}$.

due to an excited state of the BE in which the hole and one electron were excited to Γ_x and Γ_1, respectively. Note from fig. 11(a) that the oscillator strength of the $Li^{1'}$ excited BE transition is considerably less than that of the Li^1 ground state transition, which is the opposite of what is observed for the ground state and excited state α^1 and δ transitions seen for the substitutional donors. In both cases these oscillator strength ratios can be qualitatively accounted for by the relative degeneracies of the Γ_1 and $\Gamma_{3,5}$ shells.

Of the other lines shown in fig. 9, it has been argued that C and D are extraneous and probably not associated with isolated interstitial Li binding centers (Thewalt 1978). The C and D lines were not observed by Lyon et al., by the A and B lines were seen in both studies. Thewalt observed that on the basis of their excitation density dependence the A and B lines could not have the same initial states as any of the $m<5$ lines, but that their

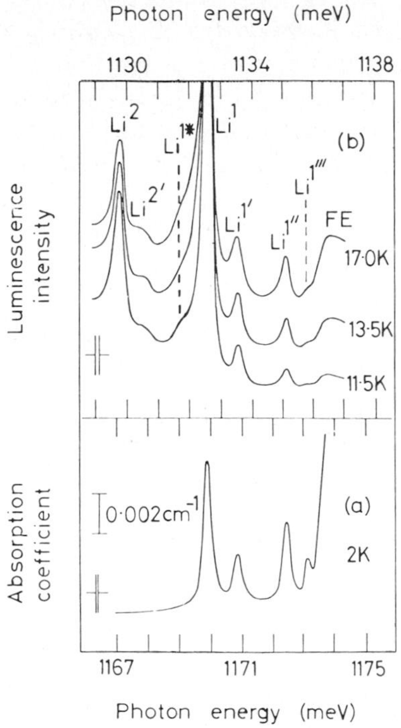

Fig. 11. Luminescence and absorption spectra of Si : Li in the TA phonon-assisted region. The absorption curve was recorded at ~2 K, while the luminescence curves were recorded at 11.5, 13.5 and 17.0 K, respectively. The absorption curve shows only BE and the FE transitions. From Henry and Lightowlers (1979).

behavior was consistent with that of Li^5 and/or Li^6. It was therefore proposed that the A and B lines were the ground state to ground state transitions of the $m = 6$ and $m = 5$ BMEC, as shown in fig. 10. The positions of the lines are quite reasonable, since the $m = 5$ and 6 complexes would be expected to be less tightly bound than the closed shell $m = 4$ BMEC. The origin of the A and B lines is a topic worthy of further study, since if the above interpretation is correct these are the only known ground state to ground state transitions of complexes having holes in more than one shell.

2.3. Shallow B acceptor (1)

Boron is the most persistent electrically active impurity found in highly-refined Si, which is probably why BMEC luminescence was first observed for Si : B (Kaminskii, Pokrovskii and co-workers 1970, 1971, 1972). One of these early spectra has already been shown in fig. 1; a more recent higher resolution spectrum is shown in fig. 12. No new B BMEC luminescence lines have been reported since 1973 when Sauer found lines up to $m = 5$. Unlike the BE's associated with the deeper acceptors, the splitting of the B BE ground state due to the coupling of the two $j = 3/2$ holes is so small as to be negligible (Thewalt 1977b) and the Si : B spectrum is therefore well described within the original shell model framework with its assumption of non-interacting single particle states. It is for this reason that B and the deeper acceptors are discussed separately.

Kirczenow (1977b) originally suggested that for acceptor complexes the electrostatic repulsion between the electrons and the acceptor ion would result in valley-orbit splittings which were small compared to those of donor complexes. An upper limit of 0.3 meV for such splittings was experimentally determined by Kulakovskii, (1978), and later reduced to 0.15 meV by Thewalt et al. (1979). The first electron shell of the acceptor complexes is therefore effectively twelve-fold degenerate. This is not to say that for acceptor complexes the shell model reduces to the earlier model proposed by Dean et al. (1976a, 1976b), since as we shall later see, the assumption of the pairing of electron spins made in that model is not borne out by Zeeman spectroscopy.

The only shell effects which can be observed in the Si : B are thus due to the closing of the first hole shell at $m = 3$. As can be seen in fig. 12 this leads to a very sharp drop in luminescence intensity between the B^3 and B^4 lines which is typical of all the acceptor spectra and does not depend on the excitation density. The B^4 and B^5 transitions are due to the recombination of holes from the inner Γ_8 shell and therefore result in excited final states. The ground state to ground state transitions of the $m = 4$ and $m = 5$ B BMEC would lie beneath the much stronger $m < 4$ lines and have not been observed. As a result the binding energies of the $m > 3$ B BMEC's cannot be determined spectroscopically.

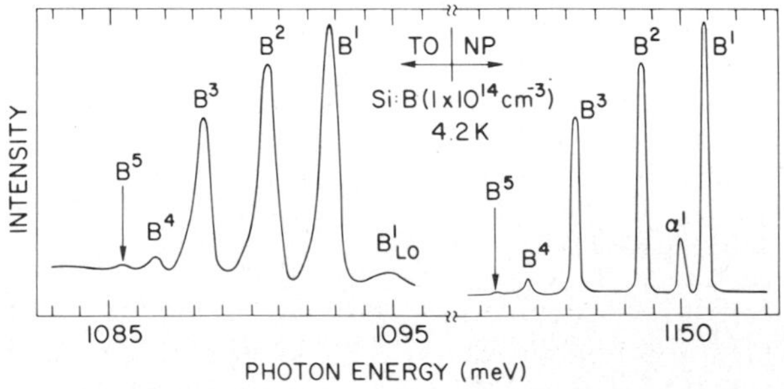

Fig. 12. TO phonon-assisted and NP photoluminescence transitions of Si : B at 4.2 K. The α^1 line resulting from P contamination is most evident in the NP region.

2.4. Deeper acceptors

For the deeper acceptors in Si the BE ground state shows a triplet splitting due to interparticle interactions (Elliot et al. 1978). It was initially thought (Lightowlers and Henry 1977, Vouk and Lightowlers 1976) that this triplet splitting originated from the j–j coupling of the two holes to form $J = 0$ and $J = 2$ states, which when coupled with the $j = 1/2$ electron gave $J = 1/2$, $3/2$ and $5/2$ levels, but it is currently accepted (Kirczenow 1977b, Elliot et al. 1978) that the splitting arises solely from the coupling of the two Γ_8 holes in the tetrahedral field of the impurity, which splits the $J = 2$ level into Γ_3 and Γ_5 states. In this notation the $J = 0$ state is Γ_1. In their reply to Sauer and Weber's (1976) rejection of BMEC models, Dean et al. (1976b) suggested that additional support for the model could be obtained if BMEC-like luminescence could be found in Si doped with the deeper acceptors, since the $m = 2$ luminescence would be expected to mirror the splittings of its BE final state.

Soon thereafter it was found that the $m = 2$ luminescence in Si:Ga consisted of a non-thermalizing doublet with a splitting exactly equal to that between the Γ_3 and Γ_5 BE states (Thewalt 1977a). Very similar results were then found by Lightowlers and Henry (1977) for Si:Al, and Lyon et al. (1978) were later able to detect the expected third component of the $m = 2$

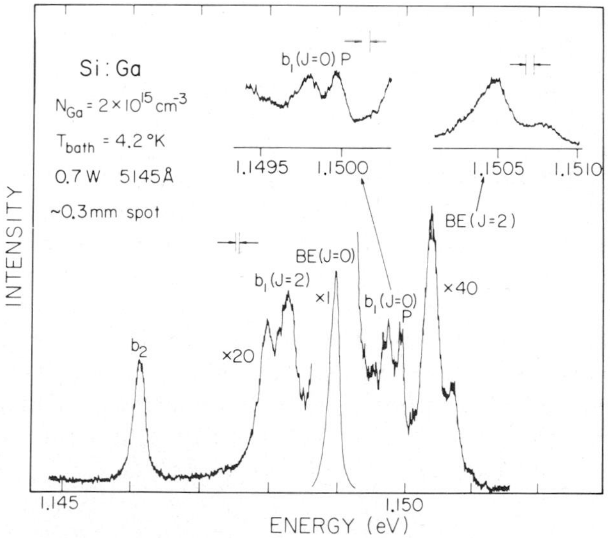

Fig. 13. Photoluminescence spectra of Si:Ga in the NP region at 4.2 K. Three BE transitions and three $m = 2$ (b_1) transitions are observed, along with a single $m = 3$ (b_2) line and an α'(P) line due to P contamination. The spacing of the $m = 2$ triplet mirrors that of the BE, indicating that the BE is the final state of the $m = 2$ transition. From Lyon et al. (1978a).

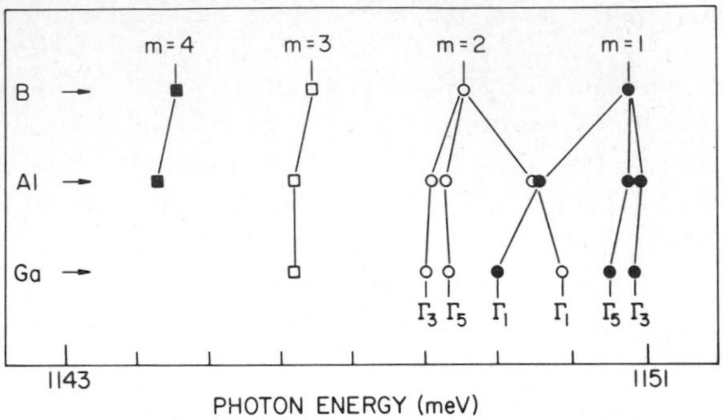

Fig. 14. Energies of the BE and BMEC NP transitions associated with the acceptors B, Al and Ga. The $m = 2\ \Gamma_1$, Γ_3 and Γ_5 lines originate from the same initial state but have the three different BE final states. The binding energy of the BE and the $m = 2$ complex are both given by the separation of the respective Γ_1 components from the FE edge at 1154.59 ± 0.1 meV.

luminescence in Si:Ga. These results verified the general concept of the BMEC model, and eliminated the previously mentioned BE-polyexciton complex model proposed by Morgan (1976). The Si:Ga spectrum is shown in fig. 13.

As for the case of B, the first electron shell of complexes associated with these deeper acceptors is taken to be twelve-fold degenerate, and shell effects are only seen at $m = 3$ as a result of the filling of the Γ_8 hole shell. Just as in the B spectrum, there is a large drop between the $m = 3$ and $m = 4$ luminescence in Si:Al (Lightowlers and Henry 1977, Thewalt 1977b, Lyon et al. 1978), while $m > 3$ luminescence has not even been observed for Si:Ga. The large drop in luminescence intensity between $m = 1$ and $m = 2$ seen for Al and Ga but not for B is due not to a shell closing effect but rather to the extra stability of the lowest energy states of the Al and Ga BE's resulting from their hole–hole coupling (only a Γ_8 shell containing *two* holes can have such interactions). This is reflected in the binding energy of the $m = 2$ complex, which in fig. 14 is seen to decrease for increasing acceptor ionization energy. This trend, along with kinetic arguments (Schmid 1977, Lyon et al. 1977), explains why no BMEC luminescence has been reported for the deeper acceptors In and Tl. Just as for all other BMEC systems, in Si:Al or Si:Ga the luminescence lines of the larger complexes showed a higher order dependence on the excitation density, or in other words, the intensity of an m line relative to an $m - 1$ line always increased with increasing excitation (Thewalt 1977b).

3. Zeeman spectroscopy

Sauer and Weber (1976) were the first to report Zeeman studies of the BMEC luminescence, providing spectra for Si:P and Si:B as well as a brief account of the results for Si:Li. In their Si:P spectrum which is shown in fig. 15 one sees that the α^1 through α^4 lines all split into six components whose spacing is independent of m. The only difference between the α lines is a progressive reduction in the thermalization of these components with increasing m, or in other words, in going from α^1 to α^4 the intensities of the higher energy Zeeman components relative to the lower energy components showed a progressively decreasing dependence on sample temperature. At the left of the diagram is a simple level scheme for donor BE which shows the origin of the six BE components. This model was introduced by Thomas et al. (1963) and later used by Cherlow et al. (1973) to interpret BE Zeeman results in Si. The behavior of the $m = 1$ through $m = 3$ B spectra was quite similar, although in that case only four components were observed.

Sauer and Weber argued that in any BMEC model the Zeeman structure of the complexes would be different from that of a simple BE, and on that basis rejected all BMEC models for the new line series. This conclusion

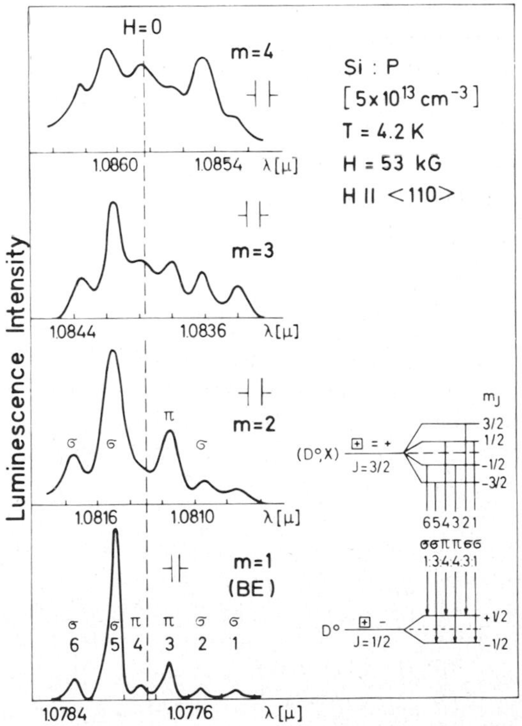

Fig. 15. Zeeman splittings of α^1 through α^4 in Si:P. The origin of the six BE (α^1) components and their predicted polarizations and oscillator strengths are given in the diagram at the right. From Sauer and Weber (1976).

was soon shown to be unfounded. Dean et al. (1976a, 1976b) demonstrated that so long as one assumed that all the electrons (holes) in the initial and final states had the same g values, and that the applied field overwhelmed any interparticle couplings (the Paschen–Back limit), the BMEC's would naturally show the same type of Zeeman splitting as did the BE. The spacings of these components and the number which can be observed are of course somewhat dependent on the direction of the field due to the anisotropy of the g factors (Cherlow et al. 1973).

In the shell model the thermalization of the substitutional donor α components is due strictly to the arrangement of the holes in the four non-degenerate levels into which a magnetic field splits the $j = 3/2$ Γ_8 shell, the electron in all cases coming from the spin-singlet Γ_1 shell. Thus the α^4 Zeeman spectrum would be expected to show no thermalization whatsoever, a result somewhat different from the earlier model of Dean et al. (1976a, 1976b) which paired off the electrons with opposite spins. Using the shell model and assuming the holes to be in thermal equilibrium, Kirczenow (1977b) was able to reproduce the relative intensities of the α^1 through α^4 Zeeman components shown in fig. 15, except that for *all* the lines the intensities of components 3 and 4 were consistently overestimated by a factor of about two. It was suggested that this might result from internal reflections coupling out the luminescence which for components 3 and 4 is also emitted parallel to the field (Kirczenow 1977b). Kulakovskii and Malyavkin (1979) contended that the discrepancy, which is also seen in the Si:Al and Si:B spectra, was too large to be accounted for by this effect. In that case the oscillator strength ratios for the Zeeman components given in fig. 15 must be incorrect, although it is not at all clear where the problem might lie.

There have been no Zeeman studies of the β line series in substitutional donor doped Si, nor any probable since, unlike the α lines, the β lines cannot be studied in the isolation of the NP replica. This is unfortunate since the thermalization of the β components would be expected to differ markedly from that of the α lines because of the ability of the $\Gamma_{3,5}$ electrons involved in β transitions to assume any total spin configuration. The same situation should obtain for the main Li^m line series, since these also originate from the recombination of $\Gamma_{3,5}$ electrons. For $m < 5$ these electrons should be free to assume any total spin configuration. Unfortunately we cannot compare the thermalization of the α^m and Li^m components since no detailed Zeeman data are yet available for Si:Li. Kulakovskii et al. (1978a) have recently studied the Si:P α spectra in the limit $g\mu H \gg kT$ (no thermal population of excited levels) and found the results to be in agreement with the shell model.

In another very recent study Kulakovskii et al. (1979) found that the $m = 1$ through 3 Zeeman splittings for Si:B could be well accounted for within the shell model, and that for $g\mu H \gg kT$ the holes simply filled the $j = 3/2$ states subject to the exclusion principle while the electrons were free to align themselves in the minimum energy (spins parallel) configuration. These authors (Kulakovskii et al. 1978a, 1979) have also studied the Si:P and Si:B Zeeman spectra in Si subjected to a uniaxial stress, which reduced the valence band degeneracy. This resulted in a considerable simplification in the spectra, which could readily be explained in terms of the shell model, but introduced the problem of the non-thermalization of the initial states.

Altukhov et al. (1977) measured the net circular polarization of the $m = 1$ through 3 B lines in the Faraday configuration, without actually resolving the components. Since the experimental resolution was not high enough to completely resolve lines having different values of m, the results cannot be regarded as definitive. On the basis of those results the authors argued that the shell model was incorrect, but that argument rested upon their misapprehension that the shell model required the Γ_1 electron shell to be lowest for acceptor complexes. A model was introduced in which various electron configurations were split by orientational effects, but this possibility is ruled out by other experimental results (Kulakovskii 1978, Thewalt et al. 1979). More recent experiments by these authors, described in ch. 6 (sect. 6.3.3) of this volume, agree with the shell model.

The Zeeman spectrum of Si doped with the deeper acceptor Al has been measured by Kulakovskii and Malyavkin (1979) whose results are shown in fig. 16. Since the Al BE ground state is split by the hole–hole interaction, the interpretation of the spectra is not quite so straightforward as for Si:B. Kulakovskii and Malyavkin considered only the large splitting between the $J = 0$ and $J = 2$ BE states, ignoring the much smaller cubic splitting of the $J = 2$ level. The resulting level scheme given in fig. 17 provided a good accounting of the observed results. Even though the experimental con-

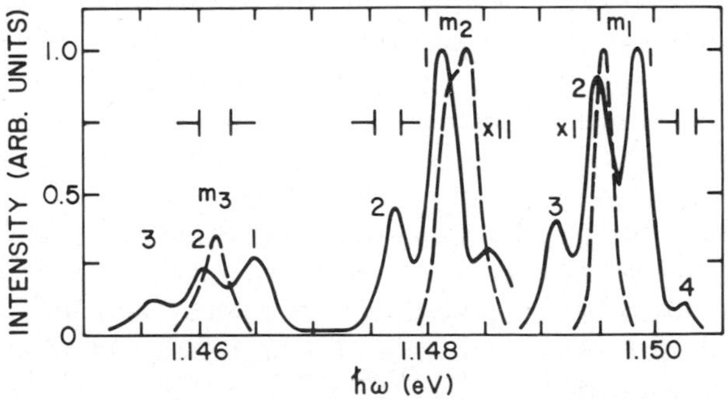

Fig. 16. Zeeman spectrum of Si:Al in the NP region at 1.8 K and 60 kG, $H \parallel \langle 111 \rangle$. The dashed lines show the zero-field spectrum. From Kulakovskii and Malyavkin (1979).

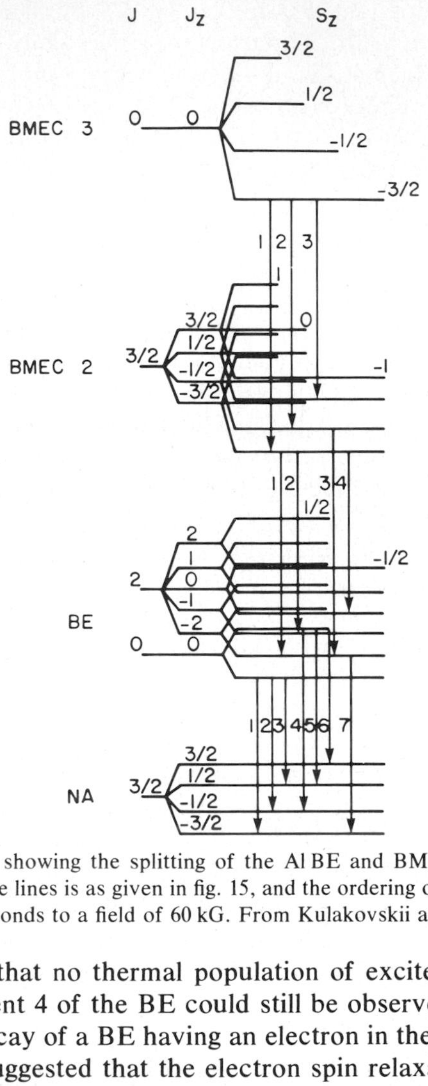

Fig. 17. Level diagram showing the splitting of the Al BE and BMEC levels in a magnetic field. The labelling of the lines is as given in fig. 15, and the ordering of the BE $J = 0$ and $J = 2$ components corresponds to a field of 60 kG. From Kulakovskii and Malyavkin (1979).

ditions were such that no thermal population of excited initial states was expected, component 4 of the BE could still be observed. This component results from the decay of a BE having an electron in the excited (+1/2) spin state, and it was suggested that the electron spin relaxation might be slow in the lowest energy BE state because the electron–hole exchange interaction would be weak when the total hole angular momentum was zero. The $m = 2$ and 3 Zeeman components were all found to originate from the lowest energy states of the $m = 2$ and 3 complexes, indicating that the Zeeman levels of these complexes were in thermal equilibrium. As in the Si:B case the electrons were found to thermalize into the lowest energy spin-parallel configuration.

Kulakovskii et al. (1979) measured the diamagnetic shifts of the main P and B lines, and from these results were able to estimate the sizes of a few of the B complexes. These estimates were in agreement with the calculations discussed in sect. 8, which show that the BMEC size does not increase with m until it becomes necessary to add particles to new shells.

4. *Uniaxial stress spectra*

The uniaxial stress studies of the BMEC luminescence spectra are considerably more extensive than the previously outlined Zeeman spectroscopy. This results both from the greater variety of degeneracy reductions which can be achieved by stress, as outlined in fig. 18, and from the fact that a relatively simple and inexpensive uniaxial stress apparatus can readily produce splittings which are far beyond the capabilities of any magnet. A compressive stress (henceforth all stresses will be considered to be compressive unless stated otherwise) of any orientation will in Si split the fourfold degenerate Γ_8 valence band into two doubly-degenerate levels, Γ_4 and $\Gamma_{5,6}$ (Γ_5 and Γ_6 being degenerate due to time-reversal symmetry), with Γ_4 lowest in energy (for holes). These two levels are often referred to in the angular momentum notation as $m_j = \pm 1/2$ and $m_j = \pm 3/2$, respectively. Although this notation is only approximate for stresses other than ⟨111⟩ and ⟨001⟩ we will nevertheless use it here for the sake of clarity, saving the group-theoretical Γ notation to label the valley-orbit electron states. As shown in fig. 18 the full conduction band degeneracy is retained for ⟨111⟩ stress, while an ⟨001⟩ stress lowers the energies of the two valleys oriented along the stress axis and raises the energies of the other four valleys. A ⟨110⟩ stress results in shifts of the conduction band valleys which are the reverse of the ⟨001⟩ case.

Before discussing the piezospectroscopic results as they are currently understood, a review of some of the earlier work will be worthwhile, since initial disagreements on both the experimental results and their interpretation led to a period of considerable controversy and confusion. Alkeev et al. (1975) published the initial stress results for Si:P and Si:B. They observed the correct doublet splittings of the P α lines, but mistook

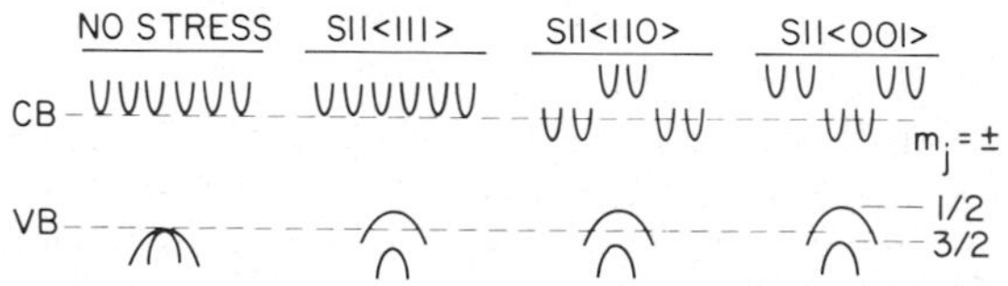

Fig. 18. Schematic representation of the effects of ⟨111⟩, ⟨110⟩ and ⟨001⟩ stress on the valence and conduction bands of Si. Each of the indicated bands is only spin degenerate, giving in unperturbed Si a total degeneracy of four at the valence band edge, and twelve for the conduction bands.

this as resulting from the conduction band splitting rather than from the valence band splitting as is now generally accepted. For the Si : B case they reported only a rigid shift of the spectrum to lower energies and failed to observe the doublet splittings reported in all later studies. In attempting to eliminate the BMEC model, Sauer and Weber (1977) reported that for large ⟨111⟩, ⟨110⟩ or ⟨001⟩ stresses the lower energy branches of the previously mentioned α^1 through α^4 doublets showed an additional splitting into two or three components. This was a very surprising result, particularly for the BE and for ⟨111⟩ stress which does not affect the electron degeneracies. No explanation was forthcoming. Since this fine structure was never observed in the many later piezospectroscopic studies it must be considered spurious, and possibly associated with inhomogeneous strain in the samples. It was soon demonstrated that shell-type BMEC models readily explained not only the doublet splittings but also the thermalizations of the doublets as a function of m (Herbert et al. 1977, Thewalt 1977d, Thewalt and Rostworowski 1978). Sauer et al. (1977) then claimed to have discredited the shell model using transient luminescence measurements, further described in sect. 5, along with a new discussion of the uniaxial stress results which was based upon a misinterpretation of how the shell model should be extended to the non-zero stress case. They also stated that, contrary to what was expected, under ⟨001⟩ stress the β^2 line showed only a doublet splitting. It will be shown in sect. 4.1 that this is not the case. These arguments and observations were refuted by Thewalt (1977d), Thewalt and Rostworowski (1978) and by Thewalt et al. (1979), and the description of the extension of the shell model to the case of non-zero stress which is given in this section was taken primarily from those three references.

Although the effects of stress on the electron levels of BE's and BMEC's is dependent upon the nature of the impurity center and the direction of stress, the effects on the Γ_8 hole shell are more general, with the specific exception of BE's associated with the deeper acceptors. Since any stress splits Γ_8 into the doubly degenerate $m_j = \pm 3/2$ and $m_j = \pm 1/2$ subshells, all lines which are due to Γ_8 hole recombination in the zero stress case should for an arbitrary stress show at least this doublet splitting, as outlined in fig. 19. For all the complexes the magnitude of the stress-induced Γ_8 shell splitting is found to be very close to that of the valence band edge. The relative population p of $m_j = \pm 3/2$ compared to $m_j = \pm 1/2$ holes as a function of the splitting Δ and the temperature can be readily calculated if one assumes that the holes are in thermal equilibrium:

$$\begin{aligned} n &= 1, \quad p = \exp(-\Delta/kT), \\ n &= 2, \quad p = [2\exp(-\Delta/kT) + \exp(-2\Delta/kT)]/[1 + 2\exp(-\Delta/kT)], \\ n &= 3, \quad p = [1 + 2\exp(-\Delta/kT)]/[2 + \exp(-\Delta/kT)], \\ n &= 4, \quad p = 1, \end{aligned} \tag{1}$$

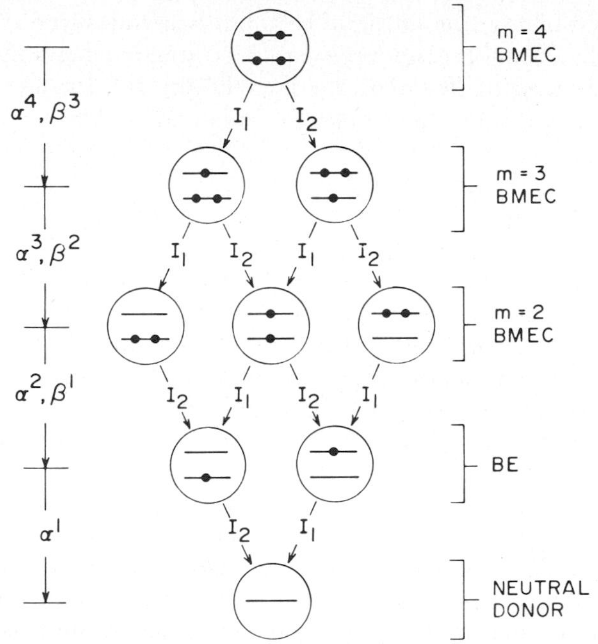

Fig. 19. The possible arrangements of holes in the $m_j = \pm 3/2$ and $m_j = \pm 1/2$ shells resulting from the stress splitting of Γ_8 is shown for donor complexes up to $m = 4$. The higher and lower energy branches of the resulting doublet transitions are labelled I_1 and I_2, respectively. From Thewalt (1977d).

where n is the number of Γ_8 holes in the complex ($n \leq 4$). These equations allow the thermalization behavior of the doublet components resulting from the valence band splitting to be predicted, although the absolute value of the intensity ratios will of course also depend on the relative oscillator strengths of transitions involving $m_j = \pm 3/2$ and $m_j = \pm 1/2$ holes. Although eqs. (1) were first given by Thewalt (1977d), the thermalization behavior had also been qualitatively described by Herbert et al. (1977) based on the earlier model of Dean et al. (1976a, 1976b) which treated the Γ_8 hole shell in the same manner as does the shell model.

The shell model description of the effects of strain on electrons in BE's and BMEC's is quite straightforward. For acceptor complexes we have already determined that valley-orbit effects are negligible, and in this limit the electrons can be thought of as occupying distinct single-valley states. For non-⟨111⟩ stress there is therefore the possibility of recombination involving electrons from lower (cold) valleys or upper (hot) valleys. This should produce a splitting identical to that of the conduction band valleys themselves. From studies of FE luminescence we know that the relaxation of electrons from the hot valleys is quite slow as a result of the large change in crystal momentum involved, but becomes very rapid once the conduction band splitting is large enough to allow intervalley scattering phonons to be emitted (Kulakovskii et al. 1978b). For the donor complexes we must consider the single-particle Hamiltonian for an electron in a BMEC:

$$H = H_{em} + H_{sc} + H_{cc} + H_s, \tag{2}$$

where H_{em} is the zero-stress effective mass Hamiltonian, H_{sc} results from the averaged fields of the other electrons and holes, H_{cc} is the central cell correction including intervalley terms and H_s represents the difference between the effective mass Hamiltonian at zero and non-zero stress. Following an approximation which is commonly made in treating the stress behavior of neutral donor states (Bir and Pikus 1972), we take the symmetry of the first three terms of eq. (2) to be unchanged by stress, and in H_s consider only the shifting of the conduction band valleys (Thewalt et al. 1979). In this simple approximation the ordering and shifts of the stress split electron shells of the complexes should be the same as those of the donors with which they are associated, remembering of course that the magnitude of the valley-orbit splitting is less in the complexes.

Under stress the symmetry group of the impurity reduces to a subgroup of T_d, and the Γ representations then belong to this subgroup and not to T_d. The splittings in Γ_1, Γ_3 and Γ_5 which are allowed by group-theoretical considerations for ⟨111⟩, ⟨110⟩ and ⟨001⟩ stresses are listed in table 1. Although a splitting of the Γ_5 level under ⟨111⟩ stress is seen to be allowed, it will not occur within the approximation we have made. In the following subsections we will label a state which transforms as Γ_x of T_d under zero stress, and Γ_y of the appropriate subgroup when stress is applied, as Γ_y^x.

Since there is now general agreement on both the structure and interpretation of the stress spectra for Si:P (Kaminskii and Pokrovskii, 1978, 1979, Kaminskii et al. 1978, Kulakovskii 1978, Thewalt et al. 1979) and Si:B (Kaminskii and Pokrovskii 1979, Kaminskii et al. 1978, Kulakovskii 1978, 1979, Thewalt et al. 1979), these results will be discussed in the

Table 1
Irreducible representations of the donor electron wave functions

Stress	Symmetry group of the impurity	Representations		
None	T_d	Γ_1	Γ_3	Γ_5
⟨111⟩	C_{3v}	Γ_1	Γ_3	$\Gamma_1+\Gamma_3$
⟨110⟩	C_{2v}	Γ_1	$\Gamma_1+\Gamma_3$	$\Gamma_1+\Gamma_2+\Gamma_4$
⟨001⟩	D_{2d}	Γ_1	$\Gamma_1+\Gamma_3$	$\Gamma_4+\Gamma_5$

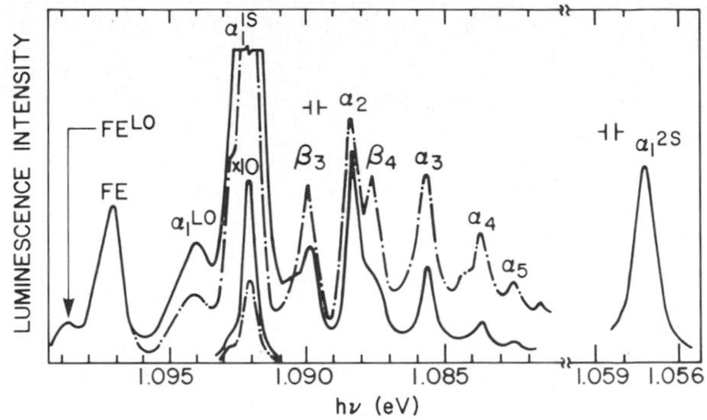

Fig. 20. Photoluminescence spectra of Si:P at 4.2 K. The solid curve shows the results for continuous excitation, while the chained curve is the synchronously detected signal resulting from a small AC modulation of the excitation beam. Note how this modulation technique enhances the higher-order lines. These authors have changed the shell model labelling scheme by denoting β^m as β_{m+1}. From Kaminskii et al. (1978).

following subsections without repeated references to the literature except for unique results. In their stress studies Kaminskii et al. (1978) used a modulated-excitation technique which takes advantage of the higher excitation density dependence of the larger m complexes to increase the relative intensities of these weak lines and thus facilitate their study, as is shown in fig. 20. This technique should find further application in BMEC spectroscopy. Kaminskii and Pokrovskii (1979) have very recently extended the piezospectroscopic studies to tensile stresses, and in addition proved that the symmetries of the initial and final states given in the shell model could accurately account for the degree of polarization of all the Si:P and Si:B stress-split components. Note that Kaminskii and coworkers (Kaminskii et al. 1978, Kaminskii and Pokrovskii 1978, Kulakovskii 1978) have altered the original shell model labelling scheme by renumbering β^m as β^{m+1}.

4.1. Substitutional donors

As in the case of unperturbed Si, the substitutional donor in the large majority of the piezospectroscopic studies has been P, although Kaminskii et al. (1978) have shown that the results for Si:Sb are almost identical. The splittings and shifts of the Si:P lines are summarized in fig. 21. For ⟨111⟩ stress only the expected identical doublet splittings resulting from the lifting of the Γ_8 degeneracy were observed. The intensity ratios of the upper component as compared to the lower component for the α^1 through α^4 doublets are shown in fig. 22 versus the splitting Δ divided by kT. The results are seen to be in excellent agreement with the predictions of eqs. (1).

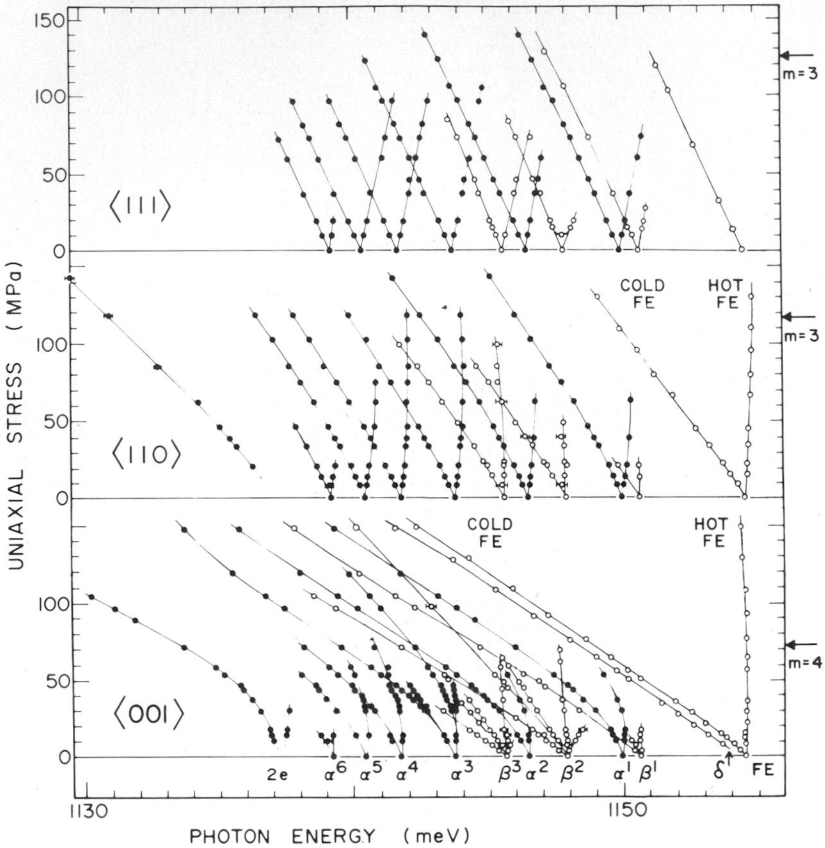

Fig. 21. Energies of the Si:P luminescence lines as a function of stress. The arrows on the right give the maximum stresses for which the named lines could still be observed. The solid circles had strong NP transitions while the open circles did not. The energy scale applies to NP transitions. From Thewalt et al. (1979).

Also to be noted is the close correspondence between the β^2 and α^3, and the β^3 and α^4 thermalization behavior. This is the expected result, since in the shell model α^m and β^{m-1} have the same initial state. The α^5 and α^6 lines showed the same non-thermalizing doublet structure as did α^4, verifying that all these transitions involve holes from the filled Γ_8 (zero stress) shell.

Since the $m > 2$ complexes necessarily have holes in the higher energy, $m_j = \pm 3/2$ subshell, their binding energy relative to the formation of a smaller complex and a FE associated with the $m_j = \pm 1/2$ valence band is reduced by stress. Above a ⟨111⟩ stress of 125 MPa (1 MPa = 10^6 Nm^{-2}), $m > 2$ luminescence could no longer be observed. The binding energy of

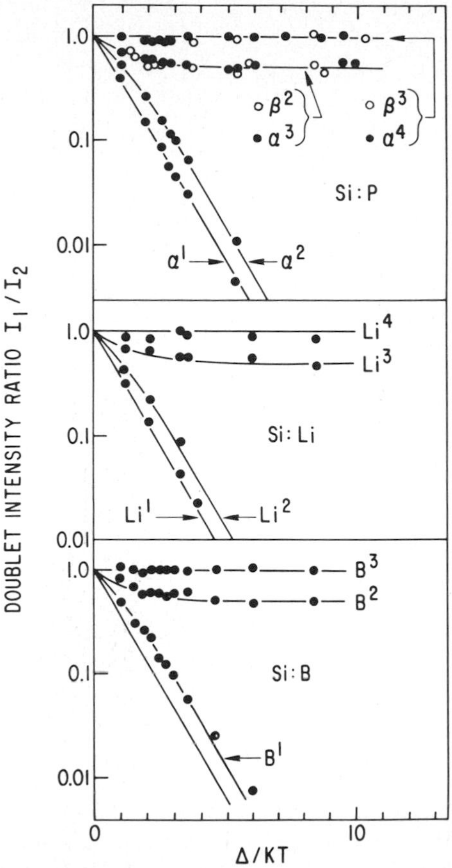

Fig. 22. Intensity ratios of the upper to lower (I_1 to I_2) components of the doublets resulting from the valence band splitting under ⟨111⟩ stress plotted as a function of the splitting Δ divided by kT. In each of the three cases the solid lines are, from top to bottom, the $n = 4, 3, 2$ and 1 cases of eqs. (1). The ratios are normalized to unity at $\Delta/kT = 0$. Note the close agreement between the β^2 and α^3, and the β^3 and α^4 behavior for Si:P. From Thewalt et al. (1979).

the $m = 3$ BMEC is given by the energy difference between the FE edge and the high energy component of the β^2 doublet. At 125 MPa this binding energy extrapolates to about 1.5 meV, which is small enough to allow rapid thermal dissociation of the $m = 3$ BMEC. The $m > 3$ luminescence must of course also disappear at that stress, since although the $m = 4$ complex still has a considerable binding energy (~3.8 meV), stable $m = 3$ complexes are a necessary step in the formation of larger BMEC's.

For ⟨110⟩ and ⟨001⟩ stresses the lifting of the electron shell degeneracies can produce extra structure in addition to the Γ_8 doublet splittings. Within

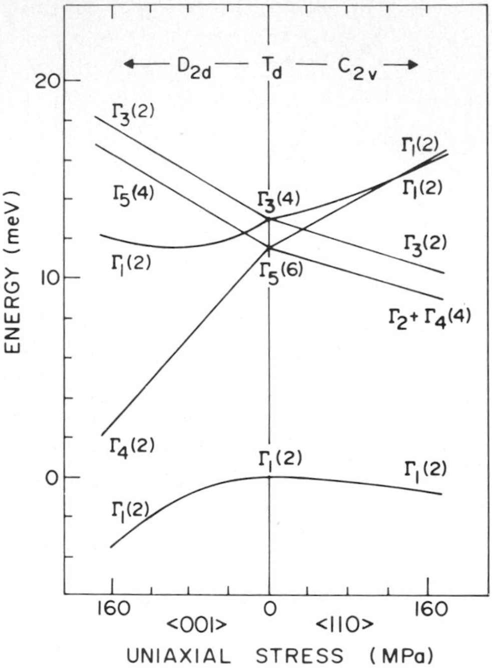

Fig. 23. The calculated energies of the P 1S neutral donor states as a function of stress. The total degeneracy of each level is given in brackets. From Thewalt and Rostworowski (1978).

the previously introduced approximation the ordering and shifts of the electron shells in the complexes will be the same as those of the donor states which are shown in fig. 23. For the ⟨110⟩ case the lowest electron subshell above Γ_1^1 (which of course cannot be split by stress) is the six-fold degenerate $\Gamma_{2,3,4}^{3,5}$, and barring thermal excitation of electrons into higher shells, extra structure resulting from electron splittings is not expected to be seen for complexes lower than $m = 8$. This was borne out by the experiments which revealed only identical doublet splittings of the α and β lines, with thermalizations as predicted by eqs. (1). Note from fig. 21 that hot FE luminescence associated with the upper conduction band valleys was seen for ⟨110⟩ stress. It was however much weaker than the hot FE luminescence seen for the ⟨001⟩ case, in agreement with earlier observations (Kulakovskii et al. 1978b). So far as could be determined, the electron populations of the donor BE and BMEC's were always in thermal equilibrium for ⟨110⟩ stresses.

As the ⟨110⟩ stress was increased a new luminescence line appeared which showed the same temperature and excitation density dependence as did the lower energy branch of the BE line α^1. This line was found to be a

BE two-electron (2e) transition in which a ground state BE recombines, leaving the donor in 1S valley-orbit excited state. Such transitions are forbidden in unstressed Si since the symmetries of the donor valley-orbit excited states, Γ_3 and Γ_5, differ from those of the two Γ_1 electrons in the ground state BE. Under ⟨110⟩ stress however, a 2e transition to the Γ_1^3 excited state would be allowed, and as shown in fig. 24 the α^1 to 2e separation as a function of stress exactly matched the calculated donor Γ_1^1 to Γ_1^3 energy difference. The 2e line should of course show the same doublet structure as the other Si:P lines, but at the temperatures involved the population of the upper hole level was too small for the higher energy 2e component to be observed. Strictly speaking, a 2e transition to the Γ_1^5 valley-orbit state is also allowed by symmetry, but the projection of Γ_1^1 onto Γ_1^3 is expected to be much larger since under stress Γ_1^1 and Γ_1^3 are both linear combinations of the zero stress Γ_1 and Γ_3 states.

Due to the large degeneracy of the $\Gamma_{2,3,4}^{3,5}$ shell the effects of stress on the binding energies of the observed complexes are the same for ⟨110⟩ stress as for ⟨111⟩, with the exception of the BE which shows a small non-linear decrease in binding energy because of the difference in the Γ_1 valley-orbit shift between unstressed Si and the high ⟨110⟩ stress limit. A similar non-linear shift in the positions of the higher α^m lines does not reflect a

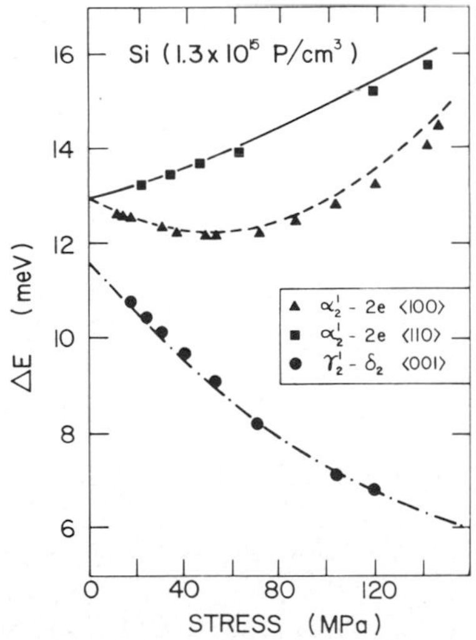

Fig. 24. The energy difference between the Si:P α_2^1 and $2e_2$ lines (the subscripts 1 or 2 here refer to recombination with $m_j = \pm 3/2$ or $\pm 1/2$ holes, respectively) as a function of ⟨110⟩ and ⟨001⟩ stresses. The solid line is the calculated Γ_1^1 to Γ_1^3 splitting of the P donor for ⟨110⟩ stress while the dashed curve gives that splitting for ⟨001⟩ stress. The observed γ_2^1 to δ_2 separation as a function of ⟨001⟩ stress is also shown to agree with the calculated (chained curve) Γ_1^1 to Γ_4^5 splitting of the P donor. From Thewalt et al. (1979).

change in the binding energies of the α^m complexes, since the α^m $(m > 1)$ transitions are not ground state to ground state. Rather these shifts reflect a stress-induced decrease of the valley-orbit splitting in the $m - 1$ final states of these transitions. As shown in fig. 25 these changes in the valley-orbit splittings of the complexes can be well described by the simple equations giving the neutral donor valley-orbit splittings as a function of stress, adjusted of course to give the observed zero stress valley-orbit energies.

For ⟨001⟩ stress the results are considerably more complicated due both to the smaller degeneracy of the next electron shell above Γ_1^1, the two-fold degenerate Γ_4^5, as well as to a reduced thermalization (we will hereinafter refer to a population as thermal if it agrees with the value calculated using the appropriate statistics at the lattice temperature) of the electron populations leading to "hot" luminescence. All the lines of course show the basic Γ_8 doublet splitting, with the thermalization of the components again agreeing with those predicted by eqs. (1). From fig. 23 we see that there are three levels which split from $\Gamma_{3,5}$ and that the β lines could therefore split into six components if these levels were all populated. The exclusion principle only requires electrons to be placed in shells above Γ_4^5 for $m > 3$. For low ⟨001⟩ stresses extra structure is seen for β^1 but this quickly

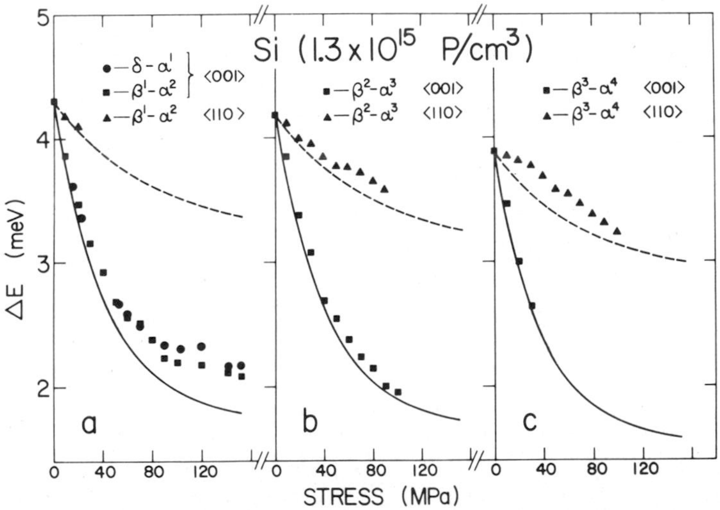

Fig. 25. The measured valley-orbit splittings of the $m = 1$, 2 and 3 complexes, as determined by the separations between the lowest energy components of β^m and α^{m+1} are plotted as a function of ⟨110⟩ and ⟨001⟩ stresses. The calculated lowest valley-orbit splittings of a simple isolated neutral donor having the observed zero-stress splittings are given for ⟨110⟩ stress (----, Γ_1^1 to $\Gamma_{2,3,4}^{3,5}$) and for ⟨001⟩ stress (—, Γ_1^1 to Γ_4^5) for comparison. Note also the close agreement between the δ to α^1 and the β^1 to α^2 splittings over the entire range of ⟨001⟩ stress. From Thewalt et al. (1979).

decreases in intensity as the stress is increased, leaving only the single β^1 component which is expected if both the electron and hole populations of the $m = 2$ complex thermalize. This is not the case for β^2, where extra structure due to electron splittings persists for splittings much larger than kT (Thewalt et al. 1979). Five β^2 components were observed and it was assumed that this was because two of the six expected components could not be resolved. Four components were observed for β^3, corresponding to the four lowest β^2 lines, with the fifth possibly lost beneath the stronger α^2 and β^2 luminescence. The β^3 results do not indicate non-thermal electron populations, since the $m = 4$ complex must have an electron in a shell above Γ_4^5.

Although it was initially thought that stress could produce no more than a doublet splitting for the α lines since the Γ_1 shell is only spin-degenerate and cannot be split by stress (Kirczenow 1977b), Thewalt et al. (1979) reported a small extra splitting of the α^3 doublet over a limited range of ⟨001⟩ stress, as is shown in figs. 21 and 26. The original prediction of only doublet splittings rested upon the implicit assumption that the energy of an α transition was independent of the arrangement of the $\Gamma_{3,5}$ electrons in the initial state. This assumption was quite reasonable for the zero stress case, but under ⟨001⟩ stress not only the Γ_1^1 but also the Γ_1^3 electrons have a non-zero amplitude in the central cell region. As a result of the same screening effects which causes such a large reduction in the valley-orbit splitting between the donor and the BE, the valley-orbit energy of a Γ_1^3

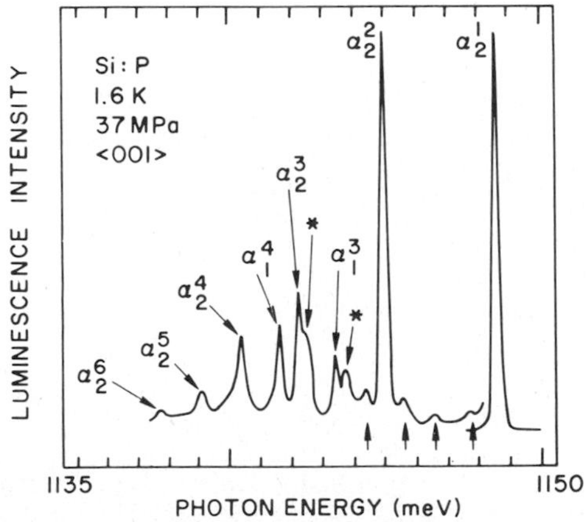

Fig. 26. Si:P spectrum at 1.6 K and an ⟨001⟩ stress of 37 MPa. The extra structure of the α^3 line is marked by asterisks. These new lines vanish as the stress is increased. The four arrows along the bottom of the figure indicate NP β^2 and β^3 components which appear under ⟨001⟩ stress and correspond to β transitions involving Γ_1^3 electron recombination. From Thewalt et al. (1979).

electron will depend on the number of Γ_1^1 electrons in the complex, which is of course reduced from two to one during an α transition. Thus even in the single-electron single-hole transition scheme, the energy of an α transition could be affected by the number of Γ_1^3 electrons present in the initial state. This explanation is supported by the fact that the α^3 fine structure components became weaker as the stress was increased and vanished at about the same point as did the hot β^2 luminescence, which also resulted from $m = 3$ complexes having electrons excited into Γ_1^3. Since there is no evidence from the β^1 and β^3 lines that the $m = 2$ and $m = 4$ BMEC have such non-thermal electron populations, the absence of the extra structure for the α^2 and α^4 lines is not surprising. There is as yet no explanation why certain complexes should have non-thermal electron populations while others do not, but as we shall see it is a very common phenomenon for ⟨001⟩ stress. Another consequence of the non-zero amplitude of the Γ_1^3 wave functions in the central cell region is the appearance of weak NP transitions for those β components which involve Γ_1^3 electron recombination, as shown in fig. 26.

The P BE also shows hot luminescence, with a δ line appearing at quite low ⟨001⟩ stresses and growing in intensity as the stress is increased. The hot BE δ luminescence is shown in figs. 27 and 28. If excited BE's are

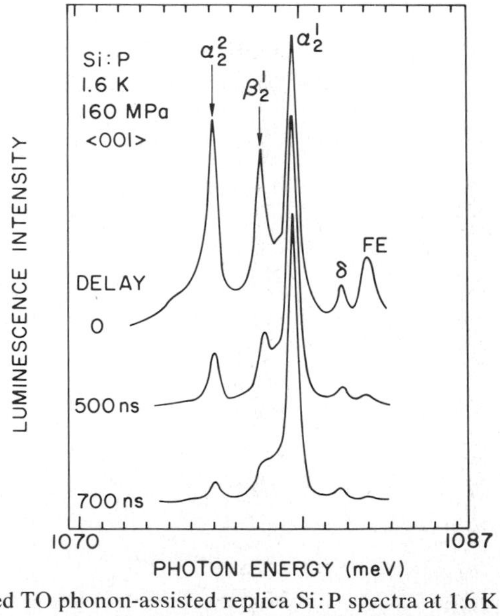

Fig. 27. Time-resolved TO phonon-assisted replica Si:P spectra at 1.6 K and an ⟨001⟩ stress of 160 MPa. The non-thermal population of excited BE's gives rise to a strong δ line. The stress-induced enhancement of α^2 and β^1 relative to α^1 can be judged by comparing the 0 delay spectrum to fig. 3. The two lower spectra were taken at delays of 500 and 700 ns after the end of the excitation pulse. From Thewalt et al. (1979).

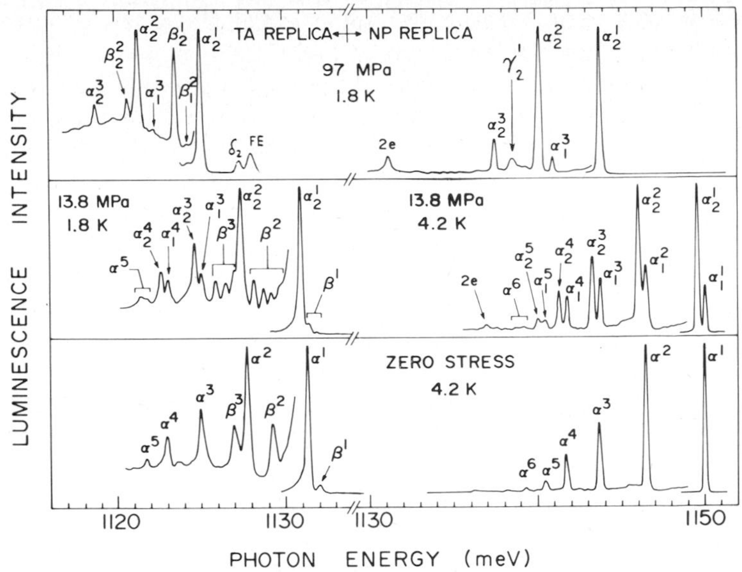

Fig. 28. NP and TA phonon-assisted photoluminescence transitions of Si:P at zero stress and two values of ⟨001⟩ stress. Note that the 13.8 MPa TA replica was taken at 1.8 K in order to suppress the high-energy components of the α^1 and α^2 lines so that the extra splittings of the β lines would be clearly visible. From Thewalt et al. (1979).

present, we would also expect from fig. 5 to see a γ^1 transition which leaves the donor in a 1S valley-orbit state identical to that of the upper electron in the excited BE. The γ^1 line is in fact observed and can be seen in the high-stress spectrum of fig. 28. From fig. 23 we would expect the excited electron in the initial and final state of the γ^1 transition to be in the Γ_4^5 shell, since only this shell could account for the observed α^1 to δ splitting as a function of stress. This is verified by the close agreement between the measured δ to γ^1 separations and the calculated P 1S Γ_1^1 to 1S Γ_4^5 splitting shown in fig. 24. The same figure shows that just as for the ⟨110⟩ case, the final state of the 2e transition induced by ⟨001⟩ stress has Γ_1^3 symmetry.

All the α transitions show strongly non-linear shifts under ⟨001⟩ stress, although only for α^1 is this shift a direct reflection of a reduction in binding energy. The valley-orbit splittings of the $m = 1$ through 3 complexes as determined from the β^m to α^{m+1} energy differences are plotted as a function of ⟨001⟩ stress in fig. 25. Once again good agreement is found with the stress behavior of the valley-orbit splittings calculated for an isolated donor having the observed zero-stress splitting, validating the simple assumption used to extend the shell model to non-zero stress.

The ⟨001⟩ stress results provide a variety of tests of the shell model assertion that β^1 and α^2 have the same initial state, the importance of which will become clearer when the thermodynamic binding energy measurements are discussed in sect. 6. We begin by noting from fig. 25 that over the entire stress range the β^1 to α^2 separation exactly matched that of δ to α^1, as expected from the shell model outlined in fig. 5. Large ⟨001⟩ stresses drastically change the relative intensities of the lines as can be seen by comparing fig. 28 with fig. 4. However, even though for large ⟨001⟩ stress the higher order lines vanish completely and the limiting α^2 (or β^1) to α^1 intensity ratio increases by at least a factor of ten due to the higher FE densities resulting from the destabilization of the electron–hole droplet (Kulakovskii and Timofeev 1977), the β^1 to α^2 intensity ratio remains equal to the zero stress value, 0.6 ± 0.1, within experimental error. The large enhancement of the α^2 and β^1 lines relative to α^1 and the shift of β^1 away from the α^1 edge can greatly facilitate the study of β^1 under high ⟨001⟩ stress. The transient luminescence results shown in fig. 27 verify that α^2 and β^1 have the same decay characteristics, which are quite distinct from those of α^1. Excitation density dependence measurements taken at large ⟨001⟩ stress also verify that α^2 and β^1 behave similarly, and quite differently from α^1 (Thewalt et al. 1979).

4.2. Interstitial Li donor

Just as in the zero-stress case, the piezospectroscopic results for Si:Li are expected to differ markedly from those of the substitutional donors because of the inverted ordering of the valley-orbit states. The results are summarized in fig. 29, and ⟨111⟩ stress is again found to produce only the identical doublet structure arising from the Γ_8 hole-shell splitting. The thermalization behavior of the doublets agreed with eqs. (1) as shown in fig. 22. The Li^5 through Li^7 doublets all showed the same lack of thermalization as did Li^4, verifying that for zero stress these three lines are due to the recombination of a hole from the filled inner Γ_8 shell and therefore result in excited final states. The Li^3 luminescence did not vanish when the $m = 3$ binding energy, as determined from the extrapolated splitting between the FE edge and the upper Li^3 component, was reduced to zero by increasing the stress. Also, the upper Li^3 component was half as intense as the lower (as expected) for stresses below the point where it became lost under the Li^1 line, but at higher stresses it was not seen to emerge on the high energy side of the Li^1 line. Similar results were found for ⟨110⟩ and ⟨001⟩ stresses. A possible explanation is that above some stress it becomes energetically favorable for the higher energy $m_j = \pm 3/2$ hole to occupy some new level, the energy of which does not increase as rapidly with stress as does that of the $m_j = \pm 3/2$ shell. Such a proposal was made by Herbert et al. (1977) to

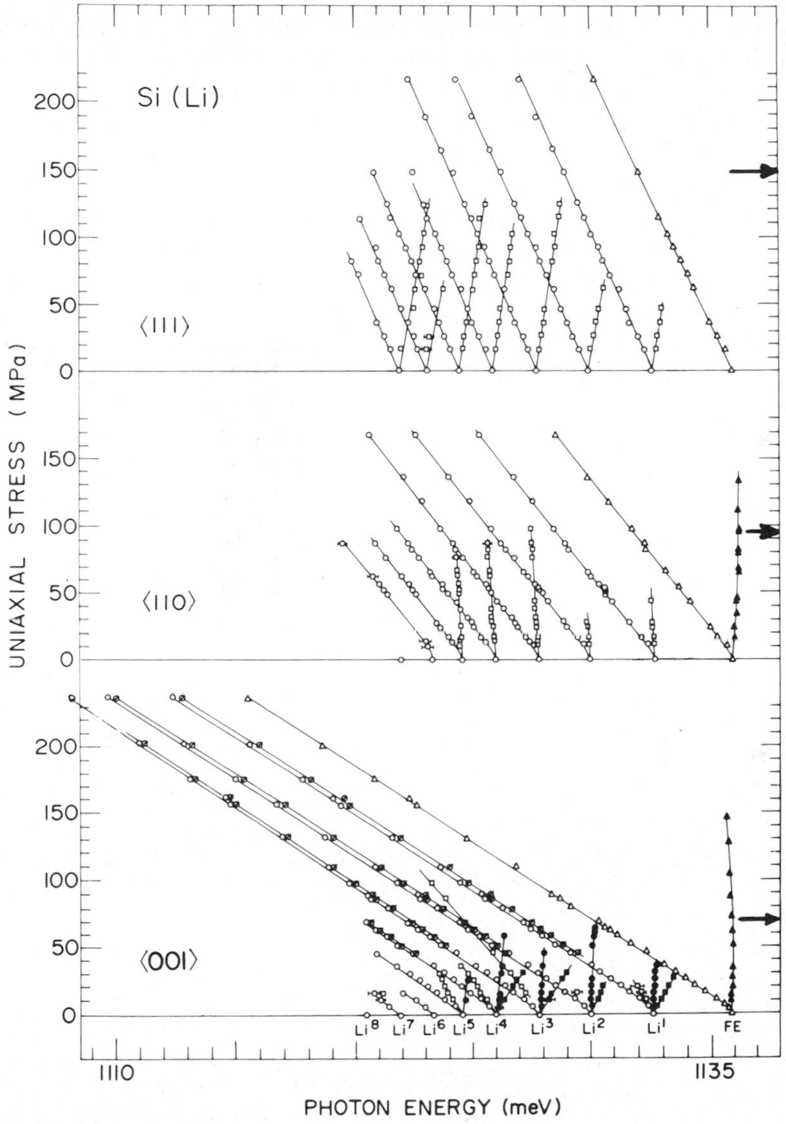

Fig. 29. The energies of the TA replica Si:Li lines as a function of stress. Transitions involving $m_j = \pm 3/2$ holes are shown as squares while those involving $m_j = \pm 1/2$ holes are shown as circles. Transitions involving electrons from the lower valleys are shown as open circles (squares) while those from the upper valleys are shown as solid circles (squares). From Thewalt et al. (1979).

explain the observation of Alkeev et al. (1975) (now known to be incorrect) that under stress the B lines shifted but did not split.

In order to discuss the ⟨110⟩ and ⟨001⟩ results we refer to the calculated behavior of the Li donor 1S valley-orbit states given in fig. 30, which in the approximation previously introduced can be used to determine the ordering of the electron shells of the complexes under stress. For the ⟨110⟩ case the lowest shell is $\Gamma^{3,5}_{2,3,4}$, and because of its six-fold degeneracy the filling of higher shells need not take place for $m < 6$. In agreement with this only the Γ_8 doublet splittings were observed for ⟨110⟩ stresses, once again showing that hot electron effects are very weak for this stress orientation.

As in the substitutional donor case, the ⟨001⟩ spectra are more complex. From fig. 30 we see that the lowest electron shell, Γ^5_4, is only doubly degenerate and that the next higher shell, Γ^3_1 also only has two-fold degeneracy. Electrons must therefore be placed in shells above Γ^5_4 for $m > 1$, and above Γ^3_1 for $m > 3$. In addition there may be hot electron luminescence resulting from the non-thermal populations which are typical for ⟨001⟩ stress. The results summarized in figs. 29 and 31 verify these predictions. Luminescence resulting from the recombination of electrons in the upper $\Gamma^{3,5}_{3,5}$ shell is seen for the main Li lines (Li^1 through Li^4), although for Li^1 through Li^3 this hot luminescence is quite weak with respect to the

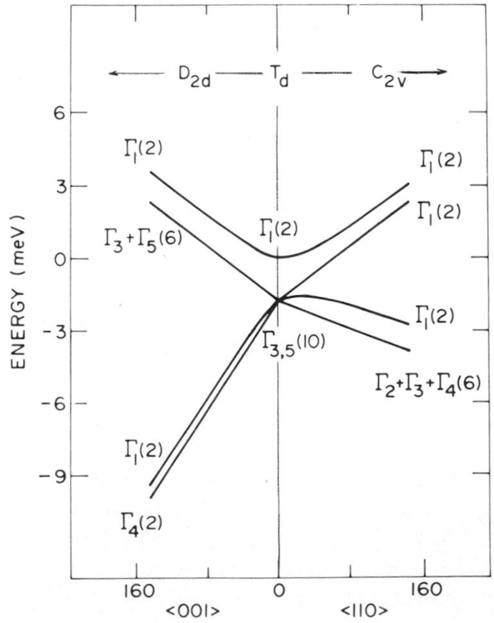

Fig. 30. Calculated energies of the Li 1S levels as a function of stress. The total degeneracies of each level are given in brackets. From Thewalt et al. (1979).

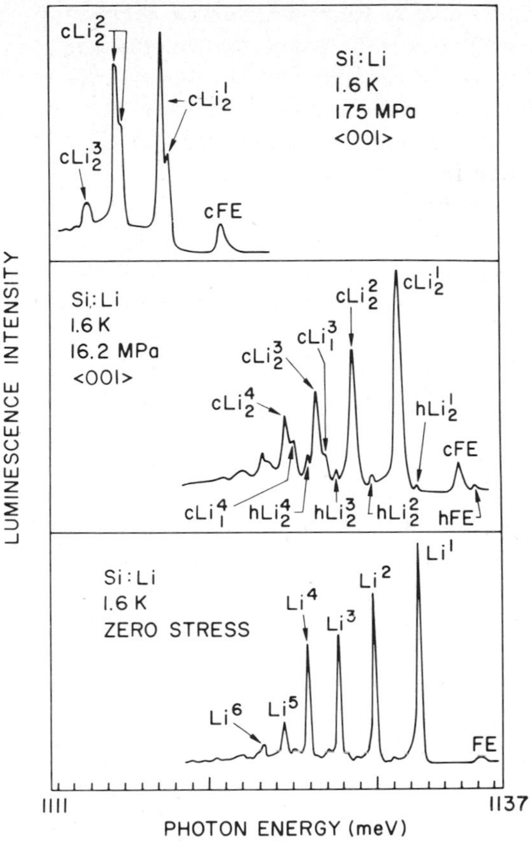

Fig. 31. Si:Li spectra at zero stress and two values of ⟨001⟩ stress. The prefixes h and c stand for hot and cold, respectively, and denote transitions involving electrons whose wave functions are associated with the upper or lower stress-split conduction band valleys. The subscripts 1 and 2 label transitions in which $m_j = \pm 3/2$ or $m_j = \pm 1/2$ holes, respectively, recombine. From Thewalt et al. (1979).

lower energy components, since for $m < 4$ it results only from the non-thermalization of the electrons. The lower energy components also show a small splitting which saturates at very low stresses, and from fig. 30 can be identified as resulting from the Γ_1^3 to Γ_4^5 shell splitting. The results for the Li^1 line shown in fig. 31 indicate a non-thermal population in Γ_1^3. There was no evidence for any of the complexes of a non-thermal electron population in the Γ_1^1 shell, which lies above $\Gamma_{3,5}$ even at zero stress.

In distinct contrast to the substitutional donor α line series, the Li^m lines show very linear shifts as a function of stress. This can be readily understood, since the lowest electron shell is excluded from the central cell region regardless of stress and therefore shifts exactly as do the lower conduction band valleys. Luminescence lines resulting from the recombination of Γ_1^3 or Γ_1^1 electrons would show some nonlinear behavior at low stresses but this would be difficult to observe, since for the Li complexes the valley-orbit energies are relatively small as compared to the substitutional donor complexes. The foregoing Si:Li piezospectroscopic results were taken from the work of Thewalt et al. (1979), which has very recently been corroborated by Henry and Lightowlers (1979). By using higher sample temperatures Henry and Lightowlers were able to thermally populate the various BE excited states and thus to observe the higher energy lines predicted by the shell model on the basis of fig. 30 but not observed in the earlier low temperature spectra.

4.3. Shallow B acceptor (2)

Of all the BE and BMEC piezospectroscopic spectra, those of Si:B are the most readily interpreted. As for all acceptor complexes, there is no valley-orbit splitting and the electrons therefore occupy single-valley states, while unlike the case of the deeper acceptors, the hole–hole interaction in the BE can be effectively ignored for B. The results are summarized in fig. 32. As before ⟨111⟩ stress results in identical doublet splittings and the thermalizations of the doublet components shown in fig. 22 agree with those predicted by eqs. (1). For ⟨110⟩ and particularly for ⟨001⟩ stresses strong hot electron luminescence resulting from the recombination of electrons associated with the upper conduction band valleys was observed. Hot electron luminescence is expected to be strong for acceptor complexes, where scattering due to the impurity short-range potential must be weak. For all three stress directions the $m > 1$ luminescence disappeared at high stresses due to the destabilization of complexes having holes in the higher energy $m_j = \pm 3/2$ shell. For ⟨111⟩ (⟨110⟩ and ⟨001⟩) stress this destabilization manifested itself when the $m = 2$ binding energy, as determined from the separation between the (cold electron) FE edge and the higher energy (cold electron) component of the B^2 line, had been reduced to roughly 1 meV.

When discussing the doublets resulting from the Γ_8 hole-shell splittings in this and the previous sections we have dealt only with the thermalization of the relative intensities of the two components, but not with the absolute intensity ratios. These involve not only the relative populations of $m_j = \pm 3/2$ and $\pm 1/2$ holes, as given by eqs. (1), but also the relative oscillator strengths of the transitions. For phonon-assisted luminescence these oscillator strength ratios were found to be in qualitative agreement with the measured (Laude et al. 1971, Capizzi et al. 1977) FE results. It was found that the oscillator strength ratio of the Si:P α doublets changed somewhat in going from zero to large ⟨001⟩ (⟨110⟩) stress, which could be understood

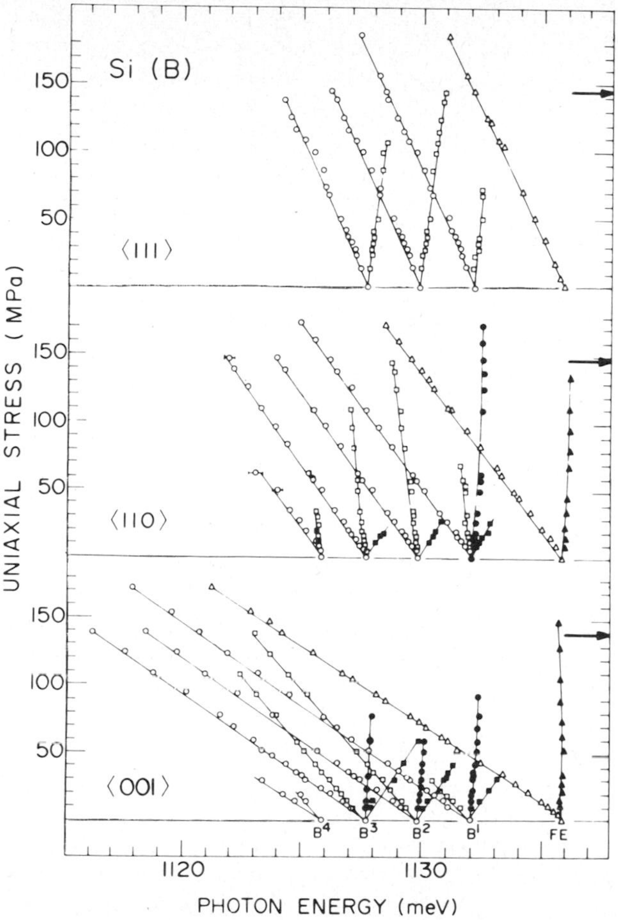

Fig. 32. Energies of the Si:B TA phonon-assisted photoluminescence lines as a function of stress. Transitions involving $m_j = \pm 3/2$ or $m_j = \pm 1/2$ holes are shown as squares and circles, respectively. Similarly, transitions involving electrons from the upper conduction band valleys are shown as solid circles (squares) while those associated with the lower valleys are shown as open circles (squares). From Thewalt et al. (1979).

in terms of the Γ_1^1 wave function gradually changing from a symmetric superposition of all six valleys to a symmetric superposition of only the two (four) valleys oriented parallel (perpendicular) to the stress axis. As seen in fig. 33 the oscillator strength ratios for Si:B differed very markedly between the phonon-assisted and the NP transitions. This difference has been explained by group-theoretical calculations of the acceptor NP transition strengths (Thewalt et al. 1979).

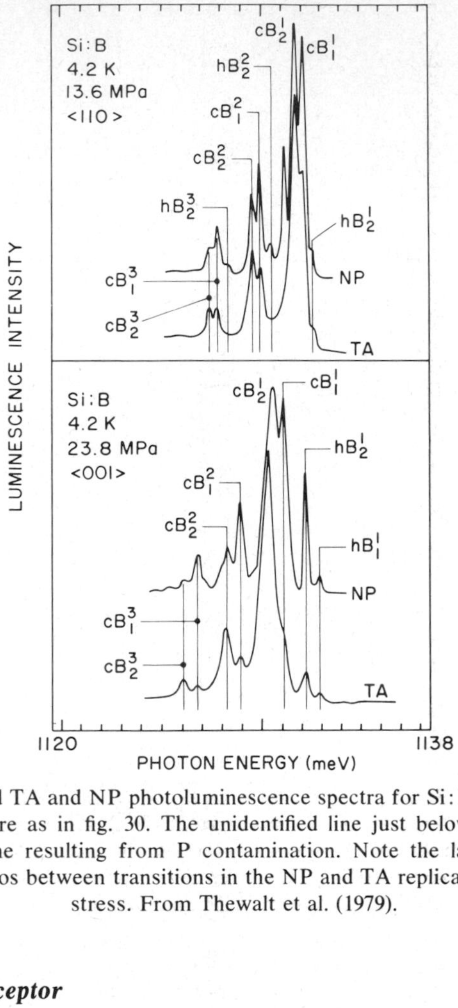

Fig. 33. Superimposed TA and NP photoluminescence spectra for Si:B under ⟨110⟩ and ⟨001⟩ stresses. The labels are as in fig. 30. The unidentified line just below cB_2^1 in the NP ⟨110⟩ spectrum is an α_2^1 line resulting from P contamination. Note the large differences in the oscillator strength ratios between transitions in the NP and TA replicas, particularly for ⟨001⟩ stress. From Thewalt et al. (1979).

4.4. Deeper Al acceptor

The only stress study of Si doped with a deeper acceptor is the Si:Al work of Kulakovskii and Malyavkin (1979). As in their Zeeman study, they considered only the large doublet splitting of the BE into $J = 0$ and $J = 2$ hole states, neglecting the much smaller cubic splitting of the $J = 2$ level. The good agreement between the calculated and measured splittings for ⟨111⟩ stress can be seen in fig. 34. Note that in the high stress limit component 3 of the $m = 2$ luminescence and components 2 and 6 of the BE luminescence would be forbidden since they are not one-hole transitions. They are however allowed at low stresses where the BE levels are still mixed states with contributions from both the $m_j = \pm 3/2$ and $\pm 1/2$ valence band edges.

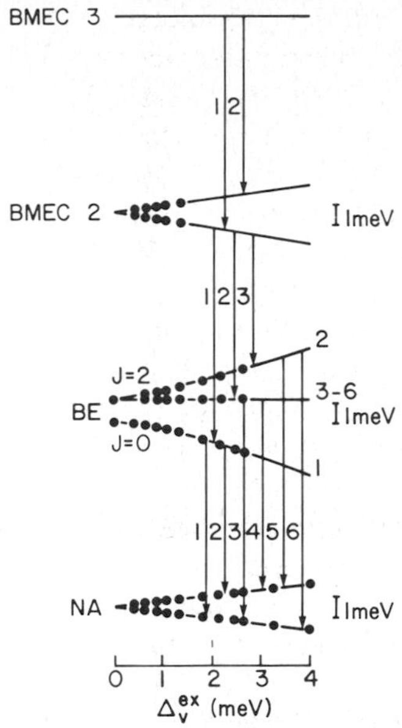

Fig. 34. Splittings of the Si:Al BE and BMEC states under ⟨111⟩ stress, and the allowed one-electron one-hole transitions. The solid lines are the calculated level splittings while the dots were experimentally determined from the energy separations of the various luminescence components. From Kulakovskii and Malyavkin (1979).

For non-⟨111⟩ stress hot electron lines very similar to those seen for Si:B were observed. An interesting energy relaxation occurred when the stress was adjusted so that the electron valley splitting equalled the splitting between the BE hole states. The intensity ratio between components 1 and 2 of the Si:Al BE luminescence was found to have a remarkably different stress dependence for recombination with either hot or cold electrons. This would seem to indicate that the admixture of $m_j = \pm 3/2$ and $\pm 1/2$ states in the $J = 0$ BE ground state is dependent upon whether the electron of the BE occupies a hot or a cold valley, probably as a result of the anisotropy of the electron effective masses.

5. Transient luminescence decay measurements

Prior to the advent of the shell model there were few measurements of the decay characteristics of the BE's and BMEC's after pulsed excitation, since at that time the transient decay data did not seem pertinent to the establishment of any model for the new luminescence processes. Kosai and Gershenzon (1974) had reported the decay characteristics of the $m = 1$

Table 2
Luminescence lifetimes of some bound excitons and bound multiexciton complexes in Si (in ns)

m	P[a]	B[b]	Li[c]
1 (BE)	245	1000	1100
2	134	417	740
3	95	256	630
4	85	172	600
5			580

[a] Hunter et al. (1979).
[b] Sauer (1974).
[c] Kosai and Gershenzon (1974).

through 6 Li lines, while Sauer (1974) described the behavior of the $m = 1$ through 4 lines in Si:B and the α^1 through α^3 lines of Si:P. Both of these early studies were hindered by background luminescence which had different decay characteristics from the lines being studied and therefore caused the decay curves to be non-exponential in certain regions. These studies revealed two general attributes of the lines: for a given dopant the lifetimes of the lines decreased with increasing m, and in comparing the decays for different impurities, the BMEC lifetimes behaved as did those of the BE, which is to say the BMEC and BE lifetimes were shorter for impurities with larger ionization energies. The first point proved that all the lines do not have the same initial state, and within the BMEC model could be taken to show that the particle densities in the complexes increased with increasing m. The second point showed that the non-radiative Auger recombination process must be just as important for BMEC's as it is known (Schmid 1977, Lyon et al. 1977) to be for BE's.

With the introduction of the shell model and the controversy which ensued, the transient decay measurements took on new significance, since in the substitutional donor shell model the β^{m-1} and α^m lines both have the same initial state and must therefore have precisely the same decay characteristics. Sauer, Schmid and Weber (1977) published plots of the log of the line intensities versus delay time after excitation for the α^1 through α^4 and β^2 and β^3 lines of Si:P which showed that β^2 and β^3 behaved almost exactly as did α^2 and α^3, respectively, and did not match the decay of α^3 and α^4. This was used as one of the grounds for rejecting the shell model. Thewalt (1977d) however argued that the experimental method used by Sauer, Schmid and Weber (1977) was faulty, since although one could certainly measure the decay characteristics of the NP α transitions with reasonable accuracy by simply monitoring the luminescence intensity at

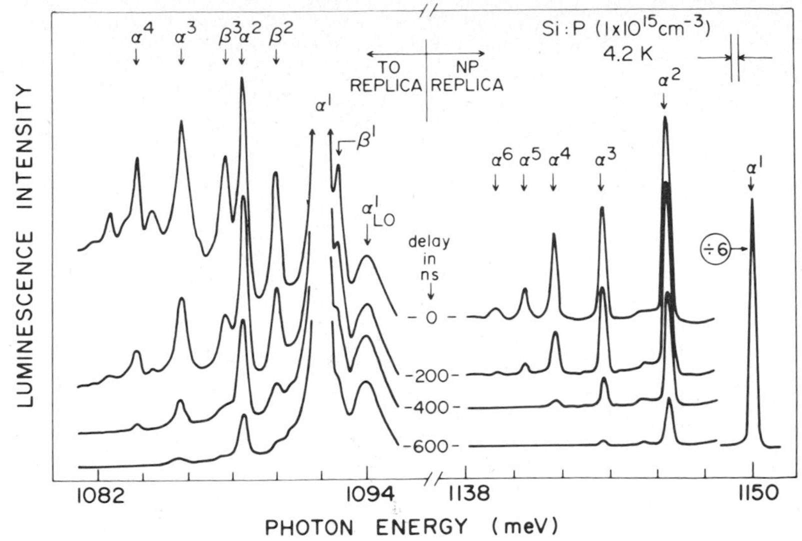

Fig. 35. TO phonon-assisted and NP photoluminescence spectra of Si:P recorded at various delays after the end of the excitation pulse. All the spectra are normalized to equal α^1 intensities. The effects of the short-lifetime background luminescence and the phonon broadened line tails can be clearly seen in the TO replica. The β^m lines are seen to decay at the same rate as the α^{m+1}. From Thewalt (1977d).

the peak wavelength as a function of delay time, this was not true of the β lines. The reasons for this can be clearly seen in fig. 4 – in the NP region the α lines are isolated and well separated, with no underlying background luminescence, while the β lines can only be studied in the phonon replicas, where they lie near the stronger α lines. Furthermore, in the phonon replicas all the lines are broadened due to the localization of the particles coupled with the dispersion of the momentum conserving phonons (Thewalt et al. 1976), and there is also a background luminescence underlying the α and β lines which may be due to the unresolved luminescence of higher order complexes. Again referring to fig. 4, the decay characteristics measured at the β^1 and β^2 positions must clearly have some α^1 component, while α^2 overlaps the β^3 position.

Thewalt (1977d) attempted to overcome these problems by recording complete spectra at various delay times so that the tailing and background effects could be directly estimated. The results are shown in fig. 35. Here we can see that the best match for β^2 is α^3, and for β^3 is α^4, as predicted by the shell model. The β^1 line is seriously obscured by α^1, but its decay is consistent with that of α^2 and is certainly much faster than that of α^1. This was also the first direct evidence that the β^1 line was not in fact merely a B BE line resulting from B contamination, since the B BE has a much longer

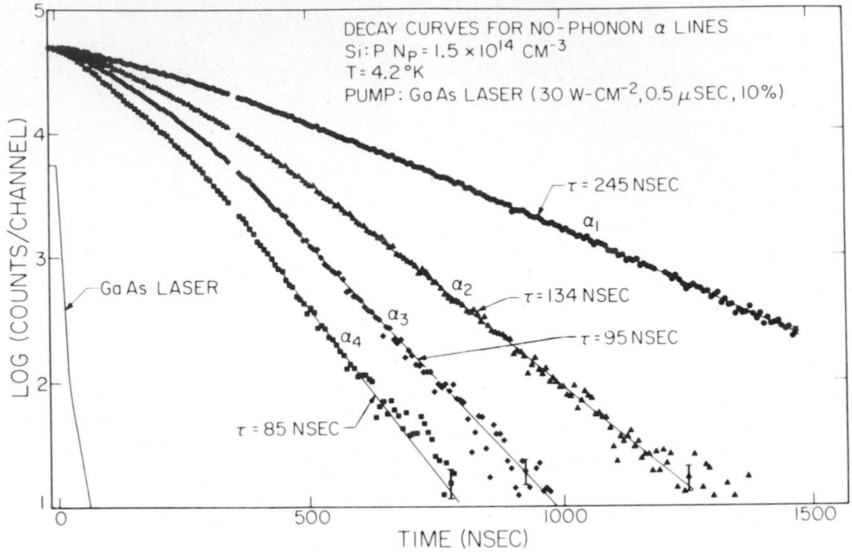

Fig. 36. Logarithm of the α^1 through α^4 NP transition intensities as a function of time after the end of the excitation pulse. The lifetimes as determined from the linear portions of the curves are also given. From Hunter et al. (1979).

($\sim 1\ \mu$s) lifetime. Transient luminescence measurements on the β^1 line in stressed Si:P were already described in sect. 4.1. Hunter et al. (1979) repeated the measurements of Sauer et al. (1977) taking care to correct for the decay of the α lines underlying the β lines. Their α decays are shown in fig. 36. Although their results regarding the $\alpha^m - \beta^{m-1}$ pairing were not definitive, they were consistent with the shell model.

6. *Thermodynamic binding energy measurements*

The BMEC binding energies which have been discussed in the previous sections were all determined spectroscopically, and are therefore model dependent since in order to use a luminescence line to determine the binding energy we must know it to be of the ground state to ground state type. Lyon et al. (1978a) attempted to circumvent this problem by directly measuring the binding energies of the complexes using thermodynamic methods. They reasoned that at sufficiently high temperatures the evaporation of excitons from complexes would dominate the recombination mechanisms, and the BMEC's and BE's would then be in thermal equilibrium with the FE gas. In this regime a plot of $\ln(T^{3/2}I_m/I_{m-1}I_{FE})$ versus $1/kT$ (where I_m is the intensity of the mth BMEC line) should yield a straight line with slope equal to the binding energy of an exciton in the mth complex: ϕ_m. Lyon et al. prepared such plots for Si:Al, Si:B and Si:P, taking the appearance of a linear behavior at elevated temperature to indicate that the system had reached the thermodynamic limit. The plot for Si:B is shown in fig. 37. Their results, and also the available spectroscopic binding energies: δ_m, are summarized in table 3. Although relaxation effects could in principle cause the optical and thermodynamic binding energies to be fundamentally different, such effects are expected to be very small for these systems. Note that spectroscopic binding energies are not available for the $m \geq 4$ lines of Si:B and Si:Al of the $m \geq 5$ line of Si:P since the ground state to ground state transitions of those BMEC's have not been observed.

For the acceptors B and Al the agreement between δ_m and ϕ_m is quite good except for a discrepancy in both $m = 3$ cases which is somewhat larger than the combined experimental errors. It has been suggested

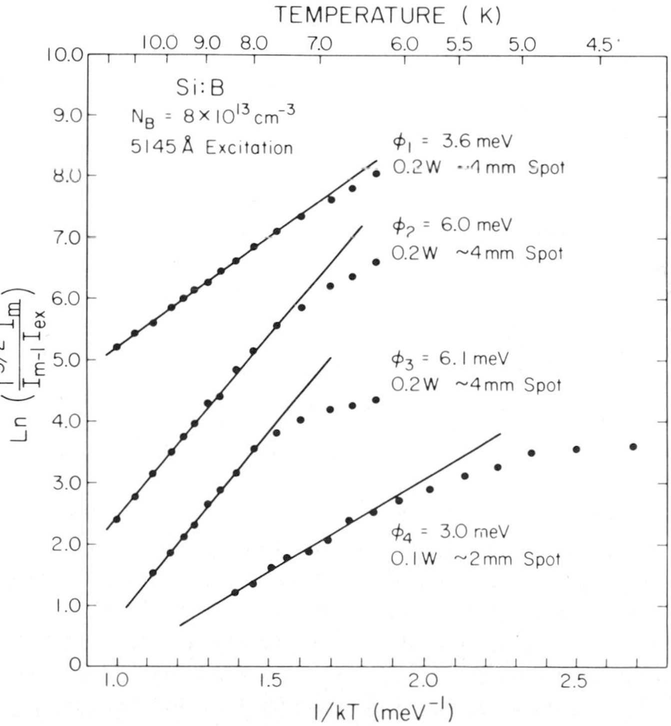

Fig. 37. Plots of the logarithm of $T^{3/2}$ times the intensity of the B^m line divided by the product of the intensities of B^{m-1} and the FE versus $1/kT$ for Si:B. The straight line segments seen at large T are taken to indicate thermodynamic equilibrium, and the thermodynamic binding energies, ϕ_m, determined from their slopes are given. From Lyon et al. (1978b).

Table 3

Binding energies (in meV) as determined thermodynamically (ϕ_m) and spectroscopically using the shell model (δ_m) for some bound excitons and bound multiexciton complexes in Si

	Si:B[a]		Si:Al[a]		Si:P[a]		Si:Li[b]	
m	δ_m	ϕ_m	δ_m	ϕ_m	δ_m	ϕ_m	δ_m	ϕ_m
1	3.9 ± 0.3	3.6 ± 0.5	5.1 ± 0.3	4.4 ± 0.5	4.6 ± 0.3	4.9 ± 0.5	3.3 ± 0.3	3.4 ± 0.5
2	6.1 ± 0.3	6.0 ± 0.5	5.2 ± 0.3	4.6 ± 0.5	3.8 ± 0.3	5.8 ± 0.5	6.0 ± 0.3	3.6 ± 0.5
3	8.3 ± 0.3	6.1 ± 0.5	8.5 ± 0.3	6.8 ± 0.5	6.7 ± 0.3	4.5 ± 0.5	8.16 ± 0.3	
4		3.0 ± 0.1			9.0 ± 0.3	3.3 ± 1	9.98 ± 0.3	
5						2.2 ± 1	7.6 ± 0.3[c]	
6							8.8 ± 0.3[c]	

[a] Lyon et al. (1978a).
[b] Lyon et al. (1978b).
[c] Using the tentative assignment given for the A and B lines in Thewalt (1978).

(Thewalt 1978) that this difference could have arisen from an underestimate of the errors inherent in determining ϕ_m since a comprehensive treatment of the possible sources of error in the thermodynamic measurements would be far from straightforward. If this is not the case, then the thermodynamic measurements indicate that the $m = 1$ and $m = 2$ B transitions result in ground state configurations while the $m = 3$ line decays into some (unspecified) excited state of the $m = 2$ complex. There is no a priori reason to expect such behavior. The large drop between ϕ_3 and ϕ_4 is in qualitative agreement with the shell model, since the $m = 3$ complex has a closed hole shell and its binding energy as a function of m should represent a local maximum. In the limit of large m the binding energy of complexes associated with any impurity must approach that of the electron–hole droplet, most recently determined to be (Schmid 1979) 9.3 ± 0.2 meV, if we assume we are in the usual light-doping limit where even a large BMEC contains only one impurity.

The discrepancies between δ_m and ϕ_m are more pronounced for Si:P, there being agreement only for the $m = 1$ (BE) line. The shell model δ_m show a drop between $m = 1$ and $m = 2$ which is interpreted as being due to the closing of the Γ_1 electron shell at $m = 1$. Thereafter the δ_m increase, reaching a maximum at $m = 4$, which has a closed Γ_8 hole shell. In the shell model δ_5 is expected to be less than δ_4, but no data is available since the $m = 5$ ground state to ground state transition has not been observed. Once again we see that the spectroscopic binding energies of the Si:P complexes do not exceed the latest estimate (Schmid 1979) of the electron–hole droplet binding energy. The thermodynamic ϕ_m values, on the other hand, show an increase in ϕ_m between $m = 1$ and $m = 2$ followed by a monotonic decrease to the rather low value of $\phi_5 = 2.2 \pm 1.0$ meV. Not even a qualitative explanation for this rather unexpected behavior has been suggested. Lyon et al. (1978a) postulated that all the $m > 1$ Si:P complexes decayed primarily into excited states of the next lower complexes whose nature could not be specified, nor could the postulated weakness of the ground state to ground state transitions for $m > 1$ be explained. It was suggested that a very weak line corresponding in the thermodynamic scheme to the ground state to ground state transition of the $m = 2$ complex had been observed, but since this line was precisely at the location of the $m = 2$ B luminescence line its identification must be considered doubtful.

In the scheme proposed by Lyon et al. (1978a) there was no explanation for the β line series, and in fact the value of ϕ_2 specifically excludes the possibility that β^1 and α^2 could have the same initial state. As we have already seen, the shell model assignment of a single initial state for β^1 and α^2 is supported by a large body of experimental evidence.

On the basis of the above arguments and the general success of the shell model in accounting for the spectra both in unperturbed Si and also in the presence of magnetic fields and uniaxial stress, the consensus of opinion currently supports the validity of the shell model. Still, the discrepancy with the thermodynamic results is troubling, but may simply be due to an incorrect assumption that over the region in which ϕ was determined the system was indeed in thermodynamic equilibrium. The known existence of bound excited states of the donor BE's and BMEC's may considerably complicate any understanding of their thermodynamic behavior. If one further considers the possibility of capture cross sections, competing processes and inhomogenious FE density profiles, *all* of which may be temperature-dependent, one might in fact be surprised by the amount of agreement between δ_m and ϕ_m.

Nevertheless, due to the discrepancy between the spectroscopic and thermodynamic binding energies, the single most crucial experiment in the BMEC field would be a determination of the excitation spectroscopy of the complexes, particularly those associated with substitutional donors such as P. The predictions of the shell model at He temperatures (where all initial states are ground states) can be readily determined from fig. 5. The α^1 luminescence should be excited either by pumping at the α^1 energy in the phonon assisted or NP region, or by pumping at the δ energy in a phonon-assisted replica, the NP δ transition having essentially zero oscillator strength since it does not involve a Γ_1 electron. Similarly, if we do a double excitation experiment in which above-band-gap light provides a population of BE's and BMEC's and a tunable light source is then used to promote m complexes into $m+1$ complexes, none of the $m > 1$ complexes should be created by pumping anywhere in the NP region since at $m = 1$ the Γ_1 shell in the initial state is full (we exclude the remote possibility of creating $2S\Gamma_1$ electrons, etc.). The α^m (and β^{m-1}) luminescence should be excited only in the phonon-assisted regions and only by pumping at β^{m-1}. There should be no absorption at all at the α^m, $m>1$ energies. These predictions are very specific and could conclusively prove the assignments given in the shell model, but the experiments are unfortunately difficult in the extreme. The main problems are the very low absorption cross sections [assumed to be similar for the BMEC's to those known for the BE's (Dean et al. 1967)], the lack of intense monochromatic pump sources in the region of interest, and the unavoidable interference of Raman scattering.

Lyon et al. (1978b) have used the same technique of measuring the relative intensities of the luminescence lines as a function of temperature to determine the thermodynamic binding energies of the $m=1$ and 2 lines of Si:Li. For the BE their ϕ_1 (3.4 ± 0.5 meV) agreed with the value of δ_1 (3.3 ± 0.3 meV) but ϕ_2 (3.6 ± 0.5 meV) was found to be considerably smaller than the spectroscopic binding energy δ_2 (6.0 ± 0.3 meV). Thus they postulated that the Li^2 line of fig. 9 was a ground state to excited state transition, and on the basis of their ϕ_2 stated that the final state of the Li^2 transition was the excited state of the Li BE responsible for the $Li^{1''}$ line shown in fig. 11. Once again the nature of the initial and final states was not described and no reason was given for the proposed dominance of this ground state to excited state transition for the $m=2$ complex. This interpretation of the Si:Li spectrum is not in agreement with that proposed by Thewalt (1978) and by Henry and Lightowlers (1979). In a more recent paper Henry and Lightowlers (to be published) argued that the interpretation given by Lyon et al. (1978b) did not agree with the stress behavior of the lines involved. Once again excitation spectroscopy, if it proves possible, should resolve this controversy.

7. Theoretical binding energy calculations

At present our knowledge of the properties of the various BMEC systems as determined by experiments, both in unperturbed semiconductors and also as a function of external perturbations, far outstrips our ability to accurately calculate any of the fundamental BMEC parameters such as the binding energies. This situation evidences the singularly difficult nature of any realistic computation of the properties of such few-body systems.

Insepov and Norman (1977) were the first to publish BMEC binding energy calculations, but their model was based on the unphysical premise that once the FE's were bound into the complexes they retained their individuality as loosely interacting FE's. The strongest argument against this assumption is the observed agreement, for all BMEC's, of the thermalizations of the doublet components resulting from the valence band splitting under applied stress with the predictions of eqs. (1). Nevertheless

Insepov and Norman obtained positive binding energies for all the complexes, although they naturally did not find any shell effects in their binding energies, and their results were not in agreement with either the spectroscopic results as determined within the shell model or with the thermodynamic (Lyon et al. 1978a) binding energy measurements.

Hartree–Fock calculations, although working well for atomic systems, will clearly fail in describing BMEC's since, as first pointed out by Dean et al. (1976a, 1976b), the main contribution to the stability of these complexes comes from correlation effects and not from interaction with the field of the impurity ion. In order to calculate BMEC binding energies a formalism which allowed the inclusion of the effects of correlation in inhomogeneous many-particle systems was needed. The density functional approach (Hohenberg and Kohn 1964, Kohn and Sham 1965, Tong and Sham 1966) is well adapted to such calculations, and was first applied to the BMEC problem by Wünsche and co-workers (1978a, 1978b, 1979), as reviewed by Wünsche et al. (1978b).

There is a very close conceptual relationship between the density functional formalism as applied to the BMEC problem and the shell model of the BMEC structure, since in the density functional approach the ground state energies and electron and hole densities of the complexes are obtained by solving single particle Schrödinger equations for electrons and holes moving in spherically symmetric self-consistent potentials which result from the field of the impurity ion and the other electrons and holes in the complex. Wünsche et al. made a number of simplifying assumptions, the most important of which were the effective mass approximation, an isotropic model band structure, and the local density approximation, Intervalley and interband matrix elements of the impurity and interaction potentials were ignored, and although the many-valley nature of the conduction band was retained, the effective masses were taken to be isotropic. Only the spherical part of the two uppermost valence band edges (as described by Baldereschi and Lipari 1973) was considered, and the light and heavy hole masses were taken to be equal. In the local density approximation the density of exchange-correlation energy at a point in the inhomogeneous system is taken to equal that of a homogeneous system having the corresponding particle densities. This is a reasonable approximation if the diameter of the exchange-correlation hole in the inhomogeneous system is small compared to the characteristic distance over which this diameter changes. Although this condition is quite often not satisfied in calculations using the local density approximation, useful results are nevertheless obtained.

The expression for the exchange energy, which was a simple sum of electron and hole terms, was taken from the work of Combescot and Nozières (1972). Corresponding simple analytical expressions of the correlation energy do not exist and it was therefore necessary to use an approximate expression, which in the limit $n_e = n_h$ agreed with the tabulated (Bhattacharya et al. 1974) values for the electron–hole droplet problem, while also giving the known (Gunnarsson and Lundqvist 1976) single-component correlation energies in the limits $n_e = 0$ or $n_h = 0$.

Rather than directly solving the coupled self-consistent Kohn–Sham equations, Wünsche et al. sought an approximate solution by using hydrogenic wave functions as a variational ansatz:

$$\phi_\lambda(\boldsymbol{r}) = \frac{1}{\sqrt{N_\lambda}} \left(\frac{|\boldsymbol{r}|}{r_\lambda}\right)^{k_\lambda/2} \exp - \left(\frac{|\boldsymbol{r}|}{r_\lambda}\right), \quad k_\lambda = 0, 1, 2, \ldots, \quad \lambda = e, h. \tag{3}$$

When substituted into the energy functional along with the interpolation formulae giving the exchange and correlation energies this yielded an analytical expression in terms of the parameters r_λ and k_λ. The energy functional was then minimized in terms of these parameters. In order to correct for systematic errors introduced by the isotropic model bandstructure, the results were scaled by the ratio of the measured electron–hole droplet binding energy compared to the electron–hole droplet binding energy as calculated within this model.

Only donor results were given, and the binding energies were found to increase monotonically in going from $m = 1$ to $m = 4$, in qualitative agreement with the observed Si:Li results. The calculation was of course not applicable to substitutional donor complexes as interpreted in the shell model, since for $m > 1$ the electrons in these complexes do not all have identical probability densities. The calculations also revealed that the size of the complexes did not change appreciably with increasing m, at least up to the point where new shells had to be populated. Thus the electron-hole density increases with increasing m, explaining the increase in binding energy. Shell effects were *not* observed, with the binding energy increasing smoothly in going from $m = 4$ to $m = 5$. This is puzzling, since the Ge calculations described below showed strong shell effects, and the difference may be related to the approximate nature of the solutions reached by using eq. (3) as an ansatz.

In a more recent paper Wünsche (1979) sought to extend these calculations by including the effect of the intervalley scattering terms of the impurity potential, and thus to explain the ground state binding energies and valley-orbit splittings of the substitutional donor complexes. A very simple model was used, in which the valley-orbit energy of a Γ_1 or $\Gamma_{3,5}$ electron was expressed in terms of a fixed valley-orbit matrix element and the Bohr radius of the electron. The radii for both Γ_1 and $\Gamma_{3,5}$ electrons were taken to be equal to the radii obtained from the previous (Wünsche and co-workers 1978a, 1978b, 1979) calculations. In this perturbative approach the valley-orbit splitting of the mth complex is just $1/(1+m)$ of the donor valley-orbit splitting. All the complexes remained bound, although the $m = 1$ excited state was above the dissociation threshold. The closing of the Γ_1 shell at $m = 1$ of course had no effect on the binding energies in this approximation. Wünsche concluded that in reality the valley-orbit energy of a Γ_1 electron would depend primarily on the number of Γ_1 electrons and not on m, and that the simple perturbation approach was not suitable for an accurate description of valley-orbit effects.

Shore and Pfeiffer (1979) have also published the results of similar density functional BMEC binding energy calculations, but for the case of Ge rather than Si. Due to the large difference between the light and heavy hole masses in Ge they decoupled the valence band edge and assumed that states associated with the light hole band would only begin to be populated for quite large complexes. Rather than use a variational ansatz as was done by the previous authors, Shore and Pfeiffer solved the coupled self-consistent Kohn–Sham equations directly. They obtained results for the $m = 1$ through $m = 10$ donor complexes in which shell effects were very evident for both electrons and holes. The binding energy dropped each time it became necessary to place an electron or a hole into a new shell. Another interesting result was their finding that for both electrons and holes the 2p-like shells filled before the 2S-like shells.

The proposal that BMEC might be the nucleation centers from which electron–hole droplets are formed is as old as the BMEC field itself (Kaminskii and Pokrovskii 1970), but it has never been clear how the discrete line spectrum and well defined states of the BMEC evolve with increasing m into the electron–hole droplet continuum spectrum given by the convolution of degenerate electron and hole populations. A recent advance in this direction has been made by Rose et al. (1979) who used density functional calculations to obtain the binding energies and luminescence spectra of small electron–hole droplets in Ge as a function of the constituent number of electron–hole pairs. The luminescence consisted of discrete lines, whose number (and therefore density) increased with increasing electron–hole droplet size. The droplet binding energies showed very distinct shell effects, as can be seen in fig. 38, with the local maxima in binding energy corresponding to closed outer electron or hole shells. It is particularly noteworthy that the binding energy for some small droplets considerably exceeded that of the bulk liquid. It would therefore not be surprising if some closed-shell BMEC's had binding energies somewhat larger than that of the electron–hole droplet (9.3 ± 0.2 meV[66]), but the only known example is the Li $m = 4$ BMEC with a spectroscopic binding energy of 9.9 ± 0.3 meV. Another bridge between the studies of BMEC's and electron–hole droplets is the density functional calculation of the energy of attachment of an electron–hole droplet to a neutral donor given by Sander et al. (1978).

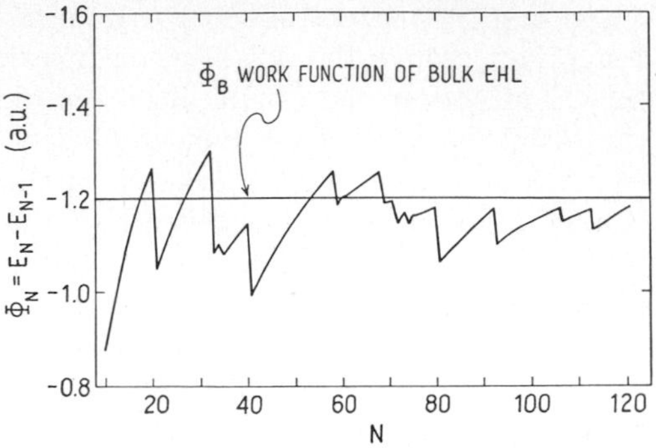

Fig. 38. The calculated work function for the removal of a separate electron–hole pair from a small droplet in Ge as a function of the number of pairs in the droplet, N. The effect of closing electron and hole shells is clearly visible, resulting in local maxima in the binding energy, and some of the small closed-shell complexes have work functions substantially larger than that of the bulk electron-hole droplet. Since the conduction band degeneracies are different for Ge and Si, the shell closing effects occur for values of N(m) different from those expected for Si BMEC's. From Rose et al. (1979).

8. BMEC spectra in other semiconductors

8.1. BMEC in Ge

The first evidence of BMEC's in a semiconductor other than Si was found by Martin (1974) in the photoluminescence spectra of Ge doped with the donors P or As. The study of BMEC's in Ge is complicated by three factors, namely the low energy of the recombination radiation, the small binding energy of the BE's and BMEC's, and the propensity for all the excitation to be channelled into electron-hole droplets as the temperature is reduced. Along with the principal BE line and a line then identified as a FE transition in which a donor electron was raised from 1S Γ_1 to 1S Γ_5, (there is no 1S Γ_3 level in Ge), Martin observed new lines which he labelled 2 through 4 for GE:P and 2 through 6 for Ge:As. On the basis of a fit of the shifts of the lines with respect to the FE edge versus line number, Martin postulated that the numbered lines corresponded to BMEC recombination in the usual labelling scheme, in other words $m = 2$ through 4 BMEC's in Si:P and $m = 2$ through 6 BMEC's in Si:As.

Thewalt (1977b) suggested that the line identified by Martin as a FE transition involving a change in the valley-orbit state of a donor was actually a shell model γ^1-type transition having as its initial state an excited BE, since the γ^1 transitions in substitutional donor doped Si had also originally been mistaken for such FE valley-orbit transitions (Dean et al. 1967). This was verified by Mayer and Lightowlers (1979, 1980) who also discovered that the donor BE excited state spectrum was considerably

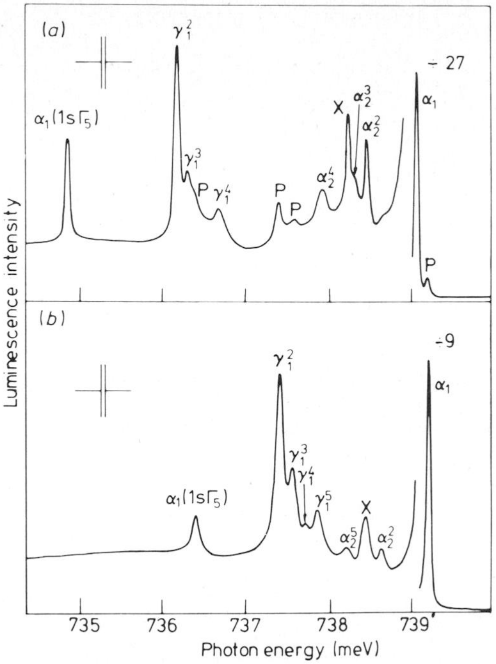

Fig. 39. Photoluminescence spectra in the NP region for Ge doped with the acceptors As (a) and P (b). Lines with subscript 1 originate from BE initial states, with α_1 being the BE ground state to ground state transition. Lines having subscript 2 have an $m = 2$ BMEC as their initial state and various excited BE final states. There are three P contamination lines in the Ge:As spectrum, and there is an unidentified X line in both spectra. From Mayer and Lightowlers to be published.

more complicated in Ge than in Si, there being four observed γ^1 transitions and three δ transitions. A detailed explanation of these excited states of the BE in Ge has not yet been forthcoming. Mayer and Lightowlers came to the conclusion that there was in fact only evidence for the $m = 2$ BMEC in the Ge luminescence, and that all of the lines previously identified as resulting from BMEC recombination by Martin were either BE γ^1 transitions or resulted from impurities. Their photoluminescence spectra for Ge:As and Ge:P are shown in fig. 39.

8.2. BMEC in SiC

Very clear BMEC photoluminescence spectra have been obtained from high-quality cubic SiC (β-SiC) crystals by Dean and co-workers (1976a, 1976b, 1977), who suggested that the impurity center involved was most likely the donor N_C (Hartman and Dean 1970). The higher photon energies

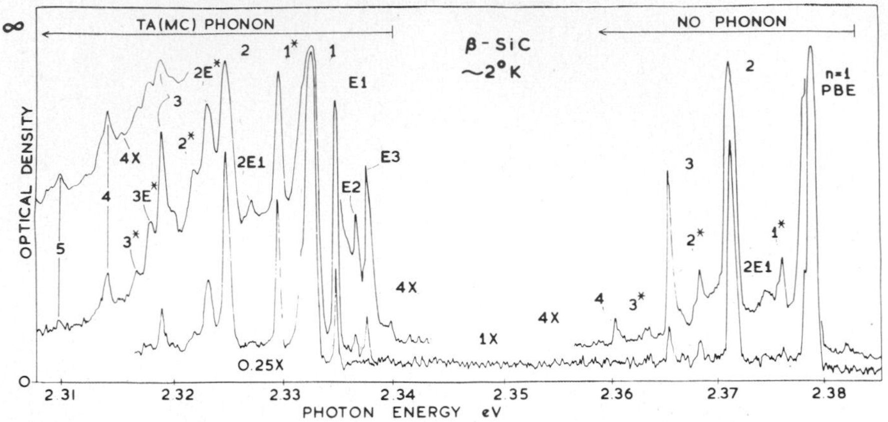

Fig. 40. Photoluminescence spectra of SiC:N recorded photographically with the indicated ratios of exposure times. The lines labelled $n = 1$, 2, 3 and 4 definitely originate from BE and BMEC recombination (with $n = m$ in our usual labelling scheme). The origin of the many other components requires further investigation. From Herbert et al. (1977).

of the SiC luminescence are a major advantage over Si from the point of view of signal detection, but unfortunately the impurity content of SiC is not nearly so well characterized or controllable as that of Si, and the origins of many of the weaker SiC luminescence lines therefore remain open to question. A typical photoluminescence spectrum is shown in fig. 40, while the Zeeman spectra can be seen in fig. 3. Also shown in that figure are transition diagrams for the $m = 1$ and 2 transitions in a magnetic field according to the model introduced by Dean et al. (1976a) to explain both the results for β-SiC as well as the Si results which originally led Sauer and Weber (1976) to reject the entire BMEC concept. The new luminescence lines seen in SiC have been firmly established as BMEC processes by studying their relative intensities as a function of excitation density (Dean et al. 1976b). As expected, the intensity of $m+1$ relative to m luminescence was always found to increase with increasing excitation.

8.3. BMEC in GaP

Sauer et al. (1979) have recently re-interpreted a number of emission lines in the photoluminescence spectra of acceptor-doped GaP as resulting from BMEC recombination. These lines had previously been interpreted as BE transitions strongly coupled to acoustical phonons (Dean et al. 1971), or more recently in terms of the camel's-back nature of the GaP conduction bands (Dean and Herbert 1976). Spectra for GaP:Zn taken at low and high excitation density are shown in fig. 41, clearly demonstrating the excitation

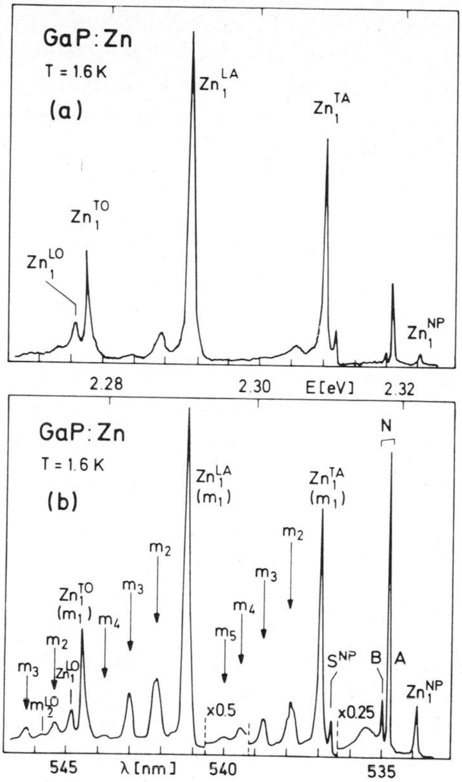

Fig. 41. NP and phonon-assisted replicas of the photoluminescence spectrum of GaP:Zn under low (a) and high (b) excitation density. The $m = 1$ through $m = 5$ lines are here labelled m_1 through m_5. From Sauer et al. (1979).

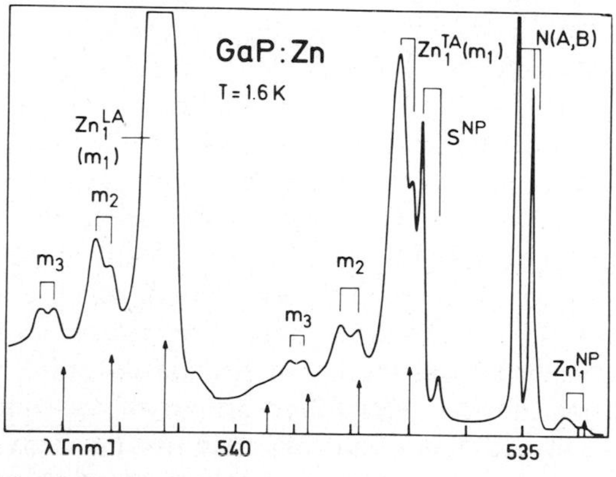

Fig. 42. High excitation luminescence spectrum of GaP:Zn under stress. The BE and BMEC lines split into identical doublets, with the relative intensity of the high energy component increasing with increasing m, as predicted by eqs. (1). From Sauer et al. (1979).

dependence typical of BMEC luminescence. Note also the large drop in intensity between $m = 3$ and $m = 4$ which results from the filling of the hole shell at $m = 3$, just as for the Si acceptor complexes.

Sauer et al. were also able to obtain uniaxial stress spectra, no mean feat considering that the samples were in the form of tiny needles. The GaP:Zn results are shown in fig. 42 and are very similar to the results for Si:B under ⟨111⟩ stress. Note the expected increase in the relative intensity of the higher energy doublet component in going from $m = 1$ to $m = 3$.

8.4. BMEC in GaAs (?)

All of the previously described systems which show BMEC luminescence have been indirect band-gap semiconductors. This is not surprising since any process which involves the binding of more than one electron–hole pair on a given site would be more probable in materials which have long lifetimes. The larger degeneracies of the band edges in indirect gap semiconductors also favor the formation of complexes. Almassy et al. (1979) have interpreted six weak luminescence lines occurring at energies below those of the three principal acceptor BE lines in GaAs resulting from BMEC luminescence. They favored a model in which all six lines resulted from $m = 2$ BMEC's, as outlined in fig. 43. Although the authors could find no other explanation for these new lines, the existence of BMEC's in GaAs must be considered questionable in the absence of any corroborating evidence. A study of the relative intensities of the lines as a function of excitation density would be most helpful. Almassy et al. did state that the new lines could be seen at low excitation, which weighs against a BMEC explanation.

9. Biexciton bound to N_P in GaP

Although it was the first BMEC to be reported in the literature (Merz et al. 1969) we have put off the discussion of the biexciton localized on the N_P impurity in GaP until the end of the chapter since unlike all the other BMEC's which have been studied this one is localized on an isoelectronic impurity. Thus this complex has an even number of mobile charges, while all the other BMEC's have an odd number. The nature of the binding of the biexciton to the impurity will therefore also be different from that of the donor or acceptor BMEC's. A thorough review of the binding of both the exciton and biexciton to N_P in GaP has recently been given by Dean and Herbert 1979. Local lattice distortions can play an important part in the binding of excitons to isoelectronic impurities, and the luminescence of the N_P isoelectronic BE in GaP shows a phonon coupling consistent with a

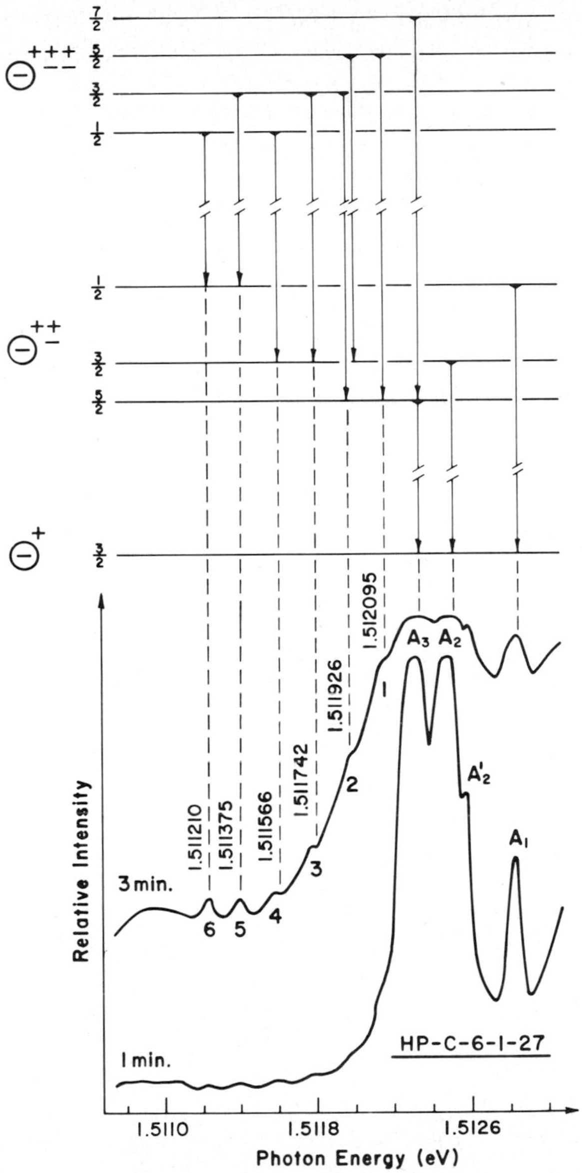

Fig. 43. Photoluminescence spectrum of high-purity GaAs at 2 K recorded photographically. The A lines result from BE recombination, while the weaker lines were interpreted as resulting from $m = 2$ BMEC states as indicated by the level diagram above the spectra, with the total J of the various states given on the left. From Almassy et al. (1979).

strong wavevector independent electron–phonon interaction. The phonon coupling of the A* and B* biexciton lines is on the other hand negligible, showing that the binding of the second exiton results in almost no change in the lattice distortion, and that correlation effects must therefore be important.

A typical spectrum showing the N_P BE and biexciton lines, along with their origin in a scheme labelling the total angular momentum of the various states, is shown in fig. 44. In this scheme the $j = 1/2$ electron and $j = 3/2$ hole of the BE can form $J = 1$ and 2 states, while the two holes of the biexciton couple to give the same threefold splitting discussed previously for the acceptor BE in Si. These assignments provided an excellent accounting of the observed Zeeman spectra. It was also found that as the excitation density was increased the A* and B* intensities increased as the square of the A and B intensities, until levels were reached where sample heating distorted the results. Also in agreement with the localized biexciton model was the short (7.5 ± 1 ns) luminescence lifetime of the A* and B* lines as compared to the A and B lines (~40 ns), which results from the possibility of Auger recombination for the biexciton, but not for the isoelectronic BE.

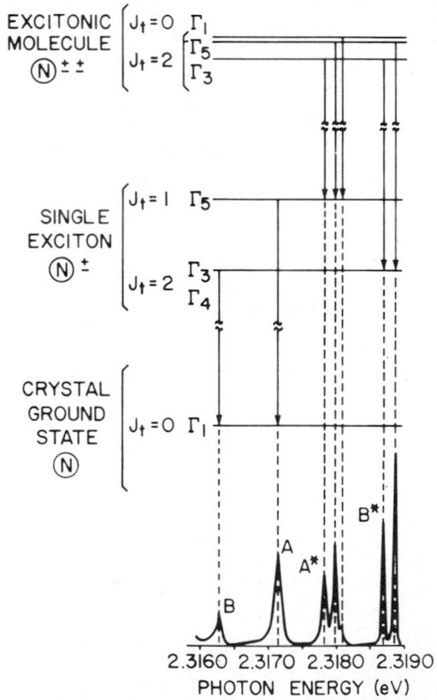

Fig. 44. Photoluminescence spectrum due to excitons and biexcitons bound to N_P in GaP. The transitions are well explained by the level diagram shown above the spectrum. From Merz et al. (1969).

10. Conclusions

In this chapter we have shown that the existence of bound multiexciton complexes is now established beyond any doubt, and that the behavior of these complexes is very well described by a remarkably straightforward model, the shell model introduced by Kirczenow 1977b. The important role assigned to electron valley-orbit splittings for donor complexes in the shell model results in very different level diagrams for substitutional-donor complexes (fig. 5) and for the interstitial Li-donor complexes (fig. 10). This explains the fundamentally different spectra seen for BE's and BMEC's associated with the two types of donor centers. Strong corroboration of the validity of the model is obtained from extensive studies of the splittings and shifts of the luminescence lines under the influence of uniaxial stresses, as was outlined in sect. 4. The available Zeeman splitting data, and particularly the thermalization of the Zeeman components, is also well accounted for.

Although much experimental work remains to be done, even in the heavily-studied Si system, our knowledge of the behavior of BMEC's already far outstrips our ability to calculate any of their fundamental properties, such as binding energies, in a convincing manner. As outlined in sect. 7, progress is being made in this direction, but any realistic calculations for the donor-based complexes will clearly have to incorporate the valley-orbit splitting in a more sophisticated manner than has until now been attempted. Another subject which needs attention is the question of why the shell model works as well as it does, or in other words, why are the many possible splittings which could arise from interparticle interactions negligible?

Finally, in sect. 8 we have seen that the study of BMEC's is expanding to semiconductors other than the original Si and Ge. It is reasonable to expect that stable BMEC's may be a general property of indirect-band-gap semiconductors, where the large band-edge degeneracies favor such multiparticle complexes. Observing BMEC luminescence in other materials may then merely be a question of purifying them to the point where the FE density can become large enough to give a reasonable probability of multiple binding at a given site. It remains to be seen whether the shell model will be applicable to the description of BMEC spectra in other semiconductors. It would of course not be surprising if modifications are necessary, since the shell model was clearly a first approximation, ignoring as it did the large number of splittings which could arise from interparticle interactions in these systems. It is this simplicity which is the beauty of the model, in that it has allowed a clear understanding of the spectroscopic properties of these systems, the complexity of which had previously seemed intractable.

References

Alkeev, N.V., A.S. Kaminskii and Ya.E. Pokrovskii, 1975, Fiz. Tverd. Tela. **17**, 843 (Sov. Phys. Solid State **17**, 535 (1975)).

Almassy, R.J., D.C. Reynolds, C.W. Litton, K.K. Bajaj and G.L. McCoy, 1979, Solid State Commun. **31**, 365 (1979).

Altukhov, P.D., K.N. El'tsov, G.E. Pikus and A.A. Rogachev, 1977, Pis'ma Zh ETF **26**, 468 (JETP Lett. **26**, 338 (1977)).

Baldereschi, A. and N.O. Lipari, 1973, Phys. Rev. **B6**, 2697.

Bhattacharya, P., V. Massida, K.S. Singwi and P. Vashista, 1974, Phys. Rev. **B10**, 5127 (1974).

Bir, G.L. and G.E. Pikus, 1972, Simmetriya i Deformatsionnye Éffekty v Poluprovodnikah (Nauka) (Symmetry and Strain-Induced Effects in Semiconductors (Wiley, New York, 1974)).

Capizzi, M., J.C. Merle, P. Fiorini and A. Frova, Solid State Commun. **24**, 451 (1977).

Cherlow, J.M., R.L. Aggarwal and B. Lax, (1973), Phys. Rev. **B7**, 4547.

Combescot, M. and P. Nozières, 1972, J. Phys. **C5**, 2369.

Dean, P.J. and D.C. Herbert, 1976, J. Lummin. **14**, 55.

Dean, P.J. and D.C. Herbert, 1979, Topics in Current Physics vol. 14, ed., K. Cho, (Springer, Berlin, Heidelberg).

Dean, P.J., J.R. Haynes and W.F. Flood, 1967, Phys. Rev. **161**, 711.

Dean, P.J., W.F. Flood and G. Kaminsky, 1967, Phys. Rev. **163**, 721.

Dean, P.J., R.A. Faulkner, S. Kimura and M. Ilegems, 1971, Phys. Rev. **B4**, 1926.

Dean, P.J., D.C. Herbert, D. Bimberg and W. J. Choyke, 1976a, Proc. 13'th Int. Conf. Phys. Semicond., Rome 1976, (a) (Tipografia Marves, Rome) p. 1298.

Dean, P.J., D.C. Herbert, D. Bimberg and W.J. Choyke, 1976b, Phys. Rev. Lett. **37**, 1635.

Elliott, K.R. and T.C. McGill, 1978, Solid State Commun. **28**, 491.

Elliott, K.R., G.C. Osbourn, D.L. Smith and T.C. McGill, 1978, Phys. Rev. **B17**, 1808.

Frenkel, J., 1931, Phys. Rev. **37**, 1276.

Gunnarsson, O. and B.I. Lundqvist, 1976, Phys. Rev. **B13**, 4274.

Hartman, R.L. and P.J. Dean, 1970, Phys. Rev. **B2**, 951.

Henry, M.O. and E.C. Lightowlers, 1977, J. Phys. **C10**, L601.

Henry, M.O. and E.C. Lightowlers, 1978, J. Phys. **C11**, L555.

Henry, M.O. and E.C. Lightowlers, 1979, J. Phys. **C12**, L485.

Henry, M.O. and E.C. Lightowlers, J. Phys. C, to be published.

Herbert, D.C., P.J. Dean and W.J. Choyke, 1977, Solid State Commun. **24**, 383.

Hohenberg, P. and W. Kohn, 1964, Phys. Rev. **136**, B 864.

Hunter, A.T., S.A. Lyon, D.L. Smith and T.C. McGill, 1979, Phys. Rev. **B20**, 2431.

Insepov, Z.A. and G.E. Norman, 1977, Zh ETF **73**, 1517 (Sov. Phys. JETP **46**, 798 (1977)).

Kaminskii, A.S. and Ya.E. Pokrovskii, 1970, Pis'ma Zh ETF **11**, 381 (JETP Lett. **11**, 255 (1970)).

Kaminskii, A.S. and Ya.E. Pokrovskii, 1978, Zh ETF **75**, 1037 (Sov. Phys. JETP **48**, 523 (1978)).

Kaminskii, A.S. and Ya.E. Pokrovskii, 1979, Zh ETF **76**, 1727.

Kaminskii, A.S., Ya.E. Pokrovskii and N.V. Alkeev, 1970, Zh ETF **59**, 1937 (Sov. Phys. JETP **32**, 1048 (1971)).

A.S. Kaminskii, V.A. Karasyuk and Ya.E. Pokrovskii, 1978, Zh ETF **74**, 2234 (Sov. Phys. JETP **74**, 1162 (1978)).

Kirczenow, G., 1977a, Solid State Commun. **21**, 713.

Kirczenow, G., 1977b, Can. J. Phys. **55**, 1787.

Kohn, W. and J.M. Luttinger, 1955, Phys. Rev. **98**, 915.

Kohn, W. and L.J. Sham, 1965, Phys. Rev. **140**, A1133.

Kosai, K. and M. Gershenzon, 1974, Phys. Rev. **B9**, 723.
Kulakovskii, V.D., 1977, Fiz. Tverd. Tela **20**, 1394 (Sov. Phys. Solid State **20**, 802 (1978)).
Kulakovskii, V.D., 1978, Pis'ma Zh ETF **27**, 217 (JETP Lett. **27**, 202 (1978)).
Kulakovskii, V.D. and A.V. Malyavkin, 1979, Phys. Stat. Sol. (b) **92**, 455.
Kulakovskii, V.D. and V.B. Timofeev, 1977, Pis'ma Zh ETF **25**, 487 (JETP Lett. **25**, 458 (1977)).
Kulakovskii, V.D., I.B. Levinson and V.B. Timofeev, 1978a, Fiz. Tverd. Tela **20**, 399 (Sov. Phys. Solid State **20**, 230 (1978)).
Kulakovskii, V.D., A.V. Malyavkin and V.B. Timofeev, 1978b, Pis'ma Zh ETF **27**, 576 (JETP Lett. **27**, 542 (1978)).
Kulakovskii, V.D., A.V. Malyavkin and V.B. Timofeev, 1979, Zh ETF **76**, 139 (Sov. Phys. JETP **49**, 139 (1979)).
Lampert, M., 1958, Phys. Rev. Lett. **1**, 450.
Laude, L.D., F.H. Pollack and M. Cardona, 1971, Phys. Rev. **B3**, 2623.
Lightowlers, E.C. and M.O. Henry, 1977, J. Phys. **C10**, L247.
Lightowlers, E.C., M.O. Henry and M.A. Vouk, 1977, J. Phys. **C10**, L713.
Lyon, S.A., D.L. Smith and T.C. McGill, 1978a, Phys. Rev. **B17**, 2620.
Lyon, S.A., D.L. Smith and T.C. McGill, 1978b, Phys. Rev. Lett. **41**, 56.
Lyon, S.A., D.L. Smith and T.C. McGill, 1978c, Solid State Commun. **28**, 317.
Lyon, S.A., G.C. Osbourn, D.L. Smith and T.C. McGill, 1977, Solid State Commun. **23**, 425.
Martin, R.W., 1974, Solid State Commun. **14**, 369.
Mayer, A.E. and E.C. Lightowlers, 1979, J. Phys. **C12**, L539.
Mayer, A.E. and E.C. Lightowlers, J. Phys. C, to be published.
Merz, J.L., R.A. Faulkner and P.J. Dean, 1969, Phys. Rev. **188**, 1228.
Morgan, T.N., 1976, Proc. 13'th Int. Conf. Phys. Semicond., Rome 1976 (Tipografia Marves, Rome) p. 825.
Mott, N.F., 1938, Proc. R. Soc. **A167**, 384.
Nishino, T., H. Nakayama and Y. Hamakawa, 1977, Solid State Commun. **21**, 327.
Parsons, R.R., 1977, Solid State Commun. **22**, 671.
Pokrovskii, Ya.E., 1972, Phys. Status Solidi (a) **11**, 385.
Pokrovskii, Ya.E., A.S. Kaminskii and K. Svistunova, 1970, Proc. 10'th Intern. Conf. Phys. Semicond., Cambridge, MA 1970 (USAEC, Springfield, VT) p. 504.
Rose, J.H., R. Pfeiffer, L.M. Sander and H.B. Shore, 1979, Solid State Comm. **30**, 697.
Sander, L.M., H.B. Shore and J.H. Rose, 1978, Solid State Commun. **27**, 331.
Sauer, R., 1973, Phys. Rev. Lett. **31**, 376.
Sauer, R., 1974, Proc. 12'th Int. Conf. Phys. Semicond., Stuttgart 1974 (Teubner, Leipzig) p. 42.
Sauer, R. and J. Weber, 1976, Phys. Rev. Lett. **36**, 48.
Sauer, R. and J. Weber, 1977, Phys. Rev. Lett. **39**, 770.
Sauer, R., W. Schmid and J. Weber, 1977, Solid State Commun. **24**, 507.
Sauer, R., W. Schmid, J. Weber and U. Rehbein, 1979, Phys. Rev. **B19**, 6502.
Schmid, W., 1977, Phys. Status Solidi (b) **84**, 529.
Schmid, W., 1979, Phys. Status Solidi (b) **94**, 413.
Shore, H.B. and R.S. Pfeiffer, 1979, Proc. 14'th Int. Conf. Phys. Semicond., Edinburgh 1978 (Institute of Physics and Physical Society, London) p. 627.
Thewalt, M.L.W., 1977a, Phys. Rev. Lett. **38**, 521.
Thewalt, M.L.W., 1977b, Can. J. Phys. **55**, 1787.
Thewalt, M.L.W., 1977e, Solid State Commun. **21**, 937.
Thewalt, M.L.W., 1977d, Solid State. Commun. **25**, 513.
Thewalt, M.L.W., 1978, Solid State Commun. **28**, 361.
Thewalt, M.L.W., 1979, Proc. 14'th Int. Conf. Phys. Semicond., Edinburgh 1978 (Institute of Physics and Physical Society, London) p. 605.

Thewalt, M.L.W. and J.A. Rostworowski, 1978, Phys. Rev. Lett. **41**, 808.
Thewalt, M.L.W., G. Kirczenow, R.R. Parsons and R. Barrie, 1976, Can. J. Phys. **54**, 1728.
Thewalt, M.L.W., J.A. Rostworowski and G. Kirczenow, 1979, Can. J. Phys. **57**, 1898.
Thomas, D.G., M. Gershenzon and J.J. Hopfield, 1963, Phys. Rev. **131**, 2397.
Tong, B.Y. and L.J. Sham, 1966, Phys. Rev. **144**, 1.
Vouk, M.A. and E.C. Lightowlers, 1976, Proc. 13'th Int. Conf. Phys. Semicond., Rome 1976 (Tipografia Marves, Rome) p. 1098.
Wannier, G.H., 1937, Phys. Rev. **52**, 191.
Wünsche, H.J., 1979, Phys. Status Solidi (b) **92**, K57.
Wünsche, H.J., V.E. Khartsiev and K. Henneberger, 1978a, Phys. Status Solidi (b) **85**, K53.
Wünsche, H.J., K. Henneberger and V.E. Khartsiev, 1978b, Phys. Status Solidi (b) **86**, 505.
Wünsche, H.J., K. Henneberger and V.E. Khartsiev, 1978, Proc. 14'th Int. Conf. Phys. Semicond., Edinburgh 1978 (Institute of Physics and Physical Society, London) p. 615.

CHAPTER 11

Biexcitons in CuCl and Related Systems

J.B. GRUN, B. HÖNERLAGE and R. LÉVY

Laboratoire de Spectroscopie
et d'Optique du Corps Solide
(associé au C.N.R.S. 232)
Université Louis Pasteur
5, rue de l'Université
67000 Strasbourg
France

Excitons
Edited by
E.I. Rashba and M.D. Sturge

Contents

1. Introduction

When excitons are created at high density in semiconductors, their envelope functions overlap, giving rise to various non-linear effects: collisions, formation of electron–hole plasmas or creation of excitonic molecules.

The existence of excitonic molecules (or biexcitons) was first predicted by Lampert and Moskalenko in 1958, and first experimental evidence has been obtained in 1968 by Mysyrowicz et al. in CuCl. Although biexcitons have been found in other crystals later on, CuCl has been studied to the largest extent because of its relatively simple bandstructure and because of the large binding energy E^{b}_{Bi} of the complex. Taking Zincblende type CuCl (T_d-point group symmetry) as an example, we will review the theoretical concepts and experimental methods used to study biexcitons.

In the next section, exciton and biexciton wave functions are constructed from electron and hole states in the effective mass approximation. The stability of the biexciton complex is discussed and the selection rules for optical transitions between the different states are established. Different non-linear spectroscopic studies of the biexciton are then described: resonant two-photon absorption and induced absorption in sect. 3, resonant two-photon Raman scattering, coherent light scattering and biexciton luminescence in sect. 4.

In sect. 5, we discuss the influence of the valence band degeneracy on the biexciton states and the optical transitions, using CuBr (T_d-point group symmetry) as an example. Then, we briefly review the situation in uniaxial crystals with C_{6v}-point group symmetry, taking CdS as an example. In the last section (sect. 6), we give a conclusion on the research work done on biexcitons.

2. Theoretical considerations

2.1. Excitons and polaritons

We start from a simple two-band model for N electrons in a periodic potential $V_{per}(\boldsymbol{r}_i)$ (Forney et al. 1974, Kittel 1967), applying the adiabatic approximation.

The valence band is filled, the conduction band empty, and the bands are assumed to be only spin degenerate. The Hamiltonian then reads:

$$H = \sum_{i=1}^{N} \left(\frac{p_i^2}{2m} + V_{\mathrm{per}}(\boldsymbol{r}_i) \right) + \frac{1}{2} \sum_{i,j}{}' \frac{e^2}{|\boldsymbol{r}_i - \boldsymbol{r}_j|}. \tag{1}$$

All other interactions are neglected for the moment. In the Hartree–Fock approximation, the N-electron ground state $|\phi_0\rangle$ is given by a Slater determinant, where all electrons i are in valence states $\psi_k(\boldsymbol{r}_i)$. An excited electron–hole pair state $|\phi_k^l\rangle$ is then constructed by replacing $\psi_k(\boldsymbol{r}_i)$ by the conduction band state $\phi_l(\boldsymbol{r}_i)$. Using this basis, an effective two-particle exciton Hamiltonian H_{ex} may be defined, which turns out to be:

$$H_{\mathrm{ex}} = \mathrm{E}_0 + H_{\mathrm{HF}}(\mathrm{e}) - \kappa H_{\mathrm{HF}}(\mathrm{h})\kappa^{+} - \frac{e^2}{|\boldsymbol{r}_{\mathrm{e}} - \boldsymbol{r}_{\mathrm{h}}|} + W(\mathrm{e}, \mathrm{h}). \tag{2}$$

Here, E_0 is the ground state energy of the crystal chosen equal to zero, $H_{\mathrm{HF}}(\mathrm{e})$ and $\kappa H_{\mathrm{HF}}(\mathrm{h})\kappa^{+}$ are the one-particle electron (e) and hole (h) operators, obtained in the Hartree–Fock approximation, and $W(\mathrm{e}, \mathrm{h})$ represents the exchange energy between electron and hole. κ is the Kramers conjugation operator.

Because of the approximation of the excited state wave function $|\phi_k^l\rangle$ (neglecting many body effects and electron–phonon coupling) the effective Coulomb interaction between electron and hole is completely unscreened in eq. (2). Since we will consider Wannier excitons, we may take these effects globally into account by screening the Coulomb interaction by the low-frequency dielectric constant ϵ_0. Furthermore, we assume that the crystal has a direct gap at the Γ point. Due to the simple band structure, the Hartree–Fock operators H_{HF} can be treated in the effective mass approximation. This leads to:

$$H_{\mathrm{ex}} = E_{\mathrm{g}} + \frac{p_{\mathrm{e}}^2}{2m_{\mathrm{e}}^*} + \frac{p_{\mathrm{h}}^2}{2m_{\mathrm{h}}^*} - \frac{e^2}{\epsilon_0 |\boldsymbol{r}_{\mathrm{e}} - \boldsymbol{r}_{\mathrm{h}}|}, \tag{3}$$

where m_{e}^* and m_{h}^* are the effective polaron masses for the electron and the hole respectively, E_{g} is the energy of the bandgap.

Now, relative and center of mass motions can be separated and the exciton dispersion $E_{\mathrm{n}}(\boldsymbol{Q})$ and exciton wave function $|\bar{\phi}_{\mathrm{ex}}^{n}(\boldsymbol{Q})\rangle$ may thus be obtained, where n stands for all exciton quantum numbers. $\boldsymbol{Q}$ is the center of mass wave vector. In the following, we consider the exciton ground state of symmetry Γ_i. In the case of Wannier excitons, the exciton wave function $|\bar{\phi}_{\mathrm{ex}}^{n}(\boldsymbol{Q})\rangle$ is given in a good approximation by

$$|\bar{\phi}_{\mathrm{ex}}^{\Gamma_i}(\boldsymbol{Q})\rangle = \frac{1}{\sqrt{V}} \mathrm{e}^{\mathrm{i}\boldsymbol{Q}\cdot\boldsymbol{R}} \bar{\psi}_{\mathrm{ex}}(\boldsymbol{r}) \cdot \bar{\phi}_{\mathrm{ex}}(\Gamma_i), \tag{4}$$

where $\bar{\psi}_{\mathrm{ex}}(\boldsymbol{r})$ is the envelope function obtained from the Schrödinger equation for the relative motion and $\bar{\phi}_{ex}(\Gamma_i)$ the Bloch part of the wave function. It is obtained from the symmetrical product of electron and hole states at wave vectors $\boldsymbol{p}_e = \boldsymbol{p}_h = 0$.

Let us now discuss the case of CuCl. CuCl is a cubic crystal with T_d-point group symmetry. Its uppermost valence band has a Γ_7 symmetry, the lowest conduction band has a Γ_6 symmetry at $\boldsymbol{Q} = 0$. From Koster et al. (1963), we find that the electron–hole product state has the symmetry:

$$\Gamma_6 \otimes \Gamma_7 = \Gamma_2 \oplus \Gamma_5. \tag{5}$$

Constructing symmetry adapted electron–hole pair states from linear combinations of conduction and valence band states, we obtain:

$$\bar{\phi}_{ex}(\Gamma_2) = \frac{1}{\sqrt{2}}(-\phi^6_{-1/2}\psi^7_{1/2} + \phi^6_{1/2}\psi^7_{-1/2}), \tag{6}$$

and for the exciton states with Γ_5 symmetry:

$$\begin{aligned}
\bar{\phi}_{ex}(x) &= \frac{i}{\sqrt{2}}(-\phi^6_{-1/2}\psi^7_{-1/2} + \phi^6_{1/2}\psi^7_{1/2}),\\
\bar{\phi}_{ex}(y) &= \frac{1}{\sqrt{2}}(\phi^6_{-1/2}\psi^7_{-1/2} + \phi^6_{1/2}\psi^7_{1/2}),\\
\bar{\phi}_{ex}(z) &= \frac{-i}{\sqrt{2}}(\phi^6_{-1/2}\psi^7_{+1/2} + \phi^6_{1/2}\psi^7_{-1/2}),
\end{aligned} \tag{7}$$

where $\phi^6_{+1/2}$, $\psi^7_{-1/2}$ are electron Bloch functions from the conduction and valence band extrema with spin $\pm\frac{1}{2}$, respectively.

Using:

$$-\psi^7_{1/2} = \kappa\psi^7_{-1/2} \quad \text{and} \quad \psi^7_{-1/2} = \kappa\psi^7_{1/2}, \tag{8}$$

eq. (7) may be expressed in terms of electron–hole states instead of two-electron states. Thus, the Bloch part of the different exciton wave functions in eq. (4) is obtained.

The analytic (short-range) part of the exchange interaction $W(\mathrm{e}, \mathrm{h})$ in eq. (2) leads to a splitting of the exciton states with Γ_2 and Γ_5 symmetry (Cho 1976). In addition, the Γ_5 exciton states are dipole active. Therefore, the non-analytic (long-range) exchange interaction leads to a splitting between the Γ_5 exciton states: we obtain the longitudinal exciton if the dipole moment of the exciton is parallel to its direction of propagation $\boldsymbol{Q}$, and the two transverse exciton states if their dipole moment is perpendicular to $\boldsymbol{Q}$. This splitting remains finite in the limit $\boldsymbol{Q} \to 0$. Including the exchange interactions, the energies are given by $E(\Gamma_2)$, $E_L(0)$ and $E_T(0)$ at $\boldsymbol{Q} = 0$.

In our starting Hamiltonian (eq. (1)), we have neglected all further interactions, such as coupling to the electromagnetic radiation field. Taking this coupling into account, we have to replace:

$$p_i \to p_i - (e/c)\,\mathbf{A}_i, \tag{9}$$

where $\mathbf{A}_i$ is the (transverse) vector potential acting on the ith electron. Consequently, some matrix elements are of the form:

$$\langle \phi_0 | (e/c) p_i \cdot \mathbf{A}_i \mid \phi^n_{\mathrm{ex}}(\mathbf{Q})\rangle \neq 0. \tag{10}$$

In eq. (10), $|\phi^n_{\mathrm{ex}}(\mathbf{Q})\rangle$ is the exciton wave function in the N-particle system corresponding to the two-particle function $|\bar{\phi}^n_{\mathrm{ex}}(\mathbf{Q})\rangle$. Equation (10) governs the dipole transitions between the crystal ground state and the dipole active exciton state. The effective two-particle Hamiltonian for the system of coupled exciton and radiation field may be diagonalized by a Bogolyubov transformation. The new quasiparticles obtained that way are called polaritons (Hopfield 1958, Pekar 1958) and their dispersion $E_{1,2}(\mathbf{Q})$ is given by the solution of:

$$n^2(Q) = \frac{Q^2c^2}{\omega^2} = \epsilon_\mathrm{b} + \frac{4\pi\beta\epsilon_\mathrm{b} E^2_\mathrm{T}(0)}{E^2_\mathrm{T}(Q) - \hbar^2\omega^2}, \tag{11}$$

where the oscillator strength β is given by:

$$4\pi\beta = \frac{E^2_\mathrm{L}(0) - E^2_\mathrm{T}(0)}{E^2_\mathrm{T}(0)}, \tag{12}$$

ϵ_b is a background dielectric constant, which takes into account the different oscillators which have been neglected in the one-oscillator model and:

$$E_\mathrm{T}(Q) = E_\mathrm{T}(0) + \hbar^2Q^2/2m^*_x \tag{13}$$

gives the dispersion of the transverse exciton, m^*_x is the exciton total mass.

2.2. Biexcitons

Let us now develop the effective Hamiltonian describing the biexciton. As in the exciton problem, we start from the N-electron Slater determinant $|\phi^{ll'}_{kk'}\rangle$, where two valence band states k, k' are unoccupied, the electrons being in the conduction band states l, l'. Then, the biexciton wave function reads:

$$|\phi^{\Gamma_i}_{\mathrm{Bi}}(\mathbf{Q})\rangle = \sum_{\substack{ll'\\kk'}} B^{ll'}_{kk'}\,\delta(\boldsymbol{l} + \boldsymbol{l}' - \boldsymbol{k} - \boldsymbol{k}', \mathbf{Q})|\phi^{ll'}_{kk'}\rangle. \tag{14}$$

The resulting effective biexciton Hamiltonian is given in the same approximations as used in the exciton problem by:

$$H_{Bi} = E_0 + H_{HF}(1) + H_{HF}(3) - \kappa(2)H_{HF}(2)\kappa^+(2) - \kappa(4)H_{HF}(4)\kappa^+(4) + \frac{e^2}{\epsilon_o}\left(\frac{1}{r_{13}} + \frac{1}{r_{24}} - \frac{1}{r_{12}} - \frac{1}{r_{14}} - \frac{1}{r_{32}} - \frac{1}{r_{34}}\right) + W_{(1,2)} + W_{(1,4)} + W_{(3,2)} + W_{(3,4)}. \tag{15}$$

It is determined from the equation:

$$\langle \phi_{kk'}^{ll'} | H | \phi_{pp'}^{mm'} \rangle = \langle \bar{\phi}_{kk'}^{ll'} | H_{Bi} | \bar{\phi}_{pp'}^{mm'} \rangle, \tag{16}$$

where $|\bar{\phi}_{kk'}^{ll'}\rangle$ is the Bloch part of the four-particle wave function

$$|\bar{\phi}_{kk'}^{ll'}\rangle = |\phi_l(1)\rangle |\phi_{l'}(3)\rangle \kappa(2) |\psi_k(2)\rangle \kappa(4) |\psi_{k'}(4)\rangle, \tag{17}$$

and the indices 1, 3 stand for electrons and 2, 4 for holes.

The Schrödinger equation:

$$H_{Bi} |\bar{\psi}_{Bi}\rangle = E_{Bi} |\bar{\psi}_{Bi}\rangle, \tag{18}$$

resulting from eq. (15) determines the biexciton energy E_{Bi} and its envelope function $|\bar{\psi}_{Bi}\rangle$ which is related to the $B_{kk'}^{ll'}$ coefficients (eq. (14)) by Fourier transformation.

Equation (18) has first been studied in the effective mass approximation for the biexciton ground state, neglecting all exchange effects by different techniques (Sharma 1968, Wehner 1969, Hanamura 1970, 1975, 1976, Adamowski et al. 1971, Akimoto and Hanamura 1972, Brinkman et al. 1973, Handel 1973, Huang 1973, Sheboul and Ekardt 1976). In this case, it turns out that the biexciton ground state is stable for all mass ratios $\sigma = m_e^*/m_h^*$†. Figure 1 gives three different theoretical curves of the biexciton binding energy E_{Bi}^b as a function of σ. E_{Bi}^b is measured in units of the binding energy of the triplet exciton E_{ex}^b. The experimental value for $\sigma = 0$ (H_2 molecule: $E_{Bi}^b/E_{ex}^b = 0.35$) is quite well reproduced by all theories. Curve (a) results from a Feynman-path integral method (Huang 1973) curves (b) and (c) from a variational calculation (Akimoto and Hanamura 1972, Brinkman et al. 1973). The experimental values are given by Klingshirn (1979). However, the quantitative agreement between theory and experiment is rather poor.

Recently, the model Hamiltonian (eq. (15)) has been completed by taking into account exchange interactions (Bassani et al. 1974, Forney et al. 1974, Ekardt and Sheboul 1976a, Quattropani and Forney 1977), coupling to phonons (Ekardt and Sheboul 1977, Hassan 1978, Ekardt and Brocksch

† In addition, it was shown from Hellmann–Feynman theorem that the structure of the Hamiltonian (eq. (15)) leads to the fact $\delta E/\delta\sigma \geq 0$ and $\delta^2 E/\delta\sigma^2 \leq 0$ in the open interval (0, 1) (Adamowski et al. 1972).

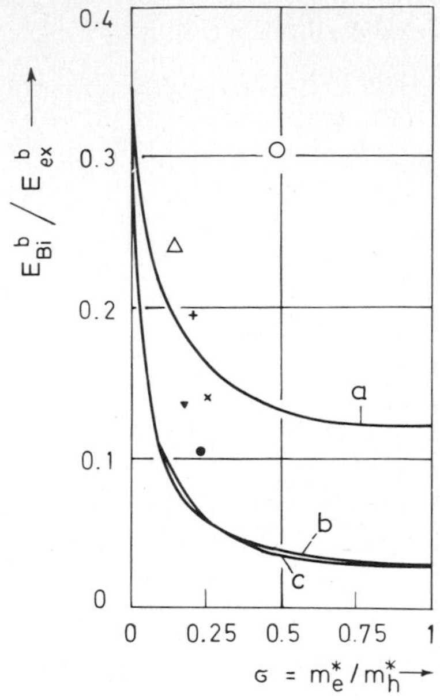

Fig. 1. Ratio of the biexciton binding energy E^b_{Bi} to the exciton binding energy E^b_{ex} as a function of the mass ratio $\sigma = m_e^*/m_h^*$. Experimental points refer to: (×) CuCl, (+) CuBr, (●) CdS, (○) ZnO, (△) InSe, (▼) ZnTe. Theoretical curves: (a) Huang (1973), (b) Brinkman et al. (1973) (c) Akimoto and Hanamura (1972) (from Klingshirn (1979).

1979) and mass-anisotropy (Akimoto 1973, Bednarek et al. 1978), and starting from better variational wave functions.

In this context, it is worthwhile pointing out that the exchange interaction always diminishes the binding energy and therefore it may be possible that the biexciton is unstable for large σ values (Bassani and Rovere 1976).

Let us now discuss the Bloch part of the four particle wave function $|\bar{\phi}^{ll'}_{kk'}\rangle$. From Pauli's principle, the total biexciton wave function (eq. (14)) has to be antisymmetric with respect to the exchange of two electrons or two holes. In analogy with the hydrogen molecule, we assume that the Bloch part of the biexciton ground state $|\bar{\phi}^{ll'}_{kk'}\rangle$ is antisymmetric with respect to the exchange of electrons and holes. A symmetric combination for the two hole states leads to a higher biexciton energy corresponding to the first rotational levels of the biexciton (Hanamura 1970, 1975, Forney et al. 1974, Ekardt and Sheboul 1976a). Consequently, the envelope function $|\bar{\psi}_{Bi}\rangle$ is a symmetric function under exchange of the electrons and holes (Γ_1 symmetry) for the ground state.

Let us study the biexciton ground state in the case $Q = l = l' = k = k' = 0$

for CuCl. The total symmetry of the biexciton state $|\bar{\phi}^{\Gamma_1}_{\mathrm{Bi}}(0)\rangle$ is given by:

$$\Gamma_{\mathrm{env}} \otimes (\Gamma_6 \otimes \Gamma_6)^- \otimes (\Gamma_7 \otimes \Gamma_7)^-. \tag{19}$$

Here the superior index (−) means that the antisymmetric product has been taken.

From Koster et al. (1963), we find:

$$(\Gamma_6 \otimes \Gamma_6)^- = (\Gamma_7 \otimes \Gamma_7)^- = \Gamma_1, \tag{20}$$

and the ground state wave function has thus a Γ_1 symmetry and its Bloch part given by (Hanamura 1970, 1975, 1976, Bassani et al. 1974, Forney et al. 1974, Quattropani and Forney 1977):

$$\bar{\phi}_{\mathrm{Bi}}(\Gamma_1) = \tfrac{1}{2}(\phi_{1/2}(1)\phi_{-1/2}(3) - \phi_{-1/2}(1)\phi_{1/2}(3))(\psi_{1/2}(2)\psi_{-1/2}(4) - \psi_{-1/2}(2)\psi_{1/2}(4)). \tag{21}$$

Changing to the exciton basis as defined in eq. (7), we find:

$$\bar{\phi}_{\mathrm{Bi}}(\Gamma_1) = \tfrac{1}{2}(\bar{\phi}^A_{\mathrm{ex}}(x)\bar{\phi}^B_{\mathrm{ex}}(x) + \bar{\phi}^A_{\mathrm{ex}}(y)\bar{\phi}^B_{\mathrm{ex}}(y) + \bar{\phi}^A_{\mathrm{ex}}(z)\bar{\phi}^B_{\mathrm{ex}}(z) + \bar{\phi}^A_{\mathrm{ex}}(\Gamma_2)\bar{\phi}^B_{\mathrm{ex}}(\Gamma_2)), \tag{22}$$

where the pairs (1, 4) and (2, 3) make up the excitons A and B respectively. In this approximation, the four-particle biexciton wave function $|\bar{\phi}^{\Gamma_1}_{\mathrm{Bi}}(Q)\rangle$ is given by

$$|\bar{\phi}^{\Gamma_1}_{\mathrm{Bi}}(\boldsymbol{Q})\rangle = \frac{1}{\sqrt{V}} \mathrm{e}^{\mathrm{i}\boldsymbol{QR}} |\bar{\psi}_{\mathrm{Bi}}\rangle |\bar{\phi}_{\mathrm{Bi}}(\Gamma_1)\rangle.$$

The corresponding wave function in the N-particle basis is denoted by $|\phi^{\Gamma_1}_{\mathrm{Bi}}(\boldsymbol{Q})\rangle$.

If we use the symmetry properties of the four-particle wave function to construct the biexciton wave function from N-electron Slater determinants, we find that the optical dipole transitions between exciton and biexciton states for different polarizations $\boldsymbol{e}$ are allowed:

$$\begin{aligned}\langle \phi^{\Gamma_1}_{\mathrm{Bi}} | e_x p_x | \phi^x_{\mathrm{ex}} \rangle &= \langle \phi^{\Gamma_1}_{\mathrm{Bi}} | e_y p_y | \phi^y_{\mathrm{ex}} \rangle \\ &= \langle \phi^{\Gamma_1}_{\mathrm{Bi}} | e_z p_z | \phi^z_{\mathrm{ex}} \rangle \neq 0,\end{aligned} \tag{23}$$

and all other transitions are equal to zero. Within this framework, we are now able to study the optical transitions between excitons, biexcitons and the crystal ground state.

3. *Non-linear absorption processes*

3.1. *Two-photon absorption*

Two-photon absorption spectroscopy (TPA) was first performed by observing the transmission of a laser beam through CuCl and CdS samples (Gale and Mysyrowicz 1974, Svorec and Chase 1976). Biexcitons are created by the simultaneous absorption of two photons of the spectrally narrow laser source, inducing the transition from the crystal ground state to the biexciton state. The absorption spectrum shows a sharp resonance at half the biexciton energy, but shows also impurity absorption levels. Therefore, it is difficult to distinguish without any doubt between one and two-photon absorption processes in this type of experiment.

We have measured the TPA transitions in CuCl from the ground state of the crystal to the biexciton state using two different light sources (dye lasers) (Vu Duy Phach et al. 1977). This technique has, compared to the study of the TPA with only one source, many interesting aspects:

(i) By varying the wavelength of the two light sources, we are able to distinguish between TPA and any one-photon absorption which is induced by the dye laser.

(ii) We are able to study directly the symmetry of the levels involved in the transition by measuring the TPA as a function of the relative polarization of the two light beams with respect to the crystal axes.

(iii) This technique enables us to measure the spectral shape of the TPA coefficient.

The samples of CuCl studied in this experiment are high-quality platelets grown by a vapour phase transport method. They are cooled down to liquid helium temperature in a double quartz cryostat.

The experimental set-up is represented in fig. 2. The sample can be simultaneously illuminated by two light beams: one, called laser, is provided by a tunable dye laser which is very intense and has a narrow emission at the energy $\hbar\omega_l$. The other, called continuum, is due to the superradiant emission of the same dye. This emission has a broad spectral width and is kept very weak in intensity. Both dye cells are pumped by the same nitrogen laser.

Concerning the laser emission, we are optically pumping a saturated solution of αNND in ethanol. The dye laser cavity is formed by an output mirror and a grating in a Littrow mount (Hänsch 1972) which does the spectral tuning. The spectral width of the emission is decreased by inserting into the cavity an inverted telescope to expand the beam section and thus to improve the efficiency of the grating. We obtain a spectral width of about 0.5 meV. If we add a Fabry–Perot, we reach a spectral width of 0.26 meV. The pulse duration is 1.2 ns, with a power of a few kilowatts. The central part of the beam is selected by a diaphragm and focussed onto the sample by a lens. The exciting light is polarized and can be attenuated by a set of neutral density filters.

The superradiant emission of another dye cell is similarly excited. We select in this broad emission a band of about 20 meV width by inserting an interference filter. This beam is also diaphragmed and polarized. It is focussed onto the sample inside the exciting spot of the intense laser. The

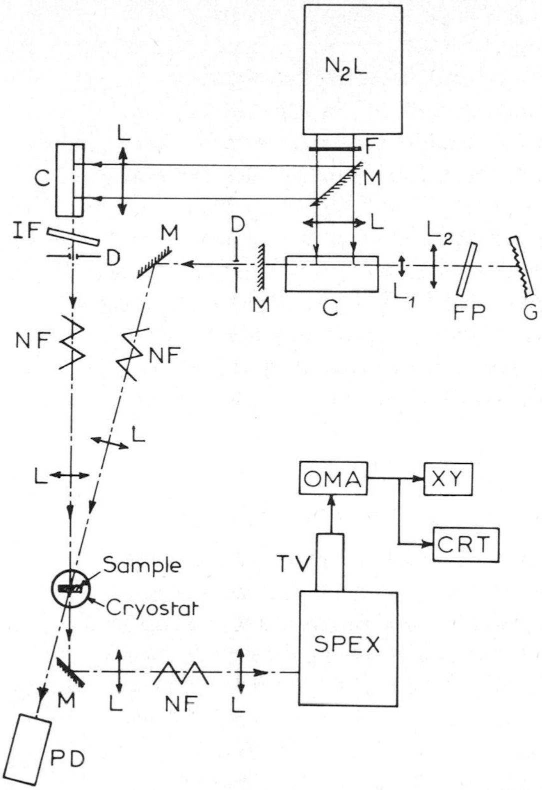

Fig. 2. Experimental set-up: C dye cells, D diaphragms, F filter, FP Fabry–Perot etalon, IF interference filter, L quartz lenses, M mirrors, NF neutral density filters, OMA optical multichannel analyzer, P polarizers, PD photocell, SPEX spectrograph, TV picture tube.

transmission of the continuum is studied through a $\frac{3}{4}$ m Spex spectrograph with a OMA television camera system and recorded by an XY plotter. The coincidence of the two spots on the surface of the sample in space and time is the most critical point in these experiments and is therefore carefully checked.

At first, the continuum and the laser are polarized parallel to each other. Figure 3 shows the observed TPA due to the transition from the crystal ground state to the biexciton level. The dotted line corresponds to the broad spectrum, $J_c(\omega)$, transmitted through the sample without laser excitation; the continuous line, $J_{c+l}(\omega)$, when both sources are exciting the crystal. The spectral position of the laser peak $\hbar\omega_l$ is reached from scattered light from the sample. A sharp absorption dip is observed with its maximum at the spectral position $\hbar_{\omega c}$. If the laser frequency is tuned, the

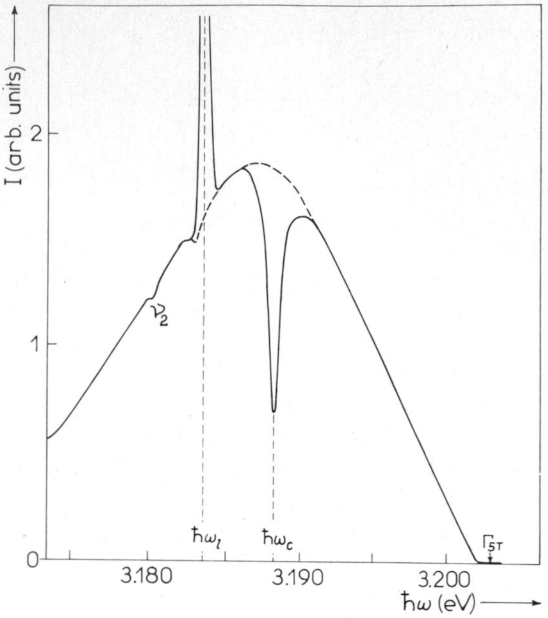

Fig. 3. Spectra of the continuum transmitted through a crystal of CuCl at liquid helium temperature in a two-photon absorption experiment. Dashed curve (---): spectrum without laser excitation. Solid curve (——): spectrum with laser excitation. $\hbar\omega_c$ absorption peak, $\hbar\omega_l$ laser diffusion peak.

spectral position of the absorption dip shifts. We shall show that this absorption is due to a TPA process corresponding to a transition from the crystal ground state to the biexciton ground state having Γ_1 symmetry.

In usual time-dependent perturbation theory, the two-photon transition probability per unit time from the crystal ground state $|\phi_0\rangle$ to the lowest biexciton state $|\phi_{\mathrm{Bi}}^{\Gamma_1}(\boldsymbol{k}')\rangle$ with momentum $\boldsymbol{k}'$ is given by:

$$W_2(\omega_l, \omega_c) \propto \frac{N_l}{\omega_l}\frac{N_c}{\omega_c}\,\delta(E_{\mathrm{Bi}} - \hbar\omega_l - \hbar\omega_c) \times \sum_{k'}\left|\sum_{k}\left[\frac{\langle\phi_{\mathrm{Bi}}^{\Gamma_1}(\boldsymbol{k}')|\boldsymbol{e}_c\cdot\boldsymbol{p}|\phi_{\mathrm{ex}}^{\Gamma_{5T}}(\boldsymbol{k})\rangle\langle\phi_{\mathrm{ex}}^{\Gamma_{5T}}(\boldsymbol{k})|\boldsymbol{e}_l\cdot\boldsymbol{p}|\phi_0\rangle}{E_{\mathrm{ex}}(\boldsymbol{k}) - \hbar\omega_l} + P_{lc}\right]\right|^2. \tag{24}$$

We consider the almost resonant transverse exciton state as intermediate state (Hanamura 1973, Bassani et al. 1974, Doni et al. 1975). Concerning the symmetry of the states involved, all transitions considered are dipole allowed. $N_i (i = l, \mathrm{c})$ denotes the number per unit volume of photons with frequency ω_i, momentum $\boldsymbol{q}_i$ and polarization vector $\boldsymbol{e}_i$. P_{lc} stands for the foregoing expression with the indices l and c of the two photons exchanged.

Using the approximations for the biexciton and exciton wave functions given by Hanamura (1973), we find for the matrix elements:

$$|\langle \phi_{\mathrm{Bi}}^{\Gamma_1}(\boldsymbol{k}')|\boldsymbol{e}_{\mathrm{c}} \cdot \boldsymbol{p}|\phi_{\mathrm{ex}}^{\Gamma_{5\mathrm{T}}}(\boldsymbol{k})\rangle|^2 = (\boldsymbol{p}_{\mathrm{cv}} \cdot \boldsymbol{e}_{\mathrm{c}})^2 |\psi_{\mathrm{ex}}(0)|^2 |g(\boldsymbol{k}')|^2 \delta(\boldsymbol{k}', \boldsymbol{k} + \boldsymbol{q}_{\mathrm{c}}), \tag{25}$$

$$|\langle \phi_{\mathrm{ex}}^{\Gamma_{5\mathrm{T}}}(\boldsymbol{k})|\boldsymbol{e}_l \cdot \boldsymbol{p}|\phi_0\rangle|^2 = (\boldsymbol{p}_{\mathrm{cv}} \cdot \boldsymbol{e}_l)^2 |\psi_{\mathrm{ex}}(0)|^2 \delta(k, q_l), \tag{26}$$

with $(\boldsymbol{p}_{\mathrm{cv}} \cdot \boldsymbol{e}_{\mathrm{i}})$ being the optical matrix element, $\psi_{\mathrm{ex}}(0)$ is the exciton envelope function and $g(\boldsymbol{k}')$ the Fourier transform of the biexciton envelope function. The first matrix element corresponds to a transition from the ground state to the intermediate state, the second one to a transition from the intermediate state to the final biexciton ground state. The only intermediate state to consider is the transverse exciton state $\Gamma_{5\mathrm{T}}$, because the two matrix elements correspond then to allowed transitions and the denominator is small ($\hbar\omega_l \simeq \hbar\omega_{\mathrm{c}} \simeq E_{\mathrm{ex}}(0)$, therefore a strong resonant effect is expected. The transition to the excited biexciton state is forbidden.

Assuming parallel polarizations of the two beams ($\boldsymbol{e}_l = \boldsymbol{e}_{\mathrm{c}}$) and $\boldsymbol{q}_l \simeq \boldsymbol{q}_{\mathrm{c}} \simeq 0$ and performing the summations over the states involved, we end up with the following transition probability:

$$W_2(\omega_l, \omega_{\mathrm{c}}) \propto \frac{N_l N_{\mathrm{c}}}{\omega_l \omega_{\mathrm{c}}} |\boldsymbol{p}_{\mathrm{cv}} \cdot \boldsymbol{e}|^4 |\psi_{\mathrm{ex}}(0)|^4 |g(0)|^2 \times \left(\frac{1}{E_{\mathrm{ex}}(0) - \hbar\omega_l} + \frac{1}{E_{\mathrm{ex}}(0) - \hbar\omega_{\mathrm{c}}} \right)^2 \delta(E_{\mathrm{Bi}} - \hbar\omega_l - \hbar\omega_{\mathrm{c}}). \tag{27}$$

$W_2(\omega_l, \omega_{\mathrm{c}})$ is related to the TPA coefficient $K_2(\omega_l, \omega_{\mathrm{c}})$ by

$$J(\omega_l) K_2(\omega_l, \omega_{\mathrm{c}}) = \frac{W_2(\omega_l, \omega_{\mathrm{c}}) \sqrt{\epsilon_{\mathrm{b}}}}{c N_{\mathrm{c}}},$$

with

$$J(\omega_l) = (c/\sqrt{\epsilon_{\mathrm{b}}}) \hbar\omega_l N_l. \tag{28}$$

If the energy of the photons of the two sources is varied, the TPA dip in the spectrum shifts. In fig. 4, we have plotted the energy of the maximum of the dip $\hbar\omega_{\mathrm{c}}$ as a function of the photon energy of the laser $\hbar\omega_l$. The sum of the two corresponds to the biexciton energy:

$$\hbar\omega_l + \hbar\omega_{\mathrm{c}} = E_{\mathrm{Bi}}. \tag{29}$$

We have evaluated from the general equations governing the variation of the intensity of the two beams travelling through the crystal, the following

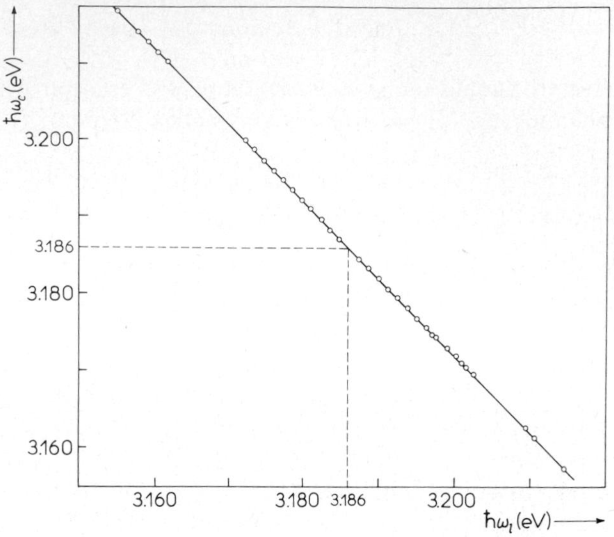

Fig. 4. Variation of the absorption peak position $\hbar\omega_c$ as a function of the laser peak position $\hbar\omega_l$.

TPA function Y for a given photon energy $\hbar\omega_l$:

$$Y = \int_0^\infty \ln \frac{J_c}{J_{c+l}} \, d(\hbar\omega)$$

$$= \frac{K_2(\omega_l, \omega_c)}{K_2(\omega_l, \omega_l)} \ln \left[1 + \frac{K_2(\omega_l, \omega_l)\, J_l(0)}{K_1(\omega_l)} (1 - \exp[-K_1(\omega_l) x_0]) \right], \tag{30}$$

x_0 is the thickness of the sample, $x_0 = 80\ \mu$m in our case.

For $\hbar\omega_l = 3.176$ eV and $K_1(\omega_l)$, the one-photon absorption coefficient, being equal to $180\ \text{cm}^{-1}$, we could determine the two-photon absorption coefficients:

$$K_2(\omega_l, \omega_l) = (2 \pm 1) \times 10^{-3}\ \text{cm W}^{-1},$$
$$K_2(\omega_l, \omega_c) = (3.0 \pm 0.7) \times 10^{-3}\ \delta(E_{Bi} - \hbar\omega_l - \hbar\omega_c)\ \text{cm W}^{-1}. \tag{31}$$

From the comparison between this coefficient and the oscillator strength f_x of the exciton per unit cell of volume Ω which is known from other measurements, we can deduce a value of the biexciton envelope function for $\boldsymbol{k}' = 0$,

$$f_x = \frac{2\Omega}{m_0} \frac{1}{E_{ex}(0)} (\boldsymbol{p}_{cv} \cdot \boldsymbol{e})^2 |\psi_{ex}(0)|^2 = 4.5 \times 10^{-3}, \quad |g(0)|^2 = 300\ \Omega. \tag{32}$$

This result is in close agreement with the theoretical prediction on the biexciton wave function given by Ekardt and Sheboul (1976b).

With this experiment, it is also possible to check the symmetry of the biexciton state involved in the transition by a study of the dependence of the TPA coefficient on the angle η between the polarization vectors of the two light sources and with respect to the crystal axis. We have shown (fig. 5) that the TPA coefficient varies as $\cos^2\eta$ and independently of the crystal axes. The corresponding symmetry of the biexciton state is Γ_1 as theoretically expected (Bassani et al. 1974).

3.2. Induced absorption

The second absorption process studied is the creation of excitons by the intense laser and the subsequent creation of biexcitons by the photons from the broad emission laser (Gogolin and Rashba 1973, Gogolin 1974, Saito et al. 1976, Bivas et al. 1977) (same experimental set-up as for the TPA).

This induced absorption is shown in fig. 6. It remains at the same spectral region when the photon energy of the laser is tuned through the exciton absorption region, contrary to the TPA absorption. Since it vanishes (dotted line) if the sample is not excited simultaneously by the intense laser, it is an induced absorption process.

This absorption is due to a transition from the transverse exciton state Γ_{5T} to the biexciton Γ_1 ground state. The transitions from the exciton level Γ_{5L} to the biexciton level Γ_1 are not observed. Transitions from the exciton states to the excited states of biexcitons are forbidden by symmetry and are not observed.

The probability for a dipole-allowed transition (creation of biexcitons

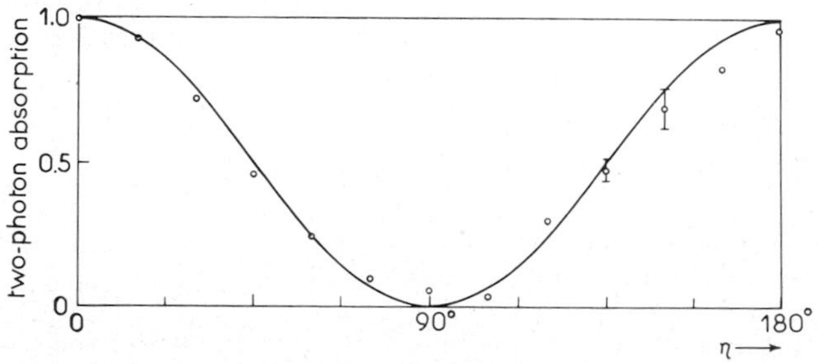

Fig. 5. Dependence of the normalized TPA coefficient on the angle η between the two polarization directions of the laser and the continuum. Experimental points (circles), $\cos^2\eta$ variation (solid curve).

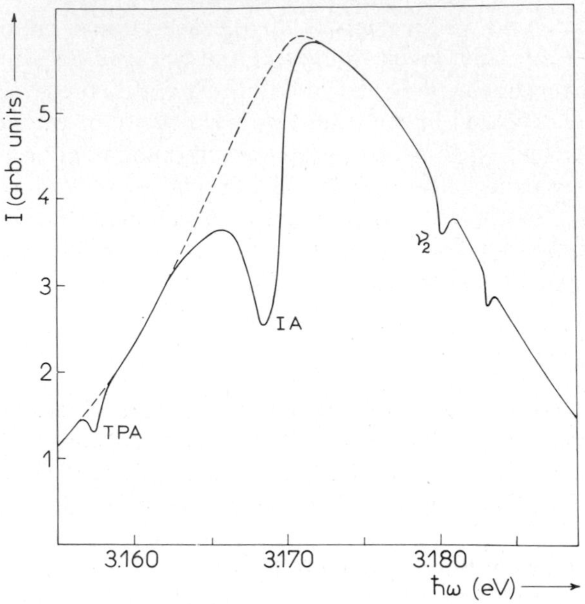

Fig. 6. Spectra of the continuum transmitted through a crystal of CuCl at 4.2 K. Dashed curve (---) without laser excitation. Solid curve (——) with laser excitation. TPA – two-photon absorption. IA – induced absorption.

from a population of excitons) is given by the following expression:

$$W_i(\hbar\omega_c) \propto \frac{N_c}{\omega_c} \sum_{kk'} |\langle\phi_{\mathrm{Bi}}^{\Gamma_1}(k')|\boldsymbol{e}_c \cdot \boldsymbol{p}|\phi_{\mathrm{ex}}^{\Gamma_{5T}}(k)\rangle|^2 \rho_x(k)\delta(E_{\mathrm{Bi}}(k') - E_{\mathrm{ex}}(k) - \hbar\omega_c), \tag{33}$$

N_c is the number of incident photons per unit volume of the probe beam. The transition from the triplet exciton state with Γ_2 symmetry to the biexciton ground state is forbidden. In our case, the initial state $|\phi_{\mathrm{ex}}^n(k)\rangle$ of the transition is the ground state Γ_{5T} of the exciton. Its energy is $E_{\mathrm{ex}}(k)$. We assume that these Γ_{5T} excitons created by the dye laser are distributed according to a Boltzmann statistics $\rho_x(k)$ with a temperature T_x different from the lattice temperature:

$$\rho_x(k) = n_x\left(\frac{2\pi\hbar^2}{m_x^* k_B T_x}\right)^{3/2} \exp - \left(\frac{\hbar^2 k^2}{2m_x^* k_B T_x}\right), \tag{34}$$

where n_x is the number of Γ_{5T} excitons per unit volume and m_x^* their effective total mass.

The final state $|\phi_{\mathrm{Bi}}^{\Gamma_1}(k')\rangle$ of the transition is the Γ_1 biexciton ground state. Its energy is $E_{\mathrm{Bi}}(k')$. Using the biexciton wave function formalism of

Hanamura (1973), the matrix element in eq. (33) is given by the following expression, the wave vector of the photon being neglected:

$$|\langle\phi_{\mathrm{Bi}}^{\Gamma_1}(k')|\boldsymbol{e}_i\cdot\boldsymbol{p}|\phi_{\mathrm{ex}}^{\Gamma_{5\mathrm{T}}}(k)\rangle|^2=(\boldsymbol{p}_{\mathrm{cv}}\cdot\boldsymbol{e})^2|\psi_{\mathrm{ex}}(0)|^2|g(k')|^2\delta(k-k'), \tag{35}$$

where $(\boldsymbol{p}_{\mathrm{cv}}\cdot\boldsymbol{e})$ is the optical matrix element, $\psi_{\mathrm{ex}}(0)$ is the exciton envelope function at $r=0$, $|\psi_{\mathrm{ex}}(0)|^2=1/\pi a_x^3$, a_x being the Bohr radius of the exciton. $g(k')$ is the Fourier transform of the relative distance between the two excitons in the biexciton wave function.

If we substitute the matrix element into eq. (33), we obtain:

$$W_i(\hbar\omega_c)\propto(\boldsymbol{p}_{\mathrm{cv}}\cdot\boldsymbol{e}_c)^2|\psi_{\mathrm{ex}}(0)|^2|g(0)|^2n_x(N_c/\omega_c)$$
$$\times(E_0-\hbar\omega_c)^{1/2}\exp\left[\frac{-2}{k_{\mathrm{B}}T_x}(E_0-\hbar\omega_c)\right], \tag{36}$$

with

$$E_0=E_{\mathrm{Bi}}(0)-E_{\mathrm{ex}}(0). \tag{37}$$

The absorption coefficient is simply given by:

$$K_i(\hbar\omega_c)=\frac{W_i(\hbar\omega_c)\sqrt{\epsilon_b}}{cN_c}. \tag{38}$$

Figure 7 represents the experimental induced absorption band (continuous

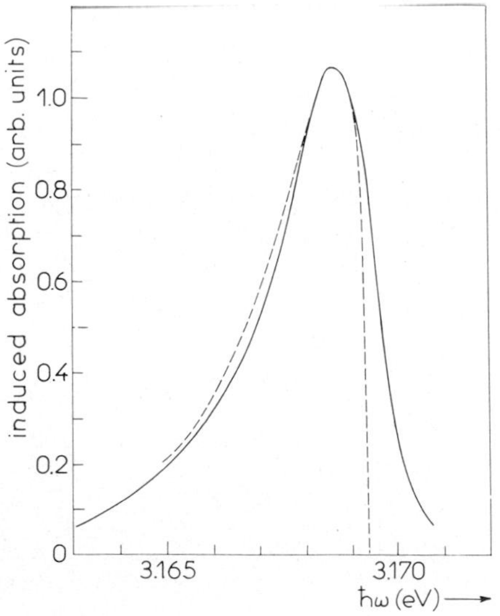

Fig. 7. Induced absorption band due to the transition exciton-biexciton. Solid curve (——) experimental band; Dashed curve (---) theoretical absorption band.

line) and the theoretical one (dotted line) at a photon energy of the laser $\hbar\omega_l = 3.2146$ eV for the following parameters (Bivas et al. 1977):

– a temperature of the exciton population $T_x = 32$ K, much higher than the lattice temperature ($T = 4.2$ K).

– the high energy cut-off $E_0 = E_{\text{Bi}}^{\Gamma_1} - E_{\text{ex}}^{\Gamma_{5T}} = 3.1694$ eV consistent with the energy of the transverse exciton $E_{\text{ex}}^{\Gamma_{5T}}$ known from hyper-Raman measurements and of $E_{\text{Bi}}^{\Gamma_1}$ known from TPA.

Since the induced absorption from the longitudinal exciton is not observed, we conclude that the excitons are mainly occupying the Γ_2 exciton and the Γ_{5T} exciton branches.

The poor fit of the two curves on the high energy side of the spectrum is probably due to the fact that polariton effects have been neglected in this analysis.

Figure 8 represents the variation of the area of the induced absorption band as a function of the energy $\hbar\omega_l$ of the photons of the dye laser. The intensity of the dye laser was kept constant, the variation of the reflection coefficient in the exciton region being taken into account.

The two arrows indicate the energetic positions of the transverse and longitudinal excitons of symmetry Γ_5. The peak observed on the low-energy side of the curve corresponds to the spectral region where it was

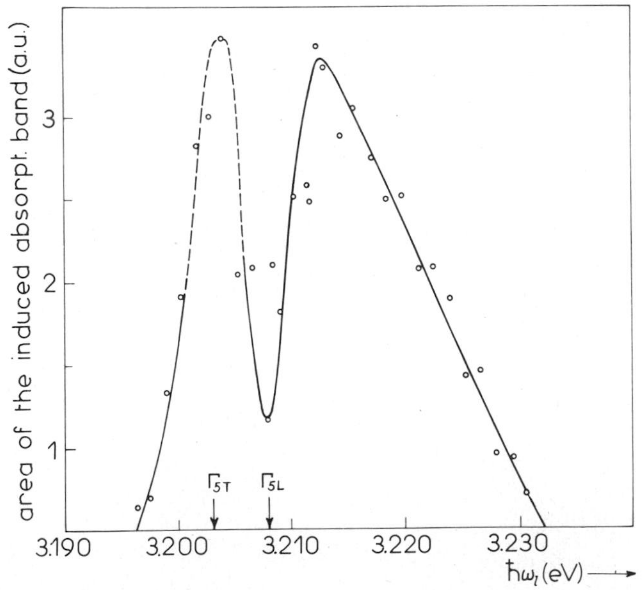

Fig. 8. Variation of the area of the induced absorption band as a function of the energy $\hbar\omega_l$ of the exciting laser photons.

difficult to distinguish between the induced absorption and the resonant two-photon absorption. A second peak is clearly observed on the high energy side. It does not correspond to any known one-photon absorption maximum.

This induced absorption due to the exciton–biexciton transition has been studied by time-resolved spectroscopy using the two times frequency doubled light of a mode-locked Yag: Nd^{3+} laser as a probe beam (Lévy et al. 1979a, 1979b).

From this study we could observe the building-up of a quasi-equilibrium of transverse excitons. Its temperature, higher than that of the lattice, decreases with time. The time behaviour of this transverse exciton population is studied giving information on relaxations and radiative decays of excitons and biexcitons.

4. Emission processes

4.1. Hyper-Raman scattering

As we have seen, biexcitons can be excited resonantly in their ground state (Γ_1) by the simultaneous absorption of two photons of the same laser when the energy of these exciting photons is equal to half the biexciton energy. These biexcitons relax on their dispersion curve by mutual collision and phonon interactions. Then, they recombine radiatively emitting a photon-likc polariton and leaving in the crystal a polariton or a longitudinal exciton (fig. 9a) (Grun et al. 1975, Nagasawa et al. 1975, Lévy et al. 1976).

If the energy of the exciting photons is detuned from the biexciton

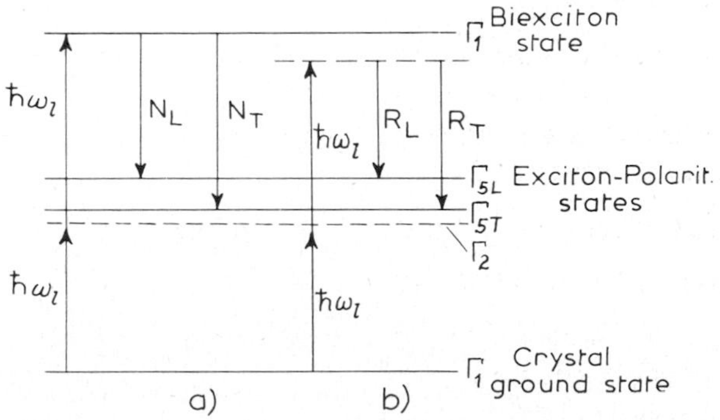

Fig. 9. Scheme of exciton and biexciton levels with the two-photon absorption, the luminescence and the hyper-Raman scattering processes related to biexcitons.

resonance, biexcitons are created only virtually, with a definite wave vector $\boldsymbol{K}$ equal to two times the wave vector $\boldsymbol{q}_l$ of the exciting polaritons. These virtual biexcitons recombine radiatively emitting a lower polariton $E_1(\boldsymbol{q})$ and leaving in the crystal either another lower polariton $E_1(\boldsymbol{k})$, a longitudinal exciton $E_L(\boldsymbol{k})$ or an upper polariton $E_2(\boldsymbol{k})$ as shown in fig. 9b.

Energies and wave vectors of the four particles involved are conserved in this two-photon Raman process or hyper-Raman scattering (HR).

$$\begin{aligned} 2\hbar\omega_l &= E_1(\boldsymbol{q}) + E_1(\boldsymbol{k}), \\ 2\hbar\omega_l &= E_1(\boldsymbol{q}) + E_L(\boldsymbol{k}), \\ 2\hbar\omega_l &= E_{1,2}(\boldsymbol{q}) + E_{2,1}(\boldsymbol{k}), \end{aligned} \tag{39}$$

and

$$\boldsymbol{K} = \boldsymbol{q}_l + \boldsymbol{q}_l = \boldsymbol{q} + \boldsymbol{k}. \tag{40}$$

This hyper-Raman scattering was first observed by Ueta's group in 1976 (Nagasawa et al. 1976a). The theory of this process was given by Bechstedt and Henneberger (1977). The importance of polariton effects was rapidly shown (Henneberger and Voigt 1976, Hönerlage et al. 1977, Itoh et al. 1977, Vu Duy Phach et al. 1978).

In the simplest experiment, the crystal is excited by a single laser beam making an angle of incidence α with the normal to the crystal surface as shown in fig. 10 (Hönerlage et al. 1977, Vu Duy Phach et al. 1978).

The laser used is the tunable dye laser described in sect. 3. The samples studied were platelets as well as large crystals, grown by vapor phase transport. They were cooled down to liquid helium temperature in a quartz cryostat (4.2 K).

The light scattered by the samples could be studied in a backward configuration, the illuminated surface of the sample being directly obser-

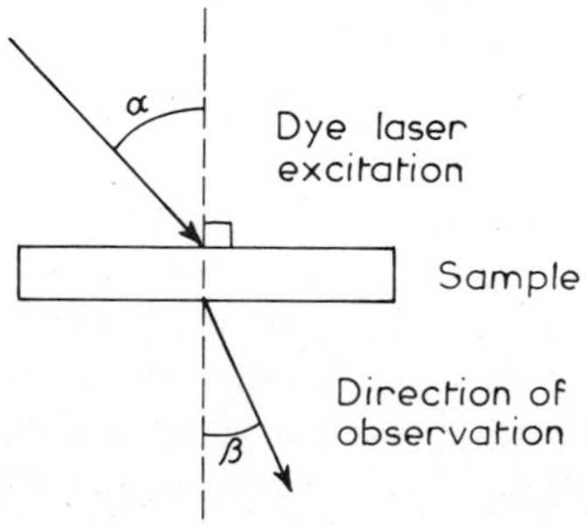

Fig. 10. Directions of the incident laser beam and of the observation relative to the sample surface.

ved, or in a forward configuration, the light transmitted through the sample being in that case recorded. The light emitted in a direction making an angle β with the normal to the crystal surface (fig. 10), analyzed through a Polaroid polarizer, was focussed on the entrance slit of a $\frac{3}{4}$m Spex spectrograph used in the second order. The spectra were recorded by an optical multi-channel analyzer (PAR). They were visualized on an oscilloscope and registered on an XY recorder.

In this scattering experiment, the energy of the exciting photons $\hbar\omega_l$ as well as the incident angle α and the observation angle β are known. To obtain the other characteristics of the scattering, we proceed as follows (Ostertag 1977, Vu Duy Phach et al. 1978): We assume that the polariton dispersion is at first known. The absolute value of the wave vector $|\boldsymbol{q}_l|$ of the incident photons is then obtained from their energy. The direction of their wave vector inside the crystal is deduced from the incident angle α and the index of refraction:

$$n(|\boldsymbol{q}_l|) = \frac{\hbar c|\boldsymbol{q}_l|}{E_1(|\boldsymbol{q}_l|)}. \tag{41}$$

The wave vectors $\boldsymbol{q}$ and $\boldsymbol{k}$ of the two polaritons scattered in the crystal and the scattering angle θ between incoming and observed polaritons are computed self-consistently by fulfilling the laws of energy and momentum conservation (eq. (39), (40)) and the index of refraction. The energies of the observed polariton and of that left in the crystal are obtained. The parameters of the polariton dispersion curve are adjusted in order to fit experimental energies of HR lines with the computed values.

The process where two lower polaritons are created in the crystal has been studied in two different experimental configurations. In the backward scattering configuration, the scattering angle θ is large ($\theta \simeq 180°$). As shown in fig. 11 by the full arrows, the polariton left in the crystal is exciton-like. Its energy remains constant and almost equal to that of the transverse exciton E_T, if the exciting laser is tuned or the angular configuration of the experiment is slightly changed. Therefore, the photon-like polariton observed gives rise to an emission line R_T which shifts in the spectrum twice as fast as the exciting photon energy, as can be seen in the spectra given in fig. 12. At an energy of the exciting photons equal to $E_{Bi}/2$, the R_T line coincides with the luminescence line N_T. The spectral positions of R_T and N_T are plotted as a function of the exciting photon energy $\hbar\omega_l$ in fig. 13. We have computed the position of the hyper-Raman line as explained previously. In this configuration, the simultaneous solution of eqs. (39) and (40) is unique. The solid line in fig. 13 represents the best fitting, obtained for the following value of $E_T = 3.2025 \pm 10^{-4}$ eV. All the other parameters of the polariton dispersion curve ϵ_b, Δ_{LT}, m_x^* have a minor influence and have been precisely determined otherwise.

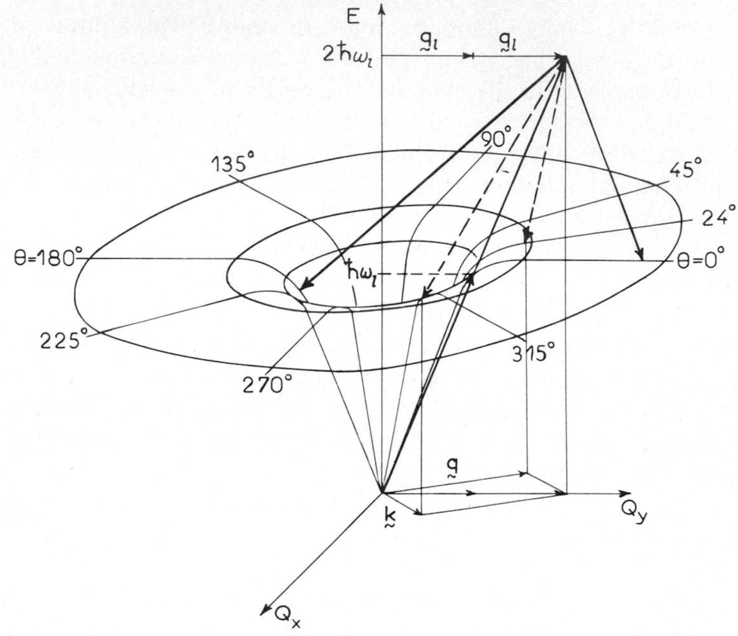

Fig. 11. Resonant two-photon Raman scattering involving the lower polariton branch.

In a forward scattering configuration, the scattering angle θ is small ($\theta < 30°$). As shown in fig. 11 by the dotted arrows, the energies of the two polaritons created in the crystal are in the bottleneck region of the polariton dispersion curve. These energies depend very sensitively on the scattering angle θ and on the energies of the exciting photons. At these small scattering angles, three simultaneous solutions of the eqs. (39) and (40) exist. Two of these solutions, corresponding to photon-like polaritons with wave vectors $|\boldsymbol{q}|$ close to that of the exciting polariton $|\boldsymbol{q}_l|$ are observed in our experiments and called R_T^- and R_T^+ respectively. The third solution corresponds to energies near the exciton absorption band and has been observed only with very thin samples. From the computed positions of the Raman lines we could deduce a precise value for the background dielectric constant $\epsilon_b \simeq 5.0 \pm 0.2$.

The second process, where a lower polariton is emitted and a longitudinal exciton left inside the crystal, has also been observed. It gives rise to the R_L emission line, showing up in backward and forward configurations. The observation of a small energy difference (0.16 ± 0.02 meV) between the spectral positions of this line in the two configurations has been related to the curvature of the longitudinal exciton band. From the computed positions of the Raman lines we have deduced directly the exciton

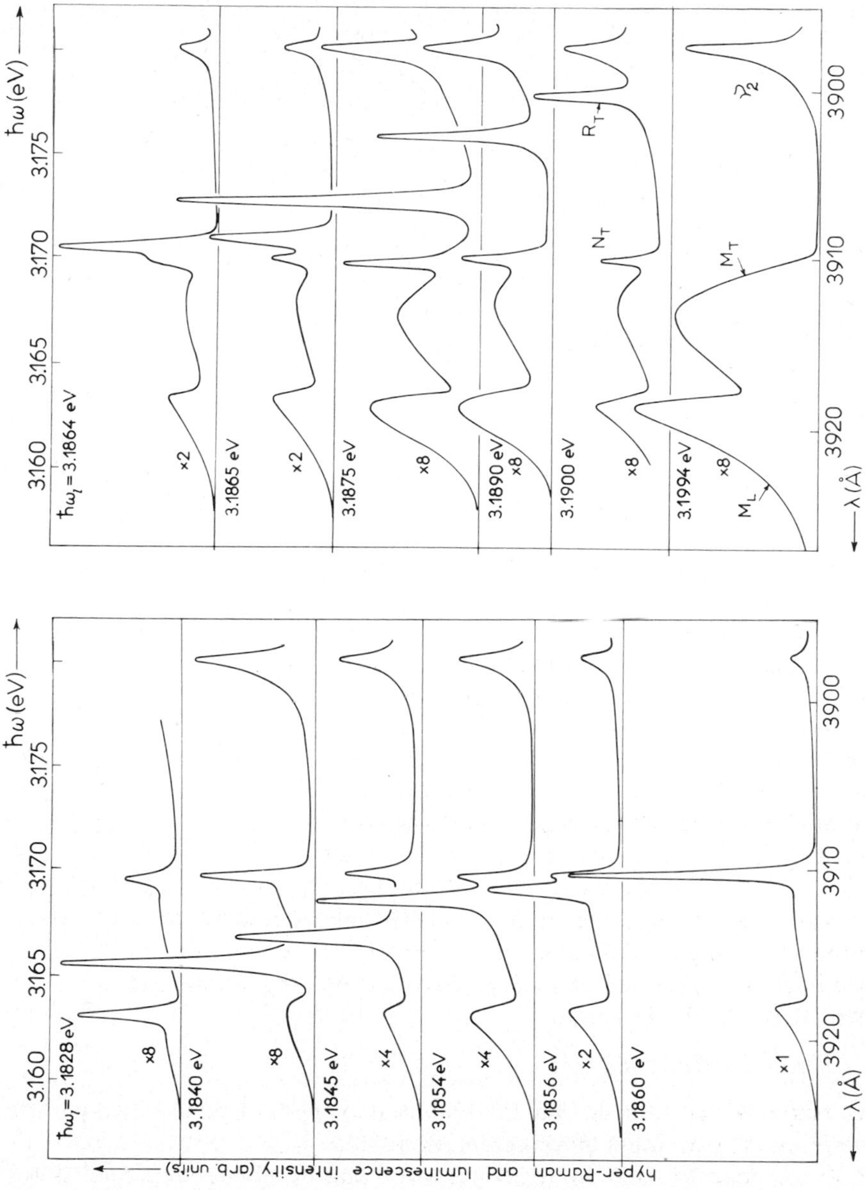

Fig. 12. Luminescence and hyper-Raman spectra of CuCl at 4.2 K for different exciting photon energies $\hbar\omega_l$.

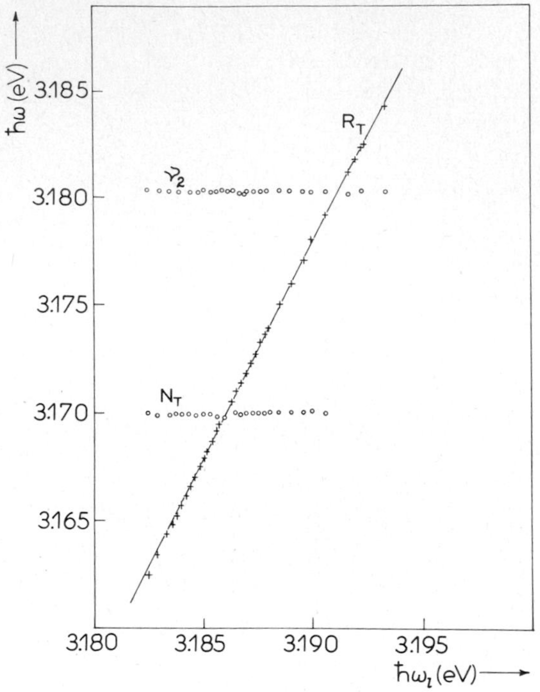

Fig. 13. Positions of the luminescence and hyper-Raman lines of CuCl at 4.2 K for different exciting photon energies.

total effective mass (Hönerlage et al. 1978b):

$$m_x^* = (2.5 \pm 0.3)\ m_0.$$

This value is in close agreement with the one deduced from electron–exciton scattering by Hönerlage et al. (1976).

We have also obtained the energy $E_L(0)$ of the longitudinal exciton at the center of the Brillouin zone

$$E_L(0) = 3.2080 \pm 10^{-4}\ \text{eV}.$$

This value is in close agreement with the one obtained by two-photon absorption measurement (Fröhlich et al. 1971).

A lower polariton and an upper polariton can be obtained as final states if the wave vector $|\boldsymbol{K}|$ of the virtual biexciton involved is smaller than $1.7|\boldsymbol{q}_l|$. In order to create such a biexciton, two separate laser beams are necessary as shown by Itoh and Suzuki (1978). They have observed the upper polariton, a lower polariton being left in the crystal. The reverse process, where a lower polariton is emitted and an upper polariton left in the crystal, has not been detected. This is due to the broad linewidth expected for this emission.

This study allows a spectroscopy in momentum space of the polariton and longitudinal exciton states, as discussed in ch. 3 of this book.

4.2. Coherent scattering

Recently, the influence of the biexciton state on the non-linear polarization of CuCl samples has been studied (Maruani et al. 1978). In this experiment, the emission from a dye laser is split into two parts: a "pump beam" characterized by the electric field vector $\boldsymbol{E}_p$ and the wave vector $\boldsymbol{k}_p$ inside the crystal and a "test-beam" ($\boldsymbol{E}_t$, $\boldsymbol{k}_t$). Both beams travel under a small angle θ through the crystal. Since the frequencies ω of the two beams are equal, the phase matching condition is very sensitive to the dispersion of the refractive index n_0. The non-linear interaction between the two beams gives rise to a non-linear polarization $\boldsymbol{P}_s$

$$\boldsymbol{P}_s(\omega) = \epsilon_0 \chi^{(3)}(-\omega, \omega, \omega, -\omega)\boldsymbol{E}_p(\omega)\boldsymbol{E}_p(\omega)\boldsymbol{E}_t^*(\omega), \tag{42}$$

which induces a coherent radiation with ($\boldsymbol{E}_s(\omega)$, $\boldsymbol{k}_s$) at a frequency ω. The wave-vector mismatch

$$\Delta \boldsymbol{k} = 2\,\boldsymbol{k}_p - \boldsymbol{k}_t - \boldsymbol{k}_s, \tag{43}$$

has its minimum value for $(\Delta k) = n_0\omega\theta'^2/c$, which is fulfilled for an angle $\theta' = -\theta$, i.e., symmetric to the test beam with respect to the pump beam. This coherent emission $\boldsymbol{E}_s$ becomes large, whenever 2ω crosses a resonance.

Apart from the emissions due to the third order non-linear susceptibility, the interaction has turned out to be so important that coherent emissions from higher order susceptibilities could be observed.

A study of the excitation spectrum of the coherent emission (Chemla et al. 1979) shows a similar structure as the excitation spectrum for non-coherent hyper-Raman scattering (Vu Duy Phach et al. 1978) in a forward scattering configuration, i.e., a minimum if the energy of the exciting photons is at half the biexciton energy. This minimum is due to the competition between coherent scattering and TPA of really created biexcitons. However, in the coherent scattering experiment (Chemla et al. 1979), a strong asymmetry of the excitation spectrum was observed. This asymmetric shape is attributed to the fact that biexcitons can decay into two polaritons so that the condition for a Fano effect (Fano 1961, Fano and Cooper 1965) is fulfilled.

4.3. Luminescence of biexcitons

The observation in 1968 of the luminescence due to the radiative decay of biexcitons was the first experimental proof of the existence of such quasiparticles (Mysyrowicz et al. 1968, Nikitine et al. 1968, Goto et al. 1970). The biexcitons were, in these experiments, indirectly created from a population of electron–hole pairs or of excitons. Since then, as we have seen, the development of tunable dye lasers has permitted the direct creation of biexcitons and the development of a complete spectroscopy of biexciton states.

4.3.1. Indirectly created biexcitons

When crystals of CuCl are excited by band-to-band absorption by the UV light of a nitrogen laser or in the exciton absorption by the light of a tunable dye laser, two broad bands called M_T and M_L are observed on the low-energy side of the free and bound exciton emission lines. The spectra obtained at different temperatures are drawn in fig. 14.

These emission bands have been explained as follows: A large number of free excitons are created directly or from thermalized free carriers. Biexcitons are then generated from this large population of excitons. They relax towards their Γ_1 ground state and they decay radiatively: a photon is emitted and a transverse or a longitudinal exciton is left in the crystal (Suga and Koda 1974).

The shape of this emission has been calculated by assuming that the biexcitons are thermalized in their ground state and characterized by a Boltzmann distribution with a definite temperature T_B. The matrix element of the transition has been assumed to be independent of $\boldsymbol{K}$ as shown by Cho (1973). The final states of the transitions have been described by the polariton and the longitudinal exciton dispersion curves. The biexciton emission spectrum of CuCl at $T = 1.8$ K and the corresponding theoretical curve are drawn in fig. 15. The fit is good at low levels of excitation (Ostertag et al. 1975).

The temperature of the biexciton gas ($T_B \simeq 20$ K) obtained here is found to be much higher than the lattice temperature. This high temperature has been related to the formation of biexcitons from the Γ_2 (and Γ_5) excitons. In order to bind together, these excitons have to get rid of the biexciton binding energy found to be 28 meV. This binding energy being very close to the LO phonon energy (26 meV), this process takes place mainly via fast LO phonon emission. Biexcitons with large wave vectors are then created. These hot biexcitons will slowly thermalize by acoustic-phonon emission. They will not reach the lattice temperature during their lifetime.

This luminescence has been studied by time-resolved spectroscopy (Lévy et al. 1979b, Ostertag and Grun 1977, Ojima et al. 1978). The CuCl

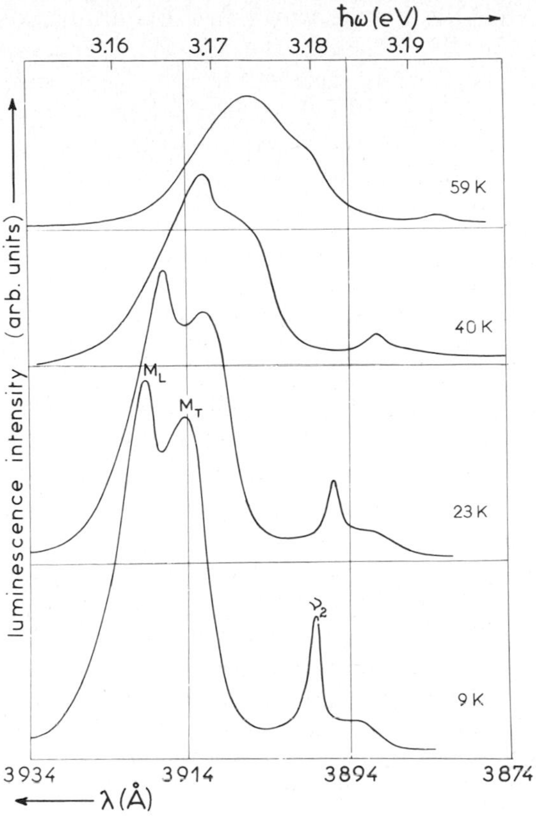

Fig. 14. Biexciton luminescence spectra of CuCl at different temperatures for an excitation in the band to band absorption.

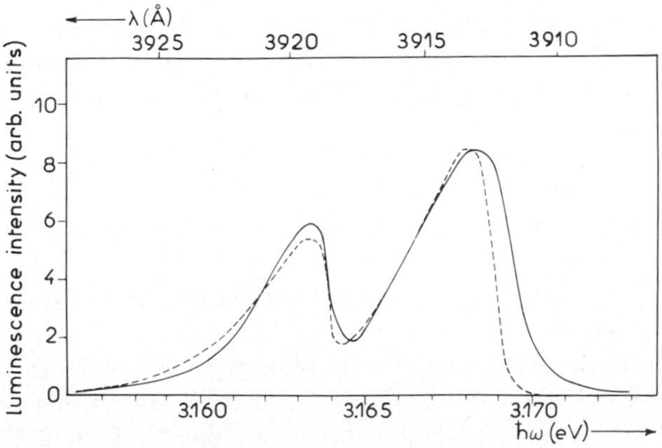

Fig. 15. Biexciton luminescence spectrum of CuCl at 1.8 K and the corresponding theoretical curve in dotted line, for an excitation in the band to band absorption.

samples were excited by the two-times frequency doubled light of a mode-locked Yag: Nd^{3+} laser. The biexciton emission was analyzed through an optical sampling gate (Kerr cell) by an optical multichannel analyzer. The building-up of a quasi-equilibrium of biexcitons was observed. A radiative lifetime of the biexcitons of the order of one nanosecond was deduced from the solutions of the kinetic equations of the different populations of quasiparticles.

4.3.2. *Directly created biexcitons*

When biexcitons are created directly by the simultaneous absorption of two photons of the same dye laser, the biexciton emission has very different characteristics (Grun et al. 1975, Nagasawa et al. 1975, Lévy et al. 1976, Hanamura and Haug 1977, Henneberger et al. 1977). As shown in figs. 12 and 16, two narrow emission lines, called N_T and N_L, appear at spectral positions corresponding to the high energy edges of the broad bands M_T and M_L. At high excitation intensities, the broad bands are observed too. The narrow lines disappear when the temperature increases. As is shown in fig. 12, the N_T and N_L lines stay at fixed spectral positions when the dye laser is tuned at the vicinity of the Γ_1 biexciton resonance. It is therefore easy to distinguish them from the hyper-Raman emission lines R_L and R_T.

These narrow emission lines have been explained by Nagasawa et al. (1975) by the existence of a Bose–Einstein condensation of biexcitons as predicted by Hanamura and Haug (1977). We have proposed a different explanation (Lévy et al. 1976): A "cold" gas of biexcitons of small wave vectors is created by two-photon absorption. It may be studied by an analysis of the polarization of its emission and by its time evolution as compared to those of the hyper-Raman scattering.

The appearance of broad emission bands M_T and M_L, even with this type of excitation, can be related to the creation of "hot" biexcitons of large wave vectors, created probably from the excitons left during the radiative recombination of cold biexcitons. Their dynamics correspond to the case discussed above.

4.4. *Polarization properties of hyper-Raman scattering and luminescence*

4.4.1. *Theoretical polarization of the light emitted by the radiative decay of virtual biexcitons*

The hyper-Raman scattering (Henneberger et al. 1977, Hönerlage et al. 1978) corresponding to the creation of virtual biexcitons and its simultaneous decay takes place in the scattering plane Q_xQ_y, as defined in fig. 17. The Q_y-axis is taken along the wave vector $\boldsymbol{q}_l$ of incident polaritons. The vector $\boldsymbol{q}$ of the observed polaritons is in this plane. It makes a scattering

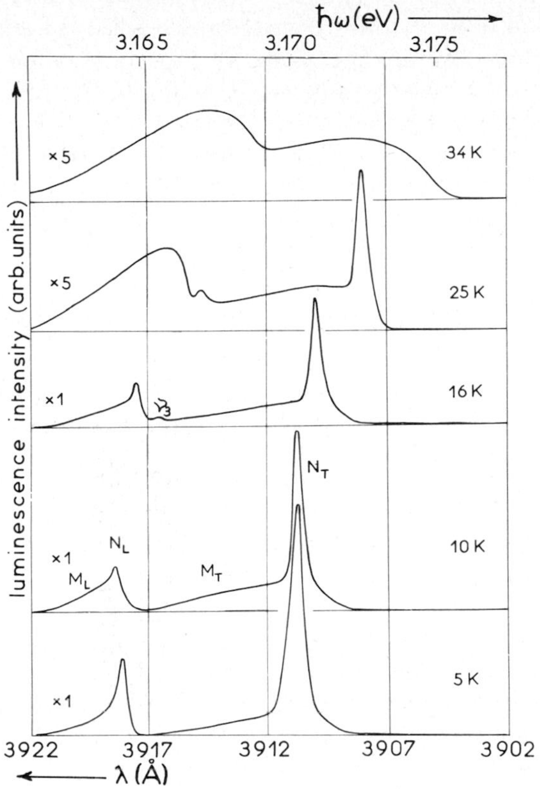

Fig. 16. Biexciton luminescence spectra of CuCl at different temperatures for an excitation by resonant two-photon absorption.

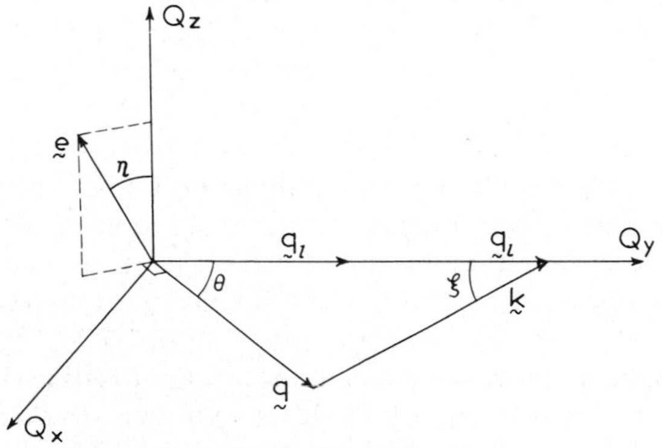

Fig. 17. Scattering configuration for the hyper-Raman process.

angle θ with q_l. Let us call ξ the angle between the wave vector k of the polariton left behind in the crystal and q. The polarization vector e of the observed polaritons makes an angle η with the Q_z-axis, chosen to be normal to the scattering plane.

The probabilities that a really or virtually created biexciton with a momentum $2q_l$ and an energy $\hbar\omega_l$ decays into a lower polariton of wave vector q and another lower polariton or a longitudinal exciton of momentum k are given by the following expressions:

$$W_T(q) = B \sum_k |1 - \sin^2\eta \sin^2(\theta + \xi)| \delta(k + q - 2q_l)\delta(E_l(k) + E_l(q) - 2\hbar\omega_l), \tag{44}$$

$$W_L(q) = B \sum_k \sin^2\eta \sin^2(\theta + \xi)\delta(k + q - 2q_l)\delta(E_L(k) + E_l(q) - 2\hbar\omega_l), \tag{45}$$

where B is a constant that does not depend on scattering angles.

Since in all our experiments $|q| \simeq |q_l|$, ξ is expressed as a function of θ by (Hönerlage et al. 1978):

$$\xi = \arc\sin \frac{\sin\theta}{\sqrt{5-4\cos\theta}}, \tag{46}$$

the function $f(\theta) = \sin^2(\theta + \xi)$ is equal to zero in a pure forward and backward configuration ($\theta = 0°$, $180°$).

4.4.2. Polarization properties of the hyper-Raman diffusion and comparison with the biexciton luminescence

In a backward scattering configuration ($\alpha = 57°$, $\beta = -37°$) we have measured the intensities of the hyper-Raman lines R_T and R_L as functions of the polarization angle η for exciting photon energies (3.1866 and 3.1863 eV) different from the biexciton resonance (3.1860 eV). These emission lines are then well separated spectrally from the luminescence lines. The experimental values obtained have been plotted in fig. 18. They fit quite well with the solid curves representing the theoretical polarizations expected for a scattering angle θ computed self-consistently as explained above. In comparison, fig. 19 represents the variation of the luminescence line N_T as a function of η at the same exciting photon energies. No polarization of this line is observed. The luminescence N_L was too small to be recorded.

Similar measurements have been done at an exciting photon energy equal to half the biexciton energy (3.1860 eV). Hyper-Raman diffusion and luminescence are then not spectrally distinguishable. The $N_T + R_T$ emission line is not polarized, as shown in fig. 20. The $N_L + R_L$ emission line is

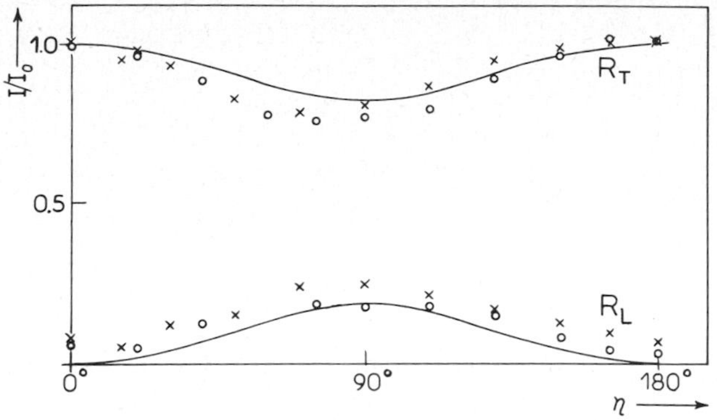

Fig. 18. R_L and R_T as a function of the polarization angle η for different exciting photon energies ((×) $\hbar\omega_l = 3.1866$ eV and (○) 3.1863 eV). The solid line represents the theoretical variation.

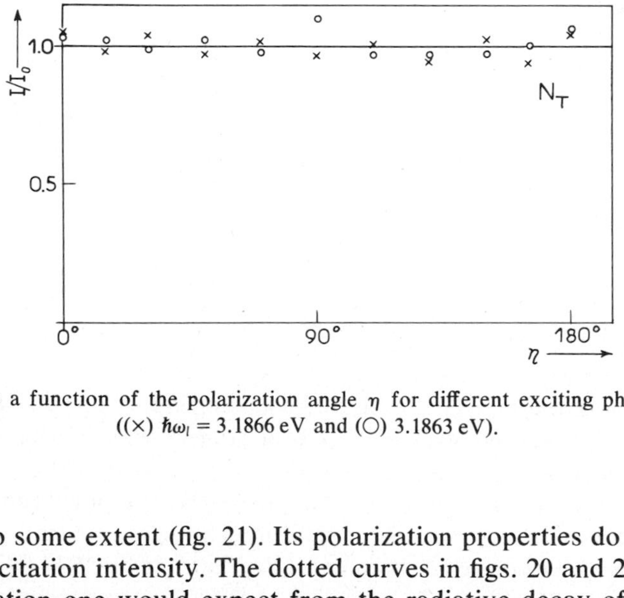

Fig. 19. N_T as a function of the polarization angle η for different exciting photon energies ((×) $\hbar\omega_l = 3.1866$ eV and (○) 3.1863 eV).

polarized to some extent (fig. 21). Its polarization properties do not change with the excitation intensity. The dotted curves in figs. 20 and 21 represent the polarization one would expect from the radiative decay of biexcitons having kept their creation wave vector $2\boldsymbol{q}_l$. This could be the case in a pure hyper-Raman scattering process or in a Bose-condensed biexciton system at wave vector $\boldsymbol{K} = 2\boldsymbol{q}_l$ (Nagasawa et al. 1975, Hanamura and Haug 1977).

In a forward scattering configuration ($\alpha = -33°$, $\beta = 57°$) we have also observed the expected polarization behaviour of the hyper-Raman emission. It was also interesting to notice that, when detuning the exciting laser from the biexciton resonance, we could not detect any luminescence lines

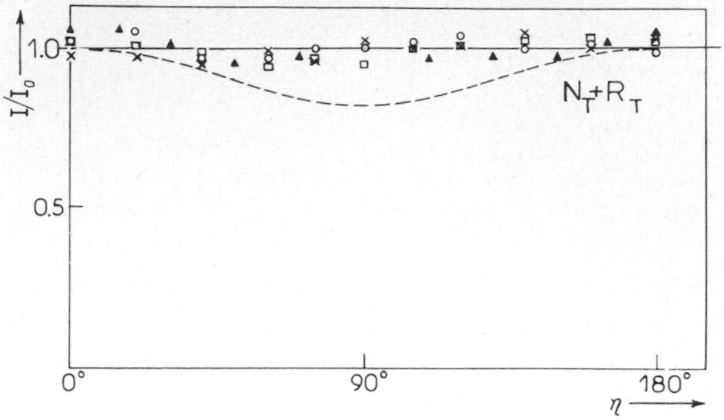

Fig. 20. $N_T + R_T$ as a function of η for an excitation at the biexciton resonance and for different excitation intensities; (□) 80%, (○) 40%, (△) 20%. The dashed line gives the polarization expected for biexcitons having kept their wave vector of creation $2\boldsymbol{q}_l$.

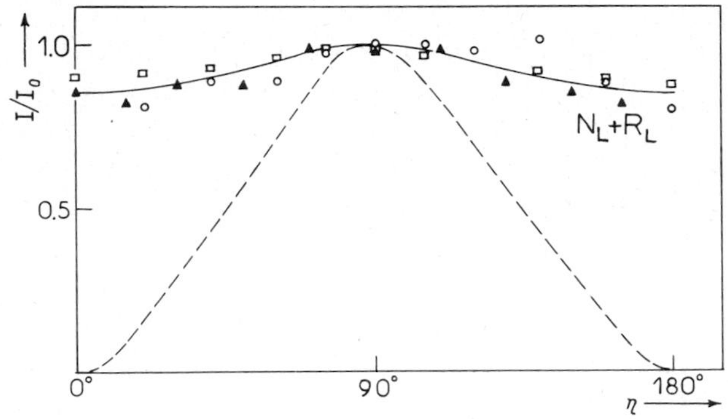

Fig. 21. $N_L + R_L$ as a function of η in the same conditions as given for fig. 20.

(N_T^+, N_T^-) corresponding to the radiative decay of biexcitons having kept the wave vector $2\boldsymbol{q}_l$ of their creation. Only a contribution to the N_T and N_L lines was observed.

This study has brought a good understanding of the polarization properties of HR emission and has also led to interesting conclusions about the relaxation of really created biexcitons (Hönerlage et al. 1978, Henneberger et al. 1979).

The luminescence of biexcitons is slightly polarized as seen for the $N_L + R_L$ emission. The fact that the similar polarization of the $N_T + R_T$ emission is not observed is simply related to the lack of precision of this measurement. No applicable luminescence lines are observed in the forward scattering configuration at the spectral positions corresponding to definite momenta of the biexcitons.

From these considerations, we can first conclude that most of the biexcitons have lost the wave vector of their creation. This can be due to elastic collisions between themselves or with acoustic phonons. Only a small part of the biexcitons decay before having lost their momentum of creation.

This biexciton distribution is also observed in the study of the time evolution of the emission from crystals excited at the biexciton resonance by a tunable dye laser pumped by a mode-locked Yag: Nd^{3+} laser. As seen in fig. 22, an emission corresponding to a hyper-Raman process follows instantaneously the laser pulse. The luminescence of relaxed biexcitons has a much longer decay time (Segawa et al. 1978, 1979).

5. *Influence of valence band degeneracy and crystal field effects*

5.1. Valence band degeneracy

Zincblende-type CuBr has the same point group symmetry as CuCl, but due to a different spin–orbit coupling, the uppermost valence band is fourfold degenerate having a Γ_8 symmetry. The exciton ground state, belonging to the series Z_{12}, is thus eight-fold degenerate. Its symmetry is

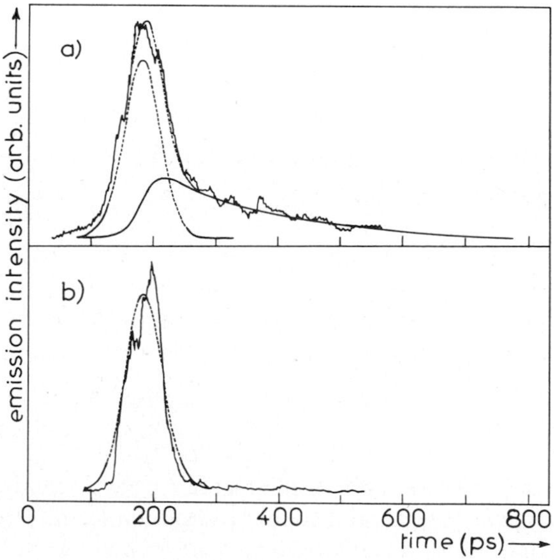

Fig. 22. (a) Time variation of the hyper-Raman diffusion and of the biexciton luminescence. (b) Time variation of the exciting laser pulse. (From Segawa et al. 1978, 1979).

given by (Koster et al. 1963):

$$\Gamma_1 \otimes \Gamma_6 \otimes \Gamma_8 = \Gamma_3 \oplus \Gamma_4 \oplus \Gamma_5. \tag{47}$$

The exciton state with a Γ_5 symmetry is dipole active.

The one-particle Hamiltonian $\kappa H_{\mathrm{HF}}(h)\kappa^+$ (Luttinger 1956) and the exchange interaction $W(\mathrm{e}, \mathrm{h})$ of eq. (2) have a different structure than for non-degenerate bands. Therefore, relative and center of mass motions cannot be separated in general. Only if the center of mass wave vector is equal to zero, the relative motion can be studied separately and the eigenvalue problem for the exciton is given by a set of coupled equations (Dresselhaus 1956, McLean and Loudon 1960, Baldereschi and Lipari 1971, Rössler 1979) yielding the exciton energies E^n_{ex} and wave functions $|\phi^n_{\mathrm{ex}}(\boldsymbol{Q})\rangle$. Concerning the exciton dispersion we are interested in, there exist different theoretical approaches (Kane 1975, Cho 1976, Altarelli and Lipari 1977). According to Cho (1976) we construct the effective exciton Hamiltonian in eq. (15) from group theoretical considerations by expanding the Hamiltonian into a power series of the total wave vector $\boldsymbol{Q}$.

In this invariant expansion, the Hamiltonian is constructed from angular momentum operators of hole states ($\mathcal{J}_{\mathrm{h}} = \frac{3}{2}$; $\mathbb{1}_{\mathrm{h}}$ is the four-dimensional unity matrix) and of electron states ($\sigma_{\mathrm{e}} = \frac{1}{2}$, $\mathbb{1}_{\mathrm{e}}$ is the two-dimensional unity matrix) and products thereof. The matrix representation of $\mathcal{J}$ operators is given in the standard basis for the holes defined as:

$$\begin{aligned}
\phi_1 &= -(x + \mathrm{i}y)\alpha/\sqrt{2}, \\
\phi_2 &= [2z\alpha - (x + \mathrm{i}y)\beta]/\sqrt{6}, \\
\phi_3 &= [2z\beta + (x - \mathrm{i}y)\alpha]/\sqrt{6}, \\
\phi_4 &= (x - \mathrm{i}y)\beta/\sqrt{2}.
\end{aligned} \tag{48}$$

They are obtained by Kramers' conjugation from the electron states:

$$\kappa\phi_1 = -v_{-3/2}, \qquad \kappa\phi_2 = v_{-1/2}, \qquad \kappa\phi_3 = -v_{1/2}, \qquad \kappa\phi_4 = v_{3/2},$$

where α and β are the spin up and down states, respectively; v_i are the electron valence states as used by Koster et al. (1963). Then, we obtain the following Hamiltonian (Cho 1976, Suga et al. 1976, Bivas et al. 1979) acting in the eight-fold product space of electron and hole as:

$$\begin{aligned}
H_{\mathrm{ex}} = {} & \Delta_0 \mathbb{1}_{\mathrm{h}} \otimes \mathbb{1}_{\mathrm{e}} + \Delta_1 \mathcal{J}_{\mathrm{h}} \cdot \sigma_{\mathrm{e}} + \Delta_2(\sigma_x \otimes \mathcal{J}^3_x + \sigma_y \otimes \mathcal{J}^3_y + \sigma_z \otimes \mathcal{J}^3_z) \\
& + C_Q[Q_x\{\mathcal{J}_x, \mathcal{J}^2_y - \mathcal{J}^2_z\} + \text{c.p.}] \otimes \mathbb{1}_{\mathrm{e}} + \text{higher orders in } Q
\end{aligned} \tag{49}$$

where $\{A, B\} = \frac{1}{2}(A\,B + B\,A)$ and c.p. means cyclic permutation. C_Q represents the Q-linear interaction. The Q dependence of the exchange interaction (Hönerlage et al. 1980) (terms that involve $\boldsymbol{Q}$, σ and $\mathcal{J}$) has been neglected as well as higher orders in the $\boldsymbol{Q}$ series expansion.

For $Q = 0$; Δ_0, Δ_1, and Δ_2 determine the energy of the exciton states, including all exchange interactions. In this case, the matrix representing eq. (49) may be diagonalized by symmetry adapted exciton wave functions $|\psi^i_{ex}\rangle$ (Cho 1976). Generally, the anisotropic exchange interaction Δ_2, which splits the states with Γ_4 and Γ_3 symmetry, is rather small and is therefore neglected.

For finite $\boldsymbol{Q}$ values, the Hamiltonian of eq. (49) has to be diagonalized numerically, yielding the exciton dispersion $E_\mu(\boldsymbol{Q})$ and the eigenfunctions $|\phi^\mu_{ex}(\boldsymbol{Q})\rangle = \Sigma_i A_{i\mu}(\boldsymbol{Q})|\psi^i_{ex}\rangle$. This approach is necessary, since the dispersive term couples the different exciton states and lifts the degeneracy present at the Γ point. This treatment corresponds to a perturbation theory for degenerate states. Due to the mixing of the different exciton states at $\boldsymbol{Q} \neq 0$, the $\mathcal{J}$th exciton state obtains some oscillator strength from the dipole-active Γ_5-exciton state. We then extend the Hopfield polariton model and we obtain the polariton dispersion $\omega(Q)$ by solving:

$$n^2(Q, \omega) = \frac{Q^2c^2}{\omega^2} = \epsilon_b + \sum_\mu \frac{\epsilon_b A^2_{5\mu}[E^2_L(0) - E^2_T(0)]}{E^2_\mu(Q) - \hbar^2\omega^2}, \tag{50}$$

in analogy to eq. (11).

Concerning the biexciton ground state, we construct the symmetry adapted biexciton wave functions as discussed in sect. 2. We obtain for the total symmetry of the biexciton ground state at $\boldsymbol{Q} = 0$ (Hanamura 1970, 1975, Comte 1974):

$$\Gamma_1 \otimes (\Gamma_6 \otimes \Gamma_6)^- \otimes (\Gamma_8 \otimes \Gamma_8)^- = \Gamma_1 \oplus \Gamma_3 \oplus \Gamma_5, \tag{51}$$

where the correct symmetry adapted wave functions may be obtained from Koster et al. (1963) in the electron–hole representation.

In order to discuss more easily the different optical transition, it is interesting to transform the biexciton wave function resulting from eq. (51) into the exciton basis (Cho 1976). Then, the selection rules for the optical transitions between exciton and biexciton states for the two-photon absorption, the hyper-Raman scattering and biexciton luminescence may be easily established (Hönerlage et al. 1980) for different directions of the wave vectors involved.

5.2. Two-photon absorption

The energies and symmetries of the biexciton ground states in CuBr have been studied by simultaneous absorption of two photons via resonant exciton intermediate state (Bivas et al. 1978, Vu Duy Phach and Lévy 1979).

The experimental set-up used is similar to the one described in sect. 3.1. Cleaved platelets of CuBr are cooled down to liquid helium temperature. Their thickness is about 300 μm and their parallel surfaces have the indices

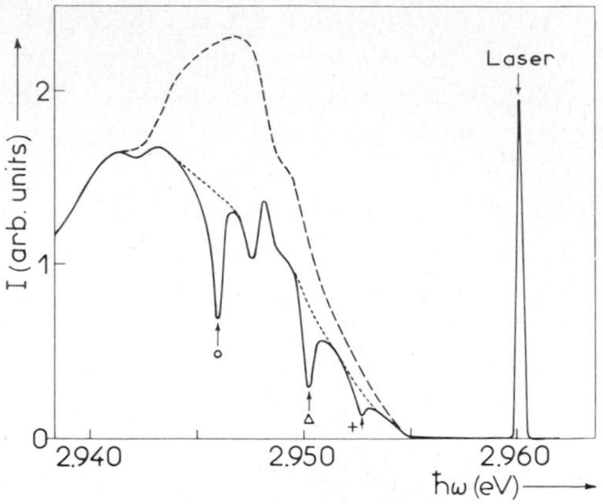

Fig. 23. Spectra of the continuum transmitted through a crystal of CuBr at liquid helium temperature. Dashed curve (---) without laser excitation. Solid curve (——) with laser excitation. The two-photon absorption dips, marked by arrows, are separated from the induced absorption band by the dotted curve.

[110]. Two antiparallel beams – a spectrally narrow and intense laser beam and a spectrally broad but weak superradiant laser beam – propagate inside the crystal almost along the [110] axis. They are set to coincide in time and space on the sample. In fig. 23, typical transmission spectra of the broad laser emission ($\boldsymbol{e}_c \| [\bar{1}10]$) through a CuBr platelet are shown when the sample is simultaneously excited by the intense laser ($\hbar\omega_l = 2.9601$ eV) ($\boldsymbol{e}_l \| [\bar{1}10]$) and when this excitation is removed. Three sharp absorption dips, marked by arrows, are observed under an induced absorption band. By tuning the laser photon energy, these dips shift in the spectrum. It is therefore possible to distinguish them from the induced absorption or from impurity absorption lines. We can thus deduce precisely the energies of the biexciton levels from the sum of the energies of the absorption dips $\hbar\omega_c$ and of the photons of the corresponding exciting laser $\hbar\omega_l$ (fig. 24).

In order to determine the symmetry of these three levels, we have measured the areas Y of the absorption dips, which are proportional to the two-photon transition probabilities (eq. (24)), as functions of the angle η between the polarization vectors $\boldsymbol{e}_l$ and $\boldsymbol{e}_c$ of the two beams, for definite direction of $\boldsymbol{e}_l$ with respect to the crystal axes.

We have obtained the following results:

For $\boldsymbol{e}_l \| [001]$, the low and high energy dips show a variation as $\cos^2 \eta$, while the intermediate dip varies as $\sin^2 \eta$ as shown in fig. 25(a).

For $\boldsymbol{e}_l \| [\bar{1}10]$, the low and high energy dips show again the same variation as $\cos^2 \eta$, while the intermediate dip remains constant, as shown in fig. 25(b).

The geometrical factors $G(\Gamma_i)$ of the transition probability for the ab-

5.3. Hyper-Raman scattering

Biexcitons may be excited resonantly by two-photon absorption into their different energy levels. Their subsequent radiative decay from their lowest Γ_1 ground state to different exciton states gives rise to luminescence lines (Grun et al. 1976, Nagasawa et al. 1976).

A hyper-Raman diffusion, where biexciton states act as almost resonant intermediate states and the different exciton and polariton states as final states, has also been observed (Bivas et al. 1979, Hönerlage et al. 1980).

As discussed above, the dispersion of multicomponent excitonic polaritons could be given theoretically and the selection rules for the hyper-Raman scattering process could be established. With this knowledge, the observed HR-emission lines could be interpreted.

We shall only consider, as an example, the results obtained for the [110] direction of wave vectors. Figure 26 shows typical emission spectra of a cleaved crystal surface [110] cooled down to liquid helium temperature. It is excited along its [110] direction by photons with a polarization vector $\boldsymbol{e}_l \| [\bar{1}10]$ and an energy $\hbar\omega_l \simeq 2.9531$ eV, close to half the Γ_1 biexciton energy.

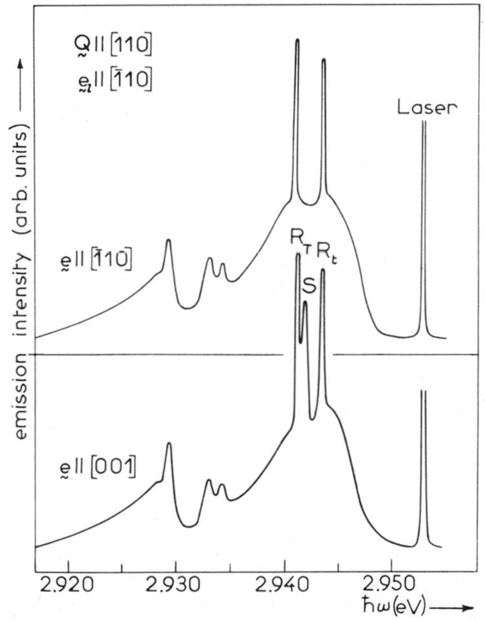

Fig. 26. Hyper-Raman spectra of CuBr at 4.2 K in a backward configuration for a cleaved surface [110]. The angles of incidence and of observation $\alpha = 23°$ and $\beta = 2°$ and the exciting photon energy $\hbar\omega_l = 2.9531$ eV are fixed.

This emission has been recorded in a backward configuration, the wave vectors of the emitted polaritons being almost parallel to the $[\bar{1}10]$ direction. Different HR lines R_T, R_L, R_t and S have been observed for two orientations of polarization of the emission, $\boldsymbol{e}\|[\bar{1}10]$ and $\boldsymbol{e}\|[001]$.

In fig. 27 are plotted the energetic positions of the HR lines obtained when the exciting photon energy $\hbar\omega_l$ is tuned.

For the [001] polarization and a backward scattering configuration, the HR line R_T corresponds to a transition to a lower polariton state A_1 (see fig. 28) observed as a photon, and to a polariton state A_3, the main contribution of which is that of a transverse Γ_5 exciton. The HR line R_t corresponds to a transition where both final states belong to the A_1 polariton branch, this last transition being allowed by the mixing of the exciton states at finite $\boldsymbol{Q}$. The line S corresponds to a stimulated HR emission.

The full lines in fig. 27 are computed in a self-consistent way by using the

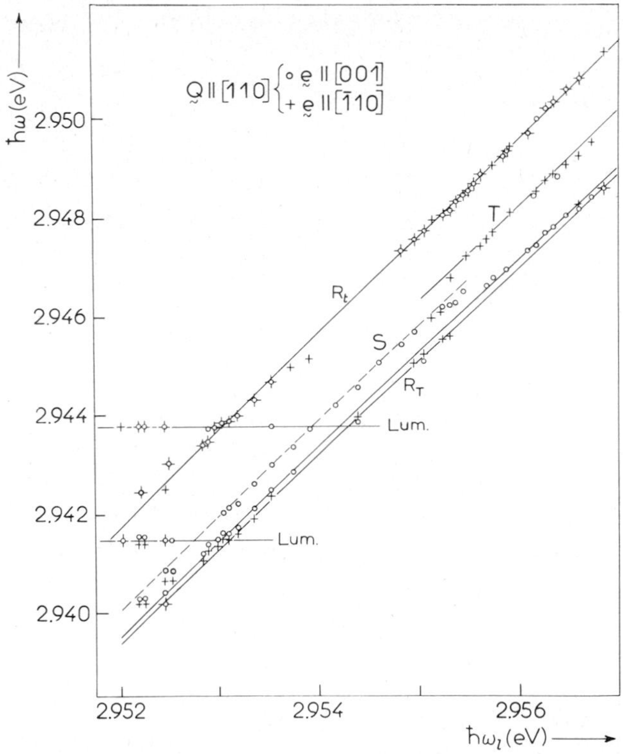

Fig. 27. Positions of hyper-Raman lines for different energies of the exciting photons and fixed angular configurations $\alpha = 23°$, $\beta = 2°$. The polarizations of the hyper-Raman lines are: $\boldsymbol{e}\|[001]$ for (○) and $\boldsymbol{e}\|[\bar{1}10]$ for (+). All wave vectors are almost parallel to the [110] direction.

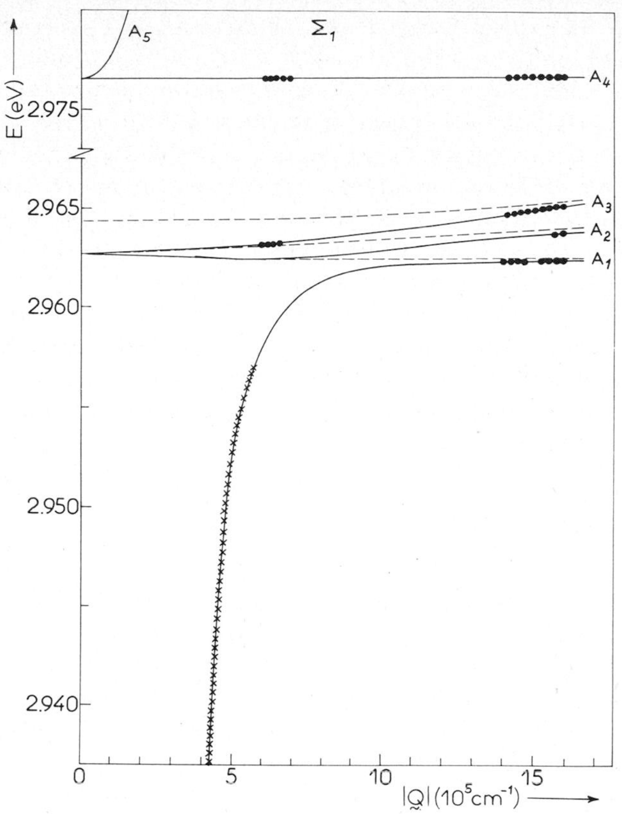

Fig. 28. Exciton (---) and polariton (——) dispersion for $\boldsymbol{Q}\|[110]$ and polarization $\boldsymbol{e}\|[001]$. The parameters used and the semi-experimental points (●) and (×) are discussed in the text.

polariton dispersion and the index of refraction, as explained before. The different parameters of the polariton dispersion have been adjusted in order to have a good fit with the backward and forward scattering in the [110] direction and $\boldsymbol{e}\|[001]$. Other measurements were then also explained (Bivas et al. 1979):

$$
\begin{aligned}
&E_{ex}(\Gamma_3) = E(\Gamma_4) = \Delta_0 + \tfrac{3}{2}\Delta_1 = 2.9633 \pm 2 \times 10^{-4}\ \text{eV}, \\
&E_{ex}(\Gamma_{5T}) = \Delta_0 - \tfrac{5}{2}\Delta_1 = 2.9646 \pm 2 \times 10^{-4}\ \text{eV}, \\
&E_{ex}(\Gamma_{5L}) = \Delta_0 - \tfrac{5}{2}\Delta_1 + \Delta_{LT} = 2.9766 \pm 10^{-4}\ \text{eV}, \qquad (53) \\
&C_Q = (69 \pm 4)10^{-11}\ \text{eV cm}, \\
&\epsilon_b = 5.4 \pm 0.2.
\end{aligned}
$$

In order to determine the exciton mass parameters, we compared the R_L

emission line in forward and backward scattering configurations. The observed energy difference, equal to 0.08 ± 0.02 meV, indicates a very weak dispersion of the longitudinal exciton, corresponding to a total exciton mass of $(11 \pm 3)\, m_0$. No anisotropy of the mass was observed.

In fig. 28, we have plotted the measured energies of the HR lines as functions of the wave vectors obtained from our calculation. These results are represented by full circles. The energies of the final polariton states not observed directly have been derived from the energies of the HR lines and the energies of exciting photons using the law of energy conservation. These results are represented by crosses. The fit of these experimental points with the polariton dispersion curve computed from the deduced parameters is quite good. In conclusion, a spectroscopy in momentum and energy space of a complicated polariton–exciton dispersion can also be achieved.

5.4. *Wurtzite structure, using CdS as an example*

In uniaxial crystal with wurtzite structure (C_{6v} point group symmetry) the uppermost valence bands are not degenerate. Due to a crystal field effect, in CdS, the splitting between the different bands is large compared with the longitudinal–transverse splitting of the dipole-active exciton states, so that the single-oscillator model for the excitonic polariton can be applied (Hopfield 1958, Pekar 1958). Uniaxiality leads, however, to a new physical feature: the mixed mode polariton (Benoit à la Guillaume et al. 1970). In a crystal direction not perpendicular to the crystallographic c-axis, a propagating exciton may have neither pure longitudinal nor transverse character.

The consequence of this effect on the biexciton luminescence has been studied in detail experimentally and theoretically (Shionoya et al. 1973, Planel and Benoit à la Guillaume 1977). The polarization properties (Planel and Benoit à la Guillaume 1977) of the biexciton luminescence confirm the theoretically predicted symmetry of the biexciton ground state (Goede 1977) to be Γ_1 (Hanamura 1970, 1975, Forney et al. 1974), and a lineshape analysis allows to determine the biexciton energy $E_B = 5.0994$ eV and the temperature of the gas.

As discussed above, the difficulty of the emission spectroscopy lies in the fact that it is very hard to distinguish between biexciton emission and non-intrinsic processes as acoustic wings of bound excitons or bound-exciton electron scattering processes (Voigt and Mauersberger 1973, Hvam 1974, Hönerlage and Rössler 1976, Klingshirn et al. 1976) from a lineshape analysis alone. A good agreement with the results from luminescence measurements was obtained in a spectroscopy where the energetic position of a reabsorption dip in a luminescence spectra was studied as a function of the excitation frequency (Schrey et al. 1979c). Similar to the two-photon absorption spectroscopy, presented in sect. 3, the intrinsic nature of the final state is thus shown and the biexciton energy determined to be 5.0983 eV.

This value is also very close to the one obtained from excitation spectroscopy of hyper-Raman scattering (Itoh et al. 1976, Nozue et al. 1978) if one corrects the energies given therein for the refractive index of air in the wavelength to energy conversion. In addition, mixed mode polaritons have been directly observed by hyper-Raman scattering via biexcitons (Schrey et al. 1979a) in a backward scattering configuration, when the crystallographic c-axis was lying inside the scattering plane (s.p.). In the case of $c \perp$ (s.p.) the excitonic polariton dispersion was obtained (Schrey et al. 1979b) from the analysis of the hyper-Raman scattering data in terms of the Hopfield model (1958). In this study, the temperature variation of the exciton states and of the L–T splitting have been obtained. A further method to distinguish intrinsic from non-intrinsic processes is the piezo-emission study of luminescence (Segawa and Namba 1973, 1975a, 1975b, Goede et al. 1975, 1976) and application of a magnetic field (Evangelisti et al. 1976) which have been performed in the recent years.

Coherent scattering experiments, as have been performed in CuCl (Maruani et al. 1978, Chemla et al. 1979), have also been done in CdS (Chemla and Maruani 1979).

6. Conclusion

The first proof of the existence of a bound state of two electrons and two holes, the biexciton, has been given by the observation of luminescence bands due to their radiative decay. Then, in the last few years, the development of tunable dye lasers has made possible detailed spectroscopy of these new quasiparticles: Two-photon absorption, induced absorption and luminescence of biexcitons have determined the main properties of biexcitons. Hyper-Raman scattering via biexciton levels has been used to study exciton and polariton dispersion curves.

The importance of biexcitons for the understanding of non-linear polarization properties of semiconductors has recently been shown. In addition, the development of tunable ultra-short light pulses has triggered a new field of research: the study of the time evolution of the properties of these quasiparticles. Both aspects have been only slightly raised in this paper. This research will certainly expand in the future.

We have considered essentially the copper halides in this paper. But many interesting results have been obtained with different materials such as II–VI compounds (only CdS has been mentioned here), indirect bandgap semiconductors (Nakahara and Kobayashi 1976, Pelant et al. 1976, 1977, Baba and Masumi 1977), layered compounds (Kurtze et al. 1979), and so on, which are not discussed here.

Acknowledgements

We would like to thank Dr. A. Bivas for a critical reading of the manuscript.

References

Adamowski, J., S. Bednarek and M. Suffczynski, 1971, Solid State Commun, **9**, 2037.
Adamowski, J., S. Bednarek and M. Suffczinski, 1972, Phil. Mag. **26**, 143.
Akimoto, O., 1973, J. Phys. Soc. Japan, **29**, 973.
Akimoto, O. and E. Hanamura, 1972, Solid State Commun. **10**, 253; 1972, J. Phys. Soc. Japan, **33**, 1537.
Altarelli, M. and N.O. Lipari, 1977, Phys. Rev. **B15**, 4898.
Baba, T. and T. Masumi, 1977, Il Nuovo Cimento, **39B**, 609.
Baldereschi, A. and N.O. Lipari, 1971, Phys. Rev. **B3**, 439. For a recent review, see: Rössler U., 1979, Festkörperprobleme XIX, J. Treusch, ed (Braunschweig FRG, Vieweg) p. 77.
Bassani, F. and M. Rovere, 1976, Solid State Commun. **19**, 887.
Bassani, F., J.J. Forney and A. Quattropani, 1974, Phys. Status Solidi b **65**, 591.
Bechstedt, F. and F. Henneberger, 1977, Phys. Status Solidi b **81**, 211.
Bednarek, S., J. Adamowski and M. Suffczinski, 1978, J. Phys. **C11**, 4515.
Benoit à la Guillaume, C., A. Bonnot and D.M. Debever, 1970, Phys. Rev. Lett. **24**, 1235.
Bivas, A., Vu Duy Phach, B. Hönerlage and J.B. Grun, 1977, Phys. Status Solidi b **84**, 235.
Bivas, A., Vu Duy Phach, R. Lévy and J.B. Grun, 1978, Proc. of the 14th Int. Conf. on Semiconductor Physics, B.L.H. Wilson, ed. (Page Brothers, Norwich) p. 497.
Bivas, A., Vu Duy Phach, B. Hönerlage, U. Rössler and J.B. Grun, 1979, Phys. Rev. **B20** 3442.
Brinkman, W.F., T.M. Rice and B. Bell, 1973, Phys. Rev. **B8**, 1570.
Chemla, D.S. and A. Maruani, 1979, private communication.
Chemla, D.S., A. Maruani and E. Batifol, 1979, Phys. Rev. Lett. **42**, 1075.
Cho, K., 1973, Opt. Comm. **8**, 412.
Cho, K., 1976, Phys. Rev. **B14**, 4463.
Comte, C., 1974, Opt. Commun. **14**, 79.
Doni, E., R. Girlanda and G. Pastori-Parravicini, 1975, Solid State Commun. **17**, 189.
Dresselhaus, G., 1956, Phys. Chem. Solids, **1**, 82.
Ekardt, W. and H.J. Brocksch, 1979, Phys. Status Solidi b **93**, 203.
Ekardt, W. and M.I. Sheboul, 1976a, Phys. Status Solidi b **73**, 475; **76**, K 89.
Ekardt, W. and M.I. Sheboul, 1976b, Phys. Status Solidi b **74**, 523.
Ekardt, W. and M.I. Sheboul, 1977, Phys. Status Solidi b **80**, 51.
Evangelisti, F., M. Capizzi, M. de Cescenti and A. Frova, 1976, Solid State Commun. **18**, 795.
Fano, U., 1961, Phys. Rev. **124**, 1866.
Fano, U. and J.W. Cooper, 1965, Phys. Rev. **137**, A1364.
Forney, J.J., A. Quattropani and F. Bassani, 1974, Il Nuovo Cimento, **22**, 153.
Fröhlich, D., E. Mohler and P. Wiesner, 1971, Phys. Rev. Lett. **26**, 554.

Gale, G.M. and A. Mysyrowicz, 1974, Phys. Rev. Lett. **32**, 727.
Goede, O., 1977, Phys. Status Solidi b **80**, 309.
Goede, O., M. Blaschke and E. Hasse, 1975, Phys. Status Solidi b **70**, K41.
Goede, O., M. Blaschke and K.H. Klohs, 1976, Phys. Status Solidi b **76**, 267.
Gogolin, A.A., 1974, Sov. Phys. Solid State, **15**, 1824.
Gogolin, A.A. and E.I. Rashba, 1973, Zh. Eksper. Teor. Fiz. Pisma, **17**, 478.
Goto, T., H. Souma and M. Ueta, 1970, J. Lumin. **1/2**, 231.
Grun, J.B., R. Lévy, E. Ostertag, Vu Duy Phach and H. Port, 1975, Proc. of Oji Seminar on the Physics of highly excited States in Solids, M. Ueta and Y. Nishina ed. (Springer-Verlag, Berlin, Heidelberg, New York) p. 49.
Grun, J.B., C. Comte, R. Lévy and E. Ostertag, 1976, J. Lumin. **12/13**, 581.
Hanamura, E., 1970, J. Phys. Soc. Japan, **29**, 50; 1975, J. Phys. Soc. Japan, **39**, 1506; **39**, 1516. For a review on biexciton binding energy calculations, see "Optical Properties of Solids. New Developments", 1976, B. Seraphin, ed. (North-Holland, Amsterdam, New York) p. 81.
Hanamura, E., 1973, Solid State Commun. **12**, 951.
Hanamura, E. and H. Haug, 1977, Physics Reports **33C**, 211 and references therein.
Handel, P.H., 1973, Phys. Rev. **B7**, 5183.
Hänsch, T.W., 1972, Appl. Opt. **11**, 895.
Hassan, A.R., 1978, Phys. Status Solidi b **87**, 31.
Henneberger, F. and J. Voigt, 1976, Phys. Status Solidi b **76**, 313.
Henneberger, F., K. Henneberger and J. Voigt, 1977, Phys. Status Solidi b **83**, 439.
Henneberger, F., J. Voigt and E. Gutsche, 1979, J. Lumin. **20**, 221.
Hönerlage, B. and U. Rössler, 1976, J. Lumin. **12/13**, 593.
Hönerlage, B., C. Klingshirn and J.B. Grun, 1976, Phys. Status Solidi b **78**, 599.
Hönerlage, B., Vu Duy Phach, A. Bivas and E. Ostertag, 1977, Phys. Status Solidi b **83**, K101.
Hönerlage, B., Vu Duy Phach and J.B. Grun, 1978a, Phys. Status Solidi b **88**, 545.
Hönerlage, B., A. Bivas and Vu Duy Phach, 1978b, Phys. Rev. Lett. **41**, 49.
Hönerlage, B., U. Rössler, Vu Duy Phach, A. Bivas and J.B. Grun, 1980, Phys. Rev. **B22**, 797.
Hopfield, J.J., 1958, Phys. Rev. **112**, 1555.
Huang, W.T., 1973, Phys. Status Solidi b **60**, 309.
Hvam, J.M., 1974, Phys. Status Solidi b **63**, 511.
Itoh, T. and T. Suzuki, 1978, J. Phys. Soc. Japan, **45**, 1939.
Itoh, T., Y. Nozue and M. Ueta, 1976, J. Phys. Soc. Japan, **40**, 1791.
Itoh, T., T. Suzuki and M. Ueta, 1977, J. Phys. Soc. Japan, **42**, 1069.
Kane, E.O., 1975, Phys. Rev. **B11**, 3850.
Kittel, C., 1967, Quantum Theory of Solids (Wiley, New York) ch. 15.
Klingshirn, C., 1979, Habilitationschrift, Karlsruhe FRG, unpublished.
Klingshirn, C., R. Lévy, J.B. Grun and B. Hönerlage, 1976, Solid State Commun. **20**, 413.
Koster, K.F., J.O. Dimmock, R.G. Wheeler and H. Statz, 1963, Properties of the Thirty-Two Point Groups (M.I.T. Press, Cambridge, Massachusetts) p. 88–101.
Kurtze, G., C. Klingshirn, B. Hönerlage, E. Tomzig and H. Scholz, 1979, J. Lumin. **20**, 151.
Lampert, M.A., 1958, Phys. Rev. Lett. **1**, 450.
Lévy, R., C. Klingshirn, E. Ostertag, Vu Duy Phach and J.B. Grun, 1976, Phys. Status Solidi b **77**, 381.
Lévy, R., B. Hönerlage and J.B. Grun, 1979a, Solid State Commun. **29**, 103.
Lévy, R., B. Hönerlage and J.B. Grun, 1979b, Phys. Rev. **B19**, 2326.
Luttinger, J.M., 1956, Phys. Rev. **102**, 1030.
Maruani, A., J.L. Oudar, E. Batifol and D.S. Chemla, 1978, Phys. Rev. Lett. **39**, 1372.
McLean, T.P. and R. Loudon, 1960, J. Phys. Chem. Solids, **13**, 1.
Moskalenko, S.A., 1958, Optics and Spectroscopy, **5**, 147.
Mysyrowicz, A., J.B. Grun, R. Lévy, A. Bivas and S. Nikitine, 1968, Phys. Lett. **26A**, 615.

Nagasawa, N., N. Nakata, Y. Doi and M. Ueta, 1975, J. Phys. Soc. Japan, **38**, 593; **39**, 987.
Nagasawa, N., T. Mita and M. Ueta, 1976a, J. Phys. Soc. Japan, **41**, 929.
Nagasawa, N., S. Koizumi, T. Mita and M. Ueta, 1976b, J. Lumin. **12/13**, 587.
Nakahara, J. and K. Kobayashi, 1976, J. Phys. Soc. Japan, **40**, 189.
Nikitine, S., A. Mysyrowicz and J.B. Grun, 1968, Helv. Phys. Acta, **41**, 1058.
Nozue, Y., T. Itoh and M. Ueta, 1978, Phys. Soc. Japan, **44**, 1305.
Ojima, M., T. Kushida, Y. Tanaka and S. Shionoya, 1978, J. Phys. Soc. Japan, **44**, 1294.
Ostertag, E., 1977, Thesis, Strasbourg, France (unpublished). 1977, Phys. Status Solidi b **84**, 673.
Ostertag, E. and J.B. Grun, 1977, Phys. Status Solidi b **82**, 335.
Ostertag, E., R. Lévy and J.B. Grun, 1975, Phys. Status Solidi b **69**, 629.
Pekar, S.I., 1958, Sov. Phys., JETP, **6**, 785.
Pelant, I., A. Mysyrowicz and C. Benoit à la Guillaume, 1976, Phys. Rev. Lett. **37**, 1708; 1977, Il Nuovo Cimento, **39B**, 655.
Planel, R. and C. Benoit à la Guillaume, 1977, Phys. Rev. **B15**, 1192.
Quattropani, A. and J.J. Forney, 1977, Il Nuovo Cimento, **39B**, 569 and literature therein.
Saito, H., A. Kurowa, S. Kuribayashi, Y. Aogaki and S. Shionoya, 1976, J. Lumin. **12/13**, 575.
Schrey, H., V.G. Lyssenko and C. Klingshirn, 1979a, Solid State Commun. **31**, 299.
Schrey, H., V.G. Lyssenko, C. Klingshirn and B. Hönerlage, 1979, Phys. Rev. **B20**, 5267.
Schrey, H., V.G. Lyssenko and C. Klingshirn, 1979, Solid State Commun. **31**, 299.
Segawa, Y. and S. Namba, 1973, Solid State Commun. **14**, 779; 1975(a), Solid State Commun. **17**, 489; 1975(b), Proc. of Oji Seminar on Physics of Highly Excited States in Solids, M. Ueta and Y. Nishina, eds., (Springer-Verlag, Berlin, Heidelberg, New York) p. 98.
Segawa, Y., Y. Aoyagi, O. Nakagawa, K. Azuma and S. Namba, 1978, Solid State Commun. **27**, 785; 1979, J. Lumin. **18/19**, 262.
Sharma, R.R., 1968, Phys. Rev. **170**, 770.
Sheboul, M.I. and W. Ekardt, 1976, Phys. Status Solidi b **73**, 165; **73**, 475.
Shionoya, S., H. Saito, E. Hanamura and O. Akimoto, 1973, Solid State Commun. **13**, 223.
Suga, S. and T. Koda, 1974, Phys. Status Solidi b **61**, 291.
Suga, S., K. Cho and M. Bettini, 1976, Phys. Rev. **B13**, 943.
Svorec, R.W. and L.L. Chase, 1976, Solid State Commun. **20**, 353.
Voigt, J. and G. Mauersberger, 1973, Phys. Status Solidi b **60**, 679.
Vu Duy Phach and R. Lévy, 1979, Solid State Commun. **29**, 247.
Vu Duy Phach, A. Bivas, B. Hönerlage and J.B. Grun, 1977, Phys. Status Solidi b **84**, 731.
Vu Duy Phach, A. Bivas, B. Hönerlage and J.B. Grun, 1978, Phys. Status Solidi b **86**, 159.
Wehner, R.K., 1969, Solid State Commun. **7**, 457.

CHAPTER 12

Free and Self-Trapped Excitons in Alkali Halides: Spectra and Dynamics

CH.B. LUSHCHIK

Institute of Physics
Academy of Sciences of the Estonian SSR
Tartu 202400
USSR

Excitons
Edited by
E.I. Rashba and M.D. Sturge

Contents

1. *Introduction*

It is fifty years now since the spectra of thin alkali halide films were measured at room temperature by Hilsch and Pohl (1930) and at 20 K by Fesefeld (1930), and narrow bands were found at the long wavelength edge of the intrinsic absorption. After Frenkel (1931, 1936) and Wannier (1937) had introduced the idea of the existence of excitons of small and large radii in solids, these bands were interpreted by Mott and Gurney (1938) and Seitz (1940) as the creation spectra of mobile currentless electronic excitations, i.e. excitons.

The first manifestations of the mobility of such excitations were the excitation of the luminescence of Tl^+ and In^+ ions in alkali halide crystals (AHC) on ultraviolet (UV) irradiation in the long-wavelength region of intrinsic absorption bands, detected by Terenin and Klement (1935) and later by Lushchik et al. (1957, 1960) as well as the external photoelectric effects in the exciton absorption band in alkali halides with F centers on UV irradiation, detected by Apker and Taft (1951).

Both the experiments by Lushchik et al. (1960), which showed that optical creation of excitons generates neither separate electrons and holes nor the characteristic recombination luminescence, as well as those by Nakai and Teegarden (1961), which did not reveal any photoconductivity in the region of the long-wavelength intrinsic absorption band, demonstrated the currentlessness of excitons in AHC. It was shown (Denks 1966) that for optical creation of separate electrons and holes, an electric field drastically decreases the emission intensity of impurity ions, while it hardly affects the energy transfer to impurity centers after an optical creation of excitons.

Thus the concept developed, and was supported experimentally, that the long-wavelength intrinsic absorption bands in AHC correspond to the creation of currentless mobile electronic excitations, i.e. excitons. However, soon the situation became complicated.

The phenomenon of the self-trapping of electrons, holes and excitons in crystals, predicted by Landau (1933) and Frenkel (1936) was experimentally detected for AHC. In 1957 Castner and Känzig found self-trapped holes (STH) in AHC by the EPR method. It turned out that soon after a hole is created in the p^6 shell of a halogen ion it interacts with the neighboring halogen ion X^-. As a result there appears a quasimolecular ion X_2^-, immobile at low temperatures which occupies two anion sites and has an effective positive charge with respect to the lattice.

Kabler (1964) and Murray and Keller (1965) detected a characteristic luminescence due to recombination of holes with STH, which Ramamurti and Teegarden (1966) and Kink et al. (1967) showed also to arise from the direct optical creation of excitons. This luminescence was interpreted as that of self-trapped excitons (STE), which can be presented as a connected pair, the electron (e) + the self-trapped hole $-X_2^-e$.

It was a paradoxical situation that arose: in some experiments the excitons in AHC showed themselves as mobile electronic excitations, in others, as immobile local excitations. For the last ten years scientists have looked for a solution to this contradictory situation.

Lushchik (1968, 1970), Blume et al. (1969), Kink and Liidja (1969), Lushchik et al. (1973) showed that the coexistence of mobile and self-trapped excitons in AHC is possible due to the existence of a barrier between the states of free excitons (FE) and STE, which makes the self-trapping of excitons difficult at low temperatures. This possibility was theoretically predicted by Landau (1933) and Toyozawa (1961) for electrons, and for excitons by Rashba (1957, 1981) and by Sumi and Toyozawa (1971).

In the sixties another peculiarity of excitons in AHC was found. By measurements of the "creation spectra" (i.e. the excitation spectra for creation) of F centers, Lushchik et al. (1961, 1965) and Ilmas et al. (1965) showed that the optical creation of excitons and electron–hole pairs leads to the efficient creation of F centres. The coloring processes of AHC in a nuclear reactor turned out to be similar to those under UV irradiation. In this connection the decay of STE into F and H (interstitial halogen atom) centers was considered by Hersh (1966), Pooley (1966), Vitol (1966), Lushchik et al. (1967, 1969).

In the last ten years investigation of the spectra and dynamics of excitons in AHC has been concerned mainly with three basic properties of excitons in AHC: STE, the coexistence of FE and STE, and the decay of excitons into the Frenkel defects.

The aim of the present review is to report the basic experimental results on the spectra and dynamics of FE and STE in AHC obtained since the reviews on excitons in AHC by Knox (1963), Kink et al. (1969), Knox and Teegarden (1968) and Lushchik (1970) were published.

Until 1970 the papers on excitons in AHC totaled about 600, today this number has probably doubled. Therefore, in this brief report we shall refer to only about 10–15 percent of generalizing and original investigations. However, some preference will be given to the investigations done in the laboratories of the Institute of Physics of the Estonian Academy of Sciences, in which the author has been taking an active part during the last 25 years.

2. *Manifestations of excitons in absorption and reflection spectra*

It is in the absorption and reflection spectra that excitons manifest themselves most conspicuously. Measurements of the intrinsic absorption spectra of AHC thin films were started by Hilsch and Pohl (1930) and continued by Eby et al. (1959) at 77 K and Teegarden and Baldini (1967) at 10 K.

In fig. 1 an absorption spectrum of a thin film of KI measured by Eby et al. (1959) is presented as an example. The narrow long-wavelength band corresponds to the creation of excitons by UV radiation. Band-to-band transitions with the creation of separate electrons and holes start in KI at 100 K in the region of photon energy $h\nu > 6.4$ eV. In this region, according to Kink et al. (1967), the slow ($\tau > 1$s) recombination luminescence of STE at 3.31 eV is excited (see fig. 1). This luminescence is also excited in the

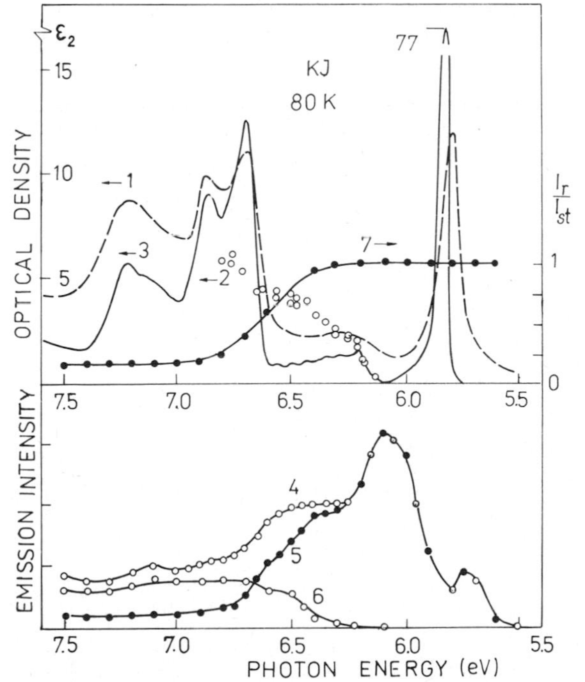

Fig. 1. Spectral and kinetic characteristics of KI at 80 K: (1) absorption spectrum of a thin film (Eby et al. 1959); (2) two-photon absorption spectrum of a crystal (Hopfield and Worlock 1965); (3) ϵ_2 spectrum (Tomiki et al. 1973); (4) excitation spectrum of the 3.31 eV luminescence for the steady state emission I_{st}; (5) excitation spectrum of the rapid component I_r with $\tau = 1.6\mu$s; (6) excitation spectrum of the slow component, with $\tau > 1$ s; (7) the ratio I_{st}/I_r (Kink et al. 1967).

region of the long-wavelength absorption bands (5.7–6.4 eV), however, not as a slow recombination emission but as a short-lived fluorescence (τ = 1.7 μs). The ratio of the intensities of the fast and steady-state components of luminescence is about 1 over the whole exciton region and decreases rapidly in the range of band-to-band transitions.

Figure 2 shows a schematic band structure of a face-centered AHC constructed in accordance with numerous theoretical calculations (e.g. the calculations by Kunz 1969), a review of which was given by Poole et al. (1975). The upper valence band is connected with a p-like hole in a p^6 shell of a halogen, and its maximum is at the point Γ of the Brillouin zone at $\boldsymbol{k} = 0$, where the valence band is split into two components due to the spin–orbit interaction. The lower component corresponds to a hole with angular momentum $j_+ = 1/2$ and the upper one to a hole with $j_+ = 3/2$. For KI the spin–orbit splitting is 1.1–1.2 eV. At the points X and L an additional splitting of the valence band into two branches occurs. The lowest conduction band corresponds to the s states of the conduction electron and has a minimum at Γ. The next conduction band corresponds to the d states of the conduction electron.

Excitons in KI and other AHC are formed at the point Γ, from an s electron with angular momentum $j_- = 1/2$ and a p hole with $j_+ = 3/2$ or $j_+ = 1/2$. In the first case the exciton has a total angular momentum $J = 1, 2,$

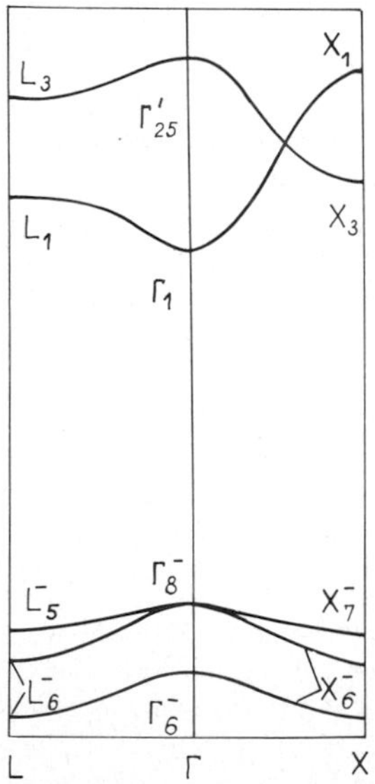

Fig. 2. A schematic band structure for AHC analogous to the band structures calculated by Kunz (1969).

in the second case $J = 0, 1$. In the dipole approximation only excitons with $J = 1$ are created optically. In the absorption spectrum these excitons give a doublet, the long-wavelength component of which is assigned to $\Gamma(3/2)$ excitons ($j_+ = 3/2$, $j_- = 1/2$, $J = 1$), and the short-wavelength one to $\Gamma(1/2)$ excitons ($j_+ = 1/2$, $j_- = 1/2$, $J = 1$).

In KI the absorption band at 5.8 eV corresponds to $\Gamma(3/2)$ excitons. Stevens and Menes (1968) studied the influence of the electric field on the absorption in the region of $\Gamma(3/2)$ excitons in KI. They managed to observe a quadratic Stark effect, by which the effective radius of an exciton with $n = 1$ (7.1 Å) was determined. For KI this radius is a little larger than the lattice constant. An exciton in KI cannot be considered as an excited state of a halogen ion, however, for an exciton with $n = 1$ the model of a large-radius exciton can also be applied approximately.

Fisher and Hilsch (1959) found a weak absorption band at 6.06 eV in the absorption spectra of the thin films of KI which was attributed to the excited state of a $\Gamma(3/2)$ exciton $n = 2$. Later Ramamurti and Teegarden (1966) and Hopfield and Worlock (1965) studied in detail the particularities of an exciton spectrum with $n = 2, 3$. In analogy with semiconductors a hydrogen-like exciton series with the energies,

$$E_n = E_g - R/n^2, \text{ where } R = \mu e^4/2\hbar^2\epsilon, \qquad \mu^{-1} = m_e^{-1} + m_h^{-1},$$

can be distinguished in AHC. Here R is a binding energy of an exciton state with $n = 1$, E_g is the bandwidth of forbidden energies, μ denotes the reduced mass, and ϵ is a dielectric constant. The excitons with $n = 2, 3$ were poorly observable in the absorption spectra of the thin films of AHC due to the imperfection of sublimated films.

When Baldini and Bosacchi (1970), Tomiki et al. (1973) and Miyata (1971) started an investigation of excitons in AHC from the reflection spectra of crystals, the task was instantly simplified. Figure 1 shows the spectrum of an imaginary part of a complex dielectric constant ϵ_2, calculated by the Kramers–Krönig relations from the reflection spectra of KI at 77 K by Tomiki et al. (1973). In this spectrum the absorption of excitons with $n = 2$ (6.2 eV) is better expressed than in the absorption spectrum of a KI film.

Hopfield and Worlock (1965) and Fröhlich and Staginnus (1967) investigated the two-photon absorption spectra of alkali iodides and bromides. The two-photon absorption spectrum of KI at 70 K is shown in fig. 1. 2p excitons arise in the region of 6.2 eV in accordance with the selection rules for two-photon processes and can be well interpreted as Wannier excitons.

Table 1 presents data by various authors on the widths of the forbidden (E_g) and valence (E_v) bands, the positions of the exciton band maxima E_e, and the values of spin–orbit splitting ΔE_{SO} for AHC.

Table 2 gives some parameters of excitons in NaI, KI, RbI, and NaBr.

Table 1
Parameters of excitons and phonons in AHC (eV). The widths of forbidden (E_g) and valence (E_v) bands. The peak positions of the exciton absorption (E_e) and emission of singlet (E_{Is}) and triplet (E_{It}) STE. (ΔE_{SO}) the spin–orbit splitting of the valence band. $h\omega_{LO}$ is the energy of a longitudinal optical phonon at the point Γ.

	E_g [a]	E_v [a]	E_e [b]	ΔE_{SO} [c]	E_{Is} [d]	E_{It} [d]	$h\omega_{LO}$ [e]
LiCl	9.4	4.5	8.65	—	—	—	—
NaCl	8.6	4.1	7.93	0.11	5.35	3.36	0.0322
KCl	8.7	2.7	7.76	0.13	—	2.31	0.0261
RbCl	8.2	2.3	7.52	0.14	—	2.23	0.0212
LiBr	7.6	4.8	7.21	—	5.3	3.95	—
NaBr	>7.0	3.8	6.72	0.48	—	4.62	0.0249
KBr	>7.3	2.6	6.76	0.44	4.42	2.28	0.0202
RbBr	>7.2	2.9	6.62	0.50	4.15	2.11	0.0160
NaI	5.86	3.3	5.62	1.12	—	4.20	0.0211
KI	6.3	2.8	5.84	1.20	4.15	3.31	0.0175
RbI	6.37	3.1	5.74	1.08	3.95	3.10	0.0132

[a] Poole et al. (1975).
[b] Teegarden and Baldini (1967). $T = 10$ K.
[c] Baldini and Bosacchi (1970). $T = 77$ K.
[d] Ikezawa and Kojima (1969). $T = 11$ K.
[e] Dolling (1974). $T = 77$ or 295 K.

The excitons with $n = 2$ in NaI have $r_2 = 33.3$ Å and can be interpreted to a good approximation as Wannier excitons with large radii.

Philipp and Ehrenreich (1963), Baldini and Bosacchi (1970), Miyata (1971) and Tomiki et al. (1973) investigated in detail the manifestations of exciton states in the spectra of reflection and optical constants (ϵ_1, ϵ_2, n, k, etc.). At low temperatures the contour of absorption bands is well des-

Table 2
Parameters of excitons in AHC.
The widths of the forbidden bands E_g(eV). The positions of the absorption maxima E_e(eV) and the binding energies R(eV) of excitons. The reduced masses μ (in the units of free electron mass). The radii (Å) of excitons with $n = 1$ (r_1) and $n = 2$ (r_2).

	E_g	E_e	R	μ	r_1	r_2
NaI [a]	5.86	5.58	0.28	0.175	8.31	33.2
KI [b]	6.31	5.85	0.36	0.21	7.1	28.4
RbI [b]	6.26	5.74	0.58	0.32	4.5	18.2
NaBr [a]	7.13	6.68	0.45	0.23	5.77	23.1

[a] Miyata (1971). $T = 78$ K.
[b] Ramamurti and Teegarden (1966). $T = 10$ K.

cribed by an asymmetric Lorentzian shape, which is characteristic of FE. In the row NaI, KI, RbI, NaBr, KBr, RbBr, and KCl the half-widths of these bands at 5 K increase from 5 to 30 meV. At low temperatures band broadening is mainly caused by the interaction with longitudinal acoustic vibrations.

For NaI and KI crystals (see, e.g. fig. 1) the half-width of the Γ band is several times smaller than for the absorption spectra of thin films. This is important, since the relatively large exciton bandwidth of films was for a long time an important argument against the existence of FE in AHC. This argument has now been shown to be invalid.

For Γ excitons with $n = 2$ in KI and RbI Baldini and Bosacchi (1970) recorded vibrational sidebands with a separation equal to the energy of longitudinal optical phonons.

Baldini and Bosacchi (1970) and Miyata (1971) found structure in the reflection bands, which correspond to Γ(3/2) excitons with $n = 1$. Narrow bands, the origin of which is difficult to understand within the framework of free exciton theory, are detected on the short-wavelength side of the exciton absorption band at a distance of 35–50 MeV from the maximum. Sumi (1975) suggested the possibility of an interference of FE and STE states.

The investigation of electronic excitations of crystals by analyzing reflection and absorption spectra contributed importantly to the understanding of energetic crystal spectra. The combination of theoretical calculation of band structures by the use of modern computers, with the employment of synchrotron radiation in experiments, has been especially fruitful. This combination was reviewed by Haensel (1969), Stephan and Robin (1974), Lushchik (1974) and Poole et al. (1975).

The optical constants characterize the electron excitations of crystals mainly at the moment of their creation. However, they do not carry full information about the subsequent behavior of the excitations: vibrational relaxation, migration and annihilation.

3. *Energy transfer by excitons*

The problem of energy transfer by excitons, first raised by Franck and Teller (1938), has been relatively less studied in AHC than in molecular crystals (Agranovich and Galanin 1978). Knox and Teegarden (1968) considered the energy transfer in AHC to be a result of dipole–dipole interaction of immobile STE with distant impurity centers. The time of exciton self-trapping was assumed to be 10^{-13} s.

The possibility of energy transfer by FE and by the jump diffusion of STE in AHC has been considered by Lushchik (1968, 1970), Kink and

Liidija (1969, 1970), Vasil'chenko et al. (1970, 1972) and Lushchik et al. (1973). However, so far the investigation of the excitation of impurity luminescence in AHC, either by coherent FE (wave packet) or by incoherent STE moving by jump diffusion, has been insufficient.

To elucidate the excitation of the luminescence of impurity centers in AHC by optically created excitons, the following processes were considered by Lushchik et al. (1973):

(1) Resonant energy transfer from thermalized or hot STE to distant impurity centers.

(2) Energy transfer by the jump diffusion of X^0e-type monohalogen STE.

(3) Energy transfer by the jump diffusion of X_2^-e-type dihalogen STE.

(4) Energy transfer by free excitons.

Resonant energy transfer from STE to the distant Eu^{2+} and Sn^{2+} centers in CaF_2–Eu, KI–Sn and KI–Eu was found by Kalder and Malysheva (1971), Kamejima et al. (1971) and Lushchik et al. (1973). For these crystals the emission spectra of STE overlap with the absorption spectra of impurity centers, as is characteristic of a strong dipole–dipole interaction and resonant energy transfer. In the crystals the decay time of STE luminescence decreases when doped by Eu^{2+} and Sn^{2+} ions. Both the energy transfer and emission are caused by thermalized STE.

Energy transfer to impurity centers due to the jump diffusion of dihalogen X_2^-e-type STE was observed for CsBr–In, and NaCl–Ag by Vasil'chenko et al. (1970, 1972) and Lushchik et al. (1972). The efficiency of this energy transfer mechanism increases rapidly when heated to the temperatures where STH X_2^- also become mobile. According to a Song's (1973) theory the activation energy of X_2^-e jump diffusion is slightly less than that for X_2^- molecules.

Vasil'chenko et al. (1970, 1972) showed that cooling of crystals to a temperature at which jump diffusion of X_2^-e-type STE is frozen out, does not cause the complete disappearance of the luminescence of impurity centers excited in the exciton absorption band. The low-temperature migration of excitons was interpreted to be a result of the migration of hot excitons before their axial relaxation.

It was shown by Lushchik et al. (1973) that in KI–Tl the excitation of Tl^+ luminescence by optically created excitons rapidly increases on cooling from 70 to 30 K, whereas the 3.31 eV emission of STE decreases simultaneously (see fig. 3). Note that after doping KI by thallium the lifetime of STE remains unchanged. Hayashi et al. (1975) detected a considerable increase in the quenching of STE emission in KI by previously created color centers (this process is characterized by the parameter *A* in fig. 3, curve 4). These experiments lead to a conclusion that before self-trapping, excitons can migrate and cause the excitation of luminescence centers.

A comparative investigation of the migration of nonrelaxed excitons in a

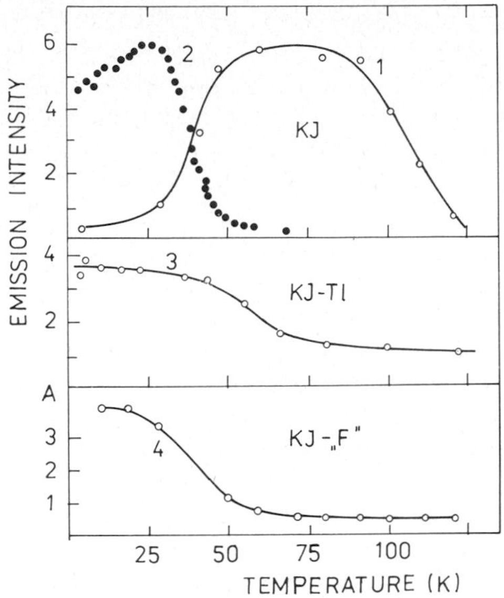

Fig. 3. Luminescence and energy transfer processes in KI: (1), (2), (3), temperature dependences of the emission intensities of self-trapped excitons at 3.31 eV (Lushchik et al. 1973), free excitons at 5.83 eV (Nishimura et al. 1979b), and Tl^+ centers (Lushchik et al. 1973); (4) temperature dependence of the parameter A characterizing the quenching degree of STE emission by previously created color centers (Hayashi et al. 1975).

homologous series of AHC at 80 K was carried out by Vasil'chenko et al. (1974). The results are presented in table 3. For each crystal the parameter $l\sigma$ is given, where l is the mean diffusion length of an exciton before self-trapping and σ the effective cross section for exciton interaction with the impurity center. The values of l given in table 3 change through the AHC series from $4000a$ in NaI to $2a$ in KCl (a = lattice constant), if we take $\sigma = a^2/4$ as an approximate estimate. In KCl an exciton becomes self-trapped almost immediately after its creation.

Table 3
The total migration path of excitons before the self-trapping (l in lattice constants) and the values $l\sigma$ (in 10^{-21} cm^{-3}) for AHC at various temperatures $T(K)$

	$l\sigma$	l	T		$l\sigma$	l	T
KCl–Tl	0.4	2	80	KI–Tl	65	235	80
NaCl–Ag	2	16	80	KI–Tl	560	2 000	5
KBr–I	30	70	80	KI–In	2600	10 000	5
NaBr–I	160	500	80	NaI–Tl	200	900	80
CsBr–In	45	235	80	NaI–Tl	900	4 000	5

Thus in certain AHC (NaI, KI) excitons are of a relatively large radius and high mobility even at 77 K, and at 4 K the mobility should increase further. On the other hand in KCl and RbCl crystals excitons have a high binding energy, a small radius and a very low mobility, and become self-trapped soon after their creation.

Kuusmann et al. (1975b) have compared the absorption spectra of FE in NaI and halogen ions in NaBr–I. There is a considerable narrowing of the absorption bands, typical of excitons in NaI. The rapid migration of excitons decreases the exciton–phonon interaction considerably. Until now no Jahn–Teller splitting of spectra, typical of impurity centers in AHC, has been observed for fast-migrating excitons with orbitally degenerate states. At the same time, the Jahn–Teller effect is pronounced for "bad" immobile cation excitons in cesium halides (Sakoda and Toyozawa 1973).

Of special interest is the migration mechanism of excitons before their axial relaxation in alkali iodides. At first, preference was given to the incoherent diffusion of monohalogen STE (Kink et al. 1969). Then, on the basis of numerous experiments, a hypothesis of the existence of FE coherent migration in alkali iodides was advanced (Kuusmann et al. 1976). It was shown by Hopfield and Thomas (1961) and Gross et al. (1961) that the displacement of exciton absorption band under reversal of an applied magnetic field should be regarded as the most convincing manifestation of coherent excitons in CdS. This effect is observed for FE in crystals with no centre of symmetry.

The existence of a longitudinal–transverse splitting is a clear manifestation of the migration of excitons with a quasimomentum. For semiconductors this effect is described by Gross (1976). In CdS $\Delta E_{LT} = 2$ meV. Baldini and Bosacchi (1970) determined ΔE_{LT} as the energy difference between the maxima in the spectra of ϵ_2 and $\mathrm{Im}(\epsilon^{-1})$. The first parameter corresponds to the creation of transverse excitons by light, the second one, to the creation of longitudinal excitons by electrons. The oscillator strength for excitons is greater in AHC than in CdS, and a value $\Delta E_{LT} \approx 50$ meV was obtained for chlorides and $\Delta E_{LT} = 100$ to 150 meV for bromides with more mobile excitons. Ejiri (1977) studied the reflection spectra of AHC by means of polarized light at various angles of incidence and also obtained $\Delta E_{LT} = 150$ meV. According to Miyata (1971) for NaI, $\Delta E_{LT} = 140$ meV. These data should be regarded as demonstrating the existence of free excitons.

4. *Emission of free excitons*

The luminescence of semiconductors in the long-wavelength edge of intrinsic absorption was first described by Kröger (1954). Further investigations showed that the edge emission of semiconductors is caused by the annihilation of free excitons and of various bound excitons, i.e. excitons, self-trapped near intrinsic and impurity defects.

Attempts to detect the edge luminescence of AHC have failed more than once (Blume et al. 1969, Kink and Liidja 1969). Kuusmann et al. (1975a) were the first to detect the edge luminescence of AHC. In fig. 4 cathodoluminescence of highly-purified NaI at 67 K is represented. Besides the well-known luminescence of X_2^-e-type STE (4.2 eV) in the region of 5.5 to 5.6 eV a distinct weak emission band peaking at 5.56 eV is visible whose half-width (70 meV) is determined by the slitwidth of the double monochromator. This emission was rapidly quenched by heating to 100 K, or by a short irradiation by an electron beam which creates radiation defects. Introduction of impurities in a concentration of 10^{16}–$10^{17}\,cm^{-3}$ also weakens the intensity of edge emission. A similar emission band at 5.75 eV was found for KI.

Kuusmann et al. (1975b) showed (see fig. 5) that the X-ray-induced edge emission of NaI increases rapidly when the temperature is lowered to 4.2 K. This process occurs in parallel with the decrease of the emission

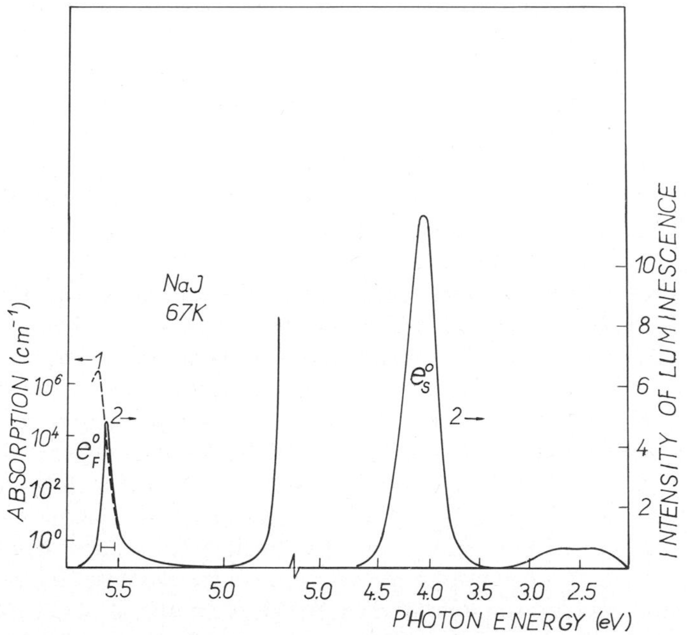

Fig. 4. The absorption spectrum (1) according to Miyata (1971) and cathodoluminescence spectrum (2) according to Kuusmann et al. (1975a) for NaI at 67 K.

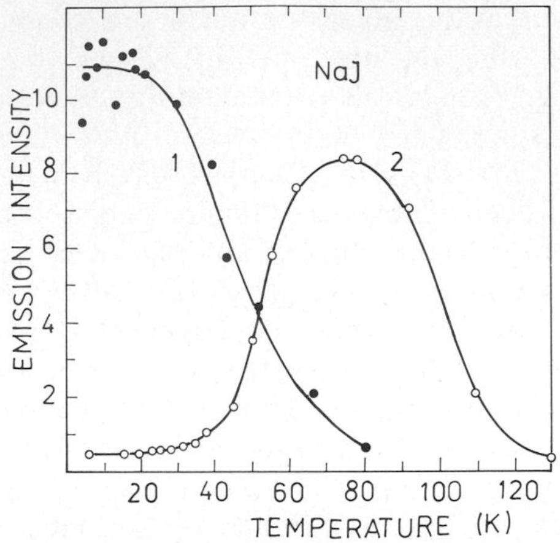

Fig. 5. Temperature dependences of the intensities of the 5.55 eV emission of free excitons (1) and 4.2 eV emission of self-trapped excitons (2) in NaI (Kuusmann et al. 1975b).

intensity of STE (4.2 eV) that was first observed by Fontana et al. (1968). The activation energies of both processes are equal to 17 meV. The phenomenon has been explained by the occurrence of a potential barrier for self-trapping of excitons.

Kuusmann et al. (1976) have reviewed Liidja's results on the decrease of STE emission for NaI, KI, RbI and NaBr excited in the long-wavelength region of intrinsic absorption. Correlation of these results with the features of edge luminescence and the data on the narrowness of exciton absorption bands as well as with the longitudinal-transverse splitting of exciton states lead to a conclusion about the coexistence of free and self-trapped excitons in AHC.

When studying the thermal ionization of Γ excitons in electric fields in NaBr and NaI, Denks et al. (1976) revealed that the Γ(3/2) exciton absorption band has a doublet structure. The intense high-energy component corresponds to excitons with $J = 1$, and the low-energy wing to the formation of metastable excitons with $J = 2$. An investigation of the edge photoluminescence of NaI and KI at 73 K disclosed an emission in the regions of 5.5 eV and 5.7 eV, respectively, which Lushchik and Soovik (1976) ascribed to the radiative decay of free Γ excitons with $J = 2$. Creation of Γ excitons with $J = 1$ by linearly polarized light gives a polarized emission of excitons with $J = 2$, whose polarization degree $p \approx 0.2$. The effect was preliminarily interpreted as an alignment of excitons in alkali halides. Obviously the basic experiments on the orientation and alignment of free excitons still lie ahead.

The experiments of Hayashi et al. (1977) on the X-ray and photon-induced edge luminescence of KI and RbI at 5 K and on its thermal quenching made possible the determination of the activation energies for exciton self-trapping.

The edge luminescence for NaI, KI, RbI and CsI (see table 4), illuminated in the exciton absorption bands at 5 K, was measured by Nishimura et al. (1977). Emission bands appeared to be very narrow and nearly in resonance with the intrinsic absorption. The positions and half-widths of the emission were practically independent of the frequency of the exciting light and the propagation direction of the emitted light. The quantum yield of the resonant emission did not exceed 10^{-3} in the best samples of NaI. The photoluminescence was quenched by heating to 70 K (see fig. 5.).

Later Nishimura et al. (1979a,b) showed that at 3 K the half-width of the resonant emission for KI and RbI is only 1.8 meV (for a band pass of 0.5 meV), which is three times less than the width of the exciton absorption band calculated by Miyata (1971) from the reflection spectrum. The emission band is of the Lorentzian shape typical of free excitons, but not of bound ones. As the temperature rises, the resonant emission of KI and RbI shifts towards the low-energy side in parallel with the exciton absorption band. In the region of 3–20 K the half-width of resonant emission increases approximately proportionally to the temperature, the increase being drastic at higher temperatures.

The half-width of the resonant emission of KI and RbI at 3 K (1.8 meV) is 10^2–10^3 times more than that predicted for FE in crystals by the theory of Segall and Mahan (1968) that ignored self-trapping and polariton effects. This discrepancy was explained by Nishimura et al. (1979a,b) by the exciton–photon interaction. If the broadening of resonant emission were due to the self-trapping of excitons by tunneling, then, according to the relation $\Delta E\tau = \hbar$, the emission bandwidth $\Delta E = 2$ meV would correspond to

Table 4
The peak positions of the absorption (E_e) and emission (E_l) bands of free excitons at 4 K. The activation energies for the self-trapping q.

	E_e (eV)	E_l (eV)	q (meV)		
NaI	5.615 [d]	5.606 [a]	17 [a]	20 [b]	
KI	5.845 [e]	5.826 [a]	33 [a]	18 [b]	18 [c]
RbI	5.74 [e]	5.724 [a]	17 [a]	14 [b]	11 [c]
NaBr	6.68 [d]			12 [b]	

[a] Nishimura et al. (1977).
[b] Kuusmann et al. (1976).
[c] Hayashi et al. (1977).
[d] Miyata (1971).
[e] Tomiki et al. (1973).

$\tau = 3 \times 10^{-13}$ s. The real tunneling time in alkali iodides is considerably greater (Kuusmann et al. 1976). The lifetime of the 5.83 eV resonant emission measured at 5 K for KI excited by pulses of synchrotron radiation was found to be $\tau < 5 \times 10^{-10}$ s (Nouailhat et al. 1979a).

It was pointed out by Fontana (1976) that a polariton-induced transparency effect can be observed in AHC. According to Fontana this phenomenon causes the intensity dips in the excitation spectra of luminescence in the region of the exciton band maximum where polaritons are created. The doublet structure of the resonant emission of FE in NaI was supposed to be related with the polariton splitting of energy states (Plekhanov and O'Konnel-Bronin 1978a). Guillot et al. (1977) and Nouailhat et al. (1978, 1979a,b) examined the resonant emission of KI irradiated by an electron beam at 5 K. Investigation of this emission as a function of the penetration depth of electrons shows that the resonant emission comes from crystal layers not thinner than 10^{-3} cm, so that the polariton effect must be taken into account (see e.g. Sumi (1976)). The creation of 10^{17} cm^{-3} color centers in the surface layer of a crystal greatly diminishes the resonant emission, whereas the luminescence of self-trapped excitons weakens by 10% only. The optical destruction of color centers increases the intensity of FE emission. A detailed examination of polariton effects has yet to be done.

It was shown by Gross (1976) and his co-workers that examination of phonon sidebands in the emission spectra of FE in semiconductors makes it possible to separate free and bound excitons and to distinguish hot FE. Plekhanov and O'Konnel-Bronin (1978a–d) and O'Konnel-Bronin et al. (1978) were the first to observe LO phonon sidebands in the spectrum of the edge luminescence of NaI, providing very important information about the dynamics of exciton processes. As follows from fig. 6, at 4.2 K the luminescence spectra of NaI excited by 5.75 eV photons reveal both resonant emission (at 5.605 eV) and phonon sidebands with an interval of 21 meV, which coincides with the energy of LO phonons. In fact for NaI at 28 K the contours of sidebands were asymmetrically broadened to the high-energy side. They reflect the Maxwellian velocity distribution of excitons within the exciton band. The latter is clearly seen by the contour of the 2LO phonon sideband at 28 K, presented in the insert of fig. 6. These experiments demonstrate quite vividly the presence of a quasimomentum of NaI excitons, i.e. prove the existence of free excitons.

At the same time the experiments carried out by O'Konnel-Bronin and Plekhanov (1979a,b) showed that excitons in NaI are not in thermal equilibrium with the lattice. This is proved, e.g. by an estimation of the effective temperature of the contour of 2LO phonon sidebands in NaI. At 28 K the effective temperature of the exciton gas is almost twice as much (48 K).

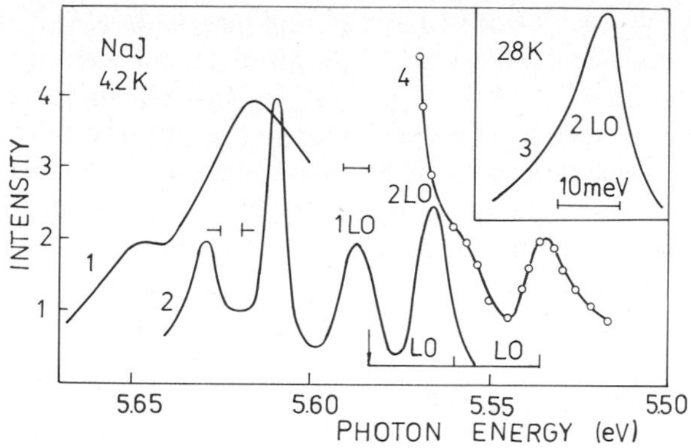

Fig. 6. Reflection (1) and photoluminescence spectra of NaI at 4.2 K (2) and 28 K (3) for excitation by 5.75 eV photons; (4) spectrum of resonance Raman scattering of 5.584 eV photons (marked by an arrow) in NaI at 4.2 K (O'Konnel-Bronin and Plekhanov, 1978, 1979a,b).

The lack of the thermal equilibrium in the exciton band due to slow vibrational relaxation through acoustic phonons is inferred from the oscillating structure in an excitation spectrum of FE emission detected in CuCl by Goto and Ueta (1967), and in CdS by Broude et al. (1958) and Gross (1976). An analogous effect was described for LiH and qualitatively for NaI by O'Konnel-Bronin and Plekhanov (1979b).

Nishimura et al. (1979a,b) investigated the phonon sidebands in the FE emission spectra of KI and RbI. In each crystal four LO phonon sidebands were observed, the intensity of which was tens of times less than that of a resonant emission. Analogously to NaI, the phonon sidebands have approximately the Maxwell contours indicating the presence of a quasimomentum.

Recently O'Konnel-Bronin and Plekhanov (1979b) observed for the first time the resonance Raman scattering of light on excitons in LiH and NaI. In fig. 6 the second-order scattering at an irradiation by 5.59 eV photons is shown for NaI. In accordance with theoretical considerations one can state that the scattering occurs with the participation of free excitons. The contribution of STE to this effect is still to be clarified. A theory of resonance Raman scattering for the systems with coexisting FE and STE was recently advanced by Hizhnyakov (1979b) and Sumi (1979).

On the basis of NaI as an example one may consider all three components of the secondary emission investigated theoretically by Hizhnyakov (1979a): the resonance Raman scattering occurring before phase relaxation, the hot luminescence arising after phase relaxation but in the course of energetic relaxation, and the thermalized ordinary luminescence observed after the phase and energetic relaxations.

In alkali halides the exciton bandwidth is considerably larger than the energy of phonons and hence the dispersion of phonons interacting with excitons is large (≈10 meV). The time of the phase relaxation, which depends on the dispersion of actual phonons, according to the estimations by Hizhnyakov and Sherman (1978), should be 10^{-12}, i.e. considerably less than that of energetic relaxation (10^{-10}–10^{-11} s). Consequently, for NaI it is possible to distinguish the resonance Raman scattering of light and hot luminescence of excitons.

5. Luminescence of self-trapped excitons

A broad-band intensive luminescence with a large Stokes shift relative to exciton absorption bands was detected for alkali iodides by Hahn and Rossel (1953), Teegarden (1957) and Morgenschtern (1959). This luminescence was excited only in the region of intrinsic absorption and was interpreted by Teegarden (1957) as an intrinsic luminescence.

Kink et al. (1967) showed that the intrinsic luminescence of KI (3.31 eV) at 80 K consists of two components: a long-lived recombination luminescence which is excited in the region of band-to-band transitions as a result of a recombination of holes with electrons released from traps, and a fast luminescence of the same spectral composition, which arises from the direct optical creation and a subsequent annihilation of excitons (see fig. 1). This luminescence is of maximum intensity in highly-purified crystals and is suppressed by the introduction of impurities.

Kabler (1964) and Murray and Keller (1965) studied the polarized intrinsic luminescence of face-centered AHC due to recombination of electrons released from the traps with self-trapped holes previously oriented by polarized light. These experiments showed that the luminescence can unambiguously be identified with that of X_2^-e-type STE. In face-centered crystals these centers are oriented along ⟨110⟩ directions. Ikezawa and Kojima (1969) investigated the excitation spectra and thermal quenching of STE luminescence.

An investigation of molecular STE luminescence in simple cubic CsBr crystals was carried out by Vasil'chenko et al. (1972). Figure 7 shows the spectral characteristics of CsBr. The low-energy bands in the absorption spectrum correspond to exciton creation. On irradiation in all these bands, independently of the energy of the exciting photons ($h\nu_e$) a broad luminescence band with the maximum at 3.5 eV appears which is shifted by more than 3 eV from the lowest-energy absorption band. This Stokes shift testifies to a post-absorption transition of the system to a new state, where ions are substantially displaced from the regular lattice sites. The same luminescence can be obtained from recombination of electrons with

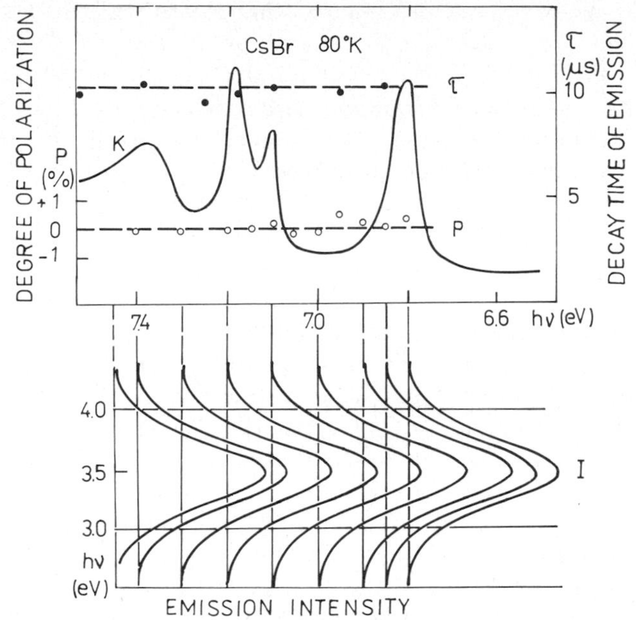

Fig. 7. Spectral and kinetic characteristics of CsBr at 80 K. Spectra of absorption k, emission *I*, emission decay time τ, and the degree of the emission polarization *p*. The dashed lines indicate the energy of exciting photons for each emission spectrum (Vasil'chenko et al. 1972).

STH. If the latter are previously oriented by a linearly polarized light, the STE luminescence is partially polarized. The degree of polarization is negative: $p \approx -0.15$. The emitters are oriented along ⟨100⟩ directions. Luminescence centers are X_2^-e molecules. The lifetime of these STE is about 10 μs and is practically independent of the frequency of the exciting radiation.

Only one characteristic of X_2^-e shows that it is not the ordinary impurity centers that are concerned. The luminescence of STE is completely unpolarized on excitation by linearly polarized radiation in the long-wavelength exciton absorption band. The absorber and the emitter are independent. The light creates a mobile exciton, the self-trapping of which leads to the formation of X_2^-e centers oriented along different ⟨100⟩ directions. The aggregate of such centers yields unpolarized luminescence.

A well-known series of investigations by Kabler and Liidja et al. (see reviews by Kabler et al. (1973), Kristoffel and Liidja (1974), Kuusmann et al. (1976)) is dedicated to the spectroscopy of molecular X_2^-e-type STE in AHC.

One of the fundamental properties of AHC is a tendency to form a molecular bond between an ionized halogen ion (X^0 atom) and one of the neighboring anions (X^-): $X^0 + X^- \rightarrow X_2^-$. The ground state of X_2^-,

$$(\sigma_g^2 \pi_u^4 \pi_g^4 \sigma_u, {}^2\Sigma_u^+),$$

is the bonding one. Such formations are self-trapped holes. The recombination of an electron with X_2^- creates an X_2^-e molecule, i.e., a molecular STE. The lowest state of such an exciton is

$$(\sigma_g np)^2(\pi_u np)^4(\pi_g np)^4(\sigma_u np)[\sigma_g(n+1)s]\,, \quad {}^{1,3}\Sigma_u^+.$$

The energy levels of free X_2^-e molecules ($D_{\infty h}$ symmetry) and for molecules in f.c.c. alkali halides (D_{2h} symmetry) are presented in fig. 8. Only odd excited states are given. On the left the levels of a monohalogen exciton (X^0e) are presented, from which the levels of X_2^-e are formed.

A metastable A_u level, from which the electron transitions in a zero magnetic field are forbidden, is the lowest level of molecular STE in alkali iodides. B_{2u} and B_{3u} levels, the distance between which is small (10^{-6} eV), are situated by 0.5–1.0 meV higher. The transitions from B_{2u} and B_{3u} to the ground state are allowed due to the spin–orbit interaction and the mixing of ${}^3\Sigma_u$ and ${}^1\Pi_u$ states.

In many AHC molecular STE yield two broad emission bands. The positions of their maxima are presented in table 1. Short-wavelength bands

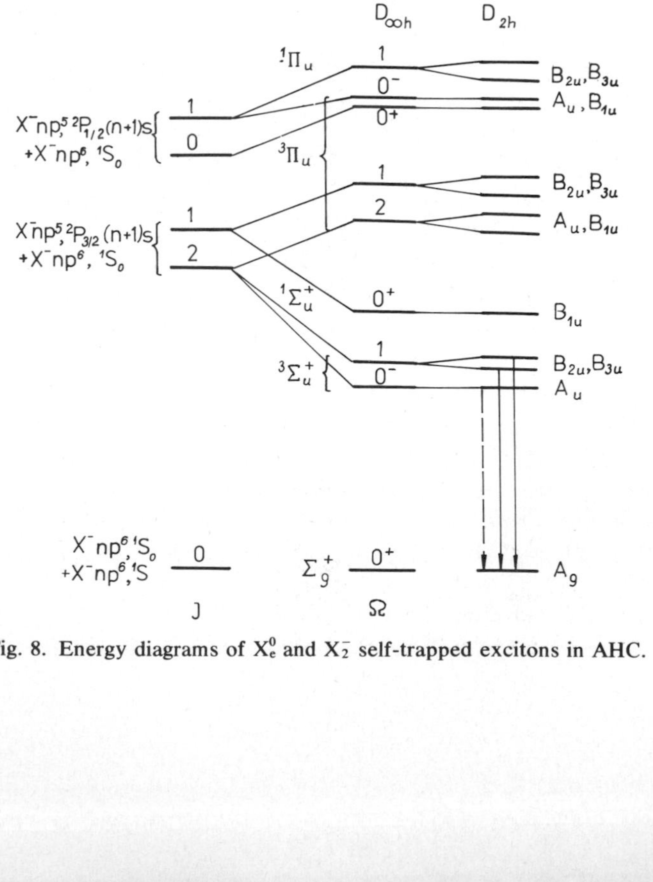

Fig. 8. Energy diagrams of X_e^0 and X_2^- self-trapped excitons in AHC.

(Blair et al. 1970) correspond to singlet-singlet $^1\Sigma_u \rightarrow {}^1\Sigma_g$ transitions and have the decay time of about 10^{-9} s. The decay time of long-wavelength bands is 10^{-6}–10^{-3} s (Kabler and Patterson 1967, Kink et al. 1967, 1971) and they correspond to triplet–singlet $^3\Sigma_u \rightarrow {}^1\Sigma_g$ transitions.

Kink and Liidja (1969), Kink et al. (1971), Kask et al. (1974) and Fishbach et al. (1973) investigated in detail the kinetics of the triplet luminescence of STE and showed that at low temperatures the luminescence consists of two components. In KI the decay time of one component is $\approx 10^{-6}$ s and does not depend on temperature, while the decay time of the other one increases by several orders of magnitude on cooling from 10 to 2 K. One of these two components corresponds to $B_{3u}, B_{2u} \rightarrow A_g$ transitions, while the second one comes from the metastable A_u level. The probability of nonradiative $A_u \rightsquigarrow B_{3u}, B_{2u}$ transitions increases rapidly with increasing temperature, bringing about a long-lived component of $B_{3u}, B_{2u} \rightarrow A_g$ luminescence. Liidja and Soovik (1978) found that the probability of the $A_u \rightarrow A_g$ forbidden transitions is 10^3–10^4 times less than that for $B_{2u} \rightarrow A_g$. The transition $A_u \rightarrow A_g$ becomes allowed due to either the interaction with nuclear magnetic moments or the asymmetric displacement of X_2^-e molecules with respect to two lattice sites.

The excitation spectrum of the triplet–singlet luminescence of KBr (2.1 eV), according to the results of Kink and Liidja (1970), spreads all over the exciton absorption region (see fig. 9). The excitation of singlet–singlet luminescence (4.2 eV) starts only in the region of $\Gamma(3/2)$ excitons with $n = 2$. This luminescence corresponds to the $^1\Sigma_u \rightarrow {}^1\Sigma_g$ transitions of the STE with an excited electronic component ($n = 2$). Yoshinary et al. (1978) observed the excitation of 4.2 eV luminescence by the laser excitation of previously created triplet excitons.

Figure 10 (Roose 1975) shows the excitation spectrum for luminescence of X_2^-e in KI (3.31 eV) at 80 K, measured over a broad spectral range. The luminescence is strongly excited in the region of direct exciton creation (5.7–6.0 eV) and at the energy of recombination of electrons with STH ($h\nu > 6.4$ eV). A rapid increase of luminescence efficiency occurs at $h\nu > 13$ eV. In this spectral region photons create hot photoelectrons, the energy of which is not sufficient to create secondary electron–hole pairs, while it suffices to create the excitons by electron impact. Ilmas et al. (1965) had previously described the creation of secondary excitons by hot photoelectrons in AHC in the course of an investigation of the excitation spectra of impurity luminescence. Beaumont et al. (1976) investigated the excitation spectra of the intrinsic luminescence of alkali chlorides and bromides at 5 K by means of synchrotron emission. The luminescence of molecular STE appeared to be unpolarized in crystals excited by linearly polarized synchrotron emission.

In AHC the model of molecular X_2^-e-type STE is well founded. Kabler et

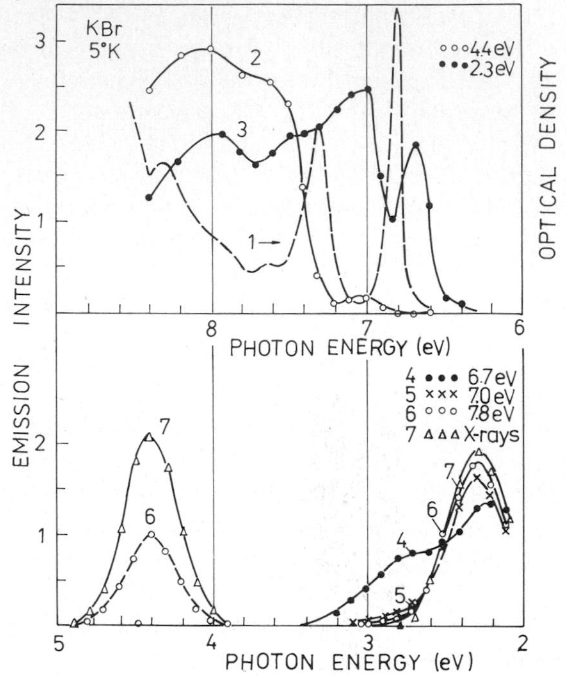

Fig. 9. Spectra of KBr at 5 K (Kink and Liidja 1970): (1) absorption spectrum; (3) and (2) excitation spectra of π luminescence (2.3 eV) and σ emission (4.4 eV) of self-trapped excitons; (4)–(7) luminescence spectra at an excitation by 6.7, 7.0, 7.8 eV photons and X-rays.

al. (1973) showed that in the transient absorption spectra of crystals containing STE bands which practically coincide with the absorption of STH (X_2^-) are distinctly seen.

Marrone et al. (1973) and Wasiela et al. (1973) studied the EPR of triplet STE in AHC, through the effect of a magnetic field on the luminescence of these excitons. The parameters of the X_2^- molecular ion included in X_2^-e are very close to those of the STH in AHC. Block et al. (1978) measured the double electron–nuclear resonance of X_2^-e in KCl. It appeared that the molecule Cl_2^- included in STE is slightly displaced with respect to the symmetric position between two anion vacancies.

Wood (1966) considered theoretically the possibility of an atomic X^0e-type STE luminescence with Stokes shift much less than for molecular X_2^-e-type STE. A search for such luminescence was carried out by Kuusmann and Lushchik (1976) and Lushchik and Soovik (1976). A band of 6.8 eV with a Stokes shift of ≈1 eV was detected in the cathodo-luminescence spectra of KCl and two luminescence bands, in the region from 5.7 to 6.4 eV in KBr, RbBr and CsBr. These bands can hypothetically be related to the lumines-

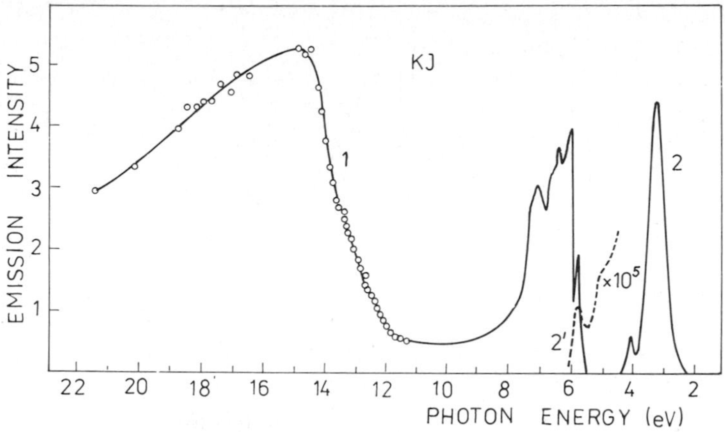

Fig. 10. Spectra of KI at 80 K (Lushchik and Kuusmann 1976). (1) excitation spectra of 3.31 eV emission of triplet self-trapped excitons; (2), (2)′ cathodoluminescence spectra.

cence of monohalogen STE. In KI and RbI analogous bands were recorded in the spectra of cathodo- and photoluminescence in the 5.0–5.5 eV region. A detailed investigation of the luminescence of atomic STE at LHeT in AHC is to be done in the future.

6. Coexistence of free and self-trapped excitons

The experimentally discovered activation barrier for the formation of molecular STE in alkali iodides was interpreted by Lushchik (1968) as the Landau (1933) barrier between the FE and STE states. Fontana et al. (1968) related the freezing of the molecular STE emission to the existence of an activation barrier between two localized states of an exciton: atomic and molecular.

The detection and detailed investigation of FE emission in alkali iodides showed that there exists a barrier between FE and STE states. With reference to excitons the barrier was examined theoretically by Rashba (1957) and by Sumi and Toyozawa (1971, 1973).

Let us consider a simplified model of an ionic crystal with coexisting FE and STE. Figure 11 shows the dependence of the energy of an excitonless crystal and of a crystal with FE and STE upon the deformation Q (the arrangement of heavy particles). The half-width of the exciton band is denoted by B, the exciton self-trapping energy by S. There is an activation barrier q between FE and STE states. For a three-dimensional AHC the local level splits off from the band only at a finite value of the deformation, the realization of which requires some energy. At small Q the parameters

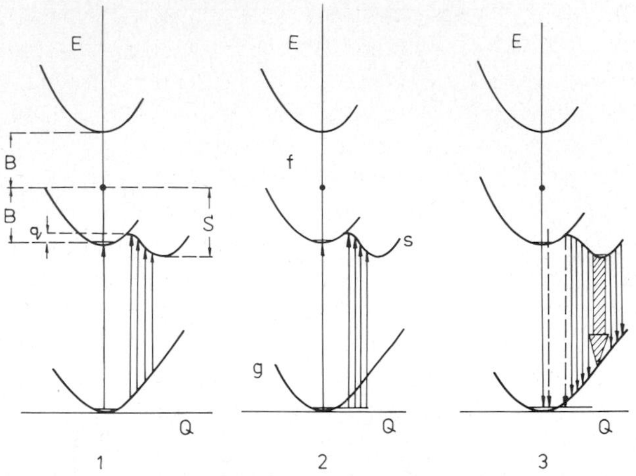

Fig. 11. Energy of a crystal without excitons (g), with free (f) and self-trapped (s) excitons, as functions of the configuration coordinates of the ions.

of FE in crystals do not differ from FE parameters at $Q = 0$. When a certain critical deformation is obtained, self-trapping of electronic excitation becomes possible, and as a result the electronic energy gained exceeds the deformation energy.

If $B \ll S$, only STE can exist in a crystal, while in the case of $B \gg S$, only FE exists. If $B \approx S$, coexistence of FE and STE is possible. For $B < S$ the STE is stable and the FE is metastable, and vice versa for $B > S$.

For $B < S$ two exciton self-trapping mechanisms are possible: thermal activation or tunneling. The theory of a tunnel self-trapping was elaborated by Iordanskii and Rashba (1977). For excitons with high kinetic energy above-barrier self-trapping is possible even at low temperatures. If $B < S$, the absorption spectrum of a crystal must contain, besides a narrow band corresponding to the creation of FE with an approximately zero momentum, a region of a continuous absorption, corresponding to the direct creation of STE.

At high temperatures, when $kT > h\omega_{\text{eff}}$ (ω_{eff} – phonon frequency effectively interacting with excitons), the continuous absorption is close to the long-wavelength side of the FE spectrum (see fig. 11). The "Urbach tail" for AHC was experimentally studied in detail by Mijata (1971), Tomiki et al. (1973), and theoretically by Sumi and Toyozawa (1971).

Hizhnyakov (1979b) analyzed the case of low temperatures. At $kT < h\omega_{\text{eff}}$ the continuous spectrum spreads not only over the long-wavelength wing of the FE spectrum but also over a small region of its short-wavelength side. In the reflection and absorption spectra of some AHC,

Baldini and Bosacchi (1970) found structure in the short-wavelength wing of the spectrum of $\Gamma(3/2)$ excitons, which may correspond to the direct creation of STE. Note, that this structure was observed in the AHC (NaI, KI, LiBr, NaBr) in which the presence of FE and STE is well established and, therefore, the activation barrier q exists.

In crystals with $B < S$ four types of emission are possible: emission of hot FE, thermalized FE, hot STE, and thermalized STE.

The emission of hot STE is the most difficult to distinguish. Being continuous it can easily be confused with an impurity emission. If the vibrational relaxation of STE is slow as compared with the oscillation period, hot luminescence is observable on the short- and long-wavelength wings of the ordinary luminescence of STE (see fig. 11, case 3). For NaI this case was described by Kuusmann et al. (1976). The detection of local vibrations for STH by Goovaerts and Schoemaker (1978) pointed out that the vibrational relaxation for molecular STE in AHC is undoubtedly slow. If the vibrational relaxation time of STE is comparable with the oscillation period, hot luminescence of STE should be observable on the short-wavelength side of STE emission. Tehver and Hizhnyakov (1975) showed that one can expect an increase of hot luminescence intensity in the region where the electronic excited state is in the vicinity of the activation barrier, as a result of the shortening of the relaxation time due to "hot migration" of excitons. Fast vibrational relaxation can be expected for atomic STE.

Kuusmann et al. (1976) and Kuusmann and Lushchik (1976) observed in a number of AHC a weak continuous cathodoluminescence, which ranges from the maxima of exciton absorption bands to the emission bands of molecular STE. Unlike the FE emission this continuous emission is not quenched even at 300 K. The quantum efficiency of this emission does not exceed 10^{-5}–10^{-6}. This emission is supposed to be as universal as the Urbach long-wavelength tail of the absorption and to correspond to the radiative annihilation of the hot states of STE.

Several attempts have been made to estimate theoretically the height of the activation barrier q for self-trapping. Rashba (1976, 1977), Toyozawa (1974) and Hizhnyakov and Sherman (1976) considered the self-trapping barrier for Frenkel excitons. Pekar et al. (1979) analyzed the case of the Wannier excitons interacting with optical polar vibrations (polarizable excitons). All these theories predict the existence of a barrier between FE and STE states, the height of which is significantly smaller than the self-trapping or the binding energies of FE.

In table 4 we summarized the values of the self-trapping barriers q, determined from the freezing of the emission of molecular STE and from the increase of FE emission at the cooling. For the excitons with $n = 1$ in NaI, KI, RbI and NaBr the experiments give the values $q = 10$–30 meV. For excitons with $n = 2$ in KI Toyozawa and Shinozuka (1979) determined

$q = 3$ meV. The general theoretical predictions agree well with the experiments.

Rashba (1957) and Toyozawa and Shinozuka (1979) predicted that in the one-dimensional case self-trapping of excitons does not require any activation energy. This case may possibly be realized for the excitons connected with linear dislocations in AHC. As is clear from experiments presence of dislocations makes the observations of FE in AHC difficult.

An investigation of excitons in semiconductors has shown that in real crystals with point defects, besides large-radius FE, bound excitons are formed, which are localized near defects. Delbecq et al. (1951) were the first to detect bound excitons in AHC. The broad absorption bands of bound excitons are shifted from the FE absorption band to the low-energy side by 0.5–1.0 eV. These absorption bands are connected with the electronic excitations arising near anion vacancies, F centers, and interstitial halogen ions. Some distinct properties of excitons bound near mercury-like ions in AHC are described, e.g. by Vasil'chenko and Lushchik (1969).

Under direct optical excitation, bound excitons in AHC are localized and interact strongly with phonons, which leads to a broadening of the absorption and emission bands and to a significant Stokes shift. No narrow resonance lines of the absorption and emission of bound excitons have been observed yet in AHC. This is possibly one more specific feature of the coexistence of FE and STE in AHC. The migration of excitons ceases near lattice defects and no longer suppresses the strong interaction of FE with phonons. Bound excitons do not prevent the identification of FE emission in AHC to such an extent as in semiconductors without STE.

7. Exciton bands and self-trapping of excitons

Hopfield and Worlock (1965) and Ramamurti and Teegarden (1966) estimated the reduced optical mass of excitons $\mu = 0.35m_0$ for a single KI crystal. Using the data on cyclotron resonance on conduction electrons, Hodby (1969) determined the effective mass of band electrons in KI as $m_e = 0.40m_0$. Combining these results, it is easy to obtain the effective mass of the valence band hole $m_h = 2.84m_0$ and the exciton translation mass $M = m_e + m_h = 3.24m_0$, which is larger than the mass of a free electron (m_0) in KI.

The exciton translation masses and the exciton bandwidths in other AHC have not been strictly determined yet. Taking into consideration that in AHC $m_e < m_h$, from the estimations made above one may take, instead of the exciton bandwidth, the experimentally determined width of the valence band E_v. As follows from table 1, which gives E_v according to Poole et al. (1975), E_v changes from 1.4 eV to 3.7 eV through the homologous series of

AHC. In alkali halides the value $\hbar\omega_{LO}$ changes from 10 to 50 meV and hence it is clear that $B \approx E_v \gg \hbar\omega_{LO}$.

According to Rashba's (1957) terminology light excitons exist in AHC. In this respect AHC differ significantly from molecular systems, in which usually $B < \hbar\omega$, the intramolecular vibration energy. Unlike AHC, semiconductors have both $B \gg \hbar\omega_{LO}$ and $B > S$.

Let us consider more thoroughly the relationship between B and S in AHC. For molecular STE in AHC $B < S_M$. Molecular STE exist in almost all AHC. An anomaly is possible only in CsF, where cations are larger than anions, that makes the approach of two halogen ions difficult.

The question of the correlation between B and S for atomic STE is not a trivial one. Since, with changing B, the probability of exciton self-trapping in the homologous series of AHC has a discontinuity, it is not excluded that in some AHC $S_A > B$, and in others $S_A < B$. A detailed investigation of this problem is still to be done. From the data available we can conclude that in KCl, RbCl, KBr and RbBr $S_A > B$, and in NaI, probably, $S_A < B$.

For KI Lushchik and Soovik (1976) supposed the coexistence of free excitons, atomic and molecular STE. Figure 12 shows the luminescence spectra of high-purity KI crystals at 67 K according to the data by Lushchik and Soovik (1976) and Kuusmann et al. (1975a). In a cathodoluminescence spectrum there is a well-resolved emission at 5.77 eV, which shifts to 5.83 eV on cooling to 5 K and corresponds to FE emission. To distinguish the emission of atomic STE a KI crystal was excited by 5.9 eV photons and its emission was measured through a crystal of 1 mm thickness. In this case intense emission from molecular STE (3.31 eV) occurs together with a 300

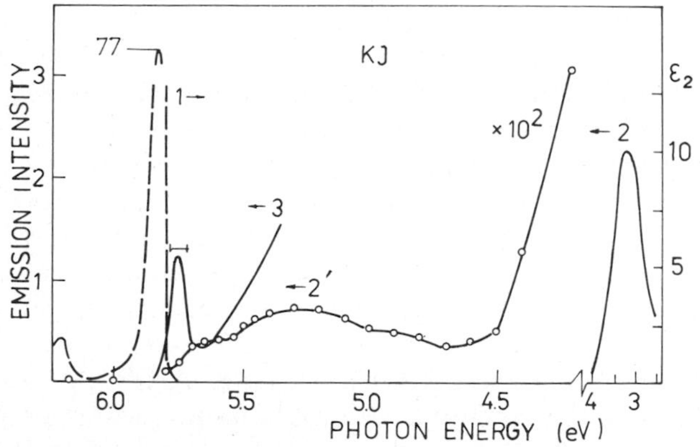

Fig. 12. Spectra of KI at 67 K: (1) ϵ_2 spectrum (Tomiki et al. 1973); (2) luminescence spectrum at an excitation by 5.9 eV photons (N. Lushchik and Soovik 1976). A correction for the emission transmission by the monochromator has been made. (3) cathodoluminescence spectrum (Kuusmann et al. 1975b).

times weaker emission in the 5.6–4.8 eV region. The excitation spectrum of the 5.4 eV emission covers the whole exciton absorption band, which is not typical of the emission of defects and impurities in a small concentration. The emission in the 5.5–5.1 eV region, with a half-width of about 0.5 eV and a Stokes shift ≈0.5 eV, is tentatively attributed to STE emission.

Kan'no et al. (1979) detected the emission of I^- ions in KCl-I. This band at 5.88 eV has the half-width of 0.23 eV and a Stokes shift relative to the absorption of I^- centers of 0.82 eV. The Stokes losses and the half-width of this localized exciton are 100 times larger than these for the emission of mobile FE in NaI.

According to Dykman and Pekar (1952) the long-range interaction of excitons with longitudinal optical vibrations is responsible for the localization of excitons in AHC. On the other hand, Sumi and Toyozawa (1971) explained the localization in terms of the short-range interaction of excitons with acoustic vibrations. This point of view is supported by many experimental facts.

Lushchik et al. (1979a,b) noted that in principle the exciton self-trapping process in alkali iodides occurs in the same way as in atomic Xe crystals. It was most clearly demonstrated by an analysis of Kink et al.'s (1978) results on the temperature dependence of the FE and STE luminescence in Xe. Since there are no optical vibrations in Xe, acoustic vibrations must be responsible for exciton self-trapping both in Xe and in alkali iodides. In the region of 20 to 60 K in NaI strong self-trapping of excitons occurs and the thermal broadening of the exciton absorption band is connected with LA-vibrations (Miyata, 1971).

The longitudinal optical vibrations should influence the process of exciton self-trapping in AHC. This problem was investigated theoretically by Pekar et al. (1979) and by Toyozawa and Shinozuka (1979). The enhancement of polarization interaction with the increase of the effective FE radius as one goes from $n = 1$ to $n = 2$, should, and in fact does, lead to a decrease of the activation barrier for self-trapping.

Lushchik et al. (1979a,b) compared the processes of self-trapping of electrons, holes and excitons in many binary ionic crystals formed by ions with different outer electron shells. The conclusion was drawn that in all reliable and supposed cases a self-trapping of *degenerate* electronic excitations occurs. Until now not a single case of the self-trapping of s holes or s electrons has been observed. However, p and d holes in AHC and AgCl, and possibly p and d electrons in $PbCl_2$ and SrO, are self-trapped. This interesting regularity is probably connected with the circumstance that orbital degeneracy and the consequent Jahn–Teller effect (Landau distortion) leads to a narrowing of the exciton band and to an increase of the energy released at the self-trapping, and hence promote the fulfilment of the self-trapping criterion $S > B$.

Unfortunately, there is as yet no strict and general theory of the coexistence of FE and STE which includes the Jahn–Teller effect. We are of the opinion that "the Landau distortion" determines the existence of the self-trapping effect, that was also predicted by Landau.

8. Exciton decay with the creation of the Frenkel defects

Lushchik, Vitol and Elango (1977), Williams (1978), Williams and Kabler (1979) have considered numerous manifestations of the creation of F centers (an electron in the field of an anion vacancy) and of H centers (interstitial halogen atom) during the radiationless decay of excitons and recombination of electrons with STH.

Figure 13(a) shows the creation spectrum of F centers in KI at 100 K with excitons being self-trapped, for the first time measured (by luminescent method) by Lushchik et al. (1961). F centers arise at an optical creation of $\Gamma(3/2)$ excitons with $n = 1$ and $n = 2$.

Taking into consideration the fact that in KI FE, atomic and molecular STE all exist, Lushchik and Kuusmann (1976) considered several possible channels of exciton decay into Frenkel defects:

(1) At a FE transition into an atomic or molecular self-trapped state.

(2) At an atomic STE decay.

(3) At an atomic STE transition into molecular STE.

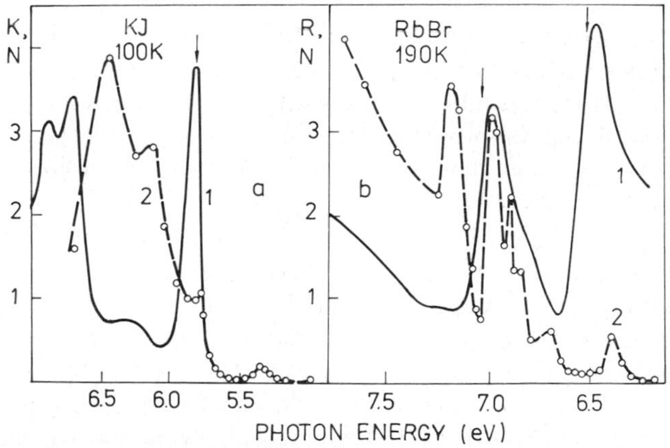

Fig. 13. An absorption spectrum for KI (1a) and a reflection spectrum for RbBr (1b). Creation spectra of color centers by UV irradiation for KI (2a) (Lushchik et al. 1961) and RbBr (2b) Vasil'chenko et al. 1980). The intensity of the luminescence photostimulated in F band is taken as a measure for the number of F centers (N) in KI. The intensity of TSL peak at 240 K is taken as a measure for the number of V_F centers (N) in RbBr.

(4) At a decay of the excited states of a molecular STE.

(5) At a decay of the lowest triplet state of a molecular STE.

Originally, the last of the possibilities was considered to be the cause for the decay of excitons into defects (Hersh 1966, Vitol 1966, Pooley 1966). However, the experiments with nanosecond pulses of an electron beam by Ueta et al. (1969) showed that in KCl and KBr the luminescence kinetics of triplet STE are slower than the rate of creation of F centers. Defects arise due to the decay of the excited states.

For defect generation it is customary to distinguish impact-recombination mechanisms, for which the displacement time of an ion from a lattice site into an interstitial one is less than the period of optical oscillation ($\tau_s < \tau_v$), from "potential displacement" mechanisms, for which $\tau_s \gg \tau_v$.

Williams et al. (1978), Yoshinari et al. (1978) and Hirai et al. (1979) showed by nanosecond laser pulse measurements that the creation time of F centers in KCl, KBr, RbBr and RbI at 5 K is respectively 11, 20, 30 and 30 ps, i.e. hundreds of times longer than τ_v. This gives preference to the mechanisms of potential displacement.

Channel 4 has been analyzed especially thoroughly. Toyozawa (1978) and Leung and Song (1978) suggested a model of decay of molecular STE with an excited electron component. The $\Sigma_u(\sigma_u\sigma_u)$ state is taken as the "pre-decaying" state (i.e. the excited state through which the defect-creating reaction takes place). Itoh and Stoneham (1977) considered the possibility of the decay of the other $\Sigma_u(\sigma_g\sigma_g)$ state with an excited hole component into F and H centers.

Williams et al. (1978) and Yoshinari et al. (1978) confirmed experimentally that optical excitation of triplet STE leads to the production of F and H centers. The creation spectrum of this effect was measured by Soda and Itoh (1979). The formation of defects occurs in the excitation regions of both the electron and hole components. However, the efficiency of the mechanism is larger in the short-wavelength region, where a decay of molecular STE is expected.

Lushchik and Kuusmann (1976), Denks et al. (1976) and Lushchik et al. (1977) considered the possibility of atomic STE decay into defects (channels 2, 3). Ortega (1979) made an analogous study of channel 1.

Vasil'chenko et al. (1980) measured the creation spectrum of color centers in RbBr crystals. As is shown in fig. 13(b), color centers are created especially effectively by an optical excitation of $\Gamma(1/2)$ excitons, but not of $\Gamma(3/2)$ excitons. A halogen atom with parallel spin and orbital momenta ($j = 3/2$) interacts strongly with a neighboring halide ion, leading to the formation of an X_2^- molecular ion. A halogen atom with antiparallel spin and orbital momenta ($j = 1/2$), however, is characterized by a repulsive interaction with the neighboring halide ion, and this may facilitate the production of Frenkel defects.

An investigation of the decay mechanisms of atomic and molecular STE with the creation of F and H centers is in progress. A detailed analysis of channels 2, 3 and 4 is especially desirable.

In AHC, besides low-energy anion excitons, the hole component of which migrates along the anion sublattice, cation excitons in the region of 12–60 eV exist, which may tentatively be interpreted as the excited states of alkali metal ions. Cation excitons in AHC have not been studied so thoroughly as the anion excitons discussed in this survey.

The creation energies of anion and cation excitons in AHC (E_{ea} and E_{ec}) as well as the equilibrium energies of the Frenkel defects' creation in anion and cation sublattices (E_{Fa} and E_{Fc}), calculated by Schulze and Hardy (1972), are presented in table 5. In many cases E_{ec} and E_{ea} exceed E_{Fa} and E_{Fc} and in some cases even $2E_{Fa}$ or $2E_{Fc}$. Anti-Schottky defect pairs – an interstitial halide ion and a neighboring interstitial metal ion – have also been detected.

Experiments have shown that in AHC one of the effective cation defect production mechanisms is exciton decay with the formation of a cation Frenkel pair (Lushchik et al. 1980). Cation defects can arise due to radiationless transitions between the excited states of STE. At these transitions vibrational instabilities can arise not only for anions but also for cations.

For a long time it was supposed that the basic channel of exciton annihilation in AHC was the transformation of exciton energy into heat (Frenkel 1931, 1936). However, Pooley and Runciman (1970) found that for NaI the basic annihilation channel is the radiative one, and this is charac-

Table 5
The energies of cation (E_{ce}) and anion (E_{ae}) excitons (eV). Formation energies of Schottky (E_{Sh}) and Frenkel cation (E_{Fc}) and anion (E_{Fa}) defects (eV) according to the calculations of Schulze and Hardy (1972).

	E_{ce}	E_{ae}	E_{Sh}	E_{Fc}	E_{Fa}
NaF	33	10.65	2.49	3.53	3.39
NaCl	33	7.93	2.26	2.88	4.60
NaBr	33	6.72	2.11	2.56	4.84
NaI	33	5.62	1.77	2.01	5.15
KF	21	9.95	1.95	4.27	3.57
KCl	21	7.76	2.20	3.46	3.73
KBr	21	7.76	2.13	3.16	4.17
KI	21	5.84	2.00	2.73	4.26
RbF	17	9.43	1.69	4.70	2.35
RbCl	17	7.52	2.05	3.71	3.51
RbBr	17	6.62	2.03	3.35	3.68
RbI	17	5.74	1.94	2.70	3.89

teristic of other alkali iodides as well. In KCl and RbCl the decay with the production of the Frenkel defects should be considered as the main annihilation channel. According to Williams (1978) in KCl the quantum yield of the exciton decay into F and H pairs is about 0.9.

The unusual and theoretically unpredicted decay of STE with the creation of Frenkel defects is a promising field for future investigations.

9. Concluding remarks; unsolved problems

During the last decade our general views on excitons in AHC have undergone a remarkable transformation.

Convincing facts have been found about the existence of free excitons in alkali iodides. These excitons have a good quasi-momentum. The investigations of coherent excitation in AHC should be continued with a special attention turned to polariton effects, interference manifestations of additional light waves, and to the orientation and alignment of excitons under excitation by linearly or circularly polarized light. The investigation of the creation of longitudinal excitons by electron beams and hot photoelectrons should be continued.

The existence of molecular X_2^-e-type STE in AHC has been proved and their structure has been elaborated. A possibility of the existence of atomic X^0e-type STE in some AHC requires further investigations.

Important results have been obtained by the study of the excited states of STE. These investigations should lead to the identification of definite predecaying states responsible for the creation of different Frenkel defects.

The use of synchrotron radiation has expanded the energy range of the electronic excitation which can be investigated. But so far the main information about high-energy excitations has been obtained by synchrotron radiation measurements of the spectra of optical constants of crystals. It is desirable to carry out a search for weak luminescence in the vacuum ultraviolet region of the spectrum, which accompanies the relaxation and annihilation processes of electronic excitations.

The experimental detection of the coexistence of FE and STE in AHC makes it possible to investigate quantitatively all stages of this most important solid state effect: the loss of a collective nature at the transformation of FE into defect-type localized excitations.

In conclusion I should like to express my gratitude to my colleagues V. Denks, M. Elango, R. Kink, I. Kuusmann, P. Liblik, G. Liidja, A. Lushchik, N. Lushchik, A. O'Konnel-Bronin, V. Plekhanov, L. Pung, T. Soovik, H. Soovik, E. Vasil'chenko and I. Vitol for the discussion of the problems considered, in the experimental investigation of which they took part directly.

References

Agranovich, V.M. and M.D. Galanin, 1978, Transfer of energy of electronic excitation in condensed media (Nauka, Moscow) (In Russian).

Apker, L. and E. Taft, 1951, Phys. Rev. **81**, 698.

Baldini, G. and B. Bosacchi, 1970, Low energy electronic transitions in the fcc alkali halides, in: Proc. Xth European congress molecular spectroscopy, Liège, 1969, eds., B. Rosen and J. Depireux. Memoires de la Societe Royale des Sciences de Liège **20**, p. 305.

Beaumont, J.H., A.J. Bourdillon and M.N. Kabler, 1976, J. Phys. C, Solid State Phys. **9**, 2961.

Blair, I.M., D. Pooley and D. Smith, 1972, J. Phys. C, Solid State Phys. **5**, 1537.

Block, D., A. Wasiela and V. Merle d'Aubigné, 1978, J. Phys. C, Solid State Phys. **11**, 4201.

Blume, H., M.P. Fontana and W.J. Van Sciver, 1969, Phys. Status Solidi **31**, 133.

Broude, V.L., V.V. Eremenko and M.K. Sheinkman, 1958, J. Techn. Phys. (USSR) **28**, 2142.

Castner, T.G. and W. Känzig, 1957, Phys. Rev. **99**, 1890.

Delbecq, C.J., J. Pringsheim and P.H. Yuster, 1951, J. Chem. Phys. **19**, 574.

Denks, V.P., 1966, Sov. Phys. Solid State (USA) **8**, 1177.

Denks, V.P., N.E. Lushchik, Ch. Lushchik and H.A. Soovik, 1976, Sov. Phys. Solid State (USA) **18**, 1254.

Dolling, G., 1974, Neutron Spectroscopy and Lattice Dynamics, in: Dynamical Properties of Solids, vol. 1, eds. G.K. Horton and A.A. Maradudin (North-Holland, Amsterdam) ch. 10.

Dykman, I.M. and S.I. Pekar, 1952, Dokl. Akad. Nauk SSSR **83**, 852.

Eby, J.E., K.J. Teegarden and D.B. Dutton, 1959, Phys. Rev. **116**, 1099.

Ejiri, A., 1977, Science of Light **26**, 1.

Fesefeld, H., 1930, Z. Phys. **64**, 623.

Fishbach, J.U., D. Fröhlich and M.N. Kabler, 1973, J. Lumin. **6**, 29.

Fischer, F. and R. Hilsch, 1959, Nachr. Akad. Wiss. Gött. II, Math.-Physik. Kl. No. 8, 241.

Fontana, M.P., 1976, Phys. Rev. Lett. **37**, 789.

Fontana, M.P., H. Blume and W.J. Van Sciver, 1968, Phys. Status Solidi **29**, 159.

Franck, J. and E. Teller, 1938, J. Chem. Phys. **6**, 861.

Frenkel, J., 1931, Phys. Rcv. **37**, 17, 1276.

Frenkel, J., 1936, Physik. Z. Sowjetunion **9**, 158.

Fröhlich, D. and B. Staginnus, 1967, Phys. Rev. Lett. **19**, 496.

Goto, T. and M. Ueta, 1967, J. Phys. Soc. Japan **22**, 488.

Goovaerts, E. and D. Schoemaker, 1978, Phys. Status Solidi **b88**, 615.

Gross, E.F., 1976, Investigations on Optics and Spectroscopy of Crystals and Liquids (Nauka, Leningrad) (In Russian).

Gross, E.F., B.P. Zakharchenya and O.V. Konstantinov, 1961, Sov. Phys. Solid State (USA) **3**, 221.

Guillot, G., E. Mercier and A. Nouailhat, 1977, J. Physique Lett. **38**, L 495.

Haensel, R., 1969, DESY-Bericht F 41–69/2.

Hahn, B. and J. Rossel, 1953, Helv. Phys. Acta **26**, 271.

Hayashi, T., T. Ohata and S. Koshino, 1975, Solid State. Commun. **17**, 945.

Hayashi, T., T. Ohata and S. Koshino, 1977, J. Phys. Soc. Japan **42**, 1647; **43**, 347.

Hersh, H.N., 1966, Phys. Rev. **148**, 928.

Hilsch, R. and R. Pohl, 1930, Z. Phys. **59**, 812.

Hirai, M., Y. Suzuki and M. Okumura, 1979, Formation of F centers and STE's in RbBr, RbI and KBr in the picosecond range, in: Third Europhysics Topical Conference Lattice Defects in Ionic Crystals, University of Kent, U.K., 1979, B 35.

Hizhnyakov, V.V., 1979a, Izv. Akad. Nauk SSSR. Ser. Fiz. **43**, 1271; Bull. Acad. Sci. USSR **43**, No. 6, 153.

Hizhnykov, V.V., 1979b, the Raman scattering and hot luminescence of self-trapped excitons, in: Proc. Conf. Light Scattering in Solids, New York, 1979, eds. J.L. Birman, H.Z. Cummins and K.K. Rebane (New York) p. 269.
Hizhnyakov, V.V. and A.V. Sherman, 1976, Trudy Inst. Fiz. Akad. Nauk Estonian SSR No. 46, 120.
Hizhnyakov, V.V. and A.V. Sherman, 1978, Phys. Status. Solidi **85**, 51.
Hodby, J.W., 1969, Phys. Rev. Lett. **23**, 1235.
Hopfield, J.J. and D.G. Thomas, 1961, Phys. Rev. **122**, 35.
Hopfield, J.J. and J.M. Worlock, 1965, Phys. Rev. **A137**, 1455.
Ikezawa, M. and T. Kojima, 1969, J. Phys. Soc. Japan **27**, 1551.
Ilmas, E., G.G. Liidja and Ch.B. Lushchik, 1965, Opt. Spectrosc. (USA) **18**, 255.
Iordanskii, S.V. and E.I. Rashba, 1978, Sov. Phys. JETP (USA) **47**, 975.
Itoh, N. and A.M. Stoneham, 1977, J. Phys. C, Solid State Phys. **10**, 4197.
Kabler, M.N., 1964, Phys. Rev. **A136**, 1296.
Kabler, M.N. and D.A. Patterson, 1967, Phys. Rev. Lett. **19**, 652.
Kabler, M.N., M.J. Marrone and W.B. Fowler, 1973, Magnetooptic effects in recombination luminescence from self-trapped excitons, in: Proc. Conf. Luminescence of Crystals, Molecules, and Solutions, Leningrad, 1972, ed. F. Williams (Plenum Press, New York, London) p. 171.
Kalder, K.A. and A.F. Malysheva, 1971, Opt. Spectrosc. (USA) **31**, 135.
Kamejima, T., S. Shionoya and A. Fukuda, 1971, J. Phys. Soc. Japan **30**, 1124.
Kan'no, K., M. Itoh and Y. Nakai, 1979, J. Phys. Soc. Japan **47**, 915.
Kask, P.A., R.A. Kink, G.G. Liidja and T.A. Soovik, 1974, Opt. Spectrosc. (USA) **30**, 149.
Kink, R.A. and G.G. Liidja, 1969, Sov. Phys. Solid State (USA) **11**, 1331.
Kink, R. and G. Liidja, 1970, Phys. Status Solidi **40**, 379.
Kink, R.A., G.G. Liidja, Ch.B. Lushchik and T.A. Soovik, 1967, Bull. Acad. Sci. USSR. Phys. Ser. (USA) **31**, 2030.
Kink, R.A., G.G. Liidja, Ch.B. Lushchik and T.A. Soovik, 1969, Trudy Inst. Fiz. Akad. Nauk Estonian SSR No. 36, 3.
Kink, R.A., G.G. Liidja and T.A. Soovik, 1971, Opt. Spectrosc. (USA) **30**, 149.
Kink, R.A., A.E. Lohmus and M.V. Selg, 1978, JETP Lett. **28**, 469.
Knox, R.S., 1963, Theory of excitons. Solid State Phys. Suppl. 5 (Academic Press, New York).
Knox, R.S. and K.J. Teegarden, 1968, Electronic excitations of perfect alkali halide crystals, in: Physics of Color Centers, ed., W.B. Fowler (Academic Press, New York) ch. 1.
Kristoffel, N. and G. Liidja, 1974, Pure and Applied Chemistry **37**, 97.
Kröger, F.A., 1954, Physica **7**, 1.
Kunz, A.B., 1969, Phys. Rev. **180**, 934.
Kuusmann, I.L. and Ch.B. Lushchik, 1976, Bull. Acad. Sci. USSR. Phys. Ser. (USA) **40**, 14.
Kuusmann, I.L., P.H. Liblik and Ch.B. Lushchik, 1975a, JETP Lett. (USA) **21**, 72.
Kuusmann, I.L., P.H. Liblik, G.G. Liidja, N.E. Lushchik, Ch.B. Lushchik and T.A. Soovik, 1975b, Sov. Phys. Solid State (USA) **17**, 2312.
Kuusmann, I.L., G.G. Liidja and Ch.B. Lushchik, 1976, Trudy Inst. Fiz. Akad. Nauk Estonian SSR No. 46, 5.
Landau, L., 1933, Phys. Z. Sowjetunion **3**, 664.
Leung, C.H. and K.S. Song, 1978, Phys. Rev. **B18**, 922.
Liidja, G.G. and T.A. Soovik, 1978, Izv. Akad. Nauk SSSR. Ser. Fiz. **42**, 3.
Lushchik, Ch., 1968, Exciton mechanisms of formation of F centers and of excitation of Tl^+ centers in alkali halide crystals, in: Proc. Conf. Color Centers in Alkali Halides, Rome, 1968, p. 189.
Lushchik, Ch., 1970, J. Lumin. **1–2**, 594.

Lushchik, Ch.B., 1974, Electronic excitations of ionic crystals in VUV region, in: Proc. Conf. Physics of Vacuum Ultraviolet Emission, Kharkov, 1972, ed., I.J. Fugol (Naukova Dumka, Kiev) p. 171.

Lushchik, Ch.B. and I.L. Kuusmann, 1976, Uspekhi Fiz. Nauk **120**, 504; Sov. Phys. Uspekhi **19**, 960.

Lushchik, N.E. and H.A. Soovik, 1976, Bull. Acad. Sci. USSR Phys. Ser. (USA) **40**, 129.

Lushchik, Ch.B., N.E. Lushchik, G.G. Liidja and L.A. Teiss, 1957, Trudy Inst. Fiz. Astr. Akad. Nauk Estonian SSR No. 6, 63.

Lushchik, Ch.B., G.G. Liidja, I.V. Jaek and E.S. Tiisler, 1960, Opt. i Spektroskopiya **9**, 70.

Lushchik, Ch.B., G. Liidja and I. Jaek, 1961, The mechanism of formation of color centres in ionic crystals by ultraviolet irradiation, in: Proc. Internat. Conf. Semiconductor Physics, Prague, 1960 (Publish. House of Czech. Acad. Sci., Prague) p. 717.

Lushchik, Ch.B., G.G. Liidja and M.A. Elango, 1965, Sov. Phys. Solid State (USA) **6**, 1789.

Lushchik, Ch.B., G.K. Vale and M.A. Elango, 1967, Izv. Akad. Nauk SSSR. Ser. Fiz. **31**, 820; Bull. Acad. Sci. USSR **31**, 826.

Lushchik, Ch.B., I.K. Vitol and M.A. Elango, 1969, Sov. Phys. Solid State (USA) **10**, 2166.

Lushchik, Ch.B., E.A. Vassil'chenko, N.E. Lushchik and L.A. Pung, 1972, Trudy Inst. Fiz. Akad. Nauk Estonian SSR No. 39, 3.

Lushchik, Ch.B., G. Liidja, N. Lushchik, E. Vassil'chenko, K. Kalder, R. Kink and T. Soovik, 1973, Exciton mechanisms of excitation of impurity centre luminescence in ionic crystals, in: Proc. Conf. Luminescence of Crystals, Molecules, and Solutions, Leningrad, 1972, ed., F. Williams (Plenum Press, New York, London) p. 162.

Lushchik, Ch.B., I. Kuusmann, P. Liblik, G. Liidja, N.E. Lushchik, V.G. Plekhanov, A. Ratas and T. Soovik, 1976, J. Lumin. **11**, 285.

Lushchik, Ch.B., I.K. Vitol and M.A. Elango, 1977, Sov. Phys. Usp. (USA) **20**, 489.

Lushchik, Ch.B., I. Kuusmann and V. Plekhanov, 1979a, J. Lumin. **18/19**, 11.

Lushchik, Ch.B., I.L. Kuusmann and V.G. Plekhanov, 1979b, Izv. Akad. Nauk SSSR, Ser. Fiz. **43**, 1162; Bull. Acad. Sci. USSR **43**, No. 6, 55.

Lushchik, Ch.B., A. Elango, R. Gindina, L. Pung, A. Lushchik, A. Maaroos, T. Nurakhmetov and L. Ploom, 1980, Semiconductors and Insulators, **5**, 133.

Marrone, M.J., F.W. Patten and M.N. Kabler, 1973, Phys. Rev. Lett. **31**, 467.

Miyata, T., 1971, J. Phys. Soc. Japan, **31**, 529.

Morgenshtern, Z.L., 1959, Opt. i Spektroskopiya **7**, 231.

Mott, N.F. and R.W. Gurney, 1938, Electronic Processes in Ionic Crystals (Oxford Univ. Press, London, New York).

Murray, R.B. and F.J. Keller, 1965, Phys. Rev. **A137**, 942.

Nakai, Y. and K. Teegarden, 1961, J. Phys. Chem. Solids **22**, 327.

Nishimura, H., C. Ohhigashi, V. Tanaka and M. Tomura, 1977, J. Phys. Soc. Japan **43**, 157.

Nishimura, H., C. Ohhigashi, V. Tanaka, H. Miyazaki and M. Tomura, 1979a, J. Lumin. **18/19**, 301.

Nishimura, H., Y. Tanaka, H. Migazaki, Ch. Ohhigashi and M. Tomura, 1979b, J. Phys. Soc. Japan **46**, 123.

Nouailhat, A., G. Guillot, E. Mercier and Truong Van Khiem, 1978, J. de Physique Lett. **39**, L223.

Nouailhat, A., G. Guillot, E. Mercier and Truong Van Khiem, 1979a, J. Lumin. **18/19**, 305.

Nouailhat, A., G. Guillot, Truong Van Khiem and J. Ortega, 1979b, J. Phys. Lett. **40**, 313.

O'Konnel-Bronin, A.A. and V.G. Plekhanov, 1979a, Sov. Phys. Solid State (USA) **21**, 356.

O'Konnel-Bronin, A.A. and V.G. Plekhanov, 1979b, Phys. Status Solidi **b95**, 75.

O'Konnel-Bronin, A.A., R.I. Gindina and V.G. Plekhanov, 1978, Sov. Phys. Solid State (USA) **20**, 356.

Ortega, J.M., 1979, Phys. Rev. **B19**, 3222.

Pekar, S.I., E.I. Rashba and V.I. Sheka, 1979, Zh. Eksperim i Teor. Fiz. **76**, 251; Sov. Phys. JETP **49**, 121.
Philipp, H.R. and H. Ehrenreich, 1963, Phys. Rev. **131**, 2016.
Plekhanov, V.G. and A.A. O'Konnel-Bronin, 1978a, JETP Lett. **27**, 27.
Plekhanov, V.G. and A.A. O'Konnel-Bronin, 1978b, Sov. Phys. Solid State (USA) **20**, 1200.
Plekhanov, V.G. and A.A. O'Konnel-Bronin, 1978c, Phys. Status Solidi **b86**, K 123.
Plekhanov, V.G. and A.A. O'Konnel-Bronin, 1978d, Sov. Phys. JETP Lett. (USA) **27**, 387.
Pooley, D., 1966, Proc. Phys. Soc. **87**, 245.
Pooley, D. and W. Runciman, 1970, J. Phys. C, Solid State Phys. **3**, 1815.
Poole, R.T., J.G. Jenkin, J. Liesegang and R.C.G. Leckey, 1975, Phys. Rev. **B11**, 5179, 5190.
Ramamurti, J. and K. Teegarden, 1966, Phys. Rev. **145**, 6986.
Rashba, E.I., 1957, Optika i Spektroskopiya **2**, 75, 88.
Rashba, E.I., 1976, Bull. Acad. Sci. USSR Phys. Ser. (USA) **40**, 20.
Rashba, E.I., 1977, Fiz. Niskikh Temperatur **3**, 524; Sov. J. Low Temp. Phys. **3**, 254.
Rashba, E.I., 1981, Ch. 13 of this volume.
Roose, N.S., 1975, Sov. Phys. Solid State (USA) **17**, 690.
Sakoda, S. and V. Toyozawa, 1973, J. Phys. Soc. Japan **35**, 172.
Schulze, P.D. and J.R. Hardy, 1972, Phys. Rev. **B5**, 3270; **B6**, 1580.
Seitz, F., 1940, Modern Theory of Solids (McGraw-Hill, New York).
Segall, B. and G. Mahan, 1968, Phys. Rev. **171**, 935.
Soda, K. and N. Itoh, 1979, Phys. Lett. **A73**, 45.
Song, K.S., 1973, J. Physique **34**, C9–495.
Stephan, G. and S. Robin, 1974, Optical properties of ionic insulators, in: Some Aspects of Vacuum Ultraviolet Radiation Physics, eds., N. Damany, J. Romand and B. Vodar (Akademie-Verlag, Berlin) ch. 3.
Stevens, M. and M. Menes, 1968, Phys. Lett. **A27**, 472.
Sumi, A., 1977, J. Phys. Soc. Japan **43**, 1286.
Sumi, A., 1979, J. Phys. Soc. Japan **47**, 1538.
Sumi, A. and Y. Toyozawa, 1973, J. Phys. Soc. Japan **35**, 137.
Sumi, H., 1975, J. Phys. Soc. Japan **38**, 825.
Sumi, H., 1976, J. Phys. Soc. Japan **41**, 526.
Sumi, H. and Y. Toyozawa, 1971, J. Phys. Soc. Japan **31**, 342.
Teegarden, K., 1957, Phys. Rev. **105**, 1222.
Teegarden, K. and G. Baldini, 1967, Phys. Rev. **155**, 896.
Tehver, I. and V. Hizhnyakov, 1975, Sov. Phys. JETP (USA) **42**, 305.
Terenin, A.N. and F.D. Klement, 1935, Acta Physicochemica USSR **2**, 941.
Tomiki, T., T. Miyata and H. Tsukamoto, 1973, J. Phys. Soc. Japan **35**, 495.
Toyozawa, Y., 1961, Progr. Theor. Phys. **26**, 29.
Toyozawa, Y., 1974, Exciton lattice interaction-fluctuation, relaxation and defects, in: Proc. Conf. Vacuum Ultraviolet Radiation Physics, Hamburg, 1974, eds., E. Koch, R. Haensel and C. Kunz (Pergamon Vieweg, Braunschweig) p. 317.
Toyozawa, Y., 1978, J. Phys. Soc. Japan **44**, 482.
Toyozawa, Y. and Y. Shinozuka, 1979, Techn. Rep. of ISSP ser. A, No. 992.
Ueta, M., Y. Kondo, M. Hirai and T. Yoshinari, 1969, J. Phys. Soc. Japan **26**, 1000.
Vasil'chenko, E.A. and N.E. Lushchik, 1969, Bull. Acad. Sci. USSR Phys. Ser. **33**, 913.
Vasil'chenko, E.A., N.E. Lushchik and Ch.B. Lushchik, 1970, Sov. Phys. Solid State (USA) **12**, 167.
Vasil'chenko, E., N. Lushchik and Ch. Lushchik, 1972, J. Lumin. **5**, 195.
Vasil'chenko, E.A., N.E. Lushchik and H.A. Soovik, 1974, Izv. Akad. Nauk SSSR. Ser. Fiz. **38**, 1267; Bull. Acad. Sci. USSR **38**, No. 6, 133.
Vasil'chenko, E.A., A.Ch. Lushchik, N.E. Lushchik and Ch.B. Lushchik, 1981, Fiz. Tverd. Tela, **22**, 1156 (Sov. Phys. Solid State, **22**, 674).

Vitol, I.K., 1966, Izv. Akad. Nauk SSSR Ser. Fiz. **30**, 564; Bull. Acad. Sci. USSR **30**, 581.
Wannier, G., 1937, Phys. Rev. **52**, 191.
Wasiela, A., G. Ascarelli and Y. Merle d'Aubigne, 1973, Phys. Rev. Lett. **31**, 993.
Williams, R.T., 1978, J. Semiconductors and Insulators **3**, 251.
Williams, R.T. and R.M. Kabler, 1979, Phys. Rev. **B9**, 1097.
Williams, R.T., J.N. Bradford and W.L. Faust, 1978, Phys. Rev. **B18**, 7038.
Wood, R.F., 1966, Phys. Rev. **151**, 629.
Yoshinari, T., K. Iwano and M. Hirai, 1978, J. Phys. Soc. Japan **45**, 1926.

CHAPTER 13

Self-Trapping of Excitons

EMMANUEL I. RASHBA

L.D. Landau Institute for Theoretical Physics
Academy of Sciences of the USSR
142432, Chernogolovka, Moscow Region
USSR

Excitons
Edited by
E.I. Rashba and M.D. Sturge

Contents

1. Introduction

The physics of self-trapped (ST) states of electrons, holes and excitons in crystals is under active investigation now. Numerous studies involve a wide class of substances (alkali halide crystals, rare gas crystals, magnetic and molecular crystals and so on) which are investigated both experimentally and theoretically. The possibility of ST of electrons was originally pointed out by Landau (1933). He showed that the states of the electron "trapped by lattice" (i.e., the states in which the lattice around an electron is strongly deformed) are more advantageous energetically as compared to the Bloch band states of the electron in the regular lattice, provided the electron–phonon interaction is sufficiently strong. Later on these states were referred to as "self-trapped" in the Western scientific literature and as "autolocalized" in the Soviet one.

The term "self-trapping" is certainly inexact to some extent. Really, the fact that a perfect crystal possesses translational symmetry results in the existence of states of the generalized Bloch wave type, corresponding to a joint motion of the electron, accompanied by a lattice deformation. But the width of the corresponding bands is small, and the effective mass of the ST states is large (it increases by a power or even by an exponential law with the coupling constant). Thus, the bands of ST electrons may be destroyed even at relatively low impurity concentrations and low temperatures ($T \neq 0$).

The consequences of the exciton–phonon coupling, if sufficiently strong, were discussed shortly after the Frenkel papers appeared, in which the concept of excitons was first introduced (Frenkel 1931). A conclusion on the possibility of self-trapping was also drawn (Peierls 1932, Frenkel 1936, Davydov 1951); however, both the arguments and the physical picture were in some sense opposite to that discussed in Landau's paper (for details see sect. 4). A synthesis of these two approaches, which made it possible to understand them as two limiting cases of the general picture, was achieved later (Rashba 1957a,d).

The problems of electron (hole) and exciton self-trapping have a great deal in common. However, there are considerable differences in them: different laws for coupling to phonons, existence of internal structure for the Wannier–Mott (WM) exciton, effect of the long-range interactions on the energy spectrum of band exciton, etc. Besides, quite different aspects of the problem under consideration are of principal physical importance for them. For electrons (holes) the problem of low and high frequency conductivity is the central one; thus, more emphasis was put on the properties of ST charge carriers at the bottom of the energy spectrum (i.e., its absolute minimum). On the contrary, for excitons, which are short-lived nonequilibrium quasi-particles (excited, e.g., by light), gross features of spectrum (structure of the energy spectrum as a whole, including highly excited states), the rate of relaxation to ST states, the possibility of coexistence of free and ST states of excitons and so on are of particular interest. Therefore, in the following we shall discuss both the problems of electron (hole) and exciton self-trapping, elucidating the question what is common and what is different in them.

In conclusion one remark should be made concerning terminology. Pekar's continuum (large) polaron was the first model of ST state, allowing consistent theoretical treatment (Pekar 1951). Before long, the problem acquired a general physical importance, and in a series of outstanding papers (Bogolyubov 1950, Lee et al. 1953, Fröhlich 1954, Feynman 1955) on polaron theory, quantum field theoretic methods penetrated extensively into solid state physics. As a result, the term polaron shortly lost its initial sense (electron interacting with long wave polar optic vibrations) and was used for a much wider class of ST states. Since the detailed mechanism of electron–phonon coupling is of significance for us, we use the term "polaron" preferably in its more narrow initial sense, in other cases using the terms "ST electron", "ST exciton" etc.

2. Basic models of self-trapped (ST) states

Although the Landau arguments concerning the possibility of ST were totally convincing, it was not confirmed experimentally for a long period of time. This was a matter even of some surprise which was expressed, e.g., in the well-known book of Mott and Gurney (1940). But subsequently the situation changed radically. Self-trapping of electrons (holes) and excitons was discovered in various classes of substances, the microscopic structure of ST states being very diverse. Therefore, while not pretending to provide a comprehensive picture, we start by demonstrating the basic models of ST states; wherever possible we mention concurrently the corresponding classes of substances. In sect. 3 we consider the Hamiltonians giving rise to these states.

2.1. Large polaron

ST of electrons (or holes) is due to their interaction with long-wave polar optic phonons. The coupling to the phonons is sufficiently strong for the characteristic electron frequencies to be much larger than the phonon frequency ω_ℓ; nevertheless, the electron cloud radius remains much larger than the lattice constant d (Pekar 1946) (for review see Pekar 1951, Appel 1968, Gross 1976).

2.2. *ST one-site electron (hole)*

The radius of the ST state is so small that the electron (hole) is concentrated practically at one of the atoms (ions). Such states may arise both in ionic and nonpolar crystals (e.g., in transition metal oxides and in sulfur), and they are usually designated as small polarons. There exists a quite definite tendency in nonpolar crystals to form states of small radius (of the order of the lattice constant). The kinetic and optical properties of small polarons have been studied extensively (for review see Austin and Mott 1969, Firsov 1975, Klinger 1979).

2.3. *Quasi-molecular ST hole*

The ST states in crystals may have symmetry which is lower than the maximal site symmetry of the lattice. For instance, ST of holes with the formation of a halide molecular ion of the X_2^- type in alkali halide crystals (Kastner and Känzig 1957) and of the R_2^+ type in rare gas crystals (Druger and Knox 1969) is very common. The structure of such states is determined by short-range interactions.

2.4. *Electron cavity*

ST of electrons may be accompanied with the formation of a vacuum cavity (bubble) around it. Such model was proposed for solvated electrons in metal–ammonia solutions (Ogg 1946) and for electrons in liquid helium (Careri et al. 1960). The formation of a cavity in helium is ascribed to the negative work function for electrons, and in metal ammonia solutions to the peculiarities of the short-range polarization interaction.

2.5. *ST Wannier–Mott exciton*

If the ratio of the hole and electron effective masses is large ($m_h/m_e \gtrsim 10$), the WM exciton may be self-trapped, the electron and the hole moving in the common potential well (Dykman and Pekar 1952) (for review see Haken 1958). If the hole is ST in the small-radius state, e.g., a quasi-molecular one, the electron will be bound to it by the usual Coulomb attraction. Such states are typical for alkali halide crystals (Tolpygo 1962, Kabler 1964; for review see Aluker et al. 1979, Lushchik 1982). Under these conditions the coupling of the electron to the lattice may remain weak due to the small electron mass.

2.6. ST Frenkel exciton

Since Frenkel excitons are electrically neutral and their internal radius is determined by the size of the molecule (ion), the ST conditions are nearly the same as for electrons in nonpolar crystals; therefore, small-radius states arise. The same picture may be expected to hold for the charge transfer exciton, if the separation between the electron and the hole does not exceed the lattice spacing.

2.7. Exciton cavity

The ST exciton may be modelled also as an excited atom or molecule (excimer) in a vacuum cavity. Both states were observed in liquid He (Dennis et al. 1969), for molecules the rotational structure is distinctly visible. In solid Ne the one-site (quasi-atomic) ST prevails (Packard et al. 1970), while in solid Ar, Kr and Xe the two-site (quasi-molecular) ST does (Jortner et al. 1965; for review see Fugol' 1978). The positronium cavities (Ferell 1957) may be also assigned to quasi-atomic states (for review see Khraplak and Yakubov 1979).

2.8. Excimers in molecular crystals

Many organic molecules are capable of forming dimers when one of the molecules is in an electronically excited state; they are known as excimers. In crystals the formation of excimers is hindered owing to the resistance to displacement (rotation) of the molecules. Therefore, excimer self-trapping usually occurs only when the lattice consists of pairs of plane-parallel molecules, which may be brought together without overcoming any appreciable barrier. The pyrene crystal (Ferguson 1958) is a well-known example (for review see Cohen 1979).

2.9. ST in quasi-one-dimensional systems

If the motion of an electron or Frenkel exciton is one-dimensional (1d), then, unlike the three-dimensional systems, "large" ST states (i.e., their size much exceeding the lattice spacing) may arise even under the nonpolar interaction conditions (Rashba 1957b,d). The polymers, some biological systems and crystals with a quasi-1d spectrum, fall into this group.

2.10. ST by phase transition

In some systems ST of electrons and excitons may be achieved by a phase transition in the surrounding area of the medium. The well-known examples are "icebergs" around positive ions in liquid helium (Atkins 1959) and magnetic polarons (de Gennes 1960) (for review see Krivoglaz 1973, Nagaev 1975).

The microscopic structure of ST states is very multiform, as can be seen from the above list of basic models, which of course is not exhaustive. Therefore, there can be no general approach to their calculation. Most appropriate for theoretical treatment are continuum models. Unfortunately, they describe the experimental situation only for certain cases. As a rule, small size states arise. Calculation of such states is always doubtful and frequently involves semi-chemical considerations, which are plausible rather than convincing.

Below we shall concentrate our attention on general qualitative regularities, rather than on numerical calculation for specific systems.

3. Hamiltonian of the exciton–phonon interaction

The exciton–phonon interaction may be reduced to the sum of the electron–phonon and hole–phonon interactions, or may be written by analogy with the electron–phonon interaction. Therefore, we write first the Hamiltonian for the electron interacting with phonons:

$$H = H_e + H_{ph} + H_{e\text{-}ph}, \tag{3.1}$$

where H_e is the electron kinetic energy,

$$H_{ph} = \sum_{qf} \omega_{qf} b^+_{qf} b_{qf} \tag{3.2}$$

is the phonon energy, and

$$H_{e\text{-}ph} = \frac{1}{N^{1/2}} \sum_{kqf} \gamma_{qf}(k) a^+_{k+q/2} a_{k-q/2} (b_{qf} + b^+_{-qf}), \quad \gamma_{qf}(k) = \gamma^*_{-qf}(k), \tag{3.3}$$

is the electron–phonon interaction. The electron band is supposed to be nondegenerate. The interaction is linear in the phonon amplitudes; the coefficients $\gamma_{qf}(k)$ are the constants of the electron–phonon coupling. The index f numerates phonon branches, N is the number of cells in the normalization volume.

After transformation from the momentum representation to the representation of electron field operators

$$a(r) = \frac{1}{V^{1/2}} \sum_k a_k \, e^{ikr}, \tag{3.4}$$

where $V = Nv$, v is the volume of a primitive cell, we get

$$H_{\text{e-ph}} = \frac{1}{N^{1/2}} \sum_{qf} \iint \mathrm{d}\boldsymbol{r}\, \mathrm{d}\boldsymbol{r}'\gamma_{qf}(\boldsymbol{r}-\boldsymbol{r}')\, \mathrm{e}^{\mathrm{i}\boldsymbol{q}(\boldsymbol{r}+\boldsymbol{r}')/2} a^{+}(\boldsymbol{r}) a(\boldsymbol{r}')(b_{qf} + b^{+}_{-qf}) \tag{3.5}$$

$$\gamma_{qf}(\boldsymbol{r}-\boldsymbol{r}') = \frac{1}{V} \sum_{\boldsymbol{k}} \gamma_{qf}(\boldsymbol{k})\, \mathrm{e}^{\mathrm{i}\boldsymbol{k}(\boldsymbol{r}-\boldsymbol{r}')}. \tag{3.6}$$

Usually the dependence of $\gamma_{qf}(\boldsymbol{k})$ on $\boldsymbol{k}$, though not on $\boldsymbol{q}$, is neglected:

$$\gamma_{qf}(\boldsymbol{k}) = \gamma_{qf}. \tag{3.7}$$

In this approximation,

$$\gamma_{qf}(\boldsymbol{r}-\boldsymbol{r}') = \gamma_{qf}\delta(\boldsymbol{r}-\boldsymbol{r}'), \tag{3.8}$$

and eq. (3.5) has the form:

$$H_{\text{e-ph}} = \frac{1}{N^{1/2}} \sum_{qf} \gamma_{qf} \int \mathrm{d}\boldsymbol{r} a^{+}(\boldsymbol{r}) a(\boldsymbol{r})\, \mathrm{e}^{\mathrm{i}\boldsymbol{q}\boldsymbol{r}}(b_{qf} + b^{+}_{-qf}). \tag{3.9}$$

When an electron is described in usual coordinate representation, rather than in a secondary quantization, eq. (3.9) is rewritten as follows:

$$H_{\text{e-ph}} = \frac{1}{N^{1/2}} \sum_{qf} \gamma_{qf}\, \mathrm{e}^{\mathrm{i}\boldsymbol{q}\boldsymbol{r}}(b_{qf} + b^{+}_{-qf}). \tag{3.10}$$

For an electron we shall use the lattice site representation, side by side with the continuum one. In the former, eq. (3.9) takes the form:

$$H_{\text{e-ph}} = \frac{1}{N^{1/2}} \sum_{nqf} \gamma_{qf}\, \mathrm{e}^{\mathrm{i}\boldsymbol{q}\boldsymbol{n}} a^{+}_{n} a_{n}(b_{qf} + b^{+}_{-qf}). \tag{3.11}$$

It is seen from eq. (3.8) that the condition (3.7) corresponds to a model in which the action of phonons on the electron is reduced to the action of an effective external field. For instance, such a description is applicable for long-wave optic phonons, which interact with the electron through their electrostatic potential: in this case the electronic band moves easily as a whole. The dependence of γ_{qf} on $\boldsymbol{k}$ arises when the lattice deformation changes the electron transfer integrals. However, it is seen from eq. (3.5) that the functions $\gamma_{qf}(\boldsymbol{r}-\boldsymbol{r}')$ decrease as exchange integrals responsible for the transfer of electrons between lattice sites, i.e., decrease exponentially, when $|\boldsymbol{r}-\boldsymbol{r}'|$ increases. Therefore, when integrated with smooth functions they may be substituted by a delta-function; so below, as a rule, we use the Hamiltonians (3.10) and (3.11).

The possible situations are determined by the type of electron–phonon coupling. The various types of this coupling are classified by the behaviour of the coefficients γ_{qf} and the vibrational frequencies ω_{qf} in the small $\boldsymbol{q}$ region ($qd \ll 1$). We list here the basic model Hamiltonians (for details see

Kittel 1963, Bir and Pikus 1972); the number of them is rather limited (below $\hbar = 1$):

(1) Optical phonons–nonpolar interaction:

$$\omega_q = \omega_0 = \text{const}, \quad \gamma_q = \gamma_0 = \text{const}. \tag{3.12}$$

(2) Optical phonons–polar interaction:

$$\omega_q = \omega_\ell = \text{const},$$

$$\gamma_q = \left(\frac{2\pi\omega_\ell}{v}\frac{e^2}{\kappa}\right)^{1/2}\frac{1}{q} = \frac{\sqrt{\alpha}}{q}\left(\frac{2\pi\omega_\ell}{v}\sqrt{\frac{2\omega_\ell}{m}}\right)^{1/2}. \tag{3.13}$$

Here ω_ℓ is the frequency of long-wave longitudinal phonons, m is the electron effective mass, α is the nondimensional electron–phonon coupling constant (Fröhlich 1954), and κ is determined by the contribution of the ionic polarization to the total dielectric polarization (Gurney and Mott 1937):

$$\alpha = \frac{e^2}{\kappa}\sqrt{\frac{m}{2\omega_\ell}}, \quad \frac{1}{\kappa} = \frac{1}{\kappa_\infty} - \frac{1}{\kappa_0}; \tag{3.14}$$

κ_0 and κ_∞ being the low and high frequency dielectric constants.

(3) Longitudinal acoustic phonons – deformation potential:

$$\omega_q = s_\ell q, \qquad \gamma_q = Cq/\sqrt{2\rho v \omega_q}. \tag{3.15}$$

Here s_ℓ is the longitudinal sound velocity, ρ is the density, C is the deformation potential.

(4) Acoustic phonons – piezoelectric interaction:

$$\omega_q = sq, \qquad \gamma_q \propto q^{-1/2}. \tag{3.16}$$

Polar interactions, (3.13) and (3.16), are long-range ones, and nonpolar interactions, (3.12) and (3.15), are short-range ones.

Although these formulas are only correct for simplified models (isotropic sound velocity, single deformation potential etc.), they nevertheless describe the main features of the phenomena.

In constructing the Hamiltonian of the exciton–phonon interaction it is useful to distinguish two limiting situations.

The ionization potential of a WM exciton is $E_{ex} \sim e^2/\kappa_\infty r_{ex} \ll E_{at}$, E_{at} is of the atomic order of magnitude (some eV). The contribution to the total energy of the electron–phonon interaction, which arises owing to modulation of the electron–hole interaction by the lattice displacements, is small under these conditions. Therefore, the Hamiltonian of the exciton–phonon interaction $H_{ex\text{-}ph}$ may be written as follows:

$$H_{ex\text{-}ph} = H_{e\text{-}ph} + H_{h\text{-}ph}, \tag{3.17}$$

i.e., as a sum of the Hamiltonians of the electron–phonon and hole–phonon interactions.

The position is quite different for small-radius excitons: for both Frenkel excitons and charge transfer excitons. Their interaction with phonons cannot be expressed directly through the electron–phonon and hole–phonon interactions in the same crystal. This is quite obvious for the Frenkel exciton. But it is correct for charge transfer excitons too. The Hippel (1936) model for excitons in the NaCl type crystals may serve as an example. According to this model, an electron transfers from a halide ion to six surrounding metal ions; the corresponding change in the Madelung energy is $\sim e^2/d$, i.e., it has the atomic order of magnitude E_{at}. As a result, the change of this energy under the lattice displacements may compete with the individual interaction of an electron and a hole with phonons, or even may be regarded as the dominating mechanism of the exciton–phonon interaction (Dykman 1954). It is correct all the more for charge transfer excitons in nonpolar or weakly-polar crystals. Such position arises, for example, in the A-PDMA crystal (anthracene-piromellitic dianhydride) which is built of mixed stacks of parallel molecules (Haarer 1974, 1977); it was discussed also for other 1d and layered systems (Agranovich et al. 1977).

The interaction of small-radius exciton with long-wave phonons is always nonpolar owing to the electrical neutrality of the exciton; thus, the laws (3.13) and (3.16) are excluded for such excitons. As a result, only nonpolar interaction laws (3.12) and (3.15) remain, their intensities being controlled by the mechanisms discussed above. The interaction with optical phonons retains its form (3.12) even in the case, when the electronic excitation of the molecule changes strongly its static dipole moment (see, e.g., Lippert 1955), i.e., the exciton possesses an electric dipole moment; but under these conditions γ_q acquires a strong angular dependence. Apparently, this case has never been investigated.

In conclusion, we consider briefly the Hamiltonian H_e. It will be written either in the lattice site representation (see eq. (4.8) below) or in the effective mass representation. In the latter case we restrict ourselves to an isotropic nondegenerate spectrum; then

$$H_e = -\frac{1}{2m}\Delta. \qquad (3.18)$$

The effect of energy spectrum anisotropy will be discussed briefly in sect. 8 in connection with low-dimensional systems. The effect of the band degeneracy has not yet been investigated, despite the fact that some experimental data apparently indicate that degeneracy favours self-trapping (Kittel 1971, Lushchik et al. 1979).

The free exciton Hamiltonian H_{ex} may differ significantly from H_e in the small $\boldsymbol{k}$ region, if the exciton transition is allowed (in the dipole approximation). Indeed, the exciton energy depends on $\boldsymbol{k}$ nonanalytically in this case because of the contribution of long-range dipole–dipole interactions, and the term with an integral operator enters into H_{ex} side by side with the term of the (3.18) type (cf. sect. 5.4).

4. *Classification of ST states and ST criteria*

The gross features of the energy spectrum of the system are determined by the ratio of the three basic parameters (in the classification of states below we follow the papers of Rashba (1957a,d)). These parameters are the half-width E_B of the electron (exciton) band in the rigid lattice, the characteristic phonon frequency ω, and the lattice deformation energy E_D in the ST state. In this section referring to excitons we have in mind the small-radius excitons only, for definiteness Frenkel excitons.

The ratio of the parameters E_B and ω determines which of the two subsystems, electrons or phonons, is the fast and which one is the slow subsystem. When

$$\omega \ll E_B, \tag{4.1}$$

the electron subsystem is the fast one, and the usual adiabatic approximation may be used to describe ST states. When

$$\omega \gg E_B, \tag{4.2}$$

phonons follow the slowly moving electron (exciton). We say that electron (exciton) is "light" in the first case and "heavy" in the second case. The Landau argument (Landau 1933) based on producing a local electron level by an instantaneous lattice field, is applicable only when the criterion (4.1) is fulfilled. On the contrary, the necessary condition of ST formulated in early papers on excitons (Peierls 1932, Frenkel 1936, Davydov 1951) is that the local lattice deformation follows the instantaneous position of exciton; this coincides with eq. (4.2).

Both the mathematical methods and the physical picture significantly differ in these two limiting cases. Therefore, we consider them separately, and begin with the case of heavy excitons, which is much easier than that of light excitons.

4.1. Heavy excitons (electrons)

The two last terms in eq. (3.1) are the largest in this case. When expressions (3.2) and (3.11) are substituted for them, the diagonalization of $H_{ph} + H_{ex\text{-}ph}$ may be done by a canonical transformation, known from the theory of impurity centers and small polarons:

$$H_{ph} + H_{e\text{-}ph} \rightarrow e^{\mathcal{U}}(H_{ph} + H_{e\text{-}ph})\, e^{-\mathcal{U}}, \tag{4.3}$$

where

$$\mathcal{U} = \frac{1}{N^{1/2}} \sum_{qfn} u_{qf} a_n^+ a_n (b_{qf} - b_{-qf}^+)\, e^{iqn}. \tag{4.4}$$

The terms in the transformed Hamiltonian which are linear in phonon amplitudes cancel at

$$u_{qf} = -\gamma_{qf}/\omega_{qf}, \tag{4.5}$$

then

$$H_{ph} + H_{e\text{-}ph} = \sum_{qf} \omega_{qf} b_{qf}^+ b_{qf} - E_D, \tag{4.6}$$

where E_D is the lattice deformation energy:

$$E_D = \frac{1}{N} \sum_{qf} \frac{|\gamma_{qf}|^2}{\omega_{qf}}. \tag{4.7}$$

It is seen from these expressions that for heavy excitons the lowering of the system energy on self-trapping, i.e., the difference between the energy of the system with ST exciton $\mathcal{H}(ST)$ and that with free exciton at the band bottom $\mathcal{H}(F)$ (rigid lattice!) equals:

$$E_{FC} \equiv \mathcal{H}(ST) - \mathcal{H}(F) \approx -E_D. \tag{4.7a}$$

From now on this energy E_{FC} will be referred to as the Franck–Condon energy (it is often denoted the lattice relaxation energy and is measured from the band center, rather than from its bottom).

The kinetic energy of electron (exciton) in the lattice site representation is

$$H_e = \sum_{n \neq m} M_{nm} a_n^+ a_m. \tag{4.8}$$

Exponential factors containing the operators b_{qf} enter into the transformed expression for H_e; they arise from the operators $\exp(\pm\mathcal{U})$. The subsequent diagonalization of the Hamiltonian may be carried out by perturbation theory in the parameter $E_B/\omega \ll 1$. Indeed, since $H_e \sim M \sim E_B$ is small, elimination of the terms which change the number of phonons gives rise to large denominators of the order of ω. Therefore, the part of the transformed Hamiltonian H_e which determines the renormalized dispersion law at $T = 0$ may be separated out by taking the mean value of H_e over the zero-phonon state. This gives the well-known result (Tjablikov 1952, Rashba 1957c, Holstein 1959):

$$(H_e)_{\text{zero-ph}} \approx \sum_{n \neq m} \tilde{M}_{nm} a_n^+ a_m, \quad \tilde{M}_{nm} = M_{nm}\, e^{-2S_{nm}}, \tag{4.9}$$

where

$$S_{nm} = \sum_{qf} \frac{|\gamma_{qf}|^2}{\omega_{qf}^2} \sin^2 \tfrac{1}{2}\boldsymbol{q}(\boldsymbol{n} - \boldsymbol{m}). \tag{4.10}$$

Equation (4.9) determines the dispersion law of the dressed heavy particle. It is designated as "localized" exciton (Davydov 1951) in some papers on the exciton theory and as nonadiabatic polaron in the small polaron theory.

It is seen that the magnitude of the energy spectrum renormalization is determined by

$$S = (1/N) \sum_{qf} |\gamma_{qf}|^2 / \omega_{qf}^2, \tag{4.11}$$

which is the usual coupling constant of impurity center theory. For nonpolar optical phonons $S = E_D/\omega_0$. The bandwidth is exponentially small under strong coupling conditions ($S \gg 1$); while if the coupling is weak ($S \ll 1$), the renormalization of the spectrum is insignificant.

It is necessary to stress that the bands of "localized" excitons, defined in such a way, are not necessarily narrow. For example, in molecular crystals of the benzene type the interaction of the exciton with high frequency ($\omega \sim 0.1$ eV) intramolecular phonons is the strongest. In this case $\omega \gg E_B$ and $S \sim 1$; but the bands, renormalized by this interaction, retain the width $\sim$0.01 eV. Such bands comprise the main subject of investigation in molecular exciton spectroscopy.

These results hold when the quantities γ_{qf} do not depend on $\boldsymbol{k}$, i.e., when the change of M_{nm} due to lattice displacements is neglected. As a rule, the coefficients M_{nm} increase when atoms become closer, and this, in principle, may result in a band broadening due to the electron–phonon interaction (effect of the "barrier preparation" type, Kagan and Klinger (1976)). However, this effect is large only for strong exponential dependence of M_{nm}'s on distances (for singlet excitons this dependence obeys a power law) and in the region of nonlinear dependence of M_{nm} on displacements.

So, the main result consists in the following: the band of heavy excitons is transformed into a band of localized (dressed) excitons, which is narrowed and shifted by E_{FC}. The spectrum of phonon excitations, including the exciton–phonon bound states, is positioned above this band (for review, as applied to molecular excitons, see Sheka 1971, Levinson and Rashba 1973).

To obtain this result we used the linearity of the operator $H_{e\text{-ph}}$ in the phonon amplitudes b_{qf}. This assumption is quite essential, as can be demonstrated on an example of the formation of excimers in molecular crystals. The gain in the energy at excimerization of plane molecules is largest when molecules are superimposed completely. In crystals of the pyrene type (fig. 1(a)) a pair of close-spaced parallel molecules is the lattice basis. These molecules move towards one another on electronic excitation – thus, the fluorescence of pyrene crystals is of excimer origin. In crystals of the anthracene type (fig. 1(b)) a single molecule is the basis. The parallel molecules are widely spaced; thus, the overlap is minor and the driving force for excimer formation is very weak. It increases only in the nonlinear region, i.e., for large displacement; these movements are, however, resisted by the environment. As a result, the excimer state is separated from the band states by a high barrier of the purely lattice origin even in the case when excimerization gives a gain in energy (fig. 1(c)). Under such conditions the band state is stable and will not be destroyed within the exciton lifetime.

4.2. *Light excitons (electrons)*

The ST states of light excitons may be described by the general method developed by Pekar (1946, 1951) in the adiabatic theory of polarons. This method is based on the following facts. In ST states (if any arise!): (i) electronic frequencies exceed significantly the characteristic phonon frequencies ω; (ii) energy of the lattice deformation is large $E_D \gg \omega$. The first fact allows one to use the adiabatic approximation, and the second, to describe the lattice vibrations quasi-classically.

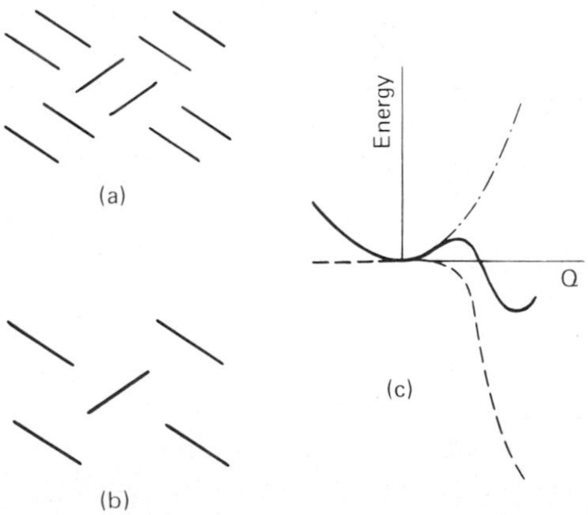

Fig. 1. Schematical arrangement of molecules in lattices of the pyrene (a) and anthracene (b) types. The dependence of the energy of two parallel molecules on the configuration coordinate in anthracene crystal – when the coordinate increases, molecules approach each other; dash-dotted curve – unexcited molecules, dashed curve – attraction energy in the excited state, solid line – total energy (c).

Under these conditions the solution of the problem may be reduced to the finding of the minimum:

$$\mathscr{H} = \min_{\psi,\{b_{qf}\}}\langle\psi|H|\psi\rangle, \tag{4.12}$$

where H is the Hamiltonian (3.1), ψ is the electronic wave function in the adiabatic approximation, and $\mathscr{H}$ is the ground state energy of the electron–phonon system. The quantities b_{qf} in eq. (4.12) should be considered as c numbers. The double variation in eq. (4.12), over ψ and lattice deformation, expresses the Pekar's self-consistency condition of ST states.

Variation over b^+_{qf} is carried out directly, e.g., for $H_{\text{e-ph}}$, chosen in the form (3.10), we have:

$$b_{qf} = -\frac{1}{N^{1/2}}\frac{\gamma^*_{qf}}{\omega_{qf}}\langle\psi|e^{iqr}|\psi\rangle. \tag{4.13}$$

The elimination of the amplitudes b_{qf} results in the following extremal problem:

$$\mathscr{H} = \min_{\psi}J[\psi], \tag{4.14a}$$

where,

$$J[\psi] = \langle\psi|H_e|\psi\rangle - \frac{1}{N}\sum_{qf}\frac{|\gamma_{qf}|^2}{\omega_{qf}}|\langle\psi|e^{iqr}|\psi\rangle|^2, \tag{4.14}$$

for the Hamiltonians (3.9) and (3.10), and

$$J[\psi] = \langle\psi|H_e|\psi\rangle - \frac{1}{N}\sum_{qf}\frac{|\gamma_{qf}|^2}{\omega_{qf}}|\langle\psi|e^{iqn}|\psi\rangle|^2, \tag{4.15}$$

for the Hamiltonian (3.11). The only difference in these expressions is that the function ψ is taken in the $\boldsymbol{r}$ representation in eq. (4.14) and in the lattice site representation in eq. (4.15). It is reasonable to use them for large-radius and small-radius states, respectively. The last term in eqs. (4.14) and (4.15) is the deformation energy E_D taken with the inverse sign. This term coincides with eq. (4.7) for the strongly localized wave function $\psi(\boldsymbol{n}) = \delta_{\boldsymbol{n}\boldsymbol{n}_0}$. The mean energy of the particle-lattice interaction is:

$$U = \langle\psi|H_{\text{e-ph}}|\psi\rangle = -2E_D. \tag{4.16}$$

4.2.1. Long-range interaction – polarons

The functional $J[\psi]$ for polarons may be found by inserting expressions (3.14) into (4.14) and extending the integration to the whole $\boldsymbol{q}$ space:

$$J[\psi] = \frac{1}{2m}\int|\nabla\psi|^2\,d\boldsymbol{r} - \frac{e^2}{2\kappa}\iint\frac{|\psi(\boldsymbol{r})|^2|\psi(\boldsymbol{r}')|^2}{|\boldsymbol{r}-\boldsymbol{r}'|}\,d\boldsymbol{r}\,d\boldsymbol{r}' \equiv K - E_D. \tag{4.17}$$

The energy of the polaron ground state (Pekar 1946) and the polaron effective mass for small momentum (Landau and Pekar 1948) are:

$$\mathcal{H} \approx -0.109\alpha^2\omega_\ell, \qquad m^* \approx 0.02\alpha^4 m. \tag{4.18}$$

The scale transformation

$$\psi(\boldsymbol{r}) \to \tilde{\psi}(\boldsymbol{r}, \nu) = \nu^{3/2}\psi(\nu\boldsymbol{r}), \tag{4.19}$$

where $\psi(\boldsymbol{r})$ is the exact extremal of $J[\psi]$, leads to the functional

$$J[\tilde{\psi}] = \nu^2 K - \nu E_D. \tag{4.20}$$

The extremum of this functional has to be achieved at $\nu = 1$. Thus, $2K = E_D$ (virial theorem), and the ratio of various terms (Pekar 1951) is:

$$|\mathcal{H}| : K : E_D : |E_e| : |U| = 1:1:2:3:4, \tag{4.21}$$

here $E_e = K + U$ is the total energy of the particle, i.e., the energy of the electronic subsystem. It follows from eq. (4.18) that the adiabatic theory is applicable for

$$\alpha^2 \gtrsim 10. \tag{4.22}$$

Under these conditions m^* exceeds m significantly; below, while discussing self-trapping, we shall assume condition (4.22) to be fulfilled. It is convenient to introduce the polaron radius r_p by the equation $K \approx (2mr_p^2)^{-1}$. Then the condition of applicability of the macroscopic description $r_p \gg d$ has the form

$$r_p/d \sim \kappa(m_0/m) \gg 1 \tag{4.23}$$

where m_0 is the free electron mass. Such a polaron is called a large polaron. Although the criteria (4.22) and (4.23) are compatible, in principle, it seems that adiabatic large polarons have not yet been discovered experimentally.

For piezopolarons, arising due to the interaction (3.16), the $J[\psi]$ coincides with eq. (4.17) (Pokatilov 1964).

4.2.2. Short-range interaction – ST of electrons and excitons

Substitution of eqs. (3.12) and (3.15) into eq. (4.14) results in the functional (Deigen and Pekar 1951, Rashba 1957b, Holstein 1959, Toyozawa 1961):

$$J[\psi] = (1/2m)\int |\nabla\psi(\boldsymbol{r})|^2\, d\boldsymbol{r} - c\int |\psi(\boldsymbol{r})|^4\, d\boldsymbol{r} = K - E_D. \tag{4.24}$$

where

$$c = C^2/2\rho s_\ell^2 \tag{4.25a}$$

$$c = \gamma^2 v/\omega_0 \tag{4.25b}$$

for acoustic and optic phonons, respectively. The scale transformation (4.19) leads to the functional,

$$J[\tilde{\psi}] = \nu^2 K - \nu^3 E_D, \tag{4.26}$$

considered as a function of ν, it is not restricted from below: $J \to -\infty$ at $\nu \to \infty$. Therefore, the absolute minimum of $J[\psi]$ is reached at singular functions $\psi(\boldsymbol{r})$, corresponding to the zero radius of the quantum state. Thus, the ST state (at least – the lowest ST state) cannot be described in the approximation used here (i.e., in the framework of the linear macroscopic theory), and the radius of the ST state has, generally speaking, the scale of d – the lattice spacing (Deigen and Pekar 1951). The radius can be larger for some numerical reasons only: if the linear approximation is violated early, if the dependence of phonon frequencies and coupling constants on $\boldsymbol{q}$ differ from the macroscopic one already in the small $\boldsymbol{q}$ region, etc. Some examples were considered by Vol and Kukushkin (1977); the radius of exciton bubbles in liquid He is relatively large only for numerical reasons, too.

It is necessary to stress that the second term in eq. (4.24) has a simple form of the "contact self-action", owing to the special choice of the interaction in the forms (3.12) and (3.15) only. For more realistic models it has the form

$$-c \iiiint K(\boldsymbol{r}_1\boldsymbol{r}_2\boldsymbol{r}_3\boldsymbol{r}_4)\psi^*(\boldsymbol{r}_1)\psi^*(\boldsymbol{r}_2)\psi(\boldsymbol{r}_3)\psi(\boldsymbol{r}_4)\, d\boldsymbol{r}_1\, d\boldsymbol{r}_2\, d\boldsymbol{r}_3\, d\boldsymbol{r}_4. \tag{4.27}$$

When the interaction is of the short-range type, the kernel K decreases as a function of any difference of arguments, not slower than $|\boldsymbol{r}_i - \boldsymbol{r}_j|^{-3}$ at $|\boldsymbol{r}_i - \boldsymbol{r}_j| \to \infty$. Such law stems, e.g., from the sound velocity anisotropy in eq. (3.15); in this way it enters into the theory for both excitons and electrons. But for excitons it may also result from the mechanism of the exciton–phonon interaction connected with modulation of the matrix elements M_{nm} (cf. eq. (4.8)), if the exciton transition is allowed; in fact, in this case the M_{nm} decreases asymptotically as $|\boldsymbol{m} - \boldsymbol{n}|^{-3}$, since they are formed by the dipole–dipole interaction. All the conclusions, which are drawn here for the simple functional (4.24) remain correct for an interaction term in the form of eq. (4.27) (Rashba 1957a,b). This follows from the fact that the functions $|\boldsymbol{r}_i - \boldsymbol{r}_j|^{-3}$ and $\delta(\boldsymbol{r}_i - \boldsymbol{r}_j)$ are transformed in the same way under the transformation $\boldsymbol{r} \to \nu \boldsymbol{r}$. If, however, the kernel K decreases faster than $|\boldsymbol{r}_i - \boldsymbol{r}_j|^{-3}$, the actual integration region has the order of magnitude $v \sim d^3$; this is effectively equivalent to the substitution of K by the product of δ functions, i.e., it leads us back to the functional (4.24).

Although the ST states cannot be described by the functional (4.24), it allows one, nevertheless, to give the sufficient criterion for their existence. Self-trapping occurs inevitably, if $J[\psi] < 0$ already in the region of applicability of the macroscopic theory, i.e., at $r_{\text{eff}} \gg d$, r_{eff} being the electron ψ-function radius. The corresponding condition is,

$$mc \gg d, \quad \text{or} \quad E_B \ll E_D. \tag{4.28}$$

The criterion (4.28) corresponds to extremely strong coupling. But the ST states correspond to the absolute minimum of the energy of the system in the whole region where $E_{FC} < 0$, which is significantly larger than the region (4.28) and may be determined by the conditions

$$mc \gtrsim d, \quad \text{or} \quad E_B \lesssim E_D. \tag{4.29}$$

It is convenient to introduce the coupling constant $\Lambda \equiv E_D/E_B$. The criteria (4.28) and (4.29) correspond to $\Lambda \gg 1$ (extremely strong coupling) and $\Lambda \gtrsim 1$ (strong coupling) respectively. The analog of eq. (4.24) written in the site representation is

$$J[\psi] = K - \frac{c}{v} \sum_n |\psi(\boldsymbol{n})|^4, \tag{4.30}$$

from this equation it follows that $E_D \sim c/v$. Since $K \sim E_B \sim (md^2)^{-1}$, the coupling constant has the order of magnitude $\Lambda \sim mc/d$, and for the interaction laws (3.13) and (3.15) we have:

$$\Lambda_{op} \sim md^2\gamma^2/\omega_0, \qquad \Lambda_{ac} \sim mC^2/\rho ds^2. \tag{4.31}$$

The constant Λ_{ac} can be estimated if we make the assumption that all the quantities entering into it have the orders of magnitude determined by expressing them through universal constants, i.e., $C \sim (m_0 d^2)^{-1} \sim E_{at}$, $s/d \sim \theta \sim E_{at}(m_0/m_{at})^{1/2}$, where $m_{at} \approx \rho v$ being the mass of atoms and θ the Debye energy: then $\Lambda_{ac} \sim 1$. It follows from this estimate that ST must be a widespread phenomenon.

Indeed, ST is observed in many wide gap materials. However, it is unknown in typical semiconductors with $E_G \lesssim 1$ eV. This may be understood (Rashba 1976) from the fact that in these materials $E_B \gg E_G$. Therefore, if ST exists, and hence the criterion $E_D \gtrsim E_B$ is fulfilled, it is probable that the energy gain on self-trapping E_{FC} is about E_B too. But then the creation of an electron–hole pair or an exciton would be associated with the expenditure of energy $\approx E_G$ and its subsequent gain $\sim |E_{FC}| \sim E_B \gg E_G$ at ST of electron and (or) hole. Under such conditions the crystal is unstable. For this reason in narrow-gap crystals either ST has to be absent, or a phase transition has to occur*. Apparently, this transition has already occurred in some crystals, so they are "ST-dielectrics". There is also a third possibility that ST occurs, but the Franck–Condon energy is small ($E_{FC} < E_G \ll E_B$). This may be possible only due to the numerical coincidence of independent parameters which results in accidental cancellation of K and E_B in eq. (4.24); it is obvious that the probability of such cancellation is small as $E_G/E_B \ll 1$. Therefore, ST in narrow-gap semicon-

* This transition resembles to some extent the Peierls transition: it is caused by the electron–phonon interaction, it is accompanied by the appearance of a charge density wave, etc. However, it arises for strong electron–phonon coupling only, unlike the Peierls transition.

ductors, though is possible in principle, actually, it should be a rare phenomenon. The probability of observation of the ST states in such crystals must increase near the phase transition curves. The ST states may be considered as precursors of phase transition, and their observation should indicate that one of the phases is an ST-dielectric.

The formulas obtained make it possible to estimate the value of the deformation in the core of the ST state. Performing the Fourier transformation of eq. (4.13), one obtains:

$$\frac{\delta_C}{d} \sim \sqrt{\frac{m}{m_{at}}} \frac{\sqrt{E_D E_B}}{\omega} \tag{4.32a}$$

$$\delta_C/d \sim C/vE. \tag{4.32b}$$

Here δ_C is the displacement in the core, E is elastic modulus. Equation (4.32a) is applicable for arbitrary nonpolar interaction, and (4.32b) for coupling to acoustic phonons. It is seen from eqs. (4.32) that $\delta_C/d \sim 1$, when estimated in terms of the universal constants, i.e., the deformation in the core is generally large, Thus, not only the macroscopic approximation, but the linear approximation as well becomes invalid in describing ST states. The well-known potential curves for Xe_2^* excimer molecule (Mulliken 1970) show, that the nonlinearity of interactions in real physical systems may be actually very strong.

4.2.3. *Mixed interaction law*

In ionic crystals the nonpolar interaction (3.12) may be present side by side with the polar interaction (3.13); this is also the case for acoustic phonons. When polar and nonpolar interactions with different groups of phonons are taken into account, the polar term E_D^p from eq. (4.17) and the nonpolar term E_D^{np} from eq. (4.24) enter into $J[\psi]$ simultaneously. But when both mechanisms of interaction (polar and nonpolar) with the same phonon branch are taken into account, a mixed term arises too (Emin and Holstein 1976, Rashba 1976);

$$-E_D^{mix} = -b \iint \frac{|\psi(\boldsymbol{r})|^2 |\psi(\boldsymbol{r}')|^2}{|\boldsymbol{r} - \boldsymbol{r}'|^2} \, d\boldsymbol{r} \, d\boldsymbol{r}'. \tag{4.33}$$

Under the scale transformation (4.19) this term transformed as $E_D^{mix} \to \nu^2 E_D^{mix}$; therefore, it may change the sign of the coefficient of ν^2 (cf. eqs. (4.20) and (4.26)). This may be of importance for the ST kinetics due to the lowering of the ST barrier (cf. sect. 5.3).

4.2.4. *ST in doped and mixed crystals*

Even if the ST criterion (4.29) is violated in a perfect crystal, it may be fulfilled in the doped crystal, if the attractive potential of the guest, Δ_i, is such that $E_D + \Delta_i \gtrsim E_B$. In this case the deformation caused by a band exciton is weak, but that by an impurity exciton is strong*. In binary mixed crystals, the excitons seen in luminescence spectra are mainly made up of the excited states of the chemical component which has the lower excitation energy. As a result, the character of the exciton spectrum may depend significantly on the dimensionality of percolation clusters. This question has been investigated recently by Shinozuka and Toyozawa (1979) in connection with the interesting results of Kobayashi's group on the spectra of the $TlCl_{1-x}Br_x$ mixed crystals (Nakahara and Kobayashi 1976, Takahei and Kobayashi 1977, 1978). In both perfect crystals no ST is observed. But introduction of Br impurity into TlCl produces an impurity band with large Stokes shift, testifying to a strong lattice deformation. The emission band of Br moves to higher frequencies as x increases. This shows quite unambiguously that the increase of impurity band width, which accompanies the increase of x, makes the conditions for ST less favourable and thus reduces the Stokes shift.

An analogous effect was also observed on trapping the carriers by some defects in $Al_xGa_{1-x}As$ (Lang and Logan 1977) and on changing the charge states of the impurity centers in the ZnS:Cr, ZnSe:Cr systems (Kaminska et al. 1979). This indicates that the electron–phonon interaction displayed in the ST effect is strong, i.e., $\Lambda \gtrsim 1$ (although the interaction of band carriers with phonons is weak in these crystals, i.e., $\lambda \ll 1$, cf. sect. 6 below).

5. *Self-consistency of band states; ST barrier*

In this section we elucidate the conditions under which the band states of excitons, similar to the exciton waves in a nondeformed lattice, will persist even when ST states exist, and establish some consequences following from the existence of two types of states.

5.1. *Self-consistency of band states*

In the previous section the extremals of the functional $J[\psi]$, corresponding to ST states, were investigated. The question arises: whether it has extremals corresponding to free (i.e., band) states:

$$\psi_k(r) = \frac{1}{V^{1/2}} e^{ikr}. \tag{5.1}$$

To solve this problem it is convenient to turn to the corresponding Euler

* Extrinsic self-trapping in narrow-gap semiconductors is not such an exotic phenomenon as intrinsic ST is (cf. sect. 4.2.2); e.g., a special type of ST has been suggested by Porowski et al. (1974) for some donor centers in InSb.

equation $\delta_{\psi^*}J[\psi]=0$:

$$-(1/2m)\Delta\psi(\boldsymbol{r})-2c\hat{K}\psi(\boldsymbol{r})=E_e\psi(\boldsymbol{r}),$$

$$\hat{K}\psi(\boldsymbol{r}_1)\equiv\iiint_{(V)} K(\boldsymbol{r}_1\boldsymbol{r}_2\boldsymbol{r}_3\boldsymbol{r}_4)\psi^*(\boldsymbol{r}_2)\psi(\boldsymbol{r}_3)\psi(\boldsymbol{r}_4)\,\mathrm{d}\boldsymbol{r}_2\,\mathrm{d}\boldsymbol{r}_3\,\mathrm{d}\boldsymbol{r}_4. \tag{5.2}$$

The interaction term is chosen here in the form (4.27); the kernel $K(\boldsymbol{r}_1, \boldsymbol{r}_2, \boldsymbol{r}_3, \boldsymbol{r}_4)$ is supposed to be symmetrized in variables $\boldsymbol{r}_1$ and $\boldsymbol{r}_2$ (as in $\boldsymbol{r}_3$ and $\boldsymbol{r}_4$, too). The eigenvalue of eq. (5.2) coincides with the electronic energy E_e, defined in sect. 4.2.1; this can be easily verified by deriving eq. (33) as the usual Schrödinger equation for electron (exiton):

$$\delta_{\psi^*}\langle\psi|H_e+H_{e\text{-ph}}|\psi\rangle=0, \tag{5.3}$$

and then substituting into it the self-consistent values (4.13) for b_{qs}.

The plane waves $\psi_k(\boldsymbol{r})$ are self-consistent states if, and only if,

$$\|f_k\|=\|\psi_k-\varphi_k\|\equiv[\textstyle\int|\psi_k(\boldsymbol{r})-\varphi_k(\boldsymbol{r})|^2\,\mathrm{d}\boldsymbol{r}]^{1/2}\rightarrow 0, \tag{5.4}$$

when $V\rightarrow\infty$. Here ψ_k is an eigenfunction of eq. (5.2) with the energy $E_e=k^2/2m$ (in the normalization region with the volume V). Making use of the Green function of the operator $(\Delta+k^2)$, eq. (5.2) can be rewritten as an integral equation:

$$\psi_k(\boldsymbol{r})=\varphi_k(\boldsymbol{r})+4mc\frac{1}{4\pi}\int\frac{\mathrm{e}^{\mathrm{i}k|\boldsymbol{r}-\boldsymbol{r}'|}}{|\boldsymbol{r}-\boldsymbol{r}'|}K\psi_k(\boldsymbol{r}')\,\mathrm{d}\boldsymbol{r}' \tag{5.5}$$

which is convenient for finding ψ_k by an iteration procedure.

The following equality,

$$\frac{1}{V^{3/2}}\iiint_{(V)} K(\boldsymbol{r}_1\boldsymbol{r}_2\boldsymbol{r}_3\boldsymbol{r}_4)\,\mathrm{d}\boldsymbol{r}_2\,\mathrm{d}\boldsymbol{r}_3\,\mathrm{d}\boldsymbol{r}_4\leqslant CV^{-\beta-1/2}, \tag{5.6}$$

is assumed to hold. The parameter β depends on the type of interaction. For long-range interactions (cf. eq. (4.17)) $\beta=\frac{1}{3}$. For short-range interactions $\beta=1$, if the interaction is chosen in the form (4.24), and $\beta=1-\eta$ ($\eta>0$, $\eta\rightarrow 0$) if it is chosen in the form (4.27); the quantity η indicates the fact that the integral in eq. (5.6) diverges logarithmically if K decreases as $|\boldsymbol{r}_i-\boldsymbol{r}_j|^{-3}$.

The substitution $\psi_k\rightarrow\varphi_k$ in the nonlinear term of eq. (5.5) leads to the following estimation of the first order correction $f_k^{(1)}$:

$$\|f_k^{(1)}\|\leqslant\frac{mc}{\pi}CV^{2/3-\beta}. \tag{5.7}$$

It may be shown that the estimated higher order corrections to f_k differ from (5.7) only by a numerical factor and contain the same power of V.

Comparing expressions (5.4) and (5.7) we arrive at a conclusion: free states φ_k are self-consistent states for short-range interactions and are not self-consistent for long-range interactions (Rashba 1957a,d). In other words, they are not self-consistent for polarons and piezopolarons, but are self-consistent for all types of excitons and for electrons (holes) under the nonpolar interaction conditions.

The origin of the free state non-self-consistency in the polaron problem may be easily understood from the fact that an electron in the quantum states φ_k in the whole volume V produces a dielectric polarization which creates a parabolic potential well for the electron. Although the depth of this well is proportional to $R^{-1} \sim V^{-1/3}$, i.e., tends to zero at $V \to \infty$, the radius of the ground quantum state ψ_0 in this well is, nevertheless, proportional to $R^{3/4}$, i.e., is vanishingly small, as compared to the initial radius $\sim R$. Therefore, the iteration procedure will result in the monotonic reduction of the quantum state radius, until the self-consistent polaron radius is reached.

5.2. Adiabatic potential surface

Here we consider the systems with short-range interaction in more detail. To understand the physical consequences, which follow from the fact that in these systems the extremals corresponding to both free and ST states may exist simultaneously (cf. sects. 4.2.2 and 5.1), it is convenient to analyze the behaviour of adiabatic potentials, i.e., the dependence of the energy of the system on atomic displacements, when the electron (exciton) is in a definite quantum state.

Lattice displacements can be described by the multidimensional coordinate Q; its components in the momentum representation are $Q_{qf} = (b_{qf} + b^{+}_{-qf})/\sqrt{2}$. The value $Q = 0$ corresponds to the undeformed lattice. The adiabatic potential is

$$\mathcal{H}(Q) = \min_{\psi} \langle \psi | H | \psi \rangle. \tag{5.8}$$

Its critical points (minima, maxima, saddle points) are achieved for functions ψ, the extremals of $J[\psi]$.

It may be stated that the extremal of $J[\psi]$ corresponding to the lowest free state always realizes the minimum of $\mathcal{H}(Q)$ (maybe the local one). To confirm the correctness of this statement, consider small deformations near $Q = 0$. The deformation will be assumed to be small, provided the deformation energy $H_{\text{ph}}(Q)$ is small. If the space distribution $Q(\boldsymbol{r})$ of the deformation is such that a localized state for an exciton (electron) exists only at some finite deformation energy, then at small Q the energy is:

$$\mathcal{H}(Q) - \mathcal{H}(Q = 0) = H_{\text{ph}}(Q) > 0.$$

This case is shown schematically in fig. 2(a) by curve 1. The case is somewhat more complicated when the spatial profile of $Q(\mathbf{r})$ is such that the level is formed already at a low deformation. The existence of such profiles can be most easily shown through the example of an exciton interacting with optical phonons in the framework of the model (3.12). Let us consider the deformation Q in the site representation ($Q \equiv \{Q_n\}$) and demand that $Q_n = Q_0 = \text{const}$ inside the region with the space scale $\sim a$ and $Q_n = 0$ outside of it. Then the condition for the formation of a local level in the potential well,

$$|\gamma Q_0| \gtrsim (ma^2)^{-1}, \quad \text{or} \quad Qa^2 \gtrsim (\gamma m)^{-1}, \tag{5.9a}$$

may be fulfilled even though the energy of lattice deformation tends to zero as $a \to \infty$:

$$H_{\text{ph}} \sim \omega_0 Q_0^2 \frac{a^3}{v} = \frac{\omega_0}{v}(Q_0 a^2)^2 / a \to 0. \tag{5.9b}$$

However, the depth of the energy level $E \sim \gamma Q_0 < 0$ is small under these conditions

$$\frac{|E|}{H_{\text{ph}}} \sim \frac{\gamma v}{\omega_0 Q_0 a^2} \frac{1}{a} \to 0 \tag{5.9c}$$

and therefore the term $H_{\text{ph}} > 0$ dominates. This case is also shown in fig. 2(a) (curve 2). Thus, the point $Q = 0$ corresponds to the minimum $H(Q)$ in both cases considered above.

The behaviour of the adiabatic potential in the region of large Q depends on the strength of the exciton–phonon interaction. Some possible situations are shown in fig. 2(b). There may be no minima at all on the adiabatic

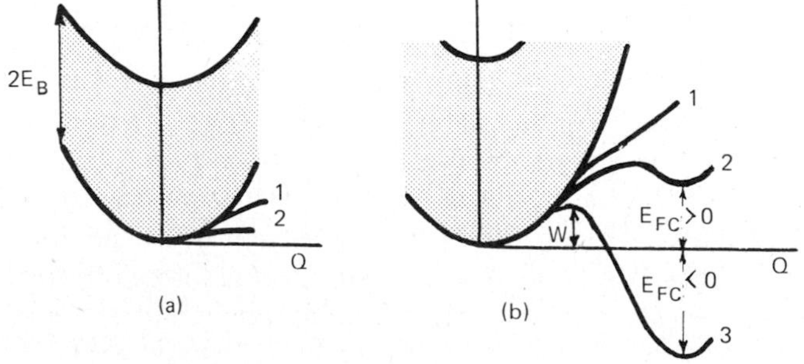

Fig. 2. Shape of adiabatic potential curves. The exciton band region is shaded. The point $Q = 0$ corresponds to free states: (a) Possible behaviour near $Q = 0$. (b) Possible behaviour at large Q.

potential surface split off the bottom of the exciton band (curve 1); the lifetime of such ST states is $\sim\omega^{-1}$. If a minimum exists, it may be located above or below the bottom of the band (curves 2 and 3). The ground state is a band state in the first case ($E_{FC} > 0$) and an ST state in the second case ($E_{FC} < 0$). The criterion (4.29) corresponds to the latter regime. The transition between both regimes occurs abruptly at the critical value of the parameter $\Lambda_{cr} \sim 1$.

5.3. *ST barrier*

The minima corresponding to free and ST states are separated by a potential barrier (Rashba 1957b,c). Below we shall designate this as an ST barrier.

The height of the ST barrier is determined by the position of the lowest saddle point on the adiabatic potential surface. The maxima on curves 2 and 3, considered as sections of such surfaces, may be put into correspondence with this saddle point. The saddle points may be found, quite analogously to eq. (4.12), from the condition that the variation $\delta\langle\psi|H|\psi\rangle = 0$, or analogously to eq. (4.14a), from the condition that $\delta J[\psi] = 0$. The height of the barrier is $W = J_{sp}$. The Euler equation for the functional (4.24) is:

$$-\frac{1}{2m}\Delta\psi - 2c|\psi|^2\psi = E\psi. \tag{5.10}$$

It was shown in sect. 4.2.2 that the absolute minimum of the functional (4.24) corresponds to singular functions $\psi(\boldsymbol{r})$ with zero radius of the quantum state. However, besides these there are regular extremals $\psi(\boldsymbol{r})$ satisfying eq. (5.10). These are the plane waves $\varphi_{\boldsymbol{k}}(\boldsymbol{r})$, describing free states, and also local type functions, corresponding to the saddle points. It is obvious from dimensionality considerations that the height of the barrier W and its spatial scale r_W are equal in order of magnitude:

$$W \sim 1/m^3c^2 \sim E_B/\Lambda^2, \qquad r_W \sim mc \sim \Lambda d. \tag{5.11a}$$

An important consequence (Rashba 1977) follows from expressions (5.11a): the space scale of the barrier is large ($r_W \gg d$) in the extremely strong coupling region ($\Lambda \gg 1$); therefore, the barrier may be described in the framework of the continuum theory. Equation (5.10) was solved numerically (Zakharov et al. 1972) in connection with the theory of the self-focusing of light. The lowest saddle point corresponds to the zero-node function $\psi(\boldsymbol{r})$, the quantum state radius being about $mc/2$, and the barrier height is:

$$W \approx \frac{11.1}{c^2m^3}, \qquad W_{op} \approx \frac{11.1}{(\gamma^2\omega_0 v)^2 m^3}, \quad W_{ac} \approx 44.4\frac{(\rho s^2)^2}{C^4 m^3}. \tag{5.11b}$$

The analog of the virial theorem (4.21) for a saddle point can be easily obtained from eq. (4.26):

$$W:|E_e|:E_D:K:|U|=1:1:2:3:4. \tag{5.12}$$

According to the data obtained by Fugol' and Tarasova (1977) the condition for the validity of the macroscopic description of the barrier, $\Lambda \gg 1$, is fulfilled quite satisfactorily in light rare-gas crystals.

It is important that at $\Lambda \gg 1$ the values of the coupling constant c entering into W and E_{FC} may differ quite considerably. Indeed, the constant of the linear exciton–phonon interaction enters into W, since the deformation in the barrier region is small; therefore, the measurement of W may serve as one of the methods for experimental determination of this constant. On the contrary, the deformation in the core of the ST state is highly nonlinear (cf. sect. 4.2.2), thus, the apparent coupling constant may be quite different in this region.

According to expressions (5.11a), the barrier height W decreases rapidly as the coupling becomes stronger (i.e., Λ increases). Although formally free states remain self-consistent at arbitrary values of Λ, physically the ST barrier is a serious restriction on the ST rate only when

$$W \gg \tfrac{1}{2}\omega_W. \tag{5.13}$$

Here ω_W is the frequency of phonons forming the ST barrier: $\omega_W=\omega_0$ for optic phonons and $\omega_W \approx s/r_W$ for acoustic phonons.

In polar crystals with a strong nonpolar exciton–phonon interaction a relatively weak polar interaction will lower the ST barrier. This may result both from the appearance of a ν-linear term in eq. (4.26), and from a decrease of the coefficient at the quadratic term owing to interference of polar and nonpolar interaction with the same phonon branch (cf. sect. 4.2.3).

Some qualitative consequences of the existence of a ST barrier have been recently discussed by Mott and Stoneham (1977).

5.4. Effect of band degeneracy and the dipole–dipole interaction

It has been supposed everywhere above that the exciton (electron) band is nondegenerate. Meanwhile, self-trapping usually proceeds from degenerate bands; this is true equally for excitons and holes. In the framework of the effective mass method the degeneracy may be taken into account in the usual fashion by introducing the matrix Hamiltonian

$$H_e=\tfrac{1}{2}k_i(\boldsymbol{m}^{-1})_{ij}k_j=-\tfrac{1}{2}\nabla_i(\boldsymbol{m}^{-1})_{ij}\nabla_j, \tag{5.14}$$

here i, j numerate Cartesian coordinates. The rank of the matrices $(\boldsymbol{m}^{-1})_{ij}$ equals the multiplicity g of the band. Expression (5.14) is substituted for

(3.18). Expression (3.3) for $H_{e\text{-ph}}$ must be also replaced by the corresponding matrix Hamiltonian. This generalization does not change the main conclusions of sect. 4.2.2 on the scale of ST states and ST conditions. However, degeneracy may turn out to be significant for the ST kinetics when $\Lambda \gg 1$, i.e., when the barrier space scale $r_W \gg d$. Indeed, the band degeneracy produces a Jahn–Teller distorted ST barrier; so, self-trapping proceeds through states of broken symmetry (Kusmartsev and Rashba 1981).

When the edge of the exciton band is at $\boldsymbol{k} = 0$ and the optical transition is allowed, there arises singular contribution to the exciton energy connected with long-range dipole–dipole interactions:

$$(H_{dd}(\boldsymbol{k}))_{\zeta\zeta'} = \frac{4\pi}{vk^2}(\boldsymbol{d}_\zeta^*\boldsymbol{k})(\boldsymbol{k}\boldsymbol{d}_{\zeta'}) \tag{5.15a}$$

Here $\boldsymbol{d}_\zeta$ is the matrix element of the dipole moment (per a primitive cell) for quantum transition to the state ζ ($\zeta = 1, 2, \ldots, g$). The expression H_{dd} is of a purely macroscopic origin and is the same for excitons of various types (phonons, magnons, electronic excitons and so on). However, it is specific just for excitons, and there is no such term in the Hamiltonians of electrons (holes). The peculiarity of H_{dd} consists in its singular behaviour at $\boldsymbol{k} \to 0$; it results in the longitudinal–transverse splitting $\epsilon_{\ell t}$ – the difference between the frequencies of longitudinal and transverse phonons in the limit $\boldsymbol{k} \to 0$. Equation (5.15a) provides an expression for H_{dd} in the momentum representation. In the coordinate representation the H_{dd} is transformed into the integral operator (Rashba 1959):

$$(H_{dd}\psi(\boldsymbol{r}))_\zeta = \frac{1}{v}\sum_{\zeta'}\int\left[(\boldsymbol{d}_\zeta^*\nabla_r)(\boldsymbol{d}_{\zeta'}\nabla_{r'})\frac{1}{|\boldsymbol{r}-\boldsymbol{r}'|}\right]\psi_{\zeta'}(\boldsymbol{r}')\,d\boldsymbol{r}'. \tag{5.15b}$$

ψ_ζ are the components of the wave function. When included into the total Hamiltonian H (see eq. (3.1)), H_{dd} introduces a term into eq. (5.10) which decreases slowly with r: $H_{dd}\psi \propto r^{-3}$. As a result, far from the localization center (i.e., at the distances exceeding the ST state radius), the dominant terms in the generalized eq. (5.10) are:

$$H_{dd}\psi \approx E\psi,$$

hence, $\psi(\boldsymbol{r}) \propto r^{-3}$ at $r \to \infty$. Therefore, the long-range interactions change the asymptotic behaviour of $\psi(\boldsymbol{r})$*, and their effect on the exciton ST barrier may turn out very significant, provided $\epsilon_{\ell t} \gtrsim W$. On the contrary, at small $\epsilon_{\ell t}$ the long-range interactions result in small corrections only, and they should be considered together with the corrections arising due to substitution of the second term in eq. (5.10) by the more general expression $\hat{K}\psi$ (cf. eq. (5.2)).

* The dipole–dipole Förster migration of ST excitons becomes possible due to this change in the asymptotic behaviour of $\psi(\boldsymbol{r})$.

5.5. ST barrier for local excitons

The ST barrier for band particles (excitons, electrons) was considered above. The question arises: whether an analogous barrier may exist for particles bound near impurities or lattice defects. In other words, whether two types of states, with weak and strong lattice deformations, may coexist in impurity centers. It may be stated that when the characteristic radius r_i of the impurity center is large for some reason, i.e., $r_i \gg r_W$, the barrier does exist and its height will be close to W, the barrier height for a band particle. When r_i decreases, the barrier becomes lower and disappears at $r_i \sim r_W$.

So far the coexistence of two types of states was not apparently observed for local excitons. But it was established lately for donor electrons in the CdF_2:In system (Langer et al. 1979).

5.6. Concluding remarks

It has been shown in sect. 5.1 that free band states are self-consistent under the nonpolar interaction. Certainly, this does not mean that these states are not perturbed by the exciton (electron)–phonon interaction. It only suggests that the adiabatic approximation is inapplicable to band states, and other methods (e.g., weak and intermediate coupling; see sect. 6) should be used. The situation is, however, profoundly complicated by the fact that when ST states exist below the bottom of the exciton band, the dependence of band parameters on the coupling constant is undoubtedly nonanalytic owing to the tunneling from free to ST states, its probability depending exponentially on the inverse coupling constant (cf. sect. 9). Therefore, the region of applicability of the perturbation theory is narrower than in the theory of polarons.

It will be shown in sect. 6 that physical meaning can be assigned to both minima on curve 3 (fig. 2(b)). They correspond to two types of excitons, actually coexisting in a crystal. In this connection the question arises, whether these two systems of functions are independent and mutually orthogonal. It may be shown that well below the barrier these conditions are fulfilled. Indeed, the overlap of wave functions, corresponding to weakly excited states near both minima, have the order of magnitude of $\exp(-S)$, i.e., it is exponentially small. Hence, the non-orthogonality integrals of electron–phonon wavefunctions are also small. It goes without saying that this "autonomy" of two types of states is absent above the barrier; in this region the spectrum is totally rearranged.

It should be stressed in conclusion that the ST barrier considered here is formed owing to the kinetic energy of an exciton (electron); its height tends to zero as $m \to \infty$. Thus, it has nothing to do with the barrier of lattice origin discussed above, as applied to heavy excitons (cf. sect. 4.1), and originates from nonlinearity of the exciton–phonon interaction. But both mechanisms may be important essentially for light excitons, since the exciton–phonon interaction is strongly nonlinear in the actual region of deformations.

The fact that self-trapping may be connected with overcoming the potential barrier was first mentioned in the initial paper of Landau (1933). Such barrier is absent for large polarons, but exists in many other cases. The question of the existence of additional barriers between different ST states (e.g., between large and small polarons, quasi-atomic and quasi-molecular excitons, etc.) cannot be solved in general form.

6. Coexistence of free and ST excitons

The results of the previous sections can be formulated briefly as follows: in three-dimensional systems with strong short-range electron–phonon interaction both free and ST states are self-consistent solutions of the adiabatic theory and have physical meaning. The exciton–phonon interaction is always of short-range character. No alternative thus exists for the formation of free and ST states. So, it should be stated that they may coexist (Rashba 1957b,d, Toyozawa 1970).

However, it remains to be elucidated what are the properties of free (i.e., band) excitons under the conditions, when the value of the coupling constant is sufficiently high ($\Lambda \gtrsim 1$) to ensure the existence of ST states lying below the free states. Won't the mean free paths of excitons be so small that the band description will lose sense? Won't the renormalization of their spectrum in the nonadiabatic approximation lead to a lowering of the band bottom comparable with E_{FC}?

To answer these questions it is sufficient to employ the well-known results of electron–phonon interaction theory in the weak coupling limit. For definiteness we restrict ourselves to the interaction with acoustic phonons; the basic results can be found, for instance, in the book of Kittel (1963, ch. 7).

The quantity

$$\lambda_{ac} \approx m^2 C^2/\rho s \tag{6.1}$$

serves as the coupling constant. The main physical parameters, such as the mean number of phonons in the cloud $\langle n \rangle$, the shift of the bottom of the band $\Delta\epsilon$ and the mean free time $\tau(\epsilon)$, relative to spontaneous phonon emission, may be expressed through λ_{ac}:

$$\langle n \rangle \sim \lambda_{ac} \ln(1/msd), \tag{6.2a}$$

$$\Delta\epsilon \sim \lambda_{ac} E_B, \tag{6.2b}$$

$$\frac{\tau^{-1}(\epsilon)}{\epsilon} \sim \lambda_{ac}, \tag{6.2c}$$

ϵ is the energy measured from the band bottom.

The interaction constants λ and Λ contain quite different combinations of the crystal parameters. Their ratio (cf. expressions (4.31)) is:

$$\lambda_{ac}/\Lambda_{ac} \sim msd = md^2(s/d) \sim \Theta/E_B \sim (m_0/m_{at})^{1/2}. \tag{6.3}$$

The last equality is based on the estimate obtained in terms of universal constants. Therefore, λ is small, as compared to Λ, in the adiabatic parameter, and the criteria $\lambda \ll 1$ and $\Lambda \gtrsim 1$ are compatible, so long as the excitons are light, i.e., $\Theta \ll E_B$.

It may be seen from (6.2a) that $\langle n \rangle$ coincides with λ_{ac} except for a logarithmic factor, i.e., $\lambda_{ac} \ll 1$ is the weak coupling criterion for free excitons. According to (6.2b), the lowering of the band bottom of free excitons at $\lambda_{ac} \ll 1$, $\Lambda_{ac} \gtrsim 1$ is small: $\Delta\epsilon \ll E_D \sim E_{FC}$; it is much less than the ST energy. And at last, according to (6.2c), the damping of free excitons is much less than their energy, i.e., the usual kinetic description holds. Thus, under the conditions of free and ST exciton coexistence the spectrum of free excitons undergoes only weak renormalization and they retain large free path lengths; the condition for this, $\lambda \ll 1$, is compatible with the ST criterion $\Lambda \gtrsim 1$*.

Under the same conditions the bands of ST excitons are exponentially narrow (Rashba 1957c, Toyozawa 1961); here formulas of the (4.9)–(4.10) type hold too, though with a slightly changed expression for S_{nm}. As a result, the kinetic regime in the band of ST excitons is easily destroyed and their diffusion is mostly of a hopping type. One may thus expect that the diffusion coefficients of free and ST excitons would show difference by many orders of magnitude. In fact, by the combined observation of exciton transport, typical of free excitons, and of the fluorescence spectrum, specific to ST excitons, the coexistence of these two types of excitons was first established in alkali iodides by Lushchik et al. (1973). Changes in kinetic properties at $\Lambda = \Lambda_{cr}$ (cf. sect. 5.2) are similar to those at a phase transition; it is from this point of view that the situation has been discussed by Sumi and Toyozawa (1973) and by Emin (1973).

The dependence of the position of the free and ST exciton band bottoms on exciton–phonon coupling is outlined in fig. 3. The lowering of the bottom of free exciton band at $\lambda \ll 1$ can be found by the perturbation

* It is of interest to establish the correspondence between the theory of dielectrics, in which there are two coupling constants λ and Λ, and the electron theory of metals. It can be verified directly that in metals the coupling constant, which determines the renormalization of the effective mass on the Fermi surface and the sound velocity (Migdal 1957), as well as electron scattering, coincides with the "large" constant Λ.

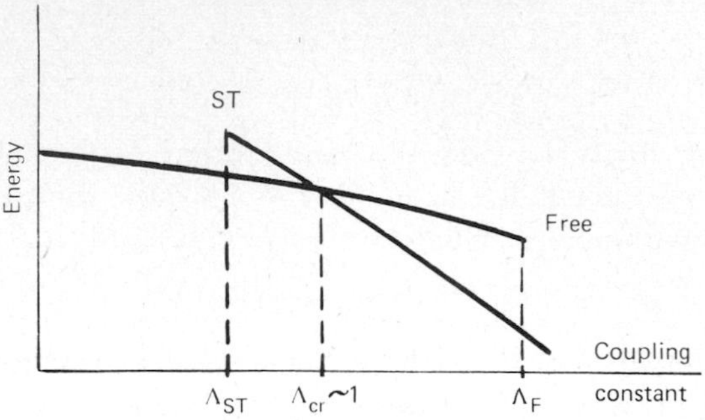

Fig. 3. Schematic dependence of the positions of free exciton and ST exciton band bottoms on the coupling constant.

theory. It is essential that the curve for ST excitons be much steeper than that for free excitons. The point Λ_F corresponds to the limit of applicability of the criterion (5.13); in this point $\Lambda_F\lambda_F \sim 1$. Metastable ST states appear first at the point Λ_{ST}. The region of $\Lambda_{ST} < \Lambda < \Lambda_F$ corresponds to the coexistence of free and ST excitons.

Since excitons are produced by light, by ionizing radiation or by some other source, their distribution is nonequilibrium and the concentration ratio of free and ST excitons is determined by the rate of relaxation transitions: overbarrier activated transitions at high temperatures, and subbarrier tunnel ones – at low temperatures (cf. sect. 9). The transport of energy in some crystals with self-trapping, observed experimentally and ascribed to free excitons, is apparently connected in all cases with non-relaxed free excitons.

However, there is another interesting possibility when free particles make a dominant contribution to the transport under equilibrium conditions, despite the fact that most of the particles are in ST states. Such a situation may arise, e.g., for small polarons, under the following conditions. At $E_{FC} < 0$ the concentration of free particles is determined by $\exp(-|E_{FC}|/T)$. The hopping conductivity in the band of ST states is proportional to $\exp(-W_{ST}/T)$, where W_{ST} is the activation energy for a particle jump to the neighbouring lattice site. It is obvious that $W_{ST} < W + |E_{FC}|$. However, at $W \gtrsim |E_{FC}|$ it may turn out that $W_{ST} > |E_{FC}|$, and then, in some temperature range at least, the contribution of free particles in the transport will dominate. This regime should be expressed most distinctly in systems with small $|E_{FC}|$, i.e., near the "phase transition".

In the foregoing we examined the cases of heavy and light excitons separately, which made it possible to reveal explicitly qualitative differences in their behaviour. Attempts to provide a unified description in the whole range of the values of parameters ω/E_B and Λ are possible only in the framework of definite interpolation schemes (see, for instance, Cho and Toyozawa (1971), Emin (1973), H. Sumi (1974), Prelovsek (1975)).

7. Wannier–Mott excitons

Self-trapping of large radius excitons (WM excitons) display a number of attractive peculiarities. In particular, from the theoretical point of view, this is the only system having both a ST barrier and a ST state, which may be described in terms of the macroscopic theory.

The adiabatic theory of large ST excitons (Dykman and Pekar 1952, Haken 1958) may be developed as natural generalization of the theory described in sect. 4. Equation (4.13) is substituted by

$$b_{qf} = -\frac{1}{N^{1/2}}\frac{1}{\omega_{qf}}\int \mathrm{d}\boldsymbol{R}\,\exp(\mathrm{i}\boldsymbol{qR})$$
$$\times \int \mathrm{d}\boldsymbol{r}\psi^*(\boldsymbol{rR})\left[\gamma_{qf}^{e*}\exp\left(\mathrm{i}\frac{m_h}{M}\boldsymbol{qr}\right)+\gamma_{qf}^{h*}\exp\left(-\mathrm{i}\frac{m_e}{M}\boldsymbol{qr}\right)\right]\psi(\boldsymbol{rR}). \qquad (7.1)$$

Here m_e, γ_{qf}^e and m_h, γ_{qf}^h are effective masses and coupling constants for electrons and holes, $M = m_e + m_h$, $\boldsymbol{r} = \boldsymbol{r}_e - \boldsymbol{r}_h$, $\boldsymbol{R} = (m_e\boldsymbol{r}_e + m_h\boldsymbol{r}_h)/M$.

For the polar interaction, $\gamma_q^e = -\gamma_q^h$, and the expression in the parentheses in eq. (7.1) vanishes at $m_e = m_h$. Therefore, the interaction is strong only when the masses differ greatly; as a result, the energy of the ST state is lower than that of the band state only when $m_h/m_e \gtrsim 10$ (here and below it is assumed for the sake of definiteness that $m_h > m_e$). The polaron radius $r_p \propto m^{-1}$; hence, under these conditions there exists a strong inequality $r_p^e \gtrsim 10 r_p^h$ for the radii of electron and hole polarons. This means that the ST exciton is suggestive of the F center with the core substituted by hole polaron. Such a state, in principle, may be macroscopic, but, since r_p^h is small, the hole polarons tend to "sit down" on the atomic space scale (cf. sect. 2.5).

At present no substances in which continual ST excitons exist are known.

Coupling of exciton to long-wave phonons is weak at an arbitrary magnitude of m_h/m_e; indeed, $b_q \propto q$ at $qr_{ex} \ll 1$, according to eq. (7.1). This is particularly evident in the multiplicative approximation $\psi(\boldsymbol{rR}) = \Psi(\boldsymbol{r})\Phi(\boldsymbol{R})$, since the effective coupling constant for the center-of-mass motion $\gamma_q^{ex} \propto q$. The physical reason for the weak coupling is the electric neutrality of the exciton, and the short-range behaviour of the interaction results in coexistence of free and ST excitons. The plot of the fig. 3 type was obtained for large excitons by Dykman and Zaslavskaya (1958), who performed the calculations by the intermediate coupling method.

The results of the calculation (Pekar et al. 1979) of the ST barrier height W and of the ST state energy $\mathscr{H}$ in the adiabatic approximation are shown in fig. 4. The curves exist in the restricted region of values of the parameter m_e/m_h only. At the edge of this region the saddle point and metastable minimum merge, and for larger values of the parameter m_e/m_h the adiabatic potential surfaces show the behaviour indicated by curve 1 (fig. 2). It is remarkable that W turns out to be small, as compared to the exciton binding energy E_{ex}. In addition, in almost the whole region where the ST state is the ground state, W is small, as compared with $|\mathscr{H}|$. The fact that W is small implies that the macroscopic approximation should be much more widely applicable for the calculation of the barrier height than of the ground state energy.

Low values of W are in a qualitative agreement with the experimental data available on alkali halide crystals (Lushchik 1982), though, apparently, the theoretical values somewhat exceed those obtained by experiment. It is natural to attribute this discrepancy, if any, to considerable nonadiabaticity in the tunneling region (see also sect. 11.2). Therefore, the development of intermediate coupling methods in the ST barrier theory seem to be very important; no work in this direction has been undertaken so far.

Equation (7.1) corresponds to the model, in which both electron and hole move in one and the same self-consistent field. There is also an alternative

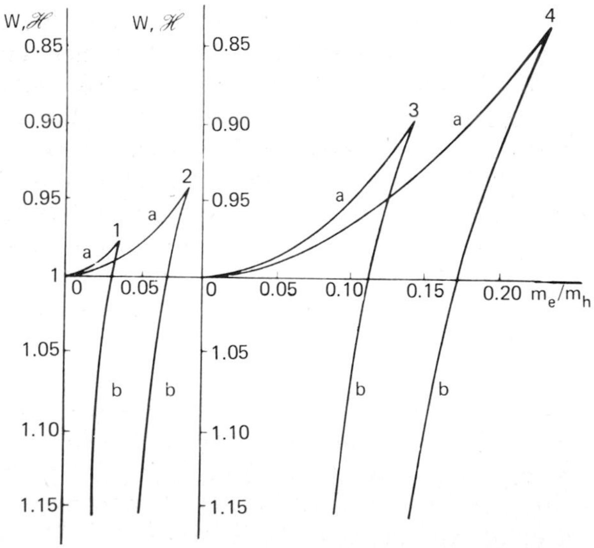

Fig. 4. Dependence of W (curves a) and $\mathscr{H}$ (curves b) on the mass ratio for ST exciton. The energy is expressed in units of the exciton binding energy $E_{ex} = \mu e^4/2\kappa_\infty^2$, $\mu = m_e m_h/M$ being the reduced mass. The numbers at the curves show the values of the parameter $4(1-\kappa_\infty/\kappa_0)$ (Pekar et al. 1979).

approach based on the model of the exciton as consisting of electron and hole polarons, bound by the Coulomb attraction (Haken 1956); it is advantageous, as applied to highly excited states. An intermediate approach between these two has been developed as well (Vinetsky and Kravchenko 1965).

In the case of the nonpolar interaction the possible types of ST states are very diversified, since both the relative magnitude and the signs of coefficients γ_{qf} for electrons and holes may be arbitrary. Depending on the values of γ and m, either one or both particles may be self-trapped. Depending on whether the signs of γ^{e} and γ^{h} coincide, electron and hole are mutually repelled or attracted through the phonon field. Accordingly, various types of ST states arise; they were considered by A. Sumi (1977). Large-radius excitons in magnetic crystals were also considered (Krivoglaz and Trushchenko 1971).

8. ST in low-dimensional systems

It is seen from the foregoing that the whole picture of self-trapping is strongly affected by the type of the electron–phonon coupling. In this section it will be shown that for the nonpolar interaction the picture is qualitatively changed when turning to the systems with low-dimensional spectrum (Rashba 1957b). All the conclusions hold when (i) the phonons are of the same dimensionality as the electron energy spectrum, (ii) both the phonon spectrum and the exciton–phonon interaction remain three-dimensional, (iii) phonon dispersion is weak, and in some other cases.

First of all, in low-dimensional systems the functions φ_k (see eq. (5.1)) are no longer self-consistent solutions of eq. (5.2). In fact, the Green function of the two-dimensional (2d) problem, which may be obtained from the Green function entering into eq. (5.5) by integration over z', diverges logarithmically at small k. The Green function of the one-dimensional (1d) problem diverges even more strongly. Therefore, the functions φ_k with $k \to 0$ are not self-consistent solutions for 1d- and 2d-systems.

8.1. Two-dimensional systems

For a two-dimensional system the expression, analogous to eq. (4.26), has the form

$$J[\tilde{\psi}] = \nu^2 K - \nu^2 E_D, \tag{8.1}$$

hence, the ratio of both terms is scale-independent. According to eq. (4.24), this ratio is determined by the value of the parameter mc. The critical value of mc may be found, if the results of numerical investigation (Chiao et al.

1964) of the problem of cylindrical light beam self-focusing, which is described by the same equation, are used. Finally,

$$(mc)_{cr} = 2.88. \tag{8.2}$$

If $mc < 2.88$, ST does not set in. If $mc > 2.88$, ST occurs without any barrier and the ST states have the space scale $\sim d$. These conclusions hold for the linear exciton–phonon coupling and isotropic phonon spectrum; the effect of the phonon spectrum anisotropy was considered by Agranovich et al. (1977).

Excitons in layered crystals with highly anisotropic energy spectrum and also surface excitons may be regarded as 2d excitons. ST of surface excitons should proceed particularly rapidly due to absence of a ST barrier: therefore, diffusion of excitons to inner cracks, boundaries of vacancy clusters, etc., may significantly intensify its self-trapping in the bulk of a crystal.

At present surface excitons are observed in the absorption spectra of Ar, Kr, Xe (Saile et al. 1976) and Ne (Saile and Koch 1979); their bands are shifted by approximately 0.1–0.6 eV below the bulk exciton absorption bands. Emission from the ST exciton states dominates in the intrinsic fluorescence spectra of all these crystals. Therefore, it would be natural to expect the experimental confirmation of surface exciton self-trapping, though it has not been observed as yet.

It has been claimed (Sugakov 1970, Brodin et al. 1971, Philpott and Turlet 1975, 1976) that surface excitons occur also in layered crystal anthracene at the free *ab*-face, which is the cleavage plane. Here the surface exciton band is shifted by ~0.02 eV above the bottom of the bulk exciton band. Apparently, surface excitons are largely autonomous and their relaxation into bulk excitons is hindered, as may be concluded from observing the surface exciton fluorescence. Since the lattice barrier resisting molecular movement is expected to be significantly lower on the surface than in the bulk (cf. the end of sect. 4.1) the formation of surface excimers and surface photodimerization seem plausible.

ST of surface excitons may be inhibited by their short radiative lifetime; it is small in the parameter $kd \ll 1$ (Agranovich 1968). But the effect of this factor may be significantly suppressed by the finite mean free path l of the exciton; when $kl \ll 1$, the radiative lifetime is determined by the parameter d/l.

8.2. One-dimensional systems

The functional $J[\psi]$ for 1d systems shows the same behaviour under the scale transformation (4.19), as does the three-dimensional polaron, this behaviour is described by eq. (4.20). Therefore, self-trapping proceeds without overcoming a barrier. The Schrödinger equation (5.10) for ST states in a one dimensional system has the exact solution (Rashba 1957b, Holstein 1959):

$$\psi(\xi) = 1/\sqrt{2}\cosh\xi, \qquad E = -mc^2/2, \qquad \xi = mcy, \tag{8.3}$$

y is the coordinate along the chain. The ground state energy is $\mathcal{H} = -mc^2/6$. This state is the only ST state – in distinction from the polaron, which possesses a sequence of ST excited states. Due to the existence of the exact solution (8.3), the theory of 1d-adiabatic exciton has been developed quite elaborately. The Landau–Pekar effective mass at small momenta is found for interaction with both acoustic (Davydov and Kislukha 1976) and optic (Mel'nikov 1977, Shaw and Whitfield 1978) phonons. The dispersion law at large momenta is asymptotic to the sound velocity* in the acoustic case (Davydov and Kislukha 1976), and tends to the phonon frequency ω_0 in the optic case (as in the weak coupling case; Suna 1964, Levinson and Rashba 1973). The spectrum of bound states of optical phonons with a ST particle was calculated too (Shaw and Whitfield 1978, Mel'nikov 1977); the lowest local frequency is $\approx 0.65\omega_0$. One dimensional excitons may arise in polymer chains (e.g., in solutions)† and in crystals with highly anistropic energy spectrum.

Spectra of polymers, formed in water solutions of cyanin dyes (Scheibe polymers; for review see Scheibe 1941) differ significantly from the spectra of monomers. This is an indication that collective effects are very important. The width of the exciton band, estimated from the oscillator strength, is high ($E_B \sim 1000\ \mathrm{cm}^{-1}$); so the formation of light excitons might be expected. For the ψ-isocyanin type substances polymerization is followed by appearance of a narrow absorption band; undoubtedly, this corresponds to free excitons (Franck and Teller 1938). On the contrary, the spectra of the substances of the pinacyanol-diethylchloride type are broadened on polymerization. It has been suggested (Rashba 1957d) that this broadening is due to ST of excitons but this interpretation remains conjectural as yet.

The most convincing are the data on charge transfer excitons in anthracene – PDMA crystal (Haarer 1974, 1977, Haarer et al. 1975, Philpott and Brillante 1979). In this crystal the width of the exciton band in a rigid lattice ($E_B \sim 1000\ \mathrm{cm}^{-1}$) is one order of magnitude larger than the frequencies of lattice phonons, i.e., excitons are light. Meanwhile, there is quite definite indication that the exciton–phonon interaction is strong: the narrow absorption band has a strong wide wing, fluorescence bands have the width of about $500\ \mathrm{cm}^{-1}$, the width of the ST exciton energy band is small (about

* Analogous to the piezo-polaron (Volovik and Edel'shtein 1974).

† Suppression of ST owing to short radiative lifetime (Möglich and Rompe 1943) is weakened by a short free path, quite analogous to 2d systems.

10 cm^{-1}), and the Franck–Condon energy is large ($S \approx 6$, cf. eq. (4.11)). The fact that the coupling is strong is thought to result from the nearly total charge transfer to neighbouring molecules in the electronically excited state (cf. sect. 3).

In Peierls dielectrics, e.g., polymer chains of polyacetylene $(CH)_n$ with conjugated bonds, the formation of neutral spin-$\frac{1}{2}$ solitons and charged spin-zero solitons is possible (Brazovskii 1978, Su et al. 1979). These solitons arise due to strong electron–phonon coupling. A new type exciton, strongly coupled to lattice, should be formed from pairs of solitons with opposite charges.

It has been claimed recently (Davydov 1974) that ST 1d excitons arise in some biological systems as well, and, particularly, are of importance for muscle action.

Finally, the trapping of excitons by dislocations turns them into 1d-excitons and should favour their self-trapping. It is possible that the suppression of exciton diffusion observed in some alkali halide crystals on introduction of dislocations (Lushchik 1982) is conditioned by this process.

8.3. Effect of anisotropy of the exciton energy spectrum

The limiting transition from 3d systems with ST barrier to nonbarrier 2d and 1d systems may be traced. For excitons, having anisotropic energy spectrum with a set of effective masses m_1, m_2 and m_3, eq. (5.11b) for the barrier height is replaced by

$$W \approx \frac{11.1}{c^2 m_1 m_2 m_3}. \tag{8.4}$$

When one or two masses $m_i \to \infty$ (2d and 1d systems, respectively) the barrier disappears.

Quasi-1d-spectra usually arise in systems built of stacks of plane molecules, the distances between molecules in a stack being considerably less (about 3 times) than the distances between stacks. Anisotropy of energy spectra of electrons, holes and triplet excitons may be very high ($\sim 10^4$) under these conditions owing to the exponential decrease of exchange integrals with distance. But for singlet excitons, especially for strong allowed transitions where the exciton band is formed due to dipole–dipole interactions ($\propto R^{-3}$), anisotropy is expected to be significantly less (~ 30). It is worth mentioning that the non-1d part of the exciton dispersion law includes the terms of the (5.15a) type, which show nonanalytic behaviour at $\boldsymbol{k} \to 0$. The angle between $\boldsymbol{k}$ and the direction of the optical transition dipole moment is an additional independent variable; therefore, the energy spectrum of such systems is effectively 4d or 5d* (depending on the polarization of the transition). The energy dependence of the density of states near the exciton band edge, the behaviour of the impurity exciton levels (Rashba 1962), etc., correspond to these higher dimensionalities; they should also show up in the ST problem. Therefore, the ST barrier may exist, its height being determined by the ratio of the singular and regular contributions to the spectrum. This problem has not been investigated so far.

9. *Self-trapping rate*

We pass now to the estimation of factors which determine the ST rate of free excitons. The ST process is illustrated in fig. 5. Excitons reach the trans-barrier region either by tunneling or by activation; arrow 2 shows the tunneling in an excited state. The mean time, until ST sets in, τ_{ST} may be roughly estimated as

$$\tau_{ST}^{-1} \sim \omega\langle D(E)\rangle, \tag{9.1}$$

where ω has the order of magnitude of the phonon frequencies, $D(E)$ is the barrier transparency which depends on the total energy of the system $E = \mathscr{H}(Q) + K_\ell$, K_ℓ is the kinetic energy of the lattice, and brackets designate the mean value over the energy distribution with temperature T.

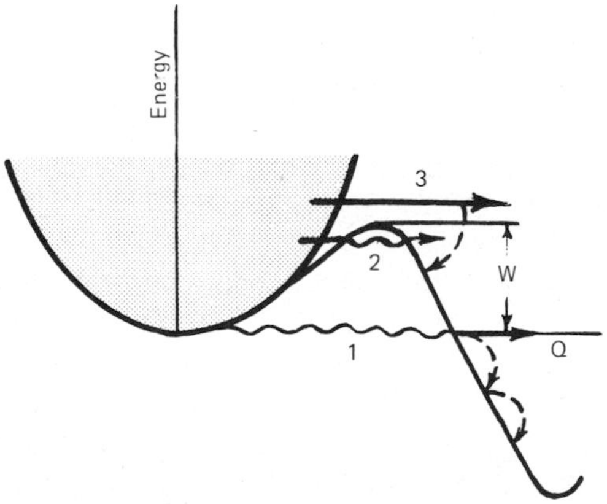

Fig. 5. Passage of the system through the ST barrier. The solid heavy line denotes the motion in classically achievable region, and wavy lines show the tunneling. Dashed lines describe the system relaxation after overcoming the barrier.

* This view of the problem was advanced by Larkin and Khmel'nitskii (1969) as applied to the theory of phase transitions.

Equation (9.1) is valid when an exciton which has penetrated into the trans-barrier region passes through the initial stages of relaxation rapidly and thus does not return to a free state. The condition that excitons relax during their lifetime τ_L is

$$\tau_{ST} \lesssim \tau_L. \tag{9.2}$$

The dominant contribution to $\langle D(E)\rangle$ comes from the vicinity of $E = E_{max}$, where the integrand reaches a maximum:

$$\frac{d}{dE}\{D(E)\exp(-E/T)\}\big|_{E=E_{max}} = 0. \tag{9.3}$$

The detailed dependence of $D(E)$ on E is at present unknown. It is obvious, however, that this dependence is strong (of the exponential type). Therefore $E_{max} < W$, and the difference between them may be very substantial. Hence, the dependence of τ_{ST}^{-1} on T may be weaker than that following from the usual activation Arrenius law. As a result, the "activation energy" q, measured from the plot $\ln(\tau_{ST}^{-1})$ vs $1/T$, may turn out to be significantly less than W.

After the barrier is passed, the system relaxes to the ground state. This process may also entail certain difficulties, so it may not come to an end for the time τ_L.

Both stages of relaxation will be discussed in this section.

9.1. Theory of the tunnel self-trapping

The theory of tunnel self-trapping may be developed only in the case $\Lambda \gg 1$, when the ST barrier is described in a continuum approximation. We restrict ourselves to the case where the total energy of the system $E = 0$; in this case tunneling proceeds along line 1, fig. 5. We follow below, the paper of Iordanskii and Rashba (1978), discussing the basic scheme and the main results only.

Let us suppose that in the barrier region the adiabatic approximation holds. This allows us to use the classical description of lattice motion. Under these conditions it is most convenient to describe the subbarrier motion by introducing imaginary time $\tau = it$. In these terms the problem is reduced in principal approximation to the calculation of minimal Hamiltonian action S for paths of the line 1 type (fig. 5), connecting the points with the same energy $\mathscr{H}(Q) = 0$, lying at opposite sides of the barrier. The transparency equals

$$D \approx \exp(-2S). \tag{9.4}$$

The extremal action for exciton, interacting with acoustic and optic phonons, in non-dimensional variables is

$$\mathscr{S}_{ac} = \min_{\psi Q} \iint \{\tfrac{1}{2}(\partial \boldsymbol{Q}/\partial \tau)^2 + \tfrac{1}{2}(\operatorname{div} \boldsymbol{Q})^2 + \tfrac{1}{2}|\nabla\psi|^2 + |\psi|^2 \operatorname{div} \boldsymbol{Q}\} \, d\boldsymbol{r} \, d\tau. \quad (9.5)$$

Here $\boldsymbol{Q}(\boldsymbol{r}\tau)$ describes the lattice, and $\psi(\boldsymbol{r}\tau)$ the exciton. The expression for action for optic phonons $\mathscr{S}_{op}$ differs from eq. (9.5) by the substitution $\operatorname{div} \boldsymbol{Q}(\boldsymbol{r}) \to Q(\boldsymbol{r})$. These expressions correspond to the model of eq. (3.15) for acoustic phonons and eq. (3.12) for optic phonons. But in distinction from sect. 4 it is more convenient to use here the coordinate representation, rather than the momentum one. The action S is related to $\mathscr{S}$ by the expressions

$$S_{ac} = \frac{\rho s_\ell}{m^2 C^2} \mathscr{S}_{ac}, \qquad S_{op} = \frac{\omega_0}{4m^3 v^2 \gamma^4} \mathscr{S}_{op}. \quad (9.5')$$

The Euler equations for lattice motion are,

$$\begin{aligned} &\partial^2 \boldsymbol{Q}/\partial\tau^2 + \Delta \boldsymbol{Q} + \nabla|\psi|^2 = 0, \quad \text{acoustic phonons} \\ &\partial^2 Q/\partial\tau^2 - Q - |\psi|^2 = 0, \quad \text{optic phonons.} \end{aligned} \quad (9.6)$$

When time derivatives are neglected, these equations determine the self-consistent deformation (4.13). After solving eq. (9.6) and eliminating $\boldsymbol{Q}(\boldsymbol{r}\tau)$ in eq. (9.5), we get

$$\mathscr{S}_{ac} = \tfrac{1}{2}\min_\psi \int [\tfrac{1}{2}|\nabla_r \psi(R)|^2 + \tfrac{1}{2}|\psi(R)|^2 \Delta_r \int G(R-R')|\psi(R')|^2 \, d^4R'] \, d^4R, \quad (9.7)$$

where

$$G(R) = G(\boldsymbol{r}\tau) = \iint \exp[i\boldsymbol{k}(\boldsymbol{r}-\boldsymbol{r}') - i\omega(\tau-\tau')] \frac{1}{\boldsymbol{k}^2 + \omega^2} \frac{d\boldsymbol{k}}{(2\pi)^3} \frac{d\omega}{2\pi}, \quad (9.8)$$

and

$$\mathscr{S}_{op} = \tfrac{1}{2}\min_\psi \int d\boldsymbol{r} \int_{-\infty}^{\infty} [\tfrac{1}{2}|\nabla_r \psi(\boldsymbol{r}\tau)|^2 + \tfrac{1}{2}|\psi(\boldsymbol{r}\tau)|^2 \int_{-\infty}^{\infty} G(\tau-\tau')|\psi(\boldsymbol{r}\tau')|^2 \, d\tau'] \, d\tau, \quad (9.9)$$

where

$$G(\tau-\tau') = -\int_{-\infty}^{\infty} \frac{\exp[-i\omega(\tau-\tau')]}{1+\omega^2} \frac{d\omega}{2\pi}. \quad (9.10)$$

The extremal problems (9.7) and (9.9) in the theory of tunnel self-trapping are analogs of the problem (4.14) in the theory of self-consistent states; however, they are four-dimensional rather than three-dimensional.

Though formulas having the sense of a generalized virial theorem may be obtained from the homogeneity of the expressions entering the right-hand sides of eqs. (9.7) and (9.9), the problem is to analyze the qualitative behaviour of extremal paths and to calculate numerically the extremal action. It is convenient to start from the qualitative analysis in the framework of a simple model, which does not use the exact results (9.7)–(9.10). We suppose that deformation produces a spherically symmetric potential well for the electron, this well being described by two parameters – the depth and radius. Then the paths of the system can be described on the depth-radius plane (fig. 6) by curves starting from the state with undeformed lattice and reaching the right edge of the ST barrier (fig. 5). The calculations show that the extremal paths are quite different for optic and acoustic phonons. In the first case they remain, for their whole extent, within a space scale no less than $\sim\Lambda d$, i.e. than the barrier scale. Therefore, they are completely within the region of applicability of the continuum theory. On the other hand, for acoustic phonons the action decreases monotonically when the path is displaced in the small space scale region, but at scales less than Λd this decrease slows down and the action approaches its stationary value. The difference in the behaviour of paths is illustrated by fig. 6.

These qualitative conclusions are supported by the analysis of eqs. (9.7)–(9.10), which were solved approximately with one-parameter function $\psi(r\tau)$. The results for transparency are

$$D_{\rm op} \approx \exp(-6.5 W_{\rm op}/\omega_0), \qquad D_{\rm ac} \approx \exp\left(-5.6\frac{\rho s_\ell}{m^2 C^2}\right), \tag{9.11}$$

$W_{\rm op}$ is determined by eq. (5.11b). The improving of the approximation should diminish the numerical coefficients in exponents. It is easy to understand the physical meaning of the exponent in the expression for $D_{\rm ac}$;

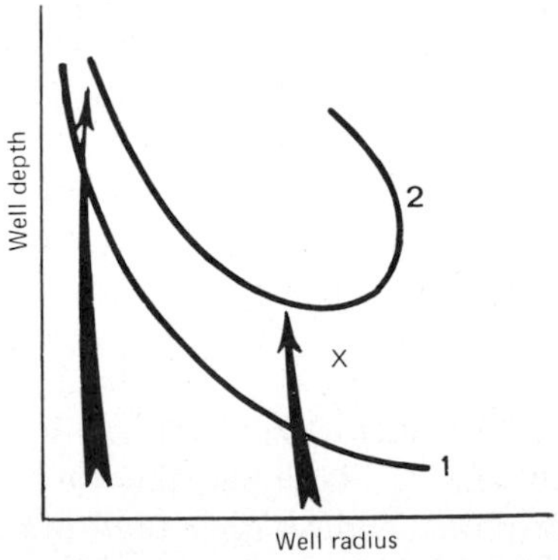

Fig. 6. Diagram of quasi-classical paths with $E = 0$ on the phase plane of a two-parameter model of self-trapping. A discrete level for exciton arises on curve 1. On curve 2 paths with $E = 0$ end. Position of the saddle point is shown by a cross. The extremal paths are shown by arrows: the right arrow denotes optic phonons, and the left one acoustic phonons.

it has the order of magnitude $\sim W_{ac}/(s/r_W)$, i.e., the ratio of W_{ac} to the frequency s/r_W of phonons with the momentum $\sim r_W^{-1}$. The similar expression

$$D \approx \exp(-4W/\omega), \tag{9.12}$$

where ω is the characteristic value of the phonon frequency, was proposed proceeding from qualitative considerations by Mott and Stoneham (1977).

The fact that the extremal paths for acoustic phonons pass near the edge of the applicability region of the continuum theory, compels us to consider the expression for D_{ac} found as a lower bound only. If the deviations from the continuum theory have such a sign that they accelerate increase of transparency at small distances, D may significantly exceed D_{ac} determined by expressions (9.11). Under these conditions the calculation of D is extremely complicated and can be performed in the framework of special models only.

It is interesting to note that so far tunneling has turned out to be the only physical process in the whole ST problem which proceeds in quite a different fashion for acoustic and optic phonons. In all other cases both groups of phonons show identical behaviour.

According to the data of Fugol' and Tarasova (1977) on solid rare gases, W increases rapidly when passing on to heavier gases, whereas the Debye frequency changes only slowly. As a result, the ST barrier, which is apparently quite inefficient in Ne, is practically opaque in Xe (cf. sect. 9.1).

9.2. Relaxation of hot ST excitons

The ST exciton is in a highly excited state after penetration through the barrier; this state should relax by emission of lattice phonons. The time needed for this process depends significantly on the particular situation. We shall discuss it here for solid rare gases, in which excimers of the R_2^* type can occur (Jortner et al. 1965). It is natural to assume that an R_2^* quasi-molecule in a strongly excited vibrational state exists as an intermediate either immediately after tunneling, or at some early stage of relaxation. The subsequent relaxation of this state may be hampered, since the vibrational frequency ω_M of quasi-molecule significantly exceeds the Debye frequency ω_D, this aspect of the problem was discussed by Jortner (1974, sects. 7 and 8). The rate of energy degradation decreases rapidly for multiphonon quantum transitions $N = \omega_M/\omega_D \gg 1$, usually as e^{-N}. According to Jortner (1974) $N = 4$, 5 and 6 for Xe, Kr and Ar, respectively. For Ne $N \approx 18$, so it is natural to assume that relaxation time becomes larger than τ_L; this explains why fluorescence from the relaxed vibrational state is not seen in the spectra of solid Ne.

9.3. ST and lattice defect production

Production of lattice defects which accompanies the ST of excitons (Lushchik et al. 1961) was most systematically studied for ionic crystals (for a review, see Lushchik et al. 1977, Williams 1978). It was already mentioned in sect. 4.2.2 that the deformation in the core of a ST state is large ($\delta_c \sim d$); production of defects is a limiting case of such highly inelastic deformation. A number of mechanisms was proposed for defect production in ionic crystals (they are summarized in the paper of Lushchik (1982)). Lattice defects are apparently created after the system overcomes the ST barrier in the course of subsequent relaxation, both in electron and vibrational levels. It has been suggested (Toyozawa 1974), that defect production, and relaxation into the ground state of ST exciton with its subsequent fluorescence, are competing channels.

However, the defect production may turn out, especially in soft lattices, to be one of the fundamental relaxation mechanisms, providing the transition of exciton to its lowest ST state. For instance, the emission bands of atomic ST excitons in solid Ne (Fugol' et al. 1975) and molecular ones in Ar (Cheshnovsky et al. 1972) are positioned very near to the corresponding gas bands. This indicates that vast cavities arise. The elastic stresses around such cavities have to be very high. Therefore, it is natural to assume that they relax due to plastic deformation including the formation of interstitials (Kusmartsev and Rashba 1980). The force acting on the cavity arises owing to quantum pressure of the electronic cloud exerted on potential wall at the vacuum–crystal boundary and to the strengthening of the Coulomb attraction of electrons to the atomic core when the screening effect of neighbours is reduced. However, the possibility that the bands observed are connected with ST of free excitons near built-in vacancy clusters and their subsequent fluorescence, cannot be totally excluded for the time being.

It has recently been proposed that ST of Frenkel excitons in diamond (cf. sect. 11.5) may be followed by production of vacancy–interstitial atom pairs (Telezhkin et al. 1980).

10. Manifestation of ST states in optical spectra

As was the case with the classification of energy spectra (see sect. 4), the optical spectra of heavy and light excitons differ essentially as well.

Under the strong exciton–phonon coupling conditions, the absorption spectrum of a heavy exciton practically coincides with the absorption spectrum of an impurity center, i.e., it should have a Gaussian shape with half-width $\sim(\omega E_D)^{1/2}$ (for a review of impurity spectra, see, e.g., Perlin

1963). Actually, the corrections to the absorption spectrum connected with the exciton motion are small at $E_B \lesssim \omega \ll E_D$; their effect may be taken into account by perturbation theory (Hizhnyakov and Sherman 1979).

On the other hand, the spectrum of light excitons under the ST conditions is as yet practically inaccessible for a rigorous analysis. The difficulties are caused both by the complicated behaviour of adiabatic potentials (curves 2 and 3, fig. 2(b)) and by the strong dependence of the exciton wave function on the deformation Q. Therefore, we shall start with qualitative analysis (following, in general, the paper of Rashba (1957c)) based on the adiabatic approximation and Franck–Condon principle. The diagram of optical transitions is shown in fig. 7.

The transitions to the free exciton state (transition 1) and its vibrational satellites dominate in the absorption spectrum at low temperatures. Optical transitions to a ST state are weak owing to the small overlap of the

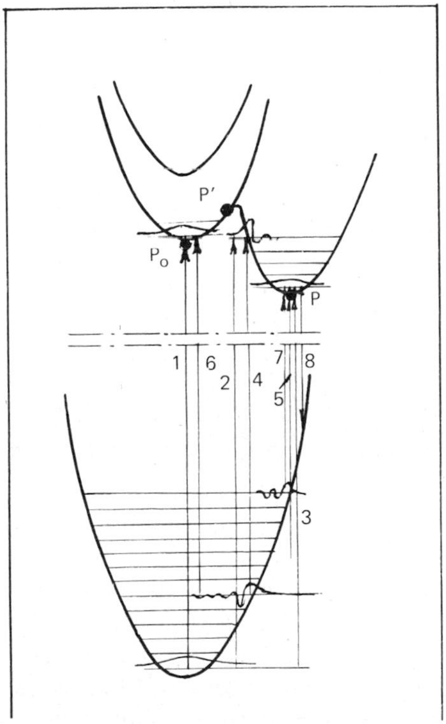

Fig. 7. Scheme of optical transitions under the ST conditions (Rashba 1957c). The case is shown when a ST state lies below the band, and the point $\mathbf{k} = 0$ to which the optical transition is allowed, is at the band bottom.

oscillator wave functions. Transitions 2 to highly excited vibrational states (near the turning points of their quasi-classical paths) will exceed in intensity transitions 3 to weakly excited states (near P). The gigant oscillator strength, inherent in transitions to shallow levels of bound excitons (for a review see Rashba 1974), will also favour the increase of the intensity of transitions when approaching the point P′. Near the barrier the absorption may have a peak, as has been shown by Hizhnyakov and Sherman (1980); it is connected with the density-of-state maximum near the saddle point existing in the quasimolecular model. The observation of this maximum might be of assistance for measuring the ST barrier height W; it has been mentioned in sect. 9 that W is only roughly determined from the temperature dependence of the ST rate.

Thus, the absorption spectrum should consist of a "narrow" principal free exciton peak, lying on a structured wide-band background, corresponding to ST excitons. In addition, the principal peak should possess a high-frequency tail within the exciton band. At the energies above P′ free and ST states are strongly mixed; as a result, at $W \lesssim \omega$ the principal peak and background are inseparable. The strength of the wide-band absorption can be estimated by the method of moments, provided that the intensity of intramolecular transition does not depend on molecular displacements. Then the zero-moment of the spectrum M_0, i.e., the total intensity of the transition, does not depend on the exciton–phonon coupling. The center of gravity of the spectrum is at the energy $\mathscr{E}_0 \equiv \mathscr{E}\,(\boldsymbol{k}=0)$, this is the energy of exciton with $\boldsymbol{k}=0$ in a rigid lattice. The first moment relative to this point is $M_1 = 0$. The second moment is:

$$M_2 = \int (\Omega - \mathscr{E}_0)^2 \sigma(\Omega)\,\mathrm{d}\Omega \Big/ \int \sigma(\Omega)\,\mathrm{d}\Omega = \frac{1}{N}\sum_{qf} |\gamma_{qf}|^2, \tag{10.1}$$

Ω being the light frequency, and σ the conductivity. Thus, M_2 does not depend on E_B and has the same magnitude, $M_2 \sim \omega^2 S$, as for impurity centers (and heavy excitons). However, the spectral distribution of the absorption is significantly changed; it is outlined in fig. 8. The absorption below $\mathscr{E}_0$ consists of the principal peak, shifted due to lowering the bottom of the exciton band caused by renormalization of the energy spectrum, and of an exponentially weak low-frequency tail. This tail should be terminated by a weak narrow zero-phonon line. The absorption above $\mathscr{E}_0$ might include an additional peak near the barrier and a high frequency tail of a mixed origin. Certainly, the drawing is highly schematic, and the scales may significantly differ from the real ones. Some conclusions on the spectrum follow also from numerical calculations based on various interpolation procedures (e.g., Cho and Toyozawa 1971, H. Sumi 1975, A. Sumi 1979). The accuracy of such calculations cannot be assessed reliably. Neverthe-

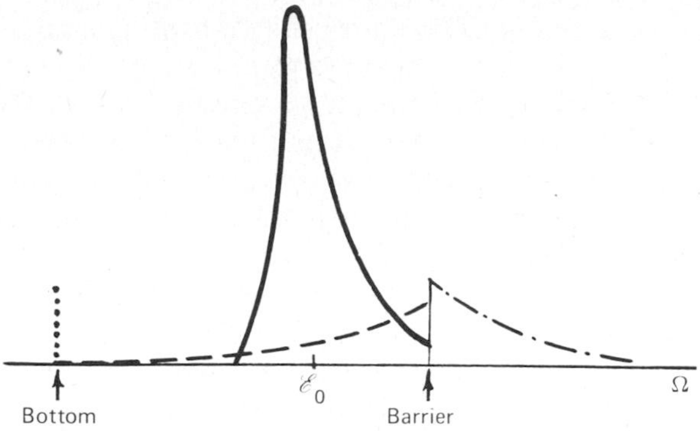

Fig. 8. Scheme of the absorption spectrum of light excitons at $T = 0$. Solid line – free exciton peak, dashed line – absorption of ST excitons between the spectrum bottom and the barrier, dash-dotted curve – absorption above the barrier. The zero-phonon peak is shown by a dotted line.

less, all the calculations indicate that the principal peak is well expressed and dominates in the spectrum at $E_B \gtrsim E_D \gg \omega$.

With increasing temperature, $T > \omega$, the motion in the initial state becomes quasi-classical too. As a result, the semi-classical Franck–Condon approximation holds, and vertical transitions between potential curves dominate. In particular, transitions 4 into highly excited states of a ST exciton near the barrier acquire considerable intensity; transitions 5 and 7 into weakly excited states are intensified too. The principal peak is broadened and weakened.

The low-frequency tail of the absorption spectrum is widely discussed in literature in connection with the Urbach–Martienssen rule. A most detailed analysis, as applied to excitons, was given by H. Sumi and Toyozawa (1971). They proceeded from the classical picture, according to which the absorption tail is ascribed to transitions to localized states of exciton in the instantaneous phonon field. It was shown that the Urbach–Martienssen rule $\ln \sigma(\Omega) = \sigma_U \Omega / T + \text{const}$ (with $\sigma_U \sim 1$) holds only in the vicinity of the inflection point in the semi-logarithmical plot $\ln \sigma$ versus T^{-1}, and that this inflection point arises from the joint effect of two factors. The first describes the decrease in the probability of lattice configurations with large deformation energy. The second is conditioned by the giant oscillator strength for shallow exciton levels near the point P' (fig. 7).

The theory of low frequency absorption, corresponding to transitions to the vicinity of point P, may be developed rigorously (Rashba 1957c). The frequency dependence of absorption is here the same as at the low frequency edge of impurity absorption. A peculiarity arises from the fact that the wavefunction of the ST state enters the electronic part of the transition matrix element. Therefore, the transition matrix element may alter significantly, as compared with its value in separate atoms, molecules etc. of an undeformed crystal, when ST is connected with the formation of quasi-molecules, excimers etc. As a result, the polarization of absorption in the tail may significantly deviate in anisotropic crystals from the polarization of the total absorption. The polarization of fluorescence from the vicinity of point P will do the same.

The fluorescence spectrum, or, from a more general point of view, the spectrum of secondary emission, depends on the frequency of exciting light and on the rate of relaxation processes. Some of them were discussed in sect. 9. At low temperatures a major part of the fluorescence, as a rule, comes from the vicinity of points P and P_0. Since in many cases the ratio of these emissions is entirely governed by the tunneling process, which is exponentially weak, the chance of obtaining a satisfactory description within the framework of interpolation procedures is less favourable, as applied to secondary emission, than in the case of the absorption spectrum. Hence, the theory may be developed either as applied to particular models, or for the excitation near special points, e.g., near the barrier (Hizhnyakov 1979, Hizhnyakov and Sherman 1980).

The above discussion was given, as applied to systems with a 3d-spectrum. In systems with a 1d-spectrum the band states are not "protected" by the ST barrier; so, the principal peak is expected to be significantly broadened.

11. ST excitons in some types of crystals

The experimental data on various materials were discussed above only as examples when the properties of ST excitons were considered. In this section we summarize the basic experimental data bearing relation to exciton self-trapping in several types of solids.

11.1. Solid rare gases (RG)

A comprehensive review on the exciton spectra of solid RG is given by Jortner (1974), Fugol' (1978) and Jortner et al. (1981). Bands of free excitons dominate in the absorption spectra. They should be attributed to the type, which is an intermediate one between the Frenkel and Wannier–Mott excitons. Several members of a hydrogen-like series are observed, but the ground state is predominantly of the Frenkel type. Fluorescence is intense; hence, solid rare gases are considered as possible candidates for UV lasers (Basov 1966). Emission with a large Stokes shift usually dominates in the solid RG fluorescence. The corresponding bands are close in position either to the nearest atomic transitions in gases, or to the emission bands of the R_2^* excimer molecules in a dense vapour. The first type of emission dominates in solid Ne; it is ascribed to quasi-atomic ST excitons – aST (Packard et al. 1970, Fugol' et al. 1972, 1975, Gedanken et al. 1973). The second type of emission dominates in solid Ar, Kr and Xe, it is ascribed to quasi-molecular ST excitons – mST. As a model for such excitons considered is an excited atom or, a quasi-molecule (excimer), confined in the cavity formed as a result of displacement of surrounding atoms, separation between atoms belonging to the excimer is about 1.5 times smaller than the interatomic spacing in perfect crystals. Calculations of the structure and energy spectrum of ST excitons are available (see, for instance, Sribnaya et al. 1979, Song and Lewis 1979, Zavt et al. 1980, Kusmartsev 1980).

The fluorescence of free and ST excitons was observed simultaneously in perfect RG crystals at low temperatures, directly proving their coexistence. It was discovered first in Xe (Debever et al. 1974, Fugol' et al. 1976, Brodmann et al. 1976), and then in Ar and Kr. Migration of excitons, corresponding to diffusion lengths up to 1000 Å, was observed too; it manifested itself in excitation energy transfer to impurities, in surface quenching of fluorescence and in the electron photoemission. The experimental data are reviewed by Fugol' (1978) and Zimmerer (1978, 1979); the data on fluorescence spectra are illustrated in fig. 9.

The width of the exciton band in solid RG is $E_B \sim 1$ eV $\gg \omega_D \sim 10$ meV. Therefore, excitons are light and an ST barrier should exist. Coexistence of free and ST excitons results from the existence of the ST barrier. The estimation of the barrier height W, performed by Fugol' and Tarasova (1977) and based on the formulas of the (5.11) type, shows that W decreases monotonically from heavy to light gases; the barrier in solid Ne is apparently too low ($W < \omega_D$) to prevent self-trapping (see also Zavt et al. 1980). Kink et al. (1978) estimated W in Xe from the temperature quenching of free exciton fluorescence (fig. 10). ST in perfect lattice is supposed to be completely suppressed by the ST barrier at $T < 60$ K; it may take place only in the vicinity of lattice defects, where the ST barrier is less. At higher T the ST barrier is overcome by activation, and the intrinsic fluorescence of mST excitons is strengthened, while the fluorescence of free and defect excitons is weakened. The barrier height found from the activation energy q is $W \approx q \approx 60$ meV. Strengthening of the defect exciton emission at $T \gtrsim 100$ K is ascribed to another process, thermal creation of defects.

It is not clear as yet whether there is a barrier between aST and mST states, and whether the degradation of the vibrational energy of the excimer (cf. sect. 9.2) may be regarded as a bottle-neck controlling the rate of conversion of aST excitons to mST excitons; it seems plausible that this is the case in Ne (Yakhot et al. 1975). At any rate the large contribution of

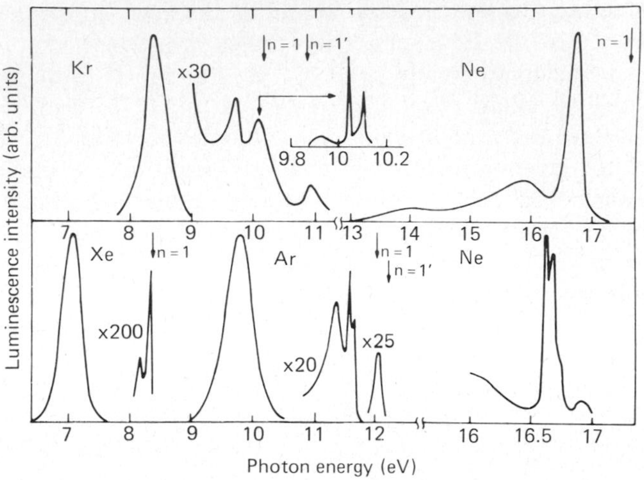

Fig. 9. Typical fluorescence spectra of solid rare gases (according to Zimmerer 1979); the data of some authors are brought together in the figure. Arrows $n,n' = 1$ correspond to the first exciton absorption bands, emission bands of free excitons in Ar, Kr and Xe coincide with them. Strong emission bands in spectra of these crystals are attributed to mST excitons. In Ne spectrum (upper curve $T = 5$ K, lower curve $T = 11$ K) the emission of free excitons is absent, the strong band is attributed to aST excitons. The low-frequency bands in Ne are assigned to nonrelaxed mST excitons in highly excited vibrational states, and the intermediate bands in Ar and Kr spectra to aST excitons.

hot fluorescence to the total fluorescence from solid Ne is a very spectacular phenomenon. Hot fluorescence was observed in the spectra of other RG as well. Since several types of excitons (free, aST and mST) may be observed in spectrum of solid RG, they provide excellent possibilities for investigation of fast relaxation processes. At present, research in this field is carried out by several different methods (see, Zimmerer 1978, 1979).

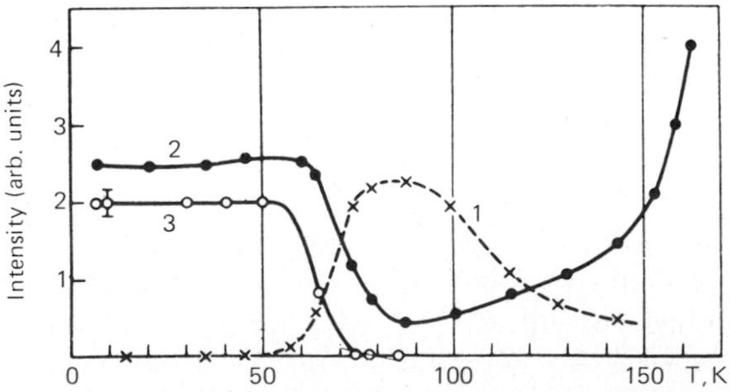

Fig. 10. Temperature dependence of the intensity of solid Xe fluorescence (Kink et al. 1978): (1) the 7.6 eV band, ascribed to mST excitons, (2) the 7.2 eV band, ascribed to mST excitons trapped by lattice defect, (3) the 8.35 eV band – free excitons.

11.2. Alkali halides

In alkali halide crystals a ST hole forms a X_2^- molecular ion (V_K center). In the V_K center the Madelung potential creates an attractive Coulomb field for electrons. The exciton is formed by the binding of an electron to X_2^- by this field (cf. sects. 2.3 and 2.5). This exciton is of the mST type, as has been shown by Kabler (1964); such an exciton is the lowest excited state of the system, so emission from the mST excited states dominates the fluorescence spectra of perfect crystals. In the spectrum of exciton absorption the bands of free excitons prevail; they are of the Wannier–Mott type. The Stokes shift of the ST exciton bands, as compared to the free exciton bands, is ~2 meV.

Coexistence of free and ST excitons was first observed by the energy transport to the activator centers; the diffusion length reached ~1000 Å in KI:Tl (Lushchik et al. 1973). The data on the observation of simultaneous emission of both mST and free excitons in alkali iodides are most convincing; the fraction of the free exciton emission is 10^{-6}–10^{-3} (Kuusmann et al.

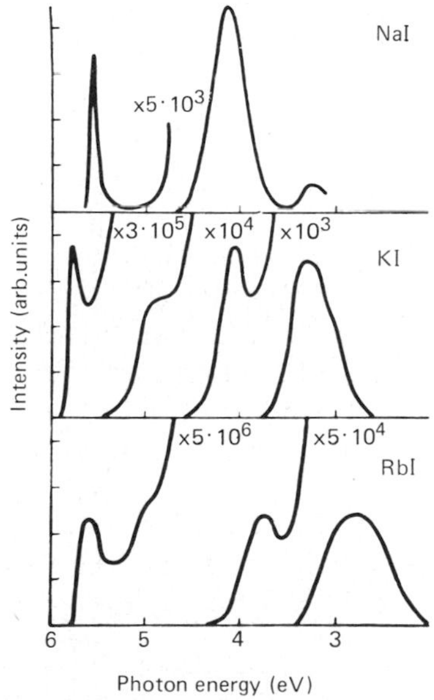

Fig. 11. Fluorescence spectrum of alkali iodides excited by an electron beam (Kuusmann et al. 1975); $T = 67$ K. The narrow high-frequency bands adjacent to intrinsic exciton absorption bands correspond to the free exciton emission. The strong low-frequency bands correspond to the ST exciton emission.

1975; see also Nishimura et al. 1977, Hayashi et al. 1977). These data are illustrated by fig. 11. No barrier is observed in the hole self-trapping; on the other hand, an activated temperature dependence of the ST rate was observed upon excitation in the exciton absorption band. This difference in the behaviour of holes and excitons is an indication of the significant effect of the polar electron–phonon interaction in initial stages of self-trapping (cf. sect. 5). This interaction is in fact the dominating one for free excitons, as follows from a comparison of the intensities of 1LO- and 2LO-satellites in the fluorescence spectrum (Plekhanov and O'Connel-Bronin 1978a). Lattice scattering of free excitons is weak; this follows from the oscillatory behaviour of the fluorescence excitation spectra with a period equal to the LO phonon frequency (Plekhanov and O'Connel-Bronin 1978b).

The activation energy for self-trapping in iodides is $q \approx 15$–$30\,\mathrm{meV}$ (Lushchik 1982; see also the review of the experimental data there). Since $E_{ex} \sim 400\,\mathrm{meV}$, and $m_e/m_h \sim 0.2$, the height of the barrier estimated from fig. 4, is $W \sim 0.1E_{ex} \sim 40\,\mathrm{meV}$. There is a satisfactory agreement between these quantities (cf. the beginning of sect. 9).

11.3. Other ionic crystals

Similar behaviour is also displayed for LiH and LiD. It follows from the Raman scattering spectrum quite definitely that the mean free path of free excitons is rather large. At the same time about one half of the fluorescence intensity should be ascribed to ST exciton emission (O'Connel-Bronin and Plekhanov 1979).

The one-site ST of holes on Ag^{2+} ions has been observed in AgCl; it is accompanied by a Jahn–Teller lattice distortion. ST excitons apparently arise when an electron is bound to such a hole (Marquard et al. 1971), and they dominate in fluorescence spectrum. Belous et al. (1978) observed emission at shorter wavelengths, which was ascribed to free anion and cation excitons. There is no ST of excitons in AgBr crystals. But in mixed $AgBr_{1-x}Cl_x$ crystals at $x \approx 0.45$ an abrupt transition from the emission of free excitons to the emission of ST excitons was observed (Kanzaki et al. 1971).

There have been reports of ST of charge carriers and excitons also in some other ionic crystals. Up till now, in all the cases when ST was observed, the carriers involved belonged to degenerate bands (Lushchik et al. 1979).

11.4. Quasi-one-dimensional systems

Convincing experimental data are available now only for the crystal anthracene – PDMA; they were discussed briefly in sect. 8.2. The spectrum

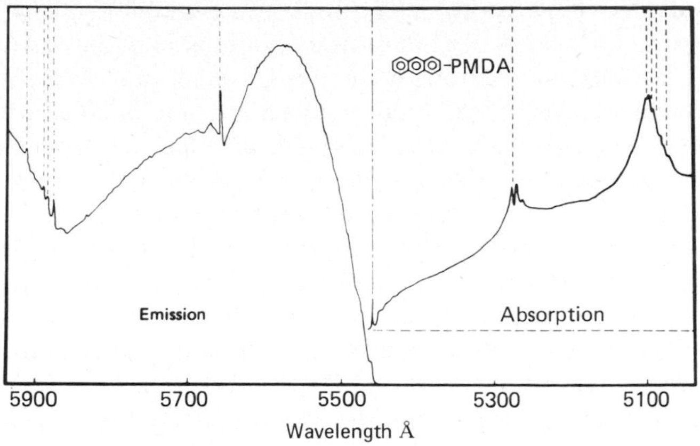

Fig. 12. Absorption and emission spectra of anthracene – PDMA; $T = 2$ K (Haarer 1977). Dash–dotted line marks the frequency of 00-transition 5459 Å, and dashed lines those of vibronic transitions.

of this crystal is shown in fig. 12. Both absorption and emission spectra develop from the narrow and very weak ($\sim 10^{-3}$ of the total oscillator strength) band of the 00 transition; it is positioned at the low frequency edge of the ST exciton absorption. Subsequent sections of $\sim 1000\,\text{cm}^{-1}$ in width correspond to vibronic satellites with intramolecular vibrations; the shape of the spectrum inside each section is determined by the interaction with intermolecular phonons. It is seen that in the absorption spectrum of anthracene – PDMA, there is no band which could be ascribed to the free exciton – in distinction from the spectra of the crystals with "three-dimensional" energy spectrum considered above.

11.5. Metastable ST states

The above considerations concerned the stable ST states, i.e., those lying below the free states. The experimental observation of metastable ST states (curve 2, fig. 2b) is more difficult.

The peculiarities of the temperature dependence of the hole mobility have been ascribed to the existence of metastable ST holes (Toyozawa and A. Sumi 1974). The observation of the extrinsic exciton self-trapping in TlCl doped with Br (Nakahara and Kobayashi 1976) makes plausible the existence of metastable ST excitons in Tl halides.

It was supposed that strong fundamental absorption peaks in the diamond and $A_{III}B_V$ type semiconductors correspond to auto-ionized

Frenkel excitons lying in the electron–hole continuum (Tolpygo 1975). According to Telezhkin et al. (1980) these excitons may be self-trapped, as a result, metastable ST excitons arise. Such model may explain the existence of hot fluorescence, which is observed in some crystals of this type.

Kagan and Kikoin (1980) have pointed out that when the electronic impurity centers possess metastable states with a strongly deformed lattice, this should result in the Fermi level "self-pinning" in heavily doped degenerated semiconductors. They applied this idea for treating the electronic properties of $Pb_{1-x}Sn_xTe$ crystals with metallic conductivity (Averkin et al. 1971, Akimov et al. 1979).

In general, the existence of metastable ST states should be a widely spread phenomenon. But the development of the methods for experimental observation of such states, including the metastable ST excitons, is a difficult problem.

12. Conclusion

The experimental data adduced above show that ST of excitons has been observed so far in several rather different classes of solids. There is no doubt that the number of such substances will increase. Upon passing to the substances with a more complicated type of chemical bond, than, for instance, in rare gas crystals and alkali halides, one may observe ST states composed after new, more sophisticated models. Such centers can be studied by traditional methods; but it seems that new experimental procedures should be developed for revealing and studying metastable ST states. Due to the small radius of the centers the theory of ST states is limited by the difficulty of making accurate quantum-chemical calculations; therefore, its role will be mainly auxiliary.

The situation will change drastically if continuum strong coupling polarons (three-dimensional) or one-dimensional continuum ST states are found experimentally. Then the theory can be used in full, and highly fascinating investigations might be forthcoming.

There is a number of problems which are already under investigation but are far from being solved. They are: the shape of optical spectra of systems with ST (and its dependence on dimensionality), transport of excitons under the coexistence conditions and so on. Here, a much better accordance between theory and experiment should be achieved than that existing at present.

Application of time-resolved techniques provides unique potentialities for tracing in stages the relaxation process accompanying self-trapping, i.e., the transition of exciton from a continuous spectrum to a discrete one, including both the tunneling and activation mechanisms. These processes are difficult for a theoretical analysis and have received yet little attention (for an analogous problem with electrons see Henry (1980)). The study of the ST barrier and the ST rate are among the most exciting problems for the near future. The potentialities of the theory are crucially dependent on whether any systems will be found in which both the ST barrier and subbarrier trajectories fall in a macroscopic region. If not, the potentialities of the theory would be very much limited; the theory will allow a description of the general qualitative regularities only. If yes, then a consistent and graceful continuum theory can be developed following the lines of sect. 9. It may happen to be rather complicated, particularly for the situations where the motion in the subbarrier region turns out to be nonadiabatic; but such mathematical difficulties seem to be surmountable. The experimental search for such systems is highly tempting.

And finally, progress in studying exciton self-trapping will be stimulated not only by the inner logic of the development of this field, but also by the increasing linkage with other fields (radiation physics, lasers, etc.).

References

Agranovich, V.M., 1968, Theory of Excitons (Nauka, Moscow Ch. III, §5 (in Russian).

Agranovich, V.M., B.P. Antonyouk, E.P. Ivanova and A.G. Mal'shukov, 1977, Zh. Eksp. Teor. Fiz. **72**, 614 (1977, Sov. Phys.-JETP **45**, 322).

Akimov, B.A., N.B. Brandt, S.A. Bogoslovskii, L.I. Ryabova and S.M. Chudinov, 1979, Pis'ma Zh. Eksp. Teor. Fiz. **29**, 11 (1978, Sov. Phys. – JETP Lett. **29**, 9).

Aluker, E.D., D.J. Lusis and S.A. Chernov, 1979, Electronic Excitations and Radioluminescence in Alkali Halide Crystals (Zinatne, Riga) [in Russian].

Appel, J., 1968, in: Solid State Physics, eds. F. Seitz, D. Turnbull and H. Ehrenreich (Acad. Press, New York and London) **21**, 193.

Atkins, K.R., 1959, Phys. Rev. **116**, 1339.

Austin, I.G. and N.F. Mott, 1969, Advan. Phys. **18**, 41.

Averkin, A.A., V.I. Kaidanov and R.B. Mel'nik, 1971, Fiz. Tekh. Poluprovodn. **5**, 91 (1971, Sov. Phys. – Semicond. **5**, 75).

Basov, N.G., 1966, IEEE QE-2, 354.

Belous, V.M., N.G. Barda, E.A. Dolbinova, I.L. Kuusmann, Ch.B. Lushchik and N.S. Roose, 1978, Zh. Nauchn. Prikl. Kinematogr. Fotograf. **6**, 460.

Bir, G.L. and G.E. Pikus, 1972, Symmetry and Strain-Induced Effects in Semiconductors (Nauka, Moscow) (English transl., 1964 (Halsted Press, New York)).

Bogolyubov, N.N., 1950, Ukr. Mat. Zh. **2**, 3 (1951, Fortschr. Phys. **4**, 1).

Brazovskii, S.A., 1978, Pis'ma Zh. Eksp. Teor. Fiz. **28**, 656 (1978, Sov. Phys. – JETP Lett. **28**, 606).

Brodin, M.S., M.A. Dudinskii and S.V. Marisova, 1971, Opt. Spektrosk. **31**, 749 (Opt. Spectrosc. (USSR) **31**, 401).

Brodmann, R., G. Tolkeihn and G. Zimmerer, 1976, Phys. Status Solidi (b)**73**, K99.

Careri, G., U. Fasoli and F.S. Gaeta, 1960, Nuovo Cimento **15**, 774.

Cheshnovsky, O., B. Raz and J. Jortner, 1972, J. Chem. Phys. **57**, 4628.

Chiao, R.Y., E. Garmire and C.H. Townes, 1964, Phys. Rev. Lett. **13**, 479.

Cho, K. and Y. Toyozawa, 1971, J. Phys. Soc. Japan **30**, 1555.
Cohen, M.D., 1979, Mol. Cryst. Liq. Cryst. **50**, 1.
Davydov, A.S., 1951, Theory of Molecular Excitons (Ukrain. Acad. of Sc. Press, Kiev) [English transl., 1962, McGraw-Hill, New York] Chs. 4 and 7.
Davydov, A.S., 1974, Biofizika **19**, 670 (1974, Biophysics (GB) **19**, 684).
Davydov, A.S. and N.I. Kislukha, 1976, Zh. Eksp. Teor. Fiz. **71**, 1090 [1976, Sov. Phys.-JETP **44**, 571).
Debever, J.M., A. Bonnot, A.M. Bonnot, F. Coletti and J. Hanus, 1974, Solid State Commun. **14**, 489.
De Gennes, P.G., 1960, Phys. Rev. **118**, 141.
Deigen, M.F. and S.I. Pekar, 1951, Zh. Eksp. Teor. Fiz. **21**, 803.
Dennis, W.S., E. Durbin, Jr., W.A. Fitzsimmons, O. Heybey and G.K. Walters, 1969, Phys. Rev. Lett. **23**, 1083.
Druger, S.D. and R.S. Knox, 1969, J. Chem. Phys. **50**, 3143.
Dykman, I.M., 1954, Zh. Eksp. Teor. Fiz. **26**, 307.
Dykman, I.M. and S.I. Pekar, 1952, Dokl. Akad. Nauk SSSR **83**, 825.
Dykman, I.M. and I.G. Zaslavskaya, 1958, Zh. Tekh. Fiz. **28**, 1959 (1959, Sov. Phys.-Tech. Phys. **3**, 1800).
Emin, D., 1973, Adv. Phys. **22**, 57.
Emin, D. and T. Holstein, 1976, Phys. Rev. Lett. **36**, 323.
Ferell, R.A., 1957, Phys. Rev. **108**, 167.
Ferguson, J., 1958, J. Chem. Phys. **28**, 765.
Feynman, R.P., 1955, Phys. Rev. **97**, 660.
Firsov, Yu.A., 1975, in: Polarons, ed. Yu.A. Firsov (Nauka, Moscow) p. 207 (in Russian).
Franck, J. and E. Teller, 1938, J. Chem. Phys. **6**, 861.
Frenkel, J., 1931, Phys. Rev. **37**, 17, 1276.
Frenkel, J.I., 1936, Phys. Zs. Sowjet. **9**, 158.
Fröhlich, H., 1954, Advan. Phys. **3**, 325.
Fugol', I.Ya., 1978, Advan. Phys. **27**, 1.
Fugol', I.Ya. and E.I. Tarasova, 1977, Fiz. Nizk. Temp. **3**, 336 (1977, Sov. J. Low Temp. Phys. **3**, 176).
Fugol', I.Ya., E.V. Savchenko and A.G. Belov, 1972, Pis'ma Zh. Eksp. Teor. Fiz. **16**, 245 (1972, Sov. Phys.-JETP Lett. **16**, 172).
Fugol', I.Ya., E.V. Savchenko and A.G. Belov, 1975, Fiz. Nizk. Temp. **1**, 750 (1975, Sov. J. Low Temp. Phys. **1**, 361).
Fugol', I.Ya., A.G. Belov, Yu.B. Poltorazkii and E.V. Savchenko, 1976, Fiz. Nizk. Temp. **2**, 400 (1976, Sov. J. Low Temp. Phys. **2**, 204).
Gedanken, A., B. Raz and J. Jortner, 1973, J. Chem. Phys. **59**, 5471.
Gross, E.P., 1976, Ann. Phys. **99**, 1.
Gurney, R.W. and N.F. Mott, 1937, Proc. Phys. Soc. **49**, 32.
Haarer, D., 1974, Chem. Phys. Lett. **27**, 91.
Haarer, D., 1977, J. Chem. Phys. **67**, 4076.
Haarer, D., M.R. Philpott and D. Morawitz, 1975, J. Chem. Phys. **63**, 5238.
Haken, H., 1956, Zs. f. Phys. **146**, 527.
Haken, H., 1958, Fortschritte der Physik **6**, 271.
Hayashi, T., T. Ohata and S. Koshino, 1977, J. Phys. Soc. Japan **43**, 347.
Henry, C.H., 1980, in: Relaxation of Elementary Excitations, eds., R. Kubo and E. Hanamura (Springer, Berlin, Heidelberg, New York) p. 19.
Hippel, A., 1936, Z. Phys. **101**, 680.

Hizhnyakov, V.V., 1979, in: Light Scattering in Solids, eds., J.L. Birman, H.Z. Cummins and K.K. Rebane (Plenum, New York, London) p. 269.

Hizhnyakov, V.V. and A.V. Sherman, 1979, Phys. Status Solidi (b)**92**, 1977.

Hizhnyakov, V.V. and A.V. Sherman, 1980, Fiz. Tverd. Tela **22**, 3254 [Sov. Phys. - Solid State **22**, 1904].

Holstein, T., 1959, Ann. Phys. **8**, 325, 343.

Iordanskii, S.V. and E.I. Rashba, 1978, Zh. Eksp. Teor. Fiz. **74**, 1872 (1978, Sov. Phys. - JETP **47**, 975).

Jortner, J., 1974, in: Vakuum Ultraviolet Radiation Physics, eds., E.E. Koch, R. Haensel and C. Kunz (Pergamon, Braunschweig) p. 263.

Jortner, J., L. Meyer, S.A. Rice and E.G. Wilson, 1965, J. Chem. Phys. **42**, 4250.

Kabler, M.N., 1964, Phys. Rev. **136**, A1296.

Kagan, Yu, and K.A. Kikoin, 1980, Pis'ma Zh. Eksp. Teor. Fiz. **31**, 367 (1980, Sov. Phys. - JETP Lett. **31**, 335).

Kagan, Yu. and M.I. Klinger, 1976, Zh. Eksp. Teor. Fiz. **70**, 255 (1976, Sov. Phys. - JETP **43**, 132).

Kaminska, M., J.M. Baranowski and M. Godlewski, 1979, in: Proc. 14-th Int. Conf. Phys. Semicond., Edinburgh, 1978 (Conf. Series N 43, Inst. of Phys., Bristol and London) p. 303.

Kanzaki, H., S. Sakuragi and K. Sakamoto, 1971, Solid State Commun. **9**, 999.

Kastner, T.G. and W.J. Känzig, 1957, J. Phys. Chem. Solids **3**, 178.

Khraplak, A.G. and I.T. Yakubov, 1979, Usp. Fiz. Nauk **129**, 45.

Kink, R., A.E. Löhmus and M.V. Selg, 1978, Pis'ma Zh. Eksp. Teor. Fiz. **28**, 505 (1978, Sov. Phys. - JETP Lett. **28**, 469).

Kittel, C., 1963, Quantum Theory of Solids (Wiley, New York, London) ch. 7.

Klinger, M.I., 1979, Problems of linear electron (polaron) transport theory in semiconductors, Intern. Ser. on Natural Philos., ed., D. ter Haar, (Pergamon Press) p. 87.

Kagan, Yu. and M.I. Klinger, 1976, Zh. Eksp. Teor. Fiz. **70**, 255 (1976, Sov. Phys. - JETP **43**, 132).

Kaminska, M., J.M. Baranowski and M. Godlewski, 1979, in: Proc. 14-th Int. Conf. Phys. Semicond., Edinburgh, 1978 (Conf. Series N 43, Inst. of Phys., Bristol and London) p. 303.

Kanzaki, H., S. Sakuragi and K. Sakamoto, 1971, Solid State Commun. **9**, 999.

Kastner, T.G. and W.J. Känzig, 1957, J. Phys. Chem. Solids **3**, 178.

Khraplak, A.G. and I.T. Yakubov, 1979, Usp. Fiz. Nauk **129**, 45 (Sov. Phys. Usp. **22**, 703).

Kink, R., A.E. Löhmus and M.V. Selg, 1978, Pis'ma Zh. Eksp. Teor. Fiz. **28**, 505 (1978, Sov. Phys. - JETP Lett. **28**, 469).

Kittel, C., 1963, Quantum Theory of Solids (Wiley, New York, London) ch. 7.

Kittel, C., 1971, Introduction to Solid State Physics (Wiley, New York, London, Sydney, Toronto) ch. 11.

Klinger, M.I., 1979, Problems of linear electron (polaron) transport theory in semiconductors, Intern. Ser. on Natural Philos., ed., D. ter Haar, (Pergamon Press) p. 87.

Krivoglaz, M.A., 1973, Usp. Fiz. Nauk **111**, 617 (1974, Sov. Phys. - Uspekhi **16**, 856).

Krivoglaz, M.A. and A.A. Trushchenko, 1971, Ukr. Fiz. Zh. **15**, 833.

Kusmartsev, F.V., 1980, Fiz. Nizk. Temp. **6**, 1046 (1980, Sov. J. Low Temp. Phys. **6**, 509).

Kusmartsev, F.V. and E.I. Rashba, 1980, in: Proc. Intern. Conf. Radiation Phys. Semiconduct. and Related Materials, Tbilisi, 1979 (Tbilisi State University Press), p. 448.

Kusmartsev, F.V. and E.I. Rashba, 1981, Pis'ma Zh. Eksp. Teor. Fiz. **33**, 164 (1981, Sov. Phys. - JETP Lett. **33**, 155).

Kuusmann, I.L., P.H. Liblik, G.G. Liidja, N.E. Lushchik, Ch.B. Lushchik and T.A. Soovik, 1975, Fiz. Tverd. Tela **17**, 3546 (1975, Sov. Phys. - Solid State **17**, 2312).

Landau, L.D., 1933, Phys. Zs. Sowjet **3**, 664.
Landau, L.D. and S.I. Pekar, 1948, Zh. Eksp. Teor. Fiz. **18**, 419.
Lang, D.V. and R.A. Logan, 1977, Phys. Rev. Lett. **39**, 635.
Langer, J.M., U. Ogonowska and A. Iller, 1979, in: Proc. 14-th Int. Conf. Phys. Semicond., Edinburgh, 1978 (Conf. Series N 43, Inst. of Phys., Bristol and London) p. 277.
Larkin, A.I. and D.E. Khmel'nitskii, 1969, Zh. Eksp. Teor. Fiz. **56**, 2087 (1969, Sov. Phys. – JETP **29**, 1123).
Lee, T.D., F. Low and D. Pines, 1953, Phys. Rev. **90**, 297.
Levinson, Y.B. and E.I. Rashba, 1973, Rep. Prog. Phys. **36**, 1499.
Lippert, E., 1955, Zs. Naturforsch. **10a**, 541.
Lushchik, Ch.B., 1982, this Volume, ch. 12.
Lushchik, Ch., G. Liidja and I. Jack, 1961, in: Proc. Internat. Conf. Semiconductor Physics, Prague, 1960 (Publish. House of Czech. Acad. Sci., Prague) p. 717.
Lushchik, Ch., G. Liidja, N. Lushchik, E. Vassilchenko, K. Kalder, R. Kink and T. Soovik, 1973, in: Luminescence of Crystals, Molecules and Solutions, ed., F.N.Y. Williams (Plenum Press, New York) p. 162.
Lushchik, Ch.B., I.K. Vitol and M.A. Elango, 1977, Usp. Fiz. Nauk **122**, 223 (1977, Sov. Phys. – Usp. **20**, 489).
Lushchik, Ch., I. Kuusmann and V. Plekhanov, 1979, J. Lumin. **18/19**, 11.
Marquard, C., R. Williams and M. Kabler, 1971, Solid State Commun. **9**, 2285.
Mel'nikov, V.I., 1977, Zh. Eksp. Teor. Fiz., **72**, 2345 (1977, Sov. Phys. – JETP **45**, 1233).
Migdal, A.B., 1958, Zh. Eksp. Teor. Fiz. **34**, 1438 (1958, Sov. Phys. – JETP **7**, N 6).
Möglich, F. and R. Rompe, 1943, Zs. f. Phys. **120**, 741.
Mott, N.F. and R.W. Gurney, 1940, Electronic Processes in Ionic Crystals (Oxford Univ. Press, Oxford) ch. 3, sect. 5.
Mott, N.F. and A.M. Stoneham, 1977, J. Phys. C.: Solid State Phys. **10**, 3391.
Mulliken, R.S., 1970, J. Chem. Phys. **52**, 5170.
Nagaev, E.L., 1975, Usp. Fiz. Nauk **117**, 437 (1975, Sov. Phys. – Uspekhi **18**, 863).
Nakahara, J. and K. Kobayashi, 1976, J. Phys. Soc. Japan **40**, 180.
Nishimura, H., G. Ohhigashi, Y. Tanaka and M. Tomura, 1977, J. Phys. Soc. Japan **43**, 157.
O'Connel-Bronin, A.A. and V.G. Plekhanov, 1979, Phys. Status Solidi (b)**95**, 75.
Ogg, R.A., 1946, Phys. Rev. **69**, 668.
Packard, R.E., F. Reif and C.M. Surko, 1970, Phys. Rev. Lett. **25**, 1435.
Peierls, R., 1932, Ann. Phys. **13**, 905.
Pekar, S.I., 1946, Zh. Eksp. Teor. Fiz. **16**, 335, 341.
Pekar, S.I., 1951, Research in the Electron Theory of Crystals (Gostekhizdat, Moscow-Leningrad) (German transl.: 1954, Untersuchungen über die Electronentheorie der Kristallen (Akademie Verlag, Berlin)).
Pekar, S.I., E.I. Rashba and V.I. Sheka, 1979, Zh. Eksp. Teor. Fiz. **76**, 251 (1979, Sov. Phys. – JETP **49**, 129).
Perlin, Yu.E., 1963, Usp. Fiz. Nauk **80**, 553 (1964, Sov. Phys. – Uspekhi **6**, 542).
Philpott, M.R. and A. Brillante, 1979, Mol. Cryst. Liq. Cryst. **50**, 163.
Philpott, M.R. and J.-M. Turlet, 1975, J. Chem. Phys. **62**, 2777.
Philpott, M.R. and J.-M. Turlet, 1976, ibid. **64**, 3852.
Plekhanov, V.G. and A.A. O'Connel-Bronin, 1978a, Pis'ma Zh. Eksp. Teor. Fiz. **27**, 30 (1978, Sov. Phys. – JETP Lett. **27**, 27).
Plekhanov, V.G. and A.A. O'Connel-Bronin, 1978b, Fiz. Tverd. Tela **20**, 2078 [1978, Sov. Phys. – Solid State **20**, 1200).
Pokatilov, E.P., 1964, Fiz. Tverd. Tela **6**, 2809 (1965, Sov. Phys. – Solid State **6**, 2233).
Porowski, S., M. Konczykowski and J. Chroboczek, 1974, Phys. Status Solidi (b)**63**, 291.
Prelovsek, P., 1975, J. Phys. C.: Solid State Phys. **8**, 2436.

Rashba, E.I., 1957a, Opt. Spektrosk. **2**, 75.
Rashba, E.I., 1957b, ibid. **2**, 88.
Rashba, E.I., 1957c, ibid. **3**, 568.
Rashba, E.I., 1957d, Izv. Akad. Nauk SSSR, Ser. Fiz. **21**, 37.
Rashba, E.I., 1959, Zh. Eksp. Teor. Fiz. **36**, 1703 (1959, Sov. Phys. – JETP **9**, 1213).
Rashba, E.I., 1962, Fiz. Tverd. Tela **4**, 3301 (1963, Sov. Phys. – Solid State **4**, 2417).
Rashba, E.I., 1974, Fiz. Tekh. Poluprovodn. **8**, 1241 (1975, Sov. Phys. – Semicond. **8**, 807).
Rashba, E.I., 1976, Izv. Akad. Nauk SSSR, Ser. Fiz. **40**, 1793 (1976, Bull. Acad. Sci. USSR **40**, No. 9, 20).
Rashba, E.I., 1977, Fiz. Nizk. Temp. **3**, 524 (1977, Sov. J. Low Temp. Phys. **3**, 254).
Saile, V. and E.E. Koch, 1979, Phys. Rev. **B20**, 784.
Saile, V., M. Skibowski, W. Steinmann, P. Gürtler, E.E. Koch and A. Kozevnikov, 1976, Phys. Rev. Lett. **37**, 305.
Scheibe, G., 1941, Z. Electrochem. **47**, 73.
Schwentner, N., E.E. Koch and J. Jortner, 1981, in: Rare Gas Solids, 2nd edition, eds. M.L. Klein and J.A. Venables (Academic Press, London) in press.
Shaw, P.B. and G. Whitfield, 1978, Phys. Rev. **B17**, 1495.
Sheka, E.F., 1971, Usp. Fiz. Nauk **104**, 593 (1971, Sov. Phys. – Uspekhi **14**, 484).
Shinozuka, Y. and Y. Toyozawa, 1979, J. Phys. Soc. Japan **46**, 505.
Song, K.S. and L.J. Lewis, 1979, Phys. Rev. **B19**, 5349.
Sribnaya, V.K., K.B. Tolpygo and E.P. Troitskaya, 1979, Fiz. Tverd. Tela **21**, 834 (1979, Sov. Phys. – Solid State **21**, 488).
Su, W.P., J.R. Schrieffer and A.J. Heeger, 1979, Phys. Rev. Lett. **42**, 1698.
Sugakov, V.I., 1970, Ukr. Fiz. Zh. **15**, 2060.
Sumi, H., 1974, J. Phys. Soc. Japan **36**, 770.
Sumi, H., 1975, ibid. **38**, 825.
Sumi, A., 1977, J. Phys. Soc. Japan **43**, 1286.
Sumi, A., 1979, J. Phys. Soc. Japan **47**, 1538.
Sumi, H., and Y. Toyozawa, 1971, J. Phys. Soc. Japan **31**, 342.
Sumi, A. and Y. Toyozawa, 1973, J. Phys. Soc. Japan **31**, 342.
Suna, A., 1964, Phys. Rev. **A135**, 111.
Takahei, K. and K. Kobayashi, 1977, J. Phys. Soc. Japan **43**, 891.
Takahei, K. and K. Kobayashi, 1978, ibid. **44**, 1850.
Telezhkin, V.A., K.B. Tolpygo and V.M. Shatalov, 1980, in: Proc. Intern. Conf. Radiation Phys. Semicond. and Related Materials, Tbilisi, 1979 (Tbilisi State University Press) p. 452.
Tjablikov, S.V., 1952, Zh. Eksp. Teor. Fiz. **23**, 381.
Tolpygo, K.B., 1962, Physics of Alkali Halide Crystals (Latvian State Univers. Press, Riga) p. 15 (in Russian).
Tolpygo, K.B., 1975, Fiz. Tverd. Tela **17**, 1769 (1975, Sov. Phys. – Solid State **17**, 1149).
Toyozawa, Y., 1961, Progr. Theor. Phys. **26**, 29.
Toyozawa, Y., 1970, J. Lumin. **1–2**, 632.
Toyozawa, Y., 1974, in: Proc. Conf. Vacuum Ultraviolet Radiation Phys., Hamburg, 1974, eds., E. Koch, R. Haensel and C. Kunz (Pergamon-Vieweg, Braunschweig) p. 317.
Toyozawa, Y. and A. Sumi, 1974, in: Proc. XII Intern. Conf. Phys. Semiconductors, Stuttgart, 1974 (B.G. Teubner, Stuttgart) p. 179.
Vinetsky, V.L. and V.Ya. Kravchenko, 1965, Ukr. Fiz. Zh. **10**, 153.
Vol, E.D. and L.S. Kukushkin, 1977, Fiz. Nizk. Temp. **3**, 222 (1977, Sov. J. Low Temp. Phys. **3**, 106).
Volovik, G.E. and V.M. Edel'shtein, 1974, Zh. Eksp. Teor. Fiz. **67**, 273 (1975, Sov. Phys. – JETP **40**, 137).
Williams, R.T., 1978, Semiconductors and Insulators, **3**, 251.

Yakhot, Y., M. Berkowitz and R.B. Gerber, 1975, Chem. Phys. **10**, 61.
Zakharov, V.E., V.V. Sobolev and V.S. Synakh, 1972, Zh. Prikl. Mekh. Tekh. Fiz. **1**, 92.
Zavt, G.S., S.P. Reifman and B.V. Shulichenko, 1980, Fiz. Tverd. Tela **22**, 841 (Sov. Phys. Solid State **22**, 490).
Zimmerer, G., 1978, in: Luminescence of Organic Solids, ed., B. DiBartolo (Plenum, New York) p. 627.
Zimmerer, G., 1979, J. Lumin. **18/19**, 875.

CHAPTER 14

Excitons in Magnetic Insulators

YUKITO TANABE

Department of Applied Physics
Faculty of Engineering
University of Tokyo, Tokyo
Japan

KIYOSHI AOYAGI

Department of Physics
Faculty of Science
Yamagata University, Yamagata
Japan

Excitons
Edited by
E.I. Rashba and M.D. Sturge

Contents

1. Introduction

In this article we are concerned with Frenkel excitons in magnetic insulators such as antiferromagnetic compounds of iron group elements. However, we do not attempt to cover all subjects of the problem here, but only try to concentrate upon the fundamental aspect of the exciton effect in these materials. (For an extensive review, see, for example, Eremenko and Petrov (1977). As a good introduction to the subject, lectures by Imbusch (1978) are recommended.)

A feature of these magnetic excitons is that both the ground and excited states involved are magnetic. In other words these states have magnetic moments due to spin and/or orbital angular momentum, which means they are degenerate or nearly so in paramagnetic states. Spectra of some typical iron group ions in paramagnetic crystals will be reviewed briefly in sect. 2 as a prerequisite to the subsequent sections.

Coherent exciton motion becomes possible in magnetically ordered states at low temperatures where long-range order is established. In this state the degeneracy mentioned above is lifted because of the internal (molecular) field arising from the exchange interactions between magnetic ions, so that we are interested in the excitations from the lowest Zeeman component of the ground state manifold to (usually the lowest) one of the split components of the excited state. The excitation can be transferred from ions to ions by the aid of multipole–multipole interaction in the case of spin allowed transition. In the case of spin forbidden transitions, however, the transfer mechanism must be spin dependent and (off-diagonal) exchange interaction will be responsible for transfers between near neighbours. This will in turn mean that dispersion as well as Davydov splittings of excitons in ordered states depend upon the relative directions of the spins and can be controlled in some cases by applying an external field on the system. These problems will be discussed in sect. 3.

Exchange interaction plays another important role in the problem of magnetic excitons. In fact, it is through this type of spin dependent electric dipole moment that it is possible to create pairs of excitons (two-magnon, exciton–magnon, two-exciton) and get information on the dispersion of excitons and magnons. These double excitations will be the subject of sect. 4.

The last section (sect. 5) will be devoted to a brief discussion of recent developments.

2. Optical spectra of paramagnetic ions in crystals

Iron group ions in compounds have several valence electrons in the 3d orbitals which form the outermost shell for them. The electronic states of the ion with N electrons in the d shell are labelled as those belonging to the "electron configuration" d^N ($N = 1, 2, \ldots, 9$). In ionic crystals, these positive ions are surrounded by neighbouring anions and further by a number of cations in the distance. In most cases, six anions occupy the nearest sites of the transition metal ion forming nearly an octahedron. In this article we shall deal only with such cases of cubic symmetry.

In an octahedral field due to anions, the five d orbitals which have equal energy in the free state are divided into a group of three orbitals (d_{yz}, d_{zx}, d_{xy}), called d_ϵ or t_2 and a group of two orbitals (d_{z^2}, $d_{x^2-y^2}$), called d_γ or e. Correspondingly, the five-fold degenerate d level splits into two, the lower one (t_2 level) being triply degenerate and the upper one (e level) doubly degenerate. The notations t_2 and e come from group theory. Separation between the two levels e and t_2 is usually denoted as $10Dq$ which is a measure of the strength of the cubic crystal (or ligand) field. The magnitude of $10Dq$ ranges from $8\,000\ \mathrm{cm}^{-1}$ to $20\,000\ \mathrm{cm}^{-1}$ depending upon the central ion and its environment. The electron configuration to describe the states of an iron group ion in crystals with N 3d electrons is now $t_2^m\, e^{N-m}$ with $m = 0, 1, \ldots, 6$ and $N - m = 0, 1, \ldots, 4$. Suppose $t_2^m\, e^{N-m}$ with largest possible value of m is the ground configuration. The lowest excited state will be reached, according to the one electron picture, by a jump of a t_2 electron into an e level corresponding to the configuration change $t_2^m e^{N-m} \rightarrow t_2^{m-1} e^{N-m+1}$ with excitation energy $10Dq$. This description, however, is much too simple. Even with t_2^m ($m = 2$–4), many electronic states, called terms, appear which belong to the same configuration but still have different total energies because of the different occupancy of the three orbital states of t_2 with different spin. This is also true with e^{N-m} ($N - m = 2$). In the case of free ions we have terms ^{2S+1}L belonging to d^N. Altogether, the states of an ion in crystals are characterized by the configuration $t_2^m e^{N-m}$ as well as the term notation $^{2S+1}\Gamma$ where S denotes the magnitude of the total electronic spin and Γ, the notation for the irreducible representations of the octahedral group, indicates transformation properties of the eigenfunction under symmetry operations. For details of the crystal or ligand field theory, the reader is referred, for example, to Ballhausen (1962) or McClure (1959b).

Excitation energies from the ground term to the excited terms of the transition metal ions in crystals are usually in the region of visible light. As well known transparent crystals containing such ions get beautifully coloured corresponding to the selective absorption in this region.

As an example, energy levels and corresponding absorption spectrum of

Cr^{3+} ion in ruby (Al_2O_3:Cr) (Tanabe and Sugano 1954, Misu 1960) are shown in fig. 1. In the left half of the figure, the energy levels of Cr^{3+} are given as a function of $10Dq$. Positions of the energy levels corresponding to $10Dq = 18\,000\ cm^{-1}$ are shown in the centre together with term notations $^{2S+1}\Gamma$ ($\Gamma = A_1(1), A_2(1), E(2), T_1(3), T_2(3)$). Note how the terms ^{2S+1}L of the free ion are split in the cubic field and the corresponding excitation energies change with the magnitude of $10Dq$. In the right half of the figure, absorption spectrum of ruby is drawn with the same ordinate as that of the left half. The abscissa has been chosen so that absorption intensity becomes stronger toward the left. The trivalent chromium ion has three d electrons. The ground term is $t_2^3\,{}^4A_2$ in which each of the three electrons occupies three different orbitals of t_2 with parallel spins. Therefore, the total spin S of this state is 3/2 (and so the left hand superscript 4). Excited states $^4T_2$ and $^4T_1$ which have the same spin multiplicity 4 as the ground state both belong to the configuration t_2^2e. Optical (spin allowed) transitions from the ground to these excited states cause comparatively strong and wide absorption bands with oscillator strength $f \sim 10^{-4}$. Weak but sharp absorption lines observed around $15\,000\ cm^{-1}$ and $21\,000\ cm^{-1}$ with $f \sim 10^{-8}$ are attributed respectively to the (spin forbidden) transitions $t_2^3\,{}^4A_2 \rightarrow t_2^3\,{}^2E$, $^2T_1$ and $t_2^3\,{}^4A_2 \rightarrow t_2^3\,{}^2T_2$ between terms with different multiplicities. These transitions are both parity forbidden in free ions because they originate from $d \rightarrow d$ transitions. In actual crystals, however, inversion symmetry around the central ions is destroyed by odd parity field of lower symmetry or gets destroyed by vibrational distortion of the surrounding crystal lattice. These effects make the electric dipole transitions possible. In the former case zero-phonon lines are observed, while in the latter one-phonon sidebands (odd-phonon assisted transitions). Vibrational motions of the surrounding ions have another important effect on the spectra. Broadness of the absorption bands accompanying the electron jump $t_2 \rightarrow e$ is in fact due to the strong dependence of their excitation energies on $10Dq$ which fluctuates with vibrations.

The cubic terms or states $^{2S+1}\Gamma$, e.g., ^2E usually further split into several components because of the field of low symmetry (in $Al_2O_3:Cr^{3+}$, actual site symmetry is trigonal, C_3 or nearly C_{3v}) as well as the spin–orbit interaction. The R_1 and R_2 lines of ruby thus correspond to the transitions to the split components $\bar{E}$ and $2\bar{A}$ (both doubly degenerate) of ^2E which has four-fold degeneracy (Sugano and Tanabe 1958). The double degeneracy (Kramers' degeneracy) in the system of odd-number electrons can only be lifted in the magnetic field.

For later discussions, we show, as another example, absorption spectrum of MnF_2 crystal (Stout 1959) and theoretical energy diagram (Tanabe and Sugano 1954) for Mn^{2+} in figs. 2 and 3, respectively. In fig. 3, both abscissa and ordinate are scaled with B as the unit. Energies of the various states

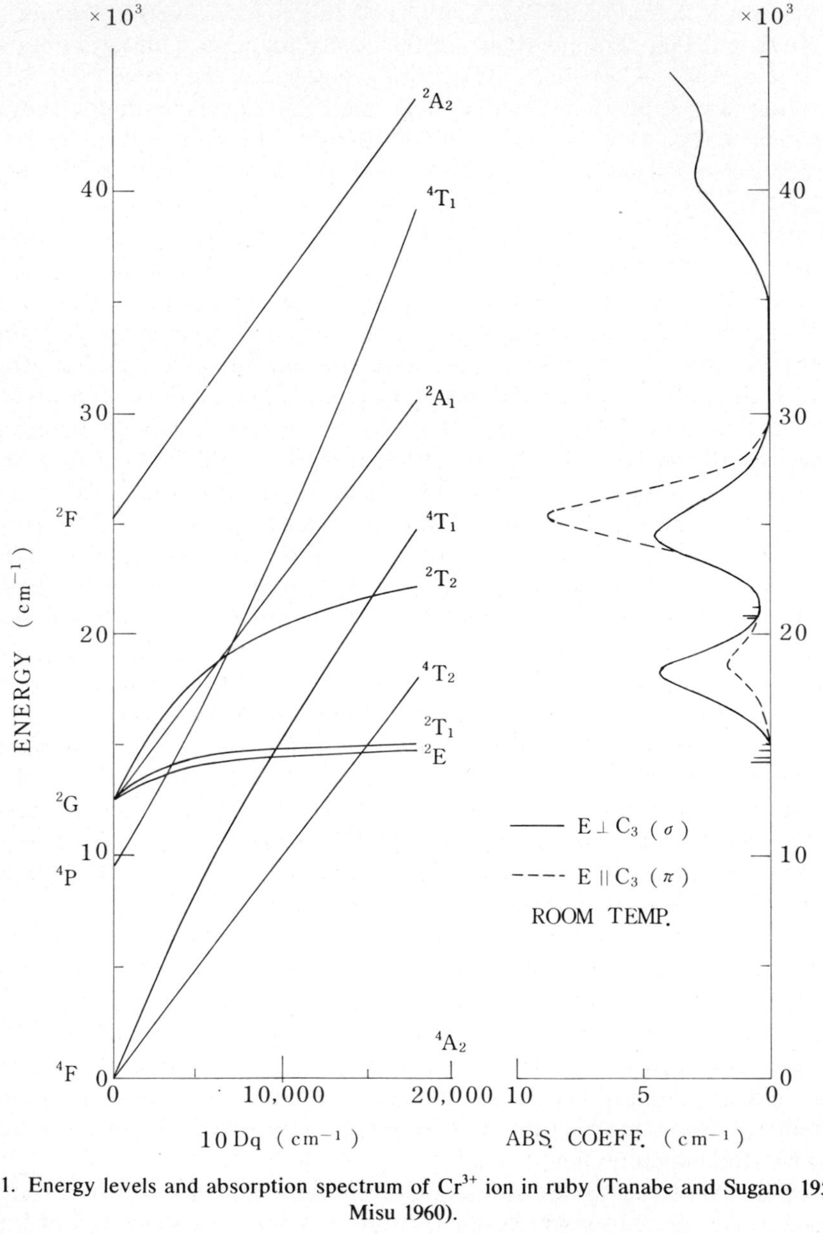

Fig. 1. Energy levels and absorption spectrum of Cr^{3+} ion in ruby (Tanabe and Sugano 1954, Misu 1960).

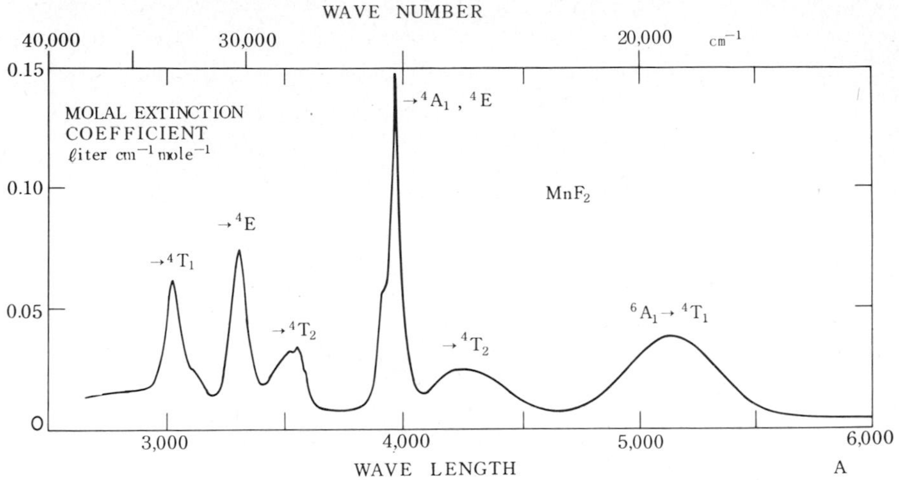

Fig. 2. Absorption spectrum of MnF_2 (Stout 1959).

relative to the ground have been calculated with $C/B = 4.48$, where B and C are the Racah parameters which gives the magnitudes of thc Coulomb (more precisely, exchange) interactions between d electrons. For free Mn^{2+} ions, $B = 860\ \text{cm}^{-1}$ and $C/B = 4.48$. In the case of MnF_2, excitation energies calculated with $10Dq = 7\,800\ \text{cm}^{-1}$, $B = 950\ \text{cm}^{-1}$ and $C = 3\,280\ \text{cm}^{-1}$ nicely agree with the observed peak positions.

We note that the ground state changes from $t_2^3e^2\,{}^6A_1$ (high spin state) to $t_2^5\,{}^2T_2$ (low spin state) because of the level crossing at $10Dq \sim 28B$. The case of MnF_2 belongs to the high spin ground state, so that the observed absorption bands and lines (fig. 2) all correspond to the spin forbidden transitions to the quartet ($S = 3/2$) states. We also note that transitions to the excited state ${}^4A_1$, ${}^4E$ with excitation energies independent of $10Dq$ (that is, those belonging to the same configuration $t_2^3e^2$ as the ground state) are observed as sharp lines, whereas those with configuration change are not, in the same way as before. Although transitions accompanying electron jumps are observed as broad bands, it is possible, under favourable conditions, e.g., in the band $t_2^3e^2\,{}^6A_1 \rightarrow t_2^4e\,{}^4T_1$ of MnF_2, to observe sharp (purely electronic, magnetic dipole) zero-phonon lines on the low energy tail of the bands. As a matter of fact, we shall be concerned only with these zero-phonon lines in later sections. From this point of view, broad bands represent an aggregate of multiphonon sidebands piled up on top of the zero-phonon line.

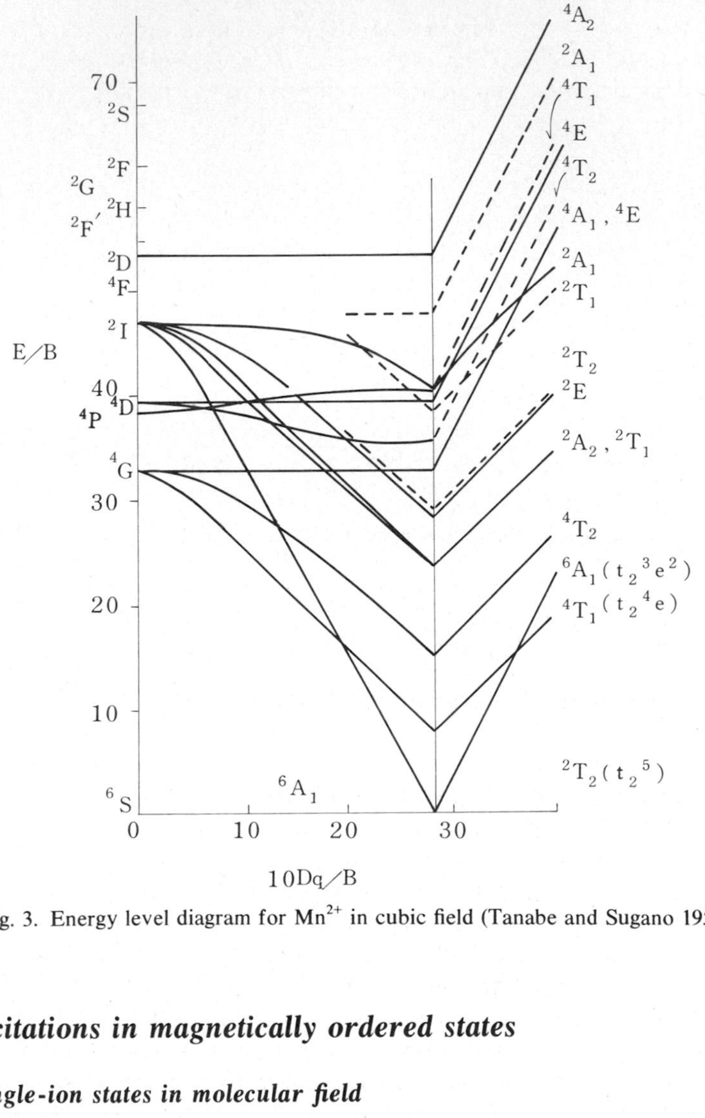

Fig. 3. Energy level diagram for Mn^{2+} in cubic field (Tanabe and Sugano 1954).

3. Excitations in magnetically ordered states

3.1. Single-ion states in molecular field

Before proceeding to the discussion of the excitation transfer, we must study the effect of magnetic interactions on the single-ion levels and states. In discussing the effect, however, we do not intend to develop a general treatment but confine ourselves here with a few typical cases, namely those of Mn and Cr compounds for which most extensive studies have been carried out. In these cases, the ground states are orbitally non-degenerate and we have pure spin multiplets $^{2S+1}A_{1\,\mathrm{or}\,2}$ with spin S. The excited states of interest are not necessarily non-degenerate. We assume, however, that the effect of spin–orbit interaction in this state is for some reason not very important except for causing only relatively small splittings, so that each orbital component may be treated separately. This is true with $^4A_1$, ^4E of Mn^{2+}, ^2E of Cr^{3+} and even with $^4T_1$ of Mn^{2+} which was referred to at the end of sect. 2.

The spin S' of the excited states in these examples is equal to $S-1$ which is less by unity than that of the ground state S. In this article, we are mainly concerned with excitons associated with such spin forbidden transitions, because excitons of spin allowed transitions do not seem to have been identified yet.

The effect of the exchange interactions between the spins

$$\mathscr{H}_{\text{exch}} = -\sum_{j>l} 2J(jl)\boldsymbol{S}_j \cdot \boldsymbol{S}_l \tag{1}$$

upon the spin $\boldsymbol{S}_j$ of the jth ion in its ground state can be treated most simply by introducing an effective magnetic (i.e., molecular) field,

$$\boldsymbol{H}_{\text{mol}}(j) = -\sum_{l} 2J(jl)\langle \boldsymbol{S}_l\rangle/2\mu_{\text{B}}, \tag{2}$$

acting upon $\boldsymbol{S}_j$, where μ_{B} is the Bohr magneton and the spin operators $\boldsymbol{S}_l$ of the neighbouring ions have been replaced by their average value $\langle \boldsymbol{S}_l\rangle$ in the lowest state. The spin degeneracy of the ground state of the jth ion is now lifted by the field $\boldsymbol{H}_{\text{mol}}(j)$ according to the magnitude of the Zeeman energy:

$$\mathscr{H}(j) = 2\mu_{\text{B}}\boldsymbol{S}_j \cdot \boldsymbol{H}_{\text{mol}}(j). \tag{3}$$

In the excited state, the exchange integral in general will have a different value, so that the Zeeman Hamiltonian for it will be

$$\mathscr{H}'(j) = E_{0j} + 2\mu_{\text{B}}\boldsymbol{S}'_j \cdot \boldsymbol{H}'_{\text{mol}}(j), \tag{4}$$

with

$$\boldsymbol{H}'_{\text{mol}}(j) = -\sum_{l} 2J'(jl)\langle \boldsymbol{S}_l\rangle/2\mu_{\text{B}}, \tag{5}$$

where E_{0j} is the excitation energy in the paramagnetic phase. When the excited state is degenerate, $\boldsymbol{S}'_j$ (as well as $J'(jl)$) is to be understood as associated with one of its orbital components we are interested in. Note that in treating each component separately, we are assuming the Zeeman energy $2\mu_{\text{B}}\boldsymbol{H}'_{\text{mol}}$ to be much larger than the effective spin–orbit interaction or the off-diagonal exchange interaction which may connect different components, and thus mixing among these components to be negligible.

The ground and excited state of the ion j will split into Zeeman components as shown in fig. 4, when both $\boldsymbol{H}_{\text{mol}}(j)$ and $\boldsymbol{H}'_{\text{mol}}(j)$ point in the same direction. We assume in this section that each spin $\boldsymbol{S}_j$ is quantized along the direction of $-\boldsymbol{H}_{\text{mol}}(j)$ unless specified otherwise.

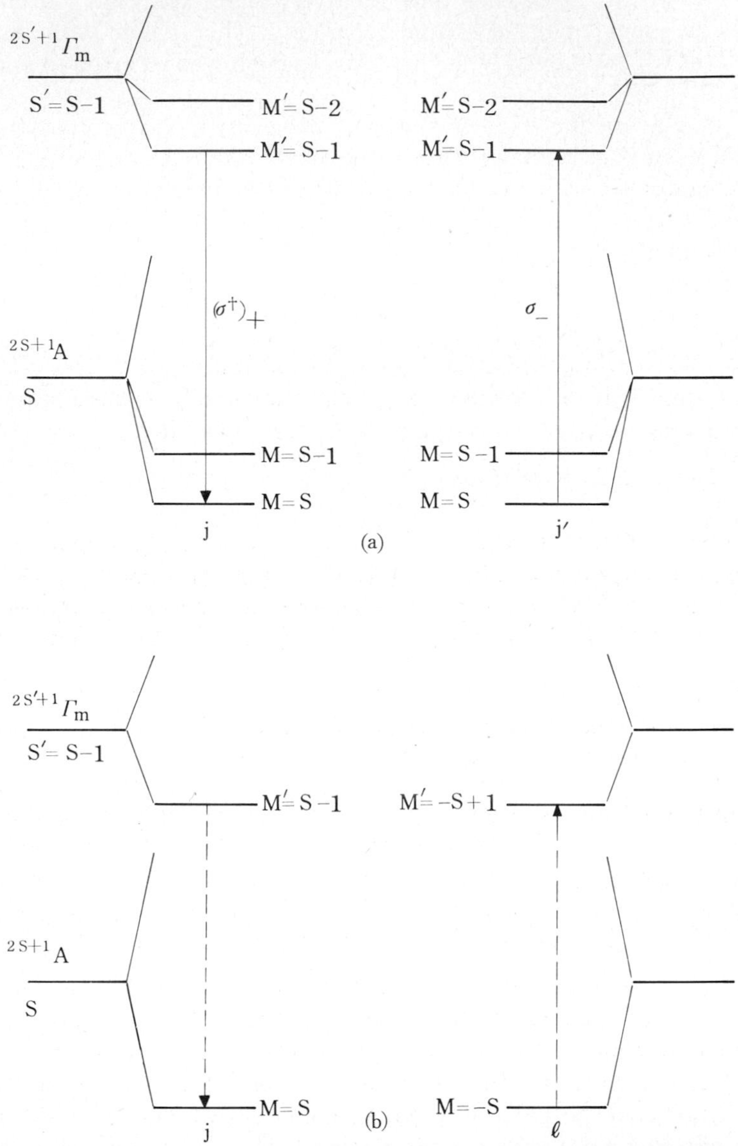

Fig. 4. Zeeman Splitting of the ground and excited levels in molecular field. (a) Excitation transfer is possible between ions of a ferromagnetic pair. (b) Excitation transfer due to eq. (6) is impossible between ions of an antiferromagnetic pair. In these figures, both spins are quantized with regard to the same axis.

3.2. Excitation transfer due to exchange

At low temperatures, only excitation from $M = S$ state to $M' = S' = S - 1$ will be important. (In the case of spin allowed transitions $M = S \rightarrow M' = S$ will be as important as the one $M = S \rightarrow M' = S - 1$.) We therefore concentrate upon the exciton associated with this transition. Suppose ion j is excited by absorption of light to $M' = S - 1$ state. This excitation cannot be kept localized at one site because of the interaction between the ions which makes transfer of excitation possible. In the present case the interaction must be spin dependent to be able to connect terms of different spin multiplicities. It is indeed possible to look for such interactions in the theory of superexchange. Instead of going into details of that theory we assume here the following form for the interaction* between the ions j and l rather phenomenologically (Freeman and Hopfield 1968, Fujiwara and Tanabe 1972):

$$V(jl) = 2K(jl)\boldsymbol{\sigma}_j \cdot \boldsymbol{\sigma}_l^{\dagger} + 2K(lj)\boldsymbol{\sigma}_j^{\dagger} \cdot \boldsymbol{\sigma}_l, \tag{6}$$

where $\boldsymbol{\sigma}_j(\boldsymbol{\sigma}_l^{\dagger})$ is a vector operator which, acting on the ground (excited) state of $j(l)$, turns it into the excited (ground) state, e.g.,

$$\langle S' = S - 1\ M' = M - 1|\sigma_-|S\ M\rangle = \sqrt{(S+M)(S+M-1)/2S(2S-1)} \tag{7}$$

with $\sigma_- = \sigma_x - i\sigma_y$. Matrix elements of the other components of $\boldsymbol{\sigma}$ automatically follow from the Wigner–Eckart theorem (Condon and Shortley 1963). The coefficient $2K(jl)$ will be called transfer integrals between j and l. For the lowest state $M = S$, we may replace σ_{j-} by $A_j^{\dagger}$, the exciton creation operator ($= |S' = S - 1\ M' = S - 1\rangle\langle S\ M = S|$) at site j.

Figure 4 shows that exciton transfer due to $V(jl)$ of (6) is possible among the ions of a ferromagnetic pair, whereas it is forbidden for an antiferromagnetic pair. This is because $M_j + M_l$ must be conserved (in case both spins are quantized with respect to the same axis) during the process of transfer due to (6). In general, effective magnitude of transfer, namely the coefficient K_{jl} of $\sigma_{j-}\sigma_{l+}^{\dagger}$ or $A_j^{\dagger}A_l$ in (6) varies as a function of the angle θ_{jl} between the spins $\boldsymbol{S}_j$ and $\boldsymbol{S}_l$:

$$K_{jl} = K(jl)\cos^2(\theta_{jl}/2), \tag{8}$$

so that for the transfer of excitation $M = S \rightarrow M' = S - 1$, we now have

$$V_{jl} = K_{jl}A_j^{\dagger}A_l + K_{lj}A_l^{\dagger}A_j \tag{9}$$

* In contrast to this we would have to consider spin independent (multipole–multipole) interaction besides the spin dependent exchange type interaction for excitons of spin allowed transitions. This would make the behaviour of the exciton $M = S \rightarrow M' = S$ different from that presented here, although $S\ M = S \rightarrow S' = S\ M' = S - 1$ exciton will have many features in common with $S\ M = S \rightarrow S' = S - 1\ M' = S - 1$ exciton.

which can be employed, of course, even in the case where interaction between the ions is not exactly of exchange type like eq. (6).

3.3 The Frenkel excitons in magnetic compounds

Putting eqs. (3), (5) and (9) together, we obtain the Hamiltonian for the exciton states:

$$\mathcal{H} = \sum_{j} \mathcal{H}_j + \sum_{j>l} V_{jl}, \tag{10}$$

where $\mathcal{H}_j$ is given by

$$\mathcal{H}_j = E_j A_j^\dagger A_j \tag{11}$$

with

$$E_j = E_{0j} + 2\mu_B(S'H'_{mol}(j) - SH_{mol}(j)), \tag{12}$$

where

$$H_{mol}(j) = -\sum_{l} 2J_{jl}\langle S_l\rangle \cos\theta_{jl}/2\mu_B, \text{ etc.} \tag{12a}$$

We note that the problem is now greatly simplified. The treatment from here proceeds exactly in the same way as in the theory of well-known Frenkel excitons. Suppose we have r magnetic ions in a magnetic unit cell of the crystal. Some of them may be equivalent, that is, they can be carried into each other by magnetic space group operations. The ions are labelled by the cell number n as well as the index α to distinguish the r magnetic ions.

The ground state wavefunction of the whole system is given by

$$\Phi_0 = \prod_{n\alpha} \phi_{n\alpha}, \tag{13}$$

where $\phi_{n\alpha}$ represents the lowest state ($M = S$) wavefunction of the ion $n\alpha$. Strictly speaking this is in general not the correct eigenfunction of the system. It actually is the ground state wavefunction in the molecular field approximation, e.g., wavefunction of the Néel state of antiferromagnets. As well known, however, this does not introduce any serious error even quantitatively in the results to be derived here. The wavefunction corresponding to the excited state where one of the ions $n\alpha$ is excited into the state $\phi'_{n\alpha}$ with $M' = S - 1$ is then given as

$$\Phi(n\alpha) = \phi'_{n\alpha} \prod_{m\beta(\neq n\alpha)}' \phi_{m\beta}. \tag{14}$$

Wavefunction of correct translational symmetry can be constructed from eq. (14):

$$\Phi(\boldsymbol{k}\alpha) = \frac{1}{\sqrt{N}} \sum_n \exp(i\boldsymbol{k} \cdot \boldsymbol{R}_{n\alpha})\Phi(n\alpha), \tag{15}$$

which represents the state in which an exciton with wave vector $\boldsymbol{k}$ is running through the sublattice α. The position of the ion $n\alpha$ is given by $\boldsymbol{R}_{n\alpha}$, and N is the number of cells in the crystal. This still is not the eigenfunction of the Hamiltonian (10). The eigenfunction can be obtained by calculating matrix elements of the Hamiltonian in terms of the r functions given above and then diagonalizing the r-dimensional matrix. The matrix elements of eq. (10) which is already diagonal with respect to $\boldsymbol{k}$ are easily calculated:

$$\langle \boldsymbol{k}\alpha | \mathcal{H} | \boldsymbol{k}\beta \rangle = E_\alpha \delta_{\alpha\beta} + V_{\alpha\beta}(\boldsymbol{k}), \tag{16}$$

where we have written E_α for $E_{n\alpha}$ which is independent of n, and

$$V_{\alpha\alpha}(\boldsymbol{k}) = \sum_{n'} \exp[i\boldsymbol{k} \cdot (\boldsymbol{R}_{n'\alpha} - \boldsymbol{R}_{n\alpha})] K_{n\alpha,n'\alpha}, \tag{17}$$

$$V_{\alpha\beta}(\boldsymbol{k}) = \sum_{m} \exp[i\boldsymbol{k} \cdot (\boldsymbol{R}_{m\beta} - \boldsymbol{R}_{n\alpha})] K_{n\alpha,m\beta}. \tag{18}$$

The eigenfunction with the energy eigenvalue $E_\nu(\boldsymbol{k})$ will be given by

$$\Phi_\nu(\boldsymbol{k}) = \sum_\alpha C_{\nu\alpha} \Phi(\boldsymbol{k}\alpha), \tag{19}$$

with the coefficients $C_{\nu\alpha}$ determined by,

$$\sum_\beta \langle \boldsymbol{k}\alpha | \mathcal{H} | \boldsymbol{k}\beta \rangle C_{\nu\beta} = E_\nu(\boldsymbol{k}) C_{\nu\alpha}. \tag{20}$$

There will of course be r exciton states $\Phi_\nu(\boldsymbol{k})$ $(\nu = 1, 2, \ldots, r)$ altogether.

Let us consider a simple case where we have two equivalent ions α and β, one with up spin and the other with down spin, that is, the case of a typical two sublattice antiferromagnet such as MnF_2 and $RbMnF_3$. We then have

$$E_\alpha = E_\beta, \qquad V_{\alpha\alpha}(\boldsymbol{k}) = V_{\beta\beta}(\boldsymbol{k}), \qquad V_{\alpha\beta}(\boldsymbol{k}) = V^*_{\beta\alpha}(\boldsymbol{k}), \tag{21}$$

so that

$$E_\pm(\boldsymbol{k}) = E_\alpha + V_{\alpha\alpha}(\boldsymbol{k}) \pm |V_{\alpha\beta}(\boldsymbol{k})|. \tag{22}$$

We see that the intrasublattice transfer leads to a dispersion of excitons, while the intersublattice transfer gives rise to a splitting. The splitting at $\boldsymbol{k} = 0$ which can be observed in optical absorption (in favourable cases) is usually called Davydov splitting. Thus the exciton nature of the excitation in magnetic crystals is exhibited in the Davydov splitting and its dispersion. The former can be observed in the ordinary single ion transitions (magnetic dipole or electric dipole assisted by the odd-parity field). The latter turns up in the line shape of the exciton–magnon transition which is made possible by the exchange dipole mechanism of sect. 4.

In the case of two sublattice antiferromagnets, $V_{\alpha\beta}(\boldsymbol{k})$ vanishes as shown in eq. (8) in case the transfer is possible only through exchange interaction (6). It is true that transfer of spin forbidden excitation can take place through multipole–multipole interaction perturbed by the spin–orbit interaction to the second order, but this effect should not be large. This is the reason for the difficulty of confirming the Davydov splitting even in the systems for which group theory predicts removal of the degeneracy for certain types of excitons*. The magnitude of $V_{\alpha\beta}(\boldsymbol{k})$ can be non-negligible when the deviation from collinearity of the two spins becomes appreciable under applied magnetic field. This has indeed been observed for ${}^6A_{1g} \to {}^4E_g$ exciton of $RbMnF_3$ (Novikov et al. 1973, see below).

3.4. Magnons as excitons

It is interesting to note that magnons can be treated within the framework of the present exciton theory with some modification. For magnons in two sublattice antiferromagnets $E_{0j} = 0$, $J'_{jl} = J_{jl}$ and exchange interaction itself takes care of excitation transfer, so that,

$$V(jl) = -2J_{jl}\boldsymbol{S}_j \cdot \boldsymbol{S}_l, \tag{23}$$

instead of eq. (6). In place of eq. (10) we now have

$$\mathcal{H} = \sum_n E_\alpha(a_n^\dagger a_n + b_n^\dagger b_n) - \sum_{n>n'} 2J_{nn'}S(a_n^\dagger a_{n'} + a_{n'}^\dagger a_n) \\ - \sum_{n>n'} 2J_{nn'}S(b_n^\dagger b_{n'} + b_{n'}^\dagger b_n) + \sum_{nm} 2J_{nm}S(a_n^\dagger b_m^\dagger + a_n b_m), \tag{24}$$

where

$$E_\alpha = g\mu_B H_A + \sum_{n'} 2J_{nn'}S - \sum_m 2J_{nm}S, \tag{25}$$

writing a_n and b_m for $A_{n\alpha}$ and $A_{m\beta}$, respectively. Note that we have included the anisotropy energy $g\mu_B H_A$, where g is the ground state g factor. In the second and third term of eq. (24) we note $-2J_{nn'}S$ plays the role of $K_{n\alpha,n'\alpha}$, the excitation transfer within the sublattice. Excitation transfer is impossible between the two sublattices as emphasized above. In the exciton theory which takes only the effect of resonance transfer the last term of eq. (24) was out of consideration, because the role of this term is to bring in the states of two more and two less excitons. The amount of mixing should be of the order of $K/2E_0$ which is negligible for the excitons in the optical

* In the case of E1 (or M1σ) exciton of MnF_2 associated with ${}^6A_{1g} \to {}^4T_{1g}$ (see below) the magnitude of $V_{\alpha\beta}(\boldsymbol{k} = 0)$ is estimated to be $\sim 10^{-5}\ \mathrm{cm}^{-1}$ (Holzrichter et al. 1971).

region. In the magnon problem, the term is not negligible, and makes the eigenvalue of the Hamiltonian (24) take the following form

$$E_m(\boldsymbol{k}) = \sqrt{(E_\alpha + V_{\alpha\alpha}(\boldsymbol{k}))^2 - |\bar{V}_{\alpha\beta}(\boldsymbol{k})|^2}, \tag{26}$$

where $V_{\alpha\alpha}(\boldsymbol{k})$ is given by eq. (17) if we replace $K_{n\alpha,n'\alpha}$ by $-2J_{nn'}S$ and

$$\bar{V}_{\alpha\beta}(\boldsymbol{k}) = -\sum_m 2J_{nm}S \exp[i\boldsymbol{k}\cdot(\boldsymbol{R}_m - \boldsymbol{R}_n)] \equiv |\bar{V}_{\alpha\beta}(\boldsymbol{k})| \exp(i\theta_k). \tag{27}$$

Diagonalization of eq. (24) can be carried out in the following way (Kittel 1963). In $\boldsymbol{k}$ representation corresponding to eq. (15), we find

$$\mathcal{H} = \sum_k [(E_\alpha + V_{\alpha\alpha}(\boldsymbol{k}))(a_k^\dagger a_k + b_{-k}^\dagger b_{-k}) - \bar{V}_{\alpha\beta}(\boldsymbol{k}) a_k b_{-k} - \bar{V}_{\alpha\beta}(\boldsymbol{k})^* a_k^\dagger b_{-k}^\dagger]. \tag{28}$$

Going over to the magnon operators,

$$\alpha_k^\dagger = u_k a_k^\dagger - v_k b_{-k},$$

$$\beta_{-k}^\dagger = -v_k a_k + u_k b_{-k}^\dagger, \tag{29}$$

the Hamiltonian is diagonalized,

$$\mathcal{H} = \sum_k [E_m(\boldsymbol{k})(\alpha_k^\dagger \alpha_k + \beta_{-k}^\dagger \beta_{-k} + 1) - (E_\alpha + V_{\alpha\alpha}(\boldsymbol{k}))], \tag{30}$$

when we determine u_k and v_k from

$$u_k = \sqrt{(E_\alpha + V_{\alpha\alpha}(\boldsymbol{k}) + E_m(\boldsymbol{k}))/2E_m(\boldsymbol{k})},$$

$$v_k = \sqrt{(E_\alpha + V_{\alpha\alpha}(\boldsymbol{k}) - E_m(\boldsymbol{k}))/2E_m(\boldsymbol{k})} \exp(i\theta_k). \tag{31}$$

Note that the eigenvalue $E_m(\boldsymbol{k})$ of eq. (26) is doubly degenerate for all wave vectors.

3.5. Symmetry considerations

Even when there are more than two ions ($r > 2$) in the magnetic cell, it is often possible to obtain eigenfunctions (19) of the matrix (16) from symmetry considerations without taking the step of direct diagonalization in case all the r ions are equivalent. This is based on the fact that the Hamiltonian of the system is invariant to the symmetry operations of the space group G which interchange the various magnetic ions while leaving the magnetic moments invariant. (Note that the group G is the unitary subgroup of the magnetic space group G′.) As is well known, the eigenfunction of such a system must behave under symmetry operations like a basis of one of the irreducible representations of the group G, and wavefunctions belonging to different irreducible representations do not mix with each other. Conversely, if we succeed in constructing such linear combinations of wavefunctions, the labour of the diagonalization is greatly reduced, because we then only have to deal with the matrix within the manifold of each irreducible representation which is of much less dimensionality.

For the details of the representation theory of the space group the reader is referred, for example, to Heine (1960). In the present problem $\Phi(\boldsymbol{k}\alpha)$ are the bases of the irreducible representation characterized by a wave vector $\boldsymbol{k}$ of the invariant subgroup T of G which consists of primitive translations of the lattice, and the eigenfunctions $\Phi_\nu(\boldsymbol{k})$ can be classified according to the irreducible representations of G($\boldsymbol{k}$), the $\boldsymbol{k}$ group of G which consists of operations of G that leave $\boldsymbol{k}$ unchanged.

For a general $\boldsymbol{k}$ in the Brillouin zone (BZ), $\Phi_\nu(\boldsymbol{k})$ all belong to one and the same irreducible representation of G($\boldsymbol{k}$), because G($\boldsymbol{k}$)/T for such $\boldsymbol{k}$ consists only of identity operation. It is for the points and lines of high symmetry in BZ that group theory can help. Especially for $\boldsymbol{k} = 0$ which is most interesting from the point of view of the Davydov splitting, we can proceed in the following way (McClure 1959a). We first study what sort of irreducible representation Γ of $\mathrm{G}(\boldsymbol{k}=0) \equiv \mathrm{G}$ are included in the reducible representation subtended by the functions $\Phi(\boldsymbol{k}=0, \alpha)$. For this purpose we calculate the character $\chi(R)$ or the trace of the matrix of representation by $\Phi(\boldsymbol{k}=0, \alpha)$ for the operation R belonging to G or rather the factor group G/T. The character will be non-vanishing only for those R that leave each sublattice invariant so that

$$\chi(R) = \sum_\alpha (\Phi(\boldsymbol{k}=0, \alpha), R\Phi(\boldsymbol{k}=0, \alpha)) = \sum_\alpha \chi^{(\alpha\sigma)}(R_\alpha), \tag{32}$$

for such R, where

$$\chi^{(\alpha\sigma)}(R_\alpha) = (\phi'_{n\alpha}, R_\alpha \phi'_{n\alpha}). \tag{33}$$

The function $\phi'_{n\alpha}$ will belong to the σth irreducible representation of the site group G_α, i.e., the set of operations which leave the site α and associated magnetic moment invariant, and $\chi^{(\alpha\sigma)}(R_\alpha)$ is the character of that representation for the operation R_α of G_α. Note that in eq. (33) R has been replaced by the operation R_α around site α which is the same rotation or reflection as R but taken with the site α as its origin.

Knowledge of the characters (32) enables us to decide whether a particular irreducible representation Γ is included or not. When we know that Γ is included, it is not difficult to construct $\Phi(\boldsymbol{k}=0, \Gamma\gamma)$ which transforms like the γth basis of the representation Γ in terms of the functions $\Phi(\boldsymbol{k}=0, \alpha)$, for example, by the method of projection operator. When Γ occurs more than twice in the reduction, we have to set up as many such functions of the same symmetry and diagonalize the Hamiltonian matrix within the manifold of these functions. When Γ occurs only once, we may say that we have found the correct eigenfunction without diagonalization.

The symmetry Γ_0 of the ground state wavefunction can of course be determined from the characters

$$\chi_0(R) = (\Phi_0, R\Phi_0). \tag{34}$$

When the product representation $\Gamma \times \Gamma_0^*$ contains the irreducible representation subtended by the electric or magnetic dipole moment operator, first order radiative process, i.e., absorption or emission is possible between Φ_0 and $\Phi(\boldsymbol{k} = 0, \Gamma\gamma)$. This is of course the selection rule that tells us whether the matrix element of the (electric or magnetic) dipole moment operator P,

$$\langle \Phi_\nu(\boldsymbol{k} = 0)|P|\Phi_0\rangle = \sum_\alpha C^*_{\nu\alpha}(\phi'_{n\alpha}, p\phi_{n\alpha})/\sqrt{N}, \tag{35}$$

with $\nu = \Gamma\gamma$, vanishes or not.

The treatment developed above is the standard procedure for excitons of molecular crystals. It can of course be applied to the excitons of magnetic crystals as it is. However, there are two points that we had better keep in mind in the present problem.

First, in the case of molecular crystals, Φ_0 usually belongs to A_{1g} or the totally symmetric representation. This is not necessarily true in the present case. Besides, double valued representations of site group inevitably appear when we treat ions of odd-number electrons. In order to avoid these complications, it is more convenient to consider as the symmetry of exciton that of the product $\Phi_\nu(\boldsymbol{k} = 0)\Phi_0^*$ or $\Gamma \times \Gamma_0^*$ instead of treating each function $\Phi_\nu(\boldsymbol{k} = 0)$ and Φ_0 separately (Loudon 1968). The symmetry of $\Phi_\nu(\boldsymbol{k} = 0)$ can be seen easily from that of exciton and of the ground state function Φ_0, because symbolically,

$$\Phi_\nu(\boldsymbol{k} = 0) \sim (\Phi_\nu(\boldsymbol{k} = 0)\Phi_0^*)\Phi_0 \tag{36}$$

as far as symmetry is concerned. Note also that the symmetry of $\Phi(\boldsymbol{k} = 0, \alpha)\Phi_0^*$ under site group symmetry operations is given by $\phi'_{n\alpha}\phi^*_{n\alpha}$.

Second, introduction of antiunitary symmetry operations, i.e., those involving time reversal as well as rotation, reflection or translation, extends G to G′, the magnetic space group. Since our Hamiltonian is invariant actually to the operations of G′, the eigenfunctions of the system are required to form bases of an irreducible co-representation of G′. This requirement leads to the degeneracy (due to time reversal) of some pairs of the eigenvalues $E_\Gamma(\boldsymbol{k} = 0)$. There is a simple criterion (Dimmock and Wheeler 1964) to decide for which irreducible representation Γ of G this occurs. We evaluate a sum of characters $\chi^{(\Gamma)}(\boldsymbol{R}'^2)$ over all the antiunitary operations $\boldsymbol{R}'$ belonging to the factor group G′/T. We have three possibilities:

$$\begin{aligned} \sum_{R'} \chi^{(\Gamma)}(R'^2) &= +g, \qquad \text{(a)}, \\ &= -g, \qquad \text{(b)}, \\ &= 0, \qquad \text{(c)}, \end{aligned} \tag{37}$$

where g is the order (number of elements) of the factor group G/T. In the case (a) there is no extra degeneracy. In the cases (b) and (c), there is extra degeneracy. The state $\Phi_{\Gamma\gamma}(\boldsymbol{k}=0)$ will be degenerate with the state $R'\Phi_{\Gamma\gamma}(\boldsymbol{k}=0)$ which are linearly independent, where R' is one of the antiunitary operations of G′ such that G′ = G + R'G. The latter functions are bases of a representation equivalent to Γ in the case (b) and those of an inequivalent irreducible representation $\bar{\Gamma}$ in the case (c). The characters of $\bar{\Gamma}$ are related to those of Γ as follows,

$$\chi^{(\bar{\Gamma})}(R) = \chi^{(\Gamma)}(R'^{-1}RR'). \tag{38}$$

The irreducible representation for $\boldsymbol{k} \neq 0$ excitons, if necessary, can be determined in a similar way, i.e., by evaluating the characters of the representation subtended by $\Phi(\boldsymbol{k}, \alpha)$. An alternative way is to start at the center $\boldsymbol{k}=0$ of BZ and work outwards along the symmetry lines using compatibility relations between the irreducible representations of symmetry lines and points (Heine 1960).

3.6. The Zeeman effect and magnetic circular dichroism

Before proceeding to the discussion of the observed examples of exciton lines in magnetic crystals, we will briefly consider the effects of external magnetic field. For simplicity, we again take up a two sublattice antiferromagnet. Suppose the magnetic field $\boldsymbol{H}$ (less than the critical field H_{cr}) is applied along the magnetization, or more exactly in the direction of the up spin, and the g values of $\phi_{n\alpha}$ and $\phi'_{n\alpha}$ are respectively g and g'. At an up spin site, both the excited and ground state will lose energy by $+g'\mu_{\mathrm{B}}S'H$ and $+g\mu_{\mathrm{B}}SH$, respectively, whereas the corresponding states of the down spin site will gain energy by the same amount (their Zeeman energy being $-g'\mu_{\mathrm{B}}S'H$ and $-g\mu_{\mathrm{B}}SH$, respectively). This means that the degeneracy of the two exciton states is lifted and we should observe a Zeeman splitting of

$$\Delta_{\mathrm{e}} = 2\mu_{\mathrm{B}}H(gS - g'S'), \tag{39}$$

as shown in fig. 5.

If we send in the light along the direction of the magnetic field in this situation, the up and down spin ion will behave oppositely for the right and left circularly polarized light. This magnetic circular dichroism (MCD) will

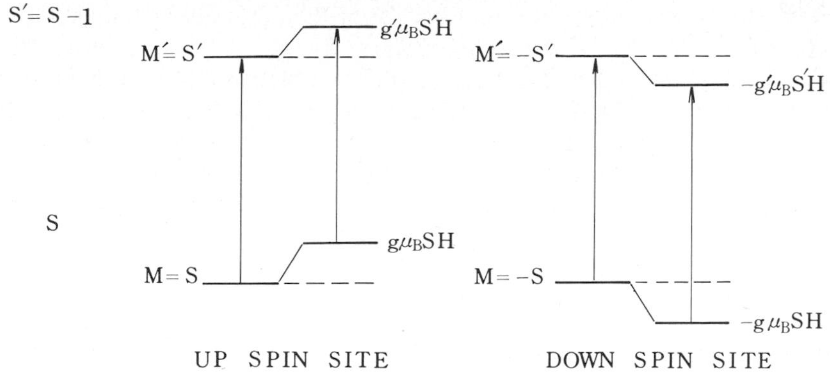

Fig. 5. Zeeman effect at up and down spin site. The same axis of quantization for both is adopted here.

be most helpful (Wong et al. 1974) for establishing the assignment of the exciton lines together with the Zeeman effect as we shall see below.

3.7. *The Davydov splitting in* Cr_2O_3

Let us now turn to the discussion of an example, the R exciton lines of Cr_2O_3 for which a remarkably large Davydov splitting is observed (Allen et al. 1969). This crystal is an antiferromagnetic insulator with its Néel temperature T_N of 308 K. It has the corundum structure with two formula units per unit cell (fig. 6). The site symmetry of the Cr^{3+} ion is C_3. Below T_N, the spins align along the crystal c-axis (chosen as z-axis below) alternately up and down, so that we have four equivalent ions ($r = 4$) in the (magnetic) unit cell. The magnetic space group G′ is $R\bar{3}'c'$, its unitary subgroup G, $R\bar{3}c$ or D_{3d}^6.

The absorption spectrum of Cr_2O_3 resembles that of Cr^{3+} in ruby shown in fig. 1. In fact, there are several sharp lines in the red to be associated with single ion transitions between $^4A_2$ and ^2E. The magnetic field dependence of the unpolarized absorption spectrum around 13 840 cm^{-1} is shown in fig. 7. The polarization character and the labelling of the lines are given in right hand columns. It is found that the lines 1 to 4 have electric dipole character, because α (axial, direction of light $\| c$) and $\sigma(\boldsymbol{E} \perp c)$ spectrum coincide, while the line 5 has a mixed electric and magnetic dipole nature. When the magnetic field applied parallel to the c-axis is increased, a satellite line 2′ appears. Lines 1 and 4 both have the same splitting factor $\Delta_e/2\mu_B H = 2.3$. In MCD, they are both polarized in the same way, with the higher and lower energy line of each split pair being, respectively, left and right circularly polarized.

The spin–orbit interaction together with the trigonal field makes the ^{2}E state split into two Kramers' doublets $\bar{\text{E}}$ and $2\bar{\text{A}}$ as in ruby. In the magnetically ordered state, each of the doublets further splits into two Zeeman components under molecular field. A schematic energy level diagram for Cr^{3+} ions with up spin (ions 1 and 3 of fig. 6) is given in fig. 8, together with the corresponding eigenfunctions. Operation C_3 of the site group $G_{\alpha=1} = C_3$ transforms, for example, the eigenfunctions according to

$$
\begin{aligned}
C_3\phi'(M' = \tfrac{1}{2}u_\pm) &= \exp[-i\tfrac{2}{3}\pi(\tfrac{1}{2} \pm 1)]\phi'(M' = \tfrac{1}{2}u_\pm), \\
C_3\phi(M = \tfrac{3}{2}a) &= e^{-i\pi}\phi(M = \tfrac{3}{2}a).
\end{aligned} \tag{40}
$$

We then find for the transition $\phi(M = \tfrac{3}{2}a) \to \phi'(M' = \tfrac{1}{2}u_-)$ the symmetry of the product $\phi'\phi^*$ is E of C_3. In fact the product transforms like $x + iy$, whereas the corresponding product at a down spin site will behave like $x - iy$. This means that the excitons associated with this transition (which we call E excitons) will be observed in σ polarization and excitons at up (down) spin sites will be created by the right (left) circularly polarized light in α spectrum. Similarly, the excitons associated with the transition $\phi(M = \tfrac{3}{2}a) \to \phi'(M' = \tfrac{1}{2}u_+)$ will be called A excitons, because the product $\phi'\phi^*$

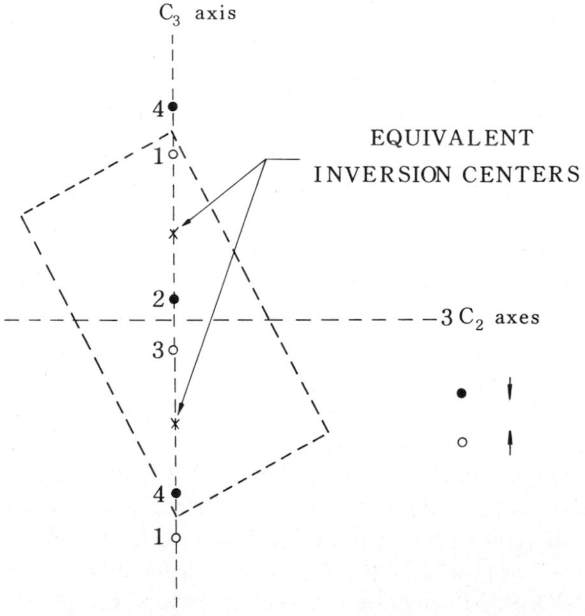

Fig. 6a. Schematic trigonal unit cell for Cr_2O_3 showing symmetry operations and labelling of the unit cell ions (Allen et al. 1969).

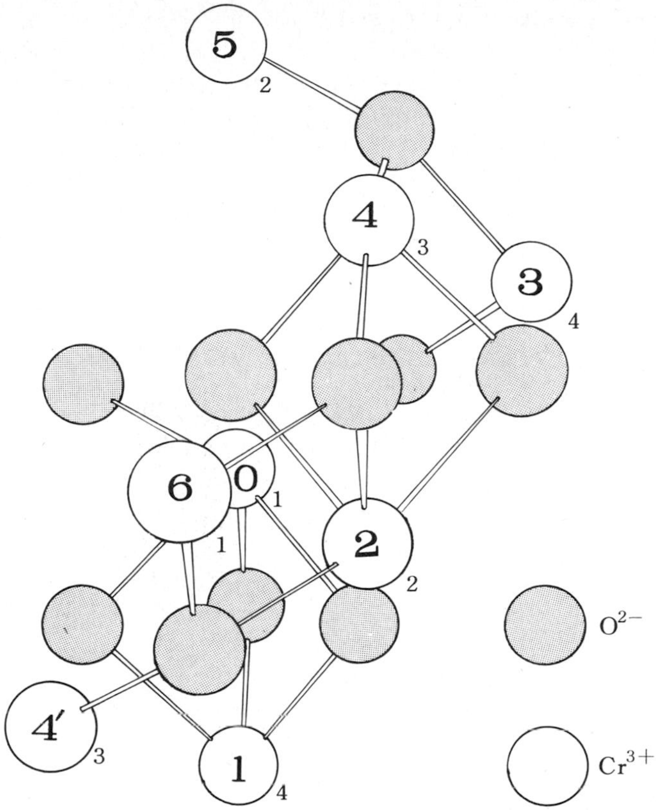

Fig. 6b. Spatial arrangement of nearest neighbours in Cr_2O_3 lattice. The number given outside each circle indicates the sublattice the Cr^{3+} ion belongs to.

transforms according to the irreducible representation A of C_3. We note they will be observed in π polarization ($\boldsymbol{E} \parallel c$).

Applying eqs. (32) and (33) to E and A excitons, we find that E exciton gives rise to two E exciton states at $\boldsymbol{k} = 0$, while A exciton leads to two A_1 and two A_2 states, where the latter E, A_1 and A_2 are the irreducible representations of D_3, i.e., G/T of the present problem. As remarked before, it is much more convenient to use $\Phi(\boldsymbol{k} = 0\alpha)\Phi_0^*$ and $\phi'_{n\alpha}\phi_{n\alpha}^*$ instead of $\Phi(\boldsymbol{k} = 0\alpha)$ and $\phi'_{n\alpha}$, respectively. The test (37) shows that the dimensions of these representations are not affected. (In the present case, R'G/T consists of $\theta\{I \mid 0\}$, $2\theta\{IC_3 \mid 0\}$, and $3\theta\{IC_2 \mid \tau\}$, where θ is the time reversal operator and τ is a translation along C_3 by half the length of the unit cell along that axis.) Since components of electric dipole moment operator transform like A_2 and E of D_3, we note only A_2 exciton states will be

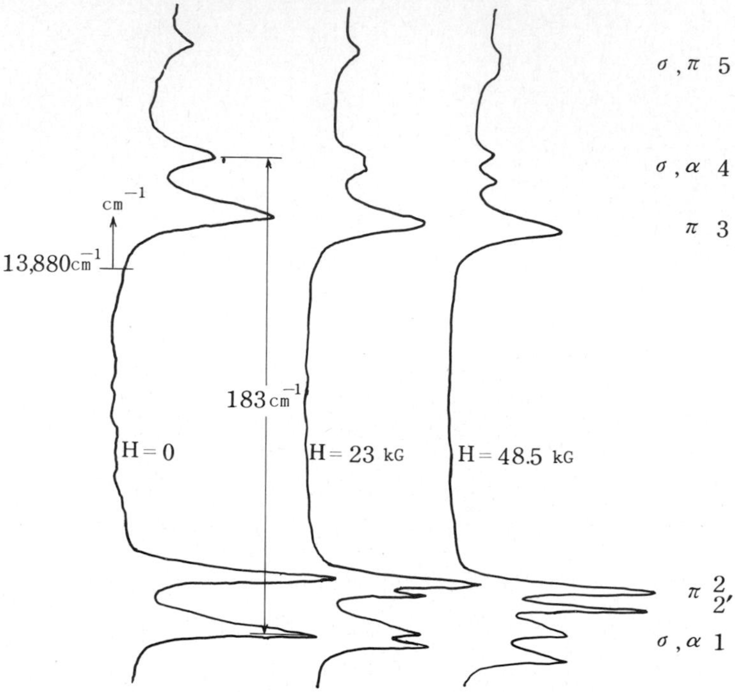

Fig. 7. Zeeman effect in unpolarized absorption spectrum of $^4A_2 \to {}^2E$ excitons in Cr_2O_3. Polarizations and labelling are given in the right hand columns (Allen et al. 1969).

observed in π spectrum. However, A_1 states can borrow intensity from A_2 states under external magnetic field applied along C_3 axis, because z component of the magnetic dipole moment operator transforms as A_2 in D_3 and the Zeeman energy can have matrix element between an A_1 and an A_2 state (or between two E states).

These conclusions can be derived of course by diagonalizing the Hamiltonian (16) for $\boldsymbol{k} = 0$ and obtaining its eigenfunctions. For E excitons, we have the following matrix:

$$(\mathcal{H}_{\alpha\beta}) = \begin{bmatrix} E_1 & 0 & V_{13} & 0 \\ 0 & E_1 & 0 & V_{13}^* \\ V_{13}^* & 0 & E_1 & 0 \\ 0 & V_{13} & 0 & E_1 \end{bmatrix}, \tag{41}$$

where V_{13} is $V_{13}(\boldsymbol{k} = 0)$ and V_{11} is absorbed into E_1. Note that V_{13} represents the coupling with fourth neighbours with parallel spins. The

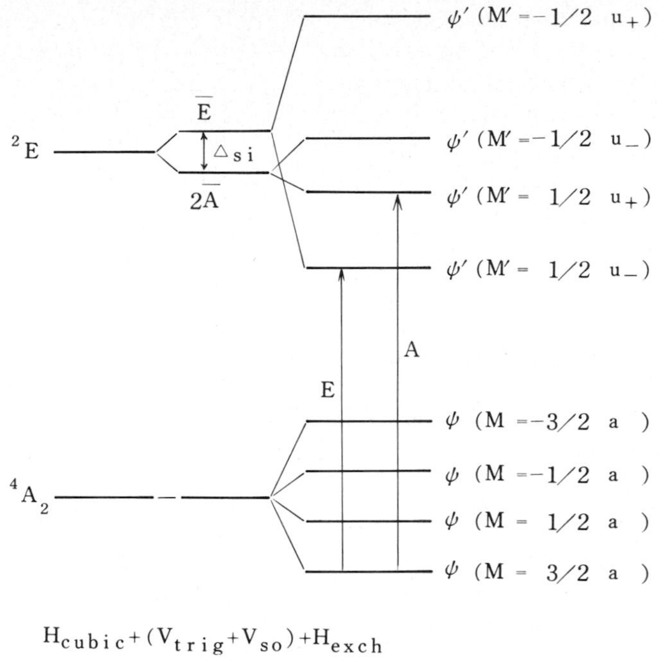

Fig. 8. Schematic energy level diagram for a Cr^{3+} ion at up spin site.

eigenvalues of matrix (41) are given by

$$E_1(\boldsymbol{k}=0)=E_1-|V_{13}|,\quad (\mathrm{E})$$
$$E_2(\boldsymbol{k}=0)=E_1+|V_{13}|,\quad (\mathrm{E}) \tag{42}$$

each doubly degenerate. This degeneracy which is essentially that of up spin and down spin state, of course, corresponds to the symmetry of E in D_3.

For A excitons, the Hamiltonian matrix is given by

$$(\mathcal{H}_{\alpha\beta})=\begin{bmatrix} E_1 & V_{12} & V_{13} & V_{14} \\ V_{12}^* & E_1 & V_{14} & V_{13}^* \\ V_{13}^* & V_{14} & E_1 & V_{12}^* \\ V_{14} & V_{13} & V_{12} & E_1 \end{bmatrix}, \tag{43}$$

where we have used the same notation as in eq. (41). However, E_1 and V_{13} in eq. (41) and those in eq. (43) are actually different parameters, because the former ones are related to the orbital state u_-, whereas the latter to u_+. Note that V_{14} which represents the coupling with the nearest (and third) neighbours with antiparallel spins is real. It is not difficult to diagonalize the matrix (43). For simplicity, however, we will put $V_{12} = 0$, because V_{12}, being the coupling with the second neighbours with antiparallel spins, is likely to be small. The eigenvalues are then easily found to be

$$E_1(\boldsymbol{k} = 0) = E_1 + V_{14} - |V_{13}|, \quad (\mathrm{A}_1)$$

$$E_2(\boldsymbol{k} = 0) = E_1 - V_{14} - |V_{13}|, \quad (\mathrm{A}_2)$$

$$E_3(\boldsymbol{k} = 0) = E_1 + V_{14} + |V_{13}|, \quad (\mathrm{A}_1)$$

$$E_4(\boldsymbol{k} = 0) = E_1 - V_{14} + |V_{13}|, \quad (\mathrm{A}_2) \tag{44}$$

where symmetry of each exciton state is given following its energy eigenvalue.

We have seen in eqs. (41) and (43) that there are several restrictions imposed on the matrix elements of the Hamiltonian to produce eigenvalues with correct degeneracy as well as eigenfunctions of correct symmetry as predicted by the group theory. It should be remarked such restrictions can be derived from the following type of general relations (Macfarlane and Allen 1971):

$$\begin{aligned}\langle \boldsymbol{k} = 0\alpha|\mathscr{H}|\boldsymbol{k} = 0\beta\rangle &= (\Phi(\boldsymbol{k} = 0\alpha), \mathscr{H}\Phi(\boldsymbol{k} = 0\beta)) \\ &= (R\Phi(\boldsymbol{k} = 0\alpha), \mathscr{H}R\Phi(\boldsymbol{k} = 0\beta)) \\ &= (R'\Phi(\boldsymbol{k} = 0\alpha), \mathscr{H}R'\Phi(\boldsymbol{k} = 0\beta))^*,\end{aligned} \tag{45}$$

which hold for any operation R belonging to G/T and for any antiunitary operation R' belonging to G′/T. Useful relations can be obtained especially by choosing R or R' which sends one of the sublattices into another.

These considerations will make the assignment in table 1 almost plausible. From the splitting factor of lines 1 and 4, a reasonable value of $g' = 1.37$ for the paramagnetic state $\bar{\mathrm{E}}$ is obtained assuming $g = 2$. Observed magnitude of the Davydov splitting of the E exciton leads to the value for $|V_{13}| = 91.5\ \mathrm{cm}^{-1}$. This in turn gives us the value $\sim 30\ \mathrm{cm}^{-1}$ for $|2K(jl)|$ of eq. (6) between the ion j and its fourth neighbour l, provided that V_{13} is mainly determined by the coupling with the six fourth neighbours, $V_{13} = 6K(jl)$. For the A exciton we have $|V_{13}| = 70\ \mathrm{cm}^{-1}$, so that $|2K(jl)|$ will be $\sim 23\ \mathrm{cm}^{-1}$. These values for $|2K(jl)|$ both seem fairly large compared with that of $2J(jl) \sim +7\ \mathrm{cm}^{-1}$ known for the fourth neighbour pairs in ruby, although $K(jl)$ should probably be compared with $-2J(jl)S$.

As remarked before, it is difficult in the case of a two sublattice antiferromagnet to observe the large Davydov splitting arising from the transfer of excitations between ions with antiparallel spins. However, in strong magnetic fields (above the critical or flop field H_{cr}) the spins will cant towards the direction of $\boldsymbol{H}$ and $|V_{\alpha\beta}(\boldsymbol{k} = 0)|$ will increase in proportion to the square of $\sin t$ where t is the tilt angle ($= (\pi - \theta_{jl})/2$). This was actually confirmed

Table 1
Exciton lines of Cr_2O_3

Lines	Observed energies	Symmetry of exciton states ($\boldsymbol{k} = 0$)	Exciton energies
1	13743.3 cm^{-1}	E	$E_1(\boldsymbol{k} = 0)^a$
2	13764.1	A_2	$E_2(\boldsymbol{k} = 0)^b$
2′	13764.1 − 3.75	A_1	$E_1(\boldsymbol{k} = 0)^b$
3	13903.4	A_2^c	$E_4(\boldsymbol{k} = 0)^b$
4	13926.3	E	$E_2(\boldsymbol{k} = 0)^a$
5	13970.5	A_2^c	

[a] Equation (42).
[b] Equation (44). In the analysis of Allen et al. (1969) non-vanishing value of V_{12} is assumed so that the theoretical splitting pattern is slightly different from that of eq. (44).
[c] In a later paper (Allen 1974) Allen assigns line 5 to A_2 and assumes line 3 to be of some other origin.

for the ^4E_g exciton (with energy 25 144.5 cm^{-1}) in $RbMnF_3$ (Novikov et al. 1973). For example, when the applied field is parallel to the [111] direction of this cubic crystal, four magnetic dipole lines (magnetic field vector of the light perpendicular to $\boldsymbol{H}$) are observed, because there are two orbital components u and v of the cubic E_g state as well as the two sublattices in this case (fig. 9). The two states $\Phi'(M' = \frac{3}{2}u)$ and $\Phi'(M' = \frac{3}{2}v)$ have different single ion energies because of the (second order) spin–orbit interaction. We do not, however, attempt further analysis here, because this problem is treated and reviewed in greater details by Eremenko and Petrov (1977). We only quote the result (Fujiwara and Tanabe 1974) that the magnitude of $2K(jl)$ (average of the transfer integrals for u and v states of ^4E_g) between the nearest neighbour ions j and l of $RbMnF_3$ is as large as 64.6 cm^{-1}, which is much larger than the value of the ground state exchange integral $2J(jl) \sim -2.36$ cm^{-1}.

3.8. Exciton dispersion in MnF_2

Our next example is the ^4E_g exciton of MnF_2 for which fairly large exciton dispersion is reported. Since symmetry consideration is helpful also for this two sublattice antiferromagnet and is going to be taken advantage of again in sect. 4, we will summarize the main results (Loudon 1968) of it here for later reference.

The crystal structure of MnF_2 is of rutile type. Its magnetic unit cell is shown in fig. 10. Below T_N (68 K), the spins of Mn^{2+} order along the c-axis antiferromagnetically, the body center ions of the unit cell forming the up

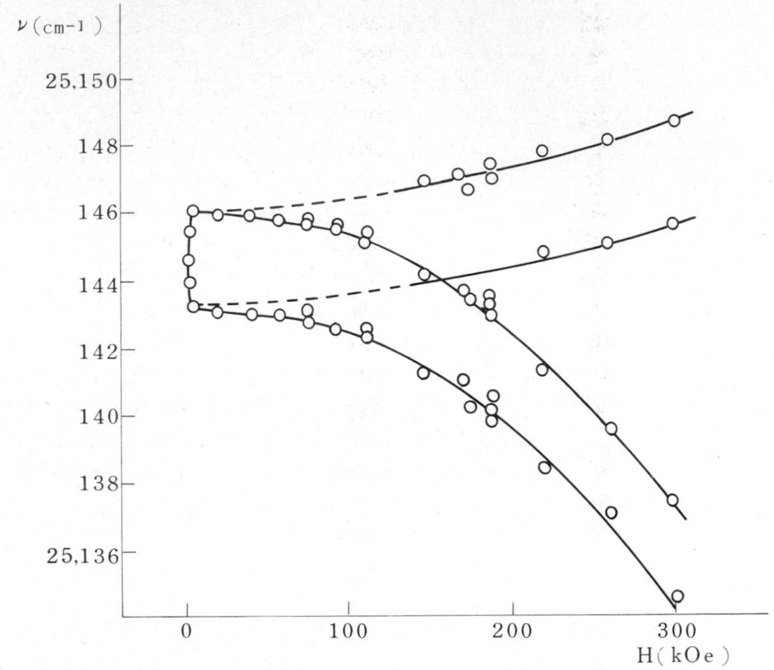

Fig. 9. Davydov splitting of ${}^6A_{1g} \to {}^4E_g$ exciton in $RbMnF_3$ as a function of external magnetic field $\boldsymbol{H} \parallel [111]$ (Novikov et al. 1973).

spin sublattice and the corner ions the down spin sublattice. The magnetic space group is $P4_2'/mnm'$ with its unitary subgroup Pnnm or D_{2h}^{12}. The site group in ordered phase is C_{2h}, while that in paramagnetic state is D_{2h}.

The cubic state ${}^4E_g$ splits into ${}^4A_g$ and ${}^4B_{3g}$ in the field of symmetry D_{2h}. In C_{2h} the latter will be ${}^4A_g$ and ${}^4B_g$, respectively. For the transition $\phi(M = \frac{5}{2}a_g) \to \phi'(M' = \frac{3}{2}a_g)$ the symmetry of the product $\phi'\phi^*$ is B_g of C_{2h}. This can easily be verified by operating the elements of C_{2h} upon this product. The exciton arising from this transition is called B_g exciton. Similarly, we have A_g exciton associated with the transition $\phi(M = \frac{5}{2}a_g) \to \phi'(M' = \frac{3}{2}b_g)$. Noting that G/T is D_{2h}, we find that B_g excitons will give rise to $\boldsymbol{k} = 0$ exciton states $B_{3g} + B_{2g}$, whereas A_g exciton will lead to those of symmetry A_g and B_{1g}. Application of the test (37) shows that the states B_{3g} and B_{2g} should be degenerate because of the time reversal symmetry. We have put a + sign to indicate this degeneracy which will be lifted only in an external magnetic field. Contrary to this, there is no need for A_g and B_{1g} state of A_g exciton to be degenerate for this reason. Since the three components m_x, m_y and m_z of the magnetic dipole moment transform like B_{3g}, B_{2g} and B_{1g} of D_{2h}, respectively, the exciton state $\boldsymbol{k} = 0$ B_{1g} will be observed as magnetic dipole transition in σ spectrum (with the direction of

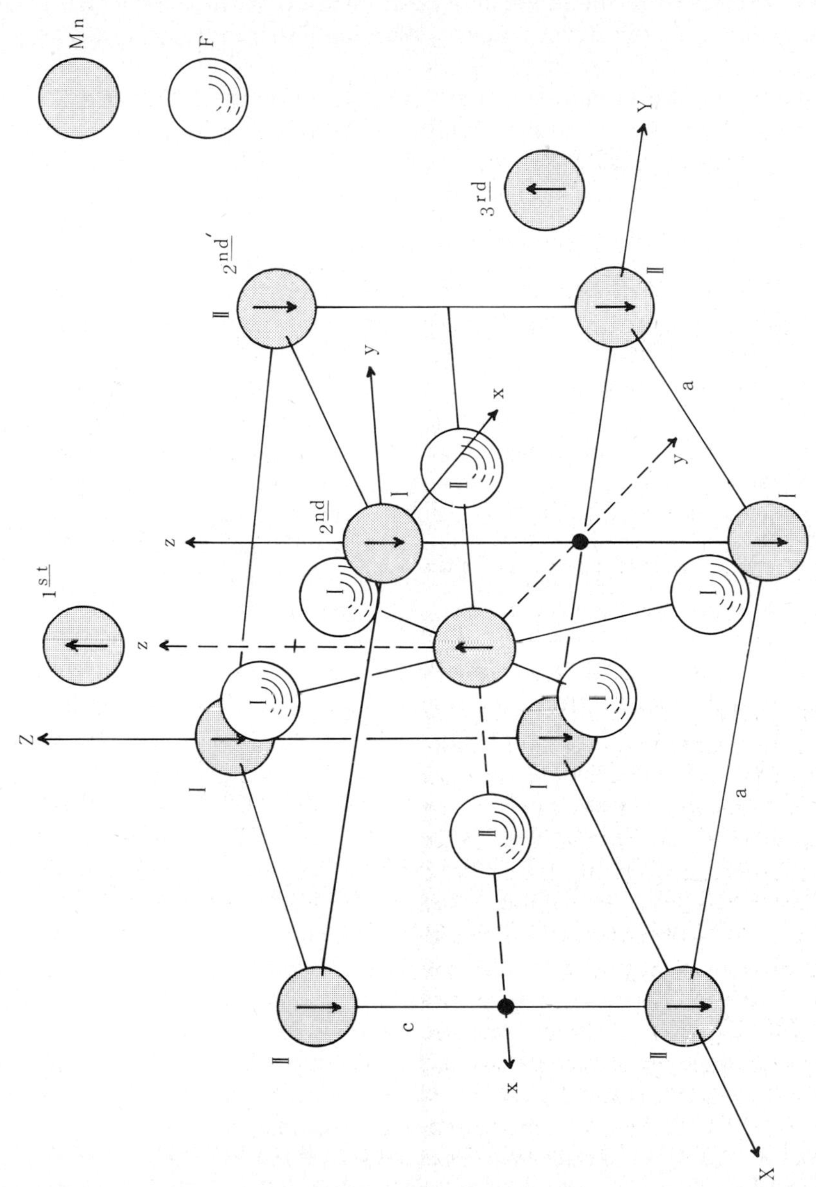

Fig. 10. Crystal structure of MnF_2. Different types of neighbouring cations of the body centre ion are designated by Roman numerals. Positions of the first, second and third neighbours are also shown.

light $\perp c$), while those of $B_{3g}+B_{2g}$ in π and α spectrum. Note that a magnon regarded as an exciton is of B_g type, because the transition creating spin deviation in the molecular field corresponds to $\phi(M=\frac{5}{2}a_g)\to\phi(M=\frac{3}{2}a_g)$. At $\boldsymbol{k}=0$, therefore, magnon state has inherent degeneracy of $B_{3g}+B_{2g}$.

The dispersion of the exciton state can be obtained from eq. (22) (and eq. (18)) with $V_{\alpha\beta}(\boldsymbol{k})=0$. If we assume excitation transfers to be appreciable only between nearest neighbours, we obtain the following expression for the dispersion:

$$V_{\alpha\alpha}(\boldsymbol{k})=2K_1\cos k_Z c+2K_3(\cos k_X a+\cos k_Y a), \tag{46}$$

where K_1 and K_3 are the effective transfer coefficients of eq. (9) for the first and third neighbour ions. (See fig. 10 for the location of the first and third neighbour of the body centre ion.)

In fig. 11(a), the absorption spectra of the transition $^6A_{1g}\to{}^4E_g$ (25 250 cm^{-1}) of Mn^{2+} in MnF_2 (Meltzer et al. 1969) are given for the two polarizations σ and π. The sharp absorption lines M1σ and M2π are magnetic dipole lines and observed only in σ and π polarization, respectively. This suggests that they are the exciton state (A_g, B_{1g}) and $B_{3g}+B_{2g}$ arising from A_g and B_g excitons of ^4E_g, respectively. The observed splitting factors $\Delta_e/2\mu_BH=1.96$ (M1σ) and 2.04 (M2π) also seem to support this assignment. Note that no appreciable Davydov splitting is observed for the former pair, i.e. A_g and B_{1g}.

As pointed out before, the dispersion of the exciton state is reflected in the line shape of the magnon sidebands to be discussed in the next section. The broad lines E1σ, π and E2σ, π in fig. 11(a) are the sidebands associated with M1σ and M2π, respectively. Leaving the theoretical interpretation of the line shape to sect. 4, we simply stress here that a satisfactory explanation of the shape of E2π is possible only by assuming a very large negative dispersion for the exciton associated with M2π or the B_g exciton originating in ^4A_g state in D_{2h}. The best choice for the values of parameters to reproduce the observed shapes are

$$2K_1=37\ \mathrm{cm}^{-1},\qquad 2K_3=0, \tag{47}$$

for the B_g exciton (M2π) and,

$$2K_1=2\ \mathrm{cm}^{-1},\qquad 2K_3=1\ \mathrm{cm}^{-1}, \tag{48}$$

for the A_g exciton (M1σ).

The result (47) shows that the transfer integral $2K(jl)$ between j and its first neighbour l in MnF_2 can be as large as 37 cm^{-1} in the $^4A_g(D_{2h})$ component of the cubic ^4E state, whereas it is rather small for its $^4B_{3g}(D_{2h})$ component. This is again very large compared to the corresponding exchange integral in the ground state $2J(jl)=0.44\ cm^{-1}$.

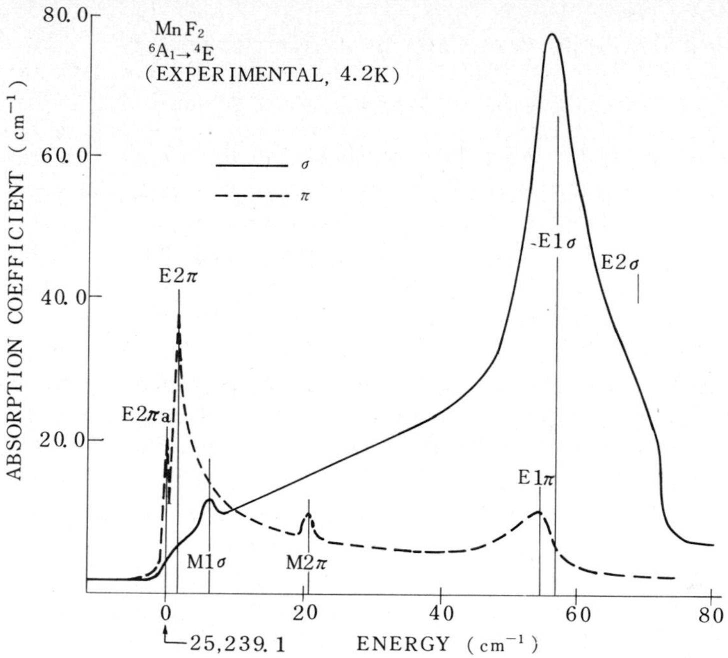

Fig. 11a. Absorption near the origin of the transition $^6A_{1g} \rightarrow {}^4E_g$ at 4.2 K (Meltzer et al. 1969).

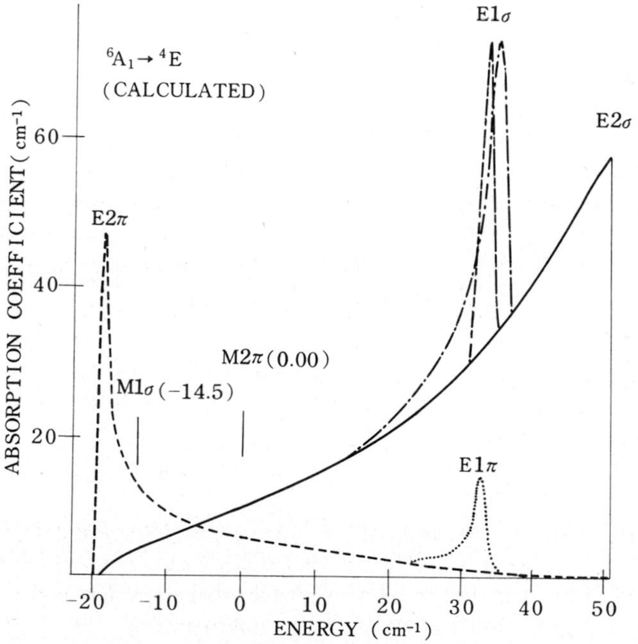

Fig. 11b. Calculated cold band absorption for the transition $^6A_{1g} \rightarrow {}^4E_g$ (Meltzer et al. 1969).

4. Exciton–magnon transitions in antiferromagnets

4.1. Mechanism of cooperative transitions

We have seen in the preceding section that the single ion transition mechanism either electric dipole or magnetic dipole can create only one exciton (including magnon) with $\boldsymbol{k} = 0$, and provides us with information on Davydov splitting. There is, however, another mechanism (Tanabe et al. 1965, Ferguson et al. 1965) involving two ions which in the case of antiferromagnets can lead to creation of two excitons (two magnons, one exciton and one magnon) on different sublattices having equal and opposite wave vectors $\boldsymbol{k}$ and $-\boldsymbol{k}$ and yields knowledge on dispersions of excitons and magnons. This is an electric dipole mechanism and its origin can be understood most easily by considering a pair of ions j and l in their ground state exchange coupled in the external electric field. In this circumstance, the exchange integral will be a function of the electric field vector $\boldsymbol{E}$. Expanding $J(jl)$ to the first order of $\boldsymbol{E}$, we have the following interaction energy between the electric field and the spin system,

$$-\boldsymbol{P}(\mathrm{j}l) \cdot \boldsymbol{E} = -2\left(\frac{\partial J(\mathrm{j}l)}{\partial \boldsymbol{E}} \boldsymbol{E}\right)(\boldsymbol{S}_\mathrm{j} \cdot \boldsymbol{S}_l), \tag{49}$$

so that we obtain an expression for the spin dependent electric dipole moment $\boldsymbol{P}(jl)$ which is associated with the ions j and l:

$$\boldsymbol{P}(\mathrm{j}l) = 2\boldsymbol{\pi}(\mathrm{j}l)\boldsymbol{S}_\mathrm{j} \cdot \boldsymbol{S}_l, \tag{50}$$

with $\boldsymbol{\pi}(jl) = \partial J(jl)/\partial \boldsymbol{E}$. Note that $\boldsymbol{\pi}(jl)$ will vanish identically when there is centre of symmetry between j and l.

Similarly, we expect the following spin dependent transition moment $\boldsymbol{P}$ for such spin forbidden excitations as considered in sect. 3:

$$\boldsymbol{P}(j^*l) = 2\boldsymbol{\pi}(j^*pl)\boldsymbol{\sigma}_{jp} \cdot \boldsymbol{S}_l, \tag{51}$$

in terms of the operator $\boldsymbol{\sigma}$ defined in eq. (7), where * has been put on the shoulder of the excited ion. The coefficient $\boldsymbol{\pi}(j^*pl)$, for example, depends of course upon the orbital state p of the excited multiplet of the ion j.

It is then natural to expect a similar transition moment operator for a double excitation:

$$\boldsymbol{P}(j^*l^*) = 2\boldsymbol{\pi}(j^*pl^*\bar{q})\boldsymbol{\sigma}_{jp} \cdot \boldsymbol{\sigma}_{l\bar{q}}, \tag{52}$$

which is also of exchange type, where p and $\bar{q}$ indicate the orbital components of the excited state. In the same way as in $\boldsymbol{\pi}(jl)$, $\boldsymbol{\pi}(j^*l^*)$ will vanish identically in case there is a centre of symmetry in the ground as well as in the doubly excited state of the pair j and l.

Let us suppose that the suffixes j and l refer to the up spin and down spin ion of an antiferromagnet, respectively, and for both j and l the same axis of spin quantization is assumed. We will keep this convention throughout this section.

A glance at fig. 12 will convince us that the dipole moment operators (50), (51) and (52) will give rise to the two magnon, exciton–magnon and two exciton absorption, respectively. (Clearly, they will not be able to create excitons when spins of the ion pair j and l are parallel.) If we introduce creation operators of spin deviations and excitons defined by

$$S_{j-} = \sqrt{2S}a_j^\dagger, \qquad S_{l+} = \sqrt{2S}b_l^\dagger, \tag{53}$$

and

$$\sigma_{jp-} = A_{jp}^\dagger, \qquad \sigma_{lq+} = -B_{lq}^\dagger, \tag{54}$$

these operators take the following forms

$$\boldsymbol{P}(jl) = 2S\boldsymbol{\pi}(jl)(a_j^\dagger b_l^\dagger + a_j b_l), \tag{55}$$

$$\boldsymbol{P}(j^*l) = \sqrt{2S}\boldsymbol{\pi}(j^*pl)A_{jp}^\dagger b_l^\dagger, \tag{56}$$

$$\boldsymbol{P}(j^*l^*) = -\boldsymbol{\pi}(j^*pl^*\bar{q})A_{jp}^\dagger B_{l\bar{q}}^\dagger, \tag{57}$$

respectively.

The transition mechanism of exchange type considered above involves mixing of odd-parity electronic state by a two ion off-diagonal exchange interaction. Several other mechanisms (Halley 1966, 1967) leading to two centre excitation have been suggested. However, they all lead to the above expression (55)–(57) for the double excitation in the final results. The treatment from now on proceeds regarding $\boldsymbol{\pi}$ coefficients as empirical parameters irrespective of the mechanism and only symmetry properties of them will be made use of in deriving the results.

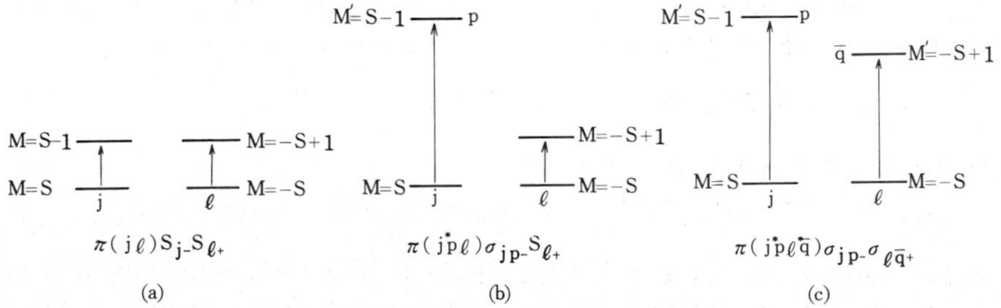

Fig. 12. Creation of (a) two magnons, (b) exciton and magnon and (c) two excitons by spin-dependent dipole moment.

4.2. Double excitations in antiferromagnets

For a typical two sublattice antiferromagnet as considered in eq. (21), the total electric dipole moment operator for the system can be put in the following form in terms of the exciton creation operators,

$$A^{\dagger}_{kp} = \frac{1}{\sqrt{N}} \sum_j \exp(i\boldsymbol{k} \cdot \boldsymbol{R}_j) A^{\dagger}_{jp}, \qquad B^{\dagger}_{kq} = \frac{1}{\sqrt{N}} \sum_l \exp(i\boldsymbol{k} \cdot \boldsymbol{R}_l) B^{\dagger}_{lq}, \tag{58}$$

for the up and down spin sublattices A and B, and magnon operators of eq. (29):

$$\boldsymbol{P}_{\rm mm} = \sum_{\boldsymbol{k}} (2S\boldsymbol{\pi}(\boldsymbol{k}) u_k^2 + 2S\boldsymbol{\pi}(\boldsymbol{k})^* v_k^{*2}) \alpha^{\dagger}_k \beta^{\dagger}_{-k}, \tag{59}$$

with

$$\boldsymbol{\pi}(\boldsymbol{k}) = \sum_l \boldsymbol{\pi}(jl) \exp[i\boldsymbol{k} \cdot (\boldsymbol{R}_l - \boldsymbol{R}_j)], \tag{60}$$

and

$$\boldsymbol{P}_{\rm em} = \sum_{kp} \{\sqrt{2S}\, \boldsymbol{\pi}_p(\boldsymbol{k}) u_k A^{\dagger}_{kp} \beta^{\dagger}_{-k} - \sqrt{2S}\, \boldsymbol{\pi}_{\bar{p}}(\boldsymbol{k}) u_k B^{\dagger}_{-k\bar{p}} \alpha^{\dagger}_k\}, \tag{61}$$

where we assume $\bar{\rm p}$ is the orbital state of the ion l corresponding to p at ion j, leading to an exciton on the B sublattice with the same energy $E_{\rm e}(\boldsymbol{k})$ as that on the A sublattice. The coefficients $\boldsymbol{\pi}_p(\boldsymbol{k})$ are defined by

$$\boldsymbol{\pi}_p(\boldsymbol{k}) = \sum_l \boldsymbol{\pi}(j^*pl) \exp[i\boldsymbol{k} \cdot (\boldsymbol{R}_l - \boldsymbol{R}_j)],$$

$$\boldsymbol{\pi}_{\bar{p}}(\boldsymbol{k}) = \sum_j \boldsymbol{\pi}(l^*\bar{p}j) \exp[i\boldsymbol{k} \cdot (\boldsymbol{R}_j - \boldsymbol{R}_l)]. \tag{62}$$

For the two exciton transition, we find

$$\boldsymbol{P}_{\rm ee} = -\sum_{kpq} \boldsymbol{\pi}_{p\bar{q}}(\boldsymbol{k}) A^{\dagger}_{kp} B^{\dagger}_{-k\bar{q}}, \tag{63}$$

with

$$\boldsymbol{\pi}_{p\bar{q}}(\boldsymbol{k}) = \sum_l \boldsymbol{\pi}(j^*pl^*\bar{q}) \exp[i\boldsymbol{k} \cdot (\boldsymbol{R}_l - \boldsymbol{R}_j)]. \tag{64}$$

The corresponding absorption coefficient for the light polarized in the direction $\boldsymbol{n}$ ($\boldsymbol{n} \parallel \boldsymbol{E}$) can be calculated in the standard way, leading to

$$\alpha^n_{\rm mm}(\bar{\nu}) = C \sum_{\boldsymbol{k}} |2S\pi^n(\boldsymbol{k}) u_k^2 + 2S\pi^n(\boldsymbol{k})^* v_k^{*2}|^2 \delta(E - 2E_{\rm m}(\boldsymbol{k})), \tag{65}$$

for two magnon absorption, and

$$\alpha^n_{\rm em}(\bar{\nu}) = C \sum_{\boldsymbol{k}} \{2S|\pi^n_p(\boldsymbol{k})|^2 u_k^2 + 2S|\pi^n_{\bar{p}}(\boldsymbol{k})|^2 u_k^2\} \delta(E - E_{\rm e}(\boldsymbol{k}) - E_{\rm m}(\boldsymbol{k})), \tag{66}$$

for the creation of p exciton and one magnon. For creating the exciton pair p and $\bar{\rm q}$ (q $\neq$ p), we obtain

$$\alpha^n_{\rm ee}(\bar{\nu}) = C \sum_k [|\pi^n_{p\bar{q}}(\boldsymbol{k})|^2 + |\pi^n_{q\bar{p}}(\boldsymbol{k})|^2] \delta(E - E_{{\rm e}p}(\boldsymbol{k}) - E_{{\rm e}\bar{q}}(\boldsymbol{k})), \tag{67}$$

but,

$$\alpha^{n}_{ee}(\bar{\nu}) = C \sum_{k} |\pi^{n}_{p\bar{p}}(\boldsymbol{k})|^2 \delta(E - 2E_e(\boldsymbol{k})), \tag{68}$$

for the pair p and $\bar{p}$.

The factor C is given by

$$C = (2\pi)^3 \bar{\nu}/hc\eta V. \tag{69}$$

The absorption coefficients (65) through (68) are for the light with wave number $\bar{\nu}(=E/hc)$ at temperature absolute zero, η is the refractive index and V is the crystal volume. In eq. (65), $E_m(\boldsymbol{k})$ is the magnon energy as given in eq. (26), $E_e(\boldsymbol{k})$ is given by eq. (22) with $V_{AB}(\boldsymbol{k})=0$, and $E_{ep}(\boldsymbol{k})$ and $E_{e\bar{q}}(\boldsymbol{k})$ in eq. (67) are also energies of excitons created on the A and B sublattice, respectively.

We now confine ourselves with the case of MnF_2 where all kinds of double excitations have been observed. The case of two magnon absorption is the simplest. If we assume $\boldsymbol{\pi}(jl)$ to be non-negligible only for the second nearest neighbour pairs of MnF_2 which is coupled antiferromagnetically, we can predict the line shape of two magnon absorption in two polarizations from symmetry considerations.

There are slightly different versions of determining the form of $\boldsymbol{\pi}(\boldsymbol{k})$ from symmetry (Loudon 1968, Sell et al. 1967, Tanabe and Gondaira 1967). We give here a method by inspection to find out how $\pi^n(jl)$ behave when we let l run over the sites of nearest neighbours of j. We first note that $\pi^n(jl)$ will change its sign like p_n, the n component of the dipole moment vector, under symmetry operations of the site group of j that will carry l into another nearest neighbour site l'. We can find in this way the relation between $\pi^n(jl)$ and $\pi^n(jl')$. Note that the site group relevant here is D_{2h}, i.e., the one in the paramagnetic phase. It is convenient to use the coordinate system (x, y, z) of fig. 10 and distinguish the two types of nearest neighbours lI and lII interacting through type I and II fluorine ions, respectively (fig. 13). Table 2 follows from this consideration together with the following relations between nonvanishing $\pi^n(jl)$ coefficients

$$\pi^z(jl\mathrm{I}) = -\pi^z(jl\mathrm{II}), \qquad \pi^y(jl\mathrm{I}) = -\pi^x(jl\mathrm{II}), \tag{70}$$

the implication of which is not difficult to see. In table 2 we have used, for example, the notation

$$\sigma(\mathrm{I})_\alpha = \mathrm{sign}(R_{l\mathrm{I}\alpha} - R_{j\alpha}), \quad \alpha = x, y, z, \tag{71}$$

to make clear how the signs of $\pi^n(jl)$ change when we let l run over the type lI neighbours of j.

The absorption coefficient for π spectrum $(n \parallel z)$ now takes the form

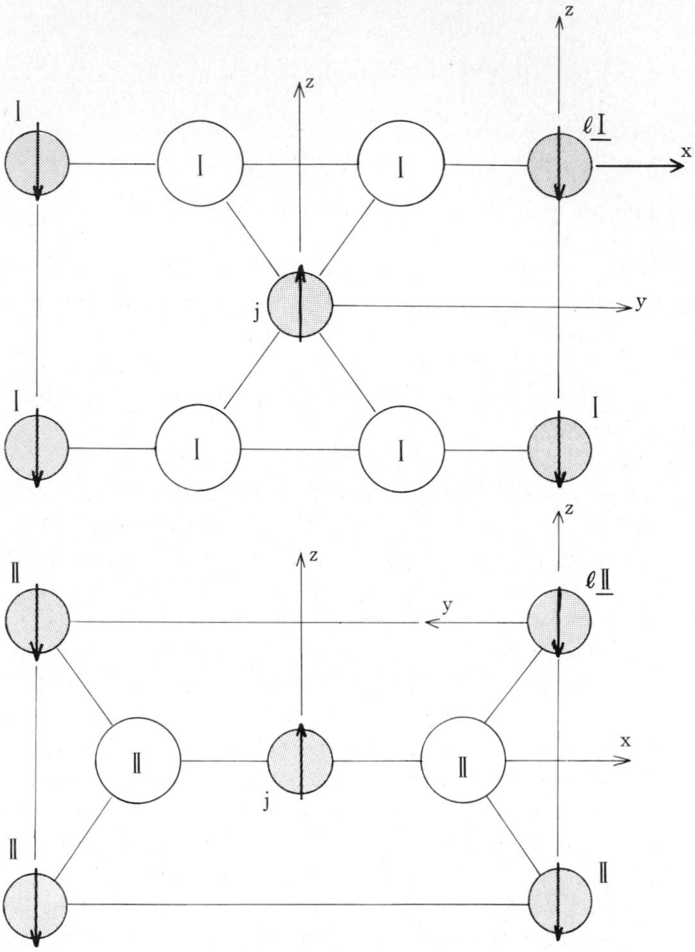

Fig. 13. Type I and Type II neighbours of the body centre ion in MnF_2.

Table 2
Symmetry of π coefficients
(two-magnon)

$\pi^n(jl)$	$l = l\text{I}$	$l = l\text{II}$
$n = z$	$\pi^z\sigma(\text{I})_z$	$-\pi^z\sigma(\text{II})_z$
$= x$	0	$\pi^x\sigma(\text{II})_x$
$= y$	$-\pi^x\sigma(\text{I})_y$	0

(Moriya 1966, Allen et al. 1966):

$$\alpha^{\pi}_{\rm mm}(\bar{\nu}) = 64(2S)^2C|\pi^z|^2 \sum_k \sin^2(\tfrac{1}{2}k_Xa)\sin^2(\tfrac{1}{2}k_Ya)\sin^2(\tfrac{1}{2}k_Zc)\delta(E-2E_{\rm m}(\boldsymbol{k})), \tag{72}$$

and that for σ spectrum ($n \parallel X, Y$) is given by

$$\alpha^{\sigma}_{\rm mm}(\bar{\nu}) = 64(2S)^2C|\pi^x|^2 \sum_k \cos^2(\tfrac{1}{2}k_Xa)\sin^2(\tfrac{1}{2}k_Ya)\cos^2(\tfrac{1}{2}k_Zc)\delta(E-2E_{\rm m}(\boldsymbol{k})). \tag{73}$$

Two magnon absorption spectra observed (Allen, Jr. et al. 1966) in MnF_2 are shown in fig. 14 (solid line) together with the theoretical curve calculated by eqs. (72) and (73) (dotted line). The agreement is not particularly good. The broken curve which shows better agreement is the result obtained by assuming for $\pi(jl)$ exponential dependence on separation between j and l and including in the summation of, eq. (60), neighbours further than the nearest. Another possibility to improve the calculated result is to take into account the effect of magnon–magnon interaction which will be discussed later.

In a similar way we can discuss the line shapes of exciton magnon absorption or the spin wave sideband. We note that $\pi^n(j^*pl)$ now changes its sign like the product $p_n\varphi'_j(p)\varphi^*_{j0}$ under the symmetry operations of D_{2h}, where $\varphi'_j(p)$ and φ_{j0} here are *orbital* functions of the excited and ground state (with symmetries p and A_{1g}, respectively) of the ion j in D_{2h}. Table 3 summarizes the result for the excited orbital states $p = B_{3g}$, B_{2g} and B_{1g}, A_g of D_{2h}. As we have seen in sect. 3, B_{3g} and B_{2g} give rise to an A_g exciton while A_g and B_{1g} lead to a B_g exciton.

The absorption coefficient for the excitation of an A_g exciton with one magnon in π polarization is given as (Meltzer et al. 1969, Loudon 1968, Tanabe and Gondaira 1967)

$$\alpha^{\pi(\sigma)}_{\rm em}(\bar{\nu}) = 128(2S)C \sum_k |\pi({\rm X})|^2 u_k^2 \cos^2(\tfrac{1}{2}k_Xa)\sin^2(\tfrac{1}{2}k_Ya)\cos^2(\tfrac{1}{2}k_Zc) \times \delta(E-E_{\rm e}(\boldsymbol{k})-E_{\rm m}(\boldsymbol{k})). \tag{74}$$

The corresponding coefficient for σ polarization is slightly more complicated:

$$\alpha^{\sigma(\pi)}_{\rm em}(\bar{\nu}) = 128(2S)C \sum_k \{|\pi({\rm Z})|^2 u_k^2 \cos^2(\tfrac{1}{2}k_Xa)\cos^2(\tfrac{1}{2}k_Ya)\sin^2(\tfrac{1}{2}k_Zc) + |\pi({\rm A})|^2 u_k^2 \sin^2(\tfrac{1}{2}k_Xa)\sin^2(\tfrac{1}{2}k_Ya)\sin^2(\tfrac{1}{2}k_Zc)\} \times \delta(E-E_{\rm e}(\boldsymbol{k})-E_{\rm m}(\boldsymbol{k})). \tag{75}$$

The expressions of $|\pi(X)|^2$, etc. are given by

$$|\pi({\rm X})|^2 = \tfrac{1}{2}(|\pi^z({\rm I})|^2 + |\pi^z({\rm II})|^2), \tag{76}$$

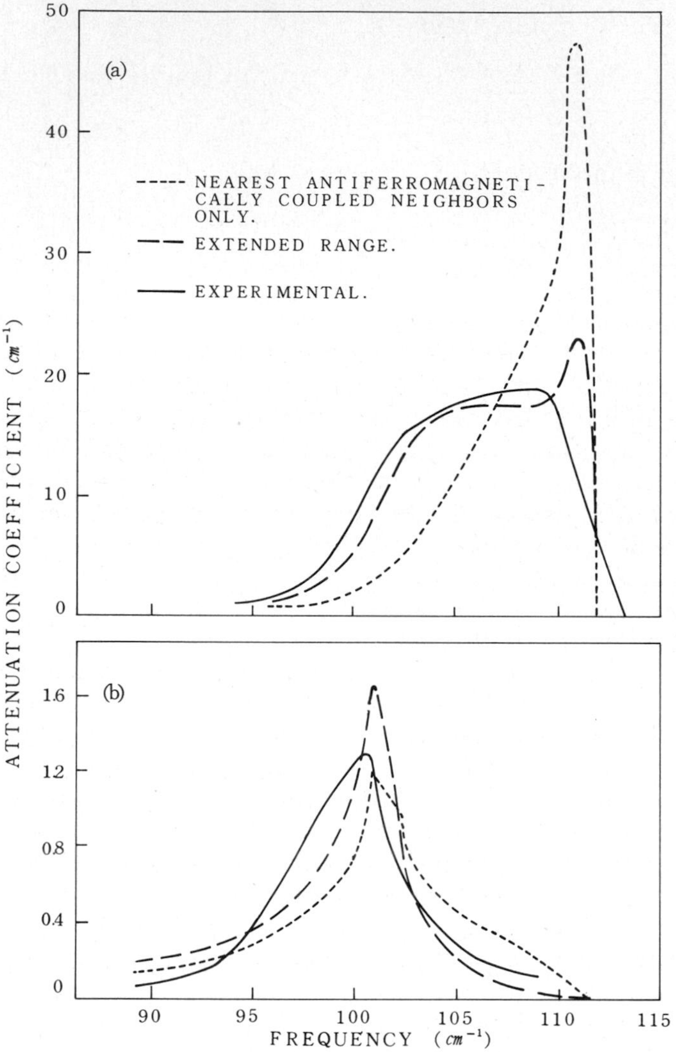

Fig. 14. Theoretical and experimental absorption coefficients for the two magnon absorption in MnF_2 at 4.2 K. (a) $\boldsymbol{E}\parallel c$, π spectrum (b) $\boldsymbol{E}\perp c$, σ spectrum (Allen et al. 1966).

$$|\pi(Z)|^2 = \tfrac{1}{8}|\pi^x(I) + \pi^x(II)|^2 + \tfrac{1}{8}|\pi^y(I) + \pi^y(II)|^2, \tag{77}$$

$$|\pi(A)|^2 = \tfrac{1}{8}|\pi^x(I) - \pi^x(II)|^2 + \tfrac{1}{8}|\pi^y(I) - \pi^y(II)|^2, \tag{78}$$

for A_g excitons. The letters X, Z, A in $\pi(X)$, etc. have been introduced to indicate that the corresponding points in BZ make significant contributions to the line shape.

Table 3
Symmetry of π coefficients (exciton–magnon)

$\pi^n(j^*pl)$	A_g exciton (B_{3g}, B_{2g}) $l = l\mathrm{I}$	$l = l\mathrm{II}$	B_g exciton (B_{1g}, A_g) $l = l\mathrm{I}$	$l = l\mathrm{II}$
$n = z$	$\pi_3^z(\mathrm{I})\sigma(\mathrm{I})_y$ [a]	$\pi_2^z(\mathrm{II})\sigma(\mathrm{II})_x$	$\pi_0^z(\mathrm{I})\sigma(\mathrm{I})_z$	$\pi_0^z(\mathrm{II})\sigma(\mathrm{II})_z$
$= x$	$\pi_2^x(\mathrm{I})\sigma(\mathrm{I})_z$	$\pi_2^x(\mathrm{II})\sigma(\mathrm{II})_z$	$\pi_1^x(\mathrm{I})\sigma(\mathrm{I})_y$	$\pi_0^x(\mathrm{II})\sigma(\mathrm{II})_x$
$= y$	$\pi_3^y(\mathrm{I})\sigma(\mathrm{I})_z$	$\pi_3^y(\mathrm{II})\sigma(\mathrm{II})_z$	$\pi_0^y(\mathrm{I})\sigma(\mathrm{I})_y$	$\pi_1^y(\mathrm{II})\sigma(\mathrm{II})_x$

[a] $\pi_p^n(\mathrm{I}) = \pi^n(j^*B_{pg}l\mathrm{I})$, $(p = 1, 2, 3)$.
For A_g, put $p = 0$. Site I and II relative to j is defined in fig. 13. When the orbital state is a mixture, for example, of B_{3g} and B_{2g}, $\phi' = a\phi'(B_{3g}) + b\phi'(B_{2g})$, $\pi^n(\mathrm{I})$ is to be obtained by calculating $a\pi_3^n(\mathrm{I}) + b\pi_2^n(\mathrm{I})$.

For B_g excitons, the coefficient (74) is to be used for σ polarization and (75) for π polarization with

$$|\pi(\mathrm{X})|^2 = \tfrac{1}{4}(|\pi^x(\mathrm{I})|^2 + |\pi^x(\mathrm{II})|^2 + |\pi^y(\mathrm{I})|^2 + |\pi^y(\mathrm{II})|^2), \tag{79}$$

$$|\pi(\mathrm{Z})|^2 = \tfrac{1}{4}|\pi^z(\mathrm{I}) + \pi^z(\mathrm{II})|^2, \tag{80}$$

$$|\pi(\mathrm{A})|^2 = \tfrac{1}{4}|\pi^z(\mathrm{I}) - \pi^z(\mathrm{II})|^2. \tag{81}$$

In fig. 15(a) we give the absorption spectrum near the origin of the transition ${}^6A_{1g} \to {}^4T_{1g}$ ($\sim$18 420 cm^{-1} at 4.2 K) of MnF_2 (Greene et al. 1965, Meltzer et al. 1969, see also Sell (1968), Eremenko and Belyaeva (1969) for early reviews). The sharp magnetic dipole lines ($f \sim 5 \times 10^{-11}$) M1σ (18 418 cm^{-1}) and M2σ (18 435 cm^{-1}) are A_g exciton lines with the splitting factor $\Delta_e/2\mu_B H = 1.79$ and 1.90, respectively. The cubic state ${}^4T_{1g}$ splits into ${}^4B_{3g}$, ${}^4B_{2g}$ and ${}^4B_{1g}$ of D_{2h}. The former two, $\phi'(M' = \tfrac{3}{2}b_{3g})$ and $\phi'(M' = \tfrac{3}{2}b_{2g})$ give rise to the two A_g excitons, because B_{3g} and B_{2g} both become B_g of C_{2h}. The sidebands E1σ, π and E2σ ($f \sim 10^{-9}$) observed as electric dipole transitions correspond to A_g exciton–magnon excitation. This was confirmed by observing the Zeeman and stress effect (Dietz et al. 1966). The sidebands do not split under magnetic field applied parallel to the c-axis. This is because the loss of energy for an exciton, say, on the up spin sublattice is balanced by the gain by a magnon on the down sublattice when $g' = g = 2$, i.e., the splitting factor Δ_{em} for magnon sideband is given by $\Delta_e - 2g\mu_B H$. Stress applied along [110] and [001] direction confirmed that sidebands E1 and E2 were associated with the exciton states M1 and M2, respectively. Another evidence to suggest that E1 and E2 are sidebands is the temperature dependent shift of the peak energy $\omega_m(T)$ relative to the zero magnon lines M1 and M2. Experimentally, the observed dependence of $\omega_m(T)/\omega_m(0)$ fairly closely follows that of $M(T)/M(0)$ up to 30 K, where

$M(T)$ is the sublattice magnetization. However, the magnitude of $\omega_m(0)$ itself was the simplest and most convincing evidence for the assignment. Observed value $\omega_m(0)$ for E1 was 57.5 cm^{-1}, whereas neutron scattering data yield 55 cm^{-1} as the maximum magnon energy at the BZ boundary. This means at the same time that the dispersion is very small for the exciton associated with the M1σ line. Indeed, calculation of the shape function (75) with appropriately chosen parameters (Sell et al. 1967) (small exciton dispersion given by eq. (46) and magnon dispersion (26) with parameters in fig. 16 (Allen, Jr. et al. 1966)) leads to the profile E1σ given in fig. 15(b) which agrees quite well with the observed. The agreement is not so good for E1π which was calculated by eq. (74). The observed peak of E1π has no sharp feature and is located much to the lower energy side, which suggests the importance of the exciton–magnon interaction.

Let us now turn to the sidebands of $^6A_{1g} \rightarrow {}^4E_g$ transition referred to at

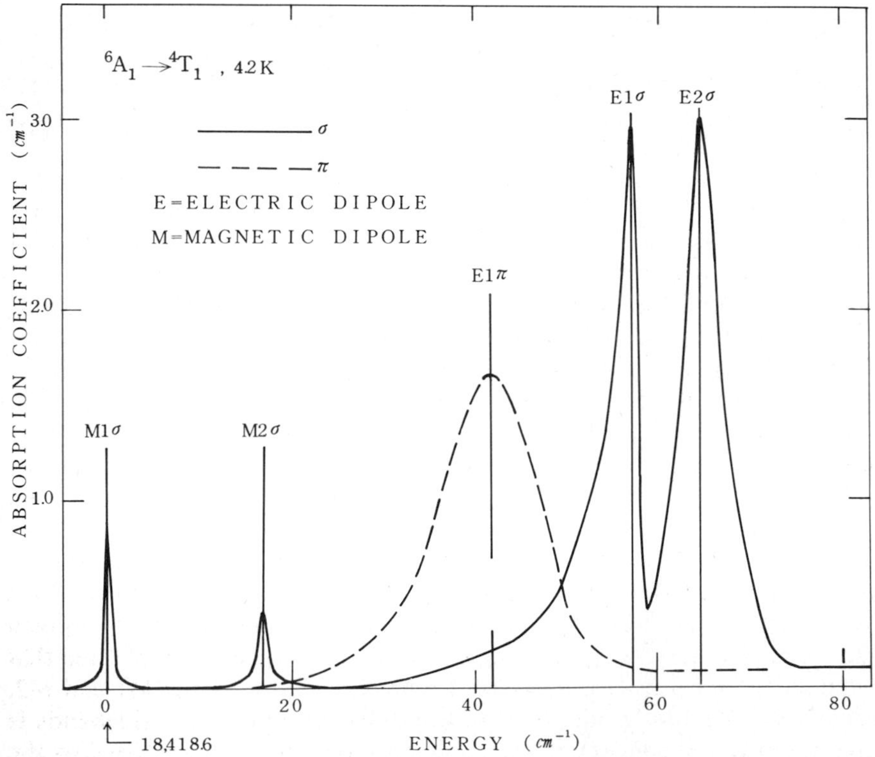

Fig. 15a. Absorption near the origin of the transition $^6A_{1g} \rightarrow {}^4T_{1g}$ (~18 420 cm^{-1}) at 4.2 K (Meltzer et al. 1969).

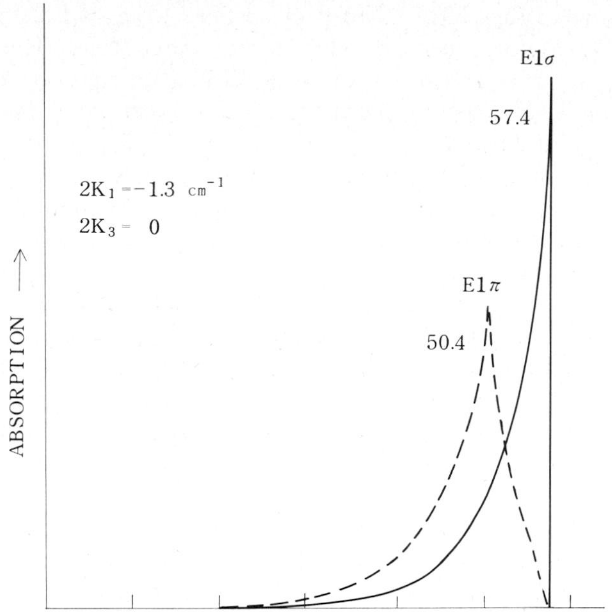

Fig. 15b. Line shape of the sidebands according to eq. (74) for π and eq. (78) for σ (Sell et al. 1967).

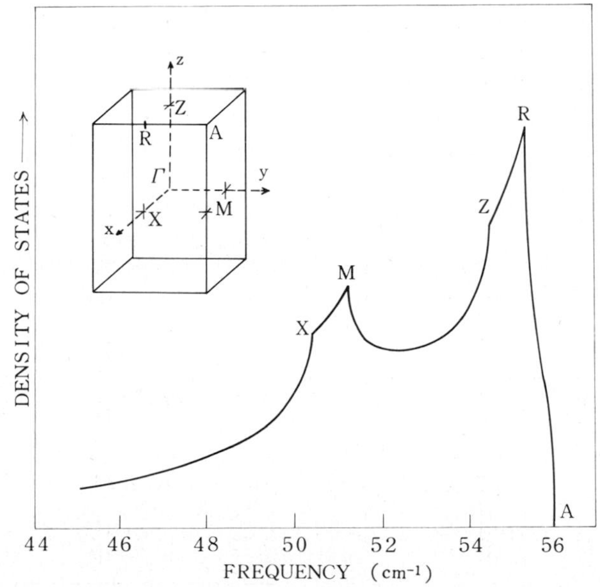

Fig. 16. BZ and relevant portion of magnon density of states in MnF_2 using exchange and anisotropy constants $J_1 = 0.22\ cm^{-1}$, $|J_2| = 1.22\ cm^{-1}$, $J_3 = 0.035\ cm^{-1}$, $2\mu_B H_A = 0.737\ cm^{-1}$ (Allen et al. 1966).

the end of sect. 3 (fig. 11(a)). The exciton lines M1σ and M2π correspond to A_g and B_g excitons, respectively, as mentioned there. The sideband of the former, i.e., E1π and σ are to be described by eqs. (74) and (75). The calculated line shape (Meltzer et al. 1969) with the parameters (48) is shown in fig. 11(b) (dot-dashed curve $\pi(Z) \neq 0$, $\pi(A) = 0$, long-dash-short-dash curve $\pi(Z) = 0$, $\pi(A) \neq 0$). The exciton dispersion is also small here as in the preceding paragraph and there is not much to comment. The shapes of E2π and σ which is the sidebands of a B_g exciton are to be given by eqs. (75) and (74), respectively. We note at once that unless we assume a very large negative exciton dispersion we shall be unable to explain the position of the zero line M2π which is dipped in the band E1π. The calculated line shape of E2π and σ were obtained with the parameters (47) and assuming $\pi(Z) \neq 0$, $\pi(A) = 0$. We find that the agreement is fairly good. The existence of the sharp line E2πa, however, cannot be understood within this theory, but can only be explained by introducing the exciton–magnon interaction.

It is now almost obvious how to work out the line shape of the two exciton absorption. The coefficient $\pi^n(j^*pl^*\bar{q})$ changes its sign like $p_n\varphi'_j(p)\varphi^*_{j0}\varphi'_l(\bar{q})\varphi^*_{l0}$ under symmetry operations of D_{2h}. The behaviours of $\pi^n(j^*pl^*\bar{q})$ upon changing site l are given in table 4. Note that table 4(b) is essentially the same as table 3. This is to be expected, because magnon is indeed one of the B_g excitons as pointed out before. According to table 4, the line shape of A_gA_g and B_gB_g exciton absorption is given by eq. (75) for π polarization, while that in σ polarization by eq. (74), if we replace the δ functions in eqs. (74) and (75) by $\delta(E - 2E_e(\boldsymbol{k}))$. For A_gB_g exciton pair, expressions (74) and (75) with the δ function appearing in eq. (67) will describe the line shape in π and σ polarization, respectively, just as in A_g exciton–magnon pair.

Table 4(a)
Symmetry of π coefficients (two-exciton)

Exciton Pair	A_gA_g		A_gB_g	
$\pi^n(j^*pl^*\bar{p})$ $\pi^n(j^*pl^*\bar{q})$	$p\bar{p} = B_{3g}B_{3g}$[a] $l = l\mathrm{I}$	$p\bar{q} = B_{3g}B_{2g}$[b] $l = l\mathrm{II}$	$p\bar{q} = B_{3g}B_{1g}$[b] $l = l\mathrm{I}$	$p\bar{q} = B_{2g}B_{1g}$[b] $l = l\mathrm{II}$
$n = z$	$\pi^z_{32}(\mathrm{I})\sigma(\mathrm{I})_z$[c]	$\pi^z_{32}(\mathrm{II})\sigma(\mathrm{II})_z$	$\pi^z_{21}(\mathrm{I})\sigma(\mathrm{I})_y$	$\pi^z_{31}(\mathrm{II})\sigma(\mathrm{II})_x$
$= x$	$\pi^x_{33}(\mathrm{I})\sigma(\mathrm{I})_y$	$\pi^x_{32}(\mathrm{II})\sigma(\mathrm{II})_x$	$\pi^x_{31}(\mathrm{I})\sigma(\mathrm{I})_z$	$\pi^x_{31}(\mathrm{II})\sigma(\mathrm{II})_z$
$= y$	$\pi^y_{32}(\mathrm{I})\sigma(\mathrm{I})_y$	$\pi^x_{33}(\mathrm{I})\sigma(\mathrm{II})_x$	$\pi^y_{21}(\mathrm{I})\sigma(\mathrm{I})_z$	$\pi^y_{21}(\mathrm{II})\sigma(\mathrm{II})_z$

[a] For $p\bar{p} = B_{2g}B_{2g}$ raplace π_{33} by π_{22}.
[b] Similarly, for $p\bar{q} = B_{2g}B_{3g}$, write π_{23} for π_{32}, etc.
[c] See footnote (c) to table 4(b).

Table 4(b)
Symmetry of π coefficients (two-exciton)

Exciton pair	A_gB_g		B_gB_g	
$\pi^n(j^*pl^*\bar{p})$ $\pi^n(j^*pl^*\bar{q})$	$p\bar{q} = B_{3g}A_g$[b] $l = lI$	$p\bar{q} = B_{2g}A_g$[b] $l = lII$	$p\bar{p} = B_{1g}B_{1g}$[a] $l = lI$	$p\bar{q} = B_{1g}A_g$[b] $l = lII$
$n = z$	$\pi^z_{30}(I)\sigma(I)_y$[c]	$\pi^z_{20}(II)\sigma(II)_x$	$\pi^z_{11}(I)\sigma(I)_z$	$-\pi^z_{11}(I)\sigma(II)_z$
$= x$	$\pi^x_{20}(I)\sigma(I)_z$	$\pi^x_{20}(II)\sigma(II)_z$	$\pi^x_{10}(I)\sigma(I)_y$	$\pi^x_{11}(II)\sigma(II)_x$
$= y$	$\pi^y_{30}(I)\sigma(I)_z$	$\pi^y_{30}(II)\sigma(II)_z$	$-\pi^x_{11}(II)\sigma(I)_y$	$\pi^y_{10}(II)\sigma(II)_x$

[a] For $p\bar{p} = A_gA_g$ replace π_{11} by π_{00}.
[b] For $p\bar{q} = A_gB_{3g}$, write π_{03} for π_{30}, etc.
[c] $\pi^n_{pq}(I) = \pi^n(j^*B_{pg}lI^*B_{qg})$ $(p, q = 1, 2, 3)$.
Put p or $q = 0$ for A_g.

Observed two exciton absorption spectra (Stokowski and Sell 1971) of MnF_2 are given in fig. 17. Three electric dipole lines are seen around 36 800 cm^{-1} which is nearly twice the energy of $^6A_{1g} \to {}^4T_{1g}$ exciton. They are E_{22} (36 879 cm^{-1}), E_{12}, E_{21} (36 917 cm^{-1}) and E'_{22} (37 012 cm^{-1}). The double exciton state E_{22} corresponds to E2(A) + E2(B), where E2(A) is the state with exciton E2 (that is M2σ of fig. 15(a)) on A sublattice. Similarly, E_{12} is E1(A) + E2(B) where E1(A) is the state with exciton E1(M1σ) on A sublattice. These assignments were made by observing their response to the uniaxial stress and comparing it with that of exciton lines M1σ and M2σ. It was found that E_{12}, E_{21} split under the stress along [110] while E_{22} and E'_{22} did not and that the shift of double exciton transition due to stress was a sum of the shifts of the constituent single excitons. (The stress behaviour of E'_{22} was the same as E_{22}, which suggests that E'_{22} probably corresponds to E_{22} plus one or more phonon or magnon excitations.) The polarization behaviour expected from the general theory is that points Z and A of BZ are accentuated in π and point X in σ, because this is the case of A_gA_g exciton.

Figure 17, however, does not tell us much about line shape. The energy of E_{22} lies 84 cm^{-1} below that of twice the energy of E2 and E_{12}, E_{21} lie 61 cm^{-1} above the sum of energies of E1 and E2. This energy shift indicates an appreciable exciton–exciton interaction compared to the exciton–magnon interaction. This seems reasonable since the magnon affects the exciton through exchange field, whereas an exciton affects its partner through ligand field or more generally vibronic effect (Chen et al. 1972, Solomon and McClure 1972, Fujiwara 1973). The large exciton–exciton interaction suggests that the observed lines are more likely to correspond

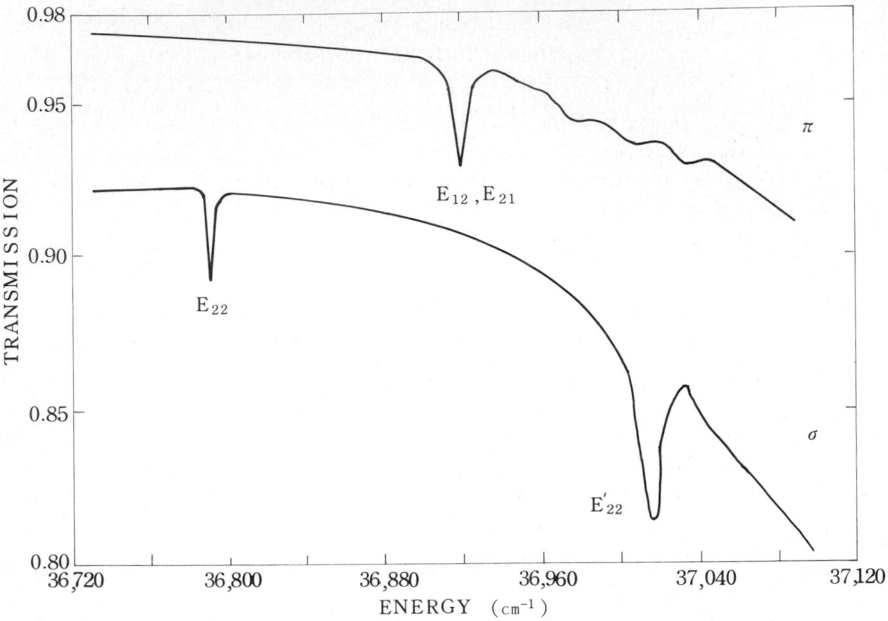

Fig. 17. Polarization properties of the two exciton transition in MnF_2 (Stokowski and Sell 1971).

to the bound states of two excitons to be discussed below than to the continuum states which we have been interested in as important in eqs. (67) and (68). It is more interesting here to try to see what the observed strong polarization of E_{22} implies. Table 4(a) then suggests that the orbital state of the constituent exciton E2(A) of E_{22} consists almost purely of either B_{2g} or B_{3g}. There are evidences supporting the former case that the orbital state of E1(A) is B_{3g} so that for this exciton the excited state wavefunction is given dominantly by $\phi_j'(M' = \frac{3}{2} b_{3g})$ while that of E2(A) is B_{2g}. This does not, however, explain why E_{11} is not observed or why π_{33} should be small. (See, however, Stokowski and Sell (1971) for a persuasive interpretation.)

4.3. Importance of the interaction between excitons

We have seen in the above that interactions between excitons (magnon–magnon, exciton–magnon, exciton–exciton) should be taken into account to attain satisfactory agreement with the experiment. For example, a very simple consideration suggests how magnon–magnon interaction will make the peak position shift in the two magnon absorption spectra.

When two spin deviations are created outside the range of J (the nearest neighbour antiferromagnetic exchange integral), their energy will be given by $2E_\alpha$ with E_α of eq. (25), i.e., the energy of a magnon at the zone boundary. But two spin deviations at the nearest neighbour site on opposite sublattices will have energy $2E_\alpha - 2|J|$ which is less than $2E_\alpha$, $2J$ being the magnon–magnon interaction energy. As seen in eqs. (72) and (73), the sharp features in the two magnon spectra correspond to the zone boundary magnons. We, therefore, expect that the peaks will shift towards lower energy by an amount $2|J|$ because of the magnon–magnon interaction.

In the case of exciton–magnon absorption, we can argue in a similar way. If we create one exciton on A sublattice and one magnon (spin deviation) on B sublattice outside the range of J, their energy will be $E_j + E_\alpha$ with E_j given by eq. (12). To excite an exciton and a magnon on nearest neighbour site will cost an amount of energy $E_j + E_\alpha + 2|J'|S' - 2|J|S = E_j + E_\alpha - 2JS\rho$, where

$$\rho = (J'S'/JS) - 1, \tag{82}$$

and J' is the exchange integral in the excited state ($S' = S - 1$). Unless $|J'|S'$ is larger than $|J|S$, the peak of exciton–magnon absorption will shift towards lower energy side. Actually, this case is a little more complicated, because the excitation transfer interaction (6) brings in a resonance term (fig. 18) of exciton–magnon pair between the two sublattices of the form $L(A_j^\dagger B_l a_j b_l^\dagger + A_j B_l^\dagger a_j^\dagger b_l)$, where L is defined by

$$L = 2K(jl)/(2S - 1). \tag{83}$$

The effect of this resonance in the case of a simple two sublattice antiferromagnet is to add or subtract L further so that the total amount of shift will be $-2JS\rho \pm L$ as the case may be. In the case of $RbMnF_3$, the

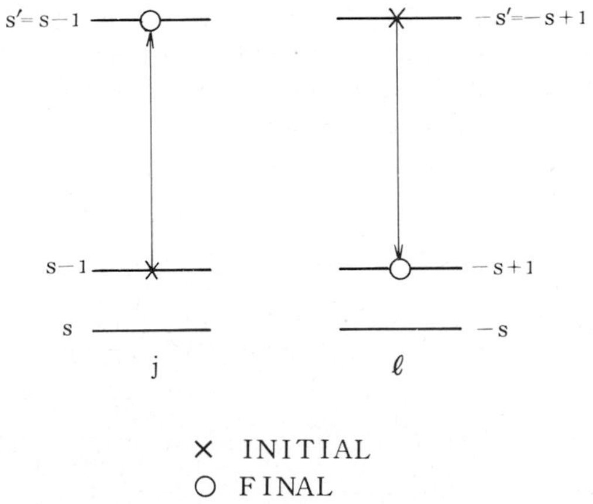

Fig. 18. Effect of resonance transfer term $LA_j^\dagger B_l a_j b_l^\dagger$ on an exciton–magnon pair.

amount of shift is $-2JS\rho + L$. In MnF_2, both signs will occur in the spectra as we shall see below. Depending upon the magnitude and sign of this quantity, bound states can appear below or above the exciton–magnon continuum.

The case of double exciton absorption may be treated with the same idea. When we create exciton p and $\bar{\mathrm{p}}$ on each sublattice, we expect that the peak will be shifted by the amount of exciton–exciton interaction as in the case of two magnon absorption. When we create different exciton p and $\bar{\mathrm{q}}$ on each sublattice, we have to consider effect of resonance with the state q and $\bar{\mathrm{p}}$ beside the interaction between the excitons p and $\bar{\mathrm{q}}$, as in the case of exciton–magnon absorption. Note that this energy transfer process need not be spin dependent.

4.4. Line shape including interaction effects

Let us step forward a little further and see how the theory including the interactions of excitons improve the line shapes of the simple theory (eqs. (72)–(75)) for MnF_2.

The absorption coefficient for the two magnon excitation is given by

$$\alpha^{n}_{\mathrm{mm}} = -NC(2S)^2/\pi \,\mathrm{Im} \sum_{\boldsymbol{\delta},\boldsymbol{\delta}'} \pi^{n}_{\delta}\pi^{n}_{\delta'} G(\boldsymbol{\delta}, \boldsymbol{\delta}')_{E+\mathrm{i}\epsilon}, \tag{84}$$

$$NG(\boldsymbol{\delta}, \boldsymbol{\delta}') = \sum_{jj'} G(a_j b_{j+\delta} + a^{\dagger}_j b^{\dagger}_{j+\delta}, a_{j'} b_{j'+\delta'} + a^{\dagger}_{j'} b^{\dagger}_{j'+\delta'}), \tag{85}$$

where we have set $\pi_\delta = \pi(j, j+\boldsymbol{\delta})$, $\boldsymbol{\delta}$ being the vector connecting the nearest neighbour on the opposite sublattice. For the definition of Green's function G on the right hand side of eq. (85), see, for example, Parkinson and Loudon (1968). Under decoupling approximation for Green's functions together with several simplifying assumptions, the equations for $G(\boldsymbol{\delta}, \boldsymbol{\delta}')$ can be solved (when the range of the exchange interaction is limited to the nearest neighbours) as the solution of an impurity problem. This is because the first spin deviation appears like an impurity to the second spin deviation. The result can be put in the following form (Elliott and Thorpe 1969, Thorpe 1970, Satoko 1970):

$$\sum_{\boldsymbol{\delta}\boldsymbol{\delta}'} \pi^{n}_{\delta}\pi^{n}_{\delta'} G(\boldsymbol{\delta}, \boldsymbol{\delta}') = 8|\pi^{z}|^2 G_{0\mathrm{A}}/(1 - 2J_2\kappa G_{0\mathrm{A}}) \quad (n = z), \tag{86}$$

$$= 8|\pi^{x}|^2 G_{0\mathrm{X}}/(1 - 2J_2\kappa G_{0\mathrm{X}}) \quad (n = x), \tag{87}$$

where

$$G_{0\mathrm{A}} = 2E/N \sum_{\boldsymbol{k}} 8 \sin^2(\tfrac{1}{2}k_{\mathrm{X}}a) \sin^2(\tfrac{1}{2}k_{\mathrm{Y}}a) \sin^2(\tfrac{1}{2}k_{\mathrm{Z}}c)/(E^2 - 4E_{\mathrm{m}}(\boldsymbol{k})^2), \tag{88}$$

$$G_{0\mathrm{X}} = 2E/N \sum_{\boldsymbol{k}} 8 \cos^2(\tfrac{1}{2}k_{\mathrm{X}}a) \sin^2(\tfrac{1}{2}k_{\mathrm{Y}}a) \cos^2(\tfrac{1}{2}k_{\mathrm{Z}}c)/(E^2 - 4E_{\mathrm{m}}(\boldsymbol{k})^2), \tag{89}$$

$$\kappa = (4Sz|J_2| - |J_2|)/E. \tag{90}$$

Equations (86) and (87) correspond respectively to the excitation of $B_{1u}(D_{4h})$ and $E_u(D_{4h})$ magnon mode around impurity or the first spin deviation. This can easily be confirmed by examining the behaviour of π^n_δ. The factor κ will be close to unity near the zone boundary where $E \sim 4Sz|J_2|$, z being the number of the second nearest neighbours.

Calculated line shapes are given in fig. 19 (Thorpe 1970). We note that the inclusion of interaction improves the agreement considerably.

For the exciton–magnon transition (Parkinson and Loudon 1968, Freeman and Hopfield 1968, Fujiwara and Tanabe 1975), we have

$$\begin{aligned}\alpha^n_{\text{em}} = -NC(2S)/\pi \,\text{Im} \sum_{\boldsymbol{\delta},\boldsymbol{\delta}'} \{&\pi^n_{\boldsymbol{\delta}}(\text{A})\pi^n_{\boldsymbol{\delta}'}(\text{A})G^{(\text{AA})}(\boldsymbol{\delta},\boldsymbol{\delta}') \\ &+ \pi^n_{\boldsymbol{\delta}}(\text{B})\pi^n_{\boldsymbol{\delta}'}(\text{B})G^{(\text{BB})}(\boldsymbol{\delta},\boldsymbol{\delta}') - \pi^n_{\boldsymbol{\delta}}(\text{A})\pi^n_{\boldsymbol{\delta}'}(\text{B})G^{(\text{AB})}(\boldsymbol{\delta},\boldsymbol{\delta}') \\ &- \pi^n_{\boldsymbol{\delta}}(\text{B})\pi^n_{\boldsymbol{\delta}'}(\text{A})G^{(\text{BA})}(\boldsymbol{\delta},\boldsymbol{\delta}')\},\end{aligned} \tag{91}$$

where

$$\begin{aligned}G^{(\text{AA})}(\boldsymbol{\delta},\boldsymbol{\delta}') &= 1/N \sum_{jj'} G(A_j b_{j+\boldsymbol{\delta}}, A^\dagger_{j'} b^\dagger_{j'+\boldsymbol{\delta}'}), \\ G^{(\text{AB})}(\boldsymbol{\delta},\boldsymbol{\delta}') &= 1/N \sum_{jl} G(A_j b_{j+\boldsymbol{\delta}}, B^\dagger_l a^\dagger_{l+\boldsymbol{\delta}'}), \quad \text{etc.,}\end{aligned} \tag{92}$$

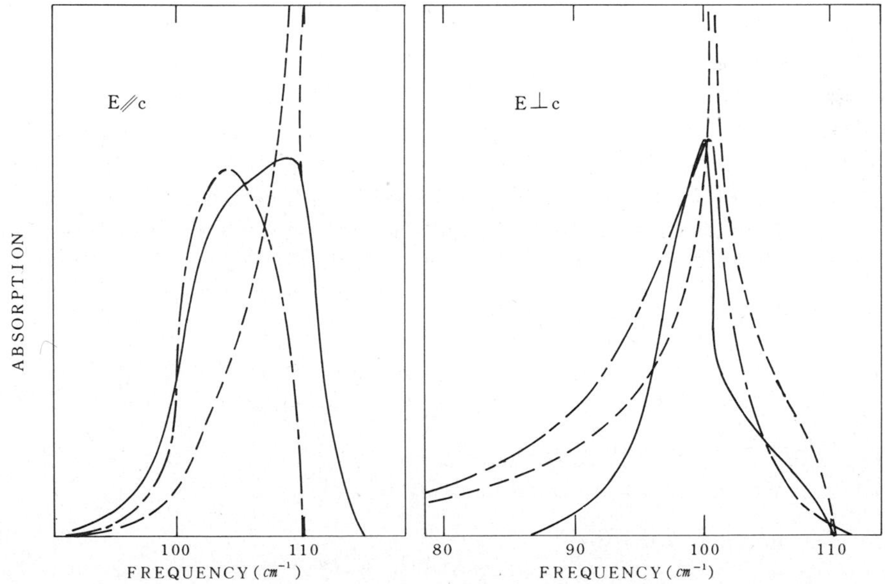

Fig. 19. The solid line is experimental: the dotted line non-interacting spin wave theory and long short dashed line includes magnon–magnon interaction effects (Thorpe 1970).

and

$$\pi_\delta(\mathrm{A}) = \pi(j^*j + \boldsymbol{\delta}), \qquad \pi_\delta(\mathrm{B}) = \pi(l^*l + \boldsymbol{\delta}). \tag{93}$$

To set up equations of motion for $G^{(\mathrm{AA})}$ etc. we need Hamiltonian for the exciton magnon interaction which reads

$$\begin{aligned}\mathcal{H}' = \sum_{jl} \{ & -2J_{jl}S[\phi_{jl}(a_j^\dagger b_l^\dagger + a_j b_l) + \rho_{jl} b_l^\dagger b_l + \epsilon_{jl} a_j^\dagger a_j] A_j^\dagger A_j \\ & + (A_j, a_j \leftrightarrow B_l, b_l) \\ & + \{L'_{jl}(a_j^2 + b_l^{\dagger 2}) + L_{jl} a_j b_l^\dagger\} A_j^\dagger B_l + (\mathrm{h.c.})\},\end{aligned} \tag{94}$$

where we have assumed the excited state is orbitally non-degenerate as usual and

$$\begin{aligned}\phi_{jl} &= J'_{jl}\sqrt{S'}/J_{jl}\sqrt{S} - 1, \\ \epsilon_{jl} &= J'_{jl}/J_{jl} - 1 + 2\mu_\mathrm{B}(H'_\mathrm{A} - H_\mathrm{A})/(-2J_{jl}Sz).\end{aligned} \tag{95}$$

Definition of ρ_{jl} is given by eq. (82) if we replace J by J_{jl}. Note that we have included uniaxial anisotropy energy $-2\mu_\mathrm{B}H_\mathrm{A}$ and $-2\mu_\mathrm{B}H'_\mathrm{A}$ in the ground and excited state energy, respectively. In the transfer term L'_{jl} is defined as

$$L'_{jl} = \tfrac{1}{2}\sqrt{(2S-1)/2S}\, L_{jl}, \tag{96}$$

with L_{jl} defined by the right hand side of eq. (83).

If we supplement $G^{(\mathrm{AA})}$, $G^{(\mathrm{AB})}$, etc., by $G^{(\mathrm{AA})}(0, \boldsymbol{\delta})$, $G^{(\mathrm{AA})}(0, 0)$, etc., defined by

$$G^{(\mathrm{AA})}(0, \boldsymbol{\delta}) = 1/N \sum_{jj'} G(A_j a_j^\dagger, A_{j'}^\dagger b_{j'+\delta}^\dagger), \quad \text{etc.,} \tag{97}$$

and assume that the range of the exchange and transfer integrals are restricted to the nearest neighbours on the opposite sublattice ($J = J_2$, $L = L_2$ in $\mathrm{MnF_2}$), we obtain the following Dyson equation for a kind of coupled impurity problem with impurity at the origin 0 and z neighbours at the sites $\boldsymbol{\delta}_i$ ($i = 1, 2, \ldots, z$):

$$\begin{aligned}\begin{bmatrix}\mathbf{G}^{(\mathrm{AA})} & \mathbf{G}^{(\mathrm{AB})} \\ \mathbf{G}^{(\mathrm{BA})} & \mathbf{G}^{(\mathrm{BB})}\end{bmatrix} &= \begin{bmatrix}\mathbf{G}_0^{(\mathrm{A})} & 0 \\ 0 & \mathbf{G}_0^{(\mathrm{B})}\end{bmatrix} \\ &+ \begin{bmatrix}\mathbf{G}_0^{(\mathrm{A})} & 0 \\ 0 & \mathbf{G}_0^{(\mathrm{B})}\end{bmatrix}\begin{bmatrix}\Delta\boldsymbol{\Lambda}^{(\mathrm{A})} & \Delta\mathbf{L}^{(\mathrm{AB})} \\ \Delta\mathbf{L}^{(\mathrm{BA})} & \Delta\boldsymbol{\Lambda}^{(\mathrm{B})}\end{bmatrix}\begin{bmatrix}\mathbf{G}^{(\mathrm{AA})} & \mathbf{G}^{(\mathrm{AB})} \\ \mathbf{G}^{(\mathrm{BA})} & \mathbf{G}^{(\mathrm{BB})}\end{bmatrix},\end{aligned} \tag{98}$$

where $\mathbf{G}^{(\mathrm{AA})}$, $\mathbf{G}^{(\mathrm{AB})}$, $\mathbf{G}_0^{(\mathrm{A})}$, $\Delta\boldsymbol{\Lambda}^{(\mathrm{A})}$, $\Delta\mathbf{L}^{(\mathrm{AB})}$, etc. are all $(z+1)\times(z+1)$ matrices whose rows and columns are labelled by $\boldsymbol{\delta}_i$ and 0:

$$\begin{aligned}\Delta\Lambda(\boldsymbol{\delta}_i, \boldsymbol{\delta}_{i'}) &= -2J_{\delta_i}S\rho_{\delta_i}\delta_{ii'}, \\ \Delta\Lambda(0, \boldsymbol{\delta}_i) &= \Delta\Lambda(\boldsymbol{\delta}_i, 0) = -2J_{\delta_i}S\phi_{\delta_i}, \\ \Delta\Lambda(0, 0) &= \sum_{\boldsymbol{\delta}} -2J_\delta S\epsilon_\delta,\end{aligned} \tag{99}$$

and

$$\Delta L(\boldsymbol{\delta}_i, \bar{\boldsymbol{\delta}}_{i'}) = L_{\boldsymbol{\delta}_i}\delta_{ii'},$$
$$\Delta L(0, \bar{\boldsymbol{\delta}}_i) = \Delta L(\boldsymbol{\delta}_i, 0) = L'_{\boldsymbol{\delta}_i}, \tag{100}$$
$$\Delta L(0,0) = \sum_{\boldsymbol{\delta}} L_{\boldsymbol{\delta}},$$

with $\bar{\boldsymbol{\delta}}_i = -\boldsymbol{\delta}_i$. We have also put $\mathbf{J}_{\delta} = \mathbf{J}_{j,j+\delta}$, $\mathbf{L}_{\delta} = \mathbf{L}_{j,j+\delta}$, etc. If it were not for $\Delta\mathbf{L}$, eq. (98) would reduce to two separate impurity problems, and each could be solved, that is, $\mathbf{G}_0^{(A)}$, $\Delta\Lambda^{(A)}$ and $\mathbf{G}_0^{(B)}$, $\Delta\Lambda^{(B)}$ diagonalized by considering the symmetry of the localized magnon modes around the exciton at the origin. The unitary transformation $\mathbf{U}_A$ and $\mathbf{U}_B$ that diagonalizes $\mathbf{G}_0^{(A)}$, $\Delta\Lambda^{(A)}$ and $\mathbf{G}_0^{(B)}$, $\Delta\Lambda^{(B)}$ with respect to irreducible representations of the site group around the localized exciton will also make $(\mathbf{U}_A)^{-1}\Delta\mathbf{L}^{(AB)}\mathbf{U}_B$ and $(\mathbf{U}_B)^{-1}\Delta\mathbf{L}^{(BA)}\mathbf{U}_A$ factorized according to the symmetry of the mode, if we correctly correlate the magnon modes on the two sublattices. For the odd-magnon mode we are interested in, only $-2\mathbf{J}_\delta S\rho_\delta$ and $\mathbf{L}_\delta$ survive in the transformed matrices, and in the case of MnF_2 we have one 4×4 (two $B_{1u}(D_{2h})$ modes on B sublattice coupled with two B_{1u} modes on A sublattice) one 2×2 ($B_{3u}(D_{2h})$ on B coupled with $B_{2u}(D_{2h})$ on A) and one 2×2 (B_{2u} on B coupled with B_{3u} on A) matrix. Note that coordinate axes xyz for an exciton on B sublattice and magnons of A sublattice surrounding it are rotated by $\pi/2$ about the z-axis compared to those (given in fig. 10) for the exciton on A sublattice and accompanying magnons of B sublattice around it. The transformed matrices of exciton–magnon interaction and exciton–magnon transfer are found to be:

$$\begin{array}{cc} & \overbrace{\quad zx^2 \qquad zy^2 \quad}^{\alpha = B_{1u}} \;\; \overbrace{\quad zx^2 \qquad zy^2 \quad}^{\beta = B_{1u}} \\ \begin{array}{l} \alpha = B_{1u}\left\{\begin{array}{l} zx^2 \\ zy^2 \end{array}\right. \\ \beta = B_{1u}\left\{\begin{array}{l} zx^2 \\ zy^2 \end{array}\right. \end{array} & \left[\begin{array}{cccc} -2JS\rho_1 & & & -L_1 \\ & -2JS\rho_2 & -L_2 & \\ & -L_1 & -2JS\rho_1 & \\ -L_2 & & & -2JS\rho_2 \end{array}\right] \end{array}, \tag{101}$$

$$\begin{array}{cc} & \begin{array}{cc} \alpha = B_{3u}x & \beta = B_{2u}y \end{array} \\ \begin{array}{l} \alpha = B_{3u}x \\ \beta = B_{2u}y \end{array} & \left[\begin{array}{cc} -2JS\rho_1 & L_1 \\ L_2 & -2JS\rho_2 \end{array}\right] \end{array} \quad \begin{array}{cc} & \begin{array}{cc} \alpha = B_{2u}y & \beta = B_{3u}x \end{array} \\ \begin{array}{l} \alpha = B_{2u}y \\ \beta = B_{3u}x \end{array} & \left[\begin{array}{cc} -2JS\rho_2 & -L_2 \\ -L_1 & -2JS\rho_1 \end{array}\right] \end{array}. \tag{102}$$

The notations α and β indicate that they are magnon modes on B and A sublattice respectively, and zx^2, zy^2 etc. following group theoretical labels show how the eigenvector transforms under the operations of D_{2h} (see fig. 20). The suffices 1 and 2 here have been introduced to distinguish the two inequivalent nearest neighbours I and II on the opposite sublattice. We can solve eq. (98) in this way for $G^{(AA)}(\boldsymbol{\delta}, \boldsymbol{\delta}')$ etc. and calculate the sum in eq. (91). This is elementary but a little tedious. If we discard the inequivalence in the excited state mentioned above and set $\rho_1 = \rho_2 = \rho$, $L_1 = L_2 = L$, the problem can be greatly simplified and we obtain the following results for the summation in eq. (91) (Freeman and Hopfield 1968):

$$\sum_{\delta\delta'} \{\ldots\} = 8|\pi(X)|^2 \sum_{s=\pm} G_{Xs} \qquad (n = z), \tag{103}$$

$$= \sum_{s=\pm} \{8|\pi_s(Z)|^2 G_{Zs} + 8|\pi_s(A)|^2 G_{As}\} \quad (n = x), \tag{104}$$

for A_g exciton, where $|\pi(X)|^2$ is given in eq. (76) and

$$|\pi_\pm(Z)|^2 = \tfrac{1}{8}|\pi^x(\mathrm{I}) + \pi^x(\mathrm{II}) \mp (\pi^y(\mathrm{I}) + \pi^y(\mathrm{II}))|^2, \tag{105}$$

$$|\pi_\pm(A)|^2 = \tfrac{1}{8}|\pi^x(\mathrm{I}) - \pi^x(\mathrm{II}) \pm (\pi^y(\mathrm{I}) - \pi^y(\mathrm{II}))|^2, \tag{106}$$

with

$$G_{P\pm} = G_{0P}/(1 - (-2JS\rho \pm L)G_{0P}), \quad (P = X, Z, A), \tag{107}$$

and the symmetry adapted Green's functions (in D_{4h})

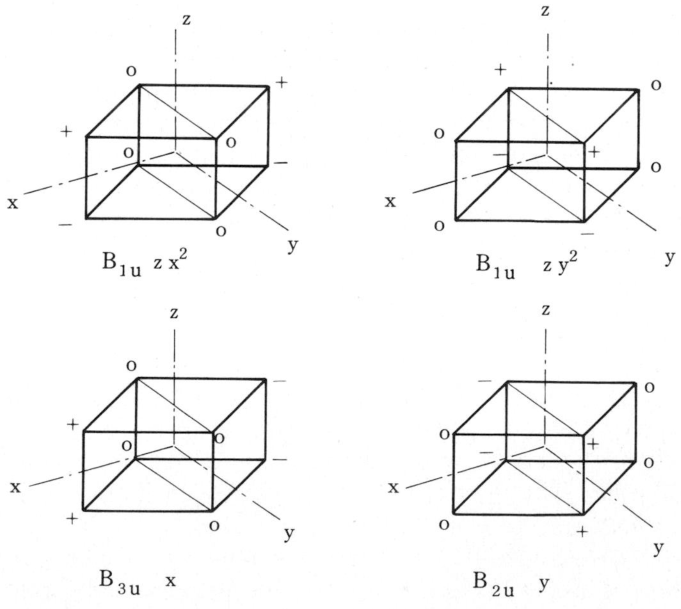

Fig. 20. Localized magnon modes around the body centre ion (D_{2h} symmetry).

$$G_{0X} = 1/N \sum_k 8\cos^2(\tfrac{1}{2}k_X a)\sin^2(\tfrac{1}{2}k_Y a)\cos^2(\tfrac{1}{2}k_Z c)u_k^2/(E - E_e(\boldsymbol{k}) - E_m(\boldsymbol{k})), \tag{108}$$

$$G_{0Z} = 1/N \sum_k 8\cos^2(\tfrac{1}{2}k_X a)\cos^2(\tfrac{1}{2}k_Y a)\sin^2(\tfrac{1}{2}k_Z c)u_k^2/(E - E_e(\boldsymbol{k}) - E_m(\boldsymbol{k})), \tag{109}$$

$$G_{0A} = 1/N \sum_k 8\sin^2(\tfrac{1}{2}k_X a)\sin^2(\tfrac{1}{2}k_Y a)\sin^2(\tfrac{1}{2}k_Z c)u_k^2/(E - E_e(\boldsymbol{k}) - E_m(\boldsymbol{k})). \tag{110}$$

The results for B_g exciton are given as,

$$\sum_{\delta\delta'} \{\ldots\} = \sum_{s=\pm} 8|\pi_s(X)|^2 G_{Xs} \quad (n = x), \tag{111}$$

$$= 16|\pi(Z)|^2 G_{Z+} + 16|\pi(A)|^2 G_{A-} \quad (n = z), \tag{112}$$

where

$$|\pi_\pm(X)|^2 = \tfrac{1}{4}|\pi^x(I) \pm \pi^y(II)|^2 + \tfrac{1}{4}|\pi^y(I) \mp \pi^x(II)|^2, \tag{113}$$

and $|\pi(Z)|^2$ and $|\pi(A)|^2$ are given in eqs. (80) and (81). In these results G_X is associated with the magnon mode $E_u(D_{4h})$ which becomes B_{3u} plus B_{2u} under D_{2h} and G_Z and G_A are associated with A_{2u} mode of $D_{4h}(B_{1u}$ of $D_{2h})$ and $B_{1u}(D_{4h})(B_{1u}(D_{2h}))$ mode, respectively.

In fig. 21 we give the calculated (Tanabe et al. 1968, Tonegawa 1969) line shape of the sidebands of ${}^6A_{1g} \to {}^4T_{1g}$ transition. The calculation corresponding to eqs. (103) and (104) includes the exciton–magnon interaction which is different for the two inequivalent neighbours ($\rho_1 \neq \rho_2$) but not the exciton–magnon pair transfer ($L = 0$). By suitable choice of parameters we obtain much better agreement with experiment than before.

Equation (112) for B_g exciton has been used to interpret the sharp line $E2\pi a$ observed below $E2\pi$ (fig. 11(a)) of the ${}^6A_{1g} \to {}^4E_g$ transition (Freeman and Hopfield 1968). This line is associated with a bound state with its energy determined from

$$1 = (-2SJ\rho - L)G_{0A}, \tag{114}$$

which will surely have a solution below continuum if $2|J|S\rho - L(<0)$ has sufficiently large absolute value. It is equally possible that G_{Z+} leads to a bound state when $2|J|S\rho + L$ is negative. No sharp line corresponding to this state, however, is observed, and it is supposed that the line is not well resolved from the peak of the continuum because of its too small binding energy and intensity.

Theoretical framework to treat exciton–exciton interaction is quite the same as developed above. Let us briefly discuss again the case of two excitons in MnF_2. For the excitation of equivalent excitons of each

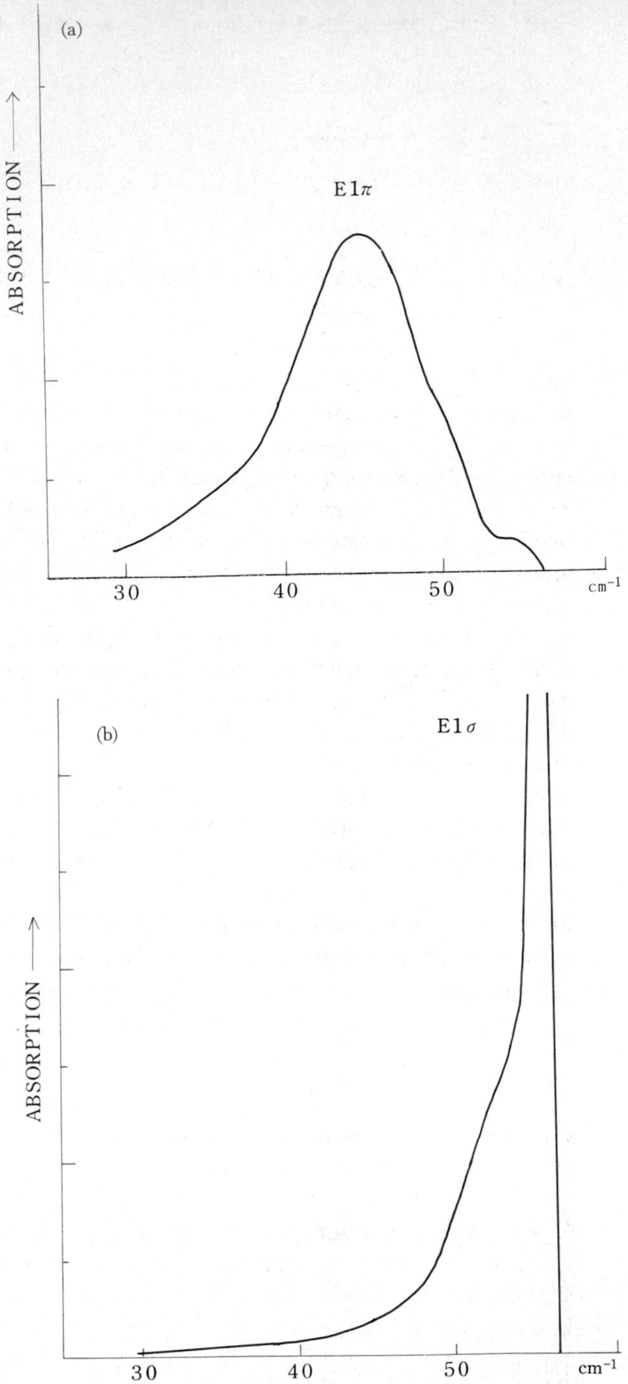

Fig. 21(a, b).

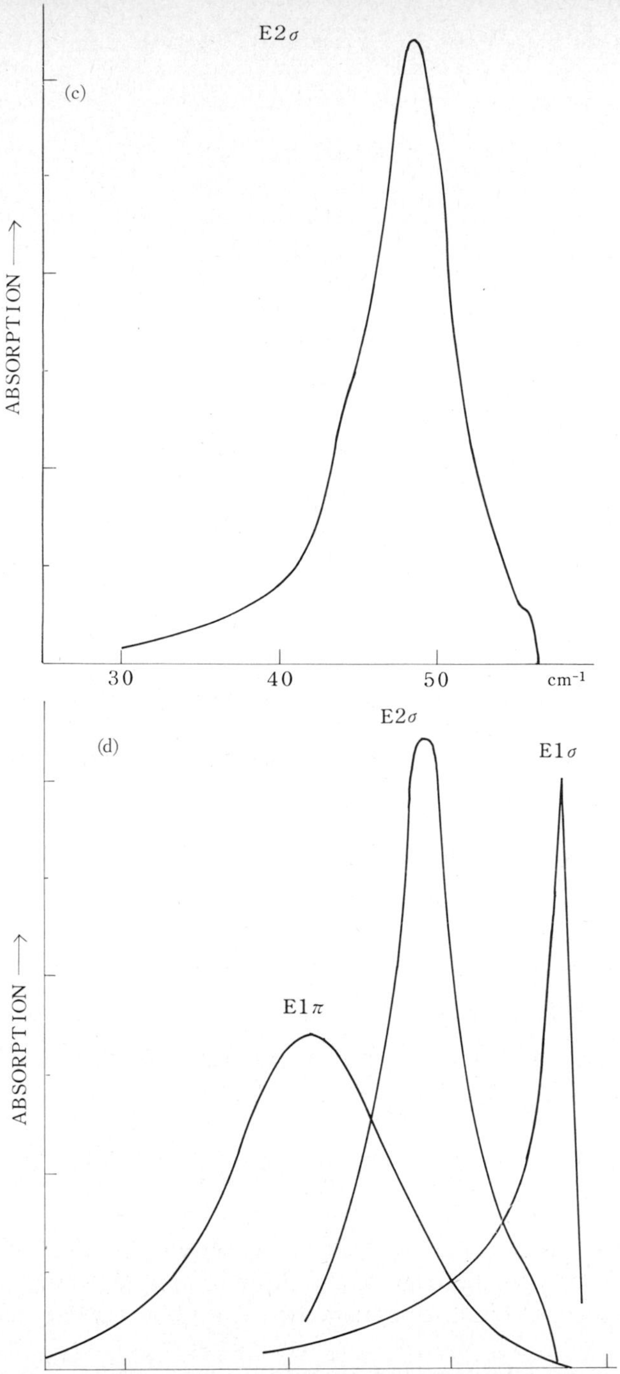

Fig. 21. Calculated line shapes (a)(b)(c) for E1π, σ and E2σ with exciton–magnon interaction (Tanabe et al. 1968) Observed shapes are given in (d).

sublattice, we have

$$\alpha^{n}_{ee} = -NC/\pi \ \mathrm{Im} \sum_{\delta\delta'} \pi^{n}_{\delta}\pi^{n}_{\delta'} G(\boldsymbol{\delta}, \boldsymbol{\delta}'), \tag{115}$$

where $\pi_\delta = \pi(j^*pj + \boldsymbol{\delta}^*\bar{p})$ and $G(\boldsymbol{\delta}, \boldsymbol{\delta}')$ is defined appropriately. We consider for simplicity the case of A_gA_g excitons ($p = \bar{p} =$ pure B_{3g} or B_{2g} of D_{2h}) in σ polarization. We then have

$$\sum_{\delta\delta'} \pi^{x}_{\delta}\pi^{x}_{\delta'} G(\boldsymbol{\delta}, \boldsymbol{\delta}') = 8|\pi^x|^2 G_X, \tag{116}$$

with

$$G_X = G_{0X}/(1 - VG_{0X}), \tag{117}$$

and

$$G_{0X} = 1/N \sum_{\boldsymbol{k}} 8\cos^2(\tfrac{1}{2}k_Xa)\sin^2(\tfrac{1}{2}k_Ya)\cos^2(\tfrac{1}{2}k_Zc)/(E - 2E_e(\boldsymbol{k}))$$

$$= \sum_{\boldsymbol{k}} p_X(\boldsymbol{k})/(E - 2E_e(\boldsymbol{k})), \quad \sum_{\boldsymbol{k}} p_X(\boldsymbol{k}) = 1, \tag{118}$$

where V is the interaction energy between the two excitons p and $\bar{p}$.

When V is sufficiently large, i.e., $|V| \gg$ (width of the two exciton continuum), bound state will appear. In such case G_{0X} may be approximated as

$$G_{0X} = 1/(E - \langle 2E_e(\boldsymbol{k})\rangle_X), \tag{119}$$

near the bound state with

$$\langle 2E_e(\boldsymbol{k})\rangle_X = \sum_{\boldsymbol{k}} 2E_e(\boldsymbol{k})p_X(\boldsymbol{k}), \tag{120}$$

so that we have a bound state at the energy

$$E_0 = \langle 2E_e(\boldsymbol{k})\rangle_X + V. \tag{121}$$

The intensity will be concentrated at this sharp line:

$$\alpha^{x}_{ee} = NC8|\pi^x|^2\delta(E - E_0). \tag{122}$$

Similarly, in the case of B_gB_g excitons, bound states will appear at

$$E_0 = \langle 2E_e(\boldsymbol{k})\rangle_A + V \tag{123}$$

in π polarization, whereas it will be found at different energy $E_0 = \langle 2E_e(\boldsymbol{k})\rangle_X + V$ in σ polarization. This kind of splitting is reported in the two exciton spectra of antiferromagnetic α-oxygen crystal (Gaididei et al. 1976).

When we consider two inequivalent excitons, we have to evaluate

$$\alpha^{n}_{ee} = -NC/\pi \ \mathrm{Im} \sum_{F,F'} \sum_{\delta,\delta'} \pi^{n}_{\delta}(F)\pi^{n}_{\delta'}(F')G^{(FF')}(\boldsymbol{\delta}, \boldsymbol{\delta}'), \tag{124}$$

where the summation over F and F′ is to be taken over A and B and,

$$\pi_\delta(\mathrm{A}) = \pi(j^*pj + \boldsymbol{\delta}^*\bar{q}), \qquad \pi_\delta(\mathrm{B}) = \pi(l^*\bar{p}l + \boldsymbol{\delta}^*q). \tag{125}$$

Green's functions $G^{(\mathrm{AA})}(\boldsymbol{\delta}, \boldsymbol{\delta}')$ etc. are defined in the same way as in eq. (92) replacing, for example, $b^\dagger_{j+\delta}$ by $B^\dagger_{j+\delta\bar{q}}$ and $A^\dagger_j$ by $A^\dagger_{jp}$, etc. We can again derive the Dyson equation (98) to determine $G^{(\mathrm{FF}')}$ as an impurity problem, with $-2J_\delta S\rho_\delta$ in $\Delta\Lambda$ and L_δ in $\Delta\mathbf{L}$ now replaced by $V_\delta(p\bar{q})$ (interaction energy between excitons p and q̄) and $L_\delta(pq)$ (exciton pair transfer integral for the resonance exchange pq̄→qp̄), respectively. The same symmetry considerations as before apply if we treat q̄ excitons around the localized p exciton like the magnons around an exciton as in exciton–magnon problem.

When V and/or L is sufficiently large the problem of obtaining (optically observable) bound states reduces to diagonalizing $\mathbf{G}$ or its inverse

$$\begin{bmatrix} (\mathbf{G}_0^{(\mathrm{A})})^{-1} & 0 \\ 0 & (\mathbf{G}_0^{(\mathrm{B})})^{-1} \end{bmatrix} - \begin{bmatrix} \Delta\Lambda^{(\mathrm{A})} & \Delta\mathbf{L}^{(\mathrm{AB})} \\ \Delta\mathbf{L}^{(\mathrm{BA})} & \Delta\Lambda^{(\mathrm{B})} \end{bmatrix} \tag{126}$$

within each (odd) symmetry species. As seen before, in the case of MnF_2, we have one 4×4 matrix for B_{1u} mode and one 2×2 matrix for each of B_{3u} and B_{2u} mode. For each symmetry adapted Green's function such as G_{0Z}, G_{0A}, and G_{0X} appearing in $(\mathbf{G}_0^{(\mathrm{A})})^{-1}$ and $(\mathbf{G}_0^{(\mathrm{B})})^{-1}$ (which are 2×2 matrices for B_{1u} mode) we can make approximation (119).

The case of double exciton E_{12}, E_{21} of MnF_2 corresponds to the case of $\mathrm{p} = B_{3g}$ and $\bar{\mathrm{q}} = B_{2g}$, and in its π spectrum we expect splittings or structures due to the inequivalence of the neighbours ($V_1 \neq V_2$) or the effect of the resonance exchange L_1 and L_2 in the 4×4 (B_{1u}) eigenvalue problem (126). However, experimentally, no splitting or structure has been detected so far. This may be due to the fact that they are actually imbedded in a phonon continuum and accordingly the line shapes are greatly influenced by it because of the resonance with the continuum states (Fano effect, Fujiwara 1973). Splitting corresponding to resonance exchange is reported in $2(^3\Sigma)\rightarrow\Sigma\Delta$ transition in the antiferromagnetic molecular crystal α-oxygen (Gaididei et al. 1975).

As an example, the case of $RbMnF_3$ is much simpler than that of MnF_2. For a pair of non-degenerate excitons p and q̄ of the same symmetry (D_{4h}) but with different energy, we have only

$$\pi^z_\delta(\mathrm{F}) = -\pi^z_{-\delta}(\mathrm{F}), \tag{127}$$

non-vanishing for the neighbours on the z-axis $\boldsymbol{\delta} = (0, 0, a)$ in π polarization. The symmetry adapted unperturbed Green's function in p_z mode is given by

$$\begin{aligned} G_{0Z} &= 1/N \sum_{k} 2\sin^2 k_z a/(E - E_{ep}(\boldsymbol{k}) - E_{e\bar{q}}(\boldsymbol{k})) \\ &= \mathrm{G}_0(\boldsymbol{\delta}, \boldsymbol{\delta}) - \mathrm{G}_0(\boldsymbol{\delta}, -\boldsymbol{\delta}). \end{aligned} \tag{128}$$

The matrix (126) is 2×2:

$$\begin{bmatrix} E-\langle E_{ep}(\boldsymbol{k})\rangle_Z-\langle E_{e\bar{q}}(\boldsymbol{k})\rangle_Z-V & L \\ L & E-\langle E_{ep}(\boldsymbol{k})\rangle_Z-\langle E_{e\bar{q}}(\boldsymbol{k})\rangle_Z-V \end{bmatrix}. \quad (129)$$

For the optically accessible state (the symmetric state) the energy of the bound state is

$$E_0 = \langle E_{ep}(\boldsymbol{k})\rangle_Z + \langle E_{e\bar{q}}(\boldsymbol{k})\rangle_Z + V - L, \quad (130)$$

with absorption intensity given by,

$$\alpha^z_{ee} = NC4|\pi^z_\delta(\mathrm{A})|^2\delta(E-E_0). \quad (131)$$

Note the minus sign before L in contrast with the exciton–magnon case.

4.5. Exciton–magnon transition in emission

It is possible to observe exciton magnon transition in emission where one exciton with wave vector $\boldsymbol{k}$ is destructed and one magnon with the same $\boldsymbol{k}$ created on the same sublattice. This process can also be described by the spin dependent (electric dipole) transition moment

$$\boldsymbol{P}(j_*j') = 2\boldsymbol{\pi}(j^*pj')\boldsymbol{\sigma}^\dagger_{jp}\cdot \boldsymbol{S}_{j'}, \quad (132)$$

which becomes, if we keep only the relevant term,

$$\boldsymbol{P}(j_*j') = \sqrt{2S}\,\boldsymbol{\pi}(j^*pj')A_{jp}a^\dagger_{j'}, \quad (133)$$

in terms of the exciton destruction operator A_{jp} and the spin deviation creation operator. As seen in fig. 22, this kind of transition can occur only in emission by a pair on the same sublattice. For a typical two sublattice antiferromagnet, the corresponding total dipole moment operator will be given by

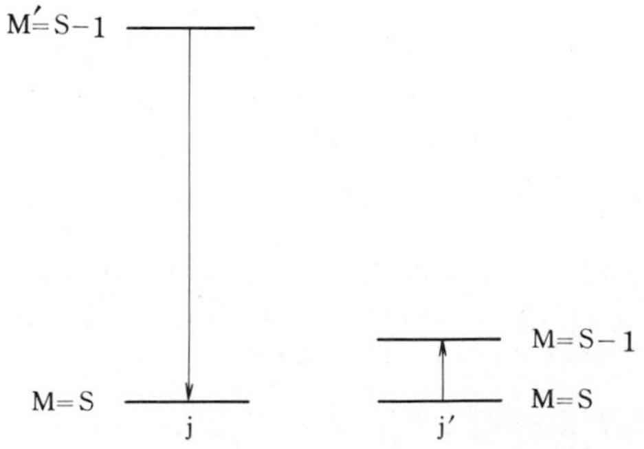

Fig. 22. Emission process described by $\pi(\sigma^\dagger_j)_+S_{j'-}$.

$$P_{\rm em} = \sum_{kp} [\sqrt{2S}\,\pi_p(k)^* u_k A_k \alpha_k^\dagger - \sqrt{2S}\,\pi_{\bar p}(k)^* u_k B_k \beta_k^\dagger], \tag{134}$$

with

$$\pi_p(k)^* = \sum_{j'} \pi(j^*pj') \exp[ik \cdot (R_{j'} - R_j)], \tag{135}$$

and a similar expression for $\pi_{\bar p}(k)$. The emission line shape will then be proportional to

$$\sum_k w(k)[2S|\pi_p^n(k)|^2 u_k^2 + 2S|\pi_{\bar p}^n(k)|^2 u_k^2]\delta(E - E_e(k) + E_m(k)), \tag{136}$$

where $w(k)$ is the initial exciton distribution over exciton states.

Such an exciton–magnon sideband corresponding to the radiative decay of free excitons is observed in the intrinsic luminescence of purest MnF_2 (less than a few ppm of impurity) at very low temperatures (e.g. at 1.32 K) (Dietz et al. 1968, 1970). A σ (and α) polarized electric dipole emission magnon sideband $\sigma 1$ (fig. 23) with a peak at 18 367 cm^{-1} is observed together with σ polarized magnetic dipole line at 18 418.6 cm^{-1} coincident with E1 (M1σ of fig. 15a) in absorption. No magnon sideband is observed in π polarization*.

If we assume only the first neighbours contribute in the sum of eq. (135) and take the symmetry of the exciton E1 (A_g, $p \approx B_{3g}$) we immediately find

$$\begin{aligned} \pi_p^y(k)^* &= 2i\pi^y(j^*pj + \delta) \sin k_z c, \\ \pi_p^z(k)^* &= 0, \end{aligned} \tag{137}$$

where $\delta = (0, 0, c)$. The solid curve in fig. 23 which shows good agreement with experiment was calculated by means of eq. (136) with eq. (137) and the parameters given in fig. 16 assuming zero dispersion (estimated to be $\sim 0.01\ cm^{-1}$) for the exciton ($w(k) = 1$).

It is interesting to note that decay rates of E1 and the peak of magnon sideband $\sigma 1$ are different at 4.2 K (though they are similar at 1.3 K) which proves that the zone centre ($k = 0$) and near boundary states are thermodynamically distinct, or that the wave vector representation for exciton is essential even though there is no energy dispersion. The decay of the intrinsic excitons is dominated by non-radiative trapping process by cen-

* Note, however, that the Hermitian conjugate of eq. (51) leads also to a dipole moment operator like (134):

$$\begin{aligned} P(j_*l) &= 2\pi(j^*pl)^*\sigma_j^\dagger \cdot S_l \to \pi(j^*pl)^*(\sigma_j^\dagger)_+ S_{l-} \\ &\to \sqrt{2S}\,\pi(j^*pl)^* A_j b_l \to \sum_k \sqrt{2S}\,\pi_p(k)^* A_k b_{-k}. \end{aligned}$$

Expressing b_{-k} in terms of $\alpha_k^\dagger$ by eq. (29) we obtain an expression like (134) with u_k replaced by v_k^*. This leads to a weak but observable π polarized sideband in emission (Chiang et al. 1978a).

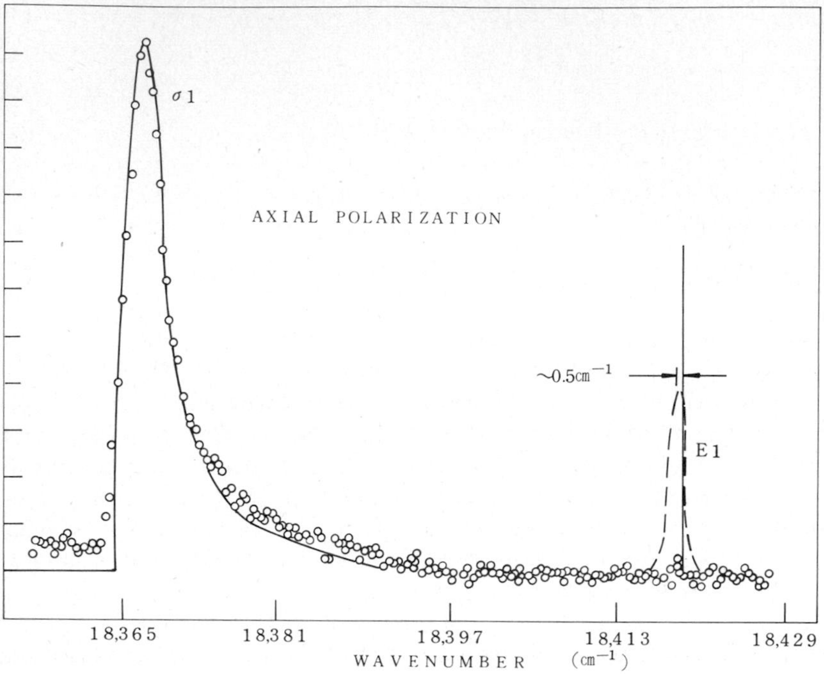

Fig. 23. Intrinsic emission near the origin of ${}^4T_{1g} \to {}^6A_{1g}$ in MnF_2. The solid curve is theoretical (Dietz et al. 1968).

tres near impurities rather than the radiative lifetime ($\tau_{rad} \sim 35$ ms), and the trapping probability is greater for the near zone boundary excitons than for the zone centre excitons at 4.2 K. That the wave vector description is necessary was also confirmed later (Macfarlane and Luntz 1973) by observing at 1.9 K the rate at which an initial non-equilibrium exciton distribution created at $\boldsymbol{k} = 0$ relaxes to an equilibrium distribution. The time $\tau(\Gamma \to Z, A)$ taken for excitons created at Γ point ($\boldsymbol{k} = 0$) to scatter from there to the Z and A point in BZ was measured by the rise time of $\sigma 1$ emission under resonant optical pumping at E1 and turned out to be $\sim 10^{-6}$ s.

As mentioned above, bulk of the visible emission from nominally pure MnF_2 (at 2 K) is known to arise from Mn levels perturbed by the impurity ions such as Mg, Zn and Ca (Greene et al. 1968). In fig. 24, observed emission line E1 and its magnon sideband $\sigma 1$ associated with impurities are shown. Roman numerals I, II and III, respectively indicate that the impurity ion is the first, second and third neighbour of the fluorescing Mn ion. The trapping levels occur some tens of cm^{-1} below the lowest intrinsic

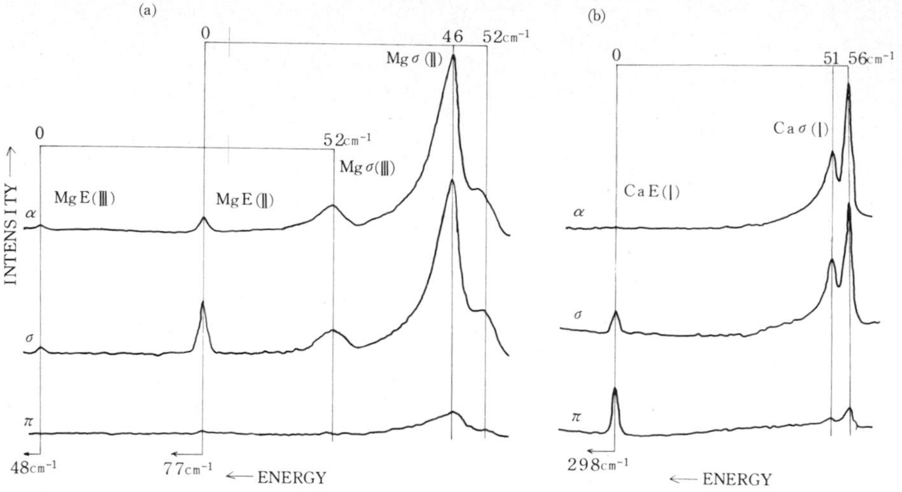

Fig. 24. The sharp structure associated with (a) Mg^{2+} and (b) Ca^{2+} impurities (Greene et al. 1968).

exciton as shown in fig. 25. This indicates that the excitation travels rapidly among the regular Mn ions until it is transferred to and trapped at a Mn ion near the impurity. The values of Δ determined from fluorescent lifetime are, for example, $\Delta = 80 \pm 5\ cm^{-1}$ ($77\ cm^{-1}$) for Mg(II), $66 \pm 5\ cm^{-1}$ ($66\ cm^{-1}$) for Zn(II) and $306 \pm 15\ cm^{-1}$ ($298\ cm^{-1}$) for Ca(I), which are in good agreement with the values in parentheses obtained from the perturbed positions of the E1 line. The MnF_2 luminescence is always dominated by such trap luminescence and this is a consequence of the efficient excitation transfer through the E1 exciton band*.

5. Recent developments

In the preceding sections we have been mainly concerned with excitons and magnons in relatively simple systems such as Cr_2O_3, MnF_2 and $RbMnF_3$. Exciton effects have been observed and studied also in other more complex systems. For example, Davydov splittings of R-line excitons of Cr^{3+} which change in company with spin reorientation have been identified in $RCrO_3$, R being rare earth elements (Meltzer 1970, Sugano et al. 1971, Satoko and Washimiya 1977). These compounds are also interesting in that some of them provide us with examples of sidebands on the low

* According to recent observation (Wilson et al. 1979) the excitation transfer above 4 K seems to occur actually through thermally activated E2 band.

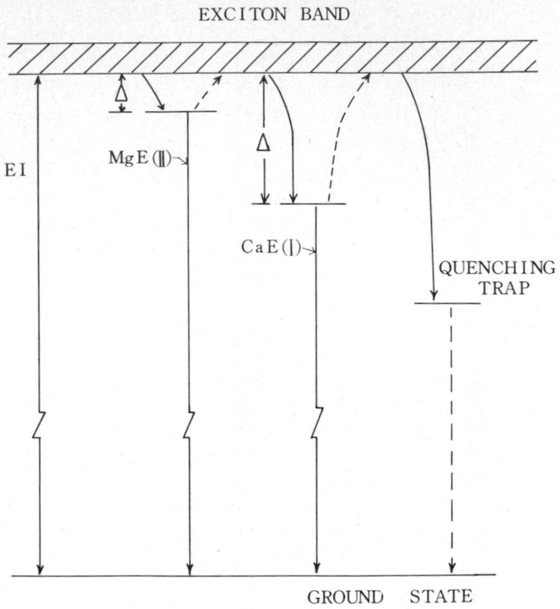

Fig. 25. Schematic representation of fluorescence from MnF_2 (Greene et al. 1968).

energy side of the exciton lines at low temperatures. The sidebands are therefore not hot bands, but they are still to be associated with simultaneous spin flip in R^{3+} and excitation ${}^4A_2 \rightarrow {}^2E$ in Cr^{3+}. The mechanism of this pair transition is not clarified yet (Aoyagi et al. 1977). This naturally brings up the subjects of excitons (and magnons) in rare earth compounds (Hüfner 1978). For example, magnon sidebands (due to single ion mechanism) in ferromagnetic $GdCl_3$ and $Gd(OH)_3$ have been identified and analyzed successfully with reasonable exciton dispersions for ${}^6P_J$ states of Gd^{3+} (Meltzer and Moos 1972). It should also be remarked that recent interest in the systems of low dimensionality has prompted optical studies of one- and two-dimensional ferro- and antiferromagnets. For this we cite only a few recent papers (Krausz et al. (1977) $AFeCl_3$, with A = K, Rb, Cs, Kojima et al. (1978) $(C_2H_5NH_3)_2MnCl_4$).

In sect. 4 we have seen that magnons in magnetic crystals play roles similar to phonons in non-magnetic crystals in many respects. It is indeed possible that many of the effects observed in the exciton magnon system can be carried over to the exciton phonon case and vice versa. (In this respect, magnetically ordered crystals are especially attractive because of the ease of interpretation of the magnons.) For example, the exciton–magnon interaction (94) can lead to multi–magnon sidebands (Moriya and Inoue 1968, Kozielski et al. 1971, Chiang et al. 1978b). Yen et al. (1967) succeeded in explaining the temperature dependent broadening of the exciton line E1 of MnF_2 by letting magnons play the part of phonons in the theory of McCumber and Sturge (1963). However, the importance of anharmonicity in the case of spin system should always be borne in mind as seen in the temperature dependence of intensities of the hot and cold bands in exciton–magnon absorption (Motizuki and Harada 1970, Shinagawa and Tanabe 1971, Fujiwara et al. 1972, Day et al. 1973, Ebara and Tanabe 1974). Spin wave approximation can never explain the decrease of intensity in the cold bands as the temperature is increased. The same remark applies to the temperature dependence of the shift and width of the sidebands (Tanaka 1971, Fujiwara and Tanabe 1972).

One of the most recent topics is the exciton dynamics (Wilson et al. 1978, Strauss et al. 1980). We have already touched upon the work by Macfarlane and Luntz (1973) (see also Cone 1978) by means of the time resolved spectroscopy to measure the decay and scattering rate in BZ for the exciton E1 of MnF_2. A similar observation was made for $^5D_4$ exciton of $Tb(OH)_3$ (Meltzer and Cone 1976, Chen and Meltzer 1980). Holzrichter et al. (1971) were able to observe magnetization induced by selective optical pumping of E1 or its sideband in MnF_2. By applying [110] stress it was possible to create excitons in one of the sublattices. Observation of the decay of magnetization enabled them to measure the intersublattice relaxation rate of $\tau_{sl}^{-1} \sim 4 \times 10^6\ s^{-1}$. We should also mention the exciton–exciton annihilation observed for $^5D_4$ exciton in $TbPO_4$ (Diggle et al. 1976) as another subject which is going to be fairly common in the exciton dynamics of magnetic crystals. Energy migration in these materials is another interesting subject. Spin barrier (eq. (8)) seems to be most effective at low temperatures in the one-dimensional antiferromagnets such as TMMC ($N(CH_3)_4MnCl_3$) and $CsMnBr_3$. There is no apparent energy migration for exciton $^6A_{1g} \rightarrow {}^4T_{1g}$ in TMMC below 50 K. The hopping time at 300 K is 10^{-12} s which is in the same region as for organic crystals (Yamamoto et al. 1977). Similar behaviour was also confirmed by observing Nd^{3+} impurity emissions in $CsMnBr_3$ (McPherson and Francis 1978).

We have no space to go into the light scattering due to the excitons and magnons in magnetic insulators. Actually, Raman spectroscopy has been one of the most powerful means of studying the excitations in these materials (Fleury and Loudon 1968). As an example we only mention here the case of CoF_2, for which Raman spectroscopy enabled us to identify all the exciton states arising from the ground multiplet $^4T_{1g}$ (Moch et al. 1971).

References

Allen, J.W., 1974, Phys. Rev. **B9**, 259.

Allen, J.W., R.M. Macfarlane and R.L. White, 1969, Phys. Rev. **179**, 523.

Allen, Jr., S.J., R. Loudon and P.L. Richards, 1966, Phys. Rev. Lett. **16**, 463.

Aoyagi, K., M. Kajiura, K. Tsushima, Y. Nakagawa and I. Tsujikawa, 1977, Physica **86–88B**, 1207.

Ballhausen, C.J., 1962, Introduction to Ligand Field Theory (McGraw-Hill, New York).

Chen, H.T. and R.S. Meltzer, 1980, Phys. Rev. Lett. **44**, 599.

Chen, M., D.S. McClure and E.I. Solomon, 1972, Phys. Rev. **B6**, 1690.

Chiang, T.C., P.R. Salvi, J. Davies and Y.R. Shen, 1978a, Solid State Commun. **26**, 217; 1978b, ibid. **26**, 527.

Condon, E.U. and G.H. Shortley, 1963, Theory of Atomic Spectra (Cambridge University Press, London) ch. 3, pp 61–64.

Cone, R.L., 1978, Proc. Conf. Dynamical Processes in the Excited States of Ions and Molecules in Solids, Georgia, F8.

Day, P., A.K. Gregson and D.H. Leech, 1973, Phys. Rev. Lett. **30**, 19.

Dietz, R.E., A. Misetich and H.J. Guggenheim, 1966, Phys. Rev. Lett. **16**, 841.

Dietz, R.E., A.E. Meixner, H.J. Guggenheim and A. Misetich, 1968, Phys. Rev. Lett. **21**, 1067; 1970, J. Lumin. **1**, **2**, 279.

Diggle, P.C., K.A. Gehring and R.M. Macfarlane, 1976, Solid State Commun. **18**, 391.

Dimmock, J.O. and R.G. Wheeler, 1964, Mathematics of Physics and Chemistry, Vol. 2, eds., H. Margenau and G.M. Murphy (Van Nostrand, New York) ch. 12, pp 725–768.

Ebara, K. and Y. Tanabe, 1974, J. Phys. Soc. Japan **36**, 93.

Elliott, R.J. and M.F. Thorpe, 1969, J. Phys. **C2**, 1630.

Eremenko, V.V. and A.I. Belyaeva, 1969, Soviet Physics-Uspekhi **12**, 320.

Eremenko, V.V. and E.G. Petrov, 1977, Adv. in Physics **26**, 31.

Ferguson, J., H.J. Guggenheim and Y. Tanabe, 1965, J. Appl. Phys. **36**, 1046.

Fleury, P.A. and R. Loudon, 1968, Phys. Rev. **166**, 514.

Freeman, S. and J.J. Hopfield, 1968, Phys. Rev. Lett. **21**, 910.

Fujiwara, T., 1973, J. Phys. Soc. Japan **34**, 36, 1180.

Fujiwara, T. and Y. Tanabe, 1972, J. Phys. Soc. Japan, **32**, 912; 1974, ibid. **37**, 1512; 1975, ibid. **39**, 7.

Fujiwara, T., W. Gebhardt, K. Petanides and Y. Tanabe, 1972, J. Phys. Soc. Japan **33**, 39.

Gaididei, Yu.B., V.M. Loktev, A.F. Prikhotko and L.I. Shanski, 1975, Fiz. Nizk. Temp. **1**, 1365; 1976, Soviet Physics-JETP **41**, 855.

Greene, R.L., D.D. Sell, R.S. Feigelson, G.F. Imbusch and H.J. Guggenheim, 1968, Phys. Rev. **171**, 600.

Greene, R.L., D.D. Sell, W.M. Yen and A.L. Schawlow, 1965, Phys. Rev. Lett. **15**, 656.

Halley, J.W., 1966, Phys. Rev. **149**, 423; 1967, ibid, **154**, 458.

Heine, V., 1960, Group Theory in Quantum Mechanics (Pergamon Press, Oxford) ch. 6, pp 265–303.

Holzrichter, J.F., R.M. Macfarlane and A.L. Schawlow, 1971, Phys. Rev. Lett. **26**, 652.

Hüfner, S., 1978, Optical Spectra of Transparent Rare Earth Compounds (Academic Press, London) p. 171.

Imbusch, G.F., 1978, Luminescence of Inorganic Solids, eds., B. DiBartolo, V. Goldberg and D. Pacheco (Plenum Press, New York) pp 115–174.

Kittel, C., 1963, Quantum Theory of Solids (Wiley, New York) ch. 4, pp 49–74.

Kojima, N., T. Ban and I. Tsujikawa, 1978, J. Phys. Soc. Japan **44**, 919, 923.

Kozielski, M., I. Pollini and G. Spinolo, 1971, Phys. Rev. Lett. **27**, 1223.

Krausz, E., S. Viney and P. Day, 1977, J. Phys. **C10**, 2685.

Loudon, R., 1968, Adv. in Physics **17**, 243.
Macfarlane, R.M. and J.W. Allen, 1971, Phys. Rev. **B4**, 3054.
Macfarlane, R.M. and A.C. Luntz, 1973, Phys. Rev. Lett. **31**, 832.
McClure, D.S., 1959a, Solid State Physics, Vol. 8, eds., F. Seitz and D. Turnbull (Academic Press, New York) pp 5–13; 1959b, ibid. Vol. 9, pp 399–525.
McCumber, D.E. and M.D. Sturge, 1963, J. Appl. Phys. **34**, 1682.
McPherson, G.L. and A.H. Francis, 1978, Phys. Rev. Lett. **24**, 1681.
Meltzer, R.S., 1970, Phys. Rev. **B2**, 2398.
Meltzer, R.S. and R.L. Cone, 1976, J. Lumin. **12/13**, 24.
Meltzer, R.S. and H.W. Moos, 1972, Phys. Rev. **B6**, 264.
Meltzer, R.S., M. Lowe and D.S. McClure, 1969, Phys. Rev. **180**, 561.
Misu, A., 1960, unpublished work.
Moch, P., J.P. Gosso and C. Dugautier, 1971, Proc. 2nd Int. Conf. Light Scattering in Solids, ed., M. Balkanski (Flammarion Sciences, Paris) p. 138.
Moriya, T., 1966, J. Phys. Soc. Japan **21**, 926.
Moriya, T. and M. Inoue, 1968, J. Phys. Soc. Japan **24**, 1251.
Motizuki, K. and I. Harada, 1970, Prog. Theor. Phys. Suppl. No 6, 40.
Novikov, V.P., V.V. Eremenko and V.V. Shapiro, 1973, J. Low Temp. Phys. **10**, 95.
Parkinson, J.B. and R. Loudon, 1968, J. Phys. **C1**, 1569.
Satoko, C., 1970, J. Phys. Soc. Japan **28**, 1367.
Satoko, C. and S. Washimiya, 1977, J. Phys. Soc. Japan **42**, 1888.
Sell, D.D., 1968, J. Appl. Phys. **39**, 1030.
Sell, D.D., R.L. Greene and R.M. White, 1967, Phys. Rev. **158**, 489.
Shinagawa, K. and Y. Tanabe, 1971, J. Phys. Soc. Japan **30**, 1280.
Solomon, E.I. and D.S. McClure, 1972, Phys. Rev. **36**, 1697.
Stokowski, S.E. and D.D. Sell, 1971, Phys. Rev. **B3**, 208.
Stout, J.W., 1959, J. Chem. Phys. **31**, 709.
Strauss, E., W.J. Miniscalco, W.M. Yen, U.C. Kellner and V. Gerhardt, 1980, Phys. Rev. Lett. **44**, 824.
Sugano, S. and Y. Tanabe, 1958, J. Phys. Soc. Japan **13**, 880.
Sugano, S., K. Aoyagi and K. Tsushima, 1971, J. Phys. Soc. Japan **31**, 706.
Tanabe, Y. and S. Sugano, 1954, J. Phys. Soc. Japan **9**, 753, 766.
Tanabe, Y. and K. Gondaira, 1967, J. Phys. Soc. Japan **22**, 573.
Tanabe, Y., T. Moriya and S. Sugano, 1965, Phys. Rev. Lett. **15**, 1023.
Tanabe, Y., K.I. Gondaira and H. Murata, 1968, J. Phys. Soc. Japan, **25**, 1562.
Tanaka, H., 1971, J. Phys. Soc. Japan **31**, 368.
Thorpe, M.F., 1970, J. Appl. Phys. **41**, 892.
Tonegawa, T., 1969, Prog. Theor. Phys. **41**, 1.
Wilson, B.A., J. Hegarty and W.M. Yen, 1978, Phys. Rev. Lett. **41**, 268.
Wilson, B.A., W.M. Yen, J. Hegarty and G.F. Imbusch, 1979, Phys. Rev. **B19**, 4238.
Wong, Y.H., F.L. Scarpace, C.D. Pfeifer and W.M. Yen, 1974, Phys. Rev. **B9**, 3086.
Yamamoto, H., D.S. McClure, C. Marzacco and M. Waldman, 1977, Chem. Phys. **22**, 79.
Yen, W.M., G.F. Imbusch and D.L. Huber, 1967, Optical Properties of Ions in Crystals, eds., H.M. Crosswhite and H.W. Moos (Interscience Publishers, New York) p. 301.

CHAPTER 15

Dynamics of Molecular Excitons: Disorder, Coherence and Dephasing

AHMED H. ZEWAIL,*† DUANE D. SMITH**
and JEAN-PIERRE LEMAISTRE***

Arthur Amos Noyes Laboratory of Chemical Physics‡
California Institute of Technology
Pasadena, California, 91125
U.S.A.

* Alfred P. Sloan Fellow and Camille and Henry Dreyfus Teacher-Scholar.
† To whom correspondence should be sent.
** Present address: The James Franck Institute, University of Chicago, Chicago, Illinois 60637.
*** National Science Foundation (USA)-Centre National de la Recherche Scientifique (France) Visiting Postdoctoral Research Fellow. Permanent address: Laboratoire d'Optique Moléculaire, LA283 de CNRS, Université de Bordeaux I, 33405 Talence, France.
‡ Contribution No. 6162.

Excitons
Edited by
E.I. Rashba and M.D. Sturge

Contents

1. *Introductory remarks*

For many years molecular crystals have been exploited to enhance our understanding of how optical molecular excitation (an exciton) "moves" in insulators. This class of solids is prototypical for at least two reasons. First, the molecules (e.g., naphthalene) stack in the crystal in unique ways making the intermolecular interactions highly anisotropic. As a result, an effective dimensionality (1-D, 2-D or 3-D) for the excitation transport may exist. Secondly, from the pioneering work of Frenkel (1931) and Davydov (1971) we expect, at least theoretically, the spectroscopy of these excitons to be rather simple and to reveal the combined properties of the molecules and the crystal structure. Armed by the Frenkel–Davydov theory, experimentalists studied molecular excitons focusing on two major problems – the *band structure* and *dynamics* of the exciton.

To probe the band structure, primarily optical and magnetic resonance spectroscopy have been used. Basically, two approaches were adopted. Either one measures the exciton density of states $\rho(E)$ or the resonance interaction matrix element, β, between molecules. Both ρ and β give information on the band structure. To illustrate the point, let us consider the simplest case of a linear chain crystal. The exciton is made of different $\boldsymbol{k}$ states, where $\boldsymbol{k}$ is the wavevector of the quasiparticle and the dispersion, energy vs. $\boldsymbol{k}$, is simply

$$E(k) = 2\beta \cos \boldsymbol{kc}. \tag{1}$$

In obtaining this expression, we assume that the molecules stack along the crystallographic c-axis and only nearest-neighbor resonance interactions between the molecules are allowed. From the equation, we see that an experimental measurement of β or $\rho(E) = \partial k/\hbar E$, will in principle, provide $E(k)$ and the exciton bandwidth, 4β. However, precisely what determines β is unclear.

The coupling β is known to depend on the vibronic state and is generally an exchange or multipolar interaction for molecular triplet and singlet states, respectively. The unique way in which molecules stack in the chain provides a comfortable way for the π and/or σ electrons on neighboring molecules to overlap, hence a nonzero β value. Believing the geometry and overlap integrals to be of primary importance, several laboratories have studied exciton band structure in a variety of crystals, hoping to establish relationships among band structure, stacking, and σ–π electron density in the excited states.

Out of the "structural studies" it may be concluded that the β's can be measured but cannot be accurately calculated. The difficulty lies, in part, with the complexity of the σ–π interactions between molecules and that, for some crystals, in the triplet state the interaction is rather small being in the micro to millivolt range.

More importantly, however, several solids with well-defined dimensionality have emerged, as we shall discuss later, from the band structural studies. But *are the band structure determinations sufficient to establish the precise nature of the exciton dynamics*? The answer is no.

Dynamically speaking, one is seeking information on the transport after the exciton is created by light absorption. Unlike the exciton structure studies, the dynamics are more intricate, requiring a detailed understanding of a large number of communication channels. It is fair to say that despite an enormous amount of work, we are still learning about the dynamics of exciton transport, especially in disordered lattices.

The theme of this article is molecular exciton dynamics, focusing on the disorder and coherent effects. We will spend little time on band structure determination, instead referring the reader to the articles of Robinson (1970) and Hochstrasser (1976) which nicely present the experimental techniques and results for a variety of crystalline solids. The recent reviews of Burland and Zewail (1979) and Silbey (1976) are also recommended for background material. Here, we shall devote our efforts to understanding the theory and experiment for *simple* systems, hoping to answer the following questions:

(a) Why can't the conventional optical spectra alone reveal dynamics of the exciton?
(b) What do we mean, in a quantum-mechanical sense, by disorder?
(c) Are excitons mobile in disordered systems?
(d) What are the experimental approaches to probe dynamics?

Rather than review all the experimental studies on coherence and disorder conducted so far, we shall draw heavily on specific work on 1-D and 2-D systems conducted in our laboratories. Using these systems as a model, we link the experimental findings with the theoretical discussion of questions pertaining to dynamics: disorder, coherence, and dephasing.

2. Definitions

2.1. Small and large excitons

Throughout the paper two types of excitons will be treated: the "small" and "large" excitons. By small and large we mean the extent to which the exciton is delocalized on the lattice. If the exciton is confined to the smallest possible *unit* (i.e., two molecules), then one has the small exciton case (SE) or what the Leiden group (Botter et al. 1976) termed the "mini" exciton. In a manner of thinking, these are dimers, but not true dimers since the two molecules are not isolated in a vacuum. Indeed, surrounding the two molecules are host molecules and the wavefunction of the dimer has finite amplitude on them (see fig. 1). On the other hand, in the large exciton case (LE), the exciton will span a large space, say, more than ten lattice sites. The physics of the SE is much simpler than that of the LE and for this reason we shall use the SE to introduce the reader unfamiliar with excitons to the problems of dynamics in *ordered* and *disordered* systems. A completely local excitation (on one molecule) will be used to illustrate some concepts regarding the origin of spectral broadening.

2.2. Disorder

Disorder in organic solids can be structural or substitutional. For the case of structural disorder, one must consider periodic or aperiodic disorder, the

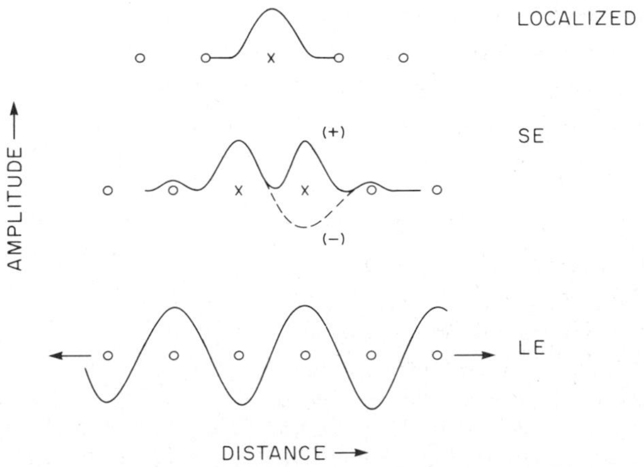

Fig. 1. Plot of the excitation amplitude for a completely localized excitation, impurity dimer states (or SE) and the large exciton case. The symbols are ○ = host × = trap and (±) to represent dimer plus and minus states.

amorphous phase and interstitial defects. For substitutional disorder the impurity (impurities) can occupy either periodic or randomly placed sites within the direct lattice. As a result of substitutional disorder (e.g., isotopically mixed organic crystals) two kinds of disorders result that are well-known in the physics literature; diagonal and off-diagonal disorder (Theodorou and Cohen 1976).

To describe diagonal and off-diagonal disorder in the present context, let us consider, for simplicity, the regular, infinite and linear array of molecules, discussed before. The localized wavefunction on site i is described by the ket $|i\rangle$. In this representation the electronic Hamiltonian for the perfect crystal has the following general matrix element:

$$\langle i|\mathcal{H}|j\rangle = \epsilon_i \delta(i, j) + \beta_{ij}\delta \quad (i, j = i \pm 1), \tag{2}$$

where ϵ_i is the local site energy, $\delta(i, j)$ is the Kronecker delta function and β_{ij} is the intersite coupling matrix element.

Assume momentarily that all the ϵ_i are the same (constant diagonal elements). One can then shift the arbitrary zero of energy to ϵ_i and recast the Hamiltonian as:

$$\mathcal{H} = \sum_i \beta_{i,i+1}[|i\rangle\langle i + 1| + |i + 1\rangle\langle i|], \tag{3}$$

where we have assumed a functional basis where all sites are orthogonal. Thus, one is left with vanishing diagonal elements and certain off-diagonal elements being nonzero and real. If the magnitude of the β's are statistically independent, one has fluctuations in the off-diagonal matrix elements. These fluctuations have a probability distribution. *A randomness in the magnitude of the off-diagonal elements is termed off-diagonal disorder.* Physically, this represents a fluctuation in the intersite interaction which may be due to optical or acoustical phonons, lattice defects, or random distributions of impurities.

Similarly, if the site energies ϵ fluctuate in some fashion then one has diagonal disorder. That is, diagonal elements of the above Hamiltonian will depart (hopefully randomly for us) from zero. It should be noted that off-diagonal disorder may or may not be independent of diagonal disorder. An opposite example of diagonal disorder is phenazine-D_8 put at random in phenazine-H_8, an isotopically mixed crystal where the protons are replaced by deuterons. In this case, the nearest-neighbor interaction is independent of isotopic identity to first order. Thus, in this rather simple picture, one has diagonal, but no off-diagonal, disorder. However, when considering longer hops between isolated impurities, one will encounter "higher-order" off-diagonal disorder in the intercluster coupling.

3. The small exciton case

3.1. The resonant and coherent limit

Here, we consider two molecules, A and B, coupled by a resonance interaction matrix element. Either molecule (or site in the crystal) could, in principle, be excited. The two site-functions are simply $\phi_A^*\phi_B$ and $\phi_A\phi_B^*$ where ϕ is the molecular wavefunction and the star denotes that the molecule is in the excited state. These product functions are degenerate in the absence of the intermolecular interaction; matrix element β. Because A and B excited molecules are assumed to be identical in zero-order, the SE states will be equally composed of $\phi_A^*\phi_B$ and $\phi_A\phi_B^*$, i.e.,

$$\Psi_\pm = (1/\sqrt{2})(\phi_A^*\phi_B \pm \phi_A\phi_B^*). \tag{4a}$$

The electronic Hamiltonian, whose eigenfunctions are given by eq. (4a) is

$$\mathcal{H} = \mathcal{H}_A + \mathcal{H}_B + V, \tag{4b}$$

where $\mathcal{H}_A$, $\mathcal{H}_B$ are the one-site (A or B) Hamiltonians and V is the intermolecular potential. Setting $\epsilon_A = \epsilon_B = 0$, the Hamiltonian, $\mathcal{H}$, can be rewritten in a second quantized form as follows:

$$\mathcal{H} = \beta(a_A^+ a_B + a_B^+ a_A). \tag{5}$$

The expression $a_A^+ a_B$ is a composite operator that creates excitation on A and annihilates excitation on B, and 2β is the total splitting with

$$\begin{aligned}\beta &= \langle \phi_A^*\phi_B | \mathcal{H} | \phi_A\phi_B^* \rangle \\ &= \langle \phi_A^*\phi_B | V | \phi_A\phi_B^* \rangle.\end{aligned} \tag{6}$$

Now we may calculate the probability of A being excited at time t with the initial condition that at $t = 0$ only A is excited. The probability $P_A(t)$ is

$$P_A(t) = |\langle \phi_A^*\phi_B | e^{-i\mathcal{H}t} | \phi_A^*\phi_B \rangle|^2 \tag{7a}$$

$$= \cos^2 \beta t, \tag{7b}$$

$$P_B(t) = \sin^2 \beta t. \tag{7c}$$

Thus, the excitation coherently oscillates between the molecules forever, with a transfer time determined by β. Such an idealized case is similar to the classical problem of coupled, resonant pendulums described, for example, in the book of Pauling and Wilson (1935). See also the article by Robinson and Frosch (1962) on intramolecular radiationless transitions.

3.2. Static diagonal disorder

If, in the above case, one puts $\epsilon_A = \Delta$, $\epsilon_B = -\Delta$, the *plus* and *minus* ($\pm$) states become

$$\Psi_+ = \cos\theta\, |\phi_A^*\phi_B\rangle + \sin\theta\, |\phi_A\phi_B^*\rangle,$$
$$\Psi_- = \sin\theta\, |\phi_A^*\phi_B\rangle - \cos\theta\, |\phi_A\phi_B^*\rangle, \tag{8a}$$

with,

$$\theta = \tfrac{1}{2}\,\mathrm{arc\,tan}(\beta/\Delta), \quad 0 \leq \theta \leq \pi/2. \tag{8b}$$

With algebra similar to that used in the derivation of eqs. (7), one obtains,

$$P_A(t) = 1 - \left(\frac{\beta^2}{\Delta^2 + \beta^2}\right) \sin^2\sqrt{\Delta^2 + \beta^2}\,t. \tag{9}$$

This is illustrated in fig. 2. Thus one arrives at the physically intuitive result that the exciton will be localized on molecule A as Δ gets larger (note that A was assumed to be prepared at $t = 0$). A practical example is the case where molecules A and B occupy different crystal field environments or are chemically distinct. The Hamiltonian (eq. (5)) now becomes:

$$\mathcal{H} = \Delta(a_A^+ a_A - a_B^+ a_B) + \beta(a_A^+ a_B + a_A a_B^+). \tag{10}$$

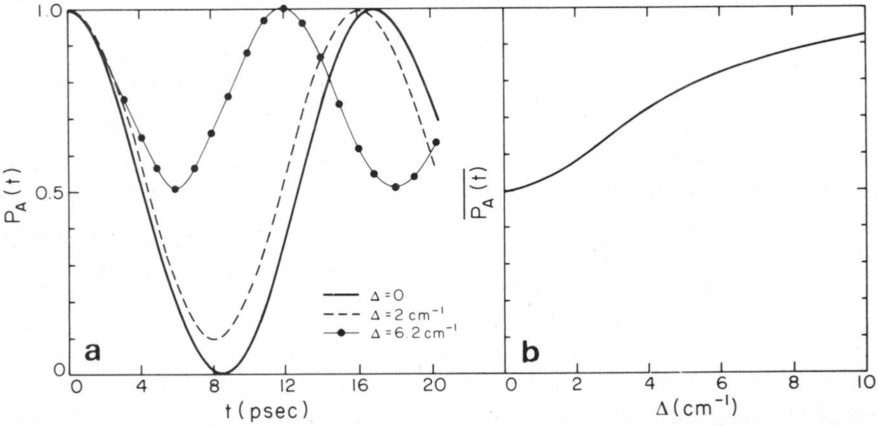

Fig. 2. (a) Plot of the time-dependent probability for finding the optical excitation on molecule A of the SE (eqs. (7b) and (9)). The plot is generated using the nearest neighbor coupling of DBN for three cases: (1) no diagonal disorder, $\Delta = 0$; (2) using the typical linewidth of the trap states in a doped DBN crystal, $\Delta = 2\ \mathrm{cm}^{-1}$; and (3) the case where the disorder is equal to the coupling $\beta = \Delta = 6.2\ \mathrm{cm}^{-1}$. (b) Plot of the root mean square probability of molecule A being excited as a function of diagonal disorder, again for the DBN dimer. Note the increase in the probability of localization on A as Δ increases.

If Δ now varies at random from one AB pair to another, eq. (10) contains a discrete diagonal disorder term in addition to the resonance term of eq. (5).

3.3. *Time-dependent disorder; the effect of phonons*

Now, suppose that A and B not only exhibit diagonal disorder, but are interacting with many of the phonons in the crystal. Such an interaction is usually assumed to be random or stochastic in nature. The phonon-induced "disorder" may cause β and/or Δ to change with time – stochastic fluctuations. Following Silbey's presentation (Rackovsky and Silbey 1973, Silbey 1976), one may then write $\mathcal{H}$ as a time-dependent Hamiltonian,

$$\mathcal{H}(t) = \Delta(t)[a_A^+ a_A - a_B^+ a_B] + \beta(t)[a_A^+ a_B + a_B^+ a_A], \tag{11}$$

thus introducing stochastic diagonal ($\Delta(t)$) and off-diagonal ($\beta(t)$) disorder. Again, one would like to solve for the time-dependent probability for molecule A or B to be excited. However, the problem is essentially intractable without invoking specific models to describe the fluctuations of β and Δ in time. Sewell (1962) and Haken and Strobl (1968) take $\Delta(t)$ and $\beta(t)$ to be Gaussian–Markov processes with δ function correlation times. This means that

$$\langle\Delta(t)\rangle = \langle\Delta + \delta\Delta(t)\rangle = \Delta, \tag{12a}$$

$$\langle\beta(t)\rangle = \langle\beta + \delta\beta(t)\rangle = \beta, \tag{12b}$$

$$\langle\delta\Delta(t)\delta\Delta(t')\rangle = \gamma_0\delta(t-t'), \tag{12c}$$

$$\langle\delta\beta(t)\delta\beta(t')\rangle = \gamma_1\delta(t-t'), \tag{12d}$$

$$\langle\delta\Delta(t)\delta\beta(t')\rangle = 0. \tag{12e}$$

Where γ_0 and γ_1, which result from the average over fluctuations, are the *diagonal* and *off-diagonal* fluctuation parameters. Note that eq. (12e) means that diagonal and off-diagonal disorder are uncorrelated. To the authors' knowledge, there is no experimental evidence to indicate that they are, or are not correlated in space or time, and it is an open issue.

A solution for $P_A(t)$-$P_B(t)$, which can be obtained in general, must invoke γ_0 and γ_1. Let us consider the quasi-coherent limit (Sewell 1962, Haken and Strobl 1968, Rackovsky and Silbey 1976, Silbey 1976) ($\gamma \ll \beta$) and further stipulate that the γ_1 mechanism is dominating. Then one obtains:

$$P_A(t) - P_B(t) = \cos 2\beta t\ e^{-2\gamma_1 t}. \tag{13}$$

It can now be seen that due to the fluctuation parameter, γ_1, damping of the probability variation will result (see fig. 3). This damping leads to line broadening of the transitions in addition to ordinary lifetime broadening via the uncertainty relation. As we shall see later, this broadening is *homogeneous* in nature and is usually masked by *inhomogeneous* effects. Fur-

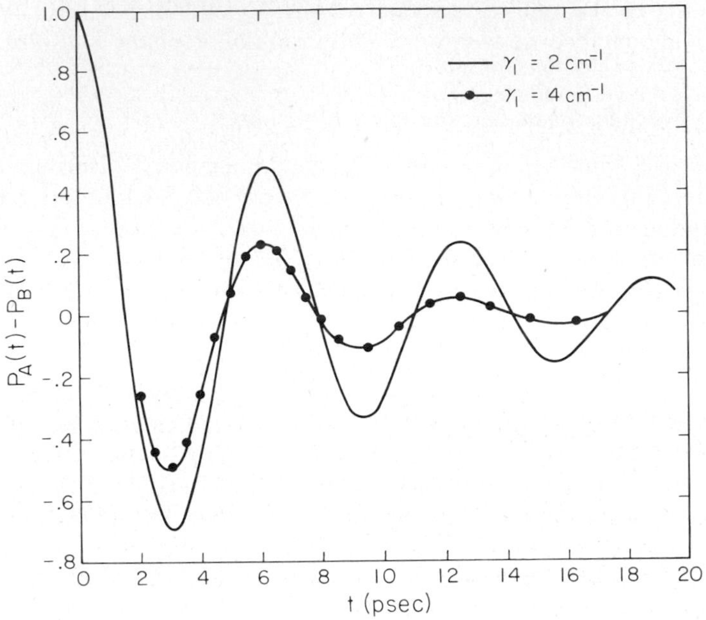

Fig. 3. The difference in the probabilities of molecule A and B being excited for the DBN dimer with no stochastic diagonal disorder and two values of off-diagonal disorder, $\gamma_1 = 2\,\mathrm{cm}^{-1}$ and $\gamma_1 = 4\,\mathrm{cm}^{-1}$.

thermore, the homogeneous broadening increases with increasing fluctuation rates which in turn are a function of the temperature of the crystal. In the coming few sections, we shall discuss the origin of this type of broadening and its relationship to dephasing.

4. Dephasing by phonons and homogeneous broadenings

4.1. An example of a completely localized exciton

To have a spatially local "exciton" requires that the excitation be on molecule A and $\beta = 0$. Hence one has two levels to consider; the ground state and the excited triplet or singlet state $(\ldots \phi_i \phi_A^* \phi_j \ldots)$. For triplet states it is also possible to learn about exciton dephasing by studying the spin transitions among the three different spin sublevels (spin angular momentum $S = 1$). But the connection between the spin resonance and optical dephasing measurements must be made depending on details of the dephasing processes (Zewail 1979).

In general, one wants to relate the dephasing of the transition to microscopic parameters. We shall follow the treatment of Jones and Zewail (1978) and Diestler and Zewail (1979) for the case of a completely local excitation on one molecule, e.g., an impurity in a host lattice (for a review, see Zewail (1980)). First let us point out the following basic approximations:

(a) The interaction among the impurity molecules is negligible.

(b) The A molecule has only two energy levels (initial i and final f), which means the transition is well isolated from all others.

(c) The coupling of the excited A molecule to all other degrees of freedom of the system (bath) is weak and the transition is homogeneously broadened.

The lineshape function can be expressed as

$$I(\omega) = \frac{1}{2\pi} \int_{-\infty}^{\infty} \langle \boldsymbol{\mu}(0)\boldsymbol{\mu}(t)\rangle e^{-i\omega t}\, dt, \tag{14}$$

where the angular brackets indicate an ensemble average of the autocorrelation function of the transition moment. An exponential decay for the correlation function yields the following lineshape:

$$I(\omega) = C[(\omega - \omega_0 - \delta\omega)^2 + \Gamma^2]^{-1}. \tag{15}$$

Where C is a normalization constant and $\delta\omega$ and Γ are, respectively, the lineshift and linewidth of the transition (frequency ω_0). With the basic approximations mentioned above, the width Γ of the $i \leftrightarrow f$ transition can be related to the *total* dephasing time, T_2, by (see Zewail 1980):

$$\Gamma = \frac{1}{T_2} = \frac{1}{2}\left(\frac{1}{T_{1i}} + \frac{1}{T_{1f}}\right) + \frac{1}{T_2'}, \tag{16}$$

where $1/T_{1i}$ and $1/T_{1f}$ are the decay rates of the initial and final states while T_2' is the rate of *pure* dephasing. The T_1 (population relaxation) and T_2' (phase relaxation) parameters can be expressed to lowest order by the following thermal averages (Jones and Zewail 1978):

$$\frac{1}{T_{1i}} = \frac{2\pi}{\hbar} \sum_{\substack{p,p' \\ i \neq i'}} W_p |\langle i'p'|V_{ep}|ip\rangle|^2 \delta(E_{i'p'} - E_{ip}), \tag{17a}$$

$$\frac{1}{T_2'} = \frac{\pi}{\hbar} \sum_{p,p'} W_p |\langle ip'|V_{ep}|ip\rangle - \langle fp'|V_{ep}|fp\rangle|^2 \delta(E_p - E_{p'}), \tag{17b}$$

where p and p' denote the phonon states, and ΣW_p contains the phonon density of states and thermal distribution for a given temperature. The operator V_{ep} describes the interaction between the localized exciton and the lattice modes.

Physically, the equations above say that the two level *homogeneous* transition width has two contributions: the "pure elastic" (T_2') and "inelastic" (T_1) scattering. For the elastic processes, the molecular or excitonic state does not change as a result of interaction with the phonons. An inelastic T_1 process, on the other hand, changes the state-of-affairs of the localized exciton from $i \leftrightarrow f$, thus leading to a lifetime broadening.

If the localized exciton is in the triplet state, then i and f could be optical or spin levels. Due to the long lifetime of the triplet state, the T_1 contribution to the optical transition linewidth is relatively minor. Similarly, for triplet state spin resonance transitions at low temperatures, spin–lattice relaxation is a slow process giving a small contribution to spin transition linewidths. Thus, one has the situation where the pure dephasing term, T_2', can be the dominant contributor to the homogeneous lineshape, particularly at higher temperatures. Very recently, one of the authors (Zewail 1979) discussed the relationship between linewidths (by pure dephasing) of optical and spin transitions using the intramolecular spin–orbit coupling mechanism (SOC). The relationship depends on the extent to which the singlet, triplet, and ground states involved in the transition are coupled by SOC. Finally, eqs. (17) contain the temperature-dependence of dephasing in the ΣW_p. Elsewhere (Jones and Zewail 1978, Diestler and Zewail 1979), we have treated the *Raman*, the *Orbach*, and other types of temperature induced dephasing using eq. (17).

4.2. The SE dephasing

Parallel to the treatment of sect. 4.1, one obtains the following expression for the SE lineshape:

$$I(\omega) = \frac{\mu^2 \pi \Gamma}{(\omega + \beta/\hbar)^2 + (\Gamma/2)^2}, \tag{18}$$

where $\Gamma = \gamma_0 + \gamma_1$ and μ is the transition moment matrix element (the two molecules are assumed to be identical, resonant and translationally equivalent). In other words, one Lorentzian lineshape for the allowed (+ state) optical transition should be observed with a linewidth determined by the sum of *both* fluctuation parameters. Several authors have dealt with the detailed nature of this model and other models (Haken 1968, Silbey 1976, Kenkre and Knox 1976) of exciton–phonon coupling (Burland and Zewail 1979). The important point here is that this type of dephasing is *homogeneous*.

Experimentally, one must establish the homogeneity of the line before extracting γ_0 and γ_1. Then γ_0 and γ_1 can be related to theoretical models which specify the nature of the coupling Hamiltonian V_{ep} (single phonon, Raman, Orbach, etc.), the phonon spectrum and lattice temperature. In sects. 7 and 8 we will discuss some spectroscopic methods used to determine Γ.

5. *Inhomogeneous broadenings and dephasings*

Recalling the discussion of sect. 3.2, we have emphasized that due to static or crystal field effects, molecule A and B might have different energies in the lattice. It is also possible that (say, due to isotopic composition) that the A molecules themselves have intrinsically different excitation energies. This inhomogeneous distribution of molecular excitation energies leads to broadening of the optical or magnetic resonance and has little to do with the dynamics belonging to the fluctuation parameters γ_0 and γ_1.

The problem of disentangling the dynamics from the observed spectra is now complicated by inhomogeneous broadening. Furthermore, there may be differences in the homogeneous width of the different molecules or packets within the inhomogeneous envelope, say, due to energy transfer or differing strain fields (Avouris et al. 1977). However, one can obtain quantitative (or nearly so) measurements of spectral and spatial transfer rates (T_1-type process) using time-resolved laser spectroscopy techniques. With the laser as a probe, one can determine the transfer rates as a function of position in the inhomogeneous line, thus determining the contribution of transport to the total dephasing.

The overall inhomogeneous width Γ_I can be related to a time constant T_2^*, which is a dephasing time, usually much shorter than the homogeneous T_2. The origin of T_2^* is complicated, but in solids it is typically assumed to be due to *random* (thus, Gaussian) distributions of excitation energies caused by strain fields, dipolar interactions, etc.

In analogy to homogeneous broadening, one might ask if the inhomogeneous broadening of spin and optical transition of local and small excitons are related. Lemaistre and Zewail (1979) have considered the question and a brief explanation of their findings follows, the reader being referred to the original paper for details.

Consider a localized exciton in its triplet state. Because inhomogeneous broadening depends on the excitation energies and other unknown parameters, one might expect the magnitude of the broadening for the singlet transition to differ from that of the triplet or spin resonance transitions. Further, since one knows that spin–orbital coupling in the molecule is finite, all three broadenings may be interrelated. But the relation necessarily depends on the degree of correlation among the inhomogeneous envelopes. By correlation, we mean the following: if a molecule in the triplet state experiences an energy shift of δ_T from the mean excitation energy and proportionally experiences a shift δ_S in the singlet state, then perfect correlation exists. Conversely, if δ_S and δ_T vary in an irregular manner with respect to one another, then the inhomogeneous broadenings are uncorrelated.

In the case where there is no correlation, the inhomogeneous broadening of the spin resonance transition, Λ_{sp}, is related to those of the optical singlet and triplet transitions, Λ_S and Λ_T, by (Lemaistre and Zewail 1979):

$$\Lambda_{sp} = |b_0|^2(\Lambda_S^2 + \Lambda_T^2)^{1/2}. \tag{19}$$

Typically, in molecular crystals, $\Lambda_S \sim \Lambda_T \sim 2\,\mathrm{cm}^{-1}$, thus giving Λ_{sp} of order Mhz when the SOC parameter b_0 (the SOC matrix element divided by the energy difference between the singlet and triplet states) is $\sim 10^{-2}$. The simple picture presented above explains several experimental results (Lemaistre and Zewail 1979) including recent optically detected magnetic resonance experiments (Tinti 1979) on NO_2^- solids.

6. The large exciton limit and localization

6.1. Homogeneous broadenings and coherent transport

In the ideal crystal limit, the Hamiltonian describing the stationary states of the crystal in the rigid lattice approximation is (Davydov 1971):

$$\mathcal{H} = \sum_n \epsilon_n a_n^+ a_n + \sum_{n,m} V_{nm} a_n^+ a_m, \tag{20a}$$

$$\mathcal{H} = E_0 + \sum_k E(\boldsymbol{k}) a_k^+ a_k. \tag{20b}$$

Where n, m label the different lattice sites, and $\boldsymbol{k}$ is the wavevector or quasi-momentum vector. There are two important differences between the LE and SE case. First, instead of *plus* and *minus* states of the SE case, one now has in the LE case N states indexed according to the quantum number $\boldsymbol{k}$. For the 1-D chain the dispersion relation, as noted in eq. (1), is $E(\boldsymbol{k}) = E_0 + 2\beta \cos \boldsymbol{k}c$. These *band states* are central to the concept of coherent transport, from which one can extract a group velocity for the exciton:

$$V_g(\boldsymbol{k}) = \frac{1}{\hbar}\nabla E(\boldsymbol{k}) = -\frac{2\beta c}{\hbar}\sin \boldsymbol{k}c, \tag{21}$$

and a density of states,

$$\rho(E) = [\nabla E(\boldsymbol{k})]^{-1}. \tag{22}$$

The second major difference between the LE and SE is the larger number of communication channels for dephasing in the LE, due in part to physical processes that do not exist in the SE case. Each $\boldsymbol{k}$ state has a homogeneous broadening, T_{2k}. The functional dependence of T_{2k} on $\boldsymbol{k}$ has not yet been measured accurately for any crystal, though such measurements are currently underway in several laboratories. It would be extremely interesting to find the $\boldsymbol{k}$ dependence of T_2 since this would serve a fundamental test for the coherence and transport models so far proposed. For instance, if the band picture is correct, one might expect that those $\boldsymbol{k}$ states with the highest group velocity ($\boldsymbol{k} = \pi/2c$) would encounter the largest number of scattering potentials and have possibly the shortest T_{2k}. On the other hand, it is arguable that those states at the boundary of the Brillouin zone ($k = \pi/c$) would be most incoherent due to Umklapp scattering processes. More on the dephasing of $\boldsymbol{k}$ states will be discussed later.

The formation of the exciton band produces unique inhomogeneous broadenings. In succinct terms, under the envelope of $\rho(E)$ vs E there are different $\boldsymbol{k}$ states which will have different singlet–triplet energy gaps (see eq. (1)). Then in much the same manner as the localized exciton case in sect. 5, each $\boldsymbol{k}$ subgroup will have its own inhomogeneous broadening. Spin and orbital inhomogeneities could be again related by SOC.

One way of modeling the homogeneous broadening of the LE is to let ϵ_n and V_{nm} fluctuate according to Gaussian–Markov statistics, as in the SE case, sect. 4.2. Thus, one replaces ϵ_n by $\epsilon_n(t)$ and V_{nm} by $V_{nm}(t)$. This problem has been dealt with by many groups; those of Haken (1968), Silbey (1976), Kenkre and Knox (1975, 1976) and others. It should be kept in mind that in the LE case, the problem is complicated again due to the large number of states. So, in effect, the fluctuations may be a combination of single and "multistate" correlations for the N level system. In a simple way of thinking, if for all $\boldsymbol{k}$ one has $T_{2k}^{-1} \ll \beta$, then one has the coherent limit. If the reverse is true, the incoherent limit prevails.

6.2. Inhomogeneous broadening and exciton localization; percolation vs. the Anderson transition

6.2.1. The impurity band

Frenkel–Davydov excitons can be localized by different means. One way is to use traps which act as a sink for the host excitation. However, at high enough dopant concentrations, intertrap or impurity "band" energy transfer occurs. Here we use the word "band" in a loose sense, since due to the disorder in the doped crystal, $\boldsymbol{k}$ is no longer a good quantum number. Consequently, several interesting questions can be raised. First, what role does disorder have on the spatial and spectral transport of optical excitation? Second, what is the explicit dependence of transfer efficiency on the impurity concentration, and in a more microscopic sense, what is the distance dependence? Third, what is the role of homogeneous and inhomogeneous broadenings on the transport? Finally, can one describe the energy transfer in these disordered systems using percolation theory, Anderson theory or neither?

Before discussing any aspect of the experimental work, we shall digress briefly to introduce the different concepts relevant to the impurity band transport in these crystals. In a broad sense, all the models predict that at some critical dopant concentration, energy transfer amongst the impurities will become efficient, resulting perhaps in delocalized or spatially extended states.

6.2.2. Anderson theory of localization

In 1958, Anderson treated the effect of diagonal disorder on electron mobility in solids. The Anderson model assumes a 3-D system, diagonal disorder with rectangular distribution of width W, and impurity–impurity coupling that falls off faster than r^{-3} and that $T = 0$ K. It is the ratio of the inhomogeneous broadening to the impurity bandwidths which determines whether or not the states at the band center will be *local* or *extended.* In 3-D, a sharp transition from localized to extended states, termed by Mott (1974, 1976) the Anderson Transition (AT), is predicted. In a perfectly 1-D disordered system, all the states are theoretically predicted to be local (Berezinskii 1973) and therefore, the AT is absent.

In quasi-1-D and 2-D systems the existence of the AT is still unsettled. Abrahams et al. (1979) argue that a 2-D system with diagonal disorder should always be localized at 0 K, while Lee's (1979) calculations suggest that it need not be.

According to Anderson, the impurity bandwidth (which is a joint function of the Hamiltonian coupling the traps and number density of the dopant) at which the AT takes place is given by

$$B_c = W/\alpha, \tag{23}$$

where α is a numerical constant and B_c is the *critical* bandwidth. The bandwidth can, in general, be written as

$$B = 2JZ, \tag{24}$$

where Z is the coordination number and J is the impurity–impurity coupling matrix element: not to be confused with β_{ij}, the nearest neighbor matrix element (see fig. 4). Thus, from eq. (23) one has

$$B \ll W/\alpha, \quad \text{(localized states)}$$

$$B \gg W/\alpha, \quad \text{(extended or partially extended states).} \tag{25}$$

The reason for using the words partially extended is that *all* the states in the band may not be extended, depending on the dimensionality and magnitude of B, W and α (see fig. 5). Recent calculations show that, at least in 3-D, off-diagonal disorder per se cannot give rise to an AT at the

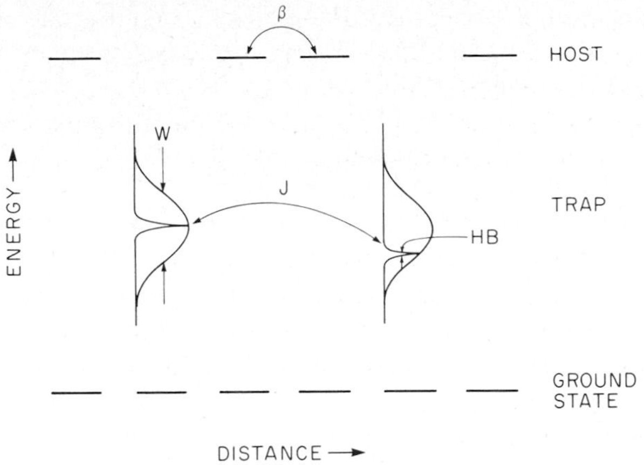

Fig. 4. A representation of a 1-D disordered chain as an ensemble of two-level systems. W = inhomogeneous broadening, J = trap-to-trap (non-nearest neighbor) coupling, HB = homogeneous broadening, β = nearest neighbor coupling.

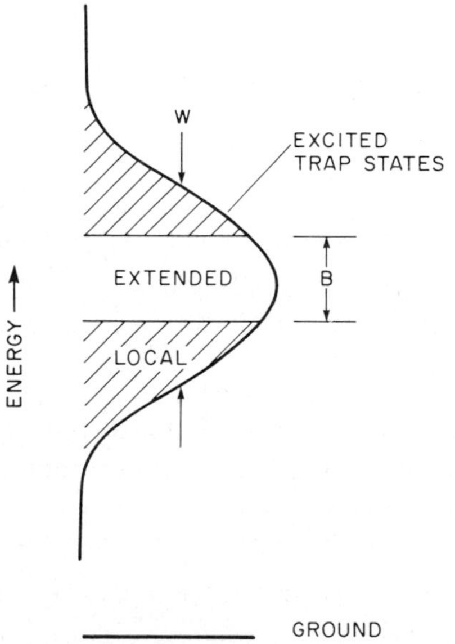

Fig. 5. Representation of important variables in the Anderson theory. Extended trap states exist if the impurity bandwidth, B, is at least W/α, where W is the inhomogeneous broadening and α is a constant.

band center, though it can contribute to the degree of localization in the tails (Klafter and Jortner 1979). Experimentally, this has not been quantitatively tested (Odagaki 1980).

Klafter and Jortner (1977, 1979) have extended the treatment to account for both diagonal *and* off-diagonal disorder. Theirs is still a low-temperature theory. Several important conclusions can be summarized:

(1) Diagonal disorder will lead to AT in organic crystals with finite inhomogeneous broadening.

(2) Off-diagonal disorder will not erode the AT, but rather shift the threshold and hence the magnitude of the critical concentration. For an exactly soluble model of a Lorentzian distribution for the impurity–impurity coupling, J and W, the AT model applies when $W/\sigma_J > 3Z$, where σ_J is the width of the J distribution.

(3) Using a continuum model for averaging the separation r of impurities in 2-D systems, they obtained the following value for J:

$$J = \beta \exp\{-a[(r/d) - 1]\}, \quad a = \ln(\Delta/\beta), \tag{26}$$

where d is the lattice constant and β is again the nearest neighbor interaction matrix element, and Δ is the trap energy depth from the host energy.

(4) The branching ratio, f, for light emission from the low-energy trap (acceptor) is given in terms of the trapping rate constant k_t and the lifetime τ of the donor:

$$f(C) = k_t(k_t + \tau^{-1})^{-1}, \tag{27}$$

where C is the donor concentration. The important point about $f(C)$ is that $f(C) = 1$, when $C = C_c$ (the critical concentration); this implies $k_t\tau \gg 1$. Thus, in this limit the threshold is independent of acceptor concentration of the energy sink and of the lifetime τ. An example will be the case of triplet impurity bands. On the other hand, if $f \ll 1$, i.e., $k_t\tau \ll 1$, then competition between excitation trapping and lifetime will take place and the so-called kinetic regime will be recovered. An example for this case is the singlet energy transfer which exhibits thresholds at much higher concentrations.

It is important to remember, in comparing theory with experiment, that Anderson's "localized" states extend over an increasing number of lattice sites as the AT is approached from below (Mott 1974, 1976). Thus efficient transfer to traps can occur even for states which are localized in the Anderson sense.

6.2.3. *Percolation theory*

The site percolation problem was discussed in the classical paper by Broadbent and Hammersley in 1957. A summary of percolation theory can be found in the review articles of Kirkpatrick (1973), Shante and Kirkpatrick (1971). Recently, Colson et al. (1977) have applied the method to exciton percolation in benzene, a lattice with an anisotropic 3-D transfer topology. The following can be summarized:

(1) If f_d is the fraction of donors, f_a is the fraction of acceptors, and P_T is the probability that an excited donor will transfer its energy to an acceptor, then $\partial P_T/\partial f_d$ will have a maximum at the critical concentration, f_d^c in ideal situations.

(2) Percolation theorists therefore relate the transport threshold to the number of sites visited by the excitation (in analogy with the original application of percolation – fluid flow). As pointed out by Colson et al. (1977), percolation theory can only give lower limits for these number due to the importance of "non-ideal" effects.

(3) Classical percolation theory cannot be applied to cases where tunneling prevails.

6.2.4. Application of Anderson and percolation theories to molecular solids

In recent work, Kopelman et al. (1977), Colson et al. (1977), and Smith et al. (1977, 1980) have shown experimentally that disordered naphthalene, benzene and phenazine exhibit an apparently critical dopant concentration for energy transfer amongst the impurities. Kopelman et al. (1977) studied the 2-D system naphthalene-D_8 (host), naphthalene-H_8 donor (trap) and betamethylnaphthalene which acted as a low-energy acceptor ("supertrap") at 1.8 K. They observed an abrupt change in the phosphorescence intensity ratio $I_d/(I_d+I_a)$ as a function of the donor concentration f_d, the "critical" value of f_d being a strong function of acceptor concentration. Kopelman and his coworkers conclude that Anderson's theory cannot explain the data and that percolation theory is more appropriate. They have used dynamic, static, and site percolation terms to explain the data and to obtain an exciton *coherence* (Kopelman et al. 1977). Recent data on naphthalene (Ahlgren and Kopelman 1980, Brown et al. 1981) show that the transfer rate is temperature dependent, and the excitons may well be localized, regardless of concentration. Thus comparison with a $T=0$ theory, such as KJ, must be done carefully.

Colson and his group (1977) have studied benzene isotopically crystals at 1.8 K, both in the triplet and singlet states. Their acceptor was the chemically distinct species, pyrazine. This is a 3-D system. Energy transfer thresholds were found at $f_d=2.8\%$ for the triplet state and 40% for the singlet state. The thresholds were discussed in terms of percolation and of Anderson localization, and it was pointed out that the exciton lifetime is a key factor in determining the difference in trap concentration for singlet and triplet percolation thresholds. The shorter singlet lifetime limits the energy transfer range despite the longer range multipolar coupling compared to the short range exchange coupling in the triplet state.

Zewail and his group studied the two-component system of phenazine-H_8 in phenazine-D_8 which has H_8 monomers and dimers in a D_8 host. For concentrations less than 5%, the number of trimers is negligible and the dimer alone serves as a trap (acceptor) for monomer (donor) excitation. The ratio of the dimer to monomer phosphorescence intensity as a function of H_8 concentration showed a threshold at ~5% *only* at low temperatures (1.13–1.3 K). This was interpreted as a transport threshold in the context of the Klafter–Jortner (KJ) model (1977) which was discussed above (see sect. 8).

7. *Experimental studies; quasi-one-dimensional excitons*

7.1. 1-D materials

Energy transfer in one dimension is extremely topical and has generated a great deal of theoretical and experimental work. This is sensible since few (if any) problems of current interest in condensed matter coherence and transport become simpler in higher dimensionalities. Only in the past several years have experiments and analytic theory (Economou and Cohen 1971, Bernasconi et al. 1978, Alexander and Bernasconi 1981) been able to penetrate into the microscopic aspects of the motion.

There are several quasi-one-dimensional systems, including organic* and inorganic (e.g., $Tb(OH)_2$) (Cone and Meltzer 1975, Meltzer and Cone 1976) in molecular crystals. The best characterized organic crystals possessing 1-D excitons are 1,2,4,5-tetrachlorobenzene and 1,4-dibromonaphthalene (DBN). In the coming sections we shall describe the experimental measurements on DBN used to determine the dimensionality, the exciton–phonon scattering rates and the effect of inhomogeneous broadening on exciton localization.

7.2. DBN crystal structure and resonance interactions

DBN ($C_{10}Br_2H_6$) crystallizes at room temperature with a $P2_1/a$ space group and eight molecules per unit cell. The crystallographic parameters are $a = 27.45$, $b = 16.62$ and $c = 4.09$ Å with $\beta = 91°51'$. The most important structural feature is that the molecules stack along the shortest crystallographic axis (c) with the planes of their rings parallel, thus favoring good

* (Francis and Harris 1971, Hochstrasser and Whiteman 1972, Schmidberger and Wolf 1972, 1974, 1975, Hochstrasser and Zewail 1974, Zewail 1974, 1975, Zewail and Harris 1974, 1975).

overlap of the π-electron systems. Molecules in a given chain are translationally equivalent. The dimensionality of DBN was first established using optical spectroscopy (Hochstrasser and Whiteman 1972, Hochstrasser and Zewail 1974). These experiments indicated that the Davydov splitting (i.e. the interaction between translationally inequivalent molecules in the lattice) was small, and that the in-chain interaction is $6.2\,\mathrm{cm}^{-1}$. Magnetic resonance studies have confirmed that the cross-chain coupling matrix element is at least 100 times less than the in-chain matrix element β_c; hence $\beta_c = -6.2\,\mathrm{cm}^{-1}$, $|\beta_{\text{cross-chain}}| \leq 0.05\,\mathrm{cm}^{-1}$. It is also known from the optical spectra that (a) the dispersion follows eq. (1) with a negative β (importantly, this places $\boldsymbol{k} = 0$ at the bottom of the band) and (b) the density of states is essentially that of a 1-D system. Further, with good oscillator strength for the triplet transition, a lifetime in the millisecond range and large excitonic bandwidth, DBN is a paradigmatic experimental system.

7.3. *The SE of DBN*

When for example, 5% DBN-H_6 is doped into DBN-D_6, impurity n-mers are formed in a statistical manner. The protonated DBN molecules form traps $65\,\mathrm{cm}^{-1}$ below the deutero host band. Due to the nearest neighbor coupling of $6.2\,\mathrm{cm}^{-1}$, the proto dimer has two states, and the lowest dimer (+) state, forming a trap below the proto monomer excitation, is the emitting state.

There is little doubt that the (±) dimeric states of DBN are at least partially coherent. This is because the separation between the states is $12.4\,\mathrm{cm}^{-1}$ while the linewidth is $\sim 2\,\mathrm{cm}^{-1}$. A lower limit to the degree of coherence is η_c = splitting/linewidth; thus $\eta_c \gtrsim 6$, for $T \lesssim 4\,\mathrm{K}$, but may be less at higher T.

Because the two molecules are translationally equivalent, the molecular transition moments are opposite in the dimer minus state and parallel in the plus state. Hence, if the states are coherent, no emission will be observed to the ground state from the minus state and the dimer plus state should have twice the radiative rate of the monomer (see fig. 6).

As well as the electronic energy being evenly distributed in the dimer, there is evidence that for some at least of the molecular modes, the vibrational as well as the electronic excitation is "shared" in the dimer. This is consistent with known vibrational excitons in DBN which play a central role in the optical spectroscopy discussed in sect. 7.4.

Magnetic resonance spectroscopy yields information on dimeric coherence on a longer time scale than the optical experiments, see the review of Burland and Zewail (1979).

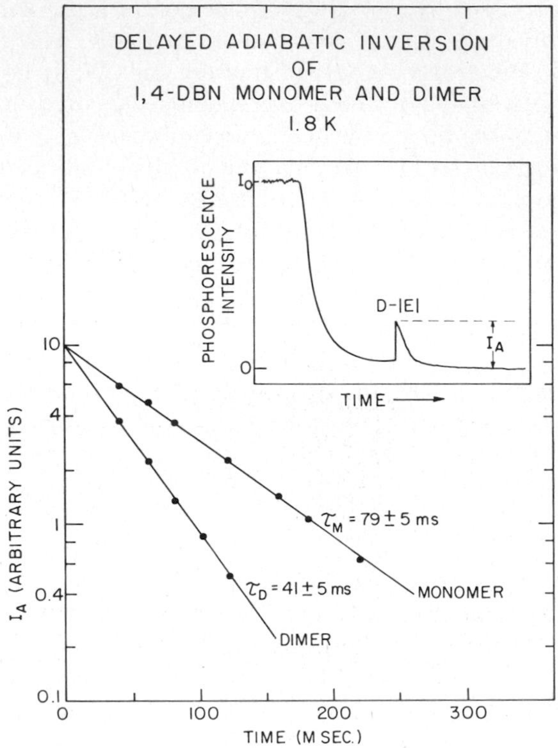

Fig. 6. Data (Zewail et al. 1979) for the determination of the DBN monomer and dimer intrinsic spin states lifetimes by delayed adiabatic spin inversion in an Optically Detected Magnetic Resonance experiment (Schmidt et al. 1969). Note that for the D − |E| spin transition, the long lifetime differs by a factor of two within the experimental error, as expected. The long-lived spin state is X, where X is the out-of-plane axis and Z is the C_2 symmetry axis in a C_{2v} point group symmetry. The other two spin states have the following lifetimes: $\tau_M^Z = 3.5 \pm 0.3$ ms, $\tau_M^Y = 5.2 \pm 0.3$ ms; $\tau_D^Z = 2 \pm 0.2$ ms, $\tau_D^Y = 4.5 \pm 0.3$ ms. The incomplete shortening of τ_D^Y is explainable by the difference in Franck–Condon factors of the (0, 0) and vibronic transitions.

7.4. Homogeneous broadening and exciton–phonon scattering in "pure" crystals

In DBN, there are eight molecules per unit cell that are grouped into two subunits. Two triplet exciton bands, separated by $50\,cm^{-1}$ have been identified (Castro and Hochstrasser 1967, 1968) and we shall refer to them as E1 and E2 (see fig. 7). In principle, there are 24 triplet bands (8 molecules × 3 spin sublevels per molecule) but only some branches are optically allowed because of symmetry. Both bands (separated by $50\,cm^{-1}$) have the same width, i.e., $4\beta = 24.8\,cm^{-1}$.

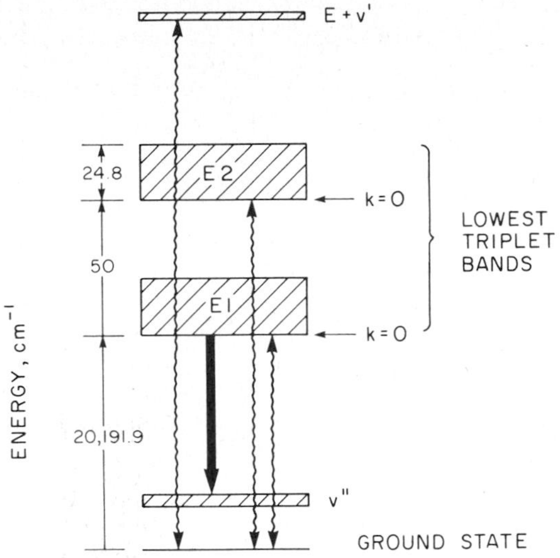

Fig. 7. Energy band diagram illustrating the position of some important states in the DBN H_6 crystal. The first triplet exciton (E1) is 50 cm^{-1} from the second triplet exciton (E2). The quantum number v is for molecular vibrational excitons and the $k = 0$ states lie at the bottom of the E1 and E2 bands. The bold arrow pointing down from E1 indicates the band-to-band transition observed in the DBN experiments. The lines with two arrow heads indicate the states pumped by the laser in our experiments.

When the pure DBN crystal is optically pumped, only the near $\boldsymbol{k} = 0$ region (hereon we shall simply call it $\boldsymbol{k} = 0$) is excited, due to the vanishingly small photon momentum. Similarly, emission to the zero vibrational level of the ground state occurs primarily from the $\boldsymbol{k} = 0$ state (neglecting phonon sidebands) resulting in one line. The absorption and emission (0, 0) spectra are displayed in fig. 8.

On the other hand, transitions to higher energy *vibrational* states in the ground state, which form narrow ($\lesssim 2$ cm^{-1}) vibrational excitons, result in an emission lineshape displaying the joint occupation and density of initial and final band states. Further, labeling the vibrational exciton wavevector as $\boldsymbol{q}$, these optical band-to-band transitions (BTBT) occur with the selection rule $\boldsymbol{k} - \boldsymbol{q} \cong 0$. Since the ground vibrational exciton has a width of the order of 2 cm^{-1} and the electronic triplet exciton band is 25 cm^{-1}, the joint density of states is essentially that of the triplet exciton. Additionally, at relatively low temperature, the high energy molecular vibrations are unpopulated, and thus, in steady state experiments, the (0, 0) emission will be ~ few cm^{-1} wide and emission to (say) the 527 cm^{-1} mode is 25 cm^{-1} wide with a

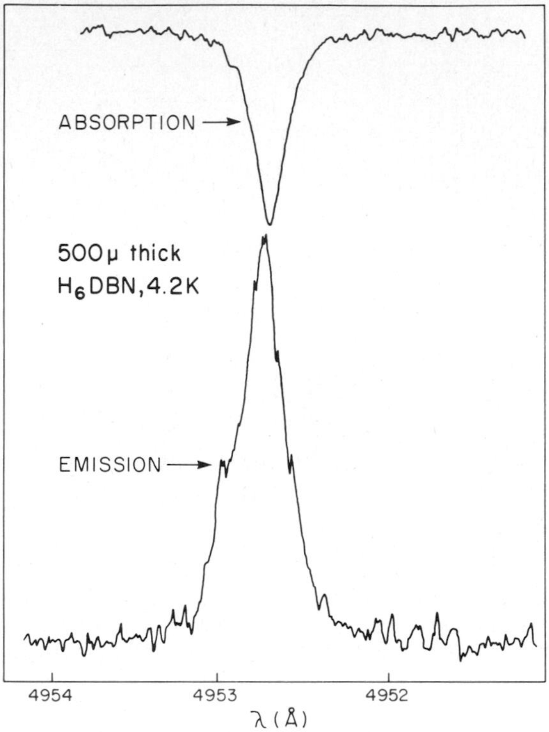

Fig. 8. The absorption and emission spectra of a thin crystal of nominally pure H_6 DBN at 4.2 K. There is a noticeable red shift and broadening of the emission with respect to the absorption as has been reported by other workers in the field. This is probably due to thermalization effects among the states. The absolute position is uncertain to $1\ \mathrm{cm}^{-1}$. The absorption width is slightly broader due to our resolution or method of crystal preparation.

lineshape, in principle, the same as the density of states given in eq. (22). However, due to homogeneous and inhomogeneous broadenings the $\boldsymbol{k}$ states have finite widths and are distributed about a mean energy such that a "double hump" spectrum is observed for the band-to-band transition (BTBT). One should note that the BTBT, for the simple 1-D cases, will generally produce a lineshape whose width is approximately equal to the difference in the dispersion of the two bands (vibrational and electronic) for a well-defined $\boldsymbol{k}$.

To optically measure the dynamics of scattering for the $\boldsymbol{k}$-states, we have recently used the technique of time-resolved laser line-narrowing spectroscopy on the BTBT. The experiments separate the inelastic and elastic scatterings and determine their contributions to the homogeneous width. The experiments can be described as follows: using a narrow-band laser, one pumps the triplet $\boldsymbol{k} = 0$ level of DBN to perform phosphores-

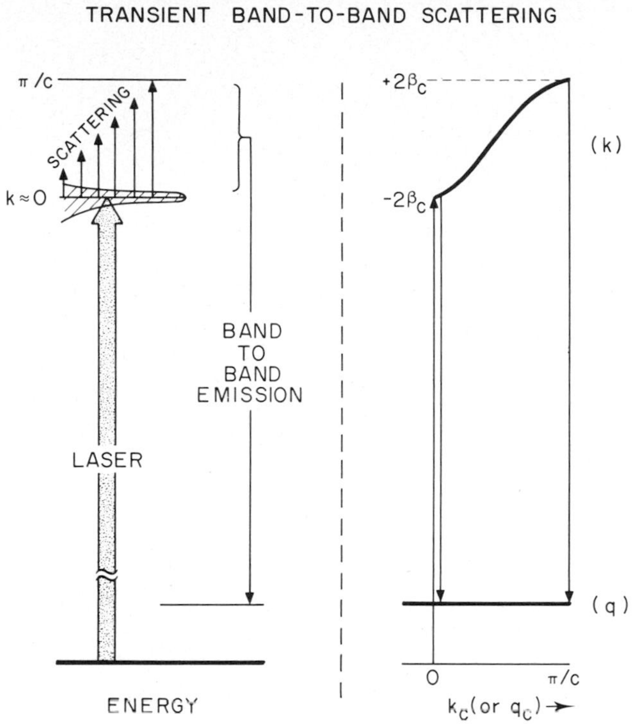

Fig. 9. A scheme for observing *transient* band-to-band optical transitions. A homogeneous exciton state with wavevector $\boldsymbol{k} = 0$ is prepared by a narrow band laser and subsequent scattering to $\boldsymbol{k} \neq 0$ states is detected by observing emission to the ground state vibrational exciton with wavevector $\boldsymbol{q}$.

cence line narrowing (PLN) following the laser excitation. The emission from the *entire* band to a ground state vibrational exciton (see fig. 9) is then detected as a function of delay time after the laser pulse (≤ 8 ns duration).

In principle, if the exciton scattering time is, say, 100 ns from $\boldsymbol{k} = 0$ to other $\boldsymbol{k}$ states, then detecting the BTBT at $t = 10$ ns after the pulse, one will only see the $\boldsymbol{k} = 0$ emission. On the other hand, observing the emission at $t = 100$ ns, or longer, one will see the entire band-to-band transition, as observed in steady state. Furthermore, if the pump laser bandwidth is narrow enough, the short time emission lineshape should give directly the total dephasing rate, T_2 of eq. (16). Knowing T_1, we can then separate T_2' processes. The experiments of Smith and Zewail (1979) have been successful in finding the scattering time this way. In these experiments one must separate real *emission* from *Raman*-type scattering. We have made a number of tests to insure the separation; one of the most convincing tests is that we observed PLN even when E2 was pumped by the laser.

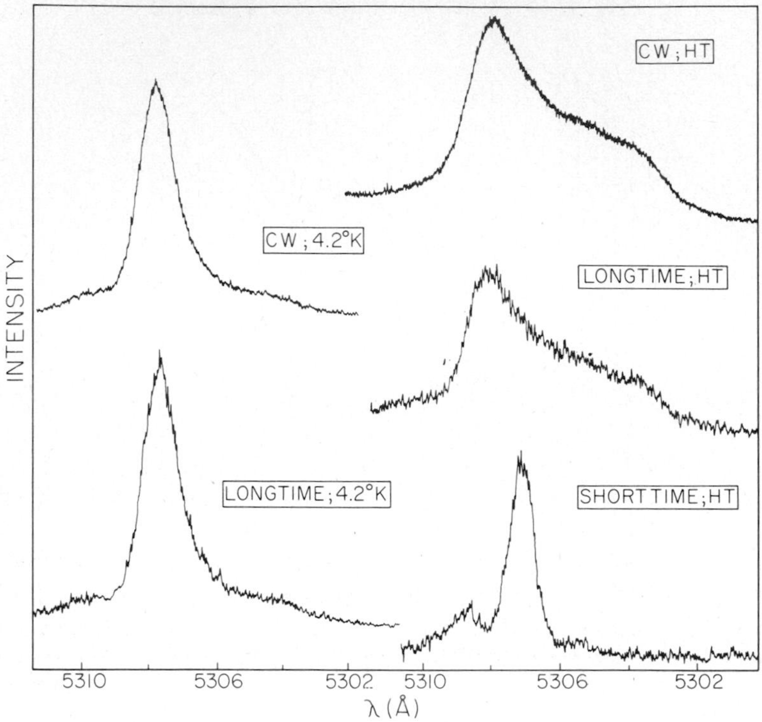

Fig. 10. Phosphorescence line narrowing and CW emission spectra (moderate resolution) of doped DBN at different temperatures and delay times. HT means high temperature (16.8 K). The most important thing to notice is that only the $\boldsymbol{k} = 0$ region is populated in the short time high temperature spectra, demonstrating the unequilibrated population among $\boldsymbol{k}$-states (see Smith and Zewail 1979).

Figure 10 shows the transient emission spectrum of DBN at different delay times. Following the time dependence of the signal we obtain a $\boldsymbol{k}$-to-$\boldsymbol{k}$ scattering time of the order of microseconds for DBN at 20 K. To obtain the *inelastic* dephasing time (see, eq. (17a)) we, at the moment, assume that the population is scattering from $\boldsymbol{k} = 0$ to π/c by a simple T_{1-} process and scattered back to $\boldsymbol{k} = 0$ by a T_{1+} process. We make a detailed balance by assuming $T_{1+}/T_{1-} = \exp -(\epsilon/KT)$ where ϵ is the energy difference. $T_{1\pm}$ are the T_1 of eq. (16) for the two channels $\boldsymbol{k} = 0$, $\boldsymbol{k} \neq 0$ and vice versa, e.g.,

$$\frac{1}{T_{1-}} = \sum_{k \neq 0} \frac{1}{T_{1k}}, \tag{28}$$

with a simplified "two-level" description. In other words, we have taken advantage of the near degeneracy of $\boldsymbol{k}$ states around $\boldsymbol{k} = \pi/c$. From the data of fig. 11, we obtained $T_{1\pm}$. Judging from the short time linewidth of

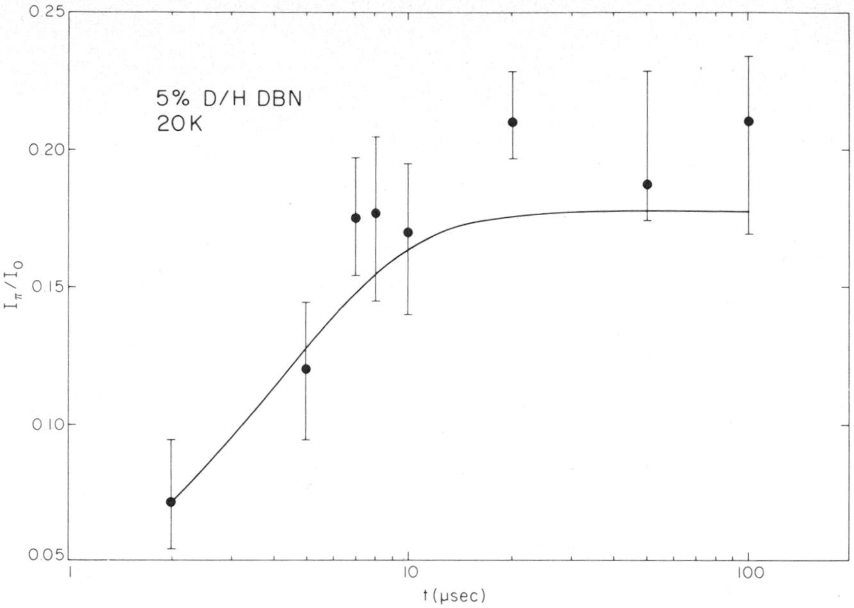

Fig. 11. The time-resolved emission intensity ratio as a function of the delay time (after the laser excitation) for $\boldsymbol{k} = 0$ and $\boldsymbol{k} = \pi/c$ for 5% D/H DBN at 20 K. The solid line is theoretical fit to a simple two level model with T_{1+} and $T_{1-} = 4$ and 22 microseconds (Smith et al., unpublished) for $\epsilon = 24\ \text{cm}^{-1}$.

the $\boldsymbol{k} = 0$ transition, we concluded that $T_2' \geqslant 5$ picoseconds (the inequality indicating an unknown inhomogeneous contribution to the linewidth).

To better our understanding of the physics of *interband* scattering, we have carried out PLN experiments on different initially prepared band states. (The reader should note that only *intraband* scattering has been discussed so far). Two interesting cases studied were (a) pumping the $\boldsymbol{k} = 0$ state of the triplet exciton in the *higher energy sublattice* (exciton E2 of fig. 7) and (b) pumping a vibrational exciton (see fig. 12) in the *excited* triplet state. When pumping the higher energy sublattice, relaxation is so rapid that, on our time scale, only emission from the lower sublattice is detectable. For case (a) we find that the time dependence of the PLN spectra is essentially the same whether the first or second exciton band is pumped. We conclude (tentatively) that a resonant $\boldsymbol{k} = 0$ phonon of approximately $50\ \text{cm}^{-1}$ is involved in the interband scattering. The exciton created in the higher energy sublattice with $\boldsymbol{k} = 0$ is selectively transferred to $\boldsymbol{k} = 0$ of the lowest band; without $\boldsymbol{k}$-to-$\boldsymbol{k}'$ scrambling! If the relaxation was a multiphonon process, the $\boldsymbol{k}$-to-$\boldsymbol{k}'$ scattering would not be $\Delta\boldsymbol{k}$ specific.

If one pumps a vibrational exciton in the triplet manifold (see fig. 7) no

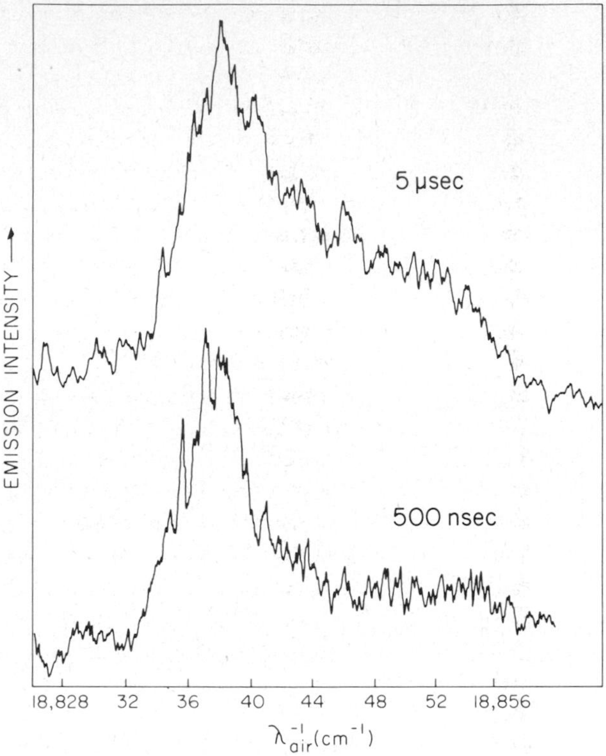

Fig. 12. The time dependence of the ($0.1350\ cm^{-1}$) band-to-band transition for 16% D/H DBN at 23 K pumping the $\mathbf{k} = 0$ state of the upper sublattice (E2). Within experimental error, the time-dependence is found to be essentially the same as pumping the lower sublattice.

large PLN is observed. That is, after rapid relaxation of the vibrational exciton, observing emission from the zero vibrational level of the E1 exciton at 500 ns after the laser pulse, all band states were occupied (within the signal to noise). This may indicate that multiphonon processes are involved which scramble the $\mathbf{k}$-to-$\mathbf{k}'$ scattering, or possibly that the $T_{1\mathbf{k}}$ of the *vibronic* exciton is so short that vibrational relaxation is from equilibrated vibrational band states. The latter is less probable because vibrational relaxation generally occurs on a picosecond time scale*.

The above results raise the following interesting question: Why is the $T_{1\mathbf{k}}$ slow? In an intuitive Fermi golden rule sense, one can see that for elastic and inelastic scattering, the number of available final states and suitable

* These experiments are still in progress.

phonons is dramatically increased for 2-D and 3-D versus 1-D*. Fox (1979) has recently considered this question from a group theoretical point of view.

From these experiments one infers that even at 20–30 K, the homogeneous broadening due to inelastic scattering of the exciton is only of the order of Mhz. At the same temperatures, the (0, 0) exciton line shape (steady state) is ~ few cm^{-1} wide. If the broadening is homogeneous, this implies that the prepared $\boldsymbol{k} = 0$ exciton packet dephases (T_2') in a pure sense in approximately 5 ps. Future theoretical treatments of exciton scattering (Davydov 1971, Silbey 1976) should attempt to separate the T_1 and T_2' processes.

7.5. *Impurity band transport: superexchange and "mobility" edges*

In the previous section, we discussed exciton motion in the nominally pure limit. In what follows, we discuss the dual: transport in the relatively dilute trap limit. At low temperature, all host excitation is trapped and transport properties are determined by energy transfer amongst the impurity clusters. The isotopically mixed DBN crystals are excellent for studying such energy transfer processes since the H_6 dimer forms a trap (acceptor) for monomer excitation and is presumably statistically distributed. That is, since the number of n-mers in a 1-D chain N long is $NC^n(1-C)^2$, C being the concentration, then the ratio of the number of dimers to monomers is always C. There is little tendency for the "acceptors" to cluster, making the experiments much easier to interpret.

We have probed impurity band transport by directly pumping the triplet state of proto-monomer (or dimer) with a pulsed laser, recording the monomer and dimer emission spectrum as a function of time after the pulse (Smith et al. 1979). By performing such experiments as a function of temperature, concentration and position of the laser in the inhomogeneous envelope, one can gain valuable information on the impurity–impurity coupling and the direct role of inhomogeneous broadening on spatial and spectral transport. Steady state experiments, on the other hand, gave us an overall view of the transport at very long times (see coming section).

For the case where the monomer is pumped, a typical set of time-resolved spectra are shown in fig. 13. As time increases, the dimer emission increases relative to the monomer. Figure 14 plots the time-dependent

* In recent unpublished experiments similar to ours, D. Hanson (private communication) found that there is no line-narrowing when the naphthalene singlet was pumped. This implies the scattering time is ≤10 ns. With the same apparatus they also observed PLN on DBN similar to ours (Smith and Zewail 1979). It should also be mentioned that recent work on TCB by Wolfrum et al. (1979) and Van Strien et al. (1980) has provided an exciton scattering time of ~μs.

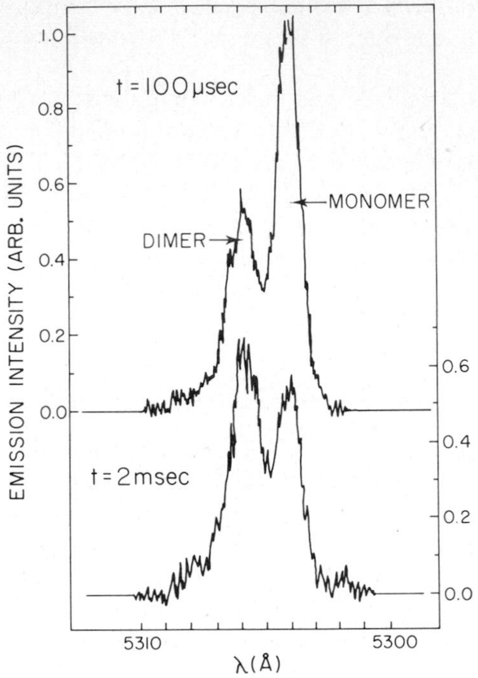

Fig. 13. Moderate resolution emission spectrum of 11.9% H/D DBN at 1.8 K for two delay times. One can clearly see that as time increases, population is transferred from the monomer to the dimer.

dimer-to-monomer phosphorescence intensity ratio (or monomer-to-dimer for pumping the dimer) for several samples at temperatures of 1.3, 1.8, and 4.2 K. At all three temperatures the 11.9 and 24.1% H_6 in D_6 samples exhibit time-dependent ratios, while in the lightly doped samples, time-dependence can be detected only at the highest temperature. In the lightly doped samples, the monomer *and* dimer lifetimes were found to be approximately 4 milliseconds and independent of temperature from 1.3 to 4.2 K.

The model used to interpret the data is as follows: if we neglect correlated motion, excitation migration among the monomers with the eventuality of being trapped at a dimer may be described as a random walk. In such a case, the rate at which distinct sites are sampled is (Soos and Powell 1972): $\omega_{MD} = C_D\dot{S}(t)$, where C_D is the fraction of dimers (and larger clusters which can also trap the monomer excitation) and $S(t)$ is the so-called sampling function (Montroll 1963). In the limit of many steps and uniform step time, $\dot{S}(t)$ has the asymptotic form in 1-D is, $\dot{S} \sim (2/\pi\tau t)^{1/2} + \cdots$ where τ is the mean hop time. The leading term in the transfer rate gives $\omega_{MD} = C_D(2/\pi\tau t)^{1/2}$.

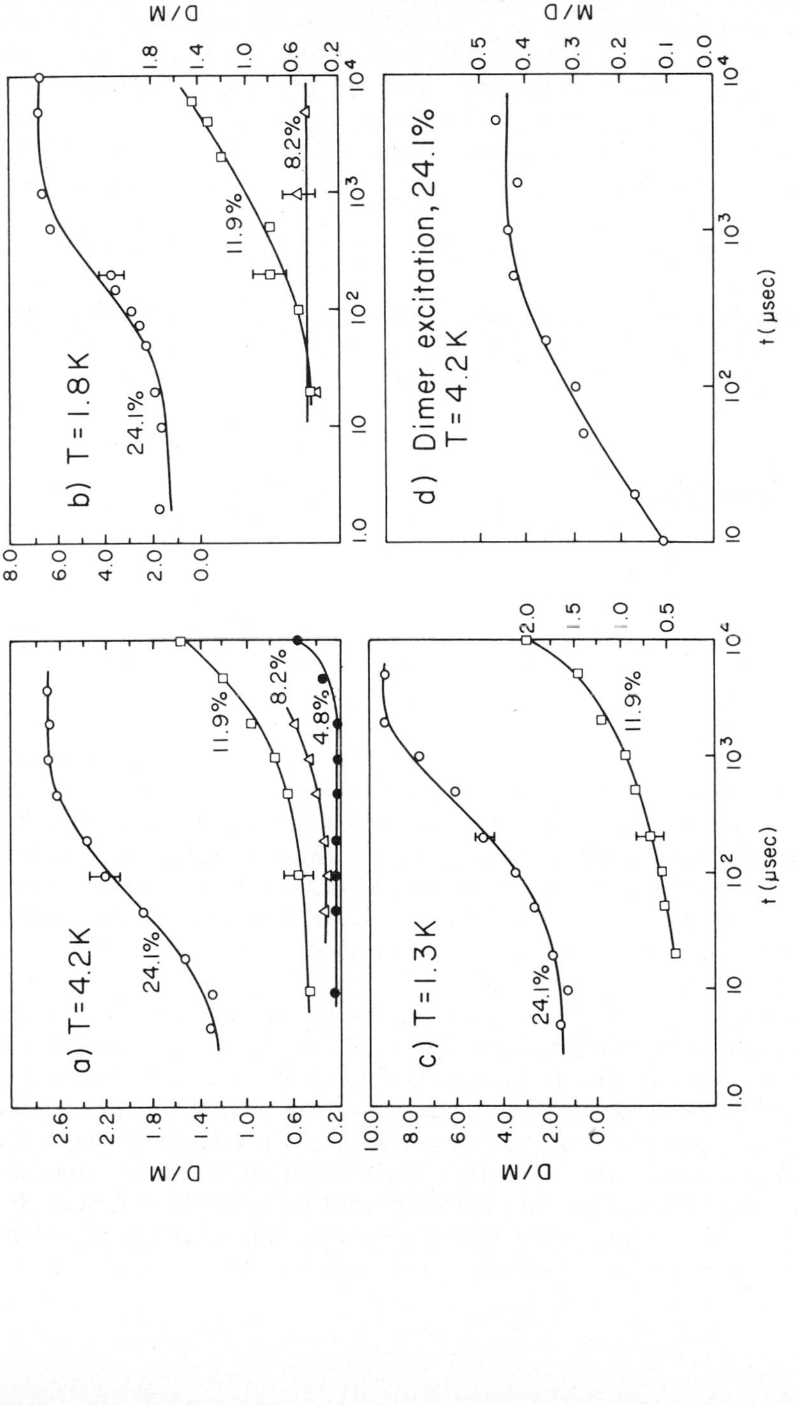

Fig. 14. Computer fit (solid lines) of the experimental D/M emission intensity ratio (squares, triangles and squares) vs. time for pumping the center of the monomer (a), (b) and (c), and the dimer (d).

Thus, using the sampling function to relate the microscopic transfer to the macroscopic observables (e.g., the dimer/monomer population ratio) one can write down a system of differential equations for the population (or probabilities) find the transient solution and fit it to the data. Smith et al. (1979) have done this, fitting the data to the solution by nonlinear regression both for time-dependent and for time-independent transfer rates. The results of the fits are the solid lines shown in fig. 14. From the fits it was discovered that the time-dependent forward transfer rate (M to D) predicted by the sampling function gives a better fit than a constant rate. Further, the concentration dependence of the energy transfer rate was found to be consistent with a Monte Carlo ensemble-averaged 1-D *superexchange* coupling.

Nieman and Robinson (1962) were the first to suggest that superexchange was the primary contributor to triplet trap-to-trap energy transfer in this type of organic crystal. However, there has not been a quantitative test of the effect. In a simplified form, the superexchange coupling matrix element for two isolated impurities in a perfectly 1-D crystal is,

$$J = \beta(\beta/\Delta)^n \chi, \tag{29a}$$

$$\langle J \rangle = \beta\chi \langle (\beta/\Delta)^n \rangle, \tag{29b}$$

where β is the nearest neighbor exchange coupling (neighboring guest and host), Δ is the trap depth, n is the number of host molecules intervening between the two impurities, and χ is a guest–host Franck–Condon overlap constant.

From the ensemble-averaged J, denoted by $\langle J \rangle$, it is clear that there are two ways to test the transport on 1-D chains by this mechanism. Either we vary n and measure the rate of energy transfer, or for fixed n, vary Δ. These two experiments have been reported in the paper by Smith et al. (1979). In varying the concentration, n and the distribution of clusters changes. To calculate the rates of energy transfer, we computed the distribution of clusters and $\langle J \rangle$ using Monte Carlo methods. Within experimental error, the agreement between theory and experiments is quite good.

The second parameter, Δ, was varied by tuning the laser within the inhomogeneously broadened line of the monomers. This selects different effective trap depths and transfer rates.

For the 11.9% H_6 in D_6DBN sample (see fig. 15) the experimental transfer rate coefficient ω has an approximate Δ^{-2} trap depth dependence. Three things are significant about the Δ dependence. First, the experiments are consistent with and quantify the super-exchange effect, so long as the averaging over n is done correctly. Secondly, the transfer rate is a decreasing function of Δ and no "mobility edges" were observed. As

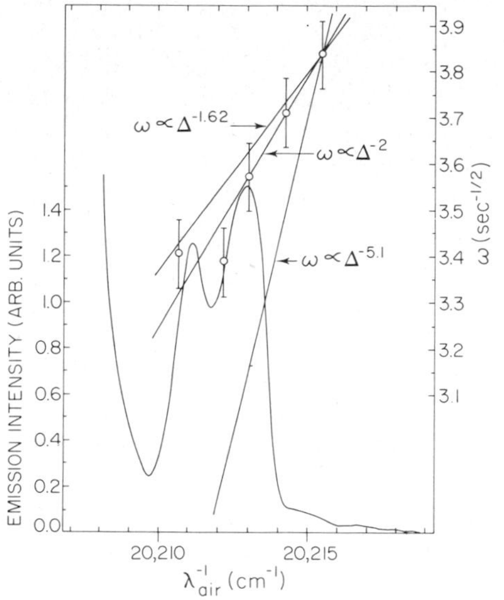

Fig. 15. The trap depth or Δ dependence of the energy transfer rate, ω. A high resolution spectrum is shown to indicate the laser position within the monomer inhomogeneous line. The peak at 20.211 cm^{-1} is probably a double monomer (see Hochstrasser and Zewail 1974). The Δ^{-2} dependence is the best fit to the data and appears linear due to the small range of Δ. The $\Delta^{-1.62}$ line is the predicted dependence from Monte Carlo computations of $\langle J\rangle$. The $\Delta^{-5.1}$ dependence is that predicted by the ensemble averaged $\langle n\rangle$ which is a poor fit as expected, since as discussed in the text, $\langle J\rangle$ cannot be accurately calculated using $\langle n\rangle$ in 1-D.

mentioned before, in perfectly 1-D systems all states are theoretically predicted to be localized (Berezinskii 1973), even with non-nearest neighbor interaction, though this has not yet been firmly established experimentally. On our experimental time scale DBN should be close to being perfectly 1-D and we do not expect abrupt changes in the transfer rate. Finally, our results clearly indicate the inhomogeneity of the monomer band.

8. *Quasi-two-dimensional systems; the phenazine crystal*

8.1. Phenazine crystalline structure and resonance interactions

Phenazine ($C_{10}N_2H_8$) crystals grown from the melt are of the monoclinic α-form with space group $P2_1/a$, two molecules per unit cell. At room temperature, the unit cell has the dimensions $a = 13.22$, $b = 5.061$, $c = 7.088$ Å, and $\beta = 109° 13'$. The most salient feature of the crystal structure

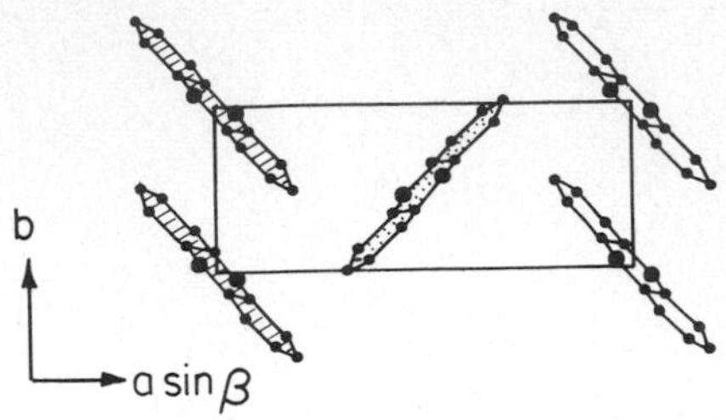

Fig. 16. Projection of the contents of a "unit" cell of phenazine onto the *ab* plane.

is chains of plane-parallel packed phenazine molecules at 45° with respect to the short axis, *b*. Neighboring chains in the *ab* plane have their long molecular axes perpendicular to one another (shown clearly in fig. 16). Intermolecular interactions are highly anisotropic for the $^3\pi\pi^*$ state, with the *b*-axis coupling (Zewail 1974, 1975) dominating at approximately 6 cm^{-1} and *ab* plane translationally inequivalent pair interaction energy (Clarke and Hochstrasser 1967) of 0.5 cm^{-1}, and coupling along the *c*-axis being immeasurably small by conventional optical techniques.

While phenazine thus has a quasi-1-D topology, excitation transfer between chains in the *ab* plane is fast on our timescale, so that phenazine should behave dynamically as a 2-D system like naphthalene.

8.2. Dependence of energy transfer on concentration and temperature: diagonal disorder and inhomogeneous broadening

Introducing phenazine-H_8 into D_8 forms trap clusters (monomer trap depth (Smith et al. 1977, 1980) is 22 cm^{-1} from the middle of the $\boldsymbol{k} = 0$ states; fig. 17) much as in the case of DBN, the major differences being the 2-D coupling and cluster statistics. Using broad band CW optical pumping, one can observe changes in the intercluster energy transfer rates by comparing the steady state populations of different impurity clusters (primarily the monomer and dimer) as a function of dopant concentration and sample temperature. Figures 18 and 19 present the experimentally measured trap phosphorescence intensities (energy resolved) as a function of impurity concentration and temperature. The data have been numerically fitted to the kinetic model and we shall interpret the results in this and coming sections.

The most conspicuous thing about fig. 19 is that at the lowest temperatures the monomer-to-dimer energy transfer shows an abrupt increase at about 4.5% donor (H_8) concentration C_M. A critical concentration of 4.5% for transport could be consistent with Klafter–Jortner application of the Anderson theory (see sect. 6.2). That is, the data indicate that the donor

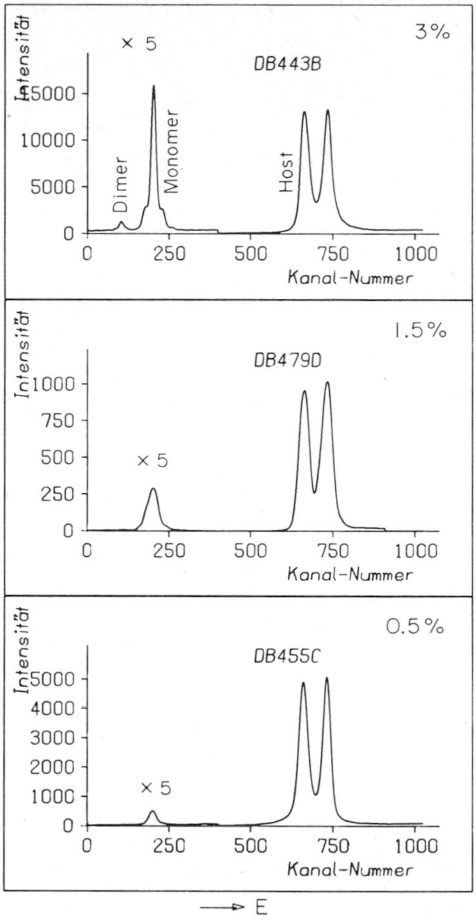

Fig. 17. The excitation spectra of isotopically-mixed phenazine crystals; 3%, 1.5% and 0.5% (Doberer and Port 1979). The doublet on the right is the Davydov $\boldsymbol{k} = 0$ states of the host (D_8). The monomer-dimer splitting is the same as that of fig. 18. The line shapes in the top spectrum are presumably because of the way the crystal was grown.

bandwidth at this concentration is sufficient to overcome some of the IB caused by diagonal disorder even at very low temperatures. In phenazine, the inhomogeneous broadening was found to be approximately 4 cm^{-1} using phosphorescence–microwave double resonance and using the known coupling parameters, a simple calculation of the impurity bandwidth yielded a value in the range of 0.1–1 cm^{-1} for $C_M \sim 4.5\%$. For more details of the calculation, see Smith et al. (1977, 1980).

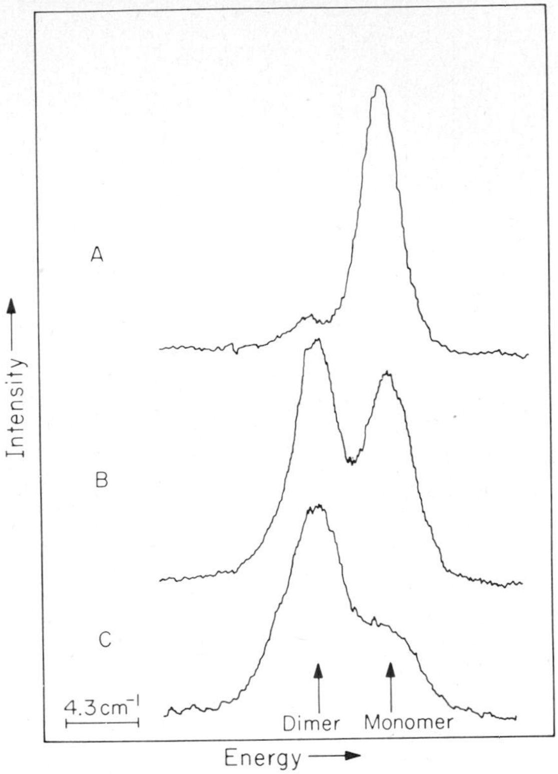

Fig. 18. The emission spectra of isotopically-mixed phenazine crystals as a function of H_8 guest concentration: A = 0.5%, B = 3.1% and C = 5.7% by weight. Note the increase in dimer intensity as C increases. The monomer dimer splitting is 4.3 cm^{-1}.

Raising the temperature, the "threshold" for energy transfer disappears (Smith et al. 1977, 1980). The KJ model does not apply, and to our knowledge, there is no theory of the Anderson transition at finite temperatures. We (Smith et al. 1977, 1980) have "bypassed" this problem by invoking a kinetic model for the flow of population among the monomer, dimer and host states. Blumen and Silbey (1979) obtained the rate equations for energy transfer among the monomers and dimers using the average transfer rates of Inokuti and Hirayama (1965) to take account of the random distribution of traps (for more details see Smith et al. (1977, 1980)). Such an "average" model cannot give true "critical" behavior or an AT. Nevertheless, we shall see that it can give a good account of the data.

Figure 20 present plots of the logarithm of the dimer-to-monomer (D/M) emission ratio as a function of temperature for several phenazine crystals.

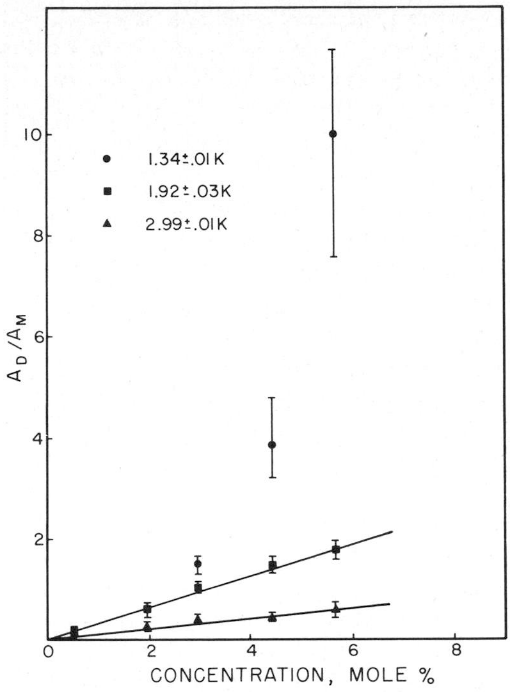

Fig. 19. The computer fit (integrated peak areas) dimer/monomer phosphorescence intensity ratio for phenazine as a function of proto trap concentration for 1.92 and 1.34 K. The data for 2.99 K are from peak heights.

Note that for most samples at high temperature (high temperature data for the 0.5% crystal are not available) a 4.4 cm^{-1} slope obtains (which is the monomer-dimer splitting) indicating complete trap-to-trap thermalization. Further note that for $C_M \leqslant 4.5\%$ there is a low temperature plateau showing that a substantial fraction of the monomer excitation cannot reach the dimer within its lifetime. For the heavily doped samples, 5.9 and 6.6%, no low temperature plateau is established at 1.3 K, though the kinetic model predicts that one should occur at still lower temperatures, corresponding to partial localization at 0 K.

Another interesting feature of fig. 20 is the negative slope only observed in the 0.5% crystal. It appears that, at this low concentration, monomer to dimer transfer is primarily via the host band, and therefore, freezes out at low temperatures.

Further support for this interpretation comes from the measurements of the monomer emission intensity as a function of temperature for a nominally pure D_8 sample (trace amounts of H_8). An approximate 18 cm^{-1}

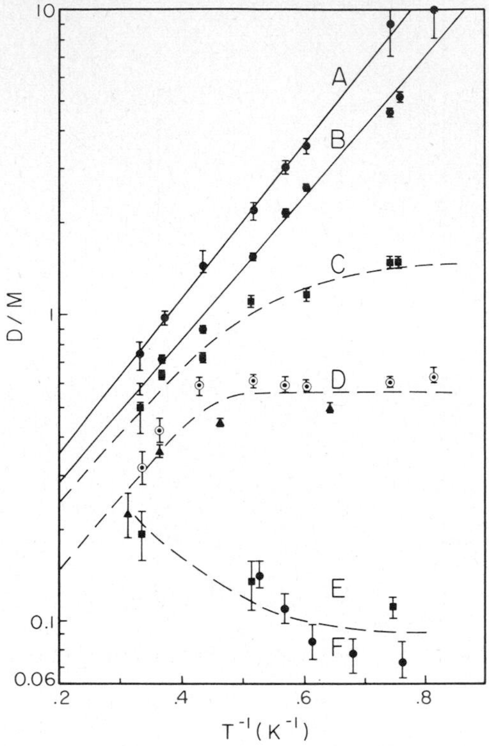

Fig. 20. Plot of log (D/M) vs. T^{-1} for (A) 5.65%, (B) 4.4%, (C) 3%, (D) 2% and (E, F) 0.5% H_8 in D_8 phenzaine. The curves E and F are for the same crystal, but different experiments. An asymptotic high temperature slope of 4.4 cm^{-1} is found for curves A through D.

activation energy for the very dilute H_8 monomer emission was found, corresponding to the monomer trap depth.

Figures 21 and 22 show the predictions of the kinetic model for the phenazine parameters. It is interesting to note how the complexion of the transport can change so dramatically over a relatively small temperature (1.3 to 4.2 K) and the concentration range (1 to 5%). This is not the case of the 1-D DBN system. Note that energy transfer amongst the monomer states cannot be separated from transfer through the host band if the temperature is high.

8.3. Dimensionality effects on "critical" behavior

In our kinetic model, we assumed that the concentration dependence of the transfer rate is given by $\beta(\beta/\Delta)^{\langle n \rangle}\chi$, rather than by eq. (29b). In 1-D the

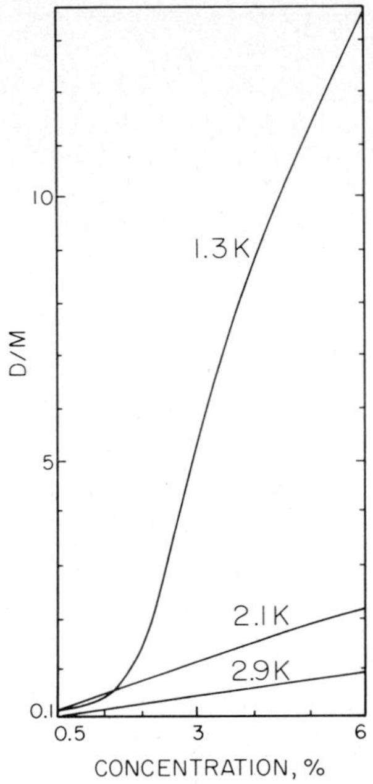

Fig. 21. Theoretical simulation of the data in fig. 19 using the rate equation model of Smith et al. (1977, 1980).

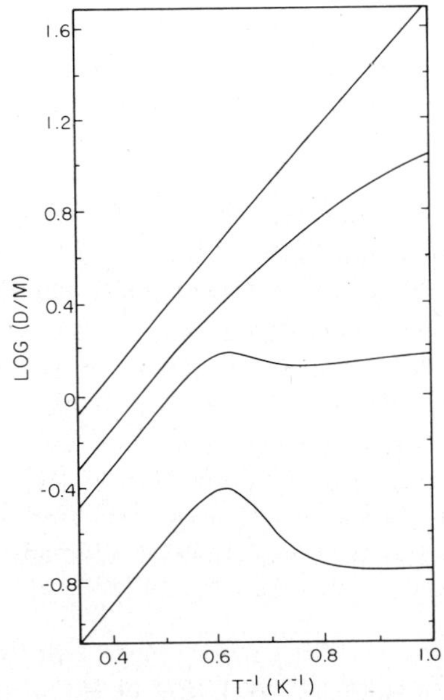

Fig. 22. Theoretical simulation of the data of fig. 20 using the rate equation model of Smith et al. (1977, 1980). Note the "knee" shaped curve for the 0.5% crystal.

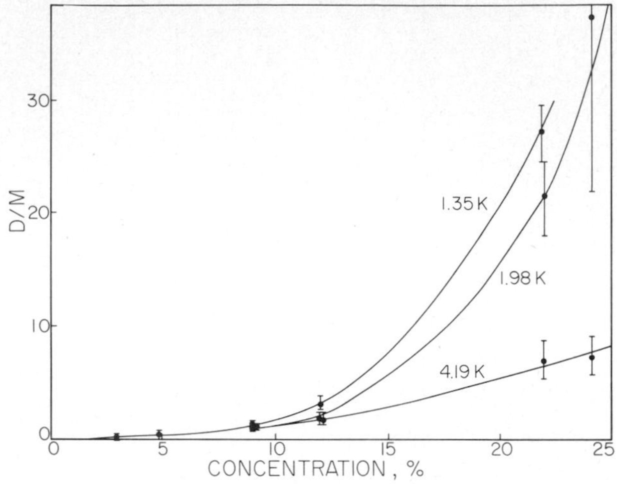

Fig. 23. The steady-state D/M ratio (determined by computer fit of the data) for DBN as a function of dopant concentration and temperature.

accuracy of this approximation is rather poor (see sect. 7.5). However, this is due to the pathological form of the 1-D continuum distribution, and using $\langle n \rangle$ in 2-D should be more accurate, especially in steady-state experiments. The continuum model explains the major features of the data well, and is as good as the other approximations made. Using

$$\langle n \rangle = \frac{0.5}{\sqrt{c}} - 1, \quad (2-\mathrm{D}), \tag{30}$$

where c is the impurity concentration, and letting χ be a parameter to account for the effect of multiple paths and Franck–Condon guest–host factors in a 2-D system, we have scaled the concentration dependence of the transfer rate, to obtain figs. 21 and 22. For the details of the modeling, the reader should consult Smith et al. (1977, 1980).

In analogy to the 2-D phenazine case, the steady state concentration and temperature dependent D/M ratio for 1-D DBN has been measured (see fig. 23). Most noticeable in the CW DBN experiments is that the "critical" concentration for transport is higher (ca. 14%) for DBN than phenazine. This is believed to be primarily a dimensionality effect (Smith et al. 1977, 1980), since among other things, the ratios (β/Δ) are comparable for DBN and phenazine.

To illustrate the effect of dimensionality and superexchange on the critical behavior, fig. 24 plots a simulation of the low-temperature phenazine case as well as for hypothetical 1-D and 3-D solids. To generate the 1-D

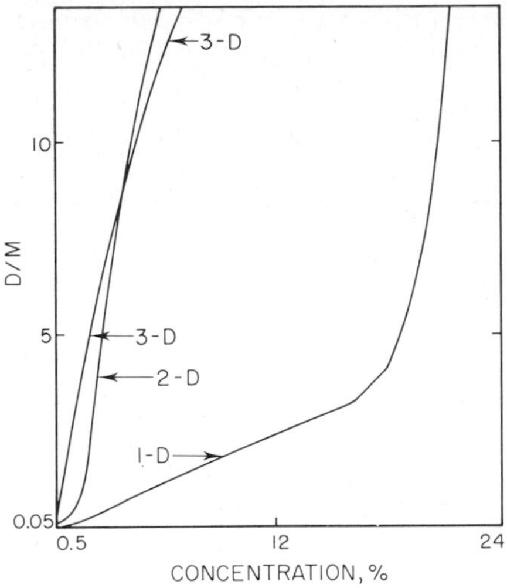

Fig. 24. The effect of dimensionality on the critical concentration for transport. The curves were generated using the communication rates found for the simulation of the 2-D phenazine data modifying only the superexchange exponents and cluster statistics.

and 3-D curves, we used the same parameter set as for the 2-D phenazine case, altering only the superexchange exponents and cluster statistics. Note the similarity between the 1-D simulation and the data of fig. 23.

9. Epilogue

In the course of the review, we have outlined some recent results from experiments and theory that are intended to determine the nature of exciton transport and coherence properties in ordered and disordered solids. We discussed some of the paramount problems in this field by choosing prototype *one*- and *two*-dimensional systems. Clearly, much remains to be explored. It is perhaps useful to end this chapter by briefly outlining the answers to the questions raised in the introduction:

(a) Conventional optical spectra do not usually reveal the dynamics because the optical transitions can be inhomogeneously broadened. The dynamics of exciton scattering by phonons leads to a homogeneous broadening that could be much smaller, especially at low temperatures, than the inhomogeneous broadening.

(b) By disorder, we mean both static and stochastic time-dependent changes in the energy of molecules in the solid. The former can lead to diagonal disorder that play a role in determining the inhomogeneous broadening. The latter, on the other hand, leads to homogeneous broadening, if the modulation of the energies is fast on the timescale of the experiment.

(c) Excitons are mobile in disordered systems if the coupling among the traps is large enough to cause transport on the time scale of the experiment. It should be kept in mind, however, that the *nature* of the transport is still not fully understood. Whether classical, quantum mechanical, coherent or incoherent modes of transport are operative is still an open question in the many systems studied.

(d) The experimental approaches we have discussed to probe dynamics involve CW lineshape analysis, magnetic resonance spectroscopy, emission techniques and laser line narrowing. In the text, we outlined the limitations and the new information which can be obtained, with specific examples from studies done on DBN and phenazine solids.

Finally, to be sure, many workers have contributed to the study of excitonic states and dynamics and it would be almost impossible to list the details of all their work here. We have chosen to pick a few illustrative cases and our only modest hope is that we have given the reader a flavor for the problems that must first be solved and the avenues that experimentalists and theorists are taking to solve those problems. (References to literature are given in the text).

Acknowledgments

This work was supported by grant number DMR77-19578 and DMR81-05034 from the National Science Foundation. We would like to thank U. Doberer and H. Port of Stuttgart University for providing us with one of the figures on phenazine. Finally, AHZ would like to thank F. Ali-OH for introducing new elements of thinking while the chapter was being written in December of 1979.

References

Abrahams, E., P.W. Anderson, D.C. Licciardello and R.V. Ramakrishnan, 1979, Phys. Rev. Lett. **42**, 673.
Ahlgren, D.C. and R. Kopelman, 1979, J. Chem. Phys. **70**, 3133; 1980, J. Chem. Phys. **73**, 1005.
Alexander, S. and J. Bernasconi (submitted J. Phys.).
Anderson, P.W., 1958, Phys. Rev. **109**, 1492.
Avouris, Ph., A. Campion and M.A. El-Sayed, 1977, in: Advances in laser spectroscopy, ed., A.H. Zewail, **113**, 57; ibid., 1977, J. Chem. Phys. **67**, 3397.
Berezinskii, V.L., 1973 Zh. Eksp. Teor. Fiz. **65**, 1251; Sov. Phys. JETP **38**, 620.
Bernasconi, J., S. Alexander and R. Orbach, 1978, Phys. Rev. Lett. **41**, 185; 1978, J. Physique, Colloque C6, Supp. au n° 8, p. C6-706.

Blumen, A. and R. Silbey, 1979, J. Chem. Phys. **70**, 3707.
Botter, B.J., C.J. Nonhof, J. Schmidt and J.H. van der Waals, 1976, Chem. Phys. Lett. **43**, 210.
Broadbent, S.R. and J.M. Hammersley, 1957, Proc. Camb. Phil. Soc. **53**, 629.
Brown, R., J. Lemaistre, J. Megel, Ph. Pec, F. Dupuy and Ph. Kottis, 1981, to be published.
Burland, D., and A.H. Zewail, 1979, Adv. Chem. Phys., eds., I. Prigogine and S. Rice (Wiley, New York) Vol. 40, p. 369.
Castro, G. and R.M. Hochstrasser, 1967, J. Chem. Phys. **47**, 1015; 1968, J. Chem. Phys. **48**, 637.
Clarke, R.H. and R. Hochstrasser, 1967, J. Chem. Phys. **47**, 1915.
Colson, S.D., S.M. George, T. Keyes and V. Vaida, 1977a, J. Chem. Phys. **67**, 4941.
Colson, S.D., R.E. Turner and V. Vaida, 1977b, J. Chem. Phys. **66**, 2187.
Cone, R.L. and R.S. Meltzer, 1975, J. Chem. Phys. **62**, 3573.
Davydov, A.S., 1971, "Theory of molecular excitons" (Plenum Press, New York).
Diestler, D.J. and A.H. Zewail, 1979, J. Chem. Phys. **71**, 3103 and 3113.
Doberer, U. and H. Port, 1979, unpublished results.
Economou, E.N. and M.H. Cohen, 1971, Rev. B**4**, 396.
Economou, E.N. and P.D. Antoniou, 1977, Phys. Rev. B**16**, 3768.
Fox, D., 1979 (private communication).
Francis, A.H. and C.B. Harris, 1971, Chem. Phys. Lett. **9**, 181, 188.
Frenkel, J., 1931, Phys. Rev. **37**, 17, 1276.
Haken, H. and G. Strobl, 1968, in: The triplet state, ed., A. Zahlan (Cambridge University Press, Cambridge).
Harris, C.B. and R.J. Hoover, 1971, Chem. Phys. Lett. **12**, 75.
Hochstrasser, R.M., 1976, Int. Rev. of Sci. Physical Chemistry, series 2, Vol. 3, Spectroscopy (Butterworths, London) p. 1.
Hochstrasser, R.M. and J.D. Whiteman, 1972, J. Chem. Phys. **56**, 5945.
Hochstrasser, R.M. and A.H. Zewail, 1974, Chem. Phys. **4**, 142.
Inokuti, M. and F. Hirayama, 1965, J. Chem. Phys. **43**, 1978.
Jones, K.E. and A.H. Zewail, 1978, Advances in Laser Chemistry, ed., A.H. Zewail (Springer-Verlag, New York) p. 196.
Jones, K.E., A.H. Zewail and D.J. Diestler, loc. cit. p. 258.
Kenkre, V.M., 1975, Phys. Rev. B**12**, 2150.
Kenkre, V.M. and R.S. Knox, 1976, J. Lumin. **12/13**, 187.
Kirkpatrick, S., 1973, Rev. Mod. Phys. **45**, 574.
Klafter, J. and J. Jortner, 1977, Chem. Phys. Lett. **49**, 410.
Klafter, J. and J. Jortner, 1979, J. Chem. Phys. **71**, 2210, and references therein.
Kopelman, R., E.M. Monberg and F.W. Ochs, 1977, Chem. Phys. **19**, 413; 1977, Chem. Phys. **21**, 373.
Lee, P.A., 1979, Phys. Rev. Lett. **42**, 1492.
Lemaistre, J.-P. and A. H. Zewail, 1979, Chem. Phys. Lett. **68**, 296, 302.
Madhukhar, A. and W. Post, 1977, Phys. Rev. Lett. **39**, 1424.
Meltzer, R.S. and R.L. Cone, 1976, J. Lumin. **12/13**, 247.
Montroll, E.W., 1963, Proc. Symp. Appl. Math. Am. Math. Soc., New York City, April, 1963, **16**, 193.
Mott, N.F., 1974, Philos. Mag. **29**, 613; 1976, Comm. Phys. **1**, 203.
Nieman, G.C. and G.W. Robinson, 1962, J. Chem. Phys. **37**, 2150.
Odagaki, T. 1980 Solid State Commun. **35**, 639.
Pauling, L., and E.B. Wilson, "Introduction to Quantum Mechanics," 1st edition (McGraw-Hill, New York) p. 314 (1935).
Rackovsky, S. and R. Silbey, 1973, Mol. Phys. **25**, 61.
Robinson, G.W., 1970, Ann. Rev. Phys. Chem. **21**, 429.

Robinson, G.W. and R.P. Frosch, 1962, J. Chem. Phys. **37**, 1962; ibid. 1963, **38**, 1187.

Schmidberger, R. and H.C. Wolf, 1972, Chem. Phys. Lett. **16**, 402; 1974, Chem. Phys. Lett. **25**, 185; 1975, Chem. Phys. Lett. **32**, 21.

Schmidt, J., W.S. Veeman and J.H. van der Waals, 1969, Chem. Phys. Lett. **4**, 341.

Sewell, G., 1962, in Polarons and Excitons, ed., G. Kuper (Plenum Press, New York) p. 233.

Shante, V.K.S. and S. Kirkpatrick, 1971, Adv. Phys. **20**, 325.

Silbey, R., 1976, Ann. Rev. Phys. Chem. **27**, 203, and references therein.

Smith, D.D. and A.H. Zewail, 1979, J. Chem. Phys. **71**, 3533.

Smith, D.D., R.D. Mead and A.H. Zewail, 1977, Chem. Phys. Lett. **50**, 358.

Smith, D.D., D.P. Millar and A.H. Zewail, 1980, J. Chem. Phys. **72**, 1187.

Smith, D.D., R.C. Powell and A.H. Zewail, 1979, Chem. Phys. Lett. **68**, 309.

Smith, D.D., D. Neikirk and A.H. Zewail, to be published.

Soos, Z.G. and R.C. Powell, 1972, Phys. Rev. B**6**, 4035.

Theodorou, G. and M.H. Cohen, 1976, Phys. Rev. B**13**, 4597.

Tinti, D. 1979 (private communication of a manuscript).

Van Strien, A.J., J. van Kooten and J. Schmidt, 1980, Chem. Phys. Lett. **76**, 7.

Wolfrum, H., K. Renk and H. Sixe, 1979, Chem. Phys. Lett. **68**, 90.

Zewail, A.H., 1974, Chem. Phys. Lett. **29**, 630; 1975, Chem. Phys. Lett. **33**, 46.

Zewail, A.H., 1979, J. Chem. Phys. **70**, 5759.

Zewail, A.H., 1980, Acc. Chem. Res. **13**, 360.

Zewail, A.H. and C.B. Harris, 1974, Chem. Phys. Lett. **28**, 8; 1975, Phys. Rev. B**11**, 935, 952.

Zewail, A.H., W. Breiland and C.B. Harris, 1979, unpublished results.

CHAPTER 16

Excitons in Strained Molecular Crystals

V.I. SUGAKOV

Institute for Nuclear Research
Academy of Sciences of the Ukrainian SSR, Kiev 252028
USSR

Excitons
Edited by
E.I. Rashba and M.D. Sturge

Contents

1. Introduction

The optical properties of molecular crystals are determined by the intermolecular interactions, which are sensitive to changes in distances and angles between molecules. Therefore, the investigation of the effect of stress on both the intermolecular distances and the orientations of molecules is an important method of studying intermolecular interactions. The first report on changes in the spectra of strained molecular crystals was given by Broude et al. (1957), who measured the light absorption in monocrystalline naphthalene and anthracene films on quartz substrates. The crystals were deposited on quartz at room temperature and then cooled down to liquid-hydrogen temperature. The difference of the thermal expansion coefficients of the molecular crystal and that of quartz, the latter being much smaller, resulted in stretching of the molecular layer. For this reason the intermolecular interaction was reduced in the cooled samples. As a result, the optical spectra were shifted to ultraviolet (towards the vapour spectra of the substances) and the Davydov splitting was strongly reduced. Later, Wiederhorn and Drickamer (1959) started their investigations of the optical spectra of molecular crystals under pressure, using specialized equipment. Now, studies of strained molecular crystals are performed by a number of research groups and appear to be an effective tool in determining the micro- and macroparameters of these substances.

2. Deformation of molecular crystals under stress

Before proceeding to the experimental and theoretical data on the optical properties of strained crystals let us consider the changes in crystals under stress. The microscopic theory of deformation in crystalline solids was developed by Born and Huang (1954). It was extended to molecular crystals by Cruickshank (1957), Powley (1967) and Walmsley (1968).

Deformation of the crystalline solid leads to translational and rotational displacements of its molecules. Assuming the molecules to be rigid one may characterize the displacements of each molecule by six quantities, three of which determine the displacement of the molecule's center of gravity and the others specify its rotations about the principal axes of inertia of the molecule. We shall denote these quantities by $U_{n\alpha i}$, where $U_{n\alpha i}$ stands for the translational displacements of the molecule at the site $n\alpha$, for $i = 1, 2, 3$, and for the angles of rotation about principal axes of the molecule, for $i = 4, 5, 6$. (The site index $n\alpha$ refers to the unit cell and labels the molecule in this unit cell.)

The deformation comprises the external and internal strain. The external strain is characterized by the variations of the shape and position of a unit cell as a whole, while the internal strain describes the changes in the intermolecular distances and rotational displacements within the cell. In the case of uniform deformation we have

$$U_{n\alpha i} = \sum_{j=1}^{3} \epsilon_{ij} R_{n\alpha j} + U_{\alpha i}, \qquad i = 1, 2, 3, \tag{1}$$

$$U_{n\alpha i} = U_{\alpha i}, \qquad i = 4, 5, 6, \tag{2}$$

where $\boldsymbol{R}_{n\alpha}$ is the position vector of the molecule $n\alpha$.

In expression (1) the first term describes external strain. It is determined by the macroscopic strain tensor ϵ_{ij}. The $U_{\alpha i}$ $(i = 1, \ldots 6)$ describe the changes of positions and orientations of molecules in the new unit cell. One obtains the values $U_{\alpha i}$ by minimizing the crystal energy at fixed external strain. As a result the internal strains $U_{\alpha i}$ are expressed in terms of the external strain.

$$U_{\alpha i} = \sum_{j,k=1}^{3} A_{\alpha,ijk} \epsilon_{jk}. \tag{3}$$

The coefficients $A_{\alpha,ijk}$ may be obtained from the microscopic calculations, taking into account the intermolecular interactions. For internal translation $(i = 1, 2, 3)$ the values $A_{\alpha,ijk}$ are components of a tensor of the third rank. For crystals with centro-symmetric site groups (e.g. naphthalene, anthracene) one has $A_{\alpha,ijk} = 0$. Therefore, $U_{\alpha i} = 0$ for $i = 1, 2, 3$.

Thus, from the relations (1)–(3) it follows that all the changes of microscopic quantities in the lattice under uniform deformation may be expressed in terms of the strain tensor. In the case of small deformation there is a linear relation between components of the strain tensor ϵ_{ij} and the stress tensor σ_{lm}

$$\epsilon_{ij} = \sum_{l,m=1}^{3} S_{ijlm} \sigma_{lm}. \tag{4}$$

The compliance tensor S_{ijlm} is inversely proportional to the elasticity tensor. The elasticity tensor for some molecular crystals has been evaluated from measurements of ultrasound velocity (Afanaseva 1967, 1968, Teslenko 1967, Huntington et al. 1969), from studies of neutron scattering (Hamamsy et al. 1977, Elnahwy et al. 1978), from calculation (Pawley 1967) within the framework of the atom–atom potential method (Kitaygorodsky 1965).

The symmetry of a uniaxially stressed crystal is lowered. For example, in crystals with two molecules per unit cell (e.g. naphthalene, anthracene) a symmetry element (such as the screw axis) may disappear under stress and the monoclinic lattice transforms into a triclinic one. This takes place when the angle of the applied external force with respect to the monoclinic axis differs from zero or 90°. Henceforth, this stress is called the uniaxial stress. In this case shear components of strain tensor U_{xy}, U_{yz} exist (where 0y coincides with the screw axis). Taking into account the nonzero components of the compliance tensor, one obtains from the relation (4)

$$U_{xy} = S_{xyxy}\sigma_{xy} + S_{xyyz}\sigma_{yz}, \tag{5}$$

$$U_{yz} = S_{yzxy}\sigma_{xy} + S_{yzyz}\sigma_{yz}. \tag{6}$$

From eqs. (5) and (6) follows that the crystal symmetry is lowered when components σ_{xy} and σ_{yz} of the stress tensor are nonzero.

When the crystal is under hydrostatic pressure P, one has $\sigma_{ik} = -P\delta_{ik}$. Therefore, in this case $\sigma_{xy} = \sigma_{yz} = 0$ and, according to eqs. (5) and (6), the shear components of the deformation tensor equal zero and the crystal symmetry is not altered.

Equations (1), (3) and (4) enable one to determine the displacements of the molecules of a crystal as a function of the applied stress. The knowledge of values of these displacements is needed for theoretical description of the dependence of the exciton spectrum on the stress.

3. Exciton Hamiltonian of strained crystals

Under uniform strain of a pure crystal translation symmetry is conserved and general exciton theory (Davydov 1968) can be applied to obtain the energy spectrum of excited states.

The Hamiltonian of a molecular crystal is given by

$$\mathcal{H} = \sum_{n\alpha} (\mathscr{E}_0 + D_\alpha) B^+_{n\alpha} B_{n\alpha} + \sum_{n\alpha m\beta} M_{n\alpha,m\beta} B^+_{n\alpha} B_{m\beta}, \tag{7}$$

here $\mathscr{E}_0$ is the free-molecule excitation energy, $D_\alpha = \sum_{m\beta} D_{n\alpha,m\beta}$ is the site shift, $M_{n\alpha,m\beta}$ is the excitation transfer matrix element between the $n\alpha$ and $m\beta$ molecules, $B^+_{n\alpha}$ and $B_{n\alpha}$ are the creation and annihilation operators of the excitation of the molecule $n\alpha$. The quantity $(\mathscr{E}_0 + D_\alpha)$ may be called the crystalline molecular term. It determines the eigenstates of the system in the absence of excitation transfer interaction.

The intermolecular interactions determine the values of the coefficients

D_α and $M_{n\alpha,m\beta}$. These coefficients comprise the information about the pressure and relate the latter to the optical properties of the crystal. It should be noted that, due to the possibility of lowering the symmetry in stressed crystals, the value of D_α may depend on α, the label of the molecule in the cell, even if D_α for unstressed crystal did not (benzene, naphthalene, anthracene, etc.). Earlier studies of spectra of strained crystals did not take this fact into account. Later, several effects due to the dependence of D_α on α in a strained crystal were predicted by Sugakov (1973).

The matrix elements D_α and $M_{n\alpha,m\beta}$ are functions of the intermolecular distances and orientation of molecules. With small deformation, D_α can be expanded in a series of displacements,

$$D_\alpha = D_0 + \delta D_\alpha, \tag{8}$$

where D_0 is the shift of the molecular term when the molecule is excited in the unstressed crystal, δD_α is the stress-induced site shift,

$$\delta D_\alpha = \sum_{m\beta} \sum_{i=1}^{6} \left(\frac{\partial D_{n\alpha,m\beta}}{\partial U_{n\alpha i}} U_{n\alpha i} + \frac{\partial D_{n\alpha,m\beta}}{\partial U_{m\beta i}} U_{m\beta i} \right). \tag{9}$$

Using relations (1)–(3) one can express δD_α in terms of the strain tensor

$$\delta D_\alpha = \sum_{i,j=1}^{3} D_{\alpha,ij} \epsilon_{ij}, \tag{10}$$

where

$$D_{\alpha,ij} = \sum_{m\beta} \left[\frac{\partial D_{n\alpha,m\beta}}{\partial U_{n\alpha i}} R_{n\alpha i} + \frac{\partial D_{n\alpha,m\beta}}{\partial U_{m\beta i}} R_{m\beta i} + \right.$$

$$\left. + \sum_{k=1}^{6} \left(\frac{\partial D_{n\alpha,m\beta}}{\partial U_{n\alpha k}} A_{\alpha,kij} + \frac{\partial D_{n\alpha,m\beta}}{\partial U_{m\beta k}} A_{\beta,kij} \right) \right]. \tag{11}$$

Similar reasoning can be applied to the excitation transfer matrix elements:

$$M_{n\alpha,m\beta} = M^0_{n\alpha,m\beta} + \sum_{i,j=1}^{3} M_{n\alpha,m\beta,ij} \epsilon_{ij}, \tag{12}$$

where $M^0_{n\alpha,m\beta}$ is the excitation transfer matrix element between the molecules $n\alpha$ and $m\beta$ in the unstressed crystal. In the following sections we shall discuss the exciton spectrum in stressed crystals for two cases: the hydrostatic pressure and the uniaxial stress.

4. Spectra of molecular crystals under hydrostatic pressure

Since the first experiments on strain effects (Broude et al. 1957, Wiederhorn and Drickamer 1959, Broude and Tomashchik 1964) it was found that the spectra of molecular crystals are shifted shorter wavelengths under expansion. Simultaneously broadening of bands is observed. The shift and broadening, on the whole, are larger for those bands for which the transition dipole moment is higher.

Most experimental studies of the strain effect on the optical spectra have been carried out under hydrostatic pressure. In these conditions the symmetry of the crystal remains unchanged ($D_\alpha = D$) and the exciton dispersion law is of the same type as that in the unstressed crystal, only the numerical values of the parameters have changed. Due to the momentum selection rules, the position of the exciton absorption bands is determined by the excitation energy of the exciton bands at the point $k = 0$. In particular, in crystals with two molecules in the unit cell, the exciton band energies are given by

$$\mathscr{E}_\mu(0) = \mathscr{E}_0 + D + L_{11}(0) - (-1)^\mu L_{12}(0), \tag{13}$$

where $\mu = 1, 2$ is the number of the exciton band,

$$L_{11}(0) = \sum_m M_{n1,m1}, \tag{14}$$

$$L_{12}(0) = \sum_m M_{n1,m2}. \tag{15}$$

$M_{n1,m1}$ and $M_{n1,m2}$ are the excitation transfer matrix elements between molecules of same and different sublattices, respectively.

The main contribution to the term D comes from the change in the Van der Waals interaction of the molecule with its environment, on excitation. This interaction can in principle be calculated by second order perturbation theory including configurational mixing of the levels (Craig 1955, Agranovich 1968). Straightforward calculation of the D value involves great difficulties and, up to the present, has not been reported. However, approximations exist which enable one to estimate the D value. The fact that the Van der Waals interaction depends on the distance between molecules as R^{-6} allows one, assuming isotropic compressibility, to write (Jones, 1968):

$$D = A(\rho/\rho_0)^2, \tag{16}$$

where A is a constant, ρ_0 is the density at $P = 1$ atm.

For deep impurity centres in solution, the effect of the excitation transfer interaction (the term $M_{n\alpha,m\beta}$) is negligible and the shift of the spectrum under stress is determined by the term D. In particular, it follows from eq. (16) that the stress-induced shift of a molecular level must be proportional to the square of the volume density. The validity of such a dependence has been confirmed in a number of experiments (Nicol 1965, Offen 1965, Jones and Nicol 1968).

A similar approximation may be employed for analyzing the excitation transfer matrix elements. In the dipolar approximation the latter descrease inversely proportionally to the cube of the distance $M_{n\alpha,m\beta} \sim R^{-3}_{n\alpha,m\beta}$. Therefore,

$$L_{\alpha\beta} \sim B_{\alpha\beta}\rho/\rho_0. \tag{17}$$

Hence, in this approximation, the Davydov splitting, which is equal to $\mathscr{E}_D = 2|L_{12}(0)|$, should vary linearly with density. However, the relation (17) was not confirmed experimentally. Otto et al. (1977) showed that Davydov splitting in anthracene crystals increases much faster with density than is expected for purely dipolar interaction. Similarly, Shirotani et al. (1974) pointed out that the Davydov splitting in tetracene varies roughly four times as rapidly with pressure as expected in the dipolar approximation.

The inapplicability of relation (17) may stem from the invalidity of the following assumptions incorporated in its derivation: (1) one should not neglect the higher multipole interactions relative to the dipolar, (2) the compressibility of the crystal may not be isotropic, (3) the orientation of molecules in the crystal may change under pressure.

Thus, at present there are no quantitative calculations of the dependence of the parameters of exciton bands on pressure. In Schipper's paper (1974) on this subject several assumptions were introduced which require further justification: the author confined himself to the dipolar approximation and neglected the angular dependence of the site shift interaction between the excited molecule and the nonexcited molecule.

Exciton spectra of molecular crystals under pressure were measured in numerous experiments (Nakashima and Offen 1968, Drickamer 1974, Shirotani et al. 1974, Otto et al. 1977, Kalinowski et al. 1978, etc.). Usually the investigations are carried out at room temperature. The bands under these conditions are broad and their structure, in particular Davydov splitting, is not easily observed experimentally. Thus clear-cut data exist only for the crystals where the splitting is large, i.e., for the crystals with strong excitation transfer interaction (table 1). Under a pressure of 6.6 kbar the Davydov splitting in tetracene is almost doubled (from $510\,cm^{-1}$ up to $1000\,cm^{-1}$). The long-wave component of the doublet always shifts further with the pressure than the short-wave one. This effect is explained as follows. The shift is governed by the terms D, L_{11}, L_{12} in formula (13). The dependence $D(P)$ gives the main contribution which leads to the long-wave shift of the both bands. At the same time, the term L_{12} grows with increasing pressure because of an increase in the excitation transfer energy. For the lower band the effect due to this term is added to that of the term D, and for the upper band they should be subtracted. Therefore, the upper band shifts more slowly with an increase of pressure.

Simultaneously with the shift of the bands a redistribution of intensities

Table 1
Dependence of Davydov splitting $\mathscr{E}_D$ on pressure for tetracene (Kalinowski et al. 1978) and pentacene crystals (Shirotani et al. 1974).

Crystal	The phase	$\mathscr{E}_D$	$\partial\mathscr{E}_D/\partial P$
Tetracene	of low pressure		46
	The phase	510	
	of high pressure		57
Pentacene		950	70

$\partial\mathscr{E}_D/\partial P$ and $\mathscr{E}_D$ values are given in cm^{-1}/kbar and cm^{-1}.

in the bands is observed. Thus, for tetracene the ratio of the intensities of Davydov doublet bands 521 nm (polarized along b) and 503 nm (polarized along a) changes by 25% at $P = 5.2$ kbar (Nakashima and Offen 1968). This may be attributed to the turning of molecules under pressure and to a modification of the configurational mixing. The latter may affect the value of the transition dipole moment.

Kalinowski and Jankowiak (1978) observed a discontinuous change in the Davydov splitting in crystal line tetracene at $P = 3$ kbar. This phenomenon was associated with the pressure-induced phase transition detected earlier (Kalinowski et al. 1976) by measuring the tetracene single crystal fluorescence versus the direction of the magnetic field.

Besides the shift of the bands with pressure, broadening of the bands was also observed. The broadening may be related to an increase of the electron–phonon interaction caused by a reduction of the intermolecular distance, by formation of defects, and by inhomogeneous deformation of the crystal. Studies of strained crystalline anthracene and phenanthrene (Tonaka et al. 1965, Jones et al. 1968) revealed a structureless emission at frequencies 3000–6000 cm^{-1} lower than the normal fluorescence. This emission was attributed to excimer states.

A number of works deal with triplet excitons in strained crystals. The study of triplet excitons in anthracene (Arnold et al. 1974) had shown that pressure shifts the exciton band to longer wavelengths at the rate of $-11.8\ cm^{-1}$/kbar while the Davydov splitting increases at the rate of $0.6\ cm^{-1}$/kbar. The latter value is minute because the excitation transfer integrals are very small for triplet excitons.

The pressure, by changing the intermolecular distances, changes the various coefficients which determine the population kinetics of the levels. Thus, Offen and Baldwin (1966) observed an increase of the naphthalene phosphorescence intensity with pressure. This result may be related to increasing intermolecular crossing between the singlet and triplet sets of the states or to a decrease of the rate of nonradiative transitions (for example, through the change in the diffusion coefficient of oxygen which causes the quenching).

An interesting effect associated with pressure-induced changing of the resonance of the levels was found in tetracene (Witten et al. 1976). The authors reported that the quantum yield of the tetracene fluorescence increases with pressure. In tetracene the energy of the lowest singlet band bottom is very close to (but somewhat lower than) twice that of the triplet band. Therefore, the radiative decay of the singlet state competes with thermally activated fission into two triplets. The probability of the latter is proportional to

$$\exp\left(-\frac{2\mathscr{E}_{\mathrm{T}}-\mathscr{E}_{\mathrm{S}}}{kT}\right), \tag{18}$$

where $2\mathscr{E}_{\mathrm{T}}$ is a doublet energy of the triplet state and $\mathscr{E}_{\mathrm{S}}$ is an energy of the singlet band bottom. So far as in tetracene $2\mathscr{E}_{\mathrm{T}} \sim \mathscr{E}_{\mathrm{S}}$ holds, the probability of decay into triplets is high, hence the quantum yield is low. These levels are shifted with pressure, $\mathscr{E}_{\mathrm{S}}$ and $\mathscr{E}_{\mathrm{T}}$ becoming smaller, and the singlet states are shifted further than the triplets. Thus, the difference $2\mathscr{E}_{\mathrm{T}}-\mathscr{E}_{\mathrm{S}}$ increases with pressure and, according to eq. (18), the singlet-to-triplet decay rate decreases, favouring growth of the fluorescence yield. Whitten et al. (1976) reported an increase of the fluorescence intensity by the factor 2.55 with growing pressure up to 4.10 kbar.

5. *Theory of exciton spectra of uniaxially stressed crystals*

We discussed above the effect of hydrostatic pressure on the crystal. Uniaxial stress may lower the symmetry of the crystal and, consequently, lead to a number of qualitatively new effects. In subsequent sections we present a short survey of the results obtained by the Kiev group in their optical studies of uniaxially stressed molecular crystals.

The energy spectrum and the exciton wavefunctions in a strained crystal with Hamiltonian (7) can be calculated in the same way as in the unstressed crystal.

For the crystal with two molecules in the unit cell the exciton wavefunction of μth band with the wave vector $\boldsymbol{k}$ is given by (Sugakov, 1973):

$$\psi^{\mu\boldsymbol{k}} = \frac{1}{\sqrt{N}} \sum_{n\alpha} a_{\alpha}^{\mu\boldsymbol{k}} \exp(\mathrm{i}\boldsymbol{k}\boldsymbol{R}_{n\alpha}) B_{n\alpha}^{+}|0\rangle, \tag{19}$$

where $|0\rangle$ is the wavefunction of the ground state of the crystal, N is the number of cells in unit volume, $\mu = 1, 2$,

$$a_\alpha^{\mu k} = \frac{1}{\sqrt{2}} \begin{bmatrix} a_+(\boldsymbol{k}) & \dfrac{|L_{12}|}{L_{12}} a_-(\boldsymbol{k}) \\ a_-(\boldsymbol{k}) & -\dfrac{|L_{12}|}{L_{12}} a_+(\boldsymbol{k}) \end{bmatrix} \tag{20}$$

$$a_\pm(\boldsymbol{k}) = \left\{ \frac{[|L_{12}(\boldsymbol{k})|^2 + (\delta(\boldsymbol{k})/2)^2]^{1/2} \pm \frac{1}{2}\delta(\boldsymbol{k})}{[|L_{12}(\boldsymbol{k})|^2 + (\delta(\boldsymbol{k})/2)^2]^{1/2}} \right\}^{1/2}, \tag{21}$$

$$\delta(\boldsymbol{k}) = D_1 - D_2 + L_{11}(\boldsymbol{k}) - L_{22}(\boldsymbol{k}), \tag{22}$$

$$L_{\alpha\beta}(\boldsymbol{k}) = \sum_m M_{n\alpha,m\beta} \exp[i\boldsymbol{k}(\boldsymbol{R}_{n\alpha} - \boldsymbol{R}_{m\beta})]. \tag{23}$$

The exciton energy of the μth band equals

$$\mathscr{E}^\mu(\boldsymbol{k}) = \mathscr{E}_0 + \frac{D_1 + D_2 + L_{11}(\boldsymbol{k}) + L_{22}(\boldsymbol{k})}{2} - (-1)^\mu \left[|L_{12}(\boldsymbol{k})|^2 + \left(\frac{\delta(\boldsymbol{k})}{2} \right)^2 \right]^{1/2}. \tag{24}$$

The factors $a_1^{\mu k}$ and $a_2^{\mu k}$ are the probability amplitudes of the electronic excitation for the molecules in the sublattices 1 and 2, respectively. It follows from eqs. (20) and (21) if $D_1 \neq D_2$ ($\delta(\boldsymbol{k}) \neq 0$) then $a_1^{\mu k} \neq a_2^{\mu k}$, and the probabilities of the electronic excitation of molecules in different sublattices do not coincide.

As a result the directions of the exciton transition dipole moments do not coincide with the direction of crystal axes. For $\delta(\boldsymbol{k}) \gg L_{12}(\boldsymbol{k})$ (i.e. when the difference between the terms of molecules of different sublattice exceeds the Davydov splitting $2L_{12}(\boldsymbol{k})$) the electronic excitation is localized on translationly equivalent molecules only: in the first band it is on one sublattice, in the second band on the other. Also, in this case the transition

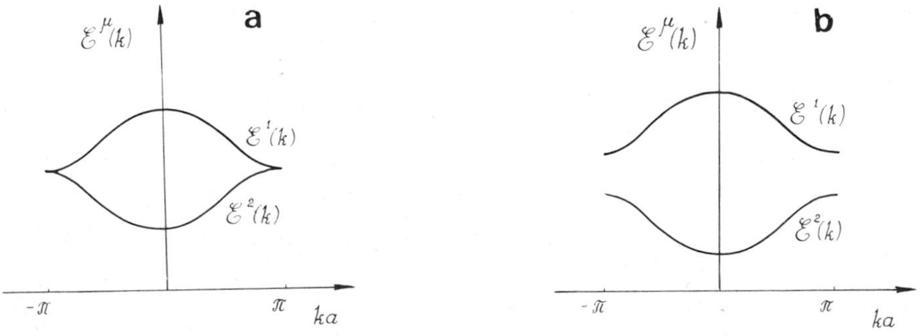

Fig. 1. The exciton dispersion law for the crystals of the C_{2h}^5 symmetry: (a) for unstrained crystals, (b) for the crystals under directional stress.

dipole moments of the first and second bands coincide with those of molecules in the respective sublattices.

There is another peculiarity of directionally stressed crystals. For example, in a crystal of C^5_{2h} symmetry, the existence of a screw axis results in the exciton bands sticking together at the boundary of the Brillouin zone. Under uniaxial stress the screw axis vanishes and a gap emerges between the bands. This phenomenon is shown in fig. 1.

6. *Theory of impurity levels in uniaxially stressed crystals*

Whenever the energy terms of molecules of different sublattices become different under uniaxial stress, impurity levels which are not split in the unstrained crystal should split under stress.

Let us consider, for example, the case of an isotopic impurity. One may assume that the intermolecular interaction matrix elements in the impurity crystal are equal to those in the perfect crystal but that the molecular term of the impurity is shifted by the value Δ. The Hamiltonian describing the electronic excitation of such a crystal with an impurity at the site 0γ is given by

$$\mathcal{H}_{\text{imp}} = \mathcal{H}_0 + \Delta B^+_{0\gamma} B_{0\gamma}, \tag{25}$$

where $\mathcal{H}_0$ is the Hamiltonian of the strained pure crystal (7).

Calculations of the eigenstates of the Schrödinger equation with Hamiltonian (25) may be carried out just as for the unstressed crystal (Rashba, 1957). The relation, which determines levels ($\mathcal{E}_\gamma$) of an impurity at the site γ, can be written as follows (Sugakov, 1973):

$$1/\Delta = \mathcal{G}_{0\gamma,0\gamma}(\mathcal{E}_\gamma), \tag{26}$$

where $\mathcal{G}_{0\gamma,0\gamma}$ is Green's function

$$\mathcal{G}_{0\gamma,0\gamma}(\mathcal{E}) = \frac{1}{N} \sum_{\mu k} \frac{a^{*\mu k}_{\gamma} a^{\mu k}_{\gamma}}{\mathcal{E} - \mathcal{E}^{\mu}(\boldsymbol{k})}. \tag{27}$$

Substituting the values of $a^{\mu k}_{\gamma}$ and $\mathcal{E}^{\mu}(\boldsymbol{k})$ from eqs. (20) and (24) into eq. (26), we obtain

$$\frac{1}{\Delta} = \mathcal{G}(\mathcal{E}_\gamma) - (-1)^{\gamma} \frac{1}{2N} \sum_{k} \frac{\delta(\boldsymbol{k})}{[\mathcal{E}_\gamma - \mathcal{E}^1(\boldsymbol{k})][\mathcal{E}_\gamma - \mathcal{E}^2(\boldsymbol{k})]}, \tag{28}$$

where

$$\mathcal{G}(\mathcal{E}) = \frac{1}{2N} \sum_{\mu k} \frac{1}{\mathcal{E} - \mathcal{E}^{\mu}(\boldsymbol{k})}. \tag{29}$$

The second term is different for nonequivalent sites of the impurity molecule and is responsible for the level splitting in a uniaxially stressed crystal. For a given strain the value of the splitting depends on the separation of the level from the bottom of the exciton band. The splitting grows when the level moves away from the band bottom and reaches its maximum value, equal to $D_1 - D_2$, if the separation is much greater than the exciton band width. The excitation transfer interaction "smears" the electronic excitation of a shallow centre around the environment of the impurity, which comprises two sublattices. Therefore, for shallow centers the inequivalence of the molecules due to the stress becomes less distinct. For deep levels, the effect of excitation transfer interaction is negligible and the excitation is completely localized on the impurity, so that the splitting coincides with the difference of the molecular terms of different sublattices.

7. *Vibronic levels in strained crystals*

The theory of the stress effect on the impurity levels can be readily extended for the case of vibronic bands generated by simultaneous excitation of a bound state of exciton and non-totally symmetric molecular vibration. In describing these states, the model of vibronic states, advanced by Rashba (1966), proved most advantageous. The phonon bands of the intramolecular vibrations are significantly narrower than the exciton bands and their width may be neglected. Therefore, following Rashba's concept, one may assume that the phonon of the bound exciton-intramolecular vibration state is "motionless" (localized on some molecule), while the exciton is moving in a certain neighbouring area. Then the Hamiltonian governing electron-vibrational excitations in a strained crystal may be written as follows

$$\mathcal{H} = \mathcal{H}_0 + \mathcal{H}_{\text{vibr}} + \mathcal{H}_{\text{int}}, \tag{30}$$

where $\mathcal{H}_0$ is the Hamiltonian of the electron subsystem (7), $\mathcal{H}_{\text{vibr}}$ is the vibration Hamiltonian,

$$\mathcal{H}_{\text{vibr}} = \sum_{n\alpha} \hbar\omega_0 b^+_{n\alpha} b_{n\alpha}, \tag{31}$$

ω_0 is the intramolecular vibration frequency, $b^+_{n\alpha}$ and $b_{n\alpha}$ are, respectively, the creation and annihilation operators of a vibration on the molecule, $\mathcal{H}_{\text{int}}$ is the exciton-intermolecular vibration interaction Hamiltonian. For non-totally symmetric vibrations the main contribution to $\mathcal{H}_{\text{int}}$ is determined by the shift Δ_ν of the vibration energy $\hbar\omega_0$ produced by the electronic excitation of the molecule. Thus,

$$\mathcal{H}_{\text{int}} = \sum_{n\alpha} \Delta_\nu B^+_{n\alpha} B_{n\alpha} b^+_{n\alpha} b_{n\alpha}. \tag{32}$$

In this approximation of "motionless" phonons the coordinate of the vibrationally excited molecule serves as one of the quantum numbers of the system. Let the vibration be on the molecule 0γ. Then the wavefunction of the system with Hamiltonian (30) can be presented in the form:

$$\psi_{0\gamma} = \sum_{n\alpha} a_{n\alpha}^{0\gamma} B_{n\alpha}^{+} b_{0\gamma}^{+}|0\rangle, \tag{33}$$

where the coefficients $a_{n\alpha}^{0\gamma}$ are calculated from the following set of equations

$$a_{n\alpha}^{0\gamma} = \Delta_\nu \mathcal{G}_{n\alpha,0\gamma}(\mathscr{E}_{\gamma M} - \hbar\omega_0) a_{0\gamma}^{0\gamma}, \tag{34}$$

$\mathscr{E}_{\gamma M}$ is a vibronic level of system, when the vibration is localized at the site γ. $\mathcal{G}_{n\alpha,0\gamma}$ is a nondiagonal Green's function (Ostapenko et al. 1979). In particular, it follows immediately from eq. (34) that the energies of the vibronic levels are determined by:

$$\frac{1}{\Delta_\nu} = \mathcal{G}_{0\gamma,0\gamma}(\mathscr{E}_\gamma - \hbar\omega_0). \tag{35}$$

Equation (35) is similar to that for impurity levels (26), but the levels obtained from eq. (35) are shifted by the intramolecular vibration energy. Just in the case of impurities, the energy levels in uniaxially stressed crystal should split because the excitation energy depends on the position of the molecule where the vibration is created. In general, there will be as many levels as there are molecules in a unit cell.

In the same way one may consider the position of vibronic levels in impure crystals. With isotopic substitution, the position of a vibronic level, when the vibration is created on the impurity molecule, is determined by the condition,

$$\frac{1}{\Delta + \Delta_\nu} = \mathcal{G}_{0\gamma,0\gamma}(\mathscr{E}_{\gamma M} - \hbar\omega_0). \tag{36}$$

When Δ and Δ_ν are of the same sign, condition (36) determines a deeper level with respect to the continuous spectrum than both the impurity level and the vibronic level in perfect crystal.

Thus, a set of levels located at different distances from the edge of the continuous spectrum can be created in the crystal.

8. *Approximate solution of exciton eigenstate problems in strained naphthalene*

Calculations of the electron impurity and vibronic levels require, apart from the values of strain tensor components, quantitative information about the dispersion law of the exciton band, but little is known on this subject. The most complete results were reported for the lowest singlet exciton bands of naphthalene crystals. (The so-called a and b bands). Using spectral data of naphthalene in deuteronaphthalene matrices, Rabinkina et al. (1970) evaluated the state density in an exciton band. In the nearest-neighbour approximation, Hong and Kopelman (1968) attempted to obtain the values of the parameters of an exciton band from the optical spectra of pair centers of isotopic impurities. However, such a reconstruction procedure did not give unique results, so Hong and Kopelman obtained three sets of exciton transfer matrix elements all of which properly accounted for the experimental data. It is noteworthy that in all the three sets the matrix element connecting nearest-neighbor molecules on different sublattices was much greater than all others. It is this property which is the basis of the calculations of a number of physical effects involving excitons in naphthalene. The matrix elements between the remoter neighbours are supposed small and are taken into account by perturbation theory. Such an approach was found useful in calculating the local and vibronic levels of impurity centers in naphthalene (Ostapenko et al. 1971 a, b, Krivenko et al. 1978). Similar simplifications can be made for calculations of bands in strained naphthalene crystals. In this case the following parameters would be considered small: (1) the difference of the terms $D_2 - D_1$ of molecules in different sublattices relative to the Davydov splitting, (2) the variation of the excitation transfer matrix element caused by the strain, relative to the matrix element itself. The effect of stress on the excitation transfer interaction between the non-nearest neighbour molecules will be neglected. Then from the relations (24) and (28), supposing indices 1 and 2 to refer to the b and a exciton bands of naphthalene respectively, we obtain the following equations for the stress-induced shift of the exciton a band $\delta\mathscr{E}^{a}$, for the change of the Davydov splitting $\delta\mathscr{E}_{D}$ and for the impurity level $\mathscr{E}_{\gamma}$ (Gorban' et al. 1978).

$$\delta\mathscr{E}^{a} = \delta D_{+} - 4\delta M - (\delta D_{-})^2/\mathscr{E}_{D}, \tag{37}$$

$$\delta\mathscr{E}_{D} = 8\delta M + 2(\delta D_{-})^2/\mathscr{E}_{D}, \tag{38}$$

$$\frac{1}{\Delta} = \mathscr{G}(\mathscr{E}_{\gamma})\left[1 - (-1)^{\gamma}\frac{\delta D_{-}}{\mathscr{E}_{\gamma}}\right] - \frac{\delta M}{M}\frac{2}{\pi\epsilon_{\gamma}}\left[\frac{E(k)}{1-k^2} - K(k)\right], \tag{39}$$

where $E(k)$ and $K(k)$ are the complete elliptic integrals, $k = 4M/\epsilon_{\gamma}$, $\epsilon_{\gamma} = \mathscr{E}_{\gamma} - \mathscr{E}_{0} - \delta D_{+}$, M is the excitation transfer matrix element between the nearest-neighbor molecules in different sublattices,

$$\delta M = \tfrac{1}{2}(\delta M_{0},(\boldsymbol{a}+\boldsymbol{b})/2 + \delta M_{0},(\boldsymbol{a}-\boldsymbol{b})/2), \tag{40}$$

$\delta M_{0,n\alpha}$ is the variation of matrix element due to strain, $\boldsymbol{a}$ and $\boldsymbol{b}$ are the basis vectors in the plane ab, $\frac{1}{2}(\pm\boldsymbol{a} \pm \boldsymbol{b})$ are the positions of nearest-neighbour molecules,

$$\delta D_{\pm} = \tfrac{1}{2}(\delta D_1 \pm \delta D_2). \tag{41}$$

In order to calculate the vibronic levels one should make a substitution $\Delta \to \Delta_\nu$, $\mathscr{E}_\gamma \to \mathscr{E}_{\gamma M} - \hbar\omega_0$. The Green's function $\mathscr{G}(\mathscr{E}_\gamma)$ can be found from the optical spectrum of naphthalene in dueteronaphthalene (Sheka 1971). Thus, the eqs. (37)–(39) enable one to calculate the position of naphthalene bands for the given parameters of strain.

9. Experimental observation of stress-induced splitting of molecular bands in naphthalene crystals

An original technique of achieving arbitrary directed stress in molecular crystals was suggested by Ostapenko et al. (1976). The deformation was created by cooling the crystal on a substrate but, in contrast to the previous experiments (Broude et al. 1957, Broude and Tomashchik 1964), the substrate was chosen to be anisotropic with different thermal expansion coefficients for different directions. Symmetry-breaking deformation

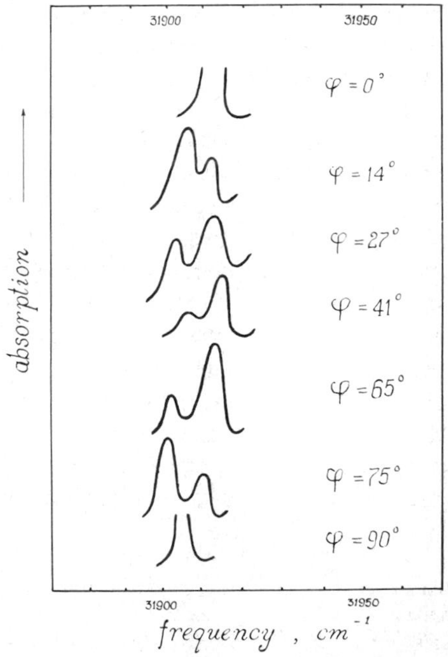

Fig. 2. Stress-induced splitting of M band of naphthalene crystal as a function of the angle φ between crystal and indole substrate axes (Ostapenko et al. 1978).

occurs if the principal axes of the investigated crystal and substrate do not coincide. By varying the angle φ between the crystal and substrate axes, arbitrarily directed stress can be achieved. In such a way Ostapenko et al. (1976) examined the spectrum of naphthalene on an indole substrate. The excitation of the bound state of the exciton with a nontotally symmetric vibration ($\hbar\omega_0 = 509\ \mathrm{cm}^{-1}$) of the naphthalene crystal gives the so-called *M* band. This band should be split in the presence of directional stress. Such a splitting of the *M* band was found. At the same time the exciton bands remained unsplit, only shifting to longer wavelengths. Figure 2 illustrates the experimentally observed splitting of *M* band versus the angle φ between the sample and substrate axes.

In order to check the accuracy of this interpretation Gorban' et al. (1978) investigated the optical spectrum of a deuteronaphthalene crystal with naphthalene impurity on an anisotropic paradibromobenzene crystal substrate. The exciton bands of deuteronaphthalene ($\mathscr{E}^a$ and $\mathscr{E}^b$), the vibronic band of naphthalene $\mathscr{E}_{M_n}$, the vibronic band of deuteronaphthalene $\mathscr{E}_{M_d}$ and the band of naphthalene impurity $\mathscr{E}_\gamma$ were analyzed. The absorption spectrum observed by Gorban' et al. (1978) is depicted in fig. 3. Since they did not measure the crystal strain, eqs. (37)–(39) contain the unknown parameters δD_+, δD_- and δM. These parameters were determined by using data on exciton bands and on the splitting of the impurity band of naphthalene in deuteronaphthalene. Then the positions of other five bands were calculated without unknown parameters. The values which were

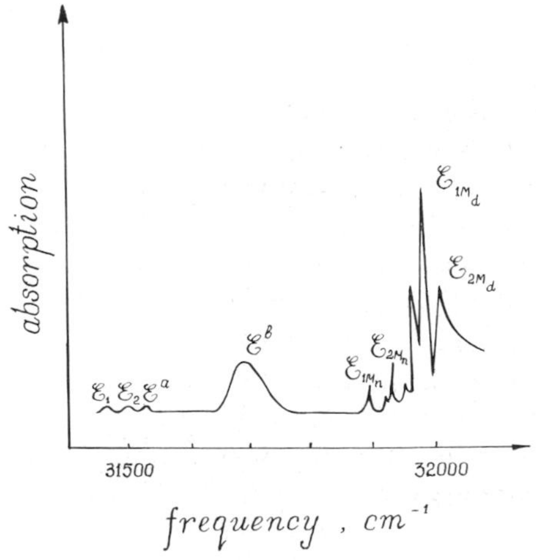

Fig. 3. Spectrum of the strained deuteronaphthalene crystal with naphthalene impurity (Gorban' et al. 1978).

Table 2.
Calculated and measured positions of exciton, impurity and vibronic levels in strained deuteronaphthalene crystal*

($\delta D_1 = 69$, $\delta D_2 = 26$, $\delta M = 0{,}5$).

Level		Theory	Experiment
$\mathscr{E}_1$		31 480	31 480
$\mathscr{E}_2$	- 118	31 511	31 510
$\mathscr{E}_2 - \mathscr{E}_1$		31	30
$\mathscr{E}_{1M_n}$		31 909	31 912
$\mathscr{E}_{2M_n}$	- 207	31 948	31 947
$\mathscr{E}_{2M_n} - \mathscr{E}_{1M_n}$		39	35
$\mathscr{E}_{1M_d}$		32 005	32 001
$\mathscr{E}_{2M_d}$	- 82	32 025	32 023
$\mathscr{E}_{2M_d} - \mathscr{E}_{1M_d}$		20	22

* All values are given in cm^{-1}.

found in such a way, together with experimental values are presented in table 2.

The difference $\delta D_2 - \delta D_1$ between the terms of molecules on different sublattices is constant for all bands registered in the same sample, but the stress-induced splittings of bands are different. As was shown in sect. 6, this should be attributed to the fact that the level position depends not only on the value of term but also on the excitation transfer interaction between

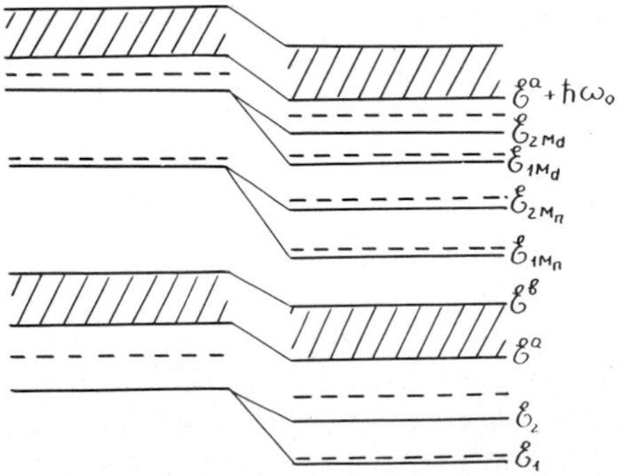

a b

Fig. 4. Energy level diagram of the impurity naphthalene ($\mathscr{E}_1$, $\mathscr{E}_2$) exciton ($\mathscr{E}^a$, $\mathscr{E}^b$) and M bands ($\mathscr{E}_{1M_n}$, $\mathscr{E}_{2M_n}$, $\mathscr{E}_{1M_d}$, $\mathscr{E}_{2M_d}$) in the unstrained deuteronaphthalene crystal (a) (solid line) and in the strained crystal (b). The dotted lines are the positions of molecular terms.

an impurity molecule and its environment. In the case of deep levels the contribution of the excitation transfer interaction is negligible. For instance, the value (35 cm^{-1}) of the splitting of the deepest M_n band is almost equal to the difference of terms ($\delta D_2 - \delta D_1 = 43\ cm^{-1}$). The splitting of shallow M_d level is half of this. These peculiarities are illustrated in fig. 4.

It is seen from the table 2, that the stress-induced change of the site shift is much greater than that of the excitation transfer matrix element. At the same time the term differences of the molecules in different sublattices are comparable to the shift of the bands.

10. *The determination of the exciton deformation potential constants*

The exciton–phonon interaction is of primary importance in considering many physical processes (Davydov 1968, Agranovitch 1968): the shape of the exciton absorption and luminescence bands, exciton migration, Raman scattering, etc. A great deal of experimental and theoretical work is devoted to the study of this interaction. Nevertheless, until recently quantitative values of the exciton–phonon interaction were not known for any molecular crystal. Recently deformation potential constants in naphthalene crystal have been evaluated by Ostapenko et al. (1978, 1979) in their experimental studies of the crystals cooled on an anisotropic substrate. The procedure of determining these constants is as follows: the stress-induced change of site shift δD_γ and excitation transfer matrix element δM is extracted from the optical data on exciton and vibronic bands of naphthalene crystal cooled on an indole substrate. Then the strain tensor components ϵ_{ij} are calculated from data on the crystal thermal expansion. Finally, using the known δD_γ, δM and ϵ_{ij} it is possible to determine the values of D_{ij} and M_{ij} from the formulae (10) and (12).

In our calculations we have used some simplifications.

(1) In naphthalene crystal the excitation transfer matrix elements between the molecules within the ab layer are known to be much greater than in the interlayer. Therefore, the matrix elements are more sensitive to the change of the intermolecular spacing in the plane ab than to that between the planes. For this reason we assume below that the site shift and excitation transfer matrix elements depend only on the tensor components ϵ_{xx}, ϵ_{xy}, ϵ_{yy}, and do not depend on the other components. Hence, for deformations of a crystal of naphthalene symmetry we have

$$\delta D_\gamma = D_{xx}\epsilon_{xx} + 2(-1)^\gamma D_{xy}\epsilon_{xy} + D_{yy}\epsilon_{yy}, \tag{42}$$

$$\delta M = M_{xx}\epsilon_{xx} + M_{yy}\epsilon_{yy}, \tag{43}$$

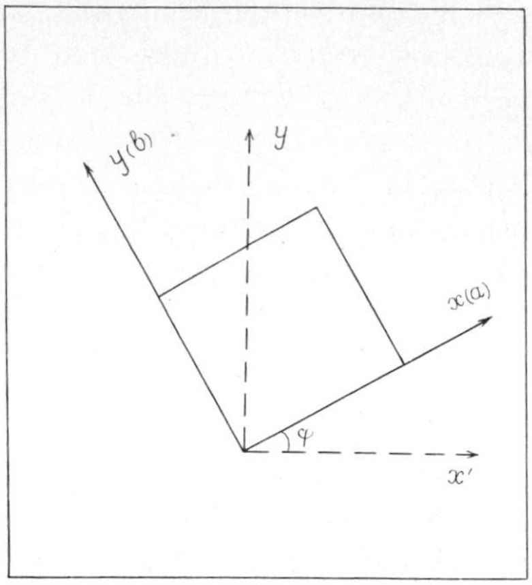

Fig. 5. The geometry of the contact of the naphthalene crystal with the substrate.

where the x- and y-axes are directed along the a- and b-axes of the crystal, respectively.

Thus, to determine the constants D_{xx}, D_{xy}, D_{yy}, M_{xx}, M_{yy} in this approximation, it is necessary to know only the strain tensor components ϵ_{xx}, ϵ_{xy}, ϵ_{yy}. In order to determine them we use a second simplification.

(2) Since the crystals under consideration were considerably thinner than the crystal substrate (fig. 5) we may assume the change of the dimensions of the naphthalene crystal in the plane ab to be equal to the temperature-induced strain of the substrate. The displacement vector in the plane ab, characterizing the deformation of naphthalene, is given by

$$\boldsymbol{U}_{\parallel} = \boldsymbol{U}_{\parallel}^{\mathrm{s}} - \boldsymbol{U}_{\parallel}^{\mathrm{n}} \tag{44}$$

where $\boldsymbol{U}_{\parallel}^{\mathrm{s}}$ and $\boldsymbol{U}_{\parallel}^{\mathrm{n}}$ are, respectively, the components of the displacement vectors of the free substrate and naphthalene on the interface when the system was cooled from room to helium temperature.

The indole crystal has rhombic symmetry, thus the x'- and y'-axes coincide with the principal axes of the thermal expansion ellipsoid. Meanwhile, the x-axis (a) in naphthalene crystal does not coincide with the principal axis. Then the components $U_{\parallel}^{\mathrm{s}}$ and $U_{\parallel}^{\mathrm{n}}$ are written in the following form

$$\begin{aligned} U_{x'}^{\mathrm{s}} &= \alpha_{11}^{\mathrm{s}} x', \\ U_{y'}^{\mathrm{s}} &= \alpha_{22}^{\mathrm{s}} y', \\ U_{x}^{\mathrm{n}} &= \alpha_{11}^{\mathrm{n}} x + \alpha_{13} z, \\ u_{y}^{\mathrm{n}} &= \alpha_{22}^{\mathrm{n}} y, \end{aligned} \tag{45}$$

where

$$\alpha_{ij}^{n(s)} = \int_{4,2^\circ}^{295^\circ} a_{ij}^{n(s)}(T)\,dT, \tag{46}$$

$a_{ij}^{n(s)}(T)$ are the thermal expansion coefficients. From eqs. (45) one can obtain the strain tensor components as a function of the angle between the x- and x'-axes,

$$\begin{aligned} \epsilon_{xx} &= \alpha_{11}^{s} - \alpha_{11}^{n} + (\alpha_{22}^{s} - \alpha_{11}^{s}) \sin^2 \varphi, \\ \epsilon_{xy} &= \tfrac{1}{2}(\alpha_{22}^{s} - \alpha_{11}^{s}) \sin 2\varphi, \\ \epsilon_{yy} &= \alpha_{22}^{s} - \alpha_{22}^{n} - (\alpha_{22}^{s} - \alpha_{11}^{s}) \sin^2 \varphi. \end{aligned} \tag{47}$$

The thermal expansion coefficients $a_{ij}^{n(s)}(T)$ of the naphthalene and indole were measured by Ostapenko et al. (1979). This enables them to obtain the strain tensor components from the relations (47). The values δD_γ and δM were calculated from the absorption spectrum data. Substituting these values in formulae (42) and (43) one finds the exciton deformation potential constants for the naphthalene crystal

$$D_{xx} = 4.2 \times 10^3\ \text{cm}^{-1}, \quad D_{xy} = D_{yx} = 1.9 \times 10^3\ \text{cm}^{-1}, \quad D_{yy} = 3.5 \times 10^3\ \text{cm}^{-1}, \tag{48}$$

$$M_{xx} = -2.7 \times 10^2\ \text{cm}^{-1}, \quad M_{yy} = -2.2 \times 10^2\ \text{cm}^{-1}. \tag{49}$$

In fig. 6 the dependence of the splitting of the M band on the angle φ between crystal and substrate axes is shown. The theoretical curve (solid line) calculated by using the constants (48) and (49) is in good agreement with the experimental data (circles).

Let us estimate the number of virtual acoustic phonons $\langle N \rangle$ which are accompanying the exciton. Using eq. (7.23) (Kittel 1963) and supposing $D_{xx} \sim D_{yy} \sim 4 \times 10^3\ \text{cm}^{-1}$ we obtain $\langle N \rangle = 0.3 < 1$. Therefore the exciton–phonon interaction in the naphthalene crystal may be considered as weak.

The parameters (48) and (49) can be used in various problems of exciton-acoustic phonon interaction effects in naphthalene.

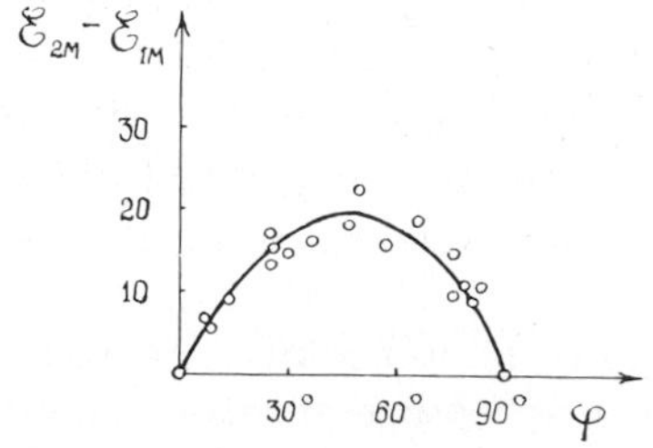

Fig. 6. M band splitting calculated by using the exciton deformation potential constants (48) and (49) (solid line).

11. Concluding remarks

The study of excitons in strained molecular crystals has been intensively pursued in recent years. The experimental studies have established the behaviour of some important physical parameters, such as the maxima and widths of exciton absorption and luminescence bands, or the decay rates of excited states with changing intermolecular distances and orientations. Unfortunately, most of these experimental investigations were carried out under hydrostatic pressure, and only a few reported on uniaxial stress effects in molecular crystals, though in semiconductors the uniaxial stress have proved to be a unique tool in analyzing the energy band structure (Bir and Pikus, 1972). In our view, it is the uniaxial stress procedures which seem to give the most perspective for optical studies of molecular crystals with the help of the stress technique. Let us consider some applications of this method.

(1) In the foregoing sections we described how the uniaxial stress procedure had enabled us to extract the values of the deformation potential of exciton– acoustic-phonon interaction in crystalline naphthalene. Until recently these values remained unknown for any molecular crystal though one needs them badly for a quantitative microscopic description of the shape of exciton absorption and luminescence bands, for an estimation of exciton diffusion coefficient or whenever the calculations of exciton eigen-states in inhomogeneously deformed crystals are attempted (Sugakov, Khotjaintsev 1976).

(2) The deformation potential constants can be also used for analyzing the dependence of the interaction between excited and nonexcited molecules on the intermolecular distances and orientations. The interaction between the nonexcited molecules can be evaluated from the calculation within framework of the atom–atom potential method, but we lack information about the interaction between the excited and nonexcited molecules. The site shift determines the interaction between the excited molecule and all the neighbouring molecules; however, from these data one can not extract, in fact, any reliable information concerning the mutual interaction in a single pair of molecules. Yet, this knowledge is required for the solution of some physical problems, such as the theory of the creation of excimers, for instance. In order to find it the intermolecular interaction ($D_{n\alpha,m\beta}$) can be approximated by some function. Then the constants of the approximation can be calculated by using the deformation potential approach.

(3) Spectroscopic study of uniaxially stressed crystals can be used to probe the exciton bands. In optical spectra of molecular crystals there are bands generated by the simultaneous excitation of dissociative states of an exciton and nontotally symmetric molecular vibrations. From these spectra

the density-of-state function of exciton bands can be obtained (Rashba 1966). Under uniaxial stress a gap between the exciton bands appears (fig. 1) and, as a result, a dip should be observed within the vibronic absorption band. The position of the dip enables one to determine the region where the exciton bands stick together.

Since the position of shallow levels depends on the parameters of exciton bands, there is another useful method of exciton band investigation by the optical study of shallow centers (Rashba 1957). The presence of the gap between exciton bands of uniaxially stressed crystals allows one to find the impurity, which generates the levels inside the gap, and to investigate the region of the band with $k \sim \pi/a,\ \pi/b$.

(4) Uniaxial stress can be used for inventing realistic models of the impurity center model in molecular crystals. The stress-induced dependence of position of impurity bands on the direction of the applied force can provide valuable information about the arrangement of molecules in the impurity center; the band splittings are informative about the symmetry of the center.

(5) Changing the direction of the applied force we can change the diffusion coefficient of the excitons. In crystals with narrow bands one can apply such a stress that the condition $|D_2 - D_1| > \mathscr{E}_D$ takes place. In this case the difference between the terms of molecules of different sublattices exceeds the Davydov splitting so the excitons move completely in one of the sublattices and the diffusion coefficient of excitons can considerably change in comparison with that in unstressed or isotropically stressed crystal, where excitons move in both sublattices.

(6) It is known that the ESR spectra of triplet excitons depend on the sublattice distribution of electronic excitation of the exciton state (Sternlicht, McConnell, 1960). Uniaxial stress changes this distribution and, hence, removes the ESR band of triplet excitons. If $|D_2 - D_1|$ changes from zero to a value greater than $\mathscr{E}_D$, the ESR spectrum changes from the spectrum of triplet excitons to that of a single molecule (Andreyev, Sugakov, 1975). Thus, from the ESR data the parameters of the exciton motion can be obtained.

It should be pointed out that further progress in the uniaxial stress spectroscopy of molecular crystals demands that a new technique be devised to create the uniaxial stress. Up to now the stress resulted from cooling the crystals on anisotropic substrates. Evidently, this procedure may be applied for studying thin samples.

Acknowledgments

The author gratefully acknowledges the valuable comments of E.I. Rashba. The author is also very grateful to M.T. Shpak, N.I. Ostapenko and M.P.

Chernomorets for constructive discussions on this article and for the use of their experimental results.

References

Afanasjeva, G.K., 1969, Sov. Phys.- Crystallogr. **13**, 892.

Afanasjeva, G.K., K.S. Alecsandrov and A.I. Kitaigorodskii, 1967, Phys. status solidi, **24**, K 61.

Agranovich, V.M., 1968, Teoriya eksitonov, Nauka, Moskva.

Andreyev, V.A., V.I. Sugakov, 1975, Ukr. Fiz. Zh. **20**, 1998.

Arnold, S., W.B. Whitten and A.S. Damask, 1974, J. Chem. Phys. **61**, 5162.

Bir, G.L., G.E. Pikus, 1972, Simmetriya i deformatsionnye effekty v poluprovodnikakh, Nauka, Moskva. (Symmetry and strain-induced effect in semiconductor, Wiley, New York. Toronto).

Born, M., and K. Huang, 1954, Dynamical theory of crystal lattices (Oxford U.P., London).

Broude, V.L. and A.K. Tomashchik, 1964, Ukr. Fiz. Zh. **9**, 39.

Broude, V.L., O.S. Pachomova and A.F. Prichot'ko, 1957, Opt. Spektr. **2**, 323.

Craig, D.P., 1955, J. Chem. Soc. **539**, 2302.

Cruickshank, D.W.J., 1957, Acta Crystallogr. **10**, 504.

Davydov, A.S., 1968, Theoriya molecularnikh excitonov, Nauka, Moskva. (1971, Theory of molecular excitons, Plenum Press, New York).

Drickamer, H.G., 1975, Pure and Appl. Chem. **43**, 379.

Elnahwy, S., M.El. Hamamsy, A.S. Damask, D.E. Cox, W.B. Daniels, 1978, J. Chem. Phys. **68**, 1161.

Gorban', K.N., N.I. Ostapenko, V.I. Sugakov, M.P. Chernomorets and M.T. Shpak, 1978, Fiz. tverd. tela **20**, 1261; Sov. Phys. Solid State **20**, 728.

Hamamsy, M.El., S. Elnahwy, A.S. Damask, 1977, J. Chem. Phys. **67**, 5501.

Hong, H.-K., R. Kopelman, 1970, Phys. Rev. Lett. **25**, 1030.

Huntington, H.B. and S.G. Gangeli, 1969, J. Chem. Phys. **50**, 3844.

Jones, P.F., 1968, J. Chem. Phys. **48**, 5448.

Jones, P.F. and M. Nicol, 1968, J. Chem. Phys. **48**, 5457.

Kalinowski, J. and R. Jankowiak, 1978, Chem. Phys. Lett. **53**, 56.

Kalinowski, J., J. Godlewski and K. Jankowiak, 1976, Chem. Phys. Lett. **43**, 127.

Kitaigorodskii, A.I., 1966, J. Chem. Phys. **63**, 9.

Kittel, C., 1963, Quantum theory of Solids (Wiley, New York, London).

Krivenko, T.A., E.I. Rashba and E.F. Sheka, 1978, Mol. Cryst. Liq. Cryst. **47**, 119.

Nakashima, T.T. and H.W. Offen, 1968, J. Chem. Phys. **48**, 4817.

Nicol, M., 1965, J. Opt. Soc. Am. **55**, 1176.

Offen, H.W., 1965, J. Chem. Phys. **42**, 2523.

Offen, H.W., 1966, J. Chem. Phys. **44**, 699.

Offen, H.W. and B.A. Baldwin, 1966, J. Chem. Phys. **44**, 3645.

Ostapenko, N.I., V.I. Sugakov and M.T. Shpak, 1971a, J. Lumin. **4**, 261.

Ostapenko, N.I., V.I. Sugakov and M.T. Shpak, 1971b, Phys. Status Solidi (b) **45**, 729.

Ostapenko, N.I., G.Yu, Khotyaintseva, M.P. Chernomorets and M.T. Shpak, 1976, Zh. exper. teor. Fiz., Pisma **23**, 377. (Sov. Phys.-JETP Lett. **23**, 341).

Ostapenko, N.I., V.I. Sugakov, M.P. Chernomorets and M.T. Shpak, 1978, Zh. exper. teor. Fiz. Pisma, **27**, 452 (1978, Sov. Phys.-JETP Lett. **27**, 423).

Ostapenko, N.I., V.I. Sugakov, M.P. Chernomorets and M.T. Shpak, 1979, Phys. Status Solidi (b) **93**, 493.

Otto, A., R. Keller and A. Rahman, 1977, Chem. Phys. Lett. **49**, 145.
Pawley, G.S., 1967, Phys. Status Solidi **20**, 347.
Rabinkina, N.V., E.I. Rashba and E.F. Sheka, 1970, Fiz. tverd. Tela **12**, 3569; Sov. Phys. Solid State **12**, 2898.
Rashba, E.I., 1957, Opt. Spectrosk. **2**, 568.
Rashba, E.I., 1966, Zh. exper. teor. Fiz. **50**, 1064 (1966, Sov. Phys.-JETP **23**, 708).
Schipper, P.E., 1974, Mol. Cryst. Liq. Cryst. **28**, 401.
Sheka, E.F., 1971, Fiz. tverd. Tela **12**, 1167; Sov. Phys. Solid State **12**, 911.
Shirotani, J., Y. Kamura and H. Inokuchi, 1974, Mol. Cryst. Liq. Cryst. **28**, 345.
Sternlicht, H and H.M. McConnell, 1960, J. Chem. Phys. **35**, 1793.
Sugakov, V.I., 1973, Fiz. tverd. Tela **15**, 2513 (1974, Sov. Phys.-Solid State **15**, 1670).
Sugakov, V.I. and V.N. Khotyaintsev, 1976, Zh. eksp. teor. fiz., **70**, 1566 (1976, Sov. Phys. JETP, **43**, 817).
Tanaka, J., T. Koda, S. Shionoya and S. Mimomuras, 1965, Bull. Chem. Soc. Japan **38**, 1159.
Teslenko, V.F., 1967, Kristallografiya, **12**, 1082; Sov. Phys. Crystall. **12**, 946.
Walmsley, S.H., 1968, J. Chem. Phys. **48**, 1438.
Whitten, W.B. and S. Arnold, 1976, Phys. Status Solidi (b) **74**, 401.
Wiederhorn, S. and H.G. Drickamer, 1959, J. Phys. Chem. Solid **9**, 330.

CHAPTER 17

Excitons in Photosynthetic and Other Biological Systems

R.M. PEARLSTEIN

Battelle Memorial Institute
Columbus Laboratories
505, King Ave
Columbus, Ohio 43201
U.S.A

Excitons
Edited by
E.I. Rashba and M.D. Sturge

Contents

1. *Introduction*

This chapter is concerned with the molecular excitons found in structures of biological interest. Among these, the chlorophyll arrays of photosynthetic organisms provide the premier example. For excitation energies below the vacuum ultraviolet, i.e., 6 or 7 eV, clear evidence for delocalization of electronic excitation energy over a set of chemically identical chromophores so far has been cited in only a few instances. Besides the photosynthetic case, the most notable examples are some of the nucleic acids and possibly the purple membrane of the salt-loving bacteria. It can be assumed that excitons in all of these systems have no charge-transfer character, that is, they are purely Frenkel excitons.

1.1. *Historical perspective*

The idea that excitons might play a role in photosynthesis was introduced only shortly after the publication of Frenkel's original papers (Frenkel 1931, 1936). Emerson and Arnold (1932) showed that chlorophyll molecules in green plants behave cooperatively in the conversion of absorbed light energy into stable chemical bond energy. A few years later, Franck and Teller (1938) discussed whether exciton transport might be the mechanism underlying this cooperativity. Because of the paucity of experimental data then available, Franck and Teller chose too simple a structural model for the arrangement of chlorophyll molecules *in vivo*, and so concluded in the negative. From the theoretical point of view the matter then lay fallow for the next quarter century, until, armed with the latest experimental findings on chloroplast structure and chlorophyll fluorescence lifetimes, Bay and Pearlstein (1963) established a theoretical model (based on Förster's theory (Förster 1948) of intermolecular excitation-energy transfer) in which hopping exciton transport is the mechanism of cooperativity.

The problem of exciton transport in photosynthetic systems reached a turning point three years later. After the 1966 Brookhaven symposium on photosynthesis (Pearlstein 1967, Robinson 1967), the existence of such transport was no longer questioned, although even now details remain to be explained and new techniques (e.g., picosecond spectroscopy) provide fresh problems for the theorist. The reason for the general acceptance of photosynthetic exciton transport at that time was simply this: In the compact chlorophyll structures found in photosynthetic organisms, exciton transport is too rapid theoretically to be a rate-limiting factor in the photochemical conversion of light energy. Put more nearly in the parlance of solid-state physics, in photosynthetic chlorophyll arrays, which are naturally "doped" with ~1 percent photochemical reaction centers, exciton trapping at the centers is not diffusion limited. Since the early 1960's much has been done to reinforce these conclusions both experimentally and theoretically, as well as to provide evidence for the existence of excitons as states of definite energy in photosynthetic systems. (See, Knox 1975, 1977, Sauer 1975, 1978, for reviews.)

1.2. Definitions

The term "exciton" is frequently used in the biophysical literature (Sybesma 1977) in a much looser sense than is customary among solid-state physicists. I shall necessarily use this biophysical definition in which an "exciton" is an elementary electronic excitation that during its lifetime resides on each of two or more *identical* molecules or chromophores. (For the sake of brevity, I deliberately exclude the subject of heterogeneous intermolecular excitation transfer, which is of considerable interest both in photosynthesis and in other parts of biology.) Thus, in biology, an exciton need not have definite momentum.

In biological systems, exciton transport usually refers to a hopping motion (see below). For this reason, the term "exciton trapping" is used to mean the localization of an exciton on a single molecular (or chromophoric) site. If the localization is completely irreversible, the term "exciton quenching" is sometimes used. Thus, in biology, exciton trapping or quenching means the removal of the exciton energy from the system of identical molecules into a localized state of a dissimilar molecule or group of molecules.

1.3. Experimental methods

Historically, the first evidence for excitons in biology was indirect. Their existence in photosynthetic organisms was inferred from the shortening of the chlorophyll fluorescence lifetime *in vivo*, i.e., by the kinetic effect of exciton quenching as an extra decay channel for the parent electronic excited state (in this case the S_1 state of chlorophyll). This is still a useful technique, especially when the quenching kinetics can be studied quantitatively. Although it cannot prove the existence of excitons without confirming evidence, the decay of luminescence intensity as a function of the state and number of exciton quenchers, and of time, can yield information about exciton transport.

2. Photosynthetic systems

The literature on excitons in photosynthetic systems is now too large to cover exhaustively in a chapter of this length. Although most topics of current interest are mentioned at least briefly, I treat in some detail only two. The first of these, the singlet exciton states of chlorophyll in certain protein complexes, is of particular interest to biochemists and biophysicists because of its possible relevance in analyzing the geometrical arrangement of chlorophyll molecules in the photochemical reaction centers. (This should not be taken to imply that exciton analysis is generally useful for deriving structural parameters.) The second topic, the transport of excitation energy as singlet excitons through the chlorophyll in the lipoprotein complexes of photosynthetic membranes, and the trapping of those excitons by the reaction centers, probably holds greater allure for the physicist. Its main outlines are clear, but much remains that is controversial.

2.1. Photosynthetic concepts

Photosynthesis is a vast subject, for a general understanding of which the reader may consult any of a number of textbooks or monographs (see, for example, Rabinowitch and Govindjee 1969, Gregory 1977, Clayton 1981). Here, I summarize some of the concepts essential in discussing excitons.

All photosynthetic organisms use the energy of light to chemically reduce carbon dioxide into carbohydrates and other products. (By this definition the salt-loving bacteria with their purple membranes are *not* photosynthetic.) All photosynthetic organisms use some form of the molecule chlorophyll as a photocatalyst in their oxidation–reduction (redox) reactions. There are two broad classes of these organisms, the photosynthetic bacteria and the green plants (including algae and cyanobacteria, or blue-green algae). The former use either bacteriochlorophyll (Bchl) *a* or Bchl *b* as photocatalyst, perform a single type of photochemical reaction, and do not evolve molecular oxygen as a by-product. The latter use chlorophyll (Chl) *a* exclusively as photocatalyst, and perform two distinct types of (electrochemically coupled) photochemical reactions, one of which splits water to evolve oxygen.

Photosynthetic chemistry involves both light and dark reactions. The former involve all steps leading from photon capture to the first chemically stable products, the latter to chemistry energized by these first stable products. In the light reactions some of the energy is used chemiosmotically. We shall be concerned here only with the physical processes leading up to and including the first step in photochemistry, i.e., light absorption, exciton transport, and quenching of the exciton energy by production of charge-separated states at the photochemical reaction centers.

It has already been mentioned that chemically distinct forms of chlorophyll appear in various organisms. There are also functional distinctions to be made regarding chlorophylls of the same chemical form. A universal distinction in all organisms is between photochemically active and photochemically inactive chlorophyll. The latter is called "antenna" chlorophyll, the former "reaction center" chlorophyll. In the green plants, there are also differences in spectroscopic and photochemical properties of the Chl *a* in each of the two photosystems, called "System 1" (the CO_2-reducing system) and "System 2" (the O_2-evolving system). Another example is provided by Bchl *a*, which has at least different spectroscopic properties in green versus purple photosynthetic bacteria. It is generally supposed that all of these differences are the result of varying physical interactions of the chlorophylls *in situ* and not of subtle chemical changes that disappear upon solvent extraction.

It is not completely clear why antenna chlorophyll exists at all. The prevalent view is that antennas increase the photon absorption cross section without incurring the metabolic cost that a proliferation of reaction centers might engender. Another possible function is the rapid transport of excitation energy away from a temporarily oxidized (i.e., "busy") reaction center, thereby lowering the probability of photodamage to its pigments (see sects. 2.2.2.2 and 2.3).

It is now recognized that, *in vivo*, all chlorophyll is bound to protein (Barber 1979, Delepelaire and Chua 1979, Markwell et al. 1979). A number of distinct chlorophyll–protein complexes have been isolated from various organisms and characterized in varying degrees. These isolated complexes are distinguished as antenna or reaction center complexes depending on whether they do not or do display photochemical activity. (The terms "antenna" and "reaction center" are more precisely used in connection with these complexes.) Reaction center complexes typically contain not only the principal chlorophyll of the parent organism, but also a derivative of that chlorophyll, called a pheophytin, in which the central magnesium atom is replaced by two hydrogens. In reaction center complexes from purple photosynthetic bacteria, chlorophylls and pheophytins have a definite 2:1 stoichiometry. Some of the antenna complexes from green plants contain the accessory pigment, Chl *b*, always a photochemically inactive pigment *in vivo*, as well as Chl *a*. Some organisms have protein complexes or other structures that contain only accessory pigments (e.g., phycobilins, Bchl *c*), but these are not considered here.

Reaction center complexes, and most antenna complexes, are found, *in vivo*, as part of lipid bilayer membranes. These membranes are very well defined in green plants, where they are called "thylakoids" and appear in the cellular organelles called "chloroplasts". Thylakoids contain a number of enzymes and other proteins in addition to the chlorophyll proteins. The structure of most chlorophyll proteins is little known and that of thylakoids even less so. However, it seems likely that the chlorophyll proteins "float" in the lipid membrane, and are capable of some lateral motion. In some cases (Park and Biggins 1964, Staehelin et al. 1980), the thylakoid displays a two-dimensional crystalline structure with a lattice spacing (10–20 nm) appropriate for protein complexes. The problem of incompletely specified structures is a critical one for exciton analysis, and is addressed frequently in what follows.

2.2. Singlet excitons in chlorophyll- and bacteriochlorophyll-protein complexes

It has already been stated that the singlet excited states are the important ones in photosynthesis, especially the S_1 which is the photochemically active state. The S_1 is also the most important state from the exciton viewpoint. In Bchl *a* the $S_0 \to S_2$ transition is well resolved from the $S_0 \to S_1$, so that S_2-exciton states may be of some interest in bacteria. Isolated Bchl-protein complexes have been studied longer and are better defined than their green plant counterparts. For this reason, they are worthy of the greater attention given them here. I consider only those complexes for which specific exciton-interaction models have been proposed.

2.2.1. Bacterial antennas

2.2.1.1. The Fenna–Matthews structure. As noted before, most chlorophyll-protein complexes are immersed in a phospholipid membrane. So far these have only been solubilized (without loss of protein tertiary structure) in the presence of detergent. However, a few complexes are water-soluble, and one of these – the Bchl *a*-protein from the green bacterium, *Prosthecochloris aestuarii* – has been structurally analyzed by single-crystal X-ray diffraction methods. In *P. aestuarii*, this protein sits atop the photochemically active membrane, and serves as an intermediate antenna, funneling energy into the Bchl *a* of the membrane's reaction center complexes from the large Bchl *c*-containing vesicles with which the Bchl *a*-protein is also in contact (Olson 1978a, 1981).

Fenna and Matthews (1975) published a structural model for the Bchl *a*-protein based on an electron density map having 0.28 nm resolution. Atomic coordinates of the Bchl *a* molecules in the protein, and other details, have since also been given (Fenna et al. 1977, Matthews et al. 1979). (The reader should be aware that the taxonomic identification of the organism from which the protein is derived has, until recently, been in question. In 1975, it was still called "*Chlorobium limicola*", rather than *P. aestuarii*. However, only the name has changed, not the material. (See, Olson 1978b.) It cannot be emphasized too strongly that, as of this writing, the Fenna–Matthews structure remains unique: it provides the only detailed information available about the state of any chlorophyll in a native protein complex. If the structure should prove atypical of chlorophyll *in vivo*, overemphasis of its properties could be quite misleading. However, as I hope to make clear here, spectroscopic properties (particularly as regards exciton aspects) that the Bchl *a* of this protein has in common with the Bchl *a* in other protein complexes strongly suggests that the interactions of Bchl with its surroundings in this protein are not at all atypical. On this basis, one can believe study of this protein ought to be most rewarding.

The Bchl *a*-protein consists of 21 molecules of Bchl *a*, three identical polypeptides each consisting of 358 amino acid residues, and possibly one or two molecules of bound water per polypeptide. These constituents are arranged in three identical subunits about an axis of three-fold symmetry. In each subunit there is a core of 7 molecules of Bchl *a* about which one of

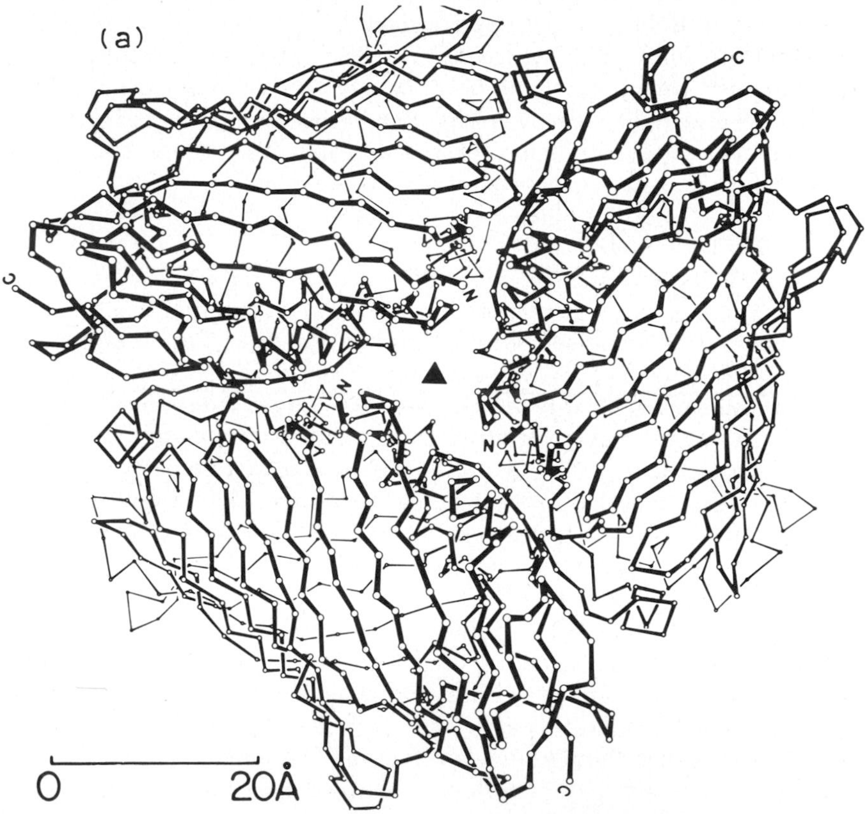

Fig. 1. Structure of the Bchl *a*-protein from *P. aestuarii* based on X-ray diffraction analysis of single crystals. (a) Projection looking along the three-fold symmetry axis, showing the polypeptide backbones of the three identical subunits.

the polypeptides is closely wrapped in a sack-like structure (fig. 1). The polypeptide "sack" is too tightly "woven" to admit even small molecules, except at the end closest to the three-fold axis. However, that end is occluded by the polypeptide of the neighboring subunit, so that in the intact protein the interior of each subunit is totally isolated in a chemical sense. It should be noted that, to date, all attempts to isolate individual subunits have resulted only in complete denaturation of the protein.

In rough outline the Bchl *a*-protein is an oblate ellipsoid of revolution of 5.7 nm minor axis (the three-fold axis) and 8.7 nm major axis. The Bchl *a* core in each subunit is circumscribed by an ellipsoidal region of dimension

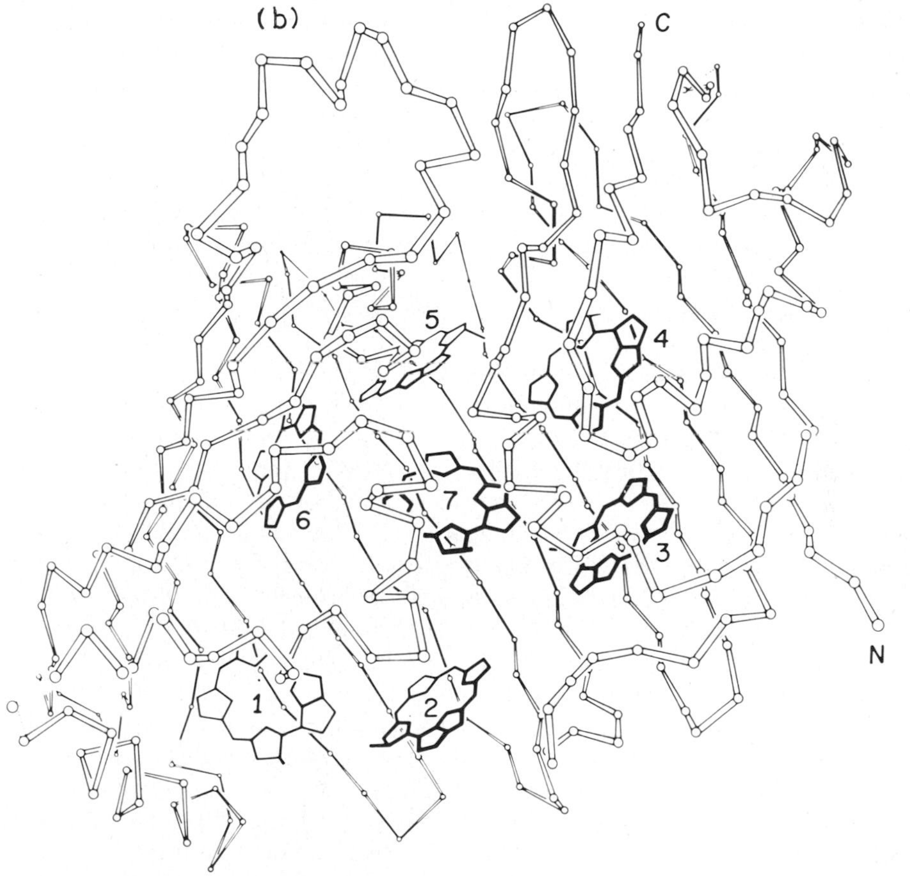

Fig. 1. (b) Expanded view of one subunit showing the polypeptide backbone and the macrocyclic rings of the seven Bchl molecules; three-fold axis parallel to the bottom edge of the page.

4.5 × 3.5 × 1.5 nm. The average center–center distance of the porphyrin ring "heads" of the Bchl *a* molecules is 1.2 nm for nearest neighbors, which is also the approximate diameter of the porphyrin ring. However, because of the relative orientations of the 7 rings there are very few close approaches (less than, say, 0.3 nm) between atoms of any two neighboring rings. Except for the fact that they lie within an angle of 40° of a mean plane, the porphyrin rings in each subunit are arrayed with no obvious symmetry. The angle between the three-fold axis and the normal to each mean plane is 62°. The distance from the three-fold axis to the center of each Bchl aggregate is 2.0 nm.

Before the X-ray structure of the Bchl *a*-protein was determined, Philipson and Sauer (1972) interpreted its low temperature absorption and CD spectra in terms of exciton interactions. Their interpretation is supported by four arguments: (1) Each of the two lowest energy electronic transitions of the Bchl in the protein displays a conservative CD spectrum with individual features of large rotational strength. (2) Each absorption band is split into subbands of unequal dipole strengths. (3) Each of the absorption subbands is much narrower than the absorption band of Bchl *a* in solution. (4) The overall width of the lowest energy electronic absorption band in the protein ($\sim 500\ \mathrm{cm}^{-1}$) is consistent with the exciton bandwidth expected for interacting dipoles about 1.2 nm apart. (Indeed, Philipson and Sauer correctly inferred this distance before it was known from the X-ray structure.) Spectroscopic data obtained since then (Olson et al. 1976, Whitten et al. 1978) have served only to strengthen Philipson and Sauer's arguments.

In spite of the success of Philipson and Sauer's interpretation, the known X-ray structure, and the simple physics of dipole–dipole interactions, the exciton states of the Bchl in the protein are still not fully characterized. The reason for this is that the exciton interactions are not the only, or even the largest, perturbation on the electronic transitions of chlorophyll *in vivo* or in native protein complexes compared to chlorophyll in solution. Quantitatively, this point has only been realized recently, from study of the Bchl *a*-protein and the LH-R26 complex (see below). The nature of this major perturbation has not yet been determined, and discussion of hypotheses regarding it are beyond the scope of this chapter. I will discuss here two effects, one of which is the defining characteristic of this perturbation. The other effect, which is probably a result of it, is the one that complicates the characterization of exciton states in the Bchl *a*-protein.

The first effect is observed as an energy shift. The centroid of the absorption band corresponding to the $S_0 \rightarrow S_1$ transition of Bchl *a* occurs at an energy lower by about $600\ \mathrm{cm}^{-1}$ in the protein than in an organic solvent (Philipson and Sauer 1972). This is an example, mentioned earlier, of the well-known but as yet not understood bathochromic absorbance shift of all chlorophylls *in vivo*.* Before the calculation of exciton effects based on the Fenna–Matthews structure, it was often assumed that such effects account for much of the shift. However, the calculations explicitly showed that at most a quarter of the shift energy (150 cm^{-1}) can be attributed to exciton effects, i.e., a shift of oscillator strength to the lower energy exciton subtransitions as a result of the geometry of the interacting transition dipoles (Pearlstein and Hemenger 1978). Thus, if one calculates the absorption band corresponding to the $S_0 \rightarrow S_1$ transition, one must shift the calculated spectrum by some 450 cm^{-1} toward lower energies to superimpose calculated and experimental centroids.

Merely shifting the calculated spectrum does not suffice to bring theory and experiment into agreement. One must also deal with the second effect, which is that the relative magnitudes and relative positions of individual peaks (i.e., individual exciton subtransitions) in both absorption and CD, and even signs of some CD features, are also given incorrectly by theory. Pearlstein and Hemenger (1978) have discussed empirical ways to deal with this problem. They found (fig. 2) that reasonably good agreement between theory and experiment can be obtained if the perturbation has at least the following two major consequences: A nonexcitonic bathochromic shift of the Bchl *a* $S_0 \rightarrow S_1$ transition energy amounting to 605 cm^{-1}, and a 90° rotation of the $S_0 \rightarrow S_1$ transition dipole direction from that predicted by molecular orbital theory for "unperturbed" Bchl. (The dipole moments of all the low-lying electronic transitions of Chl *a* or Bchl *a* are approximately directed along one or the other of two orthogonal axes in the plane of the porphyrin ring. See, Weiss 1972.) It is noteworthy, in view of findings with the LH-R26 protein complex (see below), that the theory with rotation hypothesis attributes *none* of the observed bathochromic shift to exciton effects. One must nonetheless treat the rotation hypothesis with caution pending additional experimental confirmation, such as determination of the LD spectrum of Bchl *a*-protein in a well-defined, oriented system.

Good absorption and CD spectra in the region of the $S_0 \rightarrow S_2$ transition are available for Bchl *a* both in solution and in the Bchl *a*-protein (Philipson and Sauer 1972). However, this transition has only ~10 percent of the oscillator strength of the $S_0 \rightarrow S_1$ transition, and the S_2 exciton interaction energies are correspondingly smaller. Experimentally, very little structure is resolved in the $S_0 \rightarrow S_2$ absorption and CD bands. (In addition, the relative weakness of the exciton interactions increases the importance of

* Such "bathochromic" shifts (i.e. shifts to lower energy), on going from the isolated molecule to an ordered array or complex, are common in molecular systems and are usually attributed to "crystal field" effects (see ch. 16). However, in chlorophyll the shift is largest in the noncrystalline protein complex, and the change on crystallization is small.

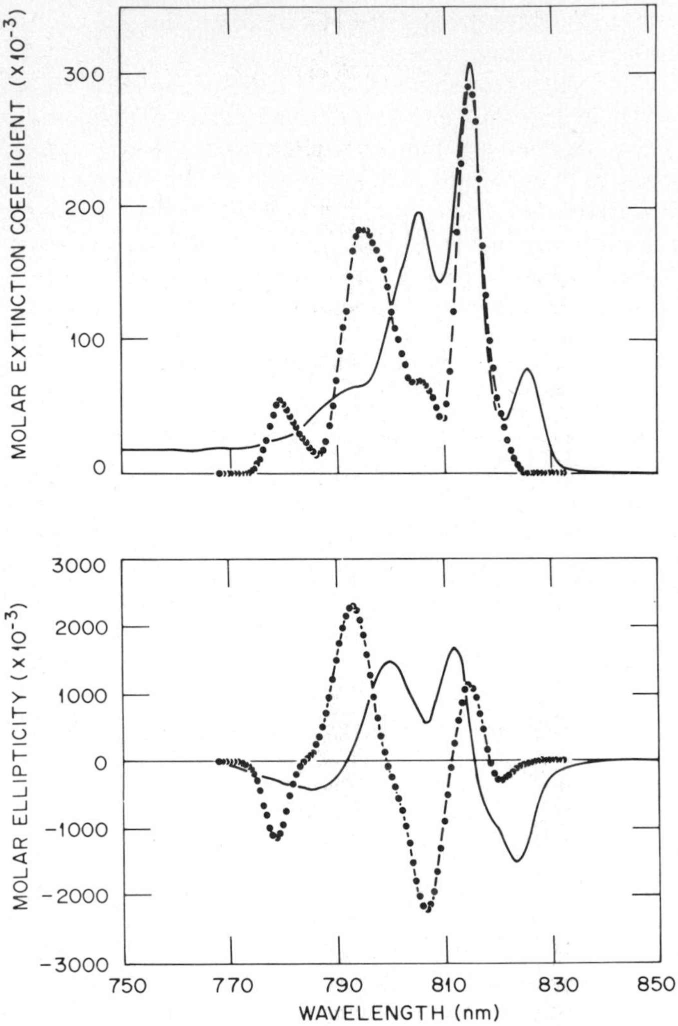

Fig. 2a. Comparison of theory (points and dashed curves) and experiment (solid curves) for absorption (labelled "molar extinction") and circular dichroism (labelled "molar ellipticity") of Bchl *a*-protein from *P. aestuarii* at low temperature (77 K, CD; 5 K, absorption). (a) Theory for standard (*y*-axis) $S_0 \rightarrow S_1$ transition dipole directions in each Bchl molecule.

perturbing effects, negligible for the S_1 band, that complicate the analysis.) For this reason, no detailed theoretical analysis of the S_2 exciton band has been given.

2.2.1.2. The LH-R26 protein. The single photosynthetic organism whose chlorophyll-protein complexes have been most intensively studied is the

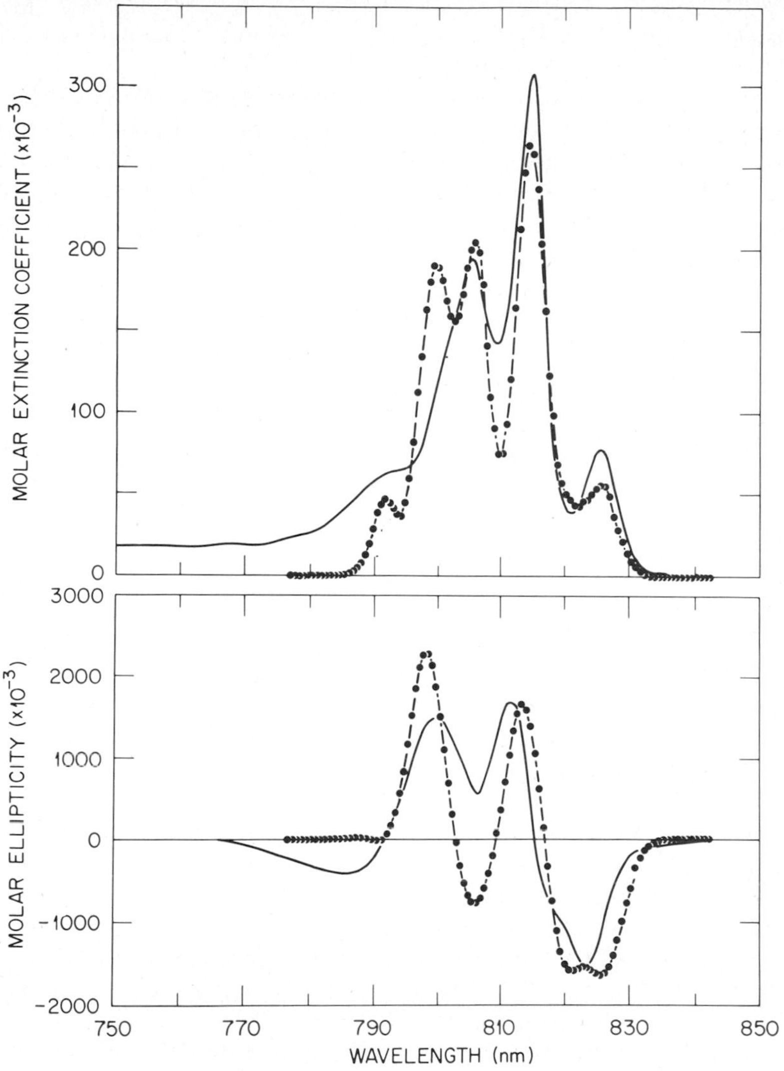

Fig. 2b. Theory for $S_0 \rightarrow S_1$ transition dipole directions rotated by 90° (parallel to molecular *x*-axes) in each Bchl molecule.

purple bacterium, *Rhodopseudomonas sphaeroides*, R26 mutant (Thornber 1975, Olson and Thornber 1979). *Rp. sphaeroides* yielded the first (Reed and Clayton 1968), and now best characterized, reaction center complex (see below). It has recently been shown (Sauer and Austin 1978) that, spectroscopically, all of the Bchl *a* (the only chlorophyll form) in *Rp. sphaeroides* R26 occurs either in those reaction center complexes or in an antenna complex, called LH-R26 ("LH" for "light-harvesting"). The latter complex is small (molecular weight, 20 000) and contains only two pigmented molecules, both Bchl *a*.

Even though its three-dimensional structure is unknown, the fact that the LH-R26 complex contains only two molecules of Bchl *a* makes it of special interest for exciton analysis. This is because for each electronic excited state of Bchl *a* there is only one exciton interaction, whose magnitude (at least for S_1) can be approximately deduced from the spectra. Ambient temperature absorption, CD, and LD (of complexes oriented in stretched polyvinyl alcohol films) spectra have been determined. Bolt and Sauer (1979) have deduced a value of 200 cm^{-1} for the S_1-exciton splitting energy ($2J$) from an exciton analysis of the LD data. The CD spectrum has the symmetric, conservative form expected for a single, dimeric exciton interaction.

From their analysis, Bolt and Sauer conclude that the two S_1-exciton transition energies are $\nu_+ = 11\,745\ cm^{-1}$ (851.4 nm) and $\nu_- = 11\,531\ cm^{-1}$ (867.2 nm), with an oscillator strength ratio, $F_+/F_- = 1.53$. From the latter, and the equation

$$F_+/F_- = \cot^2(\alpha/2), \tag{4}$$

they conclude that the angle, α, between the individual Bchl *a* $S_0 \rightarrow S_1$ transition dipoles is 78°. However, this analysis does not necessarily represent a "best fit" to absorption, CD, and LD simultaneously, so that its conclusions regarding structure are provisional.

Although the geometry of transition dipoles within LH-R26 remains to be confirmed, some aspects of the analysis –such as the magnitude of the energy J– are already fairly secure. Particularly firm is the value, ν_0, of the band-center energy for the S_1 transition, which can be estimated by two independent routes. From the LD analysis, $\nu_0 = 0.5\ (\nu_+ + \nu_-) = 11\,638\ cm^{-1}$ (859.2 nm); from the zero crossing of the CD spectrum, $\nu_0 = 11\,610\ cm^{-1}$ (861.3 nm). Thus, in the LH-R26 protein, Bchl *a* displays a nonexcitonic bathochromic absorbance shift of $12\,987\ cm^{-1} - 11\,638\ cm^{-1} = 1349\ cm^{-1}$, more than twice that for Bchl *a* in the Bchl *a*-protein from *P. aestuarii*. Moreover, because the peak of the absorption spectrum itself occurs at $11\,696\ cm^{-1}$ (855 nm), the exciton interaction actually induces a small (58 cm^{-1}) *hypsochromic* absorption shift.

There are two significant points of similarity between the S_1 exciton interactions in the LH-R26 complex and the Bchl *a*-protein from *P. aestuarii*. First, the interaction energy in the former has about the same magnitude as do the nearest-neighbor interactions in the latter. Second, none of the bathochromic absorbance shift energy, which is an order-of-magnitude larger than J in either case, can be attributed to exciton effects in the LH-R26 protein, nor in the *Prosthecochloris* protein if the transition moment rotation hypothesis is correct.

2.2.2. *Other complexes*

2.2.2.1. Chl a/b protein. Excitonic analysis of most Chl-protein complexes from the O_2-evolving organisms encounters two new difficulties beyond those already discussed for the bacterial systems. First, the occurrence of two distinct photochemical reactions in the former adds a new dimension of functionality. In addition to antenna complexes that feed excitation energy exclusively to reaction centers of one or the other reaction, there are antenna complexes that can provide excitation energy to either reaction (Butler 1978). Second, for reasons that remain obscure, green plant Chl-protein complexes are generally less well characterized than their bacterial counterparts. There is no structurally determined green-plant complex, nor is there one that has only two pigment molecules which display any sign of an exciton interaction. Reaction center complexes are particularly ill-defined; of the two types, the better defined are the Photosystem 1 (P700) complexes, which contain ~10 molecules of Chl *a* (Thornber 1975).

One of the more thoroughly studied green plant complexes is the Chl *a*/*b*-protein (also known as CP_{II}, or LHCP), usually obtained from spinach chloroplasts. The Chl *a*/*b*-protein serves as a Photosystem 2 antenna complex, and has been the object of a thoughtful and interesting study regarding exciton interactions among its pigment molecules (Van Metter 1977a, 1977b, Shepanski, 1980).

The Chl *a*/*b*-protein complex has a molecular weight of about 35 000, and apparently includes 3 molecules each of Chl *a* and Chl *b*, one carotenoid molecule, and at least one polypeptide. Van Metter (1977a and 1977b) measured ambient temperature absorption, CD, fluorescence excitation and emission, and fluorescence polarization (both excitation and emission) spectra of this complex. He also presents an analysis of this data in terms of possible exciton interactions among the $S_0 \rightarrow S_1$ transition moments of the six pigment molecules (excluding the carotenoid) in the complex. In his analysis, Van Metter makes several explicit assumptions, whose validity is admittedly open to question, but on the basis of which he reaches generally reasonable conclusions.

The absorption spectrum shows two distinct peaks in the S_1 region, each assignable uniquely to Chl *a* (672 nm) or Chl *b* (652 nm). The CD spectrum between 600 and 700 nm displays three bands of large rotational strength, the largest being a negative feature at 640 nm, the second largest a positive one at 658 nm, and the smallest a negative one at 670 nm. The ratios of rotational strengths (band areas) are 8:4:1. Thus, the two bands on either side of the Chl *b* absorption peak are much stronger than that near the Chl *a* absorption peak. The fluorescence excitation spectrum (for 730 nm, i.e. Chl *a*, emission) coincides precisely with absorption, and the fluorescence quantum yield is constant for excitation between 600 and 700 nm. The fluorescence emission spectrum shows a single peak at 680 nm, irrespective of whether Chl *a* or Chl *b* is excited. The fluorescence polarization excitation spectrum has a minimum of 0.02 near the Chl *b* absorption peak, and increases monotonically to a maximum of 0.14 at 700 nm (the longest wavelength observed). It also displays a subsidiary maximum of 0.04 at 620 nm. The fluorescence polarization emission spectrum, when Chl *b* is preferentially excited, shows a minimum of <0.01 near 665 nm, the expected Chl *b* emission maximum.

To explain these observations, Van Metter interpreted the CD data as implying that the Chl *a* molecules have relatively weak mutual exciton coupling, and at least two of the Chl *b* molecules are strongly coupled. Interactions between the two sets (*a* and *b*), it is argued, are not very strong, but large enough to effect complete transfer of electronic excitation energy from *b*'s to *a*'s in a time short compared to the *in vitro* Chl *b* fluorescence lifetime (10^{-9} s), as the fluorescence data indicate. Van Metter points out that if the three Chl *b* molecules have C_3 symmetry, the resulting degeneracy of two of the three S_1-exciton states would account for the appearance of only two features in the Chl *b* part of the CD spectrum's S_1 region. This symmetry would also provide three mutually perpendicular S_1-exciton transition dipoles, and thereby account for the Chl *b* part of the fluorescence polarization data.

Van Metter proposed a specific model of the six pigment molecules, which, he noted, is consistent with the data and is also reasonably simple. In this model, the three Chl *b*'s form an exciton-coupled trimer with C_3 symmetry. Two of the three (exciton-uncoupled) Chl *a*'s are assumed to have $S_0 \rightarrow S_1$ absorption maxima at 670 nm, and the third is shown, with the aid of a simple theoretical formalism for calculating fluorescence polarization excitation spectra, to absorb near 677 nm. In applying the formalism, Van Metter assumed that only the Chl *a*'s fluoresce (hence ignored the fluorescence emission at wavelengths shorter than 650 nm), and found quite good agreement with the observed polarization spectrum between 650 and 700 nm. From the formalism, the mutual orientations of the three Chl *a* $S_0 \rightarrow S_1$ oscillators are also obtained. The three are not mutually orthogonal, but are much closer to that configuration than to a mutually parallel one. Although his model does not allow one to deduce specific interchromophore distances, Van Metter concludes, reasonably enough, that the three Chl *b* molecules are in close proximity to one another, while the three Chl *a*'s are more widely dispersed. Thus, he supposed the former to lie in the interior of the complex, and the latter on the periphery. This agrees with a prediction made many years earlier (Pearlstein 1964), based solely on fluorescence quantum yield and lifetime data.

2.2.2.2. Purple bacterial reaction centers. The purple photosynthetic bacteria contain either Bchl *a* or Bchl *b*, never both (Olson and Thornber 1979). Organisms containing Bchl *b* are of great potential interest for the exciton analyst because features in the S_1 region of their optical spectra are better separated than in the Bchl *a*-containing purples. However, reaction center complexes from the Bchl *b*-containing organisms have become objects of intensive research only relatively recently. About their better known cousins that contain Bchl *a*, it has already been said that the most thoroughly studied are the reaction center complexes from *Rp. sphaeroides* R-26. Because the spectroscopic and photochemical properties of R-26 reaction centers also seem typical of those from among the Bchl *a*-containing purple bacteria generally, I concentrate exclusively on them here.

The R-26 reaction center complex consists of 4 molecules of Bchl *a*, 2 of bacteriopheophytin (Bph) *a* – i.e. magnesium-free Bchl, 2 of ubiquinone, a minimum of 2 distinct polypeptides, and an iron atom whose place in the complex remains undetermined (Olson and Thornber 1979). The detailed three-dimensional structure, clearly crucial for its photochemical functioning, is unknown. However, attempts have been made to deduce the orientations of the Bchl and Bph $S_0 \rightarrow S_1$ and $S_0 \rightarrow S_2$ transition moments – 12 in all, relative to a definable axis of the complex, from optical spectra (Vermeglio and Clayton 1976, Vermeglio et al. 1978, Rafferty and Clayton 1979, Shuvalov and Asadov 1979). For one of these molecules in a solvent, the two moments are orthogonal (Weiss 1972); if they are also orthogonal in each pigment molecule of the reaction center, determination of their directions would enable one to establish the orientation (within chirality) of each molecule's porphyrin ring plane – a major step toward structural elucidation. A model has evolved that so far has specified the orientations of about half of the transition moments (Vermeglio and Clayton 1976). This model is based on LD and photoselection data, and also on a particular assumption regarding the exciton interactions among the 6 pigment molecules. The biophysical importance of the problem justifies close scrutiny of this assumption.

The model in question begins with the assumption that among the 6 pigment molecules there is only one substantial S_1-exciton interaction. According to the model, that interaction is between a particular two of the four Bchl *a* molecules. This "dimeric-exciton" model identifies that pair of Bchl *a*'s as the "special pair" – a hypothetical structural arrangement of two Bchl's (or Chl's) that has been proposed as the primary photoelectron donor in both bacterial and green plant reaction centers (Boxer and Closs

1976, Shipman et al. 1976, Fong and Koester 1976). In the dimeric-exciton model the 6 pigment molecules of the reaction center are divided into three groups: Two molecules of Bchl *a*, the "special pair", that have a large S_1-exciton interaction ($|J| = 430\ cm^{-1}$); the two molecules of Bph *a*, that have negligible exciton coupling either mutually or with any of the 4 molecules of Bchl *a*; and the two remaining Bchl *a* molecules, the mutual exciton interaction of which is unspecified, but which have negligible exciton interaction with the two special pair Bchl's.

Relevant spectroscopic data on R-26 reaction centers are as follows (Reed and Ke 1973): At ~80 K, the Bchl absorption band of longest wavelength undergoes a further bathochromic shift of $360\ cm^{-1}$, from 865 nm at 300 K to 893 nm at 80 K. The CD spectrum at 80 K shows a strong positive peak at 890 nm, a weak positive inflection at ~840 nm, a very strong negative peak at 809 nm, a strong positive peak at 796 nm, a weak positive inflection at 763 nm, and a weak negative peak at 748 nm. In the photoselection experiments (Vermeglio et al. 1978) the difference in absorption of plane polarized light was observed between R-26 complexes in their reduced, or electrically neutral state, and in their oxidized state – the positively charged state that occurs upon passage of the donated electron to a secondary electron acceptor. For oxidized complexes, the S_1 region of the 80 K CD spectrum shows a weak negative inflection at 840 nm, a very strong positive peak at 800 nm, a weak positive inflection at 774 nm, and a strong negative peak at 753 nm. Neither absorption nor CD spectrum shows any band between 850 and 900 nm for oxidized complexes. In both absorption and CD of reduced complexes at 300 K, the long wavelength peak (absent in oxidized complexes) has a width ~3 times greater than that of each of the peaks in the 800 nm vicinity. Both absorption and fluorescence polarization (Ebrey and Clayton 1969) show that the 865 nm band corresponds to a single transition, i.e. either a single exciton transition, or a monomolecular (exciton uncoupled) transition. If the latter, the 865 nm absorption could be attributed to only a single Bchl *a* molecule, because the loss of this band in the oxidized complex corresponds to the loss of a single electron. Thus, there are two possibilities: Two Bchl's contributing to the 865 nm absorption, or one.

The portion of the $S_0 \rightarrow S_1$ CD spectrum ascribed to Bchl (800 nm) is decidedly *not* conservative, there being about twice as much positive as negative rotational strength. At 80 K, the positive CD bands, which occur at 800 nm and at 890 nm, are about equal in amplitude, but the latter is three times as broad as the former. Neither of the two possibilities under consideration can explain the 890 nm CD band in exitonic terms. With only one Bchl molecule absorbing at 865 nm, there obviously would be no exciton interaction; with two Bchl's that have nearly parallel $S_0 \rightarrow S_1$ oscillators (Shipman et al. 1976), neither exciton transition has much rotational strength. For the dimeric exciton model there is, however, an additional complication. With the 890 nm CD band removed from consideration, the remaining Bchl $S_0 \rightarrow S_1$ CD bands form a system that is close to conservative only if the positive shoulder in the vicinity of 840 nm is included. Such a three-featured CD spectrum (796, 809, and ~840 nm) cannot be explained excitonically with just two Bchl molecules contributing to the 800 nm absorption band. It would be difficult to explain the single, broad, strong, one-signed CD feature at 890 nm nonexcitonically. Providing an alternate explanation for the multiple, narrow, strong, two-signed bands as well would pose an even greater challenge.

The greater width of the band at 890 nm is a property shared by absorption and LD, as well as by CD, even at low temperature. The dimeric-exciton model offers at best *ad hoc* explanations for this phenomenon. If the 800 nm band is due to three exciton-coupled Bchl's, with a lone Bchl absorbing at 865 nm, there is a straightforward explanation for the width ratio. According to degenerate-ground-state theory (Hemenger 1977a, 1977b, 1978a), the spectral linewidths of an exciton-coupled N-mer are narrowed by a factor of $N^{1/2}$ relative to the monomer linewidth. This predicts a width ratio of 1.7, essentially the observed value at 80 K.

There are at least two problems for the dimeric-exciton model with respect to shifts of the long-wavelength Bchl absorption band. One of these is the ~25 nm additional bathochromic shift that occurs when the temperature is lowered from 300 K to 80 K. (The effect of lowered temperature on the shorter wavelength bands appears to be nothing more than sharpening.) This is difficult, but not impossible, to explain if the 865 nm absorption is due to only one Bchl. Explanations are more difficult in the dimeric-exciton model: A bathochromic shift of the band center would increase the 800 nm absorption at 80 K over that at 300 K; the actual absorption is less. If the cryogenic shift were to represent an increase of the exciton J with no change of band center energy, an absorbance increase would be expected at 790 nm. Again, the observation is an absorbance decrease. A precisely matched increase of J and decrease of band center energy would, of course, solve the problem.

The other shift problem possibly offers the greatest challenge to the dimeric-exciton model: what happens to 865 nm Bchl absorption upon oxidation of the reaction center? If only one Bchl absorbs at 865 nm the observed loss of the long-wavelength absorption band corresponds to the loss of one electron by the monomeric primary donor Bchl. In the dimeric-exciton model, the equivalent of one Bchl loses an electron and the exciton J vanishes. The remaining Bchl then should absorb at the band-center wavelength, ~835 nm, but no such absorption is seen. This poses two difficulties for the dimeric-exciton model. First, though the model posits a specialized form of Bchl *a* (of band-center wavelength ~835 nm), it asserts that this form is unobservable even when the exciton interaction is "turned off" (multiple hypotheses). Second, it assumes that most of the oscillator strength (~90%) of the special pair's $S_0 \rightarrow S_1$ transition lies in the long-wavelength band of the reduced complex, but makes no accounting of the 50% that should remain in the oxidized complex. It certainly does not appear in the 800 nm band, and presumably is spread over many high energy bands. The dimeric-exciton model offers no explanation for such an extreme form of dimeric hypochromism, nor why it occurs only in oxidized complexes.

From the foregoing discussion it appears that in order to explain the optical properties of the reduced reaction center, the dimeric-exciton model introduces no less than *six* hypotheses beyond those required by the assumption of a monomeric primary donor Bchl (i.e., approximately double the number). These include: (1) A nonexcitonic mechanism to explain the 800 nm CD bands. (2) A mechanism to explain the ratio of widths of the 890 nm and 800 nm absorption bands. (3) The existence of a form of Bchl *a* absorbing (in the absence of exciton interaction) at ~835 nm. (4) A mechanism to render this absorbing form unobservable in oxidized as well as reduced reaction centers. (5) A complex mechanism to explain the ~25 nm cryogenic shift of the 865 nm band, apparently involving precisely matched changes of S_1-exciton J and band center energy. (6) The existence of an extreme form of dimeric hypochromism, in oxidized reaction centers *only*, to account for the missing Bchl oscillator strength in those centers. One concludes that the existence of exciton effects in photosynthetic reaction centers is not established.

2.3. Exciton transport and trapping

As indicated in sect. 1.1, the theoretical study of exciton transport through the chlorophyll arrays of photosynthetic organisms predates the assignment of specific exciton transitions in isolated chlorophyll-protein complexes. It should be clear from sect. 2.2 that the latter effort is still in its infancy. For these reasons it is useful to begin a discussion of exciton transport with the simplest and best developed theoretical models, which are based on Förster's very weak coupling case (sect. 2.3.1). I follow that with a brief description of recent efforts to improve the theoretical base (sect. 2.3.2). I end this portion of the chapter by showing how the theoretical concepts are applied to interpret experimental fluorescence data relevant to the one-singlet-exciton migration problem.

2.3.1. Theory in the Förster limit

If $|J| \ll \tau_r^{-1}$, where τ_r is a characteristic intramolecular vibrational relaxation time, exciton coupling is said to be "very weak" (Förster 1965). In this case, electronic excitation energy transfer is completely incoherent, i.e. no memory of the phase of the exciton wave function remains. The exciton motion is then purely hopping, or diffusive, and an aggregate of identical molecules may be described by the master equation,

$$\dot{\rho}_i = \sum_{j \neq i} F_{ij}(\rho_j - \rho_i) - Q_i\rho_i. \tag{5}$$

In eq. (5), $\rho_i(t)$ is the probability that the ith molecule is excited at time t, Q_i is the rate constant for exciton quenching from molecule i, and for strongly allowed electric dipole transitions such as the $S_0 \rightarrow S_n$ ($n \geq 1$) of the chlorophylls, the symmetric rate constants F_{ij} are given by:

$$F_{ij} = \tau_0^{-1}(R_0/R_{ij})^6. \tag{6}$$

Here, τ_0 is the radiative lifetime of the excited state, R_{ij} is the distance between molecules i and j, and the characteristic distance is given (in units of nm^6) by Förster's formula,

$$R_0^6 = (8.785 \times 10^{17}\, \kappa_{ij}^2/n^4) \int F(\nu)\epsilon(\nu)\, d\nu/\nu^4. \tag{7}$$

In eq. (7), $\epsilon(\nu)$ is the molar extinction coefficient on a wave number scale, $F(\nu)$ is the normalized emission ($\int F(\nu)\, d\nu = 1$), n is the refractive index of the medium, and κ_{ij}^2 is an orientation factor:

$$\kappa_{ij}^2 = [\hat{\boldsymbol{\mu}}_i \cdot \hat{\boldsymbol{\mu}}_j - 3(\hat{\boldsymbol{\mu}}_i \cdot \hat{\boldsymbol{R}}_{ij})(\hat{\boldsymbol{\mu}}_j \cdot \hat{\boldsymbol{R}}_{ij})]^2. \tag{8}$$

In general, the orientation factor (and hence (R_0) depends on the molecular indices – $\hat{\boldsymbol{\mu}}_i$ is a unit vector in the direction of molecular transition dipole i and $\hat{\boldsymbol{R}}_{ij}$ is a unit vector in the direction of the line joining dipoles i and j. (The orientation factor is even more anisotropic if corrections to the dipolar charge distribution are included; these need not concern us here.)

For an aggregate of N molecules (i.e., N transition dipoles) the solution of eq. (5) has the form,

$$\rho_i(t) = \sum_{m=0}^{N-1} A_{im} \exp(-\gamma_m t), \tag{9}$$

where the N characteristic rate constants γ_m depend on the F_{ij}'s and the Q_i's, and the coefficients A_{im} depend on those as well as the initial condition. Of more immediate physical interest is the aggregate decay function,

$$P(t) \equiv \sum_{i=1}^{N} \rho_i(t) = \sum_{m=0}^{N-1} C_m \exp(-\gamma_m t), \tag{10}$$

(where $C_m = \Sigma_i A_{im}$), and its moments,

$$M_\alpha = \int_0^\infty t^\alpha P(t)\, dt. \tag{11}$$

M_0 is most useful, and has a particularly simple physical interpretation: it is the mean time for quenching the exciton, effectively the exciton lifetime.

In what follows, I consider lattice models first because they are simpler, give considerable physical insight, and are probably closer to reality than completely random aggregates. In a lattice model, the antenna chlorophyll molecules are assumed to occupy the sites of a regular crystal lattice and to have a definite, fixed spatial orientation. I restrict attention to simple lattices, i.e., ones in which all (nonquenching) lattice sites are equivalent. The primary donor chlorophylls of the reaction centers, which serve to quench the antenna singlet excitons, also occupy lattice sites. The quenchers may occur regularly (i.e., form a "superlattice") or at random locations – provided each quenching site has an identical set of interactions with its nearest neighboring antenna chlorophylls. The interactions of a quenching center with its nearest nonquenching neighbors in an isotropic lattice are illustrated in fig. 3. (In an isotropic lattice, all of the chlorophylls necessarily have the same spatial orientation.)

Three types of quenching centers are distinguished in fig. 3. The primary donor chlorophyll is said to interact "nondisruptively" with its nearest neighbors if the Förster transfer between the donor and its neighbors is reversible and the rate constant has the same magnitude as that between a pair of adjacent antenna chlorophylls. The quenching of the exciton by a nondisruptive center is due entirely to the rapidity of conversion of the neutral excited singlet state into the charge-separated (radical ion pair) state at the center. The reaction centers of *Rp. sphaeroides* appear to have at least approximate nondisruptive character at 300 K. A purely "disruptive" center is one whose excited state lies at a sufficiently lower energy than that of the antenna chlorophyll to "quench" the exciton (as far as the antenna is concerned) by trapping it – Förster transfer out of the center has a negligibly small rate constant. Probably no photosynthetic reaction center is purely disruptive, although the system 1 center of green plants (P700) may come close. Finally, fig. 3 shows a "generalized quencher", one which allows both conversion and partial trapping (Hemenger et al. 1972). The generalized quencher requires three parameters for its description, against the single parameter of either the disruptive or the nondisruptive quencher. It is sufficiently general to include even a case like that of the Bchl *b*-containing bacterium *Rp. viridis*, in which the reaction center, although a rapid quantum converter, lies energetically uphill of the antenna Bchl's and so is an "antitrap" (Zankel and Clayton 1969).

As written, eq. (6) applies to an aggregate of *identical* molecules, i.e., it makes no allowance for a quencher with disruptive properties. A master equation that applies in the presence of generalized quenchers is (Hemenger et al. 1972):

$$\dot{\rho}_i = \sum_{j \neq i} F_{ij}(\rho_j - \rho_i) + \sum_j S_{ij}\rho_j. \tag{12}$$

In eq. (12), F_{ij} is the reversible Förster rate constant between Chl's at sites i and j, calculated as if *every* site were occupied by an antenna Chl. For $i \neq j$, $F_{ij} + S_{ij}$ is the actual Förster rate constant (not necessarily reversible) from site j to site i with the quenching centers in place. The quantity $[(\Sigma_{j \neq i} F_{ij}) - S_{ii}]$ is the total rate constant with which excitation leaves site i in the presence of quenchers. For a lattice with a single generalized quencher at the origin ($i = 0$),

$$S_{ij} = F\{(\lambda - 1)(\delta_{i0} - \delta_{im})\delta_{jm} + (\mu - 1)\delta_{im}\delta_{j0} - [Q + q(\mu - 1)]\delta_{i0}\delta_{j0}\}. \tag{13}$$

Here, F is a reversible Förster rate constant between nearest-neighboring antenna Chl's, m is any of the nearest neighbors of $i = 0$, q is the coordination number of the lattice, and the δ's are the usual Kronecker deltas. The parameters λ, μ, and Q are just the three parameters of the generalized quencher defined in fig. 3. (The finite lifetime of the excited state in the absence of quenching is neglected here, but see below.)

Consider a lattice of N sites with periodic boundary conditions and the single quencher whose properties are defined by eq. (13). For the calculation of $P(t)$ from eq. (10) such a lattice is equivalent to an infinite lattice having regularly placed quenchers of fractional concentration $c = 1/N$. The case of a purely nondisruptive quencher ($\mu = \lambda = 1$) for which there are negligible interactions between the quencher and its nonnearest neighbors can be treated very generally. In that case, regardless of the range of the antenna Chl-antenna Chl interactions, the mean time for exciton quenching, M_0 from eq. (11), is*

$$M_0 = (N/F)(Q^{-1} + A), \tag{14}$$

where A depends on N, on lattice parameters independent of the quencher, and on the initial condition (exciton starting site). Equation (14) implies that for primitive lattices of any dimension with host–host interactions of arbitrary range and with any initial condition, the short-range purely nondisruptive quenching of incoherent excitons can be described as an additive sequence of events: There is a mean time, $(N/F)A$, to reach a quenching center, plus a mean time, $N/(QF)$, to be quenched once the exciton has reached that center.

This simple, intuitive result holds quite generally, as long as the range of direct interactions between the quencher and the antenna chlorophylls

* Note that if the decay is nonexponential, M_0 is an incomplete and therefore a possibly misleading description of the exciton dynamics (see below).

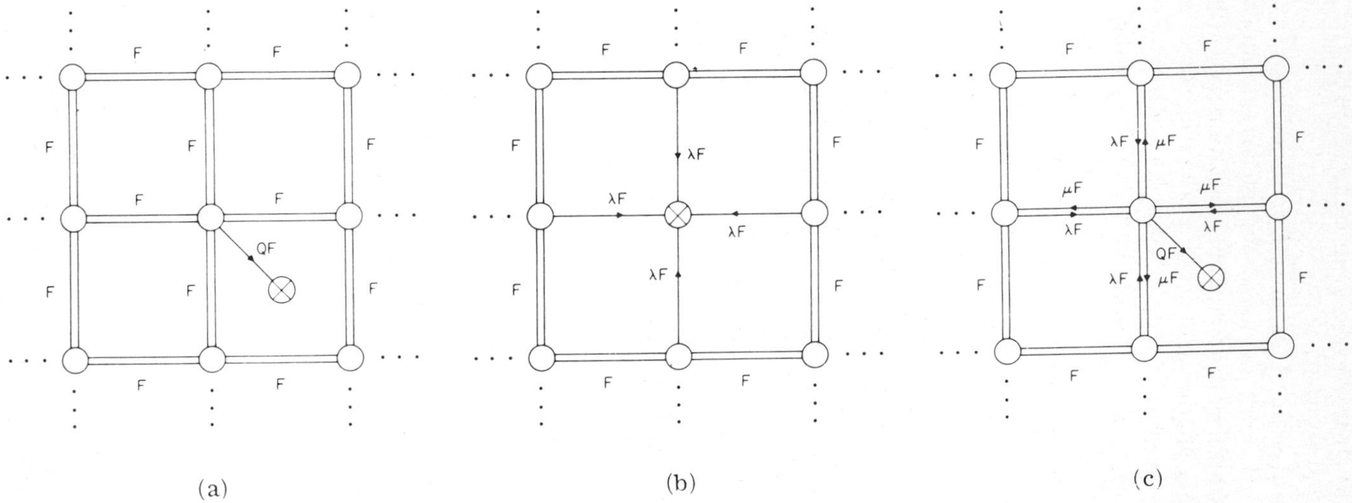

Fig. 3. Various types of quenching centers and their kinetic interactions with antenna Chl molecules in the Förster limit. For illustrative purposes, a two-dimensional square lattice with only isotropic nearest neighbor interactions is shown; results given in the text are more general. In each part, the central circle is the quenching site, other open circles represent antenna Chl molecules. Double connecting lines (without arrows) denote reversible excitation transfer pathways of Förster rate constant F. Single connecting lines with arrows denote irreversible transfer of labelled rate constants. The circle with the × (located, for clarity, in an interstitial position in parts (a) and (c)), represents the quantum-converting process, of rate constant QF, of the reaction center; in part (b), that last process is irrelevant to the exciton quenching. (a) Nondisruptive quencher. (b) Disruptive quencher. (c) Generalized quencher. The dimensionless quantities Q, λ, and μ are the three parameters that describe the generalized quencher.

remains short. In particular, for a generalized quencher (but only nearest neighbor interactions everywhere),

$$\langle M_0 \rangle = t_0 + (qF)^{-1}\{N\alpha - [(N-1)^2(\lambda - 1)/(N\lambda)]\}, \tag{15}$$

where the average is over all lattice sites, as starting sites for the exciton, weighted equally. In eq. (15), t_0 is the value of M_0 for an exciton that starts at the quenching center itself,

$$t_0 = (QF)^{-1}[1 + (\mu/\lambda)(N-1)]. \tag{16}$$

The quantity α depends on the dimensionality of the lattice:

$$\alpha = \begin{cases} (N^2 - 1)/6N, & \text{one dimension,} \\ c_1 \ln N + c_2 + o(1/N), & \text{two dimensions,} \\ c_1 + o(1/N), & \text{three dimensions.} \end{cases} \tag{17}$$

Values of c_1 and c_2 for various primitive lattices are given in table 1. The second term on the right-hand side of eq. (14) or eq. (15) is the "first passage time" of random walk theory (Montroll 1969). In eq. (15), $(qF)^{-1}$ is the step time of a constant step-time random walk, and the factor in braces is the mean number of steps required to reach the quenching center (for the first time) from an arbitrary starting site. The exciton quenching process is said to be "quencher limited" or "diffusion limited" if the t_0-term or the random-walk-term, respectively, makes the dominant contribution to M_0.

So far, I have discussed only M_0 and not the detailed shape of $P(t)$, whose mathematical form, given by eq. (10), holds also for the solution of eq. (12). For incoherent exciton transfer and quenching on two- or three-dimensional lattices, the contribution of terms in the m sum of eq. (10) is generally quite small with uniform initial excitation unless $m = 0$, i.e. $C_0 \sim 1$, $C_m \sim 0$ for $m > 0$ (Hemenger et al. 1972). When this "zero-mode dominance" (Pearlstein 1966) occurs, $P(t)$ is, of course, well-approximated

Table 1
Values of constants in eq. (17)

Lattice	No. of dimensions	q	c_1	c_2
Hexagonal	2	3	0.41350	0.06621
Square	2	4	0.31831	0.19506
Triangular	2	6	0.27566	0.23521
Simple cubic	3	6	1.51639	—
Body-centered cubic	3	8	1.39320	—
Face-centered cubic	3	12	1.34466	—

as a single exponential, $\exp(-\gamma_0 t)$, where $\gamma_0 \simeq M_0^{-1}$. The zero mode is not necessarily so dominant with other initial conditions. For example, it breaks down completely if initially the quenching center itself is excited, provided that the exciton quenching is diffusion limited. Even with uniform initial excitation, higher modes can contribute significantly if there is long-range quenching or slow back transfer from the quencher. (By "long-range quenching" I mean a situation in which exciton hopping into the quencher from its nonnearest neighbors competes with nearest neighbor hops between antenna Chls. "Slow back transfer" signifies $\lambda \gg \mu \sim Q \sim \gamma_0/F$.) None of these conditions–nonuniform initial excitation with diffusion limited quenching, long-range quenching, or slow back transfer–is easy to obtain in photosynthetic systems. For lattice models of the Chl in these, the rule is therefore zero-mode dominance of incoherent exciton quenching.

Zero-mode dominance can also break down in lattices where the exciton hopping is critically anisotropic (Hemenger et al. 1972). Critical anisotropy (defined below) is even less likely to occur in photosynthetic Chl "lattices" than, say, long-range quenching, but anisotropy generally is quite possible. The effects of anisotropy, including critical anisotropy, on the kinetics of incoherent exciton quenching are illustrated by the following example. Consider a simple cubic lattice in which the reversible Förster rate constants between nearest neighboring antenna Chl's are F, F, and βF along the a, b, and c directions, respectively ($0 \leqslant \beta \leqslant 1$). For simplicity, let $\mu = \lambda = 1$, so that from eq. (15),

$$\langle M_0 \rangle = (N/F)[(1/Q) + (\alpha/6)]. \tag{18}$$

From eq. (17) and table 1, when $\beta = 1$, $\alpha/6 = 0.25273$ (to within $o(1/N)$). As long as $\beta \gg 1/N$, the zero-mode remains dominant, and $\langle M_0 \rangle$ remains linear in N (to $o(1/N)$), although the quenching takes longer because for $1/N \ll \beta \leqslant 1$, $\alpha/6 = 0.25273 + (1/4\pi)\ln(1/\beta)$. If $\beta \ll 1/N$, the exciton quenching proceeds as if the cubic lattice had become a set of independent square lattices (the ab planes): The zero mode is again dominant (not quite as strongly) and the leading term in α is proportional to $\ln N$ (see eq. (17) and table 1). However, if the lattice is "critically anisotropic", i.e., $\beta \approx 1/N$ ($= c$, the fractional quencher concentration), a single decay mode no longer dominates and the random-walk part of the expression for $\langle M_0 \rangle$ differs from either purely two- or three-dimensional behavior. Physically, this result has a simple interpretation. The lattice becomes critically anisotropic when the time it takes the exciton to make a single nearest-neighbor hop along the c-axis ($\sim 1/\beta F$) is comparable to the diffusion-limited exciton "lifetime", $\langle M_0 \rangle$. In other words, the exciton must make at least one out-of-plane hop to sense the three-dimensionality of the lattice.

Up to this point, I have not explicitly included the finite lifetime of the exciton, in the absence of quenching, due to other decay channels, both radiative and nonradiative. If these other channels are all independent, first-order, and homogeneous over the antenna, they can be represented collectively by a single lifetime, τ. In eq. (10), each γ_m calculated as if $\tau = \infty$ becomes $\gamma_m + \tau^{-1}$. When the zero-mode is dominant, one also has

$$M_0^{-1}(\tau) = M_0^{-1}(\infty) + \tau^{-1}. \tag{19}$$

Equation (19) is a good approximation even in a one-dimensional lattice, where higher modes contribute noticeably under initial and quenching conditions for which the zero mode is strongly dominant in three dimensions (Lakatos-Lindenberg et al. 1972).

In a one-dimensional lattice, if the quenchers occur with a fractional concentration, c, but at random lattice sites, $\langle M_0 \rangle$ is generally much larger than it is with a quencher at every Nth site, periodically. Also, $P(t)$ is generally highly nonexponential with quenchers placed at random (Lakatos-Lindenberg et al. 1972). This is *not* the case for higher dimensional lattices (Hemenger et al. 1972). Both $\langle M_0 \rangle$ and $P(t)$ change little on going from periodic to random quenchers in two or three dimensions. In three dimensions, P_{random} and P_{periodic} are identical to within $o(c^{1/2})$. Thus, for lattice models of photosynthetic Chl arrays, it is virtually immaterial to the quenching kinetics of incoherent excitons whether the reaction centers themselves occur periodically or at random.

I close this section with a brief consideration of random models. If chlorophyll or other pigment molecules are placed randomly with regard to both spatial location and orientation, and there are many fewer excitation donors than acceptors (quenchers),

$$P(t) \propto \begin{cases} \exp[-kt^{1/2} - (t/\tau)], & \text{three dimensions} \\ \exp[-k't^{1/3} - (t/\tau)], & \text{two dimensions,} \end{cases} \tag{20}$$

for instantaneous excitation (Förster 1949, Wolber and Hudson, 1979). In eq. (20), k and k' are constants, τ is the donor's excited-state lifetime in the absence of quenching. If k (or k') is large, the decay is highly nonexponential (i.e., nonlinear on a $\log t$ plot), quite unlike the situation of zero-mode dominance (exponential decay) with lattice models. However, this situation does not fit the photosynthetic case, where there are many more donors (antenna chlorophylls) than acceptors (reaction centers). Although there has been recent progress (Blumen and Manz 1979, Blumen and Zumofen 1980), the problem of the kinetics of exciton quenching in a many-donor random model has not been solved analytically. A very small deviation from exponentiality is expected even in the zero-mode-dominant lattice models. Smoluchowski (1916, see also, Montroll 1946) showed that at short times the higher modes together give an additional exponential dependence on $t^{1/2}$ (in three dimensions). This is a small effect that would be very hard to discern experimentally. If experiment ultimately shows deviations from exponentiality in excess of the Smoluchowski term, one might then suspect that the principal antenna chlorophyll is not arrayed in more-or-less orderly fashion. This issue is at least as likely to be settled by structural studies as by picosecond spectroscopy.

2.3.2. Beyond the Förster limit

In the Förster limit, defined by $|J| \ll \tau_r^{-1}$ (see sect. 2.3.1), an exciton hops from one to another of a set of identical pigment molecules. In the opposite extreme, $|J| \gg \tau_r^{-1}$, the exciton motion is oscillatory. Recent theoretical approaches describe the general case in terms of a density matrix equation or a generalized master equation (Silbey 1976, Knox 1977). While different in detail, these approaches concur that, immediately following its creation, an exciton moves in a wavelike fashion (i.e., "coherently"), and after a certain period of time its motion becomes hopping ("incoherent"). Theorists differ as to the time scale on which this coherence is lost; for chlorophyll singlet excitons it is probably between 10 fs and 1 ps. Thus, except in the emerging field of subpicosecond spectroscopy, the effects of exciton coherence on energy transfer in photosynthetic systems are likely to be difficult to discern.

2.3.3. Comparison with experiment

Fluorescence is one readily available experimental monitor of the S_1-state's kinetics in photosynthetic Chl arrays. The time dependence of the fluorescence decay following excitation by a single picosecond pulse is potentially informative but it is important for the analysis of the one-exciton kinetics that such picosecond experiments be extrapolated to zero-pulse energy because of nonlinear effects. Carefully extrapolated picosecond fluorescence experiments show no apparent deviations from exponential decay (Breton and Geacintov 1980). It is of interest to compare theoretical and experimental exciton lifetimes. Experimentally, one determines the time for the fluorescence to decay to $1/e$ of its initial value. For samples whose RC's are active (i.e., not oxidized), the quantity τ^{-1} in eq. (19) is usually negligible compared to $M_0^{-1}(\infty)$, so that the experimentally determined fluorescence lifetime can be compared directly to $\langle M_0 \rangle$ from eq. (15).

An example of one such comparison is worked through here with lattice-model theory. For chromatophores of *Rp. sphaeroides* R26, which has only the single antenna pigment Bchl *a*, the extrapolated fluorescence lifetime from single-pulse picosecond excitation is 300 ± 50 ps (Campillo et al. 1977). This should be the value of $\langle M_0 \rangle$ in eq. (15). Comparison of theory and experiment thus becomes a matter of analyzing the parameters in that equation*. The t_0 term, from eq. (16), has 3 parameters, μ/λ, $(QF)^{-1}$, and N. The ratio μ/λ is the relative hopping rate, back and forth, between the RC and its nearest neighboring antenna Bchl's. It may be determined empirically as the ratio of the Förster overlap integral, from eq. (7), for RC fluorescence with antenna absorption, to the integral for antenna fluorescence with RC absorption; geometrical quantities cancel out. In the absence of such an empirical determination, one may estimate $\mu/\lambda \sim 1$ because the lowest S_1 levels of RC and antenna Bchl's are virtually monoenergetic. The quantity QF is the photoconversion rate of the isolated RC. The best current estimate (Holten et al. 1980) is $(QF)^{-1} \simeq 5$ ps. N is the number of antenna Bchl's per RC, so that $N = 60$. Thus, it appears that $t_0 \simeq 300$ ps, which is just the observed fluorescence lifetime. This is in accord with conclusions reached long ago (Pearlstein 1966, 1967), that exciton migration and conversion is not diffusion limited in photosynthetic systems.

If $\langle M_0 \rangle \simeq t_0$, the random-walk term in eq. (15) must lie within experimental uncertainty, i.e. $\leqslant 50$ ps. Campillo et al. (1977) plausibly assume a two-dimensional lattice, e.g. a square lattice, for which $q = 4$ and $\alpha = 0.32 \ln N + 0.20$ from eq. (17) and table 1. Knox (1977) has shown that for a square lattice, it is probably inconsequential whether each Bchl occupies one lattice site, or several Bchl's are bunched at each site but with the same average concentration of Bchl's. Thus, it is appropriate to take $N = 60$, as above. The parameter, $\lambda \sim 1$, for a reason similar to that why $\mu/\lambda \sim 1$ (see above and fig. 3); thus for *Rp. sphaeroides*, the λ part of the random-walk term is negligible compared to the α part. It follows that $F^{-1} \leqslant 2.2$ ps, a perfectly reasonable upper limit.

Campillo et al. (1977) use $\tau_0 = 18$ ns and estimate the antenna Bchl–Bchl $R_0 \simeq 9.0$ nm, so that from eq. (6) and the limiting value of F^{-1}, it follows that $R \leqslant 2.0$ nm. The R here is an average of the separation of dipoles within an LH-R26 complex and the separation of nearest neighboring dipoles in two adjacent complexes. The former R value is unknown but may be ~1.2 nm, which is the known separation of nearest neighboring dipoles in a subunit of the *Prosthecochloris* Bchl *a*-protein. Then linear averaging implies an upper limit of 2.8 nm for the separation of nearest neighboring dipoles in two adjacent LH-R26 complexes in the chromatophore; squared averaging implies an upper limit of 2.6 nm. These numbers can be compared with the known 2.4 nm separation of nearest neighboring dipoles in two adjacent subunits of the *Prosthecochloris* Bchl *a*-protein.

2.4. Exciton collisions and picosecond kinetics

The subject of exciton collisions, and subsequent exciton–exciton annihilation, in photosynthetic systems has been well covered in a recent

* Campillo et al. (1977) present an analysis that differs somewhat from mine.

review (Breton and Geacintov 1980), so is touched on only briefly here. With single picosecond-pulse excitation of even moderate pulse energy (10^{14} photons/cm^2), singlet exciton populations in a photosynthetic membrane can become great enough to noticeably shorten the S_1-state lifetime through exciton–exciton annihilation. While this is a complicating artifact in the study of one-exciton kinetics, it is an interesting way to study exciton dynamics, and conceivably may provide a source of structural information. For example, although the one-exciton kinetics are not very sensitive to boundary conditions (i.e., whether each reaction center is fed by a unique portion of the antenna, or all reaction centers share the entire antenna), the pair annihilation kinetics are.

An equation that has proven quite useful in describing these kinetics is (Swenberg et al. 1976):

$$\dot{S}_1 = G - KS_1 - 0.5\gamma S_1^2, \tag{21}$$

where S_1 is the number of singlet excitons per unit volume, G is their rate of generation, K is the rate constant for exciton decay in the absence of exciton collisions, and γ is the pair annihilation rate constant. Integration of eq. (21) gives the relative fluorescence quantum yield, ϕ, in terms of the pulse intensity, I, as

$$\phi = (2K/I\gamma)\ln[1 + (I\gamma/2K)]. \tag{22}$$

Equation (22) gives very good fits to experimental data; values of $\gamma \simeq 10^{-8}$ cm^3/s have been obtained for the chlorophyll arrays of green plants. More recently, equations somewhat more sophisticated than eq. (21) have been introduced (Paillotin and Swenberg 1979).

If a photosynthetic sample is excited with a train of picosecond pulses, singlet excitons created during one pulse can collide with long-lived chlorophyll triplet states created during an earlier pulse; see Monger and Parson (1977).

3. Other systems

In photosynthetic systems, exciton transport has a biological role, and certain isolated chlorophyll-protein complexes display exciton splitting in their optical spectra. In other systems, exciton transport has no established biological function, and in only one other case –the purple membrane– is there spectroscopic evidence of exciton splitting. One other class of biological systems, the nucleic acids, show some evidence of one-dimensional exciton transport along their helically stacked bases.

3.1. The purple membrane

The purple membrane occurs in salt-loving bacteria, notably *Halobacterium halobium*. The membrane contains a regular, two-dimensional array of the protein bacteriorhodopsin, which, as its name implies, has a chromophore similar to that found in the visual pigment. However, unlike the latter, bacteriorhodopsin functions to pump protons across the purple membrane in an energy-storing photoreaction (Lozier et al. 1975). The salt-loving bacteria are, therefore, solar energy converters like the photosynthetic organisms. But in a purple membrane, each pigment-protein complex is its own "reaction center." Why there should be exciton effects in the absence of a separate antenna is, at the moment, a puzzle.

The bacteriorhodopsin array in the purple membrane has a high degree of symmetry: three axes of three-fold rotational symmetry. If one chooses one of the resulting three ways to form a protein trimer, a heuristic description of resonance interactions among the three chromophores in each trimer can be given. However, because the placement of the chromophores is not known to atomic resolution, precise theoretical prediction of spectroscopic exciton effects cannot be given. Nonetheless, 560 nm absorption and CD spectra of the purple membrane have been interpreted to imply exciton interactions. One analysis of these spectra, due to Hemenger (1978b), finds $J \simeq -64\ \mathrm{cm}^{-1}$ for each pair of chromophores in the trimer. At present, all interpretations (Heyn et al. 1975, Kriebel and Albrecht 1976, Ebrey et al. 1977, Hemenger 1978b) of purple membrane spectra in terms of exciton effects must be viewed as tentative.

3.2. Nucleic acids

Nucleic acid polymers in conformations containing regularly stacked bases ought to display exciton effects because the bases are in, or almost in, Van der Waals contact (0.3–0.4 nm center–center distance) and the transition moments of the near ultraviolet electronic transitions of the bases (260 nm band) are only about ten times smaller than that of chlorophyll. Indeed, there are noticeable effects in the spectra when base stacking occurs, but so far it has not been possible to interpret these in terms of exciton bands because of the number and breadth of the electronic transitions in the 260 nm absorption band.

What evidence there is for excitons in nucleic acids comes mainly from studies of electronic excited-state quenching by exogenous agents (Hélène 1973). A model that involves a one-dimensional random walk of triplet excitons along the identical stacked bases in single-stranded polyriboadenylic acid (poly rA) appears to be the only plausible explanation of poly rA phosphorescence quenching by low concentrations of paramagnetic ions (Eisinger and Shulman 1966). Attempts to find an analogous effect with excited singlet states in poly rA have so far proved unsuccessful. This is because the metal ions quench only the triplet states, and dyes that do quench the singlet have been bound to poly rA only under conditions of imperfect base stacking (Pearlstein et al. 1979).

In poly rA all of the bases are identical (adenine). In nucleic acids where the bases are not all the same, triplet excitons are rapidly trapped on the bases whose triplet levels are lowest. The lowest excited singlet state is even less mobile because in almost all nucleic acids (poly rA is an exception) exciplex formation (i.e., a localized excited-state complex formed by two adjacent bases) competes most effectively with energy transfer (Hélène 1973). Thus, only in poly rA and one or two other simple nucleic acid polymers can excitons transport energy over 100 or more bases. For this reason, biophysical interest in excitons in nucleic acids has diminished during the past decade.

4. Conclusions

Although there may be exciton effects in nucleic acids and the bacteriorhodopsin-containing purple membrane, the chlorophyll arrays of photosynthetic organisms are the principal biological systems in which there is both a clear role and experimental evidence for excitons. Chlorophyll-containing protein complexes isolated from green plants and photosynthetic bacteria are beginning to allow the assignment of individual exciton transitions. In photosynthetic membranes, where these protein complexes are found in more-or-less regular arrays, singlet excitons undergo random walks over the antenna chlorophyll molecules to the photochemically active reaction centers. Picosecond spectroscopy has opened up new avenues for the study of exciton effects in photosynthesis, including the observation of exciton collision processes.

Note added in proof

It now appears that in some of the photosynthetic bacteria exciton quenching may come closer to the diffusion limit than previously thought; experiments are in progress to test the possibility that Bchl fluorescence from such bacteria may be nonexponential when the reaction centers themselves are preferentially excited initially. Recent evidence suggests that the actual values of $(QF)^{-1}$ and μ/λ for *Rp. sphaeroides* R26 are somewhat smaller than those given in sect. 2.3.3; precise values are still not determined.

References

Barber, J., 1979, Photochem. Photobiol. **29**, 203.
Bay, Z. and R.M. Pearlstein, 1963, Proc. Natl. Acad. Sci. U.S.A. **50**, 1071.
Blumen, A. and J. Manz, 1979, J. Chem. Phys. **71**, 4694.
Blumen, A. and G. Zumofen, 1980, Chem. Phys. Lett. **70**, 387.
Bolt, J. and K. Sauer, 1979, Biochim. Biophys. Acta **546**, 54.
Boxer, S.G. and G.L. Closs, 1976, J. Am. Chem. Soc. **98**, 5406.
Breton, J. and N.E. Geacintov, 1979, in: Chlorophyll Organization and Energy Transfer in Photosynthesis, Wolstenholme, G. and Fitzsimons, D.W., eds. (Excerpta Medica, Amsterdam) p. 217.
Breton, J. and N.E. Geacintov, 1980, Biochim. Biophys. Acta **594**, 1.
Butler, W.L., 1978, Annu. Rev. Plant Physiol. **29**, 345.
Campillo, A.J., R.C. Hyer, T.G. Monger, W.W. Parson, and S.L. Shapiro, 1977, Proc. Natl. Acad. Sci. U.S.A. **74**, 1997.
Clayton, R.K., 1981, Photosynthesis. (University Press, Cambridge).
Delepelaire, P. and N-H. Chua, 1979, Proc. Natl. Acad. Sci. U.S.A. **76**, 111.
Ebrey, T.G. and R.K. Clayton, 1969, Photochem. Photobiol. **10**, 109.
Ebrey, T.G., B. Becher, B. Mao, P. Kilbride and B. Honig, 1977, J. Mol. Biol. **112**, 377.
Eisinger, J. and R.G. Shulman, 1966, Proc. Natl. Acad. Sci. U.S.A. **55**, 1387.
Emerson, R. and W.A. Arnold, 1932, J. Gen. Physiol. **15**, 391 and **16**, 191.
Fenna, R.E. and B.W. Matthews, 1975, Nature **258**, 573.
Fenna, R.E., L.F. Ten Eyck, and B.W. Matthews, 1977, Biochem. Biophys. Res. Commun. **75**, 751.
Fong, F.K. and V.J. Koester, 1976, Biochim. Biophys. Acta **423**, 52.
Förster, Th., 1948, Ann. Physik (5) **2**, 55.
Förster, Th., 1949, Z. Naturforsch. **A4**, 321.
Förster, Th., 1965, in: Modern Quantum Chemistry, Part III, Sinanoglu, O., ed. (Academic Press, New York) p. 93.
Franck, J. and E. Teller, 1938, J. Chem. Phys. **6**, 861.
Frenkel, J.I., 1931, Phys. Rev. **37**, 17 and 1276.
Frenkel, J.I., 1936, Physik Z. Sowjet **9**, 158.
Gregory, R.P., 1977, Biochemistry of Photosynthesis (2nd ed.) (Wiley, New York).
Hélène, C., 1973, in: Physico-Chemical Properties of Nucleic Acids, Duchesne, J., ed. (Academic Press, London) p. 119.
Hemenger, R.P., 1977a, J. Chem. Phys. **66**, 1795.
Hemenger, R.P., 1977b, J. Chem. Phys. **67**, 262.
Hemenger, R.P., 1978a, J. Chem. Phys. **68**, 1722.
Hemenger, R.P., 1978b, J. Chem. Phys. **69**, 2279.
Hemenger, R.P., R.M. Pearlstein and K. Lakatos-Lindenberg, 1972, J. Math. Phys. **13**, 1056.
Heyn, M.P., P.J. Bauer and N.A. Dencher, 1975, Biochem. Biophys. Res. Commun. **67**, 897.
Holten, D., C. Hoganson, M.W. Windsor, C.C. Schenck, W.W. Parson, A. Migus, R.L. Fork and C.V. Shank, 1980, Biochim. Biophys. Acta **592**, 461.
Knox, R.S., 1975, in: Bioenergetics of Photosynthesis, Govindjee, ed. (Academic Press, New York) p. 183.
Knox, R.S., 1977, in: Topics in Photosynthesis, Vol. 2, Barber, J., ed. (Elsevier, Amsterdam) p. 55.
Kriebel, A.N. and A.C. Albrecht, 1976, J. Chem. Phys. **65**, 4575.
Lakatos-Lindenberg, K., R.P. Hemenger, and R.M. Pearlstein, 1972, J. Chem. Phys. **56**, 4852.
Lozier, R.H., R.A. Bogomolni and W. Stoeckenius, 1975, Biophys. J. **15**, 955.
Markwell, J.P., J.P. Thornber and R.T. Boggs, 1979, Proc. Natl. Acad. Sci. U.S.A. **76**, 1233.

Matthews, B.W., R.E. Fenna, M.C. Bolognesi, M.F. Schmid and J.M. Olson, 1979, J. Mol. Biol. **131**, 259.
Monger, T.G. and W.W. Parson, 1977, Biochim. Biophys. Acta **360**, 393.
Montroll, E.W., 1946, J. Chem. Phys. **14**, 202.
Montroll, E.W., 1969, J. Math. Phys. **10**, 753.
Olson, J.M., 1978a, in: The Photosynthetic Bacteria, Clayton, R.K. and Sistrom, W.R., eds. (Plenum Press, New York) p. 161.
Olson, J.M., 1978b, Int. J. Syst. Bacteriol. **28**, 128.
Olson, J.M., 1981, Biochim. Biophys. Acta **594**, 33.
Olson, J.M. and J.P. Thornber, 1979, in: Membrane Proteins in Energy Transduction, Capaldi, R.A., ed. (Marcel Dekker, New York) p. 279.
Olson, J.M., B. Ke and K.H. Thompson, 1976, Biochim. Biophys. Acta **430**, 524.
Paillotin, G. and C.E. Swenberg, 1979, in: Chlorophyll Organization and Energy Transfer in Photosynthesis, Wolstenholme, G., and Fitzsimons, D.W., eds. (Excerpta Medica, Amsterdam) p. 201.
Park, R.B. and J. Biggins, 1964, Science **144**, 1009.
Pearlstein, R.M., 1964, Proc. Natl. Acad. Sci. U.S.A. **52**. 824.
Pearlstein, R.M., 1966, Ph.D. Thesis, University of Maryland.
Pearlstein, R.M., 1967, Brookhaven Symp. Biol. **19**, 8.
Pearlstein, R.M. and R.P. Hemenger, 1978, Proc. Natl. Acad. Sci. U.S.A. **75**, 4920.
Pearlstein, R.M., F. Van Nostrand and J.A. Nairn, 1979, Biophys. J. **26**, 61.
Philipson, K.D. and K. Sauer, 1972, Biochemistry **11**, 1880.
Rabinowitch, E. and Govindjee, 1969, Photosynthesis (Wiley, New York).
Rafferty, C.N. and R.K. Clayton, 1979, Biochim. Biophys. Acta **546**, 189.
Reed, D.W. and R.K. Clayton, 1968, Biochem. Biophys. Res. Commun. **30**, 471.
Reed, D.W. and B. Ke, 1973, J. Biol. Chem. **248**, 3041.
Robinson, G.W., 1967, Brookhaven Symp. Biol. **19**, 16.
Sauer, K., 1975, in: Bioenergetics of Photosynthesis, Govindjee, ed. (Academic Press, New York) p. 115.
Sauer, K., 1978, Acc. Chem. Res. **11**, 257.
Sauer, K. and L.A. Austin, 1978, Biochemistry, **17**, 2011.
Sauer, K., E.A. Dratz and L. Coyne, 1968, Proc. Natl. Acad. Sci. U.S.A. **61**, 17.
Shepanski, J.F., 1980, Am. Soc. Photobiol. **8**, 87 (abstract).
Shipman, L.L., T.M. Cotton, J.R. Norris and J.J. Katz, 1976, Proc. Natl. Acad. Sci. U.S.A. **73**, 1791.
Shuvalov, V.A. and A.A. Asadov, 1979, Biochim. Biophys. Acta **545**, 296.
Silbey, R., 1976, Ann. Rev. Phys. Chem. **27**, 203.
Smoluchowski, M.W., 1916, Phys. Z. **17**, 557, 585.
Staehelin, L.A., J.R. Golecki and G. Drews, 1980, Biochim. Biophys. Acta **589**, 30.
Swenberg, C.E., N.E. Geacintov and M. Pope, 1976, Biophys. J. **16**, 1447.
Sybesma, C., 1977, An Introduction to Biophysics (Academic, New York).
Thornber, J.P., 1975, Ann. Rev. Plant Physiol. **26**, 127.
Tinoco, I. 1963. Radiat. Res. **20**, 133.
Van Metter, R.L., 1977a, Biochim. Biophys. Acta, **462**, 642.
Van Metter, R.L., 1977b, Ph.D. Thesis, University of Rochester.
Vermeglio, A. and R.K. Clayton, 1976, Biochim. Biophys. Acta, **449**, 500.
Vermeglio, A., J. Breton, G. Paillotin and R. Cogdell, 1978, Biochim. Biophys. Acta, **501**, 514.
Weiss, C., Jr., 1972, J. Mol. Spectrosc. **44**, 37.
Whitten, W.B., J.A. Nairn and R.M. Pearlstein, 1978, Biochim. Biophys. Acta **503**, 251.
Whitten, W.B., J.M. Olson and R.M. Pearlstein, 1980, Biochim. Biophys. Acta **591**, 203.
Wolber, P.K. and B.S. Hudson, 1979, Biophys. J. **28**, 197.
Zankel, K.L. and R.K. Clayton, 1969, Photochem. Photobiol. **9**, 7.

CHAPTER 18

Vibrational Frenkel Excitons

M.V. BELOUSOV

Institute of Physics
Leningrad State University
LENINGRAD 199164
USSR

Excitons
Edited by
E.I. Rashba and M.D. Sturge

Contents

1. Introduction

An essential feature of molecular and complex ionic crystals is the marked difference in the forces of intra- and intermolecular interactions, which makes it possible to classify the crystal phonon modes either as internal or external. Intramolecular vibrational excitations of a crystal may be considered as vibrational Frenkel excitons. Theories which describe the Davydov splitting and exciton band structure, the spectra of the doped and mixed crystals, the quasiparticle interactions for the electronic Frenkel excitons are also valid for the vibrational excitons. Historically, electronic exciton spectroscopy has developed ahead of vibrational exciton spectroscopy and to a great extent has stimulated the progress of the latter. In this paper we shall present some recent results on vibrational excitons studies which, in turn, may give a new insight on problems important for the electronic excitons as well. The success was mainly achieved through experimental studies of relatively simple physical systems for which clear theoretical models can be applied with sufficient accuracy. The high technical level of modern vibrational spectroscopy is also of great importance. This allows one to test theory quantitatively, and to develop new methods for Frenkel exciton studies.

In the present paper, we shall emphasize the effect of the exciton–exciton interaction on the two-exciton spectra in perfect crystals. Besides, we shall also consider one-exciton spectra of crystals with isotopic impurities. Attention will also be paid to two-exciton spectra of the isotopically doped crystals. The latter problem has been poorly studied up to date, whereas the former two have been investigated in detail not only for excitons and phonons but for other quasiparticles as well. The effects of the interaction between quasiparticles or between quasiparticles and an impurity are well understood now. However, a quantitative test of the accuracy of the models used meets certain difficulties. In the conventional approach the density of two- or one-particle states for a perfect harmonic crystal is first calculated and afterwards the perturbation due to the interaction of quasiparticles with each other or with impurities is taken into account. In such a case the experiment serves as a criterion for the accuracy of the models used at both steps of the calculation. On the other hand in the study of the vibronic spectra of molecular crystals Rabin'kina et al. (1970) have proposed a more efficient approach, in which the density of the exciton states is reconstructed from the experimental spectra. This approach was further developed for the spectroscopy of vibrational excitons, where it was found to be most fruitful.

The main aim of this paper is to show that comparative quantitative study of the one- and two-exciton spectra of perfect and isotopically doped and mixed crystals not only enables one to test the model of the interacting excitons but also to reconstruct the density of one-exciton states.

We shall start with a description of main theoretical results, paying special attention to the anharmonic interaction of vibrational excitons, since this problem has not been covered sufficiently by review papers. We shall next describe the methods of reconstruction of the exciton density of states. These methods will be discussed as applied to the molecular crystal of CO_2 and complex ionic crystals of $NaNO_3$ and NH_4Cl for which the most complete quantitative work have been performed. These three examples also illustrate the main effects of exciton–exciton and exciton–impurity interactions.

2. Exciton–exciton and exciton–impurity interaction: main theoretical results

2.1. Anharmonic interaction of vibrational excitons

In the Raman and IR-conductivity spectra of a harmonic perfect crystal only the narrow δ-function-like lines due to transitions with the creation of long-wavelength excitons should be observed. Broadening of the one-exciton lines and the fact that multiexciton transitions are observed are due to anharmonicity which can be taken into account in the framework of a model of the interacting quasiparticles.

In the present paper we shall dwell upon two anharmonic effects, namely, the exciton–exciton scattering and the decay of the exciton into two other excitons. These two effects influence significantly the shape of two-exciton spectra and should be taken into account when reconstructing the density of states.

The anharmonic interaction of the vibrational excitons will be considered for the model crystal with one molecule in a cell. There is no significant difference between this case and that of several molecules in a cell. The exciton state will be denoted as $|n, \boldsymbol{k}\rangle$ where the first index specifies the exciton band and the second is the exciton wave vector. The exciton energy $\omega_{n,k}$ is measured from the ground state.

Retarded Green's functions of noninteracting (free) excitons for the one-exciton $G_n(\boldsymbol{k}, \omega)$ and two-exciton $G_{nm}(\boldsymbol{k}, \omega)$ states are of the form:

$$G_n(\boldsymbol{k}, \omega) = (\omega - \omega_{n,k} + \mathrm{i}\delta_n)^{-1}, \tag{1}$$

$$G_{nm}(\boldsymbol{k}, \omega) = N^{-1} \sum_{\boldsymbol{q}} (\omega - \omega_{n,\boldsymbol{k}-\boldsymbol{q}} - \omega_{m,\boldsymbol{q}} + \mathrm{i}\delta_{nm})^{-1}. \tag{2}$$

The values δ_n and δ_{nm} represent the exciton damping which is caused by the processes which are not considered here in an explicit form. N is the number of molecules in the unit volume.

The imaginary and real parts of the Green's function are related by the Kramers–Kronig relation:

$$G(\boldsymbol{k}, \omega) = \int \frac{g(\boldsymbol{k}, z)}{\omega - z} \mathrm{d}z - \mathrm{i}\pi g(\boldsymbol{k}, \omega), \qquad \int g(\boldsymbol{k}, \omega)\, \mathrm{d}\omega \equiv 1, \tag{3}$$

where $g(\boldsymbol{k}, \omega)$ is the one- or two-exciton spectral function. The relation (3) holds for all the Green's functions used. Therefore the n or nm indices in (3) are omitted.

Consider first the interaction between excitons, resulting in their scattering on each other $|1, \boldsymbol{k} - \boldsymbol{q}; 2, \boldsymbol{q}\rangle \rightarrow |1, \boldsymbol{k} - \boldsymbol{p}; 2, \boldsymbol{p}\rangle$. This interaction is due to the fourth-order anharmonicity. Within the frame-work of the oriented gas model in which intermolecular interaction is neglected and the excitations are localized on one molecule, the interaction between excitations shifts the combination mode frequency $\omega_{1,2}^{g}$ with respect to the sum of the corresponding fundamental vibrations by $\Delta_{12} = \omega_{1,2}^{g} - \omega_{1}^{g} - \omega_{2}^{g}$. The harmonic intermolecular interaction in crystals yields the delocalization of the excitations; while the anharmonic one tends to bind two excitons into one quasiparticle (biexciton). The dynamic theory of such processes for the vibrational and rotational phonons in parahydrogen crystal has been developed by Van Kranendonk (1959, 1960). It should be stressed, however, that in these works as well as in allied works on cooperative phenomena, in the process of light absorption two quasiparticles arise at different sites and the absorption itself is allowed only due to intermolecular interaction. In vibrational excitonic spectra, as in the vibronic spectra of molecular crystals, the optical transition is intramolecular, and the role of intermolecular interaction is reduced to changing the structure of the energy spectrum. The theoretical approach used in the present paper was first introduced by Rashba (1966), who has considered the vibronic spectra of molecular crystals (see also review of Sheka 1971). A similar approach was used by Ruvalds and Zawadowski (1970) for phonons and by Agranovich (1971) for the vibrational excitons (see also review of Ruvalds (1975)). All these theories used the point interaction model, which takes into account only the intramolecular anharmonicity, thus confining the discussion to s-wave exciton–exciton scattering. The model Hamiltonian of two excitons in the momentum representation is:

$$\mathcal{H} = \sum_{\boldsymbol{k}} (\mathcal{H}_{1,\boldsymbol{k}}^{(2)} + \mathcal{H}_{2,\boldsymbol{k}}^{(2)} + \mathcal{H}_{1212,\boldsymbol{k}}^{(4)}), \tag{4}$$

$$\mathcal{H}^{(2)}_{n,k} = \omega_{n,k} b^{+}_{n,k} b_{n,k}, \tag{5}$$

$$\mathcal{H}^{(4)}_{1212,k} = \Delta_{12} \sum_{q,p} b^{+}_{1,k-q} b^{+}_{2,q} b_{1,k-p} b_{2,p}, \tag{6}$$

where $\omega_{n,k}$ denotes the exciton energy, $b^{+}_{n,k}(b_{n,k})$ define the exciton creation (destruction) operations. $\mathcal{H}^{(2)}_{n,k}$ is the harmonic term, and $\mathcal{H}^{(4)}_{1212,k}$ is the anharmonic term of the Hamiltonian. Δ_{12} is the anharmonic interaction constant, ($\Delta_{12} \ll \omega_{n,k}$). In the point interaction model Δ_{12} does not depend upon the momentum and its value is equal to the anharmonic shift of the combination mode in the isolated molecule.

The two-exciton Green's function of the interacting excitons $G^{\Delta}_{12}(\boldsymbol{k}, \omega)$ is related in the above model to the free two-exciton Green's function by:

$$G^{\Delta}_{12}(\boldsymbol{k}, \omega) = (G^{-1}_{12}(\boldsymbol{k}, \omega) - \Delta_{12})^{-1}. \tag{7}$$

The required IR-conductivity spectrum or the Raman spectrum of the two-exciton transitions are proportional, within a factor weakly dependent on the frequency, to the two-exciton spectral function, $g^{\Delta}_{12}(\boldsymbol{k} \approx 0, \omega) = -\pi^{-1} \operatorname{Im} G^{\Delta}_{12}(\boldsymbol{k} \approx 0, \omega)$.

The influence of the strength of the point interaction between excitons on the two-exciton spectral function is illustrated in fig. 1. The free exciton spectral function (curve 0) was simulated by a symmetric band of a $100\ \text{cm}^{-1}$ width which shows Van Hove singularities. In order to take into account the exciton damping, the spectral function was convoluted with Lorentz function with a width of $1.5\ \text{cm}^{-1}$, which corresponds to $\delta_{12} = 0.75\ \text{cm}^{-1}$ in expression (2).

The interaction between excitons results in a considerable alteration of the spectral function $g^{\Delta}_{12}(\boldsymbol{k}, \omega)$, and in a shift of its centre of gravity by Δ_{12}. Changes of the Van Hove singularities occur (curves 1 and 2) even for weak interactions ($\Delta_{12} = -10, -20\ \text{cm}^{-1}$). With the increase of $|\Delta_{12}|$ a sharp and asymmetric peak typical of a quasibound state arises on the low-frequency edge (curve 3, $\Delta_{12} = -30\ \text{cm}^{-1}$). Further increase of the interaction leads to the binding of the two excitons into a single-particle state (biexciton). As a result a narrow line beyond the region of the dissociated two-exciton states is revealed in the $g^{\Delta}_{12}(\boldsymbol{k}, \omega)$ spectrum (curves 4, 5 and 6, $\Delta_{12} = -40, -50$ and $-60\ \text{cm}^{-1}$).

It is worth mentioning that with the increase of the damping, δ_{12}, the biexciton line merge with the continuous spectrum of the dissociated states (fig. 1, dashed curve 4, $\delta_{12} = 3\ \text{cm}^{-1}$). As a result, the spectral function of the bound state takes a form more characteristic of a quasibound state.

The biexciton transition frequency will be designated by $\omega_{1,2,k}$ (or $\omega_{1,2}$) as distinguished from the two-exciton transitions $\omega_{1,k-q} + \omega_{2,q}$ (or $\omega_1 + \omega_2$). The frequency of the biexciton line is given by the pole of the $G^{\Delta}_{12}(\boldsymbol{k}, \omega)$ which

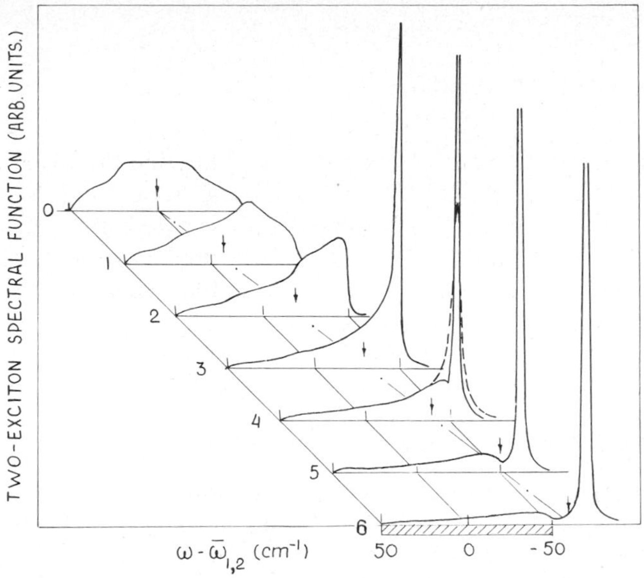

Fig. 1. Dependence of two-exciton spectra on anharmonic shift Δ_{12}. Δ_{12} varies from $0\,cm^{-1}$ (curve 0) to $-60\,cm^{-1}$ (curve 6). Solid curves are calculated for damping $\delta_{12} = 0.75\,cm^{-1}$, dashed curve for $\delta_{12} = 3\,cm^{-1}$. Region of the dissociated states is shaded. Arrows give the position of the centre of gravity of the band. The dash-dotted line shows the dependence of the combination transition frequency on the value Δ_{12} in the oriented gas approximation. Frequencies are taken from the frequency $\bar{\omega}_{1,2}$ of the centre of gravity of the initial spectrum.

is situated beyond the band of the dissociated states and can be found from:

$$\text{Re}\, G_{12}^{-1}(\boldsymbol{k}, \omega) = \Delta_{12}. \tag{8}$$

The solution of eq. (8) falls in the region of low density of the two-exciton states (e.g. on the band edge), and yields the location of the peak due to the quasibound state.

An example of a graphic solution of eq. (8) is given in fig. 2. The solid curve 1 corresponds to the left-hand side of eq. (8), whereas the horizontal line 2 represents the right-hand side. The intersection point gives the biexciton frequency $\omega_{1,2}$. The solution shown in fig. 2 corresponds to the case considered in fig. 1, curve 3. The dash-and-dot line represents the right-hand side of eq. (8) in the oriented gas approximation. All the above mentioned curves are plotted on the same scale, shown in the left of fig. 2.

As we can see the biexciton state arises if the intramolecular anharmonic interaction, resulting in the binding of the excitations, is stronger than the harmonic intermolecular interaction, responsible for the delocalization of the excitations. An approximate criterion for the existence of the biexciton

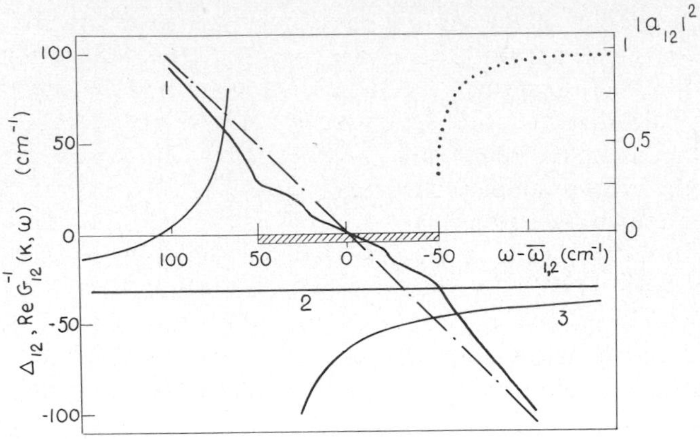

Fig. 2. Graphical solutions of eqs. (8) and (13) for the energy of the one-particle states. Re $G_{12}^{-1}(\boldsymbol{k}, \omega)$ (solid curve 1) calculated from the density of states, which is presented by curve 0 in fig. 1. Dash–dotted line corresponds to Re $G_{12}^{-1}(\boldsymbol{k}, \omega)$ in the oriented gas approximation. The intersection point of the curve 1 and the horizontal line 2 ($\Delta_{12} = -30\,\text{cm}^{-1}$) gives, according to eq. (8), the biexciton energy. The dotted curve gives the intensity of the biexciton line as a function of its frequency. Solid curves 3 are given by the right-hand side of the eq. (13) ($\Delta_{12} = -30\,\text{cm}^{-1}$, $\gamma = 40\,\text{cm}^{-1}$, $\omega_{3,\boldsymbol{k}} = \bar{\omega}_{1,2} + 50\,\text{cm}^{-1}$). Their intersections with curve 1 give the energies of the one-particle states in the case of the Fermi resonance. Frequencies are taken from the frequency $\bar{\omega}_{1,2}$ of the centre of gravity of the unperturbed two-exciton spectra. Region of the dissociated states is shaded.

state can be formulated as follows: $|\Delta_{12}| \gtrsim 0.25 T_{12}(\boldsymbol{k})$ (where $T_{12}(\boldsymbol{k})$ is the width of the two-exciton band).

Ratio of the integrated intensity of the biexciton line S_{12}^{Δ} to the total intensity of the two-exciton transitions S_{12} is given by the probability $|a_{12}|^2$ for the two excitons to be on one lattice site. This value is a measure of binding of the excitons

$$\frac{S_{12}^{\Delta}}{S_{12}} = |a_{12}|^2 = \left| \frac{\mathrm{d}}{\mathrm{d}\omega} \operatorname{Re} G_{12}^{-1}(\boldsymbol{k}, \omega) \right|^{-1}_{\omega=\omega_{1,2,\boldsymbol{k}}} . \tag{9}$$

The dependence of $|a_{12}|^2$ on the biexciton energy is shown in fig. 2 by a dotted curve. In the case of strong interaction ($|\Delta_{12}| \gtrsim T_{12}(\boldsymbol{k})$) the biexciton line contains almost all of the intensity of the two-exciton spectrum, and the intensity of the band of the dissociated states is small, which corresponds to $|a_{12}|^2 \to 1$. When $|\Delta_{12}|$ is decreased the biexciton line approaches the band edge and diminishes in intensity.

Now we shall consider the interaction between one-exciton and two-exciton states due to the third-order anharmonicity. This process was originally suggested by Fermi (1931) to explain an anomalous peak in the Raman spectrum of CO_2 molecules which was attributed to the overtone of a vibrational mode. In the molecular approximation considered by Fermi there is no dispersion of the vibrational states. Therefore the combination or the overtone mode appears as a sharp peak in the two-excitation density of states. It has been suggested that the observed spectrum results from the mixing (hybridization) of the fundamental and combination modes. It consists of two lines; its frequencies and intensities can be calculated using a simple model of two coupled oscillators.

In crystals the two-exciton states usually have a continuous spectrum. The theory of Fermi resonance in crystals has been derived by Ruvalds and Zawadowski (1970) for phonons and by Agranovich and Lalov (1971a, b) for vibrational excitons. In the point interaction model the Hamiltonian which takes into account the anharmonic interaction of $|1, \boldsymbol{k}-\boldsymbol{q}\rangle$ and $|2, \boldsymbol{q}\rangle$ excitons with each other and also their coupling with the $|3, \boldsymbol{k}\rangle$ exciton is:

$$\mathcal{H} = \sum_{\boldsymbol{k}} \left(\sum_{n=1}^{3} \mathcal{H}^{(2)}_{n,\boldsymbol{k}} + \mathcal{H}^{(3)}_{123,\boldsymbol{k}} + \mathcal{H}^{(4)}_{1212,\boldsymbol{k}} \right), \tag{10}$$

where $\mathcal{H}^{(2)}_{n,\boldsymbol{k}}$ and $\mathcal{H}^{(4)}_{1212,\boldsymbol{k}}$ are given by the expressions (5) and (6), respectively. The third order anharmonic term $\mathcal{H}^{(3)}_{123,\boldsymbol{k}}$ is given in the form:

$$\mathcal{H}^{(3)}_{123,\boldsymbol{k}} = \gamma \sum_{\boldsymbol{q}} (b^{+}_{1,\boldsymbol{k}-\boldsymbol{q}} b^{+}_{2,\boldsymbol{q}} b_{3,\boldsymbol{k}} + b_{1,\boldsymbol{k}-\boldsymbol{q}} b_{2,\boldsymbol{q}} b^{+}_{3\boldsymbol{k}}), \tag{11}$$

where γ is a coupling constant ($\gamma \ll \omega_{n,\boldsymbol{k}}$). Within the framework of the point interaction model γ does not depend upon the wave vector and its value is equal to the value of the corresponding intramolecular constant. In the molecular approximation 2γ is equal to the splitting, due to the coupling of the modes, at exact resonance.

The one-exciton Green's function of the perturbed exciton has the form:

$$G^{\gamma}_{3}(\boldsymbol{k}, \omega) = [\omega - \omega_{3,\boldsymbol{k}} + i\delta_3 - \gamma^2 G^{\Delta}_{12}(\boldsymbol{k}, \omega)]^{-1}. \tag{12}$$

At resonance the intensity of transitions into two-exciton states which are unperturbed by coupling with the one-exciton state ("unperturbed" intensity) is usually neglected since it is small compared with the "unperturbed" intensity of the one-exciton transition. In this case the one-exciton states contribution to the hybridized states governs the observed spectrum. The IR-conductivity spectrum or Raman spectrum is proportional, within a factor weakly dependent on frequency, to the one- exciton spectra function $g^{\gamma}_{3}(\boldsymbol{k} \approx 0, \omega) = -\pi^{-1} \operatorname{Im} G^{\gamma}_{3}(\boldsymbol{k} \approx 0, \omega)$. The integrated intensity of the observed spectrum and the frequency of the center of gravity coincide with the intensity and frequency of the unperturbed one-exciton transition.

Comparison between eqs. (1) and (12) indicates that the manifestation of the two-exciton states in the observed spectrum can also be understood as a result of the frequency-dependent exciton damping. The function of damping $\delta_3 - \gamma^2\,\mathrm{Im}\,G^{\Delta}_{12}(\boldsymbol{k}, \omega)$ consists of two parts. The first part describes nominally the decay and scattering of the $|3, \boldsymbol{k}\rangle$ exciton into many-particle continuum having a uniform spectrum. The second part describes the decay of the $|3, \boldsymbol{k}\rangle$ exciton into $|1, \boldsymbol{k}-\boldsymbol{q}; 2, \boldsymbol{q}\rangle$ excitons.

In a more general case, the "unperturbed" intensity of the two-exciton transitions should also be taken into account. Due to this fact the experimental spectrum results from the interference of the one- and two-exciton transitions. The situation here is similar to Fano antiresonance. The expressions for the susceptibility of the system can be taken, for instance, from Cho (1972).

We shall confine ourselves to the case which illustrates the influence of the Fermi resonance on the stability of bound states. Figure 3 shows the shape of the perturbed one-exciton spectral function $g_3^3(\boldsymbol{k}, \omega)$ calculated for different energy spacings between the unperturbed exciton and two-exciton states. The spectral function of the latter states is shown in fig. 1, curve 3. The fourth-order anharmonic interaction is not very strong in this case: as a result only quasibound states appear. The coupling with the one-exciton

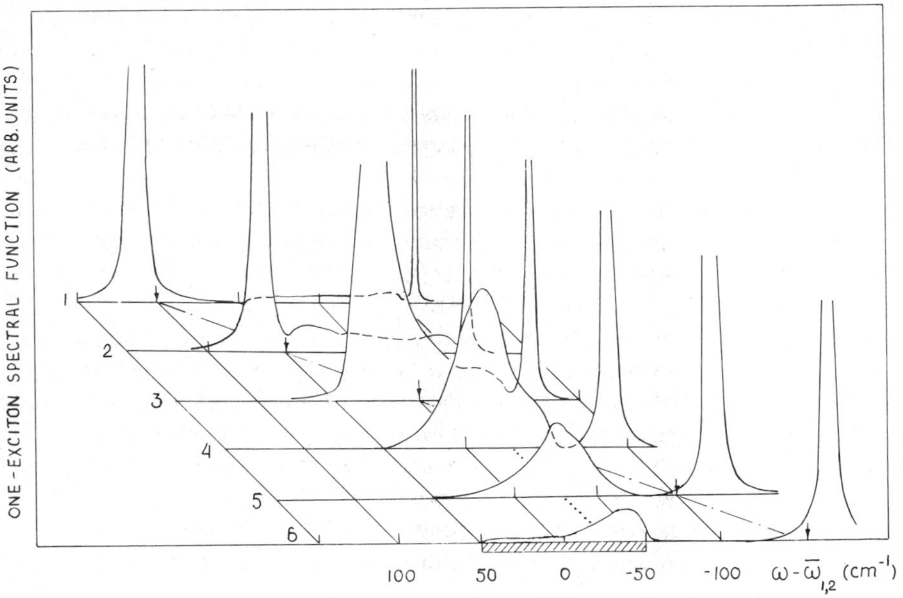

Fig. 3. Dependence of one-exciton spectra at Fermi resonance on the exciton energy (which is changed along the dash–dotted line). The region of the dissociated states is shaded. Arrows indicate the centres of gravity of one-exciton spectra. The dotted line shows the position of the centre of gravity of two-exciton spectra, unperturbed by Fermi resonance. $\gamma = 40\ \mathrm{cm}^{-1}$, $\delta_3 = \delta_{12} = 0.75\ \mathrm{cm}^{-1}$, $\Delta_{12} = -30\ \mathrm{cm}^{-1}$.

state which is located above the two-exciton band (fig. 3, curves 1 and 2) gives rise to a bound state (low-frequency peak) which is detached from the region of dissociated states. The energy of the exciton (high-frequency peak) is also changed. Apart from the narrow exciton and biexciton lines a wide band is observed which is due to the dissociated states enhanced by Fermi resonance. When the unperturbed exciton frequency is within the frequency range of the dissociated states (curve 3) the damping of the exciton increases and its band has a complex form with a steep decrease at the dissociated band edges. Near the exact resonance (curve 4) a drastic transformation of a spectrum takes place. It makes no sense to classify the observed bands into one-exciton or two-exciton bands since they are strongly hybridized. Curves 5 and 6 correspond to the case when the exciton state is below the two-exciton state. The low-frequency line in this case corresponds to the exciton, whereas the biexciton state disappears. The frequencies of the one particle lines (exciton and biexciton) are given by the poles of the $G_3^\gamma(\boldsymbol{k}, \omega)$ which are situated beyond the band of the dissociated states and can be found from:

$$\operatorname{Re} G_{12}^{-1}(\boldsymbol{k}, \omega) = \Delta_{12} + \frac{\gamma^2}{\omega - \omega_{3,\boldsymbol{k}}}. \tag{13}$$

Figure 2 gives an example of a graphic solution of eq. (13) for the case considered in fig. 3, curve 2. The energy of the unperturbed exciton falls into the high-frequency edge of the two-exciton band. As a result of the coupling two one-particle states arise: the exciton state above and the biexciton one below the band. The frequencies of these transitions are given by the intersection point of the curves 1 and 3, corresponding to the left- and right-hand side of eq. (13). The intersection of curve 3 with the dash-dotted line yields a solution in the oriented gas (molecular) approximation. In the latter case $G_{12}^{-1}(\boldsymbol{k}, \omega) = \omega - \bar{\omega}_{1,2,\boldsymbol{k}}^{\Delta}$ and eq. (13) leads to the same results as in the case of a simple model of two coupled oscillators.

In the non-resonant case ($|\omega_{1,\boldsymbol{k}-\boldsymbol{q}} + \omega_{2,\boldsymbol{q}} - \omega_{3,\boldsymbol{k}}| \ll \gamma, T_{12}(\boldsymbol{k})$) the dispersion of the right-hand side of eq. (13) can be neglected. Equation (13) is equivalent then to eq. (8). In this case the third-order anharmonicity only changes the anharmonicity shift constant and the two-exciton transition intensity.

It is worth remembering that in a harmonic crystal two-exciton transitions should not be observed. Their appearance in the anharmonic crystal results from the mixing with the one-exciton states. It should also be noted that in the nonresonant case a significant contribution to the intensity of the two-exciton transitions may also arise from remote one-exciton states, including electronic states.

2.2. Interaction of a vibrational exciton with a single isotopic impurity

Optical properties of a shallow impurity centre are strongly influenced by the interaction of the centre with all the exciton states of a host lattice. In a way the impurity serves as a probe sensitive to a density of states distribution. The isotopic impurity method has been successfully exploited in the study of the band structure of the electronic Frenkel excitons in the molecular crystals. A detailed consideration of this problem is presented in review papers by Rashba (1972) and Sheka (1972), where theoretical and experimental works are summarized, and also in the paper of Sommer and Jortner (1969). Therefore we shall only briefly review some of the aspects of the problem, necessary for the analysis of the experimental data.

For the determination of the isotopic impurity spectrum the point interaction model is usually applied. The energy of the exciton on the impurity site is changed by the value of the isotopic shift Δ_n. If the energy defect exceeds the critical value $|\Delta_n| \geqslant 0.25 T_n$ (where T_n is the exciton band width), a localized state is formed. The energy ω_n' of the latter is given by the Lifshitz equation:

$$\mathrm{Re}\, G_n^{-1}(\omega) = \Delta_n, \tag{14}$$

where $G_n(\omega)$ is a one-exciton Green's function, which is related to the density of states function $g_n(\omega)$ by the expression, similar to eq. (3):

$$G_n(\omega) = N^{-1} \sum_k G_n(k, \omega) = \int \frac{g_n(z)}{\omega - z} \mathrm{d}z - \mathrm{i}\pi g_n(\omega). \tag{15}$$

The probability of finding the exciton at the impurity site is

$$|a_n|^2 = \frac{\mathrm{d}\omega_n'}{\mathrm{d}\Delta_n} = \left| \frac{\mathrm{d}}{\mathrm{d}\omega} \mathrm{Re}\, G_n^{-1}(\omega) \right|^{-1}_{\omega=\omega_n'}. \tag{16}$$

The wave function of a localized state is a superposition of wave functions of excitons from all the band. However the intensity of the localized mode depends only on the contribution of the optically active, long-wavelength exciton. The contribution of a free exciton to the wave function of a localized exciton is resonantly dependent on their energy difference. Therefore depending on the latter the ratio of the intensities of the guest mode ($S_n'(x)$) to that of the total mode (S_n) can become either less or much larger than the relative impurity content x. Such situation is known as the Rashba effect (Rashba 1957, 1962, Broude et al. 1962) described by the relation:

$$\frac{S_n'(x)}{S_n} = x \left(\frac{\Delta_n}{\omega_n - \omega_n'} \right)^2 |a_n|^2. \tag{17}$$

Relation (17) holds for $S_n(x) \ll S_n$ in the case when the isotopic substitution does not change the intensity and symmetry of vibrations in the isolated molecule.

2.3. *Interaction of two vibrational excitons with each other and with single isotopic impurity*

The calculation of the spectrum of the two excitons interacting with each other and with an impurity is, in the general case, a three-body problem. An exact solution has been presented by Rashba (1966) for the case where of the excitations (e.g. $|m\rangle$) can be treated as immobile (i.e. having a narrow exciton band). An interesting feature of the combined excitation involving an impurity is that two types of configurations are possible: a joint configuration and a separated configuration. In the latter case the immobile exciton is on one of the molecules neighbouring the impurity. For point interaction between excitons this configuration yields a line whose frequency lies near the sum of the localized mode ω_n' and the host mode ω_m. In the case of a joint configuration the immobile exciton $|m\rangle$ is on the impurity. In this case the energy of the $|n\rangle$ exciton on the impurity site changes by $\Delta_n + \Delta_{nm}$. As a result, an additional localized mode ω'_{nm} can appear. The frequency of this mode is given by an equation similar to eqs. (7) or (14) (Rashba 1966):

$$\mathrm{Re}\, G_{nm}^{-1}(k, \omega) = \mathrm{Re}\, G_n^{-1}(\omega - \omega_m) = \Delta_{nm} + \Delta_n + \Delta_m. \qquad (18)$$

The two above-mentioned types of configurations have been observed by Broude et al. (1967) and Dolganov and Sheka (1969) in vibronic spectra.

In the case when the isotopic and anharmonic shifts are of the same sign, localized states in the two-exciton spectrum can occur even when they are absent in the one-exciton spectrum. This fact was pointed out by Agranovich et al. (1970).

2.4. *Vibrational excitons in isotopically mixed crystals*

The spectral density of vibrations of the impurity molecule and the density of states of a host crystal are related by the relation similar to eq. (7). Therefore fig. 1 can also be treated as an illustration of the dependence of the density of impurity states on the value of the isotopic shift. In this case the initial curve 0 in fig. 1 should be taken to be the density of one-exciton states. A sharp peak in the curves 4–6 corresponds to the localized mode. However it can be seen that besides this split-off peak the impurity states also have a continuous spectrum spreading over all the exciton band. This spectrum is of significant interest since it enables one to reconstruct, in principle, the density of states of the host crystal.

The manifestation of the impurity states in the observed spectrum can be understood as a result of the exciton damping due to its scattering by impurities. The presence of the impurity induces broadening of the exciton band and the appearance of the wings even when the localized mode is absent. To investigate this effect in a real crystal with anharmonic damping, one should study crystals with rather high impurity content. The inter-impurity interaction is then important and the spectrum of the mixed crystal can only be calculated using approximate methods. The conventionally used single-site approximation does not take into account the influence of clusters, which can give rise to additional structure in the spectrum of the mixed crystal. However, if the impurity vibrational shift is less than the exciton band width, and intermolecular interaction is of a long-range character, the influence of the cluster structure is not expected to be great.

For isotopic substitution, which does not change the symmetry and intensity of the vibration, the single-site approximation yields the following expression for the one-exciton Green's function of a mixed crystal $A_{1-x}B_xC$ (see the review of Elliott et al. 1974).

$$G_n(\boldsymbol{k}, \omega, x) = [\omega - \omega_{n,k} - x\Delta_n + \mathrm{i}\delta_n - \Sigma(\omega, x)]^{-1}, \tag{19}$$

where $\omega_{n,k}$ is the energy of the exciton $|n, \boldsymbol{k}\rangle$ in AC crystal. $\Sigma(\omega, x)$ is the part of self-energy and $-\mathrm{Im}\,\Sigma(\omega, x)$ is the damping function of the exciton scattered by impurities. This term can usually be calculated within the coherent potential approximation, which describes the spectrum at arbitrary impurity concentrations. For low concentrations satisfactory results can be obtained using a much simpler average T-matrix approximation (ATA). In this approximation $\Sigma(\omega, x)$ can be calculated explicitly with the help of the Green's functions $G_n(\omega)$ for the pure crystal AC, which are given by eq. (16) (Elliott et al. 1974):

$$\Sigma(\omega, x) = x(1-x)\Delta_n^2[G_n^{-1}(\omega - x\Delta_n) - (1-2x)\Delta_n]^{-1}. \tag{20}$$

At $x \to 0$ and $x \to 1$ the ATA method yields the same results as the exact solution for a single impurity.

Since there is a variety of the detailed reviews (see, for instance, Elliott et al. 1974, Barker and Sievers 1975) dealing with the spectroscopy of mixed crystals, we shall discuss only briefly the main features of their mode behaviour.

If a localized vibration exists at low concentrations, then two lines are observed in the spectrum of the mixed crystal (two-mode behaviour). The intensities of these lines have opposite dependence on concentration and are equal at a certain concentration. The disorder of the mixed crystal causes a characteristic broadening of the two modes.

One-mode behaviour is observed if no localized mode exists at $x \to 0$ and

$x \to 1$. In this case the spectrum is dominated by one band at all contentrations.

If the ratio of the isotopic shift to the exciton band width is close to a critical value of 0.25–0.35, a localized mode can exist in only one of the end crystals. In this case more complicated mode behaviour is expected. An example of such a complicated behaviour will be given in what follows for isotopically mixed $NaNO_3$ crystals (sect. 5.3).

3. Reconstruction of the density of states from two- and one-exciton spectra

The theory considered above permits us to calculate the two-exciton spectrum of a perfect crystal or the one-exciton spectrum of a crystal with isotopic impurities provided that the exciton band of perfect crystal has previously been calculated. The experiment in this case provides a criterion for accuracy of models used on both steps of calculation. This approach will be illustrated in this paper in the application to the CO_2 crystal (sect. 4).

A more efficient approach, which does not require a model description of the intermolecular interaction and the usage of the fitting parameters has been proposed by Rabin'kina et al. (1970) when considering vibronic spectra. These authors have put forward a method of reconstruction of the density of states of free excitons from spectra of vibronic absorption involving a nontotally symmetric phonon. The method proposed by them is applicable to any two-particle transition provided that the interaction between particles can be described by a point potential.

The possibility of reconstruction of the density of states of free excitons comes from the fact that the expression (7) is symmetric with respect to $G^{\Delta}_{12}(\boldsymbol{k}, \omega)$ and $G_{12}(\boldsymbol{k}, \omega)$ and can be rewritten so that to express the density of states through $G^{\Delta}_{12}(\boldsymbol{k}, \omega)$,

$$g_{12}(\boldsymbol{k}, \omega) = -\pi^{-1}\,\mathrm{Im}[(G^{\Delta}_{12}(\boldsymbol{k}, \omega))^{-1} + \Delta_{12}]^{-1}. \qquad (21)$$

$G^{\Delta}_{12}(\boldsymbol{k}, \omega)$ can easily be calculated with the help of expression (3) from the experimental spectrum of the two-exciton transitions, which is proportional to $g^{\Delta}_{12}(\boldsymbol{k}, \omega)$. If one of the excitons has a narrow band, $g_{12}(\boldsymbol{k}, \omega)$ reproduces the shape of the density of states of the other exciton.

In the case of the vibronic spectra exact reconstruction of the density of states was complicated by overlap of the absorption bands of different origin. The approach proposed by Rabin'kina et al. (1970) was realized for the case of the vibrational excitons in $NaNO_3$, KNO_3, $Ba(NO_3)_2$, SiF_4, $CaCO_3$ (Belousov et al. 1976, 1977, 1978b, 1980b). The application of this method to the $NaNO_3$ crystal will be considered in sect. 5.1.

The problem of the reconstruction of the density of states can also be solved when the two-exciton spectrum is perturbed by Fermi resonance. The works of Krauzman et al. (1974) and Fukumoto et al. (1976) give an example of the solving of such a problem. These authors have interpreted the anomolous TO spectrum in CuCl crystal as resulting from Fermi resonance between TO and (TA + LA) phonons and have performed a quantitative analysis of this spectrum with the use of the point anharmonic interaction model. The spectrum of the free two-exciton states and values of coupling constants were obtained by a fitting procedure. However, this problem can also be solved without this procedure. The two-exciton spectral function $g_{12}^{\Delta}(\boldsymbol{k}, \omega)$ sought and coupling constant γ can be obtained from the spectrum of damping which, as appears from eq. (12), can be reconstructed from $G_3^{\gamma}(\boldsymbol{k}, \omega)$:

$$\pi\gamma^2 g_{12}^{\Delta}(\boldsymbol{k}, \omega) = \mathrm{Im}(G_3^{\gamma}(\boldsymbol{k}, \omega))^{-1} - \delta_3, \tag{22}$$

$G_3^{\gamma}(\boldsymbol{k}, \omega)$ is obtained from the experimental spectrum with the use of eq. (3) which is proportional to $g_3^{\gamma}(\boldsymbol{k}, \omega)$.

Excluding by this the effect of Fermi resonance we can take into account exciton–exciton interaction with a help of the relation (21). Such a consecutive inclusion of the anharmonic interaction will be demonstrated for the case of NH_4Cl crystal in sect. 6.2. The possibility of reconstruction of the one-exciton density of states from the spectra of overtone transitions will also be illustrated.

We should remember that the spectrum observed at Fermi resonance is proportional to $g_3^{\gamma}(\boldsymbol{k}, \omega)$ only in the case when the "unperturbed" intensity of the two-exciton transitions is negligibly small. The latter assumption can be verified by altering the spectra, for instance, by isotopic substitution, hydrostatic pressure or temperature. There is a good chance of making such a test in noncentrosymmetric crystals, where the same transition can be observed both in IR and Raman spectra (Belousov et al. 1980b).

The validity of the model of the point interaction between the excitons and also between the excitons and photon should be tested. This can be done, for instance, by means of the study of the isotopically doped crystal. In the latter case there is no doubt that exciton–impurity interaction is a point one. At the same time the spectrum of a single impurity can be calculated exactly (see sect. 2.2 and 2.3). Having measured the frequencies and intensities of a guest mode we can find a real part of the Green's function (see eq. (14)) and it's derivative (see eqs. (16) and (17)) at the frequency of the localized vibration. This makes it possible to test and to correct the density of states. If we deal with a set of guest molecules with different isotopic shifts then the study of the spectra of the isotopically doped crystals enables us to reconstruct the integral characteristics of the density of states. In order to find the centre of gravity of the latter and its second moment, it is appropriate to use the dependence of the impurity mode frequency on the isotopic shift, which can be approximated as:

$$\omega_n'(\Delta_n) = \omega_n^g(\Delta_n) + M_n^{(2)}/\Delta_n, \tag{23}$$

where $M_n^{(2)} = \int (\omega - \bar{\omega}_n)^2 g_n(\omega)\,d\omega$ is the second moment $g_n(\omega)$, $\omega_n^g(\Delta_n) = \bar{\omega}_n + \Delta_n$ – the impurity vibration frequency in the oriented gas approximation, $\bar{\omega}_n$ – the frequency of the centre of gravity $g_n(\omega)$.

The relation (23) gives good accuracy for not very shallow isotopic centres ($|\Delta_n| \gtrsim T_n$) and can be obtained by means of the moment-expansion method (Sommer and Jortner 1969). If the density of states in the band is given by a model function

$$g_n(\omega) = (2\pi)^{-1} T_n^{-2} [T_n^2 - 4(\omega - \bar{\omega}_n)^2]^{1/2}, \tag{24}$$

the relation (23) is exact. The criterion for the formation of the localized state in this case is $|\Delta_n| > 0.25 T_n$. The second moment for such a band is simply related to its width: $M_n^{(2)} = (T_n/4)^2$.

In the present paper the use of the isotopical impurity method for the investigation of vibrational exciton bands will be illustrated on the CO_2 (sect. 4.3) and $NaNO_3$ (sect. 5.2) crystals.

A more detailed information on the density of states can be obtained with the use of the isotopically mixed crystal. This study is of special interest since in contrast to the case of a single impurity the spectrum of mixed crystal can only be calculated in a certain approximation. Comparison with numerical calculations by the method of Dean (see review by Elliot et al. 1974) shows that a single-site approximation yields satisfactory results in the case of a long-range interaction of vibrations. In particular, the ATA method discussed in sect. 2.4 gives the correct contour of the experimental spectrum of mixed crystals with a impurity content up to 10% (Belousov et al. 1978a). In this case a study of the isotopically mixed crystal can be used not only for testing but also for reconstruction of the density of states (Belousov et al. 1980a, b). This problem within the frames of the ATA-method is nominally equivalent to the above-considered problem of reconstruction of the spectral function of free excitons. Similar to eq. (21) we can rewrite eq. (20) so that to express the density of states through $\Sigma(\omega, x)$,

$$g_n(\omega - x\Delta_n) = -\pi^{-1}\,\mathrm{Im}[x(1-x)\Delta_n^2 \Sigma^{-1}(\omega, x) - (1-2x)\Delta_n]^{-1}, \tag{25}$$

$\Sigma(\omega, x)$ is easily obtained from $G_n(\boldsymbol{k}, \omega, x)$ calculated by eq. (3) from the experimental spectrum of mixed crystals, proportional to $g_n(\boldsymbol{k}, \omega, x)$:

$$\Sigma(\omega, x) = \omega - \omega_{n,k} - x\Delta_n + i\delta_n - G_n^{-1}(\boldsymbol{k}, \omega, x). \tag{26}$$

This approach can yield reliable results for crystals with one-mode or near-one-mode behaviour. The impurity content should not be too large, but it should be sufficient for the damping of the exciton by impurities $-\mathrm{Im}\,\Sigma(\omega, x)$ to considerably exceed the anharmonic damping δ_n. The impurity-induced spectrum should be measured for the whole region of one-exciton states. In the present paper this method is illustrated by the example of $NaNO_3$ and NH_4Cl crystals (sects. 5.3 and 6.1).

4. *Study of the ω_2 and ω_3 vibrational exciton bands in crystalline CO_2*

The carbon dioxide crystal has a space group T_h^6(Pa 3) and contains four molecules in a unit cell. Three internal vibrations of the CO_2 molecule give rise in the crystal to the exciton bands which are well separated in frequency. The totally symmetric ω_1 vibration is only Raman active, whereas ω_2 and ω_3 vibrations are only IR active.

4.1. Fermi resonance of ω_1 and $\omega_2 + \omega_2$ states

The anharmonic interaction between the fundamental and combinational vibrational modes was first introduced by Fermi (1931) to account for the anomalous peak in the Raman spectrum of CO_2. By virtue of an accidental degeneracy the totally symmetric ω_1 mode and the lowest overtone of vibration ω_2 are very close in energy. Due to the Fermi resonance their levels are repelled and the wave functions are considerably hybridized. As a result, instead of a single line in the Raman spectrum of the CO_2, there are two lines of almost equal intensity. The Raman spectrum of a gaseous CO_2 provides a typical example of the Fermi resonance in a molecule. The Raman spectrum of the same substance in the solid state reveals prominent features of Fermi resonance in crystals. Witters and Cahill (1977) have shown that in crystalline CO_2, apart from the Fermi resonance doublets (having the same shapes as in a gas), a broad band due to the $\omega_2 + \omega_2$ dissociated states is observed (fig. 4). The knowledge of the coupling constant $\gamma = 50\ \mathrm{cm}^{-1}$ (see, for instance, Herzberg 1945) enables one to evaluate the frequency of the unperturbed vibration $\bar{\omega}_1$ (shown in fig. 4 by an undulating arrow) which falls into the region of dissociated states $\omega_2 + \omega_2$. As a result of the Fermi resonance two one-particle states appear. The high-frequency line is more intense; hence it is assigned to the ω_1 exciton. The low-frequency line is ascribed to the biexciton $\omega_{2,2}$. One should bear in mind that these states in the present case are nearly completely hybridized and shifted in energy. The Fermi resonance distorts the shape of the band of the dissociated states $\omega_2 + \omega_2$, but does not shift the edge of the band. As can be seen in fig. 4, the doubled TO and LO frequencies of the ω_2 excitons (shown by arrows) are close to the edges of

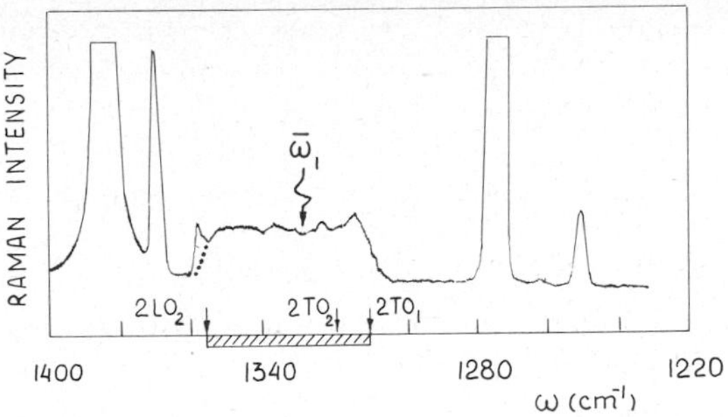

Fig. 4. Raman spectrum of natural CO_2. ($T = 20$ K). The two strong lines belong to one-particle states of the Fermi doublet. Region of the $\omega_2 + \omega_2$ dissociated states is shaded. The dotted line represents the edge of the $\omega_2 + \omega_2$ transition after ω_1 of $^{12}C\,^{16}O\,^{18}O$ has been subtracted. Two weak and narrow lines are due to $^{13}C\,^{16}O_2$ impurity. Undulating arrow shows the frequency of the $\bar{\omega}_1$ transition unperturbed by the Fermi resonance. (After Witters and Cahill (1977).)

the $\omega_2 + \omega_2$ band observed in the Raman spectrum. This is also in good agreement with the calculations of the exciton band, performed by Bogani and Schettino (1978) in the dipole–dipole approximation (the shaded region in fig. 4 corresponds to the calculated two-exciton transitions).

4.2. *Bound and dissociated states in the $\omega_3 + \omega_1$ and $\omega_3 + \omega_{2,2}$ spectra*

Dows and Schettino (1973) have studied the two-exciton IR absorption spectra in crystalline carbon dioxide. They have assigned them as resulting from the bound and dissociated states. Figure 5 shows the spectra, obtained by these authors in the region of $\omega_3 + \omega_1$ and $\omega_3 + \omega_{2,2}$ transitions. The bands due to these transitions have nearly the same intensities and their shapes are very similar, the Fermi resonance and the hybridization of ω_1 and $\omega_{2,2}$ states accounting for these facts. Apart from the isotopic structure, each of the Fermi partners consists of a very sharp line (≈ 2 cm^{-1} width) and of a broad structure on the high-frequency side extending up to the sum of $\omega_{3L} + \omega_1$ (or $\omega_{3L} + \omega_{2,2}$) with inflection at the frequency $\omega_{3T} + \omega_1$ (or $\omega_{3T} + \omega_{2,2}$) (these frequencies are shown in fig. 5 by arrows). Two sharp lines separated from the broad-band absorption can be assigned as arising from the bound states $\omega_{3,1}$ (3708 cm^{-1}) and $\omega_{3,2,2}$ (3600 cm^{-1}), while the broad feature can only be assigned as due to the dissociated states. Approximate estimates of the integrated intensities show that the biexciton and two-exciton absorption intensities are in the ratio 1:1 to each other. We note, that the shape of these two-exciton spectra is in a qualitative agreement with the model spectra calculated in the approximation of the point interaction of excitons (see fig. 1, curves 4 and 5).

A quantitative analysis of a shape of the two-exciton transition has been

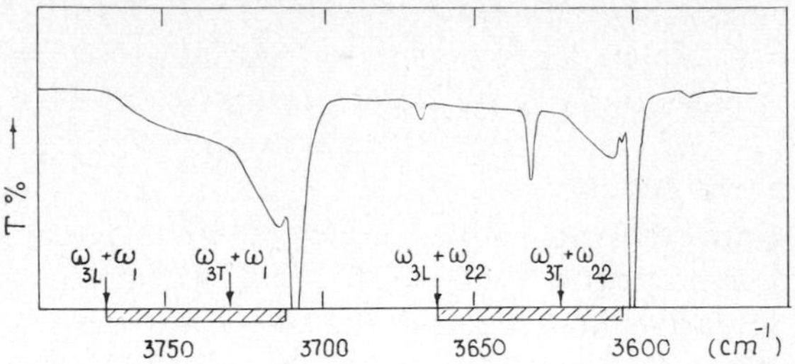

Fig. 5. IR-transmission spectra of CO_2 in the $\omega_3+\omega_1$ and $\omega_3+\omega_{2,2}$ region ($T=77$ K). The frequency region of the dissociated states is shaded. (After Dows and Schettino (1973).)

performed by Bogani (1978) who has used a point anharmonic interaction. The spectrum of the one-exciton density of states has been calculated by Bogani and Schettino (1978) in the dipole–dipole approximation allowing for the polarizability of molecules. Suggesting that the width of the band of the ω_1 vibration is small, and hence, that the free two-exciton spectral function has the same shape as the function of the density of states, Bogani (1978) has calculated the spectral function of the interacting excitons (see relation (7)). To fit the experiment the calculated spectrum had to be somewhat shifted; also a smaller value of the anharmonic shift $\Delta_{31}=-16\,\text{cm}^{-1}$ compared with the one measured in the gas phase $\Delta_{31}=-19.7\,\text{cm}^{-1}$ (Smith and Overend 1971) had to be taken. But the necessity of this correction, as will be shown below, is due to the insufficient accuracy of the model approximation of the dipole–dipole interaction of vibrations, rather than to inaccuracy of the point interaction approximation for the excitons. This is supported by the fact that the width of the ω_3 band, estimated from the width of the band due to the dissociated states and also by the spectrum of isotopic impurities (see below) is ca 10% larger than that calculated by Bogani (1978).

4.3. *Investigation of the ω_3 vibrational exciton band by means of dilute impurity modes*

An excellent example of the investigation of an exciton band by isotopical impurity method is the work of Cahill (1976). He has tested various trial densities of states for ω_3 of CO_2 by comparing experimental and calculated vibrational shifts of several guests (CO_2 and N_2O isotopes) in the $^{12}C\,^{16}O_2$ and $^{13}C\,^{16}O_2$ hosts. Cahill (1976) has concluded that the best fit to the data was obtained by utilizing the density of states, as calculated by Dows and Schettino (1973) but corrected in width. The necessity for this correction is probably due to the fact that the calculation performed by Dows and Schettino (1973) in the dipole–dipole approximation did not take into account the polarizability of the molecules. The width and second moment of the density of states, constructed by Cahill were $T_3 = 56\,\text{cm}^{-1}$ and $M_3^{(2)} = 119\,\text{cm}^{-2}$, respectively. Satisfactory results were obtained also with rectangular ($T_3 = 38.5\,\text{cm}^{-1}$, $M_3^{(2)} = 124\,\text{cm}^{-2}$) and cosine ($T_3 = 42\,\text{cm}^{-1}$, $M_3^{(2)} = 109\,\text{cm}^{-2}$) trial functions.

Figure 6 illustrates the application of the isotopic impurity method to the investigation of the exciton band. In fig. 6(a) we have represented the frequencies of impurity modes versus the reciprocal value of the isotopic shift. The experimental data obtained by Cahill (1975) for several CO_2 isotopes are given by open circles ($^{12}C\ ^{16}O_2$ host) and filled circles ($^{13}C\ ^{16}O_2$ host). The frequencies of the latter are shifted by $67.5\,\text{cm}^{-1}$, i.e. by the value of the $^{12}C\ ^{16}O_2$ host to the $^{13}C\ ^{16}O_2$ host exciton frequency shift.

The experimental data should be compared with the calculated curves Re $G_3(\omega)$, which give the frequency of the impurity mode as a function of Δ_3^{-1}. These curves and also histograms of the density of states, used for the calculation are presented in fig. 6a by solid curves (after Bogani and Schettino 1978) and by dashed curves (after Cahill 1976). The dash–dotted line gives the same dependence in the oriented gas approximation.

Figure 6(b) contains the same data as in fig. 6(a) but on the horizontal scale the gas-to-solid shifts in ω_3 are plotted. Such presentation permits more exact comparison of the calculation with the experiment. At the same time the position of the centre of gravity and the second moment of the

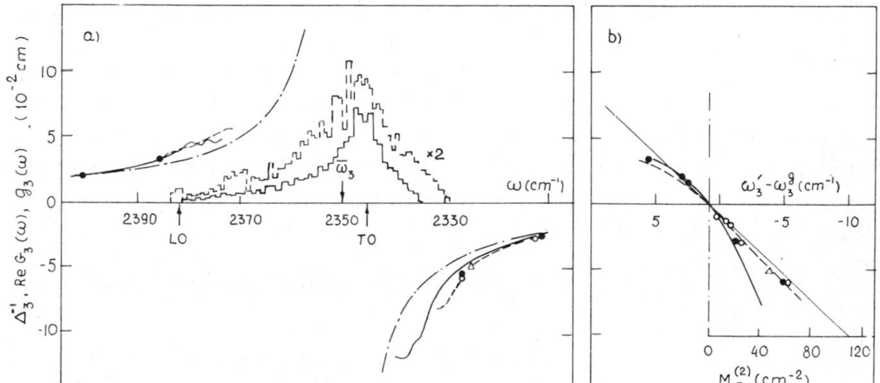

Fig. 6. Test of the calculated density of the ω_3 states (histograms) in CO_2 crystal by the isotopical impurity method. Calculated (solid and dashed curves) and experimental (open and filled circles) frequency dependencies (a) and gas-to-solid frequency shift (b) of the impurity mode as a function of the reciprocal value of the isotopic shift Δ_3^{-1}. The data of Bogani and Schettino (1978) and Cahill (1976) are presented by solid and dashed lines, respectively. By circles the results of Cahill (1975) are shown. The triangle indicates the position of $\omega_{3,1} - \omega_1$, against Δ_{31}^{-1}. The graphical method of determination of the centre of gravity $\bar{\omega}_3$ and second moment $M_3^{(2)}$ of the exciton band (see text) is also illustrated by fig. 6(b).

density of states can be obtained graphically. A fine line in fig. 6(b) is drawn through the experimental points in accordance with interpolating formula (23). It's intercept on the abscissae gives the difference between the frequency of the centre of gravity of the exciton band and the vibration frequency of the molecule in a gas. This difference for CO_2 is about 1 cm^{-1}; from this, for the $^{12}C^{16}O_2$ crystal, a value of $\bar{\omega}_3 = 2350\ \mathrm{cm}^{-1}$ is obtained. This value was used when plotting fig. 6(a). The slope of the fine line on fig. 6(b) gives the second moment of the density of states. The value of $M_3^{(2)} = 110\ \mathrm{cm}^{-1}$ is obtained numerically from the intersection of the line with a nomogram depicted at the bottom of fig. 6(b). We note that the fine line also gives the dependence $\omega_3'(\Delta) - \omega_3^{g}(\Delta)$ for the trial function of the density of states (24) with a width of $T_3 = 42\ \mathrm{cm}^{-1}$.

Data presented in fig. 6 provide evidence of the good accuracy of the function of the density of states constructed by Cahill. It is also seen that the calculation of Bogani and Schettino (1978) yields a somewhat lower value of the band width. This fact accounts for the discrepancy between the experimental $\omega_3 + \omega_1$ two-exciton spectrum and the calculation carried out by Bogani (1978). At the same time the value of $\mathrm{Re}\, G_3(\omega_{3,1} - \omega_1) = \mathrm{Re}\, G_{31}(\boldsymbol{k} \approx 0, \omega_{3,1}) = \Delta_{31}^{-1}$ (see eq. (8)) obtained from the frequency of biexciton $\omega_{3,1}$ (shown in fig. 6(a), (b) by a triangle) is in excellent agreement with the results of the experimental study of impurity modes.

5. Study of the ω_3 vibrational exciton band in the $NaNO_3$ crystal

The sodium nitrate crystal has a space group $D_{3d}^6(R\bar{3}c)$ and contains two formula units per cell. This crystal has four internal modes which are well separated in frequency. To the doubly degenerate ω_3 vibration corresponds an exciton band of 110 cm^{-1} width. The intermolecular interaction of this vibration corresponds an exciton band of this vibration is well described within the dipole–dipole approximation (Frech and Decius 1969, Plihal 1973). The bands of other vibrations are much narrower. For instance, the ω_1 and ω_4 vibrations give bands with the widths of about 1–3 cm^{-1} (as estimated from the $\omega_1 + \omega_4$ and $\omega_4 + \omega_4$ transitions).

The frequencies of one- and two-excitons transitions for three vibrational excitons in question are given in table 1. The latter also contains the values of the anharmonic and isotopic shifts, measured for the nitrate ion isolated in alkali halide crystal (after Tsuboi and Histasune 1972).

5.1. Reconstruction of the density of states from two-exciton spectra

The anharmonic shift of the $\omega_{3,1}$ and $\omega_{3,4}$ combination modes is small in comparison to the width of the ω_3 exciton band in $NaNO_3$. Therefore

Table 1
One- and two-exciton transitions in $NaNO_3$ crystal at 80 K, isotopic shifts Δ_n (at ^{14}N to ^{15}N substitution) and anharmonic shifts Δ_{nm} in NO_3^- ion

Transition	Symmetry	Wavenumber	Δ_n cm^{-1}*	Δ_{nm} cm^{-1}*
ω_4	E_g	725	−1	
	E_u(TO, LO)	727		
ω_1	A_g	1070	0	
ω_3	E_g	1389	−32	
	E_u(TO)	1355		
	E_u(LO)	1465		
$\omega_3 + \omega_4$	E_u	2080–2190	−33	−7
$\omega_3 + \omega_1$	E_u	2425–2535	−32	−12

* Tsuboi and Histasune (1972)

only broad bands due to dissociated states are revealed in the $\omega_3 + \omega_1$ and $\omega_3 + \omega_4$ two-exciton spectra. Figure 7 shows the IR-optical density curves measured in the region of the $\omega_3 + \omega_1$ and $\omega_3 + \omega_4$ transitions (Belousov et al. 1976). These curves are square-normalized. For the convenience of comparison the spectra are shifted in frequency by ω_m (where $m = 1, 4$) into the region of the one-exciton state ω_3. The sharp lines at $\omega_m + 1450\ cm^{-1}$ are due to the three-exciton transitions $\omega_m + \omega_4 + \omega_4$. One should note the antiresonant character of these lines, which indicates interaction and

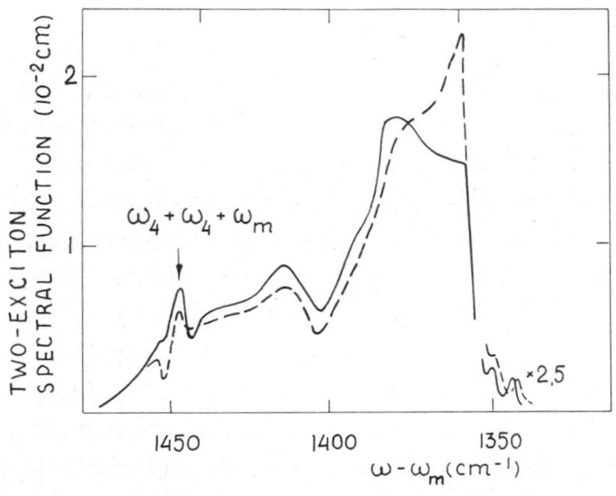

Fig. 7. IR spectra of the $\omega_3 + \omega_4$ (solid curve) and $\omega_3 + \omega_1$ (dashed curve) transitions in natural $NaNO_3$ crystal ($T = 100$ K). For convenience of comparison the spectra are square-normalized and also shifted in frequency by $\omega_m (m = 1,4)$. The localized two-exciton ^{15}N (natural abundance 0.37%) impurity modes are given on the enlarged scale. (After Belousov et al. (1976).)

interference between $|m; 4; 4\rangle$ and $|3; m\rangle$ states. The origin of the doublet, observed at the low-frequency edge of the $\omega_3 + \omega_m$ band will be discussed below (sect. 5.2). Excluding these features the optical density curves in fig. 7 reproduce the two-exciton spectral functions.

It is evident that due to small widths of the ω_1 and ω_4 bands the free two-exciton functions $g_{31}(\mathbf{k}, \omega)$ and $g_{34}(\mathbf{k}, \omega)$ should coincide and should have the same shape as the density of one-exciton ω_3 states $g_3(\omega)$. Nevertheless in fig. 7 a substantial difference between $\omega_3 + \omega_1$ and $\omega_3 + \omega_4$ can be seen. This difference arises from the exciton–exciton anharmonic interaction, the strength of which differs for the states in question (compare the values Δ_{31} an Δ_{34} in table 1).

Assuming the point interaction between the excitons, we can take into account the influence of interaction on the two-exciton spectrum (see sect. 3, relation (21) and reconstruct the spectral function of the free exciton which in the case considered coincides with the density of the ω_3 states. Figure 8 shows the densities of states, reconstructed from $\omega_3 + \omega_4$ (solid curve) and $\omega_3 + \omega_1$ (dashed curve) transitions (Belousov et al. 1976).

Close agreement between the curves indicates the applicability of the point interaction model in this case. The study of the isotopic impurity

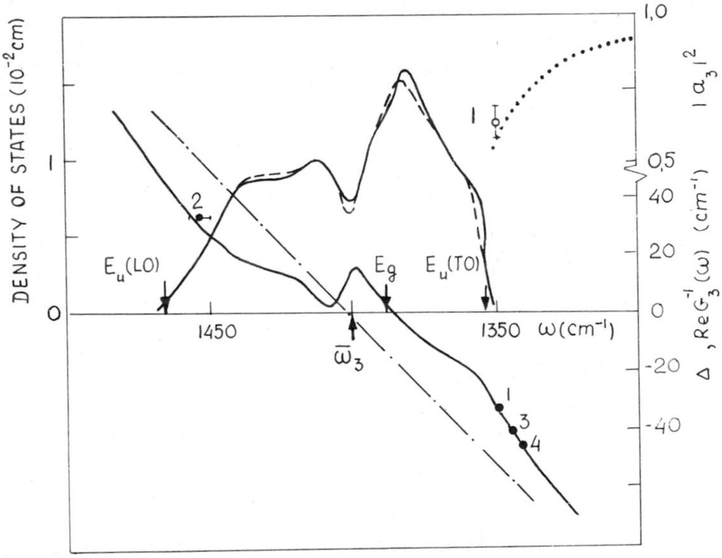

Fig. 8. Density of the ω_3 states in $NaNO_3$ crystal obtained from the $\omega_3 + \omega_4$ (solid curve) and $\omega_3 + \omega_1$ (dashed curve) two-exciton spectra. Solid curve Re $G_3^{-1}(\omega)$ calculated from the density of states, gives the dependence of the impurity mode frequency on the isotopic shift. Dashed–dotted line is the same dependence in the oriented gas approximation. The dotted curve gives the degree of localization $|a_3|^2$ of the impurity mode as a function of its frequency. The experimental data are presented by circles. (After Belousov et al. (1976).)

modes (see below) also suggest that the resultant function is the density of one-exciton states.

5.2. Study of the density of states by means of dilute impurity modes

The low-temperature spectra of the $NaNO_3$ crystal in the region of the ω_3 vibration are shown in fig. 9. The band 1450 cm^{-1} is due to $\omega_4 + \omega_4$ transitions, enhanced by the Fermi resonance with the ω_3 transition (Belousov et al. 1974). This band is not considered here. On the low-frequency edge of the reflectivity band a singularity is observed at about 1350 cm^{-1}. In Raman spectrum a weak and sharp line is revealed at 1351 cm^{-1}. The intensities of these lines increase in crystals with a ^{15}N isotope content of 2% which supports its assignment to the ω_3 exciton, localized on the $^{15}NO_3^-$ impurity ion (the natural abundance of ^{15}N is 0.37%).

The ω_3-vibration frequency of $^{15}NO_3^-$, calculated in the oriented gas approximation is $\bar{\omega}_3 + \Delta_3 = 1369\ cm^{-1}$; it falls into the region of the exciton band of $Na\,^{14}NO_3$. The interaction between the guest and host ions results in the splitting off of a localized state. The small value of the splitting (about 4–5 cm^{-1}, compared with the isotopic shift of 32 cm^{-1}) indicates that

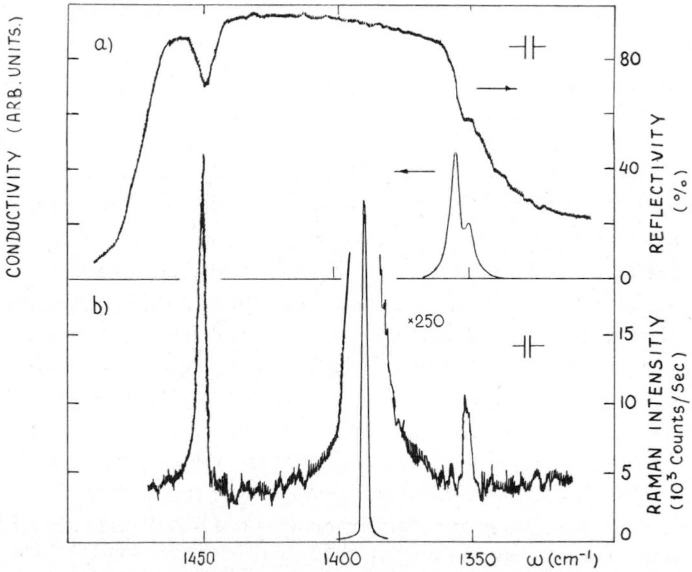

Fig. 9. Exciton transitions ω_3 in natural $NaNO_3$ crystal: (a) IR reflectivity spectrum at 100 K and calculated from IR conductivity spectrum; (b) Raman spectrum at 80 K. The line at 1350 cm^{-1} is the ^{15}N impurity mode (the natural abundance of ^{15}N 0.37%). (After Belousov et al. (1980b).)

the impurity excitation is nontotally localized allowing a strong Rashba effect. Actually two bands of comparable intensity are revealed (fig. 9a) in the conductivity spectrum, derived from the IR reflectivity spectrum. Dispersion analysis of the reflectivity shows that the relative intensity of the impurity (low-frequency) band is 25 times greater than would be expected if it were proportional to the relative impurity content. This is accounted for by the resonant contribution of the E_u(TO) exciton, located at the low-frequency band edge. In the Raman spectrum the energy spacing between the E_g exciton and localized exciton is large and the relative intensity of the impurity band is half of that expected from the relative impurity content. The degree of the localization of the vibrational exciton on the impurity $|a_3|^2$, calculated from the IR and Raman spectra according to eq. (17) is equal to 0.65 ± 0.05.

The experimental values of $\omega_3'(\Delta_3)$ and $|a_3(\omega_3')|^2$ are depicted in fig. 8 (point 1 and circle 1). Also shown by a solid curve is the calculated dependence of $\omega_3'(\Delta)$, which, according to eq. (14) represents $\mathrm{Re}\, G_3^{-1}(\omega)$. The dotted curve in fig. 8 is the calculated dependence $|a_3(\omega_3')|^2$. It is seen that the experimental data are in excellent agreement with the calculation in which the density of states reconstructed from the two-exciton spectra was used.

The presence of a ^{14}N impurity in Na $^{15}NO_3$ crystal does not produce any splitting of the localized mode. This is due to the asymmetry of the density of states, whose centre of gravity $\bar{\omega}_3 = 1401\ \mathrm{cm}^{-1}$ is shifted to the low-frequency edge of the band. Yet in the IR reflectivity spectra of the crystals with the 2% and 5% ^{15}N content a singularity arises at about $1420\ \mathrm{cm}^{-1}$. It's shape and concentrational behaviour is typical of a quasilocal mode. The frequency of this mode $\omega_3'(-\Delta_3)$ shifted by $-\Delta_3$ is represented in fig. 8 by point 2; good fit with the curve $\mathrm{Re}\, G_3^{-1}(\omega)$ is seen.

As shown by Belousov and Pogarev (1975) the ^{15}N-isotope impurity produces two lines at the low-frequency side of the $\omega_3 + \omega_4$ and $\omega_3 + \omega_1$ bands. These lines are presented in fig. 7 on an enlarged scale. The appearance of two components in the spectrum could be explained bearing in mind both the delocalization of the impurity ω_3 vibration and the anharmonic interaction (see sect. 2.3). The delocalization makes possible a transition in which the excitations are localized on the impurity and nearest host lattice ions (separated configuration). To this transition corresponds the high-frequency component of the doublet, whose position is identical – within the limits of the experimental error – with the sum of the frequencies ω_3' and ω_4 (or ω_1), whereas the intensity reflected the degree of delocalization of the impurity vibration.

The anharmonic interaction tends to localize the excitations on the impurity (joint configuration) producing the energy defect and the low-frequency $\omega_{3,m}'$ component. The splittings of the doublets ($6\ \mathrm{cm}^{-1}$ for

$\omega_3 + \omega_4$ and $9\,\mathrm{cm}^{-1}$ for $\omega_3 + \omega_1$) are smaller than the energy defect. This result is in an excellent agreement with the calculation (18) where the Green's function obtained from the two-exciton spectrum is used. The experimental values of $\omega'_{3,4}(\Delta_3 + \Delta_{34}) - \omega_4$ and $\omega'_{3,1}(\Delta_3 + \Delta_{31}) - \omega_1$ are given by points 3 and 4, respectively, in fig. 8; they fit well the calculated curve $\mathrm{Re}\, G_3^{-1}(\omega)$.

In conclusion we should state that the study of the one-exciton and two-exciton spectra in $NaNO_3$ crystal shows that the approximation of the point anharmonic interaction of the vibrational excitons is of sufficient accuracy. Similar results were obtained also for other nitrates (Belousov et al. 1977) and for the calcite crystal (Belousov et al. 1980b).

5.3. Reconstruction of the density of states from the one-exciton spectrum of mixed crystals

A reliable experimental determination of the exact density of states which has been presented above allows quantitative verification of different approximations in the theory of disordered crystals. In particular, the density of the ω_3 states of $NaNO_3$ reconstructed from the two-exciton spectra was used to calculate the vibrational spectra of the isotopically mixed $Na\,^{15}N_x\,^{14}N_{1-x}O_3$ crystals (Belousov et al. 1978a). Comparison with the experiment showed that the ATA method provides fairly good descrip-

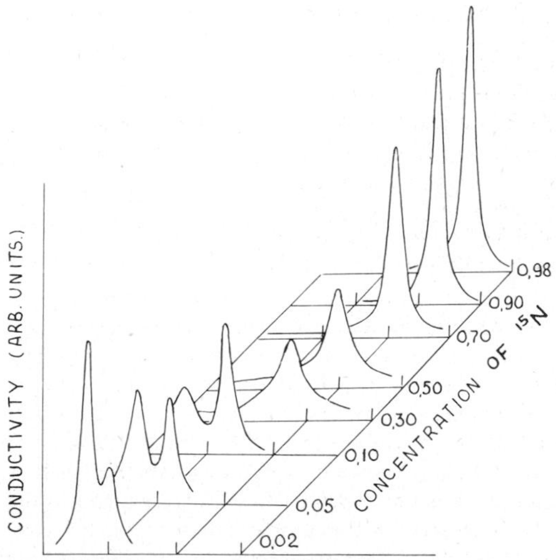

Fig. 10. Mode behaviour of ω_3 (TO) spectrum in $Na^{15}N_x\,^{14}N_{1-x}\,O_3$ crystal (100 K). (After Belousov et al. (1978a, 1980b).)

tion of the spectrum of the crystal with the impurity content up to 10%. We shall not discuss in detail these results, since in the present paper we concentrate our attention on the possibility of solving the inverse problem, i.e., the reconstruction of the density of states from the experimental spectra. However, the main features of the ω_3 mode behaviour will be considered.

In fig. 10 the IR-conductivity spectra (E_u(TO) spectra) calculated by means of Kramers–Kronig analysis of reflectivity of the mixed crystals are shown. It is seen that at $x \leqslant 0.3$ the spectrum exhibits a two-mode behaviour; a chracteristic feature of the latter is an anomalous dependence of the intensity of the modes on concentration. For example, at natural abundance of ^{15}N ($x = 0.0037$) the intensity of the guest mode amounts to about 10% of the intensity of the host mode (see fig. 9(a)). At $x = 0.05$ the modes are almost equal in intensity (fig. 10). At $x = 0.1$ the low-frequency band, originally associated with the guest mode, becomes more intense*. Further increase in x results in broading of the spectrum; the two-mode behaviour is no longer revealed. At $x = 0.98$ the band narrows again but the localized vibration is not produced.

The Raman spectra of the ω_3 exciton exhibit almost one-mode behaviour. In the natural $NaNO_3$ crystal the full width at half-maximum of the ω_3 line is 0.5 cm^{-1} at $T = 80$ K. In the mixed crystal this line is broadened up to 2.8 cm^{-1} at $x = 0.1$ and up to 3.8 cm^{-1} at $x = 0.9$ (fig. 11(a) and (b)). The spectrum arising at the wings of the line is characterized by a structure, genetically related to the localized (for $x = 0.1$) or quasilocalized (for $x = 0.9$) excitons and also to the features of the density of states. At high concentrations of the impurity ($x = 0.3$, 0.5, 0.7) broad bands of complicated shape are observed in the Raman spectrum.

The results considered above indicate that the spectrum of the ω_3 exciton is considerably influenced by the presence of ^{15}N isotope even at relatively low concentrations. As can be seen in fig. 11(a) and (b) change of the Raman spectrum induced by the impurity at $x = 0.1$ or $x = 0.9$ at $T = 80$ K can be detected for all region of exciton states.

* It should be noted that the anomalous concentration dependence of the impurity intensity (fig. 10) is a common feature of the giant enhancement of the guest band due to the Rashba effect. This effect is not widely known in vibrational spectroscopy, and its neglect may lead to the incorrect interpretation of the spectra. For instance, for the explanation of the anomalous doublet in the TO spectrum of the ω_3 mode in SiF_4 crystal, various hypotheses were put forward, all of which ignored the presence of the Si isotopes. Belousov and Pogarev (1978b) have shown that the anomaly is caused by the ^{30}Si isotope, which has the natural abundance of 3.1%. The fact that the intensities in the doublet are nearly equal is the consequence of a giant enhancement of the guest band intensity due to the resonant interaction of the impurity vibration with the TO vibration of a host crystal.

In sect. 6.3 we shall consider the case when the presence of 0.4% impurity results in the appearance of a guest line with the intensity comparable to that of host line.

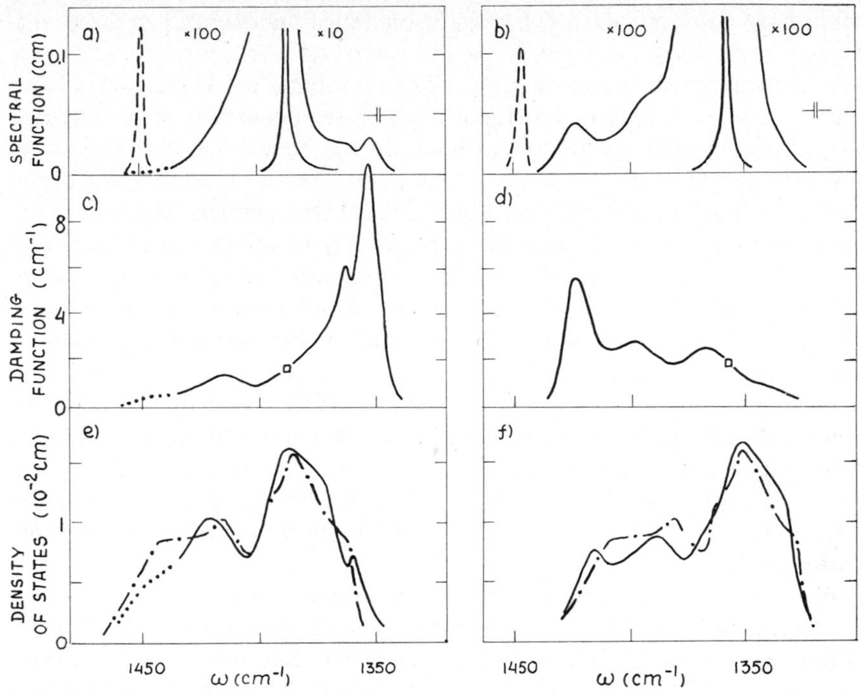

Fig. 11. Raman spectra at 80 K, (a) and (b), and the reconstructed damping functions, (c) and (d), and densities of states (e) and (f) of ω_3 exciton in $Na^{15}N_x\,^{14}N_{1-x}\,O_3$ crystal at $x = 0.1$, (a), (c), (e), and $x = 0.9$, (b), (d), (f). Dashed curves represent $\omega_4 + \omega_4$ transitions, dotted lines show data after their exclusion. By bars the values of halfwidths at half-maximum of Raman bands are given. Dash–dotted curves are $g_3(\omega - x\Delta_3)$ reconstructed from two-exciton spectrum. (After Belousov et al. (1980a, b).)

The exciton damping functions calculated from the Raman spectra of the mixed crystals in accordance with the procedure described in sect. 3 (see relation (26)), are plotted in fig. 11(c) and (d). We should note that the structure due to the singularities of the density of states function is more pronounced in the damping spectra than in the original Raman spectra.

In fig. 11(e) and (f) we plot the density of states calculated from the damping function with the help of eq. (25). For a comparison there is also shown (dash-dotted curve) the density of states as found from the two-exciton spectra. A characteristic difference between the curves is seen in the frequency regions of 1360 cm^{-1} (fig. 11(e)) and 1420 cm^{-1} (fig. 11(f)), where cluster effects are expected. However, the close general agreement between the density of states reconstructed from the spectrum of the mixed crystal, and the more exact data, obtained from the two-exciton spectra, provides evidence of the accuracy and potentialities of the mixed crystal method.

6. Study of the ω_4 vibrational exciton band in NH_4Cl crystal

The low temperature phase of NH_4Cl crystal (space group $T_d^1(P43m)$, one formula unit per cell) has been the subject of numerous spectroscopic investigations. The infrared spectra have been studied very thoroughly by Schumaker and Garland (1970) and the Raman spectra recently by Mitin et al. (1975), Gorelik et al. (1975) and Fredrickson and Decius (1977).

Ammonium chloride crystal has four internal modes which are well seperated in frequency. Totally symmetric (A_1) $\omega_1 = 3045\ cm^{-1}$ and double-degenerated (E) $\omega_2 = 1718\ cm^{-1}$ vibrations are only Raman active. Triply-degenerated (F_2), ω_4 and ω_3 vibrations are simultaneously infrared and Raman active.

From the spectrum of overtone transitions it follows that the ω_4 and ω_2 vibrations in NH_4Cl form the bands of the width of $50\ cm^{-1}$ and $80\ cm^{-1}$, respectively. In the region of the ω_3 vibration (3050–$3250\ cm^{-1}$) the spectrum has a complicated shape due to the Fermi resonance of ω_3 with $\omega_2 + \omega_4$ and $\omega_1 + \omega_{external}$ (for a quantitative analysis of this spectrum see Belousov et al. 1980b).

A problem of reconstruction of the one-exciton density of states in NH_4Cl is not simple because of the absence of transitions in which the narrow band exciton is involved. Weinstein and Cardona (1973) and Temple and Hathaway (1973) have shown that in a number of semiconductors Raman spectrum of the totally symmetric overtone transitions represents the density of the one-phonon states on a doubly expanded energy scale. Belousov et al. (1980) have demonstrated that analogous situation takes place in the spectrum of the $\omega_4 + \omega_4$ overtone transitions in NH_4Cl. However, in the latter case it was necessary to take into account the anharmonic interaction of excitons. The results of this study are presented in sect. 6.2. Since the anharmonic interaction constants in NH_4Cl are unknown we shall first consider reconstruction of the density of the ω_4 states from the one-exciton spectrum of the isotopically mixed crystal (Belousov et al. 1980b).

6.1. Reconstruction of the density of states from the one-exciton spectra of mixed crystals

The isotopic shift of the ω_4 vibration due to ^{14}N to ^{15}N substitution $\Delta_4 = -6\ cm^{-1}$ (Price et al. 1960) is small compared with the band width which is $50\ cm^{-1}$. Therefore, incorporation of the ^{15}N impurity causes only a weak perturbation of the one-exciton spectrum, which can be detected only in high precision measurements. On the other hand a small value of the perturbation increases the accuracy of the ATA method (see sect. 2.4).

The TO and LO spectral functions for natural and doped samples (fig.

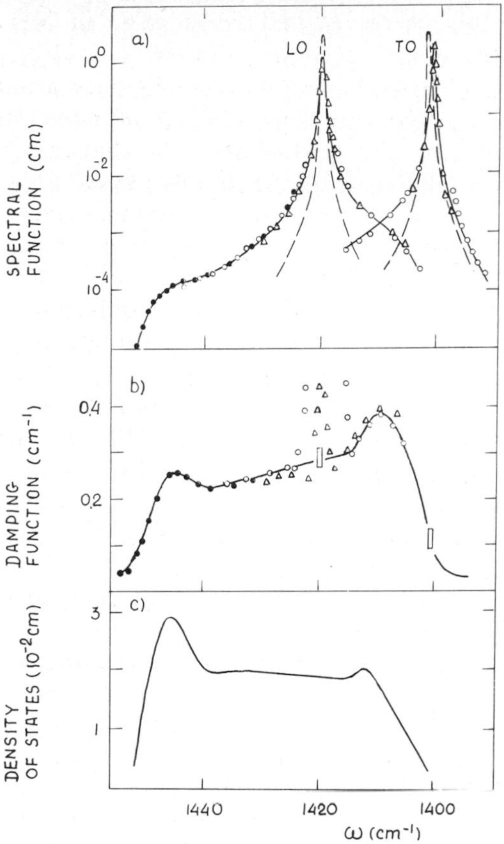

Fig. 12. Spectral function of ω_4 (TO) and ω_4 (LO) exciton in $^{15}N_{0.1}\,^{14}N_{0.9}H_4Cl$ crystal obtained from Raman spectra at 80 K (a) and the reconstructed from them damping function (b) and density of states (c). Data for various values of spectral slits are given by: (□) – 0.25, (△) – 1, (○) – 2.5, (●) – 5 cm^{-1}. Spectral functions of the ω_4 (TO) and ω_4 (LO) excitons in natural NH_4Cl crystal are presented by dashed curves (a). (After Belousov et al. (1980b).)

12(a)) were obtained by simultaneous computer analysis of Raman spectra measured for different geometries and spectral resolutions. This procedure has provided sufficient accuracy of the resultant spectral functions for both the centre and the wings of the line.

In Raman spectra of a natural NH_4Cl crystal two sharp lines at $\omega_{4T} = 1402\ cm^{-1}$ and at $\omega_{4L} = 1420\ cm^{-1}$ are observed. The full widths at half-maximum of these lines are less than $0.2\ cm^{-1}$ at $T = 80$ K. In isotopically mixed $^{15}N_{0.1}\,^{14}N_{0.9}H_4Cl$ crystal an insignificant low-frequency line shift ($\approx 0.8\ cm^{-1}$ for TO and $\approx 0.4\ cm^{-1}$ for LO) occurs which is accompanied by broadening of the LO line up to $0.6\ cm^{-1}$. The measurements with a broad slit (up to $5\ cm^{-1}$) reveal a weak spectrum occurring at the band's wings (fig. 12(a) solid curves). At the high-frequency wing of the LO-band a distinctive edge-like feature at about $1450\ cm^{-1}$ is revealed.

The exciton damping function in the mixed crystal calculated through the resultant LO-spectral function according to relations (3, 26) is given by solid curve in fig. 12(b). The density of the ω_4 states, calculated according to relation (25) from the damping function is shown in fig. 12(c).

6.2. Reconstruction of the density of states from the overtone spectrum

The forward part of fig. 13 shows the Raman spectrum of NH_4Cl which is measured in $z(xx)y$ geometry in which the transitions of A_1 and E symmetries are allowed. A very strong line at $3045\ cm^{-1}$ belongs to a totally symmetric ω_1 vibration. A broad band at 2804–2904 cm^{-1}, shown on an enlarged scale, is due to the $\omega_4 + \omega_4$ transitions. At the low-frequency edge of this band a sharp asymmetric peak is revealed; its frequency $2804\ cm^{-1}$ coincides with twice the ω_4 (TO) exciton frequency. The shape and spectral position of this peak are characteristic of a quasibound state, which indicates a strong exciton–exciton interaction. High intensity of the $\omega_4 + \omega_4$ band and also the proximity of strong ω_1 line invokes the possibility of

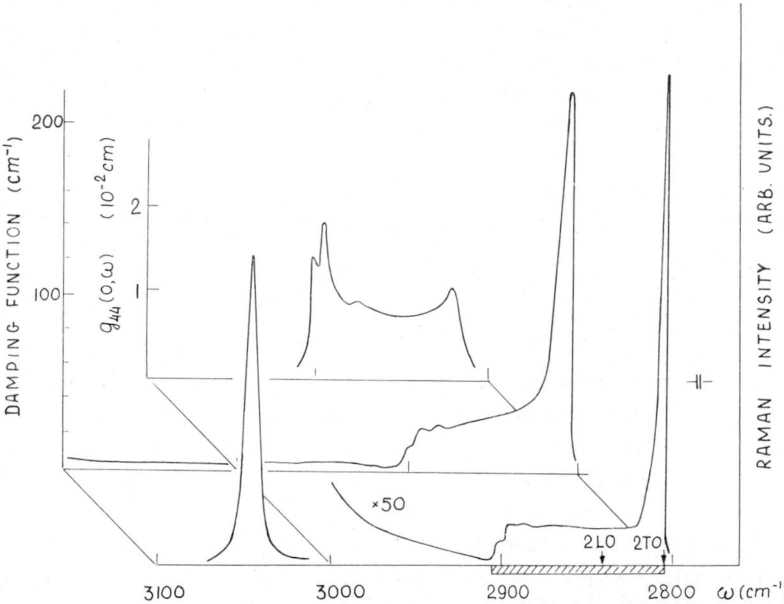

Fig. 13. Totally symmetric Raman spectrum of natural NH_4Cl crystal in the region of ω_1 and $\omega_4 + \omega_4$ transitions at 80 K and the reconstructed damping function and free two-exciton spectral function $g_{44}(0, \omega)$. Frequency region of the $\omega_4 + \omega_4$ dissociated states is shaded. (After Belousov et al. (1980b).)

hybridization of $|4;4\rangle$ and $|1\rangle$ states. This supposition is also supported by the absence of the $\omega_4+\omega_4$ band in the spectrum of E symmetry, measured for $z(y'x')y'$ geometry (where axes x' and y' are rotated by 45° with respect to x and y).

It should be stressed that in the totally symmetric $\omega_4+\omega_4$ (A_1) spectrum the selection rules allow only the transitions with creation of two excitons from the same branch (Shuvalov and Gorelick 1979). Therefore one might expect that free two-exciton spectral function $g_{44}(\boldsymbol{k}\approx 0,\omega)$ reproduces the density of states $g_4(\omega)$ on a doubly expanded energy scale.

In order to reconstruct the free two-exciton spectral function it is necessary to take into account the effects of the anharmonicity. If we can assume that the "unperturbed" $\omega_4+\omega_4$ intensity is negligible small the observed Raman spectrum is proportional to the spectral function of the ω_1 exciton $g_1^\gamma(\boldsymbol{k}\approx 0,\omega)$. Calculating the one-exciton Green's function (3) we can obtain the damping function (22) (middle part of fig. 13). A third-order anharmonicity constant can be calculated from the area under the $\omega_4+\omega_4$ curve in the damping spectrum (22); it was found to be $\gamma = 37\ \mathrm{cm}^{-1}$.

The damping spectrum in the region of the $\omega_4+\omega_4$ band is proportional to the two-exciton spectral function $g_{44}^{\Delta}(\boldsymbol{k}\approx 0,\omega)$ which is only perturbed by the fourth-order anharmonic interaction. We assume that the free two-exciton spectral function $g_{44}(\boldsymbol{k}\approx 0,\omega)$ reproduces the density of states $g_4(\omega/2)$. Proceeding from this assumption the only unknown parameter, namely, the anharmonic shift, was chosen such ($\Delta_{44} = -25\ \mathrm{cm}^{-1}$) that $g_{44}(\boldsymbol{k}\approx 0,\omega)$, calculated by means of the relation (21) would fit the density of states of fig. 12(c). The resultant curve $g_{44}(0,\omega)$ is represented in the background of fig. 13. The fact that with only one fitting parameter Δ_{44} close agreement between $g_{44}(0,\omega)$ in fig. 13 and $g_4(\omega)$ in fig. 12(c) is obtained, indicates the applicability of the point interaction approximation in this case.

6.3. Interaction between guest and host overtone states

As was indicated by Agranovich et al. (1970) strong anharmonicity may induce a localized impurity state in the two-exciton spectrum even when no localized impurity mode appears in the one-exciton spectrum. Such an effect was first observed in the spectrum of the $\omega_4+\omega_4$ overtone transitions in NH_4Cl crystal doped with ^{15}N (Belousov and Pogarev 1978c).

In fig. 14 the concentration dependence of the Raman spectra in the region of quasibound states is shown. A doublet structure is already present at natural abundance of ^{15}N ($x = 0.004$). The dip is at twice the ω_4 (TO) exciton frequency. With the increase of the impurity content the depth of the dip and the spacing between the components of the doublet increase.

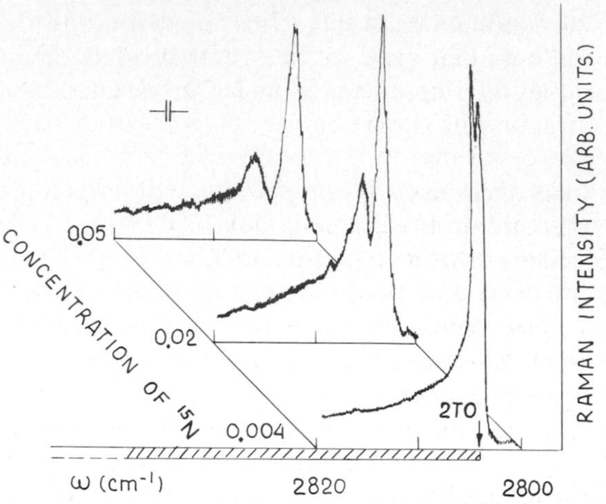

Fig. 14. Mode behaviour of totally symmetric Raman spectrum in $^{15}N_x\,^{14}N_{1-x}\,H_4Cl$ in the $\omega_4+\omega_4$ spectral region at 80 K. (After Be'ousov and Pogarev (1978c).)

The observed impurity effect was interpreted as resulting from the interaction and interference between the guest and host overtone states. We should also note, that the concentration dependence of this two-exciton spectrum resembles that of the one-exciton spectrum in the case of a strong Rashba effect (see sect. 5.3).

An interesting feature of the spectra in question is that the energies of guest and host quasibound overtone states are practically the same though the isotopic shift of the $\omega_{4,4}$ overtone of the isolated NH_4^+ ion is not too small ($2\Delta_4 = -12\,cm^{-1}$). This fact can in principle be understood only by taking into account the dispersion of the two-exciton band in $\boldsymbol{k}$ space. Agranovich et al. (1979) have performed an analysis of the two-exciton spectrum of the isotopically mixed NH_4Cl crystal by the coherent potential method using the model dispersion of the two-exciton states in $\boldsymbol{k}$ space. Their results are in qualitative agreement with experiment; particularly concentrational behaviour of the spectrum is correctly reproduced.

7. Conclusion

The works discussed in the present paper show that the comparative study of the one- and two-exciton spectra of perfect and isotopically doped and mixed crystals makes it possible to make a quantitative test of the theory of the exciton–exciton and exciton–impurity interaction. The results described here prove the high accuracy of the point (intramolecular) interaction model for the Frenkel excitons. Within the framework of this model, a simple reconstruction of the density of the one-exciton states is possible. It is important to note that such reconstruction does not require any model description of the intermolecular interaction. The approach considered enables reconstruction of the density of states with utmost reliability and precision.

The general approach discussed in this paper for the vibrational excitons is valid for the electronic Frenkel excitons. For example, reconstruction of the density of states from two-particle spectra is applicable both for the vibrational excitons (sect. 5.1) and for the vibronic transitions involving the nontotally symmetric vibrations (Rabin'kina et al. 1970). When the one-exciton electronic states are close in energy to vibronic states of the same symmetry, the coupling and hybridization of these states may be important, which is similar to Fermi resonance for the vibrational excitons (sect. 4.1 and 6.2).

The isotopical impurity method (sects. 4.3 and 5.2) is also widely used for the electronic exciton spectra.

The study of the one-exciton spectrum of an isotopically mixed crystal enables one to reconstruct the density of states even in cases when the analysis of vibronic spectra is complicated. As was shown in sects. 5.3 and 6.1, up-to-date spectroscopic technique is sufficient for the detection of very weak impurity-induced perturbations of the one-exciton spectrum. Further progress is envisaged if the problem of reconstruction of the density of states in the coherent potential approximation can be solved. In the latter case it would be possible to analyze crystals with higher impurity concentration ($>10\%$).

The problem of the two-exciton or vibronic spectra of mixed crystals seems to be also very important. Although this problem is far from being completely understood up to now, it is obvious, that these spectra can provide significant information on the density of states of a mixed crystal and also on the exciton–exciton interaction in general.

Acknowledgements

The author is very grateful to Prof. E.I. Rashba, Dr. S.A. Permogorov and Dr. D.E. Pogarev for many helpful discussions. The assistance of Dr. S.V. Pogarev in numerical calculations and experiments is also appreciated.

References

Agranovich, V.M., 1970, Fiz. Tverd. Tela **12**, 562 (1970, Sov. Phys. Solid State **12**, 430).

Agranovich, V.M. and I.I. Lalov, 1971a, Fiz. Tverd. Tela **13**, 1032 (1971, Sov. Phys. Solid State **13**, 859).

Agranovich, V.M. and I.I. Lalov, 1971b, Zh. Eksp. Teor. Fiz. **61**, 656 (1971, Sov. Phys. -JETP **34**, 656).
Agranovich, V.M., J.E. Lozovik and M.A. Mekhtiev, 1970, Zh. Eksp. Teor. Fiz. **59**, 246 (1970, Sov. Phys. -JETP **32**, 134).
Agranovich, V.M., O.A. Dubovskii and K.T. Stoichev, 1979, Fiz. Tverd Tela **21**, 3012 (1971, Sov. Phys. Solid State **21**, 1795).
Barker, Jr., A.S. and A.J. Sievers, 1975, Rev. Mod. Phys. **47**, Supplement No. 2.
Belousov, M.V. and D.E. Pogarev, 1975, Fiz. Tverd. Tela, **17**, 2832 (1975, Sov. Phys. Solid State **17**, 1896).
Belousov, M.V. and D.E. Pogarev, 1978b, Fiz. Tverd Tela **20**, 3461 (1978, Sov. Phys. Solid State **20**, 1999).
Belousov, M.V. and D.E. Pogarev, 1978c, Pis'ma Zh. Eksp. Teor. Fiz. **28**, 692 (1978, Sov. Phys. -JETP Lett. **28**, 646).
Belousov, M.V., D.E. Pogarev and A.A. Shultin, 1974, Fiz. Tverd Tela **16**, 1136 (1974, Sov. Phys. Solid State **16**, 729).
Belousov, M.V., D.E. Pogarev and A.A. Shultin, 1976, Fiz. Tverd. Tela **18**, 521 (1976, Sov. Phys. Solid State **18**, 300).
Belousov, M.V., D.E. Pogarev and A.A. Shultin, 1977, Phys. Status Solidi (b) **80**, 417.
Belousov, M.V., D.E. Pogarev and A.A. Shultin, 1978a, Fiz. Tverd Tela **20**, 1415 (1978, Sov. Phys. Solid State **20**, 814).
Belousov, M.V., D.E. Pogarev and S.V. Pogarev, 1980b, Resonant and anharmonic interaction of vibrations in perfect and isotopically mixed crystals, in: Kolebanija Okisnijkh Reshetok, eds., A.A. Lazarev and M.O. Bulanin (Nauka, Leningrad) p. 249 (in Russian).
Belousov, M.V., E.A. Ivanova, D.E. Pogarev, S.V. Pogarev, 1980a, Pis'ma Zh. Eksp. Theor. Fiz. **31**, 717 (1980, Sov. Phys.-JETP Lett. **31**, 678).
Bogani, F., 1978, J. Phys. C: Solid St. Phys. **11**, 1297.
Bogani, F. and V. Schettino, 1978, J. Phys. C: Solid St. Phys. **11**, 1275.
Broude, V.L., E.I. Rashba and E.F. Sheka, 1961, Dokl. Ak. Nauk SSSR **139**, 1085 (1962, Sov. Phys. Dokl. **6**, 718).
Broude, V.L., E.I. Rashba and E.F. Sheka, 1967, Phys. status solidi **19**, 395.
Cahill, J.E., 1975, Chem. Phys. Lett. **31**, 228.
Cahill, J.E., 1976, Chem. Phys. Lett. **39**, 98.
Cho, K., 1972, Phys. status solidi (b) **54**, 583.
Dolganov, V.K. and E.F. Sheka, 1969, Fiz. Tverd. Tela **11**, 2427 (1970, Sov. Phys. Solid State **11**, 1962).
Dows, D.A. and V. Schettino, 1973, J. Chem. Phys. **58**, 5009.
Elliott, R.J., J.A. Krumhansl and P.L. Leath, 1974, Rev. Mod. Phys. **46**, 465.
Fermi, E., 1931, Z. Phys. **72**, 250.
Frech, R. and J.C. Decius, 1970, J. Chem. Phys. **54**, 2374.
Fredrickson, L.R. and J.C. Decius, 1977, J. Chem. Phys. **66**, 2297.
Fukumoto, T., S. Nakashima, K. Tabuchi and A. Mitsushi, 1976, Phys. status solidi (b) **73**, 341.
Gorelick, V.S., G.G. Mitin and M.M. Sushchinskii, 1975, Polariton Fermi-resonance in ammonium chloride Raman spectrum, in: The Theory of Light Scattering in Condensed Matter, Proc. I-Soviet-American Symp., Moscow, 1975, eds., B. Bendow, J. Birman and V.M. Agranovich (Plenum Press, New York) p. 109.
Herzberg, G., 1945, Infrared and Raman Spectra of Polyatomic Molecules (van Nostrand, New York) part 2.5.
Krauzman, M., R.M. Pick, H. Poulet, G. Hamel and B. Prevot, 1974, Phys. Rev. Lett. **33**. 528.
Mitin, G.G., V.S. Gorelick and M.M. Sushchinskii, 1975, Fiz. Tverd. Tela **17**, 2422 (1975, Sov. Phys. Solid State **17**, 1602).
Plinal, M., 1973, Phys. status solidi (b) **58**, 315.

Price, W.C., W.F. Sherman, G.R. Wilkinson, 1960, Proc. Royal Soc. (London) **255**, 5.
Rabin'kina, N.V., E.I. Rashba and E.F. Sheka, 1970, Fiz. Tverd. Tela **12**, 3579 (1971, Sov. Phys. Solid State **12**, 2898).
Rashba, E.I., 1957, Optika i Spektroskopiya **2**, 568 [in Russian].
Rashba, E.I., 1962, Fiz. Tverd. Tela **4**, 3301 (1963, Sov. Phys. Solid State **4**, 2417).
Rashba, E.I., 1966, Zh. Eksp. Teor. Fiz. **50**, 1064 (1966, Sov. Phys. -JETP **23**, 708).
Rashba, E.I., 1972, Electronic and vibronic spectra of impurity molecular crystals I. Theory, in: Physics of Impurity Centres in Crystals, ed., G.S. Zavt (Riso AN ESSR, Tallin) p. 415.
Ruvalds, J. 1975, Quasiparticle interactions in many-body systems, in: The Theory of Light Scattering in Condensed Matter, Proc. I-Soviet-American Symp., Moscow, 1975, eds., B. Bendow, J. Birman and V.M. Agranovich (Plenum Press, New York) p. 79.
Ruvalds, J. and A. Zawadowski, 1970, Phys. Rev. **B2**, 1172.
Schumaker, N.E. and C.W. Garland, 1970, J. Chem. Phys. **53**, 392.
Sheka, E.F., 1971, Usp. Fiz. Nauk. **104**, 593 (1972, Sov. Phys. -Usp. **14**, 484).
Sheka, E.F., 1972, Electronic and Vibronic spectra of impurity molecular crystals II. Experiment, in: Physics of Impurity Centres in Crystals, ed., G.S. Zavt (Riso AN ESSR, Tallin) p. 431.
Shuvalov, A.L. and V.S. Gorelick, 1979, Preprint FIAN, No. 3 (in Russian).
Sommer, B. and J. Jortner, 1969, J. Chem. Phys. **50**, 822.
Smith, D.F. and J. Overend, 1971, J. Chem. Phys. **55**, 1157.
Temple, P.A. and C.E. Hathaway, 1973, Phys. Rev. **B7**, 3685.
Tsuboi, M. and I.S. Histasune, 1972, J. Chem. Phys. **57**, 2087.
Van Kranendonk, J., 1959, Physica (Utrecht) **25**, 1080.
Van Kranendonk, J., 1960, Canad. J. Phys. **38**, 240.
Weinstein, B.A. and M. Cardona, 1973, Phys. Rev. **B7**, 2545.
Witters, K.R. and J.E. Cahill, 1977, J. Chem. Phys. **66**, 2755.

BIBLIOGRAPHY

General theory of excitons

Agranovich, V.M., 1968, Theory of Excitons (R)* (Nauka, Moscow).

Agranovich, V.M. and M.D. Galanin, 1978, Electronic excitation energy transfer in condensed matter (R) (Nauka, Moscow) (English transl. 1982, North Holland) (in press).

Cho, K., 1979, Introduction to Excitons, in: Top. Curr. Phys. **14** (Excitons), ed. K. Cho. (Springer, Berlin) p. 1.

Collins, T.C., 1978, Excitons in solids, in: Excited States in Quantum Chemistry, ed., C.A. Nicolaides and D.R. Beck (Reidel, Dordrecht) p. 437. A brief account of the many-body theory of excitons.

Dexter, D.L. and R.S. Knox, 1965, Excitons (Interscience, New York). Still perhaps the best introduction in English. It is comprehensive (as of its date) but aims to be "useful to persons unfamiliar with the (theoretical) techniques of solid state physics".

Davydov, A.S., 1976, Theory of Solids (R) (Nauka, Moscow). Half this textbook is devoted to excitons.

Dow, J.D., 1976, Final state interactions in the optical spectra of solids: elements of exciton theory, in: Optical Properties of Solids; new developments, ed. B.O. Seraphin (North-Holland, Amsterdam) p. 33. Contains a good introduction to the effects of electric fields and disorder on the spectra of Wannier excitons.

Knox, R.S., 1963, Theory of Excitons (Academic Press, New York).

Rashba, E.I., 1974, Giant oscillator strengths associated with exciton complexes. Fiz. Tech. Polup. **8**, 1241; Sov. Phys. Semicond. **8**, 807 (1974). ("Exciton complex" here includes "bound exciton", "biexciton", etc.).

Reynolds, D.C. and T.C. Collins, 1981, Excitons: their properties and uses (Academic Press, New York).

Sham, L.J., 1980, Many body effects in the optical spectra of semiconductors, in: Physics of Semiconductors 1980: (Proc. 15th Int. Conf. on Physics of Semicond., Kyoto, 1980) eds. S. Tanaka and Y. Toyozawa (Phys. Soc. Japan, Tokyo) p. 69. Discusses the effect of the electron–hole interaction on unbound final states.

Polaritons and weak exciton–phonon interactions

Agranovich, V.M. and V.L. Ginzburg, 1966, Spatial Dispersion in Crystal Optics and the Theory of Excitons 1st ed. (Wiley, New York). 2nd ed. (R) (Nauka, Moscow, (1979)).

Loudon, R., 1975, Non-linear optics with polaritons, Proc. Int. Sch. Phys. "E. Fermi" **64**, ed. N. Bloembergen (Soc. Ital. Fisica, Bologna) p. 296.

Pekar, S.I., 1982, Crystal optics and additional light waves (R) (Naukova Dumka, Kiev) in press.

*(R) means available only in Russian.

Polivanov, Yu, N., 1978, Raman scattering of light by polaritons, Usp. Fiz. Nauk. **126**, 185; Sov. Phys. Usp. **21**, 805.

Richter, R., 1976, Resonant Raman scattering in semiconductors, in: Springer Tracts in Modern Physics **78** (Solid State Physics) ed. G. Hohler (Springer, Berlin) p. 121. A very thorough account of the theory for Wannier excitons.

Ulbrich, R.G. and C. Weisbuch, 1978, Resonant Brillouin scattering in semiconductors, Festkörperprobleme **18**, 217.

Yu, P.Y., 1979, Study of excitons and exciton–phonon interactions by resonant Raman and Brillouin spectroscopy, in: Top. Curr. Phys. **14** (Excitons) ed., K. Cho (Springer, Berlin) p. 211.

Excitons in semiconductors

Altarelli, M. and N.O. Lipari, 1977, Theory of excitons in semiconductors, in: Physics of Semiconductors (Proc. 13th Int. Conf., Rome, 1976) ed. F.G. Fumi (North-Holland, Amsterdam) p. 811. A brief account of the effective mass theory of excitons in semiconductors with degenerate band edges, with emphasis on indirect excitons.

Cho, K., 1979, Internal structure of excitons, Top. Curr. Phys. **14** (Excitons) ed. K. Cho (Springer, Berlin) p. 15. Theory of the energy levels of free Wannier excitons, including the effects of band degeneracy, exchange, and of external perturbations (stress, electric and magnetic fields).

Dean, P.J. and D.C. Herbert, 1979, Bound excitons in semiconductors, ibid p. 55. A very complete review of experiment and theory.

Permogorov, S., 1975, Hot excitons in semiconductors, Phys. Stat. Sol. **B68**, 9.

Planel, R., 1978, Spin orientation by optical pumping in semiconductors, Solid State Electron. **21**, 1437.

Rössler, U., 1979, Fine structure and dispersion of Wannier excitons, Festkörperprobleme, **19**, 77.

Schröder, U., 1973, Binding energy of excitons bound to defects: theoretical aspects, Festkörperprobleme **13**, 171.

Voos, M., R.F. Leheny and J. Shah, 1980, Radiative recombination, in: Handbook on Semiconductors, Vol. II, Ch. 6, ed. M. Balkanski (North-Holland, Amsterdam). Pays particular attention to luminescence at high pumping level: multi-exciton complexes, electron–hole liquid, hot excitons, and stimulated emission.

Zakharchenya, B.P., 1978, Spin polarization induced by optical pumping in semiconductors, in: Physics of Semiconductors, ed. B.L.H. Wilson (Inst. of Phys., London 1979) p. 31.

Magneto-optics and piezo-optics of excitons

Bimberg, D., 1977, Wannier–Mott polaritons in magnetic fields, Festkörperprobleme, **17**, 195. A comparison of recent calculations on free excitons with the latest high field experiments.

Bir, G.L. and G.E. Pikus, 1974, Symmetry and Strain-Induced Effects in Semiconductors (Nauka, Moscow, 1972; Wiley, New York, 1974).

Cho, K., 1979, Internal Structure of Excitons, loc. cit.

Dean, P.J. and D.C. Herbert, 1979, Bound excitons in semiconductors, loc. cit.

Martinez, G., 1980, Optical properties of semiconductors under pressure, in: Handbook on Semiconductors Vol. II, Ch. 4C, ed. M. Balkanski (North-Holland, Amsterdam). Deals primarily with the effect of hydrostatic pressure.

Sokolik, I.A. and E.L. Frankevich, 1973, Effect of a magnetic field on photoprocesses in organic solids, Usp. Fiz. Nauk. **111**, 261; Sov. Phys. Usp. **16**, 687 (1973).

Zakharchenya, B.P. and R.P. Seisyan, 1969, Diamagnetic excitons in semiconductors, Usp. Fiz. Nauk. **97**, 193; Sov. Phys. Usp. **12**, 70. A review of the magneto-optics of free excitons.

Multi-exciton complexes and exciton condensation

Grun, J.B., 1977, Biexcitons, experimental aspects, Nuovo Cim. **39B**, 579.

Hanamura, E., 1976, Excitonic molecules, in: Optical Properties of Solids: new developments, ed. B.O. Seraphin (North-Holland, Amsterdam) p. 81.

Hanamura, E. and H. Haug, 1977, Condensation effects of excitons, Phys. Repts. **33C**, 209. A theoretical review of the statics and dynamics of the transition from free excitons to multi-exciton complexes and to the electron–hole liquid, and of the possible Bose–Einstein condensation of excitons and biexcitons.

Hensel, J.C., T.G. Phillips and G.A. Thomas, 1977, Electron–hole liquid in semiconductors: experimental, Adv. in Solid State Phys. **32**, eds. H. Ehrenreich, F. Seitz and D. Turnbull (Academic Press, New York) p. 88.

Keldysh, L.V., 1971, Collective properties of excitons in semiconductors (R) in: Excitons in Semiconductors, ed. B.M. Vul (Nauka, Moscow).

Moskalenko, S.A., A.I. Bobrysheva, A.V. Lelyakov, M.F. Migley, P.I. Khadzhy and M.L. Shmiglink, 1974, Interactions of excitons in semiconductors (R) (Shtiintsa, Kishinev).

Pokrovskii, Ya.E. and V.B. Timofeev, 1979, Condensation of excitons and electron–hole liquids, Sov. Sci. Rev. **A1**, 191.

Rice, T.M., 1977, Electron–hole liquid in semiconductors: theoretical, Adv. in Solid State Phys. **32**, eds. H. Ehrenreich, F. Seitz, D. Turnbull (Academic Press, New York) p. 1.

Shionoya, S., 1979, Picosecond spectroscopy of biexcitons, J. Lum. **18/19**, 917.

Excitons in wide-gap inorganic insulators: self trapping

Aluker, E.D., D.J. Lusis and S.A. Chernov, 1979, Electronic excitations and radioluminescence in alkali halides (R) (Zinatne, Riga).

Fugol', I.Ya., 1978, Excitons in rare gas crystals, Adv. Phys. **27**, 1. A thorough survey of both theory and experiment.

Hayes, W., 1978, Optically detected electron paramagnetic resonance in ionic solids, Semicond. Insul. **3**, 121.

Knox, R.S. and K.J. Teegarden, 1968, Electronic excitations of perfect alkali halide crystals, in: Physics of Color Centers, ed. W. Beall Fowler (Academic Press, New York) p. 1.

Kuusmann, I.L., G.G. Liidja and Ch. B. Lushchik, 1976, Luminescence of free and self-trapped excitons in ionic crystals (R) Proc. Phys. Inst. Acad. Sci. Eston. SSR **46**, 5.

Levinson, I.B. and E.I. Rashba, 1973, Threshold phenomena and bound states in the polaron problem, Usp. Fiz. Nauk **111**, 683; Sov. Phys. Usp. **16**, 892.

Levinson, I.B. and E.I. Rashba, 1973, Electron–phonon and exciton–phonon bound states, Reps. Prog. Phys. **36**, 1499.

Rössler, U., 1976, Band structure and excitons, in: Rare Gas Solids **1**, eds. M.L. Klein and J.A. Venables, (Academic Press, London) p. 506. Concentrates on comparison of band structure calculations with experiment.

Schwentner, N., E.E. Koch and J. Jortner, 1982, Electronic excitations in condensed rare gases, in: Rare Gas Solids **3**, eds. M.L. Klein and J.A. Venables (Academic Press, New York, in press).

Toyozawa, Y., 1974, Exciton-lattice interaction: fluctuation, relaxation and defect formation, in: Vacuum UV Radiation Physics, eds. E.E. Koch, R. Hänsel and C. Kunz (Pergamon-Vieweg, Braunschweig) p. 317.

Toyozawa, Y., 1980, Self-localization of elementary excitations, in: Proc. 6th Int. Conf. VUV radiation physics, Charlottesville, Va. Appl. Opt. 19, 4101.

Williams, R.T., 1978, Photochemistry of F-center formation on halide crystals, Semicond. Insul. **3**, 251. Excited states of self-trapped excitons are precursors of defect formation.

Core excitons

These are excitons in which the hole is in an atomic core state rather than at the top of the valence band.

Altarelli, M., 1978, Theory of core excitons, J. Phys. (Paris) **39**, C4–95.

Bassani, F., 1980, Core excitons in solids, in: Proc. 6th Int. Conf. on VUV radiation physics, Charlottesville, Va. Appl Opt. **19**, 4093.

Brown, F.C., 1980, Innershell threshold spectra, in: Synchrotron Radiation Research, eds. S. Doniach and H. Winick (Plenum, New York) p. 61.

Sugano, S., 1976, Core excitation and electronic correlations in crystals, Proc. NATO Adv. Study Inst. **B12** (Spectroscopy of the Excited State) ed. B. di Bartolo (Plenum, New York) p. 279.

Excitons in magnetic insulators and metals

Eremenko, V.V., 1975, Introduction to the optical spectroscopy of magnetics (R) (Naukova Dumka, Kiev).

Eremenko, V.V. and E.G. Petrov, 1977, Light absorption in antiferromagnets, Adv. Phys. **26**, 31. Unlike most reviews of the spectra of magnetic crystals, this takes a consistently "excitonic" point of view.

Gaididei, Yu.V., V.M. Loktev, A.F. Prikhot'ko and L.I. Shanskii, 1975, Elementary excitations of α-oxygen Fiz. Nizk. Temp. (Kiev) **1**, 1365; Sov. J. Low Temp. Phys. **1**, 653. Exciton–magnon interaction and biexcitons in antiferromagnetic solid O_2.

Lovesey, S.W. and J.M. Loveluck, 1977, Magnetic Scattering, in: Top. Curr. Phys. **3**, (Dynamics of Solids and Liquids by Neutron Scattering) eds. S.W. Lovesey and T. Springer Springer, Berlin) p. 331. Includes the application of neutron scattering to exciton spectroscopy in rare earth metals.

McWhan, D.B. and C. Vettier, 1979, Neutron scattering studies at high pressure on rare earth intermetallic compounds. J. Phys. (Paris) **40**, C5-107. The spectrum of low lying excitons in such metals as PrSb can be drastically altered by hydrostatic pressure.

Excitons in molecular crystals

Agranovich, V.M. and R.M. Hochstrasser (eds.), 1982, Molecular Spectroscopy. (North-Holland, Amsterdam) (to be published in this series).

Avakian, P. and R.E. Merrifield, 1968, Triplet excitons in anthracene crystals, Mol. Cryst. **5**, 37.

Broude, V.L., E.I. Rashba and E.F. Sheka, 1981, Spectroscopy of molecular excitons (Energoizdat, Moscow); (English edition, Springer, Berlin) (in press).

Broude, V.L. and E.I. Rashba, 1974, Exciton spectra of molecular crystals, Pure Appl. Chem. **37**, 21.

Craig, D.P., 1974, Electronic energy levels of molecular solids, in: Orbital theories of molecules and solids, ed. N.H. March (Univ. Press, Oxford) p. 344.

Davydov, A.S., 1968, Theory of molecular excitons, 2nd ed. (Nauka, Moscow, 1968; Plenum, New York, 1971).
Ern, V. and M. Schott, 1976, Motion of localized excitons in organic solids, in: Localization and delocalization in quantum chemistry, eds. O. Chalvet et al. (Reidel, Dordrecht) **2**, 249.
Harris, C.B., 1974, Phosphorescence – microwave double resonance of molecular crystals and coherent states of triplet spin assemblies, Pure Appl. Chem. **37**, 73.
Hochstrasser, R.M., 1976, Triplet exciton states of molecular crystals, Int. Rev. Sci. Phys. Chem. Ser. II, **3**, ed. D.A. Ramsey (Butterworth, London) p. 1.
Kasha, M., 1976, Molecular excitons in small aggregates, in: Proc. NATO Adv. Study Inst. **B12**, (Spectroscopy of the Excited State) ed. B. di Bartolo (Plenum, New York) p. 337. Basic ideas of excitons in dimers and polymers.
Klimusheva, G.V., 1973, Exciton spectra of benzene-homologue crystals (R) in: Excitons in Molecular Crystals, ed. M.S. Brodin (Naukova Dumka, Kiev).
Kopelman, R., 1974, Excitons in pure and mixed molecular crystals, in: Excited States, ed. E.C. Lim (Academic Press, New York) **2**, 33.
Kopelman, R., 1976, Exciton percolation in molecular alloys and aggregates, in: Topics in Applied Phys. **15** (Radiationless Processes) ed. F.K. Fong (Springer, Berlin) p. 297.
Ostapenko, N.I., V.I. Sugakov and M.T. Shpak, 1973, Local excitons in molecular crystals (R) in: Excitons in Molecular Crystals, ed. M.S. Brodin (Naukova Dumka, Kiev) p. 92.
Philpott, M.R., 1973, Some modern aspects of exciton theory, Adv. Chem. Phys. **23**, 227.
Sheka, E.F., 1971, Electron-vibrational spectra of molecules and crystals, Usp. Fiz. Nauk **104**, 593; Sov. Phys. Usp. **14**, 484.

Excitons in biology

Campillo, A.J. and S.L. Shapiro, 1978, Picosecond fluorescence studies of exciton migration and annihilation in photosynthetic systems, Photochem. Photobiol. **28**, 975.
Davydov, A.S., 1979, Biology and quantum mechanics (R) (Naukova Dumka, Kiev). Discusses the possible contribution of excitonic effects to muscle action.
Guéron, M., J. Eisinger and A.A. Lamola, 1974, Excited states of nucleic acids, in: Basic Principles of Nucleic Acid Chemistry, ed. P.O.P. Ts'o (Academic Press, New York) p. 311. Discusses the evidence for triplet exciton migration in polynucleotides and DNA.
Markvart, T.J., 1978, Exciton transport in the photosynthetic unit, J. Theor. Biol. **72**, 91.
Knox, R.S., 1977, Photosynthetic efficiency, exciton transport and trapping, Top. Photosynth. **2**, 55.

Surface excitons

Agranovich, V.M. and R. Loudon eds., Surface Excitations (North-Holland, Amsterdam) (to be published in this series).
Fischer, B. and H.J. Queisser, 1975, Surface excitons, calculations of their dispersion curve, and possibilities for their experimental observation, CRC Crit. Rev. Solid State Sci. **5**, 281.
Fischer, B. and J. Lagois, 1979, Surface exciton polaritons, in: Top. Curr. Phys. **14** (Excitons) ed. K. Cho (Springer, Berlin) p. 183.
Lagois, J. and B. Fischer, 1978, Introduction to surface exciton polaritons, Festkörperprobleme, **18**, 197.

Special topics

Economou, E.N., 1978, Aspects of the theory of disordered systems, in: Excited States in Quantum Chemistry, eds. C.A. Nicolaides and D.R. Beck (Reidel, Dordrecht) p. 457. A brief and clear account of Anderson localization in a three-dimensional system, without specific reference to excitons.

Gutfreund, H. and W.A. Little, 1979, The prospects of excitonic superconductivity, in: Highly conducting one-dimensional solids, eds. J.T. Devreese, R.P. Evrard and V.E. van Doren (Plenum, New York) p. 305.

Haken, H., 1977, Non-linear interaction between excitons and coherent light, in: Proc. Int. Sch. Phys. "E. Fermi" **64**, ed. N. Bloembergen (Soc. Ital. Fisica, Bologna) p. 350.

Haken, H., 1976, Stimulated emission at high concentrations of excitons, in: Molecular spectroscopy of condensed phases, ed. M. Grosmann et al. (North-Holland, Amsterdam), p. 17.

Mott, N.F., 1978, Electronic properties of vitreous SiO_2, in: Physics of SiO_2 and its interfaces, ed. S.T. Pantelides (Pergamon, New York), p. 1. Discusses the possible existence of excitons in amorphous materials.

AUTHOR INDEX

SUBJECT INDEX

SUBSTANCE INDEX